SIGNAL PROCESSING FOR MOBILE COMMUNICATIONS

HANDBOOK

SIGNAL PROCESSING FOR MOBILE COMMUNICATIONS
HANDBOOK

Edited by
Mohamed Ibnkahla

CRC PRESS

Boca Raton London New York Washington, D.C.

Library of Congress Cataloging-in-Publication Data

Signal processing for mobile communications handbook / edited by Mohamed Ibnkahla.
p. cm.
Includes bibliographical references and index.
ISBN 0-8493-1657-X (alk. paper)
1. Signal processing. 2. Mobile communication systems. I. Ibnkahla, Mohamed.

TK5102.9.S5427 2004
621.382′2—dc22 2004042812

Direct all inquiries to CRC Press LLC, 2000 N.W. Corporate Blvd., Boca Raton, Florida 33431.

International Standard Book Number 0-8493-1657-X
Library of Congress Card Number 2004042812
Printed in the United States of America 1 2 3 4 5 6 7 8 9 0
Printed on acid-free paper

Preface

Signal processing (SP) is a key research area in mobile communications. The recent years have known a real explosion in research addressing different aspects of mobile communications signal processing. This area is continuously expanding with emerging applications and services such as interactive multimedia and Internet. SP has to meet the new challenges presented to future mobile communication systems such as very low bit error rates, very high transmission rates, real-time multimedia access, and differential quality of service (QoS).

Today's publications in this area are scattered worldwide across multiple journals and conference proceedings. Like any other discipline that seeks to reach maturity, now is the time for mobile communications signal processing to be presented to the readers in a comprehensive way and in one single book that stands by itself. This book brings together most SP techniques, delivering, for the first time in the history of SP, an in-depth survey of these techniques in a tutorial style.

The book is supported with more than 300 figures and tables, which makes it very easy to understand and accessible to students, researchers, professors, engineers, managers, and any professional involved in mobile communications.

The book investigates classical SP areas such as adaptive equalization, channel modeling and identification, multi-user detection, and array processing. It also investigates newer areas such as adaptive coded modulation, multiple-input multiple-output (MIMO) systems, diversity combining, and time-frequency analysis. It explores emerging techniques such as neural networks, Monte Carlo Markov Chain (MCMC) methods, and Chaos. It offers an excellent tutorial survey of promising approaches for future mobile communications such as cross-layer design in multi-access networks and adaptive wireless networks.

In addition to wireless terrestrial communications, the book covers most applications areas of mobile communications signal processing, such as satellite mobile communications, networking, power control and resource management, voice over IP, positioning and geolocation, cross-layer design and adaptation, etc.

I thank all the contributors for their excellent work. Thanks also to my research group at Queen's University who have dynamically contributed in writing three chapters and in the review process. Many thanks to the different reviewers (about 80) whose valuable input, remarks, and suggestions have definitely improved the technical quality of the chapters.

A special thank you to my wife, my son, and our families who have been a great support since the beginning until the final stage of this project.

Mohamed Ibnkahla
Queen's University
Kingston, Ontario, Canada

Editor

Mohamed Ibnkahla obtained an engineering degree in electronics in 1992, an M.Sc. degree in signal and image processing in 1992, a Ph.D. degree in signal processing in 1996, and an HDR (the ability to lead and supervise research) degree in digital communications and signal processing in 1998, all from the National Polytechnic Institute of Toulouse (INPT), Toulouse, France.

Dr. Ibnkahla held an Assistant Professorship at INPT (1996–1999). In 2000, he joined the Department of Electrical and Computer Engineering at Queen's University, Kingston, Ontario, Canada as Assistant Professor. He now holds the position of Associate Professor in the same department.

Since 1996, Dr. Ibnkahla has been involved in several research programs and centers of excellence, such as the European Advanced Communications Technologies and Services Program (ACTS), Communications and Information Technology Ontario (CITO), Canadian Institute for Telecommunications Research (CITR), and others. He has published a significant number of refereed journal papers, book chapters, and conference papers.

His research interests include signal processing, mobile communications, digital communications, satellite communications, and adaptive systems.

Dr. Ibnkahla received the INPT Leopold Escande Medal for the year 1997, France, for his research contributions to signal processing, and the prestigious Premier's Research Excellence Award (PREA), Ontario, December 2000, for his contributions in wireless mobile communications.

Contributors

Karim Abed-Meraim
Ecole Nationale Superieure des Telecommunications
Paris, France

Andreas Abel
ITI GmbH
Dresden, Germany

Hisham Abdul Hussein Al-Asady
Queen's University
Kingston, Ontario, Canada

Naofal Al-Dhahir
The University of Texas
Dallas, Texas

Mohamed-Slim Alouini
University of Minnesota
Minneapolis, Minnesota

Moeness Amin
Villanova University
Villanova, Pennsylvania

Hüseyin Arslan
University of South Florida
Tampa, Florida

Ghazem Azemi
Queensland University
Brisbane, Queensland, Australia

Nicholas Bambos
Stanford University
Stanford, California

A. Belouchrani
Ecole Nationale Polytechnique
Algeria

Boualem Boashash
Queensland University
Brisbane, Queensland, Australia

Helmut Bölcskei
Swiss Federal Institute of Technology (ETH)
Zurich, Switzerland

Rober Boutros
Queen's University
Kingston, Ontario, Canada

Stefano Buzzi
University of Cassino
Cassino, Italy

James J. Caffery, Jr.
University of Cincinnati
Cincinnati, Ohio

Giovanni Cherubini
IBM Research
Zurich, Switzerland

Giovanni E. Corazza
University of Bologna
Bologna, Italy

Fernando Díaz-De-María
University of Carlos III de Madrid
Madrid, Spain

Inbar Fijalkow
Université de Cergy Pontoise
Pontoise, France

Ascensio Gallardo-Antolin
University of Carlos III de Madrid
Madrid, Spain

Mounir Ghogho
University of Leeds
Leeds, England

Filippo Giannetti
University of Pisa
Pisa, Italy

Savvas Gitzenis
Stanford University
Stanford, California

Dennis L. Goeckel
University of Massachusetts
Amherst, Massachusetts

Mohamed Ibnkahla
Queen's University
Kingston, Ontario, Canada

Ming Kang
University of Minnesota
Minneapolis, Minnesota

Geert Leus
Delft University
The Netherlands

Alan Lindsey
U.S. Air Force Research Lab
Remsen, New York

Nguyen Linh-Trung
Aston University
Birmingham, England

Marco Luise
University of Pisa
Pisa, Italy

Andreas F. Molisch
Lund University
Lund, Sweden
and
Mitsubishi Electric
Research Labs
Cambridge, Massachusetts

Marc Moonen
Katholieke Universiteit Leuven
Leuven, Belgium

Massimo Neri
University of Bologna
Bologna, Italy

Raffaella Pedone
University of Bologna
Bologna, Italy

Carmen Peláez-Moreno
Universidad Carlos III
de Madrid
Madrid, Spain

Quazi Mehbubar Rahman
Queen's University
Kingston, Ontario, Canada

Atul Salhotra
Cornell University
Ithaca, New York

Anna Scaglione
Cornell University
Ithaca, New York

Wolfgang Schwarz
Dresden University of
Technology
Dresden, Germany

Noura Sellami
Université de Cergy Pontoise
Pontoise, France

Bouchra Senadji
Queensland University
Brisbane, Queensland,
Australia

Mohamed Siala
Sup'Com
El Ghazaia Ariana,
Tunisia

Wei Sun
Villanova University
Villanova, Pennsylvania

Ananthram Swami
U.S. Army Research Laboratory
Adelphi, Maryland

Lang Tong
Cornell University
Ithaca, New York

Fredrik Tufvesson
Lund University
Lund, Sweden

Jitendra K. Tugnait
Auburn University
Auburn, Alabama

Alessandro Vanelli-Coralli
University of Bologna
Bologna, Italy

Saipradeep Venkatraman
University of Cincinnati
Cincinnati, Ohio

Azadeh Vosoughi
Cornell University
Ithaca, New York

Xiaodong Wang
Columbia University
New York, New York

Hong-Chuan Yang
University of Victoria
Victoria, British Colombia,
Canada

Jun Yuan
Queen's University
Kingston, Ontario,
Canada

Qing Zhao
Cornell University
Ithaca, New York

Contents

Part I: Introduction

1 Signal Processing for Future Mobile Communications Systems: Challenges and Perspectives
Quazi Mehbubar Rahman and Mohamed Ibnkahla 1-1

Part II: Channel Modeling and Estimation

2 Multipath Propagation Models for Broadband Wireless Systems
Andreas F. Molisch and Fredrik Tufvesson 2-1

3 Modeling and Estimation of Mobile Channels
Jitendra K. Tugnait 3-1

4 Mobile Satellite Channels: Statistical Models and Performance Analysis
Giovanni E. Corazza, Alessandro Vanelli-Coralli, Raffaella Pedone, and Massimo Neri 4-1

5 Mobile Velocity Estimation for Wireless Communications
Bouchra Senadji, Ghazem Azemi, and Boualem Boashash 5-1

Part III: Modulation Techniques for Wireless Communications

6 Adaptive Coded Modulation for Transmission over Fading Channels
Dennis L. Goeckel 6-1

7 Signaling Constellations for Transmission over Nonlinear Channels
Hisham Abdul Hussein Al-Asady, Quazi M. Rahman, and Mohamed Ibnkahla 7-1

8 Carrier Frequency Synchronization for OFDM Systems
Mounir Ghogho and Ananthram Swami 8-1

9 Filter-Bank Modulation Techniques for Transmission over Frequency-Selective Channels
Giovanni Cherubini 9-1

Part IV: Multiple Access Techniques

10 Spread-Spectrum Techniques for Mobile Communications
Filippo Giannetti and Marco Luise 10-1

11 Multiuser Detection for Fading Channels
Stefano Buzzi 11-1

Part V: MIMO Systems

12 Principles of MIMO-OFDM Wireless Systems
Helmut Bölcskei 12-1

13 Space–Time Coding and Signal Processing for Broadband Wireless Communications
Naofal Al-Dhahir 13-1

14 Linear Precoding for MIMO Systems
Anna Scaglione, Atul Salhotra, and Azadeh Vosoughi 14-1

15 Performance Analysis of Multiple Antenna Systems
Ming Kang and Mohamed-Slim Alouini 15-1

Part VI: Equalization and Receiver Design

16 Equalization Techniques for Fading Channels
Geert Leus and Marc Moonen 16-1

17 Low-Complexity Diversity Combining Schemes for Mobile Communications
Hong-Chuan Yang and Mohamed-Slim Alouini 17-1

18 Overview of Equalization Techniques for MIMO Fading Channels
Noura Sellami, Inbar Fijalkow, and Mohamed Siala 18-1

19 Neural Networks for Transmission over Nonlinear Channels
Mohamed Ibnkahla, Jun Yuan, and Rober Boutros 19-1

Part VII: Voice over IP

20 Voice over IP and Wireless: Principles and Challenges
Fernando Díaz-de-María, Ascensión Gallardo-Antolín, and Carmen Peláez-Moreno **20**-1

Part VIII: Wireless Geolocation Techniques

21 Geolocation Techniques for Mobile Radio Systems
James J. Caffery, Jr. and Saipradeep Venkatraman **21**-1

22 Adaptive Arrays for GPS Receivers
Moeness Amin, Wei Sun, and Alan Lindsey **22**-1

Part IX: Power Control and Wireless Networking

23 Transmitter Power Control in Wireless Networking: Basic Principles and Core Algorithms
Nicholas Bambos and Savvas Gitzenis **23**-1

24 Signal Processing for Multiaccess Communication Networks
Qing Zhao and Lang Tong **24**-1

Part X: Emerging Techniques and Applications

25 Time–Frequency Signal Processing for Wireless Communications
Boualem Boashash, A. Belouchrani, Karim Abed-Meraim, and Nguyen Linh-Trung **25**-1

26 Monte Carlo Signal Processing for Digital Communications: Principles and Applications
Xiaodong Wang **26**-1

27 Principles of Chaos Communications
Andreas Abel and Wolfgang Schwarz **27**-1

28 Adaptation Techniques and Enabling Parameter Estimation Algorithms for Wireless Communications Systems
Hüseyin Arslan **28**-1

Index **I**-1

I

Introduction

1 **Signal Processing for Future Mobile Communications Systems: Challenges and Perspectives** *Quazi Mehbubar Rahman and Mohamed Ibnkahla* **1**-1
Introduction • Channel Characterizations • Modulation Techniques • Coding Techniques • Multiple Access Techniques • Diversity Technique • Conclusions

1

Signal Processing for Future Mobile Communications Systems: Challenges and Perspectives

1.1 Introduction 1-2
1.2 Channel Characterizations 1-2
Large-Scale Propagation Models • Small-Scale Propagation Models
1.3 Modulation Techniques 1-13
Modulation Schemes: The Classification • Different Modulation Schemes
1.4 Coding Techniques 1-20
Shannon's Capacity Theorem • Different Coding Schemes • Coding in Next-Generation Mobile Communications: Some Research Evidence and Challenges
1.5 Multiple Access Techniques 1-29
Fundamental Multiple-Access Schemes • Combination of OFDM and CDMA Systems • OFDM/TDMA • Capacity of MAC Methods • Challenges in the MAC Schemes
1.6 Diversity Technique 1-35
Classifications of the Diversity Techniques • Classifications of Diversity Combiners • Diversity for Next-Generation Systems: Some Research Evidence • Challenges in the Diversity Area
1.7 Conclusions 1-38

Quazi Mehbubar Rahman
Queen's University

Mohamed Ibnkahla
Queen's University

Abstract

This chapter briefly reviews background information on different signal processing issues of wireless mobile communications systems targeting the next-generation scenarios. The overview includes the channel characterization at the beginning of the chapter and then it steps through modulation techniques, multiple access schemes, coding, and diversity techniques. Here, along with the presentation of current research evidence, key challenges for the next-generation systems have been addressed.

0-8493-1657-X/$0.00+$1.50

1.1 Introduction

The ability to communicate on the move has evolved remarkably since Guglielmo Marconi first demonstrated radio's ability to provide continuous contact with ships sailing the English Channel. That was in 1897, and since then people throughout the world have enthusiastically adopted new wireless communications methods and services. Currently, when the telecommunications industries are deploying third-generation (3G) systems worldwide and researchers are presenting many new ideas for the next-generation wireless systems (termed 4G), several challenges are yet to be fulfilled. These include high data rate transmissions (up to 1 Gbps), multimedia communications, seamless global roaming, quality of service (QoS) management, high user capacity, integration and compatibility between 3G and next-generation components, etc. To meet these challenges, researchers are presently focusing their attentions on different signal processing issues after careful channel characterizations. This chapter will provide brief background information on these issues. It will also include some information on the current research works and challenges in these areas. The outline of the chapter follows: Section 1.2 discusses basic information on the channel characterization aspects. Section 1.3 presents an overview of the different modulation schemes that are getting the most attention in the research area. Coding techniques are discussed in Section 1.4. Section 1.5 talks about different multiple access schemes, while Section 1.6 presents different diversity scenarios. Finally, conclusions are drawn.

1.2 Channel Characterizations

The time-varying nature of the wireless mobile channel makes channel characterization and its analysis an important issue. In a mobile wireless scenario, the time-varying nature of the channel could be encountered in many different ways, e.g., a relative motion between the transmitter and the receiver, time variation in the structure of the medium, etc. All these scenarios make the channel characteristics random, and do not offer any easy analysis on the signals, transmitted through this channel. In general, as an information signal propagates through the channel, the strength of this signal decreases as the distance between the transmitter and receiver increases. The strength of the received signal depends on the characteristics of the channel and on the distance between the transmitter and the receiver. In a broad sense, the channel can be modeled in two different categories, large-scale propagation model and small-scale propagation model. These models will be discussed in the following subsections.

1.2.1 Large-Scale Propagation Models

Large-scale propagation model characterizes the received signal strength over large transmitter–receiver separation distances of several hundreds or thousands of meters. These are broadly classified in to two categories: deterministic and stochastic. Both deterministic and stochastic approaches are useful in describing a time-varying channel, even though they embrace different aspects: the stochastic model is better suited for describing global behaviors, whereas the deterministic one is more useful for studying the transmission through a specific channel realization.

1.2.1.1 Deterministic Approach

1.2.1.1.1 Free-Space Propagation Model

According to this model, the received signal power decays as a function of the distance between the transmitter and the receiver when they maintain a clear line of sight between them. In this case, the free-space signal power $P_r(d)$, received by a receiver antenna at a distance d (meters) from the transmitter, is given by

$$P_r(d) = \frac{P_t G_t G_r \lambda^2}{(4\pi)^2 d^2 L}, \quad d \geq d_0 \, (\neq 0) \geq d_f \tag{1.1}$$

TABLE 1.1 Path Loss Exponent for Different Communication Environments

Communication Environment	Path Loss Exponent
Indoor with line of sight	1.6–1.8
Free space	2
In factories with obstructions	2–3
Cellular radio in the urban area	2.7–3.5
Cellular radio in the shadowed urban area	3–5
Indoor with obstructions	4–6

where P_t represents the transmitted signal power, G_t andG_r are the transmitter and receiver antenna gains, respectively, L ($\geq$1) is the system loss factor, independent of signal propagation, λ (meters) is the wavelength, d_f is the far-field distance (also known as Fraunhofer distance), and d_0 is the received-power reference distance. The far-field distance d_f is given by

$$d_f = \frac{2D^2}{\lambda} d_f \gg D \tag{1.2}$$

where D is the largest physical linear dimension of the antenna. Using Equation 1.1, the free-space received power at a distance $d > d_0$ can be written as

$$P_r(d) = P_r(d_0)\left(\frac{d}{d_0}\right)^{-2} \tag{1.3}$$

1.2.1.1.2 Log-Distance Path Loss Model

This model shows that the average path loss[1] increases logarithmically with distance between the transmitter and the receiver of a communications system, which is given by

$$Pl_{avg}\,(\text{dB}) = Pl_{avg}\,(d_0) + 10n\log\left(\frac{d}{d_0}\right) \tag{1.4}$$

where n is the path loss exponent that indicates the rate at which the path loss for the transmitted signal increases with distance. The value of n depends on the specific propagation environment (e.g., see Table 1.1 [Rap96]). In Equation 1.4, d and d_0 hold the same definitions as in Equation 1.1.

1.2.1.2 Stochastic Approach

1.2.1.2.1 Lognormal Shadowing Model

The phenomenon that describes the random shadowing effects occurring over a large number of measurement locations having the same transmitter and receiver separation with different levels of clutter on the propagation path is referred to as lognormal shadowing. The corresponding path loss model states that the path loss *Pl*(d) at a particular location is lognormally (normal in dB) distributed about the mean distance-dependent value [Cox84] [Ber87]. The analytical expression of this model is given by

$$\begin{aligned} Pl(d) &= Pl_{avg}(d) + X_\sigma \\ &= Pl_{avg}(d_0) + 10n\log\left(\frac{d}{d_0}\right) + X_\sigma \end{aligned} \tag{1.5}$$

where X_σ (dB) is a zero-mean Gaussian distributed random variable with a variance of σ^2dB. In general, the values of n (defined earlier) and σ^2 are computed from measured data (e.g., see Table 3.6 in

[1]Path loss, expressed in dB, is defined as the difference between the effective transmitted signal power and the received signal power.

Rappaport [Rap96]), using linear regression in such a way that the contrast between the estimated and measured path losses is minimized.

Other than the general large-scale propagation models described above, there are some specific models based on the outdoor and indoor environments separately. These channel models are based on the profile of the particular area. Examples of some outdoor propagation models include the Longley–Rice model [Lon68] and Durkin's model [Dad75]. Examples of some indoor models are the Erricson multiple break-point model [Ake88] and the attenuation factor model [Sei92]. In addition to these models, Ray tracing and site-specific modeling techniques are also used for both outdoor and indoor environments.

1.2.2 Small-Scale Propagation Models

These models characterize the received signal strength of a radio signal over a short period of time or travel distance of typically 5λ to 40λ, λ being the wavelength of the signal. In this scenario, the instantaneous received signal fluctuates very rapidly and may give rise to fading, which is termed small-scale fading. In this section we will discuss different small-scale propagation models upon presenting all the relevant parameters that are required to discuss these models.

1.2.2.1 Parameters of Mobile Multipath Channel

A multipath channel is characterized by many important parameters. Among these parameters delay spread and coherence bandwidth describe the time-dispersive nature of the channel in a local area. On the other hand, Doppler spread and coherence bandwidth describe the time-varying nature of the channel in a small-scale region. Including these major parameters, here we will briefly discuss the channel parameters, which will provide a clear description of a mobile multipath channel.

1.2.2.1.1 Fading

Fading, also known as small-scale fading, is the result of interference between two or more attenuated versions of the transmitted signal arriving at the receiver in such a way that these signals are added destructively. These multiple versions of the transmitted signal result from the multiple paths present in the channel or from the rapid dynamic changes of the channel. In this case, the speed of the mobile and the transmission bandwidth of the signal also play a vital role.

1.2.2.1.2 Doppler Shift

The apparent change in frequency of the transmitted signal due to the relative motion of the mobile is known as the Doppler shift, which is given by

$$f_{ds} = \frac{v}{\lambda} \cos\theta \tag{1.6}$$

where v is the velocity of the mobile, λ is the signal wavelength, and θ is the spatial angle between the direction of motion of the mobile and the direction of arrival of the wave.

1.2.2.1.3 Excess Delay

This is the relative delay of the ith multipath signal component, compared to the first arriving component and is given by τ_i.

1.2.2.1.4 Power Delay Profile, $\Phi_c(\tau)$

This is the average output signal power of the channel as a function of excess time delay τ. In practice, $\Phi_c(\tau)$ is measured by transmitting very narrow pulses, or equivalently a wide band signal, and cross-correlating the received signal with a delayed version of itself. Power delay profile is also known as multipath intensity profile and delay power spectrum. It gets the latter name because of its frequency domain component, which gives the power spectrum density. The mean excess delay, root mean squared (rms) delay spread, and excess delay spread (XdB) are multipath channel parameters that can be determined from a power delay profile. The mean excess delay (τ_{mean}) is the first moment of the power delay profile, the rms delay spread (σ_τ) is the square root of the second central moment of the power delay profile, and the maximum excess delay (XdB) of the power delay profile is defined as the time delay during which multipath energy falls to X dB below the maximum value. τ_{mean} and σ_τ are expressed as

$$\tau_{mean} = \frac{\sum_i P(\tau_i)\tau_i}{\sum_i P(\tau_i)} \quad \text{and} \tag{1.7a}$$

$$\sigma_\tau = \sqrt{mean[(\tau)^2] - \tau_{mean}^2} \tag{1.7b}$$

where

$$mean[(\tau)^2] = \frac{\sum_i P(\tau_i)\tau_i^2}{\sum_i P(\tau_i)} \tag{1.7c}$$

1.2.2.1.5 *Delay Spread (T_m)*

Delay spread, also known as multipath spread, of the channel is the range of values of excess time delay τ, over which $\Phi_c(\tau)$ is essentially nonzero.

1.2.2.1.6 *Coherence Bandwidth (BW_{coh})*

The frequency band in which all the spectral components of the transmitted signal pass through a channel with equal gain and linear phase is known as coherence bandwidth of that channel. Over this bandwidth the channel remains invariant. BW_{coh} can be expressed in terms of rms delay spread, though there is no exact relationship between these two parameters. According to Lee [Lee89], with a frequency correlation of approximately 90%, BW_{coh} can be shown as

$$BW_{coh} \approx \frac{1}{50\sigma_\tau} \tag{1.8}$$

1.2.2.1.7 *Doppler Spread (B_d)*

Spreading of the frequency spectrum of the transmitted signal resulting from the rate of change of the mobile radio channel is known as Doppler spread. With the transmitted signal frequency f_c, the resultant Doppler spectrum has the components in the range between $(f_c - f_{d,\max})$ and $(f_c + f_{d,\max})$, $f_{d,\max}$ being the maximum Doppler frequency shift.

1.2.2.1.8 *Coherence Time (T_{coh})*

The time period during which the channel impulse response remains invariant is known as coherence time of the channel. T_{coh} is inversely proportional to the Doppler spread, and with the maximum Doppler frequency shift, $f_{d,\max}$, it is given by

$$T_{coh} \approx \frac{1}{f_{d,\max}} \tag{1.9}$$

1.2.2.2 Types of Small-Scale Fading

Small-scale fading is divided into two broad classes, which are based on the time delay spread and Doppler spread. The time delay spread-dependent class is divided into two categories, flat fading and frequency-selective fading, while the Doppler spread-dependent class is categorized as fast and slow fading. It is important to note that fast and slow fading deal with the relationship between the time rate of change of the channel and the transmitted signal, and not with propagation path loss models.

1.2.2.2.1 Flat Fading

The received signal in a mobile radio environment experiences flat fading if the channel has a constant gain and linear phase response over a bandwidth that is greater than the bandwidth of the transmitted signal. The main characteristics of a flat fading channel follow:

- Symbol period of the transmitted signal is greater than the delay spread of the channel. As a rule of thumb it should be at least 10 times greater.
- Bandwidth of the channel is greater than the bandwidth of the transmitted signal. Since the bandwidth of the transmitted signal is narrower than the channel bandwidth, the flat fading channels are also known as *narrowband channels.*
- Typical flat fading channels result in deep fades, and this requires 20 to 30 dB more transmitter power to achieve low bit error rates (BERs) during times of deep fades, compared to systems operating over nonfading channels.

1.2.2.2.2 Frequency-Selective Fading

The received signal in a mobile radio environment experiences frequency-selective fading if the channel has a constant gain and linear phase response over a bandwidth that is smaller than the bandwidth of the transmitted signal. The main characteristics of a frequency-selective fading channel follow:

- Symbol period of the transmitted signal is smaller than the delay spread of the channel. As a rule of thumb it should be at least 10 times smaller.
- Bandwidth of the channel is smaller than the bandwidth of the transmitted signal. Since the bandwidth of the transmitted signal is wider than the channel bandwidth, the frequency-selective fading channels are also known as *wideband channels.*
- Frequency-selective channel results in intersymbol interference (ISI) for the received signal.
- This type of fading channels is difficult to model compared to the flat fading channels since each multipath signal needs to be modeled individually and the channel has to be considered as a linear filter.

1.2.2.2.3 Fast Fading

The received signal, in a mobile radio environment, experiences fast fading as a result of rapidly changing channel impulse response within the symbol duration. The main characteristics of a fast fading channel follow:

- Coherence time of the channel is smaller than the symbol period of the transmitted signal. Thus this is also called *time-selective fading.*
- Doppler spread is greater than the transmitted signal bandwidth.
- Channel varies faster than the baseband signal variations.
- In *fast-flat fading* channels the amplitude of the received signal varies faster than the rate of change of the transmitted baseband signal.
- In *fast-frequency-selective* channels the amplitudes, phases, and time delays of the multipath components vary faster than the rate of change of the transmitted signal.

1.2.2.2.4 Slow Fading

The received signal, in a mobile radio environment, experiences slow fading as a result of slowly varying channel impulse response within the symbol duration. The main characteristics of a slow fading channel follow:

- Coherence time of the channel is greater than the symbol period of the transmitted signal. In this case, the channel can be assumed to be static over one or several symbol durations.
- Doppler spread is smaller than the transmitted signal bandwidth.
- Channel varies slower than the baseband signal variations.

1.2.2.3 Statistical Representation of the Small-Scale Propagation Channel

For the signal processing applications and analyses, the mobile propagation fading channels are modeled statistically in many different ways. The most popular statistical models of the fading channels are the Rayleigh, Ricean, and Nakagami fading channel models, which will be discussed briefly in this section.

1.2.2.3.1 Rayleigh Fading Channel

When the channel impulse response $c(\tau, t)$ at a delay τ and time instant t is modeled as a zero-mean complex-valued Gaussian process, the envelope $|c(\tau, t)|$ at that time instant t is known to be Rayleigh distributed. In this case the channel is said to be a Rayleigh fading channel. The Rayleigh distribution has the probability density function (PDF)

$$p(r) = \begin{cases} \frac{r}{\sigma^2} \exp\left(-\frac{r^2}{2\sigma^2}\right) & (0 \leq r \leq \infty) \\ 0 & (r < 0) \end{cases} \tag{1.10}$$

where r is the envelope of the received signal and σ^2 is the time average power of the received signal before envelope detection.

1.2.2.3.2 Ricean Fading Channel

When there are fixed scatterers or signal reflectors present in the mobile channel, in addition to the randomly moving scatterers, the channel impulse response $c(\tau, t)$ can no longer be modeled as a zero-mean complex-valued Gaussian process. In this case the envelope has a Ricean distribution and the corresponding channel is known as a Ricean fading channel. The Ricean distribution has the PDF

$$p(r) = \begin{cases} \frac{r}{\sigma^2} \exp\left(-\frac{(r^2+A^2)}{2\sigma^2}\right) I_0\left(\frac{Ar}{\sigma^2}\right) & (A \geq 0,\ r \geq 0) \\ 0 & (r < 0) \end{cases} \tag{1.11}$$

where A denotes the peak amplitude of the dominant received signal arriving at the receiver either from a fixed scatterer or through a line of sight path and $I_0\,(.)$ represents the zero-order modified Bessel function of the first kind. Ricean distribution is often described in terms of the Ricean factor K, which is defined as the ratio between the dominant signal power and the variance of the scattered power, which is given by

$$K = \frac{A^2}{2\sigma^2} \tag{1.12}$$

When $K = 0$, the channel exhibits Rayleigh fading, and when $K = \infty$, the channel remains constant.

1.2.2.3.3 Nakagami Fading Channel

Nakagami fading characterizes rapid fading in long-distance channels [Nak60]. Nakagami distribution is selected to characterize the fading channel because it provides a closer match to some experimental data than either the Rayleigh or Ricean distributions. The PDF of this distribution is given by

$$p_R(r) = \frac{2m^m x^{2m-1}}{\Gamma(m)\Omega^m} \exp\left(-\frac{mr^2}{\Omega}\right) \quad m \geq \frac{1}{2} \tag{1.13}$$

where $\Omega = E(R^2)$. The parameter m, defined as the ratio of moments, is called the fading figure, which is given by

$$m = \frac{\Omega^2}{E[(R-\Omega)^2]} \tag{1.14}$$

Some advantages of this distribution follow. This distribution can model fading conditions that are either more or less severe than Rayleigh fading. When $m = 1$, the Nakagami distribution becomes the Rayleigh distribution, when $m = 0.5$ it becomes a one-sided Gaussian distribution, and when $m \to \infty$ the distribution becomes an impulse (a constant). The Rice distribution can be closely approximated by using the following relationship between the Ricean factor (K) and the fading figure (m) [Nak60]:

$$K = \frac{\sqrt{m^2 - m}}{m - \sqrt{m^2 - m}} \quad m > 1 \tag{1.15}$$

$$m = \frac{(K+1)^2}{(2K+1)} \tag{1.16}$$

Since the Ricean distribution contains a Bessel function, while the Nakagami distribution does not, the Nakagami distribution often leads to convenient closed-form analytical expressions that are otherwise unattainable.

1.2.2.4 Statistical Models for Multipath Fading Channels

Many statistical channel models are proposed and researched for the terrestrial and satellite channel environments. Examples include Clarke's model [Cla68], the Saleh and Valenzuela model [Sal87], and the two-ray fading channel model. In this section we will discuss only the two-ray fading channel model since it gives a clear idea about the channel's fading effect. Besides, we will discuss some recently researched channel models, which are based on different types of fading channel environments.

1.2.2.4.1 Two-Ray Fading Channel Model

A commonly used multipath fading model is the Rayleigh fading two-ray channel model, as shown in Figure 1.1. Assuming that the phase of the transmitted signal does not change on both the paths, the impulse response of this channel is given by

$$h(t) = a_0\delta(t) + a_1\delta(t - t_0) \tag{1.17}$$

where $\delta(t)$ is the Kronecker delta function, defined as

$$\delta(t) = \begin{cases} 1 & \text{for } t = 0 \\ 0 & \text{otherwise} \end{cases} \tag{1.18}$$

With the input signal $x(t)$ the output of the channel $y(t)$ is expressed as

$$y(t) = a_0x(t) + a_1x(t - t_0) \tag{1.19}$$

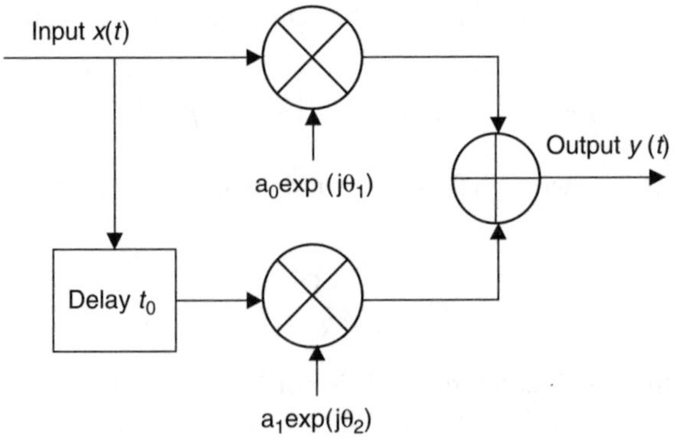

FIGURE 1.1 Two-ray fading channel model.

where a_0 and a_1 are independent and Rayleigh distributed. Letting $a_0 = 1$ and using Fourier transform on both the sides of Equation 1.19, the transfer function $H(f)$ of the channel can be found as

$$H(f) = 1 + a_1 \exp(j2\pi f t_0) \tag{1.20}$$

The amplitude response of the channel transfer function gives

$$|H(f)| = \sqrt{1 + a_1^2 + 2a_1 \cos 2\pi f t_0} \tag{1.21}$$

From Equation 1.21 it is found that the amplitude response of the channel shows frequency selectivity of the channel, and by varying t_0, it is possible to create a wide range of frequency-selective fading effects. With $a_1 = 1$, the channel results in deep fades, and with $a_1 \approx 0$, the channel becomes a flat Rayleigh fading channel.

1.2.2.4.2 Motif Model

This is a relatively new channel modeling concept [Pec00] [Pec01] [Kle02] where a semideterministic approach is developed, based on a simple ray launching technique, the Monte Carlo method, and general statistics. The model is initially developed for indoor wideband and narrowband channels. In this modeling approach an algorithm is used in which a bitmap of an indoor floor plan is utilized as a main input. This input may be obtained as a scanned blueprint with filled pixels representing walls, partitions, and obstacles. In this scanned input, different materials are distinguished from each other by different colors and textures of the pixels, where the size of a pixel is predetermined by a wavelength. For all the empty elements the prediction is calculated at once. Then the rays are launched from a transmitter antenna. Unlike the classical ray launching method, here the rays are propagated using very fast pixel graphics. When a ray hits a colored element (not empty), its neighboring elements in the bitmap are separated into a matrix called motif. To deal with all possible floor plans, many different previously generated motifs are kept stored in the database, from where the appropriate motif is selected. Upon selecting the suitable motif, a probability radiation pattern is assigned to it and a specific angle of arrival of the ray is chosen. These two components control the ray behavior in the next step. Using a random number generator and the probability radiation pattern, the next direction of the ray is chosen. A ray absorption probability is also assigned to each individual motif. A new ray is launched from the transmitter antenna when a ray reaches the boundary of the bitmap or gets absorbed in the motif.

In this model, the impulse response of the channel can easily be obtained in every empty element by recording the length of all the passing rays, each of which specifies its time delay. After dividing the time delay axis of the impulse response into discrete intervals, the incoming rays are distributed into these intervals according to their respective delays. The number of rays in each interval represents the relative power for the relevant time delay in the final impulse response. A similar procedure is carried out for calculating the angle of arrival. The main drawback of the motif concept is the requirement of computer memory, which becomes huge when motifs for many different materials are of interest.

1.2.2.4.3 Finite-State Markov Chain Model

Finite-state Markov chain (FSMC) models are widely in use in the analysis of radio channels in both the terrestrial and satellite domains [Lin02] [Hsi01] [Gua99]. The study of the finite-state Markov channel emerges from the early works of Gilbert [Gil60] and Elliott [Ell63]. They studied a two-state Markov channel known as the Gilbert–Elliott channel. Later Guan [Gua99] and Wang [Wan95] generalized FSMCs for arbitrary states.

To get an idea about this model, the example [Gua99] shown in Figure 1.2 can be taken into account. Here the model is presented for a noninterleaved fading process where all the possible fade amplitudes are divided into several nonoverlapping intervals known as channel states. In this case, the channel takes on different channel states during the transmitted symbol durations and makes transitions from one state to another according to the fading process. These transitions (Figure 1.2b) are characterized by transition probabilities between different states, while the probabilities depend on different physical channel parameters.

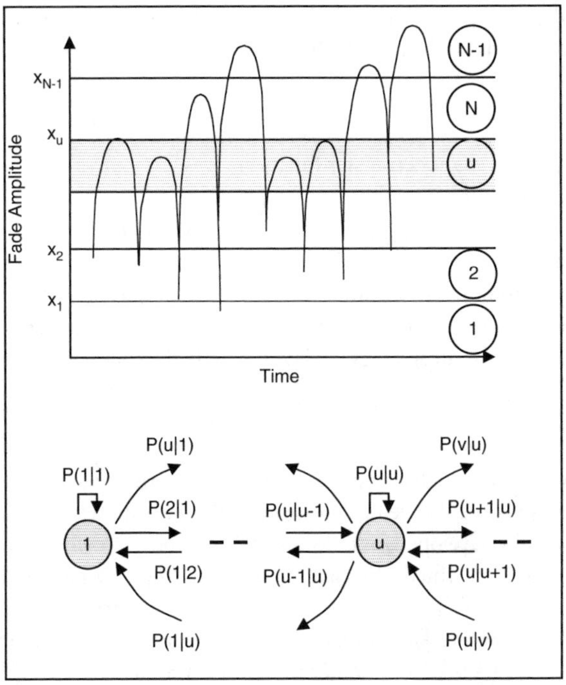

FIGURE 1.2 Finite-state Markov-chain model of a non-interleaved fading channel.

As shown in Guan [Gua99], with the aid of probabilistic theory, the equilibrium channel state probability $p(u)$ for state u, and the state transition probability $p(v|u)$ from channel state u to v, can be expressed as follows:

$$p(u) = \Pr(x_{u-1} \leq x < x_u) = \int_{x_{u-1}}^{x_u} pdf_x(x)dx \tag{1.22}$$

$$p(v|u) = \frac{p(v,u)}{p(u)} = \frac{\Pr(x_{v-1} \leq x < x_v, x_{u-1} \leq \tilde{x} < x_u)}{p(u)} == \frac{\int\limits_{x_{v-1}}^{x_v} \int\limits_{x_{u-1}}^{x_u} jpdf_{x,\tilde{x}}(x,\tilde{x})\, dx d\tilde{x}}{\int_{x_{u-1}}^{x_u} pdf_x(x)dx} \tag{1.23}$$

In the above equations, x and $\tilde{x}$ represent the fading amplitudes, with x_{u-1} and x_u being the lower and upper boundaries of the fading amplitudes, respectively; $\Pr(x)$ and $\Pr(x, y)$ represent the probability of x and joint probability of x and y, respectively; and $\text{pdf}_x(x)$ and $\text{pdf}_{x,y}(x,y)$ correspond to the PDF of x and joint PDF of x and y, respectively.

1.2.2.4.4 Loo's Satellite Channel Model

Loo [Loo85] [Loo87] [Loo94] [Loo96] [Loo98] developed some channel models for mobile satellite scenarios that represent simple and accurate probability density functions for the received signal envelope and phase. These PDFs have been shown to be dependent on the weather conditions. Loo [Loo98] has shown that for a fixed satellite Ka-band (20 to 30 GHz) channel, the signal envelope and phase can be modeled as Gaussian random variables, and their expressions are given by

$$p_w(r) = \frac{1}{\sqrt{2\pi}\sigma_r} \exp\left[-(r - m_r)^2/2\sigma_r^2\right] \tag{1.24}$$

and

$$p_w(\phi) = \frac{1}{\sqrt{2\pi}\sigma_\phi} \exp\left[-(\phi - m_\phi)^2/2\sigma_\phi^2\right] \tag{1.25}$$

where m_r, σ_r and m_ϕ, σ_ϕ are the mean and variance of the envelope and phase, respectively.

For the satellite mobile channel in the L band (1.3 to 2 GHz), Loos's model assumes that the line of sight (LOS) component under shadowing is lognormally distributed and that the multipath effect is Rayleigh distributed. The signal is then the sum of a lognormal variable z and a Rayleigh variable w (corresponding to multipath fading):

$$r \exp(j\theta) = z \exp(j\phi_0) + w \exp(j\phi) \tag{1.26}$$

where the lognormally distributed (corresponding to shadowing) random variable z has the standard deviation $\sqrt{d_0}$ and mean μ. The phases ϕ_0 and ϕ are uniformly distributed random variables in the range of 0 to 2π.

The signal envelope PDF is shown to be modeled as [Loo96, Loo98]

$$p(r) = \frac{r}{b_0\sqrt{2\pi d_0}} \int_0^{+\infty} \frac{1}{z} \exp\left[\frac{-(\ln z - \mu)^2}{2b_0\{2d_0 - (r^2 + z^2)\}}\right] \times I_0\left(\frac{rz}{b_0}\right) dz \tag{1.27}$$

where b_0 represents the average scattered power due to multipath (Rayleigh fading) and $I_0(.)$ is the zero-order modified Bessel function of the first kind.

It is clear from Equation 1.27 [Loo96] [Loo98] that when z is constant (i.e., the LOS is directly received with no shadowing), the signal envelope follows Ricean distribution:

$$p(r) = \frac{r}{b_0} \exp[-(r^2 + A^2)/2b_0] \times I_0(rA/b_0) \tag{1.28}$$

In the case where there is shadowing z, but no multipath fading (i.e., $w = 0$), the envelope PDF is lognormal, and is given by

$$p(r) = \frac{1}{r\sqrt{2\pi d_0}} \exp[-(\ln r - \mu)^2/2d_0] \tag{1.29}$$

In the case where there is no shadowing and no LOS (i.e., $z = 0$), the signal envelope PDF is Rayleigh distributed, giving

$$p(r) = \frac{r}{b_0} \exp(-r^2/2b_0) \tag{1.30}$$

1.2.2.4.5 Multiple-Input Multiple-Output Channel Models

1.2.2.4.5.1 Matrix Channel Model

The structure of this multiple-input multiple-output (MIMO) channel model, presented in Durgin [Dur03], is shown in Figure 1.3. Here the transfer functions $H_{pq}(\tau; t)$ are shown between the set of signals $\{a_p(t)\}$, sent from each of the M transmitter antennas, and the set of signals $\{b_q(t)\}$, received at the N receiver antennas. The two different time components t and τ in the channel transfer function show that these channels may be a function of time t to model a time-varying channel and a function of delay τ to model the dispersion incurred by wideband transmission.

In general a vector/matrix notation is used to keep track of all the transmitted and received signals in a MIMO system. A vector of received signals $\vec{b}(t)$at the input of the N receiver antennas may be calculated from the vector of transmitted signals $\vec{a}(t)$. The output vector is related to the input vector by the channel

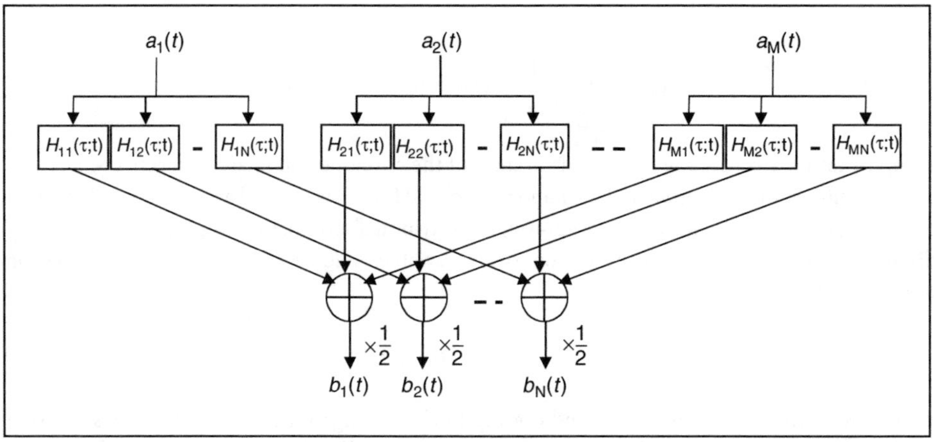

FIGURE 1.3 Matrix MIMO channel model.

transfer matrix H $(\tau; t)$ as

$$\vec{b}(t) = \frac{1}{2} \int_{-\infty}^{\infty} \mathrm{H}(\tau; t)\vec{a}(\tau)d\tau \tag{1.31}$$

where

$$\vec{a}(t) = \begin{bmatrix} a_1(t) \\ a_2(t) \\ \cdot \\ \cdot \\ a_M(t) \end{bmatrix}, \vec{b}(t) = \begin{bmatrix} b_1(t) \\ b_2(t) \\ \cdot \\ \cdot \\ b_N(t) \end{bmatrix}, \text{ and } \mathrm{H}(\tau; t) = \begin{bmatrix} H_{11}(\tau; t) & H_{21}(\tau; t) & \cdot & H_{M1}(\tau; t) \\ H_{12}(\tau; t) & H_{22}(\tau; t) & \cdot & H_{M2}(\tau; t) \\ \cdot & \cdot & \cdot & \cdot \\ \cdot & \cdot & \cdot & \cdot \\ H_{1N}(\tau; t) & H_{2N}(\tau; t) & \cdot & H_{MN}(\tau; t) \end{bmatrix} \tag{1.32}$$

In the above representation, $H_{pq}(\tau; t)$ is the channel impulse response from the pth transmitter antenna to the qth receiver antenna. For the narrowband, time-invariant MIMO channel model, the channel transfer matrix becomes a constant (H) that simplifies Equation 1.31 as

$$\vec{b}\,(t) = \frac{1}{2}\mathrm{H}\vec{a}\,(\tau) \tag{1.33}$$

where

$$\mathrm{H} = \begin{bmatrix} H_{11} & H_{21} & \cdot & H_{M1} \\ H_{12} & H_{22} & \cdot & H_{M2} \\ \cdot & \cdot & \cdot & \cdot \\ \cdot & \cdot & \cdot & \cdot \\ H_{1N} & H_{2N} & \cdot & H_{MN} \end{bmatrix} \tag{1.34}$$

1.2.2.4.5.2 Physical Scattering Model

This model [Oes03] predicts MIMO channel characteristics conforming well to experimental observations in macrocell environments. The methodology considers a predefined power delay profile valid for a specific range, system bandwidth, and antenna beam widths. A distribution of scatterers that characterizes the MIMO channel is then derived to fit the predefined power delay profile. The scattering environment is constituted by the location and scattering coefficient of each scatterer. Geometrical localization of individual antennas and scatterers is represented in an arbitrary two-dimensional coordinate system. The channel matrix is calculated using a ray-based approach, similar to geometrical optics. The proposed model is shown to be valid for any Ricean factor, including the Rayleigh fading case. This MIMO modeling approach accounts for the range dependency on a physical basis.

For more information on MIMO channel models see Chapters 13 to 15 of this book, and for further updates on other channel modeling techniques refer to Part 2 of this book.

1.3 Modulation Techniques

Digital modulation transforms digital symbols into waveforms that are attuned with the characteristics of the channel. In this section we focus on digital modulation techniques that are in use in different communication environments, some of which are being considered for the 3G and 4G mobile communications systems.

1.3.1 Modulation Schemes: The Classification

Different modulation schemes can be classified into two categories: memoryless modulation and memory modulation techniques. When a modulator maps a digital information sequence into an analog counterpart, under the constraint that an analog signal waveform at any time interval depends on one or more previously transmitted waveforms, the resultant modulation is known as the memory modulation technique. On the other hand, when mapping is performed without such constraints, the resultant modulation is known as the memoryless modulation technique. Examples include pulse amplitude modulation (PAM), phase shift keying (PSK) for memoryless modulation, and differential PSK (DPSK) for memory modulation schemes. Digital modulation schemes can also be classified as linear and nonlinear modulation techniques. In a linear modulation scheme, a modulator maps a digital information sequence into an analog counterpart by following the principle of superposition, while in the nonlinear case this principle is not followed. Examples of linear modulation schemes include PAM, PSK, etc., whereas examples of the nonlinear counterpart include continuous-phase modulation (CPM), frequency shift keying (FSK), etc. One special class of modulation technique (discussed in Section 1.3.2.9) also available in this field can use any combination of the above classes in its structure. The specialty of this modulation technique is its multiplexing capability, which can be smartly used in the area of high-data-rate applications.

1.3.2 Different Modulation Schemes

1.3.2.1 Phase Shift Keying

In this type of digital modulation technique the modulating data signals shift the phase of the constant amplitude carrier signal between M number of phase angles. The analytical expression for the mth signal waveform in PSK modulations has the general form

$$s_m(t) = g(t)\cos\left[2\pi f_c t + \theta_m\right], \quad m = 1, 2, \ldots, M \qquad 0 \le t \le T \tag{1.35}$$

where $g(t)$ is the signal pulse shape and $\theta_m = 2\pi(m-1)/M$; $m = 1, 2, \ldots, M$ are the $M(M = 2$ for binary PSK and $M = 4$ for quadrature PSK) possible phase angles of the carrier frequency f_c that convey the transmitted information for $M = 2^k$ possible k-bit (k being a positive integer) blocks or symbols. The mapping of k information bits is preferably done through Gray encoding so that the most likely errors caused by noise will result in single bit error in the k-bit symbol.

In binary phase shift keying (BPSK), the modulating data signals shift the phase of the constant amplitude carrier signal between 0 and 180 degrees, as shown in the state diagram of Figure 1.4a. A more common type of PSK modulation is quadrature phase shift keying (QPSK), where the modulating data signals shift the phase of the constant amplitude carrier signal in increments of 90 degrees, for example, from 45 to 135, −45, or −135 degrees (Figure 1.4b). QPSK ($2^2 = 4$ states) is a more spectral-efficient type of modulation than BPSK ($2^1 = 2$ states). For greater spectral efficiency in the MPSK system, we can increase the value of $M(2^x = M, x$ is an integer $> 0)$ to a higher number, but in this case we need more signal power (Figure 1.5)

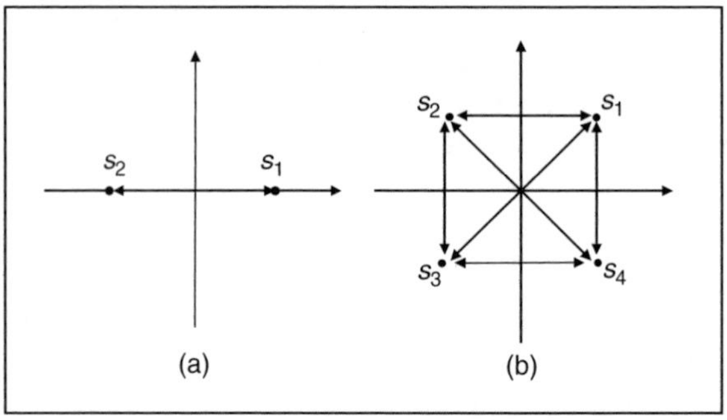

FIGURE 1.4 Phase shift keying state diagrams: (a) BSPK, (b) QPSK.

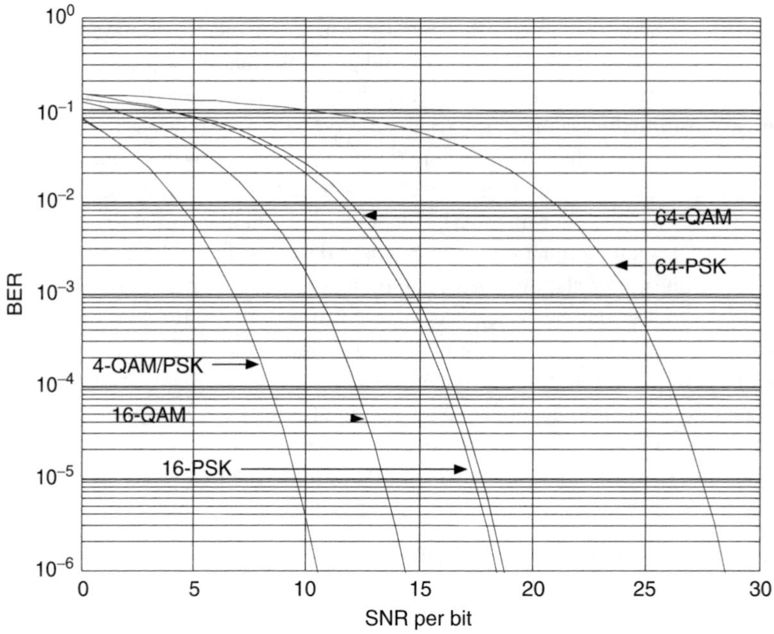

FIGURE 1.5 BER comparison between MQAM and MPSK techniques in the AWGN channel with optimum detection.

to achieve the same bit error rate performance for the MPSK system with smaller M. In other words, we gain spectral efficiency[2] at the cost of power efficiency[3] with higher-level (M) PSK. For an additive white Gaussian noise (AWGN) channel, the symbol error rate (SER) P_e for the MPSK system, using optimum detection technique, can be approximated [Pro95] for a high signal-to-noise ratio (SNR) as

$$P_e = 2Q\left(\sqrt{2\gamma_s}\sin\frac{\pi}{M}\right) \tag{1.36}$$

where γ_s is the SNR per symbol, $Q(.)$ is the Q function, and M is the level of PSK schemes.

[2]Spectral efficiency demonstrates the ability of a system (modulation scheme) to accommodate data within an allocated bandwidth.

[3]Power efficiency represents the ability of a system to reliably transmit information at the lowest practical power level.

There are many variations in the PSK modulation format that are in use because of better power and spectral efficiency requirements. Offset QPSK (OQPSK), differential QPSK (DQPSK), and $\pi/4$ DQPSK are a few examples of these PSK modulation formats. In OQPSK, the in-phase and quadrature bit streams are offset in their relative alignment by one bit period. As a result, the signal trajectories are modified in such a way that the carrier amplitude does not go through or near zero (the center of the constellation). In this case the spectral efficiency of a OQPSK-based system remains the same as that in a QPSK-based system, but the reduced amplitude variations for the former one allow a more power efficient, less linear radio frequency (RF) power amplifier to be used. For DQPSK modulation, the information is carried by the transition between states. In some cases there are also restrictions on allowable transitions. For example, in $\pi/4$ DQPSK modulation, the carrier trajectory does not go through the origin [Bur01]. The $\pi/4$ DQPSK modulation format uses two QPSK constellations offset by 45 degrees ($\pi/4$ radians). Like OQPSK, $\pi/4$ DQPSK is a power efficient modulation method, and with root cosine filtering it has better spectral efficiency than Gaussian minimum shift keying (GMSK) [Bur01] modulation.

BPSK and QPSK modulation techniques are used mostly for satellite links because of their simplified form, reasonable power and spectral efficiencies, and immunity to noise and interference. Examples include the Iridium (a voice/data satellite system) and Digital Video Broadcasting Satellite (DVB-S) systems. Besides, in both IS95 and CDMA2000[4] (also known as 3G IS-2000) cellular systems, BPSK/QPSK and OQPSK modulation techniques are used in the forward and reverse links, respectively. Eight PSK finds its application in enhanced data rate for GSM evolution (EDGE) cellular technology. $\pi/4$ DQPSK modulation is used for IS54 [North American Digital Cellular (NADC) system] and cordless personal communications services in North America, for pacific digital cellular (PDC) services [Rap96] in Japan, and for Trans European Trunked Radio (TETRA) systems in Europe. In a 3G cellular data-only system (IS856, also known as cdma2000 1xEV-DO), BPSK modulation is used in the reverse link, while QPSK and eight PSK modulations along with the quadrature amplitude modulation (QAM) technique (discussed later) are used in the forward link to support multirate data applications. In the next-generation mobile systems, researchers are still focusing on different PSK modulations as major modulation techniques. Certainly, in addition to this, coding and orthogonal frequency division multiplexing (OFDM) techniques are also considered.

1.3.2.2 Pulse Amplitude Modulation

In this type of digital modulation technique the modulating data signals shift the amplitude of the constant-phase carrier signal between M number of discrete levels. PAM is also known as amplitude shift keying (ASK) modulation. The analytical expression for the mth signal waveform in the PAM technique can be expressed in a general form as

$$s_m(t) = A_m g(t) \cos[2\pi f_c t + \theta], \quad m = 1, 2, \ldots, M \qquad 0 \le t \le T \tag{1.37}$$

where $g(t)$ is the signal pulse shape and $A_m = (2m - 1 + M)d$; $m = 1, 2, \ldots, M$ are the M possible amplitude levels of the constant-phase (θ) carrier frequency f_c that convey the transmitted information for $M = 2^k$ possible k-bit (k being a positive integer) blocks or symbols. The parameter d is related to the distance between the adjacent signal amplitudes, which is $2d$. As in the case of PSK, Gray encoding is also preferred here for mapping the k information bits into M different amplitudes. The PAM technique finds its application when it is combined with the PSK modulation technique, as shown later.

1.3.2.3 Quadrature Amplitude Modulation

QAM is simply a combination of the PAM and PSK modulation techniques. In this scheme, two orthogonal carrier frequencies (in-phase and quadrature carriers), occupying identical frequency bands, are used to transmit data over a given physical channel. The analytical expression for the mth signal waveform in the

[4] http://www.qualcomm.com/cdma/3g.html.

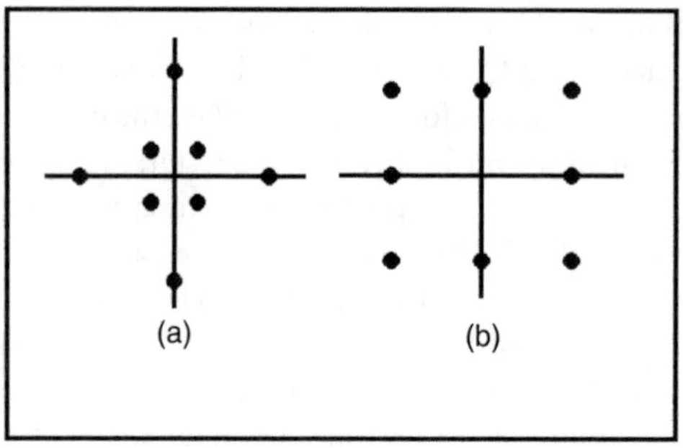

FIGURE 1.6 8-QAM Constellations: (a) 4-4; (b) rectangular.

QAM technique can be expressed in a general form as

$$s_m(t) = A_m g(t) \cos[2\pi f_c t + \theta_m], \quad m = 1, 2, \ldots, M \qquad 0 \le t \le T \tag{1.38}$$

where $A_m = \sqrt{A_{mc}^2 + A_{ms}^2}$ and $\theta_m = \tan^{-1}(A_{ms}/A_{mc}), g(t)$ is the signal pulse shape, f_c is the carrier frequency that conveys the transmitted signal information, and A_{mc} and A_{ms} are the information-bearing signal amplitudes of the quadrature carriers. In Equation 1.38, for constant θ_m, $s_m(t)$ represents the PAM signal, while for constant A_m it represents PSK signal.

By choosing the different amplitudes and phases, different constellations of QAM signals can be formed (Figure 1.6). In this case the power efficiency of the communication system will vary depending on the type of signal constellation [Gil92] used for the QAM technique. Due to the flexibility of using different amplitudes and phases, even with high-level (M > 4) QAM, the choice of decision region in QAM is not as critical as in PSK. As a result, M-QAM-based systems are more power efficient than M-PSK-based systems [Pro95]. Furthermore, the penalty in SNR for increasing M is much less in M-QAM- than in M-PSK-based systems (Figure 1.5). In an AWGN channel with optimum detection techniques, the SER of an M-QAM system with a rectangular constellation and even $k(M = 2^k)$ value is given by [Pro95]

$$P_e = 1 - \left[1 - 2\left(1 - \frac{1}{\sqrt{M}}\right) Q\left(\sqrt{\frac{3\gamma_{sa}}{M-1}}\right)\right]^2 \tag{1.39}$$

where γ_{sa} is the average SNR per symbol and M and $Q(.)$ are as defined for Equation 1.36.

QAM is used in different applications, such as microwave digital radios, Digital Video Broadcasting Cable (DVB-C) systems, and modems. In 3G cellular data-only systems (IS856), the 16-QAM technique, along with QPSK and 8QPSK modulations, is used in the forward link to support multirate data applications. These days, QAM is getting enormous attention in the field of satellite communications for both its spectral and power efficiencies [Par02].

1.3.2.4 Frequency Shift Keying

In the FSK digital modulation technique the modulating data signals shift the frequency of the constant amplitude carrier signal between M number of discrete values of the frequency components. Here, shifting between the frequency components occurs in such a way so that the phase of the shifted signal advances by an integer multiple of 2π radians, and as a result, the phase appears to be constant in the operation. The analytical expression for the mth signal waveform in FSK modulations can be expressed in a general form as

$$s_m(t) = g(t) \cos[2\pi f_m t + \theta], \quad m = 1, 2, \ldots, M \qquad 0 \le t \le T \tag{1.40}$$

where $g(t)$ is the signal pulse shape and $f_m = f_c + m\Delta f; m = 1, 2, \ldots, M$, are the M possible equal-energy orthogonal frequencies with constant-phase angle θ that convey the transmitted information for $M = 2^k$

possible k-bit (k being a positive integer) blocks or symbols. FSK is used in many applications, including cordless and paging systems.

1.3.2.5 Continuous-Phase FSK

In FSK the switching between different frequencies is carried out between M different oscillator outputs tuned in M different frequencies. When this switching is performed in successive signaling intervals, it results in spectral broadening outside the main frequency band, and consequently, an enormous bandwidth is required to transmit the signal successfully. One of the solutions to this spectral broadening problem is to use a single-frequency component whose frequency changes continuously with the change of the information-bearing signal. In this case, the resulting frequency-modulated signal is phase continuous, and hence it is known as continuous-phase FSK (CPFSK). In CPFSK the phase of the carrier is constrained to be continuous; thus it is a memory modulation technique.

To represent the CPFSK signal analytically, let us express the baseband PAM signal as

$$b_p(t) = \sum_m A_m r(t - mT) \tag{1.41}$$

where $\{A_m\}$ denotes the sequence of amplitudes obtained by mapping k-bit binary digits from the information sequence into amplitude levels $\pm 1, \pm 3, \ldots, \pm(M-1)$, and $r(t)$ is a rectangular pulse of amplitude $1/2T$ and duration T seconds. The signal $b_p(t)$ is used to frequency-modulate the carrier. Consequently, the carrier-modulated signal is expressed as

$$s(t) = g(t)\cos[2\pi f_c t + \beta(t;\tau) + \theta] \tag{1.42}$$

where $g(t)$ is the signal pulse shape, θ is the initial phase of the carrier, and $\beta(t;\tau)$ is the time-varying phase of the carrier, which is defined as

$$\beta(t;\tau) = 4\pi T f_d \int_{-\infty}^{t} b_p(\tau)d\tau \tag{1.43}$$

and solved as

$$\beta(t;\tau) = \psi_m + 2\pi h A_m v(t - mT) \tag{1.44}$$

where the parameters h, ψ_m, and $v(t)$ are defined as

$$h = 2 f_d T \tag{1.45}$$

$$\psi_m = \pi h \sum_{k=-\infty}^{m-1} A_k \tag{1.46}$$

$$v(t) = \begin{cases} 0, (t < 0) \\ t/2T, (0 \le t \le T) \\ 1/2, (t > T) \end{cases} \tag{1.47}$$

The parameter h defined in Equation 1.45 is known as the modulation index, and f_d in Equation 1.43 and Equation 1.45 is known as the peak frequency deviation. In Equation 1.43, even though $b_p(\tau)$ contains discontinuities, the integral of $b_p(\tau)$ is continuous. As a result, we have a continuous-phase signal.

1.3.2.6 Continuous-Phase Modulation

CPM is a generalized form of CPFSK where the carrier phase is given by

$$\beta(t;\tau) = 2\pi \sum_{k=-\infty}^{m} h_k A_k v(t - kT), \quad mT \leq t \leq (m-1)T \tag{1.48}$$

where $\{A_k\}$ is the sequence of M-ary information symbols selected from the alphabet of $\pm 1, \pm 3, \ldots, \pm(M-1)$; $\{h_k\}$ is a sequence of modulation indices; and $v(t)$ is some normalized waveform shape. When the modulation index varies from one symbol to another, the CPM signal is called *multi-h*.

1.3.2.7 Minimum Shift Keying

MSK is a special form of binary CPFSK in which the modulation index $h = 0.5$, which ensures the orthogonality between two signals with minimum frequency separation. In this case, the peak-to-peak frequency deviation is equal to half the bit rate. The analytical expression for this signal is given by

$$s(t) = g(t)\cos\left[2\pi f_c t + \beta(t;\tau)\right] \tag{1.49}$$

where the phase of the carrier is given by

$$\beta(t;\tau) = \psi_m + \frac{\pi}{2} A_m \left(\frac{t - mT}{T}\right), \quad mT \leq t \leq (m-1)T \tag{1.50}$$

The MSK scheme is used in the global systems for mobile communications (GSM) cellular system.

1.3.2.8 Gaussian MSK

GMSK is a derivative of MSK where the required signaling bandwidth is reduced by passing the modulating waveform through a Gaussian filter. The Gaussian filter minimizes the instantaneous frequency variations over time. GMSK is a spectrally efficient modulation scheme and is particularly useful in mobile radio systems. This constant-envelope spectrally efficient modulation technique is a self-synchronizing one and it provides good BER performance.

1.3.2.9 Orthogonal Frequency Division Multiplexing

Orthogonal frequency division multiplexing (OFDM) is a wideband modulation scheme that is specifically designed to cope with the problems of multipath reception. It achieves this by transmitting a large number of narrowband digital signals over a wide bandwidth. The idea of OFDM appeared in the literature in the sixties [Cha66] [Cha68] [Sal67]. In OFDM, the data are divided among a large number of closely spaced orthogonal carriers, which results in high spectral efficiency. In this scheme, only a small amount of data is carried on each carrier, and this significantly reduces the influence of *intersymbol interference* [She95]. Here, the parallel transmission gives the capability of supporting high-bit-rate environments. Because of the orthogonal property among the carriers, the OFDM signal can be arranged in such a way so that the sidebands of the individual carriers overlap (Figure 1.7) and the signals can still be received without adjacent carrier interference.

The OFDM signals can be easily transmitted and received using the fast Fourier transform (FFT) devices [Nee99] [Cim85] without increasing the transmitter and receiver complexities. This technique has some demerit points, too. It has a large peak-to-average power ratio (PAPR), which reduces the power efficiency and increases the cost of the power consumption of the transmitter amplifier. In this case, the operating point in the amplifiers can be backed off, but this leads to inefficient power usage. Moreover, OFDM techniques are susceptible to frequency offset and phase noise. Coding methods have been proposed in Young-Hwan [You03], Fernando [Fer98], and Cimini [Cim99] to reduce the peak-to-average power ratio.

The successful use of the OFDM technique began in the sixties for high-frequency military systems (KINEOLEX, ANDEFT, KATHRYN). In the eighties the OFDM technique found application in high-speed modems, digital mobile communications, and high-density recording. In the nineties the use of

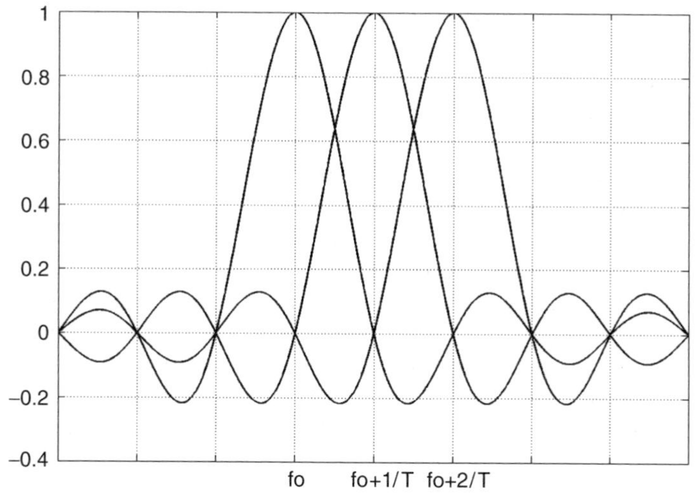

FIGURE 1.7 Spectrum associated with OFDM signal.

the OFDM technique found commercial wired applications in the digital subscriber line (DSL) [Cho91] [Sis93]. During that period, in the wireless area, OFDM became the basis for several television and radio broadcast applications, including European digital audio broadcasting (DAB) and high-definition TV (HDTV) terrestrial broadcasting [ETS97] [ETS97(2)], as well as North American digital radio broadcasting. At the beginning of the 21st century, it was adopted as a standard for new high-rate wireless local area networks (WLANs), such as IEEE 802.11, HIPERLAN II, and the Japanese Multimedia Mobile Access Communications (MMAC) [Ner99]. Currently much research is being conducted to devolve an OFDM-based system [Sam02] [Dow02] to deliver mobile broadband data service at data rates comparable to those of wired services, such as DSL and cable modems. Moreover, OFDM technology is a very attractive candidate when targeting high-quality and highly flexible mobile multimedia communications over satellite systems [Nee99].

1.3.2.10 Challenges in the Next-Generation System Concerning Different Modulation Techniques

Now that the next-generation (4G) scheme visualizes the wireless mobile communication system as a single entity by considering both the satellite and terrestrial domains, some of the challenges related to modulation issues in the satellite domain, especially in the downlink scenario, need to be tackled.

For the 4G mobile communications systems OFDM is considered to be one of the precise techniques [Vau02]. In this case the challenges include all the demerits of the OFDM technique, as discussed earlier. Although much research is being carried out to overcome these demerits, the main challenge is yet to be fulfilled. This challenge comes from the requirement of a single standard OFDM version. Currently many OFDM versions, such as vector OFDM, wideband OFDM, F-OFDM, and MIMO OFMD, are present in the application area [Vau02]. In this case OFDM needs standardization to enable its widespread use, encourage adoption, and thereby grow the market for the 4G mobile communications systems.

As shown in Costa [Cos02], Park [Par02], and Rafie [Raf89], among the spectrally efficient modulation schemes, M-QAM offers the best trade-off between implementation complexity and performance in the nonlinear channels. Consequently, for the satellite channel where both spectral and power efficiencies are the prime requirements, QAM becomes one of the strong candidates. Here one of the challenges is to come up with an optimal constellation for the QAM technique in terms of both bit-error-rate performance and complexity. Moreover, for multimedia applications where the bit rate needs to be in the gigabit range, the integration of the OFDM technique with QAM will be another challenge. In both cases, the inclusion of a fading channel scenario will be an added obstacle for the complete setting. It is worth mentioning that in the terrestrial domain the combination of OFDM and QAM is already in the application area for the IEEE 802.11 standard WLAN.

As shown in Figure 1.5, as we increase M for the M-level modulation technique, the power efficiency decreases. These results suggest the use of the adaptive modulation technique, which would be another interesting area to explore. Some of the aforementioned challenges are already being addressed in the research phase [Bou02] [Rah03] [Yua03] [Ala03] [Alo00] [Gol97].

At a high data rate, carrier acquisition and tracking of the incoming signal, which make coherent detection possible in the receiver, are extremely difficult. As a solution to this problem, a differential modulation technique (such as DQPSK) can be used with differential detection. However, it suffers from performance degradation when compared to ideal coherent detection. For a power-limited system, such as a satellite with an onboard power amplifier, this degradation cannot be tolerated. Multiple-symbol differential detection [Div90] can be used to avoid this degradation by slightly increasing the length of the observation interval. All these studies need to be extended to the fading channel scenario. Part 3 of this book provides some analytical results for some of the issues discussed in this section.

1.4 Coding Techniques

A coding technique in general is of two types: source coding and channel coding. The source coding technique refers to the encoding procedure of the source information signal into digital form. On the other hand, channel coding is applied to ensure adequate transmission quality of the signals. Channel coding is a systematic approach for the replacement of the original information symbol sequence by a sequence of code symbols in such a way as to permit its reconstruction. Here we will focus on the channel encoding technique.

Channel coding can improve the severe transmission conditions in terrestrial mobile radio communications due to multipath fading. Moreover, it can help to overcome very low SNRs for satellite communications due to limited transmit power in the downlink. The encoding process generally involves mapping every k-bit information sequence into a unique n-bit sequence, where the latter is called a code word. The amount of redundancy introduced by the encoding process is measured by the ratio k/n, whose reciprocal is known as the code rate. The output of the channel encoder is fed to the modulator, whose output is transmitted through the channel. At the receiver end, demodulation, decoding, and detection processes are carried out to decide on the transmitted signal information. In the decision process, two different strategies are used, soft decision and hard decision. When the demodulator output consists of discrete elements 0 and 1, the demodulator is said to make a hard decision. On the other hand, when the demodulator output consists of a continuous alphabet or its quantized approximation (with greater than two quantization levels), the demodulator is said to make a soft decision.

It is theoretically shown by Shannon [Sha48] that the coding technique in general improves the BER performance of a communications system. Before discussing different channel coding techniques we will take a look at this theory.

1.4.1 Shannon's Capacity Theorem

Shannon [Sha48] shows that the system capacity C bits/second of an AWGN channel is a function of the average received signal power, S, the average noise power, N, and the bandwidth, W hertz, which is given by

$$C = W \log_2\left(1 + \frac{S}{N}\right) \tag{1.51}$$

Theoretically, it is possible to transmit information with arbitrarily small BERs over such a channel at any rate, R, where $R \leq C$, using a complex coding scheme. For an information rate of $R > C$ it is not possible to find a code that can achieve an arbitrarily small error probability. Shannon's works show that the values of S, N, and W set a limit on the transmission rate, not on the error probability. In Equation 1.51, the detected noise power, N, is proportional to the bandwidth, which is given by

$$N = N_0 W \tag{1.52}$$

where N_0 is the noise power spectral density. Substituting Equation 1.52 into Equation 1.51, we get

$$\frac{C}{W} = \log_2\left(1 + \frac{S}{N_0 W}\right) \tag{1.53}$$

Now, assuming $R = C$, the ratio between the binary signal energy (E_b) and the noise spectral density (N_0) can be written as:

$$\frac{E_b}{N_0} = \frac{ST}{N_0} = \frac{S}{N_0 R} = \frac{S}{N_0 C} \tag{1.54}$$

which can be used in Equation 1.53 to yield

$$\frac{C}{W} = \log_2\left[1 + \frac{E_b}{N_0}\left(\frac{C}{W}\right)\right] \tag{1.55}$$

From Equation 1.55 it can be shown that there exists a limiting value of E_b/N_0 below which there can be no error-free communication at any information rate. This limiting value is known as *Shannon's limit*, which can be calculated by letting

$$x = \frac{E_b}{N_0}\left(\frac{C}{W}\right) \tag{1.56}$$

Substituting the above parameter in Equation 1.55 we get

$$\frac{C}{W} = \log_2[1 + x] \tag{1.57}$$

Equation 1.57 can be rewritten as

$$\frac{C}{W} = x \log_2[1 + x]^{1/x} = \frac{E_b}{N_0}\left(\frac{C}{W}\right)\log_2[1 + x]^{1/x} \tag{1.58}$$

Now, using the identity $\lim_{x \to 0}(1 + x)^{1/x} = e$ in the limit $C/W \to 0$, Equation 1.58 becomes

$$\frac{E_b}{N_{0\,Shannon}} = \frac{1}{\log_2 e} = 0.693 \tag{1.59}$$

In decibel (dB) scale it is

$$\frac{E_b}{N_{0\,Shannon\_dB}} = -1.59\ dB \tag{1.60}$$

Equations 1.59 and 1.60 give the numerical values of the Shannon's limit.

1.4.2 Different Coding Schemes

Channel coding can be classified into two major areas: waveform coding and structured sequences. The objective of waveform coding is to provide an improved waveform set so that the detection process is less subject to errors. Examples of this coding technique include M-ary signaling, antipodal, orthogonal, bi-orthogonal, and trans-orthogonal signaling. Structured sequences deal with transforming data sequences into better sequences having ordered redundancy in bits. The redundant bits can then be used for the detection and correction of errors. Examples of structured sequence coding include block and convolutional coding schemes. In this section we will mainly focus on the structured sequence type coding schemes.

Here the discussion will consider only those coding techniques that are being addressed in the recent trend of mobile and satellite communications systems. These schemes include linear block codes (e.g., Hamming codes, BCH codes, Reed–Solomon codes, etc.) and convolutional codes. Space–time codes and turbo codes will also be considered; they are getting enormous attention in the current developments of both 3G and 4G telecommunications systems. To provide improvement in power efficiency without sacrificing bandwidth efficiency, coded modulation techniques (e.g., trellis coded modulation (TCM) [Ung87]) that combine coding and modulation techniques are good choices. This section talks briefly about coded modulation techniques too.

1.4.2.1 Block Codes

In this coding scheme each k-bit information symbol block is converted to an n-bit coded symbol block with $(n-k)$ redundancy bits added to the k-bit symbols. These redundancy bits could be parity bits or check bits that do not carry any information. The resulting code is referred to as the (n,k) block code. Here the redundancy of the code is defined as the ratio between the redundant bits and k-bit symbol, i.e., $(n-k)/k$, while the code rate is defined as k/n. In block codes $2^k k$-bit message sequences are uniquely mapped into 2^k n-bit codes, out of a possible 2^n n-bit codes.

1.4.2.1.1 Vector Space and Subspace

A vector space V_n is defined as the set that contains all possible n-bit block codes. On the other hand, a subset S of the vector space V_n is called a subspace if an all-zero vector is in S and the sum (modulo-2) of any two vectors in S is also in S (closure property). The subspace properties are the basis for the algebraic characterization of linear block codes.

1.4.2.1.2 Linear Block Code

The block codes, where each of the code words can be formed by the modulo-2 sum (EX-OR) of two or more other code words, are called linear block codes. The code words are said to be linearly dependent on each other.

1.4.2.1.3 Coding Gain

Coding gain is defined as the improvement in the SNR in decibels at a specified bit-error-rate performance of an error-correcting coded system over an uncoded one with an identical system scenario.

1.4.2.1.4 Hamming Codes

This is a linear block code characterized by the following (n,k) structure:

$$(n,k) = (2^m - 1, 2^m - 1 - m) \tag{1.61}$$

where $m = 2, 3, \ldots$. These codes are capable of correcting all single bit errors and detecting all combinations of two or fewer bits in error, which can be seen from the following example. A rate 4/7, i.e., $(n,k) = (7,4)$, hamming code is shown in Table 1.2 where each output data block differs from all the other blocks by at least three bits. Hence, if a one- or two-bit error occurs in the transmission of a block, the decoder will detect that error. In the case of a single bit error, it is also possible for the receiver to match the received block to the closest valid block and thereby correct the single bit error. If a three-bit error occurs, the original block may be transformed into a new valid block and all the errors go undetected.

1.4.2.1.5 Hamming Distance

The difference in the number of bits between two coded blocks is known as the Hamming distance. A block code of a Hamming distance d can detect up to $(d-1)$ errors and correct $(d-1)/2$ errors.

1.4.2.1.6 Implementation Complexity

An increase in the coded block length results in two drawbacks in the block coding techniques, which are stated below.

TABLE 1.2 Hamming Code of Rate 4/7

Block Number	Input Data Block	Output Data Block
0	0000	000 0000
1	1000	110 1000
2	0100	011 0100
3	1100	101 1100
4	0010	111 0010
5	1010	001 1010
6	0110	100 0110
7	1110	010 1110
8	0001	101 0001
9	1001	011 1001
10	0101	110 0101
11	1101	000 1101
12	0011	010 0011
13	1011	100 1011
14	0111	001 0111
15	1111	111 1111

- *Transmission delay*: The time taken to collect k-bits to form a block increases with increasing block length, introducing delay in the transmission process, which may be unacceptable for real-time applications such as voice transmission.
- *Decoder complexity*: This increases almost exponentially with block length as the decoder searches through 2^k valid code words to find the best match with the incoming 2^n possible coded blocks. In addition to the complexity, the decoding delay can be significant.

1.4.2.1.7 BCH (Bose–Chaudhuri–Hocquenghem) Codes

These codes are generalizations of Hamming codes that allow multiple error corrections. BCH codes [Ste64] are important because at a block length of a few hundred, these codes outperform all other block codes with the same block length and code rate. For very high coding overhead with long block length, this coding scheme can be used where reliability of transmission is the key factor and data throughput is less important.

1.4.2.1.8 Reed–Solomon Codes

Reed–Solomon (RS) codes are a subclass of BCH codes that operate at the block level rather than the bit level. Here the incoming data stream is first packaged into small blocks, and these blocks are then treated as a new set of k symbols to be packaged into a supercoded block of n symbols. As a result, the decoder is able to detect and correct complete error blocks. This is a nonbinary code set that can achieve the largest possible code minimum distance for any linear code with the same encoder input and output block lengths. For nonbinary codes, the distance between two code words is defined as the number of nonbinary symbols in which the sequences differ. The code minimum distance for the RS codes is given by [Fal68]

$$d_{\text{min}} = n - k + 1 \tag{1.62}$$

These codes are capable of correcting any combination of $(n - k)/2$ or fewer symbol errors. RS codes are particularly useful for burst type error corrections, and so they are very effective with the channel with memory. They are also used in error-correcting mechanisms in CD players.

1.4.2.1.9 Interleaving

The block codes work best when errors are distributed evenly and randomly between incoming blocks. This is usually the case for AWGN channels such as landline telephone link. In a mobile radio environment, however, errors often occur in bursts as the received signal fades in and out due to the multipath propagation and the user's motion. In order to distribute these errors more evenly between coded blocks, a process

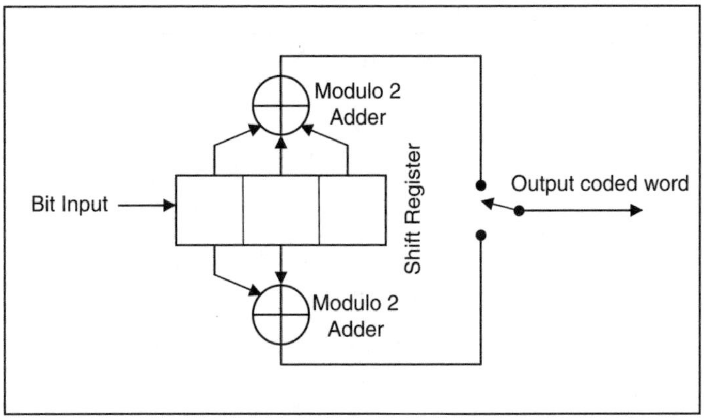

FIGURE 1.8 Rate _ convolutional encoder with constraint length of 3.

known as interleaving is used. In general, to accomplish interleaving, the encoded data blocks are read as rows into a matrix. Once the matrix is full, the data can be read out in columns, redistributing the data for transmission. At the receiver, a de-interleaving process is performed using a similar matrix filling and emptying process, reconstructing the original blocks. At the same time, the burst errors are uniformly redistributed across the blocks. The number of rows or columns in the matrix are sometimes referred to as the *interleaving depth*. The greater the interleaving depth, the greater resistance to long fades, but also the greater the latency in the decoding process as both the transmitter and receiver matrices must be full before encoding or decoding can occur.

1.4.2.2 Convolutional Codes

A convolutional code is implemented on a bit-by-bit basis from the incoming data source stream. The encoder has memory and it executes an algorithm using a predefined number of the most recent bits to yield a new coded output sequence. Convolutional codes are linear, where each branch word of the output sequence is a function of the input bits and $(k-1)$ prior bits. Since the encoding procedure is similar to the convolution operation, the coding technique is known as convolutional coding. The decoding process is usually a serial process based on present and previous received data bits (or symbols). Figure 1.8 shows an $(n, k) = (2, 1)$ convolutional encoder with a constraint length of $C_{ln} = 3$, which is the length of the shift register.

There are $n = 2$ modulo-2 adders that result in a two-bit coded word for each input bit upon EX-OR operation. The output switch samples the output of each modulo-2 adder, thus forming the two-bit code symbol associated with the single input bit. The sampling is repeated for each input bit that results in a two-bit code word. The choice of the connections between the adders and the stages of the register gives rise to the characteristics of the code. The challenge in this case is to find an optimal connection pattern that can provide codes with best distance properties. Convolutional codes have no particular block size; nonetheless, these are often forced into a block structure by periodic truncation. This requires a number of zero bits to be added at the end of the input data sequence for clearing out the data bits from the encoding shift register. Since the added zeros carry no information, the effective code rate falls below k/n. The truncation period is generally made as long as practical, to keep the code rate close to k/n.

Both the encoder and decoder can be implemented using recursive techniques, with one of the most efficient and well known being the Viterbi convolutional decoder [Bur01].

1.4.2.2.1 Pictorial Representation of Convolutional Encoder

A convolutional encoder can be represented pictorially in three different ways: state diagram, tree diagram, and trellis diagram.

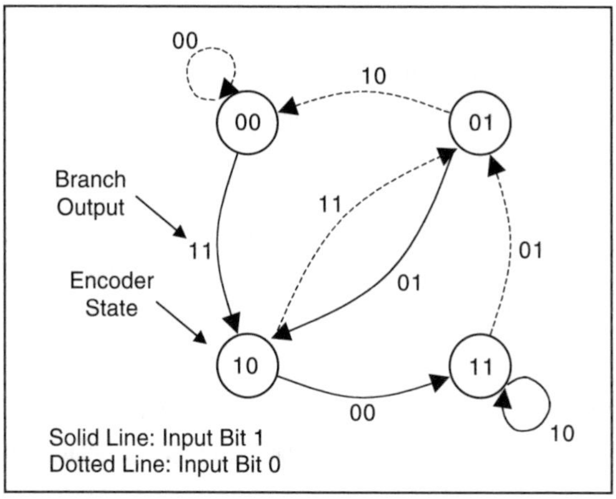

FIGURE 1.9 State diagram for the rate _ convolutional encoder with constraint length of 3 as shown in Figure 1.8.

1.4.2.2.1.1 State Diagram

In this case, the encoder is characterized by some finite number of states. The state of a rate $1/n$ convolutional encoder is defined as the contents of the rightmost $K - 1$ stages (Figure 1.8) of the shift register. The necessary and sufficient condition to determine the next output of a convolutional encoder is to have the knowledge of the current state and the next input. The state diagram for the encoder shown in Figure 1.8 can easily be drawn as shown in Figure 1.9. The states shown in the boxes of the diagram represent the possible contents of the rightmost $K - 1$ stages of the register, and the paths between the states represent the output branch words resulting from such state transitions. Table 1.3 will help to understand the state transition mechanism in Figure 1.9. Major characteristics of the state diagram follow:

- $2^{k.(K-1)}$ states
- 2^k branches entering each state while the same number of branches leave each state

1.4.2.2.1.2 Tree Diagram

The tree diagram in the convolutional encoder incorporates the time dimension in the state transition, which is not provided by the state diagram. Here the possible code sequences generated by an encoder are represented as branches of a tree. With the aid of time dimension one can easily describe the encoder as a function of a particular input sequence. For examples of tree diagrams, see Proakis [Pro95], Lee [Lee97], and Sklar [Skl88]. Since the number of branches in the tree increases as a function of 2^S, S being the sequence length, for a very long sequence this representation is not feasible. Characteristics of the tree

TABLE 1.3 State Transition Mechanism for the State Diagram

Input Bit	Register Content	Present State (Content of the Rightmost K-1 Stages)	Next State (Content of the Leftmost K-1 Stages)	Branch Output at Present State
0	000	00	00	00
1	100	00	10	11
1	110	10	11	00
1	111	11	11	10
0	011	11	01	01
0	001	01	00	10
0	010	10	01	11
1	101	01	10	01

diagram include:

- 2^k branches emanating from each node
- The whole tree repeating itself after the K*th* stage

1.4.2.2.1.3 Trellis Diagram

The trellis diagram is an intelligent pictorial representation of the tree diagram where the repetitive nature of the tree diagram is smartly utilized [Pro95] [Lee97] [Skl88]. It is called a trellis diagram because it looks like a garden trellis. Major characteristics of the trellis diagram include:

- $2^{k.(K-1)}$ states
- 2^k branches entering each state while the same number of branches leave each state

For examples of trellis diagrams, see Proakis [Pro95], Lee [Lee97], and Sklar [Skl88].

1.4.2.3 Space–Time Coding

This is basically a spatial type diversity technique (discussed in Section 1.6) where multiple antennas in the transmitter end are used with either one or more receiving antennas. This technique is known as space–time coding since it involves redundancy by transmitting the same signal using different antennas. With multiple antennas at the transmitter end, when the receiver end also uses multiple antennas, the system is known as a multiple-input multiple-output (MIMO) system. Recently, space–time coding has been getting vast recognition [Dou02] as a robust coding technique in the field of research due to its application in the 3G scenario.

1.4.2.4 Turbo Coding

Turbo coding (TC) is a specific decoding technique that was developed from two older concepts, concatenated coding and iterative decoding. These codes are built from parallel concatenation of two recursive systematic block [Bur01] or convolutional codes with nonuniform interleaving. The term *Turbo* is used to draw an analogy of this decoding process with a turbo engine in which a part of the output energy is fed back to the input to carry out its operation. Before discussing the principle of the TC technique, we will look at the concept of concatenated coding. In the concatenated coding method, two or more relatively simple codes are combined to provide much more powerful coding. In its operation, as shown in the block diagram of Figure 1.10, the output of the first encoder (outermost) is fed to the input of the second, and so on. In the decoder, the last (or innermost) code is decoded first, and then its output is fed to the next, and so on to the outermost decoder.

The principle of the decoding process of the TC technique can be explained briefly with the aid of the block diagram shown in Figure 1.11 and in terms of code array. In this case, the decoder first performs row decoding, which generates initial estimates of the data in the array. Here, for each data bit, a tentative decision and a reliability estimate for that decision are provided. The columns in the code array are then

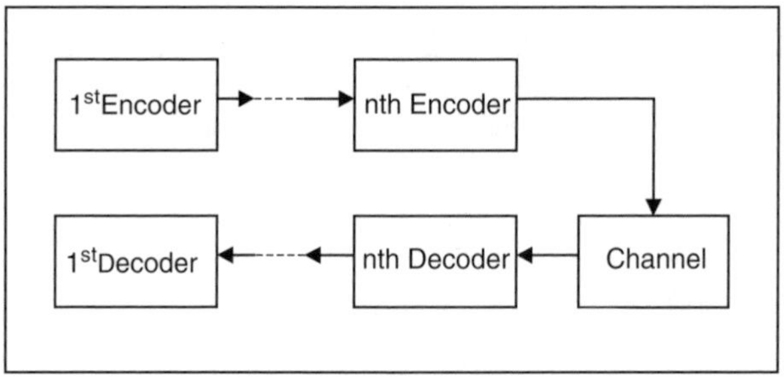

FIGURE 1.10 Concatenated coding method.

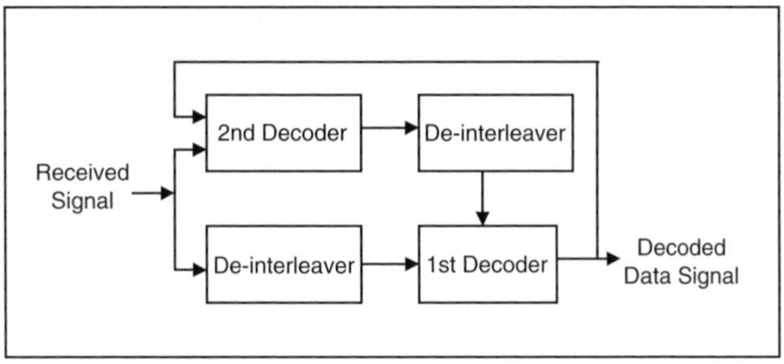

FIGURE 1.11 Turbo decoder.

decoded by taking both the original input and the previous decoder signals into consideration. In the current decoder the previous decoded signal information is known as *a priori* information on the data. This second decoding further refines the data decision and its reliability estimate. The output of this second decoding stage is fed back to the input of the first decoder. In this case, the information that was missed in the first row decoding is now decoded. The whole procedure continues until the data estimates are converged.

1.4.2.5 Coded Modulation Techniques

In both block and convolutional coding schemes, the coding gain is achieved with the price paid for the bandwidth. Since in these schemes the k-bit information signal is replaced by n-bit coded words ($n > k$), the required bandwidth gets increased, which is a major bottleneck for the band-limited channels, such as telephone channels. To overcome this problem, combined modulation and coding schemes are considered. In this case, the coding gain is achieved with the price paid for the decoder complexity. Here different coded modulation techniques are briefly addressed.

1.4.2.5.1 Trellis Coded Modulation

TCM is based on the trellis, as used in convolutional coding. In TCM, the trellis branches, instead of being labeled with binary code sequences, are represented as constellation points from the signaling constellation.

1.4.2.5.2 Block Coded Modulation

In block coded modulation (BCM) the incoming data are divided into different levels, and in each level those data streams are block coded at an equal rate.

1.4.2.5.3 Multilevel Coded Modulation

In multilevel coded modulation (MCM), which is a generalized form of BCM, the incoming data are split in different levels/branches (serial to parallel), and each of these data levels are block coded or convolutionally coded at either equal or unequal rates. Finally, the multiplexed signal results in the MCM.

1.4.2.5.4 Turbo Coded Modulation

Based on the combinations of TC and either TCM or MCM, there are many versions of turbo coded modulation techniques available in the research area, for example, turbo trellis coded modulation (T-TCM) [Gof94] [Rob98], multilevel turbo coded modulation (ML-TCM) [Wac95], etc.

1.4.3 Coding in Next-Generation Mobile Communications: Some Research Evidence and Challenges

In this section we will mention some of the current research results on coding techniques, which are getting attention in next-generation mobile communication systems, taking into account both the terrestrial and satellite domains.

The authors in Doufexi [Dou02] have utilized measured MIMO channel data to evaluate the performance of the proposed 4G space–time coded OFDM (COFDM) system. In the simulation results the authors assumed that the channel responses are constant during the period of two COFDM symbols. BERs for the half-rate convolutional coded QPSK have been presented for 2-Tx (two transmit antennas) 1-Rx (one receive antenna) and 2-Tx 2-Rx scenarios. Polarization diversity is also considered in the analysis. The results indicated that high gains can be obtained for 2-Tx 2-Rx architecture with channel correlation coefficients in the order of 0.3 to 0.5.

Turbo coded adaptive modulation and channel coding (AMC) has been examined in Classon [Cla02] for future 4G mobile systems to observe the throughput gain. Here a method of generating soft information for higher-order modulations, based on the reuse of the turbo decoding circuitry, is provided. It is shown that 3G style turbo coding can provide a 0.5- to 4-dB-link gain over 256-state convolutional codes, depending on the frame size, modulation, and channel. Here the link gains from channel coding do not directly translate into throughput gain for AMC, but they are still expected to improve throughput significantly.

In Doufexi [Dou02] space–time coded OFDM for the 4G cellular network has been proposed where the individual carriers of the OFDM techniques are modulated using BPSK, QPSK, and 16-QAM with coherent detection. The channel encoder consists of a half-rate convolutional encoder. Here the channel model considers a wide range of possible delay spreads. The results show that the space–time block codes provide diversity gain and enhance the BER performance.

In Saifuddin [Sai97], to avoid a high degree of complexity in Viterbi decoding, the authors use concatenated codes based on MCM. In this case, the outer RS code is concatenated with an MCM for high-data-rate application over satellite channels. The results show a significant coding gain in terms of BER with considerably less complexity.

Block turbo codes (BTCs) with trellis-based decoding are proposed in Vilaipornsawai [Vil02] for asynchronous transfer mode (ATM) transmission in digital video-broadcasting–return channel via satellite. In Sumanasena [Sum01] an adaptive coding and modulating transmission scheme for 3G mobile satellite systems is proposed. Here the adaptation mechanism is based on the Rice factor of the channel, which is estimated in real time using an estimation algorithm at the receiver. The transmitter, upon receiving the channel information from the receiver, determines the optimal coding and modulation scheme using a lookup table. For the coding scheme the authors use convolutional coding of rates 1/2 and 1/3, while for the modulation scheme QPSK and 8PSK modulation formats are used. The simulation results in the satellite UMTS (universal mobile transmission systems) environment show that the dynamic range of the transmission power is greatly reduced, which in turn eases the power control requirements.

Besides all the above research outputs, many investigations are in progress considering coding for both the satellite and terrestrial areas. Instead of discussing all these investigation approaches and results, we will look at some challenges laid forth in this area.

Now that the 3G system is already in use somewhat successfully in the terrestrial domain, attention in coding challenges is currently focused mainly on the satellite domain for the 4G system. In designing 4G mobile satellite systems, transmitted power is a critical concern. In this case, because of the limited satellite onboard power and the limited life span of the mobile terminal battery, the main challenge is to come up with a power-efficient coding technique. The adaptive coded modulation (ACM) technique in Sumanasena [Sum01] has already been shown to be a smart solution to this challenge. But in this case, power efficiency can be achieved at the expense of spectral efficiency. To explore more on the ACM technique, visit Chapter 6 of this book.

The use of the coded QAM technique with adaptation between different QAM constellations could be a good choice to gain both power and spectral efficiencies. It is difficult to say whether, by taking into account the huge complexity involved in this process, we can still meet our target. This doubt remains strong due to the channel, which plays an important role in the complexity issue. The use of the adaptive OFDM technique with coding can also be explored in this situation upon successfully addressing the demerits of the OFDM method discussed earlier.

1.5 Multiple Access Techniques

In general, a multiple-access (MAC) scheme offers many users in the communications system the capability to share the same spectrum resource. Different MAC schemes are either in use or in the research domain in both the terrestrial and satellite areas for providing capacity improvement in the system without significantly disturbing the system's performance. Currently, wideband code division multiple access (W-CDMA) and OFDM/time division multiple access (OFDM/TDMA) techniques are successfully in use in terrestrial mobile multimedia systems. These two MAC schemes are also getting considerable attention [Pap01] in mobile multimedia communications for nongeostationary satellite interfaces. Along with these multiple-access techniques, in this section we will briefly discuss different MAC schemes with their merits and demerits. The discussion will also cover the combination of OFDM and CDMA techniques. The capacity of the fundamental MAC schemes in the AWGN channel will also be addressed.

1.5.1 Fundamental Multiple-Access Schemes

1.5.1.1 Frequency Division Multiple Access

This is a method of combining multiple users on a given channel bandwidth using unique frequency segments. Frequency division multiple access (FDMA), by nature, is a narrowband MAC system. Here, an available frequency band (which is generally wide) is split into some smaller nonoverlapping orthogonal bands (or channels), and different information signals from different users are transmitted through these channels (Figure 1.12). In this case, each transmitter or receiver for each user uses a separate frequency band (channel) for communications. Application: Advanced mobile phone systems (AMPS).

1.5.1.1.1 Merits

In this narrowband system, the symbol duration is large compared to the average delay spread that results in low intersymbol interference (ISI). Since it is a continuous transmission scheme, system overhead in terms of bits is less than in the TDMA scheme. It is also less complex compared to the other MAC schemes.

1.5.1.1.2 Demerits

There are some cost-related demerits for the FDMA system, which vary from system to system. For example, this system has higher cell-site system costs than TDMA systems. This system is also performance limited by the nonlinear effects of the power amplifier and the stability of the system clock that generates different frequencies of interest.

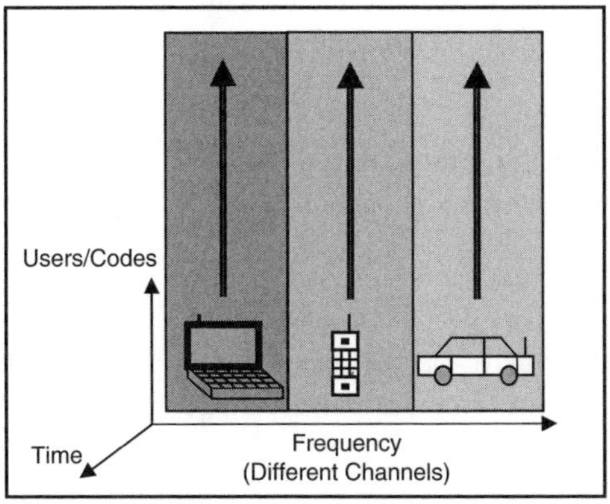

FIGURE 1.12 FDMA scheme.

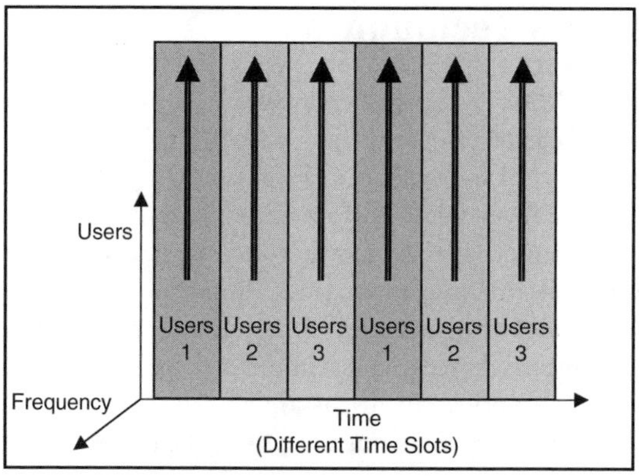

FIGURE 1.13 TDMA scheme.

1.5.1.2 Time Division Multiple Access

In this method, multiple users, using unique time segments on a given channel bandwidth, are combined. In this case, a single carrier frequency is shared between different transmitters (users), each of which is assigned a nonoverlapping time slot (Figure 1.13). Application: Global systems for mobile communications (GSM).

1.5.1.2.1 Merits

Since the data transmission occurs in burst, the transmitter or receiver can be turned off when it is not in use, resulting in low battery consumption. Due to the discontinuous nature of transmission, handoff process is simpler in the TDMA system.

1.5.1.2.2 Demerits

High synchronization overhead is required for TDMA-based systems. The system performance is limited by the stability of the digital clock that generates different time slots of interest.

1.5.1.3 Code Division Multiple Access

This is a method of combining multiple users on a given channel bandwidth using unique spreading codes or hopping patterns to distinguish any given user. In CDMA systems, several transmitters (users) simultaneously and asynchronously access a channel by modulating and spreading their narrowband information-bearing signals with preassigned wideband spreading codes. This spreading code makes it possible for the system to multiplex several users in the same time and frequency domain (Figure 1.14). Chapter 10 of this book provides more information on the CDMA technique.

1.5.1.3.1 Wideband CDMA

W-CDMA follows the same principle as the CDMA technique [Vit95]. It gets its name from its wide bandwidth (e.g., 5 MHz) requirement. Readers are cautioned here to not get this mixed up with ultrawideband[5] (UWB) technology.

1.5.1.3.2 Merits

Multipath fading in the CDMA (or W-CDMA)-based system can be substantially reduced because the signal is spread over a large spectrum. If the spread-spectrum bandwidth (BW) is greater than the coherence

[5]UWB is defined by the Federal Communications Commission (FCC) as a radio system having −10 dB of signal bandwidth greater than 500 MHz. To read more about UWB, see Chapter 28 of this book and the references therein.

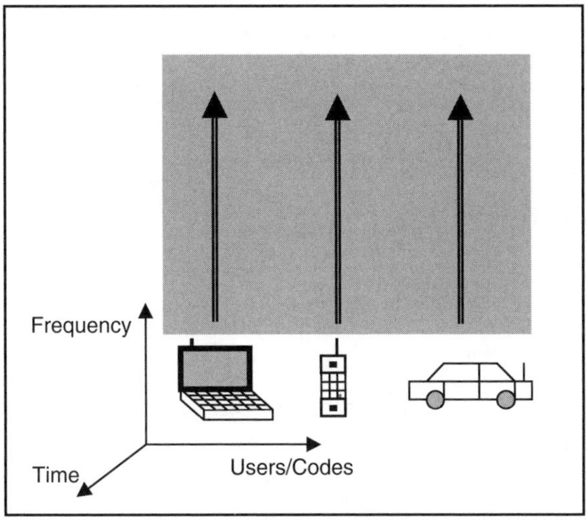

FIGURE 1.14 CDMA (W-CDMA) scheme.

BW of the channel, the inherent frequency diversity [Ada98] will mitigate the effects of small-scale fading. Additionally, the system can support multi-signaling-rate services simultaneously with a frequency reuse feature. In the satellite domain, where multiple signals from different satellites are linearly combined, W-CDMA (or CDMA) with universal frequency reuse and a Rake receiver is very efficient for soft handoff application [Vit94] [Eva98]. In this scenario, in addition to improving the received signal quality, the soft handoff technique performs much better in terms of probability of call dropping compared to the hard handoff technique; additionally, it simplifies the radio frequency (RF) interface.

1.5.1.3.3 Demerits

In the CDMA-based system, good synchronization is necessary for the spreading codes to exhibit their mutual orthogonal properties. If the mobile terminals are not synchronized, the orthogonality between the spreading codes is compromised and the overall performance is severely degraded. The capacity of a CDMA system is not a single constant number; it depends on the locations of the users as well as on their number, and it is also a function of how low a signal-to-noise ratio (SNR) is deemed to be acceptable.

Other than the basic MAC schemes stated above, there are some other versions of MAC schemes, such as space division multiple access (SDMA) [Rap96] and geographic MAC [App97]. Combinations of different basic MAC schemes are also available in the research areas, some of which are being considered for practical applications. These combinations include hybrid FDMA/CDMA (FCDMA) [Eng93], hybrid direct sequence/frequency hopped multiple access (DS/FHMA) [Dix94], time division CDMA (TCDMA) [Rap96], and time division frequency hopping (TDFH) [Gud92]. Currently, some remarkable research is going on with the combination of OFDM and CDMA techniques, and OFDM and TDMA techniques. Below, we briefly discuss these techniques.

1.5.2 Combination of OFDM and CDMA Systems

The combination of OFDM signaling and the CDMA scheme has one major advantage: it can lower the symbol rate in each subcarrier so that a longer symbol duration makes it easier to quasi-synchronize the transmission. To exploit this advantage and all the advantages of OFDM and CDMA systems discussed earlier, the combinations of these two systems have been proposed in three different ways by different researchers. These are multitone CDMA (MT-CDMA) [Van93], multicarrier CDMA (MC-CDMA) [Faz93] [Yee93] [Cho93], and multicarrier direct-sequence CDMA (MC-DS CDMA) [Das93]. A brief overview of these different schemes is given below.

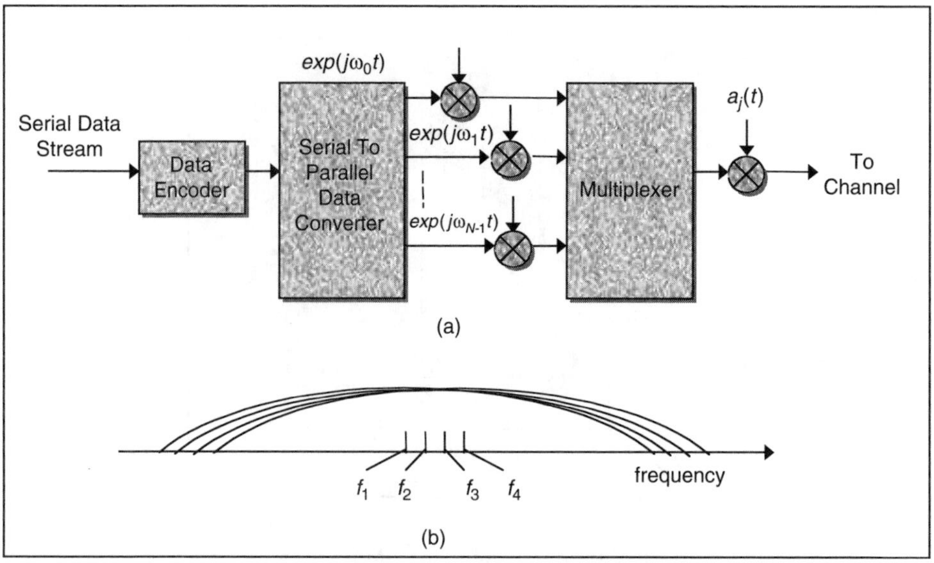

FIGURE 1.15 MT-CDMA scheme: (a) transmitter; (b) power spectrum of the transmitted signal.

1.5.2.1 MT-CDMA Scheme

In this system the transmitter spreads the OFDM signals and sends them through the channel. Figure 1.15a shows the MT-CDMA transmitter of the jth user using the PSK scheme. The transmitter spreads the *serial-to-parallel* (S/P) converted data stream after MT operation. Since before spreading it is nothing but the OFDM signals, the spectrum of each subcarrier prior to the spreading operation can satisfy the orthogonality condition with the minimum frequency separation [Van93]. After spreading, the resulting spectrum of each subcarrier no longer satisfies the orthogonality condition. When spreading is placed on top of the MT signal, as proposed in Vandendorpe [Van93], the main feature of the system is that for a constant BW, the ratio between the number of chips and the number of tones has to be constant. Hence, when the number of tones increases, the number of chips per symbol does as well. Figure 1.15b shows the power spectrum of the MT-CDMA transmitted signals for a number of tones $N_t = 4$.

1.5.2.2 MC-CDMA Scheme

In this system the transmitter spreads the input symbol first and then converts the spread symbol to OFDM signals, sending them through the channel. In other words, a fraction of the symbol corresponding to a chip of the spreading code is transmitted through a different subcarrier. In this case, the MC-CDMA transmitter spreads the original data stream over different subcarriers using a given spreading code in the frequency domain [Faz93] [Yee93] [Cho93]. The main feature of the MC-CDMA system is that as MC-CDMA spreads an information bit over many subcarriers, it can make use of information contained in good subcarriers to recover the original symbol in a deep-frequency-selective fading channel.

1.5.2.3 MC-DS CDMA Scheme

The MC-DSCDMA transmitter spreads the S/P converted data streams using a given spreading code so that the resulting spectrum of each subcarrier can satisfy the orthogonality condition with the frequency separation [Das93]. This scheme was originally proposed for an uplink communication channel because the introduction of OFDM signaling into the DSCDMA scheme is effective for the establishment of a quasi-synchronous channel.

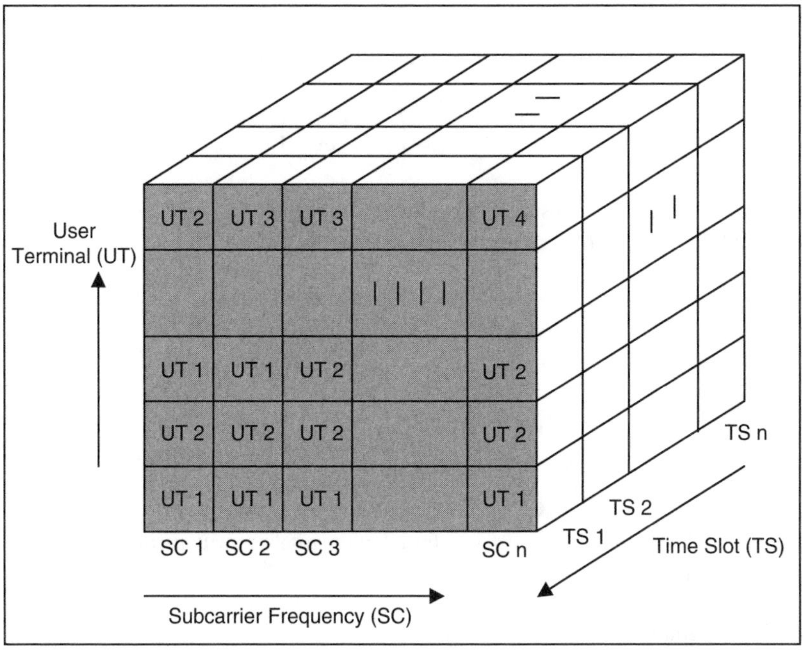

FIGURE 1.16 Block diagram of an OFDM/TDMA system.

1.5.3 OFDM/TDMA

The OFDM/TDMA-based system is a combination of OFDM transmission and TDMA techniques, which exploits all the advantages of these two techniques. Figure 1.16 illustrates the fundamental principle of this system. Here the overall channel BW is divided into a number of subcarriers, each carrying an individual bit stream with a relatively small signaling rate [Nee99] [Cim85]. In general, within a given time slot, a mobile station may use all or some of the allocated subcarriers; hence, the transmission rate of each mobile station may dynamically vary from slot to slot. This general situation actually represents the OFDM access technique [Nee99].

1.5.3.1 Merits

OFDM/TDMA technology allows the transmission of high-speed multiple-data-rates over extremely hostile channels at a relatively low complexity. Different merits of OFDM system have been discussed in the previous section. In addition to those advantages, the combination of TDMA and OFDM techniques provides the advantage of using different time slots and variable transmission rates simultaneously.

1.5.3.2 Demerits

Although the OFDM/TDMA technique provides the means for extended flexibility and multirate transmission, it requires precise synchronization between the mobile stations and thus calls for very high implementation complexity. Besides, we need to consider all the demerits of the OFDM-based system, which are discussed in the previous section.

1.5.4 Capacity of MAC Methods

Here we will address the capacity of different fundamental MAC schemes (FDMA, TDMA, and CDMA) in an ideal AWGN channel of bandwidth W accessed by K different users, each of which is transmitting an information signal with the same power P.

1.5.4.1 FDMA Capacity

In an ideal band-limited AWGN channel of bandwidth W, the capacity of a single user is given by [Pro95]

$$C = W\log_2\left(1 + \frac{P}{WN_0}\right) \tag{1.63}$$

where $N_0/2$ is the power spectral density of the additive noise. In FDMA each user occupies a bandwidth of W/K Hz. Hence the capacity of each user becomes

$$C_K = \frac{W}{K}\log_2\left(1 + \frac{KP}{WN_0}\right) \tag{1.64}$$

For K users it becomes

$$KC_K = W\log_2\left(1 + \frac{KP}{WN_0}\right) \tag{1.65}$$

Comparing Equation 1.63 and Equation 1.65 we find that the total capacity is equivalent to that of a single user with average power KP.

1.5.4.2 TDMA Capacity

In this system, since each user, with an average power of KP, transmits the information signal for $1/K$ of the total available time period, the capacity per user is given by Equation 1.64. In other words, the TDMA capacity per user is identical to the FDMA capacity per user. But in this case, there is a practical limit beyond which the transmitter power cannot be increased with the increase in the number of users.

1.5.4.3 CDMA Capacity

In a CDMA system, each user transmits a pseudorandom (PN) signal of a bandwidth W and average power P. The capacity of the system depends on the level of cooperation among K users. When the receiver for each user signal does not know the spreading waveforms of other users, the other users' signals appear as interference at the receiver of each user. The multiuser receiver consists of a bank of K single-user receivers. Assuming that each user's PN sequence is Gaussian, each user signal is corrupted by Gaussian interference of power $(K-1)P$ and additive Gaussian noise power WN_0. So, the capacity per user becomes

$$C_K = W\log_2\left(1 + \frac{P}{WN_0 + (K-1)P}\right) \tag{1.66}$$

On the other hand, if all K users are synchronous in time and the multiuser receiver performs joint demodulation and detection with all known PN sequences, it can be shown [Pro95] that the capacity of the CDMA system is identical to that of the FDMA and TDMA systems.

1.5.5 Challenges in the MAC Schemes

Many challenges are yet to be fulfilled for the MAC issue in both the terrestrial and satellite mobile communications areas. The complexity due to synchronization can be relaxed for W-CDMA-based systems using spreading coding schemes with very small cross-correlation peak values [Din98]. Profound research needs to be carried out to generate this kind of spreading coding schemes. In the case of OFDM/TDMA schemes the PAPR and frequency-offset issues of the OFDM technique, along with the synchronization issue of the TDMA scheme, need to be addressed for optimal solutions. If the shortcomings of the OFDM technique can be overcome, the combination of OFDM and CDMA techniques that results in MC-CDMA, MT-CDMA, and MC-DSCDMA schemes can be used for both MAC schemes and high-multiple-data-rate applications for the next-generation mobile communications systems.

1.6 Diversity Technique

Diversity is a family of techniques that reduces the effects of fading. In these techniques, several replicas of the same information signal transmitted over independently fading channels are supplied to the receiver. In this case, the probability of all the signal components fading simultaneously is reduced considerably. That is, if p is the probability that any signal will fade below some critical value, then p^L will be the probability that all L independently fading replicas of the same signal will fade below the critical value. When all these independently fading signals are combined, the technique is known as diversity combining. Figure 1.17 shows a block diagram of the simplest form of the diversity combining technique.

In general, a diversity system can be constructed if the following criteria are met:

- A copy of the same signal is received over two or more different paths.
- Each path fades differently.
- Some type of diversity combining on the signal replicas, received over the paths, is possible.

1.6.1 Classifications of the Diversity Techniques

There are several ways by which the receiver can be provided with L independently fading replicas of the same information-bearing signal.

In *frequency diversity* the same information-bearing signal is transmitted on L carriers, where the separation between the successive carriers equals or exceeds the coherence bandwidth (BW_{coh}) of the channel. In *time diversity* the same information-bearing signal is transmitted in L different time slots, where the separation between successive time slots equals or exceeds the coherence time (T_{coh}) of the channel. In *space diversity* (also known as *antenna diversity*) multiple antennas are used. These antennas are commonly used in the receiver section (in *transmit diversity* [Cai00], multiple transmitter antennas are used). These antennas are spaced sufficiently far apart so as to obtain signals that fade independently. The space diversity technique is one of the most popular forms of diversity used in wireless systems.

Another method of obtaining diversity is based on the use of a signal having a bandwidth much greater than the coherence bandwidth (BW_{coh}) of the channel. Such a signal with bandwidth B_s resolves the multipath components and thus provides the receiver with several independently fading signal paths. The time resolution is $1/B_s$. Consequently, with a multipath spread of T_m seconds, there are $T_m B_s$ resolvable signals components. Since $T_m = 1/BW_{coh}$, the number of resolvable signal components may also be expressed as B_s/BW_{coh}. Thus, the use of a wideband signal may be viewed as just another method for obtaining frequency diversity of the order of $L \approx B_s/BW_{coh}$. Sometimes it is called *multipath frequency diversity*.

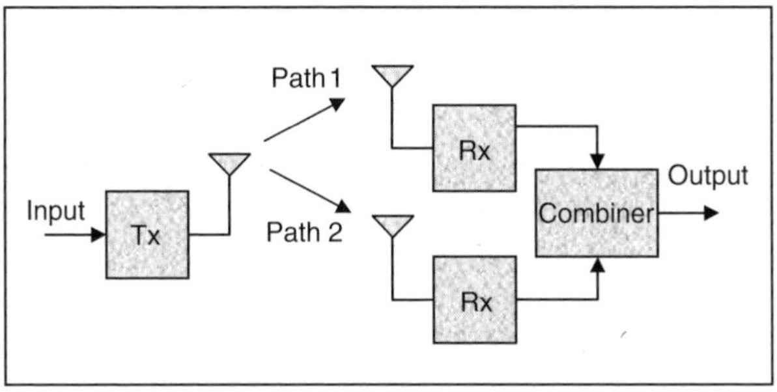

FIGURE 1.17 A simplified diversity (space) configuration.

There is one special diversity technique, known as transmit diversity, that uses multiple antennas in the transmitter end. This technique is also known as space–time coding (see the coding section) since it involves redundancy by transmitting the same signal using different antennas. It is known as the MIMO system too, as mentioned in the coding section. Currently, transmit diversity is getting notable consideration [Der02] among researchers due to its application in the 3G scenario. There are four types of transmit diversity techniques: space–time spreading (STS), also known as space–time coding; orthogonal transmit diversity (OTD); phase sweep transmit diversity (PSTD); and time delay transmit diversity (TDTD). All four techniques are being used in IS-2000 systems, while the last two are being used in IS-95 system. STS [Hoc01] performs true maximal ratio combining (discussed later) at the receiver end. In OTD [Tar99], all even numbered symbols are transmitted from one antenna, while odd numbered symbols are transmitted by the other antenna (2-Tx antennas). Thus, in the OTD technique, performance improvement is observed by filling out the long fades in the channel. In the PSPD [Kun02] technique, the same signal is transmitted by each transmitting antenna with carrier frequencies offset by f_{off} from each other, which results in artificial oscillation in the fading of each multipath component; that in turn decreases the duration of long fades and improves receiver performance. In the TDTD [Ong01] technique, one transmitting antenna transmits the same signal by adding a small delay τ_d to the signal transmitted by the second antenna (2-Tx antennas), resulting in extra multipath components in the channel by aiming to improve receiver (RAKE) diversity.

There are other diversity techniques such as *angle-of-arrival diversity* and *polarization diversity*. However, these techniques (e.g., see Popovic [Pop02] and Vaughan [Vau90]) have not been as widely used as those described above.

1.6.2 Classifications of Diversity Combiners

There are two basic types of diversity combiners, *predetection* type and *postdetection* type, which are discussed below.

1.6.2.1 Predetection Diversity Combiners

A predetection diversity combiner cophases, weights, and combines all the signals received on the different branches before signal detection. Here, the cophasing function is a difficult task to implement. The most common predetection combiners are the *selection combiner* (SC), *equal gain combiner* (EGC), and *maximal ratio combiner* (MRC). Recently, a *minimum mean squared error* (MMSE) diversity combiner specifically designed to combat ISI due to delay spread was proposed in Clark [Cla92].

1.6.2.1.1 Selection Diversity Combining

This is the simplest diversity combining technique, where the branch output having the maximum instantaneous SNR, among L diversity branches, is selected for further processing and detection.

1.6.2.1.2 Maximal Ratio Combining

Here the diversity branch signals are cophased (phase-aligned) and weighted according to their individual signal-to-noise power ratios and then summed.

1.6.2.1.3 Equal Gain Combining

EGC is based on the same principle of operation as MRC, except that all the branch weights are set to unity. For detailed information on all the above combiners, see Jakes [Jak74], Rappaport [Rap96], and Proakis [Pro95].

1.6.2.2 Postdetection Diversity Combiners

Postdetection diversity combiners weight and combine all diversity branches after signal detection and do not require the difficult-to-implement cophasing function. Since the combiner structure can be much simpler, postdetection diversity is more attractive for mobile radios. All the predetection combining techniques can also be used for postdetection combining.

1.6.3 Diversity for Next-Generation Systems: Some Research Evidence

The examples of space–time coding [Dou02] [Dou02(2)], already given in the coding section, are equally appropriate to refer to as examples of the transmit diversity technique. In those papers the transmit diversity technique is considered to be one of the strong candidates for the 4G mobile communications systems and the diversity gain in the analysis is clearly shown. Transmit diversity is also addressed in many other recent investigations, such as Zhou [Zho03] and Petré [Pet03]. In Rahman [Rah01] the authors have considered postdetection diversity combining for MT-CDMA systems, one of the probable candidates for the next-generation system. In this paper the performance of the system is analyzed in terms of BER in the Ricean fading channel. The study considers indoor environment with DQPSK modulation. The analytical results, at high SNR, show that without tracking the phase of the received signal the system performs equally well when compared with its coherent counterpart.

With the aim of improving the overall performance of next-generation communications systems, the satellite diversity combining technique is also considered in the research phase. In Caini [Cai01], the system model for the satellite diversity is based on the consideration that the coverage of the various satellites is only partially overlapping. Here, to evaluate the impact of diversity, the authors have considered the overlapping areas only. The RAKE receiver with MRC is used on the receiver side. In the analysis it is assumed that the number of fingers in the RAKE receiver is always sufficient to track all the N satellites in view, so that the kth user experiences a soft handoff of order N. In this case the ith finger of the RAKE receiver is considered to be tracking the ith satellite. Based on the numerical results presented in the paper, it is shown that satellite diversity is not only essential to achieve a satisfactory level of service availability, but also advantageous to improve the user capacity of the system, particularly in the realistic channel conditions. In Yuan [Yua03(2)], the authors have carried out some theoretical analysis on the symbol error rate (SER) over the satellite fading channel in the presence of satellite nonlinearity, using rectangular 16-QAM. The analysis is carried out over Rayliegh, Ricean, and Nakagami fading channel models for both MRC type diversity and nondiversity cases in the downlink. With close agreement with the simulation results, the theoretical analysis shows a significant improvement in terms of SER performance in the presence of diversity. In Fischer [Fis01], space diversity is used using turbo codes where the coded bits of the two coders are split between different satellites. Some low-complexity diversity combining schemes are presented in Chapter 17 of this book.

1.6.4 Challenges in the Diversity Area

Currently transmit diversity is considered to be a contender in the 3G diversity combining technique. But as researchers provide new ideas for the next-generation systems, the transmit diversity technique imposes many challenges that need further investigation. A design of a robust transmit diversity technique that exhibits satisfactory performance in various channel environments is a challenging task. In this case, optimum techniques to mitigate multipath fading effects and to suppress ISI, multiple-access interference (MAI), and other interferences are highly desirable. The trade-off between the data transmission rate and performance robustness needs to be addressed too. Some research [Gam02] [Sun02] [Yin02] is already underway to address these challenges.

For the satellite domain, in favorable channel conditions with no critical blockage, satellite diversity may be an unnecessary luxury. But when diversity combining is implemented due to the presence of hostile channels, optimum use of the available satellites to minimize the probability of call dropping becomes the main challenging task. In this case, it is necessary to have an efficient handoff algorithm for transferring control from one satellite to another and, more frequently, from one beam to another. CDMA-based systems are already offering the capability of soft handoffs [Vit94] [Eva98], but as the high-bit-rate-supporting multicarrier modulation techniques are getting attention [Ohn00] for the next-generation systems, a substantial amount of research needs to be carried out to provide satellite diversity considering these modulation techniques.

1.7 Conclusions

In this chapter we have presented an overview of different signal processing aspects for the next-generation mobile communications systems. Each section provided background information, current research evidences, and challenges on different signal processing issues. The chapter started with a presentation on the physical characteristics of the mobile channel. Then it discussed different signal processing technologies, including modulation techniques, coding schemes, multiple-access techniques, and diversity combining. Some analytical formulation on all the above issues, with different challenges, can be found in the rest of the book (e.g., Chapters 4, 7, and 17).

References

[Ada98] Adachi F., Sawahashi M. and Suda H., "Wideband DS-CDMA for next generation mobile communication systems," *IEEE Communications Magazine*, pp. 56–69, September 1998.

[Ake88] Akerberg D., "Properties of a TDMA picocellular office communication system," *IEEE GLOBECOM*, pp. 1343–1349, December 1988.

[Ala03] Al-Asady H., Boutros R. and Ibnkahla M., "Analysis and comparison between digital and analog predistortion techniques for satellite communications," *Proceedings of the IEEE Canadian Conference on Electrical and Computer Engineering*, CCECE 2003, Montreal, Canada, June 2003.

[Alo00] Alouini M.S. and Goldsmith A., "Adaptive modulation over Nakagami fading channels," *Kluwer Journal on Wireless Communications*, Vol. 13, No. 1–2, pp. 119–143, May 2000.

[App97] Digital Modulation in Communications Systems: An Introduction, application note 1298, Hewlett Packard, 1997.

[Ber87] Bernhardt R.C., "Macroscopic diversity in frequency reuse systems," *IEEE Journal on Selected Areas in Communications*, Vol. SAC 5, pp. 862–878, June 1987.

[Bou02] Boutros R. and Ibnkahla M., "New adaptive polynomial and neural network predistortion techniques for satellite transmissions," *Proceedings of the ANTEM Conference*, Montreal, Canada, 31 July–2 August 2002.

[Bur01] Burr A., *Modulation and Coding for Wireless Communications*, Prentice Hall, New York, 2001.

[Cai00] Cai X. and Akansu A.N., "Multicarrier CDMA systems with transmit diversity," *Fall 2000 IEEE Vehicular Technology Conference (VTC) Proceedings*, Boston, MA, pp. 2817–2821, September 2000.

[Cai01] Caini C. and Corazza G.E., "Satellite diversity in mobile satellite CDMA systems," *IEEE Journals on the Selected Areas in Communications*, Vol. 19, No. 7, pp. 1324–1333, July 2001.

[Cha66] Chang R.W., "Synthesis of band-limited orthogonal signals for multi-channel data transmission," *Bell Systems Technical Journal*, Vol. 45, pp. 1775–1785, December 1966.

[Cha68] Chang R.W. and Gibby R.A., "A theoretical study of performance of an orthogonal multiplexing data transmission scheme," *IEEE Transactions on Communications*, Vol. COM-16, pp. 529–540, August 1968.

[Cho91] Chow P.S. et al., "Performance evaluation of a multichannel transceiver system for ADSL and VHDSL services," *IEEE Journal in the Selected Areas in Communications*, Vol. 9, No. 6, pp. 909–919, August 1991.

[Cho93] Chouly A., Brajal A. and Jourdan S., "Orthogonal multicarrier techniques applied to direct sequence spread spectrum CDMA systems," *Proceedings of IEEE GLOBECOM'93*, Houston, pp. 1723–1728, November 1993.

[Cim85] Cimini L.J., Jr., "Analysis and simulation of a digital mobile channel using orthogonal frequency division multiplexing," *IEEE Transactions on Communications*, Vol. COM-33, No. 7, pp. 665–675, July 1985.

[Cim99] Cimini L.J. and Sollenberger R., "Peak-to-average power ratio of an OFDM signal using partial transmit sequences," *IEEE International Conference on Communications*, Vancouver, BC, Canada, June 1999, from CD-ROM.

[Cla02] Classon B. et al., "Channel coding for 4G systems with adaptive modulation and coding," *IEEE Wireless Communications*, pp. 8–13, April 2002.

[Cla68] Clarke R.H., "A statistical theory of mobile-radio reception," *Bell Systems Technical Journal*, Vol. 47, pp. 957–1000, 1968.

[Cla92] Clark M.V., Greenstein L.J., Kennedy K. and Shafi M., "MMSE diversity combining for wideband digital cellular radio," *IEEE Transactions on Communications*, Vol. 40, pp. 1128–1135, June 1992.

[Cos02] Costa E. and Puoilin S., "M-QAM OFDM system performance in the presence of a nonlinear amplifier and phase noise," *IEEE Transactions on Communications*, Vol. 50, No. 3, March 2002.

[Cox84] Cox D.C., Murray R. and Norris A., "800 MHz attenuation measured in and around suburban houses," *AT&T Laboratory Technical Journal*, Vol. 673, No. 6, July–August 1984.

[Dad75] Dadson C.E., Durkin J. and Martin E., "Computer prediction of field strength in the planning of radio systems," *IEEE Transactions on Vehicular Technology*, Vol. VT-24, No. 1, pp. 1–7, February 1975.

[Das93] DaSilva V. and Sousa E.S., "Performance of orthogonal CDMA codes for quasi-synchronous communication systems," *Proceedings of IEEE ICUPC* '93, Ottawa, Canada, pp. 995–999, October 1993.

[Der02] Derryberry R.T. et al., "Transmit diversity in 3G CDMA systems," *IEEE Communications Magazine*, pp. 68–75, April 2002.

[Din98] Dinan E.H. and Jabbari B., "Spreading codes for direct sequence CDMA and wideband CDMA cellular networks," *IEEE Communications Magazine*, pp. 48–54, September 1998.

[Div90] Divsalar D. and Simon M.K., "Multiple symbol differential detection of MPSK," *IEEE Transactions on Communications*, Vol. 38, No. 3, pp. 300–308, 1990.

[Dix94] Dixon R.C., *Spread Spectrum Systems with Commercial Applications*, 3rd ed., John Wiley & Sons Inc., New York, 1994.

[Dou02] Doufexi A., "Design considerations and initial physical layer performance results for space time coded OFDM 4G cellular network," *PIMRC 2002*, Lisbon, Portugal, September 2002.

[Dou02(2)] Doufexi A. et al., "COFDM performance evaluation in outdoor MIMO channel using space/polarization-time processing techniques," *Electronics Letters*, Vol. 38, No. 25, pp. 1720–1721, December 2002.

[Dow02] Dowler A., Doufexi A. and Nix A., "Performance evaluation of channel estimation techniques for a mobile fourth generation wide area OFDM system," *Vehicular Technology Conference, 2002, Proceedings*, VTC Fall 2002, IEEE 56th, Vol. 4, pp. 2036–2040.

[Dur03] Durgin G.D., *Space-Time Wireless Channels*, Prentice Hall, New York, 2003.

[Ell63] Elliott E.O., "Estimates of error rates for codes on burst-noise channels," *Bell Systems Technical Journal*, Vol. 42, September 1963.

[Eng93] Eng T. and Milstein L.B., "Capacities of hybrid FDMA/CDMA systems in multipath fading," *IEEE MILCOM Conference Records*, pp. 753–757, 1993.

[ETS97] ETSI, "Radio Broadcasting Systems: Digital Audio Broadcasting (DAB) to Mobile, Portable and Feed Reducers," technical report, ETS 300 401, August 1997.

[ETS97(2)] ETSI, "Digital Video Broadcasting (DVB): Framing Structure, Channel Coding and Modulation for Digital Terrestrial Television," technical report, EN 300 744, August 1997.

[Eva98] Evans J.V., "Satellite systems for personal communications," *Proceedings of the IEEE*, Vol. 86, No. 7, pp. 1325–1341, July 1998.

[Fal68] Fallager R.G., *Information Theory and Reliable Communication*, John Wiley & Sons, Inc., New York, 1968.

[Faz93] Fazel K. and Papke L., "On the performance of convolutionally-coded CDMA/OFDM for mobile communication system," *Proceedings of IEEE PIMRC* '93, Yokohama, Japan, pp. 468–472, September 1993.

[Fer98] Fernando W.A.C. and Rajatheva R.M.A.P., "Performance of COFDM for LEO satellite channels in global mobile communications," *Proceedings of IEEE ICC'98*, Atlanta, GA, pp. 412–416, 1998.

[Fis01] Fischer S. et al., "Analysis of diversity effects for satellite communication systems," *Global Telecommunications Conference, GLOBECOM 2001*, Vol. 4, pp. 2759–2763.

[Gam02] Gamal H. El, "On the robustness of space-time coding," *IEEE Transactions on Signal Processing*, Vol. 50, Issue 10, pp. 2417–2428, October 2002.

[Gil60] Gilbert E.N., "Capacity of a burst-noise channel," *Bell Systems Technical Journal*, Vol. 39, September 1960.

[Gil92] Giltin R.D., Hayes J.F. and Weinstein S.B., *Data Communications Principles*, Plenum, New York, 1992.

[Gof94] Goff S. et al., "Turbo codes and high spectral efficiency modulation," *Proceedings IEEE International Conference on Communications (ICC)*, May 1994, pp. 645–649.

[Gol97] Goldsmith A. and Chua S., "Variable-rate variable-power MQAM for fading channels," *IEEE Transactions on Communications*, Vol. 45, No. 10, pp. 1218–1231, October 1997.

[Gua99] Guan Y.L. and Turner L.F., "Generalized FSMC model for radio channels with correlated fading," *IEE Proceedings-Communications*, Vol. 146, No. 2, April 1999.

[Gud92] Gudmundson Skold B.J. and Ugland J.K., "A comparison of CDMA and TDMA systems," *Proceedings of the 42nd IEEE VTC*, Vol. 2, pp. 732–735, 1992.

[Hoc01] Hochwald B. et al., "A transmitter diversity scheme for wideband CDMA systems based on space-time spreading," *IEEE Journal on Selected Areas in Communications*, Vol. 19, Issue 1, pp. 48–60, January 2001.

[Hsi01] Hsieh J.C. et al., "A Two-Level, Multi-State Markov Model for Satellite Propagation Channels," VTC 2001, spring, Vol. 4.

[Jak74] Jakes W.C., Jr., Ed., *Microwave Mobile Communications*, Wiley, New York, 1974.

[Kle02] Klepal M. et al., "Optimizing motif models for indoor radio propagation prediction using evolutionary computation," *ISSC*, June 2002.

[Kun02] Kunnari E., "Space-time coding combined with phase sweeping transmit diversity," *Global Telecommunications Conference, 2002. IEEE GLOBECOM '02*, Vol. 2, pp. 1920–1924, November 17–21, 2002.

[Lee89] Lee W.C.Y., *Mobile Cellular Telecommunications Systems*, McGraw-Hill Publications, New York, 1989.

[Lee97] Lee L.H.C, *Convolutional Coding: Fundamentals and Applications*, Artech House,

[Lin02] Lin H.P. et al., "A Non-Stationary Hidden Markov Model for Satellite Propagation Channel Modeling," VCT 2002, fall.

[Lon68] Longley A.G. and Rice P.L., "Prediction of Tropospheric Radio Transmission Loss over Irregular Terrain; a Computer Model," ESSA technical report, ERL 79-ITS 67, 1968.

[Loo85] Loo C., "A statistical model for a land mobile satellite link,' *IEEE Transactions on Vehicular Technology*, Vol. VT-34, pp. 122–127, August 1985.

[Loo87] Loo C., "Measurements and models of a land mobile satellite channel and their applications to MSK signals," *IEEE Transactions on Vehicular Technology*, Vol. VT-36, pp. 114–121, August 1987.

[Loo94] Loo C., "Land mobile satellite channel measurement at Ka band using Olympus," *Proceedings of the IEEE Vehicular Technology Conference*, pp. 919–923, 1994.

[Loo96] Loo C., "Statistical models for land mobile and fixed satellite communications at Ka band," *Proceedings of the IEEE Vehicular Technology Conference*, pp. 1023–1027, 1996.

[Loo98] Loo C. and Butterworth J., "Land mobile satellite channel measurements and modeling," *Proceedings of the IEEE*, Vol. 86, No. 7, pp. 1442–1463, July 1998.

[Nak60] Nakagami M., "The m distribution: a general formula of intensity distribution of rapid fading," in *Statistical Methods in Radio Wave Propagation*, W.G. Hoffman, ed., 1960, pp. 3–36.

[Nee99] Nee R.V. and Prasad R., *OFDM Wireless Personal Communications*, Artech House, Norwood, MA, 1999.

[Ner99] Nee R.V. et al., "New high-rate wireless LAN standards," *IEEE Communications Magazine*, pp. 82–88, December 1999.

[Oes03] Oestges O., Erceg V. and Paulraj A.J., "A physical scattering model for MIMO macro-cellular broadband wireless channels," *IEEE Journal on Selected Areas in Communications*, Vol. 21, No. 5, June 2003.

[Ohn00] Ohnori S., Yamao Y. and Nakajima N., "The future generations of mobile communications based on broadband access technologies," *IEEE Communications Magazine*, pp. 134–142, December 2000.

[Ong01] Onggosanusi E.N. et al., "Performance analysis of closed-loop transmit diversity in the presence of feedback delay," *IEEE Transactions on Communications*, Vol. 49, Issue 9, pp. 1618–1630, September 2001.

[Pap01] Papathanassiou A. et al., "A comparison study of the uplink performance of W-CDMA and OFDM for mobile multimedia communications via LEO satellites," *IEEE Personal Communications*, pp. 35–43, June 2001.

[Par02] Park D.C. and Jeong T.J., "Complex-bilinear recurrent neural network for equalization of a digital satellite channel," *IEEE Transactions on Neural Networks*, Vol. 13, No. 3, pp. 711–725, May 2002.

[Pec00] Pechač P. and Klepal M. and Mazánek M., "Novel Approach to Indoor Semi-Deterministic Propagation Prediction," 2000 IEEE Antennas and Propagation Society International Symposium and USNC/URSI National Radio Science Meeting, Salt Lake City, July 2000.

[Pec01] Pechač P. and Klepal M., "Effective Indoor Propagation Predictions," VTC 2001, fall, Atlantic City, NJ, October 2001.

[Pet03] Petré F. et al., "Space-time block coding for single-carrier block transmission DS-CDMA downlink," *IEEE KSAC*, Vol. 21, No. 3, pp. 350–361, April 2003.

[Pop02] Popovic D. and Popovic Z., "Multibeam antennas with polarization and angle diversity," *IEEE Transactions on Antennas and Propagation*, Vol. 50, Issue 5, pp. 651–657, May 2002.

[Pro95] Proakis J.G., *Digital Communications*, 3rd ed., McGraw-Hill, New York, 1995.

[Raf89] Rafie M.S. and Shanmugan K.S., "Comparative performance analysis of M-CPSK and M-QAM over nonlinear satellite links," *GLOBECOM Conference Proceedings*, Vol. 2, Dallas, TX, pp. 1295–1302, November 1989.

[Rah01] Rahman Q.M. and Sesay A.B., "Post-detection diversity combining for non-coherent MT-CDMA system in a Rician fading channel," *IEEE Vehicular Technology Conference, 2001*, VTC 2001, fall, Vol. 2, pp. 1034–1038.

[Rah03] Rahman Q.M. and Ibnkahla M., "Performance Analysis of a QAM Based CDMA System in Mobile Satellite Rician Fading-Channel," *Proceedings of the IEEE IASTEAD Conference*, Banff, Canada, 2003.

[Rap96] Rappaport T.S., *Wireless Communications, Principles and Practice*, Prentice Hall, New York, 1996.

[Rob98] Robertson P. and Worz T., "Bandwidth efficient turbo trellis coded modulation using punctured component codes," *IEEE Journal on Selected Areas in Communications*, Vol. 16, No. 2, pp. 206–218, 1998.

[Sai97] Saifuddin A. et al., "HDR codes with concatenated multilevel codes and multiple-symbol differential detection for satellite application," *IEE Electronics Letters*, Vol. 33, No. 16, pp. 1355–1356, July 1997.

[Sal67] Saltzberg B.R., "Performance of an efficient data transmission system," *IEEE Transactions on Communications Technology*, Vol. 15, pp. 805–811, December 1967.

[Sal87] Saleh A.A.M. and Valenzula R.A., "A statistical model for indoor multipath propagation," *IEEE Journal on Selected Areas in Communications*, Vol. JSAC-5, No. 2, pp. 128–137, February 1987.

[Sam02] Sampath H. et al., "A fourth-generation MIMO-OFDM broadband wireless system: design, performance, and field trial results," *IEEE Communications Magazine*, Vol. 40, Issue 9, pp. 143–149, September 2002.

[Sei92] Seidel S.Y. and Rappaport T.S., "914 MHz path loss prediction models for indoor wireless communications in multifloor buildings," *IEEE Transactions on Antennas and Propagation*, Vol. 40, No. 2, pp. 207–217, February 1992.

[Sha48] Shannon C.E., "A mathematical theory of communication," *Bell System Technical Journal*, Vol. 27, pp. 379–423, 623–657, 1948.

[She95] Shelswell P., "The COFDM modulation system: the heart of digital audio broadcasting," *Electronics and Communications Engineering Journal*, pp. 127–136, June 1995.

[Sis93] Sistanizadeh K. et al., "Multi-tone transmission for asymmetric digital subscriber lines (ADSL)," *Proceedings of IEEE International Conference on Communications (ICC) 1993*, pp. 756–760.

[Skl88] Sklar B., *Digital Communications, Fundamental and Applications*, Prentice Hall, New York, 1988.

[Ste64] Stenbit J.P., "Tables of generators for Bose-Chadhuri codes," *IEEE Transactions on Information Theory*, Vol. IT 10, No. 4, pp. 390–391, October 1964.

[Sum01] Sumanasena M.A.K. and Evans B.G., "Adaptive modulation and coding for satellite: UMTS," *Proceedings of 54th IEEE Vehicular Technology Conference*, Fall 2001, Vol. 1, pp. 116–120.

[Sun02] Sun-Yuan K., Yunnan W. and Xinying Z., "Bezout space-time precoders and equalizers for MIMO channels," *IEEE Transactions on Signal Processing*, Vol. 50, Issue 10, pp. 2499–2514, October 2002.

[Tar99] Tarokh V., Jafarkhani H. and Calderbank A.R., "Space-time block codes from orthogonal designs," *IEEE Transactions on Information Theory*, Vol. 45, Issue 5, pp. 1456–1467, July 1999.

[Ung87] Ungerboeck G., "Trellis coded modulation with redundant signal sets, part I: introduction" *IEEE Communications Magazine*, Vol. 25, pp. 5–11, February 1987.

[Van93] Vandendorpe L., "Multitone direct sequence CDMA system in an indoor wireless environment," *Proceedings of IEEE First Symposium of Communications and Vehicular Technology*, Benelux, Delft, Netherlands, pp. 4.1.1–4.1.8, October 1993.

[Vau90] Vaughan R.G., "Polarization diversity in mobile communications," *IEEE Transactions on Vehicular Technology*, Vol. 39, pp. 177–186, August 1990.

[Vau02] Vaughan-Nichols S.J., "OFDM: back to the wireless future," *Computer*, Vol. 35, Issue 12, pp. 19–21, December 2002.

[Vil02] Vilaipornsawai U. and Soleymani M.R., "Trellis-based iterative decoding of block codes for satellite ATM," *IEEE International Conference on Communications*, ICC 2002, Vol. 5, pp. 2947–2951.

[Vit94] Viterbi A.J. et al., "Soft handoff extends CDMA cell coverage and increases reverse link capacity," *IEEE Journals on the Selected Areas in Communications*, Vol. 12, pp. 1281–1288, October 1994.

[Vit95] Viterbi A.J., *CDMA Principles of Spread-Spectrum Communications*, Addison-Wesley, Reading, MA, 1995.

[Wac95] Wachsmann U. and Huber J., "Power and bandwidth efficient digital communication using the turbo codes in multilevel codes," *European Transactions on Telecommunications*, Vol. 6, No. 5, pp. 557–567, 1995.

[Wan95] Wang H.S. and Moayeri N., "Finite-state Markov channel: a useful model for radio communication channels," *IEEE Transactions on Vehicular Technology*, Vol. 44, February 1995.

[Yee93] Yee N., Linnartz J.P. and Fettweis G., "Multicarrier CDMA in indoor wireless radio networks," *Proceedings of IEEE PIMRC '93*, Yokohama, Japan, pp. 109–113, September 1993.

[Yin02] Ying-Min W. et. al., "Joint detection for space-time block-coded TD-CDMA systems," IEEE Vehicular Technology Conference, 2002, VTC Spring 2002, Vol. 2, 2002 pp. 1007–1011.

[You03] Young-Hwan Y., "Reducing peak-to-average power ratio using simple transformed CDMA/OFDM signals," *Electronics Lett.*, Vol. 39(4), pp. 403–405, Feb 2003.

[Yua03] Yuan J. and Ibnkahla M., "Acute Theoretical Symbol Error Rate of Nonlinear Fading Channels in Satellite Communication", in the proceedings of *IEEE Canadian Conference on Electrical & Computer Engineering* (CCECE) 2003, Montreal, Canada.

[Yua03(2)] Yuan J. and Ibnkahla M., "Symbol error rate of nonlinear fading satellite communication channels", *Proc. IEEE Canadian Conference on Electrical and Computer Engineering*, CCECE 2003, Montreal Canada, June 2003 [Submitted to *IEEE Trans. Wireless Communications*]

[Zho03] Zhou Y. et. al., "Performance comparison of transmit diversity and beam-forming for the downlink of DS-CDMA system," *IEEE Trans. Wireless Commun.*, Vol. 2, No. 2, pp. 320–334, March 2003.

II

Channel Modeling and Estimation

2 **Multipath Propagation Models for Broadband Wireless Systems**
Andreas F. Molisch and Fredrik Tufvesson **2-1**
Introduction • Narrowband, Wideband, and Directional Channel Modeling • Modeling Methods for Multipath Channels • Propagation Aspects and Parameterization • Standard Models • Conclusions

3 **Modeling and Estimation of Mobile Channels** *Jitendra K. Tugnait* **3-1**
Introduction • Channel Models • Channel Estimation • Simulation Examples • Conclusions

4 **Mobile Satellite Channels: Statistical Models and Performance Analysis**
Giovanni E. Corazza, Alessandro Vanelli-Coralli, Raffaella Pedone, and Massimo Neri **4-1**
Introduction • Statistical Propagation Models • Detection Performance Analysis • Conclusions

5 **Mobile Velocity Estimation for Wireless Communications**
Bouchra Senadji, Ghazem Azemi, and Boualem Boashash **5-1**
Introduction • Received Signal Model and Statistics • Principles of Mobile Velocity Estimation • Performance Analysis of Velocity Estimators • Performance Analysis Using Simulations • Rice Factor Estimation • Application on Handover Performance • Conclusions and Perspectives

2

Multipath Propagation Models for Broadband Wireless Systems

2.1 Introduction 2-2
2.2 Narrowband, Wideband, and Directional Channel Modeling 2-3
Intuitive Description • Mathematical Description: Deterministic Case • Mathematical Description: Stochastic Case • Condensed Parameters • Directional Description
2.3 Modeling Methods for Multipath Channels 2-10
Measured Channel Impulse Responses • Deterministic Channel Computation • Tapped Delay Lines • Stochastic MIMO Models • Geometry-Based Stochastic Channel Models
2.4 Propagation Aspects and Parameterization 2-17
Amplitude Statistics • Arrival Times • Average Time Dispersion • Average Angular Dispersion at the BS • Average Angular Dispersion at the MS • MIMO Parameters • Polarization • Millimeter Wave Propagation • Ultrawideband Systems
2.5 Standard Models 2-24
The COST 207 Model • The ITU-R Models • IEEE 802.11/HIPERLAN Models • The 802.15 Ultrawideband Channel Model • The 3GPP–3GPP2 Model • The COST 259 Model
2.6 Conclusions 2-31

Andreas F. Molisch
Lund University and Mitsubishi Electric Research Labs

Fredrik Tufvesson
Lund University

Abstract

This chapter presents an overview of channel modeling for broadband communications. Because broadband communications require a large bandwidth, the channel models have to correctly reproduce the frequency selectivity of the channel; in addition, new techniques like smart antennas may also require knowledge of the directional properties of the channel. This chapter first reviews formal description methods for propagation channels, in particular the WSSUS (wide-sense stationary uncorrelated scattering) model. Next, we discuss various approaches for modeling such channels. Deterministic methods, like stored measurement results and ray tracing, are suitable for site-specific channel modeling and network planning, while stochastic methods, like tapped delay lines and geometry-based stochastic channel models,

0-8493-1657-X/$0.00+$1.50

are better for the simulation of algorithms and systems. Next, we give an overview of typical values for channel parameters, e.g., delay spread, angular spread, and arrival time statistics. Those values strongly depend on the operating environment; for example, the delay spread typically attains 10 ns in residential buildings, while it can easily reach 10 μs in urban macrocells. A summary of channel models that have been adopted by international standards organizations for systems like GSM (second-generation cellular), W-CDMA (third-generation cellular), IEEE 802.11 (wireless local area networks), and IEEE 802.15.3a (ultrawideband systems) concludes the chapter.

2.1 Introduction

Claude Shannon [203] showed that the channel determines the ultimate performance limits (information-theoretic capacity) of any communications system operating in that channel. From a more practical perspective, the performance of any transceiver, signal processing algorithm, channel code, etc., depends on the channel it is operating in [179]. For wireless systems, this channel is the *wireless propagation channel*, whose properties differ significantly from the wired channel. A correct understanding and modeling of the propagation channel is thus a vital prerequisite for understanding the performance of wireless communications systems.

The difficulties in wireless channel modeling are due to the complex propagation processes that form the basis of a wireless channel, involving reflections, scattering, diffraction, and transmission through a large number of irregular objects. For all practical purposes, it is thus necessary to derive *simplified descriptions*. The degree of admissible simplification, in turn, depends on the system for which the channel is intended.

In the past 20 years, wireless systems have changed dramatically. While the first-generation systems, like NMT (Nordic Mobile Telephony) and AMPS (Advanced Mobile Phone System) were analog and used only for speech communications, second-generation systems like GSM (Global System for Mobile Communications) used digital transmission for both speech and simple data services. Third-generation cellular systems and wireless local and personal area networks use digital transmission with much higher data rates (2 to 200 Mbit/s) to allow new applications like video telephony, web browsing, etc. Those high data rates are enabled by some new air–interface technologies, like CDMA and multiple-antenna techniques [94], [97].

As the systems have evolved, so have the channel models needed for the design and evaluation. For first-generation systems, it was sufficient to model *narrowband* channels, i.e., the attenuation between transmitter and receiver [100]. For second- and some third-generation systems, the description of the *frequency selectivity*, or its correspondence in the time domain — the *delay spread* — became an additional requirement [146]. These provide a description of the variation of the (complex) attenuation over the bandwidth of the system. Finally, recent advances in multiple-antenna systems necessitate the description of the *directional* properties of the channel [54].

There are two basic groups of propagation channel models: deterministic and stochastic. A deterministic model aims to correctly predict the channel characteristics (e.g., impulse response), in a specific location, using information about the location of the transmitter and receiver, as well as the surrounding environment. The resulting channel model is thus valid only in this specific location. Such models are commonly used for network planning, i.e., estimating how well a *given* system will work in a specific environment. Typical applications are the search for good base station sites in cellular networks and wireless local area networks (LANs).

A stochastic model, on the other hand, does not try to correctly predict each channel realization (channel at a specific location). Rather, it models the *statistical properties* of the channel. This concept can be most easily understood for narrowband channels, where the received field strength is the most important quantity. While a deterministic model would try to predict the correct field strength at, for example, each point in a room, a stochastic model would just specify that the probability density function of the field strength is Rayleigh distributed. Stochastic models can be made more refined by dividing the environments of interest into different "classes" and using different model parameters in each of those classes.

This chapter gives an overview of both deterministic and stochastic channel models for broadband wireless communications systems. The emphasis lies on frequency-selective channels and directional channel models; for narrowband modeling, we describe only the basics and otherwise refer to the literature. Also, the emphasis is on the generic structures of the models, as these influence their interplay with other aspects of signal processing. The actual parameterization in different environments and the relationship to physical propagation processes is important but will be treated only briefly, with references to the relevant literature. The remainder of the chapter is organized in the following way: In Section 2.2, we discuss the different *formal description methods* for propagation channels, stressing the difference and relation between stochastic and deterministic models. Based on this exposition, Section 2.3 develops *generic channel modeling approaches* that can approximate arbitrary propagation channels. Next, in Section 2.4, we describe how the environment influences the propagation channel, leading to the parameterization of the models of Section 2.3. Finally, in Section 2.5, we give an overview of standardized wideband and directional channel models. A list of symbols and abbreviations can be found at the end of the chapter.

As a "handbook" chapter, the purpose of this work is to give an overview of channel modeling and present a guideline through this wide area. For the many finer — but important! — points, refer to the list of references. In particular, introductory, overview, and tutorial material can be found in the papers [6], [18], [68], the textbooks [183], [213], and the books [171], [196]. References [54], [61], [148], and [242] emphasize spatial channel models, and the books [17], [20], and [227] the physical propagation effects and their impact on channel models.

2.2 Narrowband, Wideband, and Directional Channel Modeling

2.2.1 Intuitive Description

The wireless channel is characterized by the fact that there often are multiple propagation paths between transmitter and receiver (see Figure 2.1). The signals arriving via the different paths are called echoes or *multipath components* (MPCs).[1] The MPCs reach the receiver via different ways; therefore, they have different delays. The bandwidth of the system determines how many of these MPCs can be resolved, i.e., distinguished from each other. This is to say that the larger the bandwidth, the more components that can be resolved. We can interpret the effect of the different delays either in the time domain, as delay dispersion (i.e., an impulse response that is not a delta function, but has a finite support), or in the frequency domain, as frequency selectivity (i.e., a transfer function that is not constant over all frequencies) [100], [183], [196].

Consider the case where the locations of scatterers, and their radar cross sections, are known.[2] Implicitly, this gives the delays, as well as the amplitudes, of the MPCs. All scatterers lying on an ellipse (or rather an ellipsoidal annulus) result in a delay in the range $[\tau, \tau + \delta\tau]$, (see Figure 2.2), and the MPCs associated with them add up, though with different phases. As the mobile station or the scatterers move, the phases of the MPCs change, so that the MPCs in each annulus add up in a different way. This time variance of the channel is known as fading. If the number of scatterers in each annulus is large, this results in the classical Rayleigh or Rice fading described in [100]. Due to movements, also the absolute delay of scatterers can change with time; however, significant changes of the delay usually require much larger movements than significant changes of the phases.

The delay difference $\delta\tau$ that a system can resolve, often called a delay bin width, is approximately the inverse of the system bandwidth. A receiver with a small bandwidth (e.g., for a first-generation cellular

[1]The terminology MPC is mostly used when describing the different components that make up the impulse response, i.e., when the transmitted signal is a delta pulse.

[2]We assume here that only single-scattering processes occur. A discussion of that assumption is found in Section 2.3.5. We are also using *scattering* somewhat broadly, encompassing diffuse scattering as well as other interaction processes.

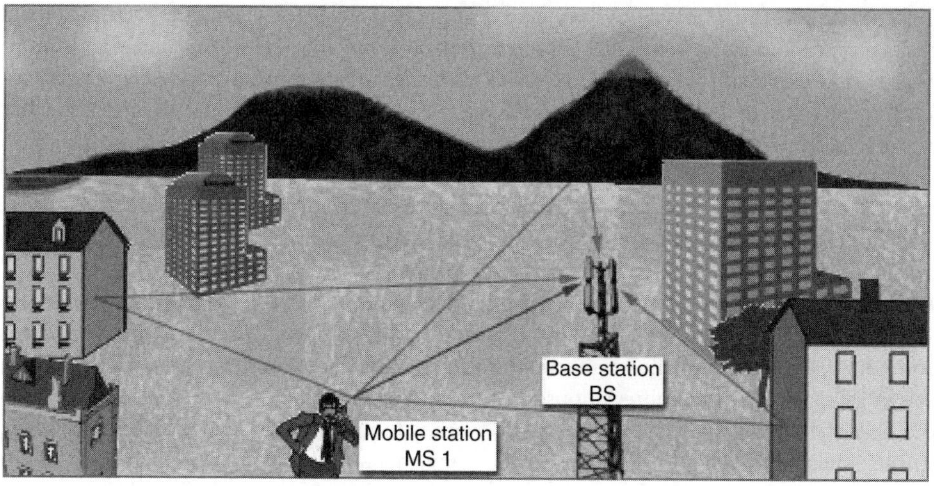

FIGURE 2.1 Multipath propagation.

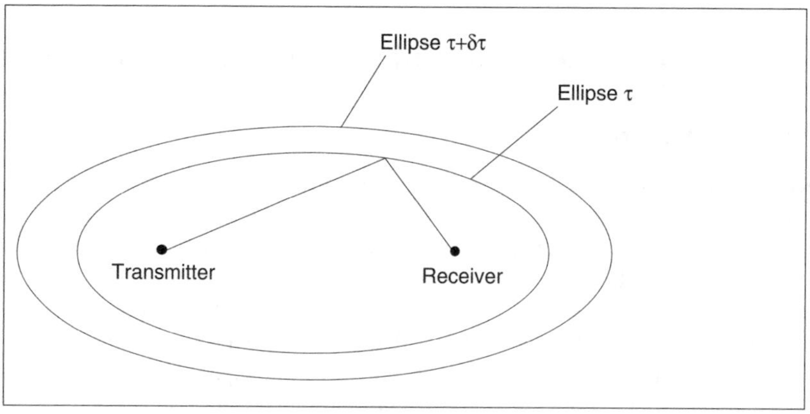

FIGURE 2.2 Scatterers on the ellipsoid result in the same delay.

system) can normally not resolve any MPCs; this is equivalent to saying that the transfer function will not vary within the bandwidth of the system.[3] In many digital wireless systems, the system has a bandwidth that is wide enough so that the channel transfer function varies noticeably within it but is small enough that several MPCs fall into each delay bin. This is the case if $\Delta\tau_{\text{spread}} > \frac{1}{BW} > \Delta\tau$ where $\Delta\tau_{\text{spread}}$ is the so-called delay spread (maximum excess delay) and $\Delta\tau$ is the delay between subsequent MPCs with significant amplitude. If both inequalities are fulfilled, the channel is wideband, but fading. If the bandwidth BW is smaller than $1/\Delta\tau_{\text{spread}}$, the channel is narrowband; if it is larger than $1/\Delta\tau$, the channel is wideband but nonfading [227].

An alternative interpretation is to consider the channel (possibly including the transmit and receive antennas) as a "black box." A signal is transmitted from the transmitter (TX), and an attenuated, distorted version of the transmitted signal arrives at the receiver (RX). (Note that the antennas are often considered

[3]This statement can be reversed: a narrowband system is *defined* as a system with a bandwidth that is so small that the transfer function does not change significantly over the system bandwidth.

part of the channel; see also Section 2.2.5.) Without requiring any knowledge of the actual propagation processes, the channel can be described as a linear filter. Due to the presence of fading, the filter is time variant and thus characterized by the *time-variant impulse response* $h(t, \tau)$ [210]. If the impulse response is known at each moment in time, the description is completely deterministic. However, in many situations, this is either not possible or not necessary. In that case, it is common to provide a *statistical* description of the channel, where the probability density functions (pdfs) of the components in the impulse response are provided [100]. For the simulation of a system, the impulse responses at different times are generated from these pdfs. This concept is basically the same for wideband and narrowband channels.

2.2.2 Mathematical Description: Deterministic Case

In this section, we make the concepts outlined in Section 2.2.1 more precise. The mathematical description of time-varying channels was derived in a seminal paper [12]; for a tutorial exposition, see [54], [171], and [227]. We will start out with the deterministic interpretation. The input $x(t)$ and output $y(t)$ of the channel are related by the time-variant impulse response $h(t, \tau)$ by $y(t) = \int_{-\infty}^{\infty} x(t-\tau)h(t,\tau)d\tau$. This relationship is analogous to the input–output relationship in linear time-invariant systems [164]; the difference lies in the fact that now the impulse response is time variant.

It is often useful to transform the impulse response into the Fourier domain. As there are two temporal variables, there are also two transform pairs. Absolute time t corresponds to Doppler frequency ν, while delay τ corresponds to frequency f. This results in the following functions [12], [105]:

- Performing the transformation $\tau \Rightarrow f$ results in the time-variant *transfer function* $H(t, f)$.
- Transformation $t \Rightarrow \nu$ gives the *delay–Doppler spread function* $s(\nu, \tau)$, also known simply as *spreading function*. This function can be interpreted physically by giving the amplitude of scatterers with specific delay and angle. If the scatterers are static and all the time variance is created by the movement of the mobile station, then the Doppler frequency is related to the angle of arrival (relative to the direction of movement) by a simple transformation [100]. Thus, the value of the spreading function for a specific (τ_i, ν_i) is related to the amplitude of the MPC with this delay and angle.
- Finally, the *bi-spreading function* $B(\nu, f)$ is defined as the Fourier transformation of $h(t, \tau)$ with respect to both t and τ.

Figure 2.3 shows the different transformations of the time-variant impulse responses, as well as the names given to them.

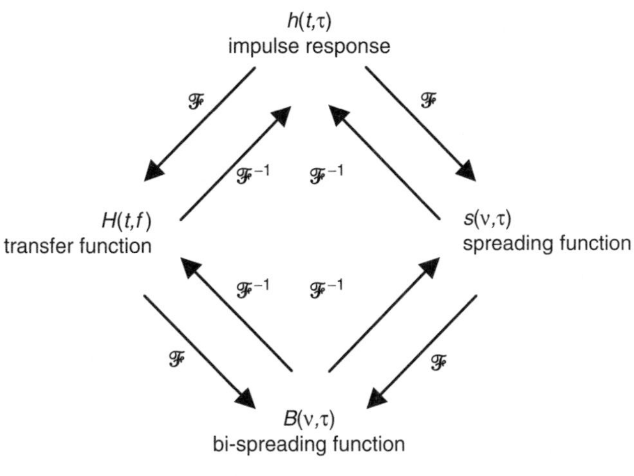

FIGURE 2.3 System functions.

We next consider the speed of the time variations and its influence on the channel characterization. Define first the quantities $\Delta\tau_{\text{spread}}$ and $\Delta\nu_{\text{spread}}$ as the *support* of the spreading function in delay and Doppler coordinates, respectively (i.e., $\Delta\tau_{\text{spread}}$ is the difference between the largest and the smallest delay of significant MPCs). Furthermore, let the *total spread* be $\Delta_{\text{total}} = \Delta\tau\Delta\nu$.

If $\Delta_{\text{total}} \ll 1$, then the interpretation of the time-varying channel impulse response is simple: the channel stays essentially time invariant during the coherence time. The coherence time, which can be approximated by the inverse of the Doppler spread $1/\Delta\nu$ [66], is then much longer than the duration of the impulse response. The channel is described for this time duration by $h(t_i,\tau)$. The other characteristic functions can be interpreted in a similar way. Also, if the symbol duration is much smaller than the coherence time, then a system "sees" a slowly time-varying channel, i.e., effectively constant for each symbol. In that case, the usual transfer function calculus, $Y(t_i, f) = H(t_i, f)X(t_i, f)$ can be used [132]. Many systems also use data "bursts," so that there is data transmission for some time, followed by a longer period of silence. If the burst duration is smaller than the coherence time, and the silence duration is larger, then we can approximate the channel as block fading, i.e., having one (time-invariant) impulse response during one burst, and a completely different one in the next.

If the total spread is *smaller* than unity, but *not much smaller*, then the characterization of the system becomes more difficult. The output is still the convolution of the input with the impulse response as described above. However, the usual transfer function calculus is not valid anymore. Interested readers should go to [12] and [112] for a more comprehensive description of this case.

Finally, a channel is called *overspread* if $\Delta_{\text{total}} > 1$ [112]. In that case, it is impossible to completely identify the channel; a determination of the time-varying impulse response (or its equivalents) is possible only if we make additional assumptions, like stipulating the validity of a parametric model. Fortunately, most practical wireless channels are significantly underspread.

2.2.3 Mathematical Description: Stochastic Case

As explained in Section 2.2.1, a purely deterministic description of the channel is often not possible or desirable. Rather, the impulse response is interpreted as a stochastic process, whose realizations are selected from an ensemble. (For a discussion on how to physically interpret ensemble in the wireless context, see [23], [105], [151].) A complete stochastic description of the impulse response would require its multidimensional pdf. This is often too complicated, so that [12] suggested a description by the autocorrelation function (ACF) only:

$$R_h(t, t'; \tau, \tau') = E\{h^*(t, \tau)h(t', \tau')\} \tag{2.1}$$

where $*$ denotes complex conjugation. Equivalently, we can define the ACFs of the Fourier transforms of the impulse response:

$$R_s(\nu, \nu'; \tau, \tau') = E\{s^*(\nu, \tau)s(\nu', \tau')\} \tag{2.2}$$

$$R_H(t, t'; f, f') = E\{H^*(t, f)H(t', f')\} \tag{2.3}$$

$$R_B(\nu, \nu'; f, f') = E\{B^*(\nu, f)B(\nu', f')\} \tag{2.4}$$

In addition, we define the mean value of the impulse response, $E\{h(t, \tau)\}$ or its equivalents.

To further simplify the autocorrelation functions, we can assume that the random process describing the channel is wide-sense stationary (WSS) and the scatterers are uncorrelated (US). In that case, the ACFs depend on only *two* instead of four variables:

$$R_h(t, t + \Delta t, \tau, \tau') = \delta(\tau - \tau')P_h(\Delta t, \tau) \tag{2.5}$$

$$R_H(t, t + \Delta t, f, f + \Delta f) = R_H(\Delta t, \Delta f) \tag{2.6}$$

$$R_s(\nu, \nu', \tau, \tau') = \delta(\nu - \nu')\delta(\tau - \tau')P_s(\nu, \tau) \tag{2.7}$$

$$R_B(\nu, \nu', f, f + \Delta f) = \delta(\nu - \nu')P_B(\nu, \Delta f) \tag{2.8}$$

where $P_h(\Delta t, \tau)$ is known as the *delay cross-power spectrum*, $R_H(\Delta t, \Delta f)$ is the *time–frequency correlation function*, $P_s(\nu, \tau)$ is the *scattering function*, and $P_B(\nu, \Delta f)$ is the *Doppler cross-power spectral density* [12], [105], [108], [155]. If the fading statistics are complex Gaussian, then the mean and the ACF completely characterize the random process. The combination of the Gaussian assumption with the WSSUS assumption is also called GWSSUS and is in widespread use. Finally, note that the stationarity of the statistics can be fulfilled only over limited areas [12], where the size of that area depends on environmental parameters as well as the considered bandwidth [54].

The function $R_H(0, \Delta f)$ is known as the frequency correlation function; $R_H(\Delta t, 0)$ is the time correlation function. The function $P(\tau) = P_h(0, \tau)$ is usually called the *power delay profile* (PDP), or *delay power density spectrum*, and it describes the expected received power for different delays τ. Assuming ergodicity, the PDP can be computed directly from measured values of the impulse response

$$P(\tau) = E_t\{|h(t, \tau)|^2\} \tag{2.9}$$

Note that in order to be physically meaningful, the expectation should be taken only over impulse responses within the area of stationarity [151].

The instantaneous PDP is defined as $P(t, \tau) = |h(t, \tau)|^2$ and is purely deterministic. An extensive discussion of the usefulness of the instantaneous PDP is given in [23].

2.2.4 Condensed Parameters

In many cases, the impulse response or the scattering function is still too complicated to quickly see the most important effects of the channel. A stronger condensation of the information is achieved by the *root mean squared (rms) delay spread*, which is the second central moment of the PDP:

$$S_\tau = \sqrt{\frac{\int_{-\infty}^{\infty} P(\tau)\tau^2 d\tau}{\int_{-\infty}^{\infty} P(\tau) d\tau} - \left(\frac{\int_{-\infty}^{\infty} P(\tau)\tau d\tau}{\int_{-\infty}^{\infty} P(\tau) d\tau}\right)^2} \tag{2.10}$$

Note that there is also an *instantaneous rms delay spread* $S_\tau(t)$ that is obtained by replacing $P(\tau)$ by $P(t, \tau)$ in Equation 2.10 [48], [151]. This quantity must not be confused with the rms delay spread [23].

A related quantity is the *coherence bandwidth*. Let us first define the normalized frequency correlation function

$$\tilde{R}_H(\Delta f) = \frac{E_t\{H^*(t, f)H(t, f + \Delta f)\}}{E_t\{|H(t, f|^2\}} \tag{2.11}$$

Then the coherence bandwidth B_c of level k is defined as the smallest number so that $|\tilde{R}_H(B_c)| < k$, where $k = 0.5, 0.75, 0.9$ has been used in the literature. The definition given in Equation 2.11 uses the complex transfer function. Similarly, coherence bandwidths for the envelope and for the phase can be defined [100].

The frequency correlation function (Equation 2.11) can also be shown to be the Fourier transform of the normalized power delay profile. This fact leads to an uncertainty relationship between rms delay spread and coherence bandwidth, namely [66],

$$B_k \geq \frac{\arccos(k)}{2\pi} \frac{1}{S_\tau} \tag{2.12}$$

Note that many papers set the coherence bandwidth simply equal to the inverse of the delay spread (times some constant), e.g., [71], [103], but this is an approximation that is valid only for specific shapes of the PDP.

The rms delay spread is the parameter that is used most frequently to describe the delay dispersion of a multipath channel. A huge number of papers report measurement campaigns whose final result is the rms delay spread. Part of the reason for its popularity is the fact that in *slightly* dispersive channels without line of sight (LOS) between transmitter and receiver, the unequalized bit error probability is proportional to S_τ^2 [13], [33], [146]. It is also popular to test a wireless system with a *prescribed* shape of the PDP (e.g., exponential), and just set S_τ (equal to the time decay constant of an exponential profile) equal to the measured S_τ. However, the rms delay spread is *not* a quantity that encompasses all (or even all important) properties of the delay dispersion. There are also other quantities that describe the delay dispersion, like the signal-to-self-interference ratio and the delay window, but they are not in widespread use. For a more extensive discussion, see [151], [183].

It must also be stressed that the extraction of the delay spread from measurements is nontrivial. Weak contributions at large delays have a considerable influence on the delay spread computed from Equation 2.9 and Equation 2.10. If noise is not eliminated, the measured delay spread can become infinite even for channels with finite-duration impulse responses. Furthermore, the windowing functions used in transformations from frequency domain to time domain also have a decisive influence on the actual value of the delay spread [189], [226].

2.2.5 Directional Description

The above descriptions of the channel model only the temporal properties of the channel. The direction of the MPCs does not enter the description; furthermore, the antenna is considered part of the channel, weighting (with the antenna pattern) and summing up the MPCs from the different directions. In other words, the impulse response describes the channel from the antenna connector at the TX to the antenna connector at the RX. This is undesirable when we want to analyze the effect of different antennas or for the case of multiple antenna elements.

The most fundamental deterministic description of the propagation channel is the double-directional impulse response [212],[4] which consists of a sum of contributions from the MPCs:

$$h(\vec{r}_T, \vec{r}_R, \tau, \Omega, \Psi) = \sum_{l=1}^{L(\vec{r})} h_l(\vec{r}_T, \vec{r}_R, \tau, \Omega, \Psi) \tag{2.13}$$

The impulse response depends on the locations of transmitter $\vec{r}_T$ and receiver $\vec{r}_R$, the delay τ, the direction of departure (DOD) Ω, the direction of arrival (DOA) Ψ, and the number of MPCs, $L(\vec{r})$, for the specific transmitter–receiver location. The $h_l(\vec{r}_T, \vec{r}_R, \tau, \Omega, \Psi)$ is the contribution of the l-th wave, which can be written as

$$h_l(\vec{r}_T, \vec{r}_R, \tau, \Omega, \Psi) = a_l e^{j\varphi_l} \delta(\tau - \tau_l) \delta(\Omega - \Omega_l) \delta(\Psi - \Psi_l)$$

Note that the absolute amplitude a, the delay, the DOA, and the DOD vary slowly (over several wavelengths) with the position, while the phase φ varies quickly.[5] However, the way the phases change with position (relative to some reference phase) is already implicit in the directional information so that it is admissible to write the double-directional impulse response as $h(\tau, \Omega, \Psi)$.

The single-directional impulse response can be obtained by integrating the double-directional impulse response (weighted by the transmit antenna pattern) over the DODs. Integrating the single-directional impulse response (weighted by the receiver antenna pattern) over the DOA results in the conventional impulse response [148].

Another representation of directional channels gives the impulse response of the channel at the elements of an antenna array. Thus, the impulse response becomes a matrix if we have arrays at both link ends, and a vector for an array at one link end [54]. We denote the transmit and receive element

[4]To be completely general, we would have to include a description of polarization as well. To avoid the cumbersome matrix notation, we omit this case here and refer interested readers to [152].

[5]To keep notation compact, we have not written these dependences on location explicitly in the above equation.

coordinates as $\vec{r}_T^{(1)}, \vec{r}_T^{(2)}, \ldots \vec{r}_T^{(N_T)}$ and $\vec{r}_R^{(1)}, \vec{r}_R^{(2)}, \ldots \vec{r}_R^{(N_R)}$, respectively, so that the impulse response from the i-th transmit to the j-th receive element[6] becomes (under the assumption of plane, narrowband waves)

$$
\begin{aligned}
h_{i,j} &= h\left(\vec{r}_T^{(i)}, \vec{r}_R^{(j)}\right) \\
&= \sum_l h_l\left(\vec{r}_T^{(0)}, \vec{r}_R^{(0)}, \tau_1, \Omega_l, \Psi_l\right) G_T(\Omega_l) G_R(\Psi_l) \exp\left(j < \vec{k}(\Omega_l), \vec{r}_T^{(i)} >\right) \exp\left(j < \vec{k}(\Psi_l), \vec{r}_R^{(i)} >\right)
\end{aligned} \tag{2.14}
$$

where G_T and G_R are the patterns of the transmit and receive antenna elements, respectively, $\vec{k}$ is the unit wave vector in the direction of the l-th DOD or DOA, and $< .,. >$ denotes the inner product. We thus see that it is always possible to obtain the impulse response matrix from a double-directional impulse response (and the knowledge of antenna positions and patterns), while the converse is not true.

If the receive arrays are uniform linear arrays, we can write Equation 2.14 as [197]

$$
\mathbf{H} = \int\int h(\tau, \Omega, \Psi) G_T(\Omega) G_R(\Psi) \vec{a}_R(\Psi) \vec{a}_T^H(\Omega) d\Psi d\Omega \tag{2.15}
$$

where we used the *steering vectors* $\vec{a}_T(\Omega) = \frac{1}{\sqrt{N_t}}[1, \exp(-j2\pi\frac{d}{\lambda}\sin(\Omega)), \ldots \exp(-j2\pi(N_t-1)\frac{d}{\lambda}\sin(\Omega))]$ and analogously defined $\vec{a}_R(\Psi)$.

The stochastic description of directional channels is analogous to the nondirectional case [67]. The autocorrelation function of the impulse response can be generalized to include the directional dependence so that it depends on six (or eight) variables. We can also introduce a generalized WSSUS condition that contributions coming from different directions are fading independently [107]. Note that the directions at the mobile station and the Doppler spreading are linked, and thus ν and Ω (or Ψ) are not independent variables anymore.

Analogously to the nondirectional case, we can define condensed descriptions of the wireless channel. The angular delay power spectrum (ADPS) and the angular power spectrum (APS) are defined (as seen from the base station antenna) via

$$
E\{s^*(\Omega, \tau, \nu) s(\Omega', \tau', \nu')\} = P_s(\Omega, \tau, \nu)\delta(\Omega - \Omega')\delta(\tau - \tau')\delta(\nu - \nu') \tag{2.16}
$$

$$
ADPS(\Omega, \tau) = \int P_s(\Omega, \tau, \nu) d\nu \tag{2.17}
$$

$$
APS(\Omega) = \int APDS(\Omega, \tau) d\tau \tag{2.18}
$$

Note also that an integration of the ADPS over Ω recovers the PDP.

The *azimuthal spread* is defined as the second central moment of the APS if all MPCs are incident in the horizontal plane. In many papers, e.g., [72], [176], it is defined in a form analogous to Equation 2.10, namely,

$$
S_\phi = \sqrt{\frac{\int APS(\phi)\phi^2 d\phi}{\int APS(\phi) d\phi} - \left(\frac{\int APS(\phi)\phi d\phi}{\int APS(\phi) d\phi}\right)^2} \tag{2.19}
$$

However, this definition is ambiguous because of the periodicity of the angle: by this definition, $APS = \delta(\phi - \pi/10) + \delta(\phi - 19\pi/10)$ would have a very different angular spread from $APS = \delta(\phi - 3\pi/10) +$

[6]If we deal with an antenna array at only one link end, the impulse responses at the antenna elements constitute a vector (instead of a matrix). However, we still write the elements of this vector as $h_{1,j}$ in order to avoid confusion with h_l, the contribution of the l-th multipath component.

$\delta(\phi - \pi/10)$, even though the two APSs differ just by a constant offset. A better definition was proposed in [67]. The first-order moment of the vector ensemble is defined as

$$\mu_\phi = \int \exp(j\phi)APS(\phi)d\phi \tag{2.20}$$

where the APS has been normalized to $\int APS(\phi)d\phi = 1$. Then the rms angular spread is

$$S_\phi = \sqrt{\int |\exp(j\phi) - \mu_\phi|^2 APS(\phi)d\phi} \tag{2.21}$$

Also note that the angular spread is only a partial description of the angular dispersion. It has been shown [10] that the correlation of signals at the elements of a uniform linear array depends only on the rms angular spread and not on the shape of the APS; however, this is valid only under some very specific assumptions [150]. Alternative condensed angular parameters were suggested in [53].

2.3 Modeling Methods for Multipath Channels

In the previous section, we have given a mathematical description of the multipath channel. In this section, we explain how it can be used to obtain *generic* simulation model structures. The actual parameter choices for those models will be discussed in Sections 2.4 and 2.5. Again, we can distinguish between deterministic and stochastic methods. The deterministic methods encompass measured impulse responses and ray tracing; the stochastic methods include tapped delay line models, frequency domain modeling, and geometry-based stochastic modeling. Deterministic models depend on the geography and morphology of the considered environment and are thus mainly suitable for site-specific modeling. This type of modeling is used for the deployment of networks. On the downside, these models do not necessarily describe the behavior in a typical environment, as they are tied to a specific location.

2.3.1 Measured Channel Impulse Responses

The most realistic deterministic models are based on direct channel measurement results. However, when using stored measurement data, it is important to know how it was measured and the limitations inherent to it. We thus briefly review the available measurement devices for impulse responses, also called *channel sounders* (for extensive reviews, see [41], [171], [172], [200]). Subsequently, we explain how directional measurements can be made.

The following techniques have been proposed to measure $h(t, \tau)$ or its equivalents:

1. Pulse generator. A direct measurement of the impulse response with short, intense excitation pulses is most straightforward conceptually, but is difficult to implement as the required peak-to-average power is very high. It has been used for ultrawideband measurements [39].
2. Correlation sounders. This is the most widespread sounding technique. The sounder transmits a pseudorandom sequence, and the receiver correlates with the same sequence [64]. Alternatively, chirp signals can be used [194]. If the ACF of the sequence approximates an impulse and if the channel is slowly time varying, then the output of the correlator approximates the channel impulse response. For quickly varying channels, corrections are required [133].
3. Swept time delay cross-correlator. This method, which is described in detail in [38], samples the received signal at a lower rate. This reduces the requirements for the analog-to-digital (A/D) converter but at the same time reduces the maximum admissible Doppler frequency for the channel to remain identifiable (i.e., underspread).

4. Network analyzer. This measurement device performs essentially a slow-frequency sweep of an exciting sine wave, thus directly measuring the transfer function. Network analyzers are readily available in most laboratories and thus very popular (e.g., [40], [87], [208]). However, they usually require a cable connection between transmitter and receiver and require the measurement environment to remain static for several seconds or minutes. Network analyzers are thus most suitable for indoor environments.

A nondirectional channel sounder can be used as a basis for constructing a directional channel sounder. The simplest way is to combine it with a directional antenna [168]. A stepper motor is used to point the antenna different directions. For each direction, the impulse response is recorded. The drawbacks of this method are the long measurement time and the fact that the resolution is limited by the beam width of the directional antenna. Alternatively, directional information can be obtained from an antenna array. In a physical array arrangement, one single-channel channel sounder receiver is connected to *each* antenna element so that the impulse responses at the different elements, $h_{i,j}$, can be measured simultaneously [143]. In a multiplexed array arrangement, there are several antenna elements connected to a single sounder via a fast radio frequency (RF) switch [222]. First, the impulse response at the first antenna element, $h_{1,1}$, is measured; then the switch connects the second element, measures $h_{1,2}$, and so on. Finally, in a virtual array arrangement, there is only a single antenna element, which is moved mechanically from one position to the next, measuring the different impulse responses, $h_{i,j}$ [116]. A basic assumption for the evaluation is that the environment does not change during the measurement procedure. Virtual arrays require between a few seconds and several minutes for one measurement run and can thus only be used in static environments; this precludes scenarios where cars or moving persons are significant scatterers. On the other hand, they avoid all problems with mutual coupling between antenna elements [99]. In nonstatic environments, multiplexed arrays are usually the best compromise between measurement speed and hardware complexity.

The measurement of double-directional channels used for multiple-input–multiple-output (MIMO) systems is even more complicated. We need to send signals from the transmit antennas that are orthogonal in either:

1. Time [221], [222] (i.e., first sending only from the first antenna, then only from the second, ...)
2. Frequency [32] (transmission of offset carriers from the different antennas)
3. Code [131] (e.g., transmission of different Walsh–Hadamard sequences from the different antennas)

In any case, each receiver antenna has to sort out the contribution from each transmit antenna element.

The channel sounders often give the impulse response at regularly spaced samples in time (and possibly space). If the channel model should follow the parametric model of Section 2.2.5, we have to extract the multipath parameters (time of arrival, direction of arrival, direction of departure, etc.) from these values. The simplest approach is a Fourier analysis; while its achievable resolution is usually sufficient in the delay domain, it is too poor in the spatial domain. For this reason, high-resolution parameter estimation approaches are often used. The most popular ones are MUSIC [198], ESPRIT [81], [191], SAGE [69], and the minimum variance method [115]. However, these algorithms might sometimes lead to numerical problems, and some of them are limited in the number of MPCs that can be estimated. The multipath parameters extracted from measurements can also be combined with the assumption of random phases for the MPCs, leading to a semideterministic channel model [153].

2.3.2 Deterministic Channel Computation

In recent years, deterministic field computation methods have developed into one of the most important methods for obtaining channel impulse responses. Progress has been driven by two factors: availability of faster computers that enabled computational tasks that were unthinkable only a few years ago and development of more efficient algorithms. This has allowed the development of a site-specific propagation analysis that, based on a database of the environment, estimates the channel characteristics.[7] For these methods, we do not need to distinguish between directional and nondirectional models; the computation methods described below inherently include directional information, and nondirectional models are created just by discarding this information.

2.3.2.1 Full Electromagnetic Description

A brute-force approach is a full solution of Maxwell's equations, through either integral or differential equation formulations. Integral equations are most often variations of the well-known method of moments [84], where the unknown currents induced in the scatterers are represented by a set of basis functions. As the number of basis functions increases with the size of the considered environment, and the computation time increases quadratically with the number of basis functions, this quickly leads to prohibitive computational complexity. Special methods that increase the efficiency of the method include natural basis sets [156], the fast multipole method [188], and the tabulated interaction method [22].

Differential equation formulations include the finite element method (FEM) [248], or the increasingly popular finite difference time domain (FDTD) method [119]. In rural outdoor environments, where essentially all scattering processes happen in one direction (neglecting back scattering), the wave equation can be approximated by a parabolic equation, which can be solved efficiently [101].

2.3.2.2 High-Frequency Approximations

By far the most widespread computation methods are based on the high-frequency approximation of the wave equation.[8] It approximates the waves as rays and establishes the (geometrical) propagation paths of those rays between TX and RX.

Let us first establish the physical mechanisms that can influence the propagation of a ray:

1. *Specular reflection*, and transmission, is conceptually the simplest process, being governed by Snell's law (angle of incidence equals angle of reflection) [181]. Most models implicitly assume that the reflectors are extended infinitely, thus limiting the accuracy in real environments (for a discussion, see [45]). Another possible source of error lies in the modeling of the actual reflection coefficients, especially for multilayer structures and for different polarization directions [93], [120].
2. *Diffraction* is a mechanism that influences the received power, especially in areas that are heavily shadowed. In microcells, diffraction around street corners is very important [58], and even the shape (chamfering) of the corner plays a significant role [218]. Similarly, diffraction plays an important role in indoor propagation. Diffraction is by nature a wave phenomenon, but can be treated for high-frequency approximations by the geometrical theory of diffraction (GTD) or the uniform theory of diffraction (UTD) [136]. Essentially, a diffraction edge serves as a secondary source for rays into all directions. The UTD is widely used, but still shows some inaccuracies when multiple edges are close together [5]. Another possible source of errors is the fact that, strictly speaking, the diffraction coefficient is frequency dependent; this can have an impact in very wideband simulations [180], [204].
3. *Diffuse scattering* arises because the surfaces involved in reflection, like house walls, furniture-lined office walls, etc., are not smooth (in the sense that the deviations from a true plane are much smaller than a wavelength). Thus, radiation can be reflected in directions other than the ones dictated by Snell's law. There are two prevalent theories of diffuse scattering, namely, the Kirchhoff theory and perturbation theory [227]. Usually only the latter one, which includes information about how fast the height of a surfaces varies, is applicable for wireless propagation. In any case, a rough surface acts

[7]Note that the accuracy of the database has an important impact on the accuracy of the resulting channel model [187].

[8]These methods are also sometimes summarily called ray tracing. However, we will avoid this notation because ray tracing is (in a convention that we follow) the name of a specific method using the imaging principle.

as a secondary source giving rise to rays in different directions. The fact that most house surfaces are not randomly rough, but rather show horizontal and vertical fine structures, can also have important effects [57].

More details about the physics and mathematical formulations of these processes can be found in [17], [227].

The way the tracking of the rays is done distinguishes ray tracing from ray launching. Ray tracing uses the imaging method for prediction of point-to-point impulse responses. Reflections at a wall are represented by an image source. The ray tracer establishes all image sources of the transmitter that can contribute at a given receiver position. This is straightforward when all involved surfaces are planes and only single reflections are taken into account. However, double reflections lead to an image of an image, and so on. This implies that the number of images increases exponentially with the number of reflection processes that are taken into account. Furthermore, since the method is inherently point to point, a new simulation has to be performed for each new TX or RX location. Acceleration methods that have been proposed in the literature include a bipartition algorithm [224] and the establishment of an image table [11]. This table includes all the various wall reflections, transmissions, and diffractions, and the areas in which they can lead to appreciable contributions. The information from this table is then used to compute the impulse response at each mobile station (MS) location.

Ray launching is an alternative method that sends off (launches) rays into all directions. The launching direction can be either random [199] or chosen from a regular, e.g., geodesic, pattern [52]. The algorithm follows the rays along their propagation path until they become too weak to be significant. This technique allows the computation of the power in a large area (for a given TX position) and is thus more efficient for comprehensive site-specific modeling. A preprocessing scheme [95] also allows fast computation for multiple transmitter locations; the environment is subdivided into tiles, and the interactions between all tiles are computed. Then, for each TX and RX position, only the interactions between the TX and the tiles that can act as first reflectors have to be computed. Ray launching is also capable of dealing with diffraction in a straightforward way; each diffraction edge serves as a secondary source of rays. Similarly, diffuse reflection (sending energy from a reflection process into all directions) can be included in the algorithm. A probabilistic scheme [173] chooses a random direction for the ray after each diffraction, with the pdf of the direction taken from measurements or fitting curves. One difficulty in ray launching is the determination of rays that actually can be received at a given location. It is usually assumed that a ray has to "hit" a so-called reception sphere in order to contribute to the received energy, but the construction of this sphere is quite tricky [52]. Alternatively, a transmitter can launch ray tubes instead of rays [114]; these alleviate the problem of determining the received power, but are more complicated to trace.

For both ray tracing and ray launching, a considerable simplification can be achieved by considering only a two-dimensional geometry; however, this might lead to exclusion of important propagation paths and thus inaccurate results. A compromise solution is 2.5-dimensional ray tracing, which adds up the contributions in the vertical and horizontal propagation planes [124]. Even more refined methods for combining results from two dimensional simulations to obtain three dimensional results are available, e.g., in [190], [245].

Note also that it is practically impossible to predict the correct phases for the different rays (in both ray tracing and ray launching) because of the inaccuracies of the databases of the geometry of reflectors and the reflection coefficients of the materials. Thus, only the average power and PDP (averaged over the small-scale fading) can be predicted deterministically, while the small-scale fading has to be treated stochastically [43]. Similarly, when comparing measurements to ray tracing results, only average powers can be compared [117]. It is remarkable that the *powers* as computed by tracing can be predicted quite well (typical standard deviation is 6 dB), but the PDPs and delay spreads usually deviate considerably [36]. The reason for this is that for average powers, errors in the ray amplitudes tend to cancel out; however, for the PDP, it is required that each resolvable ray is predicted correctly. This situation becomes even more pronounced when considering directionally resolved impulse responses.

2.3.3 Tapped Delay Lines

Most applications in signal processing require a discrete representation of the channel impulse response. Such a representation can be approached from two viewpoints: it can result from the physical modeling process or from critical or overcritical sampling of a continuous model of the channel.

The former approach follows naturally from the model of Section 2.2.5, which describes the impulse response as a sum of discrete echoes [183]:

$$h(t,\tau) = \sum_i a_i \exp(j\varphi_i)\delta(\tau - \tau_i) \tag{2.22}$$

Note that this model is a purely deterministic one if we interpret the arriving signals as completely resolvable echoes from discrete reflectors. However, in most practical cases, the resolution of the receiver is not sufficient to resolve all MPCs. In that case, the complex amplitudes $a_i \exp(j\varphi_i)$ represent the sum of several MPCs, and they thus fade; typically, the amplitudes are Rayleigh or Rice distributed, with the phases uniformly distributed. The arrival times of the MPCs are either assumed to be fixed (at equidistant intervals) or randomly distributed, e.g., according to a Poisson distribution (see Section 2.4). In the latter case, the model is usually called the Hashemi–Suzuki–Turin model [85], [216], [225].

Another problem arises from the necessity to get a discrete implementation of a continuous spreading function. When considering a band-limited system with bandwidth B, the interpolation

$$\tilde{h}_{bl}(\tau) = \sum_l A_l \,\mathrm{sinc}(B(\tau - \tau_l)) \tag{2.23}$$

with regularly spaced $\tau_l(\tau_l - \tau_{l-1} = T_s)$ is valid [227]; it establishes a relationship between the continuous function and the tap weights A_l. Note that if the *physical* scatterers fulfill the WSSUS condition, but are not equidistantly spaced, then the tap weights A_l are *not* necessarily WSSUS [217]. However, it is common to assume WSSUS also for this case.

The representation (Equation 2.23) is the correct interpolation according to the sampling theorem. For many applications, it is sufficient to have a time-discrete representation of $h(\tau)$, either because the complete simulation is time discrete or because the band limitation and interpolation are done at a different point in the simulation. In that case, a representation

$$\tilde{h}_{bl}(\tau) = \sum_l A_l\delta(\tau - \tau_l) \tag{2.24}$$

is sufficient.

Many of the standard models for wireless channels (see Section 2.5) were developed with a specific system and thus a specific system bandwidth in mind. There is often the temptation to use those models for other systems that exhibit a larger bandwidth (e.g., to test wireless LAN systems with a 20-MHz bandwidth with the International Telecommunications Union (ITU) outdoor-to-indoor model). We emphasize that this is *not* admissible and can cause completely misleading simulation results. On the other hand, it is often necessary to adjust the tap locations to a different sampling grid for a discrete simulation: in other words, a discrete simulation requires a channel representation $h(\tau) = \Sigma\delta(\tau - kT_s)$, but τ_l/T_s is a noninteger. There are three methods in widespread use:

1. Rounding to the nearest integer. This method leads to errors, but they are usually tolerable, especially if overcritical sampling is used.
2. Splitting the tap energy. The average energy is divided between the two adjacent taps $kT_s < \tau_l < (k+1)T_s$, possibly weighted by the distance to the original tap. Though this method is popular, we stress that it can give erroneous results, especially in sparse channel models. Increasing the number of energy-carrying taps by this splitting process can, for example, alter the performance of a Rake receiver in this channel.

3. Resampling. This can be done by using the interpolation formula or by describing the channel in the frequency domain and transforming it back (with a discrete Fourier transform) with the desired tap spacing.

Another interesting question is how to represent a continuous PDP by a *finite* number of taps in such a way that the representation error (for a given number of taps) is minimized [167]. This question is quite similar to the question of how to correctly approximate a Doppler spectra with a finite number of sinusoids — a topic that has drawn a lot of research and is reviewed in the monograph [166]. In general, we can distinguish the following:

1. Methods that use *fixed* tap locations and determine the tap weights according to certain criteria (method of equal distances, mean square error minimization) [167]. Note that equidistant tap spacing leads to a frequency correlation function that is periodic with the inverse of T_s; this can be a serious pitfall in the simulation of systems using frequency domain duplexing.
2. Methods that assume fixed tap weights but adjust the tap locations (method of equal areas) [167].
3. Monte Carlo methods [92], [239].
4. The Lp-norm method, which solves a nonlinear system of equations to optimize both tap location and amplitude [167].

When time variations of the channel are to be included, the temporal correlation function (or equivalently, the Doppler spectrum) also needs to be defined. Note that this has to be done for every delay tap. While some models assume a separable scattering function (i.e., the Doppler spectrum is independent of the delay), the so-called clustering effect (see Section 2.4.2) requires a more detailed model. The problem of simulating a (single) delay tap with a given Doppler spreading is equivalent to that of simulating a flat fading channel — a topic that has been extensively researched [166]. The question of nonseparable scattering functions is also addressed in [170]; an alternative method, using a Karhunen–Loeve expansion of the scattering function, is described in [240]. For large movements of the MS, also arrival times, DOAs and DODs change with time. An inheritance process can be used, where the path properties at time t_2 are the properties at time t_1, plus a (stochastic) change [247].

An alternative to the tapped delay line in the time domain is frequency domain modeling. The frequency response can, e.g., be modeled by an autoregressive process [74], [98], with the filter coefficients as parameters.

2.3.4 Stochastic MIMO Models

A generalization of the stochastic tapped delay line model is the inclusion of the directional information for single-directional [89], [134] or MIMO [236] models. Again, these models are based on the mathematical description given in Section 2.2. In all these models, it is assumed that the delay dependence is represented by the tapped delay line; thus, we are only faced with the problem of finding a representation for a flat-fading channel (for each tap). The directional model may be different for each tap; however, it is often assumed that directional and delay properties are separable.

A deterministic model again follows from the physical interpretation of the channel impulse response as a sum of resolvable echoes from a finite number of propagation paths, as given in Section 2.2.5. As in the nondirectional case, it is sometimes assumed that the wave from each direction consists of several, nonresolvable waves, so that their amplitudes are fading. However, the directional information allows the resolving of some MPCs that would not be resolvable in the delay domain alone.

Alternatively, the impulse response matrix (Equation 2.14) of a MIMO channel can be described by a stochastic model. In this case, the channel is characterized by not only the amplitude statistics of each matrix entry (which is usually Rayleigh or Ricean), but also the *correlation* between those entries [242]. The correlation matrix (for each tap) is defined by first stacking all the entries of the channel matrix in one vector $\vec{h}_{stack} = [H_{1,1}, H_{1,2}, \ldots H_{1,N_t}, H_{2,1}, \ldots H_{N_r,N_t}]^T$ and then computing the correlation matrix as $\mathbf{R} = E\{\vec{h}_{stack}\vec{h}^{T*}_{stack}\}$ where superscript $T*$ denotes the Hermitean transpose. One popular simplified model assumes that the correlation matrix can be written as a Kronecker product of the correlation matrices at the transmitter and receiver, $\mathbf{R} = \mathbf{R}_{TX} \otimes \mathbf{R}_{RX}$ [110], [241]. This model implies that the correlation matrix at the receiver is independent of the direction of transmission. While some papers have found the Kronecker model to approximate measurement data well (within a few percent relative error) [128], [138], [241], other papers found larger discrepancies [162], especially for large arrays [215]. If the model is valid, then the channel transfer function matrix can be generated as

$$\mathbf{H} = \mathbf{R}_{RX}{}^{1/2}\mathbf{G}\mathbf{R}_{TX}{}^{T/2} \tag{2.25}$$

where $\mathbf{G}$ is a matrix with independent identically distributed (i.i.d.) complex Gaussian entries. In LOS conditions, the correlation matrix should be extracted only from the diffuse non-LOS (NLOS) part of the channel, and the total channel matrix should be generated as a sum of a deterministic matrix $\mathbf{H}_{\mathrm{LOS}}$ and a stochastic matrix $\mathbf{H}_{\mathrm{NLOS}}$, which is generated according to Equation 2.25.

A more general model was proposed by [73], modeling the channel matrix as

$$\mathbf{H} = \mathbf{R}_{RX}{}^{1/2}\mathbf{G}\mathbf{R}_{TXRX}{}^{1/2}\tilde{\mathbf{G}}\mathbf{R}_{TX}{}^{T/2} \tag{2.26}$$

where $\mathbf{G}$ and $\tilde{\mathbf{G}}$ are matrices with i.i.d. Gaussian entries and $\mathbf{R}_{TXRX}$ describes the correlation of the propagation from scatterers near the transmitter to scatterers near the receiver. This model is capable of reproducing the so-called keyhole effect, where the channel transfer matrix has a low rank, even though the correlation matrices at the transmitter and receiver have full rank [4], [31], [73]. An even more general model, which allows taking the correlation between DODs and DOAs into account, is given in [231].

An alternative to the transfer function matrix model is the virtual channel representation of [197]. This model provides a simple geometrical interpretation of the channel but also captures the influence of the channel on capacity and diversity. The virtual channel is represented as

$$\mathbf{H} = \sum_m \sum_k H_V(k,m)\vec{a}_R(\tilde{\Psi}_k)\vec{a}_T^H(\tilde{\Omega}_m) = \tilde{\mathbf{A}}_R\mathbf{H}_V\tilde{\mathbf{A}}_T^H \tag{2.27}$$

where the steering vectors $\vec{a}_R(\tilde{\psi}_k)$ and $\vec{a}_T(\tilde{\Omega}_m)$ are defined in Section 2.2.5, with the virtual angles defined as $\tilde{\psi}_k = k/N_R$ and $\tilde{\Omega}_m = m/N_T$, with $-N_R \le k \le N_R$ and $-N_T \le m \le N_T$.

Interpreted physically, this is a representation of the scatterers in a beam domain, stating which power is coming from which direction. This representation has the advantage of being easily interpreted. Apart from that, it has essentially all the advantages and drawbacks of a transfer function matrix representation, as it is a Fourier transform of this quantity.

2.3.5 Geometry-Based Stochastic Channel Models

In any geometrical model, the impulse response is related to the location of scatterers. In a completely deterministic geometrical approach (for example, ray tracing), the location of the scatterers is prescribed deterministically from a database. In a geometry-based stochastic channel model (GSCM), on the other hand, the location is chosen stochastically, following a certain probability density function. The actual impulse response is then found by a simplified ray-tracing procedure, assuming that only single-scattering processes occur. The ray tracer thus needs to follow only the path from the TX to the scatterer and from there to the receiver (see Figure 2.4). This path gives the propagation delay, the direction of departure from the transmitter, and the direction of arrival at the receiver. Additional phase shifts from the reflection process can be taken into account by assigning a random phase shift (in addition to the phase shifts due to the propagation delay) to each MPC. Each path has an attenuation of its own — usually proportional to a power law (e.g., d^{-4}). The model was introduced independently in [19], [70], [126], [161], and [175]; a three-dimensional version is discussed in [144].

The above approach has a series of advantages. It emulates the physical reality and thus reproduces many effects implicitly. The small-scale fading is created by the superposition of waves from individual scatterers. The delays and directions of the waves (from which the correlations between the $h_{i,j}$ values follow) are

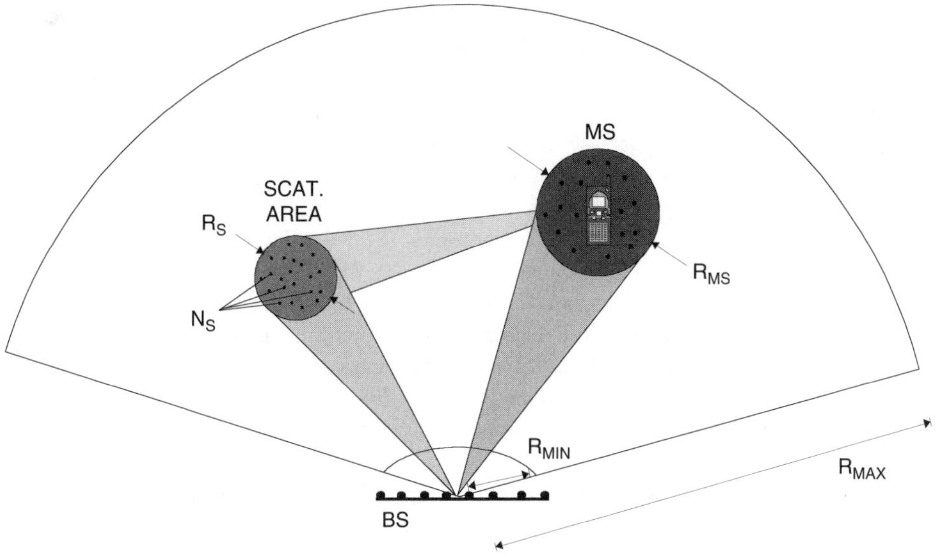

FIGURE 2.4 Geometry-based stochastic channel model (GSCM).

also implicit in the scatterer location and are thus reproduced automatically (including changes caused by large-scale movements of the MS). It is also relatively easy to parameterize the model, as many of the large-scale parameters can be derived from physical considerations. For example, the delay and DOA of signals scattered by high-rise buildings (far clusters, see below) follow immediately from the geographic position of the building, the base station (BS), and the mobile station (MS). On the downside, the model (at least in its straightforward form) relies on the assumption of single scattering. Furthermore, simulations take longer than with purely stochastic models for the same accuracy [166].

If the directional information at one link end only is required, then there exists a unique mathematical relationship between the scatterer location (using a single-scattering assumption) and the delay and angle [56], [152], [175]. This implies that it is always possible to find a scatterer geometry that corresponds to a measured angular delay power spectrum (ADPS), even though it might deviate from the actual physical scatterer arrangement. However, this equivalence is not possible anymore when the directional information at both link ends is to be taken into account. In that case, a double-scattering model must be used [147], [214]. A general model structure that takes double scattering, as well as several other propagation effects, into account is derived in [147].

2.4 Propagation Aspects and Parameterization

After all the mathematical and system-theoretic descriptions in the previous sections, we now turn to the actual parameterization of the wideband models, determining what numerical values the channel parameters take on in different environments. This section is only a very brief summary and cites only some key papers out of the hundreds of journal and conference contributions on the topic. It should also be noted that the mentioned typical values for parameters are just examples and vary widely.

The parameters of the channel, like path loss, delay spread, etc., depend on several key properties of the environment, as well as the system in question.

1. The topology and morphology of the environment have a decisive influence. For example, the delay spread and angular spread in an urban environment are significantly different from those in a rural environment. Also, finer subdivisions (e.g., into typical urban and bad urban) are widely used. Making finer differentiations is a trade-off between accuracy and complexity of the model, as well

as a question of the statistical viability of the underlying data [211]. In the following chapters, we will use a rough classification of eight different environments.

2. All parameters can vary with the carrier frequency. However, averaged over many measurements, only the average path loss shows a distinct frequency dependence. Shadowing and delay spread are sometimes reported to increase and sometimes reported to decrease with carrier frequency. In neither case is the dependence very pronounced. The frequency dependence of the narrowband parameters and the delay spread is investigated, e.g., in [2], [28], [102], [160], and [195]; for the frequency dependence of the angular spread, see [178]. Note that most of the parameter values cited in Sections 2.4.1 to 2.4.6 were measured at center frequencies of 1, 2, or 5 GHz, and with a measurement bandwidth that was *much* smaller than the center frequency.
3. The height of the base station with respect to the surrounding buildings has a strong influence. As a first consequence, it leads into a classification of macrocell (BS above rooftop), microcell (BS below rooftop), and picocell (BS indoor). But also within those classes, the height of the BS has a distinct influence on path loss, delay spread, and other parameters; see, e.g., [14], [18], [65], and [192].
4. The antenna pattern influences the parameters, especially the delay spread. Directional antennas usually lead to a smaller delay spread [141], unless they suppress the LOS [82]. In macrocells, the use of pattern downtilt at the BS decreases the delay spread [14]. Furthermore, it has been observed that the antenna polarization influences the delay spread; in particular, circularly polarized antennas reduce the delay spread [184].

2.4.1 Amplitude Statistics

The first step in any system planning is the determination of the average narrowband power. In this context, *average* means averaged over both small-scale fading and shadowing. Then the shadow fading and small-scale effects are added on to that. For the narrowband power, many models have been developed [18], [20], [183]; the most popular are presented below:

- For macrocells, the venerable Okumura–Hata model [88], [163] is still the most popular model for rural as well as urban environments. It has been modified and extended in [36] to cover higher frequencies and especially be more suitable for urban environments. The dependence of the path loss on the distance (in dB) follows $L_p[\text{dB}] = K + n_1 \log(d/d_{\text{ref}})$, where the parameters K and n depend on the environment, the carrier frequency, and the antenna heights; the reference distance d_{ref} is usually chosen to be 1 m or 1 km. Typical values for n are about 2.5 for LOS situations and 3 to 4 for NLOS situations. For other types of urban environments, the Walfisch–Bertoni model [229] and its modifications [36] are in widespread use.
- For microcells, the so-called breakpoint models are in widespread use, so that

$$L_p[dB] = \begin{cases} K + n_1 \log(d/d_{\text{ref}}) & d < d_{\text{break}} \\ K + n_1 \log(d_{\text{break}}/d_{\text{ref}}) + n_2 \log(d/d_{\text{break}}) & d > d_{\text{break}} \end{cases} \tag{2.28}$$

  One example for such a model is [65], but many other models have been published as well. Values for n_1 typically range from 1.7 (for LOS situations) to 2.5 (for NLOS); n_2 ranges from 3 to 5. Other models, which also include height dependence of the antenna, are presented in [36], [235].
- For picocells (indoor), two types of models are used. One is a simple power decay law, $K + n_1 \log(d/d_{\text{ref}})$, where exponents between 1.5 and 7 have been measured [140]. The other is the Motley–Keenan model [158], which also uses a $K + n_1 \log(d/d_{\text{ref}})$ decay law but adds an excess loss of L_{excess} for each floor or wall between transmitter and receiver. L_{excess} depends on the building materials (typically 5 to 10 dB). The excess loss is less important when there are alternative propagation paths. For example, when TX and RX are on different floors, then rays that go out of a window, are reflected off a nearby building, and enter through a window on the next floor might become the dominant propagation paths.

Superimposed on the average path loss is a lognormally distributed variation, caused by shadowing effects. The variance of this lognormal distribution varies typically between 6 and 12 dB [125]; it has also been suggested to model the variance itself as a random variable [75]. The autocorrelation function of the shadowing is usually modeled as an exponential [80] with a correlation length that ranges from 1 m in picocells [125] to some 10 m in urban cells [207] to several hundred meters in rural macrocells [211].

Finally, the small-scale statistics of the amplitude (for each tap or delay bin) is either *Rayleigh*, with a pdf

$$pdf(x) = \frac{x}{\sigma^2} \exp\left(-\frac{x^2}{2\sigma^2}\right) \tag{2.29}$$

where σ^2 is the average power; *Ricean*,

$$pdf(x) = \frac{x}{\sigma^2} \exp\left(-\frac{x^2 + A^2}{2\sigma^2}\right) I_0\left(\frac{xA}{\sigma^2}\right) \tag{2.30}$$

where A is the amplitude of the dominant component, $A^2/(2\sigma^2)$ is the Rice factor, and I_0 is the zero-order modified Bessel function of the first kind; or *Nakagami*,

$$pdf(x) = \frac{2}{\Gamma(m)} \left(\frac{m}{\Omega}\right)^m x^{2m-1} \exp\left(-\frac{m}{\Omega}x^2\right) \tag{2.31}$$

where $m \geq 1/2$ is the Nakagami m-factor, $\Gamma(m)$ is the gamma function, and Ω is the mean square value of the amplitude. The Nakagami and Rice distributions look fairly similar and can be (approximately) transformed into each other [213]. Note that in all three equations above, x is the amplitude and is understood to range from zero to infinity.

In most cases, the amplitude statistics are well described by a Rayleigh distribution. In LOS conditions, the (narrowband) Rice factor (i.e., the ratio of the power of the direct component to all other MPCs) ranges typically from 0 to about 20 dB [8], [206]; the Rice factor in a single (i.e., the first) delay bin can be appreciably higher, depending on the bin length [8]. Bins at larger delays typically obey a Rayleigh distribution; an exception to this rule occurs in ultrawideband channels [26], [246]. When considering the amplitude statistics over larger areas (combining small-scale and large-scale fading), the Suzuki distribution [216] is in widespread use; since it is somewhat complicated, it is also sometimes approximated by a Nakagami distribution.

2.4.2 Arrival Times

In land-based wireless communication systems, multipath components do not arrive in a continuum because scatterers are not continuously distributed in space. (Note that this is an important difference from ionospheric communications.) Thus, a determination of the statistics of the arrival times of the multipath components is of interest.

The simplest assumption is that scatterers are discrete but completely randomly distributed. This gives rise to a Poisson point process [225]. However, it does not agree well with measurements. An improvement is the so-called Δ-K model, which was suggested for outdoor cases in [85], [216], and [225] and for the indoor case in [86]. It defines two states: state A, where the arrival rate of paths is λ, and state B, where the rate is $K\lambda$. The model starts in state A. If a path arrives at time t, then a transition is made to state B for a minimum of time Δ. If no path arrives during that time, the model reverts to state A; otherwise, it remains in state B. This state-dependent rate leads to a clustering of MPCs. In other words, a considerable number of MPCs arrive within a shorter delay span, and some time later, another group of MPCs arrives. The clustering of MPCs is also reproduced in the Saleh–Valenzuela model [193], which uses the following discrete-time impulse response:

$$h(t) = \sum_{l=0}^{L} \sum_{k=0}^{K} a_{k,l} \delta(t - T_l - \tau_{k,l}) \tag{2.32}$$

where $a_{k.l}$ is the tap weight of the k-th component in the l-th cluster, T_l is the delay of the l-th cluster, and $\tau_{k,l}$ is the delay of the k-th MPC relative to the l-th cluster arrival time T_l. By definition, we have $\tau_{0,l} = 0$. The distributions of the cluster arrival times and the ray arrival times are given by Poisson processes, so that the interarrival times are exponentially distributed ([106] suggests Weibull distributions instead):

$$\begin{aligned} p(T_l | T_{l-1}) &= \Lambda \exp[-\Lambda(T_l - T_{l-1})], \quad l > 0 \\ p(\tau_{k,l} | \tau_{(k-1),l} &= \lambda \exp[-\lambda(\tau_{k,l} - \tau_{(k-1),l})], \quad k > 0 \end{aligned} \tag{2.33}$$

where Λ is the cluster arrival rate ranging from 20 to 300 ns and λ is the ray arrival rate ranging from 0.5 to 10 ns [193], [208] (though the higher reported values are possibly related to limited resolution bandwidth). The existence of multiple clusters has also been observed for outdoor environments. One cluster always corresponds to scattering around the mobile station, while other clusters correspond to propagation via distinct morphological or geographical features of the environment. Long-delayed echoes have been observed as an effect of mountains [50], while in urban environments, high-rise buildings or buildings surrounding large open areas cause the late-arriving clusters [202]. These clusters carry significant energy mostly when the MS has an unobstructed view to them. They thus "appear" and "disappear" as the MS moves over larger areas. A model for this process is described in [9].

Finally, the total (narrowband) shadowing of the received signal is created by the overlap of the shadowed contributions of the different clusters. It is physically reasonable to assume that shadowing for different clusters is independent (or shows, at most, a partial correlation) [78], [232]. The total shadowing distribution is thus the sum of (possibly correlated) lognormally distributed variables [25].

2.4.3 Average Time Dispersion

In many practical situations, the arrival of discrete MPCs cannot be observed; due to the limited bandwidth, they get "smeared" into a continuum. In the simplest case of a single cluster, the PDP usually follows quite well a one-sided exponential function [100]:

$$P(\tau) = P_{sc}(\tau) = \begin{cases} \exp(-\tau/S_\tau) & \tau \geq 0 \\ 0 & \text{otherwise} \end{cases} \tag{2.34}$$

For fixed wireless LANs, [59] proposed the combination of a strong (quasi-) LOS path with a profile of the form Equation 2.34. For propagation in street canyons, a modal waveguide analysis has shown good agreement with experiments, giving rise to a power law (instead of exponential) decay [177].

In the case of multiple clusters, the PDP is the sum of exponential contributions [8]:

$$P(\tau) = \sum_k \frac{P_k^c}{S_{\tau,k}^c} P_{sc}\left(\tau - \tau_{0,k}^c\right) \tag{2.35}$$

where P_k^c, $\tau_{0,k}^c$, and S_t^c are the power, delay, and delay spread of the k-th cluster, respectively. Of course the sum of all cluster powers has to add up to the narrowband power described in Section 2.4.1. For a PDP of form (Equation 2.34), the rms delay spread characterizes the delay dispersion. In the case of multiple clusters (Equation 2.35), the rms delay spread is defined mathematically, but often has a very limited physical meaning. Still, the vast majority of measurement campaigns available in the literature use only this parameter for the characterization of the delay dispersion. Typical values are presented below:

- *Indoor residential buildings*: Here the delay spreads are rather small, with 5 to 10 ns being typical values [76]; however, 30 ns has also been measured [135].

- *Indoor office environments*: Office environments show somewhat larger delay spreads than residential areas. While delay spreads of 5 ns have been measured [40], [104], both corridor-cum-office environments and "Dilbertian" environments (big halls with cubicles) typically show delay spreads between 10 and 100 ns [28], [46], [47], [87], [111], [141], [208], [220]. The room size has a clear influence on the delay spread [16].
- *Factories and airport halls*: References [83] and [182] measured delay spreads between 50 and 200 ns (roughly uniformly distributed) in different factories.
- *Microcells*: In microcells, delay spreads range from around 5 to 100 ns (for LOS situations) to 100 to 500 ns (for NLOS) [65], [113], [234], [244]. Also, measurements of delay spreads in outdoor peer-to-peer networks [90] indicate similar values. The delay spread increases with antenna height [7], [65].
- *Tunnel and mines*: Railway or car tunnels are special environments that are of interest because of the importance of coverage for emergency situations. Empty tunnels typically show a very small delay spread (on the order of 20 ns), while car-filled tunnels exhibit larger values (up to 100 ns) [243]. Reference [127] found even smaller values in mines.
- *Typical urban and suburban environments*: These show delay spreads between 100 and 800 ns [109], [165], [209], although values up to 3 μs have also been observed [103], [109], [185], while extremely low values (around 25 ns) have been measured in [244].
- *Bad urban and hilly terrain environments*: These show clear examples of multiple clusters (Section 2.4.2) that lead to much larger delay spreads. Reference [202] measured delay spreads up to 18 μs, with cluster delays of up to 50 μs, in various European cities; measurements in other European cities [15], [77], [116], [130], [223] and American cities [185] showed somewhat smaller values. References [50] and [145] have measured cluster delays of up to 100 μs in mountainous terrain.

The delay spread also shows considerable large-scale variations. Several papers [3], [79] find that the delay spread has a lognormal distribution with a variance of typically 2 to 3 dB in suburban and urban environments. Its correlation length is usually assumed to be the same as the correlation length of the shadowing (see Section 2.4.1). For indoor environments, both lognormal [26] and normal [76], [169] distributions have been suggested.

Even more noteworthy is that the delay spread shows a clear dependence on the distance between transmitter and receiver (or, equivalently, on the average path loss); for a physical interpretation, see [205]. Reference [79] reviewed the literature and found that the delay spread is proportional to d^{ε}, where $\varepsilon = 0.5$ in urban and suburban environments and $\varepsilon = 1$ in mountainous regions. The distance dependence has also been found in microcells [65], [219] and in indoor environments [24], [30], [74], [87], [238], though some of those references propose a different functional form of this dependence.

2.4.4 Average Angular Dispersion at the BS

Similarly to the delay dispersion, we can describe the angle dispersion by the angular power spectrum; in many cases, this is simplified to describe only the azimuthal power spectrum (APS). If only a single cluster occurs, the APS is found to be Laplacian [174]:

$$APS(\phi) = \exp\left[-\sqrt{2}\frac{|\phi|}{S_\phi}\right] \tag{2.36}$$

while the density of MPCs is modeled as Gaussian. In the case of multiple clusters, the APS becomes

$$APS(\phi) = \sum_k \frac{P_k^c}{\sqrt{2}S_{\phi,k}^c} APS_{sc}\left(\phi - \phi_{0,k}^c\right) \tag{2.37}$$

where superscript c again denotes "per cluster" quantities. Reference [208] proposed a generalization of the Saleh–Valenzuela model to include (clustered) DOAs, and [230] further extended it to the MIMO case, i.e., including DODs.

The *cluster* angular delay power spectrum (ADPS) can be approximated as a product of the PDP, the APS, and an elevation spectrum that has a shape similar to that of Equation 2.36. However, the *total* ADPS cannot be written in such a form; rather, it is the sum of cluster ADPSs.

As for the delay spread, many papers give only the rms angular spread as the result of measurement campaigns, though some give the cluster rms spread. While fewer papers deal with angular dispersion than with delay dispersion, the following range of values can be considered typical:

- *Indoor office environments*: Cluster angular spreads between 10 and 20 degrees have been observed [34], [208] for NLOS situations; for LOS situations, they are considerably smaller. Distribution functions for the DOAs and DODs are given in [247].
- *Industrial environments*: Reference [83] observed angular spreads between 20 and 30 degrees.
- *Microcells*: Angular spreads between 5 and 20 degrees for LOS and 10 to 40 degrees for NLOS were found in [168], while [29] measured 5 to 15 degrees, decreasing with distance. Reference [223] found about 30 degrees azimuthal spread and about 2 degrees *cluster* azimuthal spread.
- *Typical urban and suburban environments*: Reference [174] measured angular spreads on the order of 5 to 15 degrees, while [121] found 3 to 20 degrees in dense urban environments. Reference [122] found angular spreads of up to 40 degree with a distinct decrease of the angular spread with distance. In suburban environments, the angular spread was usually smaller than 5 degrees due to a frequent occurrence of LOS [121].
- *Bad urban and hilly terrain environments*: These show a considerably larger total angular spread due to the existence of multiple clusters. References [123] and [223] found around 20 degrees azimuthal spread and around 2 degrees cluster azimuthal spread.
- *Rural environments*: Angular spreads between 1 and 5 degrees have been observed [168].

In outdoor environments, the distribution of the angular spread over large areas has also been found to be lognormal and correlated with the delay spread (correlation coefficient of approximately 0.5) [3], [8]. Actually, delay spread, angular spread, and shadowing are a triplet of correlated lognormal variables, with a correlation of angular spread to shadowing around 0.75 [8], and similarly for the correlation between delay spread and shadowing. Also, for indoor peer-to-peer environments, a correlation between delay spread and angular spread has been observed [55].

The dependence of the angular spread on the distance is still a matter of discussion. Reference [174] shows examples of increasing, constant, and even decreasing azimuth spreads. The most common assumption in directional models is that the azimuthal spread is independent of distance. Furthermore, angular spread was observed to decrease with increasing antenna height [176].

The elevation spectrum is also usually modeled as Laplacian. For macrocellular environments, the elevation spread is very small, on the order of 0.3 to 3 degrees [8], [223] and around 5 degrees in microcells [223]. In indoor environments, elevation spreads can be considerably larger, and it is also more problematic to write the angular spectrum as a product of the azimuthal and elevation spectrums [247].

2.4.5 Average Angular Dispersion at the MS

A common assumption for the MS is that radiation is incident from all directions — and only in the horizontal plane; this model dates back to the 1970s [100]. However, recent studies indicate that the azimuthal spread can be considerably smaller, especially in street canyons. Cluster angular spreads on the order of 20 degree have been suggested [116], where there is either just one cluster or two (with 180 degree difference in the mean). The azimuthal power spectrum is approximated as Laplacian; another proposal is the von Mises pdf [1]:

$$pdf_{\phi}(\phi) = \frac{\exp(\kappa \cos(\phi - \bar{\phi}))}{2\pi I_0(\kappa)} \tag{2.38}$$

where the parameter κ determines the "peakiness" of the distribution and $\overline{\phi}$is the mean angle of arrival; $I_0(x)$ is the zero-order modified Bessel function of the first kind. Furthermore, the angular distribution is a function of the delay. For MSs located in street canyons without LOS, small delays are related to over-the-rooftop propagation, while later components are waveguided through the streets and thus confined to the angular range under which this canyon is seen by the MS [116]. In indoor environments with (quasi-) LOS, early components have a very small angular spread, while components with larger delays have an almost uniform distribution of the APS [39]. For the outdoor elevation spectrum, Reference [116] indicated that MPCs that propagate over the rooftops have an elevation distribution that is uniform between 0 and the angle under which the rooftops are seen; later-arriving components, which have propagated through the street canyons, show a Laplacian elevation distribution. It is worth noting that the angular distributions as measured with a channel sounder can differ greatly from those that occur in the presence of a human body close to the antenna [233].

The angular spectrum at the MS also determines the Doppler spectrum if only the MS is moving. However, moving scatterers can also introduce temporal variations of the transfer function: both pedestrians [228] and cars [142] influence this process. For fixed wireless networks, temporal Rice factors have been measured [206], ranging from –5 to 20 dB.

2.4.6 MIMO Parameters

Following the methodology of Section 2.3, measurements for MIMO channels can be represented either as joint DOA–DOD distributions or as correlation matrices for the received signal. If the DOAs are assumed to be independent of the DODs, then the Kronecker model (Section 2.3.4) is valid and the (separate) correlation matrices at the transmitter and receiver specify the model. For outdoor environments, measurement results for this case have been presented in [32] and [110]; for indoor, see [91], [137], and [138]. Equivalently, the product of the DOA and DOD distributions (Section 2.4.4 and 2.4.5) completely specifies the model. Joint DOA–DOD measurements are given in [153] and [212].

2.4.7 Polarization

Polarization has become a very important quantity for broadband wireless networks, as many recent systems exploit polarization diversity, or polarization MIMO systems. As a first approximation, parameters like delay spread, angular spread, etc., are the same for vertical and horizontal polarizations for NLOS situations. The fading of the components is independent; the *average* power leakage between the two polarizations is given by the cross-polarization discrimination (XPD). Values for the XPD in LOS indoor situations are above 10 dB; around 3 dB [60] in indoor NLOS situations; 5 dB in urban outdoor; and 12 dB in suburban outdoor environments [159]. More generally, the XPD is a function of the excess path loss: high path loss (which also implies no LOS) leads to low XPD. A model based on measurements in suburban environments is presented in [206].

2.4.8 Millimeter Wave Propagation

As mentioned at the beginning of Section 2.4, most propagation parameters are fairly invariant to changes of the operating frequency. Thus, millimeter wave propagation (referring here mostly to 17, 24, 30, and 60 GHz) does not differ too much from the propagation discussed in Section 2.4.1 to 2.4.7. However, there are a few important differences:

- Diffraction is not an efficient process for propagating energy.
- Diffuse scattering is more important, because surface roughness (relative to the wavelength) is larger.
- At 60 GHz, atmospheric gases, especially oxygen, lead to a stronger attenuation [96].

These physical facts have important consequences for the channel modeling:

- Outdoor communications tend to rely on the existence of a LOS.
- Delay spreads are typically smaller than those for centimeter wave communications. At 17 GHz, delay spreads around 20 ns are typical for indoor office environments [21], [129], whereas they are less than 10 ns at 60 GHz [42], [157], though the actual environments seem to have far larger influences than the carrier frequency [51], and higher values have been measured especially in hallways [237]. For outdoor environments, the delay spreads are somewhat larger (typically around 50 ns) [21], though smaller values have been observed for narrow streets [37]. All those measurements were done for low-positioned antennas. In point-to-multipoint distribution systems, delay spreads can reach several hundred nanoseconds [186].

2.4.9 Ultrawideband Systems

UWB channels, where the measurement bandwidth can reach several gigahertz, show many similarities to the systems we discussed above. Results for delay spread, path loss, and shadowing, based on an extensive measurement campaign [75], [76], are fairly similar to the results of Section 2.4.1 to 2.4.6. However, there are also some vital differences:

1. Due to the fine time resolution of ultrawideband systems, each resolvable delay bin contains only a few multipath components. Thus, the central limit theorem is not applicable anymore, and the amplitude statistics in each delay bin change. Both Nakagami distributions [26] and lognormal distributions [154] have been suggested.
2. Alternatively, the impulse response can also be modeled as the sum of several discrete (deterministic) components and Rayleigh-distributed clutter [49], [118].
3. For deterministic predictions, we have to take into account that the reflection and diffraction coefficients are frequency dependent [180].

2.5 Standard Models

Standard models for the mobile radio channel are important tools for the development of new radio systems. They allow the estimation of benefits for different multiple access techniques, signal processing, and other measures for enhancing the capacity and improving performance without having to build a hardware prototype for every considered system and testing it in the field.

Models for the mobile radio channel must fulfill conflicting requirements. On one hand, they should be detailed enough to reflect all relevant properties of the channel; on the other hand, they should be simple enough to allow rapid implementation and fast simulation times. The trade-off between simplicity and accuracy in the modeling is a difficult one. Furthermore, standardization of wireless systems is influenced as much by business aspects as by scientific considerations. Thus, standardized channel models — especially those used directly for commercial system design — might sometimes use oversimplifications and rather mysterious assumptions and reasoning. But even with all those caveats, standard models fulfill an important need for wireless system design, not only for industry, but also for academia. The following sections are intended mainly as lookup sections for practicing engineers, without extensive discussions of the merits and possible improvements of the models.

2.5.1 The COST 207 Model

Within the European research initiative COST 207 [63], statistical wideband models were developed. These models were the basis for defining the channel models of GSM [62]. They are very simple tapped delay line models, where each tap is independently fading. Delays, average powers, amplitude fading distributions, and Doppler spectra are time invariant.

The model specifies four different environments, each characterized by a different power delay profile.

- Rural area (RA): $P(\tau) = \begin{cases} \exp(-9.2\tau/\mu s) & 0 < \tau < 0.7\mu s \\ 0 & \text{otherwise} \end{cases}$
- Hilly terrain (HT): $P(\tau) = \begin{cases} \exp(-3.5\tau/\mu s) & 0 < \tau < 2\mu s \\ 0.1\exp(15 - \tau/\mu s) & 15 < \tau < 20\mu s \\ 0 & \text{otherwise} \end{cases}$
- Typical urban (TU): $P(\tau) = \begin{cases} \exp(-\tau/\mu s) & 0 < \tau < 7\mu s \\ 0 & \text{otherwise} \end{cases}$
- Bad urban (BU): $P(\tau) = \begin{cases} \exp(-\tau/\mu s) & 0 < \tau < 5\mu s \\ 0.5\exp(5 - \tau/\mu s) & 5 < \tau < 10\mu s \\ 0 & \text{otherwise} \end{cases}$

The fading exhibits a Rayleigh distribution, with a Doppler spectrum that is either classical (Jakes), Gaussian, or Ricean. Specifically, the Doppler categories are:

- CLASS is a Jakes spectrum $P_D(\nu) = \frac{1}{\sqrt{1-(\frac{\nu}{\nu_{max}})^2}}$ for $-\nu_{max} \leq \nu \leq \nu_{max}$. It is used for delays smaller than 0.5 μs.
- GAUS1 is the sum of two Gaussian spectra with different means and variances: $P_D(\nu) = \exp[-\frac{(\nu+0.8\nu_{max})^2}{2(0.05\nu_{max})^2}] + 0.1\exp[-\frac{(\nu-0.4\nu_{max})^2}{2(0.1\nu_{max})^2}]$. It is used for delays between 0.5 and 2 μs.
- GAUS2 is also a sum of two Gaussians, namely, $P_D(\nu) = \exp[-\frac{(\nu-0.7\nu_{max})^2}{2(0.1\nu_{max})^2}] + 0.032\exp[-\frac{(\nu+0.4\nu_{max})^2}{2(0.15\nu_{max})^2}]$ and is used for delays in excess of 2 μs.
- RICE is the sum of a classical Doppler spectrum and a delta function (corresponding to an LOS path) and is used for the first component in rural area environments: $P_D(\nu) = \frac{0.41}{2\pi\nu_{max}\sqrt{1-(\frac{\nu}{\nu_{max}})^2}} + 0.91\delta(\nu - 0.7\nu_{max})$ for $-\nu_{max} \leq \nu \leq \nu_{max}$.

Note that all Doppler spectra must be normalized so that $\int P_D(\nu)d\nu = 1$. The maximum Doppler frequency is determined from the carrier frequency and the speed of the mobile station, which implies that all scatterers can be considered stationary. Table 2.1 shows tap weights and Doppler spectra for some tapped delay line implementations, as suggested by COST 207.

The COST 207 models were derived from extensive measurement campaigns throughout Europe. It is noteworthy, however, that most of those measurements were performed with a measurement bandwidth of 200 kHz. Thus, it is problematic to use those models for an analysis of systems with larger bandwidths. This is especially true for the tapped delay line implementations described in Table 2.1.

2.5.2 The ITU-R Models

For the selection of the air interface of third-generation cellular systems, the International Telecommunications Union (ITU) developed another set of models that is available only as a tapped delay line implementation [249]. It specifies three environments: indoor, pedestrian (including outdoor to indoor), and vehicular (with high base station antennas). For each of those environments, two channels are defined: channel A (low delay spread) and channel B (high delay spread); in addition there is also a worst case scenario. The occurrence rate of the two models is also specified; all those parameters are given in Table 2.2. The amplitudes follow a Rayleigh distribution; the Doppler spectrum is uniform between $-\nu_{max}$ and ν_{max} for the indoor case and is a classical Jakes spectrum for the pedestrian and vehicular environments.

TABLE 2.1 Tapped Delay Line Implementation of COST 207 Channel Models

Tap No.	Delay/μs	Power/dB	Doppler Spectrum
RURAL AREA			
1	0	0	RICE
2	0.2	−2	CLASS
3	0.4	−10	CLASS
4	0.6	−20	CLASS
HILLY TERRAIN			
1	0	0	CLASS
2	0.2	−2	CLASS
3	0.4	−4	CLASS
4	0.6	−7	CLASS
5	15	−6	GAUS2
6	17.2	−12	GAUS2
TYPICAL URBAN			
1	0	−3	CLASS
2	0.2	0	CLASS
3	0.6	−2	GAUS1
4	1.6	−6	GAUS1
5	2.4	−8	GAUS2
6	5.0	−10	GAUS2
BAD URBAN			
1	0	−3	CLASS
2	0.4	0	CLASS
3	1	−3	GAUS1
4	1.6	−5	GAUS1
5	5	−2	GAUS2
6	6.6	−4	GAUS2

TABLE 2.2 Tapped Delay Line Implementation of ITU-R Models

Tap No.	Delay/ns	Power/dB	Delay/ns	Power/dB	Doppler Spectrum
INDOOR	CHANNEL A (50%)		CHANNEL B (45%)		
1	0	0	0	0	FLAT
2	50	−3.0	100	−3.6	FLAT
3	110	−10.0	200	−7.2	FLAT
4	170	−18.0	300	−10.8	FLAT
5	290	−26.0	500	−18	FLAT
6	310	−32.0	700	−25.2	FLAT
PEDESTRIAN	CHANNEL A (40%)		CHANNEL B (55%)		
1	0	0	0	0	CLASS
2	110	−9.7	200	−0.9	CLASS
3	190	−19.2	800	−4.9	CLASS
4	410	−22.8	1200	−8.0	CLASS
5	–	–	2300	−7.8	CLASS
6	–	–	3700	−23.9	CLASS
VEHICULAR	CHANNEL A (40%)		CHANNEL B (55%)		
1	0	0	0	−2.5	CLASS
2	310	−1	300	0	CLASS
3	710	−9	8900	−12.8	CLASS
4	1090	−10	12,900	−10	CLASS
5	1730	−15	17,100	−25.2	CLASS
6	2510	−20	20,000	−16	CLASS

In contrast to the COST 207 model, the ITU model also specifies the path loss (in dB) depending on the distance d:

- Indoor: $L_p = 37 + 30\log_{10} d + 18 N_{\text{floor}}^{\left(\frac{N_{\text{floor}}+2}{N_{\text{floor}}+1} - 0.46\right)}$.
- Pedestrian: $L_p = 40\log_{10} d + 30\log_{10} f_c + 49$; for the outdoor-to-indoor case, the additional building penetration loss is modeled as a normal variable with 12 dB mean and 8 dB standard deviation.
- Vehicular: $L_p = 40(1 - 4 \cdot 10^{-3}\Delta h_B)\log_{10} d - 18\log_{10}\Delta h_B + 21\log_{10} f_c + 80$, where f_c is the carrier frequency and Δh_B is the base station height measured from the rooftop level; the model is valid for $0 < \Delta h_B < 50$ m.

In all three cases, lognormal shadowing is imposed. The variance is 12 dB for the indoor case, 10 dB for the vehicular case, 10 dB for outdoor pedestrian users, and 12 dB for indoor pedestrian users. The autocorrelation function of the shadowing is assumed to be exponential, $R(\Delta x) = \exp(-\ln(2)|\Delta x|/d_{\text{corr}})$, where the correlation length d_{corr} is 20 m in vehicular environments, but not specified for other environments.

2.5.3 IEEE 802.11/HIPERLAN Models

The GSM and ITU models were established for the development of cellular communications systems. Wireless LANs, like the ETSI HIPERLAN and IEEE 802.11 standards, operate in a different environment and with a larger bandwidth. The standardization bodies ETSI and IEEE 802.11 thus recommended a different tapped delay line model developed by Medbo and Schramm [139]. The structure is again similar to that of the GSM model, defining five different channels (tapped delay lines):

1. Model A for a typical office environment, NLOS conditions, 50-ns rms delay spread
2. Model B for a typical large open space and office environments, NLOS conditions, 100-ns rms delay spread
3. Model C for a large open space (indoor and outdoor), NLOS conditions, 150-ns rms delay spread
4. Model D for a large open space, LOS conditions, 140-ns rms delay spread
5. Model E for a typical large open space (indoor and outdoor), NLOS conditions, 250-ns rms delay spread

The tap settings for all those cases are given in Table 2.3.

Currently, IEEE 802.11 is also developing an extension of the model to the MIMO case; details may be found in [60].

2.5.4 The 802.15 Ultrawideband Channel Model

In contrast to conventional communications systems that operate in a narrow part of the spectrum, ultrawideband signals can span the frequency range from DC to tens of GHz. Since they have a large bandwidth, the power spectral density, and thus the interference to narrowband systems, is typically very low. The large bandwidth also means that conventional channel models do not correctly describe UWB channels. For example, the number of MPCs that fall within each resolvable delay bin is small, so that the central limit theorem is no longer applicable, and the amplitude statistics are not Rayleigh anymore. Also, the channel impulse response is real, since the signals do not have a carrier frequency anymore. New channel models are thus required. A first step has been made by IEEE 802.15.3a, which standardized a channel model for indoor ultrawideband communications [35], [154].

The multipath model is defined in terms of a (Saleh–Valenzuela) model, as described in Section 2.4.2:

$$h_i(t) = X_i \sum_{l=0}^{L}\sum_{k=0}^{K} a_{k,l}^i \delta(t - T_l^i - \tau_{k,l}^i) \tag{2.39}$$

TABLE 2.3 Model for Indoor Wireless LANs (HIPERLAN, 802.11). All Taps have a Classical Doppler Spectrum, Except for the First Tap of Channel D Which Has a 10 dB Spike

	Model A		Model B		Model C		Model D		Model E	
Tap No.	Delay/ns	Power/dB	Delay/ns	Power/dB	Delay/ns	Power/dB	Delay/ns	Power/dB	Delay/ns	Power/dB
1	0	0	0	−2.6	0	−3.3	0	0	0	−4.9
2	10	−0.9	10	−3.0	10	−3.6	10	−10.0	10	−5.1
3	20	−1.7	20	−3.5	20	−3.9	20	−10.3	20	−5.2
4	30	−2.6	30	−3.9	30	−4.2	30	−10.6	40	−0.8
5	40	−3.5	50	0	50	0	50	−6.4	70	−1.3
6	50	−4.3	80	−1.3	80	−0.9	80	−7.2	100	−1.9
7	60	−5.2	110	−2.6	110	−1.7	110	−8.1	140	−0.3
8	70	−6.1	140	−3.9	140	−2.6	140	−9.0	190	−1.2
9	80	−6.9	180	−3.4	180	−1.5	180	−7.9	240	−2.1
10	90	−7.8	230	−5.6	230	−3.0	230	−9.4	320	0
11	110	−4.7	280	−7.7	280	−4.4	280	−10.8	430	−1.9
12	140	−7.3	330	−9.9	330	−5.9	330	−12.3	560	−2.8
13	170	−9.9	380	−12.1	400	−5.3	400	−11.7	710	−5.4
14	200	−12.5	430	−14.3	490	−7.9	490	−14.3	880	−7.3
15	240	−13.7	490	−15.4	600	−9.4	600	−15.8	1070	−10.6
16	290	−18.0	560	−18.4	730	−13.2	730	−19.6	1280	−13.4
17	340	−22.4	640	−20.7	880	−16.3	880	−22.7	1510	−17.4
18	390	−26.7	730	−24.6	1050	−21.2	1050	−27.6	1760	−20.9

where $a^i_{k,l}$ are the tap weights, X_i represents the lognormal shadowing, and i refers to the i-th realization. The distribution of arrival rates is given by Equation 2.33; the channel coefficients are defined as

$$a_{k,l} = p_{k,l}\xi_l\beta_{k,l} \tag{2.40}$$

where ξ_l reflects the fading associated with the l-th cluster, $\beta_{k,l}$ corresponds to the fading associated with the k-th ray of the l-th cluster, and $p_{k,l}$ is equiprobable ± 1 to account for signal inversion due to reflections. The distribution of the channel coefficients is given by

$$|\xi_l\beta_{k,l}| = 10^{(\mu_{k,l}+n_1+n_2)/20} \tag{2.41}$$

where $n_1 \propto N(0,\sigma_1^2)$ and $n_2 \propto \mathrm{N}(0,\sigma_2^2)$ are independent and correspond to the fading of each cluster and ray, respectively. Further,

$$E\left[|\xi_l\beta_{k,l}|^2\right] = \Omega_0 e^{-T_l/\Gamma} e^{-\tau_{k,l}/\gamma} \tag{2.42}$$

where T_l is the excess delay of cluster l and Ω_0 is the mean energy of the first path of the first cluster. The $\mu_{k,l}$ values are thus given by

$$\mu_{k,l} = \frac{10\ln(\Omega_0) - 10T_l/\Gamma - 10\tau_{k,l}/\gamma}{\ln(10)} - \frac{\left(\sigma_1^2+\sigma_2^2\right)\ln(10)}{20} \tag{2.43}$$

Finally, since the lognormal shadowing of the total multipath energy is captured by the term X_i, the total energy contained in the terms $\sum\sum a_{k,l}$ is normalized to unity for each realization. This shadowing term is characterized by $20\log_{10}(X_i) \propto \mathrm{N}(0,\sigma_x^2)$. Table 2.4 summarizes the parameters of the model for the four environments defined by IEEE.

Based on the work of [75], IEEE 802.15 also recommends a new modeling of the path loss. While there is still shadowing superimposed on a polynomial power decay law, now the decay exponent and the shadowing variance also become random variables, whose realizations change from building to building. Table 2.5 shows the path loss at 1-m distance, as well as the mean and standard deviation for LOS and NLOS situations. The distributions of all variables are modeled as Gaussian.

TABLE 2.4 Parameters for the Ultrawideband Channel Model

Model Parameters	CM 1[1]	CM 2[2]	CM 3[3]	CM 4[4]
Λ (1/nsec)	0.0233	0.4	0.0667	0.0667
λ (1/nsec)	2.5	0.5	2.1	2.1
Γ	7.1	5.5	14.00	24.00
γ	4.3	6.7	7.9	12
σ_1 (dB)	3.4	3.4	3.4	3.4
σ_2 (dB)	3.4	3.4	3.4	3.4
σ_x (dB)	3	3	3	3

TABLE 2.5 Parameters for UWB Path Loss Model

	LOS		NLOS	
	Mean	Standard Deviation	Mean	Standard Deviation
PL_0(dB)	47	NA	50.5	NA
n	1.7	0.3	3.5	0.97
σ (dB)	1.6	0.5	2.7	0.98

2.5.5 The 3GPP–3GPP2 Model

Up to now, we have treated single-input–single-output (SISO) standard models. However, as multiple antennas become an integral part of many wireless systems, MIMO standard models are required. These models also give a more accurate representation of the delay characteristics than the ITU models. On the other hand, those models are also more complicated, and space restrictions prevent us from giving here all the details that are necessary for implementation. Interested readers will find further resources in the references within this chapter.

The standardization organizations for third-generation cellular systems, 3GPP (third-generation partnership project) and 3GPP2, have established a double-directional channel model for the evaluation of transmit diversity and MIMO systems [27], [201], especially in the context of the high-speed data mode, HSDPA. This model not only provides directional information that is important for multiple-antenna systems, but also provides an improved model for the delay dispersion. Again, the model is a tapped delay line model [236], where the taps are specified in the angular domain as well as the delay domain. Three environments are specified: suburban macrocells, urban macrocells, and urban microcells.[9]

There are two major differences to the approaches described in Section 2.5.1 to 2.5.3:

1. Each tap (path, in the notation of 3GPP–3GPP2) is characterized by its delay and *mean* angles. However, each path consists of several subpaths that all have the same delay, but different angles of arrival and departure distributed around the mean angles.
2. The small-scale average rms delay spread, angular spread, and shadowing are not assumed to be fixed anymore. Rather, they are modeled as correlated random variables.

Consequently, the first step in the simulation is the selection of the small-scale average delay spread, angular spread, and shadowing from their joint probability density function. These spreads parameterize the ADPS. We next choose random path delays and determine the (small-scale-averaged) powers associated with these. Then the mean angle of arrival is determined for each path as a realization of a random variable. Each of the taps then consists of 20 subpaths, which all have the same delay, but *fixed* offsets from the mean

[9]In addition to the performance evaluation model discussed here, the standardization document also specifies a calibration model (sometimes erroneously called link-level model). The calibration model is extremely simplified and should be used only to test the basic correctness of simulation engines.

angle of arrival of this tap, fixed amplitudes (all identical), and random phases. These subpaths are created both at the transmitter and at the receiver. In a next step, each TX subpath is associated with *exactly one* randomly chosen RX subpath. Shadow fading is added to the total ADPS.

This procedure is valid for suburban and urban macrocells. In microcells, a slight modification is used, as the generation of the PDP and the mean angles of incidence are different. Also, several options have been defined that can be switched on to obtain a better agreement with reality:

- Polarized antennas: Following Section 2.4.7, the two polarizations are independently fading and the XPD is chosen at random from a Gaussian distribution.
- Far scatterer clusters: One far scatterer cluster has been added for the urban environment, to reflect the important bad urban situation (Section 2.4.3). Its placing within the cell is done at random and follows essentially the ideas of COST 259 (see Section 2.5.6).
- Line of sight: For urban microcells, LOS is treated in a probabilistic way, with the probability of obtaining LOS decreasing with distance.
- Urban canyons: For urban canyons the angular spectrum at the MS is modified.

A public domain implementation in MATLAB will be available at www.es.lth.se/radiocom.

2.5.6 The COST 259 Model

The European research initiative COST 259 developed a directional channel model (DCM) that has gained widespread acceptance. It is very realistic, incorporating a wealth of effects and their interplay, all for different environments; however, this makes it quite complex, and we can only point out some basic features here. A more detailed description of the first version of the model is described in [211], and a final account is found in [8] and [152].

The COST 259 DCM is even more general than the 3GPP–3GPP2 model, including small-scale as well as *continuous* large-scale changes of the channel. This is achieved efficiently by distinguishing between three different layers:

1. At the top layer, there is a distinction between different radio environments (REs), i.e., environmental classes with similar propagation characteristics (e.g., typical urban). All in all, there are 13 REs: 4 macrocellular REs (base station height above rooftop), 4 microcellular REs (outdoor, base station height below rooftop), and 5 picocellular REs (indoor).
2. Large-scale effects, which are described by their probability density functions, whose parameters differ for the different REs.
3. Small-scale fading, whose means and variances are determined by the large-scale effects.

The large-scale effects are described in a mixed geometrical-stochastic fashion, applying the concept of scatterer clusters as described in Section 2.4.2. At the beginning of a simulation, scatterer clusters (one local, around the MS, and several far scatterer clusters) are distributed at random (according to their pdfs) in the coverage area; this is the stochastic component. During the simulation, the delays and angles between the clusters are obtained deterministically from their positions and the positions of BS and MS; this is the geometrical component. Each of the clusters has a small-scale-averaged ADPS that is exponential in delay and Laplacian in azimuth and elevation. The angularly resolved complex impulse responses are then obtained from the average ADPS either directly (i.e., using a tap model; see Section 2.3.3) or by mapping it onto a scatterer distribution and obtaining the impulse responses in a geometrical way (Section 2.3.5).

In macrocells, the positions of the clusters are random, following a certain pdf. In micro- and picocells, the positions are more deterministic, using the concept of virtual cell deployment areas (VCDAs). A VCDA is essentially a map of a virtual town or office building, with a route of the MS prescribed in it. For any position of the MS, the cluster positions in delay and angle are thus known completely deterministically. This approach is somewhat similar to a ray-tracing approach but differs in two important respects: the city maps need not reflect an actual city and can thus be made more typical for many different cities. Furthermore, only the *cluster positions* are determined by ray tracing, while the behavior *within* one cluster is still treated stochastically.

2.6 Conclusions

This chapter gave an overview of channel models for broadband wireless communications. The fundamental structure of models for frequency-selective channels has been investigated for approximately 40 years and has reached a certain stage of maturity. Model structures for multiantenna systems (smart antennas and MIMO), on the other hand, are still a very active research area, as can be seen from the references cited in this chapter. For *all* channels, the model parameterization (and the channel measurements it is based on) is a work in progress. Important achievements have been made in the past, and standardized channel models, as described in Section 2.5, have played an important role in the design of second- and third-generation wireless systems. Still, the need for new operating environments and higher accuracy will necessitate additional models in the future.

From the wide variety of models, it becomes clear that there is no single master method for "best" channel modeling. Any model involves the trade-off between accuracy and complexity. When selecting (or designing) a model, it is thus important to first determine the model properties that can influence the system of interest. System design, wave propagation, and channel modeling cannot be treated as isolated disciplines, but rather have to work hand in hand.

Acknowledgment

Part of this work was supported by an INGVAR grant of the Swedish SSF.

References

[1] Abdi, A. and Kaveh, M., A space-time correlation model for multielement antenna systems in mobile fading channels, *IEEE J. Selected Areas Comm.*, 20, 550, 2002.

[2] Affandi, A., El Zein, G., and Citerne, J., Investigation on frequency dependence of indoor radio propagation parameters, in *Proceedings of the 50th IEEE Vehicular Technology Conference, VTC 1999 — Fall*, 1999, p. 1988.

[3] Algans, A., Pedersen, K.I., and Mogensen, P.E., Experimental analysis of the joint statistical properties of azimuth spread, delay spread, and shadow fading, *IEEE J. Selected Areas Comm.*, 20, 523, 2002.

[4] Almers, P., Tufvesson, F., and Molisch, A.F., Measurement of keyhole effect in wireless multiple-input–multiple-output (MIMO) channels, *IEEE Comm. Lett.*, 7, 375, 2003.

[5] Andersen, J.B., UTD multiple-edge transition zone diffraction, *IEEE Trans. Antennas Propagation*, 45, 1093, 1997.

[6] Andersen, J.B., Rappaport, T.S., and Yoshida, S., Propagation measurements and models for wireless communications channels, *IEEE Comm. Magazine*, 33, 42, 1995.

[7] Arowojolu, A.A., Turkmani, A.M.D., and Parsons, J.D., Time dispersion measurements in urban microcellular environments, in *Proceedings of the IEEE Vehicular Technology Conference*, 1994, p. 150.

[8] Asplund, H. et al., The COST 259 directional channel model II. Macrocells, *IEEE Trans. Wireless Comm.*, submitted.

[9] Asplund, H. et al., Clustering of scatterers in mobile radio channels: evaluation and modeling in the COST259 directional channel model, in *Proceedings of the IEEE International Conference Communication*, 2002, p. 901.

[10] Asztely, D., Ottersten, B., and Swindlehurst, A.L., A generalized array manifold model for local scattering in wireless communications, in *Proceedings of the IEEE International Conference on Acoustics, Speech, Signal Process*, 1997, p. 4021.

[11] Athanasiadou, G.E., Nix, A.R., and McGeehan, J.P., A microcellular ray-tracing propagation model and evaluation of its narrow-band and wide-band predictions, *IEEE J. Selected Areas Comm.*, 18, 322, 2002.
[12] Bello, P.A., Characterization of randomly time-variant linear channels, *IEEE Trans. Comm. Systems*, CS-11, 360, 1963.
[13] Bello, P. and Nelin, B.D., The effect of frequency-selective fading on the binary error probabilities of incoherent and differentially coherent matched filter receivers, *IEEE Trans. Comm.*, 11, 170, 1963.
[14] Benner, E. and Sesay, A.B., Effects of antenna height, antenna gain, and pattern downtilting for cellular mobile radio, *IEEE Trans. Vehicular Technol.*, 45, 217, 1996.
[15] Berg, J.-E. et al., Specular reflections from high-rise buildings in 900 MHz cellular systems, in *Proceedings of the 41st IEEE Vehicular Technology Conference, VTC* 1991, p. 594.
[16] Bergljung, C. and Karlsson, P., Propagation characteristics for indoor broadband radio access networks in the 5 GHz band, in *Proceedings of the IEEE International Symposium on Personal, Indoor Mobile Radio Communications*, 1999, p. 2323.
[17] Bertoni, H.L., *Radio Propagation for Modern Wireless Systems*, Prentice-Hall, New York, 2000.
[18] Bertoni, H.L. et al., UHF propagation prediction for wireless personal communications, *Proc. IEEE*, 82, 1333, 1994.
[19] Blanz, J.J. and Jung, P., A flexibly configurable spatial model for mobile radio channels, *IEEE Trans. Comm.*, 46, 367, 1998.
[20] Blaunstein, N., *Radio Propagation in Cellular Networks*, Artech House, Norwood, MA, 1999.
[21] Bohdanowicz, A., Janssen, G.J.M., and Pietrzyk, S., Wideband indoor and outdoor multipath channel measurements at 17 GHz, in *Proceedings of the 50th IEEE Vehicular Technology Conference, 1999 — Fall*, 1998.
[22] Brennan, C. and Cullen, P., Tabulated interaction method for UHF terrain propagation problems, *IEEE Trans. Antennas Propagation*, 46, 881, 1998.
[23] Bultitude, R.J.C., Estimating frequency correlation functions from propagation measurements on fading radio channels: a critical review, *IEEE J. Selected Areas Comm.*, 20, 1133, 2002.
[24] Bultitude, R.J.C. et al., The dependence of indoor radio channel multipath characteristics of transmit/receiver ranges, *IEEE J. Selected Areas Comm.*, 11, 979, 1993.
[25] Cardieri, P. and Rappaport, T.S., Statistics of the sum of lognormal variables in wireless communications, in *Proceedings of the IEEE Vehicular Technology Conference, 2000 — Spring*, p. 1823.
[26] Cassioli, D., Win, M.Z., and Molisch, A.F., The ultra-wide bandwidth indoor channel: from statistical model to simulations, *IEEE J. Selected Areas Comm.*, 20, 1247, 2002.
[27] Calcev, G. et al., A MIMO channel model for third-generation cellular systems, *IEEE Trans. Wireless Comm.*, to be submitted.
[28] Chandra, A., Kumar, A., and Chandra, P., Comparative study of path losses from propagation measurements at 450 MHz, 900 MHz, 1.35 GHz and 1.89 GHz in the corridors of a multifloor laboratory-cum-office building, in *Proceedings of the IEEE Vehicular Technology Conference, VTC 1999*, p. 2272.
[29] Chen, M. and Asplund, H., Measurements and models for direction of arrival of radio waves in LOS in urban microcells, in *Proceedings of the IEEE International Symposium Personal, Indoor Mobile Radio Communications*, 2001, p. B-100.
[30] Cheung, K.W. and Murch, R.D., Measurement, characterization and modeling of the wideband indoor channel, in *Proceedings of the 46th IEEE Vehicular Technology Conference*, 1996, p. 588.
[31] Chizhik, D., Foschini, G.J., and Valenzuela, R.A., Capacities of multi-element transmit and receive antennas: correlations and keyholes, *Electronics Lett.*, 36, 1099, 2000.
[32] Chizhik, D. et al, Multiple-input–multiple-output measurements and modeling in Manhattan, *IEEE J. Selected Areas Comm.*, 21, 321, 2003.
[33] Chuang, J.C.I., The effects of time delay spread on portable radio communications channels with digital modulation, *IEEE J. Selected Areas Comm.*, 5, 879, 1987.
[34] Chong C.-C. et al., A new statistical wideband spatio-temporal channel model for 5-GHz band WLAN systems, *IEEE J. Selected Areas Comm.*, 21, 139, 2003.

[35] IEEE 802.15.3a Channel Modeling Subcomittee, channel modeling subcommittee report 02490r0P802-15_SG3a, at http://grouper.802.org.

[36] Damosso, E. and Correia, L., Eds., Digital Mobile Radio Towards Future Generation Systems, Final Report of COST231, European Union, Brussels, Belgium, 1999.

[37] Correia, L.M. and Reis, J.R., Wideband characterisation of the propagation channel for outdoors at 60 GHz, in *Proceedings of the IEEE International Symposium on Personal, Indoor Mobile Radio Communications*, 1996, p. 752.

[38] Cox, D.C., Multipath delay spread and path loss correlation for 910 MHz urban mobile radio propagation, *IEEE Trans. Vehicular Technol.*, 25, 340, 1972.

[39] Cramer, R.J.-M., Scholtz, R.A., and Win, M.Z., Evaluation of an ultra-wide-band propagation channel, *IEEE Trans. Antennas Propagation*, 50, 561, 2002.

[40] Cuinas, I. and Sanchez, M.G., Measuring, modeling, and characterizing of indoor radio channel at 5.8 GHz, *IEEE Trans. Vehicular Technol.*, 50, 526, 2001.

[41] Cullen, P.J., Fannin, P.C., and Molina, A., Wide-band measurement and analysis techniques for the mobile radio channel, *IEEE Trans. Vehicular Technol.*, 42, 589, 1993.

[42] Dardadi, D. et al., Wideband indoor, communication channels at 60 GHz, in *Proceedings of the IEEE International Symposium on Personal, Indoor Mobile Radio Comm.*, 1996, p. 791.

[43] Degli-Esposti, V., de'Marsi, A., and Fuschini, F., A field prediction model for urban environment based on pattern recognition and statistical ray tracing, in *Proceedings of the IEEE Vehicular Technology Conference, 2001 — Spring*, p. 428.

[44] Degli-Eposti, V. et al., Wide-band measurement and ray-tracing simulation of the 1900-MHz indoor propagation channel: comparison criteria and results, *IEEE Trans. Antennas Propagation*, 49, 1101, 2001.

[45] Dersch, U. and Zollinger, E., Propagation mechanisms in microcell and indoor environments, *IEEE Trans. Vehicular Technol.*, 43, 1058, 1994.

[46] Devasirvatham, D.M.J., Time delay spread and signal level measurements of 850 MHz radio waves in building environments, *IEEE Trans. Antennas Propagation*, 34, 1300, 1986.

[47] Devasirvatham, D.M.J., A comparison of time delay spread and signal level measurements within two dissimilar office buildings, *IEEE Trans. Antennas Propagation*, 35, 319, 1987.

[48] Dimitrakopoulos, G.A. and Capsalis C.N., Statistical modeling of RMS-delay spread under multipath fading conditions in local areas, *IEEE Trans. Vehicular Technol.*, 49, 1522, 2000.

[49] Domazetovic, A. et al., A new modeling approach for wireless channels with predictable path geometries, in *Proceedings of the 56th IEEE Vehicular Technology Conference, 2002 — Fall*, p. 454.

[50] Driessen, P.E., Prediction of multipath delay profiles in mountainous terrain, *IEEE J. Selected Areas Comm.*, 18, 336, 2000.

[51] Droste, H. and Kadel, G., Measurement and analysis of wide band indoor propagation characteristics at 17 GHz and 60 GHz, in *Proceedings of the International Conference Antennas Propagation*, 1995, p. 288.

[52] Durgin, G., Patwari, N., and Rappaport, T.S., An advanced 3D ray launching method for wireless propagation prediction, in *Proceedings of the IEEE Vehicular Technology Conference* 1997, p. 785.

[53] Durgin, G. and Rappaport, T.S., Theory of multipath shape factors for small-scale fading wireless channels, *IEEE Trans. Antennas Propagation*, 48, 682, 2000.

[54] Durgin, G., *Space-Time Wireless Channels*, Cambridge University Press, Cambridge, U.K., 2003.

[55] Durgin, G.D., Kukshya, V., and Rappaport, T.S., Wideband measurements of angle and delay dispersion for outdoor and indoor peer-to-peer radio channels at 1920 MHz, *IEEE Trans. Antennas Propagation*, 51, 936, 2003.

[56] Eggers, P.C.F., Generation of base station DOA distributions by Jacobi transformation of scattering areas, *Electronics Lett.*, 34, 24, 1998.

[57] El-Sallabi, H.M. et al., Influence of diffraction coefficient and corner shape on ray prediction of power and delay spread in urban microcells, *IEEE Trans. Antennas Propagation*, 50, 703, 2002.

[58] Erceg, V., Rustako A.J., Jr., and Roman, R.S., Diffraction around corners and its effects on the microcell coverage area in urban and suburban environments at 900 MHz, 2 GHz, and 4 GHz, *IEEE Trans. Vehicular Technol.*, 43, 762, 1991.

[59] Erceg, V. et al., A model for the multipath delay profile of fixed wireless channels, *J. Selected Areas Comm.*, 17, 399, 1999.

[60] Erceg, V. et al., Indoor MIMO WLAN Channel Models, IEEE 802 standardization document 11-03-161ro, 2003.

[61] Ertel, R.B. et al., Overview of spatial channel models for antenna array communication systems, *IEEE Personal Communication*, February 1998, p. 10.

[62] ETS 300 577, 15th ed. (GSM 05.05, version 4.23.1), European Telecommunications Standards Institute, Valbonne, France, December 1999.

[63] Failli, E., Ed., *Digital Land Mobile Radio Communications: COST 207*, European Union, Brussels, Belgium, 1989.

[64] Fannin, P.C. et al., Digital signal processing techniques applied to mobile radio channel sounding, *IEE Proc.-F*, 138, 502, 1991.

[65] Feuerstein, M.J. et al., Path loss, delay spread, and outage models as functions of antenna height for microcellular system design, *IEEE Trans. Vehicular Technol.*, 43, 487, 1994.

[66] Fleury, B.H., An uncertainty relation for WSS processes and its application to WSSUS systems, *IEEE Trans. Comm.*, 44, 1632, 1996.

[67] Fleury, B.H., First- and second-order characterization of direction dispersion and space selectivity in the radio channel, *IEEE Trans. Information Theory*, 46, 2027, 2000.

[68] Fleury, B. and Leuthold P.E., Radiowave propagation in mobile communications: an overview of European research, *IEEE Communication Magazine*, February 1996, p. 70.

[69] Fleury, B.H. et al., Channel parameter estimation in mobile radio environments using the SAGE algorithm, *IEEE J. Selected Areas Comm.*, 17, 434, 1999.

[70] Fuhl, J., Molisch, A.F., and Bonek, E., A unified channel model for mobile radio systems with smart antennas, *IEEE Proc. Radar, Sonar Navigation*, 145, 32, 1998.

[71] Gans, M.J., A power-spectral theory of propagation in the mobile radio environment, *IEEE Trans. Vehicular Technol.*, 21, 27, 1972.

[72] Garcia-Pardo, J.M.M., Rodrigues, J.V.R., and Juan-Llacer, L., Angular spread at 2.1 GHz while entering tunnels, *Microwave Optical Technol. Lett.*, 37, 196, 2003.

[73] Gesbert, D. et al., Outdoor MIMO wireless channels: models and performance prediction, *IEEE Trans. Comm.*, 50, 12, 1926, 2002.

[74] Ghassemzadeh, S.S. et al., Measurement and modeling of an ultra-wide bandwidth indoor channel, *IEEE Trans. Comm.*, accepted.

[75] Ghassemzadeh, S.S. et al., UWB indoor pathloss model for residential and commercial buildings, in *Proc. IEEE Vehicular Techn. Conf., 2003, Fall*, 3115.

[76] Ghassemzadeh, S.S. et al., UWB indoor delay profile model for residential and commercial environments, *Proc. IEEE Vehicular Techn. Conf., 2003, Fall*, 3120.

[77] Glazunov, A.A., Aslund, H., and Berg, J.E., Statistical analysis of measured short-term impulse response functions of 1.88 GHz radio channels in Stockholm with corresponding channel model, in *Proceedings of the IEEE Vehicular Technology Conference, 1999 — Fall*, p. 107.

[78] Graziosi, F. and Santucci, F., General correlation model for shadow fading in mobile radio systems, *IEEE Comm. Lett.*, 6, 102, 2002.

[79] Greenstein, L.J. et al., A new path-gain/delay-spread propagation model for digital cellular channels, *IEEE Trans. Vehicular Technol.*, 46, 477, 1997.

[80] Gudmundson, M., Correlation model for shadow fading in mobile radio systems, *IEE Electronics Lett.*, 27, 2145, 1991.

[81] Haardt, M. and Nossek, J.A., Unitary ESPRIT: how to obtain increased estimation accuracy with a reduced computational burden, *IEEE Trans. Signal Processing*, 43, 1232, 1995.

[82] Hafezi, P. et al., Propagation measurements at 5.2 GHz in commercial and domestic environments, in *Proceedings of the IEEE International Symposium on Personal, Indoor Mobile Radio Communications*, 1997, p. 509.

[83] Hampicke, D. et al., Characterization of the directional mobile radio channel in industrial scenarios, based on wideband propagation measurements, in *Proceedings of the IEEE Vehicular Technology Conference, 1999 — Fall*, p. 2258.

[84] Harrington, R.F., *Field Computation by Moment Methods*, IEEE Press, Piscataway, NJ, 1993.

[85] Hashemi, H., Simulation of the urban radio propagation channel, *IEEE Trans. Vehicular Technol.*, 28, 213, 1979.

[86] Hashemi, H., The indoor radio propagation channel, *Proc. IEEE*, 81, 943, 1993.

[87] Hashemi, H. and Tholl, D., Statistical modeling and simulation of the RMS delay spread of indoor radio propagation channels, *IEEE Trans. Vehicular Technol.*, 43, 110, 1994.

[88] Hata, M., Empirical formula for propagation loss in land mobile radio services, *IEEE Trans. Vehicular Technol.*, VT-29, 317, 1980.

[89] Heddergott, R., Bernhard, U.P., and Fleury, B.H., Stochastic radio channel model for advanced indoor mobile communication systems, in *Proceedings of the IEEE International Symposium Personal, Indoor Mobile Radio Communications*, 1997, p. 140.

[90] Hendrickson, C. et al., Wideband wireless peer to peer propagation measurements, in *Proceedings Thirty-Third Asilomar Conference on Signals, Systems, Computers*, 1999, p. 183.

[91] Herdin, M. et al., Variation of measured indoor MIMO capacity with receive direction and position at 5.2 GHz, *Electronics Lett.*, 38, 1283, 2002.

[92] Hoeher, P., A statistical discrete-time model for the WSSUS multipath channel, *IEEE Trans. Vehicular Technol.*, 41, 461, 1992.

[93] Holloway, C.L., Perrini, P.L., Delyzer, R.R., and Allen, K.C., Analysis of composite walls and their effect on short-path propagation modeling, *IEEE Trans. Antennas Propagation*, 46, 730, 1997.

[94] Holma, H. and Toskala, A., *WCDMA for UMTS*, 2nd ed., Wiley, New York, 2002.

[95] Hoppe R. et al., Wideband propagation modelling for indoor environments and for radio transmission into buildings, in *Proceedings of the IEEE International Symposium on Personal, Indoor Mobile Radio Communications*, 2000, p. 282.

[96] Horikoshi, S., A study on multipath propagation modeling in millimeter wave IVC, in *Proceedings of the 5th International Symposium Wireless Personal Multimedia Communications*, 2002, p. 286.

[97] Hottinen, A. et al., *Multi-Antenna Transceiver Techniques for 3G and Beyond*, Wiley, New York, 2003.

[98] Howard, S. and Pahlavan, K., Autoregressive modeling of wide-band indoor radio propagation, *IEEE Trans. Comm.*, 40, 1540, 1992.

[99] Inoue, Y., Mori, K., and Arai, H., DOA estimation in consideration of the array element pattern, in *Proceedings of the IEEE Vehicular Technology Conference, 2002 — Spring*, 2002, p. 745.

[100] Jakes, W.C., *Microwave Mobile Communications*, Wiley, New York, 1974.

[101] Janaswamy, R. and Bach Andersen, J., Path loss predictions in urban areas sprawling over irregular terrain, in *Proceedings of the IEEE International Symposium Personal, Indoor Mobile Radio Communications*, 1998, p. 874.

[102] Janssen, G.J.M., Stigter, P.A., and Prasad, R., Wideband indoor channel measurements and BER analysis of frequency selective multipath channels at 2.4, 4.75, and 11.5 GHz, *IEEE Trans. Comm.*, 44, 1272, 1996.

[103] Kanatas, A. et al., Wideband characterization of microcellular suburban mobile radio channels at 1.89 GHz, in *Proceedings of the IEEE 56th Vehicular Technology Conference, VTC 2002 — Fall*, 2002, p. 1060.

[104] Karlsson P. et al., Wideband measurement and analysis of penetration loss in the 5 GHz band, in *Proceedings of the IEEE Vehicular Technology Conference, 1999 — Fall*, p. 2323.

[105] Kattenbach, R., *Characterisation of Time-Variant Indoor Radio Channels by Their System and Correlation Functions*, Shaker Verlag, Aachen, 1997 (in German).

[106] Kattenbach, R., Statistical distribution of path interarrival times in indoor environment, in *Proceedings of the 48th IEEE Vehicular Technology Conference*, 1998, p. 548.

[107] Kattenbach, R., Statistical modeling of small-scale fading in directional radio channels, *IEEE J. Selected Areas Comm.*, 20, 584, 2002.

[108] Kattenbach, R. and Fruechting, H., Calculation of system and correlation functions for WSSUS channels from wideband measurements, *Frequenz*, 49, 42, 1995.

[109] Kepler, J.F., Krauss, T.P., and Mukthavaram, S., Delay spread measurements on a wideband MIMO channel at 3.7 GHz, in *Proceedings of the 56th IEEE Vehicular Technology Conference, 2002 — Fall*, p. 2498.

[110] Kermoal, J.P. et al., A stochastic MIMO radio channel model with experimental validation, *IEEE J. Selected Areas Comm.*, 20, 1211, 2002.

[111] Kivinen, J., Xiongwen, Z., and Vainikainen, P., Empirical characterization of wideband indoor radio channel at 5.3 GHz, *IEEE Trans. Antennas Propagation*, 49, 1192, 2001.

[112] Kozek, W., On the transfer function calculus for underspread LTV channels, *IEEE Trans. Signal Processing*, 45, 219, 1997.

[113] Kozono, S. and Taguchi, A., Mobile propagation loss and delay spread characteristics with a low base station antenna on an urban road, *IEEE Trans. Vehicular Technol.*, 42, 103, 1993.

[114] Kreuzgruber, P. et al., Prediction of indoor radio propagation with the ray splitting model including edge diffraction and rough surfaces, in *Proceedings of the IEEE Vehicular Technology Conference*, 1994, p. 878.

[115] Krim, H. and Viberg, M., Two decades of array signal processing research: the parametric approach, *IEEE Signal Processing Magazine*, 13, 67, 1996.

[116] Kuchar, A., Rossi, J.-P., and Bonek, E., Directional macro-cell channel characterization from urban measurements, *IEEE Trans. Antennas Propagation*, 48, 137, 2000.

[117] Kuerner, T., Cichon, D.J., and Wiesbeck, W., Evaluation and verification of the VHF/UHF propagation channel based on a 3-D-wave propagation model, *IEEE Trans. Antennas Propagation*, 44, 393, 1996.

[118] Kunisch, J. and Pamp, J., Measurement results and modeling aspects for the UWB radio channel, in *Proceedings of the IEEE Conference Ultra Wideband Systems Technologies*, 2002, p. 19.

[119] Kunz, K.S. and Luebbers, R.J., *The Finite Difference Time Domain Method for Electromagnetics*, CRC Press, Boca Raton, FL, 1993.

[120] Landron, O., Feuerstein, M.J., and Rappaport, T.S., A comparison of theoretical and empirical reflection coefficients for typical exterior wall surfaces in a mobile radio environment, *IEEE Trans. Antennas Propagation*, 44, 341, 1996.

[121] Larsson, M., Spatio-temporal channel measurements at 1800 MHz for adaptive antennas, in *Proceedings of the 49th IEEE Vehicular Technology Conference*, 1999, p. 376.

[122] Laspougeas, P., Pajusco, P., and Bic, J.C., Radio propagation in urban small cells environment at 2 GHz: experimental spatio-temporal characterization and spatial wideband channel model, in *Proceedings of the 52nd IEEE Vehicular Technology Conference, 2000 — Fall*, p. 885.

[123] Laurila, J. et al., Wideband 3-D characterization of mobile radio channels in urban environment, *IEEE Trans. Antennas Propagation*, 50, 233, 2002.

[124] Jingming, Li., Wagen J.F., and Lachat, E., Propagation over rooftop and in the horizontal plane for small and micro-cell coverage predictions, in *Proceedings of the IEEE Vehicular Technology Conference*, 1997, p. 1123.

[125] Liberti, J.C. and Rappaport, T.S., Statistics of shadowing in indoor radio channels at 900 and 1900 MHz, in *Proceedings of the Military Communication Conference*, 1992, p. 1066.

[126] Liberti, J.C. and Rappaport, T.S., A geometrically based model for line-of-sight multipath radio channels, in *Proceedings of the IEEE Vehicular Technology Conference*, 1996, p. 844.

[127] Lienard, M. and Degauque, P., Natural wave propagation in mine environments, *IEEE Trans. Antennas Propagation*, 48, 1326, 2000.

[128] Lienard, M., Degauque, P., Baudet, J., and Degardin, D., Investigation on MIMO channels in subway tunnels, *IEEE J. Selected Areas Comm.*, 21, 332, 2003.
[129] Lobeira Rubio, M., Garcia-Armada, A., Torres, R.P., and Garcia, J.L., Channel modeling and characterization at 17 GHz for indoor broadband WLAN, *IEEE J. Selected Areas Comm.*, 20, 593, 2002.
[130] Martin, U., Spatio-temporal radio channel characteristics in urban macrocells, in *Proceedings of the IEE Radar, Sonar Navigation*, 145, 42, 1998.
[131] Martin, C.C., Winters, J.H., and Sollenberger, N., Multiple-input multiple-output (MIMO) radio channel measurements, in *Proceedings of the IEEE Vehicular Technology Conference, 2000 — Fall*, p. 774.
[132] Matz, G. and Hlawatsch, F., Time-frequency transfer function calculus (symbolic calculus) of linear time-varying systems (linear operators) based on generalized underspread theory, *J. Math. Phys.*, 39, 4041, 1998.
[133] Matz, G. et al., On the systematic measurement errors of correlative mobile radio channel sounders, *IEEE Trans. Comm.*, 50, 808, 2002.
[134] Mahmoudi, M. and Sousa, E.S., A statistical wideband propagation model for smart antenna systems, in *IEEE Vehicular Technology Conference*, 1998, p. 1004.
[135] MacLellan, J., Lam, S., and Lee, X., Residential indoor RF channel characterization, in *Proceedings of the IEEE Vehicular Technology Conference*, 1993, p. 210.
[136] McNamara, D.A., *Introduction to the Uniform Geometrical Theory of Diffraction*, Artech, Norwood, MA, 1990.
[137] McNamara, D.P., Beach M.A., Fletcher, P.N., and Karlsson, P., Capacity variation of indoor multiple-input multiple-output channels, *Electronics Lett.*, 36, 2037, 2000.
[138] McNamara, D.P., Beach, M.A., and Fletcher, P.N., Spatial correlation in indoor MIMO channels, in *IEEE Symposium Personal, Indoor Mobile Radio Communications*, 2002, p. 290.
[139] Medbo, J. and Schramm, P., Channel Models for HIPERLAN/2, ETSI/BRAN document 3ERI085B, European Telecommunications Standards Institute, Valbonne, France, 2000.
[140] Medbo, J. and Berg, J.-E., Simple and accurate path loss modeling at 5 GHz in indoor environments with corridors, in *Proceedings of the IEEE Vehicular Technology Conference, 2000 — Fall*, p. 30.
[141] Medbo, J., Hallenberg, H., and Berg, J.E., Propagation characteristics at 5 GHz in typical radio-LAN scenarios, in *Proceedings of the 49th IEEE Vehicular Technology Conference*, 1999, p. 185.
[142] Millott, L.J., The impact of vehicular scattering on mobile communications, in *Proceedings of the 44th IEEE Vehicular Technology Conference*, 1994, p. 1733.
[143] Mogensen, P.E. et al., Preliminary measurement results from an adaptive antenna array testbed for GSM/UMTS, in *Proceedings of the IEEE Vehicular Technology Conference*, 1997, p. 1592.
[144] Mohasseb, Y.Z. and Fitz, M.P., A 3-D spatio-temporal simulation model for wireless channels, *IEEE J. Selected Areas Comm.*, 20, 1193, 2002.
[145] Mohr, W., Wideband propagation measurements of mobile radio channels in mountainous areas in the 1800 MHz frequency range, in *Proceedings of the Vehicular Technology Conference*, 1993, p. 49.
[146] Molisch, A.F., Ed., *Wideband Wireless Digital Communications*, Prentice Hall, New York, 2000.
[147] Molisch, A.F., A generic model for MIMO wireless propagation channels in macro- and microcells, *IEEE Trans. Signal Proc.*, 52, 61, 2004.
[148] Molisch, A.F., Modeling of directional mobile radio channels, *Radio Sci. Bull.*, 302, 16, 2002.
[149] Molisch, A.F., Ultrawideband propagation channels, in *UWB Communications Systems — A Comprehensive Overview*, Kaiser, T., Ed., EURASIP Publishing.
[150] Molisch, A.F., Effect of far scatterer clusters in MIMO outdoor channel models, in *Proceedings of the IEEE Vehicular Technology Conferences, 2003 — Spring*, p. 534.
[151] Molisch, A.F. and Steinbauer, M., Condensed parameters for characterizing wideband mobile radio channels, *Int. J. Wireless Information Networks*, 6, 133, 1999.

[152] Molisch, A.F. et al., The COST 259 directional channel model. I. Philosophy and general aspects, *IEEE Trans. Wireless Comm.*, accepted pending revisions.

[153] Molisch, A.F., Steinbauer, M., Toeltsch, M., Bonek, E., and Thoma, R.S., Capacity of MIMO systems based on measured wireless channels, *IEEE J. Selected Areas Comm.*, 20, 561, 2002.

[154] Molisch, A.F., Foerster, J.R., and Pendergrass, M., Channel models for ultrawideband personal area networks, *IEEE Personal Communications Magazine*, 10, 14, 2003.

[155] Molnar, B.G., Frigyes, I., Bodnar, Z., and Herczku, Z., The WSSUS channel model: comments and a generalization, in *Proceedings of the IEEE Global Communication Conference*, 1996, p. 158.

[156] Moroney, D. and Cullen, P., A fast integral equation approach to UHF coverage estimation, in *Mobile and Personal Communications*, delRe, E., Ed., Elsevier Press, Amsterdam, Netherlands, 1995.

[157] Moraitis, N. and Constantinou, P., Indoor channel modeling at 60 GHz for wireless LAN applications, in *13th IEEE International Symposium Personal, Indoor Mobile Radio Communications*, 2002, p. 1203.

[158] Motley, A.J., and Keenan, J.M.P., Personal communication radio coverage in buildings at 900 MHz and 1700 MHz, *Electronics Lett.*, 24, 763, 1988.

[159] Nilsson, M. et al., Measurements of the spatio-temporal polarization characteristics of a radio channel at 1800 MHz, in *Proceedings of the Vehicular Technology Conference*, 1999, p. 386.

[160] Nobles, P. and Halsall, F., Delay spread and received power measurements within a building at 2 GHz, 5 GHz and 17 GHz, in *Proceedings of the 10th IEEE International Conference Antennas Propagation*, 1997, p. 319.

[161] Norklit, O. and Andersen, J.B., Diffuse channel model and experimental results for array antennas in mobile environments, *IEEE Trans. Antennas Propagation*, 46, 834, 1998.

[162] Özcelik, H. et al., Deficiencies of the 'Kronecker' MIMO radio channel model. *IEE Electronics Lett.*, 38, 1209, 2003.

[163] Okumura, Y. et al., Field strength and its variability in UHF and VHF land-mobile radio service, *Rev. Electronics Comm. Lab.*, 16, 825, 1968.

[164] Oppenheim, A.V. and Schaefer, R.W.,*Digital Signal Processing*, Prentice Hall, Upper Saddle River, NJ, 1975.

[165] Oppermann, I., Talvitie, J., and Hunter, D., Wide-band wireless local loop channel for urban and sub-urban environments at 2 GHz, in *IEEE International Conference Communication*, 1997, p. 61.

[166] Pätzold, M., *Mobile Fading Channels*, John Wiley & Sons, Chichester, U.K., 2002.

[167] Pätzold, M., Szczepanski, A., and Youssef, N., Methods for modelling of specified and measured multipath power delay profiles, *IEEE Trans. Vehicular Technol.*, 51, 978, 2002.

[168] Pajusco, P., Experimental characterization of DOA at the base station in rural and urban area, in *Proceedings of the IEEE Vehicular Technology Conference*, 1998, p. 993.

[169] Papadakis, N. et al., Wideband propagation indoor environment measurements and modeling, *Int. J. Wireless Information Networks*, 4, 101, 1997.

[170] Parra-Michel, R., Kontorovitch, V.Y., and Orozco-Lugo, A.G., Simulation of wideband channels with non-separable scattering functions, in *Proceedings of the IEEE International Conference on Acoustics, Speech, Signal Processing*, 2002, p. III-2829.

[171] Parsons, J.D., *The Mobile Radio Propagation Channel*, 2nd ed., John Wiley & Sons, Chichester, 2000.

[172] Parsons, J.D., Demery, D.A., and Turkmani, A.M.D., Sounding techniques for wideband mobile radio channels: a review, *IEE Proc.-I*, 138, 437, 1991.

[173] Pechac, P., Klepal, M., and Mazanek, M., New fast approach to wideband propagation prediction in picocells, in *Proceedings of the 11th IEEE International Conference Antennas Propagation*, 2001, p. 216.

[174] Pedersen, K.I., Mogensen, P.E., and Fleury, B., Power azimuth spectrum in outdoor environments, *IEEE Electronics Lett.*, 33, 1583, 1997.

[175] Petrus, P., Reed, J.H., and Rappaport, T.S., Geometrical-based statistical macrocell channel model for mobile environments, *IEEE Trans. Comm.*, 50, 495, 2002.

[176] Pettersen, M. et al., Characterisation of the directional wideband radio channel in urban and suburban areas, in *Proceedings of the IEEE Vehicular Technology Conference, 1999, — Fall*, p. 1454.

[177] Porrat, D. and Cox, D.C., Delay spread in microcells analysed with waveguide theory, in *Proceedings of the 55th IEEE Vehicular Technology Conference, 2002 — Spring*, p. 512.

[178] Prettie, C., Cheung, D., Rusch, L., and Ho, M., Spatial correlation of UWB signals in a home environment, in *Proceedings of the IEEE Conference Ultra Wideband Systems Technologies*, 2002, p. 65.

[179] Proakis, J.G., *Digital Communications*, 4th ed., McGraw-Hill, New York, 2000.

[180] Qiu, R.C., A study of the ultra-wideband wireless propagation channel and optimum UWB receiver design, *IEEE J. Selected Areas Comm.*, 20, 1628, 2002.

[181] Ramo, S., Whinnery, J.R., and van Duzer, T., *Fields and Waves in Communications Electronics*, 3rd ed., Wiley, New York, 1994.

[182] Rappaport, T.S., Characterization of UHF multipath radio channels in factory buildings, *IEEE Trans. Antennas Propagation*, 37, 1058, 1989.

[183] Rappaport, T.S., *Wireless Communications: Principles and Practice*, 2nd edition, Prentice Hall, New York, 2002.

[184] Rappaport, T.S. and Hawbaker, D.A., Wide-band microwave propagation parameters using circular and linear polarized antennas for indoor wireless channels, *IEEE Trans. Comm.*, 40, 240, 1992.

[185] Rappaport, T.S., Seidel, S.Y., and Singh, R., 900-MHz multipath propagation measurements for US digital cellular radiotelephone, *IEEE Trans. Vehicular Technol.*, 39, 132, 1990.

[186] Ravi, K.V. and Soma, P., Impulse response measurements of local multipoint distribution service (LMDS) channel in Singapore, in *IEEE International Conference Communications*, 2000, p. 521.

[187] Rizk, K., Wagen, J.F., and Gardiol, F., Influence of database accuracy on two-dimensional ray-tracing-based predictions in urban microcells, *IEEE Trans. Vehicular Technol.*, 49, 631, 2000.

[188] Rokhlin, V., Rapid solution of integral equations of scattering theory in two dimensions, *J. Comp. Phys.*, 96, 414, 1990.

[189] Rossi, J.P., Influence of measurement conditions on the evaluation of some radio channel parameters, *IEEE Trans. Vehicular Technol.*, 48, 1304, 1999.

[190] Rossi, J.P. and Gabillet, Y., A mixed ray launching/tracing method for full 3-D UHF propagation modeling and comparison with wide-band measurements, *IEEE Trans. Antennas Propagation*, 50, 517, 2002.

[191] Roy, R., Paulraj, A., and Kailath, T., Direction-of-arrival estimation by subspace rotation methods: ESPRIT, in *Proceedings of the IEEE International Conference Acoustics, Speech, Signal Processing*, 1986, p. 2495.

[192] Sakawa, K., Masui, H., Ishii, M., Shimizu, H., and Kobayashi, T., Microwave propagation characteristics depending on base-station antenna height in an urban area, in *Proceedings of the IEEE International Symposium Antennas Propagation*, 2001, p. 174.

[193] Saleh, A. and Valenzuela, R., A statistical model for indoor multipath propagation, *IEEE J. Selected Areas Comm.*, 5, 128, 1987.

[194] Salous, S., Nikandrou, N., and Bajj, N.F., Digital techniques for mobile radio chirp sounders, *IEE Proc. Comm.*, 145, 191, 1998.

[195] Santella, G. and Restuccia, E., Analysis of frequency domain wide-band measurements of the indoor radio channel at 1, 5.5, 10 and 18 GHz, in *Proceedings of the IEEE Global Telecommunications Conference*, 1996, p. 1162.

[196] Saunders, S.R., *Antennas and Propagation for Wireless Communications Systems*, Wiley, New York, 2002.

[197] Sayeed, A.M., Deconstructing multiantenna fading channels, *IEEE Trans. Signal Proc.*, 50, 2563, 2002.

[198] Schmidt, R.O., Multiple emitter location and signal parameter estimation, *IEEE Trans. Antennas Propagation*, 34, 276, 1986.

[199] Schoberl, T., Combined Monte Carlo simulation and ray tracing method of indoor radio propagation channel, in *Proceedings of the IEEE Microwave Symposium*, 1995, p. 1379.

[200] Schwarz, K., Martin, U., and Schüßler, H.W., Devices for propagation measurements in mobile radio channels, in *Proceedings of the IEEE Symposium Personal, Indoor Mobile Radio Communications*, 1993, p. 387.

[201] Spatial Channel Model Ad-Hoc Group (combined ad-hoc group from 3GPP and 3GPP2), Spatial channel model text description, SCM-135_Text_v7.0, available at ftp.3gpp2.org/TSGC/Working/ 2003.

[202] Seidel, S.Y, Rappaport, T.S., Jain, S., Lord, M.L., and Singh, R., Path loss, scattering and multipath delay statistics in four European cities for digital cellular and microcellular radiotelephone, *IEEE Trans. Vehicular Technol.*, 40, 721, 1991.

[203] Shannon, C.E., A mathematical theory of communication, *Bell System Tech. J.*, 27, 379 and 623, 1948.

[204] Sheikh, M.I. and Constantinou, C.C., Wideband limitations of ray-optical techniques in radiowave propagation prediction, in *Proceedings of the 10th IEEE Conference Antennas Propagation*, 1997, p. 79.

[205] Siwiak, K., Bertoni, H., and Yano, S.M., Relation between multipath and wave propagation attenuation, *IEEE Electronics Lett.*, 39, 142, 2003.

[206] Soma, P., Baum, D.S., Erceg, V., Krishnamoorthy, R., and Paulraj, A.J., Analysis and modeling of multiple-input multiple-output (MIMO) radio channel based on outdoor measurements conducted at 2.5 GHz for fixed BWA applications, in *Proceedings of the International Conference Communications*, 2002, p. 272.

[207] Sorensen, T.B., Correlation model for slow fading in a small urban macrocell, in *Proceedings of the International Symposium Personal Indoor Mobile Radio Communications*, 1998, p. 1161.

[208] Spencer, Q.H. et al., Modeling the statistical time and angle of arrival characteristics of an indoor multipath channel, *IEEE J. Selected Areas Comm.*, 18, 347, 2000.

[209] Sousa, E.S., Jovanovic, V.M., and Daigneault, C., Delay spread measurements for the digital cellular channel in Toronto, *IEEE Trans. Vehicular Technol.*, 43, 837, 1994.

[210] Steele, R. and Hanzo, L., *Mobile Radio Communications*, 2nd ed., Wiley, New York, 2000.

[211] Steinbauer, M. and Molisch, A.F., Spatial channel models, in *Wireless Flexible Personalized Communications*, Correia, L., Ed., John Wiley & Sons, Chichester, U.K., 2001.

[212] Steinbauer, M., Molisch, A.F., and Bonek, E., The double-directional mobile radio channel, *IEEE Antennas Propagation Magazine*, 43, 51, 2001.

[213] Stuber, G.L., *Principles of Mobile Communications*, Kluwer, Dordrecht, Netherlands, 1996.

[214] Svantesson, T., A double-bounce channel model for multi-polarized MIMO systems, in *Proceedings of the IEEE Vehicular Technology Conference, 2002 — Fall*, p. 691.

[215] Svantesson, T. and Wallace, J.W., Tests for assessing multivariate, normality and the covariance structure of MIMO data, in *Proceedings of the IEEE ICASSP*, 2003, p. IV-656.

[216] Suzuki, H., A statistical model for urban radio propagation, *IEEE Trans. Comm.*, 25, 673, 1977.

[217] Sykora, J., Tapped delay line model of linear randomly time-variant WSSUS channel, *Electronics Lett.*, 36, 1656, 2000.

[218] Taga, T., Furuno, T., and Suwa, K., Channel modeling for 2-GHz-band urban line-of-sight street microcells, *IEEE Trans. Vehicular Technol.*, 48, 262, 1999.

[219] Taira, K., Sekizawa, S., and Hase, Y., Wideband channel modeling for line-of-sight microcellular environment with low base station antenna height, in *Proceedings of the 48th IEEE Vehicular Technology Conference*, 1998, p. 149.

[220] Talbi, L. and Delisle, G.Y., Experimental characterization of EHF multipath indoor radio channels, *IEEE J. Selected Areas Comm.*, 14, 431, 1996.

[221] Thoma, R.S., Hampicke, D., Richter, A., Sommerkorn, G., Schneider, A., Trautwein, A.U., and Wirnitzer, W., Identification of time-variant directional mobile radio channels, *IEEE Trans. Instrum. Meas.*, 49, 357, 2000.

[222] Thoma, R.S. et al., Measurement and identification of mobile radio propagation channels, in *Proceedings of the IEEE Instrumentation and Measurement Conference*, 2001, p. 1163.

[223] Toeltsch, M. et al., Statistical characterization of urban spatial radio channels, *IEEE J. Selected Areas Comm.*, 20, 539, 2002.

[224] Torres, R.P. et al., CINDOOR: an engineering tool for planning and design of wireless systems in enclosed spaces, *IEEE Antennas Propagation Magazine*, 41, 1999.

[225] Turin, G.L., Clapp, F.D., Johnston, T.L., Fine, S.B., and Lavry, D., A statistical model of urban multipath propagation, *IEEE Trans. Vehicular Technol.*, 21, 1, 1972.
[226] Varela, M.S. and Sanchez, M.G., RMS delay and coherence bandwidth measurements in indoor radio channels in the UHF band, *IEEE Trans. Vehicular Technol.*, 50, 515, 2001.
[227] Vaughan, R. and Andersen, J.B., *Channels, Propagation and Antennas for Mobile Communications*, IEE Press, Lindon, U.K., 2003.
[228] Villanese, F., Evans, N.E., and Scanlon, W.G., Pedestrian-induced fading for indoor channels at 2.45, 5.7 and 62 GHz, in *Proceedings of the 52nd IEEE Vehicular Technology Conference, 2000 — Fall*, p. 43.
[229] Walfisch, J. and Bertoni, H.L., A theoretical model of UHF propagation in urban environments, *IEEE Trans. Antennas Propagation*, 36, 1788, 1988.
[230] Wallace, J.W., and Jensen, M.A., Modeling the indoor MIMO wireless channel, *IEEE Trans. Antennas Propagation*, 50, 591, 2002.
[231] Weichselberger, W., Özcelik, H., and Bonek, E., A novel stochastic MIMO channel model and its physical interpretation, in *6th International Symposium Wireless Personal Multimedia Communications*, 2003, accepted.
[232] Weitzen, J. and Lowe, T.J., Measurement of angular and distance correlation properties of lognormal shadowing at 1900 MHz and its application to the design of PCS systems, *IEEE Trans. Vehicular Technol.* 51, 265, 2002.
[233] Welch, T.B., Musselman, R.L., Emessiene, B.A., Gift, P.D., Choudhury, D.K., Cassadine, D.N., and Yano, S.M., The effects of the human body on UWB signal propagation in an indoor environment, *IEEE J. Selected Areas Comm.*, 20, 1778, 2002.
[234] Wiart, J., Pajusco, P., Levy, A., and Bic, J.C., Analysis of microcellular wide band measurements in Paris, in *Proceedings of the 6th IEEE International Symposium Personal, Indoor Mobile Radio Communications*, 1995, p. 144.
[235] Xia, H.H., Bertoni, H.L., Maciel, L.R., Lindsay-Stewart, A., and Rowe, R., Microcellular propagation characteristics for personal communications in urban and suburban environments, *IEEE Trans. Vehicular Technol.*, 43, 743, 1994.
[236] Xu, H., Chizhik, D., Huang, H., and Valenzuela, R., A wave-based wideband MIMO channel modeling technique, in *Proceedings of the 13th IEEE International Symposium Personal, Indoor Mobile Radio Communications*, 2002.
[237] Xu, H., Kukshya, V., and Rappaport, T.S., Spatial and temporal characteristics of 60-GHz indoor channels, *IEEE J. Selected Areas Comm.*, 20, 620, 2002.
[238] Yano, S.M., Investigating the ultra-wideband indoor wireless channel, in *Proceedings of the 55th IEEE Vehicular Technology Conference, 2002 — Spring*, p. 1200.
[239] Yip, K.W. and Ng, T.-S., Efficient simulation of digital transmission over WSSUS channels, *IEEE Trans. Comm.*, 43, 2907, 1995.
[240] Yip, K.W. and Ng, T.-S., Karhunen-Loeve expansion of the WSSUS channel output and its application to efficient simulation, *IEEE J. Selected Areas Comm.*, 15, 640, 1997.
[241] Yu, K. et al., Second order statistics of NLOS indoor MIMO channels based on 5.2 GHz measurements, in *Proceedings of the IEEE Global Telecommunications Conference*, 2001, p. 156.
[242] Yu, K. and Ottersten, B., Models for MIMO propagation channels: a review, *Wireless Comm. Mobile Comput.*, 2, 653, 2002.
[243] Zhang, Y.P. and Hwang, Y., Characterization of UHF radio propagation channels in tunnel environments for microcellular and personal communications, *IEEE Trans. Vehicular Technol.*, 47, 283, 1998.
[244] Zhao, X., Kivinen, J., Vainikainen, P., and Skog, K., Propagation characteristics for wideband outdoor mobile communications at 5.3 GHz, *IEEE J. Selected Areas Comm.*, 20, 507, 2002.
[245] Zhong, J. et al., Efficient ray-tracing methods for propagation prediction for indoor wireless communications, *IEEE Antennas Propagation Magazine*, 43, 41, 2001.

[246] Zhu, F., Wu, Z., and Nassar, C.R., Generalized fading channel model with application to UWB, in *Proceedings of the IEEE Conference Ultra Wideband Systems Technologies*, 2002, p. 13.

[247] Zwick, T., Fischer, C., and Wiesbeck, W., A stochastic multipath channel model including path directions for indoor environments, *IEEE J. Selected Areas Comm.*, 20, 1178, 2002.

[248] Zienkiewicz, O.C. and Taylor, R.L., *Finite Element Method*, Vol. 1, *The Basis*, Butterworth Heinemann, London, 2000.

[249] International Telecommunication Union, Guidelines for Evaluation of Radio Transmission Technologies for IMT-2000, Rec., ITU-RM. 1225, Geneva, Switzerland, 1997.

Definitions and Symbols

$()^*$	complex conjugation
$()^T$	transposition
<a,b>	inner product
A	amplitude of the dominant component in Rice statistics
A_i	(complex) tap weights for regularly spaced taps
$\mathbf{A}$	matrix containing steering vectors
$a_{k.l}$	tap weight of the k-th component in the l-th cluster
$a^i_{k,l}$	tap weights
a_i	path amplitude
$\vec{a}_R(\Psi)$	steering vector (of the receive antenna)
$B(\nu,\tau)$	bi-spreading function
B_c	coherence bandwidth
d_{ref}	reference distance
d_{break}	breakpoint distance
δ	Dirac delta function
Δh_B	base station height measured from the rooftop level
$\Delta\nu_{\text{spread}}$	support of the spreading function Doppler coordinates
Δ_{total}	total spread
$E\{\}$	expectation value
$E_t\{\}$	expectation over time (time average)
f	frequency
f_c	carrier frequency
ϕ	azimuth
φ_i	path phase
$\mathbf{G}$	matrix with i.i.d. complex Gaussian entries
G_R	pattern of the receive antenna elements
G_T	pattern of the transmit antenna elements
$\Gamma(m)$	gamma function
$\mathbf{H}$	transfer function matrix (for MIMO systems)
$H(t,f)$	time-variant transfer function
$h(\vec{r}_T,\vec{r}_R,\tau,\Omega,\Psi)$	double-directional impulse response
$h(t,\tau)$	time-variant impulse response
$h_{i,j}$	impulse responses at different elements
I_0	zero-order modified Bessel function
K	number of multipath components in cluster
k	unit wave vector
κ	the "peakiness" of the von Mises distribution
L	number of clusters
L_p	path loss
$L_p[\text{dB}]$	path loss in decibels

$L(\vec{r})$	number of multipath components
Λ	cluster arrival rate
λ	ray arrival rate
$N(\mu, \sigma^2)$	Gaussian distribution with mean and variance
N_{floor}	number of floors
n	decay exponents
ν	Doppler frequency
Ω	direction of departure (DOD)
$P(\tau)$	power delay profile (PDP) or delay power density spectrum
$P(t, \tau)$	instantaneous power delay profile
$P_B(\nu, \Delta f)$	Doppler cross-power spectral density
$P_h(\Delta t, \tau)$	delay cross-power spectrum
P_k^c	power of the k-th cluster
$P_s(\nu, \tau)$	scattering function
$\mathbf{R}$	correlation matrix
$\mathbf{R}_{\text{TX}}$	transmitter correlation matrix
$\mathbf{R}_{\text{RX}}$	receiver correlation matrix
$R_B(\nu, \nu'; f, f')$	bi-spreading function
$R_H(t, t'; f, f')$	autocorrelation of the transfer function
$R_H(\Delta t, \Delta f)$	time–frequency correlation function
$\tilde{R}_H(\Delta f)$	normalized frequency correlation function
$R_h(t, t'; \tau, \tau')$	autocorrelation of the impulse response
$R_s(\nu, \nu'; \tau, \tau')$	autocorrelation of the delay–Doppler spread function
$\vec{r}_R$	location of receiver
$\vec{r}_T$	location of transmitter
$s(\nu, \tau)$	delay–Doppler spread function or spreading function
$S_\tau(t)$	instantaneous rms delay spread
S_τ	rms delay spread
S_ϕ	rms angular spread
S_t^c	delay spread of the k-th cluster
σ^2	variance
T_l	delay of the l-th cluster
T_s	tap spacing
t	time
τ	delay
$\Delta\tau$	delay between subsequent MPCs with significant amplitude
$\Delta\tau_{\text{spread}}$	support of the spreading function in delay coordinates
$\tau_{0,\text{k}}^c$	delay of the k-th cluster
$\tau_{k,l}$	delay of the k-th multipath component relative to the l-th cluster arrival time T_l
X_i	lognormal shadowing
$x(t)$	input to the channel
Ψ	direction of arrival (DOA)
$y(t)$	output of the channel

Abbreviations

3GPP	third-generation partnership project
ACF	autocorrelation function
ADPS	angular delay power spectrum
APS	angular power spectrum

BS	base station
BW	bandwidth
DOA	direction of arrival
DOD	direction of departure
FDTD	finite difference time domain
FEM	finite element method
GSCM	geometry-based stochastic channel model
i.i.d.	independent identically distributed
LAN	local area network
LOS	line of sight
MIMO	multiple-input–multiple-output
MPC	multipath component
MS	mobile station
NLOS	non-line of sight
pdf	probability density function
PDP	power delay profile
RE	radio environment
RX	receiver
TX	transmitter
VCDA	virtual cell deployment area
WSS	wide-sense stationary
XPD	cross-polarization discrimination

3

Modeling and Estimation of Mobile Channels

3.1 Introduction 3-1
3.2 Channel Models 3-3
Time-Variant Channels • Time-Invariant Channels
3.3 Channel Estimation 3-8
Training-Based Channel Estimation • Blind Channel Estimation • Semiblind Approaches • Hidden Pilot-Based Approaches
3.4 Simulation Examples 3-22
Example 1 • Example 2
3.5 Conclusions 3-25

Jitendra K. Tugnait
Auburn University

Abstract

A review of various approaches to channel modeling and estimation for wireless mobile systems is presented. We begin with channel models suitable for channel estimation and equalization. Emphasis is on linear baseband equivalent models with a tapped delay line structure, and both time-invariant and time-variant models are discussed. Basis expansion modeling for time-variant channels is also presented where the basis functions are related to the physical parameters of the channel (such as Doppler and delay spreads). Channel modeling is followed by a discussion of various approaches to channel estimation, including training-based approaches, blind approaches, semiblind approaches, and hidden pilot-based approaches. In the training-based approach a sequence known to the receiver is transmitted in the acquisition mode. In blind approaches no such sequence is available (or used) and the channel is estimated based solely on the noisy received signal exploiting the statistical and other properties of the information sequence. Semiblind approaches utilize a combination of training-based and blind approaches. In the hidden pilot-based approaches a periodic (nonrandom) training sequence is arithmetically added (superimposed) at a low power to the information sequence at the transmitter before modulation and transmission.

3.1 Introduction

Propagation of signals through wireless channels (indoors or outdoors) results in the transmitted signal arriving at the receiver through multiple paths. These paths arise due to reflection, refraction, or diffraction in the channel. Multipath propagation results in a received signal that is a superposition of several delayed

0-8493-1657-X/$0.00+$1.50

and scaled copies of the transmitted signal, giving rise to frequency-selective fading. Frequency-selective fading (defined as changes in the received signal level in time) is caused by destructive interference among multiple propagation paths. The environment around the transmitter and the receiver can change over time, particularly in a mobile setting, leading to variations in the channel response with time. This gives rise to time-selective fading. Also, the channels may have a dominant path (direct path in line-of-sight channels) in addition to several secondary paths, or they may be characterized as having multiple "random" paths with no single dominant path.

Multipath propagation leads to intersymbol interference (ISI) at the receiver, which in turn may lead to high error rates in symbol detection. Equalizers are designed to compensate for these channel distortions. One may directly design an equalizer given the received signal, or one may first estimate the channel impulse response and then design an equalizer based on the estimated channel. After some processing (matched filtering, for instance), the continuous-time received signals are sampled at the baud (symbol) or higher (fractional) rate before processing them for channel estimation or equalization. It is therefore convenient to work with a baseband-equivalent discrete-time channel model. Depending upon the sampling rate, one has either a single-input single-output (SISO) (baud rate sampling) or a single-input multiple-output (SIMO) (fractional sampling) complex discrete-time-equivalent baseband channel.

Traditionally, a training sequence (known to the receiver) is transmitted during the start-up (acquisition mode). In the operational stage, the receiver switches to a decision-directed mode where the previously equalized and detected symbols are used as a (pseudo)training sequence together with the received data to update the channel or the equalizer coefficients. The various issues involved and the trade-offs among various competing approaches (linear, decision feedback, maximum likelihood sequence estimation, least mean square vs. recursive least squares, baud rate vs. fractional rate, etc.) are fairly well understood and documented; see the well-known text [30] and references therein. More recently, there has been much interest in blind (self-recovering) channel estimation and blind equalization where no training sequences are available or used and the receiver starts up without any (explicit) cooperation from the transmitter. In point-to-multipoint networks, whenever a link from the server to one of the tributary stations is interrupted, it is clearly not feasible (or desirable) for the server to start sending a training sequence to reestablish a particular link. It has also been argued [50] that a blind start-up is more straightforward to implement than a start-up that requires a training sequence; this eases interoperability issues among different manufacturers. In digital communications over fading/multipath channels, a restart is required following a temporary path loss due to a severe fade.

As in the trained case, various approaches to blind channel estimation and equalization have been developed. When sampled at the baud rate, the received signal is discrete-time stationary and typically non-minimum phase. When sampled at higher than baud rate (typically an integer multiple of baud rate), the signal is discrete-time scalar cyclostationary, and equivalently, it may be represented as a discrete-time vector stationary sequence with an underlying SIMO model. With baud rate sampling, one has to exploit the higher-order statistics (HOS) of the received signal either implicitly (as in [15] and [34], where direct design of equalizers is considered) or explicitly (as in [17] and [41]–[44], where the focus is on first estimating the channel impulse response using higher-order cumulants of the received signal). Higher-order statistics provide an incomplete characterization of the underlying non-Gaussian process. Joint channel and data estimation using maximum likelihood and related approaches ([20], [33] and references therein) exploit a complete (non-Gaussian) probabilistic characterization of the noisy signal. Computational complexity of these algorithms (explicit HOS and joint channel–data estimation) is large when the ISI spans many symbols (as in telephone channels), but they are relatively simple when the ISI span is short (as in mobile radio channels). However, they may suffer from local convergence problems.

When there is excess channel bandwidth, baud rate sampling is below the Nyquist rate, leading to aliasing and, depending upon the symbol timing phase in certain cases, causing deep spectral notches in sampled, aliased channel transfer function [14]. This renders the equalizer performance quite sensitive to symbol timing errors. Initially, in the trained case, fractional sampling was investigated to robustify the equalizer performance against timing error. More recently, in the blind context, it was discovered (see [37] and references therein) that oversampling provides some new information regarding the channel, which can be exploited for blind channel estimation and equalization provided some technical conditions are satisfied (the no-common-subchannel-zeros condition, also called channel disparity, for the underlying equivalent SIMO model). A similar SIMO model results if multiple sensors are used with or without fractional sampling. The work of [37] has spawned intense research activity in the use of second-order statistics for blind identification and equalization. It should be noted that the requisite technical conditions for applicability of these approaches are not always satisfied in practice; some examples are in [42].

In this chapter, we present a review of various approaches to channel modeling and estimation for wireless mobile systems. In Section 3.2 we present the relevant channel models, including time-variant and time-invariant models. In Section 3.3 various channel estimation methods are presented. To conclude the chapter, in Section 3.4 we illustrate the hidden pilot-based approach via simulation examples.

3.2 Channel Models

3.2.1 Time-Variant Channels

Consider a time-varying (e.g., mobile wireless) channel (linear system) with a complex baseband, a continuous-time, received signal $x(t)$, and transmitted complex baseband, and a continuous-time information signal $s(t)$ (with symbol interval T_s second) related by [30]

$$x(t) = \int_{-\infty}^{\infty} h(t;\tau)s(t-\tau)\,d\tau + w(t) \tag{3.1}$$

where $h(t;\tau)$ is the time-varying impulse response of the channel denoting the response of the channel at time t to a unit impulse input at time $t-\tau$, and $w(t)$ is the additive noise (typically white Gaussian). A delay–Doppler spread function $H(f;\tau)$ is defined as the Fourier transform of $h(t;\tau)$ [1],[30]:

$$H(f;\tau) = \int_{-\infty}^{\infty} h(t;\tau)exp^{-j2\pi ft}\,dt \tag{3.2}$$

If $|H(f;\tau)| \approx 0$ for $|\tau| > \tau_d$, then τ_d is called the (multipath) delay spread of the channel. If $|H(f;\tau)| \approx 0$ for $|f| > f_d$, then f_d is called the Doppler spread of the channel. Equation 3.1 is the most general form of a mobile channel discussed in this chapter.

In order to capture the complexity of the physical interactions characterizing the transmission through a real channel, $h(t;\tau)$ is typically modeled as a two-dimensional zero-mean random process. If $h(t;\tau)$ is wide-sense stationary in variable t, and $h(t;\tau_1)$ is uncorrelated with $h(t;\tau_2)$ for $\tau_1 \neq \tau_2$ and any t, one obtains the well-known wide-sense stationary uncorrelated scattering (WSSUS) channel [1], [30, Section 14].

In this chapter we will confine our attention to deterministic modeling of $h(t;\tau)$, which may be thought of as capturing realizations of the underlying random process.

3.2.1.1 Tapped Delay Line Model

We now consider a discrete-time channel model. If a linear modulation scheme is used, the baseband transmitted signal can be represented as

$$s(t) = \sum_{k=-\infty}^{\infty} I[k]g_T(t-kT_s) \tag{3.3}$$

where $\{I[k]\}$ is the information sequence and $g_T(t)$ is the transmit (low-pass) filter (typically a root raised cosine filter). Therefore, the baseband signal incident at the receiver is given by

$$x(t) = \sum_{k=-\infty}^{\infty} I[k]\int_{-\infty}^{\infty} h(t;\alpha)g_T(t-kT_s-\alpha)\,d\alpha + w(t) \tag{3.4}$$

After filtering with a receive filter with impulse response $g_R(t)$, the received baseband signal is given by

$$y(t) = \sum_{k=-\infty}^{\infty} I[k] \int_{-\infty}^{\infty}\int_{-\infty}^{\infty} g_R(t-\beta)h(\beta;\alpha)g_T(\beta - kT_s - \alpha)\, d\alpha\, d\beta + v(t) \tag{3.5}$$

where $v(t) = \int g_R(\tau)w(t-\tau)d\tau$. If the continuous-time signal $y(t)$ is sampled once every T_s section, we obtain the discrete-time sequence

$$y[n] := y(t)|_{t=nT_s} = \sum_{k=-\infty}^{\infty} I[k]h[n;n-k] + v[n] = \sum_{k=-\infty}^{\infty} h[n;k]I[n-k] + v[n] \tag{3.6}$$

where $h[n;k]$ is the (effective) channel response at time n to a unit input at time $n-k$ and

$$h[n;n-k] := \int_{-\infty}^{\infty}\int_{-\infty}^{\infty} g_R(nT_s - \beta)h(\beta;\alpha)g_T(\beta - kT_s - \alpha)\, d\alpha\, d\beta \tag{3.7}$$

Note that the noise sequence $\{v[n]\}$ in Equation 3.6 is no longer necessarily white; it can be whitened by further time-invariant linear filtering (see [30]). Henceforth, we assume that a whitening filter has been applied to $y[n]$, but with an abuse of notation, we will still use Equation 3.6 except for replacing $v[n]$ with whitened $w[n]$. For a causal system, $h[n;k] = 0$ for $k < 0$ ($\forall n$), and for a finite length channel of maximum length $T_s L$, $h[n;k] = 0$ for $k > L$ ($\forall n$). In this case we modify Equation 3.6 as (recall the noise-whitening filter)

$$y[n] = \sum_{k=0}^{L} h[n;k]I[n-k] + w[n] \tag{3.8}$$

Equation 3.6 represents a time- and frequency-selective linear channel. A tapped delay line structure for this model is shown in Figure 3.1. For a slowly (compared to the baud rate) time-varying system, one often simplifies Equation 3.8 to a time-invariant system as

$$y[n] = \sum_{k=0}^{L} h[k]I[n-k] + w[n] \tag{3.9}$$

where $h[k] = h[0;k]$ is the time-invariant channel response to a unit input at time 0. Equation 3.9 represents a frequency-selective linear channel with no time selectivity. It is the most commonly used model for receiver design.

Suppose that $h[n;k] = h[n]\delta[k,0]$ where $\delta[k,0]$ is the Kronecker delta located at 0, i.e., $\delta[k,0] = 1$ for $k = 0$ and $\delta[k,0] = 0$ for $k \neq 0$. Then we have the time-selective and frequency-non-selective channel whose output is given by

$$y[n] = h[n]I[n] + w[n] \tag{3.10}$$

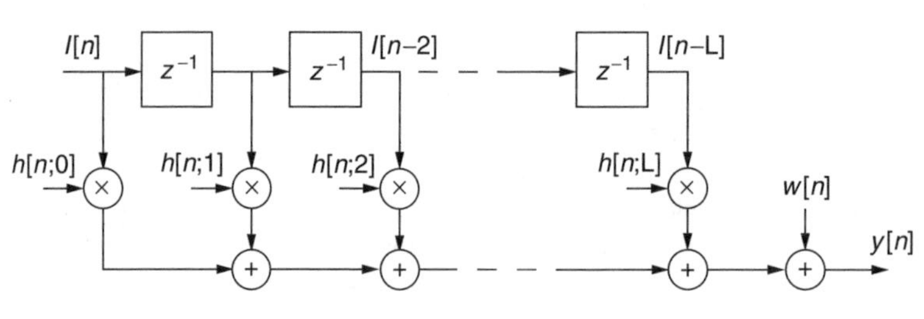

FIGURE 3.1 Tapped delay line model of frequency- and time-selective channel with finite impulse response. z^{-1} represents a unit (symbol duration) delay.

Finally, a time-non-selective and frequency-non-selective channel is modeled as

$$y[n] = hI[n] + w[n] \tag{3.11}$$

where h is a random variable (or a constant).

3.2.1.2 Basis Expansion Models

Suppose that we include the effects of transmit and receive filters in the time-variant impulse response $h(t;\tau)$ in Equation 3.1. Suppose that this channel has a delay spread τ_d and a Doppler spread f_d. It has been shown in [25] (see also [32]) that if $2f_d\tau_d < 1$ (underspread channel), $h[n;k]$ in Equation 3.6 has the (approximate) expression

$$h[n;k] = \sum_{q=1}^{Q} h_q(k)e^{j\omega_q n} \tag{3.12}$$

where

$$w_q = \frac{2\pi}{N}\left(q - \frac{1}{2} - \frac{Q}{2}\right), \quad L := \lfloor \tau_d/T_s \rfloor, \quad Q := 2\lceil f_d N T_s \rceil + 1 \tag{3.13}$$

The above representation is valid over a duration of NT_s seconds (N baud rate samples). If τ_d and f_d are known (typically true), then $h[n;k]$ is unknown up to only time-invariant quantities. Over disjoint NT_s-second long segments, only $h_q(k)$ values may change: Q, w_q, and L remain unchanged. This is the complex exponential basis expansion model (CE-BEM) representation considered in some detail in [35]. Equation 3.12 and Equation 3.13 apply to single-input single-output systems — one user and one receiver with symbol rate sampling. It is easily modified to handle multiuser, multiple transmit and receive antennas and is higher than symbol rate sampling — the basic representation remains essentially unchanged.

Let us assume that for a fixed $k=0$ (fixed tap), $h[n;k]$ follows the Jakes model [19, p. 68] and is generated by

$$h_J[n;k] = c(k)\frac{1}{\sqrt{M}}\sum_{m=0}^{M-1} \exp[j(2\pi f_d n T_s \cos(\alpha_{m,k}) + \phi_{m,k})] \tag{3.14}$$

with $c(k)=1\ \forall k$ (each tap component has the same power), $f_d =$ Doppler spread, $M=200$, T_s is the symbol interval, and random variables $\alpha_{m,k}$ and $\phi_{m,k}$ are uniformly distributed over $[0, 2\pi]$ and are mutually independent over m and k. For a fixed k, Equation 3.14 generates a random process $\{h[n;k]\}_n$ whose power spectrum approximates the Jakes spectrum [19, p. 68] as $M \uparrow \infty$ (see [35] and references therein). We consider a system with carrier frequency of 1.8 GHz, data rate of 20 kB (kB = kilo-Bauds = kilo-symbols/second), and a Doppler spread f_d of 80 Hz (corresponding to a maximum mobile velocity of 48 km/hr). We picked a block length of $N = 1000$ symbols (time duration of 50 msec). We chose $Q = 2\lceil f_d N T_s \rceil + 1 = 9$ and estimated the parameters $h_q(k)$ using a least squares approach, given Equation 3.12 and data generated via Equation 3.14. Figure 3.2(a) shows the $h[n;k]$ generated via Equation 3.14 and fitted Equation 3.12. In Figure 3.2(b) we plot the mean square error (averaged over n and 100 Monte Carlo runs) between Equation 3.14 and fitted Equation 3.12 for varying Q values. It is seen that there is little improvement beyond $Q = 9$.

The representation $h(n;l)$ in Equation 3.12 is a special case of a more general representation:

$$h[n;k] = \sum_{q=1}^{Q} h_q(k)\phi_q(n) \tag{3.15}$$

where $\{\phi_q(n)\}_{q=1}^{Q}$ is a set of orthogonal basis functions (over the time interval under consideration). Examples include wavelet-based expansions as in [26], polynomial bases as in [2], and other possibilities [29]. The CE-BEM representation is attractive because it can be related to physical parameters of the

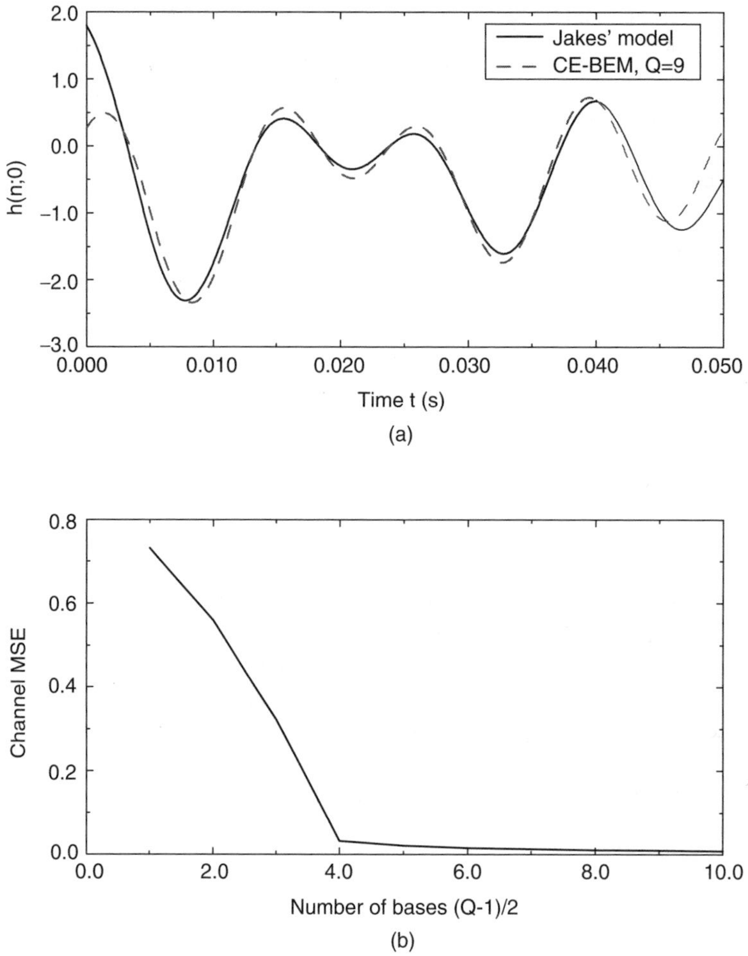

FIGURE 3.2 CE-BEM model validation for a time-selective channel. Channel MSE (mean-square error) is based on 100 Monte Carlo runs.

channel (such as delay spread, Doppler spread, and signal bandwidth, which is inversely proportional to T_s).

3.2.2 Time-Invariant Channels

After some processing (matched filtering, for instance), the continuous-time received signals are sampled at the baud (symbol) or higher (fractional) rate before processing them for channel estimation or equalization. It is therefore convenient to work with an equivalent baseband discrete-time white noise channel model [30, Section 10.1]. For a baud rate sampled system, the equivalent baseband channel model is given by

$$y[n] = \sum_{k=0}^{L} f[k]I[n-k] + w[k] \tag{3.16}$$

where $\{w[k]\}$ is a white Gaussian noise sequence with variance σ^2; $\{I[k]\}$ is the zero-mean, independent and identically distributed (i.i.d.) information (symbol) sequence, possibly complex, taking values from a finite set; $\{f[k]\}$ is finite impulse response (FIR) linear filter (with possibly complex coefficients) that

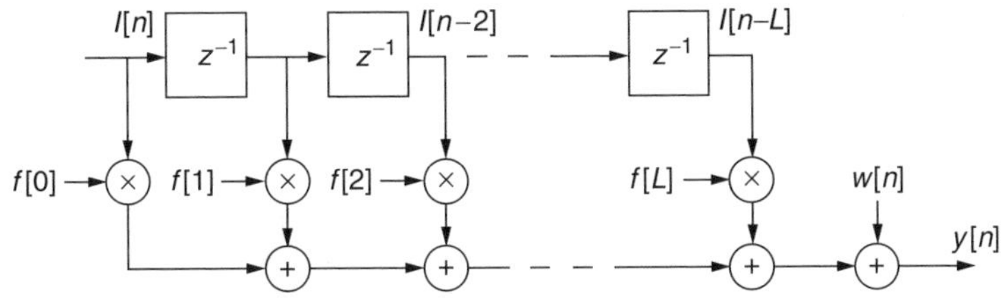

FIGURE 3.3 Tapped delay line model of the frequency-selective but time-nonselective baud-rate channel.

represents the equivalent channel, including the effects of the noise-whitening filter; and $\{y[k]\}$ is the (possibly complex) equivalent baseband received signal. A tapped delay line structure for this model is shown in Figure 3.3.

Equation 3.16 results in a SISO complex discrete-time baseband-equivalent channel model. The output sequence $\{y[n]\}$ in Equation 3.16 is discrete-time stationary. When there is excess channel bandwidth [bandwidth $> \frac{1}{2} \times$ (baud rate)], baud rate sampling is below the Nyquist rate, leading to aliasing and, depending upon the symbol timing phase in certain cases, causing deep spectral notches in sampled, aliased channel transfer functions [14]. Linear equalizers designed on the basis of the baud rate sampled channel response are quite sensitive to symbol timing errors. Initially, in the trained case, fractional sampling was investigated to robustify the equalizer performance against timing errors. Equation 3.16 does not apply to fractionally spaced samples, i.e., when the sampling interval is a fraction of the symbol duration. The fractionally sampled digital communications signal is a cyclostationary signal [10], which may be represented as a vector stationary sequence using a time series representation (TSR) [10, Section 12.6]. Suppose that we sample at P times the baud rate with signal samples spaced T_s/P seconds apart where T_s is the symbol duration. Then a TSR for the sampled signal is given by

$$y_i[n] = \sum_{k=0}^{L} f_i[k]I[n-k] + w_i[n]; \quad (i = 1, 2, \ldots, P) \tag{3.17}$$

where now we have P samples every symbol period, indexed by i. Notice, however, that the information sequence $I[n]$ is still one sample per symbol. It is assumed that the signal incident at the receiver is first passed through a receive filter whose transfer function equals the square root of a raised cosine pulse, and that the receive filter is matched to the transmit filter. The noise sequence in Equation 3.17 is the result of the fractional rate sampling of a continuous-time, filtered white Gaussian noise process. Therefore, the sampled noise sequence is white at the symbol rate, but correlated at the fractional rate. Stack P consecutive received samples in the n-th symbol duration to form a P vector $\mathbf{y}[n]$ satisfying

$$\mathbf{y}[n] = \sum_{k=0}^{L} \mathbf{f}[k]I[n-k] + \mathbf{w}[n] \tag{3.18}$$

where $\mathbf{f}[n]$ is the vector impulse response of the SIMO-equivalent channel model given by

$$\mathbf{f}[n] = \begin{bmatrix} f_1[n] & f_2[n] & \cdots & f_P[n] \end{bmatrix}^T \tag{3.19}$$

$$\mathbf{y}[n] = \begin{bmatrix} y_1[n] & y_2[n] & \cdots & y_P[n] \end{bmatrix}^T \tag{3.20}$$

$$\mathbf{w}[n] = \begin{bmatrix} w_1[n] & w_2[n] & \cdots & w_P[n] \end{bmatrix}^T \tag{3.21}$$

When $P = 2$, one way to look at the TSR model is to note that $y_1[n]$ values are odd-numbered fractionally spaced samples, $y_2[n]$ values are the even-numbered fractionally spaced samples, and n indexes the baud (symbol), similarly for $f_i[n]$. A block diagram of Equation 3.17 is shown in Figure 3.4.

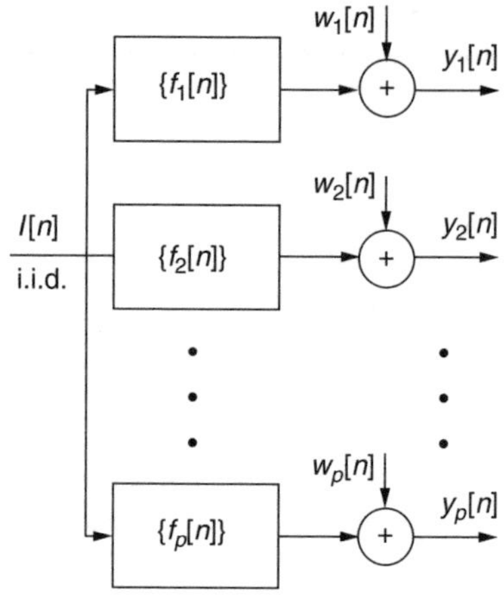

FIGURE 3.4 Block diagram of the fractionally sampled ($P \times$ baud-rate) frequency-selective but time-nonselective channel.

3.3 Channel Estimation

We first consider three types of channel estimators within the framework of maximizing the likelihood function. (Unless otherwise noted, the underlying channel model is given by the time-invariant Equation 3.18.) In general, one of the most effective and popular parameter estimation algorithms is the maximum likelihood (ML) method. The class of maximum likelihood estimators is asymptotically optimal.

Let us consider the P vector channel model given in Equation 3.18. Suppose that we have collected M samples of the observation $Y = [\mathbf{y}^T[M-1], \ldots, \mathbf{y}^T[0]]^T$. We then have the following linear model:

$$Y = \begin{pmatrix} I[M-1]\mathcal{I}_P & I[M-2]\mathcal{I}_P & \cdots & I[M-L-1]\mathcal{I}_P \\ \vdots & \textit{Block} & \textit{Hankel} & \textit{Matrix} \\ I[0]\mathcal{I}_P & I[-1]\mathcal{I}_P & \cdots & I[-L]\mathcal{I}_P \end{pmatrix} \begin{pmatrix} \mathbf{f}[0] \\ \vdots \\ \mathbf{f}[L] \end{pmatrix} + \begin{pmatrix} \mathbf{w}[M-1] \\ \vdots \\ \mathbf{w}[0] \end{pmatrix}$$

$$= \mathcal{H}(\mathbf{I})_{[MP]\times[P(L+1)]}F + W \tag{3.22}$$

where $\mathcal{I}_P$ is a $P \times P$ identity matrix, $\mathbf{I}$ and W are vectors consisting of samples of the input sequence $\{I[n]\}$ and noise $\{\mathbf{w}[n]\}$, respectively, F is the vector of the channel parameters, and a block Hankel matrix has identical block entries on its block antidiagonals.

Let θ be the vector of unknown parameters that may include the channel parameters F and possibly the entire or part of the input vector $\mathbf{I}$. Given the probability space that describes jointly the noise vector W and possibly the input data vector $\mathbf{I}$, we can then obtain, in principle, the probability density function (pdf) of the observation Y. As a function of the unknown parameter θ, the pdf of the observation $f(Y|\theta)$ is referred to as the *likelihood function*. The maximum likelihood estimator is defined by the following optimization:

$$\widehat{\theta} = \arg\max_{\theta \in \Theta} f(Y|\theta) \tag{3.23}$$

where Θ defines the domain of the optimization.

While the ML estimator is conceptually simple, and it usually has good performance when the sample size is sufficiently large, the implementation of the ML estimator is sometimes computationally intensive. Furthermore, the optimization of the likelihood function in Equation 3.23 is often hampered by the existence of local maxima. Therefore, it is desirable that effective initialization techniques are used in conjunction with the ML estimation.

3.3.1 Training-Based Channel Estimation

The training-based channel estimation assumes the availability of the input vector **I** (as training symbols) and its corresponding observation vector Y. When the noise samples are zero-mean, white Gaussian, i.e., **w** is a zero-mean, Gaussian random vector with covariance $\sigma_w^2 \mathcal{I}_{MP}$, the ML estimator defined in Equation 3.23, with $\theta = F$, is given by

$$\hat{F} = \arg\min_F \|Y - \mathcal{H}(\mathbf{I})F\|^2 = \mathcal{H}^{\dagger}(\mathbf{I})Y \tag{3.24}$$

where $\mathcal{H}^{\dagger}(\mathbf{I})$ is the Moore–Penrose pseudoinverse of the $\mathcal{H}(\mathbf{I})$ defined in Equation 3.22. This is also the classical linear least squares estimator, which can be implemented recursively, and it turns out to be the best (in terms of having minimum mean square error) among all unbiased estimators and is the most efficient in the sense that it achieves the Cramer–Rao lower bound. Various adaptive implementations can be found in [30].

3.3.1.1 Time-Variant Channels

In case of general time-varying channels represented by Equation 3.8, a simple generalization of [5] (see also [25]) is to use a periodic Kronecker delta function sequence as training:

$$I[n] = \sum_j \delta[n - j\bar{P}] \tag{3.25}$$

With Equation 3.25 as input to Equation 3.8, one obtains

$$y[n] = \sum_j h[n; n - j\bar{P}] + w[n] \tag{3.26}$$

so that if $\bar{P} > L$, we have for $0 \le i \le L$,

$$y[k\bar{P} + i] = \sum_j h[k\bar{P} + i; i] + w[k\bar{P} + i] \tag{3.27}$$

Therefore, one may take the estimate of $h[k\bar{P}; i]$ as

$$\hat{h}[k\bar{P}; i] = y[k\bar{P} + i] = h[k\bar{P} + i; i] + w[k\bar{P} + i] \tag{3.28}$$

For time samples between $k\bar{P}$ (k is an integer), linear interpolation may be used to obtain channel estimates.

If we use a CE-BEM representation Equation 3.12, then one directly estimates the time-invariant parameters $h_q(k)$. From Equation 3.8 and Equation 3.12 we have

$$y[n] = \sum_{k=0}^{L} \sum_{q=1}^{Q} \underbrace{h_q(k) e^{-j\omega_q k}}_{\tilde{h}_q(k)} I[n-k] e^{j\omega_q (n-k)} + w[n] \tag{3.29}$$

$$= \sum_{k=0}^{L} I[n-k] \underbrace{\begin{bmatrix} e^{j\omega_1 (n-k)} & \cdots & e^{j\omega_Q (n-k)} \end{bmatrix}}_{E[n-k]} \underbrace{\begin{bmatrix} \tilde{h}_1(k) \\ \vdots \\ \tilde{h}_Q(k) \end{bmatrix}}_{H[k]} + w[n] \tag{3.30}$$

Collecting M samples of the observations $Y = [y[M-1], \ldots, y[0]]^T$, we have the linear model

$$Y = \begin{pmatrix} I[M-1]E[M-1] & \cdots & I[M-L-1]E[M-L-1] \\ \vdots & Block\ \ Hankel & Matrix \\ I[0]E[0] & \cdots & I[-L]E[-L] \end{pmatrix} \begin{pmatrix} H[0] \\ \vdots \\ H[L] \end{pmatrix} + \begin{pmatrix} w[M-1] \\ \vdots \\ w[0] \end{pmatrix} \tag{3.31}$$

Now we have a model similar to that in Equation 3.22 with a solution similar to that in Equation 3.24.

3.3.2 Blind Channel Estimation

3.3.2.1 Combined Channel and Symbol Estimation

The simultaneous estimation of the input vector and the channel appears to be ill-posed; how is it possible that the channel and its input can be distinguished using only the observation? The key in blind channel estimation is the utilization of qualitative information about the channel and the input. To this end, we consider two different types of maximum likelihood techniques based on different models of the input sequence.

3.3.2.1.1 Stochastic Maximum Likelihood Estimation

While the input vector $\mathbf{I}$ is unknown, it may be modeled as a random vector with a known distribution. In such a case, the likelihood function of the unknown parameter $\theta = F$ can be obtained by

$$f(Y|F) = \int f(Y|\mathbf{I}, F) f(\mathbf{I}) d\mathbf{I} \tag{3.32}$$

where $f(\mathbf{I})$ is the marginal pdf of the input vector and $f(Y|\mathbf{I}, F)$ is the likelihood function when the input is known. Assume, for example, that the input data symbol $I[k]$ takes, with equal probability, a finite number of values. Consequently, the input data vector $\mathbf{I}$ also takes values from the signal set $\{\mathbf{I}_1, \ldots, \mathbf{I}_K\}$. The likelihood function of the channel parameters is then given by

$$f(Y|F) = \sum_{i=1}^{K} f(Y|\mathbf{I}_i, F) \text{Prob}(\mathbf{I} = \mathbf{I}_i) = C \sum_{i=1}^{K} \exp\left\{ -\frac{\|Y - \mathcal{H}(\mathbf{I}_i)F\|^2}{2\sigma^2} \right\} \tag{3.33}$$

where C is a constant, $\|Y\|^2 := Y^H Y$, Y^H is the complex conjugate transpose of the complex vector Y, and the stochastic maximum likelihood estimator is given by

$$\widehat{F} = \arg\min_F \sum_{i=1}^{K} \exp\left\{ -\frac{\|Y - \mathcal{H}(\mathbf{I}_i)F\|^2}{2\sigma^2} \right\} \tag{3.34}$$

The maximization of the likelihood function defined in Equation 3.32 is in general difficult because $f(Y|\theta)$ is nonconvex. The expectation–maximization (EM) algorithm can be applied to transform the complicated optimization to a sequence of quadratic optimizations. Kaleh and Vallet [20] first applied the EM algorithm to the equalization of communication channels with an input sequence having a finite alphabet property. By using a *hidden Markov model* (HMM), they developed a batch (off-line) procedure that includes the so-called forward and backward recursions. Unfortunately, the complexity of this algorithm increases exponentially with the channel memory.

To relax the memory requirements and facilitate channel tracking, on-line sequential approaches have been proposed in [21] for input with finite alphabet properties under an HMM formulation. Given the appropriate regularity conditions and a good initialization guess, it can be shown that these algorithms converge to the true channel value.

3.3.2.1.2 Deterministic Maximum Likelihood Estimation

The deterministic ML approach assumes no statistical model for the input sequence $\{I[k]\}$. In other words, both the channel vector F and the input source vector $\mathbf{I}$ are parameters to be estimated. When the noise is zero-mean Gaussian with covariance $\sigma_w^2 \mathcal{I}_{MP}$, the ML estimates can be obtained by the nonlinear least squares optimization:

$$\{\widehat{F}, \widehat{\mathbf{I}}\} = \arg\min \|Y - \mathcal{H}(\mathbf{I})F\|^2 \tag{3.35}$$

The joint minimization of the likelihood function with respect to both the channel and the source parameter spaces is difficult. Fortunately, the observation vector Y is linear in both the channel and the input parameters individually. In particular, we have

$$Y = \mathcal{H}(\mathbf{I})F + W = \mathcal{T}(F)\mathbf{I} + W \tag{3.36}$$

where

$$\mathcal{T}(F) = \begin{pmatrix} \mathbf{f}[0] & \cdots & \mathbf{f}[L] & & \\ & \ddots & & \ddots & \\ & & \mathbf{f}[0] & \cdots & \mathbf{f}[L] \end{pmatrix} \tag{3.37}$$

is the the so-called filtering matrix. We therefore have a separable nonlinear least squares problem that can be solved sequentially:

$$\{\widehat{F}, \widehat{\mathbf{I}}\} = \arg\min_{\mathbf{I}}\{\min_{F} \|Y - \mathcal{H}(\mathbf{I})F\|^2\} \tag{3.38}$$

$$= \arg\min_{F}\{\min_{\mathbf{I}} \|Y - \mathcal{T}(F)\mathbf{I}\|^2\} \tag{3.39}$$

If we are only interested in estimating the channel, the above minimization can be rewritten as

$$\widehat{F} = \arg\min_{F} \| \underbrace{(I - \mathcal{T}(F)\mathcal{T}^{\dagger}(F))}_{\mathcal{P}(F)} Y\|^2 = \arg\min_{F} \|\mathcal{P}(F)Y\|^2 \tag{3.40}$$

where $\mathcal{P}(F)$ is a projection transform of Y into the orthogonal complement of the range space of $\mathcal{T}(F)$, or the noise subspace of the observation, and $\mathcal{T}^{\dagger}(F)$ denotes the pseudoinverse of $\mathcal{T}(F)$. Discussions of algorithms of this type can be found in [36].

Similar to the HMM for the statistical maximum likelihood approach, the finite alphabet properties of the input sequence can also be incorporated into the deterministic maximum likelihood methods. These algorithms, first proposed by Seshadri [33] and Ghosh and Weber [12], iterate between estimates of the channel and the input. At iteration k, with an initial guess of the channel $F^{(k)}$, the algorithm estimates the input sequence $\mathbf{I}^{(k)}$ and the channel $F^{(k+1)}$ for the next iteration by

$$\mathbf{I}^{(k)} = \arg\min_{\mathbf{I}\in\mathcal{S}} \|Y - \mathcal{T}(F^{(k)})\mathbf{I}\|^2 \tag{3.41}$$

$$F^{(k+1)} = \arg\min_{F} \|Y - \mathcal{H}(\mathbf{I}^{(k)})F\|^2 \tag{3.42}$$

where $\mathcal{S}$ is the (discrete) domain of $\mathbf{I}$. The optimization in Equation 3.42 is a linear least squares problem, whereas the optimization in Equation 3.41 can be achieved by using the Viterbi algorithm [30]. Seshadri [33] presented blind trellis search techniques. Reduced-state sequence estimation was proposed in [12]. Raheli et al. [31] proposed a per-survivor processing technique.

The convergence of such approaches is not guaranteed in general. Interesting examples have been provided in [6] where two different combinations of F and $\mathbf{I}$ lead to the same cost $\|Y - \mathcal{H}(\mathbf{I})F\|^2$.

3.3.2.2 The Methods of Moments

Although the ML channel estimator discussed in Section 3.3.2.1 usually provides better performance, the computation complexity and the existence of local optima are the two major difficulties. Therefore, "simpler" approaches have also been investigated.

3.3.2.2.1 SISO Channel Estimation

For baud rate data, second-order statistics of the data do not carry enough information to allow estimation of the channel impulse response as a typical channel is non-minimum phase. On the other hand, higher-order statistics (in particular, fourth-order cumulants) of the baud rate (or fractional rate) data can be exploited to yield the channel estimates to within a scale factor.

Given the mathematical model Equation 3.16, there are two broad classes of direct approaches to channel estimation, the distinguishing feature among them being the choice of the optimization criterion. All of the approaches involve (more or less) a least squares error measure. The error definition differs, however, as follows:

- *Fitting Error:* Match the model-based higher-order (typically fourth-order) statistics to the estimated (data-based) statistics in a least squares sense to estimate the channel impulse response, as in [43] and [44], for example. This approach allows consideration of noisy observations. In general, it results in a nonlinear optimization problem. It requires availability of a good initial guess to prevent convergence to a local minimum. It yields estimates of the channel impulse response.
- *Equation Error:* This is based on minimizing an "equation error" in some equation that is satisfied ideally. The approaches of [17] and [49] (among others) fall into this category. In general, this class of approaches results in a closed-form solution for the channel impulse response so that a global extremum is always guaranteed provided that the channel length (order) is known. These approaches may also provide good initial guesses for the nonlinear fitting error approaches. Quite a few of these approaches fail if the channel length is unknown.

Further details may be found in [46] and references therein.

3.3.2.2.1.1 Indirect Channel Estimation

Here one first carries out linear blind equalization followed by channel estimation via an input–output formulation. One estimates $\{I[k]\}$ as $\{\widehat{I}[k]\}$ and then, treating $\{\widehat{I}[k]\}$ as a known input sequence to Equation 3.16, can estimate the underlying channel. Consider a linear equalizer

$$\widehat{I}[k] = \sum_{n=-N}^{N} c[n]y[k-n] \tag{3.43}$$

where $\{c[n]\}_{n=-N}^{n=N}$ are the $(2N+1)$ tap weight coefficients of the equalizer; see Figure 3.5. In the case of known training sequence transmission, the linear equalizer taps $c[n]$ are chosen to minimize the cost $E\{|\widehat{I}[k] - I[k]|^2\}$ where $\{I[k]\}$ is the training sequence. In the blind case, there is no training sequence.

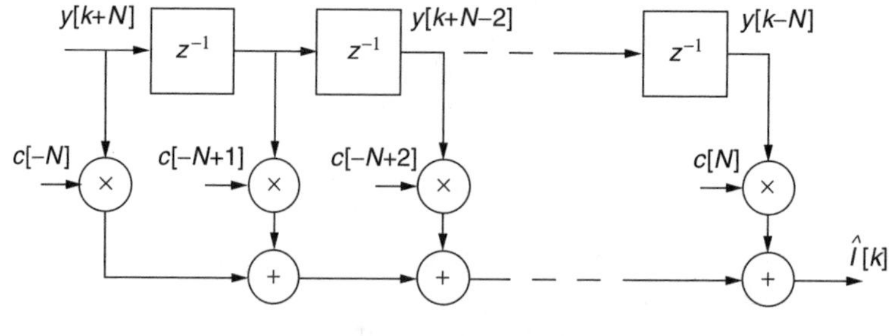

FIGURE 3.5 Structure of a baud-rate linear transversal equalizer.

The key to designing a blind equalizer is to design rules of equalizer parameter adjustment. With the lack of training sequence, the receiver does not have access to the desired equalizer output $I[k]$ to adopt the traditional minimum mean square error criterion. The blind equalizer adaptation needs to minimize some special, non-mean square error (MSE) type cost function that implicitly involves higher-order statistics of the channel output signal. The design of the blind equalizer thus translates into defining a mean cost function $E\{\Psi(\widehat{I}[k])\}$ where $\Psi(x)$ is a scalar function. Thus, the stochastic gradient descent minimization algorithm is easily determined by the derivative function $\psi(x) := \Psi'(x) := d\Psi(x)/dx$. Hence, a blind equalizer can be defined either by the cost function $\Psi(x)$ or, equivalently, by its derivative $\psi(x)$ function. Ideally, the function $\Psi(\cdot)$ should be selected such that local minima of the mean cost correspond to a significant removal of ISI in the equalizer output $\widehat{I}[k]$.

Let

$$\mathbf{C} := \begin{bmatrix} c[-N] & c[-N+1] & \cdots & c[-N] \end{bmatrix}^T \tag{3.44}$$

Let $\mathbf{C}^{(k)}$ denote the value of $\mathbf{C}$ at the k-th iteration. Then a stochastic gradient algorithm for the adaptation of $\mathbf{C}$ is given by

$$\mathbf{C}^{(k+1)} = \mathbf{C}^{(k)} - \alpha \nabla_{\mathbf{C}^{(k)}} \Psi(\widehat{I}[k]) \tag{3.45}$$

where $\nabla_{\mathbf{C}}\Psi$ denotes the gradient of Ψ w.r.t the tap vector $\mathbf{C}$ and $\alpha > 0$ is the step-size parameter [30]. The best-known blind algorithms were presented in [15] and [39] with cost functions

$$\Psi_q(x) = \frac{1}{2q}\left(|x|^q - R_q\right)^2, \quad \text{where} \quad R_q := \frac{E|I[k]|^{2q}}{E|I[k]|^q}, \quad q = 1, 2, \ldots \tag{3.46}$$

This class of *Godard algorithms* is indexed by the positive integer q. Using the stochastic gradient descent approach, equalizer parameters can be adapted accordingly.

For $q = 2$, the special Godard algorithm was developed as the constant modulus algorithm (CMA) independently by Treichler and Agee [39] using the philosophy of property restoral. For a channel input signal that has a constant modulus $|I[k]|^2 = R_2$, the CMA equalizer penalizes output samples $\widehat{I}_k$ that do not have the desired constant modulus characteristics. The modulus error is simply $e[k] = |\widehat{I}[k]|^2 - R_2$, and the squaring of this error yields the constant modulus cost function, which is identical to the Godard cost function with $q = 2$.

A deconvolution-based approach can also be found in [3]. Promising simulation results have been reported on the application of blind equalization in the popular wireless GSM (Global System for Mobile Communications) cellular system [4] using a higher-order statistical deconvolution method [3] where the estimated channel is used in conjunction with MLSE (ML sequence estimator) for symbol estimation. Boss et al. [4] report that their HOS-based approach using just 142 data samples per frame incurs an signal-to-noise ratio (SNR) loss of only 1.2 to 1.3 dB, while it saves the 22% overhead in the GSM data rate caused by the transmission of training sequences. That is, on average, the HOS-based approach of [3] requires 1.2 to 1.3 dB higher SNR than the conventional GSM system to achieve the same bit error rate. Werner et al. [50] discuss modifications of CMA, called multimodulus algorithm (MMA) and generalized MMA, for high-order QAM (quadrature amplitude modulation) and CAP (carrierless amplitude and phase) signals.

3.3.2.2.2 SIMO Channel Estimation

Here we will concentrate upon second-order statistical methods, but first let us make a few comments regarding indirect SIMO channel estimation. As noted in Section 3.1, linear equalizers designed on the basis of the baud rate sampled received signal are quite sensitive to symbol timing errors [14]. Therefore, fractionally spaced linear equalizers (typically with twice the baud rate sampling: oversampling by a factor of two) are quite widely used to mitigate sensitivity to symbol timing errors. A fractionally spaced equalizer

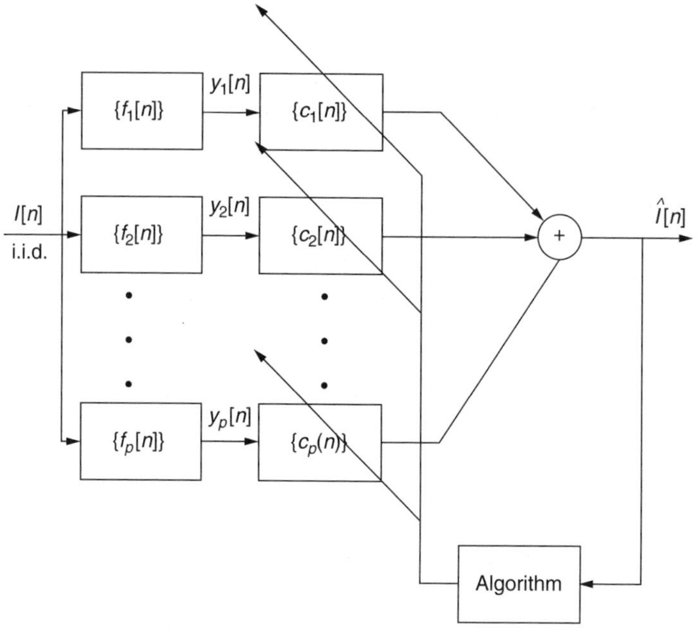

FIGURE 3.6 Block diagram of a fractionally spaced linear equalizer.

(FSE) in the linear transversal structure has the output

$$\widehat{I}[k] = \sum_{n=-N}^{N} \left(\sum_{i=1}^{P} c_i[n] y_i[k-n] \right) \tag{3.47}$$

where $\{c_i[n]\}_{n=-N}^{n=N}$ are the $(2N+1)$ tap weight coefficients of the i-th subequalizer; see Figure 3.6. Note that the FSE outputs data at the symbol rate. Similar to the SISO case, various criteria and cost functions exist to design the linear equalizers in both batch and recursive (adaptive) form.

Linear equalizers do not perform well when the underlying channels have deep spectral nulls in the passband. Several nonlinear equalizers have been developed to deal with such channels. Two effective approaches are:

- *Decision Feedback Equalizer* (DFE): A nonlinear equalizer that employs previously detected symbols to eliminate the ISI due to the previously detected symbols on the current symbol to be detected. The use of the previously detected symbols makes the equalizer output a nonlinear function of the data. DFE can be symbol spaced or fractionally spaced. Figure 3.7 shows a block diagram of a DFE.
- *Maximum Likelihood Sequence Detector*: Estimates the information sequence to maximize the joint probability of the received sequence conditioned on the information sequence.

A detailed discussion may be found in [30].

Returning to the second-order statistical methods, for single-input multiple-output vector channels the autocorrelation function of the observation is sufficient for the identification of the channel impulse response up to an unknown constant [37], [42], provided that the various subchannels have no common zeros. This observation led to a number of techniques under both statistical and deterministic assumptions of the input sequence [36]. By exploiting the multichannel aspects of the channel, many of these techniques lead to a constrained quadratic optimization:

$$\widehat{F} = \arg \min_{\|F\|=1} F^H Q(Y) F \tag{3.48}$$

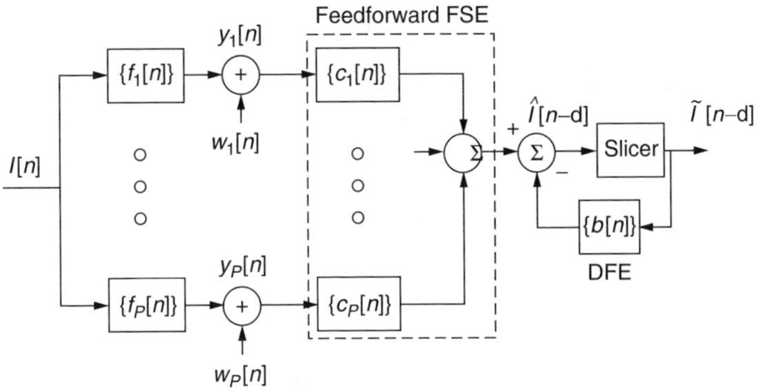

FIGURE 3.7 Feedforward and decision feedback channel equalization filter.

where $Q(Y)$ is a positive definite matrix constructed from the observation. Asymptotically (as either the sample size increases to infinity or the noise variance approaches zero), these estimates converge to true channel parameters.

3.3.2.2.2.1 The Cross-Relation Approach

Here we present a simple yet informative approach [51] that illustrates the basic idea. Suppose that we have only two channels with finite impulse responses, $f_1[n]$ and $f_2[n]$, respectively. If there is no noise, the received signals from the two channels satisfy

$$y_1[n] = f_1[n] * I[n], \quad y_2[n] = f_2[n] * I[n] \tag{3.49}$$

where $*$ is the linear convolution. Consequently, we must have

$$y_1[n] * f_2[n] = y_2[n] * f_1[n] \tag{3.50}$$

Since the convolution operation is linear with respect to the channel and $y_i[n]$ $(i = 1, 2)$ are available, the above equation is equivalent to solving a homogeneous linear equation:

$$R\tilde{F} = 0 \tag{3.51}$$

where R is constructed from the M received data samples

$$R = Y_2 - Y_1, \tag{3.52}$$

$$Y_j := \begin{pmatrix} y_j[L] & y_j[L-1] & \cdots & y_j[0] \\ y_j[L+1] & y_j[L] & \cdots & y_j[1] \\ \vdots & \vdots & \ddots & \vdots \\ y_j[M-1] & y_j[M-2] & \cdots & y_j[M-L-1] \end{pmatrix} \tag{3.53}$$

and

$$\tilde{F} := (f_1[0] \quad f_1[1] \quad \cdots \quad f_1[L] \quad f_2[0] \quad \cdots \quad f_2[L])^T \tag{3.54}$$

It can be shown that under certain identifiability conditions [36] (which include knowledge of L and no common subchannel zeros), the null space of R has dimension 1, which means that the channel can be identified up to a constant. When there is noise, the channel estimator can be obtained from a constrained quadratic optimization:

$$\widehat{\tilde{F}} = \arg \min_{\|\tilde{F}\|=1} \tilde{F}^H R^H R \tilde{F} \tag{3.55}$$

which implies that $\widehat{\tilde{F}}$ is the eigenvector corresponding to the smallest eigenvalue of $Q = R^H R$.

Hua [18] has shown that the cross-relation method combined with the ML approach offers performance close to the Cramer–Rao lower bound. The main problem with this method is that the channel length L needs to be accurately known (in addition to the no-common-subchannel-zeros condition).

3.3.2.2.2.2 Noise Subspace Approach

Alternatively, one can also exploit the subspace structure of the filtering matrix. We now consider a method proposed by Moulines et al. [28]. Define the $M \times [M + L]$ filtering matrix,

$$\mathcal{T}_{M+L}(\mathbf{f}_l) = \begin{pmatrix} f_l[0] & \cdots & f_l[L] & & \\ & \ddots & & \ddots & \\ & & f_l[0] & \cdots & f_l[L] \end{pmatrix} \tag{3.56}$$

and the $[PM] \times [M + L]$ multichannel filtering matrix,

$$\mathcal{T}_{M+L}(F) = \left(\mathcal{T}_{M+L}^T(\mathbf{f}_1) \quad \mathcal{T}_{M+L}^T(\mathbf{f}_2) \quad \cdots \quad \mathcal{T}_{M+L}^T(\mathbf{f}_P) \right)^T \tag{3.57}$$

Define $(M \geq L)$

$$\mathbf{Y}_n = \left(\mathbf{Y}_{1n}^T \quad \mathbf{Y}_{2n}^T \quad \cdots \quad \mathbf{Y}_{Pn}^T \right)^T \tag{3.58}$$

where

$$\mathbf{Y}_{in} = (\, y_i[n] \quad y_i[n-1] \quad \cdots \quad y_i[n-M+1] \,)^T \tag{3.59}$$

Then the correlation matrix $\mathbf{R} = E\{\mathbf{Y}_n\mathbf{Y}_n^H\}$ has an eigenvalue decomposition (EVD)

$$\mathbf{R} = \sum_{k=1}^{PM} \lambda_k \mathbf{q}_k \mathbf{q}_k^H \tag{3.60}$$

where λ_k values are in descending order of magnitude. It can be shown that the range space of $\mathbf{R}$ (signal subspace), also the range space of $\mathcal{T}_{M+L}(F)$, is spanned by the eigenvectors $\mathbf{q}_k$ for $k = 1, 2, \ldots, L + M$, whereas the noise subspace (orthogonal complement of the range space) is spanned by the remaining $\mathbf{q}_k$ values for $k = L + M + 1, L + M + 2, \ldots, PM$.

Define $\mathbf{g}_k = \mathbf{q}_{L+M+k+1}$ for $k = 0, 1, \ldots, PM - L - M - 1$. It then follows that

$$\mathcal{T}_{M+L}^H(F)\mathbf{g}_k = 0, \quad k = 0, 1, \ldots, PM - L - M - 1 \tag{3.61}$$

The vectors $\mathbf{g}_k$ can be estimated from data via one estimated correlation matrix $\mathbf{R}$ and its EVD. Partition the $PM-$ vector $\mathbf{g}_k$ as

$$\mathbf{g}_k = \left(\mathbf{g}_{1k}^T \quad \cdots \quad \mathbf{g}_{Pk}^T \right)^T \tag{3.62}$$

to conform to $\mathcal{T}_{M+L}(F)$, where $\mathbf{g}_{ik}$ is $M \times 1$. For a given k, define the $[L + 1] \times [L + M]$ matrix $\mathcal{T}_{M+L}(\mathbf{g}_{lk})$ just as $\mathcal{T}_{M+L}(\mathbf{f}_l)$ in Equation 3.56 except for replacing $\mathbf{f}_l$ with $\mathbf{g}_{lk}$, and similarly define $\mathcal{T}_{M+L}(\mathbf{g}_k)$ by mimicking $\mathcal{T}_{M+L}(F)$ in Equation 3.57. It has been shown by [28] that

$$\mathcal{T}_{M+L}^H(F)\mathbf{g}_k = 0 = \mathcal{T}_{M+L}^H(\mathbf{g}_k)F \tag{3.63}$$

It has been further shown in [28] that under the knowledge of L and no common subchannel zeros, the channel F can be estimated (up to a scale factor) by the optimization problem

$$\widehat{F} = \arg\min_{\|F\|=1} F^H \mathcal{Q} F, \quad \text{where} \quad \mathcal{Q} := \sum_{k=0}^{PM-L-M-1} \mathcal{T}_{M+L}(\mathbf{g}_k)\mathcal{T}_{M+L}^H(\mathbf{g}_k) \tag{3.64}$$

The solution is given by the eigenvector corresponding to the smallest eigenvalue of $\mathcal{Q}$.

As with the cross-relation approach, the noise subspace method requires that the channel length L is accurately known in addition to the channel satisfying the no-common-subchannel-zeros condition. A detailed development of this class of methods may be found in [13, Chapters 3 and 5].

3.3.2.2.2.3 *Multistep Linear Prediction*

More recently, the problem of blind channel identification has been formulated as problems of linear prediction [9], [11], [45] and smoothing [38]. We will first discuss the single-step linear prediction approach followed by its extension to multistep linear prediction. It is based on the observation that for the (noise-free) SIMO FIR channel, there exists a finite causal inverse under the no-common-subchannel-zeros condition.

Define the signal (noise-free) part of Equation 3.18 as

$$\mathbf{s}[n] = \sum_{k=0}^{L} \mathbf{f}[k]I[n-k] \tag{3.65}$$

with $s_i[n]$ denoting the i-th component of $\mathbf{s}[n]$. Consider Equation 3.17. Define the subchannel transfer function

$$F_i(z) := \sum_{n=0}^{L} f_i[n]z^{-n} \tag{3.66}$$

If $F_i(z)$ values for $1 \le i \le P$ have no common zeros (i.e., there does not exist any ρ for which $F_i(\rho) = 0$ $\forall i$), then there exists FIR subequalizers $C_i(z) := \sum_{n=0}^{M} c_i[n]z^{-n}$ with $M \ge L-1$ such that

$$\sum_{i=1}^{P} C_i(z)F_i(z) = 1 \tag{3.67}$$

Hence, there exists a causal FIR filter of length $M \le L-1$ such that

$$I[k] = \sum_{n=0}^{M}\sum_{i=1}^{P} \tilde{c}_i[n]s_i[k-n] \tag{3.68}$$

Using Equation 3.65 and Equation 3.68, we have

$$\mathbf{s}[k] = \mathbf{e}[k|k-1] + \widehat{\mathbf{s}}[k|k-1] \tag{3.69}$$

where

$$\mathbf{e}[k|k-1] := \mathbf{f}[0]I[k] \tag{3.70}$$

and

$$\widehat{\mathbf{s}}[k|k-1] := \sum_{n=1}^{L} \mathbf{f}[n]I[k-n] = \sum_{i=1}^{L_e} \mathbf{A}_i\mathbf{s}[k-i] \tag{3.71}$$

such that

$$E\{\mathbf{e}[k|k-1]\mathbf{s}[k-l]^H\} = 0 \quad \forall l \ge 1 \tag{3.72}$$

That is, by the orthogonality principle, $\widehat{\mathbf{s}}[k|k-1]$ is the one-step-ahead linear prediction (of finite length) of $\mathbf{s}[k]$, and $\mathbf{e}[k|k-1]$ is the corresponding prediction error (linear innovations). The existence of $L_e \le L-1$ in Equation 3.71 can be established. The predictor coefficients $\mathbf{A}_i$ can be estimated from data (after removal of noise effects); therefore, one can calculate $E\{\mathbf{e}[k|k-1]\mathbf{e}[k|k-1]^H\}$ from data-based correlation estimates. By Equation 3.70,

$$E\{\mathbf{e}[k|k-1]\mathbf{e}[k|k-1]^H\} = E\{|I[k]|^2\}\mathbf{f}[0]\mathbf{f}[0]^H \tag{3.73}$$

a rank one matrix. Equation 3.73 allows estimation of $\mathbf{f}[0]$ up to a scale factor (the estimate equals the eigenvector of $E\{\mathbf{e}[k|k-1]\mathbf{e}[k|k-1]^H\}$ corresponding to the largest eigenvalue). Once we have a scaled estimate of $\mathbf{f}[0]$, we can estimate the remaining channel coefficients using Equation 3.65 with $\{\mathbf{s}[k]\}$ as output and $\|\mathbf{f}[0]\|^{-2}\mathbf{f}[0]^H\mathbf{e}[k|k-1]$ $(= I[k]e^{j\alpha})$ as input (where α is arbitrary).

The above approach can be extended by using multistep linear prediction. It can be shown that

$$\mathbf{s}[k] = \mathbf{e}[k|k-2] + \widehat{\mathbf{s}}[k|k-2] \tag{3.74}$$

where

$$\mathbf{e}[k|k-2] := \mathbf{f}[0]I[k] + \mathbf{f}[1]I[k-1] \tag{3.75}$$

and

$$\widehat{\mathbf{s}}[k|k-2] := \sum_{n=2}^{L} \mathbf{f}[n]I[k-n] = \sum_{i=2}^{L_e+1} \mathbf{A}_{2i}\mathbf{s}[k-i] \tag{3.76}$$

such that

$$E\{\mathbf{e}[k|k-2]\mathbf{s}[k-l]^H\} = 0 \quad \forall l \geq 2 \tag{3.77}$$

By the orthogonality principle, $\widehat{\mathbf{s}}[k|k-2]$ is the two-step-ahead linear prediction (of finite length) of $\mathbf{s}[k]$, and $\mathbf{e}[k|k-2]$ is the corresponding prediction error. Define

$$\mathbf{E}_k := ((\mathbf{e}[k+1|k-1] - \mathbf{e}[k+1|k])^T, \quad \mathbf{e}[k|k-1]^T)^T \tag{3.78}$$

so that we have

$$\mathbf{E}_k = \begin{pmatrix} \mathbf{f}[1] \\ \mathbf{f}[0] \end{pmatrix} I[k] \tag{3.79}$$

By Equation 3.79,

$$E\left\{\mathbf{E}_k\mathbf{E}_k^H\right\} = E\{|I[k]|^2\} \begin{pmatrix} \mathbf{f}[1] \\ \mathbf{f}[0] \end{pmatrix} (\mathbf{f}[1]^H \quad \mathbf{f}[0]^H) \tag{3.80}$$

a rank one matrix. That is, we can estimate $\mathbf{f}[0]$ and $\mathbf{f}[1]$ simultaneously up to the same scale factor. By adding larger step predictors, one can estimate the entire channel impulse response simultaneously. An advantage over the one-step predictor approach is that the results are not unduly influenced by any estimation errors in estimating the leading coefficient $\mathbf{f}[0]$.

The multistep linear prediction approach was proposed by [9] in a different form and by [11] in the form given above. Both of them assumed FIR channels with known channel length and no common subchannel zeros. Tugnait [45] extended the approach of [11] by allowing common subchannel zeros, infinite impulse response (IIR) channels, and unknown channel length. It is shown in [45] that minimum-phase common subchannel zeros pose no problems for the multistep linear prediction approach, and in the presence of non-minimum-phase common subchannel zeros, the multistep linear prediction approach yields a minimum-phase equivalent version of these zeros. It is also worth noting that linear prediction approaches (both single step and multistep) are robust against overdetermination of channel length, unlike the cross-relation and noise subspace approaches.

Extensions to CE-BEM models of the linear prediction and multistep linear prediction approaches may be found in [47] and [24], respectively.

3.3.3 Semiblind Approaches

Semiblind approaches utilize a combination of training-based and blind approaches. Here we present a brief discussion about the idea and refer the reader to a recent survey [8] for details. The objective of semiblind channel estimation (and equalization) is to exploit the information used by blind methods as well as the information exploited by the training-based methods. Semiblind channel estimation assumes additional knowledge of the input sequence. Specifically, part of the input data vector is known. Both the statistical and deterministic maximum likelihood estimators remain the same except that the likelihood function needs to be modified to incorporate the knowledge of the input. However, semiblind channel estimation may offer significant performance improvement over either the blind or the training-based methods, as demonstrated in the evaluation of the Cramer–Rao lower bound in [8].

There are many generalizations of blind channel estimation techniques to incorporate known symbols. Cirpan and Tsatsanis [7] extended the approach of Kaleh and Vallet by restricting the transition of the hidden Markov model. In [23], the knowledge of the known symbol is used to avoid the local maxima in the maximization of the likelihood function. A popular approach is to combine the objective function used to derive the blind channel estimator with the least squares cost in the training-based channel estimation. For example, a weighted linear combination of the cost for the blind channel estimator and that for the training-based estimator can be used [16], [22].

3.3.4 Hidden Pilot-Based Approaches

In the hidden pilot (or superimposed training)-based approach, one takes

$$I[n] = c[n] + b[n] \tag{3.81}$$

where $\{b[n]\}$ is the information sequence and $\{c[n]\}$ is a nonrandom periodic training (pilot) sequence. Exploitation of the periodicity of $\{c[n]\}$ allows identification of the channel without allocating any explicit time slots for training, unlike traditional training methods. There is no loss in information rate. On the other hand, some useful power is wasted in superimposed training, which could have otherwise been allocated to the information sequence. This lowers the effective SNR for the information sequence and affects the bit error rate (BER) at the receiver.

Suppose that the superimposed training sequence $c[n] = c[n + m\bar{P}]\ \forall m; n$ is a nonrandom periodic sequence with period $\bar{P}$. Reference [27] uses the second-order statistics of the received signal to estimate the channel, whereas [48] and [52] exploit the first-order statistics. References. [27] and [52] deal with time-invariant models, whereas [48] is also applicable to time-variant basis expansion models.

Consider Equation 3.18. Define the i-th length $\bar{P}$ subrecord of $\mathbf{y}[n]$ as

$$\mathbf{y}_i[n] := \mathbf{y}[n + \bar{P}], \quad 0 \le i \le R-1, \quad 0 \le n \le \bar{P}-1 \tag{3.82}$$

The number of subrecords is $R = N/\bar{P}$, where N is the data record length in symbols (assume R is an integer). It then follows that the (synchronized) average of $\mathbf{y}_i[n]$ is

$$\begin{aligned}\hat{\mathbf{m}}_y[n] &:= \frac{1}{R}\sum_{i=0}^{R-1} \mathbf{y}_i[n] \\ &= \sum_{k=0}^{L} \mathbf{f}[k]c[n-k] + \frac{1}{R}\sum_{i=0}^{R-1} \mathbf{u}_i[n]\end{aligned} \tag{3.83}$$

where $\mathbf{u}_i[n] := \mathbf{u}[n + \bar{P}]$ and

$$\mathbf{u}[n] := \sum_{k=0}^{L} \mathbf{f}[k]b[n-k] + \mathbf{w}[n] \tag{3.84}$$

If both the information-bearing signal $b[n]$ and the additive noise $\mathbf{w}[n]$ are zero mean, then

$$E\{\hat{\mathbf{m}}_y[n]\} := \sum_{k=0}^{L} \mathbf{f}[k]c[n-k] \tag{3.85}$$

In [52], the superimposed training is picked to be $c[n] = \sum_k a\delta[n - k\bar{P}]$, where $\delta[n]$ is the Kronecker delta function. This immediately leads to $E\{\hat{\mathbf{m}}_y[n]\} = \mathbf{f}[n], 0 \le n \le L \le \bar{P} - 1$. Therefore, one may view $\hat{\mathbf{m}}_y[n]$ as an estimate of $\mathbf{f}[n]$:

$$\hat{\mathbf{f}}[n] = \frac{1}{a}\hat{\mathbf{m}}_y[n] = \frac{1}{aR}\sum_{i=0}^{R-1} \mathbf{y}_i[n], \quad 0 \le n \le \bar{P} - 1 \tag{3.86}$$

The choice of [52] leads to a poor peak-to-average power ratio of the transmitted signal, which is undesirable if the transmit power amplifier has some nonlinearity. Also, in practice, linear systems arise because of linearization about some operating (set) point — e.g., bias in amplifiers. These set points are typically unknown (at least not known precisely) *a priori*, and one does not normally worry about them since unknown means are estimated and removed before processing (blocked by capacitor coupling, etc.) and they are not needed in any processing. However, if (time-varying) mean $E\{\mathbf{y}(n)\}$ is what we wish to use (as in [52]), then we must include a dc offset term $\mathbf{m}$ in Equation 3.18 and modify it as

$$\mathbf{y}[n] = \sum_{k=0}^{L} \mathbf{f}[k]\,(b[n-k] + c[n-k]) + \mathbf{w}[n] + \mathbf{m} \tag{3.87}$$

In this case, the method of [52] does not apply. A fix, and an alternative approach, has been provided in [48], which we discuss next.

Under the above assumptions, we have

$$E\{\mathbf{y}[n]\} = \sum_{l=0}^{L} \mathbf{f}[l]c[n-l] + \mathbf{m} \tag{3.88}$$

Since $\{c[n]\}$ is periodic, we have

$$c[n] = \sum_{m=0}^{\bar{P}-1} c_m e^{j\alpha_m n} \quad \forall n \tag{3.89}$$

where

$$\alpha_m := 2\pi m/\bar{P}, \quad c_m := \frac{1}{\bar{P}}\sum_{n=0}^{\bar{P}-1} c[n]e^{-j\alpha_m n} \tag{3.90}$$

The coefficients c_m are known at the receiver since $\{c[n]\}$ is known. We have

$$E\{\mathbf{y}[n]\} = \sum_{m=0}^{\bar{P}-1} \underbrace{\left[\sum_{l=0}^{L} c_m \mathbf{f}[l]e^{-j\alpha_m l}\right]}_{=:\mathbf{d}_m} e^{j\alpha_m n} + \mathbf{m} \tag{3.91}$$

The sequence $E\{\mathbf{y}[n]\}$ is periodic with cycle frequencies α_m, $0 \leq m \leq \bar{P}-1$. A mean square (ms) consistent estimate $\hat{\mathbf{d}}_m$ of $\mathbf{d}_m$, for $\alpha_m \neq 0$, follows as [48]

$$\hat{\mathbf{d}}_m = \frac{1}{N}\sum_{n=1}^{N} y[n]e^{-j\alpha_m n} \tag{3.92}$$

As $T \to \infty$, $\hat{\mathbf{d}}_m \to \mathbf{d}_m$ m.s. if $\alpha_m \neq 0$ and $\hat{\mathbf{d}}_0 \to \mathbf{d}_0 + \mathbf{m}$ m.s. if $\alpha_m = 0$.

We now establish that given $\mathbf{d}_m$ for $1 \leq m \leq \bar{P}-1$, we can (uniquely) estimate $\mathbf{h}(l)$ if $\bar{P} \geq L+2$, $\alpha_m \neq 0$, and $c_m \neq 0\ \forall m \neq 0$. Since $\mathbf{m}$ is unknown, we will omit the term $m=0$ for further discussion. Define

$$\mathbf{V} := \begin{bmatrix} 1 & e^{-j\alpha_1} & \cdots & e^{-j\alpha_1 L} \\ 1 & e^{-j\alpha_2} & \cdots & e^{-j\alpha_2 L} \\ \vdots & \vdots & \vdots & \vdots \\ 1 & e^{-j\alpha_{\bar{P}-1}} & \cdots & e^{-j\alpha_{\bar{P}-1} L} \end{bmatrix}_{(\bar{P}-1)\times(L+1)} \tag{3.93}$$

$$\mathcal{H} := \begin{bmatrix} \mathbf{f}^H[0] & \mathbf{f}^H[1] & \cdots & \mathbf{f}^H[L] \end{bmatrix}^H \tag{3.94}$$

$$\mathcal{D} := \begin{bmatrix} \mathbf{d}_1^H & \mathbf{d}_2^H & \cdots & \mathbf{d}_{\bar{P}-1}^H \end{bmatrix}^H \tag{3.95}$$

$$\mathcal{C} := \underbrace{\left(\text{diag}\{c_1, c_2, \ldots, c_{\bar{P}-1}\}\mathbf{V}\right)}_{=:\mathcal{V}} \otimes \mathcal{I}_{\bar{P}} \tag{3.96}$$

where $\mathcal{C}$ is $[P(\bar{P}-1)] \times [P(L+1)]$ and $\otimes$ denotes the Kronecker product. Omitting the term $m=0$ and using the definition of $\mathbf{d}_m$ from Equation 3.88, it follows that

$$\mathcal{C}\mathcal{H} = \mathcal{D} \tag{3.97}$$

In Equation 3.93 $\mathbf{V}$ is a Vandermonde matrix with a rank of $L+1$ if $\bar{P}-1 \geq L+1$ and the α_i values are distinct. Since $c_m \neq 0\ \forall m$, rank$(\mathcal{V})$ = rank$(\mathbf{V}) = L+1$. Finally, rank$(\mathcal{C})$ = rank$(\mathcal{V}) \times$ rank$(\mathcal{I}_{\bar{P}}) = P(L+1)$. Therefore, we can determine $\mathbf{f}[l]$ values uniquely. Define $\hat{\mathcal{D}}$ as in Equation 3.97 with $\mathbf{d}_m$ replaced with $\hat{\mathbf{d}}_m$. Then we have the channel estimate

$$\hat{\mathcal{H}} = (\mathcal{C}^H\mathcal{C})^{-1}\mathcal{C}^H\hat{\mathcal{D}} \tag{3.98}$$

Note that precise knowledge of the channel length L is not required; an upper bound L_u suffices. Then we estimate $\mathbf{f}[i]$ for $0 \leq i \leq L_u$ with $\hat{\mathbf{f}}[i] \to 0$ m.s. for $i \geq L+1$ (true channel length) as record length $N \to \infty$. Also, we do not need $c_m \neq 0$ for every m. We need at least $L+2$ nonzero c_m values. This can be accomplished by picking a large $\bar{P}$ and a suitable $\{c[n]\}$ (e.g., picked to satisfy a peak-to-average power constraint). Implicit in this approach (also in [27] and [52]) is the need at the receiver for synchronization with the transmitter's superimposed training sequence.

Time-variant channels using CE-BEM representation are considered in [48] with periodic non-random hidden training sequences.

3.3.4.1 Equalization

With $\hat{\mathbf{f}}[i]$ denoting the estimated $\mathbf{f}[i]$, define

$$\tilde{\mathbf{y}}[n] := \mathbf{y}[n] - \sum_{i=0}^{L}\hat{\mathbf{f}}[i]c[n-i] - \hat{\mathbf{m}} \approx \sum_{i=0}^{L}\mathbf{f}[i]b[n-i] + \mathbf{w}[n] \tag{3.99}$$

where $\hat{\mathbf{m}} := (1/N)\sum_{n=1}^{N}[\mathbf{y}[n] - \sum_{i=0}^{L}\hat{\mathbf{f}}[i]c[n-i]]$. That is, $\tilde{\mathbf{y}}[n]$ is obtained by removing the (estimated) contribution of the superimposed training and the dc offset from the noisy data. Equation 3.99 with the estimated channel is used to equalize the channel and to detect the information sequence. If we use a linear MMSE (minimum mean square error) equalizer, then it also requires the knowledge of the correlation function of $\tilde{\mathbf{y}}[n]$. We estimate the noise variance σ_v^2 as (tr$\{A\}$ denotes trace of matrix A)

$$\hat{\sigma}_v^2 := N^{-1}\text{tr}\left\{\left[(1/N)\sum_{n=1}^{N}\tilde{\mathbf{y}}[n]\tilde{\mathbf{y}}^H[n]\right] - \sum_{i=0}^{L}\hat{\mathbf{f}}[i]\hat{\mathbf{h}}^H[i]\right\} \tag{3.100}$$

If Equation 3.100 yields a negative result, we set it to zero. The correlation function of $\tilde{\mathbf{y}}[n]$ can then be estimated using the estimated channel (instead of the less reliable sample averaging); only the zero lag correlation requires $\hat{\sigma}_v^2$.

3.4 Simulation Examples

In this section we present simulation examples to illustrate some of the approaches to channel estimation.

3.4.1 Example 1

Consider a continuous-time channel $\tilde{h}(t)$ given by $\tilde{h}(t) = \sum_{i=1}^{4} a_i\, p_{4T_s}(t - \tau_i; 0.2)$ where T_s is the symbol interval; $p_{4T_s}(t; 0.2)$ denotes the raised cosine pulse with a roll-off factor of 0.2 and a length truncated to $4T_s$ (i.e., $p_{4T_s}(t; 0.2) = 0$ for $|t| > 2T_s$); the amplitudes a_i are mutually independent, zero-mean, complex Gaussian with same variance for all i values; and delays τ_i are mutually independent, uniformly distributed over $[0, 4T_s]$. The continuous-time channel $\tilde{h}(t)$ is sampled once every T_s second to yield the discrete-time channel $f[n] := \tilde{h}((n-1)T_s)$. Thus we have $P = 1$ in Equation 3.87 leading to

$$y[n] = \sum_{l=0}^{7} f[l][b[n-l] + c[n-l]] + w[n] + m \tag{3.101}$$

Let L_u be the upper bound on channel length $L = 7$. We take $L_u = 10$. The channel is randomly generated in each Monte Carlo. The input information sequence $\{b[n]\}$ is i.i.d. equiprobable 4-QAM (quadrature amplitude modulation), taking values $(\pm 1 \pm j)/\sqrt{2}$. The training sequence was chosen to have a period $\bar{P} = 15$ with $c[n] = \sum_k \sqrt{\alpha}\delta(n - 15k)$, as in [2]; α is picked to yield a particular training-to-information sequence power ratio (TIR)

$$\alpha = \sigma_c^2/\sigma_b^2 \tag{3.102}$$

where σ_b^2 and σ_c^2 denote the average power in the information sequence $\{b[n]\}$ and training sequence $\{c[n]\}$, respectively. Complex white zero-mean Gaussian noise was added to the received signal and scaled to achieve an SNR at the receiver (relative to the contribution of $\{I[n]\}$). A mean value m was added to the noisy received signal to achieve a specified dc offset-to-signal ac component (DCAC) power ratio $\frac{m^2}{E\{|y[n]-w[n]-m|^2\}}$. The normalized mean square error in estimating the channel impulse response averaged over 100 Monte Carlo runs was taken as the performance measure for channel estimation. It is defined as (before Monte Carlo averaging)

$$\text{NCMSE} := \left[\sum_{l=0}^{L_u} \|f[l] - \hat{f}[l]\|^2\right]\left[\sum_{l=0}^{L_u} \|f[l]\|^2\right]^{-1} \tag{3.103}$$

The results of averaging over 100 Monte Carlo runs are shown in Figure 3.8 for various SNRs and DCAC power ratios for a record length of $T = 150$ symbols and a TIR of -2.33 dB ($\alpha = 0.585$). The methods of [48] and [52] were simulated. It is seen that the method of [48] is insensitive to the presence

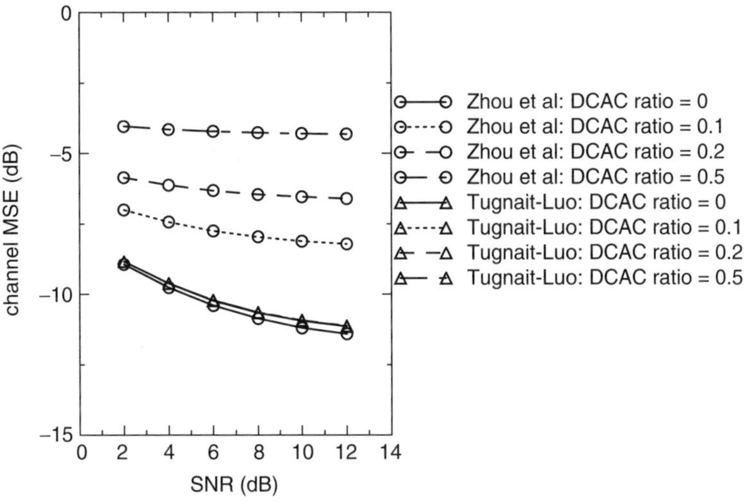

FIGURE 3.8 Example 1: Normalized channel MSE (3.103) based on $N = 150$ symbols per run, 100 Monte Carlo runs, $\bar{P} = 15$, TIR $\alpha = 0.585$. DCAC ratio $= \frac{m^2}{E\{|y[n]-w[n]-m|^2\}}$. The curves for the proposed method for different DCAC ratios are overlaid (very close). Zhou et al. is [52] and Tugnait-Luo is [48].

of the unknown mean m, whereas the method of [52] is very sensitive. For $m = 0$, the performance of the method of [48] is slightly inferior to that of [52]. In the method of [52], $\hat{f}[l]$ values are estimated directly from data for $1 \leq l \leq L_u + 1 = 11$, whereas in [48], we first estimate $\hat{\mathbf{d}}_m$ values for $1 \leq m \leq \bar{P} - 1 = 14$ and then use Equation 3.98. Since we estimate more variables (14 vs. 11), this may account for the slightly inferior performance of our method for $m = 0$.

3.4.2 Example 2

This example is exactly as Example 1 except for the training sequence, which was taken to be an m sequence (maximal-length pseudorandom binary sequence) of length 15 ($= \bar{P}$), $c[n] = \sqrt{\alpha}\tilde{c}(n)$, and

$$\{\tilde{c}(n)\}_{n=0}^{15} = \{-1, -1, -1, 1, 1, 1, 1, -1, 1, -1, 1, 1, -1, -1, 1\} \tag{3.104}$$

The peak-to-average power ratio for this sequence is 1 (the best possible). The results of averaging over 100 Monte Carlo runs are shown in Figure 3.8 for various SNRs and DCAC power ratios for a record length of T = 150 symbols and a TIR of −2.33 dB ($\alpha = 0.585$). Only the method of [48] was simulated since the method of [52] does not apply to this model. It is seen that as in Example 1, the method of [48] is insensitive to the presence of the unknown mean m. Equalization performance (BER) of a linear MMSE equalizer based on the estimated channel (Example 2) is shown in Figure 3.10 for two different record lengths of T = 150 and 300 symbols. The linear equalizer was designed as noted earlier with an equalizer length of 10 symbols and a delay of 5 symbols. Also shown is the performance of a linear equalizer based upon perfect knowledge of the channel and noise variance. The results are based on 100 Monte Carlo runs. It is seen that the performance improves with record length. Note that for our choice of $\alpha = 0.585$, the SNR relative to $\{b[n]\}$ would be 2 dB less than the SNR shown in Figure 3.9, which is relative to $\{I[n]\}$. To reflect this loss in SNR due to inclusion of the superimposed training, we redraw Figure 3.10 as Figure 3.11 with the SNR for the curve for the known channel linear MMSE equalizer adjusted by 2 dB; the other two curves remain unchanged.

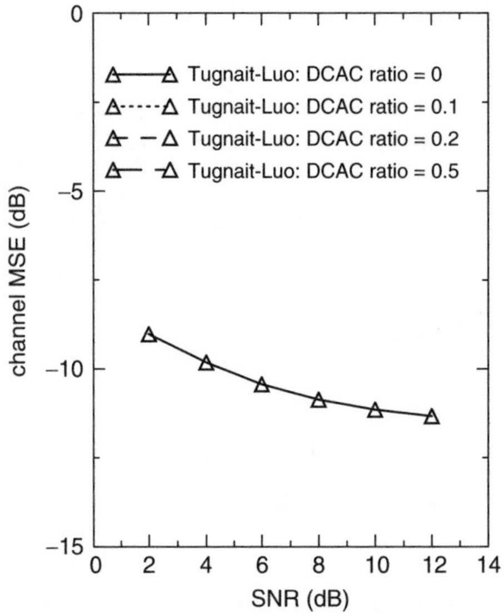

FIGURE 3.9 Example 2: Normalized channel MSE (3.103) based on $N = 150$ symbols per run, 100 Monte Carlo runs, $\bar{P} = 15$. DCAC ratio $= \frac{m^2}{E\{|y[n]-w[n]-m|^2\}}$, TIR $\alpha = 0.585$. The curves for the proposed method for different DCAC ratios are overlaid (very close). Tugnait-Luo is [48].

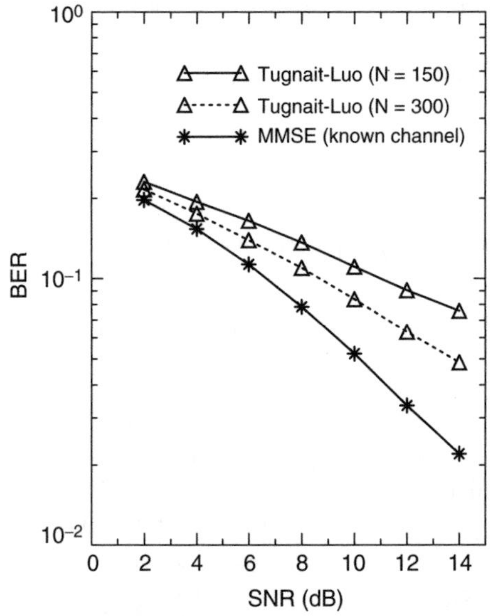

FIGURE 3.10 Example 2: Equalization performance using linear MMSE equalizers based on $N = 150$ or 300 symbols per run, 100 Monte Carlo runs, $\bar{P} = 15$. DCAC ratio $= 0$, TIR $\alpha = 0.585$. Tugnait-Luo is [48].

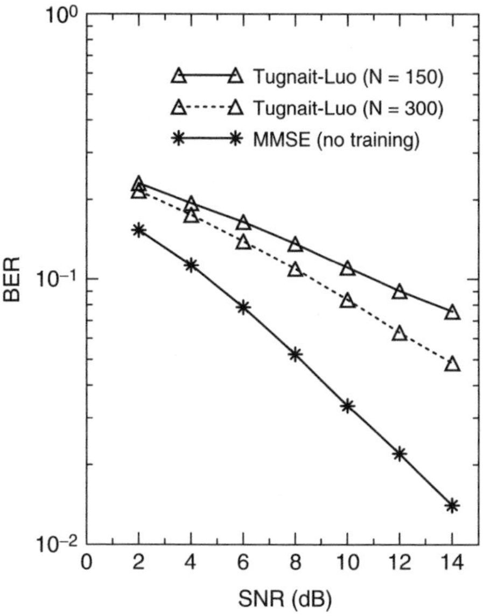

FIGURE 3.11 Example 2: Figure 3.10 redrawn with the curve for the known-channel linear MMSE equalizer adjusted by 2dB – no power is wasted in training in this case.

3.5 Conclusions

A review of various approaches to channel modeling and estimation for wireless mobile systems was presented. Emphasis was on linear baseband equivalent models with a tapped delay line structure, and both time-invariant and time-variant models were discussed. Basis expansion modeling for time-variant channels was considered where the basis functions are related to the physical parameters of the channel (such as Doppler and delay spreads). Channel modeling was followed by a discussion of various approaches to channel estimation, including training-based approaches, blind approaches, semiblind approaches, and hidden pilot-based approaches. In the training-based approach a sequence known to the receiver is transmitted in the acquisition mode. In blind approaches no such sequence is available (or used) and the channel is estimated based solely on the noisy received signal, exploiting the statistical and other properties of the information sequence. Semiblind approaches utilize a combination of training-based and blind approaches. In the hidden pilot-based approaches a periodic (nonrandom) training sequence is arithmetically added (superimposed) at a low power to the information sequence at the transmitter before modulation and transmission. Finally, simulation results were presented to illustrate some of the approaches.

Acknowledgment

This work was prepared in part under the support of the U.S. Army Research Office under grant DAAD19-01-1-0539.

References

[1] P.A. Bello, "Characterization of randomly time-variant channels," *IEEE Trans. Comm. Syst.*, CS-11, 360–393, 1963.

[2] D.K. Borah and B. Hart, "Receiver structures for time-varying frequency-selective fading channels," *IEEE J. Selected Areas Commun.*, SAC-17, 1863–1875, 1999.

[3] D. Boss, B. Jelonek, and K.D. Kammeyer, "Eigenvector algorithm for blind MA system identification," *Signal Process.*, 66, 1–26, 1998.

[4] D. Boss, K.-D. Kammeyer, and T. Petermann, "Is blind channel estimation feasible in mobile communication systems? A study based on GSM," *IEEE J. Selected Areas Commun.*, SAC-16, 1480–1492, 1998.

[5] J.K. Cavers, "Pilot symbol assisted modulation and differential detection in fading and delay spread," *IEEE Trans. Commun.*, COM-43, 2206–2212, 1995.

[6] K.M. Chugg, "Blind acquisition characteristics of PSP-based sequence detectors," *IEEE J. Selected Areas Commun.*, SAC-16, 1518–1529, 1998.

[7] H.A. Cirpan and M.K. Tsatsanis, "Stochastic maximum likelihood methods for semi-blind channel estimation," *IEEE Signal Proc. Lett.*, SPL-5, 21–24, 1998.

[8] E. de Carvalho and D.T.M. Slock, "Semi-blind methods for FIR multichannel estimation," chapter 7 in *Signal Processing Advances in Wireless and Mobile Communications*, Vol. 1, G.B. Giannakis, Y. Hua, P. Stoica, and L. Tong, Eds., Upper Saddle River, NJ: Prentice Hall, 2001.

[9] Z. Ding, "Matrix outer-product decomposition method for blind multiple channel identification," *IEEE Trans. Signal Process.*, 45, 3054–3061, 1997.

[10] W.A. Gardner, *Introduction to Random Processes: With Applications to Signals and Systems*, 2nd ed., New York: McGraw-Hill, 1989.

[11] D. Gesbert and P. Duhamel, "Robust blind channel identification and equalization based on multistep predictors," in *Proceedings of the IEEE International Conference Acoustics, Speech, Signal Processing*, Seattle, WA, April 1997, pp. 3621–3624.

[12] M. Ghosh and C.L. Weber, "Maximum-likelihood blind equalization," *Opt. Eng.*, 31, 1224–1228, 1992.

[13] G.B. Giannakis, Y. Hua, P. Stoica, and L. Tong, Eds., *Signal Processing Advances in Wireless and Mobile Communications*, Vol. 1, *Trends in Channel Estimation and Equalization*, Upper Saddle River, NJ: Prentice Hall, 2001.

[14] R.D. Gitlin and S.B. Weinstein, "Fractionally-spaced equalization: an improved digital transversal equalizer," *Bell Syst. Tech. J.*, 60, 275–296, 1981.

[15] D.N. Godard, "Self-recovering equalization and carrier tracking in two-dimensional data communication systems," *IEEE Trans. Commun.*, COM-28, 1867–1875, 1980.

[16] A. Gorokhov and P. Loubaton, "Semi-blind second order identification of convolutive channels," in *Proceedings of the IEEE International Conference Acoustics, Speech, Signal Processing*, Munich, Germany, 1997, pp. 3905–3908.

[17] D. Hatzinakos and C.L. Nikias, "Blind equalization using a tricepstrum based algorithm," *IEEE Trans. Commun.*, COM-39, 669–681, 1991.

[18] Y. Hua, "Fast maximum likelihood for blind identification of multiple FIR channels," *IEEE Trans. Signal Process.*, SP-44, 661–672, 1996.

[19] W.C. Jakes, *Microwave Mobile Communications*, New York: Wiley, 1974.

[20] G.K. Kaleh and R. Vallet, "Joint parameter estimation and symbol detection for linear or nonlinear unknown dispersive channels," *IEEE Trans. Commun.*, COM-42, 2406–2413, 1994.

[21] V. Krishnamurthy and J.B. Moore, "On-line estimation of hidden Markov model parameters based on Kullback-Leibler information measure," *IEEE Trans. Signal Process.*, SP-41, 2557–2573, 1993.

[22] S. Lasaulce, P. Loubaton, and E. Moulines, "A semi-blind channel estimation technique based on second-order blind method for CDMA systems," *IEEE Trans. Signal Process.*, SP-51, 1894–1904, 2003.

[23] J. Laurila, K. Kopsa, and E. Bonek, "Semi-blind signal estimation for smart antennas using subspace tracking," in *Processing 1999 IEEE Workshop on Signal Processing Advances in Wireless Communications*, Annapolis, MD, May 1999.

[24] W. Luo and J.K. Tugnait, "Blind identification of time-varying channels using multistep linear predictors," in *Proceedings of the 2002 IEEE International Conference on Acoustics, Speech, Signal Processing*, Orlando, FL, May 13–17, 2002.

[25] X. Ma, G.B. Giannakis, and S. Ohno, "Optimal training for block transmissions over doubly-selective fading channels," in *Proceedings of the 2002 ICASSP*, Orlando, FL, April 2002, pp. II-1509–II-1512.

[26] M. Martone, "Wavelet-based separating kernels for array processing of cellular DS/CDMA signals in fast fading," *IEEE Trans. Commun.*, COM-48, 979–995, 2000.

[27] F. Mazzenga, "Channel estimation and equalization for M-QAM transmission with a hidden pilot sequence," *IEEE Trans. Broadcasting*, 46, 170–176, 2000.

[28] E. Moulines, P. Duhamel, J. Cardoso, and S. Mayrargue, "Subspace-methods for the blind identification of multichannel FIR filters," *IEEE Trans. Signal Process.*, SP-43, 516–525, 1995.

[29] M. Niedzwiecki, *Identification of Time-Varying Processes*, New York: Wiley, 2000.

[30] J.G. Proakis, *Digital Communications*, 4th ed., New York: McGraw-Hill, 2001.

[31] R. Raheli, A. Polydoros, and C.K. Tzou, "Per-survivor processing: a general approach to MLSE in uncertain environments," *IEEE Trans. Commun.*, COM-43, 354–364, 1995.

[32] A.M. Sayeed and B. Aazhang, "Joint multipath-Doppler diversity in mobile wireless communications," *IEEE Trans. Commun.*, COM-47, 123–132, 1999.

[33] N. Seshadri, "Joint data and channel estimation using fast blind trellis search techniques," *IEEE Trans. Commun.*, COM-42, 1000–1011, 1994.

[34] O. Shalvi and E. Weinstein, "Super-exponential methods for blind deconvolution," *IEEE Trans. Inf. Theory*, IT-39, 504–519, 1993.

[35] C. Tepedelenlioglu and G.B. Giannakis, "Transmitter redundancy for blind estimation and equalization of time- and frequency-selective channels," *IEEE Trans. Signal Process.*, SP-48, 2029–2043, 2000.

[36] L. Tong and S. Perreau, "Multichannel blind channel estimation: from subspace to maximum likelihood methods," *Proc. IEEE*, 86, 1951–1968, 1998.

[37] L. Tong, G. Xu, and T. Kailath, "A new approach to blind identification and equalization of multipath channels," *IEEE Trans. Inf. Theory*, IT-40, 340–349, 1994.

[38] L. Tong and Q. Zhao, "Joint order detection and blind channel estimation by least squares smoothing," *IEEE Trans. Signal Process.*, SP-47, 1999.

[39] J.R. Treichler and M.G. Agee, "A new approach to multipath correction of constant modulus signals," *IEEE Trans. Acoustics Speech Signal Process.*, ASSP-31, 349–472, 1983.

[40] J.R. Treichler, M.G. Larimore, and J.C. Harp, "Practical blind demodulators for high-order QAM signals," *Proc. IEEE*, 86, 1907–1926, 1998.

[41] J.K. Tugnait, "Identification of linear stochastic systems via second- and fourth-order cumulant matching," *IEEE Trans. Inf. Theory*, IT-33, 393–407, 1987.

[42] J.K. Tugnait, "On blind identifiability of multipath channels using fractional sampling and second-order cyclostationary statistics," *IEEE Trans. Inf. Theory*, IT-41, 308–311, 1995.

[43] J.K. Tugnait, "Blind estimation and equalization of digital communication FIR channels using cumulant matching," *IEEE Trans. Commun.*, COM-43, 1240–1245, 1995.

[44] J.K. Tugnait, "Blind equalization and estimation of FIR communications channels using fractional sampling," *IEEE Trans. Commun.*, COM-44, 324–336, 1996.

[45] J.K. Tugnait, "Multistep linear predictors-based blind equalization of FIR/IIR single-input multiple-output channels with common zeros," *IEEE Trans. Signal Process.*, SP-47, 1689–1700, 1999.

[46] J.K. Tugnait, "Channel estimation and equalization using higher-order statistics," chapter 1 in *Signal Processing Advances in Wireless and Mobile Communications*, Vol. 1, *Trends in Channel Estimation and Equalization*, G.B. Giannakis, Y. Hua, P. Stoica, and L. Tong, Eds., Upper Saddle River, NJ: Prentice Hall, 2001, pp. 1–39.

[47] J.K. Tugnait and W. Luo, "Linear prediction error method for blind identification of periodically time-varying channels," *IEEE Trans. Signal Processing*, SP-50, 3070–3082, 2002.

[48] J.K. Tugnait and Weilin Luo, "On channel estimation using superimposed training and first-order statistics," in *Proceedings of the 2003 IEEE International Conference on Acoustics, Speech, Signal Processing*, Hong Kong, April 6–10, 2003.

[49] J. Vidal and J.A.R. Fonollosa, "Adaptive blind equalization using weighted cumulant slices," *Int. J. Adaptive Control and Signal Processing*, 10, No. 2-3, 213–238, 1996.

[50] J.-J. Werner, J. Yang, D.D. Harman and G.A. Dumont, "Blind equalization for broadband access," *IEEE Communications Magazine*, 37, 87–93, 1999.

[51] G. Xu, H. Liu, L. Tong, and T. Kailath, "A Least-Squares Approach to Blind Channel Identification," *IEEE Trans. Signal Processing*, SP-43, 2982–2993, 1995.

[52] G.T. Zhou, M. Viberg, and T. McKelvey, "A first-order statistical method for channel estimation" *IEEE Signal Processing Letters*, SPL-10, 57–60, 2003.

4

Mobile Satellite Channels: Statistical Models and Performance Analysis

Giovanni E. Corazza
University of Bologna, Italy

Alessandro Vanelli-Coralli
University of Bologna, Italy

Raffaella Pedone
University of Bologna, Italy

Massimo Neri
University of Bologna, Italy

4.1 Introduction 4-1
4.2 Statistical Propagation Models 4-2
Narrowband Statistical Models • Wideband Statistical Models
4.3 Detection Performance Analysis 4-20
Uncoded Transmission over LMS Channels • Coded Transmission over LMS Channels
4.4 Conclusions 4-31

Abstract

In this chapter, the problem of modeling electromagnetic propagation and its effects on land mobile satellite communications is addressed. A thorough review of the most accepted statistical models proposed in the scientific literature is presented, considering large- and small-scale fading, single- and multiple-state structures, narrowband and wideband channels, and first- and second-order statistics. Building upon a thorough characterization of propagation effects, the attention is turned to the assessment of digital transmission error probability for both uncoded and coded signals. Performance analysis based on closed-form expressions, upper bounds, and numerical simulation is included. An extensive bibliography is offered to the interested reader, for further insights and investigation.

4.1 Introduction

Mobile satellite systems are an essential part of the global communication infrastructure, providing a variety of services to several market segments, such as aeronautical, maritime, vehicular, and pedestrian. In particular, the two last cases are jointly referred to as the land mobile satellite (LMS) segment and constitute a very important field of application, development, and research, which has attracted the interest of numerous scientists in the last few decades. One fundamental characteristic of an LMS system is the necessity to be designed for integration with a terrestrial mobile network counterpart, in order to optimize the overall benefits from the point of view of the users and network operators. In essence,

0-8493-1657-X/$0.00+$1.50

satellite and terrestrial mobile systems share the market segment along with many technical challenges and solutions, although they also have their own peculiar characteristics. A classic and central problem in any mobile communication system is that of modeling electromagnetic propagation characteristics. In LMS communications, as for terrestrial networks, multipath fading and shadowing are extremely important in determining the distribution of the received power level. In addition, it is common to also have a strong direct or specular component from the satellite to the user terminal, which is essential to close the link budget, and which modifies significantly the statistics with respect to terrestrial outdoor propagation. Another peculiarity is the dependence of the propagation channel on the satellite orbital configuration, which can be geostationary, at an altitude H of 36,000 km on the equatorial plane, or nongeostationary. In the latter case, it is possible to have orbital planes with arbitrary inclination and to adopt low-altitude Earth orbits (LEOs, $H = 500$ to 1700 km), medium-altitude Earth orbits (MEOs, $H = 5000$ to 10,400 km), or highly elliptical orbits (HEOs, apogee around 48,000 to 50,000 km). All of these nongeostationary orbits entail that, as seen from the mobile user, the satellite changes its elevation angle continuously, effectively varying the propagation channel conditions. In terms of frequency bands, the spectrum allocation for LMS is generally adjacent or nearly contiguous with respect to the terrestrial network allocation, so that no major differences are observed. In particular, at L- or S-band, atmospheric effects are normally negligible with respect to electromagnetic interaction with orography, buildings, trees, and other obstacles.

In terms of modeling the LMS propagation channel, there are three basic alternatives: geometric analytic, statistical, and empirical. Generally speaking, the statistical modeling approach is less computationally intensive than a geometric analytic characterization, and is more phenomenological than an empirical regression model. The most remarkable advantage of statistical models is that they allow flexible and efficient performance predictions and system comparisons under different modulation, coding, and access schemes. For these reasons, in the first part of this chapter we focus our attention on a thorough review of statistical LMS propagation models, considering large- and small-scale fading, single-state and multistate models, first- and second-order characterization, and narrowband and wideband propagation.

Accurate propagation channel modeling is the fundamental prerequisite for being able to perform realistic quality of service predictions. As the reader will notice, there is an astonishingly large number of proposed models, all of which are able to properly fit the available data from measurement campaigns. As a consequence, the selection of a particular model should be tightly linked to its use in performance prediction, which is the focus of the second part of this chapter. There, we address a review of performance evaluation methodologies over the LMS channel, considering both uncoded and coded transmission, as well as coherent, pseudocoherent (pilot-aided), and noncoherent detection. In particular, coded performance is reported in terms of bounding techniques for both convolutional and turbo codes, along with sample simulation results in uncorrelated and correlated channels, possibly considering satellite diversity.

4.2 Statistical Propagation Models

This section is devoted to the review of statistical models for the land mobile satellite fading channel. When considering a typical LMS link, the relevant propagation phenomena are distance-dependent attenuation, diffraction, absorption, and scattering. Atmospheric attenuation phenomena can be safely disregarded at L- and S-band frequencies, typical for LMS. In general, the received signal contains a *direct component* and a number of multiple paths, which are usually identified as the *diffuse component*. Two concurrent effects can be observed: *large-scale* and *small-scale* fading. Large-scale fading, commonly identified as *shadowing*, models the attenuation caused by the orography and large obstacles, such as hills, buildings, trees, etc., through absorption and diffraction mechanisms. In order to measure a significant power level variation due to large-scale fading, the mobile terminal must typically travel several hundred wavelengths. Instead, small-scale fading models variations in the signal amplitude due to constructive and destructive interference in the sum of multiple rays, mainly caused by reflections over surrounding surfaces. In this case, power fluctuations are measurable over distances comparable to the wavelength.[1]

The terminal motion induces time-varying channel characteristics. When the environment changes significantly (e.g., moving from urban to suburban areas), fading becomes nonstationary and is hardly tractable. Therefore, the assumption of quasi-stationarity is typically accepted, modeling the real phenomena as single or multiple wide-sense stationary propagation states. As a consequence, statistical models can be classified into *single-state* and *multi-state*. More precisely, a single-state model describes the propagation channel through a single, spatially invariant statistical distribution. A multi-state model contains a collection of single-state models[2] and accounts for macro-variations of the channel through probabilistic state transitions.

Considering a specific propagation state, the transfer function of the channel, $H(f, t)$, is modeled as a wide-sense stationary uncorrelated scattering (WSSUS) random process, i.e., WSS in both f and t. From the transfer function, both first- and second-order fading statistics can be extracted. First-order statistics contain the probability density function (pdf) and the cumulative density function (cdf) of the received signal envelope and phase for each resolvable path. Second-order statistics account for the autocorrelation properties of the fading random process occurring in the time domain (resulting in a specific Doppler spectrum, level crossing rate, and fade duration) and frequency domain (leading to the multipath intensity profile and delay spread). More precisely, considering a couple of frequencies (f, f') and time instants (t, t'), the process autocorrelation function satisfies

$$R_H(f, f'; t, t') = R_H(\Delta f; \Delta t) \tag{4.1}$$

where $\Delta f = f' - f$ and $\Delta t = t' - t$.

The domain supporting the function $R_H(\Delta f) \doteq R_H(\Delta f; 0)$ is defined as the channel *coherence bandwidth* $(\Delta f)_c$ and represents an indication of the channel memory in the frequency domain. Similarly, the domain supporting the function $R_H(\Delta t) \doteq R_H(0; \Delta t)$ is defined as the channel *coherence time* $(\Delta t)_c$ and gives an indication of the channel memory in the time domain. These two memory indicators are inversely related to a pair of very useful parameters: the *delay spread*,

$$T_s \approx \frac{1}{(\Delta f)_c} \tag{4.2}$$

and the *Doppler spread*,

$$B_d \approx \frac{1}{(\Delta t)_c} \tag{4.3}$$

Based on the relationship between the transmitted signal data rate, $1/T$, and these parameters, selectivity in time and frequency can be defined. Namely, the fading process is *frequency nonselective* or *flat* when $T \gg T_s$, i.e., the delay spread produces negligible time dispersion on the signal. Assuming that the signal bandwidth $B \approx 1/T$, frequency nonselectivity translates into

$$B \ll (\Delta f)_c \tag{4.4}$$

i.e., the signal bandwidth is much less than the coherence bandwidth, and fading is multiplicative. When $B > (\Delta f)_c$, fading is *frequency selective* and introduces significant time dispersion and frequency distortion, and the multiple paths can be resolved. Depending on frequency flatness or selectivity, statistical models are classified as *narrow-band* or *wide-band*, respectively. The majority of LMS models in the

[1]As a consequence, in order to counteract large- and small-scale fading, it is possible to resort to macro- and microdiversity, respectively.

[2]Note that the multiple states could also share the same single-state distribution with different parameter values.

literature are narrowband, in line with the fact that in present satellite environments the direct component is generally strong, and multipath power at resolvable delay is relatively very low and can be neglected with no accuracy loss. However, the trend for broadband services and the technological improvements in the space and ground segments are justifying more and more the use of wideband models, which so far have received less attention in the literature.

Fading is identified as *time nonselective* when

$$T \ll (\Delta t)_c \tag{4.5}$$

i.e., the symbol period is significantly smaller than the coherence time. This implies that $B_d \ll B$, i.e., the Doppler spread is much smaller than the signal bandwidth. If this condition is not verified, fading is *time selective*. In addition to the multipath Doppler spectrum, in LMS communications it is possible to have a Doppler offset, essentially due to the relative satellite motion, which is significant for non-geostationary orbits.

The mobile terminal moves in an environment with specific space coherence characteristics. We define the *shadowing coherence distance*, $(\Delta s)_c$, as the distance that must be traveled to have uncorrelated large-scale fading events. Obviously, the coherence distance of shadowing is much larger than that of small-scale fading. Given a specific terminal speed, this means that small-scale fading events are much faster than large-scale fading events.

Finally, similarly to time, frequency, and space correlation, an angular coherence span can be defined to identify the angular separation needed to ensure uncorrelated fading events or, in other words, to measure the channel memory in the angular domain. Again, a distinction between small- and large-scale fading is necessary. This notion is useful in the design of antenna arrays and beam-forming networks and for macro-diversity analysis.

4.2.1 Narrowband Statistical Models

In this section, the main narrowband models proposed in the literature are presented, for both single-state and multistate processes. We begin with a first-order characterization, which is instrumental for the evaluation of average error probability in uncoded transmissions or in coded transmission assuming ideal interleaving (see Section 4.3), evaluation of outage probability and link budget margins, etc. Then we will address second-order statistics.

All model parameters depend on the environment surrounding the mobile terminal (i.e., rural, suburban, urban) and on the geometrical characteristics of the link, e.g., the satellite elevation angle. Note that the link geometry is time variant in nongeostationary systems.

4.2.1.1 Single-State Statistical Models

Considering the problem of modeling a single-state propagation environment, various distributions have been proposed that model separately either large-scale or small-scale fading: lognormal, Rayleigh [1], [2], Rice [3], Nagakami [4], and Norton [5]–[7]. On the other hand, several authors have defined composite distributions that are able to model at the same time both large- and small-scale fading, such as Suzuki [8], Loo [9], RLN (Rice–lognormal) [10], [11], Hwang et al. [12], GRLN (generalized RLN) [13], Xie and Fang [14], Pätzold et al. [15], International Telecommunications Union (ITU) model for LMS systems [16], and Adbi et al. [17]. We review all of these models below. The large number of alternatives testifies to the fact that the problem of characterizing LMS propagation has spurred the interest of many researchers. However, it is fair to say that the differences between the above models are not dramatic in terms of capability of fitting measurement campaigns. Therefore, the relative merits pertain to physical meaningfulness and ease of applicability to the performance evaluation problem.

Let the baseband-equivalent received signal sample in the presence of nonselective fading be defined as

$$y = gx + n \tag{4.6}$$

where x is the complex transmitted signal, g is the multiplicative complex fading channel coefficient, and n is a complex Gaussian disturbance with zero mean and variance equal to the one-sided power spectral density of the additive white Gaussian noise (AWGN), N_0. In particular, $g = re^{j\theta}$, where r is the fading envelope and θ the phase rotation introduced by the channel. The average power associated with the fading envelope is given by

$$\mathcal{E}\{r^2\} = \Omega \tag{4.7}$$

where $\mathcal{E}(\cdot)$ is the expectation operator. A useful normalization amounts to setting $\Omega = 1$, although this is not always adopted in the literature.

4.2.1.1.1 Single-State First-Order Characterization

The first-order characterization amounts to expressing the relevant probability density functions for the fading envelope, and possibly for the fading phase. Note that the following distributions all refer to $r \geq 0$, and are understood to be null for $r < 0$.

4.2.1.1.1.1 Lognormal Distribution

Large-scale shadowing can be described through a Gaussian distribution in decibels. Converting decibels in the linear scale, the lognormal distribution is obtained:

$$p_{\text{Lognormal}}(r) = \frac{1}{\sqrt{2\pi}r\delta} \exp\left\{-\frac{(\ln r - \mu)^2}{2\delta^2}\right\} \tag{4.8}$$

where δ and μ indicate the standard deviation and mean of the associated normal variate, respectively. The commonly used *dB-spread* is defined as the ratio δ/h, where $h = (ln10)/20$. No phase rotation is usually associated with shadowing.

4.2.1.1.1.2 Rayleigh Distribution

Following the studies by Ossanna [18], the Rayleigh distribution was derived by Gilbert [1] and Clarke [2] under the assumption that no multipath component is prevalent, and applying the central limit theorem. The channel coefficient becomes a zero-mean complex Gaussian variate, with uniform phase distribution in $[0, 2\pi]$ and Rayleigh-distributed envelope according to the pdf

$$p_{\text{Rayleigh}}(r) = \frac{r}{\sigma^2} \exp\left[-\frac{r^2}{2\sigma^2}\right] \tag{4.9}$$

where $2\sigma^2 = \Omega$ is the variance of the associated complex Gaussian random variable. Under these assumptions, the overlaying fading random process is referred to as GWSSUS (Gaussian WSSUS). The Rayleigh distribution is not used frequently in LMS because it does not contain a direct component, and therefore it models situations where the link budget is often not closed.

4.2.1.1.1.3 Rice Distribution

In rural and suburban environments, where a direct component exists, fading can be modeled according to the Rice distribution [3], which assumes a constant mean power for both direct and diffuse components. The corresponding pdf is

$$p_{\text{Rice}}(r) = \frac{r}{\sigma^2} \exp\left[-\frac{r^2 + s^2}{2\sigma^2}\right] I_0\left(\frac{rs}{\sigma^2}\right) \tag{4.10}$$

where σ^2 has the same meaning as for the Rayleigh distribution, s is the magnitude of the direct component, and $I_0(\cdot)$ is the zero-order modified Bessel function of the first kind, defined as

$$I_0(z) = \frac{1}{\pi} \int_0^{\pi} e^{z \cos\varphi} d\varphi \tag{4.11}$$

It follows that

$$\Omega = \mathcal{E}\{r^2\} = s^2 + 2\sigma^2 \tag{4.12}$$

A frequently used parameter is the power ratio between the direct and the diffuse components, identified as the *Rice factor*, K:

$$K = \frac{s^2}{2\sigma^2} \tag{4.13}$$

When $K = 0$, the Rice distribution becomes the Rayleigh pdf. Normalizing $\Omega = 1$, the Rice distribution can be expressed in terms of a single parameter, the Rice factor, as

$$p_{\text{Rice}}(r) = 2r(K+1)\exp[-r^2(K+1) - K]I_0(2r\sqrt{K(K+1)}) \tag{4.14}$$

The normalized Rice distribution, with a single parameter, is very simple and surprisingly good in fitting measurement campaigns with satisfactory accuracy.

4.2.1.1.1.4 Nakagami Distribution

An alternative to Rayleigh is the Nakagami distribution [4], also identified as m-distribution. It contains many other distributions as special cases. The envelope pdf is given by

$$p_{\text{Nakagami}}(r) = \frac{2}{\Gamma(m)}\left(\frac{m}{\Omega}\right)^m r^{2m-1}\exp\left[-\frac{mr^2}{\Omega}\right] \tag{4.15}$$

where $\Gamma(\cdot)$ is the Gamma function, m is the shape factor, $m = \Omega^2/\mathcal{E}\left\{(r^2-\Omega)^2\right\} \geq 1/2$, and $\Omega = \mathcal{E}\{r^2\}$. When the envelope is Nakagami distributed, the corresponding instantaneous power is Gamma distributed. In the special case where $m = 1$, Rayleigh fading is obtained, with an exponentially distributed instantaneous power. For $m > 1$, the signal strength fluctuations reduce, compared to Rayleigh fading. For $m = 1/2$, Nakagami reduces to the one-sided Gaussian distribution.

4.2.1.1.1.5 Norton Distribution

Another possible alternative for modeling small-scale fading is represented by the Norton distribution [5], [6], [7], which is a combination of the Rice and Nakagami distributions. The pdf is

$$p_{\text{Norton}}(r) = \frac{r^m m}{\sigma^2 s^{m-1}}\exp\left[-\left(\frac{r^2+s^2}{2\sigma^2}\right)m\right] I_{m-1}\left(\frac{rsm}{\sigma^2}\right) \tag{4.16}$$

where $s \geq 0$ is the direct component amplitude and $I_{m-1}(\cdot)$ is the modified Bessel function of the first kind of order $m-1$, $m \geq 1/2$ and σ^2 have a significance similar to that for the Rice and Nakagami distributions. With three parameters, the Norton distribution is able to produce better fitting than other distributions. However, the additional complexity may not be justified in most situations.

4.2.1.1.1.6 Suzuki Distribution

In order to describe jointly large- and small-scale fading, a composite pdf can be employed. The Suzuki pdf [8] is widely accepted for urban terrestrial mobile channels; it combines Rayleigh and lognormal statistics as follows:

$$p_{\text{Suzuki}}(r) = \frac{r}{\sqrt{2\pi}\,\delta}\int_0^\infty \frac{1}{\sigma^3}\exp\left[-\frac{1}{2}\left(\frac{r^2}{\sigma^2} + \frac{(\ln\sigma - \mu)^2}{\delta^2}\right)\right] d\sigma \tag{4.17}$$

where $2\sigma^2$ is the average received power for the Rayleigh process, and δ and μ are the standard deviation and the mean value of the normal variable associated with the lognormal distribution, respectively. Similar results have been obtained in the concomitant studies by Hansen and Meno [19].

Starting from the Suzuki pdf, many other models have been introduced that employ compositions of Rayleigh, Rice, and lognormal distributions to describe the peculiarities of satellite mobile channels.

4.2.1.1.1.7 Loo Distribution

Loo [9] proposed a model suitable for rural environments, specifically accounting for shadowing due to roadside trees. The Loo model assumes that the received signal is affected by nonselective Rice fading with lognormal shadowing on the direct component only, while the diffuse scattered component has a constant average power level. The total complex fading coefficient g is the sum of the direct component contribution with lognormal envelope S and the diffuse component with Rayleigh envelope R:

$$g = Se^{j\phi_0} + Re^{j\phi} \tag{4.18}$$

where the phases ϕ_0 and ϕ are uniformly distributed. For the fading envelope, the resulting pdf conditioned to a certain S value is Rice distributed. So the envelope pdf can be written as

$$\begin{aligned} p_{\text{Loo}}(r) &= \int_0^{\infty} p_{\text{Rice}}(r\,|\,S)\, p_{\text{Lognormal}}(S) dS = \\ &= \frac{r}{\sigma^2\sqrt{2\pi}\delta} \int_0^{\infty} \frac{1}{S} \exp\left(-\frac{(\ln S - \mu)^2}{2\delta^2} - \frac{(r^2 + S^2)}{2\sigma^2}\right) I_0\left(\frac{rS}{\sigma^2}\right) dS \end{aligned} \tag{4.19}$$

where $2\sigma^2$ is the average scattered power due to multipath, and δ and μ are the standard deviation and the mean value of the normal variate associated with the lognormal distribution, respectively. It can be shown that the fading envelope pdf corresponds to a lognormal distribution for large envelope values and to a Rayleigh distribution for small values [9]:

$$p_{\text{Loo}}(r) = \begin{cases} p_{\text{Lognormal}}(r) & \text{for } r \gg \sigma \\ p_{\text{Rayleigh}}(r) & \text{for } r \ll \sigma \end{cases} \tag{4.20}$$

4.2.1.1.1.8 RLN Distribution

The Rice–lognormal model [10], [11] is a composition of Rice and lognormal statistics, with shadowing affecting both direct and diffuse components, and not only the direct path, as in the Loo model. As a matter of fact, the diffuse component power is no longer constant, since it suffers the same variations as the direct component. This is based on the observation that large-scale fading is caused by major obstacles that are likely to affect both direct and multipath components. The fading envelope is thus factored as the product of two independent variates, i.e.,

$$r = SR \tag{4.21}$$

The shadowing S is lognormal with pdf

$$p_S(S) = \frac{1}{\sqrt{2\pi}h\delta_{\text{dB}}S} \exp\left[-\frac{1}{2}\left(\frac{\ln S - h\mu_{\text{dB}}}{h\delta_{\text{dB}}}\right)^2\right] \tag{4.22}$$

where $h = (ln10)/20$, and μ_{dB} and δ^2_{dB} are the mean and variance of the associated normal variate expressed in dB, respectively. δ_{dB} is the shadowing dB spread. R is a Rice process normalized in power (i.e., $\mathcal{E}\{R^2\} = 1$) whose pdf can be expressed as a function of the Rice factor, K:

$$p_R(R) = 2(K+1)R\exp[-R^2(K+1) - K]I_0(2R\sqrt{K(K+1)}) \tag{4.23}$$

Letting

$$p(r|S) = 2(K+1)\frac{r}{S^2}\exp\left[-(K+1)\frac{r^2}{S^2} - K\right] I_0\left(2\frac{r}{S}\sqrt{K(K+1)}\right) \tag{4.24}$$

it follows that

$$p_r(r) = \int_0^\infty p(r|S)p_S(S)dS \tag{4.25}$$

Depending on the combination of K, μ_{dB}, and δ_{dB}, many of the previous nonselective statistical models can be derived as limiting cases of the RLN channel model:

- when $K = 0$, r is described by a Rayleigh–lognormal pdf (Suzuki);
- when $K \to \infty$, $p_R(R)$ tends to a Dirac pulse located at $R = 1$ and $p_r(r)$ tends to $p_S(r)$; i.e., the channel is lognormal;
- in the limit for $\delta_{dB} \to 0$, $p_S(S)$ tends to $\delta(S - e^{h\mu_{dB}})$, a Dirac pulse located at its mean value; therefore, $p_r(r) \to p(r|e^{h\mu_{dB}})$ and the channel is Ricean. When also $K = 0$, the channel is Rayleigh.

This asymptotic behavior of the RLN model gives a flavor of generality to performance analysis, because several channel models are considered at the same time.

The RLN model [10] has been fitted and validated against experimental data collected by ESA [20] in four different environments (rural tree-shadowed, urban, suburban, and open) and in a vast range of elevation angles α (20 to 80 degrees). This makes it also suitable for modeling the communication channel of a global satellite system adopting a nongeostationary orbit. Fitting can be performed by determining the optimum triplet $(K, \mu_{dB}, \delta_{dB})$ by means of the following polynomial empirical model:

$$\begin{aligned} K(\alpha) &= K_0 + K_1\alpha + K_2\alpha^2 + K_3\alpha^3 \\ \mu_{dB}(\alpha) &= \mu_0 + \mu_1\alpha + \mu_2\alpha^2 + \mu_3\alpha^3 \\ \delta_{dB}(\alpha) &= \delta_0 + \delta_1\alpha + \delta_2\alpha^2 + \delta_3\alpha^3 \end{aligned} \tag{4.26}$$

where the fitting coefficients K_i, μ_i, and δ_i ($i = 0, 1, 2, 3$) are provided in Table 4.1 for the considered environmental conditions. In this form, the RLN model can be identified as a hybrid statistical-empirical model, whereby the statistical model parameters for any elevation angle are obtained in a rapid empirical manner.

4.2.1.1.1.9 Pätzold et al. Distribution

Pätzold et al. [21] proposed an extended RLN model, where small-scale fading is modeled again through a Rice variate R, but with cross-correlated in-phase and quadrature components and a possible Doppler shift on the direct component:

$$R = |\rho e^{j(2\pi f_\rho t + \theta_\rho)} + R_p + jR_q| \tag{4.27}$$

TABLE 4.1 Empirical Coefficients for RLN Parameters in Different Environments

	Rural Tree Shadowed Area	Urban Area	Suburban Area	Open Area
K_0	2.731	1.750	−13.60	26.43
K_1	$-1.074\cdot10^{-1}$	$6.700\cdot10^{-2}$	$9.650\cdot10^{-1}$	−2.644
K_2	$2.744\cdot10^{-3}$	0.0	$-1.663\cdot10^{-2}$	$8.337\cdot10^{-2}$
K_3	0.0	0.0	$1.187\cdot10^{-4}$	$-4.111\cdot10^{-4}$
μ_0	−20.25	−52.12	−1.998	3.978
μ_1	$9.919\cdot10^{-1}$	2.758	$-9.919\cdot10^{-3}$	$-1.742\cdot10^{-1}$
μ_2	$-1.684\cdot10^{-2}$	$-4.777\cdot10^{-2}$	$1.520\cdot10^{-3}$	$2.647\cdot10^{-3}$
μ_3	$9.502\cdot10^{-5}$	$2.714\cdot10^{-4}$	$-1.266\cdot10^{-5}$	$-1.367\cdot10^{-5}$
δ_0	4.500	7.800	8.000	0.0
δ_1	$-5.000\cdot10^{-2}$	$-3.542\cdot10^{-1}$	$-3.741\cdot10^{-1}$	0.0
δ_2	0.0	$6.5\cdot10^{-3}$	$6.125\cdot10^{-3}$	0.0
δ_3	0.0	$-3.958\cdot10^{-5}$	$-3.333\cdot10^{-5}$	0.0

where ρ is the direct component amplitude, f_ρ is the direct component Doppler shift, θ_ρ is the direct component initial phase, and $R_p + jR_q$ is the complex diffuse component. The large-scale fading S is still lognormally distributed and multiplies the Rice variate, as in the RLN distribution. The two variables are statistically uncorrelated, and the corresponding envelope pdf is the same as that for the RLN model, the differences lying in the second-order characterization, visibly in the Doppler spectrum.

4.2.1.1.1.10 GRLN Distribution

The generalized RLN model [13] contains the RLN and Loo models as particular cases. The diffuse multipath component is subdivided into two parts, shadowed and unshadowed respectively. The complex fading coefficient is written as

$$g = re^{j\theta} = x + jy = RSe^{j\phi} + x_1 + jy_1 \tag{4.28}$$

where the variate R is Ricean with parameters s and σ^2 and is multiplied by the lognormal variable S with parameters μ_{dB} and δ_{dB}, and the resulting product is added to the zero-mean complex Gaussian variate $x_1 + jy_1$, with Rayleigh envelope and variance $2\sigma_1^2$. S, x_1, and y_1 are mutually uncorrelated. Introducing the Rice factor, K, and the mean power ratio between the shadowed and unshadowed diffuse components, $\xi = \sigma^2/\sigma_1^2$, the normalized envelope distribution conditioned on shadowing can be written as

$$p(r|S) = \frac{2r\xi(K+1)}{1+\xi S^2} e^{-\xi \frac{KS^2+(K+1)r^2}{1+\xi S^2}} I_0\left(2rS\frac{\xi\sqrt{K(K+1)}}{1+\xi S^2}\right) \tag{4.29}$$

This $p(r|S)$ expression must be substituted into Equation 4.25 to have the GRLN distribution. Note that for $\xi \to 0$ the GRLN tends to the Loo distribution, while for $\xi \to \infty$ the GRLN tends to the RLN distribution. The conditional phase pdf, $p(\theta|S)$, is given by

$$p(\theta|S) = \frac{1}{2\pi} e^{-\frac{\xi KS^2}{1+\xi S^2}} + \sqrt{\frac{\xi KS^2}{\pi(1+\xi S^2)}} \cos\theta \times \left[1 - \frac{1}{2}\mathrm{erfc}\left(\sqrt{\frac{\xi KS^2}{\pi(1+\xi S^2)}} \cos\theta\right)\right] . e^{-\frac{\xi KS^2}{1+\xi S^2}\sin^2\theta} \tag{4.30}$$

4.2.1.1.1.11 Xie and Fang Distribution

Xie and Fang [14] derived another extension of the RLN model based on propagation scattering theory. As before, the fading envelope is factored as the product of two independent variables related to large-scale and small-scale fading:

$$r = RS \tag{4.31}$$

where S is lognormally distributed with parameters μ and δ, and R is modeled through a generalization of the Rice distribution. The generalization allows the diffuse component to result from the composition of generic amplitude/phase scattering contributions. As a result, the real and imaginary small-scale fading components are still Gaussian but with a different mean and variance (α, σ_1^2 and β, σ_2^2, respectively). Therefore,

$$p_R(R) = \frac{R}{\sigma_1\sigma_2} \exp\left(-\frac{\sigma_1^2R^2 + \sigma_2^2\alpha^2 + \sigma_1^2\beta^2}{2\sigma_1^2\sigma_2^2}\right)\frac{1}{2\pi} \times \int_0^{2\pi} \exp\left(\frac{2\sigma_2^2\alpha R\cos\theta + 2\sigma_1^2\beta R\sin\theta + (\sigma_1^2 - \sigma_2^2)R^2\cos^2\theta}{2\sigma_1^2\sigma_2^2}\right) d\theta \tag{4.32}$$

The channel envelope pdf can be determined as the double integral

$$p_r(r) = \int_0^{\infty} \frac{1}{S} p_R\left(\frac{R}{S}\right) p_S(S)dS \tag{4.33}$$

This model contains several particular cases. When $\delta = 0$, and $\sigma_1^2 = \sigma_2^2$, large-scale fading vanishes and Equation 4.33 reduces to the Rice distribution. Furthermore, when $\alpha = \beta = 0$, it reduces to the Rayleigh distribution. If $\delta \neq 0$, $\sigma_1^2 = \sigma_2^2$, and $\alpha = \beta = 0$, Equation 4.33 reduces to the Suzuki pdf. Finally, when $\sigma_1^2 = \sigma_2^2$ and $\alpha^2 = \beta^2 = (1 - 2\sigma_1^4)/2$, Equation 4.33 reduces to the RLN model.

4.2.1.1.1.12 Other Models

Hwang et al. [12], [22] proposed an extended Loo model where shadowing on the diffuse component is introduced. This is different from RLN because shadowing on the direct component is independent from shadowing on the diffuse component. The complex fading coefficient is therefore

$$g = re^{j\theta} = sS_1e^{j\phi} + RS_2e^{j(\varphi+\phi)} \tag{4.34}$$

where s is the direct component amplitude, R is the Rayleigh-distributed diffuse component, and S_1 and S_2 are lognormally distributed with parameters μ_1, δ_1 and μ_2, δ_2, respectively. When $\delta_2 = 0$, the Hwang et al. model coincides with the Loo model.

Tjhung and Chai [23], [24] proposed a Nakagami–lognormal (NLN) statistical model to characterize multipath fading, shadowing, and path loss, using the same formulation as the RLN model. The only difference stands in modeling the variate R with a Nakagami m-distribution in place of the Rice distribution.

Abdi et al. [17] propose a model describing the direct component S through a Nakagami distribution and the diffuse component R by means of a Rayleigh distribution. The complex fading coefficient is expressed as

$$g = Se^{j\phi_0} + Re^{j\phi} \tag{4.35}$$

where ϕ is uniformly distributed over $[0, 2\pi)$ and ϕ_0 is the deterministic phase associated with the direct component. S and R are independent variates. Overall, this model can also be interpreted as a Rice distribution with Nakagami-distributed amplitude for the direct component.

The comparison among the different first-order single-state statistical models is summarized in Table 4.2, where the last column describes the statistical correlation between direct and diffuse components.

4.2.1.1.2 Single-State Second-Order Characterization

The first-order statistics are unable to characterize events occurring at specific pairs of time samples, frequency samples, or spatial samples. Therefore, second-order characterization becomes mandatory when memory effects are present in various domains. For frequency-non-selective channels, time and space domain correlations are the only relevant quantities. Time-related second-order characteristics (induced by the relative motion between the terminal and the satellite) are important for applications such as coding, interleaving, automatic repeat–request, etc. Below, Doppler spectrum, level crossing rate, fade duration, and shadowing correlation are considered.

TABLE 4.2 Comparison among First-Order Single-State Statistical Models

Model	Year	Direct Component	Diffuse Component	Correlation between Direct and Diffuse
Rice	1945	Constant	Rayleigh	Zero
Loo	1985	Lognormal	Rayleigh	Zero
RLN	1994	Lognormal	Lognormal–Rayleigh	Unity
GRLN	1995	Lognormal	Part 1: Rayleigh Part 2: Rayleigh–lognormal	Variable
Xie and Fang	2000	Lognormal	Lognormal–Generalized Rayleigh	Unity
Pätzold et al.	1998	Lognormal	Lognormal–Rayleigh	Unity
Hwang et al.	1997	Lognormal	Lognormal–Rayleigh	Zero
Tjhung and Chai	1998	Lognormal	Lognormal–Nakagami	Unity
Abdi et al.	2003	Nakagami	Rayleigh	Zero

4.2.1.1.2.1 Doppler Spectrum

The Doppler spectrum arises due to the different angles of arrival of the the various multipath components to or from a moving terminal. Therefore, the characterization of the Doppler spectrum requires a specific statistical model for the multipath angle of arrival and intensity, and for the terminal motion. Clarke [2] assumes two-dimensional horizontal propagation with uniform angles of arrival over the azimuth, equal multipath amplitude, and constant terminal speed and direction. Under these assumptions, the channel autocorrelation function at the central frequency f_0 is given by

$$R_H(t, t'; f_0, f_0) = R_H(\Delta t) = 2\sigma^2 J_0(2\pi f_{d,m} \Delta t) \tag{4.36}$$

where $J_0(\cdot)$ is the zero-order Bessel function of the first kind, $2\sigma^2$ is the mean power of the Rayleigh-distributed diffuse component, and $f_{d,m}$ is the maximum Doppler shift associated with the terminal motion, which is a function of the terminal speed v according to

$$f_{d,m} = \frac{v}{c} f_0 \tag{4.37}$$

where c is the speed of light. By taking the Fourier transform of the above correlation function, the two-dimensional (2D) *isotropic Doppler spectrum*, also known as Jakes Doppler spectrum [25], is obtained:

$$S_d(f_d) = \begin{cases} \dfrac{2\sigma^2}{\pi f_{d,m}\sqrt{1-\left(\frac{f_d}{f_{d,m}}\right)^2}} & \text{for } |f_d| \le f_{d,m} \\ 0 & \text{elsewhere} \end{cases} \tag{4.38}$$

The spectrum is continuous and symmetrical, tending to infinity for $|f_d| = f_{d,m}$. Two modifications on the two-dimensional isotropic spectrum are necessary for LMS application. First, if the satellite is nongeostationary with apparent velocity v_{sat}, then an additional Doppler frequency shift is present over the entire spectrum and equals $f_{d,sat} = f_0 v_{sat}/c$. Second, if a direct component with amplitude s is present, the resulting Doppler spectrum consists of a continuous and a discrete component. The continuous component is again the shifted two-dimensional isotropic spectrum, and the discrete contribution is a Dirac delta centered in $f_{d,sat} + f_{d,d}$, where $f_{d,d}$ is the additional Doppler shift dependent on the terminal motion and direct component angle of arrival. In summary, the two-dimensional isotropic LMS Doppler spectrum can be written as follows:

$$S_{d,2D}(f_d) = \begin{cases} \dfrac{2\sigma^2}{\pi f_{d,m}\sqrt{1-\left(\frac{f_d - f_{d,sat}}{f_{d,m}}\right)^2}} + s^2\delta(f_d - f_{d,sat} - f_{d,d}) & \text{for } |f_d - f_{d,sat}| \le f_{d,m} \\ 0 & \text{elsewhere} \end{cases} \tag{4.39}$$

The two-dimensional LMS Doppler spectrum is plotted in Figure 4.1.

The extension from two-dimensional to three-dimensional (3D) modeling was tackled by various researchers in the literature. Two remarkable examples are given by the works of Aulin [26] and Parsons [27], who presented two different distributions for the angles of arrival in the vertical plane, resulting in the Doppler spectra represented in Figure 4.2 [27]. Note that the second-kind discontinuities at $|f_d| = f_{d,m}$ are eliminated, although the Aulin spectrum maintains a pair of first-kind discontinuities. More impressing is the result that is obtained by assuming an isotropic angle of arrival distribution over an hemisphere. In this case, it can be rigorously shown that the corresponding Doppler spectrum is simply rectangular [28]:

$$S_{d,3D}(f_d) = \frac{\sigma^2}{f_{d,m}} \text{rect}\left\{\frac{f_d - f_{d,sat}}{2 f_{d,m}}\right\} + s^2\delta(f_d - f_{d,sat} - f_{d,d}) \tag{4.40}$$

where $\text{rect}\{x\} = 1$ for $x \in [-1/2, 1/2]$ and zero otherwise. Evidently, this Doppler spectrum shape is much more amenable to numerical simulation, in addition to being more physically meaningful for LMS channels, where three-dimensional propagation is certainly the case.

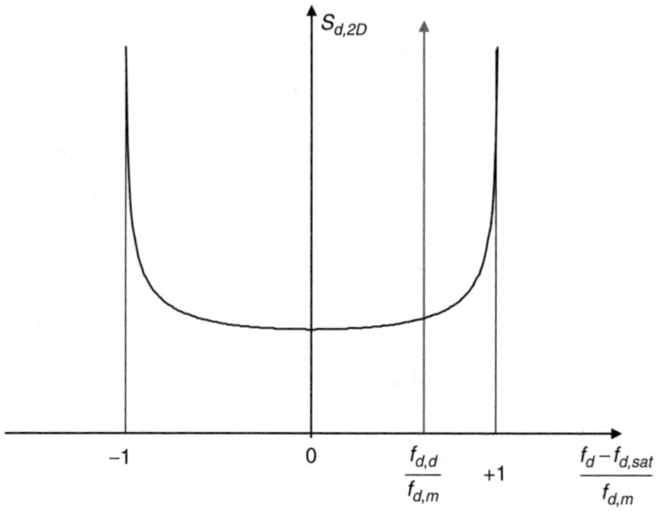

FIGURE 4.1 Two-dimensional isotropic LMS Doppler spectrum.

Another possible way to model the Doppler spectrum is given by the frequency-shifted Gaussian power spectral density [15].

4.2.1.1.2.2 *Level Crossing Rate and Average Fade Duration*

The *level crossing rate* (LCR), N_X, was defined by Rice [29], [30] as the expected rate at which the envelope r crosses a specific level X with positive slope:

$$N_X = \int_0^{\infty} \dot{r} p(X, \dot{r}) d\dot{r} \tag{4.41}$$

where $\dot{r}$ is the time derivative of the signal envelope and $p(X, \dot{r})$ is the joint pdf of the envelope r and its derivative $\dot{r}$ computed when r equals the level X.

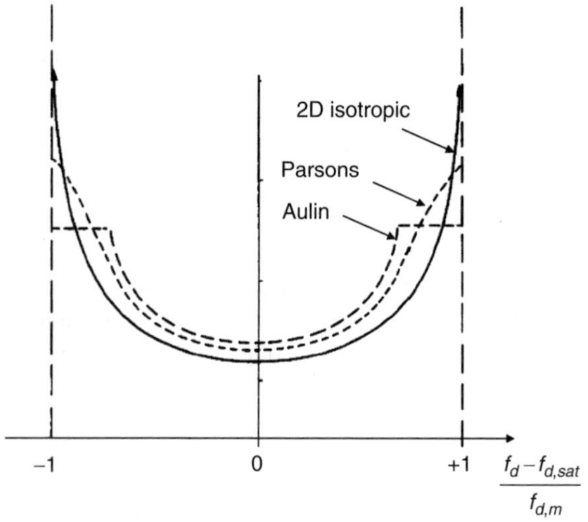

FIGURE 4.2 Comparison among two-dimensional isotropic, Aulin, and Parsons Doppler spectra.

The *average fade duration* (AFD), L_X, is defined as the average length of the time intervals corresponding to the envelope falling below a certain level X. Indicating with τ_i the duration of the i-th fade time interval, the probability to have $r \leq X$ during an observation time T can be written as

$$P(r \leq X) = \frac{\sum_i \tau_i}{T} \tag{4.42}$$

Thus L_X becomes

$$L_X = \frac{P(r \leq X)}{N_X} = \frac{\sum_i \tau_i}{T N_X} \tag{4.43}$$

The LCR and AFD parameters can be used, for example, in the selection of the optimum interleaving span in forward error correction, to break channel correlation.

For the two-dimensional isotropic Doppler spectrum, the LCR can be expressed as

$$N_X = \sqrt{2\pi}\, f_{d,m} \bar{X} \exp(-\bar{X}^2) \tag{4.44}$$

where $\bar{X} = X/(\sigma\sqrt{2})$, and the associated AFD results in

$$L_X = \frac{\exp(\bar{X}^2) - 1}{\bar{X} f_{d,m}\sqrt{2\pi}} \tag{4.45}$$

In [9] expressions for LCR and AFD are provided for the Loo model:

$$N_X = \frac{\sqrt{1-\rho^2}}{\sqrt{2\pi}} \cdot \frac{\sigma_1^2\left(\sigma_1^2 + 2\rho\sigma_1\delta_1 + \delta_1^2\right)^{1/2}}{\sigma_1^2(1-\rho^2) + 4\rho\sigma_1\delta_1} p_{\text{Loo}}(X) \tag{4.46}$$

$$L_X = \frac{1}{N_X}\int_0^X p_{\text{Loo}}(r)dr \tag{4.47}$$

where the parameters $\rho, \sigma_1, and\, \delta_1$ are respectively the mutual correlation and the variances of the small-scale and large-scale rate-of-change distributions.

In [11] the LCR is computed for the RLN model, resulting in

$$N_X = \sqrt{2\pi}\,\beta\sigma_r\, p_{\text{RLN}}(X) \tag{4.48}$$

where σ_r is the variance of the envelope (function of σ, δ, and μ) and β is the Doppler effective bandwidth, expressed as

$$\beta = \frac{\sigma_{\dot{r}}^2}{4\pi^2\sigma_r^2} \tag{4.49}$$

$\sigma_{\dot{r}}^2$ being the variance of the envelope derivative.

4.2.1.1.2.3 Fade and Non-fade Duration Statistics

The average fade duration characterization can be augmented by considering the full statistical behavior (cdf) of fade durations (d_f) and non-fade durations (d_{nf}). The fade state is defined as the set of time intervals during which the signal level falls below a specified threshold, while the nonfade state comprises all the remaining periods of time. An example of this approach is given in [31], where the desired statistical characterization is obtained by fitting measurement data resulting from a campaign conducted in Australia. By expressing fade and non-fade durations in units of distance (meters), the cdf of fade durations can be modeled by a lognormal law, while the non-fade duration is well described by a power law, for a fade level

X in the range of 2 to 8 dB. The cdfs are respectively

$$P(d_f > D_X) = \frac{1}{2}\left\{1 - \text{erf}\left(\frac{\ln D_X - \ln a}{\sigma_F\sqrt{2}}\right)\right\} \tag{4.50}$$

and

$$P(d_{nf} > D_X) = bD_X^{-c} \tag{4.51}$$

where erf(·) is the error function, $\ln a$ is the mean value of $\ln D_X$, σ_F is the standard deviation of $\ln D_X$, and b and c are constants. The parameters a, b, c, and σ_F are obtained through regression methods from measurement data.

4.2.1.1.2.4 Shadowing Correlation Function

Modeling spatial correlation for large-scale fading in LMS is important for the design of power control and handoff schemes. Assuming lognormally distributed shadowing, and indicating with S_x(dB) and S_{x+d}(dB) the shadowing values in dB at the generic points x and $x+d$, the dependence of the shadowing autocovariance $C_S(\cdot)$ on the distance d can be modeled through an exponential distribution as follows:

$$C_S(d) = \mathcal{E}\{S_x(\text{dB})S_{x+d}(\text{dB})\} - \mathcal{E}\{S_x(\text{dB})\}\mathcal{E}\{S_{x+d}(\text{dB})\} = \delta_{\text{dB}}^2\gamma^d \tag{4.52}$$

where δ_{dB} is the dB spread and $\gamma \in [0,1]$ is an empirical parameter determining the correlation decay with increasing distances d. In [32], this model is used with $\gamma = (1/e)^{(1/(\Delta_s)_c)}$, where $(\Delta_s)_c$ is the effective coherence distance in meters. The model was fitted on measurements in L and S-bands for elevation angles of 60 and 80 degrees. The effective coherence distance $(\Delta_s)_c$ was found to be between 16 and 20 m.

4.2.1.2 Multistate Statistical Models

Single-state models describe wide-sense stationary fading conditions and are thus unable to model nonstationary transitions, e.g., due to the terminal moving from an urban to suburban environment or due to a change in elevation angle for a nongeostationary link. In order to be able to describe these variations in the statistical nature of the channel, multistate statistical models have been introduced, assuming in general an underlying Markov process. Each state in the Markov chain is described by one of the single-state statistical models from the previous section. A Markov chain model requires two matrices: the state probability array, Π $(1 \times M)$, and the state transition probability matrix, P $(M \times M)$, where M is the overall number of states. Each element in the matrix P, P_{ij}, represents the transition probability from state i to state j. The generic element of the array Π, Π_i, represents the total probability of being in state i. In general, letting $p_i(r)$ be the pdf describing the i-th state, the application of the total probability theorem leads to the total pdf being written as

$$p_{tot}(r) = \sum_{i=1}^{M} p_i(r)\Pi_i \tag{4.53}$$

The various multistate models differ in the number of states in the Markov chain and the statistics describing the single discrete states. In the following, the Lutz et al. [33], Barts and Stutzman [34], Vucetic and Du [35], Rice and Humprays [36], Karasawa et al. [37], Fontan et al. [38], and Wakana [39] models are presented.

4.2.1.2.1 Lutz Model

Lutz et al. [33] introduced a two-state statistical model based on an extensive measurement campaign over European areas at elevation angles in the range of 13 to 43 degrees. Channel states are identified as either *good* or *bad*. The good channel state corresponds to areas with unobstructed direct components from the satellite, whereas the bad channel state corresponds to shadowed areas. In both cases, the diffuse signal component is modeled by a Rayleigh distribution with mean power $2\sigma^2$. In the good state where the direct component is present with amplitude s, the overall received envelope follows a Rice distribution, with Rice factor K. Instead, the bad state is conditionally Rayleigh distributed, with mean power varying according to a lognormal distribution, due to shadowing. The bad state is therefore Suzuki distributed. The most significant parameter of this model is the shadowing time-share A, which determines the probability to be in either the bad or good state. Lutz et al. model the instantaneous received power $W = r^2$ by combining the pdfs of the two states according to A as

$$p(W) = (1 - A)p_{\chi_2^2(\mathrm{d})}(W) + A \int_0^{\infty} p_{\chi_2^2(0)}(W|W_S)p_{\mathrm{LogNormal}}(W_S)dW_S \tag{4.54}$$

where W_S is the instantaneous shadowing power variable and $p_{\mathrm{LogNormal}}(W_S)$ is the corresponding log-normal distribution with parameters μ_{dB} and δ_{dB}:

$$p_{\mathrm{LogNormal}}(W_S) = \frac{10/\ln 10}{\sqrt{2\pi}\,\delta_{\mathrm{dB}} W_S} \exp\left\{-\frac{(10\mathrm{Log}W_S - \mu_{\mathrm{dB}})^2}{2\delta_{\mathrm{dB}}^2}\right\} \tag{4.55}$$

Further, $p_{\chi_2^2(\mathrm{d})}(W)$ is the noncentral χ^2 pdf with two degrees of freedom and noncentrality parameter $d = s^2$, associated with the square of the Rice-distributed envelope with Rice factor $K = d/2\sigma^2$:

$$p_{\chi_2^2(\mathrm{d})}(x) = \frac{1}{2\sigma^2} \exp\left(-\frac{x+d}{2\sigma^2}\right) I_0\left(\frac{\sqrt{xd}}{\sigma^2}\right) \tag{4.56}$$

Finally, $p_{\chi^2(0)}(S)$ is the central χ^2 pdf with two degrees of freedom associated with the square of a Rayleigh-distributed envelope with mean power $2\sigma^2$. For different satellite elevations, environments, and antennas, the parameters A, K, μ_{dB}, and δ_{dB} have been evaluated from the statistics of the recordings through a least squares curve-fitting procedure [33]. Transitions between states are described by the first-order Markov chain, as reported in Figure 4.3. Transition probabilities from state i to state j are indicated as P_{ij}, with $i, j = g, b$ according to good or bad states. It results in

$$P_{gb} = \frac{v}{RD_g} \tag{4.57}$$

$$P_{bg} = \frac{v}{RD_b} \tag{4.58}$$

where D_g and D_b are the average distances in meters over which the good and bad states tend to persist, respectively; v denotes the terminal speed; and R is the transmission data rate. Further, $P_{gg} = 1 - P_{gb}$ and $P_{bb} = 1 - P_{bg}$. The time-share of shadowing can be consequently determined as

$$A = \frac{D_b}{D_b + D_g} \tag{4.59}$$

As can be noticed, A is independent of the speed and data rate of the user. *Saunders and Evans* [40] estimate the time-share of shadowing A through geometrical-statistical principles in terms of physical parameters such as street width and building height distributions. This allows predictions to be made for systems

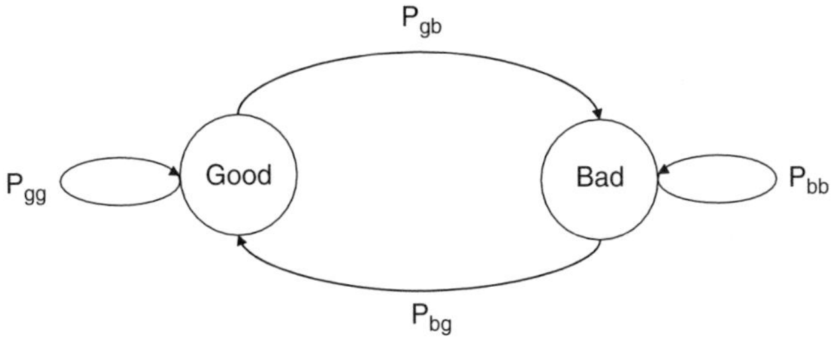

FIGURE 4.3 Markov chain for Lutz model.

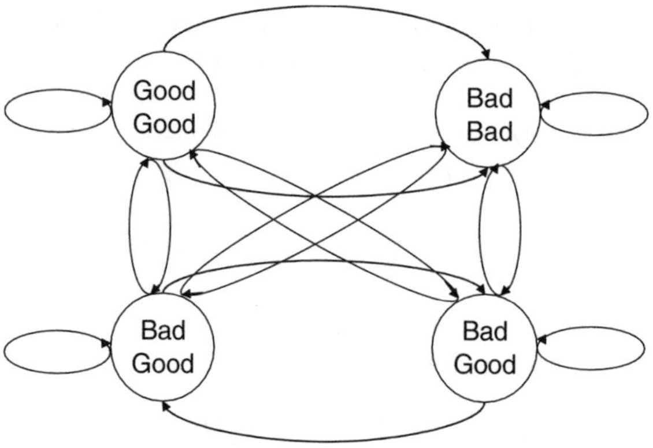

FIGURE 4.4 Four-state model for satellite diversity.

operating in areas where no direct measurements are available, and permits existing measurements to be scaled for new parameter ranges. A generalization of the Lutz model to include satellite diversity was proposed in [41], accounting for the statistical interdependence among the different link fading processes at different elevation and azimuth angles. The model has four states for a twofold satellite diversity, representing all possible combinations between the good and bad conditions. The corresponding Markov chain is reported in Figure 4.4. For the definition of the correlation coefficient, the channel envelope for each single satellite link is expressed as

$$r_i(t) = \begin{cases} 0 & \text{bad channel state} \\ 1 & \text{good channel state} \end{cases} \tag{4.60}$$

with $i = 1, 2$ for the two satellites, respectively. The correlation coefficient ρ is computed as

$$\rho = \frac{\mathcal{E}\{(r_1(t) - \bar{r}_1)(r_2(t) - \bar{r}_2)\}}{\sigma_1 \sigma_2} \tag{4.61}$$

where $\bar{r}_i = \mathcal{E}\{r_i(t)\} = 1 - A_i$ and $\sigma_i^2 = \mathcal{E}\{(r_i(t) - \bar{r}_i)^2\} = A_i(1 - A_i)$, for $i = 1, 2$, A_i being the shadowing time-share of the relevant satellite link. In [41], the parameter ρ is computed as a function of the transition and states probabilities of the related Markov process, resulting in

$$\rho = \frac{1}{\sigma_1 \sigma_2}(p_0 - A_1 A_2) \tag{4.62}$$

p_0 being the equilibrium state probability for the bad–bad state.

4.2.1.2.2 *Barts and Stutzman Model*

Barts and Stutzman [34] proposed a two-state model similar to that of Lutz et al., taking into account the vegetative shadowing phenomenon of mobile satellite systems. The model comprises a shadowed state and an unshadowed state, which are opportunely combined through the shadowing time-share factor. The only difference from Lutz is that in the bad state a Loo distribution is used in place of the Suzuki distribution.

4.2.1.2.3 *Vucetic and Du Model*

Vucetic and Du [35] introduced a general M-state model, in which each state can have either Rayleigh, Rice, or Loo distribution. An experimental application for $M = 4$ is also reported, where two states are Rice distributed, corresponding to unshadowed channel conditions, while the remaining states are described by linear combinations of Rayleigh and lognormal distributions, corresponding to shadowed channels.

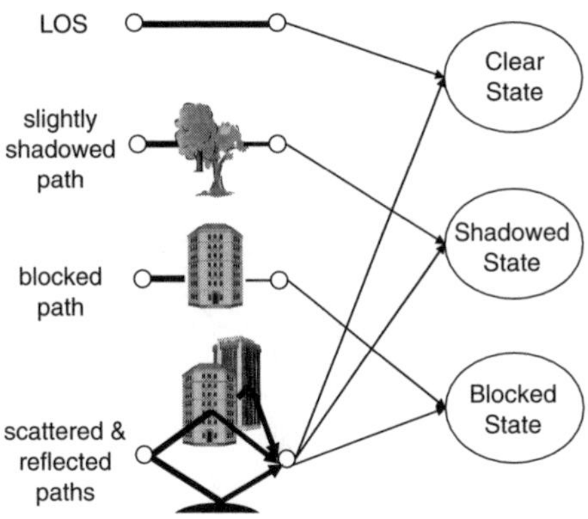

FIGURE 4.5 Karasawa three state propagation channel model.

4.2.1.2.4 Rice and Humphreys Model

Rice and Humphreys [36] proposed a variation on the Lutz et al. model, where the direct component amplitude s in the good state is modeled through a bimodal distribution. As a consequence, a three-state model is obtained. Fitting with experimental results is provided for a number of cases.

4.2.1.2.5 Karasawa et al. Model

Karasawa et al. [37] proposed a three-state statistical model, represented pictorially in Figure 4.5. The clear state, with time-share C, is described by the Rice distribution, corresponding to an unblocked direct component; the shadowed state, with time-share S, is described by the Loo pdf, corresponding to a shadowed direct component and a Rayleigh-distributed diffuse component; and the blocked state, with time-share B, is described by a Rayleigh-distributed fading channel, referring to a complete blockage of the direct component by buildings. Thus, the resulting total pdf is a weighted linear combination of Rice, Loo, and Rayleigh distributions, as follows:

$$p_r(r) = Cp_{\text{Rice}}(r) + Sp_{\text{Loo}}(r) + Bp_{\text{Rayleigh}}(r) \tag{4.63}$$

where the distribution parameters and the associated weights are determined by fitting measurement campaigns. Note that when $B = 0$ the model reduces to the Barts and Stutzman model. In [42] an interesting method for computing the weighting factors is proposed, based on photogrammetry and image processing, to derive the time-share of each state.

4.2.1.2.6 Fontan et al. Model

Fontan et al. [38], [43] also proposed a three-state statistical model. State 1 corresponds to the line-of-sight (LOS) conditions, state 2 corresponds to moderate shadowing conditions, and state 3 describes strong shadowing situations. The Loo model was used to describe all three states with different parameters defined by fitting measurement campaigns.

4.2.1.2.7 Wakana model

Wakana [39] proposed a model comprising two main states: the fade and nonfade states. The propagation channel is assumed to be in a specific main state according to the comparison of the signal level to a threshold. The transition between fade and nonfade states is described through a five-state Markov chain, which is reported in Figure 4.6. The main fade state is divided into two Rayleigh fade substates, corresponding to short and long fade duration, respectively, and the main nonfade state is divided into three Ricean

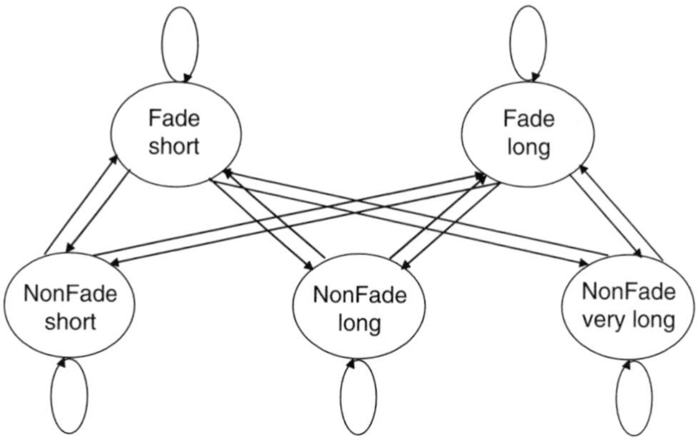

FIGURE 4.6 Markov chain for Wakana model.

sub-states, corresponding to short, long, and very long duration, respectively. The comparison among the different multi-state models is summarized in Table 4.3.

4.2.2 Wideband Statistical Models

When the signal bandwidth is larger than the coherence bandwidth of the channel, frequency selectivity leads to wideband channels. By discretizing the delay of the multipath components, the channel impulse response can be written as

$$h(t) = \sum_{i=0}^{N-1} A_i(t)\delta[t - \tau_i(t)]e^{j[2\pi f_{d,i}t+\theta_i(t)]} \tag{4.64}$$

where $A_i(t)$, $\tau_i(t)$, $f_{d,i}$, and $\theta_i(t)$ are the amplitude, delay, Doppler shift, and phase of the various paths, respectively; $\delta(t)$ is the Dirac delta function; and N is the number of paths. This model is also referred to as the tapped delay line model, where N is the number of taps. Normally, the path of index $i = 0$ corresponds to the direct component, while the terms for $i \neq 0$ correspond to multipath echoes. Each path can be described by a narrowband model. The multiple paths are combined according to appropriate delays and relative amplitudes, depending on the peculiarities of the local environment, and are assumed to be mutually uncorrelated.

4.2.2.1 DLR Wideband Model

In [44], the DLR model is presented. It consists of a tapped delay line model with a random number of taps with randomly varying tap delays. The parameters of the model are obtained through fitting procedures

TABLE 4.3 Comparison among Multistate Statistical Models

Model	Year	Number of States	Single-State Distributions
Lutz et al.	1991	2 (diversity: 4)	Rice, Suzuki
Barts and Stutzman	1992	2	Rice, Loo
Vucetic and Du	1992	M	Rayleigh, Rice, Loo
Rice and Humphreys	1997	3	Rice (2 states), Suzuki
Karasawa et al.	1997	3	Rice, Loo, Rayleigh
Fontan et al.	1997	3	Loo
Wakana	1997	5 (2 main)	Rayleigh (2 states), Rice (3 states)

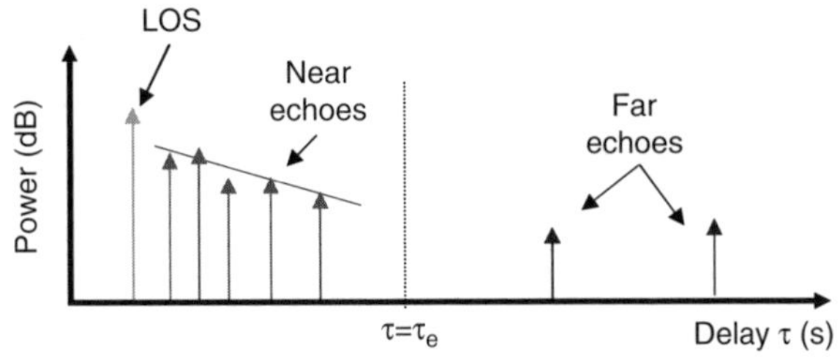

FIGURE 4.7 Echoes region definition for DLR model.

with measured data for different elevation angles and environments (open, rural, suburban, urban, etc.). The impulse response with N echoes is divided into three parts with different behaviors: direct path, near echoes, and far echoes. This subdivision is shown in Figure 4.7. The direct path is modeled according to the Lutz narrowband model. A number of N_{NE} *near echoes* appears in the close vicinity of the receiver with delays $0 < \tau_i \leq \tau_e$, where τ_e is a function of the environment and elevation angle [44], and is in the order of a few hundred nsec. The remaining $N_{FE} = N - N_{NE}$ echoes are *far echoes*, and they appear with delays $\tau_e < \tau_i \leq \tau_{max}$, where τ_{max} depends on the environment and elevation and is in the order of 10 to 15 μsec. The number of near and far echoes is a Poisson-distributed random variable

$$p_{Poisson}(N_i) = \frac{\lambda_i^{N_i}}{N_i!} e^{-\lambda_i} \quad i = NE, FE \tag{4.65}$$

where λ_{NE} and λ_{FE} are the mean values of the number of near and far echoes, respectively. The delays of the near echoes follow an exponential distribution, while the far echo delays are uniformly distributed in $[\tau_e, \tau_{max}]$. In [16], each tap is again modeled according to the Lutz narrowband model, but the number of paths is time varying, following a birth–death Markov process.

4.2.2.2 Saunders et al.

An alternative method is proposed by Saunders et al. [45]. The satellite mobile channel comprises two cascaded processes: the satellite process, associated with the satellite-to-Earth path, and the terrestrial process, associated with the effects of the mobile motion relative to terrestrial scatterers. The satellite process is characterized by a propagation path loss (including free space, antenna radiation and pattern, and atmospheric absorption effects), a time delay associated with the satellite-to-ground path length, and a Doppler shift due to satellite motion relative to the ground. The terrestrial process comprises a direct part and a multipath part. The latter is modeled according to a tapped delay line where the taps are Rice distributed, in accordance with the Doppler shift induced by terminal motion.

4.2.2.3 IMR Channel Model

In [46], a wideband characterization of the LMS channel is proposed in the presence of Intermediate Module Repeaters (IMRs), or gap fillers, introduced to avoid link obstruction in built-up urban areas, for extended coverage. Interestingly, the presence of IMRs in the system architecture consistently increases the multipath propagation phenomenon, and wideband modeling becomes mandatory. For each IMR, a tapped delay line model is introduced in [46] for two different environments: the vehicular (with six paths) and the outdoor-to-indoor and pedestrian ETSI (with four paths) [47]. A hexagonal cellular layout model is assumed, considering the paths coming from the nearest IMR (reference) and the six IMRs in the first tier. For the vehicular environment, the corresponding multipath intensity profile as a function of relative delay is reported in Table 4.4. The delays are computed assuming a geostationary satellite and an IMR coverage radius of 400 m at a frequency of 2 GHz. Lognormal shadowing should also be superimposed on this small-scale characterization.

TABLE 4.4 Intensity–Delay Profile for Vehicular Environment

Reference IMR		IMR 1		IMR 2		IMR 3	
Intensity (dB)	Delay (μsec)	Intensity (dB)	Delay (μsec)	Intensity (dB)	Delay (μsec)	Intensity (dB)	Delay (μsec)
0.0	1.99	−3.7	0.32	−13.2	2.44	−17.5	5.18
−1.0	2.30	−4.7	0.63	−14.2	2.75	−18.5	5.49
−9.0	2.70	−12.7	1.03	−22.2	3.15	−26.5	5.89
−10.0	3.08	−13.7	1.41	−23.2	3.53	−27.5	6.27
−15.0	3.72	−18.7	2.05	−28.2	4.17	−32.5	6.91
−20.0	4.50	−23.7	2.83	−33.2	4.95	−37.5	7.69

IMR 4		IMR 5		IMR 6	
Intensity (dB)	Delay (μsec)	Intensity (dB)	Delay (μsec)	Intensity (dB)	Delay (μsec)
−17.5	6.16	−13.2	4.41	−3.7	1.30
−18.5	6.47	−14.2	4.72	−4.7	1.61
−26.5	6.87	−22.2	5.12	−12.7	2.01
−27.5	7.25	−23.2	5.50	−13.7	2.39
−32.5	7.89	−28.2	6.14	−18.7	3.03
−37.5	8.67	−33.2	6.92	−23.7	3.81

4.3 Detection Performance Analysis

This section presents a review of techniques for the calculation of error probability over the LMS channel. Detection performance is analyzed here in wide-sense stationary conditions, i.e., using a single-state model, which is the most sensible approach for average measures. The percentile performance in multistate fading conditions can be readily obtained by proper composition of the single-state results. The selection of a specific single-state LMS channel model from the previous section depends both on the capability to fit empirical data and on the corresponding amenability to derivation of closed-form expressions or bounds for error probability or other quality measures. This is a fundamental requirement in order to be able to optimize complex system trade-offs with limited use of time-consuming numerical simulations, and also to validate the simulation results. From this point of view, the RLN model appears to be very suitable, also because it contains as particular cases Rice, Rayleigh, Suzuki, lognormal, and AWGN. Closed-form analytical expressions for the symbol error probability are obtained for uncoded transmission. In the case of coded performance, upper bounding techniques are presented, along with numerical simulation results.

4.3.1 Uncoded Transmission over LMS Channels

This section tackles the performance analysis in the case of uncoded transmission, considering nonselective single-state LMS models. From Equation 4.6, the baseband-equivalent received signal is $y = re^{j\theta}x + n$, where r and θ are the fading envelope and phase, respectively. The phase rotation is resolved both in the case of coherent detection, whereby ideal phase estimation is assumed, and in the case of noncoherent detection. In these cases, only the fading envelope plays a role in determining the receiver performance. Furthermore, under the assumption of time nonselective fading, i.e., $T \ll (\Delta t)_c$, the fading envelope operates as a multiplicative constant on the instantaneous received energy per symbol, so that the channel is conditionally AWGN. As a consequence, the average symbol error probability, P_e, can be obtained by averaging the conditional AWGN error probability, $P(e|r)$, with respect to the envelope pdf as follows:

$$P_e = \mathcal{E}_r\{P(e|r)\} = \int_0^{\infty} P(e|r)p_r(r)\mathrm{d}r \tag{4.66}$$

Although Equation 4.66 is formally simple, its evaluation may not be straightforward in some cases, especially when shadowing is considered. Over the years, a large extent of scientific literature has been devoted to this problem for given modulation schemes and fading models. Here, we characterize the error probabilities for M-ary coherent and noncoherent modulation schemes. Considering large- and small-scale fading, P_e can be reexpressed as

$$P_e = \mathcal{E}_S\left\{\mathcal{E}_{R|S}\{P(e|R,S)\}\right\} = \mathcal{E}_S\left\{P(e|S)\right\} \tag{4.67}$$

$$= \mathcal{E}_S\left\{\int_0^\infty P(e|R,S)p_{R|S}(R|S)\mathrm{d}R\right\} \tag{4.68}$$

$$= \int_0^\infty p_S(S)\left\{\int_0^\infty P(e|R,S)p_{R|S}(R|S)\mathrm{d}R\right\}\mathrm{d}S \tag{4.69}$$

We will apply this equation set to the case of the RLN model, for which $\mathcal{E}_{R|S}\{\cdot\}$ can be substituted by $\mathcal{E}_R\{\cdot\}$ because of the independence between small-scale and large-scale effects. Note that averaging over the shadowing distribution may be substituted by the evaluation of a specific percentile, $x_{\%}$, to derive $P_e(x_{\%})$, the error probability not exceeded for $x_{\%}$ of the time.

4.3.1.1 *M*-ary Coherent Detection

Considering the case of coherent detection, we assume the following expression for the conditional symbol error probability [48], which is rigorous for binary transmission ($M = 2$) and is a very good approximation for M-ary transmission under high signal-to-noise ratio (SNR) conditions:

$$P(e|r) = \frac{1}{2}\bar{N}_a\,\mathrm{erfc}(r\sqrt{\rho}) \tag{4.70}$$

where $\rho = \frac{d^2}{4N_0}$, N_0 is the one-sided noise power spectral density (PSD), $\bar{N}_a$ and d are the average number of adjacent points and the minimum Euclidean distance within the signal constellation, respectively, and erfc $(\cdot)$ is the complementary error function.

Averaging over the Rice pdf, we have the small-scale average error probability for the M-ary constellation and coherent detection:

$$P(e|S) = \frac{1}{2}\bar{N}_a\int_0^\infty \mathrm{erfc}(RS\sqrt{\rho})p_R(R)\mathrm{d}R \tag{4.71}$$

where $p_R(R)$ is given by Equation 4.23. Equation 4.71 can be integrated in closed form [49], leading to the symbol error probability in the Rice–lognormal channel:

$$P_e = \bar{N}_a\mathcal{E}_S\left\{Q(U,V) - \frac{1}{2}\left[1+\sqrt{\frac{p}{1+p}}\right]e^{-\frac{U^2+V^2}{2}}I_0(UV)\right\} \tag{4.72}$$

where

$$\left.\begin{matrix}V\\U\end{matrix}\right\} = \sqrt{K\left[\frac{1+2p}{2(1+p)} \pm \sqrt{\frac{p}{1+p}}\right]} \tag{4.73}$$

and $p = p_0/(1+K)$ with $p_0 = S^2\rho$. In the case of Rayleigh–lognormal fading, i.e., $K = 0$, Equation 4.72 simplifies to the more familiar result:

$$P_e = \frac{\bar{N}_a}{2}\left[1 - \mathcal{E}_S\left\{\sqrt{\frac{p_0}{1+p_0}}\right\}\right] \tag{4.74}$$

4.3.1.2 *M*-ary Orthogonal Noncoherent Detection

Consider the case of an M-ary orthogonal constellation, with points at Euclidean distance d. Let E_s be the average energy per symbol and E_b the average energy per bit. Assuming optimum noncoherent detection, the conditional error probability is provided by [48]

$$P(e|r) = \frac{1}{M} e^{-2r^2\rho} \sum_{i=2}^{M} (-1)^i \binom{M}{i} e^{\frac{2r^2\rho}{i}} \tag{4.75}$$

For frequency shift keying (FSK) modulation it holds $d = \sqrt{2E_s}$. In addition, this includes as a special case differential phase shift keying (DPSK) performance with differential demodulation, by simply substituting $M = 2$ and $d = 2\sqrt{E_b}$.

Considering M-ary noncoherent transmission over the RLN channel, the small-scale average probability of error is

$$P(e|S) = \frac{1}{M} \sum_{i=2}^{M} (-1)^i \binom{M}{i} \int_0^{\infty} e^{-2\frac{i-1}{i} S^2 R^2 \rho} p_R(R) dR \tag{4.76}$$

and the integral can be solved in closed form [50], so that the large-scale average probability of error in the RLN channel is

$$P_e = \frac{K+1}{M} e^{-K} \sum_{i=2}^{M} (-1)^i \binom{M}{i} \mathcal{E}_S \left\{ \frac{\exp\left[\frac{K(K+1)}{K+1+2\rho_0 \frac{i-1}{i}}\right]}{K+1+2\rho_0 \frac{i-1}{i}} \right\} \tag{4.77}$$

For the Rayleigh–lognormal channel, i.e., $K = 0$, Equation 4.77 simplifies to

$$P_e = \frac{1}{M} \sum_{i=2}^{M} (-1)^i \binom{M}{i} \mathcal{E}_S \left\{ \frac{1}{1+2\rho_0 \frac{i-1}{i}} \right\} \tag{4.78}$$

In the literature, further results for the symbol error probability over other LMS channels can be found. In [51] the interested reader can find symbol error probabilities over the Loo channel for binary noncoherent FSK demodulation and binary coherent PSK demodulation, also including the phase error induced by fading (pseudocoherent detection). For an additional extensive collection of results over small-scale and large-scale fading, refer to the book by Simon and Alouini [52]. The key idea there is that most error probability expressions contain either the Gaussian Q-function, $Q(\alpha)$, or the Marcum Q-function, $Q(\alpha; \beta)$. By providing particular formulations for the Q-functions, Simon and Alouini reformulate the integral in Equation 4.66 in a tractable form for a large number of cases. As far as multistate models are concerned, performance analysis for a DPSK system with differential demodulation over the Lutz LMS channel has been addressed in [53].

4.3.2 Coded Transmission over LMS Channels

Satellite communications have played a fundamental role in driving and stimulating the development of efficient forward error correction techniques and analytical tools for their performance evaluation, because of the great impact of the achievable coding gain and the possibility of affording somewhat high decoding complexity in the receiver. In particular, performance evaluation often entails the use of analytical bounding techniques assisted by computer simulation for the validation and analysis of performance in nonideal channel conditions. Since the publication of the pioneering work by Gallager [54], many authors have contributed to this activity by proposing different solutions to the bounding problem [55]–[76]. This great variety of different solutions and their analytical complexity hinder the possibility to report here an exhaustive overview of all approaches. The interested reader is therefore referred to the original works and to the references therein. Here we focus on convolutional and turbo coding techniques, with the intent to show the underpinning principles and some LMS applications.

Upper bounding the decoder performance usually involves the use of union bounds, which entail the evaluation of a pairwise error probability, $P(\mathbf{a} \rightarrow \tilde{\mathbf{a}})$, i.e., the probability that the decoder chooses an incorrect sequence $\tilde{\mathbf{a}}$. Pairwise error probabilities can be generally expressed as

$$P(\mathbf{a} \rightarrow \tilde{\mathbf{a}}) = \text{Prob}\{m(z|\tilde{\mathbf{a}}) > m(z|\mathbf{a}) \mid \mathbf{a}\} \tag{4.79}$$

where $m(z|\mathbf{a})$ and $m(z|\tilde{\mathbf{a}})$ are the metrics corresponding to the hypotheses $\mathbf{a}$ and $\tilde{\mathbf{a}}$, respectively, and z is the soft detector output. For maximum likelihood (ML) decoding the decision metrics are in the form $m(z|\mathbf{a}) = \ln p_{z|\mathbf{a}}(z|\mathbf{a})$, where $p_{z|\mathbf{a}}(z|\mathbf{a})$ is the conditional detector output pdf. Notably, this pdf is the common building block of all the proposed bounding techniques and incorporates the effects of the channel and the detector structure, whereas the code properties are reflected in the form of the union bounds.

In the presence of an LMS channel, or in general of any time-variant channel, the major difficulty is the proper consideration of second-order effects, such as time correlation, described in Section 4.2.1.1.2. To break this correlation, which has harmful effects on the code performance, interleaving is usually introduced, with a span that should be larger than the channel coherence time. Although residual correlation may still be present, for simplicity most of the analytical approaches assume either uncorrelated fading, i.e., ideal interleaving, or block fading correlation, i.e., constant fading on a block of coded symbols and uncorrelated fading between consecutive blocks. The task of quantifying the actual correlation effects is therefore left to numerical simulations. One of the most effective countermeasures to correlated fading is the adoption of diversity, which can take on various forms, such as satellite, space, or multipath diversity. To include this important feature, diversity of order L is assumed in the following.

Finally, in an LMS channel the effects of phase noise cannot be neglected. To this end, it is customary either to use noncoherent receivers or to perform carrier recovery, possibly with a pilot-aided scheme, and proceed to pseudocoherent detection. In the following, the detector output pdf is reported with reference to pseudocoherent binary PSK (BPSK) and noncoherent M-ary orthogonal demodulation, respectively, over the Rice and RLN channels. Before doing so, we note that when ideal coherent demodulation is assumed, a more direct approach is followed: The pairwise error probability can in fact be evaluated by direct computation of Equation 4.79. It has been shown in [56] that for the conditional pairwise error probability of ideal coherent demodulation, it holds

$$P(\mathbf{a} \rightarrow \tilde{\mathbf{a}}|\mathbf{r}) \le \exp\left\{-\frac{E_s}{4N_0} d^2(\mathbf{a}, \tilde{\mathbf{a}})\right\} \tag{4.80}$$

where

$$d(\mathbf{a}, \tilde{\mathbf{a}}) = \sqrt{\sum_{a_k \neq \tilde{a}_k} r_k^2 |a_k - \tilde{a}_k|^2} \tag{4.81}$$

is the weighted Euclidean distance between the two sequences $\mathbf{a} = (a_1, \ldots, a_k, \ldots)$ and $\tilde{\mathbf{a}} = (\tilde{a}_1, \ldots, \tilde{a}_k, \ldots)$, and r_k is the fading process sample. The upper bound on the coherent pairwise error probability is then achieved by averaging $P(\mathbf{a} \rightarrow \tilde{\mathbf{a}}|\mathbf{r})$ with respect to the fading pdf.

4.3.2.1 Pseudo-coherent BPSK Detection

Pseudo-coherent detection can be aided through the insertion of a pilot tone or pilot symbols in both the forward and return links of satellite networks [62]. The pilot and traffic streams are usually associated with orthogonal channels (e.g., a different spreading code, carrier frequency, or time slot) with a specific pilot-to-traffic power ratio,[3] ρ_p. In each diversity branch, the traffic and pilot signals are matched filtered, and sampled on the in-phase and quadrature rails. The resulting pilot samples are further filtered to reduce

[3] The pilot-to-traffic power ratio has a particular significance in code division multiple access (CDMA) based networks where the fraction of power assigned to the pilot channel is directly related to a capacity reduction.

the noise effect on the reference signal, which is used to derotate the data stream.[4] This filtering operation introduces a distortion on the reconstructed pilot. This effect is included in the analysis by introducing the coefficient ρ, which represents the correlation between the filtered pilot and the ideal pilot samples. The output of the ℓ-th diversity branch, z^{ℓ}, is therefore

$$z^{\ell} = \Re\{s\ p^*\} \tag{4.82}$$

where s and p are the complex samples of the data and pilot branches, defined respectively as $s = ud + n$ and $p = v + \eta$, where d is the BPSK data symbol, u and v are the fading process samples operating on the data and the filtered pilot branch, respectively, and n and η are the corresponding noise samples. The LMS channel model influence is therefore contained in the statistics of u and v, which are here taken to be complex Gaussian with a non-zero mean (Rice fading).

Assuming, without loss in generality, that a +1 symbol was transmitted, the conditional soft output may be written as

$$z_{+1}^{\ell} = \mathbf{V}^T \mathbf{A} \mathbf{V} \tag{4.83}$$

where

$$\mathbf{A} = \frac{1}{2}\begin{bmatrix} 0 & 1 & 0 & 0 \\ 1 & 0 & 0 & 0 \\ 0 & 0 & 0 & 1 \\ 0 & 0 & 1 & 0 \end{bmatrix} \tag{4.84}$$

and

$$\mathbf{V}^T = [u^p + n^p \quad v^p + \eta^p \quad u^q + n^q \quad v^q + \eta^q] \tag{4.85}$$

Therefore, z_{+1}^{ℓ} is a quadratic form (q.f.) in Gaussian random variables. Assuming independent identical distributions for the statistics in the L diversity channels and ideal maximum ratio combining (MRC), the resulting combined signal, z_{+1}, is simply a sum of L independent q.f.s with identical distribution. By exploiting the property that any quadrature form in Gaussian random variables can be transformed into a linear function of independent noncentral χ^2 random variables [77], it has been demonstrated [62] that for the pdf of the detector output, it holds that

$$p_{z|+1}(x) = A \exp\left(\frac{x}{2\beta}\right) \sum_{k=0}^{\infty} \sum_{i=0}^{L-1+k} \sum_{j=0}^{i} C_{ijk} D_{ijk} (-x)^{L-1+k-i} g_{ij}(x) \tag{4.86}$$

where A, C_{ijk}, D_{ijk}, and $g_{ij}(x)$ are defined as

$$A = \frac{1}{[2(\alpha+\beta)]^L} \exp\left[-\frac{1}{2}\left(\frac{\alpha\delta_1}{\alpha+\beta} + \delta_3\right)\right] \tag{4.87}$$

$$C_{ijk} = \binom{L-1+k}{i}\binom{i}{j} \frac{2^{i-j-2k}\Gamma(L+i)}{k!\Gamma(L+k)\Gamma(L+j)} \tag{4.88}$$

$$D_{ijk} = \left(\frac{\alpha\beta}{\alpha+\beta}\right)^{i+j} \left(\frac{\delta_1}{\alpha}\right)^{j} \left(\frac{\delta_3}{\beta}\right)^{k} \tag{4.89}$$

$$g_{ij}(x) = \begin{cases} Q_{L+i+j}\left(\sqrt{\frac{\beta\delta_1}{\alpha+\beta}}, \sqrt{\frac{\alpha+\beta}{\alpha\beta}x}\right) & x > 0 \\ 1 & x \leq 0 \end{cases} \tag{4.90}$$

[4]Note that the filter bandwidth must be chosen so as to strike a good balance between noise rejection and fading tracking accuracy.

and $Q_M(a, b)$ is the generalized Marcum Q-function. The parameters α, β, δ_1, and δ_3 appearing in the above equations are defined as

$$\alpha \triangleq \frac{1}{2}(\sigma_1\sigma_2 + \rho) \tag{4.91}$$

$$\beta \triangleq \frac{1}{2}(\sigma_1\sigma_2 - \rho) \tag{4.92}$$

$$\delta_1 \triangleq 2L\left[\frac{\bar{u}\sigma_2 + \bar{v}\sigma_1}{\sqrt{2\sigma_1\sigma_2(\sigma_1\sigma_2 + \rho)}}\right]^2 \tag{4.93}$$

$$\delta_3 \triangleq 2L\left[\frac{\bar{u}\sigma_2 - \bar{v}\sigma_1}{\sqrt{2\sigma_1\sigma_2(\sigma_1\sigma_2 - \rho)}}\right]^2 \tag{4.94}$$

where

$\bar{u} = \sqrt{\frac{K}{2(K+1)}}$ is the mean value of the normalized Rice fading distribution.
$\bar{v} = \sqrt{\rho_p}$ is the mean value of the filtered pilot component.
σ_1^2 is the total variance (fading plus noise) in the data branch.
σ_2^2 is the total variance (fading plus noise) in the pilot branch.

Finally, due to evident symmetry, it holds that

$$p_{z|-1}(x) = p_{z|+1}(-x) \tag{4.95}$$

Equation 4.86 can be extended to the RLN case by introducing the shadowing variate.

4.3.2.2 Non-coherent *M*-ary Detection

In M-ary orthogonal detection, all of the signal space dimensions must be observed in all L diversity branches. Therefore, a noncoherent receiver consists of a bank of ML parallel square envelope detectors (SEDs) followed by the combining and soft decision sections [61]. The output of the generic n-th detector (i.e., the subsystem formed by the collection of L SEDs) is given as

$$z_n = \sum_{\ell=1}^{L} \left|z_n^\ell\right|^2 \tag{4.96}$$

where $z_n^\ell = z_n^{\ell,p} + jz_n^{\ell,q}$ is the output of the ℓ-th SED of the n-th detector. In RLN fading, and conditioning on shadowing, the detector output z can be shown [61] to have:

- For the correct detector (i.e., the signal is present), a noncentral χ^2 distribution with $2L$ degrees of freedom and a noncentrality parameter λ given by

$$\lambda = \frac{\gamma LK^2}{K+1} \tag{4.97}$$

  where γ is the symbol energy-to-noise PSD ratio
- For the $M-1$ incorrect detectors (i.e., the signal is absent), a central χ^2 distribution with $2L$ degrees of freedom

By using the hypergeometric function ${}_0F_1$, the central and noncentral χ^2 pdfs can be expressed in a unified form [61]. Inserting the average on shadowing, the unconditional unified detector output pdf is given by

$$p_z(x) = \mathcal{E}_S\left\{\frac{x^{L-1}}{(L-1)!}\left(\frac{K+1}{K+1+\gamma S^2}\right)^L \right.$$
$$\left. \times \exp\left(-\frac{x(K+1)+\gamma LKS^2}{K+1+\gamma S^2}\right) {}_0F_1\left(L; \frac{x\gamma LK(K+1)S^2}{(K+1+\gamma S^2)^2}\right)\right\} \tag{4.98}$$

Notably, Equation 4.98 contains the following as particular cases:

Rice fading channel: $p_S(S) \to \delta(S-1)$

$$p_z(x) = \frac{x^{L-1}}{(L-1)!}\left(\frac{K+1}{K+1+\gamma}\right)^L \exp\left(-\frac{x(K+1)+\gamma LK}{K+1+\gamma}\right) {}_0F_1\left(L; \frac{x\gamma LK(K+1)}{(K+1+\gamma)^2}\right) \quad (4.99)$$

AWGN channel: $p_S(S) \to \delta(S-1)$ and $K \to \infty$

$$p_z(x) = \frac{x^{L-1}}{(L-1)!} e^{-(x+\gamma L)} {}_0F_1(L; x\gamma L) \quad (4.100)$$

Rayleigh fading channel: $p_S(S) \to \delta(S-1)$ and $K = 0$

$$p_z(x) = \frac{x^{L-1}}{(L-1)!} \frac{e^{-x/(1+\gamma)}}{(1+\gamma)^L} \quad (4.101)$$

Incorrect detector: $\gamma = 0$,

$$p_{\tilde{z}}(x) = \frac{x^{L-1}}{(L-1)!} e^{-x} \quad (4.102)$$

These expressions do not depend on the particular decoding strategy, but only on the employed detector, and can be used to evaluate upper bounds for different coding schemes.

Below, we will report the analysis of convolutional and turbo coding schemes, which are widely employed in current and planned satellite networks. In addition, low-density parity check (LDPC) coding schemes are also of interest for their powerful correcting performance, and are being considered in the design of new satellite systems. The interested reader is referred to [72–74] for further details.

4.3.2.3 Convolutional Code Performance Analysis

The detector output pdfs from the preceding section can be used to evaluate the performance of a convolutional coded transmission scheme, employing either a pseudo- or noncoherent detector. The soft decisions at the detector output are usually deinterleaved and fed into a Viterbi decoder. The union bound on the average error probability is given by

$$P_e \le \sum_{\mathbf{a}} \sum_{\tilde{\mathbf{a}} \ne \mathbf{a}} P(\mathbf{a}) P(\mathbf{a} \to \tilde{\mathbf{a}}) \quad (4.103)$$

where $P(\mathbf{a})$ is the *a priori* probability of sequence **a**. When Equation 4.103 is applied to linear convolutional codes, it can be simplified due to the uniform error property. In this case, the bound can be written in terms of the weight-enumerating function (WEF), which can be efficiently evaluated by standard transfer function techniques [48]:

$$P_e \le \sum_{d \ge d_{min}} A_d P_d(\mathbf{a} \to \tilde{\mathbf{a}}) \quad (4.104)$$

where d is the Hamming distance between sequences (defined as the number of corresponding different symbols), d_{min} is the code minimum distance, A_d enumerates the error events at distance d, and $P_d(\mathbf{a} \to \tilde{\mathbf{a}})$ is the pairwise error probability pertaining to any pair of sequences at distance d. Similarly, the bit error probability is upper bounded as

$$P_b \le \sum_{d \ge d_{min}} B_d P_d(\mathbf{a} \to \tilde{\mathbf{a}}) \quad (4.105)$$

where B_d describes the contribution of the error events at Hamming distance d to the bit error probability.

The computation of the pairwise error probability can be complex in some cases, and it can be performed in correlated fading conditions. Under the assumption of ideal interleaving (i.e., memoryless channel), the pairwise error probability itself can be simplified through upper bounds in various forms, e.g., using the Gallager bound [54], the Bhattacharyya bound,[5] or the Chernoff bound [55]. To be able to use transfer function techniques, the result must be expressed in the form

$$P_d\left(\mathbf{a} \to \tilde{\mathbf{a}}\right) \leq \prod_{\tilde{a}_k \neq a_k} Z = Z^d \tag{4.106}$$

where Z is the *code performance parameter*, which depends on LMS channel conditions and detector structure. Here, we adopt the Bhattacharyya bound for the pseudocoherent case, for which

$$Z = \int_{-\infty}^{\infty} \sqrt{p_{z|+1}(x)p_{z|-1}(x)}dx \tag{4.107}$$

where $p_{z|\pm1}(x)$ are the detector output pdfs reported in Equation 4.86 and Equation 4.95.

For the noncoherent case, a tighter upper bound is obtained by using the Chernoff bound [48], for which the code performance parameter is expressed as a function of an optimization parameter, $\rho \in [0, 1]$, as

$$\begin{aligned} Z(\rho) = {} & \frac{M}{2}\int_0^{\infty} e^{[\rho x]}\, p_{\tilde{z}}(x)\,[F_{\tilde{z}}(x)]^{\frac{M}{2}-1}\,\mathrm{d}x \\ & \times \int_0^{\infty} e^{[-\rho x]}\left\{p_z(x)\,[F_{\tilde{z}}(x)]^{\frac{M}{2}-1} + \left(\frac{M}{2}-1\right)p_{\tilde{z}}(x)F_z(x)\,[F_{\tilde{z}}(x)]^{\frac{M}{2}-2}\right\}\mathrm{d}x \end{aligned} \tag{4.108}$$

where $p_z(x)$ and $p_{\tilde{z}}(x)$ are given for the Rice channel in Equation 4.98 and Equation 4.102, respectively, and $F_z(x)$ and $F_{\tilde{z}}(x)$ are the corresponding cdfs.

In Figure 4.8 to Figure 4.11 the bit error probability union bounds for the pseudo- and noncoherent cases are reported. To corroborate the analytical evaluation and show the effects of nonideal conditions, simulation results are also reported. The analysis is carried out on a LMS channel characterized by Rice fading and diversity order L equal to either 1 or 2. The upper bounds have been evaluated under the assumptions of infinite decoding depth in the Viterbi algorithm, infinite quantization, and memoryless channel. The reference system is a CDMA satellite network. This reference case is of interest for second-and third-generation satellite networks, e.g., for Globalstar [78] and for the satellite wideband CDMA (SW-CDMA) air interface, designed for the satellite component of UMTS (Universal Mobile Telecommunications System) [79].

In the pseudo-coherent CDMA case, the pilot tone is multiplexed in the code domain, i.e., via a reserved spreading code, and is therefore identified as code division multiplexed pilot (CDMP) [62]. In the analysis, the pilot-to-traffic power ratio, ρ_p, has been set to −10 dB, and the following system parameters have been adopted: convolutional coding with rate 1/4, constraint length of 9, generator polynomials 235, 275, 313, and 357 (octal), bit reversal interleaving (800 rows × 64 columns), bit rate R_b = 9.6 kbit/sec, chip rate R_c = 1.2288 kchip/sec, carrier frequency f_0 = 1.618 GHz, and Rice fading channel. Two mobile terminal speeds have been simulated: v = 5 mph, identified as *slow fading*, and v = 80 mph, identified as *fast fading*. In Figure 4.8, we report the results for the slow fading channel, including coded and uncoded cases, for various values of the Rice factor, K, with and without diversity. The interleaving depth (800 × 64) is large enough to break fading correlation, which is the condition under which the analysis holds. The agreement between analysis and simulations is very satisfactory (for the uncoded case the two curves are indistinguishable). It is worthwhile to note the very small performance loss when K drops from 10 to 3 dB.

[5]The Bhattacharyya bound is a particular case of the Gallager bound.

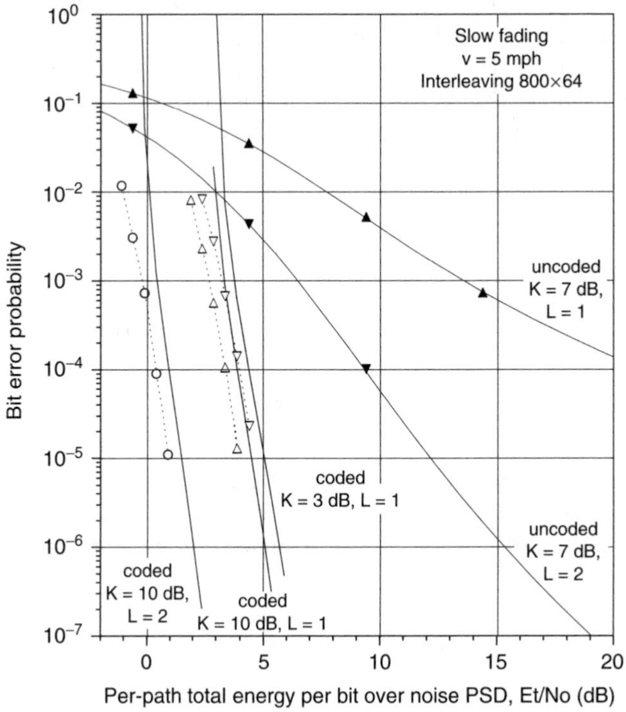

FIGURE 4.8 Error probability performance of the CDMP technique in slow Rice fading (interleaver, 800 × 64; $v = 5$ mph; $\rho_p = -10$ dB). Lines with markers are from simulations. Lines without markers are upper bounds.

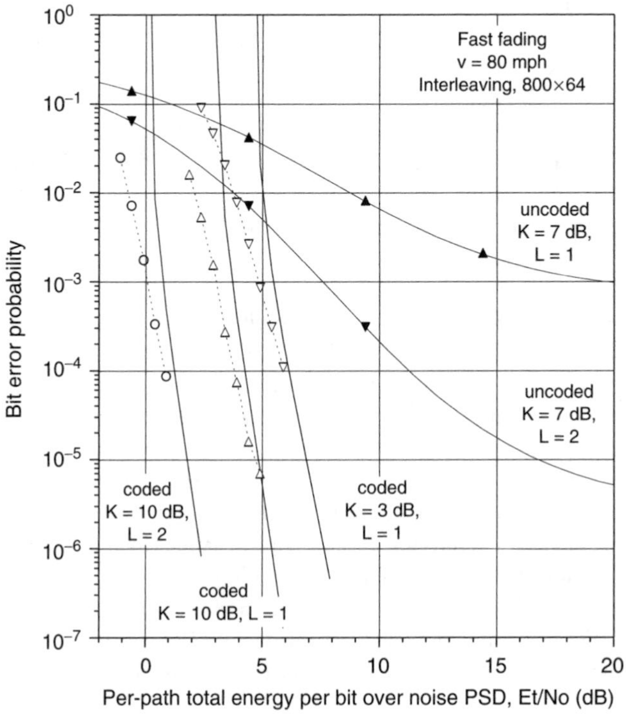

FIGURE 4.9 Error probability performance of the CDMP technique in fast Rice fading (interleaver, 800 × 64; $v = 80$ mph; $\rho_p = -10$ dB). Lines with markers are from simulations. Lines without markers are upper bounds.

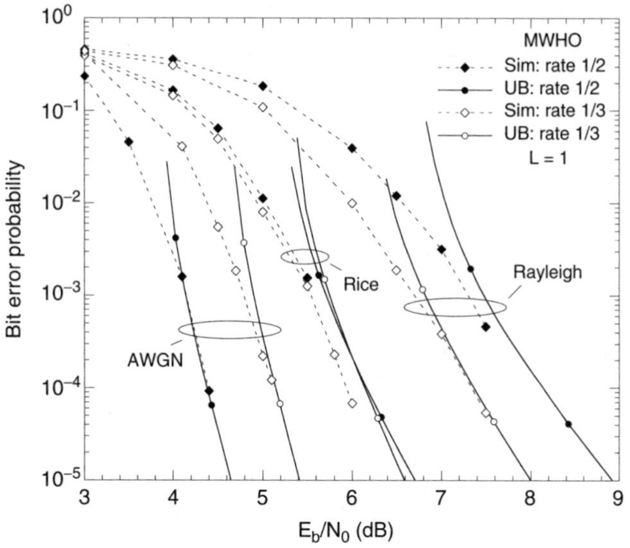

FIGURE 4.10 Upper bounds and simulations for MWHO on AWGN, Rice ($K = 3$), and Rayleigh channels (rate = 1/2 and 1/3, $L = 1$).

This is due to the fact that pilot-aided fading estimation is extremely good under these conditions. Similarly, Figure 4.9 contains the analysis and simulation for the fast fading conditions. In this case, the performance loss when K drops from 10 to 3 dB is larger, due to the difficulty in estimating the rapid phase fluctuations with a small direct component. The agreement between analysis and simulations confirms the correctness of the described analysis in accounting for the imperfect channel estimation due to pilot filtering.

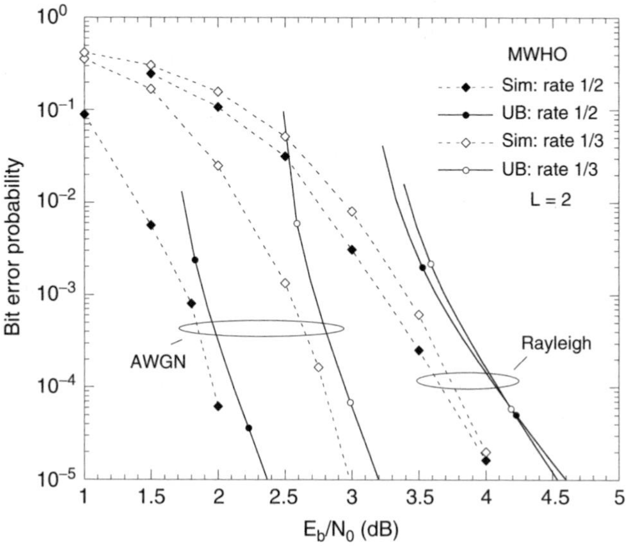

FIGURE 4.11 Upper bounds and simulations for MWHO on AWGN and Rayleigh channels (rate = 1/2 and 1/3, $L = 2$).

TABLE 4.5 Pilot-Aided Coherent vs. Non-coherent (NC) Performance in Rice Fading

L	R_b (kb/s)	v (mph)	K (dB)	E_b/N_0 NC MWHO (dB)	E_b/N_0 CDMP (dB)	Gain (dB)
1	9600	5	10	8.9	8.0	0.9
1	9600	80	10	5.2	4.2	1.0
2	9600	5	10	7.5	5.7	1.8
2	9600	80	10	5.5	4.3	1.2

For the noncoherent detection case, the performance of M-ary Walsh–Hadamard orthogonal (MWHO) convolutionally coded direct-sequence CDMA is reported [61], [78]. The number of orthogonal waveforms, M, has been set to 64, and the following system parameters have been adopted: convolutional coding with rate $r = 1/2$ and $r = 1/3$, constraint length of 9, and generator polynomials 753 and 561 (octal) for rate 1/2 and 711, 663, and 557 (octal) for rate 1/3. Figure 4.10 shows upper bounds and simulation results for MWHO in AWGN, Rice, and Rayleigh channels without diversity, $L = 1$, for $r = 1/2$ and $r = 1/3$. First, we note that the agreement between simulations and upper bounds confirms that all the realistic assumptions for the simulation (finite truncation depth, finite interleaving, etc.) do not bring noticeable performance degradation. However, the performance degrades significantly when smaller fading bandwidths are considered [61]. For an error probability of $P_b = 10^{-4}$ in the Rayleigh channel, the gain of coded MWHO with respect to uncoded MWHO amounts to 27.8 dB. It can be noted that in the AWGN channel, rate 1/2 is better than rate 1/3 (due to the higher non-coherent combining loss for rate 1/3), while the opposite is true for the Rayleigh channel (due to the higher intrinsic time diversity achieved by bit-interleaving a three-bit symbol, which more than overcomes the non-coherent combining loss). Note also that the performance of the two coding rates is about the same on the Rice channel, with $K = 3$.

In Figure 4.11 the case of two satellite diversity, $L = 2$, is reported for the AWGN and Rayleigh channels. In general, bounds are somewhat looser when diversity is considered. Note that in this case performance of the two coding rates is about the same in the Rayleigh channel, while the advantage of rate 1/2 in the AWGN channel is increased. For rate 1/2, the gain in single-link E_b/N_0 in going from $L = 1$ to $L = 2$ amounts to about 4.4 dB in the Rayleigh channel at $P_b = 10^{-4}$, while it is 3.6 dB for rate 1/3.

Finally, in Table 4.5 the CDMP pseudocoherent scheme is compared with the rate 1/2 noncoherent MWHO scheme adopted in [78]. The comparison is based on a frame error rate specification of 10^{-2}. Notably, the largest CDMP gain, amounting to 1.8 dB, is obtained for dual-satellite diversity with slow-moving users, which are likely to represent the largest portion of the user population. Moreover, the slow-moving users will likely possess a low-gain handheld phone. Consequently, the lower E_b/N_0 requirement will be instrumental in ensuring a longer battery life or higher link margin.

4.3.2.4 Turbo Code Performance Analysis

Performance evaluation of turbo coding schemes can resort to similar bounding techniques, essentially based on the evaluation of the WEF and pairwise error probability. However, due to the presence of the internal turbo interleaver, the evaluation of WEF is not straightforward. A solution is to consider an abstract interleaver, namely, the *uniform interleaver*,[6] i.e., the ensemble of all possible interleavers [59]. Under this assumption, the WEF can be evaluated [80] and the error probability is computed starting from Equation 4.104. Unfortunately, the union upper bound for turbo codes diverges for E_b/N_0 values below

[6]The uniform interleaver refers to the internal turbo interleaver of the turbo coder and not to the channel interleaver, which is still considered ideal; i.e., it ensures a perfect decorrelation of the channel.

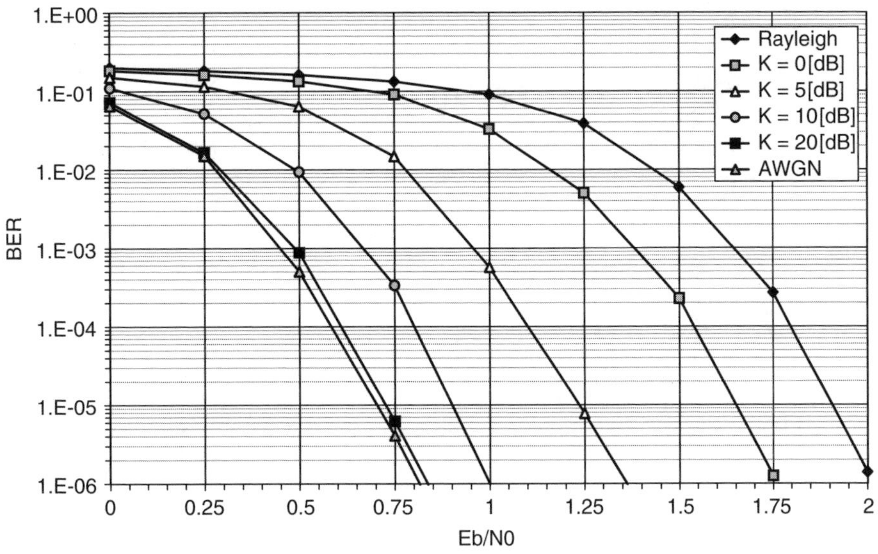

FIGURE 4.12 Turbo code bit error rate (BER) over uncorrelated Rice channel (rate = 1/3).

the cutoff rate threshold. Various approaches have been proposed to overcome this problem, based on variants of the Gallager bound and on the tangential sphere bound, see [63],[65],[66],[68],[75], [76]. For more details, the interested reader is referred to the original papers and the references therein. Whatever the form of the bounds, the detector output pdfs reported in Equation 4.86 and Equation 4.98 are still instrumental for the computation of the pairwise error probability.

Bounding techniques are hardly applicable when correlated fading is considered, and numerical simulation becomes the predominant instrument. Here, we report on simulation results for the cases of uncorrelated and correlated Rice fading. A rate 1/3 turbo code is considered, with generator polynomials for the constituent codes 013 and 015 (octal), and an internal interleaver span equal to 3000 bits. Log-MAP iterative decoding with six decoding iterations is used, assuming ideal channel estimation. In Figure 4.12, bit error rate results over uncorrelated Rice channel are reported for six different Rice factors, including Rayleigh and AWGN channels. Note the very small performance loss in Rayleigh fading, testifying to the powerful error correcting capabilities of turbo codes in memoryless fading channels. In Figure 4.13, the bit error rate is reported in the case of a correlated Rice channel, with Rice factor $K = 5$ dB, as a function of E_b/N_0, for different values of the normalized maximum Doppler shift, $\mathrm{ND} = f_{d,m}T$. As correlation increases, performance degrades significantly because the internal interleaver becomes unable to break the channel memory. Again, diversity is the most effective countermeasure in this case.

4.4 Conclusions

The problem of modeling the propagation channel and evaluating the performance of digital transmission schemes for LMS applications will continue to spur the interest of a number of researchers, in a strive for the achievement of excellent modeling accuracy and near-Shannon bound performance. In this chapter we have toured the state of the art in this exciting field of knowledge in an effort to let the reader grasp the abundance of intellectual endeavors that have been produced within it. Far from claiming exhaustiveness, we apologize for those contributions that have escaped our analysis. Possible avenues left for future exploration include the exploitation of new frequency bands, the consideration of ultra-wideband communications, and the derivation of new upper bounds for LMS coded transmission.

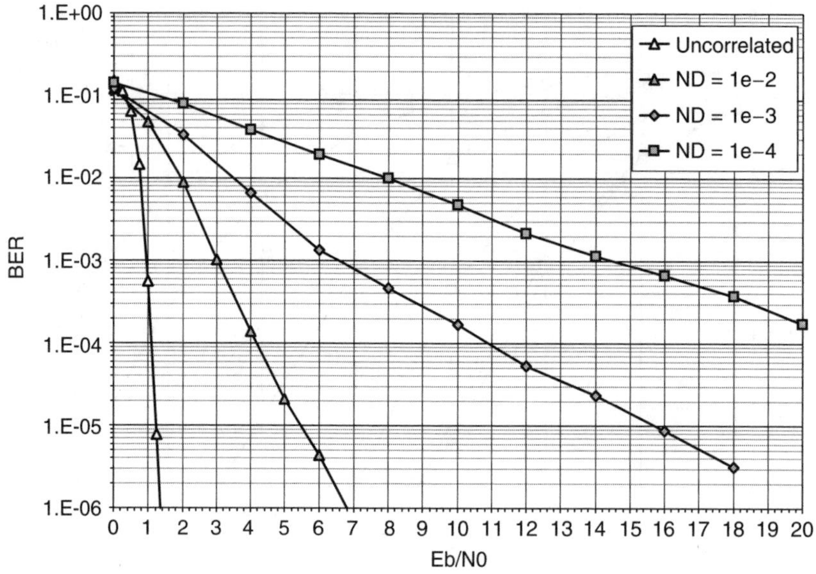

FIGURE 4.13 Turbo code bit error rate (BER) over correlated Rice channel (rate = 1/3, $K = 5$ dB).

References

[1] E.N. Gilbert, "Energy Reception for Mobile Radio," *Bell System Technol. J.*, 44, 1779–1803, 1965.

[2] R.H. Clarke, "A Statistical Theory of Mobile-Radio Reception," *Bell System Technol. J.*, 47, 957–1000, 1968.

[3] S.O. Rice, "Mathematical Analysis of a Sine Wave Plus Random Noise," *Bell System Tech. J.*, 27, 109–157, 1948.

[4] M. Nakagami, "The m-Distribution: A General Formula of Intensity Distribution of Rapid Fading," in *Statistical Methods in Radio Wave Propagation*, W.G. Hoffman, Ed., Pergamon, Oxford, 1960.

[5] K.A. Norton, L.E. Vogler, W.V. Mansfield, and P.J. Short, "The Probability Distribution of the Amplitude of a Constant Vector Plus a Rayleigh Distributed Vector," *Proc. IRE*, 43, 1354–1361, 1955.

[6] V.Y. Iskam and V.A. Shaptsev, "Properties of the Nakagami-Rice Distribution as a Model for Signal Fading," *Telecommun. Radio Eng.*, 2, 129–132, 1985.

[7] A.L. Martin and T. Vu-Dinh, "A Statistical Characterization of Point-to-Point Microwave Links Using Biased Rayleigh Distributions," *IEEE Trans. Antennas Propagation*, 45, 806–821, 1997.

[8] H. Suzuki, "A Statistical Model for Urban Radio Propagation," *IEEE Trans. Commun.*, 25, 673–680, 1977.

[9] C. Loo, "A Statistical Model for a Land Mobile Satellite Link," *IEEE Trans. Vehicular Technol.*, 34, 122–127, 1985.

[10] G.E. Corazza and F. Vatalaro, "A Statistical Model for Land Mobile Satellite Channels and Its Application to Nongeostationary Orbit Systems," *IEEE Trans. Vehicular Technol.*, 43, 738–742, 1994.

[11] G.E. Corazza, A. Jahn, E. Lutz, and F. Vatalaro, "Channel Characterization for Mobile Satellite Communications," in *Procedure of the First European Workshop on Mobile/Personal Satcoms, EMPS*, 1994, pp. 225–262.

[12] S.H. Hwang, K.J. Kim, J.Y. Ahn, and K.C. Whang, "A Channel Model for Nongeostationary Orbiting Satellite System," in *IEEE Vehicular Technology Conference*, May 4–7, 1997, pp. 41–45.

[13] F. Vatalaro, "Generalized Rice-Lognormal Channel Model for Wireless Communications," *IEEE Electron. Lett.*, 31, 1899–1900, 1995.

[14] Y. Xie and Y. Fang, "A General Statistical Channel Model for Mobile Satellite Systems," *IEEE Trans. Vehicular Technol.*, 49, 744–752, 2000.
[15] M. Pätzold, U. Killat, F. Laue, and Y. Li, "On the Statistical Properties of Deterministic Simulation Models for Mobile Fading Channels," *IEEE Trans. Vehicular Technol.*, 47, 254–269, 1998.
[16] ESA, "Robust Modulation and Coding for Personal Communication Satellite Systems: Phase I," Final Report, April 1999.
[17] A. Abdi, W.C. Lau, M.S. Alouini, and M. Kaveh, "A New Simple Model for Land Mobile Satellite Channels: First- and Second-Order Statistics," *IEEE Trans. Wireless Commun.*, 2, 519–528, 2003.
[18] J.F. Ossanna, Jr., "A Model for Mobile Radio Fading Due to Building Reflections: Theoretical and Experimental Fading Waveform Power Spectra," *Bell System Technol. J.*, 43, 2935–2971, 1964.
[19] F. Hansen and F.I. Meno, "Mobile Fading-Rayleigh and Lognormal Superimposed," *IEEE Trans. Vehicular Technol.*, 26, 332–335, 1977.
[20] M. Sforza and S. Buonomo, "Characterization of the Propagation Channel for Nongeostationary LMS Systems at L- and S-Bands: Narrow Band Experimental Data and Channel Modelling," in *Procedure XVII NAPEX Conference*, Pasadena, CA, June 14–15, 1993.
[21] M. Pätzold, U. Killat, and F. Laue, "An Extended Suzuki Model for Land Mobile Satellite Channels and Its Statistical Properties," *IEEE Trans. Vehicular Technol.*, 47, 617–630, 1998.
[22] M.S. Karaliopoulos and F.N. Pavlidou, "Modelling the Land Mobile Satellite Channel: A Review," *Electron. Commun. Eng. J.*, 11, 235–248, 1999.
[23] T.T. Tjhung and C.C. Chai, "Bit Error Rate Performance of $\pi/4$ DQPSK Nakagami-Lognormal Channels," *IEEE Electron. Lett.*, 34, 625–627, 1998.
[24] T.T. Tjhung and C.C. Chai, "Fade Statistics in Nakagami-Lognormal Channels," *IEEE Trans. Commun.*, 47, 1769–1772, 1999.
[25] W.C. Jakes, *Microwave Mobile Communications*, IEEE Press, New York, 1994.
[26] T. Aulin, "A Modified Model for the Fading Signal at a Mobile Radio Channel," *IEEE Trans. Vehicular Technol.*, 28, 182–203, 1979.
[27] J.D. Parsons, *The Mobile Radio Propagation Channel*, John Wiley & Sons Ltd., New York, 2000.
[28] F. Vatalaro and A. Forcella, "Doppler Spectrum in Mobile-to-Mobile Communications in the Presence of Three-Dimensional Multipath Scattering," *IEEE Trans. Vehicular Technol.*, 46, 213–219, 1997.
[29] S.O. Rice, "Mathematical Analysis of Random Noise," *Bell System Tech. J.*, 23, 282–332, 1944.
[30] S.O. Rice, "Mathematical Analysis of Random Noise," *Bell System Tech. J.*, 24, 46–156, 1945.
[31] Y. Hase, W.J. Vogel, and J. Goldhirsh, "Fade-Durations Derived from Land-Mobile-Satellite Measurements in Australia," *IEEE Trans. Commun.*, 39, 664–668, 1991.
[32] P. Taaghol and R. Tafazolli, "Correlation Model for Shadow Fading in Land-Mobile Satellite Systems," *IEEE Electron. Lett.*, 33, 1287–1289, 1997.
[33] E. Lutz, D. Cygan, M. Dippold, F. Dolainsky, and W. Papke, "The Land Mobile Satellite Communication Channel: Recording, Statistics, and Channel Model," *IEEE Trans. Vehicular Technol.*, 40, 375–386, 1991.
[34] R.M. Barts and L. Stutzman, "Modeling and Simulation of Mobile Satellite Propagation," *IEEE Trans. Antennas Propagation*, 10, 1209–1218, 1992.
[35] B. Vucetic and J. Du, "Channel Modeling and Simulation in Satellite Mobile Communication Systems," *IEEE J. Selected Areas Commun.*, 10, 1209–1218, 1992.
[36] M. Rice and B. Humphreys, "Statistical Models for the ACTS K-Band Land Mobile Satellite Channel," in *Procedure IEEE Vehicular Technology Confrence*, Phoenix, AZ, 1997, pp. 46–50.
[37] Y. Karasawa, K. Kimura, and K. Minamisono, "Analysis of Availability Improvement in LMSS by Means of Satellite Diversity Based on Three-State Propagation Channel Model," *IEEE Trans. Vehicular Technol.*, 46, 1047–1056, 1997.
[38] F. Pérez-Fontan, M. Vásquez-Castro, C.E. Cabado, J.P. García, and E. Kubista, "Statistical Modeling of the LMS Channel," *IEEE Trans. Vehicular Technol.*, 50, 1549–1567, 2001.

[39] H. Wakana, "Propagation Model for Simulating Shadowing and Multipath Fading in Land Mobile Satellite Channel," *IEEE Electron. Lett.*, 33, 1925–1926, 1997.

[40] S.R. Saunders and B.G. Evans, "Physical Model of Shadowing Probability for Land Mobile Satellite Propagation," *IEEE Electron. Lett.*, 32, 1548–1549, 1996.

[41] E. Lutz, "A Markov Model for Correlated Land Mobile Satellite Channels," *Int. J. Satellite Communi.*, 14, 333–339, 1996.

[42] R. Akturan and W.J. Vogel, "Photogrammetic Mobile Satellite Service Prediction," *IEEE Electron. Lett.*, 31, 165–166, 1995.

[43] F. Pérez-Fontan, M. Vásquez-Castro, S. Buonomo, J.P. Poiares-Baptista, and B. Arbesser-Rastburg, "S-Band LMS Propagation Channel Behaviour for Different Environments, Degrees of Shadowing and Elevation Angles," *IEEE Trans. Broadcasting*, 44, 40–76, 1998.

[44] A. Jahn, H. Bischl, and G. Heiß, "Channel Characterization for Spread Spectrum Satellite Communications," *IEEE 4th Int. Symp. Spread Spectrum Tech. Appl. Proc*, 3, 1221–1226, 1996.

[45] M.A.N. Parks, S.R. Saunders, and B.G. Evans, "A Wideband Channel Model Applicable to Mobile Satellite Systems at L-Band and S-Band," in *IEE Colloqium on Propagation Aspects of Future Mobile Systems*, October 1996, pp. 12/1–12/6.

[46] SATIN, "Simulation Results and Evaluation," *Deliverable, no. 7*, 2003.

[47] ETSI TR 101 112 (UMTS 30.03), "Universal Mobile Telecommunications System (UMTS); Selection Procedures for the Choice of Radio Transmission Technologies of the UMTS," Version 3.2.0, April 1998.

[48] A.J. Viterbi and J.K. Omura, *Principles of Digital Communication and Coding*, McGraw-Hill, New York, 1979.

[49] W.C. Lindsey, "Error Probabilities for Rician Fading Multichannel Reception of Binary and N-ary Signals," *IEEE Trans. Commun.*, 10, 339–350, 1964.

[50] I.S. Gradshteyn and I.M. Ryzhik, *Table of Integrals, Series, and Products*, Academic Press, London, 1980.

[51] C. Loo, "Digital transmission through a Land Mobile Satellite Channel," *IEEE Trans. Commun.*, 38, 693–697, 1990.

[52] M.K. Simon and M.S. Alouini, *Digital Communication over Fading Channels: A Unified Approach to Performance Analysis*, John Wiley & Sons, New York, 2000.

[53] D. Cygan, "Analytical Evaluation of Average Bit Error Rate for the Land Mobile Satellite Channel," *IEEE Trans. Commun.*, 7, 99–102, 1989.

[54] R.G. Gallager, "A Simple Derivation of the Coding Theorem and Some Applications," *IEEE Trans. Inf. Theory*, 11, 3–18, 1965.

[55] D. Divsalar and M. Simon, "Trellis Coded Modulation for 4800–9600 bits/s Transmission over a Fading Mobile Satellite Channel," *IEEE J. Selected Areas Commun.*, 5, 162–175, 1987.

[56] D. Divsalar and M.K. Simon, "The Design of Trellis Coded MPSK for Fading Channels: Performance Criteria," *IEEE Trans. Commun.*, 36, 1004–1012, 1988.

[57] G. Poltyrev, "Bounds on the Decoding Error Probability of Binary Linear Codes via Their Spectra," *IEEE Trans. Inf. Theory*, 40, 1284–1292, 1994.

[58] L.C. Perez, J. Seghers, and D.J. Costello, Jr., "A Distance Spectrum Interpretation of Turbo Codes," *IEEE Trans. Inf. Theory*, 42, 1698–1709, 1996.

[59] S. Benedetto and G. Montorsi, "Unveiling Turbo Codes: Some Results on Parallel Concatenated Coding Schemes," *IEEE Trans. Inf. Theory*, 42, 409–428, 1996.

[60] T.M. Duman and M. Salehi, "New Performance Bounds for Turbo Codes," *IEEE Trans. Commun.*, 46, 717–723, 1998.

[61] G.E. Corazza and R. De Gaudenzi, "Analysis of Coded Noncoherent Transmission in DS-CDMA Mobile Satellite Communications," *IEEE Trans. Commun.*, 46, 1525–1535, 1998.

[62] G.E. Corazza and R. De Gaudenzi, "Pilot-Aided Coherent Uplink for Mobile Satellite CDMA Networks," *IEEE Trans. Commun.*, 47, 773–783, 1999.

[63] T.M. Duman and M. Salehi, "The Union Bound for Turbo-Coded Modulation Systems over Fading Channels," *IEEE Trans. Commun.*, 47, 1495–1502, 1999.

[64] T.M. Duman and M. Salehi, "Performance Bounds for Turbo-Coded Modulation Systems," *IEEE Trans. Commun.*, 47, 511–521, 1999.

[65] I. Sason and S. Shamai, "Improved Upper Bounds on the ML Decoding Error Probability of Parallel and Serial Concatenated Turbo Codes via Their Ensemble Distance Spectrum," *IEEE Trans. Inf. Theory*, 46, 24–47, 2000.

[66] I. Sason and S. Shamai, "Improved Upper Bounds on the ML Performance of Turbo Codes for Interleaved Rician Fading Channels, with Comparison to Iterative Decoding," in *Procedure IEEE International Conference on Communications*, vol. 2, 2000, pp. 591–596.

[67] I. Sason and S. Shamai, "On Gallager Bounding Technique with Applications to Turbo-Like Codes over Fading Channels," in *The 21st IEEE Convention of the Electrical and Electronic Engineers in Israel*, 2000, pp. 431–434.

[68] F. Babich, G. Montorsi, and F. Vatta, "Improved Union Bounds on Turbo Codes Performance," *IEE Proc. Commun.*, 147, 337–344, 2000.

[69] I. Sason and S. Shamai, "On Improved Bounds on the Decoding Error Probability of Block Codes over Interleaved Fading Channels, with Applications to Turbo-Like Codes," *IEEE Trans. Inf. Theory*, 47, 2275–2299, 2001.

[70] S. Shamai and I. Sason, "Variations on the Gallager Bounds, Connections, and Applications," *IEEE Trans. Inf. Theory*, 48, 3029–3051, 2002.

[71] Lizhong Zheng and D.N.C. Tse, "Diversity and Multiplexing: A Fundamental Trade-Off in Multiple-Antenna Channels," *IEEE Trans. Inf. Theory*, 49, 1073–1096, 2003.

[72] I. Sason and S. Shamai, "Improved Upper Bounds on the Ensemble Performance of ML Decoded Low Density Parity Check Codes," *IEEE Commun. Lett.*, 4, 89–91, 2000.

[73] Jilei Hou, P.H. Siegel, and L.B. Milstein, "Performance Analysis and Code Optimization of Low Density Parity-Check Codes on Rayleigh Fading Channels," *IEEE J. Selected Areas Commun.*, 19, 924–934, 2001.

[74] Guosen Yue, Xiaodong Wang, and K.R. Narayanan, "Design of Low Density Parity Check Codes for Turbo Multiuser Detection," *Proc. IEEE Int. Conf. Commun.*, 4, 2703–2707, 2003.

[75] A.M. Viterbi and A.J. Viterbi, "Improved Union Bound on Linear Codes for the Input-Binary AWGN Channel, with Applications to Turbo Codes," in Information Theory, 1998. Proceedings. 1998 IEEE International Symposium on, 1998, p. 29.

[76] F. Vatta, G. Montorsi, and F. Babich, "Achievable Performance of Turbo Codes over the Correlated Rician Channel," *IEEE Trans. Commun.*, 51, 1–4, 2003.

[77] S. Kotz, N.L. Johnson, and D.W. Boyd, "Series Representations of Distributions of Quadratic Forms in Normal Variables. I. Central Case," *Ann. Math. Stats.*, 38, 823–837, 1967.

[78] P. Monte and S. Carter, "The Globalstar Air Interface," *Proc. AIAA Satellite Commun. Conf.* 4, 1614–1621, 1994.

[79] ITU-R M.1457, "Detailed Specifications of the Radio Interfaces of International Mobile Telecommunications—2000 (IMT-2000)," International Telecommunications Union, April 2001.

[80] B. Vucetic and J. Yuan, *Turbo Codes: Principles and Applications*, Kluwer Academic Publishers, Dordrecht, Netherlands, 2000.

5

Mobile Velocity Estimation for Wireless Communications

5.1 Introduction 5-2
Importance of Velocity Estimation • Existing Velocity Estimators • Structure of the Chapter

5.2 Received Signal Model and Statistics 5-5
Received Signal Model • Multipath Component Model • The Scattering Distribution • Statistics of the Multipath Fading

5.3 Principles of Mobile Velocity Estimation 5-9
Examples of Derivations of the Velocity • Examples of Velocity Estimators

5.4 Performance Analysis of Velocity Estimators 5-12
Effect of Shadowing • Effect of AWGN and Nonisotropic Scattering

5.5 Performance Analysis Using Simulations 5-17
Simulations of the Received Signal • Simulation Results

5.6 Rice Factor Estimation 5-21
Existing Methods • Envelope-Based Estimators • Simulation Results

5.7 Application on Handover Performance 5-25
Handover Decision Algorithms • Effect of an Error in Velocity Estimation on the System's Quality of Service

5.8 Conclusions and Perspectives 5-28

Bouchra Senadji
Queensland University of Technology

Ghazem Azemi
Queensland University of Technology

Boualem Boashash
Queensland University of Technology

Abstract

In this chapter we explain the principles of velocity estimation and review both analytically and using simulations the properties of existing estimators. We then discuss the importance of accurate velocity estimation in the context of handover algorithm design. We show how an error in velocity estimation can significantly increase the probability of dropped calls, leading to a poorer quality of service of the system. Some velocity estimators require prior estimation of the Rician factor. This chapter also presents two Rice factor estimators that have been shown to have better performance than existing estimators.

0-8493-1657-X/$0.00+$1.50

5.1 Introduction

This chapter presents the principles of mobile velocity estimation and illustrates its importance in the context of handoff algorithm design for mobile communications systems. Current mobile communications systems are under constant pressure to increase their capacity while maintaining a high quality of service. This is due to an ever-growing population of mobile phone users and an increasing demand for multimedia services. The main constraints system designers have to face are the distortions introduced by the mobile communications channel. These can be classified as Rayleigh or Rician fading, shadowing, and path loss [62, p. 16].[1] These distortions limit the performance of mobile communications systems in terms of efficiency and quality of service. Many standard functions of the system are significantly enhanced by *a priori* knowledge or estimation of the *mobile velocity*. The next section provides examples where mobile velocity plays a key role in the overall performance of the communications system.

5.1.1 Importance of Velocity Estimation

In microcellular systems, the cell size is smaller than in macrocellular systems (Figure 5.1). As a result, microcellular systems require faster and more reliable handovers.[2] Also, since the base station (BS) is at lamppost level, if the mobile station (MS) rounds a corner, the signal power received by an MS can drop rapidly by 20 to 30 dB over distances as small as 10 m (Figure 5.2 and Figure 5.3) [6]. In that case, an emergency handover needs to be processed toward a target BS if the call is to be maintained. The drop in signal strength can be detected by applying short temporal window averaging in the received signal. However, the window size is velocity dependent and is optimal only when an accurate estimation of the mobile velocity is available [7].

In multitier systems (Figure 5.1), cells of different sizes coexist in a two-layer structure, i.e., microcells on the lower layer and macrocells (umbrella) in the upper layer. Within these systems, different types of handover have to be managed between the umbrellas and the microcells. In order to minimize the number of handovers, the MS velocity can be used in a cell layer assignment strategy, in which an MS is allocated to different hierarchical layers according to its velocity. The umbrella cells are used for fast-moving users and

[1]The distortions of the mobile communications channel are reviewed in Section 5.2.

[2]Handover or handoff is the process of transferring the control of a mobile station from one base station or channel to another. It is an essential component of mobile communications systems.

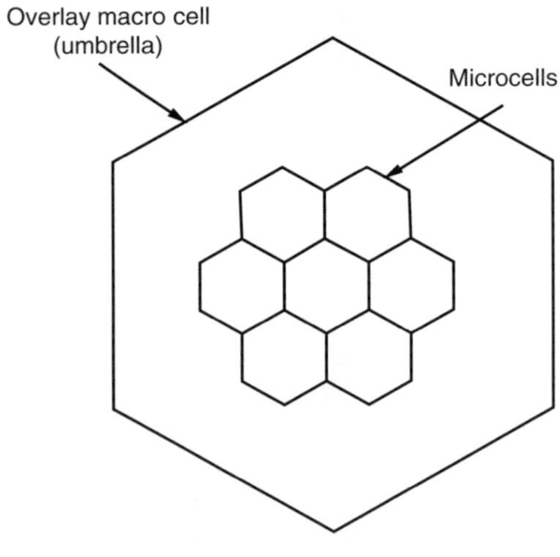

FIGURE 5.1 A multitier (microcell/macrocell overlay) system.

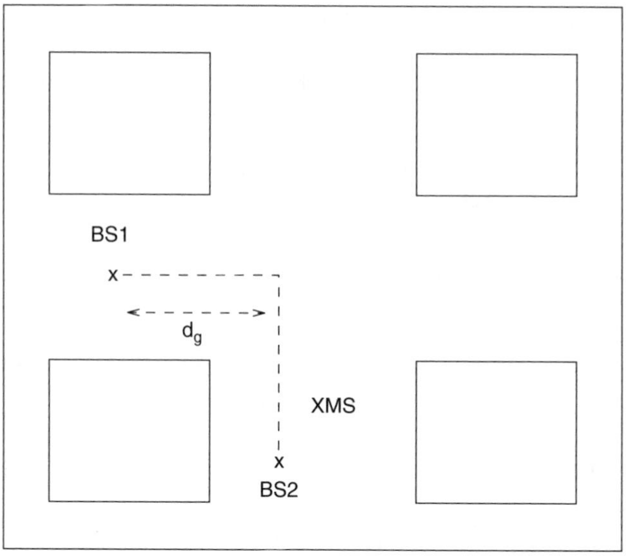

FIGURE 5.2 Corner effect: the MS turns round a corner, the LOS from the current BS (BS1) is lost, and an LOS is established between the MS and the target BS (BS2).

microcells for slow-moving users. This results in a reasonable grade of service (GoS) for both microcellular and multitier systems [33].

Reliable estimates of the MS velocity are also useful for effective dynamic channel assignment and the optimization of adaptive multiple-access wireless receivers [26, 69]. Since the performance of many receiver techniques depends on the fading rate of the received signal, an adaptive communications receiver can improve the performance and reduce the complexity of current systems by using Doppler information to control the receiver parameters. These parameters include the pilot filter bandwidth, the automatic gain control

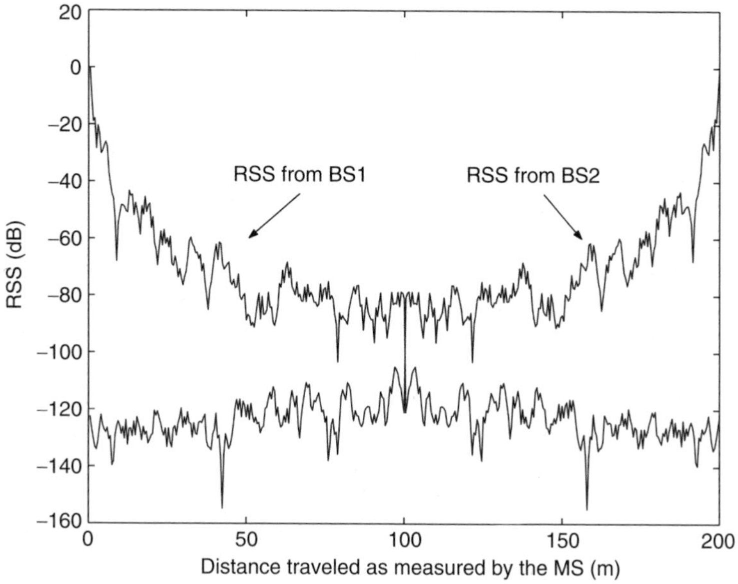

FIGURE 5.3 Received signal strength (RSS) at the MS from two neighboring BSs as a function of distance.

loop bandwidth, the phase tracker bandwidth, and the size of the interleaver [19, 52]. Based on this idea, Lee and Cho [42] proposed a power control scheme for codedivision multiple-access (CDMA) systems, which selects the power control step size based on the MS velocity. The MS velocity also affects the performance of a communications system operating in a pseudonoise (PN) tracking in CDMA systems [20, 22].

Mobile velocity estimation can occur at both BS and MS sites. For example, in the current CDMA 2000 system, the BS is responsible for power control [4, 41], and velocity is therefore estimated by the BS. However, a decision to hand off is made by the MS, and the MS is then responsible for estimating the velocity. Velocity estimation is performed by exploiting the statistics of a received pilot signal. For example, in CDMA 2000 the pilot signal consists of a constant-amplitude, zero-phase signal, spread by an all-one code sequence. It is then added to the information signal spread by an orthogonal sequence. The combination of the two signals is modulated by a carrier frequency and transmitted [21]. This process is equivalent to transmitting a pure carrier of constant amplitude alongside the information signal. At the reception, the combination of the two signals is demodulated, then correlated by each spreading sequence. Assuming no loss of orthogonality, the correlation of the received signal with the all-one sequence cancels the information signal and reveals the distortions introduced by the channel on the pilot signal.

5.1.2 Existing Velocity Estimators

Several mobile velocity estimators have been proposed in the literature, and some have been reviewed in [65]. Among the existing estimators are the zero-crossing rate of the in-phase or quadrature component [7], level crossing rate [7], and the autocovariance of the envelope of the received signal [5]. In [30, 31], the received signal samples are used to estimate the autocovariance function of the received faded envelope, from which the velocity information is extracted. Similarly, in [61], two methods for approximating the mobile velocity are proposed, based on the autocovariance function of the envelope and quadrature components of the received signal. The method proposed in [55] is based on the squared deviations of the logarithmically compressed signal envelope. In [1], the rate of maxima of the envelope of the received signal is used as a velocity estimator. The method in [49] is based on the estimator proposed in [1] and applies the continuous wavelet transform to locate the extrema of the received signal envelope. In [66], velocity estimators are derived, based on the spectral moments of the received signal. The method described in [36] uses the switching rate of diversity branches in a selection diversity combiner to estimate the MS velocity. In [68], mobile velocity is estimated by applying an eigenmatrix pencil method to the received signal samples. In [71], two methods are proposed for estimating mobile velocities for Global System for Mobile Communications (GSM) radios. The methods are based on estimating the deviation of the received signal strength. The method proposed in [50] uses the local stationarity of the received signal and expands it on a basis of smooth local complex exponentials. Velocity estimates have also been obtained by estimating the maximum Doppler frequency using eigenspace methods [6]. The method is designed under specific assumptions of the angular distribution of the incident power. Other velocity estimators have been proposed that require prior estimation of the channel [40] and covariance function of the channel power [48]. In [28], multiple BS and multidimensional scaling are used to estimate the velocity. This method requires knowledge of the average signal strength for all locations. In [26], a maximum likelihood estimator is derived and requires prior estimation of the channel parameters. A few estimators have also been proposed, which do not provide an explicit formula for the MS velocity, but rather classify the MS velocity as being fast, medium, or slow [39, 73, 74].

It was shown in [10, 11] that since all the above-mentioned techniques are based on the statistics of either the envelope or quadrature components of the received signal, they are not robust to shadowing. Other estimators have been proposed in [8, 9] based on the instantaneous frequency (IF) of the received signal. Recent work has already exploited the concept of IF estimation and time–frequency signal processing in wireless communications, e.g., in blind source separation, channel coding and capacity, interference excision, multiuser detection, code design, multicarrier transmission, and synchronization [2, 3, 12, 13, 17, 27, 35, 37, 38, 56, 57, 58, 59, 60, 72], and led to significant improvement in the systems [16, Chapter 13]. As shown in [9], the main advantage of IF-based velocity estimators is that they are robust to shadowing.

5.1.3 Structure of the Chapter

The chapter is structured as follows. Section 5.2 focuses on the received signal model in a mobile communications environment. In Section 5.3, we explain the principles behind velocity estimation and provide examples selected from the above-mentioned list of velocity estimators. The examples are selected on a basis of simplicity and performance. Section 5.4 provides the framework for analyzing the performance of velocity estimators with application to one specific estimator. In Section 5.5, we present simulation results of a selection of velocity estimators. Because some velocity estimators require prior knowledge of the Rice factor, we derive in Section 5.6 two Rice factor estimators. Section 5.7 discusses the importance of accurate velocity estimation in the context of handover decision algorithms, and Section 5.8 concludes this chapter.

5.2 Received Signal Model and Statistics

5.2.1 Received Signal Model

Propagation, in a mobile communications environment, is subject to three phenomena: path loss, shadowing, and multipath fading [62, p. 16]. Path loss affects all radio communications and refers to the loss of received power as the MS travels away from the BS. Path loss is completely characterized by the distance between the BS and the MS, the operating wavelength, the antenna height, and the surrounding terrain [62, p. 16]. Shadowing, or shadow fading, is caused by terrain configurations between the BS and MS and is produced by variations of the average of the envelope of the received signal over a few wavelengths. It is a random process and follows a lognormal distribution, i.e., when expressed in decibel, its distribution is Gaussian [62, p. 87]. Multipath fading, also referred to as fast fading, is due to the constructive and destructive superposition of many reflected, scattered, and diffracted plane waves arriving at the MS with different time delays and phase shifts. Figure 5.4 represents a typical received pilot signal envelope.

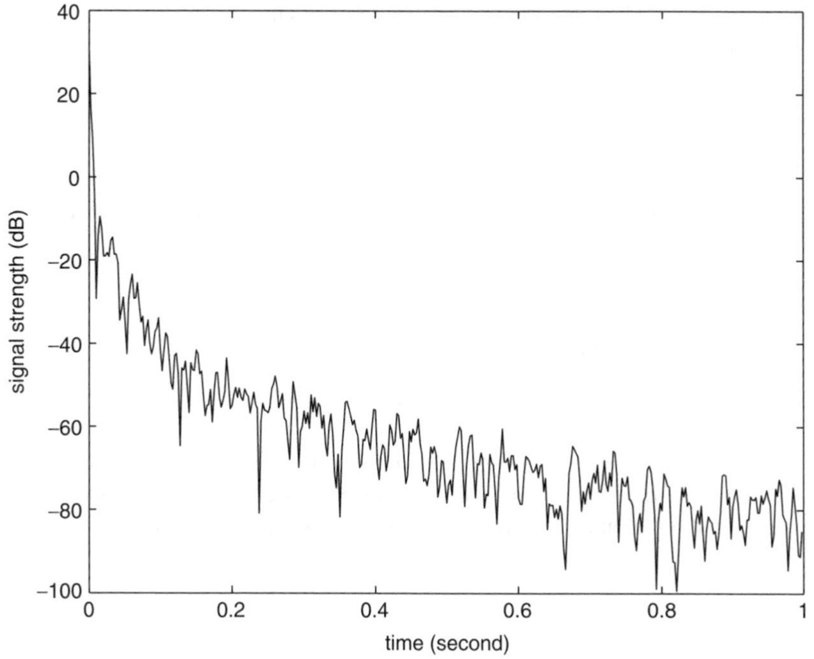

FIGURE 5.4 Typical received signal strength at an MS traveling at $v = 50$ km/h.

We assume, in what follows, that the distortions introduced by the channel are described by the model proposed by Lee and Yeh [44]. In this model, the real signal $y(t)$ received by an MS (or a BS) is the product of path loss, shadowing, and multipath fading with some additive noise:

$$\begin{aligned} y(t) &= m(t)s(t) + n(t) \\ &= p(t)m_o(t)s(t) + n(t) \end{aligned} \tag{5.1}$$

where $s(t)$ represents the multipath fading, $m_o(t)$ is the lognormal shadowing given by [49],

$$m_o(t) = 10^{L(t)/20} \tag{5.2}$$

$L(t)$ is a zero-mean Gaussian process, and $p(t)$ represents the path loss given by [49]:

$$p(t) = P_o(vt + d_o)^{-\alpha/2} \tag{5.3}$$

P_0 accounts for antenna parameters, transmitted power, and other relevant system parameters; vt represents the distance between the BS and an MS traveling at velocity v at time t; d_o is the distance between the BS and MS at time $t = 0$; and α, the exponent of the distance dependence, reflects the amount of power loss as a function of distance. The additive noise, $n(t)$, is assumed bandpass Gaussian with one-sided power spectral density (PSD) $S_N(f)$ given by [7]:

$$S_N(f) = \begin{cases} \frac{N_0}{2} & : \quad |f - f_c| < \frac{B_0}{2} \\ 0 & : \quad |f - f_c| > \frac{B_0}{2} \end{cases} \tag{5.4}$$

where B_0 represents the system bandwidth and is chosen equal to $\frac{2v_{max}}{\lambda}$, where v_{max} is the maximum mobile velocity and λ is the operating wavelength. $\frac{2v_{max}}{\lambda}$ represents the maximum expected Doppler frequency over the range of velocities.

5.2.2 Multipath Component Model

The multipath component, $s(t)$, is a result of the superposition of a number N of incoming waves, including a possible line-of-sight (LOS) component. Each incident wave $w_k(t)$ is modeled as a constant-amplitude, time-varying phase signal:

$$w_k(t) = \alpha_k e^{j(2\pi f_c t + \Phi_k(t))} \tag{5.5}$$

where α_k represents the amplitude of wave k and $\Phi_k(t)$ is the phase shift caused by the moving vehicle. If $w_k(t)$ arrives with angle of incidence θ_k with respect to the direction of travel of the MS, the Doppler shift created is

$$f_k = \frac{v}{\lambda} cos(\theta_k) \tag{5.6}$$

and the subsequent phase shift introduced by that wave is

$$\Phi_k(t) = 2\pi f_k t + \Psi_k = 2\pi \frac{v}{\lambda} cos(\theta_k)t + \Psi_k \tag{5.7}$$

If $w_k(t)$ is a LOS component, θ_k and Ψ_k are deterministic and depend on the position of the vehicle. Otherwise, θ_k and Ψ_k are random and are generally modeled as uniformly distributed between $(0, 2\pi)$ [34]. Combining Equation 5.5 and Equation 5.6, the multipath fading, $s(t)$, can be written as

$$s(t) = \Re\left\{\sum_{k=1}^{N} w_k(t)\right\} \tag{5.8}$$

$$= \Re\left\{e^{j2\pi f_c t}\left[\sum_{k=1}^{N} \alpha_k e^{j2\pi \frac{v}{\lambda}\cos(\theta_k)t + \Psi_k}\right]\right\} \tag{5.9}$$

where $\Re\{\}$ refers to the real-part operator, $w_1(t)$ is a possible LOS component, and $\{w_2(t), w_3(t) \ldots, w_N(t)\}$ are the result of $N-1$ independent scatterers and, as a consequence, are assumed independent and identically distributed (i.i.d). Signal $s(t)$ can be further written as

$$s(t) = s_i(t)\cos 2\pi f_c t - s_q(t)\sin 2\pi f_c t \quad (5.10)$$

where $s_i(t)$ and $s_q(t)$ are the respective in-phase and quadrature phase components of $s(t)$, defined as

$$s_i(t) = \sum_{k=1}^{N} \alpha_k \cos\left(2\pi \frac{v}{\lambda}\cos(\theta_k)t + \Psi_k\right) = x_i(t) + m_i \quad (5.11)$$

$$s_q(t) = \sum_{k=1}^{N} \alpha_k \sin\left(2\pi \frac{v}{\lambda}\cos(\theta_k)t + \Psi_k\right) = x_q(t) + m_q \quad (5.12)$$

where $E\{x_i(t)\} = E\{x_q(t)\} = 0$, $m_i = E\{s_i(t)\}$, $m_q = E\{s_q(t)\}$, and $E\{\}$ represents the expectation operator.

In the presence of a LOS component, the means m_i and m_q are nonzero [34] and

$$s(t) = (x_i(t) + m_i)\cos 2\pi f_c t - (x_q(t) + m_q)\sin 2\pi f_c t \quad (5.13)$$

$$= x(t) + m_i \cos 2\pi f_c t - m_q \sin 2\pi f_c t \quad (5.14)$$

where $x_i(t)$ and $x_q(t)$ are the respective in-phase and quadrature phase components of signal $x(t)$, i.e.,

$$x(t) = x_i(t)\cos 2\pi f_c t - x_q(t)\sin 2\pi f_c t \quad (5.15)$$

5.2.3 The Scattering Distribution

An important parameter in mobile communications system design is the distribution of incoming waves around a particular MS. It is referred to as scattering distribution. For example, in a typical macrocellular environment, the MS is usually uniformly surrounded by local scatterers, so that the plane waves arrive from many directions without an LOS component. As a consequence, the scattering distribution is usually considered isotropic. However, in a microcellular environment, the antennas of the BSS are only moderately elevated above the local scatterers. As a result, an LOS component may or may not exist and the scattering distribution is usually nonisotropic. The von Mises density provides a model for the distribution of the angle of arrival of the incoming waves [1, 45]. It is a function of one parameter, χ , which determines the directivity of the incoming waves. It is given by

$$p(\theta) = \frac{1}{2\pi I_0(\chi)} e^{\chi \cos\theta}, \; \chi \geq 0, \; -\pi \leq \theta \leq \pi \quad (5.16)$$

where $I_n(.)$ is the modified Bessel function of order n and θ is the angle of incidence of the incoming waves. Figure 5.5 shows the polar plots of $p(\theta)$ against the angle of arrival of the plane waves for three different values of χ. It can be seen that for $\chi = 0$ the scattering distribution is isotropic and that it becomes more directive as χ increases.

5.2.4 Statistics of the Multipath Fading

When the number of incoming waves is sufficiently large (generally greater than six, [34, p. 69]), the in-phase and quadrature components of $x(t)$, defined in Equations 5.11 and 5.12, respectively, tend to be independent zero-mean Gaussian processes, with variance σ^2. As a consequence, the envelope of $s(t)$, defined as

$$|s(t)| = \sqrt{(x_i(t) + m_i)^2 + (x_q(t) + m_q)^2}$$

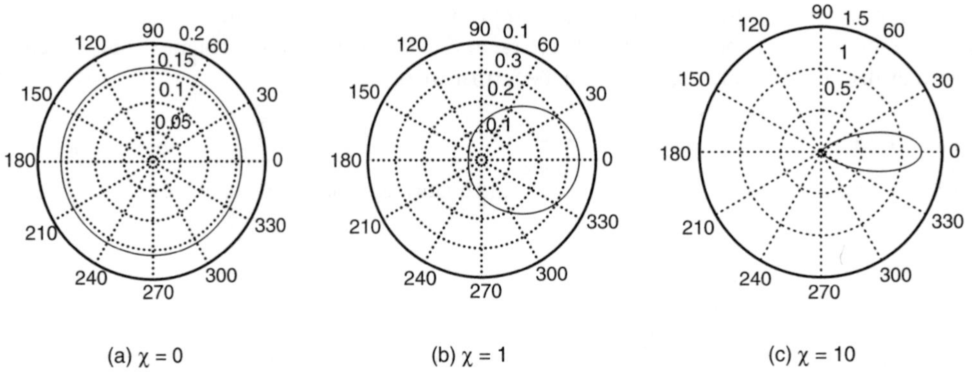

FIGURE 5.5 Polar plots of $p(\theta)$ in terms of θ for $\chi = 0$, $\chi = 1$, and $\chi = 10$. For $\chi = 0$ the scattering distribution is isotropic and becomes more directive as χ increases.

follows a Ricean distribution with Rice factor $K = \frac{\eta^2}{2\sigma^2}$, where $\eta^2 = m_i^2 + m_q^2$ [62, p. 518]. The Rice factor K represents the ratio of the power in the LOS component and scatter components of the signal $s(t)$.

The PSD of $x(t)$ is given in [62, p. 42] as

$$S_X(f) = \begin{cases} \frac{\sigma^2}{\sqrt{f_m^2-(f-f_c)^2}}[G(\theta)p(\theta) + G(-\theta)p(-\theta)] & : \ |f - f_c| < f_m \\ 0 & : \ |f - f_c| > f_m \end{cases} \tag{5.17}$$

where $f_m = \frac{v}{\lambda}$ is the maximum Doppler frequency shift, $p(\theta)$ is the scattering distribution, and $G(\theta)$ is the gain of the MS antenna. We will assume in the remainder of the chapter that a vertical mono-pole antenna with $G(\theta) = \frac{3}{2}$ is used. Note that the PSD $S_X(f)$ is centered about the carrier frequency with a spectral width of $2f_m = \frac{2v}{\lambda}$.

Assuming the scattering distribution of equation 5.16, it can be shown that the nth spectral moment of $x(t)$ is given by [1]

$$a_n = a_0(2\pi f_m)^n q_n(\chi) \tag{5.18}$$

where $a_0 = \frac{3}{2}\sigma^2$ and

$$q_n(\chi) = \frac{1}{\pi I_0(\chi)} \int_0^{\pi} e^{\chi \cos\theta} \cos^n \theta \, d\theta \tag{5.19}$$

Specifically, based on Equation 5.18 and Equation 5.19, we obtain

$$a_1 = a_0(2\pi f_m)\frac{I_1(\chi)}{I_0(\chi)} \tag{5.20a}$$

$$a_2 = a_0(2\pi f_m)^2 \left(\frac{I_0(\chi) + I_2(\chi)}{2I_0(\chi)}\right) \tag{5.20b}$$

In the case of isotropic scattering ($\chi = 0$), the first two spectral moments reduce to

$$a_1 = 0 \quad \text{and} \quad a_2 = 2a_0(\pi f_m)^2 \tag{5.21}$$

These statistics will be used further in the design of velocity estimators and in the evaluation of their performance.

5.3 Principles of Mobile Velocity Estimation

We can see from Equation 5.1 and Equation 5.9 that the information on mobile velocity is contained in both the envelope and phase of the received signal. The abundance of existing velocity estimators is a reflection of the numerous ways of extracting this information. The velocity of a mobile unit is generally estimated by exploiting the statistics of the envelope, phase, in-phase, and quadrature phase components or any other feature of the received signal. The approach for deriving a velocity estimator is based on the principle that there exists an exact, nonrandom relationship between the mobile velocity and the statistics of the Rician component $s(t)$ in the presence of isotropic scattering. Several exact and equivalent expressions of the velocity can be derived, depending on which feature of $s(t)$ is being considered, i.e., envelope, phase, in-phase, or quadrature phase, and which order of the statistic is chosen, i.e., first order, second order, etc. Below are some examples of these exact expressions of velocity.

5.3.1 Examples of Derivations of the Velocity

Below are some examples of how the mobile velocity can be calculated from the statistics of various features of $s(t)$.

Example 5.1 Level Crossing Rate Method

The envelope phase description of the Rician fading $s(t)$ is

$$s(t) = r(t)\cos(2\pi f_c t + \psi(t)) \tag{5.22}$$

where $r(t)$ and $\psi(t)$ represent the envelope and phase of $s(t)$, respectively.

The level crossing rate (LCR) is defined as the average number of upcrossings per second the envelope $r(t)$ makes of a predetermined level R_o. It is obtained as [43, p. 77]

$$LCR_r(R_0) = \int_0^{\infty} \dot{r}p(r = R_0, \dot{r})d\dot{r} = \frac{R_0}{a_0}\sqrt{\frac{a_2}{2\pi}}\exp\left(-\frac{R_0^2}{2a_0}\right) \tag{5.23}$$

where $p(r, \dot{r})$ denotes the joint probability density function (pdf) of $r(t)$ and its time derivative $\dot{r}(t)$, $a_0 = \frac{3}{2}\sigma^2$, and a_2 is the second spectral moment of $x(t)$, given in Equation 5.20b. If the level R_o is chosen equal to $\sqrt{2a_0}$, the LCR of level R_o becomes

$$LCR_r(R_0) = \sqrt{\frac{a_2}{\pi a_0}}\, e^{-1} \tag{5.24}$$

In the case of isotropic scattering, $a_2 = 2a_0(\pi f_m)^2$ (Equation 5.21). It follows that Equation 5.24 reduces to

$$LCR_r(R_0) = \sqrt{2\pi}\frac{v}{\lambda}\, e^{-1} \tag{5.25}$$

The mobile velocity can then be exactly expressed in terms of the average number of upcrossings the envelope $r(t)$ makes of level $R_o = \sqrt{2a_0}$ as

$$v = \frac{\lambda e}{\sqrt{2\pi}} LCR_r(R_0) \tag{5.26}$$

Example 5.2 Rate of Maxima Method

The rate of maxima (ROM) of a given process $r(t)$ is defined as the average number of maxima per second of $r(t)$. It can be derived as [53]

$$ROM_r = \int_0^{\infty} \ddot{r}\, p(\dot{r} = 0, \ddot{r})d\ddot{r} \tag{5.27}$$

where $p(\dot{r}, \ddot{r})$ denotes the joint (pdf) of the first and second derivatives of the process $r(t)$.

Using the results in [53], the ROM of the envelope $r(t)$ of signal $s(t)$ in Equation 5.22 can be obtained as

$$ROM_r = \frac{1.5651}{\pi}\sqrt{\frac{a_2}{2a_0}} \tag{5.28}$$

which, in the case of isotropic scattering, and following the approach of Example 1, reduces to

$$ROM_r = 1.5651\frac{v}{\lambda} \tag{5.29}$$

From Equation 5.29, the mobile velocity can then be exactly expressed in terms of the rate of maxima of the envelope of $s(t)$ as

$$v = 0.6389\lambda ROM_r \tag{5.30}$$

Example 5.3 Zero Crossing Rate Method

The zero crossing rate (ZCR) of a given process $a(t)$ is defined as the average number of positive-going zero crossings per second of that process. It is given by

$$ZCR_a = \int_0^{\infty} \dot{a}\, p(a = 0, \dot{a}) d\dot{a} \tag{5.31}$$

In [54], the ZCR definition was applied to the in-phase component of signal $x(t)$ defined in Equation 5.11 and Equation 5.15 and was shown to be equal to

$$ZCR_{x_i} = \frac{1}{2\pi}\sqrt{\frac{a_2}{a_0}} \tag{5.32}$$

which in the case of isotropic scattering reduces to

$$ZCR_{x_i} = \frac{v}{\sqrt{2}\lambda} \tag{5.33}$$

The mobile velocity can then be obtained using the following simple expression:

$$v = \sqrt{2}\lambda ZCR_{x_i} \tag{5.34}$$

The same result can be obtained using the quadrature component of $x(t)$ instead of the in-phase component.

Example 5.4 Covariance-Based Method

For the covariance (COV)-based method, consider a process $r_1(t) = r^2(t)$ where $r(t)$ is the envelope of the Rician fading $s(t)$. We can show that for a given time lag τ

$$E\{(r_1(t+\tau) - r_1(t))^2\} = 2R_{r_1}(0) - 2R_{r_1}(\tau) \tag{5.35}$$

where $R_{r_1}(\tau) = E\{r_1(t+\tau)r_1(t)\}$ is the autocorrelation function of process $r_1(t)$.

In [7], it was proved that in the presence of isotropic scattering

$$E\{(r_1(t+\tau) - r_1(t))^2\} \simeq R_{r_1}(0)\left(\frac{2\pi\tau_t v}{\lambda}\right)^2 \tag{5.36}$$

where τ_t is the sample spacing in seconds per sample and $R_{r_1}(0)$ is the variance of the squared envelope.

The mobile velocity can therefore be extracted from Equation 5.36 as

$$v = \frac{\lambda}{2\pi\tau_t}\sqrt{\frac{2R_{r_1}(0) - 2R_{r_1}(\tau)}{R_{r_1}(0)}} \tag{5.37}$$

where $R_{r_1}(\tau)$ is the autocorrelation of the envelope square $|s(t)|^2$ for a given time lag τ.

Example 5.5 Instantaneous Frequency-based Method

The IF of $s(t)$ is defined as [14]

$$f_{i,s}(t) = \frac{1}{2\pi}\frac{d\psi(t)}{dt} \tag{5.38}$$

where $\psi(t)$ is the phase of $s(t)$. In Appendix A, we show that the first moment of $|f_{i,s}|$ is

$$E\{|f_{i,s}|\} = \frac{v}{\sqrt{2}\lambda}\, I_0\left(\frac{K}{2}\right)\, e^{\frac{-K}{2}} \tag{5.39}$$

where | | stands for the absolute value operator, $f_{i,s}$ is the IF of the multipath component $s(t)$, and K is the Rice factor. The velocity of a mobile unit can therefore be expressed in terms of the first-order moment of the absolute value of the IF of the Rician fading component $s(t)$ as

$$v = \sqrt{2}\lambda\, I_0^{-1}\left(\frac{K}{2}\right)\, e^{\frac{K}{2}}\, E\{|f_{i,s}|\} \tag{5.40}$$

5.3.2 Examples of Velocity Estimators

The expressions of the velocity in all the above examples are exactly equivalent. If we were to calculate the velocity using the LCR of the envelope of $s(t)$ (Equation 5.26) or the first-order moment of the IF of $s(t)$ (Equation 5.40), we would obtain exactly the same result. These exact expressions become estimators for two main reasons. The most important reason is that, in reality, we do not have access to $s(t)$, and the scattering distribution is not necessarily isotropic. The exact expressions obtained from exploiting the statistics of features of $s(t)$ in an isotropic environment become approximate in a nonisotropic environment, and in the presence of shadowing and additive noise. The second reason is that some parameters required to compute the exact expression of v must also be approximated. For example, in Equation 5.40, since exact values for $E\{|f_{i,s}|\}$ or K are not available, estimated values must be used, resulting in another source of error.

In the presence of shadowing, additive noise, and nonisotropic scattering, the previous exact expressions become the following estimators.

Example 5.1 LCR Estimator

From the expression of the velocity in Equation 5.26, we can derive an estimator

$$\hat{v}_{LCR} = \frac{\lambda e}{\sqrt{2\pi}} LCR_r(R_0) \tag{5.41}$$

where $LCR_r(R_0)$ represents the number of level of upcrossings of the envelope of $y(t)$ and where $\hat{R}_0 = \sqrt{2\hat{a}_0}$, with

$$\hat{a}_0 = \frac{1}{T}\int_T y_i^2(t)dt - \left(\frac{1}{T}\int_T y_i(t)dt\right)^2 \tag{5.42}$$

where $y_i(t)$ is the in-phase component of the received signal.

Example 5.2 ROM Estimator

From the expression of the velocity in Equation 5.30, we can derive an estimator

$$\hat{v}_{ROM} = 0.6389\lambda ROM_r \tag{5.43}$$

where ROM_r is the rate of maxima of the envelope of the received signal $y(t)$.

Example 5.3 ZCR Estimator

From the expression of the velocity in Equation 5.34, we can derive the following estimator:

$$\hat{v}_{ZCR} = \sqrt{2}\lambda ZCR_{z_i} \tag{5.44}$$

where ZCR_{z_i} is the zero crossing rate of $z_i(t) = y_i(t) - \bar{y}_i(t)$, where $y_i(t)$ is the in-phase component of the received signal $y(t)$ and $\bar{y}_i(t)$ an estimate of its mean.

Example 5.4 COV-Based Estimator

From the expression of the velocity in Equation 5.37, we can derive the following estimator:

$$\hat{v}_{COV} = \frac{\lambda}{2\pi\tau_t}\sqrt{\frac{\overline{V}}{R(0)}} \tag{5.45}$$

where τ_t is the sample spacing in seconds per sample, $R(0)$ is the variance of the envelope square of the received signal, and

$$\overline{V} = 2(R(0) - R(\tau)) \tag{5.46}$$

Example 5.5 IF-Based Estimator

The IF-based estimator is obtained from Equation 5.40 as

$$\hat{v}_{IF} = \lambda\sqrt{2}I_0^{-1}\left(\frac{\hat{K}}{2}\right)e^{\frac{\hat{K}}{2}} < |\hat{f}_{i,y}(t)| > \tag{5.47}$$

where

$$< |\hat{f}_{i,y}(t)| > = \frac{1}{T}\int_T |\hat{f}_{i,y}(t)| dt$$

is the time average of the IF estimate of the received signal over the duration T of the recorded signal, and $\hat{K}$ is the estimated Ricean factor.

All the above-mentioned and other existing estimators exhibit different performances in the presence of additive noise, shadowing, and nonisotropic scattering. For example, in the presence of shadowing, the envelope of $s(t)$ is distorted while its IF is not. As a consequence, any estimator based on the IF of the received signal would be expected to be more robust to shadowing than estimators based on the envelope of the received signal. The approach for evaluating the performance of velocity estimators is decribed in the next section.

5.4 Performance Analysis of Velocity Estimators

The examples of the previous section showed that several exact expressions of the velocity of a mobile unit can be obtained from exploiting different statistics and different features of the Rician component $s(t)$. In the presence of additive noise, shadowing, and a nonisotropic environment, these exact expressions are used as velocity estimators with different properties and performances. Any performance analysis of a velocity estimator should evaluate its behavior in the presence of shadowing, additive noise, and nonisotropic scattering. This approach is illustrated below for the IF-based estimator but can be applied to any other estimator.

5.4.1 Effect of Shadowing

In this section, the effect of shadowing is studied in the absence of additive noise. The model for the received signal is therefore $y(t) = m(t)s(t)$. We observe that $m(t)$ causes a distortion in the amplitude and quadrature components of the received signal. Thus, the performance of all velocity estimators that are based on the statistics of the envelope and quadrature components of the received signal deteriorates in the presence of shadowing. However, since the presence of shadowing changes only the amplitude of the received signal [43, p. 203; 62, p. 91], it produces no phase distortion. Hence, the IF of the received signal is not affected by $m(t)$; i.e., both signals $s(t) = r(t)\cos(2\pi f_c t + \psi(t))$ and $y(t) = m(t)r(t)\cos(2\pi f_c t + \psi(t))$ have the same IF, $\frac{1}{2\pi}\dot{\psi}(t)$. The IF-based estimator is therefore robust to shadowing and to any other amplitude distortion, such as path loss.

5.4.2 Effect of AWGN and Nonisotropic Scattering

5.4.2.1 Derivation of the Normalized Bias

In this section, the performance of the IF-based estimator is investigated in the presence of additive white Gaussian noise (AWGN), nonisotropic scattering, and the absence of shadowing. The performance criterion is the normalized bias of the estimator. Ideally, the variance of the estimator should also be derived. However, at the time of publication, there is no closed-form expression of the variance. This is also the case for all existing estimators. The variance will be investigated in the next section using Monte Carlo simulations.

The normalized bias of estimator $\hat{v}_{IF}$ in Equation 5.47 is defined as

$$\begin{aligned} \varepsilon(\chi, \gamma, K) &= \frac{\hat{v}_{IF}}{v} - 1 \\ &= \frac{\sqrt{2}\lambda}{v} I_0^{-1}\left(\frac{K}{2}\right) e^{\frac{K}{2}} E\{|f_{i,y}|\} - 1 \end{aligned} \tag{5.48}$$

where γ is the signal-to-noise ratio (SNR) and

$$E\{|f_{i,y}|\} = \frac{1}{2\pi}\int_{-\infty}^{\infty} |\dot{\psi}|\, p(\dot{\psi})\, d\dot{\psi} \tag{5.49}$$

$p(\dot{\psi})$ is given by [54]

$$\begin{aligned} p(\dot{\psi}) = {} & \frac{1}{8a}\sqrt{\frac{2}{ab_0\mathcal{B}}} \exp\left(\frac{b_1^2\rho + c\mathcal{B}}{2\mathcal{B}} - \frac{b_0 b_2 \rho}{\mathcal{B}}\right) \\ & \times \left[(c+1)\, I_0\left(\frac{c\mathcal{B} - b_1^2\rho}{2\mathcal{B}}\right) + c\, I_1\left(\frac{c\mathcal{B} - b_1^2\rho}{2\mathcal{B}}\right)\right] \end{aligned} \tag{5.50}$$

where

$$\rho = \frac{Q^2}{2b_0} \tag{5.51}$$

with $Q^2 = \frac{3}{2}M^2\eta^2$

$$\mathcal{B} = b_0 b_2 - b_1^2 \tag{5.52a}$$

$$a = \frac{b_2 - 2b_1\dot{\psi} + b_0\dot{\psi}^2}{2\mathcal{B}} \tag{5.52b}$$

$$c = \frac{Q^2(b_2 - b_1\dot{\psi})^2}{4a\mathcal{B}^2} \tag{5.52c}$$

and b_n is the nth spectral moment of $y_0(t) = m(t)x(t) + n(t)$. $y_0(t)$ is the received signal in the absence of an LOS component.

In the absence of shadowing, $m(t)$ is a constant M. Since both $x(t)$ and $n(t)$ are zero-mean, independent processes, the nth spectral moment of $y_0(t)$ can be written as

$$\begin{aligned} b_n &= (2\pi)^n \int_0^{\infty} (f - f_c)^n (M^2 S_X(f) + S_N(f))df \\ &= M^2 a_n + (1 - (-1)^{n+1}) \frac{N_0 \pi^n}{4(n+1)} B_0^{n+1}, \\ & n \geq 0 \end{aligned} \tag{5.53}$$

5.4.2.2 Performance in the Presence of AWGN and Isotropic Scattering

In the case of isotropic scattering ($\chi = 0$, Figure 5.5(a)), $E\{|f_{i,y}|\}$ is given by (Equation 5.A5)

$$E\{|f_{i,y}|\} = \frac{1}{2\pi} \sqrt{\frac{b_2}{b_0}} \exp\left(\frac{-\rho}{2}\right) I_0\left(\frac{\rho}{2}\right) \tag{5.54}$$

The normalized bias of Equation 5.48 can then be written as

$$\varepsilon(0, \gamma, K) = \frac{\lambda}{\sqrt{2}\pi v} \sqrt{\frac{b_2}{b_0}} I_0\left(\frac{\rho}{2}\right) I_0^{-1}\left(\frac{K}{2}\right) e^{\frac{K-\rho}{2}} - 1 \tag{5.55}$$

where ρ is defined in Equation 5.51. After using Equation 5.18 and Equation 5.53, Equation 5.55 becomes

$$\varepsilon(0, \gamma, K) = \sqrt{\frac{\gamma + \frac{1}{6}(\frac{\lambda B_0}{v})^2}{\gamma + 1}} I_0\left(\frac{K\gamma}{2(\gamma+1)}\right) I_0^{-1}\left(\frac{K}{2}\right) e^{\frac{K}{2(\gamma+1)}} - 1 \tag{5.56}$$

Figure 5.6 and Figure 5.7 show the effect of AWGN on the IF-based estimator with respect to γ and v, for two different values of K. The bandwidth B_0 is assumed to be $B_0 = \frac{2v_{max}}{\lambda} = 170$ Hz, which allows

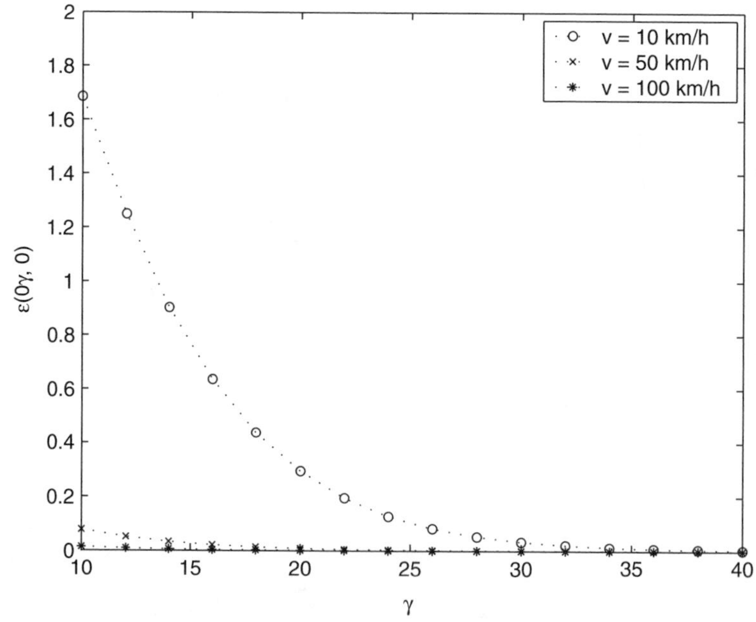

FIGURE 5.6 The normalized bias of the IF-based velocity estimator as a function of SNR for three different MS velocities, assuming $K = 0$.

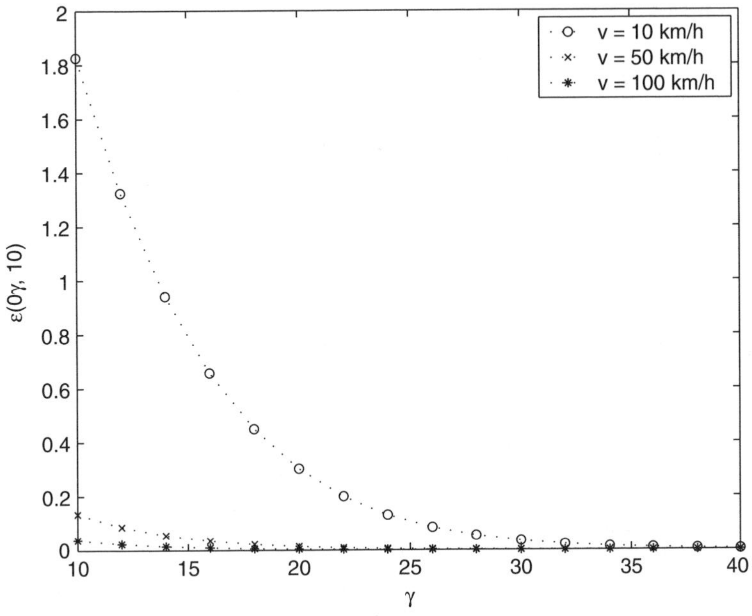

FIGURE 5.7 The normalized bias of the IF-based velocity estimator as a function of SNR for three different MS velocities, assuming $K = 10$.

for velocities up to $v_{max} = 100$ km/h at $f_c = 900$ MHz. We observe that for low velocities and low SNR, the bias is significant. The normalized bias is also seen to increase slightly as K increases. This is also the case for some existing estimators [7]. However, Appendix C shows that the IF estimator allows for an improvement in SNR by a factor of $3 \ln \frac{B_0}{2f_1}$, where f_1 is the lower cutoff frequency of the output low-pass filter at the output of the IF estimator. For typical values, $B_0 = 170$ Hz and $f_1 = 10$ Hz, the improvement is about 8 dB. This means that the normalized bias of the IF-based estimator for an input SNR of 10 dB is the same as that of the ZCR method for an input SNR of 18 dB. Thus, even for small velocities, the IF-based estimator will still exhibit a negligible bias at low SNRs due to the improvement of SNR at the output of the IF estimator.

5.4.2.3 Performance in the Presence of Nonisotropic Scattering and in the Absence of AWGN

The effect of the scattering distribution is investigated here in the absence of noise ($\gamma \to \infty$). In this case, using Equation 5.50, the normalized bias in Equation 5.48 in the absence of additive noise, $\varepsilon(\chi, \infty, K)$, can be computed numerically. Figure 5.8 shows the effect of the scattering distribution on the IF-based estimator for different values of χ. Because the estimator in Equation 5.47 was derived in the case of isotropic scattering, the error in estimating the MS velocity increases as χ increases. Comparing Figure 5.8(a) and (b), we observe that there is a negligible increase in normalized bias when K increases. This suggests that $\varepsilon(\chi, \infty, K)$ can be approximated by $\varepsilon(\chi, \infty, 0)$. Using Equation 5.48 and Equation 5.A5, we obtain

$$\begin{aligned} \varepsilon(\chi, \infty, K) &\simeq \varepsilon(\chi, \infty, 0) \\ &= \sqrt{2\, q_2(\chi)} - 1 \end{aligned} \tag{5.57}$$

Appendix B shows that the ZCR-based velocity estimator (Equation 5.44) has the same normalized bias as the IF-based estimator (Equation 5.47) in the presence of nonisotropic scattering. The ZCR-based velocity estimator is generally more robust than the LCR and covariance-based methods in the presence of nonisotropic scattering [7]. It then follows that the IF-based velocity estimator also outperforms the above-mentioned estimators in the presence of nonisotropic scattering.

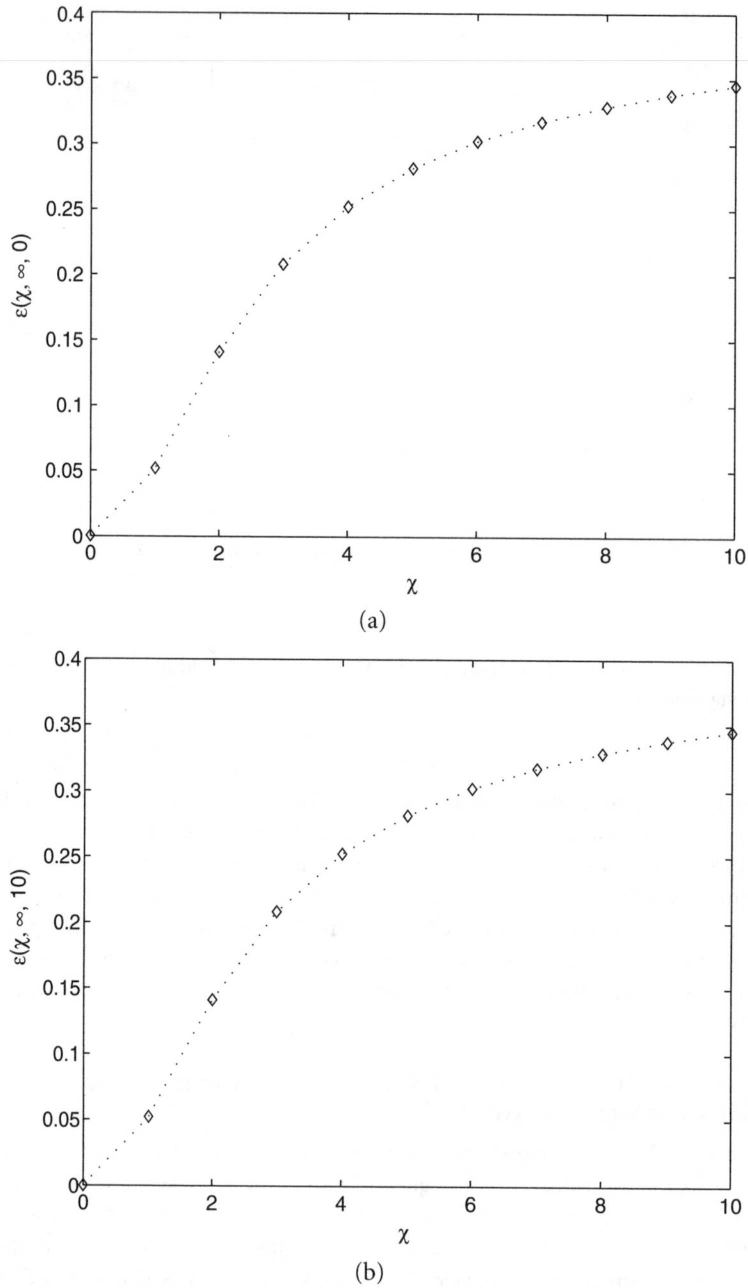

FIGURE 5.8 The normalized bias of the IF-based velocity estimator as a function of χ: (a) for $K = 0$ and $v = 40$ km/h, (b) for $K = 10$ and $v = 40$ km/h.

In summary, unlike the ZCR, LCR, and COV-based methods, or any method based on the envelope, in-phase, or quadrature phase of the received signal, the IF-based velocity estimator is robust to shadowing. It also exhibits superior performance in the presence of AWGN due to the improvement in SNR introduced by the IF estimator. In the presence of nonisotropic scattering, the IF-based estimator has the same performance as the ZCR and is generally more robust than LCR and COV-based methods. The effects of additive noise, nonisotropic scattering, and shadowing combine in a nonlinear way and are easier to evaluate using simulations. This is the object of the next section.

5.5 Performance Analysis Using Simulations

In Section 5.4, the performance of the IF-based estimator was evaluated analytically in the case of restrictive assumptions. For example, the IF-based estimator was evaluated in terms of the bias introduced in the presence of additive noise, shadowing, and nonisotropic scattering considered separately. It is also important to investigate the performance of an estimator when all these effects are present at once, which is the case in real life. Analytical derivations can, in that case, become intractable, and it is often easier to use simulations. Also, the IF-based estimator was evaluated in terms of bias only. However, the variance of an estimator is another important measure of performance. At time of publication, in the case of the IF-based estimator or any other velocity estimator, there is no closed-form expression for the variance. In this case, Monte Carlo simulations are used. Finally, the analytical derivations of Section 5.4 assumed perfect knowledge of the first-order moment of the IF of the received signal and of the Rice factor. A more accurate performance study would require taking into account the statistics of the estimators used for the first order of the IF and for the Rice factor. However, the derivations would again become too complex and simulations are a better alternative.

5.5.1 Simulations of the Received Signal

Using Jakes method for simulating the multipath component of the received signal, the in-phase and quadrature components of $x(t)$ defined in Equation 5.15 can be modeled as [34]

$$x_i(t) = \sum_{n=1}^{N} \cos\left(2\pi f_m t \cos\left(\frac{2\pi n}{N}\right) + \Psi_n\right) \tag{5.58a}$$

$$x_q(t) = \sum_{n=1}^{N} \sin\left(2\pi f_m t \cos\left(\frac{2\pi n}{N}\right) + \Psi_n\right) \tag{5.58b}$$

where N is the number of waves arriving at the MS antenna and $\Psi_n(n = 1, 2, \ldots, N)$ are assumed to be i.i.d. uniformly on $(0, 2\pi)$. The process $L(t)$ of Equation 5.2 is modeled as a zero-mean Gaussian process with PSD $S_L(\nu)$ given by

$$S_L(\nu) = \frac{2d_0\sigma_L^2}{1 + (2\pi\nu d_0)^2} \tag{5.59}$$

where d_0 and σ_L^2 are the correlation length and variance of L, respectively. Let ν_{max} denote the maximum spatial frequency of $S_L(\nu)$, and $D = vT$, with T the total duration of the simulated signal. A model for L is [49]

$$L(t) = \sum_{k=-J}^{J-1} \left[\frac{2}{CD} S_L\left(\frac{k+1/2}{D}\right)\right]^{\frac{1}{2}} \times \cos\left(\frac{2\pi t}{T}(k+1/2) + \theta_k\right) \tag{5.60}$$

where $\theta_k(k = -J, -J+1, \ldots, J)$ are assumed to be i.i.d. uniformly on $(0, 2\pi)$, $J = D\nu_{max}$, and

$$C = \frac{1}{D\sigma_L^2} \sum_{k=-J}^{J-1} S_L\left(\frac{k+1/2}{D}\right) \tag{5.61}$$

In these simulations, the carrier wavelength is $\lambda = 1/3$ ($f_c = 900$ MHz), and the exponent of the distance dependence is $\alpha = 2$. The standard deviation of lognormal shadowing ranges from 6.5 to 8.2 dB at 900 MHz in urban areas for microcells [47]. Here, the standard deviation and correlation length of the lognormal shadowing are $\sigma_L = 8$ dB and $d_0 = 50$ m, respectively. Parameter P_0 is assumed equal to 1 W.

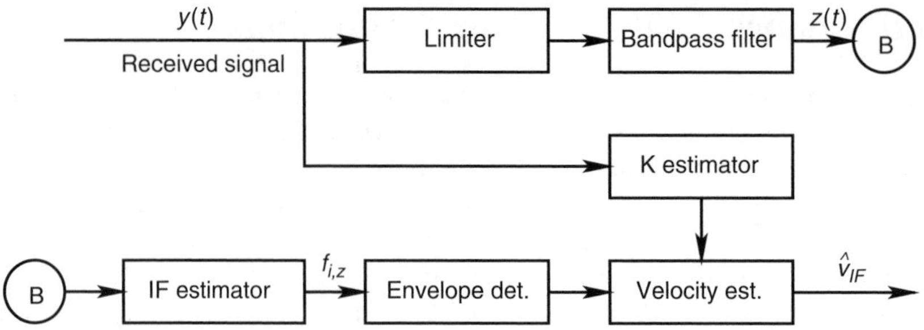

FIGURE 5.9 Block diagram of the IF-based MS velocity estimator.

5.5.2 Simulation Results

We consider that only a finite duration of the received signal is observed, and that the IF and moments of the signal are estimated using a finite duration of the received signal. The performance of the IF-based estimator is then compared with that of other estimators.

For each value of the MS velocity, the estimator $\hat{v}_{IF}$ in Equation 5.47, is computed by using the approach illustrated in Figure 5.9. A limiter is used to remove spurious amplitude variations from the received signal without destroying the information in the phase. It is followed by a bandpass filter of suitable bandwidth B_0, in order to extract the constant-amplitude signal. A value of $B_0 = 170$ Hz is chosen for the system bandwidth. This process is widely used in FM receivers [18, pp. 205–207]. The signal $z(t)$ is a constant-amplitude frequency-modulated signal and can be expressed as

$$z(t) = A\cos(2\pi f_0 t + \phi(t)) + n_o(t) \tag{5.62}$$

where A is the maximum amplitude at the limiter output and $n_o(t)$ is the bandpass noise at the output of the filter. The IF of the signal $z(t)$ can be estimated using [15]

$$f_{i,z}(t) = \frac{1}{2\pi}\frac{z_i(t)z_q'(t) - z_i'(t)z_q(t)}{z_i^2(t) + z_q^2(t)} \tag{5.63}$$

with $z_i(t) = A\cos(\phi(t)) + n_{oi}(t)$ and $z_q(t) = A\sin(\phi(t)) + n_{oq}(t)$. In practice, however, this method is difficult to implement as coherent detectors are needed to recover the in-phase and quadrature components of $z(t)$. In order to overcome this problem, a balanced discriminator can be deployed to estimate the IF of $z(t)$ [18, p. 216]. This is the method used in these simulations. The samples of the signal $z(t)$ are passed through an finite impulse response (FIR) differentiator. The envelope of the output of the differentiator is proportional to $f_{i,z}(t)$. Therefore, unlike other estimators such as the ZCR estimator, the IF-based estimator does not require coherent detectors. The Ricean K factor is estimated using the estimator of Equation 5.70 derived in the next section. The IF-based estimator of Equation 5.47 can then be calculated using the envelope of the estimated IF and the estimated K factor.

The performance of the IF-based estimator in the presence of shadowing and additive noise was compared with that of the ZCR, LCR, ROM, and COV-based methods by computing the average normalized bias and variance over 100 realizations of the received signal. The sensitivity of the COV-based method to additive noise can be reduced significantly by choosing a large time lag. However, a large time lag reduces the accuracy of the estimator itself. Here, the time lag in the COV-based method was chosen to be 2.5 msec, which according to our simulations gives an accurate estimate of the MS velocity in the ideal case. In order to limit the delay in obtaining velocity estimates for real-time implementation, we used a window of 1 sec of the simulated signal to estimate the unknown velocity. As the window length increases, the bias introduced by estimating $E\{|f_{i,y}|\}$ using the time average of the estimated IF decreases, and the delay in obtaining velocity estimates increases.

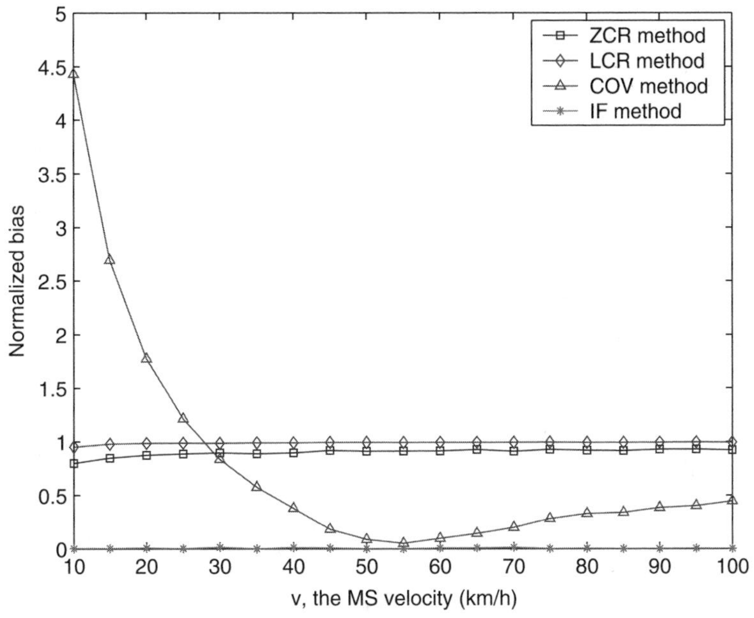

FIGURE 5.10 Normalized bias as a function of the MS velocity in the presence of shadowing.

Figure 5.10 shows the effect of shadowing on each of the velocity estimators in the absence of additive noise and in the case of isotropic scattering. We observe that in the presence of shadowing, the normalized bias in estimated velocity for the IF-based estimator is negligible compared to the other estimators. This confirms that the IF-based estimator is robust to shadowing.

Figure 5.11 shows the effect of additive noise in the absence of shadowing, assuming isotropic scattering. We observe that the IF-based estimator is generally more robust than the other estimators. It exhibits a

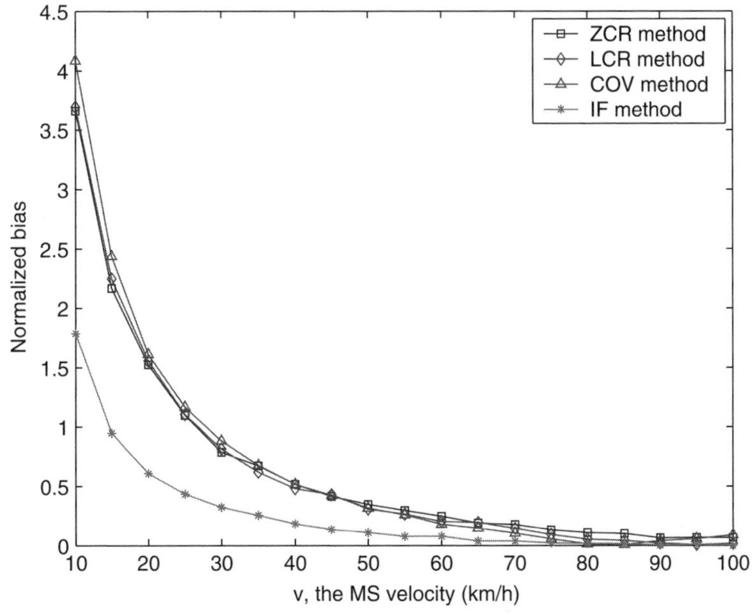

FIGURE 5.11 Normalized bias as a function of the MS velocity in the presence of AWGN for SNR = 10 dB.

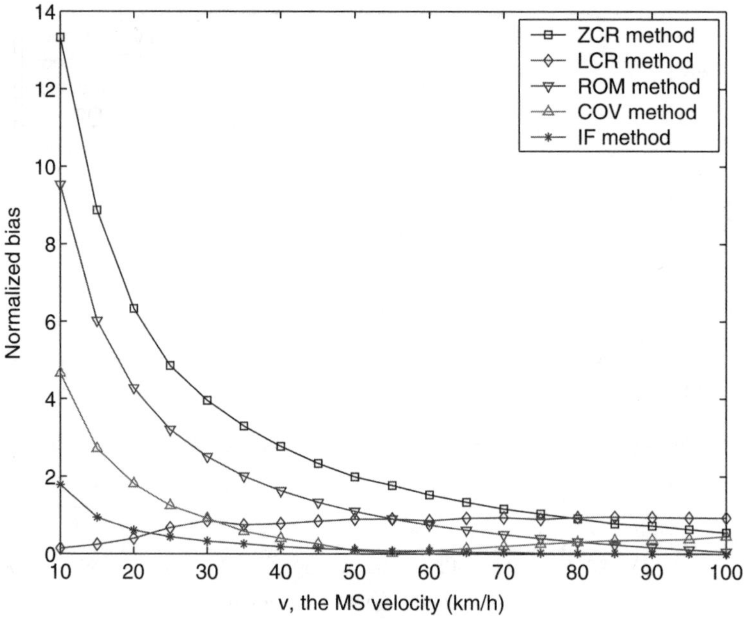

FIGURE 5.12 Normalized bias as a function of the MS velocity in the presence of AWGN, shadowing, and isotroic scattering for SNR = 10 dB.

smaller bias than the other estimators for small velocities. This is due to the improvement in SNR at the output of the IF estimator (see Appendix C). As the MS velocity increases, the system bandwidth approaches the actual Doppler shift, and therefore, the difference between the normalized biases of all the estimators becomes particularly small. Finally, Figure 5.12 to Figure 5.14 represent the respective normalized bias,

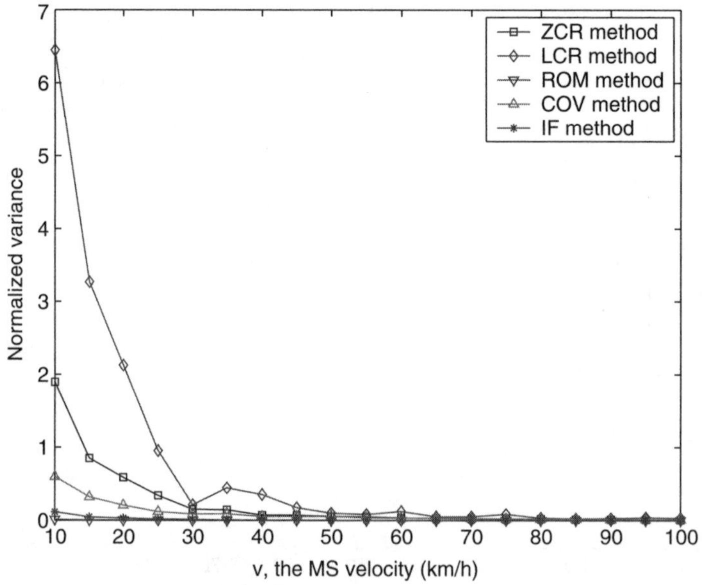

FIGURE 5.13 Variance as a function of the MS velocity in the presence of AWGN, shadowing, and isotroic scattering for SNR = 10 dB.

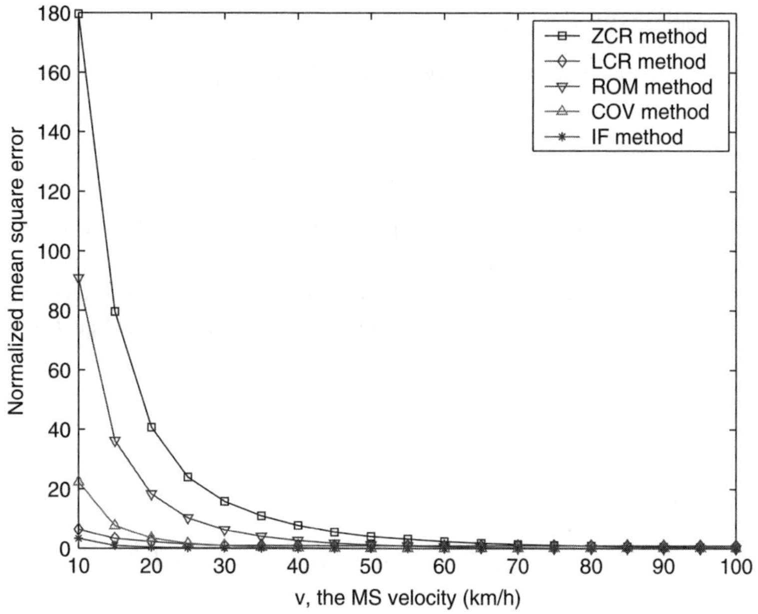

FIGURE 5.14 Mean square error as a function of the MS velocity in the presence of AWGN, shadowing, and isotropic scattering for SNR = 10 dB.

the variance, and the mean square error of the above-considered estimators in the presence of additive noise, shadowing, and isotropic scattering. They indicate that the ROM estimator exhibits slightly lower variance than the IF-based estimator, but higher bias and mean square error.

5.6 Rice Factor Estimation

5.6.1 Existing Methods

The value of K influences most local power and mobile velocity estimators [7]. It is also a measure of the severity of the fading, $K = 0$ being the most severe Rayleigh fading and $K = \infty$ indicating that there is no fading. Hence, knowledge of the Ricean K factor is a good indicator of the channel quality [25].

Traditional methods for estimating the Ricean K factor are based on measurements of the received power. In [25], the probability distribution of the measured data is computed and compared to a set of hypothesis distributions using a suitable goodness-of-fit test. In [46], a maximum likelihood estimate of K is obtained using an expectation–maximization (EM) algorithm. Both methods have a high degree of complexity, which makes them relatively time-consuming [24]. A simple and rapid method for K estimation uses two estimated moments of the received power [24]. The same estimator has been derived in [66] using an approach based on the covariance of the received power. In [63], the Ricean K factor is estimated using the first two moments of the envelope of the received signal. However, this method did not receive much attention, as it did not provide a closed-form expression for K. However, Tepedelenlioĝlu et al. [64] recently proposed a general class of moment-based estimators for K, which included the estimator of [63]. They derived the asymptotic variance of each member of the family and showed that the asymptotic variance of the estimator in [63] is the closest to the Cramer–Rao lower bound. In this section, we present two explicit and simple estimators for K based on an approximation of the method in [63]. We also show, using simulations, that one of the proposed estimators has a lower mean square error than the estimator in [24] and [66].

5.6.2 Envelope-Based Estimators

The ratio of the first two moments of the envelope, $r(t)$, of the Rician fading is [53]

$$E_r \triangleq \frac{E\{r\}}{\sqrt{E\{r^2\}}} = \frac{\sqrt{\pi}e^{-\frac{K}{2}}}{2\sqrt{K+1}}\left[(1+K)I_0\left(\frac{K}{2}\right)+KI_1\left(\frac{K}{2}\right)\right] \tag{5.64}$$

Using Equation 5.64, $\hat{K}$ can be obtained by first estimating the first two moments of $r(t)$, and then solving Equation 5.64 numerically for K. This was the method originally proposed in [63]. However, since Equation 5.64 does not provide an explicit formula for K, it has not been used as an estimator.

To overcome the aforementioned drawback, we approximate the right side of Equation 5.64, and derive two explicit estimators for K. We first rewrite Equation 5.64 as

$$E_r = \frac{g(K)}{K+1} \tag{5.65}$$

where $g(K)$ is defined as

$$g(K) \triangleq \frac{\sqrt{\pi}}{2}e^{-\frac{K}{2}}\sqrt{K+1}\left[(1+K)I_0\left(\frac{K}{2}\right)+KI_1\left(\frac{K}{2}\right)\right] \tag{5.66}$$

Figure 5.15 plots $g(K)$ for $K \in [0, 100]$. It suggests that $g(K)$ can be well approximated with a low-order polynomial function $g_N(K)$, where

$$g_N(K) = \sum_{i=0}^{N} p_{iN}K^i \tag{5.67}$$

The coefficients p_{iN} $(i = 0, 1, \ldots, N)$ are computed by fitting $g_N(K)$ to $g(K)$ in a least squares sense. It can be shown that for $N > 2$, $p_{iN}, i > 2$ is of order 10^{-5}. We choose to approximate $g(K)$ with a linear and quadratic function, $g_1(K)$ and $g_2(K)$, respectively.

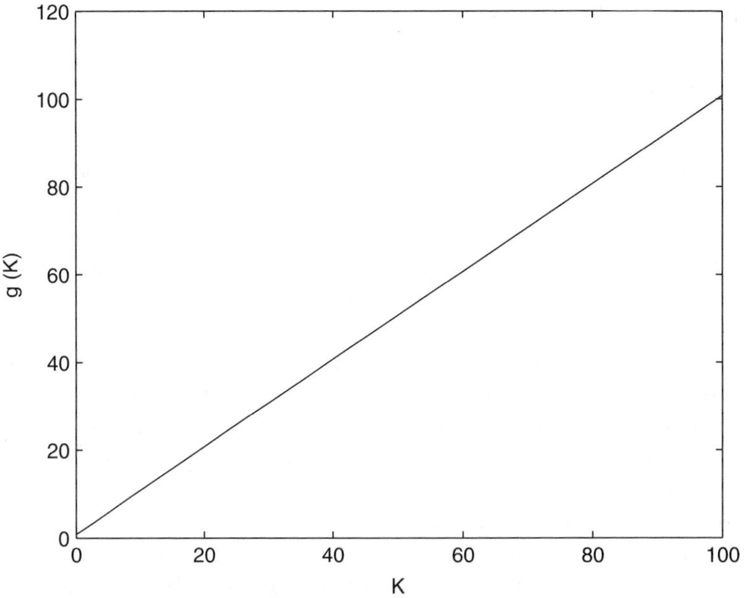

FIGURE 5.15 $g(K)$ in Equation 5.66 vs. K. We observe that $g(K)$ can be well approximated by a low-order polynomial.

Replacing $g(k)$ by its linear approximation in Equation 5.65 leads to the following estimator:

$$\widehat{K_1} = \frac{\widehat{E_r} - p_{01}}{p_{11} - \widehat{E_r}} \tag{5.68}$$

In Equation 5.68, $\widehat{E_r}$ is the ratio of the estimated first and second moments of the envelope of the received signal, i.e., $\widehat{E_r} = \overline{R}/\sqrt{\overline{R^2}}$ where, assuming local ergodicity,

$$\overline{R^n} = \frac{1}{T}\int_T R^n(t)dt, \quad n = 1, 2$$

and T is the duration of the observed signal. By fitting $g_1(K)$ to $g(K)$ in a least squares sense, we find

$$p_{01} = 0.7967, \; p_{11} = 0.9969 \tag{5.69}$$

Following the same approach, $g(k)$ is replaced by its quadratic approximation in Equation 5.65, which leads to the following estimator:

$$\widehat{K_2} = \frac{\widehat{E_r} - p_{12} + \sqrt{(\widehat{E_r} - p_{12})^2 + 4p_{22}(\widehat{E_r} - p_{02})}}{2p_{22}} \tag{5.70}$$

where

$$p_{02} = 0.8293, \; p_{12} = 0.9866, \; p_{22} = 0.0005 \tag{5.71}$$

Parameter E_r in Equation 5.64 can be approximated by

$$E_r \simeq E_{r1} = \frac{p_{01} + p_{11}K}{K+1} \tag{5.72}$$

or

$$E_r \simeq E_{r2} = \frac{p_{02} + p_{12}K + p_{22}K^2}{K+1} \tag{5.73}$$

depending on the degree of accuracy required.

Figure 5.16 plots Equation 5.64 and its approximations (Equation 5.72 and Equation 5.73). It verifies that E_{r2} is a better approximation than E_{r1}.

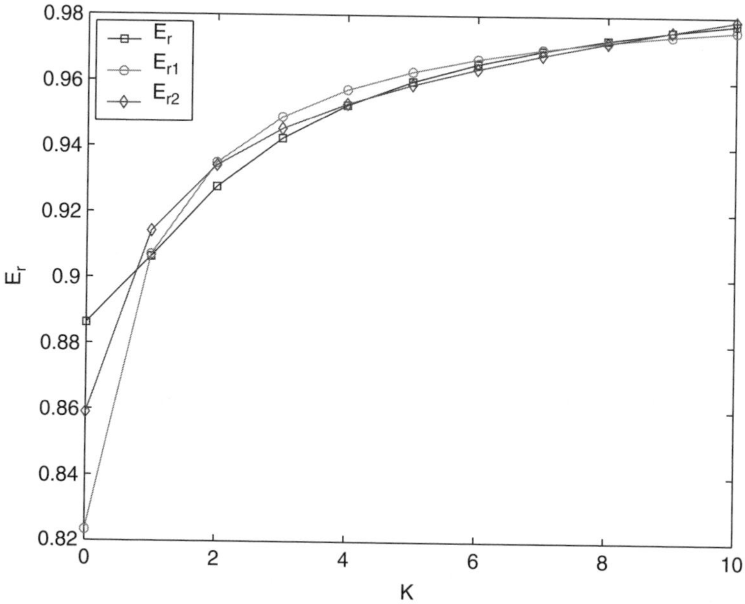

FIGURE 5.16 E_r vs. K (Equation 5.64) and its approximations (Equation 5.72 and Equation 5.73).

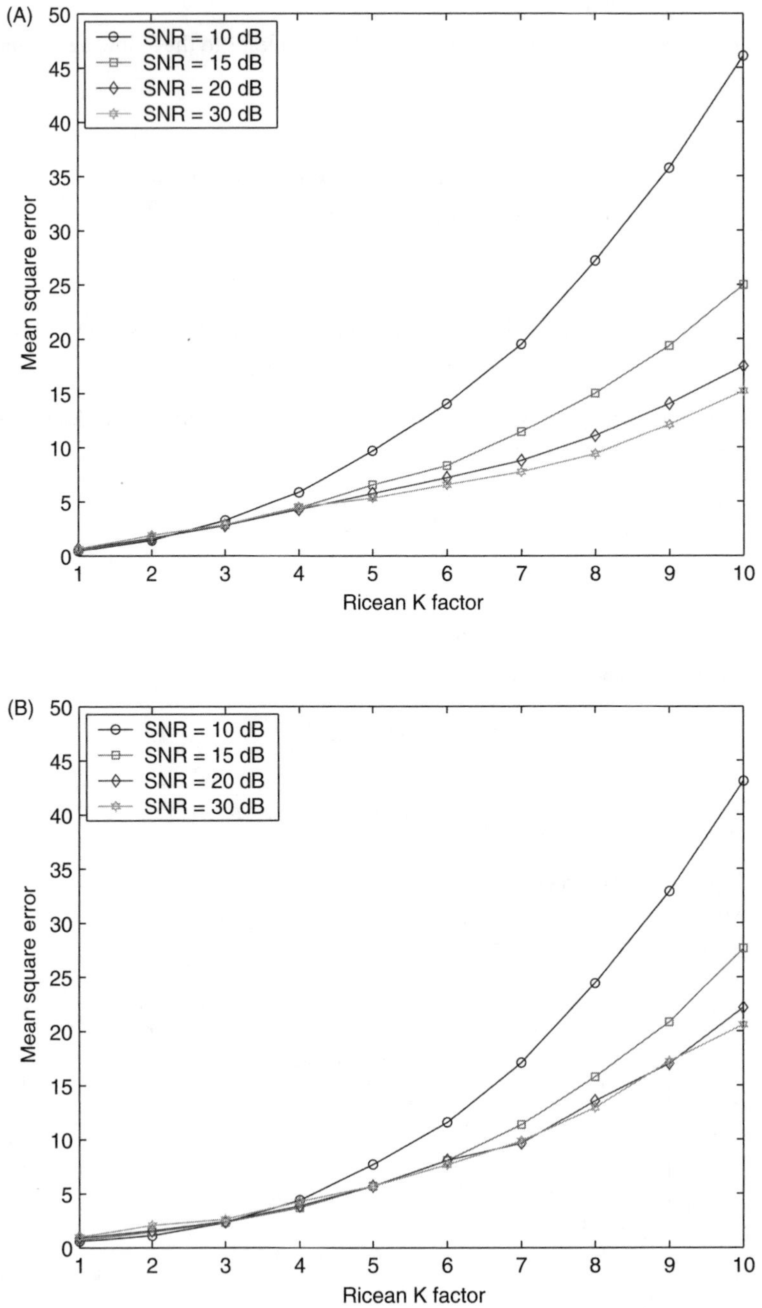

FIGURE 5.17 The MSE of the (A) estimator in Equation 5.70, (B) estimator in Equation 5.74, and (C) estimator in Equation 5.68, for different values of the SNR. The estimator in Equation 5.70 outperforms the other estimators for SNRs larger than 15 dB.

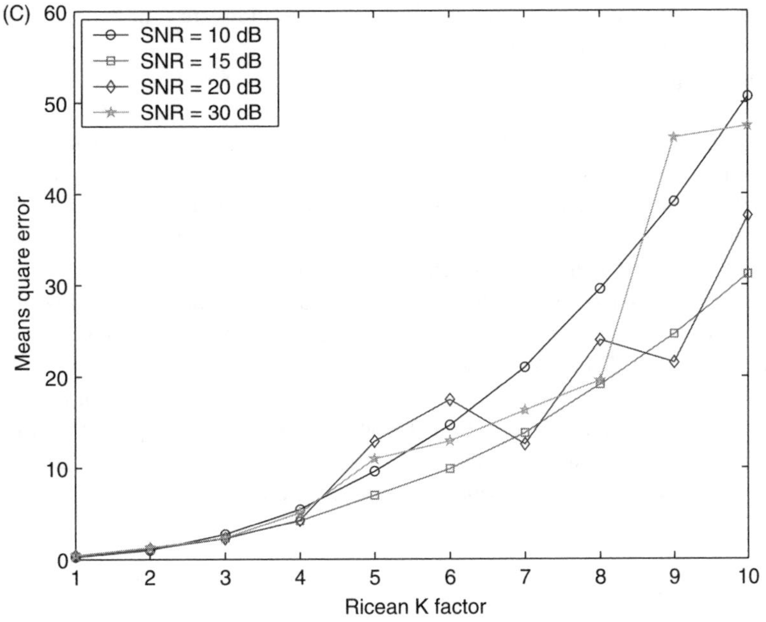

FIGURE 5.17 (*Continued*).

5.6.3 Simulation Results

In this section, the performance of the proposed estimators is compared to that of the COV-based estimator [24], [66]. The performance criterion is the mean square error of the estimators. The COV-based estimator is defined as [66]

$$\hat{K}_{COV} = \frac{\sigma_y^4 - c_{|y|^2}(0) + \sigma_y^2\sqrt{\sigma_y^4 - c_{|y|^2}(0)}}{c_{|y|^2}(0)} \tag{5.74}$$

where $c_{|y|^2}(\tau)$ refers to the covariance of $|y(t)|^2$ and σ_y^2 is the power of the received signal.

The mean square error (MSE) of the three estimators was computed over 1000 realizations of the received signal. Figure 5.17 shows the MSE of the estimators as a function of SNR. We observe that for SNRs larger than 15 dB, the estimator given by Equation 5.70 is superior to the estimators given by Equation 5.68 and Equation 5.74.

5.7 Application on Handover Performance

5.7.1 Handover Decision Algorithms

In cellular, microcellular, and picocellular systems, as a vehicule crosses the cell boundary between two BSs, control has to be transferred from the current BS to the target BS. This process is referred to as handover or handoff and is generally categorized into soft handoff and hard handoff processes [70, 75]. In hard handover processes, only one BS is in control of an MS at a given time. Soft handover processes require two or more BSs to be allocated to a single MS, in order to prevent the received signal power from dropping below a given threshold level. Soft handover processes are currently specific to CDMA systems where power control is critical [75]. The aim of this section is to study the effect of a biased velocity estimation on the performance of handover decision algorithms. The derivations are done in the case of hard handoff processes but can also be adapted to soft handoff processes.

Consider an MS traveling from its current base station, BS1, to a target base station, BS2 (Figure 5.18). The decision to hand off is based on the received signal strength from both base stations. In the absence of shadowing, fading, and neglecting the effect of additive noise, the received signal envelopes are only affected by path loss. The path loss is inversely proportional to d^α, where d is the distance to the BS and α is a parameter determined by the environment. At mid-distance from the two BSs, the received strengths are identical, or equivalently, the difference between the two received strengths is zero. A decision to hand off from the current BS to the target BS could then be made when the sign of the difference between the two received strengths changes. However, in the presence of shadowing or fading, the sign can change several times around the cell boundary, which could result in a large number of unnecessary handovers. This is called the ping-pong effect. The effects of fast fading are significantly reduced through time-averaging the received signals over a window of size L. In [62], it was recommended that the window size corresponds to a distance of 40λ, hence to a period of time $T = 40\lambda/v$. Assuming the signal strengths are measured every T_s seconds, the time averaging should occur over $L = [\frac{40\lambda}{vT_s}]$ samples, where $[x] \leq x < [x] + 1$. The window size is therefore velocity dependent and is smaller for higher velocities than for smaller velocities.

5.7.2 Effect of an Error in Velocity Estimation on the System's Quality of Service

The quality of service of a system requires that the probability of a call being dropped must be less than p_{out}, where p_{out} is referred to as outage probability. Here we investigate how an error in velocity estimation δv would impact the probability of lost calls.

Every sampling time T_s, the MS of Figure 5.18 records measurements of the received signal strengths from BS1 and BS2 and computes a time average of the two received signal envelopes based on current and past measurements. Let $\Delta y_1(k)$ and $\Delta y_2(k)$ be the kth sample of the time-averaged signals from BS1 and BS2, respectively. Assuming a measurement time $k \geq L$, $\Delta y_1(k)$ and $\Delta y_2(k)$ can be expressed as

$$\Delta y_i(k) = \frac{1}{L} \sum_{j=k-L+1}^{k} y_i(j), \quad i = \{1,2\} \tag{5.75}$$

where $y_i(j)$ is the signal strength received from BSi at time j.

An error, δv, in velocity estimation leads to the choice of a temporal window

$$L' = \left[\frac{40\lambda}{(v + \delta v)T_s}\right] = L + \delta L \tag{5.76}$$

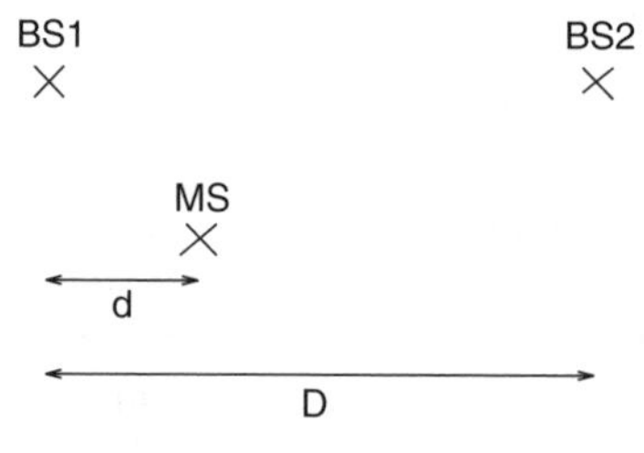

FIGURE 5.18 We consider an MS traveling from its current base station, BS1, toward a target base station, BS2.

where δL is a positive integer when δv is negative and a negative integer when δv is positive. It follows that the averaged signal strengths in Equation 5.75 become

$$\Delta' y_i(k) = \frac{1}{L + \delta L} \sum_{j=k-L-\delta L+1}^{k} y_i(j), \quad i = \{1, 2\} \tag{5.77}$$

Assuming an error δL such that $\delta L / L \ll 1$, Equation 5.77 can be written as

$$\Delta' y_i(k) = \frac{1}{L}\left(1 - \frac{\delta L}{L}\right)\left[L \Delta y_i(k) + \sum_{j=k-L-\delta L+1}^{k-L} y_i(j)\right] \tag{5.78}$$

$$\delta L \geq 0 \tag{5.79}$$

$$\Delta' y_i(k) = \frac{1}{L}\left(1 - \frac{\delta L}{L}\right)\left[L \Delta y_i(k) - \sum_{j=k-L+1}^{k-L-\delta L} y_i(j)\right] \tag{5.80}$$

$$\delta L < 0 \tag{5.81}$$

where $\Delta y_i(k), i = \{1, 2\}$, is defined in Equation 5.75. An error δL in window size leads to an error δe_i in average signal strength, where δe_i is defined as

$$\delta e_i(k) = \Delta' y_i(k) - \Delta y_i(k) \tag{5.82}$$

$$= \frac{1}{L} \sum_{k-L-\delta L+1}^{k-L} y_i(j) - \frac{\delta L}{L^2}\left[L \Delta y_i(k) + \sum_{k-L-\delta L+1}^{k-L} y_i(j)\right] \tag{5.83}$$

$$\delta L \geq 0 \tag{5.84}$$

$$\delta e_i(k) = -\frac{1}{L} \sum_{j=k-L+1}^{k-L-\delta L} y_i(j) - \frac{\delta L}{L^2}\left[L \Delta y_i(k) - \sum_{j=k-L+1}^{k-L-\delta L} y_i(j)\right] \tag{5.85}$$

$$\delta L < 0 \tag{5.86}$$

where $i = \{1, 2\}$.

It is assumed, in handover decision algorithm design, that the process of time averaging has smoothed the effect of fast fading but not that of shadowing [51]. As a consequence, merely handing over from BS1 toward BS2 and back to BS1 when the difference $\Delta y_1(k) - \Delta y_2(k)$ changes sign would still result in a high number of unnecessary handovers. A more appropriate handover decision algorithm consists of handing over from BS1 toward BS2 when $\Delta y_1(k) - \Delta y_2(k) < -h$, and back to BS1 when $\Delta y_1(k) - \Delta y_2(k) > h$, where h is a hysteresis value determined by the system and by the environment (see Figure 5.19) [70]. The choice of h can be critical in terms of minimizing two conflicting criteria. If h is too small, shadowing is still dominant, resulting in a high number of unnecessary handovers. If h is too large, the handover process is delayed and a call could be lost.

A call is lost when the base station is BS1 and $\Delta y_1(k)$ drops below a given threshold th determined by the sensitivity of the receiver, or similarly, when the base station is BS2 and $\Delta y_2(k)$ is below th. The probability of a call being dropped at time k is then [70]

$$p_{LC}(k) = prob(BS1(k) \& \Delta y_1(k) < th) + prob(BS2(k) \& \Delta y_2(k) < th) \tag{5.87}$$

If an error δv occurs in the estimation of the velocity, a decision to hand off from BS1 toward BS2 would now be made when $\Delta' y_1(k) - \Delta' y_2(k) < -h$, and back to BS1 when $\Delta' y_1(k) - \Delta' y_2(k) > h$. Equivalently, a hand off toward BS2 would occur when $\Delta y_1(k) - \Delta y_2(k) < -h + \delta h$, and back to BS1 when $\Delta y_1(k) - \Delta y_2(k) > h + \delta h$, where $\delta h = \delta e_1(k) - \delta e_2(k)$. This could be interpreted as the decision to hand off in a system where the mobile velocity is known or is estimated in a very accurate way, but

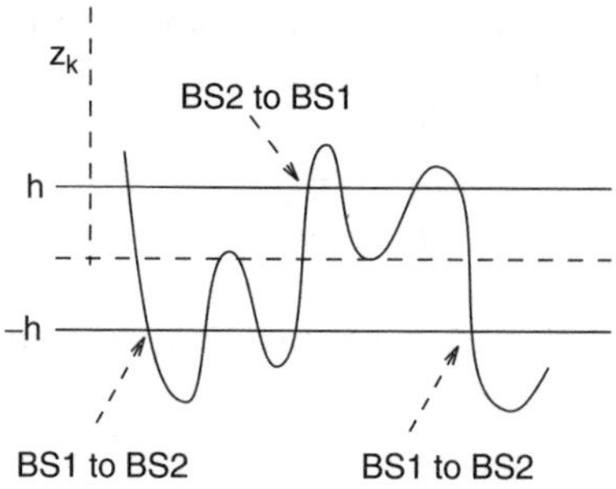

FIGURE 5.19 A handover is processed toward BS2 when $\Delta y_1(k) - \Delta y_2(k) < -h$, and back to BS1 when $\Delta y_1(k) - \Delta y_2(k) > h$, where h is a hysteresis value determined by the system and by the environment.

where two distinct hysteresis values, $h_1 = -h + \delta h$ and $h_2 = h + \delta h$, are used for handing over from BS1 to BS2 and from BS2 to BS1, respectively, where $h_1 \neq h_2$. It is assumed that the hysteresis value h was optimized so that the probability of lost calls resulting from a delay in handover is below the outage probability of the communications system. However, in the presence of an error in velocity estimation, if δh is negative, a delay in handing over from BS1 to BS2 occurs, and if δh is positive, a delay in handing over back to BS1 occurs. Either way, a delay in handover occurs, and depending on the value of δh, this delay could contribute to increasing the probability of losing a call. There is another way an error in velocity estimation can contribute to an increase in the probability of lost calls. Because of the error in velocity estimation, the probability of losing a call at time k is now

$$p_{LC}(k) = prob(BS1(k)\&\Delta y_1(k) < th - \delta e_1(k)) + prob(BS2(k)\&\Delta y_2(k) < th - \delta e_2(k)) \quad (5.88)$$

Equation 5.88 can be interpreted as the probability of lost calls in a system where the mobile velocity is known or is estimated in a very accurate way, but where the sensitivity of the receiver is different. If $\delta e_1(k)$ and $\delta e_2(k)$ are positive, the receiver can be considered as having better sensitivity. However, if either $\delta e_1(k)$ or $\delta e_2(k)$ is negative, the sensitivity of the receiver could be considered worse, which could contribute to an increase in the probability of lost calls. In [29], it was found that an increase of 20% in velocity estimation error results in the probability of lost calls increasing by 50%.

5.8 Conclusions and Perspectives

This chapter outlined the principles and methods of velocity estimation, provided examples of velocity estimators, and gave a framework for analyzing the performance of velocity estimators. It also discussed the implications of erroneous velocity estimation on the wireless telecommunications systems and, more specifically, in the context of handover algorithm design. It was found that an error in velocity estimation could significantly deteriorate the quality of service of the system in terms of probability of dropped calls. The CDMA 2000 system is currently being implemented in Australia. Although the service provider, TELSTRA, sets the standards to be met by mobile phone manufacturers, it has no control over which or even if a velocity estimator is to be used by the telephone hand set manufacturer. As a result, two mobile phone sets from different manufacturers used at the same location can have different performances. This suggests the need, in the future, for more control over which standards are to be used by mobile phone manufacturers.

Acknowledgments

Prof. B. Boashash thanks the editor for his invitation to write this chapter. The material in the chapter is mostly extracted from the Ph.D. thesis of G. Azemi, whose supervisors were B. Boashash and B. Senadji. The authors also thank Dr. Sylvie Perreau and TELSTRA for technical discussions related to this chapter.

Appendix A: Derivation of Equation 5.39

In this appendix, we derive the first moment of the envelope of $|f_{i,s}|$. Using Equation 5.38, the expected value of $|f_{i,y}|$ can be written as

$$E\{|f_{i,y}|\} = \frac{1}{2\pi}\int_0^{\infty}\int_{-\pi}^{\pi}\int_{-\infty}^{\infty}\int_{-\infty}^{\infty} |\dot{\psi}|\ p(r,\psi,\dot{r},\dot{\psi})\ d\dot{\psi}\ d\dot{r}\ d\psi\ dr \tag{5.A1}$$

where $p(r,\psi,\dot{r},\dot{\psi})$ is the joint pdf of r, ψ, $\dot{r}$, $\dot{\psi}$. In the case of isotropic scattering, the joint pdf is given by [54]

$$\begin{aligned} p_1(r,\psi,\dot{r},\dot{\psi}) = {} & \frac{r^2}{4\pi^2 b_0 b_2} \\ & \times \exp\left(\frac{-1}{2b_0}(r^2 - 2Qr\cos\psi + Q^2)\right) \\ & \times \exp\left(\frac{-1}{2b_2}(\dot{r}^2 + r^2\dot{\psi}^2)\right) \end{aligned} \tag{5.A2}$$

Substituting Equation 5.A2 into Equation 5.A1, and since $p_1(r,\psi,\dot{r},\dot{\psi})$ is an even function of $\dot{\psi}$, Equation 5.A1 can now be written as

$$E\{|f_{i,y}|\} = \frac{1}{\pi}\int_0^{\infty}\int_{-\pi}^{\pi}\int_{-\infty}^{\infty}\int_{0}^{\infty} \dot{\psi}\ p_1(r,\psi,\dot{r},\dot{\psi})\ d\dot{\psi}\ d\dot{r}\ d\psi\ dr \tag{5.A3}$$

Using the following identities [23],

$$\int_{-\infty}^{\infty} \exp\left(\frac{-1}{2b_2}\dot{r}^2\right) d\dot{r} = \sqrt{2\pi b_2} \tag{5.A4a}$$

$$\int_0^{\infty} \dot{\psi}\ \exp\left(\frac{-1}{2b_2}r^2\dot{\psi}^2\right) d\dot{\psi} = b_2 r^{-2} \tag{5.A4b}$$

$$\int_{-\pi}^{\pi} \exp\left(\frac{1}{b_0}Qr\cos\psi\right) d\psi = 2\pi\, I_0\left(\frac{Q}{b_0}r\right) \tag{5.A4c}$$

$$\int_0^{\infty} \exp(-\mu^2x^2)\, I_0(\nu x)dx = \frac{\sqrt{\pi}}{2\mu}\exp\left(\frac{\nu^2}{8\mu^2}\right) I_0\left(\frac{\nu^2}{8\mu^2}\right) \tag{5.A4d}$$

in Equation 5.A3, we obtain

$$E\{|f_{i,y}|\} = \frac{1}{2\pi}\sqrt{\frac{b_2}{b_0}}\exp\left(\frac{-\rho}{2}\right) I_0\left(\frac{\rho}{2}\right) \tag{5.A5}$$

where $\rho = \frac{Q^2}{2b_0}$. In the absence of noise ($\gamma \to \infty$) and in the presence of isotropic scattering, b_0 and b_2 are equal to a_0 and $2\pi^2 f_m^2 a_0$, respectively, and since $K = \frac{\eta^2}{2\sigma^2}$, Equation 5.A5 becomes Equation 5.39.

Appendix B: Effect of the Scattering Distribution on the ZCR Method

In this appendix, we study the effect of nonisotropic scattering on the estimator in Equation 5.44, in the absence of shadowing and additive noise. In this case, ZCR_{x_i} in Equation 5.44 is given by Equation 5.32. From Equation 5.53, in the absence of shadowing and additive noise, b_0 and b_2 become M^2a_0 and M^2a_2, respectively. In this case, Equation 5.44 reduces to

$$N_{\mathrm{ZCR}} = \frac{v}{\lambda}\sqrt{q_2(\chi)} \tag{5.B1}$$

It follows that the normalized bias for the ZCR-based velocity estimator is derived as

$$\varepsilon(\chi, \infty, K) = \sqrt{2q_2(\chi)} - 1 \tag{5.B2}$$

which is the same as Equation 5.57. Equation 5.B2 proves that both ZCR and IF-based estimators have the same performances in the presence of nonisotropic scattering.

Appendix C: SNR Improvement in the IF Estimator

In this appendix, we prove that the IF estimator improves the SNR. If we assume that the power of the noise at the input of the IF estimator is $N_0 B_0$, the SNR at the input of the IF estimator is derived as

$$\gamma_R = \left(\frac{S}{N}\right)_R = \frac{A^2}{2N_0 B_0} \tag{5.C1}$$

Based on the results in [18, Section 10.3], when γ_R is more than a threshold level[3], the signal $y_{out}(t)$ at the output of the IF estimator, can be expressed as

$$y_{out}(t) = f_{i,y}(t) = f_{i,s}(t) + \xi(t)$$

where $\xi(t)$ represents the IF noise. The PSD of the IF noise is given by [18, p. 414]

$$S_\xi(f) = \frac{N_0 f^2}{A^2} \prod\left(\frac{f}{B_0}\right)$$

where $\prod(\,.\,)$ represents the rectangular pulse function. Thus, the power of the IF noise in the frequency band $(f_1, \frac{B_0}{2})$ can then be computed as

$$N_D = 2\int_{f_1}^{\frac{B_0}{2}} S_\xi(f)df = \frac{2N_0}{3A^2}\left(\frac{B_0^3}{8} - f_1^3\right)$$

The power of $f_i(t)$ in the frequency band $(f_1, \frac{B_0}{2})$ can also be computed as [43, p. 256]

$$S_D = \left(\frac{1}{4\pi^2}\right) 2\int_{f_1}^{\frac{B_0}{2}} S_\phi(f)df = \frac{B_0^2}{8}\ln\frac{B_0}{2f_1}$$

Therefore, assuming $f_1 \ll B_0$, the SNR at the output of the IF estimator can be derived as

$$\gamma_D = \left(\frac{S}{N}\right)_D \simeq 3\gamma_R \ln\frac{B_0}{2f_1} \tag{5.C2}$$

It follows from Equation 5.C2 that the IF estimator improves the SNR by a factor $3\ln\frac{B_0}{2f_1}$.

[3] This threshold is found to occur in the vicinity of 10-dB input SNR and varies in the approximate range of 6 to 13 dB [67, p. 458].

References

[1] A. Abdi and M. Kaveh. A new velocity estimator for cellular systems based on higher order crossings. In *Proceeding of the Asilomar Conference on Signals, Systems, and Computers*, Vol. 2, 1998, pp. 1423–1427.

[2] M.G. Amin. Interference mitigation in spread spectrum communication systems using time-frequency distributions. *IEEE Trans. Signal Process.*, 45, Jan. 1997.

[3] M.G. Amin, C. Wang, and A.R. Lindsey. Optimum interference excision in spread spectrum communications using open-loop adaptive filters. *IEEE Trans. Signal Process.*, 47,1966–1976, 1999.

[4] M. Anderson and S. Perreau. Robust power control for cdma networks subject to modelisation errors. In *Proceedings of the IEEE International Conference on Acoustics, Speech, and Signal Processing, ICASSP'03*, Hong Kong, 2003, pp. 612–619.

[5] K.D. Anim-Appiah. On generalized covariance-based velocity estimation. *IEEE Trans. Veh. Technol.*, 48, 1546–1557, 1999.

[6] M.D. Austin and G.L. Stüber. Eigen-based doppler estimation for differentially coherent CPM. *IEEE Trans. Veh. Technol.*, 43, 781–785, 1994.

[7] M.D. Austin and G.L. Stüber. Velocity adaptive handoff algorithms for microcellular systems. *IEEE Trans. Veh. Technol.*, 43, 549–561, 1994.

[8] G. Azemi. Mobile Velocity Estimation Using a Time-Frequency Approach. Ph.D. thesis, Queensland University of Technology, Brisbane, Australia, 2003.

[9] G. Azemi, B. Senadji, and B. Boashash. Mobile unit velocity estimation based on the instantaneous frequency of the received signal. Submitted.

[10] G. Azemi, B. Senadji, and B. Boashash. Mobile unit velocity estimation in micro-cellular systems using the ZCR of the instantaneous frequency of the received signal. In *Proceedings of the International Symposium on Signal Processing and Its Applications, ISSPA'03*, Vol. 2, Paris, France, 2003, pp. 289–292.

[11] G. Azemi, B. Senadji, and B. Boashash. A novel estimator for the velocity of a mobile station in a microcellular system. In *Proceedings of the International Symposium on Circuits and Systems, ISCAS'03*, Vol. 2, Bangkok, Thailand, May 2003, pp. 212–215.

[12] S. Barbarossa and A. Scaglione. Optimal precoding for transmissions over linear time-varying channels. In *Seamless Interconnection for Universal Services, GLOBECOM'99*, Vol. 5, Piscataway, NJ, 1999, pp. 2545–2549.

[13] A. Belouchrani and M.G. Amin. Blind source separation based on time–frequency signal representations. *IEEE Trans. Signal Process.*, 46, 2888–2897, 1998.

[14] B. Boashash. Estimating and interpreting the instantaneous frequency of a signal. Part 1. Fundamentals. *Proc. IEEE*, 80, 519–538, 1992.

[15] B. Boashash. Estimating and interpreting the instantaneous frequency of a signal. Part 2. Algorithms and applications. *Proc. IEEE*, 80, 539–568, 1992.

[16] B. Boashash, Ed. *Time-Frequency Signal Analysis and Processing: A Comprehensive Reference*. Elsevier, Amsterdam, 2003.

[17] B. Boashash, A. Belouchrani, K. Abed-Meraim, and L.-T. Nguyen. Time-frequency signal processing for wireless communications. Chapter 25 in *Signal Processing for Mobile Communications Handbook*, CRC Press, Boca Raton, FL, M. Ibnkahla, Ed. 2003.

[18] A.B. Carlson, P.B. Crilly, and J.C. Rutledge. *Communication Systems: An Introduction to Signals and Noise in Electrical Communication*, 4th ed. McGraw-Hill, New York, 2002.

[19] A. Chockalingam, P. Dietrich, L.B. Milstein, and R.R. Rao. Performance of closed-loop power control in DS-CDMA cellular systems. *IEEE Trans. Veh. Technol.*, 47, 774–789, 1998.

[20] M.J. Chu and W.E. Stark. Effect of mobile velocity on communications in fading channels. *IEEE Trans. Veh. Technol.*, 49, 202–210, 2000.

[21] Standard Document, Ed. *Physical Layer Standard for CDMA 2000 Spread Spectrum Systems*. 3GPP, 1999.

[22] H.S.H. Gombachika and O.K. Tonguz. Influence of multipath fading and mobile unit velocity on the performance of PN tracking in CDMA systems. In *Proceedings of the Vehicular Technology Conference, IEEE 47th*, Vol. 3, 1997, pp. 2206–2209,

[23] I.S. Gradshteyn and I.M. Ryzhik. *Table of Integrals, Series, and Products*. Academic Press, New York, 1994.

[24] D.G. Greenstein, L.J. Michelson, and V. Erceg. Moment-method estimation of the Ricean K-factor. *IEEE Commun. Lett.*, 3, 175–176, 1999.

[25] D. Greenwood and L. Hanzo. Characterization of mobile radio channels. In *Mobile Radio Communications*, R. Steele, Ed., Pentech, London, 1992, pp. 163–185.

[26] H. Hansen, S. Affes, and P. Mermelstein. A Rayleigh Doppler frequency estimator derived from maximum likelihood theory. In *IEEE Workshop on Signal Processing Advances in Wireless Communications, SPAWC'99*, 1999, pp. 382–386.

[27] R. Hass and J.C. Belfiore. A time–frequency well-localized pulse for multiple carrier transmission. *Wireless Personal Commun.*, 5, 1–18, 1997.

[28] M. Hellebrandt, R. Mathar, and M. Scheibenbogen. Estimating position and velocity of mobiles in a cellular radio network. *IEEE Trans. Veh. Technol.*, 46, 65–71, 1997.

[29] E. Holmbakken, G. Azemi, V. Karawalevu, and B. Senadji. The importance of accurate velocity estimation in designing handover algorithms for microcellular systems. In *Workshop on Signal Processing and Applications, WoSPA*, Brisbane, Australia, December 2002.

[30] J.M. Holtzman. Adaptive measurement intervals for handoffs. In *Proceedings of the IEEE International Conference on Communications, ICC*, Vol. 2, June 1992, pp. 1032–1036.

[31] J.M. Holtzman and A. Sampath. Adaptive averaging methodology for handoffs in cellular systems. *IEEE Trans. Veh. Technol.*, 44, 59–66, 1995.

[32] M. Ibnkahla, Ed. *Signal Processing for Mobile Communications Handbook*. CRC Press, Boca Raton FL, 2003.

[33] K. Ivanov and G. Spring. Mobile speed sensitive handover in a mixed cell environment. In *Proceedings of the Vehicular Technology Conference, IEEE 45th*, Vol. 2, 1995, pp. 892–896.

[34] W.C. Jakes, Ed. *Microwave Mobile Communication*. IEEE Press, Washington, DC, 1974.

[35] Z.M. Kamran, A.R. Leyman, and K. Abed-Meraim. Techniques for blind source separation using higher-order statistic. In *IEEE Workshop on Statistical Signal and Array Processing, Pennsylvania, USA*, August 2000, pp. 334–338.

[36] K. Kawabata, T. Nakamura, and E. Fukuda. Estimating velocity using diversity reception. In *Proceedings of the Vehicular Technology Conference, IEEE 44th*, Vol. 1, 1994, pp. 371–374.

[37] T. Keller and L. Hanzo. Adaptive multicarrier modulation: a convenient framework for time-frequency processing in wireless communications. *Proc. IEEE*, 88, 611–640, 2000.

[38] J.E. Kleider and M.E. Humphrey. Robust time-frequency synchronization for OFDM mobile applications. In *Proceedings of the International Symposium on Signal Processing and Its Applications, ISSPA'99*, Vol. 1, 1999, pp. 423–426.

[39] Y.C. Ko and G. Jeong. Doppler spread estimation in mobile communication systems. In *Proceedings of the Vehicular Technology Conference, IEEE 55th*, Vol. 4, 2002, pp. 1941–1945.

[40] L. Krasny, H. Arslan, D. Koilpillai, and S. Chennakeshu. Doppler spread estimation in mobile radio systems. *IEEE Commun. Lett.*, 5, 197–199, 2001.

[41] A. Kurniawan, S. Perreau, J. Choi, and K. Lever. Power control and diversity antenna arrays for cdma systems. In *Proceedings of the International Conference on Information, Communications and Signal Processing*, Singapore, 2001.

[42] H. Lee and D. Cho. A new adaptive power control scheme based on mobile velocity in wireless mobile communication systems. In *Proceedings of the Vehicular Technology Conference, IEEE 53rd*, Vol. 4, 2001, pp. 2878–2882.

[43] W.C.Y. Lee. *Mobile Communication Engineering: Theory and Applications*. McGraw-Hill, New York, 1998.

[44] W.C.Y. Lee and Y.S. Yeh. On the estimation of the second-order statistics of log-normal fading in mobile radio environment. *IEEE Trans. Commun.*, 22, 809–873, 1974.

[45] J. Lin and J.G. Proakis. A parametric method for Doppler spectrum estimation in mobile radio channels. In *27th Conference on Information Systems and Sciences*, March 1993, pp. 875–880.

[46] T.L. Marzetta. EM algorithm for estimating the parameters of a multivariate complex Rician density for polarimetric SAR. In *Proceedings of the IEEE International Conference on Acoustics, Speech, and Signal Processing, ICASSP'95*, Vol. 5, May 1995, pp. 3651–3654.

[47] P.E. Mogensen, P. Eggers, C. Jensen, and J.B. Andersen. Urban area radio propagation measurements at 995 and 1845 MHz for small and micro-cells. In *Proceedings of the Global Communication Conference*, Vol. 2, December 1991, pp. 1297–1302.

[48] D. Mottier and D. Castelain. A Doppler estimation for UMTS-FDD based on channel power statistics. In *Proceedings of the Vehicular Technology Conference, IEEE 50th*, Vol. 5, Fall 1999, pp. 3052–3056.

[49] R. Narasimhan and D.C. Cox. Speed estimation in wireless systems using wavelets. *IEEE Trans. Commun.*, 47, 1357–1364, 1999.

[50] R. Narasimhan and D.C. Cox. Estimation of mobile speed and average received power in wireless systems using best basis methods. *IEEE Trans. Commun.*, 49, 2172–2183, 2001.

[51] G.P. Pollini. Trends in handover design. *IEEE Commun. Mag.*, 34, 82–89, 1996.

[52] J.G. Proakis. *Digital Communications*, 3rd ed. McGraw-Hill, New York, 1995.

[53] S.O. Rice. Mathematical analysis of random noise. *Bell System Tech. J.*, 24, 46–156, 1945.

[54] S.O. Rice. Statistical properties of a sine wave plus random noise. *Bell System Tech. J.*, 109–157, 1948.

[55] A. Sampath and J.M. Holtzman. Estimation of maximum Doppler frequency for handoff decisions. In *Proceedings of the Vehicular Technology Conference, IEEE 43rd*, 1993, pp. 859–862.

[56] A.M. Sayeed. Canonical time-frequency processing for broadband signaling over dispersive channels. In *Proceedings of the IEEE-SP International Symposium on Time-Frequency and Time-Scale Analysis*, New York, 1998, pp. 369–372.

[57] A.M. Sayeed and B. Aazhang. Communication over multipath fading channels: a time-frequency perspective. Chapter 3 in *Wireless Communications: TDMA versus CDMA*, S.G. Glisic and P.A. Leppanen, Eds. Kluwer Academic, Dordrecht, Netherlands, 1997.

[58] A.M. Sayeed and B. Aazhang. Joint multipath-Doppler diversity in mobile wireless communications. *IEEE Trans. Commun.*, 47, 123–132, 1999.

[59] A.M. Sayeed, A. Sendonaris, and B. Aazhang. Multiuser detection in fast-fading multipath environments. *IEEE J. Selected Areas Commun.*, 16, 1691–1701, 1998.

[60] A. Scaglione, S. Barbarossa, and G.B. Giannakis. Optimal adaptive precoding for frequency-selective Nakagami-m fading channels. In *Proceedings of the Vehicular Technology Conference, IEEE 52nd*, Vol. 3, 2000, pp. 1291–1295.

[61] W. Sheng and S.D. Blostein. SNR-independent velocity estimation for mobile cellular communications systems. In *Proceedings of the IEEE International Conference on Acoustics, Speech, and Signal Processing, ICASSP'02*, Vol. 3, 2002, pp. 2469–2472.

[62] G.L. Stüber. *Principles of Mobile Communication*. Kluwer Academic Publishers, Dordrecht, Netherlands, 1996.

[63] K.K. Talukdar and W.D. Lawing. Estimation of the parameters of the Rice distribution. *J. Acoust. Soc. Am.*, 89, 1193–1197, 1991.

[64] C. Tepedelenlioğlu, A. Abdi, and G.B. Giannakis. The Ricean *K* factor: estimation and performance analysis. 2, 799–810, 2003.

[65] C. Tepedelenlioğlu, A. Abdi, G.B. Giannakis, and M. Kaveh. Review: estimation of Doppler spread and signal strength in mobile communications with applications to handoff and adaptive transmission. *Wireless Commun. Mobile Comput.*, 1, 221–242, 2001.

[66] C. Tepedelenlioğlu and G.B. Giannakis. On velocity estimation and correlation properties of narrowband mobile communication channels. *IEEE Trans. Veh. Technol.*, 50, 1039–1052, 2001.

[67] J.B. Thomas. *An Introduction to Statistical Communication Theory*. John Wiley & Sons, New York, 1967.

[68] M. Türkboylari and G.L. Stüber. Eigen-matrix pencil method-based velocity estimation for mobile cellular radio systems. In *Proceedings of the IEEE International Conference on Communications, ICC*, Vol. 2, 2000, pp. 690–694.

[69] R. Vijayan and J.M. Holtzman. Analysis of handoff algorithms using nonstationary signal strength measurements. In *Proceedings of the IEEE Global Telecommunications Conference, GLOBCOM'92*, Vol. 3, December 1992, pp. 1405–1409.

[70] R. Vijayan and J.M. Holtzman. A model for analyzing handoff algorithms (cellular radio). *IEEE Trans. Veh. Technol.*, 42, 351–356, 1993.

[71] L. Wang, M. Silventoinen, and Z. Honkasalo. A new algorithm for estimating mobile speed at the TDMA-based cellular system. In *Proceedings of the Vehicular Technology Conference, IEEE 46th*, Vol. 2, 1996, pp. 1145–1149.

[72] Y. Wang, L. Gao, M. Zhao, J. Chen, Z. Zhang, and Y. Yao. Time-frequency code for multicarrier DS-CDMA systems. In *Proceedings of the Vehicular Technology Conference, IEEE 55th*, Vol. 3, 2002, pp. 1224–1227.

[73] C. Xiao. Estimating velocity of mobiles in EDGE systems. In *Proceedings of the IEEE International Conference on Communications, ICC*, Vol. 5, 2002, pp. 3240–3244.

[74] C. Xiao, K.D. Mann, and J.C. Oliver. Mobile speed estimation for TDMA-based hierarchical cellular systems. *IEEE Trans. Veh. Technol.*, 50, 981–991, 2001.

[75] N. Zhang and J.M. Holtzman. Analysis of cdma soft-handoff algorithm. *IEEE Trans. Veh. Technol.*, 47, 710–718, 1998.

III

Modulation Techniques for Wireless Communications

6 **Adaptive Coded Modulation for Transmission over Fading Channels**
Dennis L. Goeckel 6-1
Introduction • Adaptive System Model • Adaptivity in Single-Input Single-Output Systems • Adaptivity in Multiantenna Systems • Conclusions

7 **Signaling Constellations for Transmission over Nonlinear Channels**
Hisham Abdul Hussein Al-Asady, Quazi Mehbubar Rahman, and Mohamed Ibnkahla 7-1
Introduction • System Model • Craig's Method • Probability of Symbol Error for 16-ary QAM Format • Probability of Symbol Error for the Circular 32-ary QAM Format • Conclusions

8 **Carrier Frequency Synchronization for OFDM Systems** *Mounir Ghogho and Ananthram Swami* 8-1
Introduction • Basics of OFDM • Effect of CFO on System Performance • Carrier Frequency Offset Estimation • Repetitive Slots-Based CFO Estimation • Null-Subcarrier-Based CFO Estimation • Identifiability • Performance Analysis • Simulation Results • Conclusions

9 **Filter-Bank Modulation Techniques for Transmission over Frequency-Selective Channels** *Giovanni Cherubini* 9-1
Introduction • Critically Sampled Filter Banks • Discrete Multitone Modulation • O-QAM OFDM Modulation • Discrete Wavelet Multitone Modulation • Filtered Multitone Modulation • Conclusion

6

Adaptive Coded Modulation for Transmission over Fading Channels

Dennis L. Goeckel
University of Massachusetts

6.1 Introduction 6-2
6.2 Adaptive System Model.. 6-4
Model for a Wireless Link • Adaptation in Response to Path Loss/Shadowing • Analytic Model for Fine-Scale Adaptation
6.3 Adaptivity in Single-Input Single-Output Systems 6-8
Information Theoretic Bounds • Design for Uncoded Systems • Coded Modulation Structures • Designing with a Given Coded Modulation Structure
6.4 Adaptivity in Multiantenna Systems 6-14
Information Theoretic Considerations • Adaptive Coded Modulation for MIMO Systems
6.5 Conclusions 6-16

Abstract

Adaptive signaling, where transmission parameters are adjusted based on information about the current state of the channel, has the capability to significantly increase the throughput of wireless communication systems. Such adaptation can take many forms (power, code rate, modulation, etc.) and can be performed at many different timescales, depending on the system capabilities and the rate at which the channel state varies. After a general introduction to the adaptive signaling methodology, this chapter focuses on symbol-by-symbol adaptation of coded modulation in response to measurements of the multipath fading. In such systems, there are two key design decisions: (1) choosing the structure of the coded modulation on which adaptation will be performed, and (2) deciding how to adapt that structure based on available measurements. These two design problems are discussed in detail for systems where the transmitter and receiver each have a single antenna. Extensions to systems with multiple antennas at the transmitter or receiver, which is a problem of great current interest in the research community, are also discussed. Performance results confirm the significant gains achievable through adaptation of the coded modulation in wireless systems, and future directions in this important area are discussed.

0-8493-1657-X/$0.00+$1.50

6.1 Introduction

The main goal of modern communications systems is to make universal high-speed access to information a reality. Under nearly any realizable scenario, the end user in such a system is free of tethers, thus making robust high-speed data transfer across the wireless channel a topic of extreme importance. However, the wireless channel presents a number of challenges. In particular, the wireless signal experiences [57]: (1) path loss — the signal attenuates with distance from the transmitter; (2) shadowing — large objects between the transmitter and receiver can obstruct the radio signal; and (3) multipath fading — reflections from objects in the environment can add constructively or destructively at the receiver. Since all three of these effects change with time or position, they cannot be known at the time that the system is designed. Thus, if no form of adaptation is performed while the system is operating, the system must be designed to in some sense deal with the worst case, which can be very expensive in terms of system resources. For example, if all users in a cellular system have to assume worst-case path loss and shadowing (e.g., behind a large building at the very edge of the cell) regardless of their location, they will transmit at maximum power, thus maximizing battery usage and the interference to other users. Therefore, wireless system adaptation has been a topic of critical interest in recent years.

Wireless link adaptation can be defined as shown in Figure 6.1: any altering of the parameters at the transmitter based on information about the current link state. Note that *link state* will be defined very generally; in particular, in addition to wireless channel conditions (path loss, shadowing, multipath fading), user data requirements will be included in the definition. Methods of adaptation can be classified by the type of adaptation performed (power, code rate, modulation, etc.) and the timescale of that adaptation.

The timescale of the adaptation depends on the link state phenomena for which measurements are provided to the adaptation algorithm. User data rate requirements, path loss, and shadowing change at a timescale that is long relative to the symbol rate. This makes measurements of such phenomena relatively robust [28, 66], particularly at high signal-to-noise ratios (SNRs), and has resulted in widespread penetration into current and pending systems of slow adaptations, such as power control [47, 68, 70] and data rate adaptation through variable spreading, code rate, or code aggregation [48]. Thus, throughout this chapter, it will be largely assumed that the measurements of the path loss and shadowing are accurate and known at both the transmitter and receiver, thus yielding a wireless system with a known given average received signal power but experiencing variable multipath fading.

Multipath fading is caused by the arrival at the receiver of many signal reflections, the superposition of which causes the instantaneous received signal power to vary widely [57, Chapter 4], as described in Section 6.2 below. Since significant nulls can occur, this signal fading is one of the most difficult problems to deal with in wireless communications systems. In particular, when the received power drops too low, a burst of bit errors can occur, and such bursts tend to dominate the error probability — even if the occurrence of such system power drops is relatively unlikely. This results in a significantly higher required average received SNR for a given level of performance relative to systems operating over additive white Gaussian

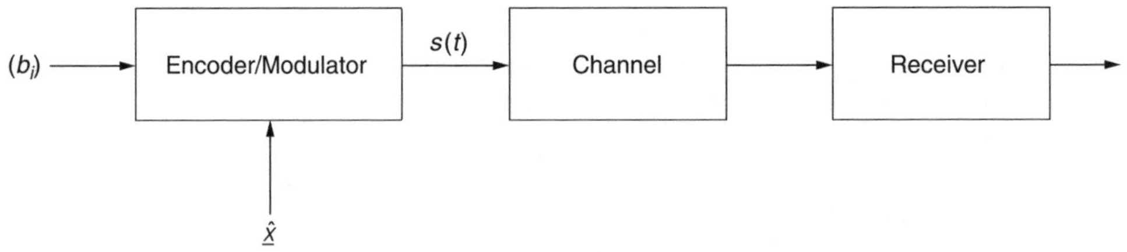

FIGURE 6.1 General adaptation framework: The transmitter sets parameters based on $\underline{\hat{X}}$, which contains information about the channel state, while forming the transmitted signal $s(t)$ from the information bit sequence (b_i). Note that $\underline{\hat{X}}$ can take many forms: path loss/shadowing estimates [47, 68, 70], number of errors corrected in previous packets [54, 55], explicit multipath fading estimates [31, 32], etc.

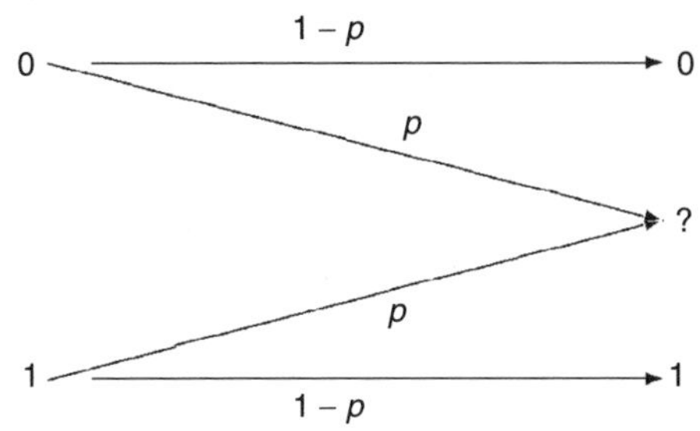

FIGURE 6.2 The binary erasure channel [16, p. 188] to represent a discrete-valued fading channel with channel state information available at the receiver.

noise (AWGN) channels [53, p. 820]. However, unlike the effects of path loss and shadowing, which vary slowly and thus limit the ability of the system designer to average their effects over time, multipath fading varies relatively rapidly with time, position, and frequency. Thus, in many scenarios, well-designed systems achieve diversity, allowing them to average effectively over the effects of the multipath fading and thus significantly reduce its impact [53, p. 821]—even if there is no channel knowledge at the transmitter. For many systems, such nonadaptive solutions come at the cost of system latency or complexity, which motivates the consideration of transmission schemes that employ measurements of the multipath fading values.

Since this chapter will largely focus on the design of techniques for adaptation based on measurements of the multipath fading, it is important to understand the applicability of such adaptation. Thus, the gains from having knowledge of the channel at the transmitter will be a key topic discussed, and as motivated in the previous paragraph, it is often a question of the system complexity and latency allowable. To make this more concrete, consider the simple information theoretic example drawn from [16, p. 188], which is shown in Figure 6.2. First, to see how this represents a fading channel, consider a binary transmission system for which there are essentially no errors when the signal is transmitted over an AWGN channel (i.e., no fading); a simple example is coherently detected binary phase-shift keying (BPSK) with a relatively high SNR [53, p. 820]. Now assume that a BPSK system is operating over a discrete-valued fading channel described as follows. First, the state of the channel is independent for separate channel uses, which implies that a deep interleaver [53, p. 467] is employed. For a given channel use:

1. With probability p, the transmitted signal is multiplied by zero (hence disappears).
2. With probability $1 - p$, the transmitted signal is multiplied by $\alpha = \frac{1}{1-p}$ (hence amplified, which keeps the average received SNR identical to the AWGN case).

Assuming that channel state information (CSI), which throughout this paper will be the value of the multiplicative factor (in this case 0 or α), is available at the receiver, this yields the model shown in Figure 6.2. Consider signaling over the channel shown in Figure 6.2 with and without channel state information at the transmitter. With CSI at the transmitter, transmission is halted when $\alpha = 0$ and a single bit is transmitted whenever $\alpha = \frac{1}{1-p}$. Thus, with a very simple receiver identical to that for the BPSK system operating over the AWGN channel, the system reliably transmits $1 - p$ information bits per channel use. Next consider the case where there does not exist channel state information at the transmitter. Using information theoretic results [16, p. 188], the capacity of the channel without transmitter CSI is still $1 - p$ information bits per channel use, but now it requires very long code words and the typical sequence decoding employed for the achievability statement of Shannon's capacity [59]. This simple example captures the key idea to adaptive signaling in response to the multipath fading in many cases—it will often not make sense from a Shannon capacity, but it can greatly simplify system design in practical systems [30].

Thus, adaptation in response to transmitter knowledge of the multipath fading has the promise of greatly simplifying the system design or, for a fixed system complexity, has the promise of greatly improving system performance (such as average data rate) [31, 32]. However, it is the very property that makes adaptation fruitful that also complicates its implementation; in particular, adaptation can be exploited because of the time-varying nature of user needs, path loss, shadowing, and multipath fading. But this time-varying nature makes that adaptation difficult; in particular, although changes in user needs and path loss generally happen over a long enough timescale that they can be reliably estimated, the time-varying nature of the shadowing [66] and the multipath fading [24, 27] make channel measurements outdated by the time that they are ready to be used. In other words, the channel has changed since the measurements were performed, and thus the utility of such measurements in representing the current state of the channel can be questioned. This will be particularly exacerbated, of course, in systems that seek to adapt to the multipath fading [48].

The consideration of the design of signaling schemes that employ inherently outdated or noisy measurements is best done by carefully considering the channel characteristics *conditioned* on the measurements available. Naturally, if the support of the probability density function of the conditional channel given the measurements is very narrow, indicating that the system is fairly certain of the channel value, one can design coded modulation structures and rules for adapting those structures based on the assumption that the channel is fully known [31, 32] and suffer only mild degradations. However, such schemes can be very sensitive, even if the probability density function only shows a little spread around the estimated value [24, 27]. In such cases, not only must the rules of adaptation consider such spread, but it often will affect the types of coding and modulation structures that are effective, as demonstrated in Section 6.3.3.

This chapter is organized as follows. In Section 6.2, the system model that will be used throughout this work is presented. Section 6.3 provides a detailed derivation of the key issues in adaptive signaling using the simplest case of a system where there is only a single antenna employed at each the transmitter and receiver. Section 6.4 discusses recent extensions of these results to systems with multiple antennas at the transmitter and receiver, and Section 6.5 presents conclusions and avenues for future work in the field.

6.2 Adaptive System Model

6.2.1 Model for a Wireless Link

The transmitted signal in a wireless communications system is affected by three factors: path loss, shadowing, and multipath fading. In complex baseband notation [53], the signal $r(t)$ that is received when the signal $s(t)$ is transmitted can be written as

$$r(t) = L(t)X(t)s(t) + n(t) \tag{6.1}$$

where $L(t)$ is a real-valued random process that represents the combined effect of the path loss and shadowing, $X(t)$ is a complex random process representing the effect of the multipath fading, and $n(t)$ is a stationary complex Gaussian random process with (two-sided) power spectral density $S_N(f) = \frac{N_0}{2}$, representing additive noise. In Equation 6.1, the fading has been assumed to be frequency-non-selective [53, p. 816]; this is appropriate for a narrowband single-carrier system or a single subcarrier of a wideband orthogonal frequency division multiplexing (OFDM) system [6, 71]. Extensions of the concepts presented in this chapter to frequency-selective channels are conceptually straightforward, although such channels offer inherent natural diversity with little system latency, and hence often reduce the gain available through adaptive signaling.

Understanding the characteristics of the processes $L(t)$ and $X(t)$ in Equation 6.1 is crucial in determining methods of adaptation based on such. The random process $L(t)$ is caused by path loss, which is determined by the distance the receiver is from the transmitter, and shadowing, which is determined by the existence of large objects between the transmitter and receiver. For a stationary user, the path loss and shadowing are generally modeled as constant, despite the fact that it could be argued that the movement of large objects can affect the shadowing. For a user in motion, the shadowing will be the more variable of the two effects, and the distance over which it is highly correlated can be roughly modeled as 100 m in a macrocellular suburban environment [35]. For a user at walking speed (say, 2 m/sec), this implies that the shadowing correlation time is on the order of 50 sec; for a user in a vehicle (say, 88 km/h), this implies that the shadowing correlation time is on the order of 4 sec. This suggests that it is quite plausible to make estimates of the path loss and shadowing and to employ such in wireless communications systems. In fact, this is very often done in current and next-generation cellular system implementations [48]. Also, since $L(t)$ varies at a relatively long timescale, it will be assumed throughout the remainder of this chapter that it is measured accurately and known at the transmitter and receiver.

In contrast, consider the random process $X(t)$, which represents the multipath fading. In a wireless environment, the signal $s(t)$ is reflected to the receiver from many objects. Because the propagation distance is different for each of these reflections, the reflected signals will arrive at slightly different times at the receiver. For a narrowband system, which has a relatively long symbol interval, there will not be appreciable intersymbol interference (ISI) [53, p. 817]. However, because of the large carrier frequencies typically employed in modern wireless communications systems, even a small difference in arrival times for two paths can result in a large phase difference between those paths. For example, a path-length difference of only 1 ft results in the signal being delayed by 1 nsec, which causes a full 2π rotation in phase when the carrier frequency is 1 GHz. Hence, the phase of any given arriving path is generally modeled as uniformly distributed. Since the process $X(t)$ is caused by the sum of very many roughly independent paths, projected onto each of the in-phase (real) and quadrature (imaginary) components, the central limit theorem [52, p. 214] motivates its modeling as a complex Gaussian random process [3].

By considering the genesis of the multipath fading as described above, it is easy to observe that the phase of a given path will change greatly for each movement of the reflecting object, the receiver, or the transmitter by one wavelength. Hence, even with only walking speed mobility (say, 2 m/sec), a system with a 1-GHz carrier will yield a process $X(t)$ that changes independently six times per second (or, as commonly stated, with a 6-Hz Doppler frequency) [51, 56], which makes adaptation challenging, since feedback of the channel characteristics provided to the transmitter at some delay must accurately model the current channel fading for adaptation to be effective. Note that this problem will be exacerbated at higher mobilities and higher carrier frequencies.

Mathematically, $X(t) = X_R(t) + jX_I(t)$ will be assumed to be a zero-mean stationary Gaussian random process with an autocorrelation function of the real part $X_R(t)$ (or imaginary part $X_I(t)$) defined as

$$R_X(\tau) = E[X_R(t)X_R(t+\tau)] = E[X_I X_I(t+\tau)]$$

and the real part $X_R(t)$ and imaginary part $X_I(t)$ will be assumed to be independent of one another. The zero-mean assumption implies that a line-of-sight path is not present, which corresponds to the most pessimistic case — Rayleigh fading. Throughout this chapter, the popular Jakes model [41] will generally be adopted, which is characterized by $R_X(\tau) = J_0(2\pi f_d \tau)$, where $J_0(\cdot)$ is the zero-order Bessel function of the first kind and f_d is the Doppler frequency, which is defined as the number of wavelengths of motion of an object per second. It will be assumed that although the Doppler frequency might be large, it will not approach the symbol rate, and thus the channel $X(t)$ can be assumed to be constant over the support of a single signaling pulse $p(t)$, which is termed the *slowly fading* assumption in most digital communications texts [53, p. 816]. We hasten to emphasize, however, that the use of the word *slow* in this context is with reference to the symbol interval — not the amount of time between channel estimation and signal transmission, where even such "slow" multipath fading can have a significant effect.

6.2.2 Adaptation in Response to Path Loss/Shadowing

There are many forms of adaptation currently employed in response to path loss/shadowing in wireless communications systems. In fact, even the base station selection process, where a mobile generally decides to associate with the base station from which it sees the largest average received signal strength, can be viewed as a form of adaptation. Such adaptations will be called "slow" adaptations throughout this work, and they will be characterized by schemes that adapt the transmitter at an interval on the order of (at least) many (hundreds of) symbols. A good tutorial on slow adaptations, particularly in current standards, is provided in [48].

First, consider wireless system adaptations that adapt depending on user needs. In particular, one of the key features of third-generation cellular systems is supporting users with high data rates. This is often done by simply allocating more of the time/bandwidth/code space to the users. For example, in Enhanced Data Rates for GSM Evolution (EDGE) systems, which are built on a time-division multiple-access (TDMA) framework, users with high data rate needs are allocated more time slots. In the code-division multiple-access (CDMA)-based IS-95 Revision B, high data rate users are allocated multiple spreading codes, which is termed "code aggregation" [48].

Next, consider adaptations based on the current channel conditions for a given user. In fixed-rate systems, such as first- and second-generation cellular telephone systems, where the rate of the vocoder is generally fixed, the key is to adapt the system such that acceptable performance is maintained at this fixed rate. The transmission technology and channel assumptions fix a minimum average received SNR γ_0 required for acceptable operation — the goal of adaptation is to maintain γ_0, which can be done by adapting the transmitted power in response to measurements of the path loss and shadowing. Methods of performing such adaptation include channel inversion, where the transmitted power is set proportional to the channel loss, and truncated channel inversion (e.g., [17, 66]). Truncated channel inversion is defined by a threshold L_0, which breaks the policy into two cases:

1. $L(t_0) \geq L_0$: The transmitted power is set to $\gamma_0/L(t_0)$, which results in an average received SNR of γ_0.
2. $L(t_0) < L_0$: The transmitted power is set to zero, which results in an outage.

Using this policy, the required average received SNR (and no more) is obtained whenever possible, but excessive power is not wasted by inverting the channel when there are large losses on the channel.

Outside of current standards, the setting of the rate (coding, modulation, spreading factor) of the system to the current average received SNR has clearly emerged as a critical topic. In particular, turbo codes [4] and low-density parity-check (LDPC) codes [23, 46] are approaching channel capacities on a variety of channels. Thus, assuming enough receiver complexity for the decoding of such codes and enough latency to allow perfect interleaving of the coded symbols, the rate of nearly error-free systems should approach the Shannon capacity [59] of the independent and identically distributed (IID) discrete-time Rayleigh fading channel, defined by:

$$Y_i = \alpha_i X_i + n_i$$

where Y_i is the received sequence, α_i is the IID sequence of Rayleigh channel fading values, X_i is the transmitted sequence, and n_i is the noise sequence. The Shannon capacity for such a channel without CSI at the transmitter is shown as the lower curve in Figure 6.3, where the SNR on the horizontal axis is the average received SNR (i.e., the SNR after the path loss and lognormal shadowing are considered). The only requirement for the highly efficient operation of such as system is the knowledge of this average received SNR at the transmitter so that the rate of the transmitter can be set appropriately, and this can be obtained by feedback of the path loss and shadowing. We emphasize that approaching the curves in Figure 6.3 still requires high decoding complexity and significant latency, which motivate whether adaptation with the additional knowledge of the values of the multipath fading can improve on the performance in Figure 6.3 in terms of performance vs. system complexity.

6.2.3 Analytic Model for Fine-Scale Adaptation

The main portion of this chapter will be dedicated to the design and analysis of adaptive systems that use explicit measurements of the multipath fading to perform system adaptation. This is a topic that was considered in the 1970s [11, 37, 38] and then became popular again in the early 1990s (e.g., [1, 6, 14, 29, 30, 65]).

A block diagram of the typically-employed system is shown in Figure 6.4. Given the model shown in Figure 6.4 and channel model given in Section 6.2.1, the key to designing adaptive signaling systems is considering signaling for the *conditional* channel for the symbol of interest (call it s_k) given the outdated

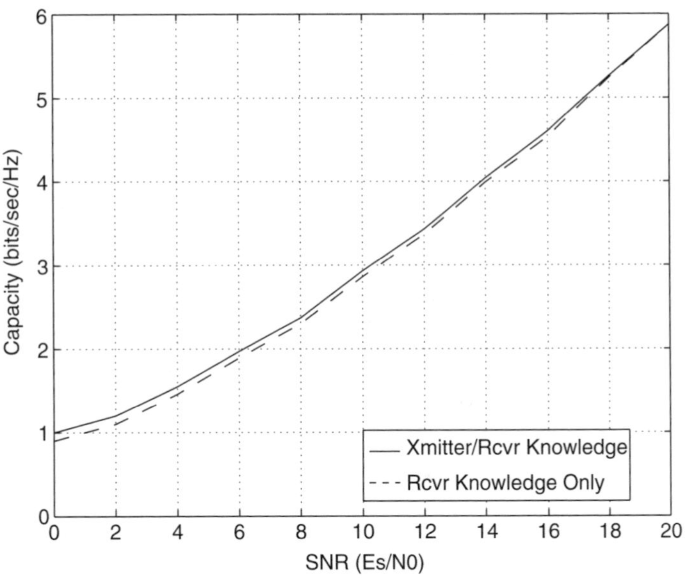

FIGURE 6.3 The Shannon capacity of an independent and identically distributed Rayleigh fading channel, assuming (1) perfect CSI available only at the receiver, and (2) perfect CSI available at both the transmitter and receiver [30]. Note that the gain in Shannon capacity resulting from having perfect CSI available only at the transmitter is only slight, as discussed in Section 6.3.1.

measurement $\underline{\hat{X}} = (\hat{X}(t-\tau_1), \hat{X}(t-\tau_2), \ldots, \hat{X}(t-\tau_N))^T$. It will be assumed that a measurement $\hat{X}(t-\tau_i)$ is equal to the true value $X(t-\tau_i)$ plus additive Gaussian noise of variance $\sigma_\epsilon^2 = \frac{1}{\frac{E_p}{N_0}}$ in each of the in-phase and quadrature components. For example, such would be the case in an adaptive system employing pilot symbol assisted modulation (PSAM) [12] with a pilot symbol energy of E_p [7, 25].

Note that the model in Figure 6.4 captures the critical issue of delay in the feedback path from the receiver to the transmitter, since the most recent estimate is assumed to have been made τ_1. seconds ago. In other words, the "outdated" nature of the estimates takes into account this key implementation issue in adaptive communications systems.

Denoting Y as the magnitude of the fading that multiplies s_k in the matched filter output for the kth symbol and using the fact that linear functionals of a Gaussian random process are jointly Gaussian, Y is Rician when conditioned on the vector $\underline{\hat{X}}$, with probability density function [25]

$$p_{Y|\underline{\hat{X}}}(y|\underline{x}) = \frac{y}{\sigma^2} e^{-\frac{y^2+s^2}{2\sigma^2}} I_0\left(\frac{ys}{\sigma^2}\right), \qquad y \geq 0 \tag{6.2}$$

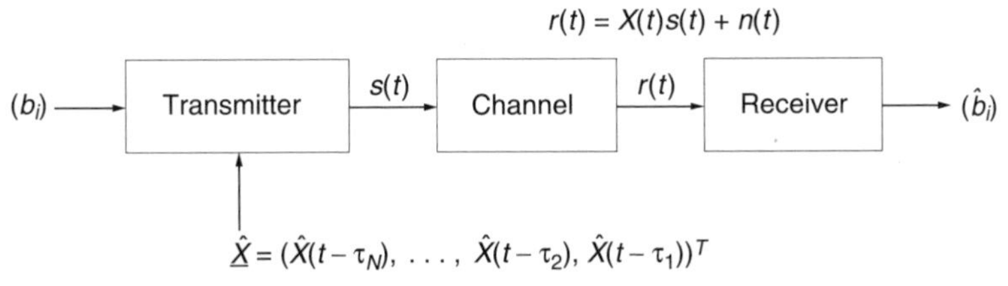

FIGURE 6.4 A block diagram of the system, where (b_i) is the sequence of information bits to be transmitted across the channel, $s(t) = \sum_{k=-\infty}^{\infty} s_k p(t - kT_s)$ is the transmitted signal, $r(t)$ is the received signal, $n(t)$ is additive white Gaussian noise, $\underline{\hat{X}}$ is the vector of outdated channel measurements, and $(\hat{b}_i)$ is the sequence of information bit estimates output from the receiver.

where $I_0(\cdot)$ is the zero-order modified Bessel function. Using the assumption that $X(t)$ can be assumed constant over the support of $p(t)$ and normalizing the fading such that $E[(X_R(kT_s))^2] = E[(X_I(kT_s))^2] = 1$ (note that this simplification will make the average received energy twice that provided by simply the path loss and shadowing, which will be accounted for below), the noncentrality parameter in Equation 6.2 is given by

$$s^2 = \left(\underline{\rho}^T\left(\Sigma_{\underline{X}} + \sigma_\epsilon^2 I_N\right)^{-1}\underline{x}_R\right)^2 + \left(\underline{\rho}^T\left(\Sigma_{\underline{X}} + \sigma_\epsilon^2 I_N\right)^{-1}\underline{x}_I\right)^2 \tag{6.3}$$

where I_N is an N by N identity matrix. The $(m, n)^{th}$ element of $\Sigma_{\underline{X}}$, the N by N autocorrelation matrix of the in-phase component of $\hat{\underline{X}}$ when the channel estimates are noiseless, is given by $R_X(\tau_{N-m+1} - \tau_{N-n+1})$, and the correlation vector of the in-phase component of $\hat{\underline{X}}$ with the in-phase component of the fading of interest is given by $\underline{\rho}$, where $\rho_i = R_X(\tau_{N-i+1})$. The parameter σ^2 in Equation 6.2 is the mean square error of a minimum mean square error (MMSE) estimator [72, p. 54] of the in-phase (or quadrature) fading of interest, and is given by

$$\sigma^2 = 1 - \underline{\rho}^T(\Sigma_{\underline{X}} + \sigma_\epsilon^2 I_N)^{-1}\underline{\rho} \tag{6.4}$$

Understanding the Rician density in Equation 6.2 and the expression for the Rician noncentrality parameter s in Equation 6.3 is key to designing effective adaptive coded modulation schemes. In particular, for $s = 0$, the Rician probability density function in Equation 6.2 is equivalent to a Rayleigh probability density function, indicating that coded modulation structures designed for Rayleigh fading channels are pertinent for application when s is small; likewise, as $s \to \infty$, the (properly normalized) Rician density function approaches a delta function, thus indicating that the effective channel approaches an AWGN channel. Since coded modulation schemes for Rayleigh fading channels differ greatly from AWGN schemes, the interpretation of Equation 6.3 is used extensively in the design of structures, as demonstrated in Section 6.3.3.

There is one limitation to directly employing the result in Equation 6.2. In particular, it presumes that the autocorrelation function $R_X(\tau)$ of the random process $X(t)$ is known at the transmitter; however, this autocorrelation function can vary greatly in wireless systems [51, p. 88–89]. Thus, it must generally be estimated [18, 19], either implicitly or explicitly, or uncertainties in it must be worked into system design [24, 27]. To address both possibilities in one framework, the autocorrelation function will be assumed to lie in some uncertainty class $\mathcal{R}$, which matches the approach taken in [27] directly. If it can be accurately estimated through techniques as described in [18, 19], this class can be shrunk accordingly (in the limit to a single autocorrelation function).

Thus, given the model for the system measurements and the measurements of the autocorrelation function, one should be able to ascertain (1) how much predictor error σ^2 will generally be in the system, and (2) what is the uncertainty class $\mathcal{R}$ over which some sort of robustness will be maintained. Understanding both of these for a given system configuration will be the key to understanding the design of the coded modulation. In particular, the former will allow the choice of the coded modulation structure, while the latter will allow one to design on that structure.

6.3 Adaptivity in Single-Input Single-Output Systems

6.3.1 Information Theoretic Bounds

Before considering the derivation of practical signaling schemes that adapt to the multipath fading, it is instructive to consider the improvement in Shannon capacity that is available when CSI is made available to the transmitter. Assume that the sequence of zero-mean complex Gaussian channel fading coefficients affecting the transmitted symbols forms an IID sequence; in other words, an IID Rayleigh fading channel is assumed. If the criterion is to maximize the average data rate under an average power constraint, Goldsmith [30] has demonstrated that the information theoretic capacity when perfect CSI is available at both the transmitter and receiver is achieved with variable-power Gaussian codebooks, where the power depends on the current value of the channel fading. A comparison of the Shannon capacity when perfect CSI is available at both the transmitter and receiver with the capacity when CSI is available only at the receiver is shown in Figure 6.3. Note that, somewhat surprisingly, the gain in channel capacity is only slight. However, as with the example of the erasure channel in Section 6.1, it should be remembered that this assumes very long code words and large decoding complexities. In particular, the minimum distance of the codes becomes very large, and since IID Rayleigh fading is assumed, the diversity achieved by a given code is very large—even for the case of CSI only at the receiver.

The analysis in the previous paragraph and Figure 6.3 applies to the Shannon capacity [59], which is generally appropriate if the average rate of a system is being considered. Recently, however, there has been significant interest in whether a system can transmit a fixed amount of information within a given time constraint—such approaches lead to measures such as the outage capacity [22] or the delay-limited capacity [36]. Such analyses are generally still done under the assumption of infinite-length code words, which are required for the random coding arguments invoked, but now under the assumption that a given code word will only see some small number of fades. Thus, in essence, this yields a view at system operation in a diversity-limited context. The metric is based on the probability that the system experiences a set of fading values for which it can communicate at the desired rate. In contrast to Figure 6.3, such analyses [8, 49] have demonstrated the significant gains possible when knowledge of the channel fading values is provide to the transmitter in addition to the receiver. This has motivated work in the design of practical adaptation schemes that focus on outage probability [43]. These results lead to the preliminary conclusion that the gain from having estimates of the channel fading at the transmitter is highly reliant on the decoder complexity and system latency allowed.

6.3.2 Design for Uncoded Systems

In this section, adaptive uncoded modulation will be designed. Unlike coded schemes, where the design is complicated by memory in the trellis and questions about the proper structure, uncoded schemes present a simple framework to demonstrate many of the key issues.

As description in Section 6.2.3, there is a key issue of robustness to uncertainties in the autocorrelation function $R_X(\tau)$, which can be captured by designing for an uncertainty class $\mathcal{R}$ that shrinks to a single point when the autocorrelation is known or can be accurately estimated. If the class $\mathcal{R}$ is a single point as is often considered for prediction-based methods [18, 19, 44, 45], the application of Equation 6.2 is identical regardless of the number of outdated estimates N employed. This is observed by noting that, for a vector of random variables drawn as samples from a stationary Gaussian random process, (1) the marginal probability density function is independent of the sampling time, and (2) the conditional probability density function for any one of the variables is Gaussian when conditioned on the others, with variance given by σ^2 in Equation 6.4. Thus, given the predicted value and σ^2, the choice of the signal set is independent of N. Hence, when the autocorrelation function $R_X(\tau)$ is known exactly, the design for $N = 1$ with the appropriate σ^2 is sufficient to characterize performance of a given scheme.

When $\mathcal{R}$ is made larger to capture uncertainties in the knowledge of $R_X(\tau)$, the design becomes greatly complicated [26]. Thus, robust design with general $\mathcal{R}$ with only a single outdated estimate ($N = 1$) will be considered; however, we hasten to note that, per the previous paragraph, designing adaptive coded modulation for known $R_X(\tau)$ and *any* N is a simplification of this case.

Since the case $N = 1$ will be considered, the quantity $\rho = R_X(\tau_1)$, which, since $X_R(t)$ and $X_I(t)$ are normalized to have unit energy, is the correlation coefficient of the in-phase (or quadrature) components of the multipath fading process between the time of channel estimation and the time of data transmission, will be important. Assuming that the estimates are noiseless ($\sigma_\epsilon^2 = 0$) implies that $\sigma^2 = 1 - \rho^2$, and it is observed that the mean square prediction error increases rapidly with decreasing correlation between the estimate and the current value, as expected. Throughout much of this chapter, systems will be designed for a given ρ, which captures the amount of information in the channel estimate about the current fading value. A conversion to mean square predictor error, if desired, can be obtained by the transformation $\rho = \sqrt{1 - \sigma^2}$.

Designing robustly using a single outdate estimate requires performance to be guaranteed for all $\rho \in [\rho_{\min}, 1]$, where $\rho_{\min}$ is the minimum value of $R_X(\tau_1)$.

6.3.2.1 Design Rules

The design rules for uncoded systems have been well established by a number of authors [7, 27, 31]. The signal sets considered in this section will be 0-QAM (quadrature amplitude modulation) (no data transmitted), 2-QAM, 4-QAM, 16-QAM, and 64-QAM with two-dimensional Gray mapping, although the extension to any set of signal sets is immediate.

Following [27], let P_b be the target bit error probability for the system, which operates at the average received SNR $\frac{E_s}{N_0}$, where E_s is the average received energy per QAM symbol. For now, it will be assumed that the average energy E_s is not varied over time; energy adaptation will be discussed in detail below. Specification of the adaptive transmitter requires finding $\tilde{M}(h), \forall h$, where $\tilde{M}(h)$ is the number of signals in the QAM signal set employed when $|\hat{X}(kT_s - \tau_1)| = h$. If $\tilde{M}(h)$ is chosen such that P_b is maintained for each h,

$$\tilde{M}(h) = \max\left\{M : \sup_{\rho_{min} \leq \rho \leq 1} \tilde{P}_M\left(\frac{E_s}{N_0}, h, \rho\right) \leq P_b\right\} \tag{6.5}$$

where $\tilde{P}_M(\frac{E_s}{N_0}, h, \rho)$ is defined as the bit error probability of the M-QAM signal set at average received SNR $\frac{E_s}{N_0}$ when $R_X(\tau_1) = \rho$ and $|\hat{X}(kT_s - \tau_1)| = h$. Assume that maximum likelihood symbol detection, given the current channel fading amplitude, is employed on the samples of the matched filter output at the receiver. A tight approximation to the bit error rate of M-QAM modulations is given by [31]

$$P_M\left(y^2 \frac{E_s}{N_0}\right) \approx 0.2 \exp\left(-\frac{3}{4(M-1)} \frac{E_s}{N_0} y^2\right) \tag{6.6}$$

which will be employed for all M for much of the design work for uncoded systems. If errors in channel estimation *at the receiver* are considered, the right side of Equation 6.6 will increase, of course, but it will often fit into the same functional form [7], which is convenient, since the same optimization will apply. Using Equation 6.6 yields

$$\tilde{P}_M\left(\frac{E_s}{N_0}, h, \rho\right) = E\left[P_M\left(Y^2 \frac{E_s}{N_0}\right) \Big| |\hat{X}(kT_s - \tau_1)| = h\right]$$
$$\approx \begin{cases} \dfrac{0.2 \exp\left[-\frac{h^2 \rho^2}{2(1-\rho^2)}\left(1 - \frac{1}{1 + \frac{3}{2}\frac{E_s}{N_0}\frac{(1-\rho^2)}{(M-1)}}\right)\right]}{1 + \frac{3}{2}\frac{E_s}{N_0}\frac{(1-\rho^2)}{(M-1)}} & \rho < 1 \\ 0.2 \exp\left(-\frac{3}{4}\frac{E_s}{N_0}\frac{h^2}{(M-1)}\right) & \rho = 1 \end{cases} \tag{6.7}$$

where the second line is obtained by substituting Equation 6.2 and Equation 6.6 into the first line and evaluating the expectation over Y using [34, 6.614.3].

From Equation 6.5, Equation 6.7 must be evaluated at its supremum on $\rho \in [\rho_{min}, 1]$. Since the right side of Equation 6.7 is a continuous function on this closed interval, it achieves its maximum on this interval at a point that will be denoted ρ^*. The following solution is found by standard calculus techniques. Let

$$\tilde{\rho} = \begin{cases} 0 & h \geq \sqrt{2} \\ \sqrt{\left(1 + \frac{2(M-1)}{3}\frac{N_0}{E_s}\right)\frac{(2-h^2)}{2}} & 0 \leq h \leq \sqrt{2} \end{cases}$$

The worst-case autocorrelation is then given by

$$\rho^* = \begin{cases} \rho_{min} & \tilde{\rho} \leq \rho_{min} \\ \tilde{\rho} & \rho_{min} < \tilde{\rho} < 1 \\ 1 & 1 \leq \tilde{\rho} \end{cases} \tag{6.8}$$

The signal set is specified using Equation 6.7 and Equation 6.8 in $\tilde{M}(h) = \max\{M : \tilde{P}_M(\frac{E_s}{N_0}, h, \rho^*) \leq P_b\}$. Note that $\tilde{M}(h)$ is nondecreasing in h. Thus, the adaptive scheme can be specified by the values $h_m, m = 2, 4, 16, 64$, where h_m is defined as the threshold such that for $h \geq h_m$, m-QAM can be employed.

The discrete nature of the set of rates for any finite collection of signal sets hurts the performance of the system; in particular, for all h such that $h_m < h < h_{m+1}$, the estimate is better than that required to use m-QAM but not good enough to use $(m+1)$-QAM. Energy adaptation provides a means to solve this problem [31]. Rather than employing the method of [31], an alternate method, which is analogous to truncated channel inversion and the power pruning of [20], is described here. The advantage of this method is that, with very little loss of optimality, it is easily extended to coded modulation structures, where the overall optimization problem of [31] is not easily framed when channel prediction is not perfect [27]. Once a signal set has been chosen, the system is essentially a fixed-rate system; thus, the goal changes from maximizing average rate to attempting to allow communication at this fixed rate with the least amount of power. Thus, after the signal set is chosen, Equation 6.7 and Equation 6.8 are used to decide the minimum energy required to maintain P_b given the channel estimate h, and this energy is employed rather than the average energy. Any excess energy is put into a "bank" on which successive symbols can draw.

6.3.2.2 Numerical Results

As discussed in Section 6.3.1, systems with a significant amount of decoding complexity and allowable latency only have the potential for a small amount of improvement when CSI is provided to the transmitter. As might be expected, uncoded systems, which have the least decoder complexity and essentially no latency, benefit the most when transmitter CSI is available. In particular, uncoded systems operating over frequency-non-selective Rayleigh fading channels perform very poorly, because they do not achieve diversity. Because of this, coherently decoded quadrature phase-shift keying (QPSK) with only receiver CSI requires an SNR of 34 dB to achieve a bit error rate of 10^{-4} on a frequency-non-selective Rayleigh fading channel [53, p. 829], whereas the same technique requires an SNR of less than 10 dB to achieve the same bit error rate on an AWGN channel. The reason for this discrepancy is that the QPSK system operating over the Rayleigh fading channel is extremely susceptible to deep signal fades. Although the occurrence of such is relatively uncommon, the error rate during a bad fade can be orders of magnitude above that occurring when the average received SNR is observed, and thus these bad fades dominate the error rate.

In adaptive signaling, CSI is available at the transmitter. Arguably, the greatest utility of such information is that signaling can be avoided when bad fades are present. In particular, with perfect transmitter CSI [31], average rates in excess of 2 bits per symbol are possible at bit error rates of 10^{-5} for average received SNRs under 20 dB. Thus, there is a significant gain in system performance when transmitter CSI is available in uncoded systems.

However, as pointed out in [24, 27], the assumption of perfect CSI is dangerous when channel estimates are outdated or noisy, as would be the case with realistic delay in the feedback path from the receiver to the transmitter. In particular, the conditional density function given in Equation 6.2 becomes Rician (rather than a delta function), and hence the conditional channel acts like a fading channel. For example, for the example described in Section 6.3.2.1, adaptive signaling assuming perfect channel estimation can miss its target bit error rate by two orders of magnitude—even for the relatively high correlation coefficients of $\rho = 0.96$. In this case, bad predictions, which are relatively uncommon, lead to instantaneous error rates that are orders of magnitude above the target and thus dominate system performance. Using the design method of Equation 6.2 reveals that there are still significant gains in adaptive signaling vs. nonadaptive signaling when transmitter CSI is not perfect—even when the correlation coefficient drops as low as

$\rho = 0.85$. We conclude from this section that adaptive signaling is particularly effective for simple, low-latency systems such as adaptive uncoded QAM systems [24, 27].

6.3.3 Coded Modulation Structures

As discussed in Section 6.2.3, the conditional channel given an outdated measurement can vary from almost Rayleigh to almost AWGN—depending on both the channel estimate and the mean square prediction error σ^2. For small σ^2, the conditional channel is nearly always Rician with a large noncentrality component [53, p. 811] (hence, approaching AWGN), whereas for large σ^2, the channel often approaches Rayleigh. It is well known that coded modulation structures optimized for AWGN channels (e.g., [63]) are not well-matched to Rayleigh channels [58]. Thus, the characterization of the mean square prediction error in a given system determines the types of coded modulation structures to be employed.

For systems where the mean square prediction error is anticipated to be nearly zero, coded modulation structures designed for the AWGN channel can be employed without interleaving [31]. For systems with a moderate amount of channel prediction error, structures designed for a Rayleigh fading channel can have aspects of structures designed for an AWGN channel embedded in them [27]. Finally, for adaptive systems where the mean square prediction error is expected to be large, adaptive bit-interleaved coded modulation (BICM) [50] is preferable. These three types of structures are presented below.

6.3.3.1 Coding Structures with (Nearly) Perfect Prediction

If the current channel fading $X(t)$ is known accurately at the transmitter (i.e., σ^2, the prediction error of an MMSE predictor, is small), the effective channel given the outdated estimate is roughly AWGN. Thus, coding structures designed for AWGN channels [63] should be employed [32]. In particular, a base trellis-coded modulation scheme [63] tuned to the average received SNR can be selected and then uncoded bits can be added or deleted based on the channel estimate. This structure is shown in Figure 6.5. As described in the legend to Figure 6.5, the interleaver can be removed in this case, since the Euclidean distance between two possible paths at the receiver can be precisely controlled. This structure allows symbol-by-symbol adaptation (unlike changing the rate of a convolutional encoder) and, because parallel branches are generally effective when communicating over AWGN channels, is an efficient coding structure over a wide range of instantaneous rates of the system.

6.3.3.2 Coding Structures with Moderate Prediction Error Statistics

When there is a moderate amount of predictor error power (i.e., moderate values of σ^2), the use of uncoded bits can be detrimental, since there will be a high number of channels that are not strongly Rician per Equation 6.2, and it is well known that the use of uncoded bits on fading channels is problematic, per Section 6.3.2.2. However, note that the channels become more Rician *as the predicted value increases.* This is fortuitous, because it allows the retention of the adaptive coded modulation structure shown in

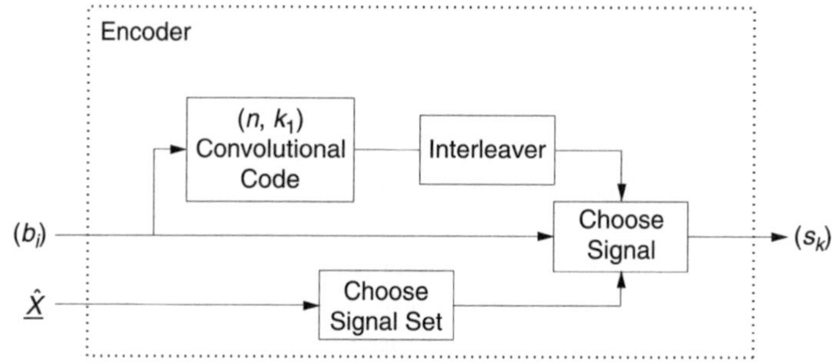

FIGURE 6.5 The adaptive trellis coding diagram—to be employed when performing adaptive signaling with low to moderate prediction errors. When the prediction error is low, the interleaver can be removed.

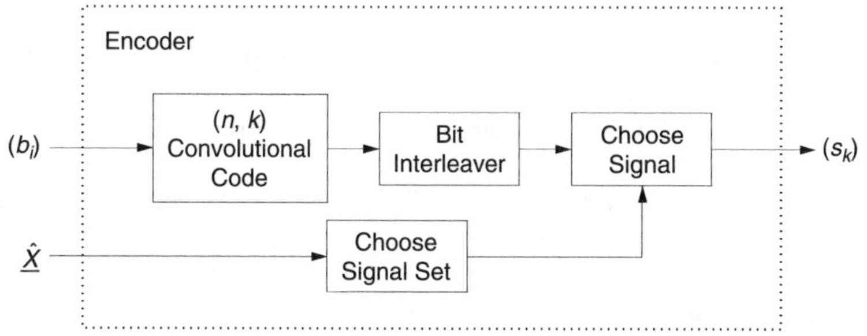

FIGURE 6.6 The adaptive bit-interleaved coded modulation scheme—to be employed when performing adaptive signaling with relatively frequent large prediction errors.

Figure 6.5, except that the base convolutional code is chosen to have no parallel branches [27]. Thus, when the estimate is small (and the channel nearly Rayleigh), a code appropriate for such a fading channel is employed [58]. When the channel estimate is large (and thus the channel strongly Rician), the structure in Figure 6.5 adds uncoded bits, which are appropriate in such a situation.

6.3.3.3 Coding Structures with Large Prediction Error Statistics

When there is a significant amount of prediction error (i.e., large values of σ^2), the use of parallel branches (i.e., uncoded bits) is not possible under almost any channel measurement, since from Equation 6.2, it can be seen that the channel will be nearly Rayleigh with very high probability. Thus, symbol-by-symbol rate adaptation is desirable, but parallel branches are not allowable. A structure that allows such was presented in [50] and is shown in Figure 6.6. Note that the instantaneous rate is adapted, but all bits are coded and thus the scheme retains diversity (in this case, against bad predictions) equal to the minimum Hamming distance of the convolutional code.

6.3.4 Designing with a Given Coded Modulation Structure

6.3.4.1 Design Rules

Unlike uncoded systems, which do not possess memory and thus allow simple symbol-by-symbol adaptation, as demonstrated in Section 6.3.2.1, the memory in coded modulation schemes complicates design. The main techniques that have been developed for coded modulation systems are described in [27] and [32]. Since the techniques in the later work of [27] include those in [32], the techniques of [27] will be briefly described. In particular, it is important to protect both the coded and uncoded information bits. This is done by maintaining the intersubset and intrasubset differences, which, roughly stated (see [27] for details), is the pairwise error probability between the two signal points in different and in the same subsets [63], respectively. When the prediction error power is small, the intrasubset and intersubset differences for the coded modulation structures described by Figure 6.5 can be maintained by simply preserving the *received* Euclidean distance between adjacent points in the signal set [32]. When the prediction error power is moderate, the intersubset and intrasubset differences must be maintained separately [27]. For the adaptive BICM structure of Figure 6.6, there are only intersubset differences to be maintained [50].

6.3.4.2 Performance Results

As demonstrated in Sections 6.3.1 and 6.3.2.2, the gains when CSI is available at the transmitter should decrease as system decoding complexity and latency are increased. Thus, it is anticipated that the gains described in this section will be smaller than those shown in Section 6.3.2.2, and this is indeed observed. In particular, nonadaptive systems employing coded modulation and interleaving over frequency-non-selective Rayleigh fading channels have an enormous potential for gain, as evidenced by the vast difference between the performance of nonadaptive uncoded systems [53, p. 829] and the channel capacity shown in Figure 6.3. In contrast, adaptive signaling schemes with perfect prediction are signaling for a channel that is conditionally AWGN, which implies that the gains between uncoded systems [31] and channel capacity (see Figure 6.3) for systems with CSI at the transmitter are not so vast; in fact, they are similar to those attainable for AWGN channels [32].

The performance loss when perfect channel predictions are not available at the transmitter can be mitigated by employing the techniques of Sections 6.3.3.2 and 6.3.3.3. Doing such with eight-state trellis codes provides gains over nonadaptive schemes that are on the order of 25 to 75% in data rate [27, 50], which, as expected, do not match the exorbitant gains seen in the uncoded case. One would expect the gains to decrease even further for more complicated codes and, in the complexity/latency limit, almost disappear as prescribed by Figure 6.3.

6.4 Adaptivity in Multiantenna Systems

Wireless systems employing multiple antennas at the transmitter or receiver have demonstrated both the theoretical [21, 62] and practical [21] ability to greatly increase system capacities far beyond those previously imagined. In particular, the disparate fading values between different pairs of antennas in multiple-input multiple-output (MIMO) systems lead to a large increase in the number of degrees of freedom of the system, and capacities can even dwarf those attainable for the AWGN channel at the same average transmitted SNR—even if channel state information is not available at the transmitter [21, 62].

Throughout this section, a multiple-antenna system will be referred to as an (M, N) system if it employs M transmit and N receive antennas. The system model generally employed for a narrowband MIMO system is given by

$$\underline{Y} = H\underline{X} + \underline{Z} \tag{6.9}$$

where $\underline{X}$ is an $M \times 1$ vector whose j^{th} component represents the signal transmitted by the jth antenna. Similarly, the received signal and received noise are represented by $N \times 1$ complex vectors, $\underline{Y}$ and $\underline{Z}$, respectively. Generally, it is assumed that the entries of the $N \times M$ matrix H, whose entry (i, j) represents the fading from transmitter j to receiver i, are identically distributed zero-mean jointly complex Gaussian random variables. For many of the early results, the entries of H were considered to be independent [21, 62], although the impact of correlation of the entries has been widely considered in recent years [13, 60, 67].

First, a review of recent information theoretic results for systems that employ some form of channel information at the transmitter is considered. Although this topic is relatively new, single-user MIMO information theory has progressed very rapidly, and interesting results in multiuser MIMO information theory are starting to appear. Next, work concerned with the adaptation of practical structures built on Equation 6.9 with knowledge of the values of H at the transmitter is considered.

6.4.1 Information Theoretic Considerations

An excellent recent tutorial of information theoretic considerations for MIMO systems, including adaptation-based ones on transmitter knowledge of the channel, can be found in [33].

6.4.1.1 MIMO Single-User Systems

It was established in early work on MIMO systems [62] that the Shannon capacity of an (N, N) MIMO system operating over a block-fading channel with CSI available at the transmitter and receiver is obtained by decomposing the channel into its eigenmodes, and then performing water filling [16, p. 349] on the eigenmodes based on the corresponding eigenvalues, where an eigenvalue indicates the SNR of the corresponding eigenmode. The system takes the convenient form of a single codebook designed for the AWGN channel followed by a beam former that is adapted to each block [5].

As noted throughout this work, the assumption of perfect channel state information at the transmitter is problematic on wireless communication channels due to their time-varying nature. This is particularly true in the case of MIMO channels, since there are far more numerous coefficients to estimate than in the single-input single-output case. There has been a recent set of papers [40, 42, 61, 64] that consider the Shannon capacity of MIMO systems when there is mean and covariance feedback. The results in [40, 42, 61, 64] demonstrate the trade-offs in Shannon capacity associated with having channel state information available at the transmitter; in particular, they generalize the results in [62] and reveal when Shannon capacity can be obtained by beam forming — only a scalar codebook followed by a beam former [33] rather than vector coding. In all cases, the Shannon capacity with knowledge of the channel at the transmitter only grows linearly with the number of antennas — only the leading constant and the simplicity of the system are possibly improved [2, 21], thus echoing the result of Section 6.3 with respect to gains in ergodic capacity when CSI is available at the transmitter.

In [5], information theoretic measures based on the notion of outage are considered when there is perfect CSI available at both the transmitter and receiver. Recall that such measures (see Section 6.3.1) attempt to capture the notion of system latency by limiting the number of fading blocks (and hence diversity) that a given code word experiences. In this case, a large number of antennas allows spatial diversity to be exploited, and in the limit, the effective channel can be made to look AWGN by employing a beam-forming approach. Hence, an error control code designed for an AWGN channel concatenated with a beam former is the optimal approach [5].

6.4.1.2 Multiuser MIMO Systems

The information theory of multiuser MIMO systems has only recently been explored. The most striking result in this context is the key gains that CSI at the transmitter [9, 10, 69] can provide, and this has generated a lot of interest in "dirty paper coding" methods [15]. In particular, unlike single-user MIMO systems, where generally only capacity-multiplying factors are improved or system complexity is decreased, CSI at the transmitter can allow the number of degrees of freedom to be increased [10, 33]. The facilitation of the penetration of such results into practical multiuser systems, where obtaining CSI at the transmitter can be complicated (and, at best, noisy and outdated), is an area that promises to be fruitful for research in the future.

6.4.2 Adaptive Coded Modulation for MIMO Systems

As demonstrated in Section 6.4.1, there has been significant recent work establishing information theoretic bounds and methods for achieving those bounds for MIMO systems. It is now incumbent upon the communication theory community to translate those gains into practice. In particular, it will be important to consider the nature of obtaining channel state information at the transmitter. If the CSI provided to the transmitter is reliable and the channel is relatively constant over a long block, the path will be clear — standard scalar coding followed by beam forming matched to the current CSI. Code rate (and power) adaptation will only need to be done on a block-by-block basis. However, if the CSI is noisy or outdated, it will be interesting to consider whether the results of [40, 42, 61, 64] apply; that is, it will be interesting to consider the robustness of beam forming when practical coded modulation schemes are employed.

Recent work on adaptive coded modulation for multiple-antenna systems has followed the basic tenets above. In particular, there has been some consideration about how to perform adaptation in practical systems in the MIMO environment. In [2], consideration is given to a scheme that tracks the eigenspace of the system so that water-filling type schemes can be employed — the parallel channel idea of [62]. Recent papers (e.g., [39]) have considered different forms of adaptation — antenna selection, beam forming, and space–time coding. Because such papers generally rely on (nearly) perfect prediction assumptions, there has been little consideration of the effects of imperfect channel state information in these works. Recent work [73] has considered the impact of imperfect channel state information. In particular, robustness under imperfect channel state information is obtained by adding an Alamouti scheme over an inner beam former (which would be ideal with perfect channel state information), and the optimality of such is shown for a system needing to transmit two information bits across the channel. These papers represent the start of research in a very important area that will bring the information theoretic gains of Section 6.4.1 to application.

6.5 Conclusions

Adaptive signaling, where the transmitted signal in a wireless system is adjusted based on channel state information available to the transmitter, has been an area of significant research interest for over a decade. For single-user systems, the gain from having channel state information at the transmitter is generally a reduction in complexity and latency. Conversely, the gain of adaptive signaling is generally a function of the system complexity — large gains for very simple uncoded systems and almost no gain in the system Shannon capacity. However, for systems characterized by outage capacity or those employing multiple antennas, channel state information at the transmitter can have a significant role, particularly in systems with multiple users. This indicates the importance of the design of practical adaptive coded modulation for multiple-input multiple-output systems. As in single-antenna systems, variability of the channel coefficients between the time of channel estimation and the time of data transmission will be a key concern that will need to be addressed.

References

[1] S. Alamouti and S. Kallel, "Adaptive trellis-coded multiple-phase-shift keying for Rayleigh fading channels," *IEEE Transactions on Communications*, 42, 2305–2314, 1994.

[2] B. Bannister and J. Zeidler, "Feedback assisted transmission subspace tracking for MIMO systems," to appear in the *IEEE Journal on Selected Areas in Communications.*

[3] P. Bello, "Characterization of randomly time-variant linear channels," *IEEE Transactions on Communications Systems*, COM-11, 360–393, 1963.

[4] C. Berrou, A. Glavieux, and P. Thitimajshima, "Near Shannon limit error-correcting coding and decoding: turbo-codes," in *Proceedings of International Conference on Communications*, May 1993.

[5] E. Biglieri, G. Caire, and G. Taricco, "Limiting performance of block-fading channels with multiple antennas," *IEEE Transactions on Information Theory*, 47, 1273–1289, 2001.

[6] J. Bingham, "Multicarrier modulation for Data Transmission: an idea whose time has come," *IEEE Communications Magazine*, May 1990, pp. 5–14.

[7] X. Cai and G. Giannakis, "Adaptive modulation with adaptive pilot symbol assisted estimation and prediction of rapidly fading channels," in *Proceedings of the Conference on Information Sciences and Systems*, 2003.

[8] G. Caire, G. Taricco, and E. Biglieri, "Optimal power control for the fading channel," *IEEE Transactions on Information Theory*, 45, 1468–1489, 1999.

[9] G. Caire and S. Shamai, "On achievable rates in a multi-antenna broadcast downlink," in *Proceedings of the 38th Annual Allerton Conference on Communication, Control, and Computing*, October 2000, pp. 1188–1193.

[10] G. Caire and S. Shamai, "On the achievable throughput of a multi-antenna Gaussian broadcast channel," to appear in *IEEE Transactions on Information Theory.*

[11] J. Cavers, "Variable-rate transmission for Rayleigh fading channels," *IEEE Transactions on Communications*, 20, 15–22, 1972.

[12] J. Cavers, "An analysis of pilot symbol assisted modulation for Rayleigh fading channels," *IEEE Transactions on Vehicular Technology*, 40, 686–693, 1991.

[13] C. Chuah, D. Tse, J. Kahn, and R. Valenzuela, "Capacity scaling in MIMO wireless systems under correlated fading," *IEEE Transactions on Information Theory*, 48, 637–651, 2002.

[14] L. Cimini, Jr., "Performance studies for high-speed indoor wireless communications," *Wireless Personal Communications*, 2, 67–85, 1995.

[15] M. Costa, "Writing on dirty paper," *IEEE Transactions on Information Theory*, 48, 637–650, 2002.

[16] T. Cover and J. Thomas, *Elements of Information Theory*, New York: Wiley, 1991.

[17] L. Ding and J.S. Lehnert, "Performance analysis of an uplink power control using truncated channel inversion for data traffic in a cellular CDMA system," in *Proceedings of the IEEE Vehicular Technology Conference*, 2000.

[18] A. Duel-Hallen, S. Hu, and H. Hallen, "Long-range prediction of fading signals: enabling adaptive transmission for mobile radio channels," *IEEE Signal Processing Magazine, Special Issue on Advances in Wireless and Mobile Communications*, 17, 62–75, 2000.

[19] T. Eyceoz, A. Duel-Hallen, and H. Hallen, "Deterministic channel modeling and long range prediction of fast fading mobile radio channels," *IEEE Communication Letters*, 2, 254–256, 1998.

[20] E. Feig, "Practical aspects of DFT-based frequency division multiplexing for data transmission," *IEEE Transactions on Communications*, 38, 929–932, 1990.

[21] G. Foschini, "Layered space-time architecture for wireless communication in a fading environment when using multi-element antennas," *Bell Labs Technical Journal*, 41–59, Autumn 1996.

[22] G. Foschini and M. Gans, "On limits of wireless communications in a fading environment when using multiple antennas," *Wireless Personal Communications*, 6, 311–335, 1998.

[23] R. Gallager, *Low Density Parity-Check Codes*, Cambridge, MA: MIT Press, 1963.

[24] D. Goeckel, "Robust adaptive coded modulation for time-varying channels with delayed feedback," in *Proceedings of the Thirty-Fifth Annual Allerton Conference on Communication, Control, and Computing*, 1997, pp. 370–379.

[25] D. Goeckel, "Adaptive coding for fading channels using outdated fading estimates," in *Proceedings of the IEEE Vehicular Technology Conference*, 1998, pp. 1925–1929.

[26] D. Goeckel, "Strongly robust adaptive signaling for time-varying channels," in *Proceeding of the 1998 International Conference on Communications*, June 1998, pp. 454–458.

[27] D. Goeckel, "Adaptive coding for time-varying channels using outdated fading estimates," *IEEE Transactions on Communications*, 47, 844–855, 1999.

[28] A. Goldsmith, L. Greenstein, and G. Foschini, "Error statistics of real-time power measurements in cellular channels with multipath and shadowing," *IEEE Transactions on Vehicular Technology*, 43, 439–446, 1994.

[29] A. Goldsmith, "Variable-rate coded MQAM for fading channels," in *Proceedings of the IEEE Global Communications Conference: Communication Theory Miniconference*, 1994, pp. 186–190.

[30] A. Goldsmith, "Capacity and dynamic resource allocation in broadcast fading channels," in *Proceedings of the Allerton Conference on Communications, Control, and Computing*, 1995, pp. 915–924.

[31] A. Goldsmith and S. Chua, "Variable-rate variable-power MQAM for fading channels," *IEEE Transactions on Communications*, 45, 1218–1230, 1997.

[32] A. Goldsmith and S. Chua, "Adaptive coded modulation for fading channels," *IEEE Transactions on Communications*, 46, 595–602, 1998.

[33] A. Goldsmith, S. Jafar, N. Jindal, and S. Vishwanath, "Capacity limits of MIMO channels," to appear in *IEEE Transactions on Wireless Communications.*

[34] I. Gradshteyn and I. Rhyzhik, *Table of Integrals, Series, and Products*, New York: Academic Press, 1980.

[35] M. Gudmundson, "Correlation model for shadow fading in mobile radio systems," *Electronics Letters*, 27, 2145–2146, 1991.

[36] S. Hanly and D. Tse, "Multi-access fading channels. Part II. Delay-limited capacities," *IEEE Transactions on Information Theory*, 44, 2816–2831, 1998.

[37] J. Hayes, "Adaptive feedback communications," *IEEE Transactions on Communications Technology*, 16, 29–34, 1968.

[38] V. Hentinen, "Error performance for adaptive transmission on fading channels," *IEEE Transactions on Communications*, 22, 1331–1337, 1974.

[39] S. Hu and A. Duel-Hallen, "Combined adaptive modulation and transmit diversity using long range prediction for flat mobile radio channels," in *Proceedings of the Global Communications Conference*, 2001.

[40] S. Jafar and A. Goldsmith, "Transmitter optimization and optimality of beamforming for multiple antenna systems with imperfect feedback," submitted to *IEEE Transactions on Wireless Communications.*

[41] W. Jakes, Jr., *Microwave Mobile Communications*, New York: Wiley, 1974.

[42] E. Jorswieck and H. Boche, "Channel capacity and capacity-range of beamforming in MIMO wireless systems under correlated fading with covariance feedback," submitted to *IEEE Journal on Selected Areas in Communications.*

[43] K. Kamath and D. Goeckel, "Adaptive modulation schemes for minimum outage probability in wireless systems," submitted to *IEEE Transactions on Communications.*

[44] V. Lau and M. Macleod, "Variable rate adaptive trellis coded QAM for high bandwidth efficiency applications in Rayleigh fading channels," in *Proceedings of the Vehicular Technology Conference*, 1998, pp. 348–352.

[45] V. Lau and M. Macleod, "Variable-rate trellis coded QAM for flat-fading channels," *IEEE Transactions on Communications*, 49, 1550–1560, 2001.

[46] D. MacKay, "Good error-correcting codes based on very sparse matrices," *IEEE Transactions on Information Theory*, 45, 399–431, 1999.

[47] A. Monk and L. Milstein, "Open-loop power control error in a land mobile satellite system," *IEEE Journal on Selected Areas in Communications*, 13, 205–212, 1995.

[48] S. Nanda, K. Balachandran, and S. Kumar, "Adaptation techniques in wireless packet data services," *IEEE Communications Magazine*, 54–64, 2000.

[49] R. Negi and J. Cioffi, "Delay-constrained capacity with causal feedback," *IEEE Transactions on Information Theory*, 48, 2478–2494, 2002.

[50] P. Örmeci, X. Liu, D. Goeckel, and R.D. Wesel, "Adaptive bit-interleaved coded modulation," *IEEE Transactions on Communications*, 49, 1572–1581, 2001.

[51] K. Pahlavan and A. Levesque, *Wireless Information Networks*, New York: John Wiley & Sons, 1995.

[52] A. Papoulis, *Probability, Random Variables, and Stochastic Processes*, 3rd ed., New York: McGraw-Hill, 1991.

[53] J. Proakis, *Digital Communications*, 4th ed., New York: McGraw-Hill, 2001.

[54] M. Pursley and C. Wilkins, "Adaptive-rate coding for frequency-hop communications over Rayleigh fading channels," *IEEE Journal on Selected Areas in Communications*, 17, 1224–1232, 1999.

[55] M. Pursley and J. Shea, "Channel quality estimation with channel error counters for adaptive signaling in wireless communications," in *Proceedings of the International Symposium on Information Theory*, 2000.

[56] T. Rappaport and C. McGillen, "UHF fading in factories," *IEEE Journal on Selected Areas in Communications*, 7, 40–48, 1989.

[57] T. Rappaport, *Wireless Communications*, New York: Prentice Hall, 1996.

[58] C. Schlegel and D. Costello, Jr., "Bandwidth efficient coding for fading channels: code construction and performance analysis," *IEEE Journal on Selected Areas in Communications*, 7, 1356–1368, 1989.

[59] C. Shannon, "A mathematical theory of communications," *Bell Systems Technical Journal*, 27, 379–423, 623–656, 1948.

[60] D. Shiu, G. Foschini, M. Gans, and J. Kahn, "Fading correlation and its effect on the capacity of multielement antenna systems," *IEEE Transactions on Communications*, 48, 502–513, 2000.

[61] S. Simon and A. Moustakas, "Optimizing MIMO antenna systems with channel covariance feedback," to appear in *IEEE Journal on Selected Areas in Communications.*

[62] I. Telatar, "Capacity of multi-antenna Gaussian channels," *European Transactions on Telecommunications*, 10, 586–595, 1999.

[63] G. Ungerboeck, "Channel coding with multilevel/phase signals," *IEEE Transactions on Information Theory*, 28, 55–67, 1982.

[64] E. Visotsky and U. Madhow, "Space-time transmit precoding with imperfect feedback," *IEEE Transactions on Information Theory*, 47, 2632–2639, 2001.

[65] B. Vucetic, "An adaptive coding scheme for time-varying channels," *IEEE Transactions on Communications*, 39, 653–663, 1991.

[66] S. Wei and D. Goeckel, "Error statistics for average power measurements in wireless communication systems," *IEEE Transactions on Communications*, 50, 1535–1546, 2002.

[67] S. Wei, D. Goeckel, and R. Janaswamy, "On the asymptotic capacity of MIMO systems with antenna arrays of fixed length," submitted to *IEEE Transactions on Wireless Communications.*

[68] R. Yates, "A framework for uplink power control in cellular radio systems," *IEEE Journal on Selected Areas in Communications*, 13, 1341–1347, 1995.

[69] W. Yu and J. Cioffi, "Trellis precoding for the broadcast channel," in *Proceedings of the Global Communications Conference*, 2001, pp. 1344–1348.

[70] J. Zander, "Performance of optimum transmitter power control in cellular radio systems," *IEEE Transactions on Vehicular Technology*, 41, 57–62, 1992.

[71] R. van Nee and R. Prasad, *OFDM for Wireless Multimedia Communications*, Norwood, MA: Artech House, 2000.

[72] H. Van Trees, *Detection, Estimation, and Modulation Theory*, Vol. I, New York: Wiley, 1968.

[73] S. Zhou and G. Giannakis, "Optimal transmitter eigen-beamforming and space-time block coding based on channel mean feedback," in *IEEE Transactions on Signal Processing*, October 2002.

7

Signaling Constellations for Transmission over Nonlinear Channels

Hisham Abdul Hussein Al-Asady
Queen's University

Quazi M. Rahman
Queen's University

Mohamed Ibnkahla
Queen's University

7.1 Introduction 7-2
7.2 System Model 7-3
7.3 Craig's Method 7-4
7.4 Probability of Symbol Error for 16-ary QAM Format 7-5
The (8,8) Constellation Format • The (4,12) Constellation Format • The (5,11) Constellation Format • The (6,10) Constellation Format • 16-Rectangular Constellations with a Circular Format • The Effect of Nonlinearity and the Application of a Predistortion Technique • Total Degradation • Results and Discussions
7.5 Probability of Symbol Error for the Circular 32-ary QAM Format 7-20
The (4,11,17) Constellation Format • The (5,11,16) Constellation Format • 32-Rectangular Constellations with a Circular Format • The Effect of Nonlinearity and the Application of a Predistortion Technique • Results and Discussions
7.6 Conclusions 7-30

Abstract

This chapter presents a simplified mathematical model to evaluate the performance of any given circular constellation of the M-level quadrature amplitude modulation (M-QAM) technique in terms of probability of error (POE) or symbol error rate (SER). With the aim to work on memoryless nonlinear satellite channels, the model is derived as a generalized form for both linear and nonlinear channels in the presence of additive white Gaussian noise (AWGN). The analysis provides the means to calculate the optimal ring ratio (RR) and phase difference (PD) for several possible candidates of 16- and 32-QAM circular constellations. The effects of RR and PD on the POE performance are investigated in the analysis. In the 16-QAM system, the analytical formulation has been extended for total degradation (TD) performance measure as a function of input back-off (IBO) of the nonlinear amplifier. To overcome the nonlinear distortion, data predistortion is taken into account. A POE performance comparison between different constellations for both 16- and 32-QAM systems has also been presented in this chapter. The analytical results are validated by simulation.

0-8493-1657-X/$0.00+$1.50

7.1 Introduction

The rapid evolution of global information technology demands high-data-rate transmission via satellites in the presence of available bandwidth, which in turn requires a spectrally efficient modulation technique. In this case, high-level modulation techniques (M-ary) are the favorable candidates. Among all the existing M-ary modulation techniques M-level quadrature amplitude modulation (M-QAM) offers the maximum power/spectral efficiency [Pro02] and appears to be a potentially attractive modulation scheme for satellite communications [Ibn04].

Unlike terrestrial wireless systems, where power is not a major concern, satellite systems are severely limited in power for the payload and need nonlinear onboard high-power amplifiers (HPAs), such as the traveling wave tube (TWT), to operate efficiently (i.e., close to their maximum power). As a result, these HPAs introduce nonlinear distortions [Web95] [Gol97]. Since M-QAM signals ($M > 4$) lack a constant envelope, they become highly sensitive to these nonlinearities and this results in making a simple tractable probability of error (POE) evaluation technique difficult.

The objective of our work is to provide a tractable mathematical model for evaluating the POE of M-QAM ($M > 4$) systems with different constellations, in the presence of a nonlinear satellite channel. We focus our study on 16- and 32-QAM systems.

Currently the 16-QAM scheme is used in the forward link of the third-generation (3G) cellular data-only system (IS856) and is being widely investigated (e.g., see [Che00]) in satellite domains. In [Dim90], the effect of nonlinear distortion on the error performance of the 16-QAM rectangular signaling scheme, operating in a typical satellite channel, was investigated by means of computer simulation. In [Cra91], Craig provided a simple formulation technique for evaluating the POE of two-dimensional signals in the linear additive white Gaussian noise (AWGN) channel, which was applied and extended later by many researchers for POE evaluation. In [Chi98], to find the POE expression, the author followed Craig's method by assuming that the receiver knows the saturation level of the nonlinear amplifier (NA) through which it can modify the decision boundary. Xiaodai et al. in [Dng99a] and [Dng99b] have extended Craig's method for linear fading channels.

Even though Craig's method provides a simplified technique, it requires a clear knowledge about the decision boundaries of the constellations. For nonlinear systems, this becomes a difficult task unless the decision boundaries of the constellations at the HPA output are assumed to be known. To overcome this bottleneck, here we propose a simpler approach than Craig's method. Instead of considering the decision regions we take into account the distances between the symbols for different constellations to calculate the POE and its lower and upper bounds [Pro02]. Our approach not only eliminates the need to find the decision regions for each constellation, but also avoids any cumbersome numerical integration, which is required by all the above-mentioned analyses.

In an AWGN channel with the 16-QAM system we present the POE analysis for the (8,8), (4,12), (5,11), (6,10), and (4,8,4) circular constellations, while with the 32-QAM system the analysis is presented for the (5,11,17), (4,12,16), and (4,8,4,8,8) circular constellations. Among these, the (4,8,4) and (4,8,4,8,8) constellations represent 16- and 32-rectangular constellations, respectively, in circular format. For each case, we present a generalized form for the POE expression. Discussion on the nonlinear environment is also included in the manuscript. Although any type of HPA will fit with the analysis, our discussion considers the TWT amplifier due to its major application in the satellite communications domain. These systems have also been studied with a data predistorter. The predistorter calculates the optimal ring ratio (RR) and the optimal phase difference (PD) that need to be used at the transmitter to overcome the nonlinear effects of the TWT amplifier. For the 16-QAM system, total degradation[1] (TD) performance measure in terms of input back-off[2] (IBO) is also included in the analysis.

[1]Total degradation is a performance figure that describes the difference between the maximum power of the HPA and the output power of a linear amplifier required to guarantee a predefined POE.

[2]Input back-off is defined as the ratio between maximum input power and average input power of the NA [Kar89].

In summary, the contribution of this chapter includes the formulation of a simplified mathematical model to calculate the POE for any circular constellations of M-QAM systems in an AWGN channel. This modeling technique is equally applicable for both linear and nonlinear channels. With the availability of both the upper and lower bounds of the POE, deriving the best-approximated POE expressions is an important contribution to this study. The analysis also presents the optimal RRs and PDs for all the studied constellations. The analytical results, which agree with the well-known closed-form POE expression [Pro02], are validated with computer simulations. A final contribution is made to this study by the extension of the mathematical model for the data predistortion technique that results in a POE performance improvement in a nonlinear channel. The chapter provides a fine comparison between POE performances for different constellations (both for 16- and 32-QAM), taking into consideration different parameters such as RRs, PDs, and IBO (for the HPA).

The remainder of this chapter is organized as follows. Section 7.2 briefly discusses the system model that we have used in our analysis. In Section 7.3, Craig's method [Cra91] for evaluating the POE of two-dimensional signals is presented. The reason behind presenting this method is to provide readers with a clear picture of its limitations, which in later sections are shown to be avoided through our simplified method for evaluating the POE. In Section 7.4, the POE and its bounds for different 16-QAM circular constellations are derived. Analytical and simulation results along with a brief comparison between different 16-QAM constellations are presented in this section. Also presented are discussions on the effect of nonlinearity on the system's performance on the predistortion techniques used to improve the POE performance in the presence of nonlinearity. Finally, a TD performance measure for different 16-QAM circular constellations is presented. Section 7.5 covers the same areas as Section 7.4, but for a 32-QAM system. Conclusions are drawn in Section 7.6.

7.2 System Model

Figure 7.1 shows a block diagram of the nonlinear communications system, which we have considered in our analysis. In this figure, the data source is a random data generator. These random data are fed into the QAM modulator, which modulates different signal constellations with different values of RRs and PDs. The TWT amplifier provides two types of distortions that affect the overall system performance. These distortions are the results of AM/AM and AM/PM conversion of the incoming signal (see Appendix).

In the system model we have considered an AWGN channel, which is represented by a random Gaussian noise generator, as shown in the block diagram. The demodulator demodulates different M-QAM (M = 16 or 32) signal constellations with different values of RRs and PDs.

The detection process incorporates an optimum detector (MLSE [maximum likelihood sequence estimation] type) that takes the nonlinear effect into account to get a minimum POE.

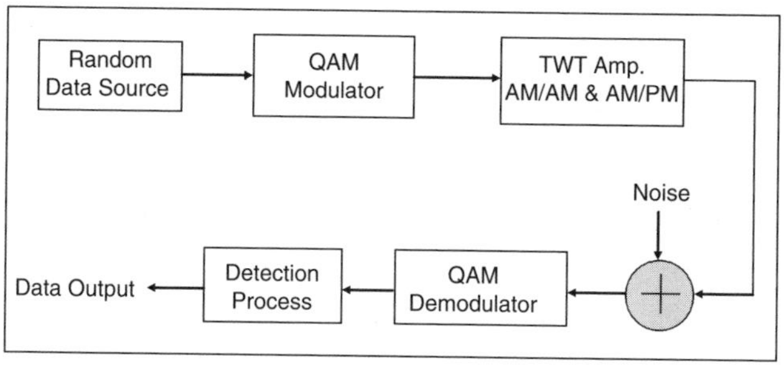

FIGURE 7.1 Nonlinear communication system.

7.3 Craig's Method

In [Cra91], Craig provided a general method to calculate the POE for any two-dimensional signal constellations based on the boundaries of the decision regions. Figure 7.2 illustrates a possible received noiseless symbol x_i in two-dimensional signal space and its decision boundaries. The transmitted symbol x_i is chosen from a constellation signal set. In general, the decision boundaries are either closed (Figure 7.2a) or open (Figure 7.2b) regions. If the two-dimensional Guassian noise, superimposed on x_i, results in the received signal falling outside the appropriate decision boundaries, a symbol decision error occurs. The decision region, given that x_i is sent in, can be divided into trilaterals (triangle or triangles with a vertex at infinity). These trilaterals are formed by straight lines, originating at the symbol x_i and terminating at the intersection of two decision boundaries. The erroneous decision region given x_i is therefore composed of disjoint subregions $1, 2, \ldots, 6$, as illustrated in Figure 7.6a.

The joint probability density function (PDF) of Gaussian noise $n = n_x + jn_y$ is given by

$$p_n(n_x, n_y) = \frac{1}{\pi N_0} \exp\left\{-\frac{n_x^2 + n_y^2}{N_0}\right\} \tag{7.1}$$

In polar coordinates, the PDF of Gaussian noise is expressed as

$$p_n(r, \theta) = \frac{r}{\pi N_0} \exp\left\{-\frac{r^2}{N_0}\right\} \tag{7.2}$$

For example, the probability that the received signal falls into an erroneous area of subregion 1 in Figure 7.2a is given by

$$P_{AWGN-1} = \int_0^{\eta_0} \int_{R(\theta)}^{\infty} p_n(r, \theta) dr\, d\theta = \frac{1}{2\pi} \int_0^{\eta_0} \exp\left\{-\frac{R^2(\theta)}{N_0}\right\} d\theta \tag{7.3}$$

where $R(\theta) = \frac{l_0' \sin \psi_0}{\sin(\theta + \psi_0)}$ and $l_0' = \sqrt{b_0 E_0}$ is the distance shown in Figure 7.2. The subscript 0 means all the parameters are for the AWGN channel. Hence the average signal-to-noise ratio (SNR) per symbol is given by $\gamma = \frac{E_0}{N_0}$, and E_0 is the signal energy before interrupted by noise. $N_0$is the one-sided power spectrum density of Gaussian noise. Parameters η_0, b_0, and ψ_0 are determined by the geometry of the trilaterals corresponding to the subregion, and θ is a dummy variable of integration. In summary, the probability of

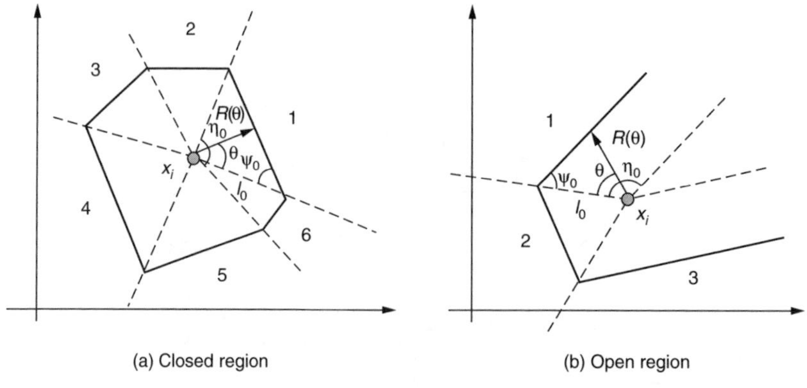

FIGURE 7.2 Decision boundaries of 2-D signal constellation.

error for subregion 1 can be rewritten as a function of SNR:

$$P_{AWGN-1}(\gamma) = \frac{1}{2\pi}\int_{0}^{\eta_0} \exp\left\{-\frac{b_0\gamma\sin^2\psi_0}{\sin^2(\theta+\psi_0)}\right\} d\theta \tag{7.4}$$

The above equations are well suited to numerical evaluation in that they have finite integration limits and an exponential integrand [Dng99]. The detection error probability, given that x_i is sent, is then the sum of probabilities that the received signal falls into each erroneous subregion (i.e., subregions 1 to 6 in Figure 7.2a):

$$P(e|\, x_i) = \sum_{j=1}^{6} P_{AWGN-j}(\gamma) \tag{7.5}$$

The probability of error, given any other symbol in the constellation is transmitted, can be obtained similarly. Then the exact average probability of symbol error for M-ary data in AWGN is the weighted sum of probabilities for all subregions of every possible signal point, which can be calculated as

$$Pe_{AWGN} = \sum_{i=1}^{M_s} P(x_i)\sum_{j=1}^{G_i} P_{i,j}(\gamma) = \sum_{i=1}^{M_s}\sum_{j=1}^{G_i} \frac{P(x_i)}{2\pi}\int_{0}^{\eta_{0,i,j}} \exp\left\{-\frac{b_{0,i,j}\gamma\sin^2\psi_{0,i,j}}{\sin^2(\theta+\psi_{0,i,j})}\right\} d\theta \tag{7.6}$$

where $P(x_i)$ is the prior probability of transmitted symbol x_i; $b_{0,i,j}$, $\psi_{0,i,j}$, and $\eta_{0,i,j}$ are parameters corresponding to x_i, subregion j, and are determined only by input constellation; M_s means M-ary symbols; and G_i is the total number of subregions for symbol x_i.

7.4 Probability of Symbol Error for 16-ary QAM Format

From the previous section we see that Craig's method calls for a clear knowledge of the decision boundaries of all the constellations. Moreover, it needs numerical integration to find out the average POE. To overcome these shortfalls, in this section and the following one we propose a simplified approach to evaluate the POE, which does not need either the numerical integration or the knowledge of the decision regions. We discuss this simplified methodology for finding out the POEs of different 16-QAM circular constellations.

In this study we consider the (8,8), (4,12), (5,11), and (6,10) constellations as well as the 16-rectangular constellation with circular format (4,8,4). Here, after presenting the structures of each constellation, we discuss the analytical technique used to derive the POE expression and then talk about the POE results in the presence of a predistortion technique. The analysis, which is dependent on the geometry of the different constellations, provides optimal RRs and PDs for different constellations. Moreover, we briefly discuss the TD performance as a function of IBO. Finally, the section concludes with results and discussions on the above findings, including a performance comparison between different constellations. Here the analytical results are verified with computer simulations.

7.4.1 The (8,8) Constellation Format

7.4.1.1 The Structure

One of the most popular 16-QAM constellations, shown in Figure 7.3, is the circular (8,8) constellation. It is constructed by two concentric circles, each of which has eight symbols. In its structure the distances between any two neighboring symbols in each circle are kept equal. The ratio between the outer circle radius (R) and the inner circle one (r) is defined as a ring ratio (α), i.e.,

$$\alpha = \frac{R}{r} \tag{7.7}$$

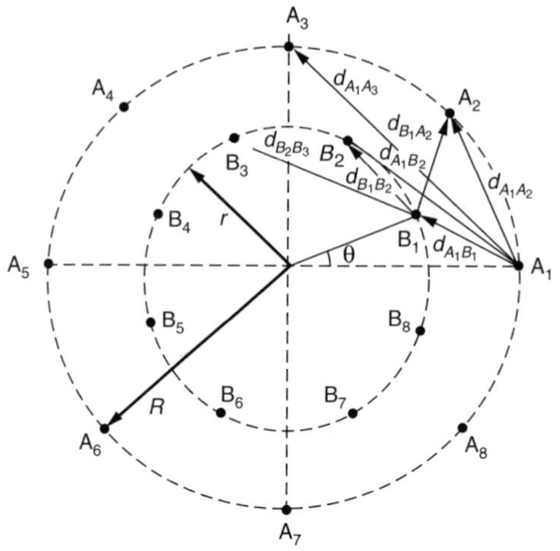

FIGURE 7.3 Circular (8,8) constellation.

The angle θ between the first outer circle symbol and the nearest inner circle symbol is defined as a phase difference.

7.4.1.2 Probability of Error Analysis

The analytical technique to find the POE, presented in this section, is dependent on the geometry of the constellation diagram. Here, as a result of symmetry of the symbol orientations on both circles, we can arbitrarily choose any two symbols, one from each circle, for the analysis. In this case, we choose symbols A_1 and B_1 and define:

$d_{A_1A_i}$ = distance between symbols A_1 and A_i $(i \neq 1)$, in the outer circle
$d_{A_iB_i}$ = distance between any two closest inner and outer circle symbols
$d_{A_1B_i}$ = distance between the outer circle symbol A_1 and any other inner circle symbol B_i $(i \neq 1)$
$d_{B_1A_{10-i}}$ = distance between the inner circle symbol B_1 and any other outer circle symbol A_{10-i} $(i \neq 1)$
$d_{B_1B_i}$ = distance between symbols B_1 and B_i $(i \neq 1)$, in the inner circle

With the aid of Figure 7.3, all the above distances can easily be equated as

$$d^2_{A_1A_i} = 4E_{s_2}\sin^2\left[\frac{(i-1)\pi}{8}\right], \quad i = 2, 3, \ldots, 8 \tag{7.8}$$

$$d^2_{A_iB_i} = E_{s_1}[1 + \alpha^2 - 2\alpha\cos(\theta)], \quad i = 1, 2, \ldots, 8 \tag{7.9}$$

$$d^2_{A_1B_i} = E_{s_1}\left[1 + \alpha^2 - 2\alpha\cos\left\{\theta + \frac{(i-1)\pi}{4}\right\}\right], \quad i = 1, 2, 3, \ldots, 8 \tag{7.10}$$

$$d_{B_1A_{10-i}} = d_{A_1B_i}, \quad i = 2, 3, \ldots, 8 \tag{7.11}$$

$$d^2_{B_1B_i} = 4E_{s_1}\sin^2\left[\frac{(i-1)\pi}{8}\right], \quad i = 2, 3, \ldots, 8 \tag{7.12}$$

where α and θ have already been defined, and E_{s_1} and E_{s_2} are the energies of the inner and outer circle symbols, respectively, which are defined as

$$E_{s_1} = r^2 \tag{7.13a}$$

$$E_{s_2} = R^2 \tag{7.13b}$$

Now, if we transmit any arbitrary symbol from the transmitter in the presence of AWGN of one-sided power spectral density (PSD) N_0, the *lower bound* of the POE could be given as

$$P_{e_{Lower}} \geq Q\left(\sqrt{\frac{d_{\min}^2}{2N_o}}\right) \tag{7.14}$$

where $d_{\min}$ represents the minimum distance among all the distances calculated in Equation 7.8 to Equation 7.12. From the expressions of the distance parameters it is clear that $d_{\min}$ is either a function of RR, PD, and symbol energy or a function of symbol energy only. It is worthwhile to mention that the symbol energy itself is a function of the RR, α.

We can also express the upper bound for the POE in terms of $d_{\min}$, which will be given as

$$P_{e_{weakupper}} \leq 15Q\left(\sqrt{\frac{d_{\min}^2}{2N_o}}\right) \tag{7.15}$$

where the factor 15 represents the number of neighboring symbols around the transmitted one.

For calculating the best-approximated POE for the transmitted symbol in the presence of AWGN, we can carry out the following procedure. First, assuming that one of the outer circle symbols (e.g., A_1) is transmitted, we calculate the corresponding POE, and then assuming that one of the inner circle symbols (e.g., B_1) is transmitted, we calculate the corresponding POE. Finally, we average the above POE expressions over the PDF of the symbols. Assuming all the symbols are equally likely, we find that the PDF of any symbol in either the outer or inner circle is equal to 8/16 = 0.5 (i.e., $P(A_1) = P(B_1) = 0.5$). As a result, the best-approximated POE expression becomes

$$P_{e(8,8)} \approx \sum_{i=1}^{8} Q\left(\sqrt{\frac{d_{A_1B_i}^2}{2N_0}}\right) + \frac{1}{2}\sum_{i=2}^{8} Q\left(\sqrt{\frac{d_{A_1A_i}^2}{2N_0}}\right) + \frac{1}{2}\sum_{i=2}^{8} Q\left(\sqrt{\frac{d_{B_1B_i}^2}{2N_0}}\right) \tag{7.16}$$

Equation 7.14 to Equation 7.16 provide generalized POE expressions for the circular (8,8) constellation, which are equally applicable for both linear and nonlinear environments. The expressions can easily be evaluated by calculating the constellations' distances.

7.4.2 The (4,12) Constellation Format

7.4.2.1 The Structure

Figure 7.4 shows the circular (4,12) constellation whose structure is the same as that of the (8,8) one, with the only exception being that in this case the inner and outer circles have 4 and 12 symbols, respectively. As before, RR (α) and PD (θ) hold the same definitions.

7.4.2.2 Probability of Error Analysis

Again, as a result of the symmetry of the symbol orientations on both circles, we can arbitrarily choose any two symbols, one from each circle, for presenting the POE analysis. In this case, choosing symbols A_1 and B_1 as before, we define $d_{A_1B_1}$ and $d_{B_1A_i}$ as the distance between symbols A_1 and B_1, and between the inner circle symbol B_1 and any other outer circle symbol A_i ($i \neq 1$), respectively. We use the same definitions for $d_{A_1A_i}$, $d_{A_1B_i}$, and $d_{B_1B_i}$ as stated in Section 7.4.1.2.

Here, from the geometry of Figure 7.4, all the above distances can easily be equated as

$$d_{A_1B_1}^2 = E_{s_1}[1 + \alpha^2 - 2\alpha\cos(\theta)] \tag{7.17}$$

$$d_{B_1A_i}^2 = E_{s_1}\left[1 + \alpha^2 - 2\alpha\cos\left\{\theta + \frac{(i-1)\pi}{6}\right\}\right], \quad i = 1, 2, 3, \ldots, 12 \tag{7.18}$$

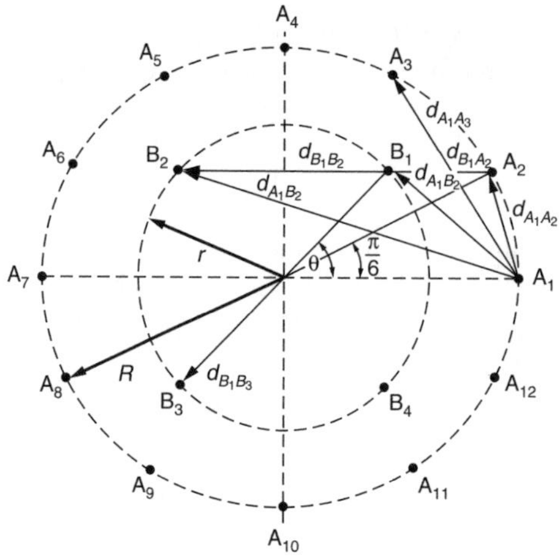

FIGURE 7.4 (4,12) Circular constellation.

$$d^2_{A_1 A_i} = 4E_{s_2} \sin^2\left[\frac{(i-1)\pi}{12}\right], \quad i = 2, 3, \ldots, 12 \tag{7.19}$$

$$d^2_{A_1 B_i} = E_{s_1}\left[1 + \alpha^2 - 2\alpha \cos\left\{\theta + \frac{(i-1)\pi}{2}\right\}\right], \quad i = 1, 2, 3, 4 \tag{7.20}$$

$$d^2_{B_1 B_i} = 4E_{s_1} \sin^2\left[\frac{(i-1)\pi}{4}\right], \quad i = 2, 3, 4 \tag{7.21}$$

where E_{s_1}, E_{s_2}, α, and θ hold the same definitions as before. In this case, in transmitting an arbitrary symbol from the transmitter in the presence of AWGN of one-sided PSD N_0, the lower and upper bounds of the POE will hold the same expressions of Equation 7.14 and Equation 7.15, respectively, while for the best-approximated POE, the expression would be

$$P_{e(4,12)} \approx \frac{3}{4}\sum_{i=1}^{4} Q\left(\sqrt{\frac{d^2_{A_1 B_i}}{2N_0}}\right) + \frac{3}{4}\sum_{1=2}^{12} Q\left(\sqrt{\frac{d^2_{A_1 A_i}}{2N_0}}\right) + \frac{1}{4}\sum_{i=2}^{4} Q\left(\sqrt{\frac{d^2_{B_1 B_i}}{2N_0}}\right) + \frac{1}{4}\sum_{i=1}^{12} Q\left(\sqrt{\frac{d^2_{B_1 A_i}}{2N_0}}\right) \tag{7.22}$$

where the factors ¾ and ¼ come from the fact that $P(A_1) = 3/4$ and $P(B_1) = 1/4$ with equally likely transmitted symbols. Equation 7.22 provides a generalized POE expression for the circular (4,12) constellation.

7.4.3 The (5,11) Constellation Format

In the (5,11) circular constellation format, as shown in Figure 7.5, the inner and outer circles accommodate 5 and 11 symbols, respectively. From the geometry of this circular constellation, by following a similar approach to the one stated in the two previous sections, we can easily find out the best-approximated POE expression in transmitting an arbitrary symbol from the transmitter in the presence of the same previous channel environment. This expression is given by

$$\begin{aligned} P_{e(5,11)} \approx{} & \frac{11}{16}\sum_{i=1}^{5} Q\left(\sqrt{\frac{d^2_{A_1 B_i}}{2N_0}}\right) + \frac{11}{16}\sum_{i=2}^{11} Q\left(\sqrt{\frac{d^2_{A_1 A_i}}{2N_0}}\right) \\ & + \frac{5}{16}\sum_{i=2}^{5} Q\left(\sqrt{\frac{d^2_{B_1 B_i}}{2N_0}}\right) + \frac{5}{16}\sum_{i=1}^{11} Q\left(\sqrt{\frac{d^2_{B_1 A_i}}{2N_0}}\right) \end{aligned} \tag{7.23}$$

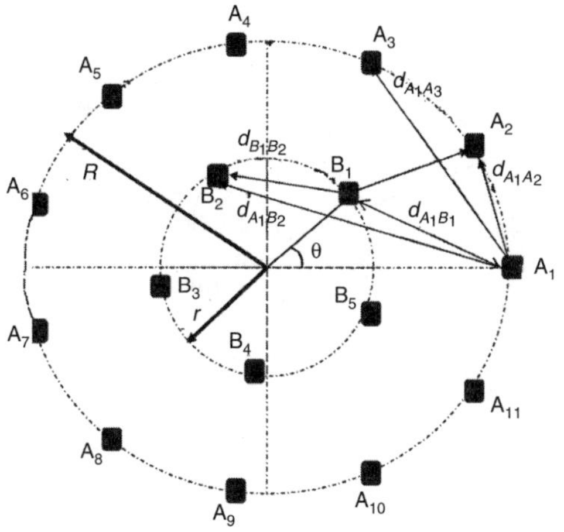

FIGURE 7.5 (5,11) Circular constellation.

where

$$d^2_{A_1B_i} = E_{s_1}\left[1 + \alpha^2 - 2\alpha\cos\left\{\theta + \frac{2(i-1)\pi}{5}\right\}\right], \quad i = 1,2,\ldots,5 \tag{7.24}$$

$$d^2_{A_1A_i} = 4E_{s_2}\sin^2\left[\frac{(i-1)\pi}{11}\right], \quad i = 2,3,\ldots,11 \tag{7.25}$$

$$d^2_{B_1B_i} = 4E_{s_1}\sin^2\left\{\frac{(i-1)\pi}{5}\right\}, \quad i = 2,3,4,5 \tag{7.26}$$

$$d^2_{B_1A_i} = E_{s_1}\left[1 + \alpha^2 - 2\alpha\cos\left\{\theta + \frac{2(i-1)\pi}{11}\right\}\right], \quad i = 1,2,3,\ldots,11 \tag{7.27}$$

In Equation 7.24 to Equation 7.27, E_{s_1}, E_{s_2}, α, and θ hold the same definitions as before. For upper- and lower-bound POE expressions we can always refer to Equation 7.14 and Equation 7.15.

7.4.4 The (6,10) Constellation Format

In this circular constellation format (Figure 7.6), the inner and outer circles contain 6 and 10 symbols, respectively.

Here, for transmitting an arbitrary symbol from the transmitter with the same channel scenario as before, the best-approximated POE expression becomes

$$P_{e(6,10)} \approx \frac{10}{16}\sum_{i=1}^{6} Q\left(\sqrt{\frac{d^2_{A_1B_i}}{2N_0}}\right) + \frac{10}{16}\sum_{i=2}^{10} Q\left(\sqrt{\frac{d^2_{A_1A_i}}{2N_0}}\right) + \frac{6}{16}\sum_{i=2}^{6} Q\left(\sqrt{\frac{d^2_{B_1B_i}}{2N_0}}\right) + \frac{6}{16}\sum_{i=1}^{10} Q\left(\sqrt{\frac{d^2_{B_1A_i}}{2N_0}}\right) \tag{7.28}$$

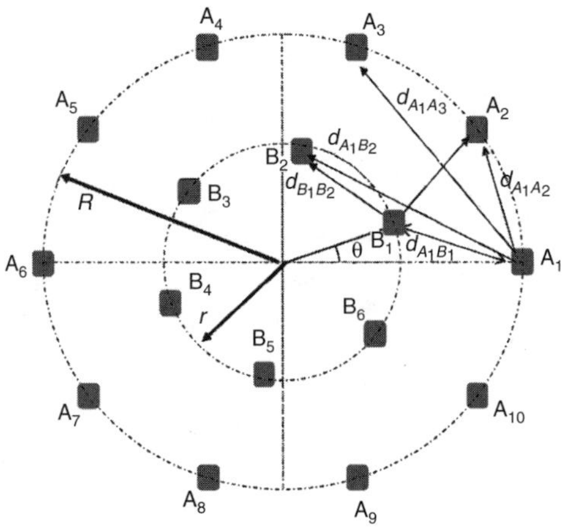

FIGURE 7.6 (6,10) Circular constellation.

where

$$d^2_{A_1B_i} = E_{s_1}\left[1+\alpha^2 - 2\alpha\cos\left\{\theta + \frac{(i-1)\pi}{3}\right\}\right], \quad i = 1, 2, \ldots, 6 \tag{7.29}$$

$$d^2_{A_1A_i} = 4E_{s_2}\sin^2\left[\frac{(i-1)\pi}{10}\right], \quad i = 2, 3, \ldots, 10 \tag{7.30}$$

$$d^2_{B_1B_i} = 4E_{s_1}\sin^2\left[\frac{(i-1)\pi}{6}\right], \quad i = 2, 3, \ldots, 6 \tag{7.31}$$

$$d^2_{B_1A_i} = E_{s_1}\left[1+\alpha^2 - 2\alpha\cos\left\{\theta + \frac{(i-1)\pi}{5}\right\}\right], \quad i = 1, 2, 3, \ldots, 10 \tag{7.32}$$

In the above equations E_{s_1}, E_{s_2}, α, and θ have already been defined in the previous sections. For evaluating the upper- and lower-bound POE expressions, we can use Equation 7.14 and Equation 7.15.

7.4.5 16-Rectangular Constellations with a Circular Format

7.4.5.1 The Structure

The 16-QAM rectangular constellation can be represented by a circular format with three concentric circles, as shown in Figure 7.7. In this case, the first, second, and third circles hold 4, 8, and 4 symbols, respectively. Due to the presence of three circles, here we need to define two RRs, α_1 and α_2, as

$$\alpha_1 = \frac{R}{r_1} \tag{7.33a}$$

$$\alpha_2 = \frac{R}{r_2} \tag{7.33b}$$

where r_1, r_2, and R are the inner, middle, and outer circle radii, respectively. Here we also define θ_1 and θ_2 as the PDs between the first outer circle symbol and the nearest center circle symbol, and between the first center circle symbol and the nearest inner circle symbol, respectively. From the geometry it is clear that $\theta_1 = \theta_2 =$ angle between symbols C_1 and B_2.

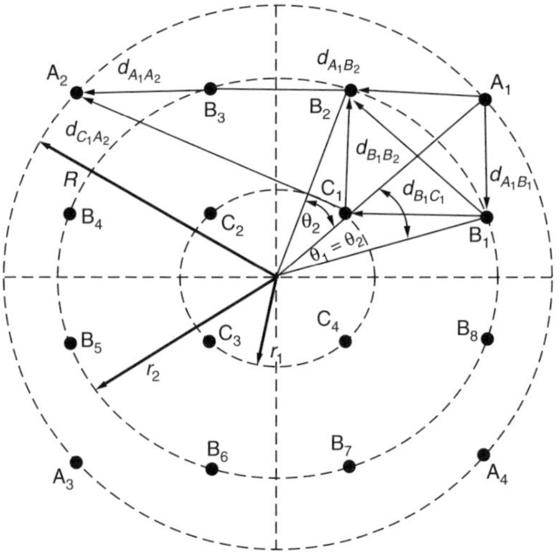

FIGURE 7.7 (4,8,4) Circular format for 16-rectangular constellation.

7.4.5.2 Probability of Error Analysis

As before, depending on the geometry of the constellation diagram, the POE analytical technique will be presented in this section. As a result of symmetry of the symbol orientations on all circles, we can arbitrarily choose any three symbols, one from each circle, for presenting the analysis. In this case, we choose symbols A_1, B_1, and C_1 and define:

$d_{A_1A_i}$ = distance between symbols A_1 and A_i $(i \neq 1)$, in the outer circle
$d_{A_1B_i}$ = distance between the outer circle symbol A_1 and any other center circle symbol B_i
$d_{A_1C_i}$ = distance between the outer circle symbol A_1 and any other inner circle symbol C_i
$d_{C_1C_i}$ = distance between symbols C_1 and C_i $(i \neq 1)$, in the inner circle
$d_{C_1B_i}$ = distance between the inner circle symbol C_1 and any other center circle symbol B_i
$d_{B_1B_{i_{even}}}$ = distance between the center circle symbol B_1 and any even-numbered center circle symbol $B_{i_{even}}$ $(i_{even} = 2, 4, 6, 8)$
$d_{B_1B_{i_{odd}}}$ = distance between the center circle symbol B_1 and any odd-numbered center circle symbol $B_{i_{odd}}$ $(i \neq 1 \text{ and } i_{odd} = 3, 5, 7)$

By following the geometry of Figure 7.6, all the above distances can easily be equated as

$$d^2_{A_1A_i} = 4E_{s_3}\sin^2\left[\frac{(i-1)\pi}{4}\right], \quad i = 2,3,4 \tag{7.34}$$

$$d^2_{A_1B_i} = E_{s_2}\left[1+\alpha_2^2 - 2\alpha_2\cos\left\{\theta_1 + \frac{(i-1)\pi}{2}\right\}\right], \quad i = 1,2,\ldots,8 \tag{7.35}$$

$$d^2_{A_1C_i} = E_{s_1}\left[1+\alpha_1^2 - 2\alpha_1\cos\left\{\frac{(i-1)\pi}{2}\right\}\right], \quad i = 1,2,3,4 \tag{7.36}$$

$$d^2_{C_1C_i} = 4E_{s_1}\sin^2\left[\frac{(i-1)\pi}{4}\right], \quad i = 2,3,4 \tag{7.37}$$

$$d^2_{C_1B_i} = E_{s_1}\left[1+\left(\frac{\alpha_1}{\alpha_2}\right)^2 - 2\left(\frac{\alpha_1}{\alpha_2}\right)\cos\left\{\theta_2 + \frac{(i-1)\pi}{2}\right\}\right], \quad i = 1,2,\ldots,8 \tag{7.38}$$

$$d^2_{B_1 B_{i_{even}}} = 4E_{s_2} \sin^2\left[\frac{\theta_1 + \theta_2}{2} + \frac{(i_{even} - 2)\pi}{8}\right], \quad i_{even} = 2, 4, 6, 8 \tag{7.39}$$

$$d^2_{B_1 B_{i_{odd}}} = 4E_{s_2} \sin^2\left[\frac{(i_{odd} - 1)\pi}{8}\right], \quad i_{odd} \neq 1 \text{ and } i_{odd} = 3, 5, 7 \tag{7.40}$$

where E_{s_1}, E_{s_2}, and E_{s_3} are the energies of the inner, center, and outer circle symbols, respectively, which are defined as

$$E_{s_1} = r_1^2 \tag{7.41a}$$

$$E_{s_2} = r_2^2 \tag{7.41b}$$

$$E_{s_3} = R^2 \tag{7.41c}$$

In this case, in transmitting an arbitrary symbol from the transmitter in the presence of AWGN of one-sided PSD N_0, the lower and upper bounds of the POE expressions of Equation 7.14 and Equation 7.15 can be applied by following the previous approach to evaluate the best-approximated POE as

$$\begin{aligned} P_{e(16\text{Rec})} \approx\ & \frac{1}{4}\sum_{i=2}^{4} Q\left(\sqrt{\frac{d^2_{A_1 A_i}}{2N_0}}\right) + \frac{1}{4}\sum_{i=1}^{8} Q\left(\sqrt{\frac{d^2_{A_1 B_i}}{2N_0}}\right) + \frac{1}{4}\sum_{i=1}^{4} Q\left(\sqrt{\frac{d^2_{A_1 C_i}}{2N_0}}\right) \\ & + \frac{1}{4}\sum_{i=2}^{4} Q\left(\sqrt{\frac{d^2_{C_1 C_i}}{2N_0}}\right) + \frac{1}{4}\sum_{i=2}^{8} Q\left(\sqrt{\frac{d^2_{C_1 B_i}}{2N_0}}\right) + \frac{1}{4}\sum_{i=1}^{4} Q\left(\sqrt{\frac{d^2_{C_1 A_i}}{2N_0}}\right) \\ & + \frac{1}{2}\sum_{i=2}^{8} \left[Q\left(\sqrt{\frac{d^2_{B_1 B_{i_{even}}}}{2N_0}}\right) + Q\left(\sqrt{\frac{d^2_{B_1 B_{i_{odd}}}}{2N_0}}\right)\right] + \frac{1}{2}\sum_{i=1}^{4} Q\left(\sqrt{\frac{d^2_{B_1 C_i}}{2N_0}}\right) \\ & + \frac{1}{2}\sum_{i=1}^{4} Q\left(\sqrt{\frac{d^2_{B_1 A_i}}{2N_0}}\right) \end{aligned} \tag{7.42}$$

where the seventh term on the right-hand side of Equation 7.42 works as follows: when i is even, $d_{B_1 B_{i_{odd}}} = 0$, and when i is odd, $d_{B_1 B_{i_{even}}} = 0$. Equation 7.42 provides the general form for the best-approximated POE expressions, which can be equally applicable to both linear and nonlinear channels.

7.4.6 The Effect of Nonlinearity and the Application of a Predistortion Technique

Due to the presence of the HPA in the communication link, the distances between the signal constellation points will be distorted according to the model and back-off values of the HPA. Back-off of an HPA represents the backing off of its operating point away from the saturation region, which in turn results in reduced nonlinear distortion with the expense of power efficiency of the amplifier. The back-off could be input or output type. The IBO of an HPA is defined (in decibel scale) as the difference between its input saturation power and the average input signal power, while the output back-off (OBO) in decibel scale is defined in the same way for output powers. As an example, we can refer to Figure 7.8a and 7.8b, where (8,8) circular constellations at the input and output of the TWT, respectively, are shown in the presence of minimum back-off (i.e., the amplitudes of outer circle symbols are on the saturation point of the HPA). The resulting distortion due to these HPAs affects the POE performance. One way to overcome this problem is to use a data predistortion technique, where the transmitted data is predistorted in such a way that the output of the HPA provides the optimum constellation with respect to POE performance.

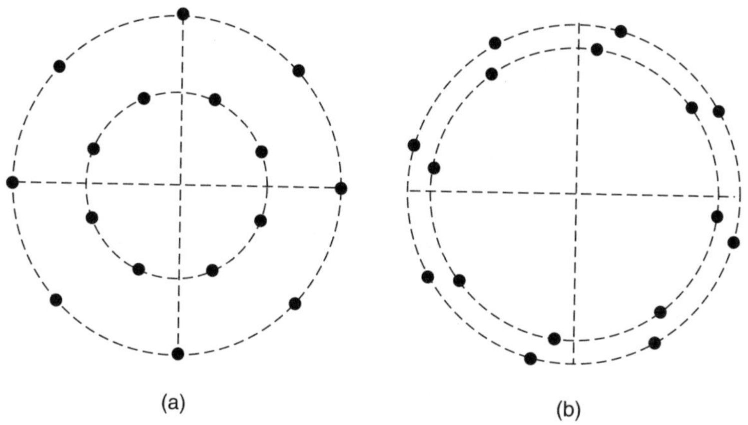

FIGURE 7.8 (8,8) circular constellation at the input (a), and at the output (b) of TWT.

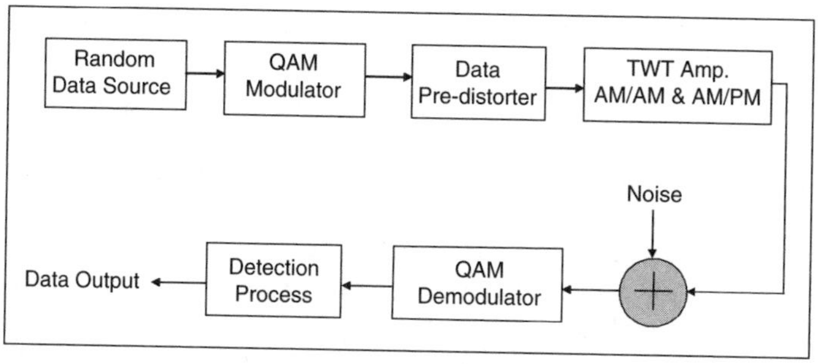

FIGURE 7.9 Nonlinear communication system with data predistortion.

Figure 7.9 shows the block diagram of the nonlinear communication system with data predistortion. In this system the data predistorter distorts the outgoing data in such a way so that the nonlinear amplifier, which follows the predistorter, keeps the original constellation intact.

There are many techniques (e.g., see [Muh00] [Ibn98] [Ber97]) available in the research and application domains that can be used to achieve data predistortion. Since the major focus of this chapter is not on searching for the optimal data predistortion technique, we will consider a simple technique for our analysis. With this technique data predistortion for the HPA can be achieved by using two different methods: the direct method and the optimal method. When using the direct method for a circular constellation with two circles, we first calculate the RR α' and PD θ' at the output of a linear channel. Now for the HPA we calculate the radius R' of the outer circle from the operating point of the HPA. This operating point is set according to the back-off value of the HPA. With the aid of α' and R' we then find the value of the radius of the inner circle r' for the nonlinear case. Subsequently, by using the model of the amplifier, the predistorted inner circle radius can be found. Here, for the TWT, the AM/AM model (see Appendix) is given by

$$A(r) = \frac{2r}{1+r} \tag{7.43}$$

With $A(r) = r'$ and $r = r_p$, the predistorted inner circle radius would be

$$r_p = \frac{1}{r'} - \sqrt{\frac{1}{r'^2} - 1} \tag{7.44}$$

while the predistorted RR α_p would be

$$\alpha_p = \frac{R}{r_p} \tag{7.45}$$

The predistorted PD θ_p can be equated as

$$\theta_p = P(R') - [\theta' - P(r_p)] \tag{7.46}$$

In Equation 7.40, the TWT AM/PM model (see Appendix), given by

$$P(r) = \frac{\pi}{3}\frac{r^2}{1+r^2} \tag{7.47}$$

can be used for calculating the phases $P(R')$ and $P(r_p)$ for outer and inner circles, respectively.

It is interesting to note here that with a small rotation of the symbols in the inner circle (not true for the 16-rectangular constellation with a circular format), sometimes we get better performance with the use of a predistortion technique. This better performance can only be ensured through the use of an exhaustive search method, which we will call the optimal method. In this case, the best possible RR α' and PD θ' are extracted for the linear case from the set of POE performance curves, after the exhaustive search, by varying the RRs and PDs one at a time. After that, the same procedure used for the direct method is followed.

For the three-circle circular constellation, the same two methods can be applied. In this case for the direct method, we first calculate the RR α_1' and PD θ_1' between the outer and center circles at the output of a linear channel, and then we follow the same procedure for the two-circle case. Later we calculate the RR α_2' and PD θ_2' between the center and inner circles at the output of the same linear channel and then repeat the two-circle procedure. For the optimal case we follow the search method. The 16-rectangular constellation with a circular format provides the same results for both methods.

Unlike other exhaustive search methods, due to the simplified forms of the POE expressions, this optimal method can easily be applied with a small price paid for the search time. Since the performance improvement in the second case is not very significant, the necessity of an exhaustive search is totally dependent on the requirement of the application and on the individual's choice.

7.4.7 Total Degradation

Total degradation is a performance figure in the nonlinear environment that describes the difference between the maximum power of the HPA and the output power of a linear amplifier required to assure a predefined POE. The TD of a system, which can be a function of either IBO or OBO of an HPA, is dependent on the operating point of the HPA. In terms of IBO, TD is defined as

$$TD(dB) = IBO(dB) + SNR_{NL}(dB) - SNR_L(dB) \tag{7.48}$$

where SNR_{NL} and SNR_L are the required SNRs for nonlinear and linear amplifiers, respectively, at a targeted POE.

For multilevel signals, the operating point of the HPA cannot be chosen in such a way that the average signal power is at the saturation point, because in this case the high-level symbols will pass beyond saturation. For this reason, while working with multilevel signals, we need to know a minimum IBO (IBO_{min}) that will guarantee the highest level of the signal symbols at the saturation point of the amplifier. The value of IBO_{min} is dependent on the number of symbols in each level of a multilevel signal constellation. For each constellation the IBO_{min} can be calculated as the decibel difference between the average power of the symbols and the maximum symbol power in that constellation.

The plot for TD performance, with different constellations as a function of IBO, gives a set of convex curves from which the minimum IBO and the corresponding total degradation can be calculated for each constellation.

The optimum IBO (which corresponds to the minimal TD) can be found by plotting the TD curves as a function of different IBOs, which take into account the optimum RRs and PDs.

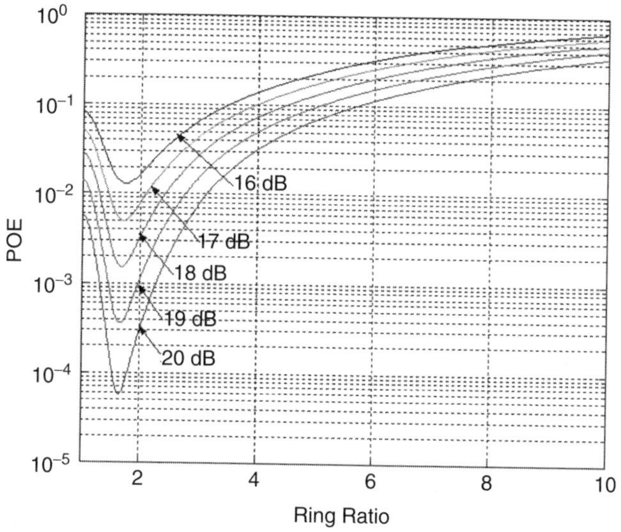

FIGURE 7.10 Effect of RR on the POE performance of circular (8,8) constellations in the presence of a TWT.

7.4.8 Results and Discussions

In this section we will present some analytical results based on the findings in the previous subsections, taking into account the TWT as part of the system. Some supporting simulation results are also presented here. Moreover, this section provides a brief performance comparison between the different 16-QAM circular constellations presented above.

Figure 7.10 shows the effect of various RRs on the POE performance on the circular (8,8) QAM constellation in the presence of the TWT and a predicted optimum PD of 22.5°. With this value, any inner circle symbol positions itself at an equidistant length from its two neighboring outer circle symbols providing the best possible structure for detecting any one of the symbols. In the figure, the effects of the RR are observed for different SNRs in decibel. From the figure it is found that the POE performance continues to improve with continuous increase in the RR up to its optimal value, while an increase in the RR beyond its optimal value results in worse POE performance. There is an explanation behind this particular observation. In the beginning (RR = 1) all the constellation points lie on the same circle providing a very small decision region for detection. With the assumption that the maximum amplitude of the symbols is normalized to 1 as the RR is increased, the decision region becomes wider for detection with a decrease in the radius of the inner circle. As a result, the performance keeps improving until the RR reaches its optimum value. After the optimal RR is reached, the constellation points in the inner circle become so crowded that the decision region in this circle becomes very small for detection purposes. Thus, the system performance starts to deteriorate. The optimal RR is found to be almost invariant with the SNR, as shown in the plots. They show that the POE performance is asymmetrical with respect to the optimal RR. This results in a large degradation for a small deviation in the RR from its optimal value.

Figure 7.11 presents the effects of various PDs on the POE performance for the circular (8,8) QAM constellation. Here, the effects of PDs are observed on different SNRs in decibel, while the optimal RR is 1.645. In this figure the optimal PD is shown to be 22.5°, which is obvious since the optimal RR, based on which the PDs are plotted, was achieved by predicting the above value. In this case, it is found that if we vary the optimal RR to some extent, the optimal PD does not vary, which proves the validity of its value. In the figure, this optimal PD is shown to be independent of the SNRs. The figure also shows that the POE performance is symmetrical with respect to the optimal PD.

Table 7.1 presents different optimal RRs and optimal PDs. These are found through an exhaustive search method by plotting different best-approximated POE expressions for different 16-QAM circular

TABLE 7.1 Optimal RRs and PDs at the TWT Output

16-QAM Circular Constellations	Optimal Ring Ratio	Optimal Phase Difference (Degrees)
(8,8)	1.645	22.5
(4,12)	2.732	45
(5,11)	2.17	46
(6,10)	2.03	21
16-rec (4,8,4)	3 and 1.34126	26.565 for both

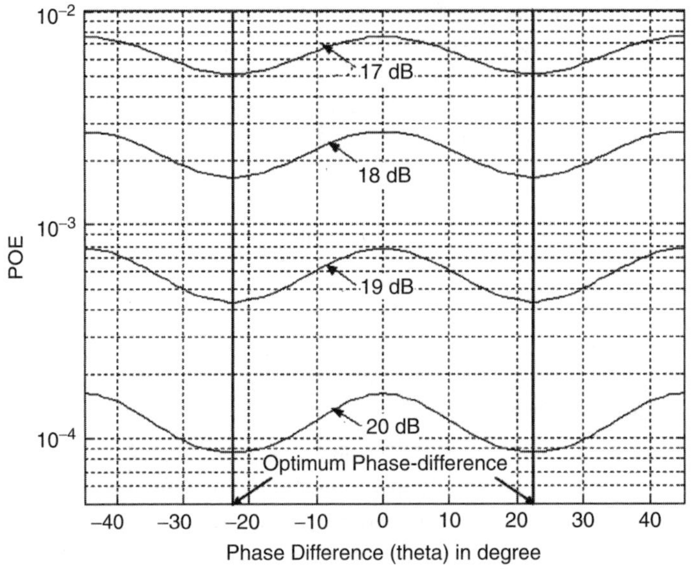

FIGURE 7.11 Effect of PD on the POE performance of circular (8,8) constellations in the presence of a TWT.

constellations. Due to the presence of three circles for the 16-rectangular constellation with a circular format, we find two optimal RRs and PDs for this case.

To verify the accuracy of our approach, we present some simulation results in Figure 7.12. In this case we consider circular (8,8) QAM constellations with predistorted data sequences. We predistort the transmitted data in such a way that the TWT's output gives the optimal RR and optimal PD. Here, the communications system model, shown in Figure 7.9, is used for the purpose of this simulation. The simulation results presented in this scenario are based on the Monte Carlo simulation. In the transmitter, the information data bits are generated using random data sources. At the receiver, the detected symbols are compared with the original transmitted data on a symbol-by-symbol basis. Figure 7.12 shows that the simulation results are in congruence with theoretical results. Figure 7.13 compares our rectangular 16-QAM constellation best-approximated POE results with the existing exact [Pro02], Craig [Cra91], and approximate [Alu00] counterparts. It is shown that at reasonable POEs of interest ($>10^{-1}$) our results agree more with the existing "exact" result than with Craig's. The slight disagreement in the latter case is due to the presence of the numerical integration, which is avoided in the method discussed in this chapter.

For data predistortion, Table 7.2 presents the optimal RRs and PDs at the predistorter output for different constellations. These optimal values were obtained using the results of Section 7.4.6. Figure 7.14 illustrates the concept of data predistortion for the rectangular 16-QAM constellation.

In Figure 7.15 we present the POE performance as a function of the SNR for different circular 16-QAM constellations with their individual optimal RRs and PDs. The results are presented for both predistortion and nonpredistortion techniques. The figure also includes the POE performance curve for a 16-PSK (phase shift keying) system.

TABLE 7.2 Optimal RRs and PDs at the Predistorter Output

16-QAM Circular Constellations	Optimal Ring Ratio	Optimal Phase Difference (Degrees)
(8,8)	2.8284	45.8241
(4,12)	5.27426	72.918
(5,11)	4.0958	72.3256
(6,10)	3.732	46.98
16-rec (4,8,4)	5.82842 and 2.23607	54.84927 and 46.5651

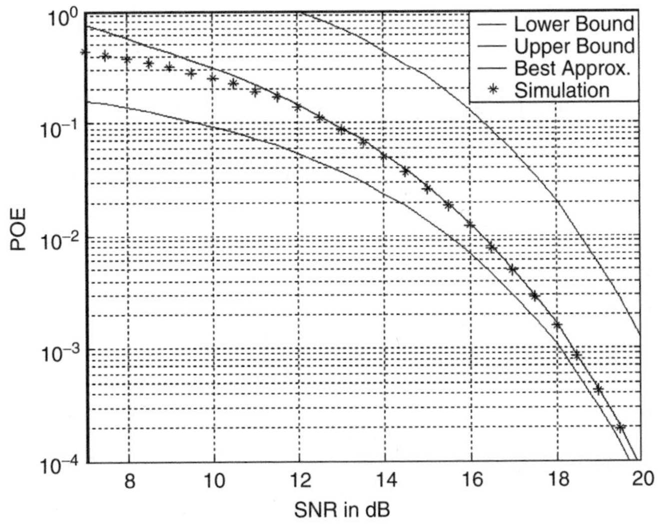

FIGURE 7.12 Theoretical and simulation results of POE performance for the (8,8) circular constellation in the presence of a TWT.

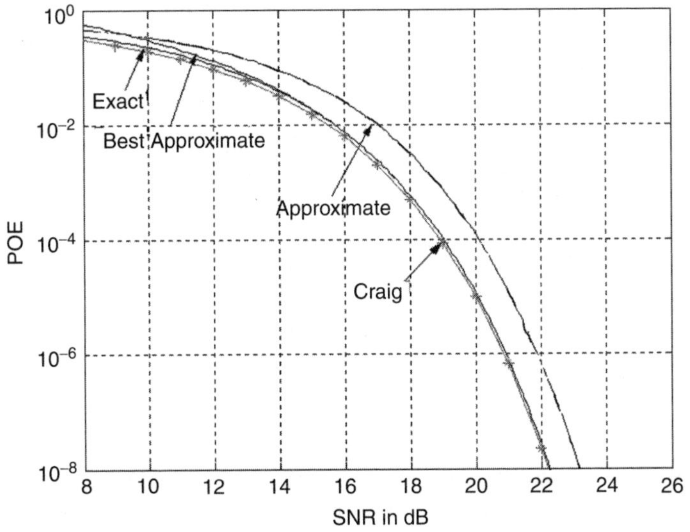

FIGURE 7.13 16-Rectangular POE performance in the AWGN channel with exact, approximate, Craig and best approximate representations.

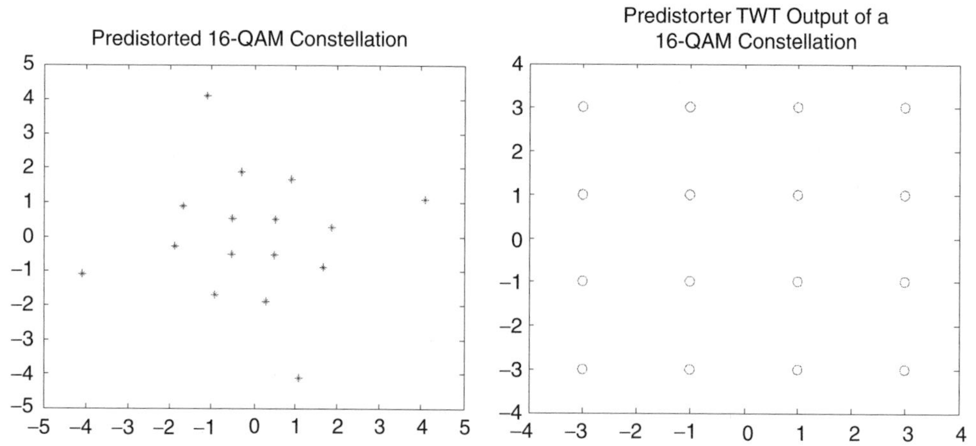

FIGURE 7.14 Illustration of data predistortion, the rectangular 16-QAM case: constellation at the predistorter output (left), constellation at the amplifier output (right).

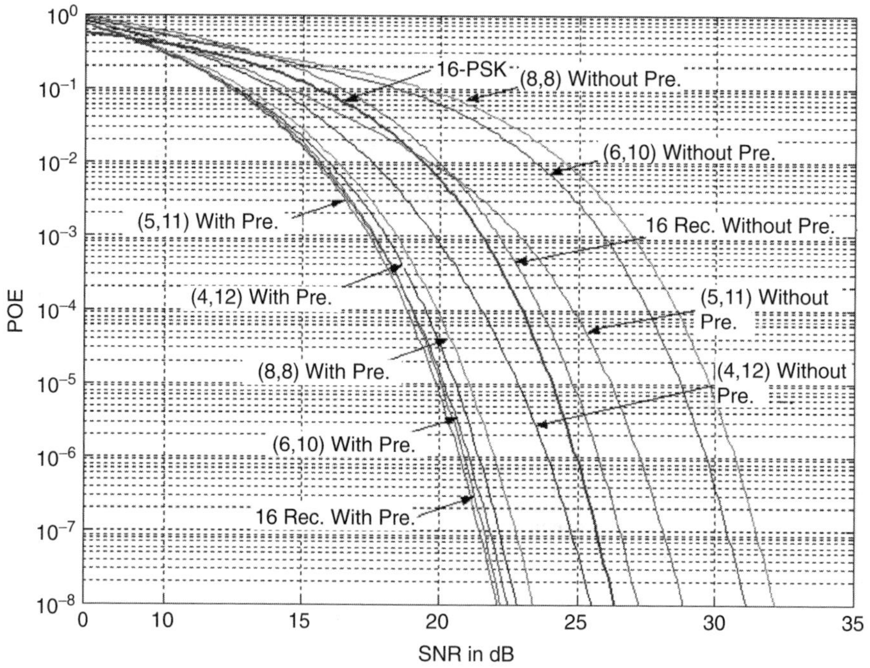

FIGURE 7.15 Different 16-QAM circular constellations with and without the use of a predistortion technique in the presence of a nonlinear HPA. For each constellation, the minimal IBO was used.

The figure shows that without predistortion, the circular (4,12) 16-QAM constellation performs the best in terms of POE for a nonlinear channel. However, the application of a predistorter results in the best performance for the circular (5,11) 16-QAM constellation. From the figure it is clear that the application of a predistorter in a nonlinear system improves the system performance.

TABLE 7.3 Minimum IBO, Optimum IBO, and Optimum TD for Different 16-QAM Circular Constellations

	Star (8,8)	Star (6,10)	Star (5,11)	Star (4,12)	Rec. 16-QAM
IBO_{min}(dB)	1.8017	1.4508	1.2270	1.0596	2.5527
IBO_{opt}(dB)	5.400	4.6509	3.4271	1.4596	2.5527
TD_{opt}(dB)	8.9345	7.6949	6.6282	3.0980	7.2181

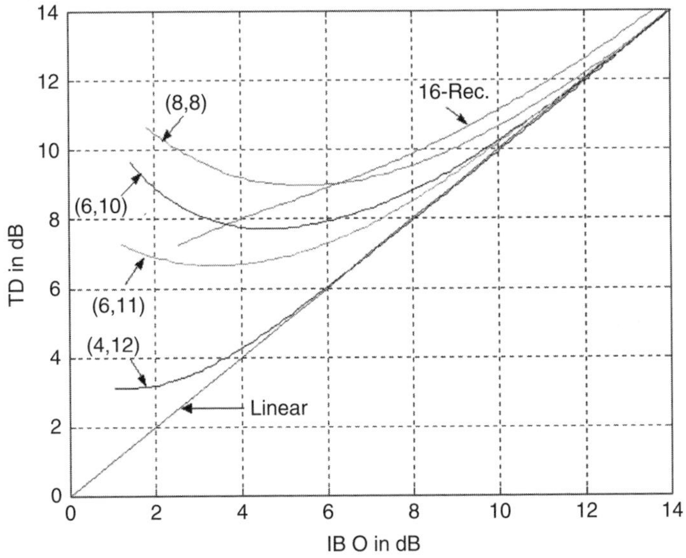

FIGURE 7.16a TD performance for different 16-QAM circular constellations in the presence of a TWT.

It is worth noting that with the use of a predistortion technique the rectangular 16-QAM constellation gives a nearly identical performance to that of the circular (5,11) constellation. However, the predistortion requirements for the circular (5,11) constellation have a lower complexity than those required for the rectangular 16-QAM constellation. Therefore, it can be concluded that among different 16-QAM constellations, the circular (5,11) constellation is of practical interest for a nonlinear channel in the presence of a data predistorter.

In Figure 7.16a we present the TD performance with a predefined POE of 10^{-4} as a function of different IBOs for different circular 16-QAM constellations (the optimal RRs and PDs have been used, and no data predistortion is considered here). This figure provides the optimal IBO and the corresponding optimal TD for each constellation. The figure also displays the TD performance of a linear constant modulation scheme (16-PSK) for comparison. It is observed that for two-circle circular constellations, the TD is higher for the constellations having more symbols in the inner circle. Table 7.3 summarizes the optimum IBO, optimum TD, and minimum IBO values.

Finally, Figure 7.16b presents the POE performance as a function of SNR for different circular 16-QAM constellations with their individual optimal RRs and PDs when the HPA is forced to work with optimal IBO for each constellation. POE performances of all the constellations with data predistortion are also displayed, which show their superiority over the nonpredistortion technique with optimal IBO.

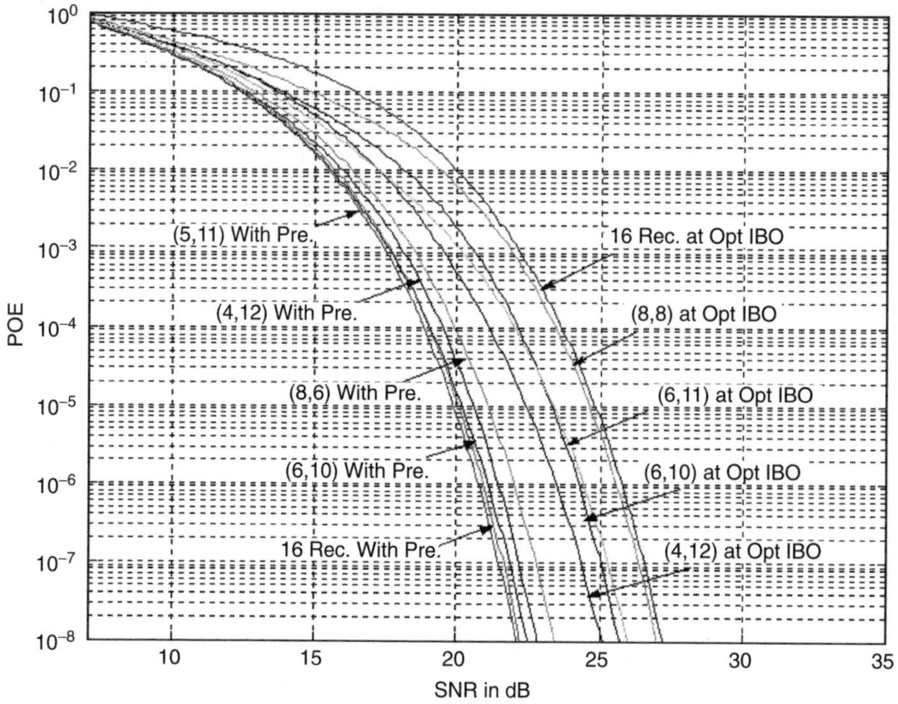

FIGURE 7.16b Different 16-QAM circular constellations with and without the use of a pre distortion technique at their optimum IBO in the presence of a TWT.

7.5 Probability of Symbol Error for the Circular 32-ary QAM Format

In this section we will study three different circular 32-ary QAM constellations: the (4,11,17), (5,11,16), and the 32-rectangular constellation with the circular format (4,8,4,8,8). To find out the POE expressions, the same methodology detailed in Section 7.4.5 will be followed for the first two constellations. However, for the 32-rectangular constellation, where five concentric circles represent the structure, the method will be extended by taking into consideration four different RRs and PDs. This section will include a discussion on the predistortion technique used in this study and its application.

7.5.1 The (4,11,17) Constellation Format

This constellation, as shown in Figure 7.17, is represented with three concentric circles where the inner, center, and outer circles contain 4, 11, and 17 symbols, respectively. For the three circles the resultant RRs, α_1 and α_2, and PDs, θ_1 and θ_2, hold the same definitions as seen in Section 7.4.5. Here in transmitting any arbitrary symbol of the 32-QAM signal, the lower-bound, upper-bound, and best-approximated POE expressions in the presence of AWGN with a PSD of N_0 can be expressed as

Lower bound:

$$P_{e_{Lower}} > Q\left(\sqrt{\frac{d_{\min}^2}{2N_o}}\right) \tag{7.49}$$

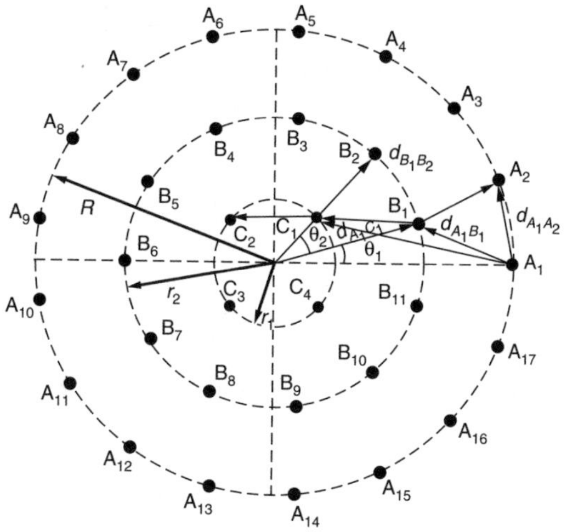

FIGURE 7.17 32-QAM with a (4,11,17) circular constellation.

Upper bound:

$$P_{e_{upper}} < 31Q\left(\sqrt{\frac{d_{\min}^2}{2N_o}}\right) \tag{7.50}$$

Best-approximated POE:

$$\begin{aligned} P_{e(4,11,17)} \approx{}& \frac{17}{32}\sum_{i=2}^{17} Q\left(\sqrt{\frac{d_{A_1A_i}^2}{2N_0}}\right) + \frac{17}{32}\sum_{i=1}^{11} Q\left(\sqrt{\frac{d_{A_1B_i}^2}{2N_0}}\right) + \frac{17}{32}\sum_{i=1}^{4} Q\left(\sqrt{\frac{d_{A_1C_i}^2}{2N_0}}\right) \\ &+ \frac{11}{32}\sum_{i=2}^{11} Q\left(\sqrt{\frac{d_{B_1B_i}^2}{2N_0}}\right) + \frac{11}{32}\sum_{i=1}^{17} Q\left(\sqrt{\frac{d_{B_1A_i}^2}{2N_0}}\right) + \frac{11}{32}\sum_{i=1}^{4} Q\left(\sqrt{\frac{d_{B_1C_i}^2}{2N_0}}\right) \\ &+ \frac{1}{8}\sum_{i=2}^{4} Q\left(\sqrt{\frac{d_{C_1C_i}^2}{2N_0}}\right) + \frac{1}{8}\sum_{i=1}^{11} Q\left(\sqrt{\frac{d_{C_1B_i}^2}{2N_0}}\right) + \frac{1}{8}\sum_{i=1}^{17} Q\left(\sqrt{\frac{d_{C_1A_i}^2}{2N_0}}\right) \end{aligned} \tag{7.51}$$

In Equation 7.49 and Equation 7.50, $d_{\min}$ represents the minimum distance among all possible distances between the symbols, as shown in Figure 7.17. The factor 31 in Equation 7.50 represents the number of neighboring symbols around the transmitted symbol. With the help of the geometrical structure presented in Figure 7.17, the distances in Equation 7.51 can easily be equated as

$$d_{A_1A_i}^2 = 4E_{s_3}\sin^2\left[\frac{(i-1)\pi}{17}\right], \quad i = 2,3,\ldots,17 \tag{7.52}$$

$$d_{A_1B_i}^2 = E_{s_2}\left[1+\alpha_2^2 - 2\alpha_2\cos\left\{\theta_1 + \frac{2(i-1)\pi}{11}\right\}\right], \quad i = 1,2,\ldots,11 \tag{7.53}$$

$$d^2_{A_1C_i} = E_{s_1}\left[1+\alpha_1^2 - 2\alpha_1\cos\left\{(\theta_1+\theta_2)+\frac{(i-1)\pi}{2}\right\}\right], \quad i=1,2,3,4 \tag{7.54}$$

$$d^2_{B_1B_i} = 4E_{s_2}\sin^2\left[\frac{(i-1)\pi}{11}\right], \quad i=2,3,\ldots,11 \tag{7.55}$$

$$d^2_{B_1A_i} = E_{s_2}\left[1+\alpha_2^2 - 2\alpha_2\cos\left\{\theta_1-\frac{2(i-1)\pi}{17}\right\}\right], \quad i=1,2,\ldots,17 \tag{7.56}$$

$$d^2_{B_1C_i} = E_{s_1}\left[1+\left(\frac{\alpha_1}{\alpha_2}\right)^2 - 2\left(\frac{\alpha_1}{\alpha_2}\right)\cos\left\{\theta_2+\frac{(i-1)\pi}{2}\right\}\right], \quad i=1,2,3,4 \tag{7.57}$$

$$d^2_{C_1C_i} = 4E_{s_1}\sin^2\left[\frac{(i-1)\pi}{4}\right], \quad i=2,3,4 \tag{7.58}$$

$$d^2_{C_1B_i} = E_{s_1}\left[1+\left(\frac{\alpha_1}{\alpha_2}\right)^2 - 2\left(\frac{\alpha_1}{\alpha_2}\right)\cos\left\{\theta_2-\frac{2(i-1)\pi}{11}\right\}\right], \quad i=1,\ldots,11 \tag{7.59}$$

$$d^2_{C_1A_i} = E_{s_1}\left[1+\alpha_1^2 - 2\alpha_1\cos\left\{(\theta_1+\theta_2)-\frac{2(i-1)\pi}{17}\right\}\right], \quad i=1,2,\ldots,17 \tag{7.60}$$

where E_{s_1}, E_{s_2}, and E_{s_3} are the energies of the inner, center, and outer circle symbols, respectively, which hold the same definitions as seen in Equation 7.41.

7.5.2 The (5,11,16) Constellation Format

In this constellation (Figure 7.18), the inner, center, and outer circles contain 5, 11, and 16 symbols, respectively. Here we use the same definitions for RRs, PDs, and symbol energies as mentioned in Section 7.5.1. In this case, with a similar channel scenario as before, the lower- and upper-bound POE can be expressed

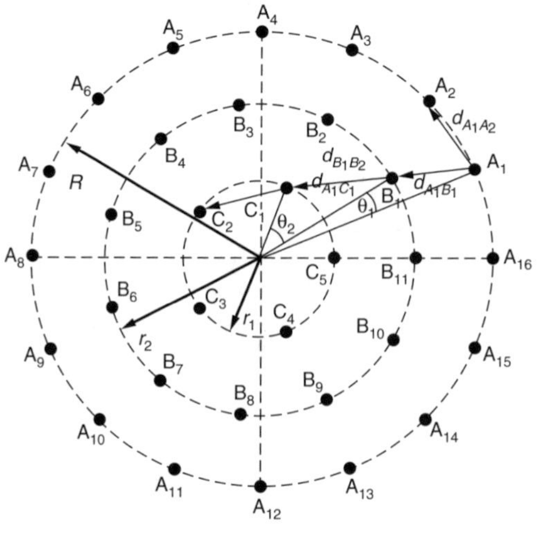

FIGURE 7.18 32-QAM with a (5,11,16) circular constellation.

with the help of Equation 7.49 and Equation 7.50, while the best-approximated bound can be equated as

$$P_{e(5,11,16)} \approx \frac{1}{2}\sum_{i=2}^{16} Q\left(\sqrt{\frac{d^2_{A_1A_i}}{2N_0}}\right) + \frac{1}{2}\sum_{i=1}^{11} Q\left(\sqrt{\frac{d^2_{A_1B_i}}{2N_0}}\right) + \frac{1}{2}\sum_{i=1}^{5} Q\left(\sqrt{\frac{d^2_{A_1C_i}}{2N_0}}\right)$$
$$+ \frac{11}{32}\sum_{i=2}^{11} Q\left(\sqrt{\frac{d^2_{B_1B_i}}{2N_0}}\right) + \frac{11}{32}\sum_{i=1}^{16} Q\left(\sqrt{\frac{d^2_{B_1A_i}}{2N_0}}\right) + \frac{11}{32}\sum_{i=1}^{5} Q\left(\sqrt{\frac{d^2_{B_1C_i}}{2N_0}}\right) \quad (7.61)$$
$$+ \frac{5}{32}\sum_{i=2}^{5} Q\left(\sqrt{\frac{d^2_{C_1C_i}}{2N_0}}\right) + \frac{5}{32}\sum_{i=1}^{11} Q\left(\sqrt{\frac{d^2_{C_1B_i}}{2N_0}}\right) + \frac{5}{32}\sum_{i=1}^{16} Q\left(\sqrt{\frac{d^2_{C_1A_i}}{2N_0}}\right)$$

where

$$d^2_{A_1A_i} = 4E_{s_3}\sin^2\left[\frac{(i-1)\pi}{16}\right], \quad i = 2,3,\ldots,16 \quad (7.62)$$

$$d^2_{A_1B_i} = E_{s_2}\left[1 + \alpha_2^2 - 2\alpha_2\cos\left\{\theta_1 + \frac{2(i-1)\pi}{11}\right\}\right], \quad i = 1,2,\ldots,11 \quad (7.63)$$

$$d^2_{A_1C_i} = E_{s_1}\left[1 + \alpha_1^2 - 2\alpha_1\cos\left\{(\theta_1+\theta_2) + \frac{2(i-1)\pi}{5}\right\}\right], \quad i = 1,2,\ldots,5 \quad (7.64)$$

$$d^2_{B_1B_i} = 4E_{s_2}\sin^2\left[\frac{(i-1)\pi}{11}\right], \quad i = 2,3,\ldots,11 \quad (7.65)$$

$$d^2_{B_1A_i} = E_{s_2}\left[1 + \alpha_2^2 - 2\alpha_2\cos\left\{\theta_1 - \frac{(i-1)\pi}{8}\right\}\right], \quad i = 1,2,\ldots,16 \quad (7.66)$$

$$d^2_{B_1C_i} = E_{s_1}\left[1 + \left(\frac{\alpha_1}{\alpha_2}\right)^2 - 2\left(\frac{\alpha_1}{\alpha_2}\right)\cos\left\{\theta_2 + \frac{2(i-1)\pi}{5}\right\}\right], \quad i = 1,2,3,4,5 \quad (7.67)$$

$$d^2_{C_1C_i} = 4E_{s_1}\sin^2\left[\frac{(i-1)\pi}{5}\right], \quad i = 2,3,4,5 \quad (7.68)$$

$$d^2_{C_1B_i} = E_{s_1}\left[1 + \left(\frac{\alpha_1}{\alpha_2}\right)^2 - 2\left(\frac{\alpha_1}{\alpha_2}\right)\cos\left[\theta_2 - \frac{2(i-1)\pi}{11}\right]\right], \quad i = 1,2,\ldots,11 \quad (7.69)$$

$$d^2_{C_1A_i} = E_{s_1}\left[1 + \alpha_1^2 - 2\alpha_1\cos\left\{(\theta_1+\theta_2) - \frac{(i-1)\pi}{8}\right\}\right], \quad i = 1,2,\ldots,16 \quad (7.70)$$

7.5.3 32-Rectangular Constellations with a Circular Format

7.5.3.1 The Structure

The 32-QAM rectangular constellation can be represented by a circular format with five concentric circles, as shown in Figure 7.19. In this case, the circles A, B, C, D, and E accommodate 4, 8, 4, 8, and 8 symbols, respectively. These five circles provide four RRs, α_1, α_2, α_3, and α_4, defined respectively as

$$\alpha_1 = \frac{R}{r_1} \quad (7.71a)$$

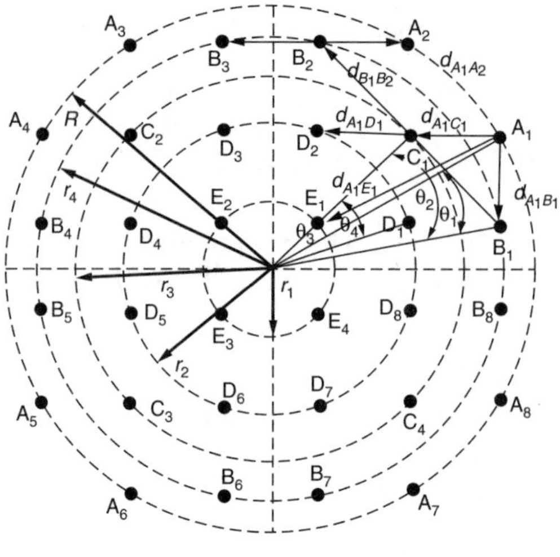

FIGURE 7.19 Circular representation (4,8,4,8,8) of 32-rectangular QAM.

$$\alpha_2 = \frac{R}{r_2} \tag{7.71b}$$

$$\alpha_3 = \frac{R}{r_3} \tag{7.71c}$$

$$\alpha_4 = \frac{R}{r_4} \tag{7.71d}$$

where r_1, r_2, r_3, r_4, and R are the radii of circles E, D, C, B, and A, respectively. Here we also define θ_1, θ_2, θ_3, and θ_4 as the PDs between the symbols A_1 and B_1, B_1 and C_1, C_1 and D_1, and D_1 and E_1, respectively.

7.5.3.2 Probability of Error Analysis

In this case, as a result of the symmetry of the symbol orientations on all the circles, we will arbitrarily choose any five symbols, one from each circle, for presenting the analysis. Here, we choose symbols A_1 B_1, C_1, D_1 and E_1 and define:

$d_{A_1A_i}$ = distance between symbols A_1 and A_i $(i \neq 1)$, in circle A
$d_{B_1B_i}$ = distance between symbols B_1 and B_i $(i \neq 1)$, in circle B
$d_{C_1C_i}$ = distance between symbols C_1 and C_i $(i \neq 1)$, in circle C
$d_{D_1D_i}$ = distance between symbols D_1 and D_i $(i \neq 1)$, in circle D
$d_{E_1E_i}$ = distance between symbols E_1 and E_i $(i \neq 1)$, in circle E
$d_{B_1A_i}$ = distance between symbols B_1 and A_i
$d_{C_1A_i}$ = distance between symbols C_1 and A_i
$d_{C_1B_i}$ = distance between symbols C_1 and B_i
$d_{D_1A_i}$ = distance between symbols D_1 and A_i
$d_{D_1B_i}$ = distance between symbols D_1 and B_i
$d_{D_1C_i}$ = distance between symbols D_1 and C_i
$d_{E_1A_i}$ = distance between symbols E_1 and A_i
$d_{E_1B_i}$ = distance between symbols E_1 and B_i
$d_{E_1C_i}$ = distance between symbols E_1 and C_i
$d_{E_1D_i}$ = distance between symbols E_1 and D_i

According to the structure presented in Figure 7.19, these distances can easily be expressed as

$$d^2_{A_1A_i} = 4E_{s_5}\sin^2\left[(\theta_2-\theta_1)+\frac{(i-2)\pi}{8}\right],\quad i=2,4,6,8 \tag{7.72a}$$

$$d^2_{A_1A_i} = 4E_{s_5}\sin^2\left[\frac{(i-1)\pi}{8}\right],\quad i=3,5,7 \tag{7.72b}$$

$$d^2_{B_1B_i} = 4E_{s_4}\sin^2\left[\theta_2+\frac{(i-2)\pi}{8}\right],\quad i=2,4,6,8 \tag{7.73a}$$

$$d^2_{B_1B_i} = 4E_{s_4}\sin^2\left[\frac{(i-1)\pi}{8}\right],\quad i=3,5,7 \tag{7.73b}$$

$$d^2_{C_1C_i} = 4E_{s_3}\sin^2\left[\frac{(i-1)\pi}{4}\right],\quad i=2,3,4 \tag{7.74}$$

$$d^2_{D_1D_i} = 4E_{s_2}\sin^2\left[\theta_3+\frac{(i-2)\pi}{8}\right],\quad i=2,4,6,8 \tag{7.75a}$$

$$d^2_{D_1D_i} = 4E_{s_2}\sin^2\left[\frac{(i-1)\pi}{8}\right],\quad i=3,5,7 \tag{7.75b}$$

$$d^2_{E_1E_i} = 4E_{s_1}\sin^2\left[\frac{(i-1)\pi}{4}\right],\quad i=2,3,4 \tag{7.76}$$

$$d^2_{B_1A_i} = E_{s_4}\left[1+\alpha_4^2-2\alpha_4\cos\left\{28.072+\theta_1+\frac{(i-2)\pi}{4}\right\}\right],\quad i=2,4,6,8 \tag{7.77a}$$

$$d^2_{B_1A_i} = E_{s_4}\left[1+\alpha_4^2-2\alpha_4\cos\left\{\theta_1+\frac{(i-1)\pi}{4}\right\}\right],\quad i=1,3,5,7 \tag{7.77b}$$

$$d^2_{C_1A_i} = E_{s_3}\left[1+\alpha_3^2-2\alpha_3\cos\left\{28.072-(\theta_2-\theta_1)+\frac{(i-2)\pi}{4}\right\}\right],\quad i=2,4,6,8 \tag{7.78a}$$

$$d^2_{C_1A_i} = E_{s_3}\left[1+\alpha_3^2-2\alpha_3\cos\left\{(\theta_2-\theta_1)-\frac{(i-1)\pi}{4}\right\}\right],\quad i=1,3,5,7 \tag{7.78b}$$

$$d^2_{C_1B_i} = E_{s_3}\left[1+\left(\frac{\alpha_3}{\alpha_4}\right)^2-2\left(\frac{\alpha_3}{\alpha_4}\right)\cos\left\{67.38-\theta_2+\frac{(i-2)\pi}{4}\right\}\right],\quad i=2,4,6,8 \tag{7.79a}$$

$$d^2_{C_1B_i} = E_{s_3}\left[1+\left(\frac{\alpha_3}{\alpha_4}\right)^2-2\left(\frac{\alpha_3}{\alpha_4}\right)\cos\left\{\theta_2-\frac{(i-1)\pi}{4}\right\}\right],\quad i=1,3,5,7 \tag{7.79b}$$

$$d^2_{D_1A_i} = E_{s_1}\left[1+\alpha_2^2 - 2\alpha_2\cos\left\{28.072+\theta_3-(\theta_2-\theta_1)+\frac{(i-2)\pi}{4}\right\}\right], \quad i=2,4,6,8 \tag{7.80a}$$

$$d^2_{D_1A_i} = E_{s_1}\left[1+\alpha_2^2 - 2\alpha_2\cos\left\{\theta_3-(\theta_2-\theta_1)+\frac{(i-1)\pi}{4}\right\}\right], \quad i=1,3,5,7 \tag{7.80b}$$

$$d^2_{D_1B_i} = E_{s_2}\left[1+\left(\frac{\alpha_2}{\alpha_4}\right)^2 - 2\left(\frac{\alpha_2}{\alpha_4}\right)\cos\left\{67.38-\theta_2+\theta_3+\frac{(i-2)\pi}{4}\right\}\right], \quad i=2,4,6,8 \tag{7.81a}$$

$$d^2_{D_1B_i} = E_{s_2}\left[1+\left(\frac{\alpha_2}{\alpha_4}\right)^2 - 2\left(\frac{\alpha_2}{\alpha_4}\right)\cos\left\{\theta_2-\theta_3-\frac{(i-1)\pi}{4}\right\}\right], \quad i=1,3,5,7 \tag{7.81b}$$

$$d^2_{D_1C_i} = E_{s_2}\left[1+\left(\frac{\alpha_2}{\alpha_3}\right)^2 - 2\left(\frac{\alpha_2}{\alpha_3}\right)\cos\left\{\theta_3+\frac{(i-1)\pi}{2}\right\}\right], \quad i=1,2,3,4 \tag{7.82}$$

$$d^2_{E_1A_i} = E_{s_1}\left[1+\alpha_1^2 - 2\alpha_1\cos\left\{28.072-(\theta_2-\theta_1)+\frac{(i-2)\pi}{4}\right\}\right], \quad i=2,4,6,8 \tag{7.83a}$$

$$d^2_{E_1A_i} = E_{s_1}\left[1+\alpha_1^2 - 2\alpha_1\cos\left\{(\theta_2-\theta_1)-\frac{(i-1)\pi}{4}\right\}\right], \quad i=1,3,5,7 \tag{7.83b}$$

$$d^2_{E_1B_i} = E_{s_1}\left[1+\left(\frac{\alpha_1}{\alpha_4}\right)^2 - 2\left(\frac{\alpha_1}{\alpha_4}\right)\cos\left\{67.38-\theta_2+\frac{(i-2)\pi}{4}\right\}\right], \quad i=2,4,6,8 \tag{7.84a}$$

$$d^2_{E_1B_i} = E_{s_1}\left[1+\left(\frac{\alpha_1}{\alpha_4}\right)^2 - 2\left(\frac{\alpha_1}{\alpha_4}\right)\cos\left\{\theta_2-\frac{(i-1)\pi}{4}\right\}\right], \quad i=1,3,5,7 \tag{7.84b}$$

$$d^2_{E_1C_i} = E_{s_1}\left[1+\left(\frac{\alpha_1}{\alpha_3}\right)^2 - 2\left(\frac{\alpha_1}{\alpha_3}\right)\cos\left\{\frac{(i-1)\pi}{2}\right\}\right], \quad i=1,2,3,4 \tag{7.85}$$

$$d^2_{E_1D_i} = E_{s_1}\left[1+\left(\frac{\alpha_1}{\alpha_2}\right)^2 - 2\left(\frac{\alpha_1}{\alpha_2}\right)\cos\left\{53.13-\theta_3+\frac{(i-2)\pi}{4}\right\}\right], \quad i=2,4,6,8 \tag{7.86a}$$

$$d^2_{E_1D_i} = E_{s_1}\left[1+\left(\frac{\alpha_1}{\alpha_2}\right)^2 - 2\left(\frac{\alpha_1}{\alpha_2}\right)\cos\left\{\theta_3-\frac{(i-1)\pi}{4}\right\}\right], \quad i=1,3,5,7 \tag{7.86b}$$

As seen from the geometry of the constellation diagram, the equality $d_{X_iY_j} = d_{Y_jX_i}$ is satisfied in all the above equations. In these equations,E_{s_1}, E_{s_2}, E_{s_3}, E_{s_4}, and E_{s_5} are the energies of the circles E, D, C, B, and A, respectively, which are defined as

$$E_{s_1} = r_1^2 \tag{7.87a}$$

$$E_{s_2} = r_2^2 \tag{7.87b}$$

$$E_{s_3} = r_2^2 \tag{7.87c}$$

$$E_{s_4} = r_2^2 \tag{7.87d}$$

$$E_{s_5} = R^2 \tag{7.87e}$$

Now $d_{\min}$ can be found from the distances defined above. Then we can apply Equation 7.49 and Equation 7.50 to find the lower and upper bounds of the POE expressions when transmitting a symbol in the presence of AWGN with a PSD of N_0. With the same scenario the best-approximated upper bound can be derived as

$$P_{e(32Rec)} \approx 0.25P_{e/A_1} + 0.25P_{e/B_1} + 0.125P_{e/C_1} + 0.25P_{e/D_1} + 0.125P_{e/E_1} \tag{7.88}$$

where the conditional probabilities are given by

$$P_{e/A_1} \approx \sum_{i=1}^{4} Q\left(\sqrt{\frac{d^2_{A_1E_i}}{2N_0}}\right) + \sum_{i=1}^{8} Q\left(\sqrt{\frac{d^2_{A_1D_i}}{2N_0}}\right) + \sum_{i=1}^{4} Q\left(\sqrt{\frac{d^2_{A_1C_i}}{2N_0}}\right) + \sum_{i=1}^{8} Q\left(\sqrt{\frac{d^2_{A_1B_i}}{2N_0}}\right) + \sum_{i=2}^{8} Q\left(\sqrt{\frac{d^2_{A_1A_i}}{2N_0}}\right)$$

$$P_{e/B_1} \approx \sum_{i=1}^{4} Q\left(\sqrt{\frac{d^2_{B_1E_i}}{2N_0}}\right) + \sum_{i=1}^{8} Q\left(\sqrt{\frac{d^2_{B_1D_i}}{2N_0}}\right) + \sum_{i=1}^{4} Q\left(\sqrt{\frac{d^2_{B_1C_i}}{2N_0}}\right) + \sum_{i=2}^{8} Q\left(\sqrt{\frac{d^2_{B_1B_i}}{2N_0}}\right) + \sum_{i=1}^{8} Q\left(\sqrt{\frac{d^2_{B_1A_i}}{2N_0}}\right)$$

$$P_{e/C_1} \approx \sum_{i=1}^{4} Q\left(\sqrt{\frac{d^2_{C_1E_i}}{2N_0}}\right) + \sum_{i=1}^{8} Q\left(\sqrt{\frac{d^2_{C_1D_i}}{2N_0}}\right) + \sum_{i=2}^{4} Q\left(\sqrt{\frac{d^2_{C_1C_i}}{2N_0}}\right) + \sum_{i=1}^{8} Q\left(\sqrt{\frac{d^2_{C_1B_i}}{2N_0}}\right) + \sum_{i=1}^{8} Q\left(\sqrt{\frac{d^2_{C_1A_i}}{2N_0}}\right)$$

$$P_{e/D_1} \approx \sum_{i=1}^{4} Q\left(\sqrt{\frac{d^2_{D_1E_i}}{2N_0}}\right) + \sum_{i=2}^{8} Q\left(\sqrt{\frac{d^2_{D_1D_i}}{2N_0}}\right) + \sum_{i=1}^{4} Q\left(\sqrt{\frac{d^2_{D_1C_i}}{2N_0}}\right) + \sum_{i=1}^{8} Q\left(\sqrt{\frac{d^2_{D_1B_i}}{2N_0}}\right) + \sum_{i=1}^{8} Q\left(\sqrt{\frac{d^2_{D_1A_i}}{2N_0}}\right)$$

$$P_{e/E_1} \approx \sum_{i=2}^{4} Q\left(\sqrt{\frac{d^2_{E_1E_i}}{2N_0}}\right) + \sum_{i=1}^{8} Q\left(\sqrt{\frac{d^2_{E_1D_i}}{2N_0}}\right) + \sum_{i=1}^{4} Q\left(\sqrt{\frac{d^2_{E_1C_i}}{2N_0}}\right) + \sum_{i=1}^{8} Q\left(\sqrt{\frac{d^2_{E_1B_i}}{2N_0}}\right) + \sum_{i=1}^{8} Q\left(\sqrt{\frac{d^2_{E_1A_i}}{2N_0}}\right) \ldots$$

7.5.4 The Effect of Nonlinearity and the Application of a Predistortion Technique

Equations 7.49 to 7.51, 7.61, and 7.88 provide a generalized form of the POE expressions for 32-QAM circular constellations. These equations are equally applicable for the nonlinear environment in the presence of a HPA, where the distances would vary depending on the model of the nonlinear amplifier. To overcome the nonlinearity in a 32-QAM system with circular constellations, we can follow the same predistortion approach as seen in Section 7.4.6.

7.5.5 Results and Discussions

In this section we will briefly discuss some results that provide a clear comparison between all the addressed 32-QAM constellations with and without the application of a predistortion technique in the presence of the TWT. We also present a comparative analysis between the addressed 16-QAM and 32-QAM constellations with the aid of POE vs. SNR curves. In the graphs we add the POE curves for 16- and 32-PSK systems in the linear AWGN channel just to reflect the popularity of 16- and 32-QAM systems over their PSK counterparts.

Here, Figure 7.20 shows POE plots for all the addressed 32-QAM circular constellations with and without the presence of a data predistortion technique in the nonlinear AWGN channel (in the presence of a nonlinear amplifier); the minimal IBO is used for each constellation. The figure also includes the POE plots for the 16- and 32-PSK systems in the linear AWGN channel. From the curves it is clearly noticeable that in the nonlinear environment, with a data predistortion technique, all the addressed 32-QAM constellations perform better than both the 16- and 32-PSK systems. As seen from the figure, the (5,11,16) circular constellation performs best among all the addressed constellations.

Figure 7.21 illustrates data predistortion for a 32-QAM rectangular constellation. Figure 7.22 shows some interesting results that help us to see the POE performance comparison between the addressed 16- and 32-QAM constellations, with data predistortion, and the 16- and 32-PSK based on linear system performance. From the figure we can comment that with a little sacrifice in the spectral efficiency, we can achieve a significant gain in the power efficiency by using the (5,11) constellation, with data predistortion,

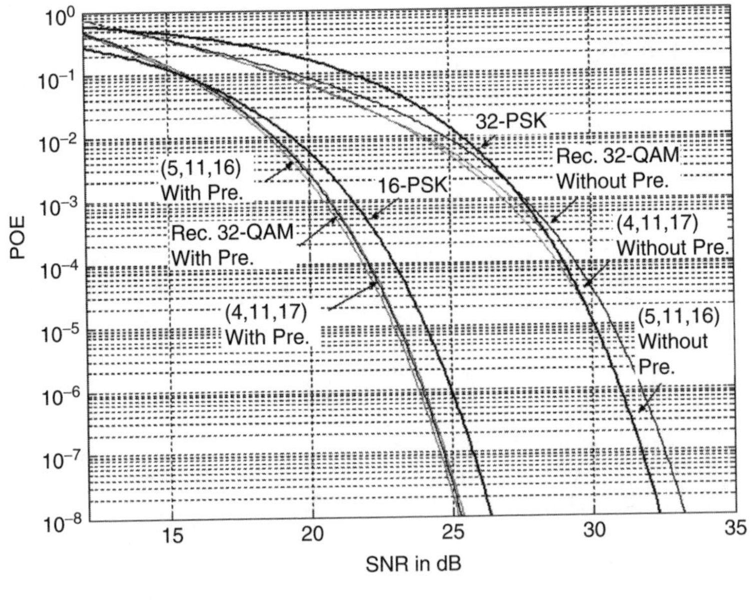

FIGURE 7.20 Performance comparison between different 32-QAM circular constellations with and without the application of a data predistortion technique in the presence of a nonlinear HPA.

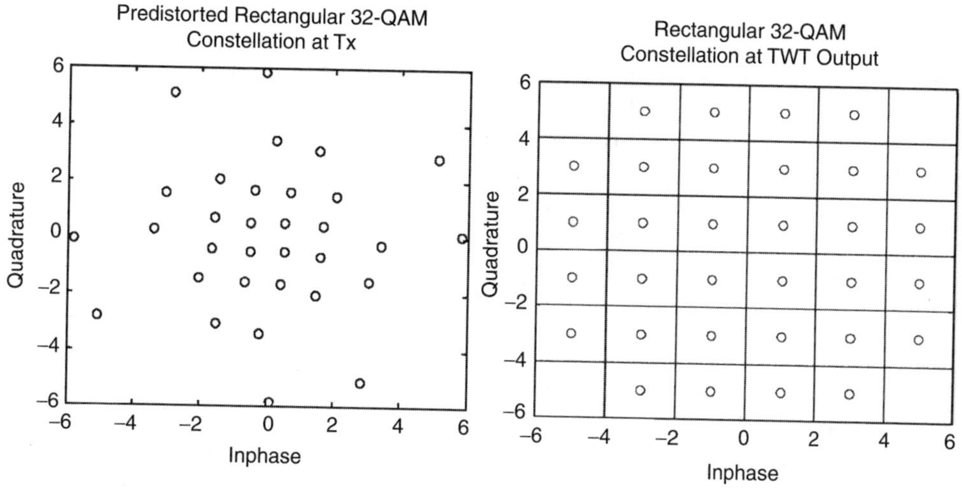

FIGURE 7.21 Illustration of data predistortion, the rectangular 32-QAM case: constellation at the predistorter output (left), constellation at the amplifier output (right).

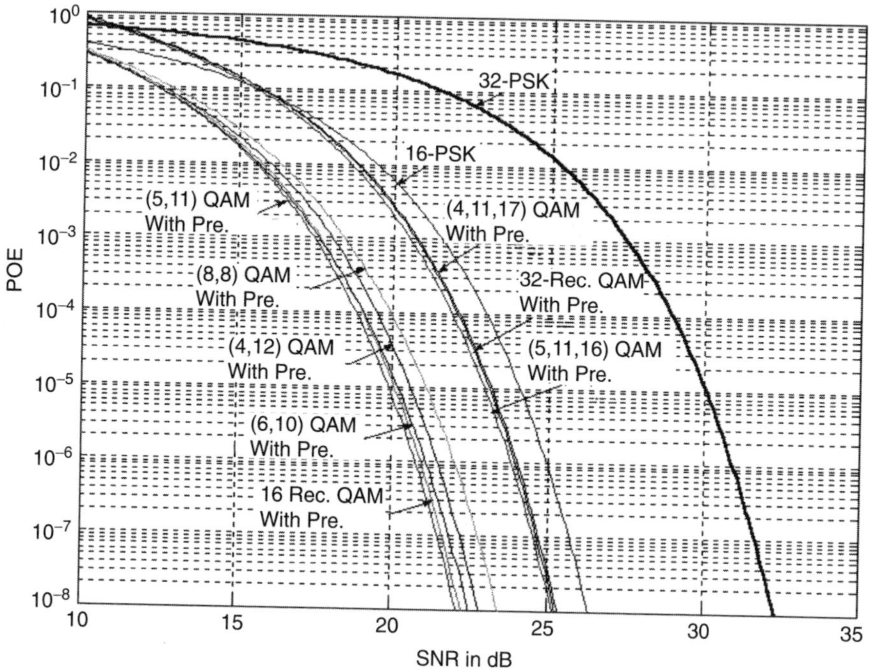

FIGURE 7.22 Performance comparison between different 16 and 32-QAM circular constellations with the use of a data predistortion technique in the presence of an NA.

over the other 16- and 32-QAM (and PSK) constellations. The above figures also show that the 16- and 32-QAM systems, with the use of a data predistortion technique, could be great candidates in designing a communications system, such as a satellite communications system, where nonlinearity is a big challenge.

7.6 Conclusions

This chapter presents a simple and efficient method to calculate the POE of any 16- and 32-QAM circular constellation over a memoryless satellite channel. This method, which can be extended for any M-QAM circular constellation, provides a way to find the optimal RRs and PDs for all the addressed QAM constellations. A data predistortion technique is introduced into the system to overcome the nonlinear distortions that result from the TWT amplifier. The results show that the RR has more influence than the PD on the variation of the POE performance. Computer simulations support the theoretical results. The analytical results provide a comparison between different POE performances for different constellations, which helps in finding the right QAM constellation for the right system.

Appendix

TWT Based on the Saleh Model [Sal81]

TWT amplifiers are commonly used in mobile satellite communication systems for high-power application. When operated near the maximum power region, these amplifiers cause nonlinear distortions. These distortions are modeled in terms of amplitude-to-amplitude (AM/AM) and amplitude-to-phase (AM/PM) conversions. Many researchers have provided different models on these conversions, among which the Saleh model [Sal81] is most widely used:

$$A(r) = \frac{\alpha_a r}{1 + \beta_a r^2} \tag{7.A1}$$

$$P(r) = \frac{\pi}{3} \frac{\alpha_p r^2}{1 + \beta_p r^2} \tag{7.A2}$$

Equation 7.A1 and 7.A2 represent the AM/AM and AM/PM conversions, respectively, and α and β are the arbitrary constants. Figure 7.A(a) and (b) show the normalized AM/AM and AM/PM conversions, respectively, with $\alpha_a = 2$ and $\beta_a = 1$, and $\alpha_p = \beta_p = 1$ (these values are used in this chapter). From these figures it is clear that as the operating point of the amplifier moves close to the saturation region, the distortion of the output increases due to the nonlinearity.

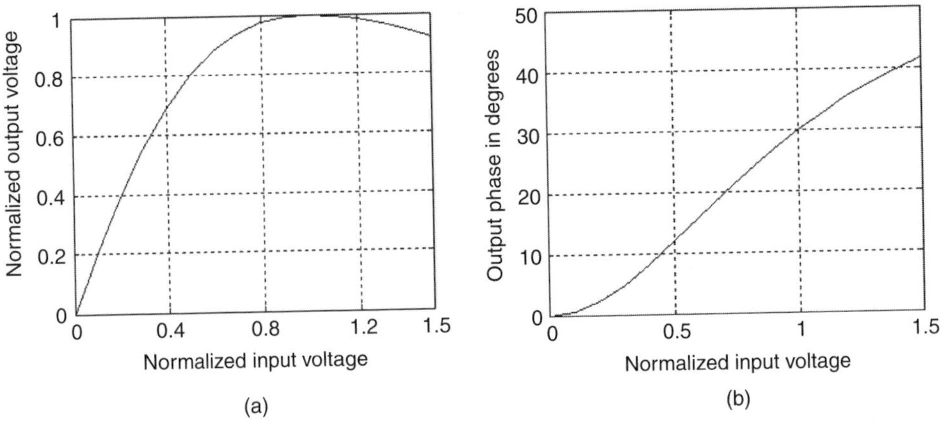

FIGURE 7.A (a) AM/AM conversion and (b) AM/PM conversion.

References

[Alu00] M.S. Alouini and A. Goldsmith, "Adaptive modulation over Nakagami fading channels," *Kluwer J. Wireless Personal Commmun.*, 13, 119–143, 2000.

[Ber97] A. Bernardini and S. De Fina, "Analysis of different optimization criteria for IF predistortion in digital radio links with nonlinear amplifiers," *IEEE Trans. Commun.*, 45, 421–428, 1997.

[Che00] L. Chee-Siong, S. Vlahoyiannatos, and L. Hanzo, "Satellite based turbo-coded, blind-equalized 4-QAM and 16-QAM digital video broadcasting," *IEEE Trans. Broadcasting*, 46, 23–33, 2000.

[Chi98] G. Chrisikos, "Analysis of 16-QAM over a nonlinear channel," in *IEEE Symposium on Personal, Indoor and Mobile Radio Communications (PIMRC)*, Vol. 3, September 1998, pp. 1325–1329.

[Cra91] J.W. Craig, "A new, simple and exact result for calculating the probability of error for two-dimensional signal constellations," in *Proceedings of the IEEE MILCOM'91*, Vol. 2, November 1991, pp. 571–575.

[Dim90] C.E. Dimakis, S.S. Kouris, and S.A. Kosmopoulos, "Performance evaluation of 16-QAM signalling through nonlinear channels in an AWGN and interference environment: a simulation approach," in *Proceedings of the IEEE*, Vol. 137, October 1990, pp. 315–322.

[Dng99a] D. Xiaodai, N.C. Beaulieu, and P.H. Wittke, "Signaling constellations for fading channels," *IEEE Trans. Commun.*, 47, 703–714, 1999.

[Dng99b] D. Xiaodai, N.C. Beaulieu, and P.H. Wittke, "Error probabilities of two-dimensional M-ary signaling in fading," *IEEE Trans. Commun.*, 47, 352–355, 1999.

[Gol97] A.J. Goldsmith and C. Soon-Ghee, "Variable-rate variable-power MQAM for fading channels," *IEEE Trans. Commun.*, 45, 1218–1230, 1997.

[Ibn98] M. Ibnkahla, N.J. Bershad, J. Sombrin, and F. Castanie, "Neural network modeling and identification of nonlinear channels with memory: algorithms, applications, and analytic models," *IEEE Transactions Signal Process.*, 46, 1208–1220, 1998.

[Ibn04] M. Ibnkahla, Q. Rahman, A. Sulyman, H. Al-Asady, Y. Jun, and A. Safwat, "High speed satellite mobile communications: Technologies and challenges," *Proc. IEEE*, Special issue on Gigabit wireless communications: Technologies and challenges, pp. 312–339, February 2004.

[Kar89] G. Karam and H. Sari, "Analysis of predistortion, equalization, and ISI cancellation techniques in digital radio system with nonlinear transmit amplifiers," *IEEE Trans. Commun.*, 37, 1245–1254, 1989.

[Muh00] K.J. Muhonen, M. Kavehrad, and R. Krishnamoorthy, "Look-up table techniques for adaptive digital predistortion: a development and comparison," *IEEE. Trans. Vehicular Technol.*, 49, 1995–2002, 2000.

[Pro02] J.G. Proakis, *Digital Communications*, 4th ed., New York: McGraw-Hill, 2002.

[Sal81] A. Saleh, "Frequency-independent and frequency-dependent nonlinear models of TWT amplifiers," *IEEE Trans. Commun.*, 29, 1715–1720, 1981.

[Web95] W.T. Webb and R. Steele, "Variable rate QAM for mobile radio," *IEEE Trans. Commun.*, 43, 2223–2230, 1995.

8

Carrier Frequency Synchronization for OFDM Systems

8.1 Introduction 8-1
8.2 Basics of OFDM 8-2
OFDM Modulation • Demodulation
8.3 Effect of CFO on System Performance 8-6
8.4 Carrier Frequency Offset Estimation 8-7
8.5 Repetitive Slots-Based CFO Estimation 8-8
Nonlinear Least Squares Method • Computationally Simpler Estimators
8.6 Null-Subcarrier-Based CFO Estimation 8-10
Deterministic Maximum Likelihood Estimation • Special Case: Repetition of Identical Slots
8.7 Identifiability 8-13
8.8 Performance Analysis 8-15
The Conditional CRB • The Unconditional CRB: Rayleigh Channel • Optimal Choice of Null-Subcarriers
8.9 Simulation Results 8-17
8.10 Conclusions 8-19

Mounir Ghogho
University of Leeds

Ananthram Swami
Army Research Laboratory

Abstract

Orthogonal frequency division multiplexing (OFDM) is, with substantial progress in digital signal processing, becoming an important part of the telecommunications arena. The most appealing feature of OFDM is the simplicity of the receiver design due to the efficiency with which OFDM can cope with the effects of frequency-selective multipath channels. Multicarrier systems such as OFDM are, however, more sensitive to carrier frequency offset (CFO) than are single-carrier systems. We first review the basics of OFDM communications systems and then address the CFO estimation problem. Both pilot-based and blind estimation techniques are presented. The Cramér–Rao bound is established, and an optimal null carrier placement scheme is derived.

8.1 Introduction

Orthogonal frequency division multiplexing (OFDM) has become the standard of choice for wireless local area networks (LANs) such as HIPERLAN/2 and IEEE 802.11a; it has been adopted in Europe for

0-8493-1657-X/$0.00+$1.50

Digital Audio Broadcasting (DAB) and Digital Video Broadcasting (DVB), Multimedia Mobile Access Communications (MMAC) in Japan, and fixed wireless; and is being considered for several IEEE 802.11 and 802.16 standards, including wideband metropolitan area networks (MANS) [1]. The popularity of OFDM stems from its ability to transform a wideband frequency-selective channel to a set of parallel flat-fading narrowband channels, which substantially simplifies the channel equalization problem. Because of the time–frequency granularity that it offers, OFDM appears to be a natural solution when the available spectrum is not contiguous, for overlay systems, and to cope with issues such as narrowband jamming. In the multiuser context, this granularity also accommodates variable quality-of-service (QoS) requirements and bursty data.

But carrier synchronization turns out to be a critical issue for multicarrier systems. In this chapter, we review the basics of OFDM and study the frequency synchronization problem in detail. Clearly, issues such as timing recovery and channel estimation are important, but will not be treated here. Synchronization usually consists of an acquisition phase followed by a tracking phase. In the acquisition phase, initial (often coarse) estimates of the synchronization parameters are obtained. Finer estimates, as well as tracking of small time variations, of these parameters are acquired during the tracking phase. Here we assume perfect frame and time synchronization and focus on carrier frequency offset (CFO) estimation.

The origins of OFDM can be traced back to a 1957 multicarrier HF analog modem (the Kineplex), but it was only with digital implementations via fast Fourier transform (FFT) and the introduction of the cyclic prefix, both by Weinstein and Ebert [2], and seminal analyzes by Cimini [3], that OFDM became practical. Several recent books and papers on OFDM cover different aspects of signal and system design [4], [5], [6], [7], [8].

Notation: We will let $x(i,m)$ denote the mth element of the vector $\boldsymbol{x}(i)$. Superscripts $\mathcal{H}$ and T will denote conjugate transposition and transposition. $\mathbf{F}$ will denote the FFT matrix with (k,m)th entry $\frac{1}{\sqrt{M}}\exp(-j2\pi km/M)$. We use the Matlab notation $A(1:m,1:n)$ to denote the submatrix formed by the first m rows and first n columns of the matrix A; when all columns are included '$1:n$' is replaced simply by ':'. $\mathcal{R}[z]$, $\mathcal{I}[z]$, and $\arg\{z\}$ will denote the real part, the imaginary part, and the argument of z, respectively. The subscript 2π denotes modulo-2π operation, and Tr denotes the trace operator. For a positive scalar Q, $\lfloor Q \rceil$ is the nearest integer to Q. If $Q = m+0.5$ with m being a positive integer, $\lfloor Q \rceil = m$.

8.2 Basics of OFDM

Assume that the information-bearing symbols are to be transmitted at the rate of R symbols per second over a multipath propagation channel. The duration of each symbol is therefore $T_s = 1/R$. If the delay spread,[1] τ_{max}, of the channel is larger than about 10% of the symbol duration, then the received signal may suffer from significant intersymbol interference (ISI), which can drastically increase the symbol error rate (SER)[2] unless countermeasures are undertaken. Such a channel is said to be dispersive or frequency selective. There are two main approaches to cope with such channels. The first approach is to use a single-carrier system with an equalizer at the receiver to compensate for the ISI, which spans $\lceil \tau_{max}/T_s \rceil$ symbols. The implementation of the equalizer may become very challenging for channels with large delay spreads and at higher data rates. The second approach is based on multicarrier modulation, such as orthogonal frequency division multiplexing. Here, we focus on the latter approach.

The operational principle of an OFDM system is that the available bandwidth is divided into a large number of subbands, over each of which the wireless channel can be considered nondispersive or flat fading. The original data stream at rate R is split into M parallel data streams, each at rate R/M. The symbol duration, T, for these parallel data streams is therefore increased by a factor of M, i.e., $T = MT_s$. Conceptually, each of the data streams modulates a carrier with a different frequency and the resulting signals are

[1]The delay spread is a measure of the delay of the longest path (or last echo) with respect to that of the earliest path (or first arrival); typically the root mean square (RMS) delay spread is used.

[2]The symbol error rate is the rate of errors made during the symbol detection process at the receiver.

transmitted simultaneously (in reality, a single modulator is used, as we discuss later). Correspondingly, the receiver consists of M parallel receiver paths. Due to the increased symbol duration, the ISI over each channel is reduced to $\lceil \tau_{max}/(MT_s) \rceil$ symbols. Thus, an advantage of OFDM is that, for frequency-selective fading channels, the OFDM symbols are less affected by channel fades than are single-carrier transmitted symbols. This is due to the increased symbol duration in an OFDM system. While many symbols during a channel fade might be lost in a single-carrier system, the symbols of an OFDM system can still be correctly detected since only a fraction of each symbol might be affected by the fade. On the other hand, if the channel is time selective, i.e., the channel impulse response varies significantly within the OFDM symbol period, then the channel matrix is no longer Toeplitz and conventional OFDM would fail.

Since multicarrier modulation is based on a block transmission scheme, measures have to be taken to avoid or compensate for interblock interference (IBI), which contributes to the overall ISI. OFDM systems can be categorized by the way they handle IBI. In the most popular system, a guard time is introduced between consecutive OFDM symbols as a *cyclic prefix* (CP); i.e., the tail end of the OFDM symbol is prefixed. The length of the cyclic prefix is chosen to be larger than the expected delay spread; after proper time synchronization, the receiver discards the CP and thus the IBI is eliminated. Time guarding by zero-padding the OFDM symbols has also been proposed in [9], [10]. The issue here is one of turning the transmitter on and off and increased receiver complexity vs. the increased signal-to-noise ratio (SNR) and decreased SER. Comparisons between cyclic-prefixing and zero-padding OFDM systems may be found in [11].

To achieve high resilience against channel dispersion, a large number of subcarriers is required. However, the implementation of a large number of modulators and demodulators can be very complex, in terms of both the physical size of the radio, and the difficulty of locking in multiple oscillators. This complexity can be significantly reduced by digitally performing the modulation and demodulation using the discrete Fourier transform (DFT) and its inverse (IDFT) [2]. An efficient implementation of the DFT may be obtained by any of the available FFT algorithms.

The choice of the OFDM parameters is a trade-off between various, often conflicting requirements. The length of the CP is dictated by the delay spread of the channel. Introduction of the CP entails a reduction in rate (or wasted bandwidth), as well as an SNR loss; to minimize these inefficiencies, the number of subcarriers, M, should be large. However, a large number of subcarriers induces high implementation complexity, increased sensitivity to frequency offset and phase noise (since the subcarriers get closer to each other as M increases), and increased peak-to-average power ratios (PAPRs). M is dictated by concerns regarding practical FFT sizes as well as the coherence time of the channel. We will not address the issue of practical choice of OFDM parameters here; refer to [3], [6], [8]. In this chapter, we address the crucial issue of CFO estimation.

We confine our attention to OFDM; generalized schemes that also convert frequency-selective channels into a bank of flat-fading channels exist [12], see also [13]. In [12], the OFDM scheme is shown to be the optimal precoder, which uses a cyclic prefix. In a multiple-user setting, OFDM has been shown to be the optimal scheme in the sense of maximizing the SNR for each user [14].

8.2.1 OFDM Modulation

OFDM modulation consists of M (usually a power of 2) subcarriers, equi-spaced at a separation of $\Delta f = B/M$, where B is the total system bandwidth. All subcarriers are mutually orthogonal over a time interval of length $T = 1/\Delta f$. Each subcarrier is modulated independently with information-bearing symbols (this does not preclude coding across the subcarriers). Each OFDM block is preceded by a CP whose duration is usually longer than the delay spread of the propagation channel, so that IBI can be eliminated at the receiver without affecting the orthogonality of the subcarriers. Practical OFDM systems are not fully loaded in order to avoid interference between adjacent OFDM systems: some of the subcarriers at the edges of the OFDM block are not modulated; these subcarriers are referred to as virtual subcarriers (VSCs). The number of these VSCs is dictated by system design requirements and in general is about 10% of M. Some subcarriers, other than the VSCs, may also be deactivated. For example, when channel state information

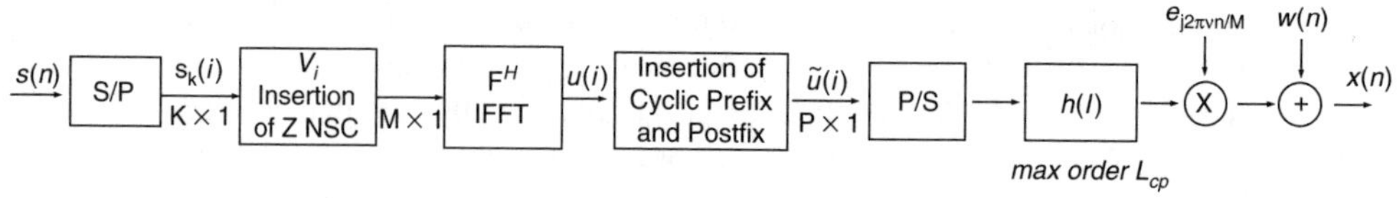

FIGURE 8.1 Discrete-time complex baseband representation.

(CSI) is available to the transmitter, subcarriers experiencing deep fades will be left unmodulated. Further, synchronization preambles are often made by nulling a large number of subcarriers. Indeed, a preamble consisting of a repetition of two identical slots is obtained by nulling all the odd subcarriers [15]. Here, deactivated subcarriers will be referred to as null subcarriers (NSCs). The set of NSCs includes the VSCs, whose placement and number are imposed by system design; the number and placement of the remaining NSCs are controlled by the system user and could vary across the OFDM symbols. Let $\mathcal{M} = \{0, \ldots, M-1\}$ denote the entire set of subcarriers, and let $\mathcal{K}_i$ (respectively $\mathcal{Z}_i$) denote the subset of $\mathcal{M}$ that contains the K_i (respectively Z_i) modulated (respectively null) subcarriers during the ith OFDM symbol or block.

The discrete-time block diagram of a standard OFDM system is depicted in Figure 8.1. The vector modulating the entire set of subcarriers during the ith block can then be expressed as $\boldsymbol{s}(i) := \mathbf{V}_i \boldsymbol{s}_{\mathcal{K}}(i)$, where $\boldsymbol{s}_{\mathcal{K}}(i)$ is the K_i element vector of symbols transmitted on the activated subcarriers and $\mathbf{V}_i$ is the $M \times K_i$ matrix whose (m, n)th entry is 1 if the nth symbol is transmitted on the mth subcarrier during the ith OFDM block, and is zero otherwise. Matrix $\mathbf{V}_i$ is a full-rank submatrix of an $(M \times M)$ permutation matrix. We assume without loss of generality that the symbols are zero mean and have unit variance, i.e., $E|s(i, m)|^2 = 1$. The $(M \times 1)$ data block $\boldsymbol{s}(i)$ is first precoded by the IFFT matrix $\mathbf{F}^{\mathcal{H}}$. The resulting $(M \times 1)$ vector $\boldsymbol{u}(i) = \beta_i \mathbf{F}^{\mathcal{H}} \boldsymbol{s}(i)$ is called the time-domain block vector, or the time-domain OFDM symbol. We have also introduced a normalization parameter, $\beta_i := \sqrt{M/K_i}$, to ensure that the transmitted power is kept constant regardless of K_i, the number of active subcarriers. Next, a CP of length L_{cp} is inserted by replicating the last L_{cp} elements of each block in the front. The redundant block vector can be expressed as

$$\tilde{\boldsymbol{u}}(i) = [u(i, M - L_{cp}), u(i, M - L_{cp} + 1), \ldots, u(i, M-1), u(i, 0), \ldots, u(i, M-1)]^T$$

The $P (= M + L_{cp})$ samples of each block are then pulse shaped, upconverted to the carrier frequency, and transmitted sequentially through the channel.

8.2.2 Demodulation

We model the frequency-selective channel as a finite impulse response (FIR) filter with channel impulse response (CIR) $\boldsymbol{h} = [h_0, \ldots, h_L]^T$, where L is the channel order. In practice, the system is *usually* designed such that $L \leq L_{cp} \leq M$. We assume that the CIR is time invariant over $N \geq 1$ consecutive symbol blocks, but could vary from one set of N blocks to the next.

The received signal is downconverted to baseband and sampled at the rate of P samples per extended OFDM symbol. We will index these samples by $[-L_{cp}, \ldots, M-1]$. We will assume that time synchronization has been achieved. Discarding the samples $n = -L_{cp}, \ldots, -1$ is known as discarding the cyclic prefix. The noise-free received signal corresponding to the ith OFDM symbol, $\tilde{\boldsymbol{u}}(i)$, can be written as

$$x(i, n) = \sum_{\ell=0}^{L} h_\ell \tilde{u}(i, n-\ell)$$

for $n = 0, \ldots, M-1$. Recall that with the insertion of CP, we have $\tilde{u}(i, \ell) = u(i, M + \ell)$ for $\ell = -L_{cp}, \ldots, -1$, and $\tilde{u}(i, \ell) = u(i, \ell)$ for $\ell = 0, \ldots, M-1$. Then, collecting samples, $n = 0, \ldots, M-1$, of $x(i,n)$, we obtain

$$\begin{bmatrix} x(i,0) \\ \vdots \\ x(i,M-1) \end{bmatrix} = \begin{bmatrix} h_0 & 0 & \dots & 0 & 0 & \dots & 0 \\ h_1 & h_0 & \dots & 0 & 0 & \dots & 0 \\ \vdots & \vdots & \ddots & \vdots & \vdots & \ddots & \vdots \\ h_{L-1} & h_{L-2} & \dots & h_0 & 0 & \dots & 0 \\ h_L & h_{L-1} & \dots & h_1 & h_0 & \mathbf{0} & 0 \\ \vdots & \vdots & \ddots & \vdots & \vdots & \ddots & \vdots \end{bmatrix} \begin{bmatrix} u(i,0) \\ u(i,1) \\ \vdots \\ u(i,M-1) \end{bmatrix}$$

$$+ \begin{bmatrix} h_1 & h_2 & \dots & h_{L-1} & h_L & 0 & \cdots & 0 \\ h_2 & h_3 & \dots & h_L & 0 & 0 & \cdots & 0 \\ \vdots & \vdots & \ddots & \vdots & \vdots & \vdots & \ddots & \vdots \\ h_L & 0 & 0 & 0 & 0 & 0 & \cdots & 0 \\ \mathbf{0} & \mathbf{0} & \mathbf{0} & \mathbf{0} & \mathbf{0} & 0 & \cdots & 0 \end{bmatrix} \begin{bmatrix} u(i,M-1) \\ u(i,M-2) \\ \vdots \\ u(i,M-L_{cp}) \end{bmatrix}$$

$$= \begin{bmatrix} h_0 & 0 & \dots & 0 & 0 & \mathbf{0} & h_L & h_{L-1} & \dots & h_1 \\ h_1 & h_0 & \dots & 0 & 0 & \mathbf{0} & 0 & h_L & \dots & h_2 \\ \vdots & \vdots & \ddots & \vdots & \vdots & \ddots & \vdots & \vdots & \ddots & \vdots \\ h_{L-1} & h_{L-2} & \dots & h_0 & 0 & \mathbf{0} & 0 & 0 & \dots & h_L \\ h_L & h_{L-1} & \dots & h_1 & h_0 & \mathbf{0} & 0 & 0 & \dots & 0 \\ \vdots & \vdots & \ddots & \vdots & \vdots & \ddots & \vdots & \vdots & \ddots & \vdots \end{bmatrix} \begin{bmatrix} u(i,0) \\ u(i,1) \\ \vdots \\ u(i,M-1) \end{bmatrix}$$

$$\boldsymbol{x}(i) = \mathbf{H}_c \boldsymbol{u}(i) \tag{8.1}$$

Notice that the matrix $\mathbf{H}_c$ is circulant with first column, $[h_0, h_1, \dots h_L, 0, \dots, 0]^T$.

Let ν (a real number) denote the CFO normalized to the subcarrier spacing; i.e., the actual frequency offset is $\nu\Delta f$ Hz. In the presence of CFO, the mth sample of the ith received OFDM symbol will experience a phase shift equal to $2\pi\nu(iP+m)/M$. After discarding the CP, the mth sample will have phase shift $2\pi\nu(iP+L_{cp}+m)/M$, $m=0,\dots,M-1$. Under the narrowband assumption, the noise and CFO corrupted counterpart of Equation 8.1 becomes

$$\boldsymbol{x}(i) = \beta_i e^{j2\pi\nu(iP+L_{cp})/M}\mathbf{D}(\nu)\mathbf{H}_c\mathbf{F}^H\boldsymbol{s}(i) + \boldsymbol{n}(i) \tag{8.2}$$

where

$$\mathbf{D}(\nu) := \text{diag}\,(1, \exp(j2\pi\nu/M), \dots, \exp(j2\pi\nu(M-1)/M))$$

$\mathbf{H}_c$ is the $(M \times M)$ circulant matrix defined in Equation 8.1, and $\boldsymbol{n}(i)$ is the $M \times 1$ noise vector, which is assumed to be zero-mean circular Gaussian with covariance matrix $\sigma^2\mathbf{I}$.

Demodulation is based on the well-known property that any circulant matrix can be diagonalized by premultiplication by the FFT matrix and postmultiplication by the IFFT matrix. Let $H_k := \sum_{l=0}^{L} h_l e^{-j2\pi kl/M}$ denote the frequency response of the channel at frequency $2\pi k/M$, and let $\mathbf{D}(H) := \text{diag}\,(H_0, \dots, H_{M-1})$. Then, the signal in Equation 8.2 can be rewritten as

$$\begin{aligned} \boldsymbol{x}(i) &= \beta_i e^{j2\pi\nu(iP+L_{cp})/M}\mathbf{D}(\nu)\mathbf{F}^H\mathbf{D}(H)\mathbf{F}\mathbf{F}^H\boldsymbol{s}(i) + \boldsymbol{n}(i) \\ &= \beta_i e^{j2\pi\nu(iP+L_{cp})/M}\mathbf{D}(\nu)\mathbf{F}^H\mathbf{D}(H)\boldsymbol{s}(i) + \boldsymbol{n}(i) \end{aligned} \tag{8.3}$$

Therefore, after FFT processing, the so-called frequency-domain received symbol blocks are obtained as

$$y(i) = \mathbf{F}x(i) = \beta_i e^{j2\pi\nu(iP+L_{cp})/M}\mathbf{F}\mathbf{D}(\nu)\mathbf{F}^H\mathbf{D}(H)s(i) + \eta(i) \quad (8.4)$$

where $\eta(i) = \mathbf{F}n(i)$ is again additive white Gaussian, noise (AWGN) with covariance matrix $\sigma^2\mathbf{I}$.

In the absence of CFO (i.e., $\nu = 0$), the frequency-domain blocks are obtained as

$$y(i) = \beta_i\mathbf{D}(H)s(i) + \eta(i) \quad (8.5)$$

The effect of the frequency-selective channel on the OFDM signal is completely captured by scalar multiplications of the data symbols by the frequency responses of the channel at the subcarrier frequencies. Further, demodulation at the receiver does not color the additive noise. If none of the channel zeros coincide with an activated subcarrier, maximum likelihood detection of the symbols is straightforward. Zero-forcing and minimum mean square error (MMSE) equalizers can be applied on a per-carrier basis. From Equation 8.5, we see that under the constraint of constant transmitted power, the presence of NSCs (i.e., $\beta_i = M/K_i > 1$) implies a higher local SNR[3] at the modulated subcarriers at the expense of bandwidth efficiency. But for frequency-selective channels, the local SNR can vary significantly across the subcarriers. Information transmitted on a subcarrier that is experiencing a deep fade (i.e., low SNR) could be lost, i.e., frequency selectivity of the channel could degrade bit error rate (BER). There are two major techniques to mitigate this problem. The first approach is to code across the subcarriers, typically by using a convolutional code. This improves SER at the expense of reduced rate, but does not require channel state information (CSI) at the transmitter. The second approach is based on power or bit loading techniques and assumes that CSI is available at the transmitter. System capacity can be maximized by adapting the powers or bit loads of the different subcarriers to the channel. In the case of bit loading, the constellation sizes of the symbols transmitted on the different subcarriers could be adjusted according to the corresponding SNRs using a water-filling method [16]. This is the typical scenario in discrete multitone (DMT), as the wired version of OFDM is called [17].

In the presence of CFO, the orthogonality between the subcarriers is destroyed since $\mathbf{F}\mathbf{D}(\nu)\mathbf{F}^H = \mathbf{I}_M$ if $\nu = 0$. Thus CFO causes intercarrier interference (ICI). Next, we review the effects of CFO on system performance.

8.3 Effect of CFO on System Performance

The CFO can be several times as large as the subchannel spacing. It is usually divided into an integer part and a fractional part. The integer part causes a circular shift of the transmitted symbols, but does not cause ICI; i.e., the orthogonality of the subcarriers is maintained. The fractional part, however, causes ICI. In a seminal paper, Pollet et al. [18] showed that the SNR degradation in decibel due to CFO is proportional to SNR and to $(M\nu)^2$; i.e., the SNR loss is quadratic in the number of carriers. Further, in the single-carrier case, the loss term proportional to SNR vanishes. For 16-QAM (quadrature amplitutde modulation), at a nominal per-carrier SNR of 14 dB (corresponding to an SER of 10^{-5}), and $M = 64$ subcarriers, a CFO offset of $\nu = 0.002$ causes a 5-dB degradation in SNR and increases the SER to 10^{-2}. More precise techniques for computing the SNR loss may be found in [19]. Earlier works on quantifying the impact of CFO include [20], [21]. In this chapter, we invoke the narrowband assumption; a discussion of the wideband case may be found in [22]. For a tutorial treatment of synchronization, refer to [23], [24], [25], [26].

From Equation 8.4, it is clear that performing the FFT before correcting the CFO causes ICI or ISI (the CP takes care of the IBI). The impulse response of this ISI can be readily expressed in terms of the CFO. However, treating this as an equalization problem negates one of the main advantages of OFDM: low-complexity equalization. Further, it increases the number of parameters to be estimated.

[3] The local SNR at subcarrier k is defined as $(M/K_i)(|H_k|^2/\sigma^2)$.

8.4 Carrier Frequency Offset Estimation

CFO estimation techniques may be classified as time-domain (pre-FFT) or frequency-domain (post-FFT) techniques. The latter are usually used to estimate the integer part of the CFO after the fractional part has been identified and corrected. Time-domain methods are used to estimate the fractional part of the CFO, although some of these techniques can also estimate the integer part of the CFO. Here, we focus on time-domain methods.

Time-domain methods can be classified into those that exploit the time diversity provided by the cyclic prefix (see [27], [28] and references therein) or oversampling [29], those that ignore the cyclic prefix and rely on pilots or null subcarriers in the OFDM block [30]–[43], and semiblind approaches that exploit the constant-modulus (CM) [44], [45], the non-Gaussianity, or the finite alphabet (FA) property [46] of the symbols. Maximum likelihood (ML) approaches are reported in [49], [50], [51], [52], although some assume that the channel is flat fading. Pilot design issues are considered in [53], [54]. Comparisons of CFO estimation algorithms may be found in [55], [56]. Techniques based on single-carrier clock recovery algorithms are described in [57].

Estimation of carrier frequency offset, in the presence of carrier modulation and channel fading, can be cast as a harmonic retrieval (HR) problem in multiplicative and additive noise. This is best appreciated from the time series version of Equation 8.4:

$$
\begin{aligned}
x(i,m) &= \frac{\beta_i}{\sqrt{M}} e^{j2\pi\nu(iP+L_{cp})/M} e^{j2\pi\nu m/M} \sum_{k\in\mathcal{K}} s_k H_k e^{j2\pi mk/M} + n(i,m) \\
&= g(i,m) e^{j2\pi\nu m/M} + n(i,m) \\
&= \delta_i \sum_{k\in\mathcal{K}} \alpha_k e^{j2\pi m f_k/M} + n(i,m)
\end{aligned}
$$

The second equation leads to approaches that do not exploit the structure of the OFDM signal (placement of NSCs, pilots if any). The third equation converts the problem into a standard HR problem in AWGN, with constraint $f_k = \nu + k$. With this interpretation, eigenmethods arise naturally; in particular, MUSIC- and ESPRIT-based approaches were first developed for CFO estimation in [32, 34, 37], with identifiability conditions being established in [42].

In the literature on synchronization and channel estimation, it is typical to classify algorithms as pilot-based (or data-aided), blind, and semiblind methods. In pilot-based methods, a known pilot symbol is transmitted and channel parameters are estimated given the channel model and the known input; in blind techniques, the input is not known, but some statistical properties, such as independent and identically distributed (iid) inputs, may be known. The semiblind algorithms refer both to those that use both pilots and statistical properties of the unknown data and to those that exploit additional features (such as finite alphabet) of the unknown symbol stream.

Some CFO estimation methods are classified as being data aided, although they do not use the known pilot block. This is the case for some estimation methods that are based on structuring the OFDM symbol as a repetition of $J \geq 2$ identical slots [15], [33], [35]. These methods are, in fact, NSC-based techniques, since a repetition of $J \geq 2$ identical slots can be generated by nulling all subcarriers whose frequencies are not multiples of $J\,\Delta f$. In [30], two identical OFDM symbols were used to estimate the CFO. Even though this method is not NSC based, it can be mathematically described as such by considering the two symbols as a double-size OFDM symbol (i.e., $2M$ subcarriers) generated by setting the subcarriers with odd frequencies to zero. These repetitive slot-based techniques require the number of NSCs to be larger than or equal to half the total number of subcarriers. Since the number of training zeros is large, these methods have been called data aided, although no data are transmitted on those carriers. When the only NSCs are those imposed by system design requirements, i.e., the VSCs, the method is referred to as blind [32]. If in addition to these NSCs, extra NSCs are inserted, the technique is classified as semiblind since the receiver knows the locations of the NSCs [47]. Here, we avoid this confusion and refer to the approach simply as the NSC-based approach. We give a general framework for this approach and show that many of the existing time-domain methods are special cases of it. A nice property of the NSC-based methods is that CFO and channel estimations can be decoupled, which implies that no model is required for the channel during CFO estimation. Joint CFO and channel estimation techniques have recently been proposed to improve estimation accuracy at the expense of increased computational complexity (see, e.g., [45], [48] and references therein). The above-mentioned CM- and FA-based methods fall in this category.

8.5 Repetitive Slots-Based CFO Estimation

In OFDM systems, pilot symbols are usually transmitted prior to the information frame. For example, in IEEE 802.11a, the preamble is a series of identical slots. Using this preamble structure, a nonlinear least squares (NLS) CFO estimator is proposed in [40]. Correlation-based estimation methods were proposed in [30], [33], [35]. In this section, we review this approach and show the links between the NLS and correlation-based techniques.

In order to compare the repetitive slots-based approach with the NSC-based approach, we assume that the preamble is a single OFDM block made of J identical subblocks of length $Q = M/J$ each; we assume that Q is an integer. A cyclic prefix is also used with this preamble. The case where the preamble is made up of a sequence of identical OFDM blocks can be treated similarly. For example, two identical OFDM symbols can be thought of as two half symbols of a $2M$-point OFDM block. In this case, a guard interval is not needed between the identical blocks.

In this section, we do not specify how the repetitive structure of the OFDM block is generated. Let this time-domain (post-IFFT) preamble be denoted by the $(M \times 1)$ vector $\boldsymbol{u}$. Using the results in Section 8.2.2 and dropping the $'i'$ index, the received signal can be written as

$$\begin{aligned} x(i,m) &= \frac{\beta_i}{\sqrt{M}} e^{j2\pi\nu(iP+L_{cp})/M} e^{j2\pi\nu m/M} \sum_{k\in\mathcal{K}} s(i,k) H_k e^{j2\pi mk/M} + n(i,m) \\ &= g(i,m) e^{j2\pi\nu m/M} + n(i,m) \\ &= \delta_i \sum_{k\in\mathcal{K}} \alpha(i,k) e^{j2\pi m f_k/M} + n(i,m) \end{aligned} \tag{8.6}$$

where $z(k) = e^{j2\pi\nu(ML_{cp}+k)/M} H_c(k,:)\boldsymbol{u}$; the last equality follows from the slot structure $u(k+\ell Q) = u(k)$, where $k = 0,\ldots,Q-1$ and $\ell = 0,\ldots,J-1$. [4] Although $z(k)$ depends upon ν, the ν-dependent factor can be absorbed into $\boldsymbol{u}$ and hence ignored. Estimating ν is now in a standard form: harmonic retrieval in additive and multiplicative noise. The multiplicative noise, the vector $\boldsymbol{z} = [z(0),\ldots,z(Q-1)]^T$, may be modeled as an unknown nonrandom vector. Note that the acquisition range increases with J and is given by $(-J/2, J/2)$. If $J = 2$, then only the fractional part of the CFO can be identified.

8.5.1 Nonlinear Least Squares Method

The NLS estimators of $\boldsymbol{z}$ and ν are obtained by minimizing the following criterion (see [40]):

$$\sum_{\ell=0}^{J-1}\sum_{k=0}^{Q-1} \left| x(k+\ell Q) - z(k)e^{j2\pi\nu\ell/J} \right|^2 \tag{8.7}$$

This criterion is quadratic in the $z(k)$ values, and the ML estimate of $z(k)$ conditioned on ν is given by $\hat{z}(k) = \sum_\ell x(k+\ell Q) exp(-j2\pi\nu\ell/J)$. Plugging this back into the criterion in Equation 8.7, and simplifying, the NLS of ν is found to be[5]

$$\hat{\nu}_{REP} = \arg\max_{\nu} \sum_{k=0}^{Q-1} \xi_\nu(k) \tag{8.8}$$

[4]The circulant matrix $\mathbf{H}_c$ is also block circulant with $Q \times Q$ blocks; hence, $z(k)$ is not a function of ℓ.

[5]The subscript *REP* indicates that the estimate is based on the repetitive property of the preamble.

where

$$\xi_\nu(k) = \frac{1}{J}\left|\sum_{\ell=0}^{J-1} e^{-j2\pi\ell\nu/J} x(k+\ell Q)\right|^2 \tag{8.9}$$

Thus, the estimator proposed in [40] can also be expressed as

$$\hat{\nu}_{REP} = \arg\max_{\nu} \sum_{m=1}^{J-1} \mathcal{R}\left[r(mQ)e^{-j2\pi m\nu/J}\right] \tag{8.10}$$

where $r(\tau)$ is the correlation

$$r(\tau) = \sum_{k=0}^{M-\tau-1} x^*(k)x(k+\tau)$$

When $J = 2$, the estimator can be given in closed form as

$$\hat{\nu}_{REP} = \frac{1}{\pi}\arg\{r(M/2)\} \tag{8.11}$$

If $J > 2$, no closed-form solution is available for the optimization problem in Equation 8.10. The NLS estimator can be initialized or even replaced by the following simpler estimators.

8.5.2 Computationally Simpler Estimators

The expected value of $r(mQ)$ is given by[6]

$$E\{r(mQ)\} = (J-m)\|z\|^2 e^{j2\pi m\nu/J}, \qquad m = 1,\ldots,J-1$$

where $\|z\|$ is the l_2 norm of $\mathbf{z}$. Therefore, the phase of any correlation coefficient $r(mQ)$ can be used to estimate ν. This implies that the estimator in Equation 8.11 is valid even when $J > 2$. In order to improve value accuracy, the phases of the $r(mQ)$ values, $m = 0,\ldots,J-1$, can be judiciously combined. Next, we present two ways of combining these phases.

8.5.2.1 Approximate NLS Estimator

Let ϕ_m denote the unwrapped phase of the correlation estimate $r(mQ)$ for $m = 1,\ldots,J-1$. The NLS criterion in Equation 8.10 can be rewritten as

$$\sum_{m=1}^{J-1} |r(mQ)|\cos(\phi_m - 2\pi m\nu/J) \tag{8.12}$$

---

[6]With $k = q + \ell Q, 0 \le q < Q, 0 \le \ell < J$ we can rewrite $r(mQ)$ as

$$r(mQ) = \sum_{\ell=0}^{J-1-m}\sum_{q=0}^{Q-1} x^*(q+\ell Q)x(q+\ell Q+mQ)$$
$$= \sum_{\ell=0}^{J-1-m}\sum_{q=0}^{Q-1} [z(q)e^{j2\pi\nu\ell/J} + n(q+\ell Q)]^*[z(q)e^{j2\pi\nu(\ell+m)/J} + n(q+\ell Q+mQ)].$$

Treating the channel as nonrandom, and given $\boldsymbol{u}$, the expected value of $r(mQ)$ follows immediately; the $m = 0$ term is independent of ν.

Setting the derivative of this criterion with respect to ν to zero, we obtain

$$\sum_{m=1}^{J-1} m|r(mQ)|\sin(\phi_m - 2\pi m\nu/J) = 0 \tag{8.13}$$

The phase of $E\{r(mQ)\}$ is $2\pi m\nu/J$; hence, under the small error approximation, i.e., $\sin(\phi_m - j2\pi m\nu/J) \approx (\phi_m - j2\pi m\nu/J)$, the approximate NLS estimator is obtained as

$$\tilde{\nu}_{REP} = \frac{J}{2\pi}\frac{\sum_{m=1}^{J-1} m|r(mQ)|\phi_m}{\sum_{m=1}^{J-1} m^2|r(mQ)|} \tag{8.14}$$

This technique requires phase unwrapping. This task is not too demanding given the fact that the lags of the few correlations to be computed are quite far apart from each other. In our simulations, this phase unwrapping has always been carried out successfully.

8.5.2.2 BLUE Estimator

A technique of combining the individual phases without phase unwrapping was developed in [39]. It is based on the best linear unbiased estimator (BLUE) concept. Let

$$\varphi(m) = [\arg\{r(mQ)\} - \arg\{r((m-1)Q)\}]_{2\pi}$$

The estimator was developed by first calculating the statistics of these phase estimates. In order for the weighting coefficients to be channel independent, the estimation errors were averaged over the channel, which was assumed to be Rayleigh distributed. The BLUE estimator of ν was then expressed as

$$\check{\nu}_{REP} = \frac{J}{2\pi}\sum_{m=1}^{p} w(m)\varphi(m) \tag{8.15}$$

where p is a design parameter and the weighting coefficients are given by

$$w(m) = 3\frac{(J-m)(J-m+1) - p(J-p)}{p(4p^2 - 6pJ + 3J^2 - 1)}$$

It was shown in [39] that the variance of the above estimator is minimum when $p = J/2$.

8.6 Null-Subcarrier-Based CFO Estimation

Here, we give a general framework for the NSC-based approach. The number of NSCs and their placement are arbitrary. We adopt a deterministic maximum likelihood approach for CFO estimation. In the next section, we derive necessary and sufficient conditions on the number of NSCs and their placement to ensure identifiability of the CFO. In Section 8.8, for a given number of NSCs, we derive the best placement in terms of the performance of the estimator. We follow the development in [41].

Let $\mathbf{D}(H_{\mathcal{K}_i}) = \operatorname{diag}(H_n, n \in \mathcal{K}_i)$, and denote the ith block of CFO rotated and faded symbols by

$$\boldsymbol{\alpha}(i) = e^{j2\pi\nu(iP+L_{cp})/M}\mathbf{D}(H_{\mathcal{K}_i})\boldsymbol{s}_{\mathcal{K}}(i)$$

Recall from Section 8.2.1 that $\boldsymbol{s}(i) := \mathbf{V}_i\boldsymbol{s}_{\mathcal{K}}(i)$, where $\mathbf{V}_i$ is the $K_i \times M$ subcarrier selection matrix for the ith OFDM symbol. Since $\mathbf{D}(H)$ is diagonal and $\mathbf{V}_i$ is a (tall) permutation matrix, it follows that $\mathbf{D}(H)\mathbf{V}_i = \mathbf{V}\mathbf{D}(H_{\mathcal{K}_i})$. Then, the signal model in Equation 8.4 can be rewritten as

$$\boldsymbol{x}(i) = e^{j2\pi\nu(iP+L_{cp})/M}\beta_i\mathbf{D}(\nu)\mathbf{F}^H\mathbf{V}_i\boldsymbol{\alpha}(i) + \boldsymbol{n}(i) \tag{8.16}$$

8.6.1 Deterministic Maximum Likelihood Estimation

Frequency synchronization is often required prior to channel estimation. Thus, at this stage, the channel coefficients, the H_k values, may be regarded as unknown nonrandom parameters. Therefore, the $\boldsymbol{\alpha}(i)$ values will be modeled as unknown nonrandom vectors; further, the exponential term in Equation 8.16 can be absorbed into the $\boldsymbol{\alpha}(i)$ values.

Our objective is to estimate the CFO using N symbols, $\mathbf{x}(i), i = 1, \ldots, N$. If a reference symbol is used for frequency synchronization, $N = 1$ and a large number of NSCs are usually deployed (see previous section). For blind and semiblind methods, N is usually larger and the number of NSCs significantly lower. For example, in the blind CFO estimation proposed in [32], the only NSCs are the VSCs, which are imposed by system design. The semiblind approach is based on the insertion of extra NSCs into the OFDM symbols [41]. Clearly, the number of symbols N that can be used in an estimation scheme depends upon whether the channel can be expected to remain static over N symbols, as well as the delay/complexity trade-offs.

Since the additive noise is assumed Gaussian and white, the maximum likelihood estimates of ν and $\boldsymbol{\alpha} = \{\boldsymbol{\alpha}(1), \ldots, \boldsymbol{\alpha}(N)\}$ are obtained by minimizing the cost function

$$J(\nu, \boldsymbol{\alpha}) = \sum_{i=1}^{N} \|\mathbf{x}(i) - \beta_i \mathbf{D}(\nu)\mathbf{F}^H \mathbf{V}_i \boldsymbol{\alpha}(i))\|^2 \tag{8.17}$$

where $\|.\|$ is the ℓ_2 norm. This criterion is quadratic in the $\boldsymbol{\alpha}(i)$ values. $\mathbf{V}_i$ is a tall permutation matrix with pseudoinverse $\mathbf{V}_i^T$. Now, if ν is known, the ML estimate of $\boldsymbol{\alpha}(i)$ is given by

$$\hat{\boldsymbol{\alpha}}(i) = \left(1/\beta_i^2\right)\mathbf{V}_i^T \mathbf{F}\mathbf{D}(-\nu)\mathbf{x}(i) \tag{8.18}$$

Substituting these estimates into the cost function, the ML estimate of ν is obtained as

$$\begin{aligned}\hat{\nu}_{NSC} &= \arg\max_{\nu} \sum_{i=1}^{N} \mathbf{x}^H(i)\mathbf{D}(\nu)\mathbf{F}^H \mathbf{V}_i \mathbf{V}_i^T \mathbf{F}\mathbf{D}^H(\nu)\mathbf{x}(i) \\ &= \arg\max_{\nu} \sum_{i=1}^{N} \left\|\mathbf{V}_i^T \mathbf{F}\mathbf{D}^H(\nu)\mathbf{x}(i)\right\|^2\end{aligned} \tag{8.19}$$

It is instructive to rewrite this estimator as

$$\hat{\nu}_{NSC} = \arg\max_{\nu} \sum_{i=1}^{N} \sum_{k \in \mathcal{K}_i} |X(i, \nu + k)|^2 = \arg\min_{\nu} \sum_{i=1}^{N} \sum_{k \in \mathcal{Z}_i} |X(i, \nu + k)|^2 \tag{8.20}$$

where $X(i, f)$ denotes the discrete-time Fourier transform (DTFT)

$$X(i, f) = \sum_{\ell=0}^{M-1} x(i, \ell) \exp(-j2\pi f \ell / M)$$

We can therefore interpret the above estimator as follows: in the absence of CFO, the subcarriers are orthogonal and the energy of the received signal at the NSCs should be zero. We estimate the CFO as the frequency shift that minimizes the energy at the NSCs or maximizes the energy at the active carriers (see also [32], [41], [42]). We note that the estimators developed in [32], [42] are thus ML.

The estimator can also be written in terms of the time-domain correlation function as in [41]:

$$\hat{\nu}_{NSC} = \arg\max_{\nu} \sum_{\tau=1}^{M-1} \mathcal{R}\left[\sum_{i=1}^{N} \left[r_i(\tau)\psi_{\mathcal{K}_i}^*(\tau)\right] e^{-j2\pi\tau\nu/M}\right] \tag{8.21}$$

where

$$\psi_{\mathcal{K}_i}(\tau) = \frac{1}{M} \sum_{k \in \mathcal{K}_i} e^{j2\pi k\tau/M}$$

and

$$r_i(\tau) = \sum_{\ell=0}^{M-1-\tau} x^*(i,\ell)x(i,\ell+\tau)$$

$x(i,\ell)$ was defined earlier as the ℓth entry of $\boldsymbol{x}(i)$.

8.6.2 Special Case: Repetition of Identical Slots

Here, we assume $N = 1$ and we drop the time index i from all vectors and matrices defined above. As mentioned previously, an OFDM symbol structured as a repetition of J identical slots can be generated by nulling the subcarriers whose normalized frequencies are not multiples of J. The active subcarriers are equi-spaced. Next, we consider how the presence of VSCs impacts the estimator.

8.6.2.1 Virtual Subcarriers Absent

The elements of $\mathcal{K}$ are now $n_m = mJ$ $m = 0, \ldots, Q-1$, where $Q = M/J$ is assumed an integer and $J \geq 2$. Now $\psi_{\mathcal{K}}(\tau)$ is nonzero only if τ is a multiple of Q, i.e.,

$$\psi_{\mathcal{K}}(\tau) = \frac{K}{M}\delta(\tau - mQ), \qquad m = 0, \pm 1, \pm 2, \ldots \tag{8.22}$$

The estimator in Equation 8.21 thus reduces to the repetition slot-based NLS estimator in Equation 8.10. Therefore, we have shown that the latter estimator is also the NSC-based ML estimator when the odd subcarriers are deactivated and no virtual subcarriers are present, and the channel (which may be frequency selective) is unknown.

8.6.2.2 Virtual Subcarriers Present

In the more realistic case where some of the subcarriers at the edge of the spectrum are nulled to avoid interference between adjacent OFDM systems, the ML estimator is different from the estimator in Equation 8.10. Let the number of active subcarriers be $K = (2I + 1)$. The useful subcarriers are then $\{0, 1, \ldots, I, M-I, \ldots, M-2, M-1\}$ with $I < M/2$.[7] The subcarriers of this set, whose frequencies are not multiples of J are also nulled in order for the OFDM symbol to have a repetitive structure. The function $\psi_{\mathcal{K}}(\tau)$ is still real-valued, but different from that in Equation 8.22. A sketch of this function when $M = 64$, $J = 4$, and $I = 24$ is displayed in Figure 8.2. Here, most of the correlation coefficients contribute to the ML estimator. The estimator in Equation 8.10 is still consistent[8] but is no longer ML. The estimator in Equation 8.10 now consists of using only the $(J-1)$ highest correlation coefficients and could therefore be seen as an approximate ML estimator. In the simulation section, we investigate the difference in performance of these two estimators, as well as that of two others. We will see that the performances are very similar for practical values of J and for moderate to high SNRs.

[7] Because of the modulo-M notation, the $M - K$ band-edge VSCs appear in the middle in the usual DFT style.

[8] We mean consistency in the mean square sense: an estimate is consistent if it is unbiased and its variance goes to zero as the number of samples increases.

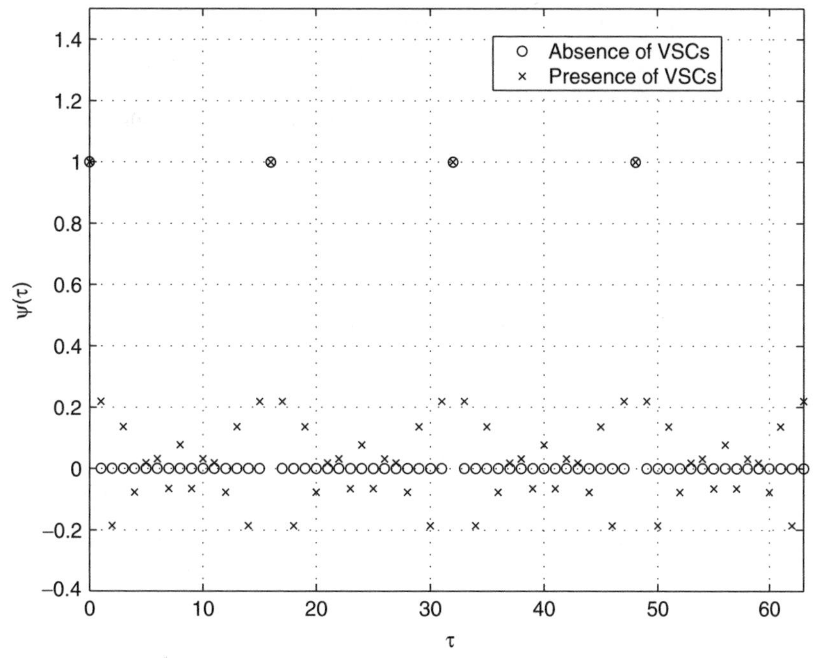

FIGURE 8.2 Plot of $\psi_{\mathcal{K}}(\tau)$ when $M = 64$, $J = 4$, and $I = 24$.

8.7 Identifiability

Here we address the problem of loss of identifiability due to the placement of the NSCs and the channel zeros when the channel is unknown. We consider the noise-free case. We will distinguish between two types of nonidentifiability: (1) due to the location of the channel zeros (LOCZ) and (2) due to ambiguity, arising solely from the number and placement of the NSCs. The distinction between the two cases will become clear in what follows. The main result is given in Result 2 (see also remark 1).

The identifiability results when $N > 1$ are the same as in the case of $N = 1$ if the number and placement of the NSCs remain the same across the symbols. However, if the placements of the NSCs are shifted from one symbol to another, it was shown in [43] that the LOCZs have no effects on identifiability. Here, we focus on the case of one OFDM symbol, $N = 1$. We therefore drop the time index 'i' from the above variables.

An obvious necessary condition to avoid LOCZ-induced nonidentifiability is

$$\boldsymbol{\alpha}^H \boldsymbol{\alpha} \neq 0 \quad \Leftrightarrow \quad \boldsymbol{\alpha} \neq \mathbf{0} \tag{8.23}$$

which ensures that the energy of the *received* signal is not zero. If the channel can be modeled as an FIR fading channel with $L + 1$ taps,[9] then at most L of the H_n values, $n \in \mathcal{K}$, would be zero, corresponding to the case where all the channel zeros are fortuitously at the subcarrier frequencies. The condition in Equation 8.23 can be satisfied, independently of the channel, if at least $L + 1$ subcarriers are activated, i.e., $K > L$, or equivalently, if the number of NSCs, Z, satisfy $Z < M - L$. This condition guarantees that the true ν maximizes the criterion in Equation 8.19, but does not guarantee that the maximizer is unique.

[9]This implies that the discrete-time CIR $h_k = 0$ for $k > L$. Since the channel is random, the h_k values for $k \leq L$ can take on arbitrary values, including zeros. However, we assume that the norm of the CIR is nonzero; otherwise, the energy of the received signal would be zero regardless of the transmitted signal.

Assume that the desired acquisition range for the CFO is $\mathcal{R}_\nu = (-Q, Q)$, with Q being an integer smaller than $M/2$. The corresponding range for the nonnormalized frequency is $(-Q\Delta f, Q\Delta f)$. The value of Q depends on *a priori* information about the frequency offset experienced by the studied system. In general, Q must be less than 20% of M to ensure that the spectrum of the received signal is not significantly distorted by the receive matched filter.

For any scalar f, let

$$\mathbf{G}_{\mathcal{K}}(f) = \mathbf{F}_{\mathcal{K}}\, \mathbf{D}^H(f)\Psi_{\mathcal{K}}\mathbf{D}(f)\mathbf{F}_{\mathcal{K}}^H$$

where $\mathbf{F}_{\mathcal{K}} := \mathbf{V}^T\mathbf{F}$ is the $K \times M$ matrix obtained by removing the rows of $\mathbf{F}$ corresponding to the NSCs, and $\Psi_{\mathcal{K}} = \mathbf{F}_{\mathcal{K}}^H\mathbf{F}_{\mathcal{K}}$ is a Toeplitz matrix created from the $\psi_{\mathcal{K}}(\tau)$ values. According to Equation 8.19, the objective function (recall that we are considering the noise-free case)

$$J(f) = \boldsymbol{\alpha}^H\, \mathbf{G}_{\mathcal{K}}(f - \nu)\, \boldsymbol{\alpha}$$

should be maximized over $f \in \mathcal{R}_\nu$. The following observations follow immediately:

1. $J(\nu) = \boldsymbol{\alpha}^H\boldsymbol{\alpha}$.
2. For $f - \nu$, not an integer, $J(f) < J(\nu)$, provided that Equation 8.23 is satisfied.
3. The maximizers of $J(f)$, $f \in \mathcal{R}_\nu$, are of the form $f = \nu + m$, where m is an integer.
4. For m, an integer, the (i, j) entry of $\mathbf{G}_{\mathcal{K}}(m)$ is

$$\sum_{\ell=1}^{K} \delta(n_\ell - n_i - m, \bmod M)\, \delta(n_\ell - n_j - m, \bmod M)$$

   where $n_\ell, n_i, n_j \in \mathcal{K}$. Hence, $\mathbf{G}_{\mathcal{K}}(m)$ is a diagonal matrix; further, the diagonal entries are either zero or unity. Let $\mathbf{g}_{\mathcal{K}}(m) = [g_{n_1}(m), \ldots, g_{n_K}(m)]^T$ denote the diagonal of $\mathbf{G}_{\mathcal{K}}(m)$.
5. Hence, we have

$$J(m + \nu) = \sum_{n_\ell \in \mathcal{K}} |\alpha_{n_\ell} g_{n_\ell}(m)|^2$$

Notice that $g_{n_\ell}(m) = 1$ only if the subcarriers n_ℓ and $n_\ell + m$ are both active. Thus, with $m = 0$, $g_{n_\ell}(0) = 1$, $\forall n_\ell \in \mathcal{K}$. Identifiability is clearly lost if every active subcarrier has an active neighbor m to its right (cyclically). If the channel has a zero at one of the active subcarriers, say n_i, then the ith entry of $\boldsymbol{\alpha}$, $\alpha_{n_i} = 0$. If there exists an m such that $g_{n_i}(m) = 0$ and $g_{n_j}(m) = 1, \forall j \neq i$, then $J(\nu) = J(\nu + m)$ and identifiability is lost. Identifiability can be restored either by restricting the acquisition range $\mathcal{R}_\nu$ or by selecting $\mathcal{K}$ appropriately, as we show next.

Let $P_{\mathcal{K}}(m)$ denote the number of zero entries in the vector $\mathbf{g}_{\mathcal{K}}(m)$; we note that $P_{\mathcal{K}}(-m) = P_{\mathcal{K}}(m)$. If the underlying channel is FIR$(L + 1)$, it can null out at most L of the activated subcarriers. Hence, if the set $\mathcal{K}$ is chosen such that $P_{\mathcal{K}}(m) > L$ for $m = 1, \ldots, \lfloor Q \rfloor$, then identifiability is assured in $(-Q, Q)$ regardless of channel zeros. We summarize these results in the following:

Result 1 Suppose that the channel has nulls at H_n, $n \in \mathcal{K}$. Then the CFO is not uniquely identifiable in $(-Q, Q)$ if for some integer $m \neq 0$, such that $m + \nu \in (-Q, Q)$, $g_n(m) = 0$, $\forall n \in \mathcal{K}$, and $g_n(m) \neq 0$, $n \notin \mathcal{K}$.

Result 2 The CFO is uniquely identifiable in $(-Q, Q)$ for any FIR channel of order L if $K > L$ and

$$P_{\mathcal{K}}(m) > L, \qquad m = 1, \ldots, \lfloor Q \rfloor$$

The following remarks and special cases are in order (we drop the subscript $\mathcal{K}$ on P for convenience).

1. $P(m)$ is the number of active subcarriers that do not have a neighbor at distance m and is a function of the number of NSCs and their placement. Thus, Result 2 says that at least $L + 1$ active

subcarriers must have nearest neighbors no closer than Q, in order to ensure identifiability within $(-Q, Q)$.

2. Let us rewrite the normalized CFO as the sum of a fractional part $\tilde{\nu}$ $(-0.5 \leq \tilde{\nu} < 0.5)$ and an integer, m, i.e., $\nu = \tilde{\nu} + m$. From Result 2 (with $Q = 0.5$ and $\lfloor Q \rfloor = 0$), it suffices to have $K > L$ in order for $\tilde{\nu}$ to be uniquely identifiable regardless of the channel zeros.
3. If $Q \geq 1$, we can show that $P(m) \leq \min(K, Z)$. Thus, another necessary (but not sufficient) condition is $\min(K, Z) > L$.
4. For consecutive NSCs as in [32], a case where the only NSCs are the VSCs imposed by system design, we have that $P(m) = \min(M - K, K, m)$. Hence, with $m = 1$, we need $1 > L$ to ensure identifiability; in other words, the scheme of [32] guarantees channel-independent identifiability only for an AWGN channel; see also [42]. Note, however, that this scheme is ambiguity-free in $(-M/2, M/2)$.
5. For equi-spaced NSCs $(Z = M - K \leq M/2)$, $P(m) = Z$ if $m \neq iM/Z$ (multiple of M/Z) and $P(iM/Z) = 0$ (i.e., an ambiguity), $i = 1, \ldots$ Therefore, the CFO is uniquely identifiable in $(-M/2Z, M/2Z)$, provided $L < Z < M - L$.
6. For equi-spaced active subcarriers $(K < M/2)$, $P(m) = K$ if $m \neq iM/K$ and $P(iM/K) = 0$, $i = 1, \ldots$ The CFO can be uniquely identified in $(-M/2K, M/2K)$, provided $K > L$ (or $Z < M - L$).
7. For distinctly spaced NSCs [42], we have that $P(m) \geq (Z - 1)$. Hence, identifiability is ensured over $(-M/2, M/2)$ if $L + 1 < Z < M - L$. The scheme of [42] uses the smallest allowed number of NSCs, $Z = L + 2$.
8. The above remarks show that the scenario that is most robust to LOCZ is the equi-spaced NSCs (or active subcarriers). Indeed, for a fixed $Z \leq M/2$, a longer delay spread can be tolerated than in the other scenarios. However, it is ambiguity-free only if the acquisition range is $(-Q, Q)$ with $Q \leq M/(2Z)$. The consecutive NSC scenario is the most vulnerable to LOCZ, but it is ambiguity-free. The scenario of NSCs with distinct spacing is ambiguity-free, and it is slightly more vulnerable to LOCZ than the equi-spaced NSC case. Note also that the maximum Z for a given M in the case of NSCs with distinct spacing [42] is given by $\sqrt{2M}$. This may be restrictive if a large number of NSCs is needed to achieve a certain performance. In the next section, we show that the equi-spaced NSC scenario leads to the best performance in terms of accuracy of the CFO estimates.
9. Results 1 and 2 should be useful for designing a reference symbol for frequency synchronization.

8.8 Performance Analysis

In this section, we investigate the effects of the number of activated subcarriers and their placements on the performance of frequency offset estimation. Again we focus on the case of one OFDM symbol, $N = 1$. We find the placement that, for a fixed K minimizes the Cramér–Rao bound (CRB). This criterion is independent of the estimation algorithm; further, the ML estimator, derived earlier, achieves the CRB for large M. In this section we assume that the identifiability conditions discussed in the previous section are satisfied.

8.8.1 The Conditional CRB

The ML estimator in the Section 8.7 was derived by assuming that the faded symbol vector $\boldsymbol{\alpha}$ was unknown and nonrandom. The CRB in this case will be referred to as the conditional CRB (CCRB), since the $\boldsymbol{\alpha}$ is treated as a deterministic vector. The parameter vector describing the signal model is $\boldsymbol{\theta} = [\nu, \boldsymbol{\alpha}, \sigma^2]^T$. The CCRB is the inverse of the conditional Fisher information matrix (CFIM). The final expression for the CCRB of the CFO (see the Appendix for the detailed derivations) is

$$CCRB_{\mathcal{K}}(\nu) = \frac{\sigma^2}{8\pi^2\beta^2 M}\left[\boldsymbol{\alpha}^H \mathbf{F}_{\mathcal{K}} \mathbf{W} (\mathbf{I} - \Psi_{\mathcal{K}}) \mathbf{W} \mathbf{F}_{\mathcal{K}}^H \boldsymbol{\alpha}\right]^{-1} \tag{8.24}$$

where $\mathbf{W} = M^{-3/2}\text{diag}(0, \ldots, M-1)$ and $\mathbf{I}$ is the $M \times M$ identity matrix. We recall that $\mathbf{F}_{\mathcal{K}} := \mathbf{V}^T\mathbf{F}$ and $\Psi_{\mathcal{K}} = \mathbf{F}_{\mathcal{K}}^H\mathbf{F}_{\mathcal{K}}$. If $K = M$, $\Psi_{\mathcal{K}} = \mathbf{I}_M$ and $CCRB_{\mathcal{K}}(\nu) = \infty$, i.e., the CFO is nonidentifiable if all the subcarriers are active. The CCRB is useful to predict the performance of the CFO estimation for a particular channel and a particular data sequence. To predict the performance under the assumption of a random channel, a channel-independent bound must be derived.

8.8.2 The Unconditional CRB: Rayleigh Channel

Since our objective is to evaluate and compare bounds on the performance of CFO estimators for different placements of the NSCs, we need to derive a CRB that is channel-independent. We will use the unconditional CRB (sometimes called the stochastic CRB), which considers the vector $\boldsymbol{\alpha}$ as a random vector. The approach of averaging the CFIM (i.e., treating the H_n values as nuisance parameters) is not useful since this results in a quantity that is independent of the NSC location.

Here, we consider the Rayleigh channel; i.e., we assume that $\{H_n\}$ is a stationary sequence of zero-mean complex Gaussian random variables. Let $\mathbf{R}_H$ and $\mathbf{R}_\alpha$ denote the covariance matrices of $\{H_n, n \in \mathcal{K}\}$ and $\{\alpha_n, n \in \mathcal{K}\}$, respectively. We have that $\mathbf{R}_\alpha = \mathbf{S}\mathbf{R}_H\mathbf{S}^H$ where $\mathbf{S} = \text{diag}(s_n, n \in \mathcal{K})$. The covariance matrix of $\boldsymbol{x}$ in Equation 8.16 is then given by

$$\mathbf{R}_x = \beta^2\mathbf{D}(\nu)\mathbf{F}_{\mathcal{K}}^H\mathbf{R}_\alpha\mathbf{F}_{\mathcal{K}}\mathbf{D}^H(\nu) + \sigma^2\mathbf{I} \tag{8.25}$$

The unconditional CRB (UCRB) is (see [58])

$$UCRB_{\mathcal{K}}(\nu) = \frac{1/(8\pi^2M)}{\text{Tr}\{\mathbf{R}^{-1}\mathbf{W}\mathbf{R}\mathbf{W} - \mathbf{W}^2\}} \tag{8.26}$$

where $\mathbf{R} = \beta^2\gamma\mathbf{F}_{\mathcal{K}}^H\bar{\mathbf{R}}_\alpha\mathbf{F}_{\mathcal{K}} + \mathbf{I}$, $\gamma = E\{|H_n|^2\}/\sigma^2$ is the average SNR, and $\bar{\mathbf{R}}_\alpha = \mathbf{R}_\alpha/E\{|H_n|^2\}$ is the normalized covariance.

In the blind scenario, $\mathbf{S}$ is unknown, and it is reasonable to assume that $\bar{\mathbf{R}}_\alpha = \mathbf{I}$ (the random symbols will decorrelate the α_n values even if the H_n values are correlated). Then, $\mathbf{R} = \beta^2\gamma\Psi_{\mathcal{K}} + \mathbf{I}$ and the UCRB in Equation 8.26 simplifies accordingly.

In the data-aided case, $\mathbf{R} = \beta^2\gamma\Psi_{\mathcal{K}} + \mathbf{I}$ if the H_n values are mutually uncorrelated and the s_n values have constant amplitude. Note that the correlation between the H_n values decreases with the delay spread of the channel. In the sequel, we consider the case where $\mathbf{R}_\alpha = \mathbf{I}$.

8.8.3 Optimal Choice of Null-Subcarriers

The UCRB derived in Equation 8.26, with $\mathbf{R} = \beta^2\gamma\Psi_{\mathcal{K}} + \mathbf{I}$, is channel independent but is a function of the subset of the activated subcarriers $\mathcal{K}$, i.e., the number of activated subcarriers *and* their placement. The minimization of the UCRB with respect to $\mathcal{K}$ for a fixed K will provide the optimal choice of $\mathcal{K}$.

The matrix inversion in Equation 8.26 can be avoided since $\Psi^2 = \Psi$, so that

$$(\beta^2\gamma\Psi_{\mathcal{K}} + \mathbf{I})^{-1} = \mathbf{I} - \frac{\beta^2\gamma}{1+\beta^2\gamma}\Psi_{\mathcal{K}} \tag{8.27}$$

The UCRB can be written as (since $\psi_{\mathcal{K}}(k,k) \equiv 1$)

$$UCRB_{\mathcal{K}}(\nu) = \frac{1/(8\pi^2M\eta)}{\beta^2\text{Tr}\{\{\}\mathbf{W}^2\} - \beta^4\text{Tr}\{\Psi_{\mathcal{K}}\mathbf{W}\Psi_{\mathcal{K}}\mathbf{W}\}} \tag{8.28}$$

where $\eta = \gamma^2/(1+\beta^2\gamma)$ is independent of the channel set.

Result 3 For an uncorrelated Rayleigh fading channel, the optimal (in the sense of minimum UCRB) placement of a fixed number of active subcarriers, K, is given by

$$\mathcal{K}^* = \arg\min_{\mathcal{K}} \sum_{k,\ell=0}^{M-1} k\ell|\psi_{\mathcal{K}}(k,\ell)|^2 \tag{8.29}$$

If $K = M/J$ where M is the total number of carriers, and K and $J > 1$ are positive integers, the performance is best when the activated subcarriers are equi-spaced (by J). In the complementary scenario where the number of null subcarriers $Z = M/J$, the performance is best when the null subcarriers are equi-spaced.

Since the optimality result above is channel independent, the optimal placements of the active (or the null) subcarriers can be derived off-line.

From Result 3, we deduce that if $K \leq M/2$ (respectively $K > M/2$), the UCRB is minimized when the activated (respectively null) subcarriers are as equi-spaced as possible. Further, the UCRB is maximized when the activated (or null) subcarriers are adjacent.

As mentioned earlier, some subcarriers at the edges of the signal spectrum are deactivated. The placement of these consecutive NSCs is fixed and their number is dictated by system design requirements. When extra NSCs are inserted to improve CFO estimation performance, their optimal placement can be derived using Result 3. The minimization of the UCRB in this case can be achieved by an exhaustive search over all possible placements of the NSCs.

8.9 Simulation Results

Here, we compare the performances of the various techniques developed in this paper. We assume a preamble is available for CFO estimation. This preamble consists of one OFDM symbol structured as a repetition of $J \geq 2$ identical slots.

We consider an OFDM system with a total of 64 subcarriers. There are 11 virtual subcarriers at the edges of the spectrum. The useful part of an OFDM symbol that contains 64 samples is preceded by a cyclic prefix of length 16. Quadrature phase shift keying (QPSK) modulation is used. The channel has 15 paths, with path delays $0, 1, 2, \ldots, 14$ samples. The magnitudes of the channel coefficients, the h_i values, are Rayleigh distributed with exponential power delay profile, with decay parameter 1/5, while their phases are uniformly distributed over $(-\pi, \pi)$. Further, the channel coefficients are independent of each other. We consider the scenario where the channel is static over an OFDM symbol. The comparison is based on the mean square errors (MSEs) on CFO estimation, which are calculated using 2000 Monte Carlo runs.

Figure 8.3 and Figure 8.4 display the MSEs vs. SNRs for different values of J. It is seen that the proposed approximate NLS (ANLS) estimator has nearly the same accuracy as the NLS and ML estimators. This accuracy is close to the UCRB. This suggests that ANLS should be preferred to the latter estimators, as it is computationally simpler. Indeed, no numerical optimization is required for the ANLS method. Furthermore, the ANLS estimator performs better than the BLUE estimator developed in [39] for low SNRs. For high SNRs, all the techniques have identical accuracy.

Figure 8.5 illustrates MSEs vs. J the number of repetitions in the preamble, for two values of the SNR, 0 and 10 dB. It is seen that at high SNRs, all techniques have similar performance. For low SNRs, the performance first improves as J increases and then deteriorates. There seems to be an optimal value for J this value varies with total number of subcarriers. In our simulations, $J = 8$ seems to be the optimal choice when $M = 64$. Performance deteriorates when J is large because of the decrease in frequency diversity when a large number of NSCs are present. For example, if $J = 32$ and $M = 64$, only one subcarrier is modulated. Because the channel is random, the single active carrier could experience a deep fade, which would cause a very low SNR. A simultaneous fade on several subcarriers is less likely than a fade on one subcarrier. However, if all subcarriers are modulated, the CFO becomes unidentifiable. Therefore, there

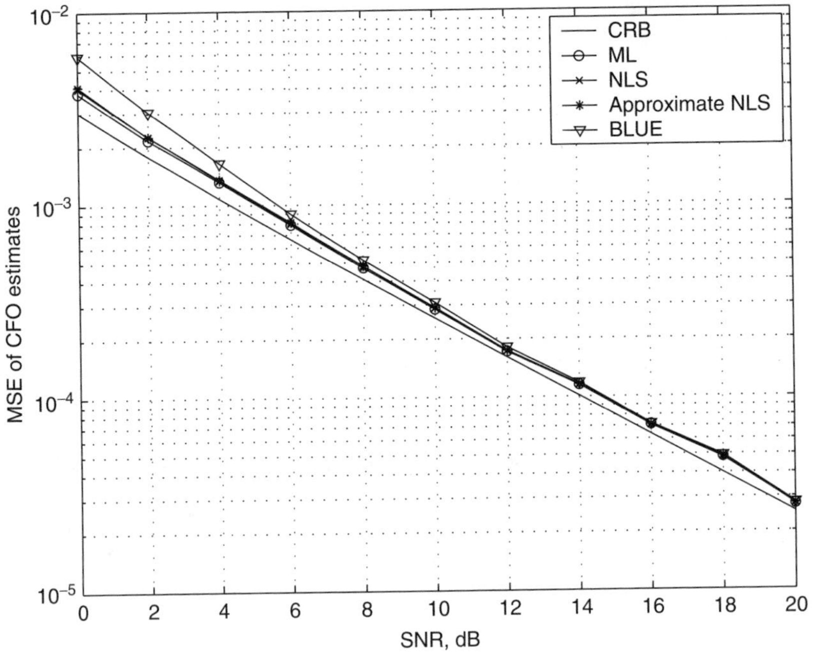

FIGURE 8.3 MSE of CFO estimates; $J = 4$, $M = 64$, $L = 15$.

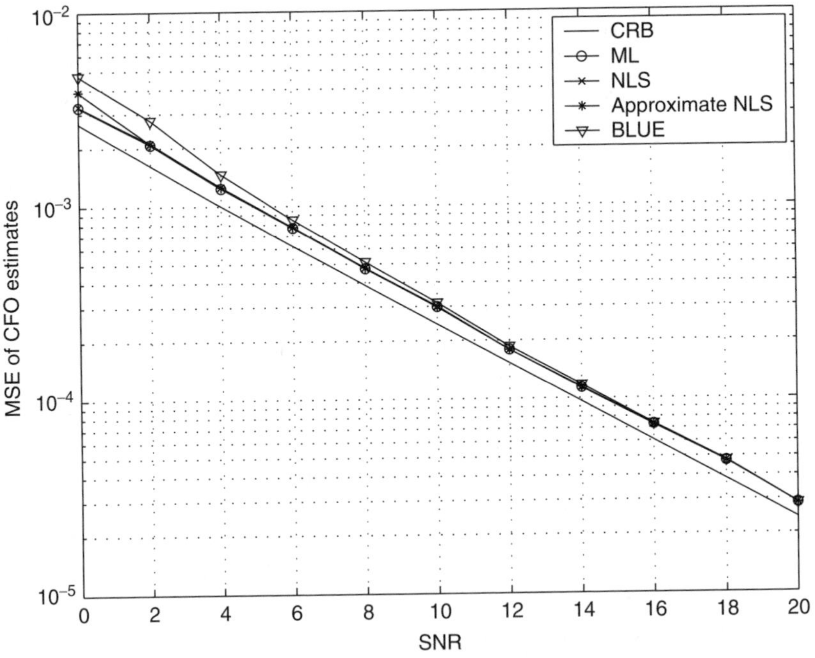

FIGURE 8.4 MSE of CFO estimates; $J = 8$, $M = 64$, $L = 15$.

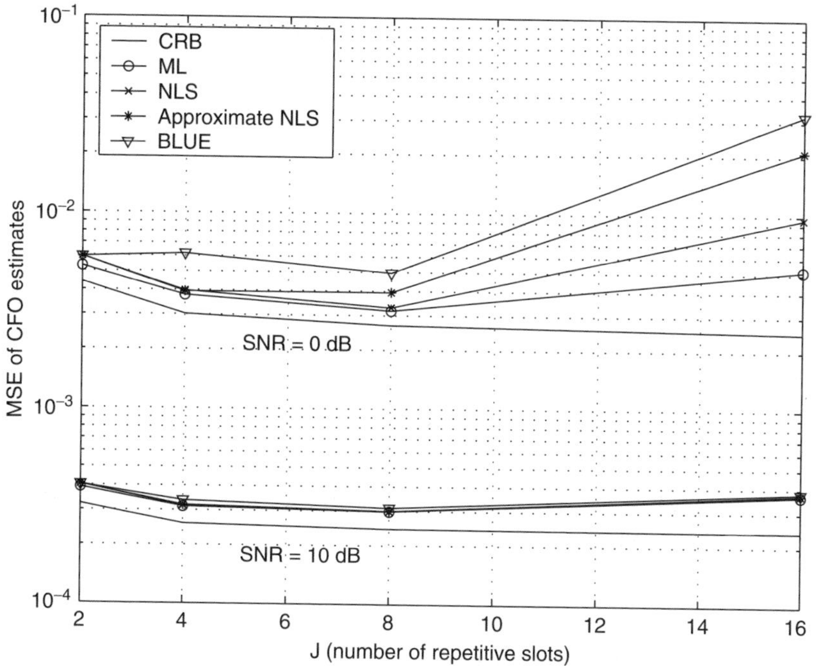

FIGURE 8.5 MSE of CFO estimates vs. the number of repetitive slots, J; $M = 64$, $L = 15$.

must be a trade-off between these two phenomena. This explains why there is an optimal value for J in terms of CFO estimation performance.

8.10 Conclusions

In this chapter we have addressed CFO estimation in OFDM systems. Due to the high sensitivity of multi-carrier systems to CFO, accurate estimation algorithms are required in order to achieve high performance with such systems. Here we have reviewed several existing CFO estimation methods. We have shown that many of these methods share the same underlying approach, which is the exploitation of null subcarriers. Based on the same approach, a new computationally simple estimator, the approximate nonlinear squares estimator, was developed. It was shown that the performance of the ANLS estimator is very close to that of the computationally more demanding NLS estimator. Further, the ANLS estimator outperforms existing computationally simple estimators.

Identifiability conditions in the case of random multipath channels were established. For a given channel order, these conditions (on the number and placement of the null subcarriers) guarantee CFO identifiability in a specified acquisition range regardless of the channel. Moreover, closed-form expressions for both the conditional (with respect to the channel) and the unconditional CRB were derived. Using the latter, results on optimal placement of null subcarriers in terms of estimation performance were presented.

Appendix: Conditional CRB

The parameters describing the noiseless signal are ν and $\boldsymbol{\alpha}$. These parameters are considered deterministic. The conditional CRB (CCRB) is given by the inverse of the conditional Fisher information matrix (CFIM). Since the additive noise is white, Gaussian, and circular, the entries of the CFIM are given by

$$J_{\nu,\nu} = \frac{8\pi^2\beta^2 M}{\sigma^2}\boldsymbol{\alpha}^H \mathbf{F}_{\mathcal{K}} \mathbf{W}^2 \mathbf{F}_{\mathcal{K}}^H \boldsymbol{\alpha}$$

$$J_{\nu,\alpha^R} = \frac{4\pi\beta^2\sqrt{M}}{\sigma^2}\mathcal{I}\left[\boldsymbol{\alpha}^H\mathbf{F}_{\mathcal{K}}\mathbf{W}\mathbf{F}_{\mathcal{K}}^H\right]$$

$$J_{\nu,\alpha^I} = \frac{4\pi\beta^2\sqrt{M}}{\sigma^2}\mathcal{R}\left[\boldsymbol{\alpha}^H\mathbf{F}_{\mathcal{K}}\mathbf{W}\mathbf{F}_{\mathcal{K}}^H\right]$$

$$J_{\alpha^R,\alpha^R} = J_{\alpha^I,\alpha^I} = \frac{2\beta^2}{\sigma^2}\mathbf{I}_K; \; J_{\sigma^2,\sigma^2} = \frac{2M}{\sigma^4}$$

where $\mathbf{W} = M^{-3/2}\text{diag}(0,\ldots,M-1)$ and $\mathbf{I}_K$ is the $K \times K$ identity matrix. The other CFIM entries are identically zero. Inverting this matrix, followed by some manipulations, leads to the expression for the CCRB given in Equation 8.24.

Acknowledgments

The authors thank Sergio Barbarossa and Saswat Misra for their feedback on an early version and for many discussions related to this chapter.

References

[1] R. van Nee et al., "New high-rate wireless LAN standards," *IEEE Commun. Mag.*, December 1999.

[2] S. Weinstein and P. Ebert, "Data transmission by frequency division multiplexing using the discrete Fourier transform," *IEEE Trans. Commun. Technol.*, 19, 628–634, 1971.

[3] L.J. Cimini, Jr., "Analysis and simulation of a digital mobile channel using orthogonal frequency domain multiplexing," *IEEE Trans. Commun.*, 3, 665–675, 1985.

[4] J.A.C. Bingham, "Multicarrier modulation for data transmission: an idea whose time has come," *IEEE Commun. Mag.*, 28, 5–14, 1990.

[5] F. Khaled and G.P. Fettweis, Eds., *Multi-Carrier Spread-Spectrum*, Kluwer Academic Publishers, Dordrechts, Netherlands, 1997.

[6] R. van Nee and R. Prasad, *OFDM for Wireless Multimedia Communications*, Artech House, Norwood, MA, 2000.

[7] Z. Wang and G.B. Giannakis, "Wireless multicarrier communications: where Fourier meets Shannon," *IEEE Signal Process. Mag.*, 17, 29–48, May 2000.

[8] L. Hanzo, M. Muenster, B.-J. Choi and T. Keller, *OFDM and MC-CDMA for Broadband Multi-User Communications*, John Wiley & Sons, New York, 2003.

[9] G.B. Giannakis, "Filterbanks for blind channel identification and equalization," *IEEE Signal Process., Lett.*, 4, 184–187, 1997.

[10] A. Scaglione, G.B. Giannakis, and S. Barbarossa, "Redundant filter-bank precoders and equalizers. Part I. Unification and optimal designs" and "Part II. Blind channel estimation, synchronization and direct equalization," *IEEE Trans. Signal Process.*, 47, 1988–2022, 1999.

[11] B. Muquet, Z. Wang, G.B. Giannakis, M. de Courville, and P. Duhamel, "Cyclic prefixing or zero padding for wireless multicarrier transmissions?" *IEEE Trans. Signal Process.*, 50, 2136–2148, 2002.

[12] A. Scaglione, S. Barbarossa, and G.B. Giannakis, "Filterbank transceivers optimizing information rate in block transmissions over dispersive channels," *IEEE Trans. Inf. Theory*, 45, 1019–1032, 1999.

[13] X.-G. Xia, "New precoding for intersymbol interference cancellation using nonmaximally decimated multirate filterbanks with ideal FIR equalizers," *IEEE Trans. Signal Process.*, 45, 2431–2441, 1997.

[14] A. Scaglione, G.B. Giannakis, and S. Barbarossa, "Lagrange/Vandermonde MUI eliminating user codes for quasi-synchronous CDMA in unknown multipath," *IEEE Trans. Signal Process.*, 48, 2057–2073, 2000.

[15] T.M. Schmidl and D.C. Cox, "Robust frequency and timing synchronization for OFDM," *IEEE Trans. Commun.*, 45, 1613–1621, 1997.

[16] T.M. Cover and J.A. Thomas, *Elements of Information Theory*, Wiley InterScience, New York, 1991.

[17] P.S. Chow, J.M. Cioffi, and J.A.C. Bingham, "A practical discrete multitone transceiver loading algorithm for data transmission over spectrally shaped channels," *IEEE Trans. Commun.*, 43, 773–775, 1995.

[18] T. Pollet, M. van Bladel, and M. Moeneclaey, "BER sensitivity of OFDM systems to CFO and Wiener phase noise," *IEEE Trans. Commun.*, 43, 1995.

[19] K. Sathananthan and C. Tellambura, "Probability of error calculation of OFDM systems with frequency offset," *IEEE Trans. Commun.*, 49, 1884–1888, 2001.

[20] T. Pollet and M. Peeters, "Synchronization with DMT modulation," *IEEE Commun. Mag.*, 37, 80–86, 1999.

[21] L. Wei and S. Schegel, "Synchronization requirement for multi-user OFDM on satellite mobile and two path Rayleigh fading channel," *IEEE Trans. Commun.*, 43, 887–895, 1995.

[22] A.-B. Salberg and A. Swami, "Doppler and frequency-offset synchronization in wideband OFDM: estimators and performance analysis," in *Proceedings of the IEEE NORSIG'02*, Tromso-Trondheim, Norway, October 2002.

[23] U. Mengali and A.N. D'Andrea, *Synchronization Techniques for Digital Receivers*, Plenum Publisher, New York, 1997.

[24] H. Meyr, M. Moeneclaey, and S.A. Fechtel, *Synchronization, Channel Estimation, and Signal Processing*, Vol. 2, *Digital Communication Receivers*, Wiley InterScience, New York, 1997.

[25] G. Vazquez and J. Riba, "Non-data-aided digital synchronization," in *Signal Processing Advances in Wireless and Mobile Communications*, Vol. 2, G.B. Giannakis, Y. Hua, P. Stoica, and L. Tong, Eds., Prentice Hall, New York, 2000, chap. 9, pp. 357–402.

[26] C. Georghiades and E. Serpedin, "Synchronization," in *The Handbook of Communications*, J. Gibson, Ed., CRC Press, Boca Raton, FL, 2002.

[27] N. Lashkarian and S. Kiaei, "Class of cyclic-based estimators for frequency-offset estimation of OFDM systems," *IEEE Trans. Commun.*, 48, 2139–2149, 2000.

[28] X. Ma and G.B. Giannakis, "Exploiting the cyclic-prefix for blind frequency-offset estimation in OFDM," in *Proceedings of 35th Asilomar Conference on Signals, Systems, and Computers*, Pacific Grove, CA, November 4–7, 2001.

[29] B. Park et al., "A blind OFDM synchronization algorithm based on cyclic correlation," *IEEE Signal Proc. Lett.*, 11, 2004.

[30] P.H. Moose, "A technique for orthogonal frequency division multiplexing frequency offset correction," *IEEE Trans. Commun.*, 42, 2908–2914, 1994.

[31] M. Luise and Reggianini, "Carrier frequency offset acquisition and tracking for OFDM systems," *IEEE Trans. Commun.*, 44, 1996.

[32] H. Liu and U. Tureli, "A high efficiency carrier estimator for OFDM communications," *IEEE Commun. Lett.*, 2, 104–106, 1998.

[33] M. Morelli and U. Mengali, "An improved frequency offset estimator for OFDM applications," *IEEE Commun. Lett.*, 3, 75–77, 1999.

[34] H. Bölcskei, "Blind high-resolution uplink synchronization of OFDM-based multiple access schemes," in *Proceedings of the IEEE SPAWC'99*, Annapolis, MD, 1999, pp. 166–169.

[35] H.-K. Song, Y.-H. You, J.-H. Paik, and Y.-S. Cho, "Frequency-offset synchronization and channel estimation for OFDM-based transmission," *IEEE Commun. Lett.*, 4, 95–97, 2000.

[36] M. Morelli, A.N. D'Andrea, and U. Mengali, "Frequency ambiguity resolution in OFDM systems," *IEEE Commun. Lett.*, 4, 134–136, 2000.

[37] U. Tureli, H. Liu, and M. Zoltowski, "OFDM blind carrier offset estimation: ESPRIT," *IEEE Trans. Commun.*, 48, 2000.

[38] M. Morelli and U. Mengali, "Carrier-frequency estimation for transmission over selective channels," *IEEE Trans. Commun.*, 48, 1580–1589, 2000.

[39] M. Morelli, A.N. D'Andrea, and U. Mengali, "Feedback frequency synchronization for OFDM applications," *IEEE Commun. Lett.*, 5, 28–30, 2001.

[40] J. Li, G. Liu, and G.B. Giannakis, "Carrier frequency offset estimation for OFDM-based WLAN's," *IEEE Signal Process. Lett.*, 8, 80–82, 2001.

[41] M. Ghogho, A. Swami, and G.B. Giannakis, "Optimized null-subcarrier selection for CFO estimation in OFDM over frequency-selective fading channels," in *GLOBECOM*'2001, San Antonio, TX, November 2001.

[42] X. Ma, C. Tepedelenlioglu, G.B. Giannakis, and S. Barbarossa, "Non-data aided carrier-offset estimators for OFDM with null subcarriers: identifiability, algorithms, and performance," *IEEE J. Selected Areas Commun.*, 19, 2504–2515, 2001.

[43] S. Barbarossa, M. Pompili, and G.B. Giannakis, "Channel-independent synchronization of OFD multiple access systems," *IEEE J. Selected Areas Commun.*, 20, 474–486, 2002.

[44] M. Ghogho and A. Swami, "A blind frequency offset synchronization for OFDM transmitting constant-modulus symbols," *IEEE Commun. Lett.*, 5, 2002.

[45] M. Ghogho and A. Swami, "Blind frequency-offset estimation for OFDM and multicarrier systems," in *2nd IEEE International Symposium on Signal Processing and Information Technology, ISSPIT2002*, Marrakech, Morocco, December 2002.

[46] M. Ghogho, "On blind frequency-offset synchronization for OFDM communications," in *XI European Signal Processing Conference, EUSIPCO'2002*, Toulouse, France, September 3–6, 2002.

[47] M. Ghogho and A. Swami, "Semi-blind frequency offset synchronization for OFDM," in *IEEE International Conference on Acoustics, Speech and Signal Processing (ICASSP'02)*, Orlando, FL, May 2002.

[48] M. Ghogho and A. Swami, "Blind channel identification for OFDM systems with receive antenna diversity," in *Proceedings of the IEEE SPAWC'03*, Rome, June 2003.

[49] F. Daffara and A. Chouly, "Maximum likelihood frequency detectors for orthogonal multicarrier systems," in *Proceedings of the ICC*, May 1993, pp. 766–771.

[50] J.-J. van de Beek, M. Sandell, and P.O. Borjesson, "ML estimation of time and frequency offset in OFDM systems," *IEEE Trans. Signal Process.*, 45, 1800–1805, 1997.

[51] M.G. Hebley and D.P. Taylor, "The effect of diversity on a burst mode carrier-frequency estimator in the frequency-selective multipath channel," *IEEE Trans. Commun.*, 46, 553–560, 1998.

[52] A.J. Coulson, "Maximum likelihood synchronization for OFDM using a pilot symbol: algorithms," *IEEE J. Selected Areas Commun.*, 19, 2486–2494, 2001.

[53] S. Kapoor, D.J. Marchok, and Y.-F. Huang, "Pilot assisted synchronization for wireless OFDM using pilot carriers," in *Proceedings of the VTC'98*, 1998, pp. 2077–2080.

[54] U. Lambrette, M. Speth, and H. Meyr, "OFDM robust frequency synchronization by single carrier training data," *IEEE Commun. Lett.*, 1, 46–48, 1997.

[55] T. Keller et al., "OFDM synchronization techniques for frequency-selective fading channels," *IEEE J. Selected Areas Commun.*, 19, 999–1008, 2001.

[56] S. Patel, L.J. Cimini, Jr., and B. McNair, "Comparison of frequency offset estimation techniques for burst OFDM," in *Proceedings of the IEEE VTC 2002*, 2002, pp. 772–776.

[57] M. Luise, M. Marselli, and R. Reggiannini, "Low-complexity blind carrier frequency recovery for OFDM signals over frequency-selective radio channels," *IEEE Trans. Commun.*, 50, 1182–1188, 2002.

[58] M. Ghogho, A. Swami, and T. Durrani, "Frequency estimation in the presence of Doppler spread: performance analysis," *IEEE Trans. Signal Process.*, 49(4), 2001.

[59] M.H. Hsieh and C.H. Wei, "A low-complexity frame synchronization and frequency offset compensation scheme for OFDM systems and fading channels," *IEEE Trans. Vehicular Technol.*, 48, 1596–1609, 1999.

9

Filter-Bank Modulation Techniques for Transmission over Frequency-Selective Channels

Giovanni Cherubini
IBM Research

9.1 Introduction 9-2
9.2 Critically Sampled Filter Banks 9-4
Orthogonality Conditions • Efficient Implementation • Example of Critically-Sampled Filter Bank
9.3 Discrete Multitone Modulation 9-9
9.4 O-QAM OFDM Modulation 9-11
9.5 Discrete Wavelet Multitone Modulation 9-16
9.6 Filtered Multitone Modulation 9-17
Filter-bank Design • Per-Subchannel Adaptive Equalization and Precoding
9.7 Conclusion 9-27

Abstract

Filter-bank modulation techniques are well suited for data transmission over channels that exhibit high signal attenuation at frequencies within the passband. The task of modulating several carriers in parallel is efficiently accomplished in the digital domain by employing filter banks. Symbol–vector sequences are presented at the filter-bank input, where the symbol period is typically much longer than that of a single-carrier system transmitting at the same bit rate. The narrowband subchannel signals obtained at the filter-bank output are transmitted over the channel. At the receiver, demodulation is also accomplished by filter-bank techniques. The orthogonality conditions and the efficient implementation of critically sampled filter banks are addressed first. Then the implementation and various system aspects of filter-bank modulation techniques are discussed, with emphasis on discrete multitone (DMT) modulation, offset quadrature amplitude modulation (O-QAM) orthogonal frequency division multiplexing (OFDM) modulation, and filtered multitone (FMT) modulation.

0-8493-1657-X/$0.00+$1.50

9.1 Introduction

The tenet of multicarrier transmission is to divide the complex problem of wideband transmission over channels that exhibit high signal attenuation at frequencies within the passband into a set of simply resolved narrowband transmission problems. Shannon's classic paper (Shannon, 1948), which marked the beginning of the information age, clearly identifies multicarrier transmission as the optimum method to solve complex transmission problems. Although Shannon's work provided nonconstructively proved bounds on transmission in every other way, it did point to one method for handling a linear channel with intersymbol interference constructively as the method of choice: multicarrier transmission. It is interesting to observe that recent extensions of Shannon's work indicate that multicarrier frequency division multiplexing is also optimum for the multiuser channel (Cheng and Verdu, 1993).

The idea of subdividing a signal frequency band into a set of contiguous bands was recognized very early in the fields of signal processing and data communications as a powerful technique for achieving efficient system realizations. The spectral partitioning can generally be realized in the form of overlapping or nonoverlapping subbands, as illustrated in Figure 9.1. In signal processing, subband partitioning was introduced to perform short-time spectrum analysis of speech signals, initially in analog and then in digital form (Schafer and Rabiner, 1971). For the digital representation of speech signals, it was later found (Crochiere et al., 1976) that if the subbands are individually quantized with possibly different accuracy (subband coding), it is possible to achieve, for the same total bit rate, a signal quality better than that obtained by quantizing the full-band signal. This work motivated the development of the quadrature mirror filter (QMF) (Esteban and Galand, 1977) as a fundamental building block for spectral splitting. The QMF structure allows spectral decomposition into two contiguous low-pass/high-pass overlapping subbands in such a way that all the aliasing incurred in the initial analysis stage is eliminated during signal reconstruction by the synthesis stage. Later, the technique was extended (Smith and Barnwell, 1984) by introducing the so-called perfect reconstruction QMF bank that also allowed complete elimination of amplitude and phase distortion in the reconstructed signal. Multirate filter banks have been comprehensively studied (e.g., in Vetterli, 1987; Vaidyanathan, 1993; Fliege, 1994).

In data communications, the motivation for dividing the spectrum of the communication channel into a plurality of subchannels and performing parallel data transmission over the subchannels was to increase system robustness against amplitude and phase distortion introduced by the communication channel, impulse noise, etc. An early reference of a commercial system employing this concept was the Collins Radio Co. Kineplex system (Doelz et al., 1957) designed for HF transmission, which used four-phase differential modulation for parallel data transmission over 20 subchannels within the voice band and achieved a data rate of 3 kbit/s. For at least a decade, this data rate was 10 times faster than that achieved by other single-carrier commercial modems.

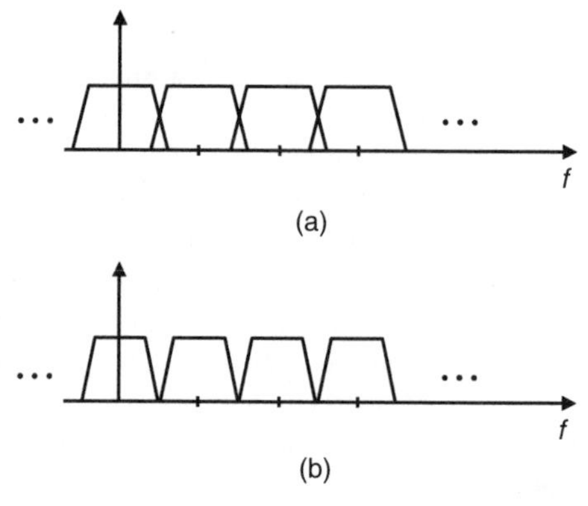

FIGURE 9.1 Spectral subdivision into contiguous (a) overlapping and (b) nonoverlapping bands.

For good performance, signal reception on each subchannel should ideally be free of intersymbol interference (ISI) as well as intersubchannel interference (ICI). Although the latter objective can be achieved by avoiding spectral overlap between the subchannels, as illustrated in Figure 9.1b, this approach was usually discarded on the premise that because rather sharp filters are not easily realizable, filters with smooth characteristics would lead to an unacceptable loss of spectral efficiency. In Hartmut (1960) time-limited sinusoids resulting in sinc(f)-type overlapping spectra were used as a set of orthogonal functions to achieve parallel data transmission. Subsequently, certain orthogonality conditions were introduced in Chang (1966) to derive filter transfer functions to limit the spectral overlap of any subchannel to the two adjacent subchannels, whereas both ISI and ICI are avoided at the output of a *known* transmission channel. Motivated by this work, an analog transmission scheme was studied in Saltzberg (1967), where orthogonality between the overlapping subchannels was achieved through offset quadrature amplitude modulation (O-QAM).

Soon thereafter, it was recognized in Weinstein and Ebert (1971) that digital parallel data transmission systems can be realized efficiently if the discrete Fourier transform (DFT), implemented by the fast Fourier transform (FFT), is employed. In the approach of Weinstein and Ebert (1971) modulation is achieved by a DFT operation on consecutive blocks of complex transmit symbols, and a similar DFT operation is employed at the receiver. In later designs, signal modulation and demodulation are usually performed by inverse DFT (IDFT) and DFT operations, respectively. Because the serial transmission of consecutive DFT output blocks corresponds to a time-windowing operation with a rectangular pulse, a sinc(f)-shaped spectral characteristic is obtained for each subchannel. To cope with ISI and ICI at the receiver, two important techniques were described in Weinstein and Ebert (1971): if the DFT blocks are cyclically extended prior to transmission and a sufficient number of subchannels are chosen, each subchannel can essentially be equalized by a complex-valued single-tap equalizer, the so-called frequency-domain equalizer. The significance of this cyclic extension was later recognized in Peled and Ruiz (1980) as being a method to ensure that circular convolution holds. Furthermore, if a nonrectangular time-windowing operation is performed at the DFT output prior to signal transmission, then the effect of ICI can be significantly mitigated at the receiver.

The now popular multicarrier techniques known as orthogonal frequency division multiplexing (OFDM) and discrete multitone (DMT) follow the DFT-based modulation principle described above. These techniques also incorporate significant extensions and refinements brought about during the intensive implementation phase that started in the early 1980s (Bingham, 1990). For example, the European telecommunications standards for digital audio and video broadcasting employ OFDM as physical-layer technology (ETS, 1994, 1997).

The gap between the approaches of Saltzberg (1967) and Weinstein and Ebert (1971) was bridged in Hirosaki (1981), where a computationally efficient digital parallel data transmission scheme was proposed based on a combination of O-QAM and DFT processing. The structure of Hirosaki (1981) paralleled the one obtained in the 1970s by Bellanger et al. (1976) with work on time division multiplexing (TDM) to frequency division multiplexing (FDM) transmultiplexers. The latter marked a turning point in the field of multirate filter banks by showing that uniform digital filter banks can be efficiently realized by a combination of polyphase filtering and discrete Fourier transformation.

The conditions for perfect reconstruction of the transmit signal at the output of the receive filter bank are satisfied if the transmission channel is ideal. Therefore, orthogonality between the subchannels is destroyed at the receiver whenever amplitude and phase distortion are introduced by the transmission channel, causing unacceptable performance degradation in most cases. To maintain orthogonality of signals transmitted over different subchannels, various methods can be used. One approach consists of cyclically extending the transmit frames prior to transmission, as mentioned above in the case of DMT modulation. Filter-bank transceivers that employ filter-bank precoders to introduce redundancy in the transmit signal so that the information rate over dispersive channels is optimized have been treated in Scaglione et al. (1999a, 1999b, 2002) and Stamoulis et al. (2001). Another approach is to employ filtering as a means to minimize ISI and ICI (Kammeyer et al., 1992; Haas and Belfiore, 1994; Vallet and Taieb, 1995; Matheus et al., 2000). In any case, spectral efficiency must be sacrificed. Hence, if the loss of efficiency due to cyclic extensions is comparable with that experienced by multicarrier systems where orthogonality is achieved through nonoverlapping subchannel characteristics, then the latter may be preferred if unique benefits are derived from a high level of subchannel spectral containment. A filter-bank modulation technique with the aforementioned spectral properties is the filtered multitone (FMT) modulation (Cherubini et al., 2000, 2002; Benvenuto et al., 2000).

In multiple-access systems, significant benefits can be gained from multicarrier modulation techniques that exhibit negligible spectral overlap between subchannels. Consider, for example, the problem of ranging and power adjustment of a station joining a network. During the initial registration process, the station has no knowledge of important transmission parameters such as correct transmit power setting and round-trip delay. If multicarrier schemes are employed, whereby adjacent subchannel spectral characteristics exhibit significant overlap, signals that are received with improper timing phase cause severe ICI (Jacobsen et al., 1995; Sari et al., 1997). A high level of spectral containment results in negligible ICI, independent of the timing phase of the received signals, and allows a straightforward solution for determining transmission parameters (Cherubini, 2000).

In this chapter, the orthogonality conditions and the efficient implementation of critically sampled filter banks, which typically yield nonnegligible overlap of adjacent subchannel spectral characteristics, are reviewed first. DMT, O-QAM OFDM, and discrete wavelet multitone (DWMT) modulation techniques are then discussed. Finally, noncritically sampled filter banks, which allow transmission over individual subchannels with excess bandwidth, are discussed, with emphasis on FMT modulation. In particular, joint trellis coding and precoding is described for a system employing FMT modulation for transmission over slowly time-varying frequency-selective channels.

9.2 Critically Sampled Filter Banks

9.2.1 Orthogonality Conditions

Figure 9.2 shows the baseband equivalent of a communications system employing filter-bank modulation with critical sampling. The input symbols at the n-th modulation interval are represented by the vector $\mathbf{A}(nT) = [A_0(nT), A_1(nT), \ldots, A_{M-1}(nT)]^{\mathrm{T}}$, where $\mathbf{A}^{\mathrm{T}}$ denotes the transpose of $\mathbf{A}$. The symbols in $\mathbf{A}(nT)$ are taken from the two-dimensional constellations $\mathcal{A}_m$, $m = 0, 1, \ldots, M-1$, and are transmitted in parallel over M subchannels. Therefore, the symbol rate of each subchannel is equal to $1/T$, where T denotes the modulation interval. After upsampling by a factor M, each sequence is filtered by a band-limited filter properly allocated in frequency. The filter on the m-th branch is characterized by the impulse response $\{h_m(kT/M)\}$, with transfer function $H_m(z) = \sum_{k=-\infty}^{+\infty} h_m(kT/M)\, z^{-k}$ and frequency response

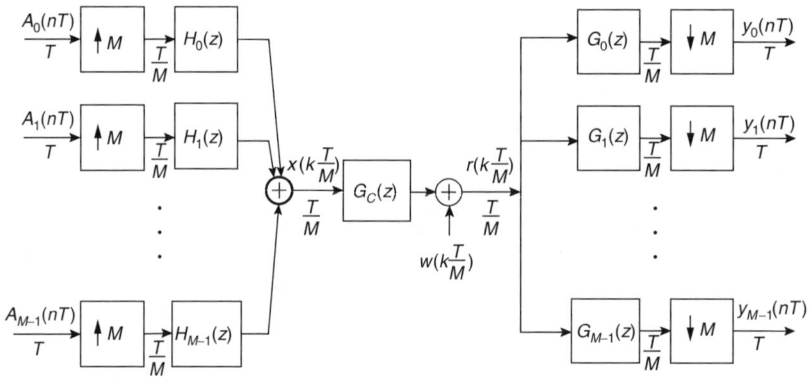

FIGURE 9.2 Block diagram of a communications system employing critically sampled filter banks.

$\mathcal{H}_m(f) = H_m(z)|_{z=e^{j2\pi fT/M}}$. The transmitter output signal is given by the sum of the filter output signals,

$$x\left(k\frac{T}{M}\right) = \sum_{m=0}^{M-1}\sum_{n=-\infty}^{+\infty} A_m(nT)\, h_m\left((k-nM)\frac{T}{M}\right) \tag{9.1}$$

and is transmitted over a noisy channel at the transmission rate M/T. The channel transfer function is denoted by $G_c(z)$. The sequence $\{w(kT/M)\}$ is assumed to be a sequence of additive white Gaussian noise samples.

The received signal $\{r(kT/M)\}$ is filtered by M filters in parallel having impulse responses $\{g_m(kT/M)\}$, $m = 0, \ldots, M-1$, and transfer functions $G_m(z) = \sum_{k=-\infty}^{+\infty} g_m(kT/M)\, z^{-k}$, $m = 0, \ldots, M-1$. The vector sequence $\{\mathbf{y}(nT) = [y_0(nT), \ldots, y_{M-1}(nT)]^{\mathrm{T}}\}$ at the symbol rate $1/T$, which is employed to detect the vector sequence $\{\mathbf{A}(nT)\}$ of transmitted symbols, is obtained by downsampling the output signals of the receive filters by the factor M.

Let us consider as transmit filters finite impulse response (FIR) filters of length γM, with $h_m(kT/M) = 0$ for $k < 0$ and $k > \gamma M - 1$. The inherent signal shaping capability of filter-bank systems offers several degrees of freedom to the communications system designer. As discussed above, the selection of the set of filters has traditionally been performed under the constraint of complete elimination of ISI and ICI. Assuming the channel is ideal and noiseless, if the transmit and receive filter banks are designed such that certain orthogonality conditions are satisfied, the subchannel output signals are delayed versions of the transmitted symbol sequences at the corresponding subchannel inputs.

With reference to the scheme of Figure 9.2, assuming matched receive filters with impulse responses given by $g_m(kT/M) = h_m^*((\gamma M - k)T/M)$, $\forall k$, $m = 0, \ldots, M-1$, where $*$ denotes complex conjugation, the impulse response relative to the m-th input at the output of the j-th subchannel is given before downsampling by

$$\sum_{p=0}^{\gamma M-1} h_m\left(p\frac{T}{M}\right) g_j\left((k-p)\frac{T}{M}\right) = \sum_{p=0}^{\gamma M-1} h_m\left(p\frac{T}{M}\right) h_j^*\left((\gamma M + p - k)\frac{T}{M}\right) \tag{9.2}$$

Therefore, transmission in the absence of ISI over a subchannel and in the absence of ICI between subchannels is achieved if orthogonality conditions are satisfied that in the time domain are expressed as

$$\sum_{p=0}^{\gamma M-1} h_m\left(p\frac{T}{M}\right) h_j^*\left((p + M(\gamma - n))\frac{T}{M}\right) = \delta_{m-j}\delta_{n-\gamma}, \quad m, j = 0, \ldots, M-1 \tag{9.3}$$

where δ_m denotes the Kronecker delta. Hence, in the ideal channel case considered here, the vector sequence at the output of the receive filter bank is a replica of the transmitted vector sequence with a delay of γ modulation intervals, that is, $\{\mathbf{y}(nT)\} = \{\mathbf{A}((n-\gamma)T)\}$. Sometimes the elements of a set of orthogonal impulse responses that satisfy Equation 9.3 are called wavelets.

In the frequency domain, the conditions in Equation 9.3 are expressed as

$$\sum_{\ell=0}^{M-1} \mathcal{H}_m\left(f - \ell\frac{1}{T}\right) \mathcal{H}_j^*\left(f - \ell\frac{1}{T}\right) = \delta_{m-j}, \quad m, j = 0, \ldots, M-1 \tag{9.4}$$

9.2.2 Efficient Implementation

For large values of M, a direct implementation of a filter-bank modulation system, as shown in Figure 9.2, would require an exceedingly large computational complexity, as all filtering operations are performed at a high transmission rate equal to M/T. A reduction in the number of operations per unit of time is obtained by resorting to the polyphase representation of the various filters (Benvenuto and Cherubini,

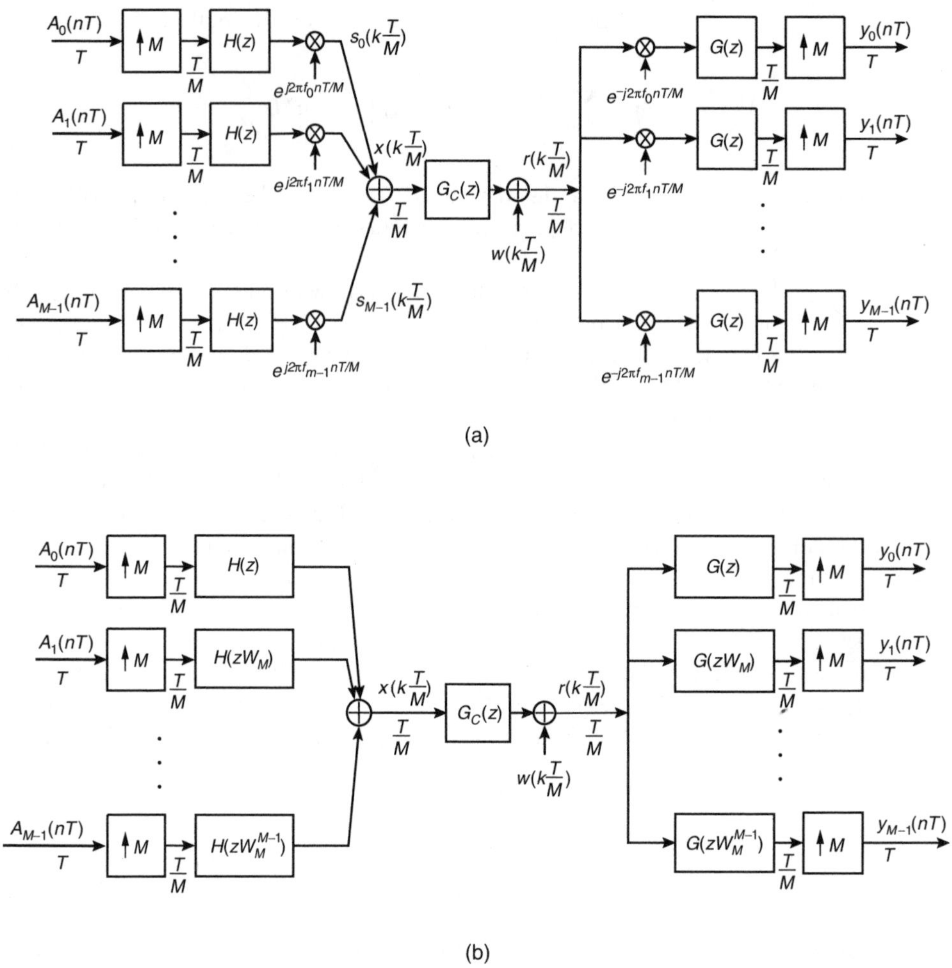

FIGURE 9.3 Block diagram of a communications system employing critically sampled uniform filter banks: (a) general scheme and (b) equivalent scheme for $f_m = m/T, m = 0, \ldots, M-1$.

2002). System complexity can be further reduced by resorting to uniform filter banks. In this case the frequency responses of the various filters are obtained by shifting the frequency response of a prototype filter with impulse response $\{h(kT/M)\}$ and transfer function $H(z)$ around carrier frequencies given by $f_m = \frac{m}{T}$, $m = 0, \ldots, M-1$, as illustrated in Figure 9.3a. The spacing in frequency between the subcarriers is $\Delta f = 1/T$. Bellanger et al. (1976) have shown that for the critically sampled case the transmit and receive filter-bank transmitter and receiver can equivalently be realized, as depicted in Figure 9.4. The M-branch filters with transfer functions $H^{(\ell)}(z), \ell = 0, \ldots, M-1$, and corresponding impulse responses $\{h^{(\ell)}(nT)\}$, $\ell = 0, \ldots, M-1$, are baseband filters that represent the M polyphase components of the prototype filter. The impulse responses of the polyphase filter components are obtained by extracting regularly T-spaced samples from the prototype filter impulse response, which has T/M-spaced samples. Each polyphase filter component is then applied to a different time-domain point of the IDFT output. Polyphase filter-bank structures are attractive because the required filtering operations are performed at the symbol rate $1/T$ instead of the transmission rate M/T. A trade-off between processing speed and parallelism is generally allowed in systems that employ filter banks.

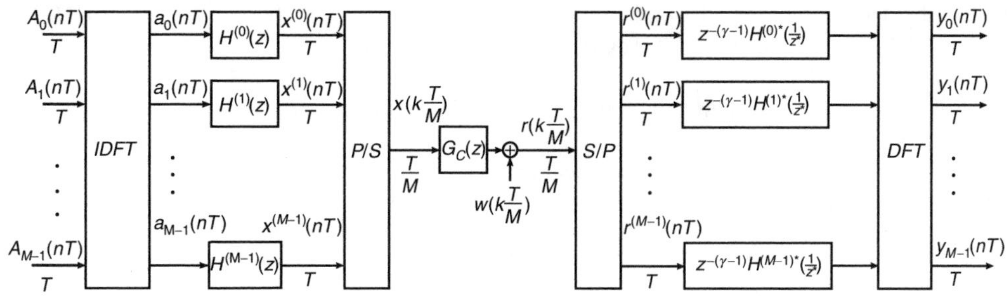

FIGURE 9.4 Efficient realization of critically sampled filter banks.

To derive the efficient implementation of uniform filter banks, the scheme represented in Figure 9.3a is considered. The m-th subchannel signal at the channel input is given by

$$x_m\left(k\frac{T}{M}\right) = e^{j2\pi\frac{mk}{M}} \sum_{n=-\infty}^{+\infty} A_m(nT)h\left((k-nM)\frac{T}{M}\right) \tag{9.5}$$

Since $e^{j2\pi\frac{mk}{M}} = e^{j2\pi\frac{m(k-nM)}{M}}$, we obtain

$$x_m\left(k\frac{T}{M}\right) = \sum_{n=-\infty}^{\infty} A_m(nT)h\left((k-nM)\frac{T}{M}\right)e^{j2\pi\frac{m(k-nM)}{M}} = \sum_{n=-\infty}^{+\infty} A_m(nT)h_m\left((k-nM)\frac{T}{M}\right) \tag{9.6}$$

where $h_m(kT/M) = h(kT/M)e^{j2\pi\frac{mk}{M}}$. Recalling the definition $W_M = e^{-j2\pi/M}$, the z-transform of $\{h_m(kT/M)\}$ is expressed as $H_m(z) = H(zW_M^m)$, $m = 0, \ldots, M-1$. Observing Equation 9.6, the equivalent scheme of Figure 9.3b is obtained. The scheme of Figure 9.3b may be considered as a particular case of the general scheme represented in Figure 9.2. In particular, the transfer functions of the filters can be expressed using the polyphase representation. Observing Equation 9.5, the overall signal $x(kT/M)$ is expressed as

$$x\left(k\frac{T}{M}\right) = \sum_{m=0}^{M-1} e^{j2\pi\frac{mk}{M}} \sum_{n=-\infty}^{+\infty} h\left((k-nM)\frac{T}{M}\right)A_m(nT) \tag{9.7}$$

With the change of variables $k = pM + \ell$, for $p = -\infty, \ldots, +\infty$ and $\ell = 0, 1, \ldots, M-1$, we get

$$x\left((pM+\ell)\frac{T}{M}\right) = \sum_{m=0}^{M-1} e^{j2\pi\frac{m}{M}(pM+\ell)} \sum_{n=-\infty}^{+\infty} h\left(((p-n)M+\ell)\frac{T}{M}\right)A_m(nT) \tag{9.8}$$

Observing $e^{j2\pi mp} = 1$, setting $x^{(\ell)}(pT) = x((pM+\ell)T/M)$, $h^{(\ell)}(pT) = h((pM+\ell)T/M)$, and interchanging the order of summations, the ℓ-th polyphase component of $x(kT/M)$ is expressed as

$$x^{(\ell)}(pT) = \sum_{n=-\infty}^{+\infty} h^{(\ell)}((p-n)T) \sum_{m=0}^{M-1} W_M^{-\ell m} A_m(nT) \tag{9.9}$$

The sequences $\{h^{(\ell)}(pT)\}, \ell = 0, 1, \ldots, M-1$, denote the polyphase components of the prototype filter impulse response, with transfer functions given by

$$H^{(\ell)}(z) = \sum_{p=0}^{\gamma-1} h^{(\ell)}(pT)z^{-p}, \quad \ell = 0, 1, \ldots, M-1 \tag{9.10}$$

Recalling the definition of the DFT operator as an $M \times M$ matrix, $\mathbf{F}_M = [(e^{-j2\pi/M})^{\ell m}]$, $\ell, m = 0, 1, \ldots, M-1$, the IDFT of the vector $\mathbf{A}(nT)$ is expressed as

$$\mathbf{F}_M^{-1}\mathbf{A}(nT) = \mathbf{a}(nT) = [a_0(nT), \ldots, a_{M-1}(nT)]^{\mathrm{T}} \tag{9.11}$$

Then the inner summation in Equation 9.9 yields

$$\frac{1}{M}\sum_{m=0}^{M-1} W_M^{-\ell m} A_m(nT) = a_\ell(nT), \quad \ell = 0, 1, \ldots, M-1 \tag{9.12}$$

and

$$x^{(\ell)}(pT) = M\sum_{n=-\infty}^{+\infty} h^{(\ell)}((p-n)T)a_\ell(nT) = M\sum_{n=-\infty}^{+\infty} h^{(\ell)}(nT)a_\ell((p-n)T) \tag{9.13}$$

Including the factor M in the definition of the prototype filter impulse response, the efficient implementation of a uniform transmit filter bank given by an IDFT, a polyphase network with M branches, and a parallel-to-serial (P/S) converter are obtained, as illustrated in Figure 9.4.

With reference to Figure 9.3a, it is interesting to derive the polyphase representation of the receive filter bank by observing the relation between the received sequence $r(kT/M)$ and the output of the m-th subchannel $y_m(nT)$, given by

$$y_m(nT) = \sum_{k=-\infty}^{+\infty} g\left((nM-k)\frac{T}{M}\right) e^{-j2\pi\frac{m}{T}k\frac{T}{M}}\, r(kT/M) \tag{9.14}$$

With the change of variables $k = pM + \ell$, for $p = -\infty, \ldots, +\infty$, $\ell = 0, 1, \ldots, M-1$, and recalling the expression of the matched prototype filter impulse response, we obtain

$$y_m(nT) = \sum_{p=-\infty}^{+\infty}\sum_{\ell=0}^{M-1} h^*\left(((\gamma - n + p)M + \ell)\frac{T}{M}\right) e^{-j2\pi\frac{m}{M}(pM+\ell)}\, r\left((pM+\ell)\frac{T}{M}\right) \tag{9.15}$$

Observing that $e^{-j2\pi\frac{m}{M}pM} = 1$, setting $r^{(\ell)}(pT) = r((pM+\ell)T/M)$ and $h^{(\ell)^*}(pT) = h^*((pM+\ell)T/M)$, and interchanging the order of summations, we get

$$y_m(nT) = \sum_{\ell=0}^{M-1} e^{-j\frac{2\pi}{M}m\ell} \sum_{p=-\infty}^{+\infty} h^{(\ell)^*}((\gamma - n + p)T)\, r^{(\ell)}(pT) \tag{9.16}$$

Using the relation $e^{-j2\pi m\ell/M} = W_M^{m\ell}$, we finally find the expression

$$y_m(nT) = \sum_{\ell=0}^{M-1} W_M^{m\ell} \sum_{p=-\infty}^{+\infty} h^{(\ell)^*}((\gamma - n + p)T)\, r^{(\ell)}(pT) \tag{9.17}$$

Hence an efficient implementation of a uniform receive filter bank, also illustrated in Figure 9.4, is given by a serial-to-parallel (S/P) converter, a polyphase network with M branches, and a DFT. Note that the filter of the m-th branch at the receiver is matched to the filter of the corresponding branch at the transmitter.

9.2.3 Example of Critically-Sampled Filter Bank

The perfect reconstruction conditions in Equation 9.3 are satisfied by OFDM systems for transmission over an ideal channel. The transmit and receive filter banks use a prototype filter whose respective impulse

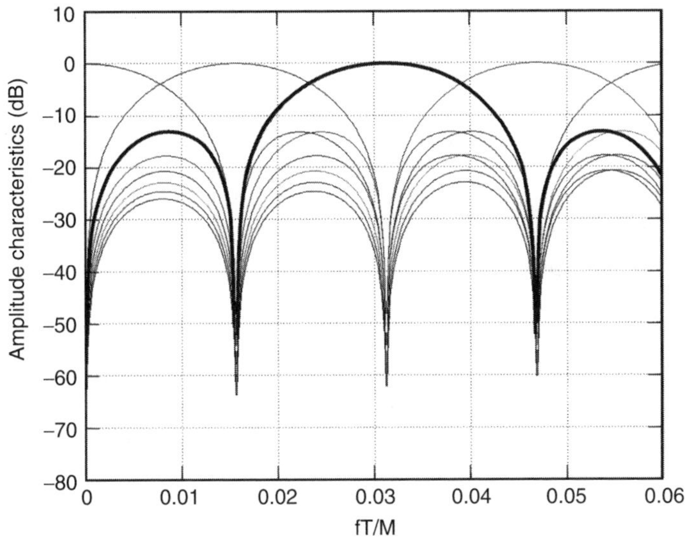

FIGURE 9.5 Amplitude characteristics of the frequency responses of adjacent subchannel filters in a DMT system for $f \in (0, 0.06\,M/T)$ and $M = 64$. (From Cherubini et al., *IEEE Commun. Mag.*, 38, 98–104, 2000. © 2000 IEEE.)

response and frequency response are given by

$$h\left(k\frac{T}{M}\right) = \begin{cases} 1 & \text{for } 0 \leq k \leq M-1 \\ 0 & \text{otherwise} \end{cases} \tag{9.18}$$

and

$$\mathcal{H}(f) = \frac{\sin(\pi f T)}{\sin(\pi f T/M)} e^{-j2\pi f \frac{M-1}{2}\frac{T}{M}} \tag{9.19}$$

The impulse responses of the polyphase components of $\{h(kT/M)\}$ are trivially given by $\{h^{(\ell)}(nT)\} = \delta_n, \ell = 0, \ldots, M-1$.

Because the frequency responses of the polyphase components are constant, the transmit signal is directly obtained by applying a P/S conversion at the output of the IDFT. Assuming an ideal channel, at the receiver a S/P converter forms blocks of M samples, with boundaries between blocks placed such that each block at the output of the IDFT at the transmitter is presented unchanged at the input of the DFT. At the DFT output, the input blocks of M symbols are reproduced without distortion with a delay equal to T. Note, however, that the orthogonality conditions are satisfied only if the channel is ideal. Figure 9.5 shows the amplitude of the frequency response of adjacent subchannel filters obtained in the frequency interval $(0, 0.06\,M/T)$ for $M = 64$. The spectra of adjacent subchannels approximately cross at the -3 dB point, and the first sidelobe is as high as -13 dB. Note that the choice of a rectangular window of length M as impulse response of the baseband prototype filter leads to a significant overlapping of the spectral components of transmitted signals in adjacent subchannels.

9.3 Discrete Multitone Modulation

Let us consider the baseband equivalent system shown in Figure 9.4, assuming the impulse response of the channel $\{g_c(kT/M)\}$ has a length equal to $N_c > 1$, with $g_c(kT/M) = 0$ for $k < 0$ and $k > N_c - 1$. In this case the orthogonality conditions for the system described in Section 9.2.3 are no longer satisfied. However, ISI and ICI can be eliminated by the following method.

For the same channel bandwidth and hence for a given transmission rate M/T, the modulation must be carried out at the rate $1/T' = M/((M + N_c - 1)T) < 1/T$. After modulation, each block of samples is cyclically extended by copying the $N_c - 1$ samples $a_{M-N_c+1}(nT), \ldots, a_{M-1}(nT)$ in front of the block. After P/S conversion, where the $N_c - 1$ samples of the cyclic extension, also known as cyclic prefix, are the first to be sent, the $M + N_c - 1$ samples are transmitted over the channel. At the receiver, blocks of samples of length $M + N_c - 1$ are taken. The boundaries between blocks are set such that the last M samples depend only on the elements of one cyclically extended block of samples. The first $N_c - 1$ samples of a block are then discarded. The vector $\mathbf{r}(nT')$ of the last M samples of the block received at the n-th modulation interval can be expressed as

$$\mathbf{r}(nT') = \mathbf{X}(nT')\mathbf{g}_c + \mathbf{w}(nT') \tag{9.20}$$

where $\mathbf{g}_c = [g_c(0), \ldots, g_c(N_c - 1), 0, \ldots, 0]^{\mathrm{T}}$ is the M-component vector of the channel impulse response extended with $M - N_c$ zeros, $\mathbf{w}(nT')$ is a vector of additive white Gaussian noise samples, and $\mathbf{X}(nT')$ is an $M \times M$ circulant matrix given by

$$\mathbf{X}(nT') = \begin{bmatrix} a_0(nT') & a_{M-1}(nT') & \cdots & a_1(nT') \\ a_1(nT') & a_0(nT') & \cdots & a_2(nT') \\ \vdots & \vdots & & \vdots \\ a_{M-1}(nT') & a_{M-2}(nT') & \cdots & a_0(nT') \end{bmatrix} \tag{9.21}$$

Note that the matrix $\mathbf{X}(nT')$, being circulant, satisfies the relation

$$\mathbf{F}_M \mathbf{X}(nT')\mathbf{F}_M^{-1} = \begin{bmatrix} A_0(nT') & 0 & \cdots & 0 \\ 0 & A_1(nT') & \cdots & 0 \\ \vdots & \vdots & & \vdots \\ 0 & 0 & \cdots & A_{M-1}(nT') \end{bmatrix} = diag\,\{\mathbf{A}(nT')\} \tag{9.22}$$

where $diag\{\mathbf{A}\}$ denotes the diagonal matrix with elements on the diagonal given by the elements of the vector $\mathbf{A}$. Defining the DFT of the vector $\mathbf{g}_c$ as

$$\mathbf{G}_c = \mathbf{F}_M\ \mathbf{g}_c = [\mathcal{G}_c(0), \mathcal{G}_c(1), \ldots, \mathcal{G}_c(M-1)]^{\mathrm{T}} \tag{9.23}$$

and using Equation 9.22, it turns out that the demodulator output is given by

$$\mathbf{x}(nT') = \mathbf{F}_M\ \mathbf{r}(nT') = diag\{\mathbf{A}(nT')\}\mathbf{G}_c + \mathbf{W}(nT') \tag{9.24}$$

where $\mathbf{W}(nT')$ is given by the DFT of the vector $\mathbf{w}(nT')$. Recalling the properties of $\mathbf{w}(nT')$, it turns out that $\mathbf{W}(nT')$ is a vector of independent Gaussian random variables.

Equalizing the channel using the zero-forcing criterion, the signal $\mathbf{x}(nT')$ is multiplied by the diagonal matrix $\mathbf{K}$, whose elements on the main diagonal are given by

$$\mathcal{K}_m = \frac{1}{\mathcal{G}_c(m)}, \quad m = 0, \ldots, M-1 \tag{9.25}$$

Therefore, the input to the data detector is given by

$$\mathbf{y}(nT') = \mathbf{K}\mathbf{x}(nT') = \mathbf{A}(nT') + \mathbf{K}\mathbf{W}(nT') \tag{9.26}$$

If the sequence of input symbol vectors $\mathbf{A}(nT')$ is a sequence of independent, identically distributed random vectors, Equation 9.26 shows that the sequence $\mathbf{A}(nT')$ can be detected by assuming transmission over M independent and orthogonal subchannels in the presence of additive white Gaussian noise. A disadvantage of this simple equalization scheme is the reduction in the modulation rate by a factor $(M + N_c - 1)/M$. Therefore, it is essential that the length of the channel impulse response is much smaller than the number of subchannels, so that the reduction of the modulation rate due to the cyclic prefix can be considered negligible. One approach to reduce the length of the overall channel impulse response is to equalize the channel before demodulation by a so-called time-domain equalizer (Chow et al., 1991; Melsa et al., 1996; Baldemair and Frenger, 2001). Imperfectly shortened channel impulse responses, however, give origin to ISI and ICI. To mitigate this problem, alternative equalization methods have been proposed. In Vandendorpe et al. (1998) fractionally spaced linear and decision-feedback multiple-input multiple-output (MIMO) detectors were developed for multitone systems without cyclic prefix. More recently, a per-tone equalization structure has been proposed, whereby the time-domain equalizer operations are transferred to the frequency domain (Van Acker et al., 2001; Leus et al., 2003). Another low-complexity frequency-domain equalization method for DMT systems without the time-domain guard interval provided by the cyclic prefix is presented in Trautmann and Fliege (2002).

Performance of a multicarrier transmission system is usually measured in terms of achievable bit rate for given channel characteristics. The number of bits per modulation interval that can be loaded on the m-th subchannel is given by (Cioffi, 1997)

$$\beta_m = \log_2\left(1 + \frac{\text{SNR}_m\,\gamma_{\text{code}}}{\Gamma\,\gamma_{\text{margin}}}\right) \tag{9.27}$$

where SNR_m is the signal-to-noise ratio at the m-th subchannel output, γ_{code} denotes the coding gain, Γ denotes the SNR gap between the minimum SNR required for reliable transmission of β bits per modulation interval and the SNR required by 2^{β}-ary QAM to achieve a bit error probability of 10^{-7}, $\beta \gg 1$, and γ_{margin} denotes a margin that is usually required against additional noise sources that may be introduced by the channel and are not accounted for by the white Gaussian noise model. The achievable bit rate is therefore obtained by summing the values given by Equation 9.27 over the subchannels allocated for transmission and by multiplying the result by the modulation rate, and is given by

$$\text{R} = \frac{1}{T}\sum_{m\in\mathcal{M}} \beta_m \quad \text{[bit/s]} \tag{9.28}$$

where T is the modulation interval and $\mathcal{M}$ denotes the set of subchannels allocated for transmission. In practice, a technique called bit loading is employed to determine the number of bits to be transmitted over each subchannel per modulation interval. Bit loading maximizes the number of transmitted bits under the constraints that the input symbols, into which the information bits are mapped, can take only values from the two-dimensional constellations $\mathcal{A}_m$, $m = 0, 1, \ldots, M - 1$, and that the power of the transmitted signal is fixed (Chow et al., 1995; Leke and Cioffi, 1997; Campello, 1999).

In transmission systems that employ filter-bank modulation techniques, for a given bandwidth of the transmission channel, the modulation interval increases as the number of subchannels increases. Therefore, to reduce the delay in the recovery of the information, coding is usually applied across the subchannels. An example of the application of trellis coding across the subchannels for FMT modulation will be given in Section 9.6.

9.4 O-QAM OFDM Modulation

An O-QAM OFDM system with M subchannels may be regarded as a system of M ordinary O-QAM subchannels operating in parallel at different frequencies, whereby the spectra of the M subchannel signals are overlapping. The carrier frequencies are separated by $1/T$, where T denotes the modulation interval. Moreover, the carriers employed in adjacent subchannels have a phase difference of $\pi/2$. The baseband equivalent of an analog O-QAM OFDM system with M subchannels is illustrated in Figure 9.6. The real and imaginary parts of the complex symbol $A_m(nT)$ transmitted over the m-th subchannel at the n-th modulation interval are separated in time by $T/2$. We assume uniform transmit and receive filter banks

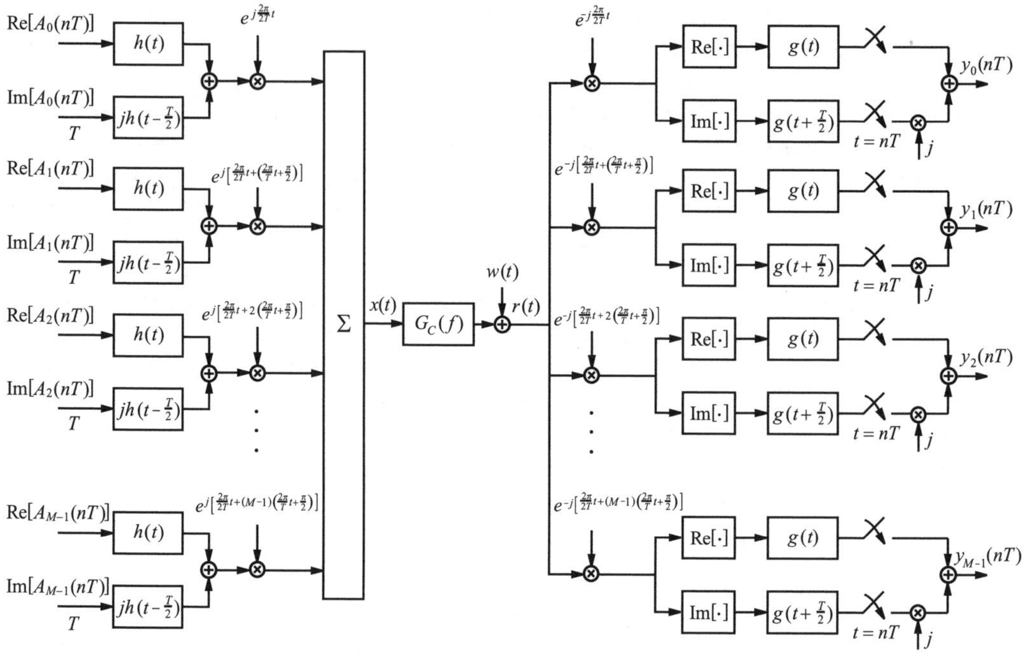

FIGURE 9.6 Block diagram of a communications system employing O-QAM OFDM modulation.

with impulse responses of the transmitter and receiver filters given by the real functions $h(t)$ and $g(t)$, respectively. Because the system is time invariant and the filter banks are uniform, to derive the conditions for the absence of ICI and ISI it is sufficient to consider the interference on a single symbol due to symbols transmitted on the other subchannels and on the same subchannel at other modulation intervals. Let us consider the signal path from subchannel $\ell + m$ to subchannel ℓ of the O-QAM OFDM system. The conditions for the absence of ISI and ICI can be obtained directly from the block diagram of Figure 9.6. The real and imaginary parts of the received sample at the output of subchannel ℓ at time $t = 0$, as well as the interference from the real and imaginary parts of symbol $A_{\ell+m}(nT)$, are considered separately. The conditions that the functions $h(t)$ and $g(t)$ must satisfy can then be expressed as

$$\left[\mathrm{Re}\left\{h(t-nT)e^{j(\frac{2\pi}{T}t+\frac{\pi}{2})m} \otimes g(t)\right\}\right]_{t=0} = \delta_m \, \delta_n \tag{9.29}$$

$$\left[\mathrm{Re}\left\{jh\left(t-nT-\frac{T}{2}\right)e^{j(\frac{2\pi}{T}t+\frac{\pi}{2})m} \otimes g(t)\right\}\right]_{t=0} = 0 \tag{9.30}$$

$$\left[\mathrm{Im}\left\{h(t-nT)e^{j(\frac{2\pi}{T}t+\frac{\pi}{2})m} \otimes g\left(t+\frac{T}{2}\right)\right\}\right]_{t=0} = 0 \tag{9.31}$$

$$\left[\mathrm{Im}\left\{jh\left(t-nT-\frac{T}{2}\right)e^{j(\frac{2\pi}{T}t+\frac{\pi}{2})m} \otimes g\left(t+\frac{T}{2}\right)\right\}\right]_{t=0} = \delta_m \, \delta_n \tag{9.32}$$

where $\otimes$ denotes convolution. Assuming that the function $h(t)$ is real and symmetric, i.e., $h(t) = h(-t)$, and that the receiver filters are matched to the transmitter filters, i.e., $g(t) = h^*(-t) = h(-t)$, Equation 9.30 and Equation 9.31 are satisfied. Moreover, Equation 9.32 is equivalent to Equation 9.29, and

Equation 9.29 is satisfied if m is an odd number. Therefore, the conditions for absence of ISI and ICI reduce to

$$\int_{-\infty}^{+\infty} h(t-nT)h(t)\cos\left(\frac{2\pi}{T}2mt\right)dt = \delta_m\,\delta_n \tag{9.33}$$

The optimization, in the sense of minimum out-of-band energy, of finite-duration pulses that satisfy Equation 9.33 is presented in Vahlin and Holte (1996).

Let us now consider the digital implementation of an O-QAM OFDM system. The sampling rate of the transmit signal is assumed equal to N/T, where $N = 2M$. The prototype for the transmit filter bank can be chosen as a real FIR filter of length $2\gamma M$ that approximates the square root of a Nyquist filter. The impulse response and the transfer function of the prototype filter are denoted by $\{h(kT/M)\}$ and $H(z)$, respectively. Then the orthogonally multiplexed complex O-QAM sequence at the transmitter output can be expressed in z-transform notation as

$$X(z) = \sum_{m=0}^{N/2-1} \tilde{A}_m\left(e^{-j\left[\frac{2\pi}{2T}\frac{T}{N}+m\frac{2\pi}{T}\frac{T}{N}\right]\frac{N}{2}}z^{\frac{N}{2}}\right)H\left(e^{-j\left[\frac{2\pi}{2T}\frac{T}{N}+m\frac{2\pi}{T}\frac{T}{N}\right]}z\right) \tag{9.34}$$

where

$$\tilde{A}_m\left(z^{\frac{N}{2}}\right) = j^m\left[\mathrm{Re}\{A_m(z^N)\} + j\mathrm{Im}\{A_m(z^N)\}z^{-\frac{N}{2}}\right] \tag{9.35}$$

Recalling that $e^{-j\left[\frac{2\pi}{2T}\frac{T}{N}+m\frac{2\pi}{T}\frac{T}{N}\right]} = W_N^{m+\frac{1}{2}}$ and defining

$$X(z) = X_{\mathrm{Re}}(z) + jX_{\mathrm{Im}}(z) \tag{9.36}$$

we have

$$X_{\mathrm{Re}}(z) = \mathrm{Re}\left\{\sum_{m=0}^{N/2-1}\tilde{A}_m\left(-j(-1)^m z^{\frac{N}{2}}\right)H\left(W_N^{m+\frac{1}{2}}z\right)\right\} \tag{9.37}$$

and

$$X_{\mathrm{Im}}(z) = \mathrm{Im}\left\{\sum_{m=0}^{N/2-1}\tilde{A}_m\left(-j(-1)^m z^{\frac{N}{2}}\right)H\left(W_N^{m+\frac{1}{2}}z\right)\right\} \tag{9.38}$$

Let us assume that N is a multiple of four. Then the real sequence $X_{\mathrm{Re}}(z)$ can be decomposed into two terms as

$$X_{\mathrm{Re}}(z) = \mathrm{Re}\left\{\sum_{m=0}^{\frac{N}{4}-1}\tilde{A}_{2m}\left(-jz^{\frac{N}{2}}\right)H\left(W_N^{2m+\frac{1}{2}}z\right)\right\} + \mathrm{Re}\left\{\sum_{m=0}^{\frac{N}{4}-1}\tilde{A}^*_{2m+1}\left(jz^{\frac{N}{2}}\right)H^*\left(W_N^{2m+\frac{3}{2}}z\right)\right\} \tag{9.39}$$

Note that as the impulse response of the prototype filter is real, $H^*(W_N^{2m+\frac{3}{2}}z) = H(W_N^{-(2m+\frac{3}{2})}z)$. Hence the second term on the right-hand side of Equation 9.39 is

$$\mathrm{Re}\left\{\sum_{m=0}^{\frac{N}{4}-1}\tilde{A}^*_{2m+1}\left(jz^{\frac{N}{2}}\right)H\left(W_N^{2m+\frac{3}{2}}z\right)\right\} \tag{9.40}$$

With the change of variables $\ell = N/2 - m - 1$, Equation 9.40 becomes

$$\mathrm{Re}\left\{\sum_{\ell=\frac{N}{4}}^{\frac{N}{2}-1}\tilde{A}^*_{N-2\ell-1}\left(jz^{\frac{N}{2}}\right)H\left(W_N^{2m+\frac{1}{2}}z\right)\right\} \tag{9.41}$$

Therefore, the real part of the transmit signal can be expressed as

$$X_{\mathrm{Re}}(z) = \mathrm{Re}\left\{\sum_{m=0}^{\frac{N}{2}-1} V'_m\left(z^{\frac{N}{2}}\right) H\left(W_N^{2m+\frac{1}{2}} z\right)\right\} \tag{9.42}$$

where

$$V'_m\left(z^{\frac{N}{2}}\right) = \begin{cases} \tilde{A}_{2m}(-jz^{N/2}) & \text{for } 0 \leq m \leq N/4-1 \\ \tilde{A}^*_{N-2m-1}(jz^{N/2}) & \text{for } N/4 \leq m \leq N/2-1 \end{cases} \tag{9.43}$$

Recalling now the polyphase decomposition $H(z) = \sum_{\ell=0}^{N/2-1} z^{-\ell} H^{(\ell)}(z^{N/2})$, we get

$$\begin{aligned} X_{\mathrm{Re}}(z) &= \mathrm{Re}\left\{\sum_{m=0}^{\frac{N}{2}-1} V'_m\left(z^{\frac{N}{2}}\right) \sum_{\ell=0}^{\frac{N}{2}-1} \left(W_N^{2m+\frac{1}{2}} z\right)^{-\ell} H^{(\ell)}\left(-jz^{\frac{N}{2}}\right)\right\} \\ &= \mathrm{Re}\left\{\sum_{\ell=0}^{\frac{N}{2}-1} z^{-\ell}\, W_{2N}^{-\ell}\, H^{(\ell)}\left(-jz^{\frac{N}{2}}\right) \sum_{m=0}^{\frac{N}{2}-1} W_{N/2}^{-\ell m}\, V'_m\left(z^{\frac{N}{2}}\right)\right\} \\ &= \sum_{\ell=0}^{\frac{N}{2}-1} z^{-\ell}\, \mathrm{Re}\left\{v'_m\left(z^{\frac{N}{2}}\right) H^{(\ell)}\left(-jz^{\frac{N}{2}}\right) W_{2N}^{-\ell}\right\} \end{aligned} \tag{9.44}$$

where

$$v'_\ell\left(z^{\frac{N}{2}}\right) = \sum_{m=0}^{\frac{N}{2}-1} W_{N/2}^{-\ell m}\, V'_m\left(z^{\frac{N}{2}}\right) \tag{9.45}$$

is obtained by the IDFT of $V'_m(z^{N/2})$, $m = 0, \ldots, N/2 - 1$. In an analogous way, for the imaginary part of the transmit signal we obtain

$$X_{\mathrm{Im}}(z) = \sum_{\ell=0}^{\frac{N}{2}-1} z^{-\ell}\, \mathrm{Im}\left\{v''_\ell\left(z^{\frac{N}{2}}\right) H^{(\ell)}\left(-jz^{\frac{N}{2}}\right) W_{2N}^{-\ell}\right\} \tag{9.46}$$

where

$$v''_\ell\left(z^{\frac{N}{2}}\right) = \sum_{m=0}^{\frac{N}{2}-1} W_{N/2}^{-\ell m}\, V''_m\left(z^{\frac{N}{2}}\right) \tag{9.47}$$

and

$$V''_m\left(z^{\frac{N}{2}}\right) = \begin{cases} \tilde{A}_{2m}(-jz^{N/2}) & \text{for } 0 \leq m \leq N/4-1 \\ -\tilde{A}^*_{N-2m-1}(jz^{N/2}) & \text{for } N/4 \leq m \leq N/2-1 \end{cases} \tag{9.48}$$

The efficient implementation of the transmit filter bank is illustrated in Figure 9.7. The input symbols are first preprocessed to generate the sequences $V'_m(z^{N/2})$ and $V''_m(z^{N/2})$. Observing the relationship between $V'_m(z^{N/2})$ and $V''_m(z^{N/2})$ obtained from Equation 9.43 and Equation 9.48, two $N/2$-point IDFT processors that perform Equation 9.45 and Equation 9.47 can be combined into one IDFT processor. The IDFT is then followed by a polyphase network with $N/2$ branches, where the $N/2$ filters have transfer functions $H^{(\ell)}(-jz^{N/2})$, $\ell = 0, \ldots, N/2 - 1$, and by a P/S converter.

The efficient implementation of the receive filter bank for an O-QAM OFDM is obtained as follows. A uniform filter bank with a prototype filter having an impulse response given by $g(kT/N) = h((\gamma M - k)T/N)$,

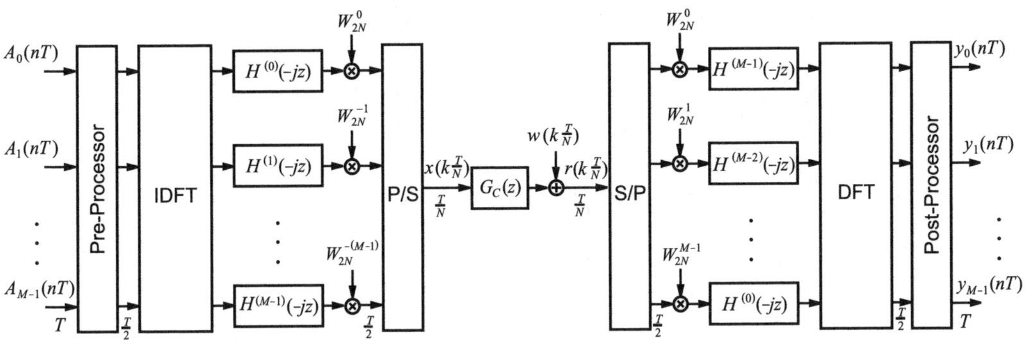

FIGURE 9.7 Efficient realization of filter banks for O-QAM OFDM modulation.

matched to the prototype filter of the transmit filter bank, is assumed. The output of the m-th subchannel at instants that are multiples of half the modulation interval, $y_m(nT/2)$, can be expressed as

$$
\begin{aligned}
y_m\left(n\frac{T}{2}\right) &= \sum_{k=-\infty}^{+\infty} g\left(\left(n\frac{N}{2}-k\right)\frac{T}{N}\right) e^{-j\left(\frac{2\pi}{2T}k\frac{T}{N}+m\left(\frac{2\pi}{T}k\frac{T}{N}+\frac{\pi}{2}\right)\right)} r\left(k\frac{T}{N}\right) \\
&= \sum_{p=-\infty}^{+\infty}\sum_{\ell=0}^{\frac{N}{2}-1} h^*\left(\left((\gamma-n+p)\frac{N}{2}+\ell\right)\frac{T}{N}\right) e^{-j\left(\frac{2\pi}{2N}\left(p\frac{N}{2}+\ell\right)+m\left(\frac{2\pi}{N}\left(p\frac{N}{2}+\ell\right)+\frac{\pi}{2}\right)\right)} r\left(\left(p\frac{N}{2}+\ell\right)\frac{T}{N}\right) \\
&= e^{-j\frac{\pi}{2}m}\, e^{-j\left(\frac{2\pi}{2}m+\frac{\pi}{2}\right)n} \sum_{\ell=0}^{\frac{N}{2}-1} e^{-j\left(\frac{2\pi}{N}m\ell+\frac{2\pi}{2N}\ell\right)} \sum_{p=-\infty}^{+\infty} h^{(\ell)*}\left((\gamma-n+p)\frac{T}{2}\right) e^{-j\left(\frac{2\pi}{2}m+\frac{\pi}{2}\right)(p-n)}\, r^{(\ell)}\left(p\frac{T}{2}\right)
\end{aligned}
\tag{9.49}
$$

Let us define $\tilde{y}_{\text{Re},m}(nT/2)$ and $\tilde{y}_{\text{Im},m}(nT/2)$ by

$$
y_m\left(n\frac{T}{2}\right) = e^{-j\frac{\pi}{2}m}\left(\tilde{y}_{\text{Re},m}\left(n\frac{T}{2}\right) + j\tilde{y}_{\text{Im},m}\left(n\frac{T}{2}\right)\right) \tag{9.50}
$$

Note that $\tilde{y}_{\text{Re},m}(nT/2)$ can be expressed for even and odd subchannel indexes as

$$
\tilde{y}_{\text{Re},2m}\left(n\frac{T}{2}\right) = \text{Re}\left\{(-j)^n \sum_{\ell=0}^{\frac{N}{2}-1} W_{N/2}^{m\ell}\, W_{2N}^{\ell} \sum_{p=-\infty}^{+\infty} h^{(\ell)*}\left((\gamma-n+p)\frac{T}{2}\right) e^{-j\frac{\pi}{2}(p-n)}\, r^{(\ell)}\left(p\frac{T}{2}\right)\right\} \tag{9.51}
$$

and

$$
\tilde{y}_{\text{Re},2m+1}\left(n\frac{T}{2}\right) = \text{Im}\left\{j^n \sum_{\ell=0}^{\frac{N}{2}-1} W_{N/2}^{m\ell}\, e^{-j\frac{2\pi}{2N}3\ell} \sum_{p=-\infty}^{+\infty} h^{(\ell)*}\left((\gamma-n+p)\frac{T}{2}\right) e^{-j\frac{3\pi}{2}(p-n)}\, r^{(\ell)}\left(p\frac{T}{2}\right)\right\} \tag{9.52}
$$

respectively, for $m = 0, \ldots, N/4 - 1$. Recalling that the impulse response of the prototype filter is real, Equation 9.52 can be written as

$$
\text{Re}\left\{(-j)^n \sum_{\ell=0}^{\frac{N}{2}-1} W_{N/2}^{m\ell}\, e^{j\frac{2\pi}{2N}3\ell} \sum_{p=-\infty}^{+\infty} h^{(\ell)*}\left((\gamma-n+p)\frac{T}{2}\right) e^{-j\frac{\pi}{2}(p-n)}\, r^{(\ell)*}\left(p\frac{T}{2}\right)\right\} \tag{9.53}
$$

With the change of variables $q = N/2 - m - 1$, Equation 9.53 becomes

$$\text{Re}\left\{(-j)^n \sum_{\ell=0}^{\frac{N}{2}-1} W_{N/2}^{q\ell}\, W_{2N}^{\ell} \sum_{p=-\infty}^{+\infty} h^{(\ell)*}\left((\gamma - n + p)\frac{T}{2}\right) e^{-j\frac{\pi}{2}(p-n)}\, r^{(\ell)*}\left(p\frac{T}{2}\right)\right\} \tag{9.54}$$

for $q = N/4, \ldots, N/2 - 1$. In an analogous way we get

$$\tilde{y}_{\text{Im},2m}\left(n\frac{T}{2}\right) = \text{Im}\left\{(-j)^n \sum_{\ell=0}^{\frac{N}{2}-1} W_{N/2}^{ml}\, W_{2N}^{\ell} \sum_{p=-\infty}^{+\infty} h^{(\ell)*}\left((\gamma - n + p)\frac{T}{2}\right) e^{-j\frac{\pi}{2}(p-n)}\, r^{(\ell)}\left(p\frac{T}{2}\right)\right\} \tag{9.55}$$

for $m = 0, \ldots, N/4 - 1$, and

$$\tilde{y}_{\text{Im},2m+1}\left(n\frac{T}{2}\right) = -\text{Im}\left\{(-j)^n \sum_{\ell=0}^{\frac{N}{2}-1} W_{N/2}^{ql}\, W_{2N}^{\ell} \sum_{p=-\infty}^{+\infty} h^{(\ell)*}\left((\gamma - n + p)\frac{T}{2}\right) e^{-j\frac{\pi}{2}(p-n)}\, r^{(\ell)*}\left(p\frac{T}{2}\right)\right\} \tag{9.56}$$

for $q = N/2 - m - 1$ and $q = N/4, \ldots, N/2 - 1$. Equations 9.51 and 9.54 to 9.56 suggest that the efficient realization of the receive filter bank comprises a S/P converter, a polyphase network with $N/2$ branches, where the $N/2$ filters have transfer functions $H^{(N/2-1-\ell)}(-jz^{N/2}), \ell = 0, \ldots, N/2 - 1$, and an $N/2$-point DFT processor, as illustrated in Figure 9.7. For the derivation of the orthogonality conditions for O-QAM OFDM filter banks with respect to the polyphase components of an FIR prototype filter, refer to Siohan et al. (2002). A group band data modem based on the O-QAM OFDM scheme, where the per-subchannel equalization method introduced in Hirosaki (1980) is employed, is described in Hirosaki et al. (1986).

9.5 Discrete Wavelet Multitone Modulation

Discrete Wavelet Multitone (DWMT) modulation is a filter-bank modulation scheme where the elements of the vector of input symbols are real-valued, and the impulse responses of the filtering elements in the transmit and receive filter banks are also real-valued. The orthogonality conditions are satisfied, assuming an ideal channel without noise and distortion (Sandberg and Tzannes, 1995). In DWMT modulation, all signal processing operations involve real signals. Therefore, for the same number of dimensions per modulation interval of the transmitted signal, the minimum bandwidth of a subchannel for DWMT is half the minimum bandwidth of a subchannel for DMT or O-QAM OFDM modulation. As illustrated by the subchannel frequency responses in Figure 9.8, obtained for $f \in (0, 0.06\, M/T)$, and a number of subchannels $M = 64$, in general DWMT has a higher spectral containment of individual subchannel signals than DMT.

Although DWMT is characterized by higher spectral containment than DMT, each subchannel requires a bandwidth larger than the minimum bandwidth of $1/(2T)$. In fact, note that the first sidelobe of the amplitude characteristic of a subchannel frequency response in Figure 9.8 is more than 45 dB below the maximum value of the characteristic. Consider, however, the amplitude characteristics at the band-edge frequencies, which are the mid-frequencies between the center frequencies of adjacent subchannels. At the band-edge frequencies, where the characteristics of adjacent subchannels cross, the amplitude characteristic is only 3 dB below the maximum. The efficient implementation of filter banks for DWMT is described in Benvenuto and Cherubini (2002). It is worth mentioning that the filtering elements for DWMT modulation can be efficiently implemented using cosine-modulated filter banks (Vaidyanathan, 1993).

Note that DWMT modulation is not as well suited for transmission over channels that exhibit a passband characteristic as the other filter-bank modulation schemes discussed in this chapter. In fact, a DWMT signal

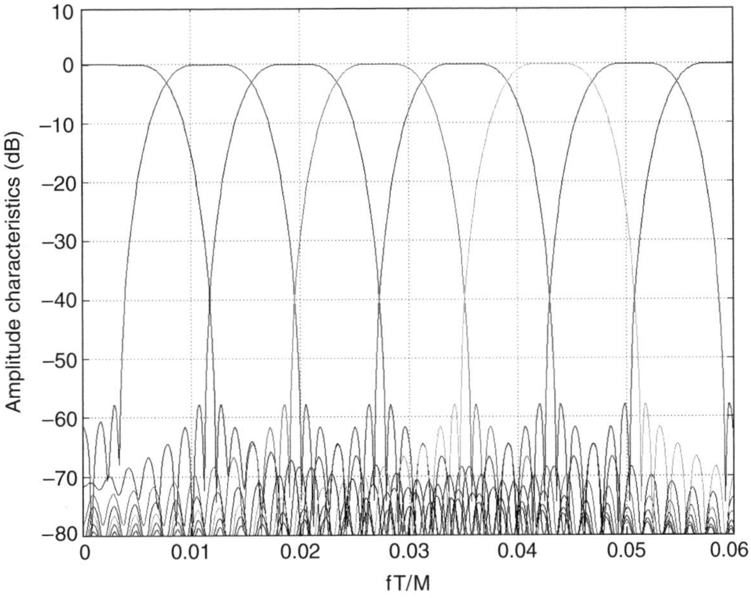

FIGURE 9.8 Amplitude characteristics of the frequency responses of adjacent subchannel filters in a DWMT system for $f \in (0, 0.06\,M/T)$ and $M = 64$. (From Cherubini et al., *IEEE Commun. Mag.*, 38, 98–104, 2000. © 2000 IEEE.)

is generated by real-valued input symbols and real-valued filter impulse responses, so that the spectrum of the baseband equivalent signal has Hermitian symmetry around the frequency $f = 0$. The passband signal can be obtained in this case by single sideband (SSB) or vestigial sideband (VSB) modulation. It turns out that SSB and VSB modulation schemes for digital passband transmission are characterized by inferior performance compared to double sideband amplitude and phase modulation (DSB-AM/PM), which can be applied in a straightforward manner to the other filter-bank modulation schemes considered here, owing to the difficulties that these schemes present for carrier-phase recovery. To obviate this problem, pilot tones can be employed to provide carrier-phase information. Transmission of pilot tones, however, is usually not practical in several applications, as it reduces the power efficiency of the system and introduces one or more spectral lines in the signal spectrum, and should be avoided if not necessary.

9.6 Filtered Multitone Modulation

High spectral containment is more easily achieved by relaxing the perfect reconstruction constraint or by resorting to noncritically sampled uniform filter banks, as shown in the equivalent baseband block diagram of Figure 9.9 for the general case $K > M$. FMT is a filter-bank modulation technique where the M-branch filters are frequency-shifted versions of a prototype filter that achieves a high level of spectral containment, such that the ICI is negligible compared with noise signals. Let us consider linear-phase FIR prototype filters of length γM, with $h(kT/K) = 0$ for $k < 0$ and $k > \gamma M - 1$. In general, larger values of γ allow a better approximation of filters with transfer functions that exhibit sharp spectral roll-off and high attenuation of out-of-band energy, but lead to an increase in system latency. The choice of the prototype filter allows various trade-offs among the number of subchannels, the level of spectral containment, signal latency, transmission efficiency, and complexity.

An efficient implementation of uniform filter-bank modulation systems also exists for the noncritically sampled case of $K > M$. With reference to Figure 9.9, assuming subchannel center frequencies given by

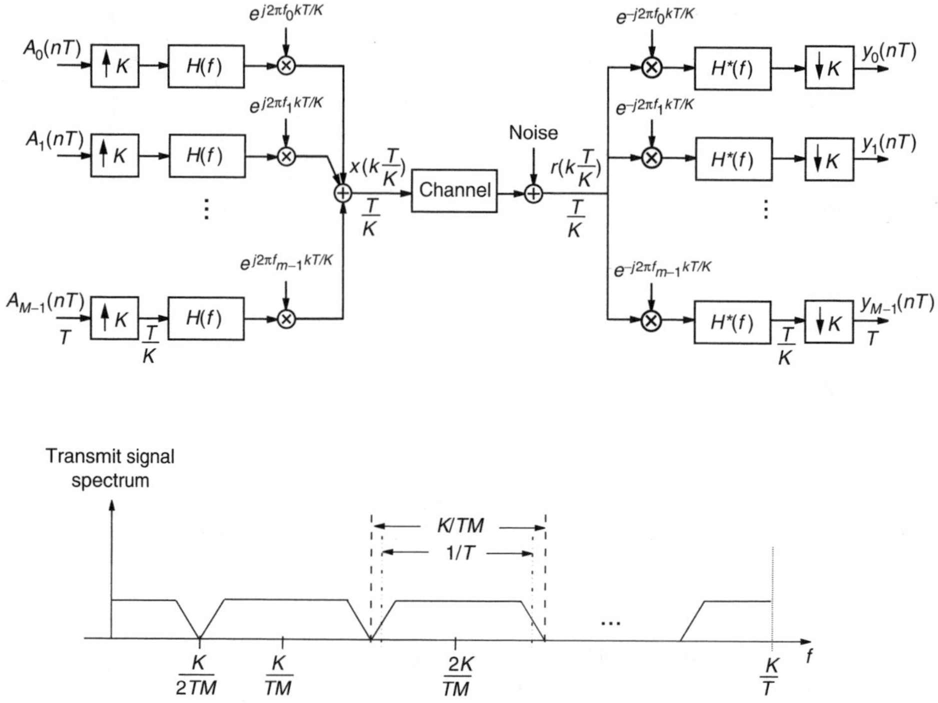

FIGURE 9.9 Block diagram of a communications system employing noncritically sampled filter banks. (From Cherubini et al., *IEEE J. Selected Areas Commun.*, 20, 1016–1028, 2002. © 2002 IEEE.)

$f_m = mK/(MT)$, $m = 0, \ldots, M-1$, the signal $x(kT/K)$ input to the channel at time kT/K is given by

$$
\begin{aligned}
x\left(k\frac{T}{K}\right) &= \sum_{m=0}^{M-1}\sum_{n=-\infty}^{+\infty} A_m(nT)h\left[(k-nK)\frac{T}{K}\right]e^{j2\pi m\frac{K}{MT}k\frac{T}{K}} \\
&= \sum_{n=-\infty}^{+\infty}\sum_{m=0}^{M-1} A_m(nT)e^{j2\pi m\frac{K}{MT}k\frac{T}{K}}\, h\left[k\frac{T}{K}-nT\right]
\end{aligned}
\tag{9.57}
$$

With the change of variables $k = pM + \ell, \ell = 0, 1, \ldots, M-1$, we get

$$
x\left(pM\frac{T}{K}+\ell\frac{T}{K}\right) = \sum_{n=-\infty}^{+\infty}\sum_{m=0}^{M-1} A_m(nT)e^{j2\pi m\frac{K}{MT}\ell\frac{T}{K}}h\left(pM\frac{T}{K}+\ell\frac{T}{K}-nT\right) \tag{9.58}
$$

By introducing $a_\ell(nT) = \sum_{m=0}^{M-1} A_m(nT)e^{j2\pi\frac{m\ell}{M}}$, Equation 9.58 is expressed as

$$
x\left(pM\frac{T}{K}+\ell\frac{T}{K}\right) = \sum_{n=-\infty}^{+\infty} a_\ell(nT)h\left(pM\frac{T}{K}+\ell\frac{T}{K}-nT\right) \tag{9.59}
$$

Clearly, $a_\ell(nT), \ell = 0, \ldots, M-1$, are obtained from $A_m(nT), m = 0, \ldots, M-1$, via an IDFT, as discussed in the preceding sections. Furthermore, by adopting the general expression for signal interpolation where a filter index $q = \lfloor\frac{pM+\ell}{K}\rfloor - n$, a basepoint index $\eta_{p,\ell} = \lfloor\frac{pM+\ell}{K}\rfloor$, and a fractional index $\nu_{p,\ell} = \frac{pM+\ell}{K} - \eta_{p,\ell}$

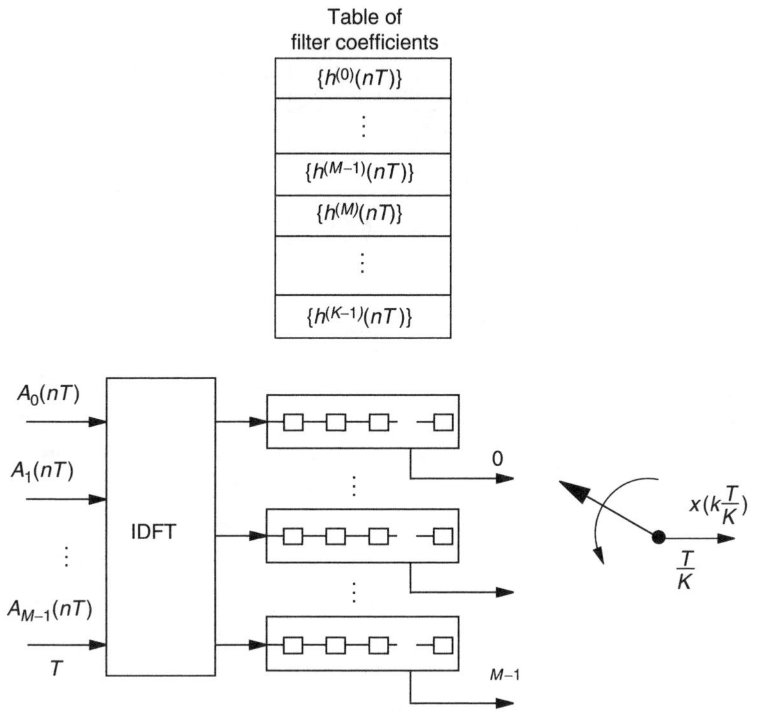

FIGURE 9.10 Efficient realization of an FMT modulator. (From Cherubini et al., *IEEE J. Selected Areas Commun.*, 20, 1016–1028, 2002. © 2002 IEEE.)

are introduced (Gardner, 1993), the transmit signal can be expressed as

$$\begin{aligned} x\left(pM\frac{T}{K}+\ell\frac{T}{K}\right) &= \sum_{q=-\infty}^{+\infty} a_\ell[(\eta_{p,\ell}-q)T]\,h[(\nu_{p,\ell}+q)T] \\ &= \sum_{q=-\infty}^{+\infty} a_\ell[(\eta_{p,\ell}-q)T]\,h^{(\nu_{p,\ell}K)}(qT), \quad \ell=0,1,\ldots,M-1 \end{aligned} \tag{9.60}$$

where $0 \le \nu_{p,\ell} < 1$ and $\nu_{p,\ell}K = (pM+\ell) \bmod K$. Hence, we find that the transmit signal at time kT/K is computed by convolving the signal samples stored in the $(k \bmod M)$-th delay line at the IDFT output with the $(k \bmod K)$-th polyphase component (with respect to K) of the prototype filter. In other words, the integer number $\nu_{p,\ell}K$ provides the address of the polyphase component that needs to be applied at the $(k \bmod M)$-th output of the IDFT to generate the transmitted signal $x(kT/K)$. Therefore, each element of the IDFT output frame is filtered by a periodically time-varying filter with a period equal to $[lcm(M,K)]T/K$, where $[lcm(M,K)]$ denotes the least common multiple of M and K. This transmitter structure is depicted in Figure 9.10.

Note that the m-th subchannel, $m = 0,\ldots, M-1$, can be considered a prototype baseband channel that is translated in frequency by $f_m = mK/(MT)$ Hz, as shown in Figure 9.9. By resorting to noncritically sampled filter banks, modulation with an excess bandwidth of $\alpha = K/M - 1$ within each subchannel is feasible and ensures total spectral containment within a subchannel. By letting $K \to M$, the penalty in bandwidth efficiency becomes vanishingly small at the price of an increase in implementation complexity because filters with increasingly sharper spectral roll-offs must then be realized. For a critically sampled system with $M = K$, the efficient realization of the transmit filter bank shown in Figure 9.10 becomes equivalent to that in Figure 9.4. In that case each element of the IDFT output frame is processed by a filter that is no longer periodically time varying.

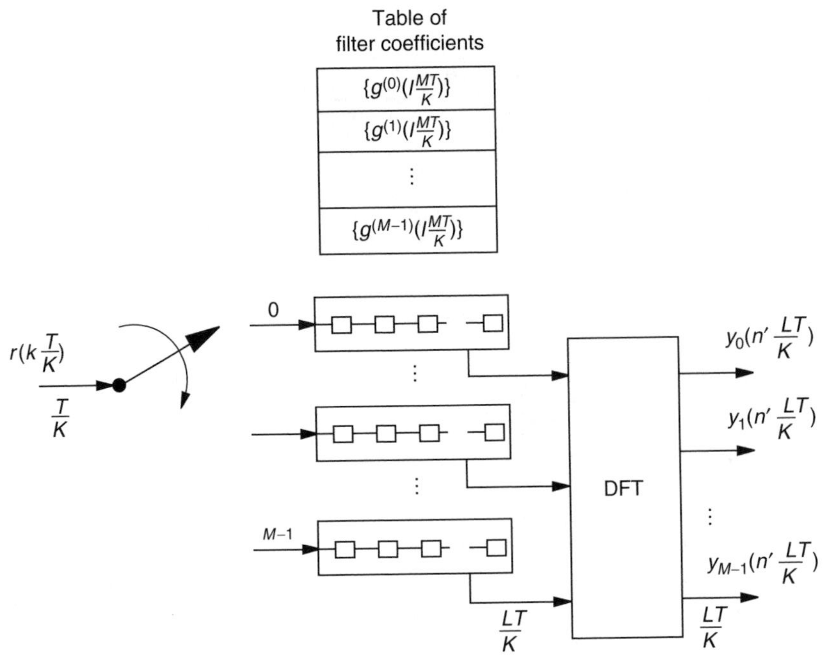

FIGURE 9.11 Efficient realization of an FMT demodulator. (From Cherubini et al., *IEEE J. Selected Areas Commun.*, 20, 1016–1028, 2002. © 2002 IEEE.)

We now turn to the efficient implementation of the FMT demodulator, where we assume the same rate for the received signals as for the transmit signals and consider in general a downsampling factor $L \leq K$ (i.e., in Figure 9.9 replace $\downarrow K$ with $\downarrow L$). The block diagram of an FMT demodulator is depicted in Figure 9.11. The received signal is denoted by $r(kT/K)$, and the filtering elements on the M branches are given by polyphase components (with respect to M) of a prototype filter $\{g(kT/K)\}$ with T/K-spaced coefficients, defined as $\{g^{(\ell)}(pMT/K)\} = \{g((pM+\ell)T/K)\}$, $\ell = 0, 1, \ldots, M-1$.

The m-th output signal of the FMT demodulator at time $n'LT/K$ is given by

$$y_m\left(n'\frac{L}{K}T\right) = \sum_{k=-\infty}^{+\infty} r\left(k\frac{T}{K}\right) e^{-j2\pi m \frac{K}{MT} k \frac{T}{K}} g\left[(Ln'-k)\frac{T}{K}\right] \tag{9.61}$$

Letting $k = pM + \ell$, $\ell = 0, 1, \ldots, M-1$, we obtain

$$y_m\left(n'\frac{L}{K}T\right) = \sum_{\ell=0}^{M-1} \sum_{p=-\infty}^{+\infty} r\left[(pM+\ell)\frac{T}{K}\right] g\left[(Ln'-pM-\ell)\frac{T}{K}\right] e^{-j2\pi \frac{m\ell}{M}} \tag{9.62}$$

which can be expressed as

$$y_m\left(n'\frac{L}{K}T\right) = \sum_{\ell=0}^{M-1} z_\ell\left(n'\frac{L}{K}T\right) e^{-j2\pi \frac{m\ell}{M}} \tag{9.63}$$

where

$$z_\ell\left(n'\frac{L}{K}T\right) = \sum_{p=-\infty}^{+\infty} r\left[(pM+\ell)\frac{T}{K}\right] g\left[(Ln'-pM-\ell)\frac{T}{K}\right] \tag{9.64}$$

Clearly, $y_m(n'LT/K), m = 0, 1, \ldots, M-1$, are obtained from $z_\ell(n'LT/K), \ell = 0, 1, \ldots, M-1$, via a DFT. Furthermore, if the polyphase components (with respect to M) of the received signal are defined as

$r^{(\ell)}(pMT/K) = r((pM+\ell)T/K), \ell = 0, 1, \ldots, M-1$, and a filter index $q' = \lfloor \frac{Ln'-\ell}{M} \rfloor - p$, basepoint index $\eta'_{n',\ell} = \lfloor \frac{Ln'-\ell}{M} \rfloor$, and fractional index $\nu'_{n',\ell} = \frac{Ln'-\ell}{M} - \eta'_{n',\ell}$ are introduced, we obtain

$$z_\ell\left(n'\frac{L}{K}T\right) = \sum_{q'=-\infty}^{+\infty} r^{(\ell)}\left[(\eta'_{n',\ell} - q')\frac{MT}{K}\right] g^{(\nu'_{n',\ell}M)}\left(q'\frac{MT}{K}\right) \tag{9.65}$$

Note that if the receive prototype filter is causal and matched to the transmit prototype filter, i.e., $\{g(kT/K)\} = \{h^*(\gamma M - kT/K)\}$, where γM denotes the filter length assumed to be a multiple of M, then Equation (9.64) becomes

$$z_\ell\left(n'\frac{L}{K}T\right) = \sum_{p=-\infty}^{+\infty} r\left[\left(p + \frac{\ell}{M}\right)\frac{MT}{K}\right] h^*\left[\left(p + \gamma + \frac{\ell - Ln'}{M}\right)\frac{MT}{K}\right] \tag{9.66}$$

In general, a new DFT output frame at time $kT/K = n'LT/K$ is obtained by the method illustrated in Figure 9.11. The commutator is circularly rotated L steps from its position at time $(n'-1)LT/K$, allowing a set of L consecutive received signals $r(kT/K)$ to be input into the M delay lines. The content of each delay line is then convolved with a polyphase component (with respect to M) of the receive prototype filter. The integer number $\nu'_{n',\ell}M$ provides the address of the polyphase component that needs to be applied at the ℓ-th branch. The resulting signals are then input to the DFT to finally yield the signals $y_m(n'LT/K), m = 0, 1, \ldots, M-1$, that are used for the detection of the transmitted symbols. Note that the DFT output frames are obtained at the rate of $(K/L)/T$.

9.6.1 Filter-bank Design

For the realization of transmit and receive filter banks, a linear-phase FIR prototype filter of length γM is considered that approximates a filter with a suitable frequency response characteristic $H_{\text{ideal}}(e^{j2\pi fT})$. Then each of the polyphase filter components (with respect to M) has γ coefficients. The parameter γ is defined as the (time-domain) *overlap factor* of the system.

The first example of prototype filter for FMT systems has the target frequency response given by

$$H_{\text{ideal},1}(e^{j2\pi fT}) = \begin{cases} \left|\frac{1+e^{-j2\pi fT}}{1+\rho e^{-j2\pi fT}}\right| & \text{if} - 1/(2T) \le f \le 1/(2T) \\ 0 & \text{otherwise} \end{cases} \tag{9.67}$$

where the parameter $0 \le \rho \le 1$ controls the spectral roll-off of the filter. The frequency response $H_{\text{ideal},1}(e^{j2\pi fT})$ exhibits spectral nulls at the band edges and, when used as the prototype filter characteristic, leads to transmission free of ICI but with ISI within a subchannel. For $\rho \to 1$, the frequency characteristic of each subchannel is characterized by steep roll-off toward the band-edge frequencies. On the other hand, for $\rho \to 0$ one obtains the partial-response class I characteristic. Figure 9.12 shows the spectral characteristics of adjacent subchannel filters in the frequency interval $(0, 0.06\, M/T)$ obtained by using a prototype FIR filter designed for $M = 64, \rho = 0.1$, and $\gamma = 10$. For transmission with 0, 6.25, or 12.5% excess bandwidth within a subchannel, a modulator with upsampling parameter values of $K = 64$, $K = 68$, or $K = 72$ would be chosen, respectively. Note that the spectral energy outside of a subchannel is suppressed by more than 67 dB, with further suppression possible by increasing the length of the prototype filter.

Another choice for the target filter characteristic is obtained if the prototype filter design is based on a square root raised cosine Nyquist filter with excess bandwidth α, i.e.,

$$H_{\text{ideal},2}(e^{j2\pi fT}) = \begin{cases} 1 & \text{if } |f| \le \frac{1-\alpha}{2T} \\ \frac{1}{\sqrt{2}}\sqrt{1 - \sin\left(\frac{\pi}{\alpha}\left(fT - \frac{1}{2}\right)\right)} & \text{if } \frac{1-\alpha}{2T} < |f| \le \frac{1+\alpha}{2T} \\ 0 & \text{otherwise} \end{cases} \tag{9.68}$$

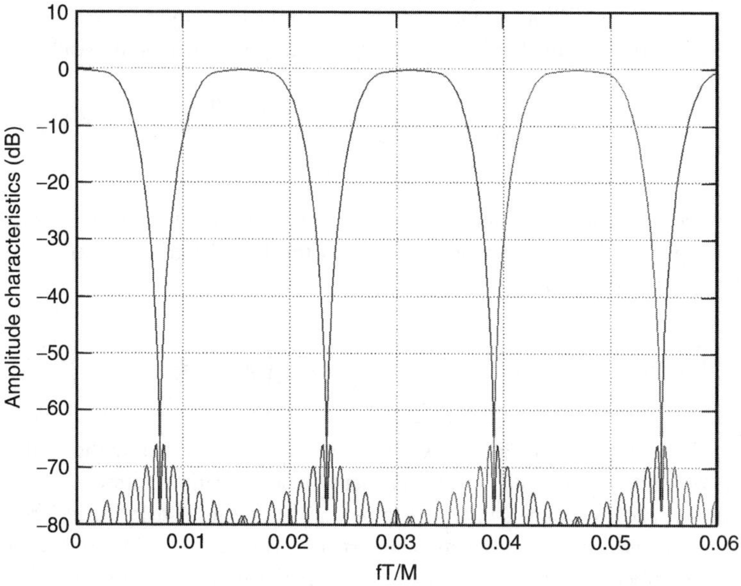

FIGURE 9.12 Amplitude characteristics of the frequency responses of adjacent subchannel filters in an FMT system for $f \in (0, 0.06M/T)$ and $M = 64$. (From Cherubini et al., *IEEE Commun. Mag.*, 38, 98–104, 2000. © 2000 IEEE.)

The frequency response $H_{\text{ideal},2}(e^{j2\pi fT})$ leads to transmission free of ICI and ISI within a subchannel if the channel is ideal. In other words, the perfect reconstruction conditions are, in this case, satisfied.

To develop a prototype filter design algorithm for FMT, we determine the conditions for ICI suppression in a multicarrier system employing critically sampled uniform filter banks. For the special case $K = M$, the modulator and demodulator structures derived in Section 9.2 take the simple form depicted in Figure 9.4. The prototype filter transfer functions at the transmitter and receiver are $H(z)$ and $H^*(1/z^*)$, respectively. Note that the commutators are replaced by the equivalent P/S and S/P converters, and that the filtering elements on the M branches of the transmit and receive filter banks are not time varying. A normalized transmission interval $T/M = 1$ is assumed. Let us express the transfer function of the prototype filter as

$$H(z) = \sum_{\ell=0}^{M-1} z^{-\ell} H^{(\ell)}(z^M) \tag{9.69}$$

where $H^{(\ell)}(z)$ denotes the transfer function of the ℓ-th polyphase component (with respect to M) of $H(z)$. The transfer function of the noiseless channel is expressed in terms of its polyphase components as

$$G_c(z) = \sum_{\ell=0}^{M-1} z^{-\ell} G_c^{(\ell)}(z^M) \tag{9.70}$$

By repeatedly applying the noble identities (Vaidyanathan, 1993, pp. 119–120) and the polyphase identity (Vaidyanathan, 1993, p. 133), the transfer matrix of the filter-bank communications system shown in Figure 9.4 is expressed as

$$\Lambda(z) = \mathbf{F}_M \, \Gamma(z) \mathbf{F}_M^{-1} \tag{9.71}$$

where the matrix $\Gamma(z)$ is given by

$$\Gamma(z) = \tag{9.72}$$

$$\begin{bmatrix} H^{(0)}(z) & 0 & \cdots & 0 \\ 0 & H^{(1)}(z) & \cdots & 0 \\ \vdots & \vdots & & \vdots \\ 0 & 0 & \cdots & H^{(M-1)}(z) \end{bmatrix} \times \begin{bmatrix} z^{-1}G_c^{(0)}(z) & z^{-1}G_c^{(1)}(z) & \cdots & z^{-1}G_c^{(M-1)}(z) \\ z^{-2}G_c^{(M-1)}(z) & z^{-1}G_c^{(0)}(z) & \cdots & z^{-1}G_c^{(M-2)}(z) \\ \vdots & \vdots & & \vdots \\ z^{-2}G_c^{(1)}(z) & z^{-2}G_c^{(2)}(z) & \cdots & z^{-1}G_c^{(0)}(z) \end{bmatrix}$$

$$\times \begin{bmatrix} \left[H^{(0)}\left(\frac{1}{z^*}\right)\right]^* & 0 & \cdots & 0 \\ 0 & \left[H^{(1)}\left(\frac{1}{z^*}\right)\right]^* & \cdots & 0 \\ \vdots & \vdots & & \vdots \\ 0 & 0 & \cdots & \left[H^{(M-1)}\left(\frac{1}{z^*}\right)\right]^* \end{bmatrix} = z^{-1}$$

$$\times \begin{bmatrix} H^{(0)}(z)\,G_c^{(0)}(z)\,H^{(0)*}\left(\frac{1}{z^*}\right) & H^{(0)}(z)\,G_c^{(1)}(z)\,H^{(1)*}\left(\frac{1}{z^*}\right) & \ldots & H^{(0)}(z)\,G_c^{(M-1)}(z)\,H^{(M-1)*}\left(\frac{1}{z^*}\right) \\ z^{-1}H^{(1)}(z)\,G_c^{(M-1)}(z)\,H^{(0)*}\left(\frac{1}{z^*}\right) & H^{(1)}(z)\,G_c^{(0)}(z)\,H^{(1)*}\left(\frac{1}{z^*}\right) & \ldots & H^{(1)}(z)\,G_c^{(M-2)}(z)\,H^{(M-1)*}\left(\frac{1}{z^*}\right) \\ \vdots & \vdots & & \vdots \\ z^{-1}H^{(M-1)}(z)\,G_c^{(1)}(z)\,H^{(0)*}\left(\frac{1}{z^*}\right) & z^{-1}H^{(M-1)}(z)\,G_c^{(2)}(z)\,H^{(1)*}\left(\frac{1}{z^*}\right) & \ldots & H^{(M-1)}(z)\,G_c^{(0)}(z)\,H^{(M-1)*}\left(\frac{1}{z^*}\right) \end{bmatrix}$$

Observing Equation 9.69 and recalling that $\Lambda(z)$ is diagonal if and only if $\Gamma(z)$ is circulant, with elements on the diagonal given by the IDFT of the first row of $\Gamma(z)$, we find that a set of sufficient conditions for ICI suppression is given by

$$\begin{aligned} &H^{(0)}(z)H^{(0)*}\left(\frac{1}{z^*}\right) = H^{(1)}(z)H^{(1)*}\left(\frac{1}{z^*}\right) = \cdots = H^{(M-1)}(z)H^{(M-1)*}\left(\frac{1}{z^*}\right) \\ &H^{(0)}(z)H^{(1)*}\left(\frac{1}{z^*}\right) = H^{(1)}(z)H^{(2)*}\left(\frac{1}{z^*}\right) = \cdots = z^{-1}H^{(M-1)}(z)H^{(0)*}\left(\frac{1}{z^*}\right) \\ &\vdots \\ &H^{(0)}(z)H^{(M-1)*}\left(\frac{1}{z^*}\right) = z^{-1}H^{(1)}(z)H^{(0)*}\left(\frac{1}{z^*}\right) = \cdots = z^{-1}H^{(M-1)}(z)H^{(M-2)*}\left(\frac{1}{z^*}\right) \end{aligned} \tag{9.73}$$

Note that in general the conditions in Equation 9.73 cannot be satisfied by a FIR prototype filter, as they require strict band limitation within a subchannel. Furthermore, for an ideal channel with transfer function $G_c(z) = 1$, it can readily be seen that the conditions for ICI suppression are given by the first row of Equation 9.73, which requires that the correlation functions of the polyphase components are all equal. Therefore, if a linear-phase FIR prototype filter is chosen that approximates a filter with strictly band-limited frequency response and polyphase components that satisfy the first row of Equation 9.73, then the system will exhibit no ICI if the channel is ideal. If the channel is nonideal, the level of ICI will depend on the stopband attenuation. Let us define $\tilde{H}_{\text{ideal}}(z)$ as a filter that satisfies the equality

$$\tilde{H}_{\text{ideal}}(z)\tilde{H}_{\text{ideal}}{}^*\left(\frac{1}{z^*}\right) = H^{(0)}(z)H^{(0)*}\left(\frac{1}{z^*}\right) = \cdots = H^{(M-1)}(z)H^{(M-1)*}\left(\frac{1}{z^*}\right) \tag{9.74}$$

Note that for an ideal transmission channel, Equation 9.74 gives the transfer function of each subchannel. As mentioned above, we consider FIR filters of length γM. Then each of the polyphase filter components has γ coefficients. The approximation problem is formulated first as an unconstrained optimization problem (Princen, 1995) and then reduced to a form that requires finding the vector of filter coefficients

$\mathbf{h} = [h(\gamma M/2), \ldots, h(\gamma M - 1)]^{\mathrm{T}}$ that minimizes the quadratic objective function

$$\mathbf{h}^{\mathrm{T}}\mathbf{Q}^{\mathrm{T}}\mathbf{Q}\mathbf{h} + \frac{\kappa}{2}(\mathbf{J}\mathbf{h} - \mathbf{v})^{\mathrm{T}}(\mathbf{J}\mathbf{h} - \mathbf{v}) \tag{9.75}$$

where the first term reflects the stopband mean energy of the prototype filter computed by dividing the stopband interval $(\pi/M, \pi)$ into $N_s - 1$ subintervals, and the second term denotes the L_2 norm of the approximation error of Equation 9.74. In Equation 9.75 κ is a positive constant, and the elements of the matrices **S**, **J**, and the vector **v** are given by

$$S(i,j) = \left[2\cos\left(\frac{2j+1}{2}(\omega_s + i\Delta\omega)\right)\right], \; 0 \le i \le N_s - 1, 0 \le j \le \frac{\gamma M}{2} - 1$$

with $\omega_s = \pi/M, \Delta\omega = (\pi - \omega_s)/(N_s - 1)$,

$$J(k + i\gamma, j) = [\tilde{J}(k + i\gamma, \gamma M/2 + j) + \tilde{J}(k + i\gamma, \gamma M/2 - 1 - j)], \; 0 \le k \le \gamma - 1,$$
$$0 \le i \le M - 1, \; 0 \le j \le \gamma M/2 - 1$$

with $\tilde{J}(k + i\gamma, l) = \begin{cases} h(l - kM)\delta_{(l-i) \bmod M} & \text{if } l \ge kM \\ 0 & \text{otherwise} \end{cases}, \; 0 \le l \le \gamma M - 1$, and

$$v(n) = \frac{1}{2\pi M}\int e^{j\omega[n \bmod \gamma]}\left|\tilde{H}_{\mathrm{ideal}}(e^{j\omega})\right|^2 d\omega, \; 0 \le n \le \gamma M - 1$$

The minimization of Equation 9.75 can be performed, e.g., by an iterative least squares algorithm (Rossi et al., 1996). If the duration of the correlation function of $H_{\mathrm{ideal}}(z)$ is greater than $2\gamma - 1$, windowing of the correlation may be used to avoid the Gibbs phenomenon.

9.6.2 Per-Subchannel Adaptive Equalization and Precoding

In FMT systems, per-subchannel signal equalization is employed at the receiver in the form of symbol or fractionally spaced linear or decision-feedback equalizers. If the transmission channel is slowly time varying, precoding techniques can also be applied. We recall that the frequency responses of FMT subchannels are characterized by steep roll-off toward the band-edge frequencies, where they exhibit near spectral nulls. This suggests that per-subchannel decision-feedback equalization be performed to recover the transmitted symbols. In this section, we address this topic and also consider the application of precoding techniques to FMT modulation for transmission over slowly time-varying frequency-selective channels (Cherubini et al., 2002).

In an FMT receiver employing per-subchannel equalization, the signals $y_m(n'LT/K), m = 0, 1, \ldots, M - 1$, at the FMT demodulator output are input for symbol detection to M adaptive decision-feedback equalizers having feed-forward linear sections with LT/K-spaced coefficients. If the coefficient vectors of the feed-forward linear section and of the feedback section of the m-th equalizer at time nT are defined as $\mathbf{c}_m(nT) = \{c_m^{(l)}(nT), l = 0, \ldots, N_f - 1\}$ and $\mathbf{d}_m(nT) = \{d_m^{(i)}(nT), i = 1, \ldots, N_b\}$, respectively, the equalizer output on the m-th subchannel at time nT is given by

$$s_m(nT) = \sum_{l=0}^{N_f-1} y_m\left[\left(n\frac{K}{L} - l\right)\frac{LT}{K}\right]c_m^{(l)}(nT) - \sum_{i=1}^{N_b}\hat{A}_m[(n-i)T]d_m^{(i)}(nT) \tag{9.76}$$

where $\hat{A}_m(nT)$ denotes the symbol decision on the m-th subchannel at time nT, which is provided by a memoryless decision element. Note that the choice of $L = K/2$ results in fractionally $T/2$-spaced coefficients of the linear feed-forward equalizer section.

Error propagation inherent in decision-feedback equalization can be avoided by resorting to precoding techniques. For example, per-subchannel Tomlinson–Harashima precoding (Tomlinson, 1971; Harashima and Miyakawa, 1972) can be applied in a straightforward manner. The application of precoding techniques

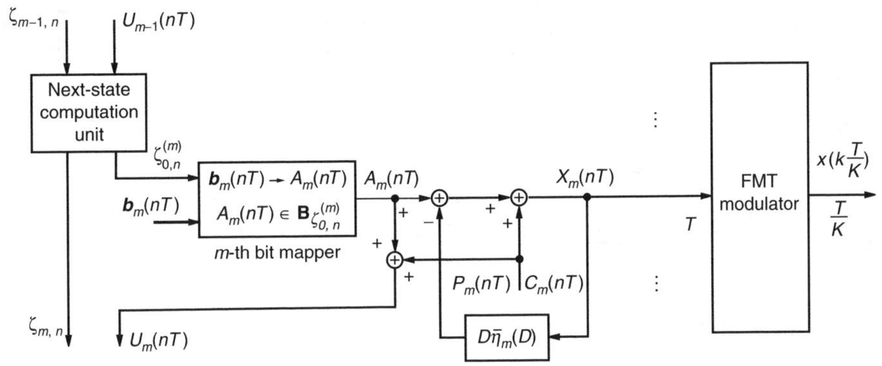

FIGURE 9.13 FMT transmitter employing trellis-augmented precoding. (From Cherubini et al., *IEEE J. Selected Areas Commun.*, 20, 1016–1028, 2002. © 2002 IEEE.)

in conjunction with trellis-coded modulation (TCM) for FMT transmission requires further discussion. *Flexible precoding* (Eyuboglu and Forney, 1992) or *trellis-enhanced precoding* (Laroia, 1996) for trellis-coded transmission over an ISI channel allows coding gains to be achieved for arbitrary constellations provided that the ISI channel is linearly invertible. However, if the channel exhibits spectral nulls, as is usually the case for an FMT subchannel characteristic, infinite error propagation can occur within the inverse precoder at the receiver. In that case, feedback trellis encoding can still be achieved by *trellis-augmented precoding* (Cherubini et al., 1997), but then only certain constellations are allowed. With trellis-augmented precoding, error propagation in the FMT receiver is completely avoided.

The block diagram of an FMT transmitter that employs trellis-augmented precoding is illustrated in Figure 9.13. For transmission over the m-th subchannel, an $N_m \times N_m$ point signal constellation is assumed, with N_m even, $\forall m$. The transmitter consists of M units chained together to allow trellis coding across the subchannels. This arrangement is chosen to reduce decoding delay and is possible because transmission over M independent subchannels may be assumed. The m-th subchannel response in D-transform notation is denoted by $\eta_m(D) = 1 + D\tilde{\eta}_m(D) = 1 + \eta_{1,m}D + \eta_{2,m}D^2 + \cdots$. We consider a trellis code with 2^ν states and conventional encoding based on a systematic convolutional encoder. For the remainder of this section, we assume that the reader is familiar with precoding techniques, such as those described in Eyuboglu and Forney (1992) and Laroia (1996).

To explain transmitter operations, it is assumed that the subchannels with indices from 1 to $M-1$ are used for transmission. Let us consider the m-th trellis precoding element. At the n-th modulation interval, the inputs to this element are the binary information vector $\mathbf{b}_m(nT)$, which may take one of $N_m^2/2$ different values, and the binary value $\zeta_{0,n}^{(m)}$, which represents the least significant bit of the TCM state $\zeta_{m,n} = (\zeta_{\nu-1,n}^{(m)}, \ldots, \zeta_{0,n}^{(m)})$ at the m-th subchannel during the n-th modulation interval. The mapping of information bits into symbols operated by the m-th signal mapper follows the usual rules of trellis coding (Ungerboeck, 1982). In particular, note that $A_m(nT) \in \mathbf{B}_{\zeta_{0,n}^{(m)}}, \zeta_{0,n}^{(m)} \in \{0, 1\}$, where $\mathbf{B}_0$ and $\mathbf{B}_1$ are the two sets obtained at the first partitioning level of the $N_m \times N_m$-point signal constellation of the m-th subchannel.

The m-th precoder operates on the symbol $A_m(nT)$ and determines the m-th subchannel input signal $X_m(nT)$. The sequence of the m-th subchannel input signals is given by

$$X_m(D) = A_m(D) + C_m(D) - P_m(D) \tag{9.77}$$

where

$$P_m(D) = [\eta_m(D) - 1]X_m(D) = D\tilde{\eta}_m(D)X_m(D) \tag{9.78}$$

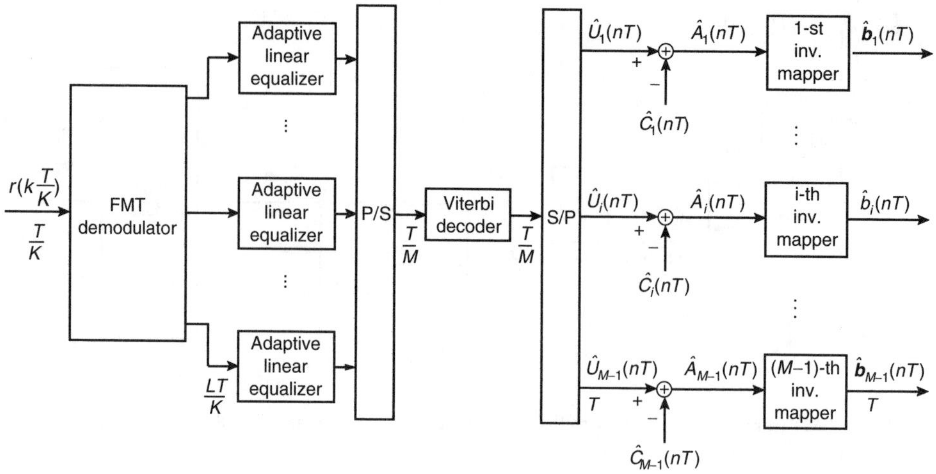

FIGURE 9.14 FMT receiver employing trellis-augmented precoding. (From Cherubini et al., *IEEE J. Selected Areas Commun.*, 20, 1016–1028, 2002. © 2002 IEEE.)

and

$$C_m(D) = -Q_{\Lambda_X^{(m)}}\{A_m(D) - P_m(D)\} \tag{9.79}$$

where $Q_{\Lambda_X^{(m)}}\{\cdot\}$ denotes quantization to the closest point of the lattice $\Lambda_X^{(m)}$ underlying the power-efficient modulo extension of the fundamental region of the m-th signal constellation. Note that by defining the signal $U_m(D) = A_m(D) + C_m(D)$ and observing the expression of $P_m(D)$, the signal $X_m(D)$ can be expressed as

$$X_m(D) = \frac{A_m(D) + C_m(D)}{1 + D\tilde{\eta}_m(D)} = \frac{U_m(D)}{\eta_m(D)} \tag{9.80}$$

Therefore, $U_m(D)$ represents the output of the m-th subchannel in the absence of noise. To allow correct decoding operations with trellis coding performed across the subchannels, the symbol $U_m(nT)$ must represent a valid continuation of the code sequence $(\ldots, U_{m-2}(nT), U_{m-1}(nT))$, assuming the TCM state $\zeta_{m,n}$. Recalling that the m-th signal constellation is an $N_m \times N_m$-point constellation with N_m even, $A_m(nT) \in \mathbf{B}_{\zeta_{0,n}^{(m)}}$, and $C_m(nT) \in \Lambda_X^{(m)}$, we find that $U_m(nT) = A_m(nT) + C_m(nT) \in \mathbf{B}_{\zeta_{0,n}^{(m)}}$. This implies that $U_m(nT)$ represents a valid continuation of the code sequence. To obtain a valid continuation of the code sequence also at the $(m+1)$-th trellis precoding element, the information about the current TCM state $\zeta_{m,n}$ and about the code symbol $U_m(nT)$ chosen is sent to the $(m+1)$-th next-state computation unit, which determines the state $\zeta_{m+1,n} = (\zeta_{\nu-1,n}^{(m+1)}, \ldots, \zeta_{0,n}^{(m+1)})$. After the $(M-1)$-th trellis precoding element, a delay of one modulation interval is provided to allow the information about the TCM state and the code symbol chosen to be presented at the beginning of the next modulation interval at the input of the first element.

The block diagram of the receiver of an FMT system with trellis-augmented precoding is illustrated in Figure 9.14. The m-th subchannel output signal is given by $V_m(D) = U_m(D) + W_m(D)$, where $W_m(D)$ represents a sequence of additive Gaussian noise samples. In this receiver, per-subchannel linear adaptive equalization is included to compensate for mismatches between the subchannel frequency responses assumed at the transmitter and the actual responses. The M subchannel output signals are converted from parallel to serial, and the resulting sequence is input to a Viterbi decoder. The decoder output sequence $\hat{U}_m(D)$ is converted from serial to parallel. The detected sequences $\hat{A}_m(D)$, $m = 0, 1, \ldots, M-1$, are then given by the memoryless operation

$$\hat{A}_m(D) = \hat{U}_m(D) - Q_{\Lambda_X^{(m)}}\{\hat{U}_m(D)\} = \hat{U}_m(D) - \hat{C}_m(D) \tag{9.81}$$

Note that error propagation in the receiver is completely avoided. The sequences of binary information vectors $\hat{\mathbf{b}}_m(D)$ are then recovered from the sequences $\hat{A}_m(D)$.

9.7 Conclusion

Filter-bank modulation techniques represent interesting solutions for difficult transmission environments, as they provide some unique advantages in terms of spectral efficiency, robustness against impulse noise and narrowband interference, and flexibility in spectrum allocation in multiple-access systems.

Noncritically sampled filter-bank modulation schemes digitally generate a set of tightly packed single-carrier-like multiband signals with minimum analog filtering requirements and offer filtering-based implementation alternatives to time-domain cyclic extensions.

In multiple-access systems, application of noncritically sampled filter-bank modulation does not require synchronization between transmissions taking place simultaneously, thus allowing simpler system operations. Compared with schemes using cyclic extensions, enhanced performance is obtained at the cost of per-subchannel filtering, which corresponds to functions performed at a significantly lower rate than the transmission rate.

References

Baldemair, R. and Frenger, P., A time-domain equalizer minimizing intersymbol and intercarrier interference in DMT systems, in *Proceedings of the IEEE GLOBECOM '01*, IEEE, Piscataway, NJ, 2001.

Bellanger, M.G., Bonnerot, G. and Coudreuse, M., Digital filtering by polyphase network: application to sample-rate alteration and filter banks, *IEEE Trans. Acoustics Speech Signal Process.*, ASSP-24, 109–114, 1976.

Benvenuto, N. and Cherubini, G., *Algorithms for Communications Systems and Their Applications*, John Wiley & Sons, Chichester, U.K., 2002.

Benvenuto, N., Cherubini, G. and Tomba, L., Achievable bit rates of DMT and FMT systems in the presence of phase noise and multipath, in *Proceedings of the 51st IEEE Annual Vehicular Technology Conference "VTC 2000,"* IEEE, Piscataway, NJ, 2000.

Bingham, J.A.C., Multicarrier modulation for data transmission: an idea whose time has come, *IEEE Commun. Mag.*, 28, 5–14, 1990.

Campello, J., Practical bit loading for DMT, in *Proceedings of the 1999 IEEE International Conference on Communications, "ICC 1999,"* IEEE, Piscataway, NJ, 1999, pp. 801–805.

Chang, R.W., Synthesis of band-limited orthogonal signals for multichannel data transmission, *Bell System Tech. J.*, 1775–1796, 1966.

Cheng, R.S. and Verdu, S., Gaussian multiaccess channels with ISI: capacity region and multiuser water-filling, *IEEE Trans. Inf. Theory*, 39, 773–785, 1993.

Cherubini, G., A hybrid TDMA/CDMA system based on filtered multitone modulation for upstream transmission in HFC networks, in *Multi-Carrier Spread Spectrum and Related Topics*, Fazel, K. and Kaiser, S., Eds., Kluwer Academic Publishers, Boston, MA, 2000, pp. 79–86.

Cherubini, G., Ölçer, S. and Ungerboeck, G., Trellis precoding for channels with spectral nulls, in *Proceedings of the IEEE International Symposium on Information Theory, "ISIT '97,"* IEEE, Piscataway, NJ, 1997, p. 464.

Cherubini, G., Eleftheriou, E., Ölçer, S. and Cioffi, J., Filter bank modulation techniques for very high-speed digital subscriber lines, *IEEE Commun. Mag.*, 38, 98–104, 2000.

Cherubini, G., Eleftheriou, E. and Ölçer, S., Filtered multitone modulation for very high-speed digital subscriber lines, *IEEE J. Selected Areas Commun.*, 20, 1016–1028, 2002.

Chow, J.S., Tu, J.C. and Cioffi, J.M., A discrete multitone transceiver system for HDSL applications, *IEEE J. Selected Areas Commun.*, 9, 895–908, 1991.

Chow, P.S., Cioffi, J.M. and Bingham, A.C., A practical discrete multitone transceiver loading algorithm for data transmission over spectrally shaped channels, *IEEE Trans. Commun.*, 43, 773–775, 1995.

Cioffi, J.M., Asymmetrical digital subscriber lines, in *The Communications Handbook*, Gibson, J.D., Ed., CRC Press Inc., Boca Raton, FL, 1997, pp. 450–479.

Crochiere, R.E., Weber, S.A. and Flanagan, J.L., Digital coding of speech in sub-bands, *Bell System Tech. J.*, 1069–1085, 1976.

Doelz, M.L., Heald, E.T. and Martin, D.L., Binary data transmission techniques for linear systems, *Proc. IRE*, 656–661, 1957.

Esteban, D. and Galand, C., Application of quadrature mirror filters to split band voice coding schemes, in *Proceedings of the IEEE ICASSP '77*, IEEE, Piscataway, NJ, 1977, pp. 191–195.

ETS (European Telecommunication Standard), Radio broadcast systems; digital audio broadcasting (DAB) to mobile, portable and fixed receivers, ETS 300 401, November 1994.

ETS (European Telecommunication Standard), Digital broadcasting systems for television, sound and data services; framing structure, channel coding and modulation for digital terrestrial television, ETS 300 744, February 1997.

Eyuboglu, M.V. and Forney, G.D., Trellis precoding: combined coding, precoding and shaping for intersymbol interference channels, *IEEE Trans. Inf. Theory*, 38, 301–314, 1992.

Fliege, N.J., *Multirate Digital Signal Processing*, John Wiley & Sons, Chichester, U.K., 1994.

Gardner, F., Interpolation in digital modems. Part I. Fundamentals, *IEEE Trans. Commun.*, 41, 501–507, 1993.

Haas, R. and Belfiore, J.-C., Multiple carrier transmission with time-frequency well-localized impulses, in *Proceedings of the IEEE 2nd Symposium on Communications and Vehicular Technology in the Benelux*, IEEE, 1994, pp. 187–193.

Harashima, H. and Miyakawa, H., Matched transmission technique for channels with intersymbol interference, *IEEE Trans. Commun.*, 20, 774–780, 1972.

Harmuth, H.F., On the transmission of information by orthogonal time functions, *AIEE Trans.*, Pt. I (Communications and Electronics), 248–255, 1960.

Hirosaki, B., An analysis of automatic equalizers for orthogonally multiplexed QAM systems, *IEEE Trans. Commun.*, COM-28, 73–83, 1980.

Hirosaki, B., An orthogonally multiplexed QAM system using the discrete Fourier transform, *IEEE Trans. Commun.*, COM-29, 982–989, 1981.

Hirosaki, B., Hasegawa, S. and Sabato, A., Advanced groupband data modem using orthogonally multiplexed QAM technique, *IEEE Trans. Commun.*, COM-34, 587–592, 1986.

Jacobsen, K.S., Bingham, J.A.C. and Cioffi, J.M., Synchronized DMT for multipoint-to-point communications on HFC networks, in *Proceedings of the IEEE GLOBECOM '95*, IEEE, Piscataway, NJ, 1995, pp. 963–966.

Kammeyer, K.D., Tuisel, U., Schulze, H. and Bochmann, H., Digital multicarrier-transmission of audio signals over mobile radio channels, *Eur. Trans. Telecommun.*, 3, 243–253, 1992.

Laroia, R., Coding for intersymbol interference channels: combined coding and precoding, *IEEE Trans. Inf. Theory*, 42, 1053–1061, 1996.

Leke, A. and Cioffi, J.M., A maximum rate loading algorithm for discrete multitone systems, in *Proceedings of the IEEE GLOBECOM '97*, IEEE, Piscataway, NJ, 1997, pp. 1514–1518.

Leus, G., Barhumi, I. and Moonen, M., Per tone equalization for MIMO-OFDM systems, in *Proceedings of the 2003 IEEE International Conference on Communications, "ICC 2003,"* IEEE, Piscataway, NJ, 2003, p. ASP05-1.

Matheus, K., Kammeyer, K.D. and Tuisel, U., Implementation of multicarrier systems with polyphase filterbanks, in *Multi-Carrier Spread Spectrum and Related Topics*, Fazel, K. and Kaiser, S., Eds., Kluwer Academic Publishers, Boston, MA, 2000, pp. 329–336.

Melsa, P.J.W., Younce, R.C. and Rohrs, C.E. Impulse response shortening for discrete multitone transceivers, *IEEE Trans. Commun.*, 44, 1662–1672, 1996.

Peled, A. and Ruiz, A., Frequency domain data transmission using reduced computational complexity algorithms, in *Proceedings of the IEEE ICASSP'80*, IEEE, Piscataway, NJ, 1980, pp. 964–967.

Princen, J., The design of nonuniform modulated filter banks, *IEEE Trans. Signal Process.*, 43, 2550–2560, 1995.

Rossi, M., Zhang, J.-Y. and Steenaart, W., Iterative least squares design of perfect reconstruction QMF filter banks, in *Proceedings of the IEEE CCECE'96*, IEEE, Piscataway, NJ, 1996, pp. 762–765.

Saltzberg, B., Performance of an efficient parallel data transmission system, *IEEE Trans. Commun. Technol.*, COM-15, 805–811, 1967.

Sandberg, S.D. and Tzannes, M.A., Overlapped discrete multitone modulation for high speed copper wire communications, *IEEE J. Selected Areas Commun.*, 13, 1571–1585, 1995.

Sari, H., Levy, Y. and Karam, G.S., An analysis of orthogonal frequency-division multiple access, in *Proceedings of the IEEE GLOBECOM '97*, IEEE, Piscataway, NJ, 1997, pp. 1635–1639.

Scaglione, A., Barbarossa, S. and Giannakis, G.B., Filterbank transceivers optimizing information rate in block transmissions over dispersive channels, *IEEE Trans. Inf. Theory*, 45, 1019–1032, 1999a.

Scaglione, A., Giannakis, G.B. and Barbarossa, S., Redundant filterbank precoders and equalizers. Part I. Unification and optimal designs, *IEEE Trans. Signal Process.*, 47, 1988–2006, 1999b.

Scaglione, A., Stoica, P., Barbarossa S., Giannakis, G.B. and Sampath, H., Optimal designs for space-time linear precoders and decoders, *IEEE Trans. Signal Process.*, 50, 1051–1064, 2002.

Schafer, R.W. and Rabiner, L.R., Design of digital filter banks for speech analysis, *Bell System Tech. J.*, 3097–3115, 1971.

Shannon, C.E., A mathematical theory of communication, *Bell System Tech. J.*, 379–423, 623–656, 1948.

Siohan, P., Siclet, C. and Lacaille, N., Analysis and design of OFDM/OQAM systems based on filterbank theory, *IEEE Trans. Signal Process.*, 50, 1170–1183, 2002.

Smith, M.J.T. and Barnwell, T.P., A procedure for designing exact reconstruction filterbanks for tree structured sub-band coders, in *Proceedings of the IEEE ICASSP'84*, IEEE, Piscataway, NJ, 1984, pp. 27.1–27.4.

Stamoulis, A., Giannakis, G.B. and Scaglione, A., Block FIR decision-feedback equalizers for filterbank precoded transmissions with blind channel estimation capabilities, *IEEE Trans. Commun.*, 49, 69–83, 2001.

Tomlinson, M., New automatic equalizer employing modulo arithmetic, *Electron. Lett.*, 7, 138–139, 1971.

Trautmann, S. and Fliege, N.J., Perfect equalization for DMT systems without guard interval, *IEEE J. Selected Areas Commun.*, 20, 987–996, 2002.

Ungerboeck, G., Channel coding with multilevel/phase signals, *IEEE Trans. Inf. Theory*, IT-28, 55–67, 1982.

Vahlin, A. and Holte, N., Optimal finite duration pulses for OFDM, *IEEE Trans. Commun.*, 44, 10–14, 1996.

Vaidyanathan, P.P., *Multirate Systems and Filter Banks*, Prentice Hall, Englewod Cliffs, NJ, 1993.

Vallet, R. and Taieb, K.H., Fraction spaced multi-carrier modulation, *Wireless Personal Commun.*, 2, 97–103, 1995.

Van Acker, K., Leus, G., Moonen, M., van de Viel, O. and Pollet, T., Per tone equalization for DMT-based systems, *IEEE Trans. Commun.*, 49, 109–119, 2001.

Vandendorpe, L., Cuvelier, L., Deryck, F., Louveaux, J. and van de Wiel, O., Fractionally spaced linear and decision-feedback detectors for transmultiplexers, *IEEE Trans. Signal Process.*, 46, 996–1011, 1998.

Vetterli, M., A theory of multirate filter banks, *IEEE Trans. Acoustics Speech Signal Process.*, ASSP-35, 356–372, 1987.

Weinstein, S.B. and Ebert, P.M., Data transmission by frequency-division multiplexing using the discrete Fourier transform, *IEEE Trans. Commun. Technol.*, COM-19, 628–634, 1971.

IV

Multiple Access Techniques

10 Spread-Spectrum Techniques for Mobile Communications *Filippo Giannetti and Marco Luise* **10**-1
A Brief History of Wireless Communications • Fundamentals of Digital Spread-Spectrum Signaling • Code-Division Multiple Access • A Review of 2G and 3G Standards for CDMA Mobile Communications • Synchronization for Spread-Spectrum and CDMA Signals • Architecture of DSP-Based DS/SS and CDMA Receivers for Wireless Mobile Communications • Multiuser Detection • Perspectives and Conclusions

11 Multiuser Detection for Fading Channels *Stefano Buzzi* **11**-1
Introduction • Signal and Channel Model • Multiuser Detection with Known CSI • Multiuser Detection with Unknown CSI • Conclusions

10

Spread-Spectrum Techniques for Mobile Communications

10.1 A Brief History of Wireless Communications........ 10-2
The Wireless Revolution • 2G and 3G Cellular Systems • DSP Components for Wireless Communications

10.2 Fundamentals of Digital Spread-Spectrum Signaling 10-4
Narrowband • Spread-Spectrum • Frequency-Hopping Spread Spectrum • Direct-Sequence Spread-Spectrum • DS/SS Signal Model • Real Spreading • Complex Spreading • DS/SS Bandwidth Occupancy • Spreading Factor • Short Code • Long Code • Processing Gain • Pseudorandom Sequence Generators • Basic Architecture of a DS/SS Modem

10.3 Code-Division Multiple Access 10-10
Frequency-, Time-, and Code-Division Multiplexing • Multirate Code-Division Multiplexing • Multiple Access Interference • Capacity of a CDMA System • Cellular Networks and the Universal Frequency Reuse

10.4 A Review of 2G and 3G Standards for CDMA Mobile Communications 10-14
IS-95 • UMTS/UTRA • cdma2000

10.5 Synchronization for Spread-Spectrum and CDMA Signals.. 10-18
Synchronization Functions • Code Synchronization • Carrier Frequency and Phase Synchronization

10.6 Architecture of DSP-Based DS/SS and CDMA Receivers for Wireless Mobile Communications 10-21
IF vs. Baseband Sampling • Correlation Receiver • Rake Receiver

10.7 Multiuser Detection............................... 10-24
Multiuser Detection in the UL • The Decorrelating and MMSE Detectors

10.8 Perspectives and Conclusions...................... 10-27
The Challenge to Mobile Spread-Spectrum Communications • 4G Wireless Communications Systems • Concluding Remarks

Filippo Giannetti
University of Pisa

Marco Luise
University of Pisa

0-8493-1657-X/$0.00+$1.50

Wireless mobile communications has greatly benefited from the adoption of advanced signal processing techniques and components. In particular, the advent of low-cost very large scale integrated (VLSI) chips has allowed the design and implementation of efficient communications systems based on spread-spectrum signals, namely, the different code-division multiple-access (CDMA) radio networks currently in operation or being deployed. The aim of this chapter is to provide the reader with a basic understanding of the diverse digital signal processing (DSP) algorithms that lie at the foundation of such systems. After a short introduction to wireless communications (Section 10.1), the chapter develops into a description of the basic characteristics of digital spread-spectrum signals (Section 10.2). This sets the framework for the subsequent general study of CDMA techniques (Section 10.3) and description of current standards (Section 10.4). Next, the focus is shifted to the functions of signal synchronization for optimum data detection (Section 10.5) and to the design of a digital CDMA receiver (Section 10.6), highlighting the main relevant DPS-intensive tasks. The final part of the chapter is devoted to the introduction of an advanced topic related to the detection of digital information in a CDMA signal, namely, multiuser detection (Section 10.7).

10.1 A Brief History of Wireless Communications

10.1.1 The Wireless Revolution

In the developed countries, the number of *wireless* access connections between the user terminals of a telecommunication network (phones, laptops, palmtops, etc.) and the fixed, high-capacity transport network has already exceeded the number of *wired* ones. Untethered communications and computing have become sort of a lifestyle, and the trend will undoubtedly grow further in the near future, with the development of *wireless fidelity* (Wi-Fi) hot spots and the commercialization of low-cost wireless *local area networks* (LANs) for the home. The picture we have just depicted is what we may call the *wireless revolution* [Rap91]. Started in Europe in the early 1990s, with the American countries lagging a little bit, it will likely come to its full evolution within the next few years, to ramp up again when the Asian developing countries catch up [Sas98]. Driving elements of this revolution are the advances in components and techniques for *digital signal processing* (DSP), which represent the background, we could say the *humus*, upon which the whole of the wireless network is developed. The fundamental contributions of DSP to mobile communications lie, of course, in the relevant, traditional, physical-layer issues, namely, source and channel coding, modulation, signal synchronization and equalization, and multiple access. In particular, modern mobile cellular systems have benefited from the adoption of *spread-spectrum* (SS) signaling techniques that will be treated in detail in the following sections.

10.1.2 2G and 3G Cellular Systems

The real start of the revolution was the advent of the so-called *second-generation* (2G) digital pan-European *time-division multiple-access* (TDMA) cellular communications systems, the well-known GSM (*Global System for Mobile Communications*) [Pad95]. The growth of cellular communications had already started with earlier analog systems, the so-called *first-generation* (1G) systems, but the real breakthrough was marked by the initially slow, then exponential, diffusion of GSM terminals, fostered by continent-wide compatibility through international roaming. In the U.S., the advent of 2G digital cellular telephony was somewhat slowed down by the coexistence of incompatible systems and by the consequent lack of a nationwide accepted unique standard [Pad95]. The two competing 2G American standards were the 'digital' AMPS (*Advanced Mobile Phone System*) IS-54/136, whose technology was developed with the specific aim of being back-compatible (as far as the assigned radio channels are concerned) with the preexisting 1G analog AMPS system, and the highly innovative code-division multiple-access (CDMA) system IS-95 [Koh95, Gil91]. Although slower to gain momentum, 2G systems in the Americas had the merit of introducing the technology of SS signals and CDMA that in a few years would be adopted worldwide for *third-generation* (3G) systems. A similar story about 2G digital cellular telephony could be told about Japan and its *Personal*

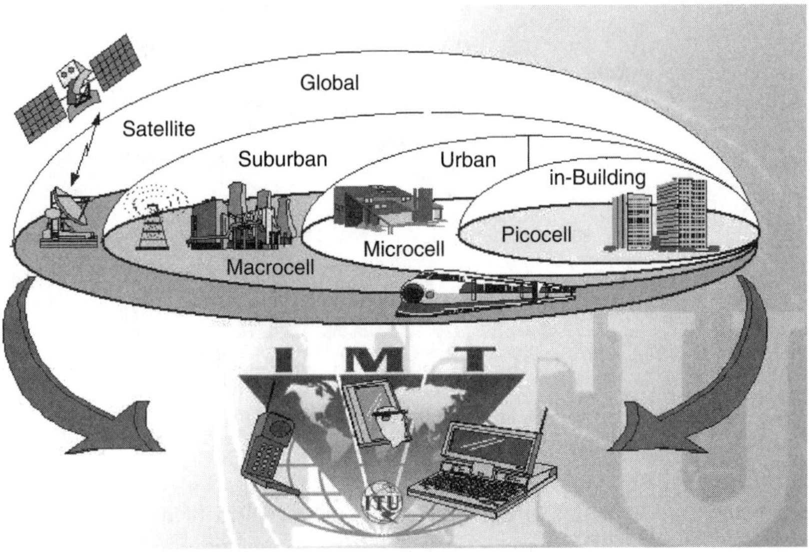

FIGURE 10.1 Architecture of an IMT-2000-compliant system (http://www.itu.org).

Digital Cellular (PDC) system, which is actually quite similar from a technical standpoint to the American IS-54/136.

At the dawn of the third millennium, the ITU (*International Telecommunications Union*), based in Geneva, Switzerland, took the initiative to promote the development of a universal 3G mobile/personal wireless communication system with high capacity and a high degree of interoperability among the different network components, as depicted in Figure 10.1. Under initiative IMT-2000 (*International Mobile Telecommunications for the year 2000*) [Chi92], a call for proposal was issued in 1997 to eventually set up the specifications and the technical recommendations for a universal system. At the end of the selection procedure, and in response to the different needs of the national industries, operators, and PTTs, two different noncompatible standards survived: UMTS (*Universal Mobile Telecommunications System*) for Europe and Japan [Dah98], [Ada98] and cdma2000 for the U.S. [Kni98]. Both are based on a different mixture of TDMA and CDMA technologies, but the real common innovative factor is the universal adoption of CDMA for a maximum of flexibility and multimedia support.

A more detailed description of 2G and 3G CDMA-based wireless systems will be presented in Section 10.4.

10.1.3 DSP Components for Wireless Communications

The celebrated *Moore's law* states that the number of transistors on a chip with a fixed area roughly doubles every year and a half. Hence, the price of microelectronics components (and microelectronics-based equipment) halves in eighteen months, or in other words, the power of VLSI (*very large scale integrated*) circuits doubles past the same period. Over the last few decades Moore's empirical prediction was remarkably prescient. The minimum size of the CMOS (*complementary metal-oxide semiconductor*) transistor has decreased on average by 13% per year from 3 μm in 1980 to 0.13 μm in 2002, while die areas have increased by 13% per year and design complexity (as measured by the number of transistors on a chip) has increased at an annual growth rate of 50% for *dynamic random access memories* (DRAMs) and 35% for microprocessors. Performance enhancements have been equally impressive. For example, clock frequencies for leading-edge general-purpose microprocessors have increased by more than 30% per year. An example related to Intel microprocessor transistor count is shown in Figure 10.2. The remarkable progress in semiconductor technology has fueled the growth of commercial wireless communications

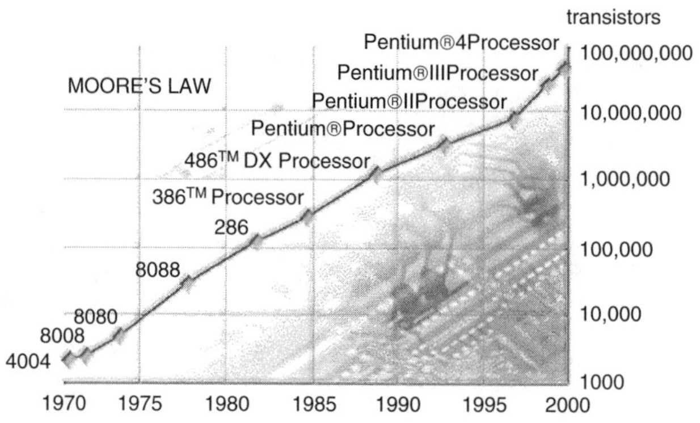

FIGURE 10.2 Moore's law and Intel microprocessors. (Courtesy of Intel.)

systems through the advances of the DSP algorithms that can be implemented on a low-power, small-size, low-cost wireless terminal [Rab00]. The efficiency of wireless communications is significantly increasing with the application of more sophisticated multiple-access and digital modulation/processing techniques. This allows accommodation of the dramatic growth in the number of subscribers experienced by 2G and 3G systems, and at the same time offers increased functionality with better quality of service.

In the following, we will briefly outline the different SS techniques that are currently adopted (year 2003) in the main wireless cellular systems to establish a multimedia (i.e., voice, video, and data) full-duplex link between a *mobile terminal* (MT) and the *base station* (BS) of a cellular radio communications access network. We will also show the different constraints placed on the *down-link* (DL, or *forward-link*, from the BS to the MT) and the *up-link* (UL, or *return-link*, from the MT to the BS) by the different degrees of affordable complexity of the two pieces of equipment.

10.2 Fundamentals of Digital Spread-Spectrum Signaling

10.2.1 Narrowband

By this term we indicate those data-modulated signals whose bandwidths around their carrier frequencies is comparable to the information bit rate. Most popular narrowband modulated signals are *phase shift keying* (PSK), *quadrature amplitude modulation* (QAM), and a few nonlinear modulations like *Gaussian minimum shift keying* (GMSK), which belongs to the broader class of *continuous-phase modulations* (CPMs). For the sake of simplicity, we will restrict our attention in the sequel to linearly modulated PSK and QAM signals. The generic expression of a bandpass modulated signal $s_{BP}(t)$ is

$$s_{BP}(t) = s_I(t) \cdot \cos(2\pi f_0 t) - s_Q(t) \cdot \sin(2\pi f_0 t) \tag{10.1}$$

where f_0 is the carrier frequency and $s_I(t)$ and $s_Q(t)$ are two baseband signals that represent the *in-phase* (I) and *quadrature* (Q) components, respectively, of the I/Q modulated signal. A more compact representation of the modulated signal (Equation 10.1) is its *complex envelope* (or *baseband equivalent*), defined as $s(t) \triangleq s_I(t) + \mathrm{j}s_Q(t)$, where $\mathrm{j} \triangleq \sqrt{-1}$. The relationship between the bandpass modulated signal and its own complex envelope is straightforward: $s_{BP}(t) = \Re\{s(t) \cdot \exp(\mathrm{j}2\pi f_0 t)\}$, where $\Re\{\cdot\}$ takes the 'real part' of the argument. In the case of a linear modulation, the complex envelope is made of uniformly spaced data pulses as follows:

$$s(t) \triangleq A \cdot \sum_{i=-\infty}^{\infty} d[i] \cdot g_T(t - iT_s) \tag{10.2}$$

where $d[i] \triangleq d_I[i] + jd_Q[i]$ is the i-th transmitted symbol taken from a *constellation* of W points in the complex plane,[1] T_s is the signaling interval (i.e., the reciprocal of the symbol rate R_s), and $g_T(t)$ is the basic T_s-energy pulse shape. Denoting with P_s the average power of the modulated signal (Equation 10.1), the average energy per data symbol is $E_s = P_s \cdot T_s$, and the amplitude coefficient in (Equation 10.2) is equal to $A \triangleq \sqrt{2P_s/\mathrm{E}\{|\,d[i]\,|^2\}}$, where $\mathrm{E}\{\cdot\}$ is the statistical expectation operator. Each symbol $d[i]$ in the constellation is addressed by taking $\log_2(W)$ binary symbols (bits) from a binary information source running at rate $R_b = R_s \log_2(W)$.

The *power spectral density* (PSD) of Equation 10.2 is proportional to $|G_T(f)|^2$, where $G_T(f)$ is the Fourier transform of the pulse shape $g_T(t)$. As a result, the spectral occupancy B of a narrowband modulated signal is strictly related to the bandwidth of the pulse shape $g_T(t)$. For instance, in the case of a transmission with *no-return-to-zero* (NRZ) rectangular shaping, the baseband occupancy, evaluated at the first spectral null, is $B = R_s$, while in the case of Nyquist's *square root raised cosine* (SRRC) band-limited shaping, the baseband occupancy is $B = (1+\alpha)R_s/2$, where $0 \leq \alpha \leq 1$ is the pulse *roll-off factor*. Notice that the bandwidth occupancy of the modulated signal is twice that of the baseband signal, i.e., $B_{RF} = 2B$.

10.2.2 Spread-Spectrum

With this term we address any modulated signal whose bandwidth around the carrier frequency is much larger than its information bit rate. In SS signaling, the bandwidth occupancy of the transmitted signal is intentionally increased well beyond the value required for conventional narrowband transmission, the transmitted power being the same. This is done to enhance signal robustness against interference, be it intentional (i.e., hostile jamming) or unintentional (e.g., man-made noise, narrow- and wideband transmitters operating on the same carrier frequency), and distortions caused by frequency-selective radio channels. Spectrum spreading also reduces the level of radiated power density (i.e., watts per bandwidth unit) and allows the simultaneous utilization of the same (wide) bandwidth by multiple users (this is called *multiple access*; see Section 10.3). Spectrum spreading can be essentially accomplished in two ways: *frequency hopping* and *direct sequence.*

10.2.3 Frequency-Hopping Spread Spectrum

Frequency-hopping spread spectrum (FH/SS) is the first SS technique, at least from a historical perspective, and was conceived in 1941 to provide secure military communications during World War II [Sch82]. The basic idea underlying this early SS concept is relatively simple: just (rapidly) change the transmit carrier frequency every T_{hop} seconds. The signal spectrum then "hops" from frequency to frequency following a periodic *pseudorandom* (i.e., apparently random) sequence of carrier values to escape hostile jamming or eavesdropping by unauthorized listeners. In so doing, the transmitted spectrum 'spans' a large *radio frequency* (RF) bandwidth and implements a spread-spectrum modulation [Sim85], [Dix94]. The (periodic) frequency pattern is also made known to the authorized receiver, which is then capable of tracking the transmitted carrier. However, due to strict requirements concerning oscillator frequency stability and switch rate, FH/SS does not find significant applications in multiple-access commercial systems for mobile and cellular communications, and therefore it will not be further considered in the following.

10.2.4 Direct-Sequence Spread Spectrum

Direct-sequence spread spectrum (DS/SS) is at the moment the main technique to generate an SS signal. In particular, spectral spreading is accomplished by direct multiplication in the time domain of the information-bearing symbols, running at rate $R_s = 1/T_s$, and a sequence made of binary symbols (*chips*)

[1]The usual assumption is that the i-th data symbol $d[i]$ is a random variable with uniform probability distribution over the W-point constellation alphabet, and that $d[i]$ is statistically independent of $d[k]$, $i \neq k$.

$c[k]$ running at the much higher rate $R_c \triangleq 1/T_c \gg R_s$, with T_c the chip interval [Dix94], [Pic82], [Sim85]. The result of this product (which is carried out before transmit pulse shaping takes place) is a stream of 'chipped' high-rate symbols running at the chip rate R_c. The resulting signal to be transmitted turns out to have a bandwidth occupancy wider than that of conventional modulation schemes, and in particular comparable to the chip rate. The binary sequence $c[k]$ is usually periodic with period L chips and is referred to as *spreading sequence* or *spreading code*. DS/SS is widely used in commercial communications systems, and for this reason, in the following we will restrict our attention to the description of the DS/SS signal format and to its relevant features.

10.2.5 DS/SS Signal Model

Assuming (as is always the case) that one symbol interval T_s spans an integer number M of chip intervals T_c of the spreading sequence (i.e., $M \triangleq T_s/T_c = R_c/R_s$ is an integer), a DS/SS signal can be represented as

$$s^{(SS)}(t) \triangleq A \cdot \sum_{k=-\infty}^{\infty} d[k//M] \cdot c[|k|_L] \cdot g_T(t - kT_c) \tag{10.3}$$

where the operators $k//M \triangleq \text{int}(k/M)$ and $|k|_L \triangleq k \bmod L$. The spreading sequence $c[k]$, either real- or complex-valued, runs at rate $R_c = 1/T_c$ and therefore a new T_c-energy chip shaping pulse $g_T(t)$ is generated every T_c seconds. The index k "ticks" at the rate R_c and the spreading code index $|k|_L$ cycles over the interval $0, 1, \ldots, L-1$ every L chip intervals, scanning the whole code length. The data index $k//M$ runs at the symbol rate $R_s = R_c/M$; i.e., it is updated every M ticks of the index k. The normalization constant is again $A \triangleq \sqrt{2P_s/\mathrm{E}\{|d[i]|^2\}}$, and the average energy per chip is now $E_c = P_s \cdot T_c$.

10.2.6 Real Spreading

In real spreading (RS) DS/SS, the spreading sequence $c[k]$ is real-valued and the chips usually belong to the alphabet $\{-1, +1\}$. In the particular case of complex-valued data $d[i]$, such a DS/SS format is named *quadrature real spreading* (Q-RS).

10.2.7 Complex Spreading

In complex spreading (CS) DS/SS, the spreading sequence $c[k]$ is complex-valued, i.e., $c[k] \triangleq c_I[k] + jc_Q[k]$, where both the real and imaginary parts of the sequence are as in RS above. Two spreading codes are thus needed for the transmission of a CS DS/SS signal. In IS-95, CS is used with real-valued data symbols $d[i]$, while UMTS/cdma2000 features CS with complex data (see Section 10.4 for more details).

10.2.8 DS/SS Bandwidth Occupancy

The bandwidth occupancy of the DS/SS signal is dictated by the chip rate R_c (as is apparent from Equation 10.3) and by the shape of the basic pulse $g_T(t)$. For instance, in the case of a DS/SS transmission with NRZ rectangular chip shaping, the baseband occupancy of the SS signal (Equation 10.3), evaluated at the first spectral null, is $B^{(SS)} = R_c$, and thus $B^{(SS)} \gg B$. However, apart from some particular applications (mainly military communications and Global Positioning System (GPS) radiolocation), the spectrum of the DS/SS signal shall be strictly limited due to regulatory issues. This leads to the definition of a *band-limited spread-spectrum* (BL-SS) signal (which sounds like a paradox). As in the case of conventional narrowband modulations, spectrum limitation can be achieved by resorting to Nyquist SRRC shaping of the chip pulses so that the bandwidth occupancy becomes $B^{(SS)} = (1+\alpha)R_c/2$ for the baseband signal and $B_{RF}^{(SS)} = (1+\alpha)R_c$ for the modulated signal.

10.2.9 Spreading Factor

This is defined as $M \triangleq T_s/T_c = R_c/R_s$ and represents the ratio between the bandwidth occupancy of the SS signal $B^{(SS)}$ and the bandwidth B of the corresponding narrowband modulated signal. Typical values of the spreading factor commonly adopted in commercial wireless communications systems range from 4 to 256. If channel coding is used, and the chip and bit rates are kept constant, the spreading factor is reduced by a factor r, where $r < 1$ is the coding rate. This is easily understood since channel coding has the effect of increasing the symbol rate by a factor $1/r$, thus expanding the signal bandwidth by the same factor *before* spectrum spreading takes place. On the other hand, keeping the chip rate constant means keeping the same RF bandwidth, and so the net result is a decrease of the spreading factor.

10.2.10 Short Code

A spreading sequence is called *short code* when one data symbol interval exactly spans a (small) integer number n of the spreading sequence repetition periods, i.e., $T_s = nLT_c$. In this case, the spreading factor is $M = nL$. The simplest case of short code spreading is the case with $n = 1$, whereby the code repetition period is equal to one symbol interval and $M = L$. An example of signal spreading by a short code, with $M = L = 8$ and NRZ chip shaping, is shown in Figure 10.3. Short code spreading is specified as an option for the UL of UMTS to enable multiuser detection (see Section 10.7).

10.2.11 Long Code

A spreading sequence is called *long code* when its repetition period is much longer than the data symbol duration, i.e., $LT_c \gg T_s$. In this case, the spreading factor is $M \ll L$. An example of signal spreading by a long code, with $M = 8$ and $L = 24$, is shown in Figure 10.4. Long code spreading is specified for the DL of UMTS and for both the DL and UL of IS-95 and cdma2000.

10.2.12 Processing Gain

This is defined as $G_p \triangleq T_b/T_c = R_c/R_b$ and is related to the spreading factor as $G_p = M/\log_2(W)$. The processing gain has to do with the antijamming capability of the DS/SS signal and is often confused with the spreading factor. The two are coincident only in the case of binary modulation with no channel coding (see above).

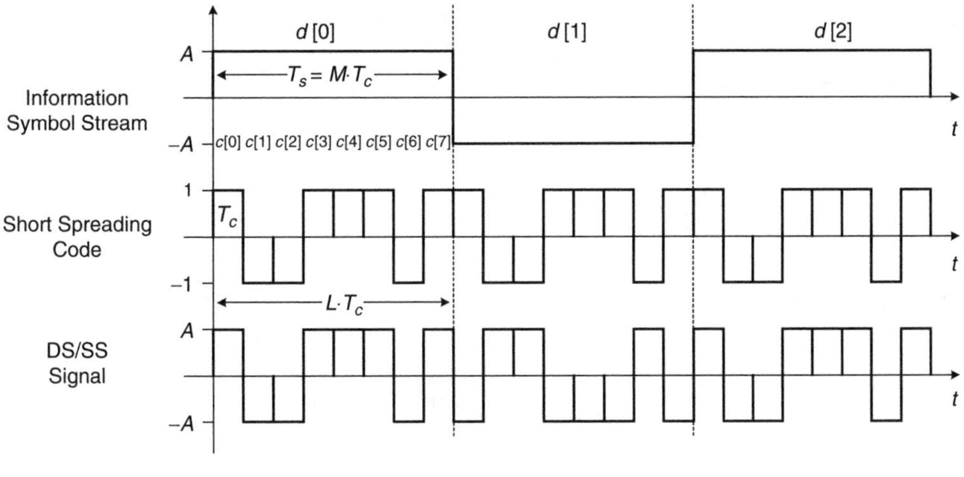

FIGURE 10.3 DS/SS signal with short code, $M = L = 8$.

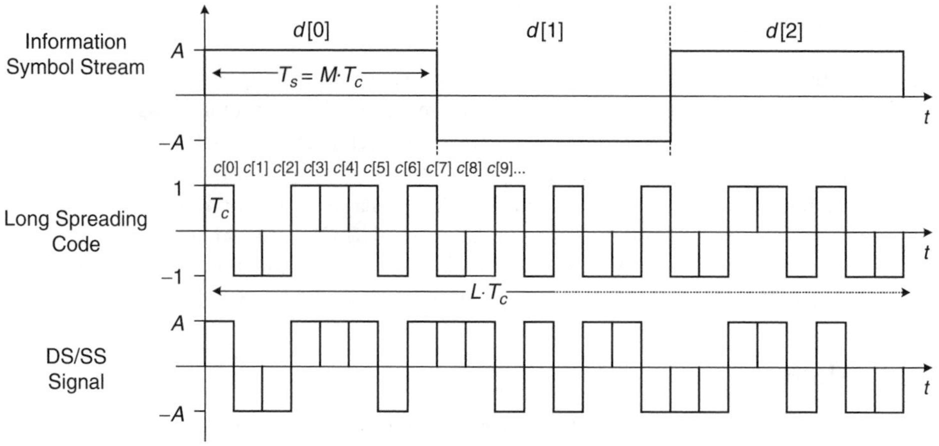

FIGURE 10.4 DS/SS signal with long code, $M = 8$, $L = 24$.

10.2.13 Pseudorandom Sequence Generators

A real-valued binary spreading sequence $c[k]$ with *pseudonoise* (PN) properties is also called *pseudorandom binary sequence* (PRBS) and can be generated by an m-stage *linear feedback shift register* (LFSR) whose taps are properly set according to the polynomial theory [Pet72]. In particular, it is possible to design sequence generators based on one or more LFSR structures [Dix94], [Pic82], [Sim85], [Din98] to obtain spreading codes with random-like appearance and special features as far as the repetition period and correlation properties are concerned [Sar80]. Refer to [Dix94] for further details on the main code sets used in the practice (Gold codes, Kasami codes, Walsh–Hadamard and OVSF codes, ML codes).

10.2.14 Basic Architecture of a DS/SS Modem

Figure 10.5 shows the general outline of the modulator section of a digital DS/SS modem, where thick lines denote complex-valued signals. The information-bearing binary data undergo channel encoding and are mapped onto a complex constellation, thus yielding the data symbol stream $d[i]$ running at symbol rate R_s. $d[i]$ is then oversampled (repeated) by a factor M, and spectrum spreading is eventually accomplished by direct multiplication with the spreading code $c[k]$ running at chip rate $R_c = MR_s$. Next, chip pulse shaping is performed in the digital domain by a *finite impulse response* (FIR) filter featuring SRRC response $g_T[m]$ and operating at sample rate $R_{sa} \triangleq 1/T_{sa}$ (for instance, $R_{sa} = 4R_c$). Notice that, the resulting baseband DS/SS signal at the filter output is a digital version of Equation 10.3 with a sampling

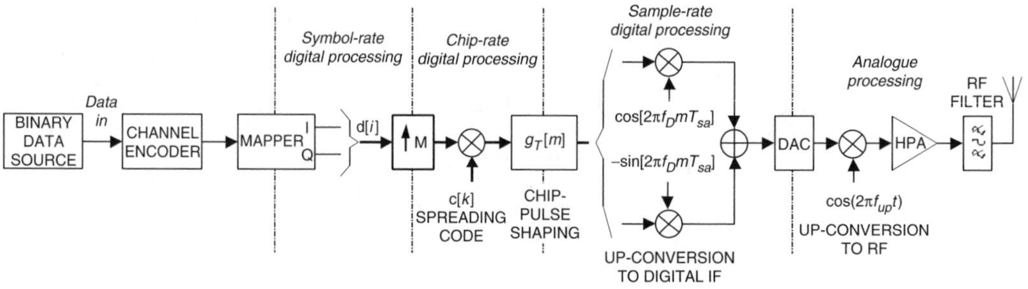

FIGURE 10.5 Block diagram of the transmit section of a digital DS/SS modem.

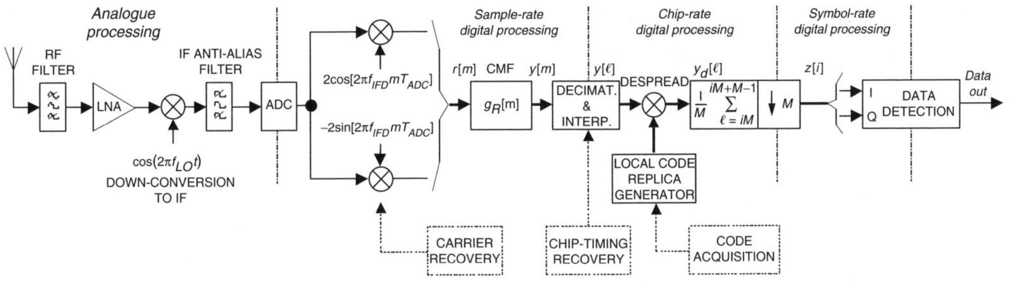

FIGURE 10.6 Block diagram of the receive section of a digital DS/SS modem.

frequency R_{sa}. Such a DS/SS signal is up-converted to a digital *intermediate frequency* (IF) f_D, then *digital-to-analog conversion* (DAC) takes place, followed by conventional analog processing (final up-conversion to RF, power amplification, and band limiting) before the signal is sent on air. The basic outline of the twin demodulator section[2] is sketched in Figure 10.6. The analog received signal undergoes RF amplification and filtering, followed by down-conversion to IF. In many state-of-the-art implementations, *analog-to-digital conversion* (ADC) takes place directly at IF with bandpass sampling techniques, as shown. The conversion rate R_{ADC} is invariably faster than the chip rate R_c to perform subsequent baseband filtering with no aliasing issues (e.g., $R_{ADC} = 4R_c$). The digital signal is then converted to baseband by an I/Q demodulator driven by a digital oscillator under the control of the carrier synchronization unit (see Section 10.5). Each component of the resulting complex signal $r[m]$ is fed to a filter $g_R[m]$ matched to the chip pulse (*chip-matched filter*, CMF). On both I/Q rails, a decimator/interpolator changes the clock rate of the digital signal $y[m]$ and takes the optimum samples $y[\ell]$, at chip rate, from the digital stream available at the filter output. Assuming SRRC chip pulse shaping in the transmitter, ideal CMF in the receiver, and perfect chip timing and carrier synchronization, the chip rate signal $y[\ell]$ at the decimator/interpolator output is

$$y[\ell] = A \cdot d[\ell//M] \cdot c[|\ell|_L] + n[\ell] \tag{10.4}$$

where $n[\ell]$ is a random process (which represents the contribution of the filtered channel noise) whose I/Q components are independent and identically distributed zero-mean Gaussian random variables with variance $\sigma_n^2 = N_0/T_c$. Subsequently, the spreading code is removed from the received signal by performing the so-called *despreading* operation:

$$y_d[\ell] = y[\ell] \cdot c[|\ell|_L] = A \cdot d[\ell//M] \cdot c^2[|\ell|_L] + n[\ell]c[|\ell|_L] = A \cdot d[\ell//M] + n'[\ell] \tag{10.5}$$

where we took into account that $c^2[|\ell|_L] = 1$ and where, due to the properties of the spreading code, $n'(\ell)$ is a noise component statistically equivalent to $n[\ell]$. It is seen that despreading requires a local replica $c[\ell]$ of the same spreading code used by the transmitter, and also requires perfect *synchronicity* of the remote and local codes. After despreading, we are left with a (digital) narrowband rectangular-pulse signal plus *additive white Gaussian noise* (AWGN). The optimum processing for data detection is thus (digital) integration over a one-symbol period to yield the following complex-valued *sufficient statistics*:

$$z[i] \triangleq \frac{1}{M} \cdot \sum_{\ell=iM}^{iM+M-1} y_d[\ell] = \frac{1}{M} \cdot \sum_{\ell=iM}^{iM+M-1} y[\ell] \cdot c[|\ell|_L] \tag{10.6}$$

[2]This is meant to be a sample receiver architecture taken as the baseline for discussion. More details on possible DS/SS receiver architectures can be found in Section 10.6.

Recalling Equation 10.4 and Equation 10.5, we obtain

$$z[i] = \frac{1}{M} \cdot \sum_{\ell=iM}^{iM+M-1} A \cdot d[\ell//M] + \frac{1}{M} \cdot \sum_{\ell=iM}^{iM+M-1} n'[\ell] = A \cdot d[i] + \nu[i] \tag{10.7}$$

where $\nu[i]$ is a noise term whose I/Q components are independent and identically distributed zero-mean Gaussian random variables with variance $\sigma_\nu^2 = \sigma_n^2/M = N_0/MT_c = N_0/T_s$. From Equation 10.6, we see that despreading accumulation is equivalent to the operation of *cross-correlation* between the received signal $y[\ell]$ and the local code replica $c[\ell]$, computed on the i-th symbol period to yield the i-th sufficient statistics $z[i]$. This is why we also speak of *correlation receiver* (CR).

The decision strobe (Equation 10.7) is eventually passed to the final detector, which may be a slicer to regenerate the transmitted digital data stream or, if channel coding is adopted, a soft-input channel decoder such as a Viterbi algorithm [Vit67, Vit79b]. Looking at Equation 10.7, we also see that the sufficient statistics are exactly the same as those we would obtain in the case of conventional narrowband transmission of the symbol stream $d[i]$ over an AWGN channel with matched-filter detection. The conclusion is that spreading and despreading are completely transparent to the end user as far as the *bit error rate* (BER) performance of the link is concerned.

As already mentioned, the block diagram in Figure 10.6 also shows some ancillary functions, namely carrier frequency/phase and chip timing recovery, together with spreading code time alignment, which will be dealt with in detail in Section 10.5.

10.3 Code-Division Multiple Access

10.3.1 Frequency-, Time-, and Code-Division Multiplexing

In the DS/SS schemes discussed above, the data stream generated by an information source is spread over a wide frequency band using one or two spreading codes. Starting from this consideration, we can devise an access system allowing multiple users to share a common radio channel. This can be achieved by assigning each user a different spreading code and letting all of the signals *simultaneously* access, in DS/SS mode, the *same frequency spectrum*. All the user signals are therefore transmitted at the same time and over the same frequency band, but they can nevertheless be identified thanks to the particular properties of the spreading codes. The users are kept separated in the *code domain*, instead of the time or frequency domain, as in conventional *time-* or *frequency-division multiple access*, respectively (TDMA, FDMA). Such a multiplexing/multiple-access technique, based on DS/SS transmission, is called *code-division multiple access* (CDMA), and the particular spreading sequence identifying each user is known as a *signature*.

In the DL of a wireless network, the N user signals in DS/SS format are obtained from a set of N traffic-bearing channels that are physically co-located into a single site, for instance, the radio base station. The BS keeps all N tributary streams $d_n[k//M]$, $n = 1, 2, \ldots, N$ synchronous and performs spectrum spreading on each channel using a set of N different signature codes $c_n[k]$, $n = 1, 2, \ldots, N$, all having the same start epoch (i.e., the start instant of the repetition period). The resulting *synchronous code-division multiplexing* (S-CDM) signal is

$$s^{(CDM)}(t) \triangleq A \cdot \sum_{k=-\infty}^{\infty} \left(\sum_{n=1}^{N} d_n[k//M] \cdot c_n[|k|_L] \right) g_T(t - kT_c) \tag{10.8}$$

The most popular class of binary spreading codes commonly used for S-CDM in the DL of wireless networks is the *Walsh–Hadamard* (WH) set [Ahm75], [Din98], which is made of $L = 2^\zeta$ sequences (with ζ an integer) with repetition period L each (in practical systems L = 32, 64, 128, or 256). The WH

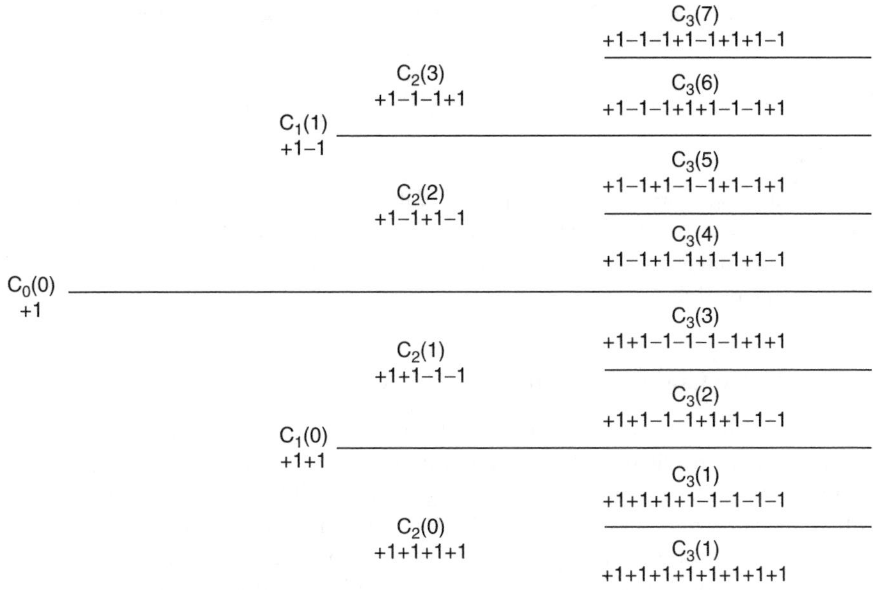

FIGURE 10.7 OVSF codes tree — each layer is the complete set of WH codes of a particular length.

sequences are *orthogonal*, that is

$$\frac{1}{L}\sum_{k=0}^{L-1} c_n[k]\cdot c_m[k] = \begin{cases} 0 & n \neq m \\ 1 & n = m \end{cases} \tag{10.9}$$

S-CDM with WH codes is also known as *synchronous orthogonal* CDM (S-O-CDM). Demultiplexing of channel i out of the multiplex signal (Equation 10.8) is easily accomplished by the simple correlation receiver described in Section 10.2, provided, of course, that it uses the signature code $c_i[k]$. Thanks to the orthogonality property of the WH codes, no interference from the other channels is experienced (see also Section 10.3.3 on multiple access interference). Fast techniques to perform *multiplexing/demultiplexing* (MUX/DEMUX) of a large number of channels with WH functions are also available (Walsh–Hadamard transform [Ahm75]).

In the UL of a wireless network, the DS/SS signals are originated by N spatially separated MTs that access the same physical medium (the same RF bandwidth), and their superposition is collected at the receiver antenna. Synchronization of the different, sparse MTs is very difficult to achieve, and so the access is *asynchronous* — in this case, *asynchronous CDMA* (A-CDMA). The resulting received signal at the BS receiver is[3]

$$r(t) = \sum_{n=1}^{N} e^{j(2\pi\Delta f_n t+\theta_n)} s_n^{(SS)}(t-\tau_n) + w(t) \tag{10.10}$$

where τ_n, Δf_n, and θ_n are the propagation delay, carrier frequency, and phase offsets, respectively, of the generic n-th signal, and $w(t)$ is an AWGN contribution. Recalling the DS/SS signal description presented

[3]We neglect here multipath propagation that will be dealt with in Section 10.6.3 on the *Rake receiver*.

in Section 10.2, the generic n-th traffic channel in DS/SS format can be expressed as

$$s_n^{(SS)}(t) \triangleq A_n \cdot \sum_{k=-\infty}^{\infty} d_n[k//M] \cdot c_n[|k|_L] \cdot g_T(t - kT_c) \quad (10.11)$$

where A_n, $d_n[i]$, and $c_n[k]$ are the signal amplitude, data symbols, and signature sequence, respectively. The signal received by an MT in the DL can actually be seen as a special case of Equation 10.10, with $\tau_n = \tau$, $\Delta f_n = \Delta f$, and $\theta_n = \theta$, for every n; i.e., all signals are synchronous and experience the same carrier frequency/phase shift.

10.3.2 Multirate Code-Division Multiplexing

Advanced communications systems (2G and 3G cellular) support different kind of services (e.g., voice, video, data), and so the required bit rate per user can be variable from a few kilobits per second up to a few megabits per second. Therefore, the spreading scheme shall be flexible enough to easily allocate signals with different bit rates on the same bandwidth. This can be achieved basically by three different techniques, or a mixture of the three: (1) multicode, (2) symbol repetition, and (3) variable spreading factor. In the *multicode* option, if the user requires a higher bit rate than the one of the basic channel, he or she is simply allocated more than one signature code, and the high-rate bit stream is split into a number of parallel lower-rate streams, each occupying a single (basic rate) channel. This is done in UMTS and cdma2000. With symbol repetition, the basic rate is considered the highest in the network. If a lower rate is needed, it is obtained by generating a "fake" high-rate stream obtained by repetition a number of time of the same low-rate datum. This is the technique used in IS-95. Using a *variable spreading factor*, the chip rate R_c (and therefore SS bandwidth $B^{(SS)}$) is kept constant, but the spreading factor M is varied according to the bit rate of the signal to be transmitted. This also has to be done without altering the property of mutual code orthogonality outlined above. The solution to this issue, applied in both UMTS and cdma2000, is the special class of codes named *orthogonal variable spreading factor* (OVSF) [Ada97], [Din98]. The OVSF code set is a reorganization of the WH codes into *layers*. The codes on each layer have twice the length of the codes in the previous layer. Also, the codes are organized in a tree, wherein any two "children" codes on the layer underneath a "parent" code are generated by repetition and repetition with sign change, respectively (see Figure 10.7), of the parent. The peculiarity of the tree is that any two codes are not only orthogonal within each layer (each layer is just the complete set of the WH codes of the corresponding length), but also orthogonal *across* layers, provided that the shorter is not an "ancestor" of the longer one. Orthogonality across layers means that the two codes are orthogonal both on the length of the longer code (after extension by repetition of the shorter code to make it the same length as the other) and on the length of the shorter code (for all of the longer code sections having the length of the shorter one). As a consequence, we can use the shorter code for a higher-rate transmission with a smaller spreading factor, and the longer code for a lower-rate transmission with a higher spreading factor (recall that the chip rate is always the same). The two codes will not give rise to any channel cross talk.

10.3.3 Multiple Access Interference

In the DL of a cellular network each BS sends out an N-channel S-O-CDM signal, so that any MT located within a cell experiences no interference from the other channels in the multiplex thanks to the orthogonality property of the *channelization* WH codes. To see this, assume that the generic mobile receiver intends to detect the useful traffic channel #h. The output of the relevant correlation receiver is (see Equation 10.6)

$$z_h[i] = A_h \cdot d_h[i] + \mu[i] + \nu[i] \quad (10.12)$$

where

$$\mu[i] \triangleq \sum_{\substack{n \neq h \\ n=1}}^{N} \left\{ \frac{1}{M} \cdot \sum_{\ell=iM}^{iM+M-1} A_n \cdot d_n[\ell // M] \cdot c_n[|\ell|_L] \cdot c_h[|\ell|_L] \right\} = \sum_{\substack{n \neq h \\ n=1}}^{N} \left\{ \frac{A_n \cdot d_n[i]}{M} \cdot \sum_{\ell=0}^{M-1} c_n[\ell] \cdot c_h[\ell] \right\} \tag{10.13}$$

represents a sort of cross-talk term caused by the presence of the other user signals and is referred to as *multiple-access interference* (MAI). Recalling Equation 10.9, if WH codes are used, the MAI term $\mu[i]$ vanishes and Equation 10.12 collapses to Equation 10.7.

Without entering into further detail, we can easily argue that MAI is on the contrary *intrinsic* to detection of an UL channel. In this case, the N aggregate DS/SS signals (Equation 10.10) are received by the BS in A-CDMA mode. The output of the correlation receiver for channel #h can still be put in a form similar to those in Equation 10.12 and Equation 10.13, with a different, slightly more involved expression of the MAI term $\mu[i]$. But now the *asynchronous* access that is peculiar of the UL precludes the use of orthogonal codes to cancel MAI. The usual choice of the channelization codes is thus a set of (long) PN sequences that, thanks to their randomness, makes the MAI term similar to additional Gaussian noise, uncorrelated with respect to the useful traffic signal. This approach is pursued in the UL of all of the 2G and 3G commercial systems currently in use. A more advanced (and more complex) solution to the issue of the MAI, i.e., multiuser detection, will be described in Section 10.7.

10.3.4 Capacity of a CDMA System

In the DL of a wireless network, S-O-CDM is commonly adopted. Therefore, any BS can radiate a CDM signal containing up to L traffic channels (i.e., the size of the orthogonal code set) without introducing any *quality-of-service* (QoS) degradation due to the presence of multiple users. With WH codes, the spreading factor M is equal to the code length L (short codes), and this represents the key to understanding the interplay between spectrum spreading and multiplexing. Using DS/SS transmission, the signal bandwidth is expanded by a factor M, and this is a potential bandwidth waste. But using S-O-CDM, we can put M signals in the same bandwidth with no interference, thus regaining the original bandwidth efficiency. This is exactly what happens with FDMA (where we increase the total bandwidth by the number of channels M that we put on to adjacent carriers) and TDMA (where the channel signaling rate is increased by a factor M to accommodate the M tributary signals). The spreading factor thus represents the ultimate capacity of the DL in a CDMA network, and in this respect CDMA is no better or worse than FDMA and TDMA.

In the UL on the contrary, synchronization of the MTs is no longer feasible and A-CDMA with long spreading codes (e.g., PN sequences) is adopted. In this case, the receiver may experience a significant amount of MAI (as is apparent from Equation 10.12 and Equation 10.13 that may cause significant degradation of the QoS. The nice feature of A-CDMA, also known as *graceful degradation* property, is that the QoS degradation can be traded off with a capacity increase [Gil90, Gil91]. To see this, observe that MAI can be seen as an additional, independent noise term summing up to the usual thermal and antenna noise contributions. If N traffic channels are active (i.e., the useful plus $N-1$ interferers) and all the interfering channels are received with the same power level[4] P_s, the MAI can be modeled as a white Gaussian noise process, whose equivalent PSD is found to be

$$I_0 \triangleq \frac{(N-1) \cdot P_s}{1/T_c} = (N-1) \cdot E_c \tag{10.14}$$

As a consequence, the BS correlation receiver experiences an equivalent total noise PSD given by $N_0 + I_0$.

[4]All 2G and 3G CDMA systems implement specific *power control* strategies to ensure that this condition is attained within a small error margin.

Assuming *quadrature PSK* (QPSK) modulation, the BER of the link is therefore

$$P(e) = \mathrm{Q}\left(\sqrt{\frac{2E_b}{N_0 + I_0}}\right) = \mathrm{Q}\left(\sqrt{\frac{2E_b}{N_0 + (N-1)\cdot E_c}}\right) = \mathrm{Q}\left(\sqrt{\frac{2E_b}{N_0}\cdot\frac{1}{1+(N-1)/G_p\cdot E_b/N_0}}\right) \tag{10.15}$$

where $\mathrm{Q}(x) \triangleq \sqrt{1/2\pi}\int_{-\infty}^{x} \exp(-y^2/2)\mathrm{d}y$ represents the Gaussian integral function, and according to the definition of the processing gain, we let $G_p = E_c/E_b$. It is apparent that the number N of concurrently active users depends on the specification about the QoS of the link, expressed in terms of a target BER: the lower the QoS (i.e., the larger the target BER), the higher the capacity will be in terms of number of active users. From this standpoint, it is easily seen that the UL capacity of CDMA can be further boosted up by appropriate use of channel coding [Vit79a].

10.3.5 Cellular Networks and the Universal Frequency Reuse

As outlined above, multiple access can be granted to DS/SS signals by assigning different spreading codes to all of the different active users within a given cell. A problem arises when we run out of codes in the DL and more users ask to access the network. With reasonable spreading factors (up to 256), the number of concurrently active channels is too low to serve a large user population like we have in a large metropolitan area or a vast suburban area. This also applies to conventional FDMA or TDMA radio networks where the number of channels is equal to the number of carriers in the allocated bandwidth or the number of time slots in a frame, respectively. The solution to this issue lies in the well-known notion of *cellular network with frequency reuse.* In a nutshell, the area to be served by the wireless system is split into cells, and the channels within the RF allocated bandwidth are distributed over the different cells. In so doing, the same channels can be allocated more than once to different cells (they can be reused), provided that the relevant cells are sufficiently far apart, so that the interference necessarily caused by the use of the same channels on the same frequency is sufficiently low. The distribution of cells on the territory is characterized by the so-called frequency reuse factor Q. Of course, frequency reuse has an impact on the overall network efficiency in terms of users per cell (or users per km^2) since the number of channels allocated to each cell is a fraction $1/Q$ of the overall channels allocated to the communications service provider. For instance, in GSM network, Q is equal to 7 or 9.

In CDMA-based networks, the spectral efficiency is maximized by *universal frequency reuse* whereby the same carrier frequency is used in each cell, and the same WH orthogonal code sets (i.e., the same channels) are used within each cell on the same carrier. Of course, something has to be done to prevent neighboring users at the edge of two adjacent cells and using the same WH code to heavily interfere with each other. The solution is to use a different *scrambling code* on different cells to cover the channelization (traffic) WH codes [Fon96], [Din98]. This represents a sort of code reuse technique, where *code* refers to the (orthogonal) channelization codes in each cell. Denoting with $c_m[k]$, $m = 1, \ldots, M$ the set of M WH codes to be used by the generic cell #u, and with $p_u[k]$ the overlaying sequence, the resulting *composite spreading signature code* is $c'_{u,m}[k] \triangleq p_u[k]\cdot c_m[k]$. The orthogonality between any pair of composite sequences is preserved, as in the original set of WH codes. Sequences typically used as overlay are the maximal-length PN and codes belonging to the Gold set [Gol67], [Gol68], [Sar80], [Din98].

10.4 A Review of 2G and 3G Standards for CDMA Mobile Communications

Although this handbook is focused on DSP components and techniques, we think that the reader may be interested in a short presentation from a communications systems perspective of the main features of the current (2003) major wireless CDMA communications systems that we have already mentioned. The mother of all commercial systems is the American standard IS-95 issued in 1993, which later evolved

TABLE 10.1 The cdmaOne Family

System Name	System Kind	UL Frequency Band (MHz)	DL Frequency Band (MHz)
TIE/EIA IS-95	Cellular	824–849	869–894
ANSI J-STD-008	Personal Communications System, PCS	1850–1910	1930–1990
Globalstar™	Satellite	1610–1626.5	2483.5–2500

into the well-established family of different systems, namely, cdmaOne™, whose main representatives are summarized in Table 10.1.

10.4.1 IS-95

This is the most widespread CDMA standard to date (2003). The chip rate is the same in both the UL and DL and is equal to 1.2288 Mchip/s for a nominal bandwidth occupancy of 1.25 MHz (roll-off factor of 0.2). In the DL, S-CDM with length 64 WH codes is used, for up to 64 multiplexed channels. The maximum net bit rate on each channel is 9600 bit/s; also, some channels are reserved for control. In particular, the channel corresponding to the WH function that is constant throughout the code period bears no data modulation (the *pilot* channel) and is used as a reference for synchronization and channel estimation (see Section 10.5). The traffic channels are all protected against transmission errors by a powerful forward-error correcting code whose coding rate is $r = 1/2$, so that the actual maximum signaling rate is 19,200 symbols/s. Considering the spreading factor 64 of the WH codes, we end up with the chip rate $19.2 \times 64 = 1228.8$ kchip/s, as above. The modulation and spreading format is "real data, complex code" since two different codes are used on the I and Q rails on the same copy of the digital datum. The I and Q codes are long (although they are called *short codes* in the standard), and they also serve as scrambling code to perform BS identification. If a lower data rate than 9600 bit/s has to be used, the channel encoder just reclocks the slower output data to make them enter the DS/SS modulator at the same invariable rate of 19.2 kbit/s. This is called *symbol repetition* and allows the support of bit rates that are integer submultiples of the maximum basic rate (in this case, 4800, 2400, and 1200 bit/s).

The arrangement for the UL of IS-95 is a little bit more complicated. We have here asynchronous CDMA with very long codes (periodicity of $2^{42} - 1$ chips). The chip rate is the same as in the DL (1.2288 Mchip/s), as is the basic maximum data rate (9600 bit/s). But the coding rate is now 1/3, and so the symbol rate at the modulator input is $3 \times 9.6 = 28.8$ ksymbols/s. At the other end of the modulator we also have a conventional "real data, complex code" section with scrambling codes.[5] In between, we have multilevel 64-ary modulation with a set of 64 orthogonal WH functions [Pro95]. Here, the WH functions are not used as orthogonal codes to perform multiplexing, as they are in the DL. Rather, they are used as symbols in a signal constellation to implement the robust process of *orthogonal functions modulation*, which is suited to efficient *noncoherent* demodulation. As a consequence, the signaling rate of the orthogonal-modulated stream raises to $28.8/\log_2(64) \times 64 = 307.2$ ksymbols/s. After four times of symbol repetition, scrambling with the very long channelization code and I/Q modulation/scrambling take place.

10.4.2 UMTS/UTRA

Although the 3G Euro-Japanese standard UMTS is substantially different in all details from IS-95, many of the good ideas that were introduced by the latter are incorporated into the former. Table 10.2 summarizes the main parameters of the two versions of UMTS, namely, UTRA-FDD and UTRA-TDD, where UTRA

[5]In the UL, the two I/Q components of the modulated signal are staggered in time (i.e., delayed one with respect to the other) by half a chip period. This reduces the envelope fluctuations of the modulated signal, thus preventing distortion by the subsequent stages of nonlinear amplification in the MT radio frequency transmitter.

TABLE 10.2 UMTS UTRA-FDD and -TDD Main System Parameters

Parameter	UTRA-FDD	UTRA-TDD
Frequency band	UL: 1920–1980 DL: 2110–2170	1900–1920 and 2010–2025
Maximum data rate	2.048 Mbit/s	2.048 Mbit/s
Framing	10 ms	10 ms
Chip rate	3.840 Mchip/s	3.840 or 1.280 Mchip/s
Pulse shaping	SRRC roll-off factor of 0.22	SRRC roll-off factor of 0.22
Carrier spacing	5 MHz	5 MHz
Spreading factor	4–256	Either 1 (no spreading, pure TDMA) or 16
DL multiplexing	S-O-CDM Optional short codes	S-O-CDM Short codes
UL multiple access	A-CDMA Long codes	S-O-CDM Short codes
UL pilot channel	Dedicated physical control channel running in parallel with the data channel	"Midamble": short series of pilot symbols at the center of the data burst
DL pilot channel	Common pilot channel plus dedicated pilot symbol in the traffic channels	"Midamble": short series of pilot symbols at the center of the data burst

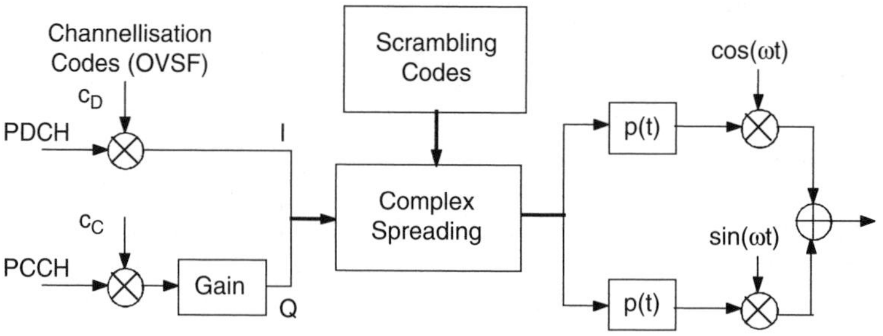

FIGURE 10.8 UTRA-FDD up-link spreading and modulation.

stands for *UMTS terrestrial radio access* and F/TDD stands for *frequency/time division duplexing*. In both versions, UL and DL, the chip rate is 3.840 Mchip/s with a roll-off factor of 0.22 for a nominal bandwidth occupancy of 5 MHz. This explains the qualification of wideband CDMA of UTRA-FDD. The latter offers symmetrical links with supported bit rates ranging from 16 (the basic, slowest rate) to 2048 kbit/s, with a variable spreading factor (OVSF codes; see "Multirate Code-Division Multiplexing" in Section 10.3.2). As is seen in Figure 10.8, each user has a UL *physical dedicated control channel* (PCCH) in addition to the customary traffic channel (*physical dedicated data channel*, PDCH). The two channels are locally multiplexed and spread with channelization OVSF codes (they may have different bit rates), and the two of them are further scrambled via complex spreading — the product between the complex-valued spread I/Q PDCH/PCCH signal and the complex-valued I/Q scrambling code.

The arrangement for modulation and spreading in the DL (traffic and control channels) is similar but a little more complicated (see Figure 10.9). The presence of the *serial-to-parallel* (S/P) converters on the data fluxes indicates that modulation is QPSK (complex binary data), but spreading with the channelization codes (synchronous OVSFs) is real (single code per channel). After multiplexing is accomplished, base station-unique real scrambling is carried out.

UTRA-TDD is different from UTRA-FDD for the duplexing mode: instead of placing the UL and DL on two different carrier frequencies as in FDD, the UL and DL share the same carrier and simply alternate in time. This gives total symmetry of signal format between UL and DL but, on the contrary, allows asymmetry of capacity by simply "moving" the boundary between the UL and DL slots every 10 ms, as shown in Figure 10.10. The signal format is actually somewhat simpler than its FDD counterpart (fixed spreading factor, option for a factor-of-three slower chip rate, etc.), and so UTRA-TDD is left for less demanding applications than FDD.

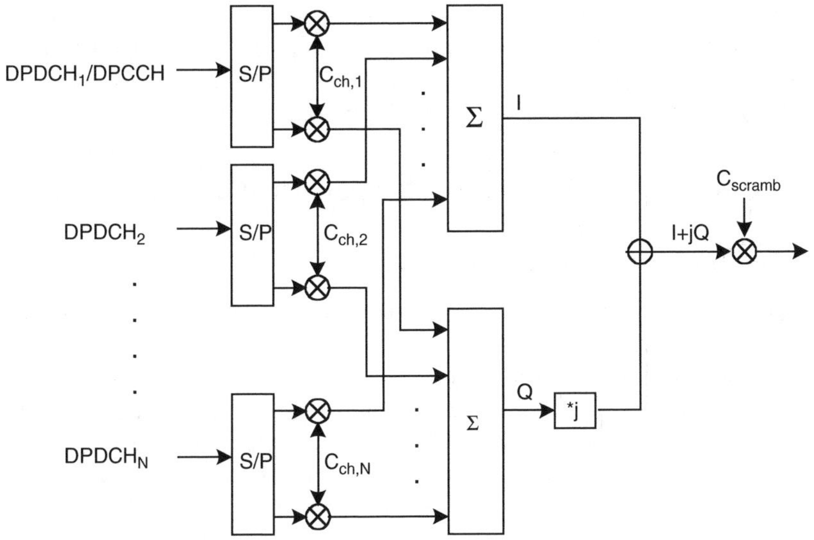

FIGURE 10.9 UTRA-FDD down-link modulation, spreading, multiplexing, and scrambling.

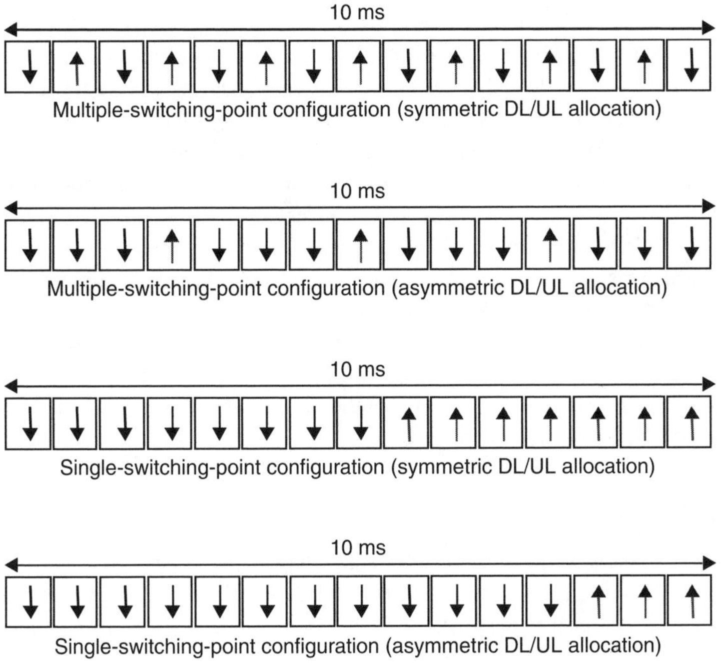

FIGURE 10.10 Asymmetric UL and DL capacity in UTRA-TDD.

10.4.3 cdma2000

As far as the general architecture is concerned (at least from a physical-layer perspective), the differences with UTRA-FDD are not very relevant. The main differentiation lies in a sort of multicarrier mode of cdma2000 that allows best back-compatibility with IS-95. The chip rates of cdma2000 are integer multiples of the IS-95 chip rate (for instance, $1.2288 \times 3 = 3.6864$ Mchip/s), so that one cdma2000 carrier fits exactly into the bandwidth of an integer number of cdmaOne channels. Many features like OVSF codes, dedicated control channels in the UL, common pilot channels in the DL, coherent demodulation in both the UL and DL, packet access, fast power control, etc., are shared with UTRA-FDD, and the relevant end-user performance and services are similar. For real worldwide roaming, the design of dual-mode cdma2000/UMTS terminals is being pursued by most equipment manufacturers.

10.5 Synchronization for Spread-Spectrum and CDMA Signals

10.5.1 Synchronization Functions

Broadly speaking, the word *synchronization* (often abbreviated *sync*) refers to those signal processing functions that in digital communication receivers are carried out to achieve correct time alignment of the incoming waveform with certain locally generated references. For instance, in a baseband pulse amplitude modulation system the signal samples must be taken at proper instants to minimize the intersymbol interference. To this purpose the receiver generates clock ticks indicating the location of the optimum sampling times. As a second example, in bandpass transmissions a coherent receiver needs *carrier* synchronization (or recovery), which means that the demodulation sinusoid must be locked in phase and frequency to the incoming carrier. Clock and carrier recovery are instances of *signal* synchronization, which is carried out within the physical layer of the system. As is apparent from this short discussion, sync functions are crucial for proper demodulation and data recovery and often represent a critical issue in the design of a communication link.

Figure 10.11 shows the general outline of (the baseband equivalent of) a conventional coherent correlation receiver for DS/SS user #h, where we have highlighted the relevant synchronization functions. For the sake of simplicity, we used a pseudoanalog description whereby ADC takes place at the CMF output. After frequency and phase offset correction, the received waveform $r(t)$ is fed to the CMF and then sampled (interpolated) at chip rate. Spectral despreading is performed by multiplying the samples $y_h[\ell]$ by a locally generated replica of the user's code $c_h[\ell]$, and finally, the resulting sequence is accumulated over a symbol period $T_s = MT_c$ (see also Section 10.2). Correct receiver operation requires accurate recovery of the signal time offset τ_h to ensure that the local code replica is properly aligned with the signature sequence in the received signal. This goal is usually achieved in two steps: (1) a coarse alignment is obtained first, and (2) it is used as a starting point for fine code tracking.

From the discussion above it appears that the synchronization problem in CDMA systems is similar to that encountered with narrowband signals, with two main differences. One is the presence of MAI, and the other is the broader signal bandwidth that makes the time offset compensation (code synchronization) much more complex, as discussed in the next section.

10.5.2 Code Synchronization

As mentioned earlier, data detection relies on the availability at the receiver of a time-aligned version of the spreading code. The delay τ_h must therefore be estimated and tracked to ensure such an alignment. The main difference with respect to narrowband modulations is the estimation accuracy. In narrowband

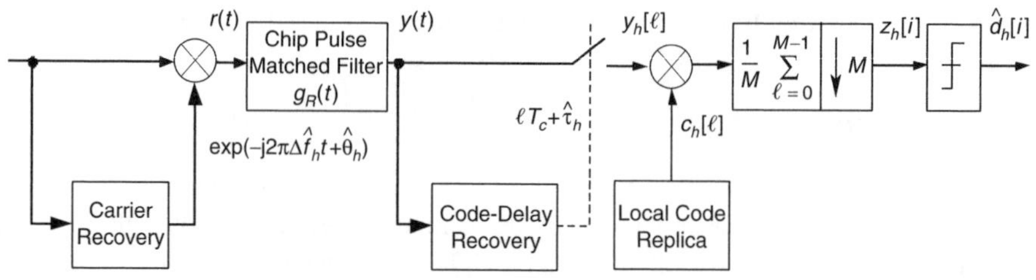

FIGURE 10.11 Coherent DS/SS CDMA receiver.

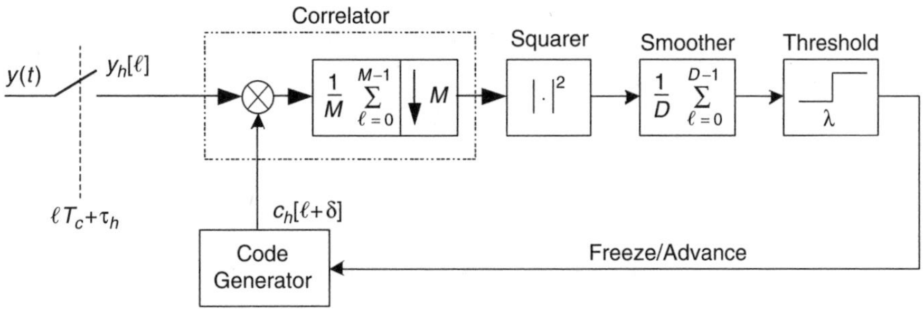

FIGURE 10.12 Serial code acquisition of a DS/SS CDMA signal.

systems, timing errors must be small compared to the *symbol time*, whereas in spread-spectrum systems, they must be small compared with the *chip* time, which is M times smaller. Assume again an AWGN channel and, for simplicity, that the spreading factor M equals the code repetition length L (the *short-code* DS/SS format). Then code acquisition amounts to finding the start of the symbol interval and is usually performed by correlating the input signal with the local replica of the code sequence (*sliding correlator*). The magnitude of the correlation is used in two ways: it is compared to a threshold to decide whether the intended user is actually transmitting and, if this is the case, to locate the position of the maximum. This location is taken as a coarse estimate of the delay τ_h.

The search for the maximum of the correlation may be performed with either serial or parallel schemes [De98a]. Serial schemes test all the possible code delays in sequence until the threshold is crossed. They are simple to implement, but are inherently time-consuming and their acquisition time cannot be established *a priori* (it can only be predicted statistically). Parallel schemes look for all the possible code epochs in parallel and choose the one corresponding to the maximum correlation. They guarantee short acquisitions but are computationally intensive.

A simplified scheme of a serial detector is shown in Figure 10.12. The code generator provides a spreading code with a tentative initial epoch δ, which is correlated over a symbol interval with the (digitized) received signal. The result is squared to cancel out any possible phase offset (noncoherent processing) and data dependence, then it is smoothed on a D-symbol dwell time, and finally it is compared to a threshold λ. If the threshold is crossed, the code epoch is frozen and fine timing recovery is started. Otherwise, the code generator is stepped forward by (a fraction of) one chip and a new trial acquisition is performed. The value of the threshold λ is a design parameter to be set according to the following considerations: low values of λ yield high probability of false acquisition events caused by large noise peaks; and vice versa, high threshold values produce occasional acquisition failures (i.e., missed detections). The false alarm probability and missed detection probability depend on λ and on the signal-to-noise ratio. The threshold value is therefore chosen based on the operating conditions.

Conventional correlation-based methods have satisfactory performance in a power-controlled system, i.e., when signals from different users arrive at the receiver with comparable amplitudes. However, they fail in a *near–far* situation where strong-powered users interfere with weaker ones. In these cases MAI becomes a serious impairment to achieve accurate code synchronization. Improvements are obtained by taking into account the statistical properties of the MAI. For example, near–far resistant code acquisition is achieved by modeling MAI as colored Gaussian noise [Ben98].

Initial code acquisition provides the receiver with a coarse estimate of the delay τ_h. Fine timing recovery (also called *code tracking*) is then needed to locate the optimum chip rate sampling instants. Code tracking is typically performed by means of a feedback loop, where a suitable timing error signal $e[i]$ is used to update the timing estimate at a symbol rate according to a recursive equation:

$$\hat{\varepsilon}[i+1] = \hat{\varepsilon}[i] - \gamma e[i] \tag{10.16}$$

where $\hat{\varepsilon}[i]$ is the i-th estimate of the normalized chip delay $\varepsilon \stackrel{\Delta}{=} \tau_h/T_c$ and γ is a design parameter. Figure 10.13 depicts the architecture of a *digital delay-lock loop* (DDLL) [Deg93] performing such a

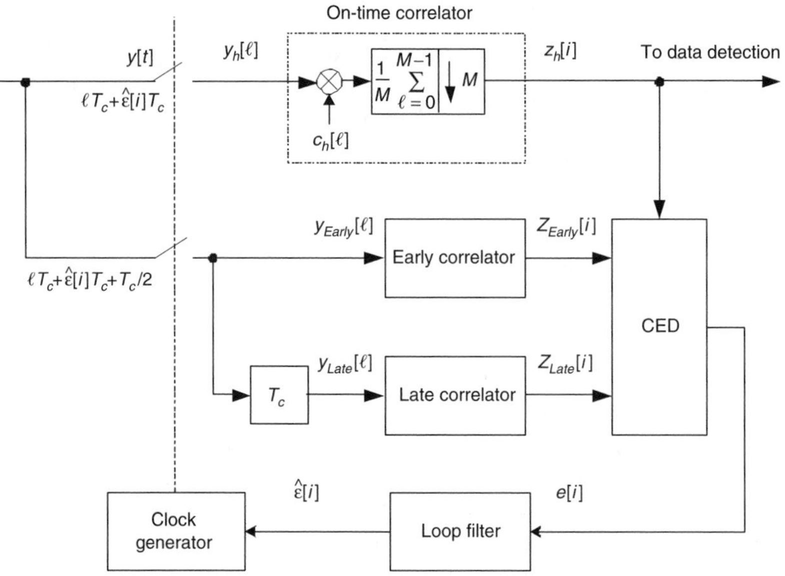

FIGURE 10.13 Digital delay-lock loop for fine timing recovery of a DS/SS CDMA signal.

function. It is seen that data demodulation relies on the so-called *on-time* samples, i.e., those taken at the optimum sampling instants, denoted as $y_h[\ell] \triangleq y(t)|_{t=\ell T_c+\tau_h}$. The DLL also needs the so-called *early–late* samples, $y_{Early}[\ell] \triangleq y(t)|_{t=\ell T_c+T_c/2+\tau_h}$ and $y_{Late}[\ell] \triangleq y(t)|_{t=\ell T_c-T_c/2+\tau_h}$, i.e., those taken in between two consecutive on-time samples. Two possible *chip-timing error detectors* (CEDs) insensitive to the phase offset (not requiring prior carrier phase recovery) are the following [Moe91]:

$$e[i] = \Re\{(z_{Early}[i] - z_{Late}[i]) \cdot z_h^*[i]\} \quad \text{(algorithm 1)} \tag{10.17}$$

$$e[i] = |z_{Early}[i]|^2 - |z_{Late}[i]|^2 \quad \text{(algorithm 2)} \tag{10.18}$$

where

$$z_h[i] \triangleq \frac{\sum_{\ell=iM}^{iM+M-1} y_h[\ell]c_h[|\ell|_L]}{M} \qquad z_{Early}[i] \triangleq \frac{\sum_{\ell=iM}^{iM+M-1} y_{Early}[\ell]c_h[|\ell|_L]}{M} \qquad z_{Late}[i] \triangleq \frac{\sum_{\ell=iM}^{iM+M-1} y_{Late}[\ell]c_h[|\ell|_L]}{M} \tag{10.19}$$

Equation 10.17 is reminiscent of the early–late timing detector for narrowband transmissions, while Equation 10.18 is typical of spread-spectrum signals. Their performance is satisfactory in the absence of channel distortion and near–far effects.

10.5.3 Carrier Frequency and Phase Synchronization

Carrier phase recovery for coherent demodulation can be carried out at the output of the despreader–accumulator in the correlation receiver as it is in conventional narrowband modulations, and so we will not further pursue this issue. On the contrary, carrier frequency acquisition may reveal an issue in DS/SS receivers, especially when the frequency offset Δf to be recovered is comparable with the symbol rate R_s. This situation is typically encountered at receiver start-up in the low-cost MT, wherein the local oscillator has an inherent instability that may be that large. In this case, spreading code acquisition becomes unreliable since the frequency offset decorrelates the signal within the integration window T_{ACQ} of the code acquisition unit ($T_{ACQ} = T_s$ in Figure 10.13). The result of this decorrelation is an equivalent signal-to-noise ratio loss in code acquisition that amounts to

$$\Lambda = \left[\frac{\pi \Delta f T_{ACQ}}{\sin(\pi \Delta f T_{ACQ})} \right]^2 \tag{10.20}$$

For example, a frequency offset of half the symbol rate (with $T_{ACQ} = T_s$) causes a loss of more than 6 dB. On the other hand, frequency offset cannot be reliably estimated unless the code sequence is coarsely acquired. This problem can be effectively approached with joint estimation of the frequency offset and code phase. This leads to a bi-dimensional grid search in which the frequency uncertainty range is partitioned into a number of bins and the code acquisition test is repeated for all of the bins [De98a] after frequency precompensation with the hypothesized offset. At the end of the process, coarse estimates of code phase and frequency offset are obtained.

Once coarse frequency acquisition is attained, fine frequency tracking can be performed by means of conventional feedback schemes based on a quadri-correlator or a dual-filter detector [Men97]. Frequency recovery for CDMA transmissions on frequency-selective channels is still an open problem. Current research investigates methods to alleviate the combined effects of MAI and multipath on the acquisition process.

10.6 Architecture of DSP-Based DS/SS and CDMA Receivers for Wireless Mobile Communications

10.6.1 IF vs. Baseband Sampling

The current trend in modem design, both in a low-cost MT and in an expensive BS, is to perform as much signal processing in the digital domain as possible. Considering the receiver section of the modem (the most challenging to implement) we have basically two possible alternatives as far as the location of the ADC stage is concerned. The first approach, shown in Figure 10.14.a, is mainly pursued in low-power MTs. The RF received signal is converted to baseband using a conventional I/Q analog tuner followed by twin I/Q ADC and baseband digital signal processing for data detection. The critical points of this arrangement are the possible amplitude imbalance of the two I and Q analog rails, as well as the imperfect quadrature between the two I/Q carriers used for IF-to-baseband conversion. A precision receiver with no power consumption or cost constraint is likely to be implemented according to the IF sampling architecture shown in Figure 10.14.b, where the ADC stage is shifted toward the antenna. Incidentally, we observe that this is just the general scheme of the so-called *software radios* [Mit95], where all of the signal processing, apart from the initial RF-to-IF conversion, is performed in the digital domain, and the DSP components are reprogrammable to a certain extent. With a software radio, changing the signal format to be treated just amounts to changing the software that drives the (programmable) DSP components, instead of changing a piece of dedicated hardware (a card, or the whole modem), as in conventional equipment.

Coming back to Figure 10.14.b, the ADC operates directly on the IF signal, and the relevant digital output is still a bandpass signal on a different, digital, intermediate frequency, $f_{IFD} = f_{IF} \pm kf_{sa}$ (k is an integer), as in Figure 10.6. The analog front end is simplified, and the task of baseband conversion is deferred to the digital section (with no issues of amplitude imbalance and imperfect quadrature). The main drawback of the IF sampling approach is the tighter requirements for the ADC, which must now handle faster IF signals (instead of baseband signals, as in Figure 10.14.a). In particular, the converter rise and fall times, i.e., the time needed to open and close the gate of the sampling device, will be commensurate to the analog IF frequency and turn out to be much shorter than those of the converters operating on the baseband signal. Also, the IF sampling ADC is in general more expensive and power-consuming than the two baseband converters in Figure 10.14.a.

Concerning the bandwidth of the receiver front end, notice that in conventional narrowband modulation supporting multirate transmissions, the bandwidth of the data signal varies according to the data rate. In the design of a digital receiver for such signals, the bandwidth of the analog front end shall be set to accommodate the widest possible spectrum (i.e., that relevant to the highest symbol rate). Therefore, a

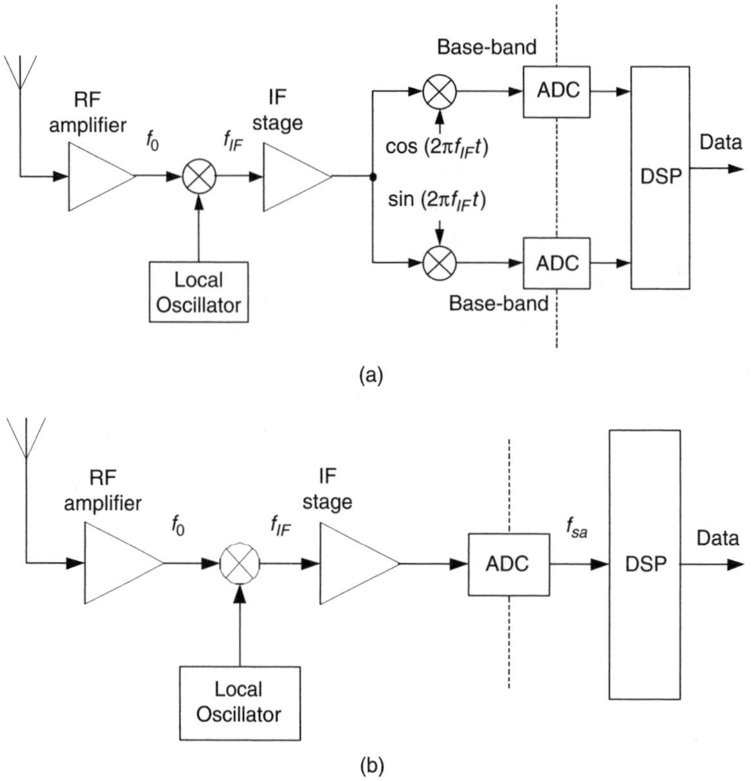

FIGURE 10.14 (a) Baseband sampling implementation of a digital receiver. (b) IF sampling implementation of a digital receiver.

digital decimator/interpolator stage must be implemented after the ADC in order to match the processing rate to the symbol rate (the receiver usually operates at a fixed number of samples/symbol interval). On the contrary, in the case of DS/SS transmissions, the bandwidth occupancy is determined by the chip rate, which is usually a fixed system parameter, independent of the actual symbol rate. This particular feature avoids the need of a dedicated decimator stage like that mentioned above, since the despreading/accumulation function (Equation 10.6) carried out in a DS/SS receiver implicitly performs decimation from chip rate down to symbol rate.

10.6.2 Correlation Receiver

This is the conventional receiver for DS/SS transmissions and it has already been described in detail in Section 10.2. The correlation receiver is optimum in each of the following conditions:

1. Single-user transmission over the AWGN channel
2. Multiuser transmission over the AWGN channel in S-O-CDM mode (no MAI is experienced by the receiver)
3. Multiuser transmission over the AWGN channel in A-CDMA mode with long PN codes and a large number of equi-powered active users (allowing the modeling of MAI as an additional Gaussian noise contribution)

10.6.3 Rake Receiver

The radio signal in a wireless mobile communications system operating in an urban or indoor environment experiences the phenomenon of *multipath propagation* [Tur80]. In such scenarios, both the transmit and the receive antennas have usually moderate to weak directivity features, so that the signal can propagate from the transmitter to the receiver through a number of different *propagation paths*. The different paths (or *rays*, as they are called in electromagnetism) are due to reflection and scattering on walls, buildings, trees, furniture, and other surfaces or obstacles. A simple model for the received signal at the receiving end is thus

$$r(t) \triangleq \sum_{k=1}^{K} a_k \cdot s^{(SS)}(t - \tau_k) + w(t) \tag{10.21}$$

where $s^{(SS)}(t)$ is the baseband equivalent of the transmitted DS/SS signal, $w(t)$ is the usual channel noise term, K is the number of paths, $a_k \triangleq |a_k| \cdot \exp(j\angle a_k)$ is a complex random variable representing the amplitude and phase shift experienced by the signal traveling on the generic k-th path, and τ_k is the relevant propagation delay. In Equation 10.21 we assumed that the channel parameters (amplitudes, phase shifts, and delays) are constant in time, but this may not be the case for mobile communications at high speed. In an urban environment, the amplitude coefficient $|a_k|$ is usually modeled as a Rayleigh random variable, while the phase shift $\angle a_k$ is assumed to be uniformly distributed over $(0, 2\pi)$. From Equation 10.21, the frequency response of the multipath channel turns out to be

$$H(f) = \sum_{k=1}^{K} a_k \cdot e^{-j2\pi f \tau_k} \tag{10.22}$$

Typical values of the delays τ_k are $1 \div 10$ μs for rural environments, $0.1 \div 1$ μs for urban scenarios, and $1 \div 100$ ns for indoor propagation. In general, the channel amplitude and phase responses reveal considerably variable over the signal bandwidth, thus causing nonnegligible distortion on the received signal: we have a *frequency-selective* channel.

On the multipath channel, the CR in no longer optimal since the additional rays may cause intersymbol interference. A popular way to deal with DS/SS signal detection consists of resorting to the so-called Rake receiver, whose block diagram is depicted in Figure 10.15. The Rake is a kind of time diversity receiver made of a bank of N_R identical conventional DS/SS detectors, operating in parallel and called fingers (just like the fingers of a gardener's rake). In the code acquisition phase, DS/SS demodulator 1 tries to locate the strongest path within the multipath signal (Equation 10.21) — by finding the strongest correlation peak of the local code replica with the received signal — and "tunes" onto it. After this, the other detectors will try to find the second and third strongest paths and lock onto them. To achieve this, they carry out a continuous search around the previously found locations, in particular looking for secondary correlation

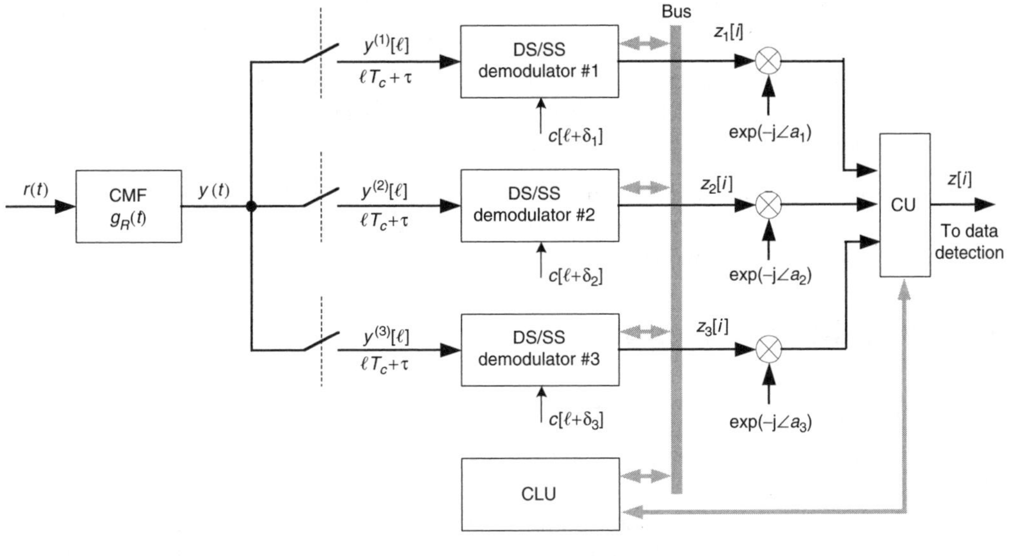

FIGURE 10.15 Outline of a Rake receiver.

peaks generated by weaker signal echoes. The search range is limited to a few chip intervals, depending on the maximum channel delay the receiver has to cope with. Eventually, each finger of the Rake locks onto a different time-delayed replica of the incoming signal, correlates it with a properly shifted version of the local code replica, and yields a decision strobe $z_n[i]$ at symbol rate. This operation mode presumes that the different signal replicas (echoes) can be resolved; i.e., they can be singled out by a correlation procedure. This is true only when the difference between the various signal delays is greater than one chip time. In this case, the different signal echoes on the different propagation paths turn out to be uncorrelated (due to the properties of the spreading code), and they can be resolved via correlation processing.

Assume now for simplicity that finger 1 locks onto path 1, finger 2 onto path 2, and so forth. The different partial-decision variables $z_n[i]$ are phase-corrected by $\exp(-\mathrm{j}\angle a_n)$ to perform coherent detection. Then, a *combination logic unit* (CLU) controls a *combination unit* (CU), which is in charge of selection or combining of the finger outputs to provide a final decision strobe $z[i]$ for data detection. The selection/combining criterion may be one of the following:

1. *Selection combining* (SC), where the CLU just selects the maximum-amplitude strobe:

$$z[i]|_{SC} \stackrel{\Delta}{=} z_{\bar{n}}[i] \;\Leftrightarrow\; \bar{n} : |z_{\bar{n}}[i]| = \max_n \{|z_n[i]|\} \tag{10.23}$$

2. *Equal gain combining* (EGC), where the CLU sums up all strobes at the finger outputs (upon carrier phase counterrotation):

$$z[i]|_{EGC} \stackrel{\Delta}{=} \sum_{n=1}^{N_R} z_n[i] \cdot \mathrm{e}^{-\mathrm{j}\angle a_n} \tag{10.24}$$

3. *Maximal ratio combining* (MRC), where the CLU performs a weighted sum of all the strobes at the fingers outputs:

$$z[i]|_{MRC} \stackrel{\Delta}{=} \sum_{n=1}^{N_R} |a_n| \cdot z_n[i] \cdot \mathrm{e}^{-\mathrm{j}\angle a_n} = \sum_{n=1}^{N_R} a_n^* \cdot z_n[i] \tag{10.25}$$

It can be shown that MRC maximizes the signal-to-noise ratio at the Rake output, since Equation 10.25 implements a channel-matched filter. In fact the Rake represents the optimum receiver for DS/SS single-user reception over a multipath channel, provided that all the following conditions are met:

1. The number of Rake fingers is exactly equal to that of the propagation paths.
2. The Rake can resolve all the paths; i.e., each finger locks onto a different path.
3. *Interchip interference* (ICI) is negligible; i.e., perfect fine time recovery is carried out.
4. The spreading code is Delta correlated; i.e., it has null off-zero autocorrelation.

According to conditions 1 and 2, the number of fingers in the Rake should be equal to the number of resolvable paths in the channel. In practice, a reasonable trade-off between complexity and multipath robustness is represented in many cases by the choice $N_R = 3$ [Deg94], [De98b].

10.7 Multiuser Detection

10.7.1 Multiuser Detection in the UL

The discussion in the previous sections has outlined the main issue that affects CDMA systems as far as capacity and QoS are concerned — interference. This especially applies to the (asynchronous) UL wherein interference in the form of MAI is generated within one's own cell (intracell interference).

The seminal work by Verdù [Ver86] and almost two decades of related research have shown how to cope with such issues [Mos96, Ver98]. MAI in the UL can be counteracted by the adoption at the BS of a suited

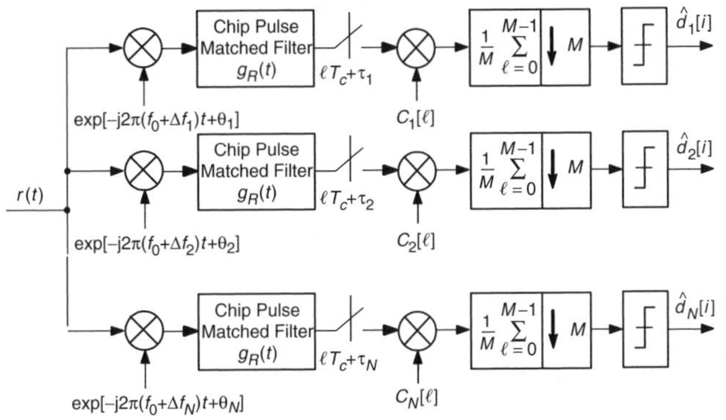

FIGURE 10.16 Bank of conventional single-user detectors in the UL.

joint or *multiuser data detection* (MUD) scheme that performs demodulation of *all* the UL data streams in a single, centralized signal processing unit. With centralized detection, the presence of a certain amount of MAI coming from interfering channels can be accounted for when demodulating the useful, intended channel. This standpoint applies reciprocally to *all* the channels, leading to the above-mentioned notion of multiuser demodulation. The conventional arrangement of the BS channel demodulator is shown in Figure 10.16. The data demodulation unit is constructed as an array of independent single-channel demodulators in the form of conventional correlation detectors, like that depicted in Figure 10.6. Those detectors are optimum in the AWGN environment — they simply ignore the issue of MAI and lead to strong suboptimality. On the contrary, a multiuser detector is a centralized data demodulation unit whose general scheme is depicted in Figure 10.17. By concurrently observing all of the correlators outputs, MAI can be taken into account when detecting each channel, and it can be mitigated or cancelled by suited signal processing. This leap forward in the performance of the CDMA receiver, not surprisingly, comes at the expense of a substantial increase in the complexity (signal processing power) of the demodulator. The front end of the centralized MUD scheme in Figure 10.17 consists of the same bank of N correlators with the individual N user signature waveforms that we find in the multiple conventional correlation receiver in Figure 10.16. The real core of MUD lies in the elaborate postprocessing of the array of matched filter outputs (sufficient statistics) [Mos96]. The optimum AWGN MUD scheme originally proposed and analyzed by Verdù [Ver86] encompasses a *Viterbi algorithm* (VA) to perform parallel maximum likelihood sequence estimation. Since the number of states in the trellis of the VA is equal to 2^N, a receiver with this optimum structure has a complexity that is exponential in the number of users N, and therefore it does

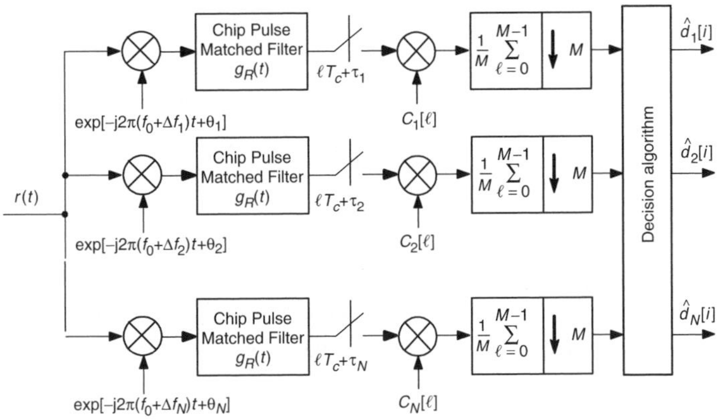

FIGURE 10.17 General scheme of a multiuser detector.

not easily lend itself to practical implementations. Several different suboptimum MUD structures have therefore appeared since then, all of them attempting to reduce the estimation complexity by replacing the Viterbi decoder with a different device (decorrelating detector, successive cancellations, and so on [Mos96]).

A further constraint that has to be satisfied in order for MUD to be applicable is that the CDMA signals to be used in the multiple-access network bear *short codes*, i.e., $L = M$, so that the symbol period cross-correlations between the different spreading codes are well defined and *stationary in time*. This condition is not met in the case of long spreading codes, as is unfortunately the case for the UL of IS-95.

10.7.2 The Decorrelating and MMSE Detectors

Let us now assume that we intend to demodulate a three-channel CDMA signal with simple, real spreading of real-valued symbols. To simplify matters, we will also assume that the three channels (codes) are synchronous, but they are not orthogonal. Although this is not fully representative of the situation encountered in the UL (asynchronous nonorthogonal), it leads to similar results. Assuming ideal carrier and code timing synchronization, the symbol rate sampled outputs of the code matched filters in the front end of the MUD in Figure 10.17 are (see Equation 10.13)

$$\begin{aligned}
z_1[i] &= A_1 \cdot d_1[i] + \rho_{1,2} \cdot A_2 \cdot d_2[i] + \rho_{1,3} \cdot A_3 \cdot d_3[i] + \nu_1[i] = A_1 \cdot d_1[i] + \mu_1[i] + \nu_1[i] \\
z_2[i] &= \rho_{2,1} \cdot A_1 \cdot d_1[i] + A_2 \cdot d_2[i] + \rho_{2,3} \cdot A_3 \cdot d_3[i] + \nu_2[i] = A_2 \cdot d_2[i] + \mu_2[i] + \nu_2[i] \\
z_3[i] &= \rho_{3,1} \cdot A_1 \cdot d_1[i] + \rho_{3,2} \cdot A_2 \cdot d_2[i] + A_3 \cdot d_3[i] + \nu_3[i] = A_3 \cdot d_3[i] + \mu_3[i] + \nu_3[i]
\end{aligned} \tag{10.26}$$

where $\rho_{n,m} = \rho_{m,n} \triangleq (1/L)\sum_{k=0}^{L-1} c_n[k] \cdot c_m[k]$ is the cross-correlation coefficient between the two spreading codes of channel n and channel m, A_n is the amplitude of signal on channel n, $d_n[i]$ is the i-th data symbol on channel n, and $\nu_n[i]$ is the noise component affecting the n-th detector rail. The effect of MAI, which is represented by the samples $\mu_n[i]$, is apparent, and it is also clear that it can be potentially destructive. Assume, for instance, that $\rho_{1,2} \neq 0$ (i.e., codes 1 and 2 are not orthogonal) and that $A_2 \gg A_1$. Then the MAI contribution $\rho_{1,2}A_2d_2[i]$ in Equation 10.26 may overwhelm the useful term $A_1d_1[i]$ for data detection of channel 1. This phenomenon is called the *near–far effect*: user 2 can be considered as located near the receiver in the BS, thus received with a large amplitude, while user 1 (the one we intend to demodulate) is the far user and is weaker than user 2. Generalizing to N users, Equation 10.26 can easily be cast into a simple matrix form. If we arrange the cross-correlation coefficients $\rho_{i,k}$ into the (symmetric) correlation matrix

$$\mathbf{R} \triangleq \begin{bmatrix} 1 & \rho_{1,2} & \cdots & \rho_{1,N} \\ \rho_{2,1} & 1 & & \rho_{2,N} \\ \vdots & & \ddots & \vdots \\ \rho_{N,1} & \rho_{N,2} & \cdots & 1 \end{bmatrix} \tag{10.27}$$

and we introduce the diagonal matrix of the user amplitudes $\mathbf{A} \triangleq \text{diag}\{A_1, A_2, \ldots, A_N\}$, we have

$$\mathbf{z} = \mathbf{RAd} + \boldsymbol{\nu} \tag{10.28}$$

where the N-dimensional vectors $\mathbf{z}$, $\mathbf{d}$, and $\boldsymbol{\nu}$ simply collect the samples of received signal ($z_n[i]$), data ($d_n[i]$), and noise ($\nu_n[i]$), respectively (we have dropped the time index, i for simplicity). A simple multiuser detector is the *decorrelating detector* that applies a linear transformation to vector $\mathbf{z}$ to provide N soft decision variables relevant to the N data bits to be estimated. Collecting such N decision variables $g_n[i]$ into vector $\mathbf{g}$, the linear *joint* transformation on the correlator output vector $\mathbf{z}$ is just

$$\mathbf{g} = \mathbf{R}^{-1}\mathbf{z} \tag{10.29}$$

so that

$$\mathbf{g} = \mathbf{R}^{-1}(\mathbf{RAd} + \boldsymbol{\nu}) = \mathbf{Ad} + \mathbf{R}^{-1}\boldsymbol{\nu} = \text{diag}\{A_1 d_1[i], A_2 d_2[i], \ldots, A_N d_N[i]\} + \boldsymbol{\nu}' \tag{10.30}$$

and

$$g_n[i] = A_n \cdot d_n[i] + \nu_n'[i] \tag{10.31}$$

where $\nu_n'[i]$ is just a noise component. It is apparent that the MAI has been completely cancelled (provided that the correlation matrix is invertible) or, in other words, the different channels have been *decorrelated*. The drawback is an effect of noise enhancement due to the application of the decorrelating matrix $\mathbf{R}^{-1}$: the variance of the noise components in ν' is in general larger than that of the components in $\boldsymbol{\nu}$. Therefore, the decorrelating detector works properly only when the MAI is largely dominant over noise.

A different approach is pursued in the design of the *minimum mean square error* (MMSE) multiuser detector: the linear transformation is now with a generic $N \times N$ matrix $\mathbf{Z}$ whose components are such that the MSE between the soft output decision variables in $\mathbf{g}$ and the vector of the data symbols is minimized:

$$\mathbf{g} = \mathbf{Zz}, \quad \text{with } \mathbf{Z} \text{ such that } \mathrm{E}\{|\mathbf{g} - \mathbf{d}|^2\} = \min \tag{10.32}$$

where E{·} denotes statistical expectation. Solving for $\mathbf{Z}$, we get

$$\mathbf{Z} = \left(\mathbf{R} + \sigma_\nu^2 \mathbf{A}^{-2}\right)^{-1} \tag{10.33}$$

with σ_ν^2 indicating the variance of the noise components in Equation 10.8. The MMSE detector tries to optimize the linear transformation with respect to both MAI and noise. If noise is negligible with respect to MAI, the matrix in Equation 10.33 collapses into the decorrelating matrix $\mathbf{R}^{-1}$. Vice versa, if the MAI is negligible, matrix $\mathbf{Z}$ is diagonal and collapses into a set of scaling factors on the correlator outputs that do not affect data decisions at all (and, in fact, in the absence of MAI, the outputs of the correlators are the optimum decision variables without any need of further processing).

From this short discussion on MUD algorithms, it is clear that in general such techniques are quite challenging to implement, both because they require nonnegligible processing power (for instance, to invert the decorrelating or MMSE matrices) and because they call for *a priori* knowledge or real-time estimation of signal parameters, such as the correlation matrix. But the potential performance gain of MUD also had an impact on the standardization of 3G systems (in particular, UMTS) in that an option for short codes in the UL was introduced just to allow for the application of such techniques in the BS [Ada98], [Oja98].

Interference mitigation techniques can also be used in the MT of the end user to counteract intercell interference coming from other cells in the network. This would require the adoption of a short-code format in the DL (see Section 10.2), which at the moment of this writing is not supported by any of the existing standards. Therefore, we will not dwell any further on this subject of active research [Fan04].

10.8 Perspectives and Conclusions

10.8.1 The Challenge to Mobile Spread-Spectrum Communications

At the moment of this writing (2003) the present SS systems being deployed all over the world are the 3G cellular networks, as already discussed in Sections 10.1 and 10.4. But we have to say that in some areas (notably Europe and the U.S.) 3G systems have been relented by a number of factors, both technical and economical. The uncertainities about the perspective of a successful diffusion of 3G systems favor the advent and development of some serious competitors. For instance, many forecasts envisage a coexistence of copper *wired LANs* to link fixed PCs within an office or building to *wireless networks* characterized by high bit rate and a certain support of mobility and handovers. Such wireless networks, called *wireless LANs* (WLANs) [Nee99], actually represent a viable alternative to 3G systems in some applications and not just a replacement of a traditional wired LAN. In fact, with WLANs, laptops, palmtops, possibly portable mp3 players, and video terminals are all linked together, either via a central access point in a star topology (with immediate provision of connectivity with the fixed network) or directly with each other in an ad hoc, decentralized network. The former architecture is typical of IEEE 802.11a-b-g networks [ie802], which are at the moment gaining more and more popularity; the latter is the paradigm of Bluetooth piconets/scatternets [Teg02] and IEEE 802.15 ad hoc communications [ie80a]. WLANs are becoming the standard untethered connection for nomadic computing and Internet access, and may seriously challenge 3G systems when full mobility is not a fundamental requirement. To this respect, so-called hot-spots, that is, high-traffic areas served by 802.11 WLAN public access points, are being created in most public buildings and communication-demanding areas (railway stations, airports, university campuses, etc.). Still to come are WLANs for the home (such as home RF and similar products currently being developed) that belong more to the field of consumer electronics than to telecommunications.

We will not enter here into the technical details of the 802.11 and 802.15 standards, because they do not properly fall into the category of mobile communications. Suffice it to say that both are in some ways based on FH/SS, so they still belong in a sense to the big family of SS techniques. On the contrary, we want to say more about the future of SS communications, which seems to be intertwined with that of another technology that has marked the 1990s: *multicarrier* (MC) *modulation* [Sar95]. In MC transmission, the output of a high-rate data source is split into many low-rate streams modulating adjacent subcarriers within the available bandwidth. If N is the number of subcarriers, the symbol rate on each of them is accordingly reduced by a factor N with respect to the source rate, and this squeezes the signal bandwidth around the subcarrier to a point that the transmission channel looks locally flat. Correspondingly, the channel distortion on each subcarrier is reduced to a multiplicative factor that can be compensated for by a simple one-tap equalizer. The possibility of easing the equalization function has motivated the adoption of MC transmission as a standard in a number of current applications, for example, European *Digital Audio Broadcasting* (DAB) and terrestrial *Digital Video Broadcasting* (DVB), IEEE 802.11a-g WLANs, and *Asymmetric Digital Subscriber Line* (ADSL) and its high-speed variant VDSL, just to mention a few.

The idea for coming to a highly efficient and flexible wireless signaling technique for next-generation systems is to combine SS and MC communications. On one hand, SS signaling warrants easy and efficient multiple access, and on the other, MC modulation yields robustness against channel distortions, as explained above. The resulting technique is referred to as *multicarrier CDMA*, or MC-CDMA for short [Faz03]. Such a modulation/multiple-access method appears the most credited signal format for the *fourth-generation* (4G) systems.

10.8.2 4G Wireless Communications Systems

Although the end user may not be aware of it, at the same time that 3G systems are being commercially developed, R&D laboratories worldwide are working hard to devise prototypes of 4G wireless networks. At the moment, a great many 4G architectures exist, according to the different visions of producers and research labs. The main ambition of 4G is to combine traditional cellular communications, WLANs, broadband *wireless local loops* (WLLs), and possibly broadcasting into a ubiquitous, universal, broadband, flexible wireless access network that can accommodate business as well as residential users, multimedia communications as well as Internet access. In a word, 4G will go where 3G cannot go.

In this ambitious picture, a variety of signal processing techniques will be integrated, and in the writers' opinion, SS will no longer be the fundamental one, as in the previous generation. For instance, many proposals for 4G systems or even 3½G systems (that is, an intermediate generation between current 3G and long-term future 4G) are based on *multiple-input multiple-output* (MIMO) signal transmission technologies as the factor to boost system capacity [Tel99]. With 3G, we can say that SS has attained its full maturity and will be either integrated or replaced by different, at times more advanced, techniques. We have already mentioned MC-CDMA, which tries to capitalize from the advantages of both SS and MC modulations, as a good candidate for 4G. But again, competitors are in view. Some envisage the comeback of pure FDMA for multiple access, in its modern variant called *orthogonal FDMA* (OFDMA) [ie80b]. In a sense, OFDMA, which has recently been standardized for WLL applications by the IEEE 802.16 committee [ie80b], is a by-product of OFDM technology. Also, the lesson of intentionally broadening the signal spectrum well beyond the Nyquist bandwidth has been completely taken up by those supporting *ultrawideband* (UWB) signaling for short-range communications [IEE02]. In UWB systems, the information conveyed by a single bit is coded into a sequence of ultrashort pulses (less than 1 ns each) whose spectrum is apparently spread over an ultrawide bandwidth. Admittedly, this is not conventional DS/SS or FH/SS, but is indeed a form of wide-spectrum signaling. The debate in the different laboratories or standardization bodies about the pros and cons of such technologies is really hot, so insisting on one instead of another would currently be unfair or unmotivated.

10.8.3 Concluding Remarks

One thing can be said for sure: the heritage of SS techniques will extend well beyond 3G systems and 802.11x WLANs, and well into 4G systems. But what 4G wireless communications will actually be, nobody knows at the moment. Developing, testing, and deploying a worldwide and such ambitious standard is something that takes at least 5 years, even with the current fast pace of technology. Take, for instance, into consideration the history of 3G: first commercial deployment of UMTS networks started in Japan in 2002, while the first international wide-scale research projects on wideband CDMA (that later developed into UMTS) date back to 1995. Shortening this cycle a little bit leads to a 5-year period. During the same time it may happen, as it happened with UMTS/cdma2000, that boundary conditions change and new, "leaner" technologies emerge (as Bluetooth/WLANs) that may change the scenario and perspectives of the standard as well. One of the goals of 4G is coming to a unified signal format for low-rate as well as wide-bandwidth access, to make cellular network coexistent with WLANs, WLLs, and broadcasting. But on the contrary, traditional broadcasting networks are developing standards that are encompassing some form of return channel from the user terminals to the transmitter, just to give the end user some interactivity and to provide, in a sense, the same kind of multimedia interactive services provided nowadays by 3G cellular. So the question is, Will 4G really be the integrated universal standard system as it is envisaged above, or will it only be something that looks like a bigger, smarter, wider-bandwidth cellular network, interoperating with a number of heterogeneous networks as today happens with 3G? We personally fear that the WLAN syndrome may be round the corner (and in particular we think of decentralized, ad hoc networks [ietfw] as an unpredictable factor that may menace the ubiquity of 4G), but this is just an opinion. On the contrary, DSP will have more and more relevance to keep at a lower cost the promises of increased efficiency and flexibility of next-generation wireless communications, whatever their architecture might be.

Acknowledgments

The authors express their gratitude to Dr. Vincenzo Lottici of the University of Pisa, Italy, for his helpful comments and suggestions.

References

[Ada97] F. Adachi, M. Sawahashi, K. Okawa, "Tree-Structured Generation of Orthogonal Spreading Codes with Different Lengths for Forward Link of DS CDMA Mobile Radio," *IEE Electronics Letters*, 33(1), 27–28, 1997.

[Ada98] F. Adachi, M. Sawahashi, H. Suda, "Wideband DS-CDMA for Next-Generation Mobile Communications Systems," *IEEE Communications Magazine*, September 1998, pp. 56–69.

[Ahm75] N. Ahmed, K.R. Rao, *Orthogonal Transforms for Digital Signal Processing*, Springer-Verlag, New York, 1975.

[Ben98] S.E. Bensley, B. Aazhang, "Maximum Likelihood Synchronization of a Single-User for Code-Division Multiple-Access Communication Systems," *IEEE Transactions on Communications*, 46(3), 392–399, 1998.

[Chi92] S. Chia, "The Universal Mobile Telecommunication System," *IEEE Communications Magazine*, December 1992, pp. 54–62.

[Dah98] E. Dahlman, B. Gudmundson, M. Nilsson, J. Sköкld, "UMTS/IMT-2000 Based on Wideband CDMA," *IEEE Communications Magazine*, September 1998, pp. 70–80.

[De98a] R. De Gaudenzi, F. Giannetti, M. Luise, "Signal Synchronization for Direct-Sequence Code-Division Multiple Access Radio Modems," *European Transactions on Telecommunications*, 9(1), 73–89, 1998.

[De98b] R. De Gaudenzi, F. Giannetti, "DS-CDMA Satellite Diversity Reception for Personal Satellite Communication: Satellite-to-Mobile Link Performance Analysis," *IEEE Transactions on Vehicular Technology*, 47(2), 658–672, 1998.

[Deg93] R. De Gaudenzi, M. Luise, R. Viola, "A Digital Chip Timing Recovery Loop for Band-Limited Direct-Sequence Spread-Spectrum Signals," *IEEE Transactions on Communications*, 41(11), 1993.

[Deg94] R. De Gaudenzi, F. Giannetti, "Analysis of an Advanced Satellite Digital Audio Broadcasting System and Complementary Terrestrial Gap-Filler Single Frequency Network," *IEEE Transactions on Vehicular Technology*, 43(2), 194–210, 1994.

[Din98] E.H. Dinan, B. Jabbari, "Spreading Codes for Direct Sequence CDMA and Wideband CDMA Cellular Networks," *IEEE Communications Magazine*, September 1998, pp. 48–54.

[Dix94] R.C. Dixon, *Spread Spectrum Systems with Commercial Applications*, Wiley Interscience, New York, 1994.

[Fan04] L. Fanucci, F. Giannetti, M. Luise, M. Rovini, *Experimenting with CDMA and Interference Mitigation: From System Architecture to Hardware Testing through VLSI Design*, Kluwer Academic Publisher, New York, 2004.

[Faz03] K. Fazel, S. Kaiser, *Multi-Carrier and Spread Spectrum Systems*, Wiley, Europe, 2003.

[Fon96] M.-H. Fong, V.K. Bhargava, Q. Wang, "Concatenated Orthogonal/PN Spreading Sequences and Their Application to Cellular DS-CDMA Systems with Integrated Traffic," *IEEE Journal on Selected Areas in Communications*, 14(3), 547–558, 1996.

[Gil90] K.S. Gilhousen et al., "Increased Capacity Using CDMA for Mobile Satellite Communication," *IEEE Journal on Selected Areas in Communications*, 8(4), 503–514, 1990.

[Gil91] K.S. Gilhousen et al., "On the Capacity of a Cellular CDMA System," *IEEE Transactions on Vehicular Technology*, 40(5), 303–312, 1991.

[Gol67] R. Gold, "Optimal Binary Sequences for Spread Spectrum Multiplexing," *IEEE Transactions on Information Theory*, 13, 619–621, 1967.

[Gol68] R. Gold, "Maximal Recursive Sequences with 3-Valued Recursive Cross-Correlation Functions," *IEEE Transactions on Information Theory*, 14, 154–156, 1968.

[ie802] http://www.ieee802.org/11/.

[ie80a] http://www.ieee802.org/15/.

[ie80b] http://www.ieee802.org/16/.

[IEE02] *IEEE Journal on Selected Areas in Communications, Ultra Wide Band Radio in Multiaccess Wireless Communications,* special issue, December 2002, Vol. 20.

[ietfw] http://www.ietf.org/html.charters/manet-charter.html

[Kni98] D. Knisely et al., "Evolution of Wireless Data Services: IS-95 to cdma2000," *IEEE Communications Magazine*, October 1998, pp. 140–149.

[Koh95] R. Kohno, R. Meidan, L.B. Milstein, "Spread Spectrum Access Methods for Wireless Communications," *IEEE Communications Magazine*, January 1995, pp. 58–67.

[Men97] U. Mengali, A.N. D'Andrea, *Synchronization Techniques for Digital Receivers*, Plenum Press, New York, 1997.

[Mit95] J. Mitola, "The Software Radio Architecture," *IEEE Communications Magazine*, 33(5), 26–38, 1995.

[Moe91] M. Moeneclaey, G. De Jonghe, "Tracking Performance of Digital Chip Synchronization Algorithms for Bandlimited Direct-Sequence Spread-Spectrum Communications," *IEE Electronics Letters*, 1147–1149, 1991.

[Mos96] S. Moshavi, "Multi-User Detection for DS-CDMA Communications," *IEEE Communications Magazine*, October 1996, pp. 124–136.

[Nee99] R. van Nee, G. Awater, M. Morikura, H. Takanashi, M. Webster, K.W. Halford, "New High-Rate Wireless LAN Standards," *IEEE Communications Magazine*, December 1999, pp. 82–88.

[Oja98] T. Ojanperä, R. Prasad, "An Overview of Air Interface Multiple Access for IMT-2000/UMTS," *IEEE Communications Magazine*, September 1998, pp. 82–95.

[Pad95] J.E. Padgett, C.G. Günther, T. Hattori, "Overview of Wireless Personal Communciations," *IEEE Personal Communications Magazine*, January 1995, pp. 28–41.

[Pet72] W.W. Peterson, E.J. Weldon, Jr., *Error-Correcting Codes*, 2nd ed., MIT Press, Cambridge, MA, 1972.

[Pic82] R.L. Pickholtz, D.L. Schilling, L.B. Milstein, "Theory of Spread-Spectrum Communications: A Tutorial," *IEEE Transactions on Communications*, 30, 855–884, 1982.

[Pro95] J.G. Proakis, *Digital Communications*, 3rd ed., McGraw-Hill, New York, 1995.

[Rab00] J.M. Rabaey, "Low-Power Silicon Architectures for Wireless Communications," *Proceedings of the IEEE ASP-DAC 2000*, Asia and South Pacific, pp. 377–380.

[Rap91] T.S. Rappaport, "The Wireless Revolution," *IEEE Communications Magazine*, November 1991, pp. 52–71.

[Sar80] D.P. Sarwate, M.B. Pursley, "Crosscorrelation Properties of Pseudorandom and Related Sequences," *Proceedings of the IEEE*, 68(5), 593-619, 1980.

[Sar95] H. Sari, G. Karam, J. Janclaude, "Transmission Techniques for Digital Terrestrial TV Broadcasting," *IEEE Commununications Magazine*, 36, 100–109, 1995.

[Sas98] A. Sasaki, M. Yabusaki, S. Inada, "The Current Situation of IMT-2000 Standardization Activities in Japan," *IEEE Communications Magazine*, September 1998, pp. 145–153.

[Sch82] R.A. Scholtz, "The Origins of Spread-Spectrum Communications," *IEEE Transactions on Communications*, 30, 822–854, 1982.

[Sim85] M.K. Simon, J.K. Omura, R.A. Scholtz, B.K. Levitt, *Spread Spectrum Communications*, Computer Science Press, Rockville, MD, 1985.

[Teg02] S. Teger, D.J. Waks, "End-User Perspectives on Home Networking," *IEEE Communications Magazine*, 40(4), 114–119, 2002.

[Tel99] I.E. Telatar, "Capacity of Multi-Antenna Gaussian Channels," *European Transactions on Telecommunications*, 10, 585–595, 1999.

[Tur80] G.L. Turin, "Introduction to Spread-Spectrum Antimultipath Techniques and Their Application to Urban Digital Radio," *Proceedings of the IEEE*, 68(3), 328–353, 1980.

[Ver86] S. Verdù, "Minimum Probability of Error for Asynchronous Gaussian Multiple Access Channels," *IEEE Transactions on Information Theory*, 32(1), 85–96, 1986.

[Ver98] S.Verdù, *Multiuser Detection*, University Press, Cambridge, MA, 1998.

[Vit67] A.J. Viterbi, "Error Bounds for Convolutional Codes and an Asymptotically Optimum Decoding Algorithm," *IEEE Transactions on Information Theory*, 13, 260–269, 1967.

[Vit79a] A.J. Viterbi, "Spread Spectrum Communications: Myths and Realities," *IEEE Communications Magazine*, May 1979, pp. 11–18.

[Vit79b] A.J. Viterbi, J.K. Omura, *Principles of Digital Communciation and Coding*, McGraw-Hill, New York, 1979.

11

Multiuser Detection for Fading Channels

11.1 Introduction **11**-2

11.2 Signal and Channel Model **11**-2
The Fading Channel Model • Transmitted Signal Model • Continuous-Time Received Signal Model

11.3 Multiuser Detection with Known CSI **11**-6
Conventional Single-User Detection • Optimum Multiuser Detection • Linear Multiuser Detection • Approximate MMSE Detection: Linear Serial Interference Cancellation • Constrained ML Detection: Nonlinear Serial Interference Cancellation • Performance Results • Some Numerical Results • Sliding-Window One-Shot Multiuser Receivers

11.4 Multiuser Detection with Unknown CSI **11**-25
Signal Representation in Unknown CSI • Available Strategies to Cope with Missing CSI • Channel Estimation Based on the Least Squares Criterion • Trained Adaptive Code-Aided Symbol Detection • Code-Aided Joint Blind Multiuser Detection and Equalization • Subspace-Based Blind MMSE Detection • Minimum Variance Blind Detection • Two-Stage Blind Detection • Performance Results

11.5 Conclusions **11**-38

Stefano Buzzi
Università degli Studi di Cassino

Abstract

This chapter is an introduction to multiuser detection for code division multiple-access systems operating over fading channels. It mainly focuses on the relevant case of slow frequency-selective fading channels, which are frequently encountered in multiuser high-data-rate communications over wireless links. First, the problem of multiuser detection with known channel state information is considered, i.e., assuming knowledge of the channel impulse response, either for all active users or for the user of interest only. In particular, the optimum multiuser detector, the linear multiuser detectors, and interference cancellation receivers are reviewed and compared. Then, the focus of the chapter is shifted toward the problem of adaptive and blind multiuser detection with no channel state information. First, multiuser channel estimation techniques based on the least squares criterion are reviewed and then several linear detectors, requiring knowledge of the spreading code of the user of interest only, are illustrated and compared. In particular, the chapter emphasis is on recursive least squares (RLS) trained adaptive multiuser detection and on the exposition of blind techniques for joint suppression of multiuser interference and intersymbol interference.

0-8493-1657-X/$0.00+$1.50

11.1 Introduction

Multiuser detection deals with the simultaneous detection of multiple information streams that are transmitted over a common channel through digitally modulated signals overlapping in both time and frequency. Multiuser detection finds its main field of application in the design of reception structures for code division multiple-access (CDMA) systems, which are nowadays widely in use in worldwide third-generation (3G) cellular systems. Conversely, it does not apply to orthogonal multiple-access schemes such as frequency division multiple-access (FDMA) and time division multiple-access (TDMA) systems, which, by virtue of the orthogonality of the transmitted signal, just require a conventional matched filter to achieve optimum performance.

In a CDMA channel, the communication reliability is thus impaired not only by the additive thermal noise, but also by the multiple-access interference (MAI), which is originated by the other users simultaneously accessing the channel. Until the early 1980s, the conventional approach to deal with multiple-access channels was to consider MAI as an additional Gaussian noise source, so that the conventional matched filter would be the optimal receiver. This approach, however, was shown to be wrong by Sergio Verdú, the pioneer of multiuser detection, who derived the optimum minimum error probability multiuser receiver in [20, 21] and showed that the near–far problem could be circumvented by resorting to detection algorithms explicitly taking into account the structure of the MAI.

Since that time, multiuser detection has become an intense area of research, attracting the interest of many researchers, from both industry and academia, and expanding at an explosive rate in the scientific literature. This chapter contains a tutorial survey of multiuser detection techniques for fading channels. The extension of multiuser detection techniques, originally conceived by Verdú for the nonfaded Gaussian channel, to fading channels was initiated by Zvonar and Brady in the 1990s [24–29] under some simplifying assumption such as the perfect knowledge of the channel impulse response. Several later works have then relaxed these hypotheses, considering the issues of channel estimation in multiuser systems and of (possibly blind) data detection in multiple-access systems in the presence of unknown frequency-selective fading channels.

The focus of this chapter is on the direct-sequence CDMA (DS/CDMA) technique and on frequency-selective fading channels, which are frequently encountered in mobile communications applications. It is also worth pointing out that another important multiple-access strategy is the multicarrier CDMA (MC/CDMA) technique [5], which can be thought of as a mixture between DS/CDMA and orthogonal frequency division multiplexing (OFDM), and which is currently employed in some 3G cellular wireless network standards. Even though this chapter focuses on DS/CDMA, most of the techniques presented here can be extended with straightforward modifications to MC/CDMA systems as well.

In the following sections, a review of the basic multiuser detection techniques for fading channels assuming perfect channel state information (CSI) is provided (Section 11.3). Then, the focus of the chapter is shifted to the design of multiuser detectors for the case that the channel is not known at the receiver (Section 11.4). In particular, multiuser channel estimation techniques based on the least squares criterion are presented, along with trained and blind procedures for data detection in the presence of unknown multipath distortion and unknown multiple-access interference. Finally, concluding remarks end this review (Section 11.5).

11.2 Signal and Channel Model

11.2.1 The Fading Channel Model

It is well known [1, 16] that a fading channel may be in general represented as a linear time-variant (LTV) system whose (equivalent low-pass) impulse response $h_C(t, \tau)$ is modeled as a complex, possibly Gaussian, random process with respect to the variable t. Accordingly, upon transmission of a certain signal, whose complex envelope is denoted by $p(t)$, the baseband equivalent of the received signal, say $r(t)$, can be written as

$$r(t) = \int_0^{\infty} h_C(t, \tau) p(t - \tau) d\tau \quad (11.1)$$

Starting upon the second-order statistics of the channel impulse response $h_C(t, \tau)$, it is possible to define some ensemble parameters that are useful to specialize the fading channel model and to determine the type of distortion that the fading imposes on the transmitted signal. These parameters are given by:

- The multipath delay spread T_m
- The channel coherence bandwidth $(\Delta f)_c$
- The Doppler spread B_d
- The channel coherence time $(\Delta t)_c$

The multipath delay spread T_m represents an average measure of the time spread that, due to the multipath, a strictly time-concentrated signal experiences during its propagation. The inverse of T_m is referred to as the channel coherence bandwidth $(\Delta f)_c$, and its value is an estimate of the frequency range in which the channel behavior may be considered constant. Otherwise stated, if we transmit two sinusoidal tones with frequencies spaced apart less than $(\Delta f)_c$, these tones will experience the same (random) attenuation.

The Doppler spread B_d, on the contrary, represents the frequency spread that a sinusoidal tone undergoes when transmitted on a fading channel. Such a phenomenon is due to the relative motion between the transmitter and the receiver. The value of B_d provides a measure of how rapidly the channel impulse response varies in time. Finally, the channel coherence time $(\Delta t)_c$, which is approximately equal to the inverse of B_d, represents the time during which the channel behavior is stationary or, alternatively, the fading process is highly correlated.

Now, depending on the values of these parameters with respect to the bandwidth W and to the signaling interval of the digital transmission that we are considering, Equation 11.1 can be specialized in that the channel impulse response $h_C(t, \tau)$ assumes particular simplified structures. In the following, we concentrate on the model of (possibly time-selective) frequency-selective fading channels, which are of major importance when dealing with transmission of wideband signals (as DS/CDMA signals are) on cellular indoor and outdoor channels. According to this model, the received signal can be deemed as the superposition of randomly delayed and scaled replicas of the transmitted signals [16]; i.e., the channel impulse response can be written as

$$h_C(t, \tau) = \sum_{l=0}^{L-1} c_l(t) \delta(\tau - \tau_l) \quad (11.2)$$

where $c_l(t)$ is a complex Gaussian random process, representing the complex attenuation of the l-th path; τ_l is the random delay of the l-th path, with $\tau_l W$ not much smaller than 1;[1] and the integer number L, the number of multipath replicas, depends on the product WT_m and on the scattering environment. Hereafter, a Rayleigh fading channel model is assumed, namely the fading processes $c_l(t)$ are complex Gaussian zero-mean random variates at any instant t and $\forall l$; it is also assumed that their variance is normalized so that

$$\sum_{l=0}^{L-1} E\{|c_l(t)|^2\} = 1 \qquad \forall t \quad (11.3)$$

[1]Note that if for any l, $\tau_l W \ll 1$, then Equation 11.2 represents the impulse response of a frequency-flat channel.

with $E\{\cdot\}$ denoting statistical expectation. Furthermore, the channel coefficients are assumed independent, i.e.,

$$E\{c_l(t)c_{l'}^*(t)\} = E\{|c_l(t)|^2\}\delta_{l,l'} \tag{11.4}$$

with $(\cdot)^*$ denoting conjugate, and $\delta_{l,l'}$ the discrete Kronecker delta function; this hypothesis, usually referred to as uncorrelated scattering (US), can be physically justified by noticing that paths propagating at different delays undergo independent attenuations. Another common assumption is that of wide-sense stationary (WSS) fading; i.e., the quantity $E\{c_l(t)c_l^*(u)\}$ is a function of the time difference $(t-u)$ only. Obviously, the way that the channel coefficients $\{c_l(t)\}_{l=0}^{L-1}$ change with time is tied to the channel coherence time, which for a base-to-mobile communication is given by

$$(\Delta t)_c \approx \frac{1}{B_d} = \frac{\lambda_c}{v} \tag{11.5}$$

with v the mobile terminal speed and λ_c the wavelength corresponding to the signal carrier frequency. If the transmitted signal symbol interval, say T, is such that $T \ll (\Delta t)_c$, then the fading is *slow* or *highly correlated*, and the channel gains $\{c_l(t)\}_{l=0}^{L-1}$ are approximately constant over a certain number of signaling intervals; if instead $T \approx (\Delta t)_c$, then the fading is *fast* or *independent*, and the channel gains change at any signaling interval. Thus, from the above considerations, we may conclude that low mobile speeds and high signaling rates lead to slower and slower fading, whereas high mobile speeds and lower signaling rates lead to the opposite extreme of fast fading. As an example, in indoor environments we usually have to deal with slow fading, while for outdoor channels and speech services the slow fading condition $T \ll (\Delta t)_c$ may not be valid.

11.2.2 Transmitted Signal Model

Consider an asynchronous DS/CDMA system employing a BPSK modulation format.[2] The complex envelope of the signal, say $x_k(t)$, transmitted by the k-th active user can thus be written as

$$x_k(t) = A_k e^{j\phi_k} \sum_{m=-P}^{P} b_k(m) s_k(t - \tau_k - mT_b) \qquad k = 0, \ldots, K-1 \tag{11.6}$$

The meanings of the symbols in the above expression follows:

- $A_k e^{j\phi_k}$ is a complex gain accounting for the transmitted signal strength and the phase offset of the local oscillator.
- $(2P+1)$ is the transmitted frame length.
- K is the number of the active users.
- $\{b_k(m)\}_{m=-P}^{P}$ is the sequence of the $(2P+1)$ information symbols from the k-th user. Since we are dealing with a binary modulation, $b_k(m)$ is a random variate taking on values in the set $\{+1, -1\}$ with uniform distribution $\forall m, k$.
- $\tau_k \in (0, T_b)$ is the relative delay of the k-th user.
- T_b is the common network signaling interval.
- $s_k(\cdot)$ is the baseband equivalent of the *signature waveform* assigned to the k-th user, i.e., the unique waveform that enables extraction of the k-th user information stream from the whole received signal.

[2]Extending the subsequent derivations to complex signaling constellations is straightforward.

For the DS modulation format herein considered the complex envelope $s_k(\cdot)$ of the signature waveform is expressed as

$$s_k(t) = \sum_{n=0}^{N-1} \beta_{n,k} \psi_{T_c}(t - nT_c) \tag{11.7}$$

wherein

- $\{\beta_{n,k}\}_{n=0}^{N-1}$ is the pseudonoise sequence of the k-th user, with $\beta_{n,k} \in \{+1, -1\}, \forall n, k$.
- N is the processing gain.
- T_c is the chip interval, related to the bit interval T_b by the relation $T_b = NT_c$.
- $\psi_{T_c}(\cdot)$ is the chip pulse, which we assume to be a unit-height rectangular pulse supported on the interval $(0, T_c)$. The results derived in the following, however, apply with almost no modifications to the alternative customary model of raised-cosine pulse that, at the price of introduction of a certain amount of interchip interference in the time domain, exhibits the desirable property of being strictly band limited.

11.2.3 Continuous-Time Received Signal Model

Since the signals $x_k(\cdot)$ are assumed to propagate through a frequency-selective fading channel, the signal at the receiver is finally given by

$$r(t) = \sum_{m=-P}^{P} \sum_{k=0}^{K-1} \sum_{l=0}^{L-1} c_{k,l}(m) A_k b_k(m) s_k(t - \tau_k - \tau_{k,l} - mT_b) + w(t) \tag{11.8}$$

In Equation 11.8 the term $w(t)$ represents the additive thermal noise, modeled as a sample function from a zero-mean, white, complex Gaussian process with a power spectral density (PSD) $2\mathcal{N}_0$, while $c_{k,l}(m)$ is the complex Gaussian random gain affecting the l-th replica of the k-th user signal in the m-th signaling interval. Notice that the continuous-time variable t has been replaced by the discrete variable m, in that we assume that the fading coefficients are constant at least for a time duration equal to the network signaling interval T_b; obviously, the dependence on the temporal index m may be either effective or fictitious depending on the values of the packet length $(2P + 1)$ and of the mobile terminal speed. Notice also that the phase terms $\{\phi_k\}_{k=0}^{K-1}$ have disappeared in that they have been included in the circularly symmetric channel coefficients. The $(L \times L)$-dimensional channel covariance matrix will be denoted by $\boldsymbol{M_C}$ and is defined as

$$\boldsymbol{M_C} = E\left\{\boldsymbol{c}_0(m)\boldsymbol{c}_0^H(m)\right\} = \ldots = E\left\{\boldsymbol{c}_{K-1}(m)\boldsymbol{c}_{K-1}^H(m)\right\} \qquad \forall m \tag{11.9}$$

In the above expression, $\boldsymbol{c}_k(m)$ denotes the k-th user fading vector in the m-th signaling interval, that is,

$$\boldsymbol{c}_k(m) = [c_{k,0}(m), \ldots, c_{k,L-1}(m)]^T \qquad \forall k = 0, \ldots, K-1,$$

the superscripts $(\cdot)^H$ and $(\cdot)^T$ denote conjugate transpose and transpose, respectively, and the matrix $\boldsymbol{M_C}$ is time invariant by virtue of the WSS assumption. Furthermore, for notation simplicity we have assumed that $\boldsymbol{M_C}$ is independent of the user index k, and under the US hypothesis it assumes a diagonal structure; in particular, the normalization condition in Equation 11.3 implies that $\text{tr}(\boldsymbol{M_C}) = 1$, with $\text{tr}(\cdot)$ denoting the trace operator. We also explicitly notice that since we are considering an asynchronous system and have assumed that the fading coefficients $c_{k,l}(\cdot)$ depend on the user index k, the signal in Equation 11.8 refers to the uplink model in a cellular system. However, the downlink model is simply obtained by particularizing Equation 11.8, letting τ_k = constant, $\forall k$ (i.e., we have a synchronous transmission), and $c_{k,l}(m) = c_{k',l}(m), \forall k, k'$ (i.e., all the signals propagate through the same fading channel). Finally, we also assume that since in the uplink model each signal $x_k(t)$ experiences its own fading distortion, the channel vectors $\boldsymbol{c}_k(m)$ are independent with respect to the index k; i.e., the matrices $E\{\boldsymbol{c}_k(m)\boldsymbol{c}_{k'}^H(q)\}$ have all-zero entries $\forall m, q, k \neq k'$. Of course, this assumption no longer holds if we consider the downlink model, wherein, as anticipated, we have $\boldsymbol{c}_k(m) = \boldsymbol{c}_{k'}(m)\ \forall m, k, k'$.

11.3 Multiuser Detection with Known CSI

We first focus on the case that the receiver has full CSI; i.e., the channel vectors $\boldsymbol{c}_k(m), \forall k = 0, \ldots, K-1$ and $\forall m = -P, \ldots, P$ are perfectly known to the receiver. Before proceeding to the derivation of the multiuser detectors, it is convenient to express the signal in Equation 11.8 through some matrix-based notations. Indeed, upon defining

$$\boldsymbol{C}(m) = \text{Diag}(\boldsymbol{c}_0(m), \boldsymbol{c}_1(m), \ldots, \boldsymbol{c}_{K-1}(m)), \qquad LK \times K \tag{11.10}$$

$$\widetilde{\boldsymbol{C}} = \text{Diag}(\boldsymbol{C}(-P), \boldsymbol{C}(-P+1), \ldots, \boldsymbol{C}(P)), \qquad (2P+1)LK \times (2P+1)K \tag{11.11}$$

$$\boldsymbol{s}_k(t) = \begin{bmatrix} s_k(t - \tau_k - \tau_{k,0}) \\ s_k(t - \tau_k - \tau_{k,1}) \\ \vdots \\ s_k(t - \tau_k - \tau_{k,L-1}) \end{bmatrix}, \qquad L \times 1 \tag{11.12}$$

$$\boldsymbol{s}(t) = \left[\boldsymbol{s}_0^T(t), \boldsymbol{s}_1^T(t), \ldots, \boldsymbol{s}_{K-1}^T(t)\right]^T, \qquad KL \times 1 \tag{11.13}$$

$$\widetilde{\boldsymbol{s}}(t) = \begin{bmatrix} \boldsymbol{s}(t + PT_b) \\ \boldsymbol{s}(t + (P-1)T_b) \\ \vdots \\ \boldsymbol{s}(t - PT_b) \end{bmatrix}, \qquad (2P+1)KL \times 1 \tag{11.14}$$

$$\boldsymbol{A}_k = A_k \boldsymbol{I}_L, \qquad \boldsymbol{A} = \text{Diag}(\boldsymbol{A}_0, \boldsymbol{A}_1, \ldots, \boldsymbol{A}_{K-1}), \tag{11.15}$$

$$\widetilde{\boldsymbol{A}} = \boldsymbol{I}_{2P+1} \otimes \boldsymbol{A}, \qquad (2P+1)KL \times (2P+1)KL \tag{11.16}$$

$$\boldsymbol{b}(m) = [b_0(m), b_1(m), \ldots, b_{K-1}(m)]^T, \qquad K \times 1 \tag{11.17}$$

and

$$\widetilde{\boldsymbol{b}} = \begin{bmatrix} \boldsymbol{b}(-P) \\ \boldsymbol{b}(-P+1) \\ \vdots \\ \boldsymbol{b}(P) \end{bmatrix}, \qquad (2P+1)K \times 1 \tag{11.18}$$

the received signal can be written as

$$r(t) = \widetilde{\boldsymbol{s}}^T(t)\widetilde{\boldsymbol{A}}\widetilde{\boldsymbol{C}}\widetilde{\boldsymbol{b}} + w(t) \tag{11.19}$$

From the above equation, it is seen that the thermal noise-free received waveform $r(t) - w(t)$ is a linear combination of the entries of the $KL(2P+1)$-dimensional vector $\widetilde{s}(t)$, whereby the signal $r(t)$ can be discretized with no loss of information by considering the following projection:

$$\widetilde{\boldsymbol{r}} = \int_{-\infty}^{+\infty} r(t)\widetilde{\boldsymbol{s}}(t)dt = \widetilde{\boldsymbol{R}}\widetilde{\boldsymbol{A}}\widetilde{\boldsymbol{C}}\widetilde{\boldsymbol{b}} + \underbrace{\int_{-\infty}^{+\infty} n(t)\widetilde{\boldsymbol{s}}(t)dt}_{\widetilde{\boldsymbol{z}}} \tag{11.20}$$

In Equation 11.20, we have

$$\widetilde{\boldsymbol{R}} = \int_{-\infty}^{+\infty} \widetilde{\boldsymbol{s}}(t)\widetilde{\boldsymbol{s}}^T(t)dt = \begin{bmatrix} \boldsymbol{R}(0) & \boldsymbol{R}(1) & \dots & \boldsymbol{R}(2P) \\ \boldsymbol{R}(-1) & \boldsymbol{R}(0) & \dots & \boldsymbol{R}(2P-1) \\ \vdots & \vdots & \ddots & \vdots \\ \boldsymbol{R}(-2P) & \dots & \dots & \boldsymbol{R}(0) \end{bmatrix}, \quad (2P+1)KL\times(2P+1)KL \tag{11.21}$$

with

$$\boldsymbol{R}(\ell) = \int_{-\infty}^{+\infty} \boldsymbol{s}(t)\boldsymbol{s}^T(t+\ell T_b)dt = \boldsymbol{0} \quad \forall|\ell| > \nu + 1 \tag{11.22}$$

and

$$\nu = \left\lfloor \frac{\max_{k,l} \tau_{k,l}}{T_b} \right\rfloor$$

while the $(2P+1)KL$-dimensional vector $\widetilde{\boldsymbol{z}}$ is a zero-mean complex Gaussian random vector with covariance matrix

$$E\{\widetilde{\boldsymbol{z}}\widetilde{\boldsymbol{z}}^H\} = 2\mathcal{N}_0\widetilde{\boldsymbol{R}} \tag{11.23}$$

11.3.1 Conventional Single-User Detection

The simplest detection strategy for detecting the transmitted symbols is to model the MAI contribution as additional white Gaussian noise and to resort to a plain RAKE detector. The information symbols are thus detected according to the following linear detection rule:

$$\widetilde{\boldsymbol{b}}_{\text{SU}} = \text{sgn}[\Re\{\widetilde{\boldsymbol{C}}^H\widetilde{\boldsymbol{A}}^H\widetilde{\boldsymbol{r}}\}] \tag{11.24}$$

with $\Re(\cdot)$ denoting real part and $\text{sgn}(\cdot)$ the signum function. A block scheme of this detector is reported in Figure 11.1. Note that the entries of the $(2P+1)K$-dimensional vector $\widetilde{\boldsymbol{C}}^H\widetilde{\boldsymbol{A}}^H\widetilde{\boldsymbol{r}}$ can be obtained by properly combining the sampled (at rate $1/T_b$) outputs of a bank of LK filters matched to the delayed replicas of the users' signatures. Although the above detection rule is very simple and can be implemented with little effort (note that if the data symbols of only one user are to be detected, this receiver requires knowledge of the fading coefficients for only that user), it is well known that its performance is heavily limited by the multiuser interference, and in particular by the presence of unbalanced CDMA signal amplitudes. While in the absence of fading MAI can be contrasted through the adoption of power control, in a fading channel the random amplitude fluctuations make the situation even more critical, and the conventional single-user receiver exhibits an error floor in the high signal-to-noise ratio region. The bad performance behavior of the single-user receiver has thus fostered the design of alternative detection structures with improved performance.

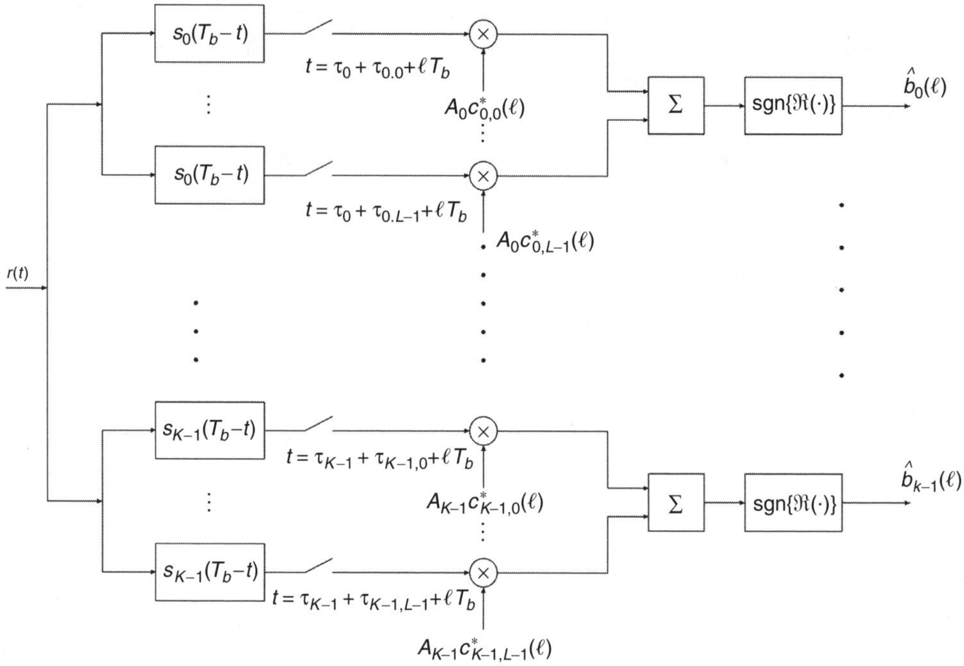

FIGURE 11.1 Block scheme of the conventional single-user receiver.

11.3.2 Optimum Multiuser Detection

If complete CSI is available, then the receiver has perfect knowledge of the entries of the matrices $\widetilde{\boldsymbol{A}}$ and $\widetilde{\boldsymbol{C}}$, and the probability density function (pdf) of the received vector, conditioned on a certain transmitted symbol sequence $\widetilde{\boldsymbol{b}}$, is written as

$$f(\widetilde{\boldsymbol{r}}|\widetilde{\boldsymbol{b}}) = \gamma \exp[-(\widetilde{\boldsymbol{r}} - \widetilde{\boldsymbol{R}}\widetilde{\boldsymbol{A}}\widetilde{\boldsymbol{C}}\widetilde{\boldsymbol{b}})^H \widetilde{\boldsymbol{R}}^{-1} (\widetilde{\boldsymbol{r}} - \widetilde{\boldsymbol{R}}\widetilde{\boldsymbol{A}}\widetilde{\boldsymbol{C}}\widetilde{\boldsymbol{b}})] \tag{11.25}$$

with γ a proper normalization constant. Assuming that all the possible transmitted symbol sequences are equally likely, the minimum error probability receiver implements the maximum likelihood (ML) decision rule; i.e., it selects as the transmitted sequence the one that maximizes the pdf (25):

$$\widetilde{\boldsymbol{b}}_{\mathrm{ML}} = \arg \max_{\widetilde{\boldsymbol{b}} \in \{+1,-1\}^{(2P+1)K}} f(\widetilde{\boldsymbol{r}}|\widetilde{\boldsymbol{b}}) = \arg \max_{\widetilde{\boldsymbol{b}} \in \{+1,-1\}^{(2P+1)K}} \Omega(\widetilde{\boldsymbol{b}}) \tag{11.26}$$

with

$$\Omega(\widetilde{\boldsymbol{b}}) = [2\Re\{\widetilde{\boldsymbol{b}}^H \widetilde{\boldsymbol{C}}^H \widetilde{\boldsymbol{A}}^H \widetilde{\boldsymbol{r}}] - \widetilde{\boldsymbol{b}}^H \widetilde{\boldsymbol{C}}^H \widetilde{\boldsymbol{A}}^H \widetilde{\boldsymbol{R}}\widetilde{\boldsymbol{A}}\widetilde{\boldsymbol{C}}\widetilde{\boldsymbol{b}}] \tag{11.27}$$

From the above equation, it is seen that the functional $\Omega(\widetilde{\boldsymbol{b}})$ depends on the $(2P + 1)K$-dimensional vector

$$\widetilde{\boldsymbol{x}} = \widetilde{\boldsymbol{C}}^H \widetilde{\boldsymbol{A}}^H \widetilde{\boldsymbol{r}} \tag{11.28}$$

whose entries can be obtained as the outputs of K filters matched to the channel-modified signatures[3] sampled at rate $1/T_b$. Since the maximization is to be performed over a discrete-valued set, an exhaustive search over all the $2^{(2P+1)K}$ possible transmitted symbol sequences is to be done. At first sight, it might thus seem that implementing the optimum ML receiver entails a computational complexity exponential in the number of transmitted symbols $(2P+1)K$. On the other hand, since the blocks of the matrix $\widetilde{\boldsymbol{R}}$ are nonzero only in proximity of the principal diagonal (see Equation 11.22), the Viterbi algorithm can be applied in order to achieve some complexity reduction. To elaborate, note that it is easily seen that upon defining

$$\boldsymbol{H} = \widetilde{\boldsymbol{C}}^H \widetilde{\boldsymbol{A}}^H \widetilde{\boldsymbol{R}} \widetilde{\boldsymbol{A}} \widetilde{\boldsymbol{C}}$$

and denoting by $\boldsymbol{H}(i,j)$ the (i,j)-th block of the matrix $\boldsymbol{H}$ of dimension $KL \times KL$, it is seen that $\boldsymbol{H}(i,j)$ is nonzero only if $|i-j| \leq (\nu+1)$. As a consequence, we have

$$\widetilde{\boldsymbol{b}}^H \boldsymbol{H} \widetilde{\boldsymbol{b}} = \sum_{i,j=-P}^{P} \boldsymbol{b}^H(i) \boldsymbol{H}(i+P+1, j+P+1) \boldsymbol{b}(j) =$$

$$\sum_{i=-P}^{P} \left[\boldsymbol{b}^H(i) \boldsymbol{H}(i+P+1, i+P+1) \boldsymbol{b}(i) + 2\Re \left\{ \sum_{j=\max(i-\nu-1,-P)}^{i-1} \boldsymbol{b}^H(i) \boldsymbol{H}(i+P+1, j+P+1) \boldsymbol{b}(j) \right\} \right] \tag{11.29}$$

Moreover, if we partition the vector $\widetilde{\boldsymbol{x}}$ in $(2P+1)K$-dimensional vectors, i.e.,

$$\widetilde{\boldsymbol{x}} = [\boldsymbol{x}^H(-P), \boldsymbol{x}^H(-P+1), \ldots, \boldsymbol{x}^H(P)]^H$$

we also have

$$\widetilde{\boldsymbol{b}}^H \widetilde{\boldsymbol{C}}^H \widetilde{\boldsymbol{A}}^H \widetilde{\boldsymbol{r}} = \widetilde{\boldsymbol{b}}^H \widetilde{\boldsymbol{x}} = \sum_{i=-P}^{P} \boldsymbol{b}^H(i) \boldsymbol{x}(i) \tag{11.30}$$

As a consequence, if we define the $(\nu+1)K$-dimensional state vector at the i-th symbol interval as

$$\boldsymbol{g}(i) = [\boldsymbol{b}^H(i-1), \boldsymbol{b}^H(i-2), \ldots, \boldsymbol{b}^H(i-\nu-1)]^H \tag{11.31}$$

the functional $\Omega(\widetilde{\boldsymbol{b}})$ is finally expressed as

$$\Omega(\widetilde{\boldsymbol{b}}) = \sum_{i=-P}^{P} \zeta_i(\boldsymbol{b}(i), \boldsymbol{g}(i)) \tag{11.32}$$

with

$$\zeta_i(\boldsymbol{b}(i), \boldsymbol{g}(i)) = -\boldsymbol{b}^H(i) \boldsymbol{H}(i+P+1, i+P+1) \boldsymbol{b}(i)$$
$$-2\Re \left\{ \sum_{j=\max(i-\nu-1,-P)}^{i-1} \boldsymbol{b}^H(i) \boldsymbol{H}(i+P+1, j+P+1) \boldsymbol{b}(j) \right\} + 2\Re\{\boldsymbol{b}^H(i) \boldsymbol{x}(i)\} \tag{11.33}$$

[3]Note that these matched filters are time invariant if the fading is constant over the whole received frame, while they are time varying in the opposite situation where the channel impulse response is not constant.

Given the decomposition Equation 11.32, it is now straightforward to apply the Viterbi algorithm in order to obtain a procedure with a computational complexity exponential in $[(\nu + 2)K - 1]$.

11.3.3 Linear Multiuser Detection

Although the Viterbi algorithm enables a substantial complexity reduction for the ML optimum detectors, its exponential complexity cannot be afforded in many practical applications, and it is thus of interest to devise alternative lower-complexity solutions. The adoption of linear multiuser detection structures is thus a viable means to obtain reception structures that exhibit satisfactory performance levels without incurring the overwhelming complexity of the optimum ML receiver.

When considering linear multiuser detection, two possible alternative detection strategies can be considered. More precisely, multiuser detection can take place either before or after multipath combining. In the former approach, first MAI is suppressed and then the replicas of the signal of interest are properly combined, while in the second approach MAI is suppressed after the multipath combining has been carried out. In this section we will explore both approaches with reference to the two most popular linear multiuser detectors: the decorrelating detector and the minimum mean square error (MMSE) detector.

11.3.3.1 The Decorrelating Detector

To begin with, let us first consider the case that MAI is suppressed prior to multipath combining. Given Equation 11.20, consider the following decorrelating linear transformation:

$$\widetilde{\boldsymbol{y}} = \widetilde{\boldsymbol{R}}^{-1}\widetilde{\boldsymbol{r}} = \widetilde{\boldsymbol{A}}\widetilde{\boldsymbol{C}}\widetilde{\boldsymbol{b}} + \underbrace{\widetilde{\boldsymbol{R}}^{-1}\widetilde{\boldsymbol{z}}}_{\widetilde{\boldsymbol{z}}_d} \tag{11.34}$$

Note that the vector $\widetilde{\boldsymbol{y}}$ has $(2P + 1)KL$ entries, and it contains L MAI-free elements for each transmitted data symbol, while $\widetilde{\boldsymbol{z}}_d$ is a complex Gaussian vector with covariance matrix

$$\widetilde{\boldsymbol{Q}} = 2\mathcal{N}_0\widetilde{\boldsymbol{R}}^{-1} \tag{11.35}$$

The decorrelating transformation eliminates the MAI at the price of noise enhancement and correlation. In order to detect the symbol $b_k(m)$, with $m \in \{-P, \ldots, P\}$ and $k \in \{0, \ldots, K-1\}$, we define the following selection matrix

$$\boldsymbol{S}_k(m) = [\boldsymbol{0}_{L,(m+P)KL+kL}, \boldsymbol{I}_L, \boldsymbol{0}_{L,(P-m+1)KL-(k+1)L}] \quad L \times (2P+1)KL \tag{11.36}$$

and consider the L-dimensional subvector[4]

$$\begin{aligned}\boldsymbol{y}_k(m) &= \widetilde{\boldsymbol{y}}((m+P)KL + kL + 1 : (m+P)KL + (k+1)L) = \boldsymbol{S}_k(m)\widetilde{\boldsymbol{y}} \\ &= A_k\boldsymbol{c}_k(m)b_k(m) + \boldsymbol{z}_{d,k}(m)\end{aligned} \tag{11.37}$$

with

$$\boldsymbol{z}_{d,k}(m) = \boldsymbol{S}_k(m)\widetilde{\boldsymbol{z}} = \widetilde{\boldsymbol{z}}((m+P)KL + kL + 1 : (m+P)KL + (k+1)L) \tag{11.38}$$

a complex Gaussian L-dimensional random vector with covariance matrix

$$\boldsymbol{Q}_k(m) = E\left\{\boldsymbol{z}_{d,k}(m)\boldsymbol{z}_{d,k}^H(m)\right\} = \boldsymbol{S}_k(m)\widetilde{\boldsymbol{Q}}\boldsymbol{S}_k^H(m) \tag{11.39}$$

Given the decorrelated subvector in Equation 11.37, the optimal detection rule amounts to the cascade of a noise-whitening filter and a filter matched to the whitened signal of interest. It is easily seen that given

[4]Note that due to the noise correlation, considering only the L entries corresponding to the bit to be detected is suboptimal.

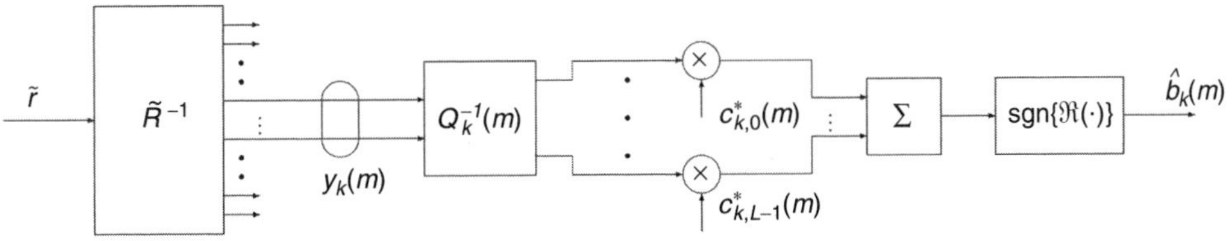

FIGURE 11.2 Block scheme of the decorrelating receiver operating prior to multipath combining.

the following eigendecomposition,

$$Q_k(m) = U_k(m)\Lambda_k(m)U_k^H(m) \tag{11.40}$$

the whitened observables are

$$y_{w,k}(m) = \Lambda_k^{-1/2}(m)U_k^H(m)y_k(m) = A_k\Lambda_k^{-1/2}(m)U_k^H(m)c_k(m)b_k(m) + \Lambda_k^{-1/2}(m)U_k^H(m)z_k(m) \tag{11.41}$$

and the final decision rule for the symbol $b_k(m)$ is written as

$$\begin{aligned}\widehat{b}_k(m) &= \text{sgn}\left[\Re\left(c_k^H(m)U_k(m)\Lambda_k^{-1/2}(m)y_{w,k}(m)\right)\right] \\ &= \text{sgn}\left[\Re\left(c_k^H(m)Q_K^{-1}(m)y_k(m)\right)\right]\end{aligned} \tag{11.42}$$

A block scheme of this detector is reported in Figure 11.2. A simplified form of the detector (Equation 11.42) can be obtained by skipping the noise-whitening step for the decorrelated subvector in Equation 11.37 and by direct application of the multipath combining step. In this situation the decision rule is written as

$$\widehat{b}_k(m) = \text{sgn}\left[\Re\left(c_k^H(m)y_k(m)\right)\right] \tag{11.43}$$

Numerical results show that the performance loss incurred by the suboptimal decision rule in Equation 11.43 with respect to the optimal one in Equation 11.42 is quite limited.

The alternative form of the decorrelating detector suppresses MAI after multipath combining has taken place. To better illustrate, let us consider the $(2P+1)K$-dimensional vector $\widetilde{x}$, which is defined in Equation 11.28, and rewrite it here:

$$\widetilde{x} = \widetilde{C}^H\widetilde{A}^H\widetilde{r} = \underbrace{\widetilde{C}^H\widetilde{A}^H\widetilde{R}\widetilde{A}\widetilde{C}}_{H}\widetilde{b} + \widetilde{C}^H\widetilde{A}^H\widetilde{z} \tag{11.44}$$

Under the assumption that H is nonsingular,[5] the transmitted signals can be decorrelated by simply premultiplying the vector $\widetilde{x}$ by the inverse of the matrix H. We thus obtain the following detection rule:

$$\widehat{\widetilde{b}}_{\text{DEC}} = \text{sgn}[\Re\{H^{-1}\widetilde{x}\}] = \text{sgn}[\Re\{H^{-1}\widetilde{C}^H\widetilde{A}^H\widetilde{r}\}] \tag{11.45}$$

Note that this detector can also be obtained by minimizing the objective function $\Omega(\widetilde{b})$ over the set $\mathcal{R}^{(2P+1)K}$, with $\mathcal{R}$ the real field. A block scheme of this detector is reported in Figure 11.3.

[5]This assumption is usually fulfilled for carefully chosen spreading codes and when the user's number does not exceed the processing gain.

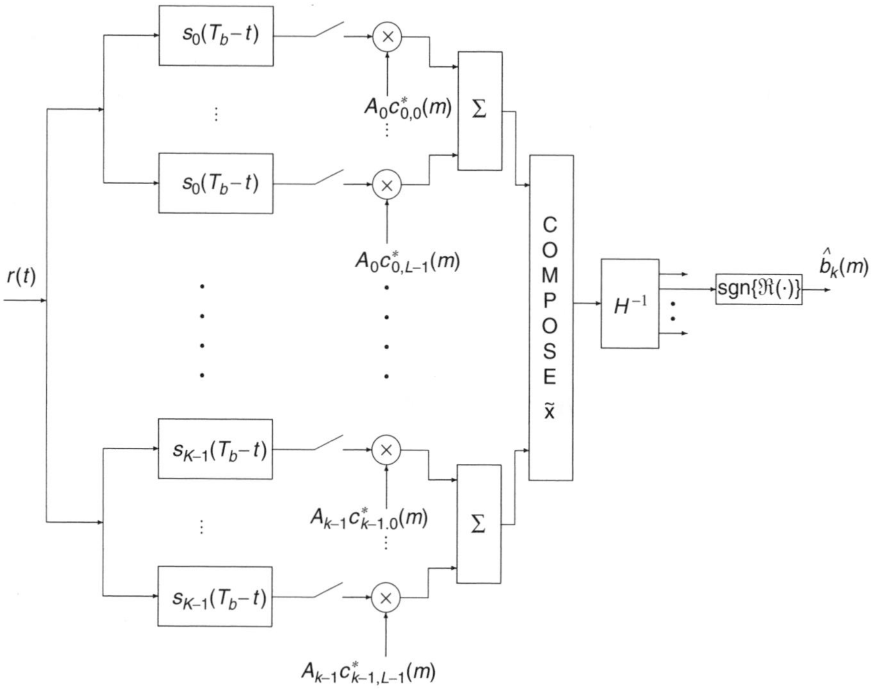

FIGURE 11.3 Block scheme of the decorrelating receiver operating after multipath combining.

11.3.3.2 Remarks

Some remarks on the illustrated decorrelating detection rules in Equations 11.42, 11.43, and 11.45 are now in order. First of all, we note that Equation 11.42 and Equation 11.43 can be thought of as two-stage receiving structures. Indeed, the first stage (i.e., the $\widetilde{\boldsymbol{R}}^{-1}$ transformation) performs decorrelation of each path with respect to MAI and multipath interference, and the second stage of the receiver takes care of multipath combining. Interestingly, it is seen that the receiver's first stage is independent of the channel coefficients (indeed, multiuser detection is performed prior to multipath combining), and moreover, the second stage of the receiver requires knowledge of only the vector $\boldsymbol{c}_k(m)$ to detect the bit $b_k(m)$. As a consequence, if the receiver is interested in decoding the information stream transmitted from a subset of the active users, only the channel coefficients of the users of interest are to be estimated.[6]

Conversely, the decision rule in Equation 11.45 depends, through the matrix $\boldsymbol{H}^{-1}$, on all the channel coefficients for all the active users. Accordingly, even if the receiver is interested in detecting only one user, the CSI for all the active users must be available. Moreover, it is also worth pointing out that in a fast fading environment, the matrix $\boldsymbol{H}^{-1}$ is to be recomputed at each channel variation, while the first stage of the receivers in Equations 11.42 and 11.43 is independent of the channel. On the other hand, results will show that the receiver in Equation 11.45 exhibits superior performance over the other two detectors, especially when the condition $K \ll N$ is not fulfilled. In particular, since the decorrelating stage $\widetilde{\boldsymbol{R}}^{-1}$ treats each multipath replica as a virtual independent signature, receivers in Equations 11.42 and 11.43 are affected by a larger noise enhancement effect and, for increasing K, are much more sensitive to the near–far effect than the receiver in Equation 11.45 [8, 9].

[6]Note, however, that the delays of the multipath replicas of all the active users are needed in order to implement the receiver's first stage.

11.3.3.3 The MMSE Detector

Parallel to the exposition of the decorrelating detectors, two alternative forms of the MMSE detector can be derived. We start by considering the case that the MMSE interference suppression is made prior to multipath combining. Let

$$\widetilde{\boldsymbol{d}} = \widetilde{\boldsymbol{C}}\widetilde{\boldsymbol{b}} \tag{11.46}$$

and assume that an estimate of the $(2P+1)KL$-dimensional vector $\widetilde{\boldsymbol{d}}$, say $\widehat{\widetilde{\boldsymbol{d}}}$, is obtained through the following linear processing:

$$\widehat{\widetilde{\boldsymbol{d}}} = \boldsymbol{D}^H\widetilde{\boldsymbol{r}} \tag{11.47}$$

with $\boldsymbol{D}$ a square matrix of order $(2P+1)KL$ minimizing the following objective function:

$$\mathcal{J}(\boldsymbol{D}) = E\{\|\widetilde{\boldsymbol{d}} - \boldsymbol{D}^H\widetilde{\boldsymbol{r}}\|^2\} \tag{11.48}$$

Letting

$$\Sigma_c = E\{\widetilde{\boldsymbol{C}}\widetilde{\boldsymbol{C}}^H\} = \boldsymbol{I}_{(2P+1)K} \otimes \boldsymbol{M_C}$$

direct application of the orthogonality principle leads to the solution

$$\boldsymbol{D} = \left(\underbrace{\widetilde{\boldsymbol{R}}\widetilde{\boldsymbol{A}}\Sigma_c\widetilde{\boldsymbol{A}}^H\widetilde{\boldsymbol{R}}^H + 2\mathcal{N}_0\widetilde{\boldsymbol{R}}}_{E\{\widetilde{\boldsymbol{r}}\widetilde{\boldsymbol{r}}^H\}}\right)^{-1} \underbrace{\widetilde{\boldsymbol{R}}\widetilde{\boldsymbol{A}}\Sigma_c}_{E\{\widetilde{\boldsymbol{r}}\widetilde{\boldsymbol{d}}^H\}} \tag{11.49}$$

In order to detect the information symbol $b_k(m)$, we now consider the L-dimensional subvector

$$\boldsymbol{d}_k(m) = \boldsymbol{S}_k(m)\boldsymbol{D}^H\widetilde{\boldsymbol{r}} \tag{11.50}$$

Strictly speaking, the vector $\boldsymbol{d}_k(m)$ can be decomposed in the sum of three different contributions: the useful signal, the MAI, and the thermal (colored) noise. On the other hand, it may be reasonably assumed that the MAI contribution to the vector $\boldsymbol{d}_k(m)$ is negligible, since the MAI has been filtered by the matrix $\boldsymbol{D}$, which has the capability of forcing to zero the MAI components with increasingly large strength. Accordingly, assuming that the MAI has been nullified by the filter $\boldsymbol{D}$, the optimum processing rule for detecting the bit $b_k(m)$ is a noise-whitening filter followed by a filter matched to the useful whitened signal.[7] It can be easily shown that the covariance matrix of the thermal noise contribution in Equation 11.50 is

$$2\mathcal{N}_0\boldsymbol{S}_k(m)\boldsymbol{D}^H\widetilde{\boldsymbol{R}}\boldsymbol{D}\boldsymbol{S}_k^H(m) = \boldsymbol{U}\Lambda\boldsymbol{U}^H$$

thus implying that the whitened observables are written as

$$\boldsymbol{d}_{w,k}(m) = \Lambda^{-1/2}\boldsymbol{U}^H\boldsymbol{S}_k(m)\boldsymbol{D}^H\widetilde{\boldsymbol{r}} \tag{11.51}$$

[7]Note also that an alternative approach is to model the residual MAI at the output of the filter $\boldsymbol{D}$ as a colored Gaussian process [14] and to resort to a transformation that whitens the sum of the output MAI and thermal noise contribution. In our approach, instead the MAI contribution is neglected so that the design of the whitening filter and of the combining stage does not depend on the MAI. Notice that this simplification comes at the expense of a negligible performance loss and that, moreover, in the limiting cases of either vanishingly small thermal noise or increasingly large MAI strength, the two approaches, under mild conditions, end up equivalent.

which leads to the decision rule

$$\widehat{b}_k(m) = \text{sgn}\left[\Re\left\{\widetilde{C}(:, (P+m)K+k+1)^H \widetilde{A}^H \widetilde{R} D S_k^H(m) U \Lambda^{-1} U^H S_k(m) D^H \widetilde{r}\right\}\right] \tag{11.52}$$

A simplified form of the above decision rule is obtained by skipping the noise-whitening step, which leads to

$$\widehat{b}_k(m) = \text{sgn}\left[\Re\left\{c_k^H(m) S_k(m) D^H \widetilde{r}\right\}\right] \tag{11.53}$$

Finally, an MMSE multiuser detector operating after the multipath combining scheme can be obtained by linearly estimating the information symbols $\widetilde{b}$ conditioned on the channel coefficient realizations. More precisely, given the decision rule

$$\widehat{\widetilde{b}} = \text{sgn}[\Re\{D^H \widetilde{r}\}] \tag{11.54}$$

the $(2P+1)KL \times (2P+1)K$-dimensional matrix D solves the following minimization problem:

$$D = \arg\min_{Y} E\{\|\widetilde{b} - Y^H \widetilde{r}\|^2 | \widetilde{C}\} \tag{11.55}$$

Direct application of the orthogonality principle leads to the solution

$$\begin{aligned} D &= [E\{\widetilde{r}\widetilde{r}^H | \widetilde{C}\}]^{-1} E\{\widetilde{r}\widetilde{b}^H | \widetilde{C}\} \\ &= (\widetilde{A}\widetilde{C}\widetilde{C}^H \widetilde{A}^H \widetilde{R} + 2\mathcal{N}_0 I_{(2P+1)KL})^{-1} \widetilde{R}\widetilde{A}\widetilde{C} \end{aligned} \tag{11.56}$$

with $I_{(2P+1)KL}$ the identity matrix of order $(2P+1)KL$. Applying the matrix inversion lemma,[8] it can easily be seen that the solution D can also be written as

$$D = \widetilde{A}\widetilde{C}(\widetilde{C}^H \widetilde{A}^H \widetilde{R}\widetilde{A}\widetilde{C} + 2\mathcal{N}_0 I_{(2P+1)K})^{-1} \tag{11.57}$$

Substituting the above solution into Equation 11.54 finally yields the decision rule

$$\begin{aligned} \widehat{\widetilde{b}} &= \text{sgn}[\Re\{(\widetilde{C}^H \widetilde{A}^H \widetilde{R}\widetilde{A}\widetilde{C} + 2\mathcal{N}_0 I_{(2P+1)K})^{-1} \widetilde{C}^H \widetilde{A}^H \widetilde{r}\}] \\ &= \text{sgn}[\Re\{(\widetilde{C}^H \widetilde{A}^H \widetilde{R}\widetilde{A}\widetilde{C} + 2\mathcal{N}_0 I_{(2P+1)K})^{-1} \widetilde{x}\}] \end{aligned} \tag{11.58}$$

which shows that the observables are first fed to a multipath combining stage and then are forwarded to an interference suppression stage that takes care of MAI and thermal noise reduction.

11.3.3.4 Remarks on the MMSE Detector

As for the three implementations of the decorrelating detector, the same remarks also apply to the receivers in Equations 11.52, 11.53, and 11.58. In particular, it is easily seen that the first two receivers can be interpreted as two-stage receivers, wherein the first stage is independent of the channel coefficients and the second stage depends only on the channel experienced by the users of interest. The decision rule in Equation 11.58, instead depends in an involved way on all the channel coefficients, but on the other hand, it outperforms the other two receivers, whose first stage treats the multipath replicas as additional fictitious users.

Finally, it is also worth pointing out that in the limit $\mathcal{N}_0 \to 0$, i.e., for vanishingly small thermal noise level, the MMSE detectors converge to the decorrelating detectors. More precisely, applying standard

[8] The matrix inversion lemma states the following identity: $(A+BCD)^{-1} = A^{-1} - A^{-1}B(C^{-1} + DA^{-1}B)^{-1}DA^{-1}$.

techniques for matrix analysis, it can be shown that the receivers in Equations 11.52, 11.53, and 11.58 converge, as $\mathcal{N}_0 \to 0$ and under mild conditions, to the decision rules in Equations 11.42, 11.43, and 11.45, respectively. The proof of this result, which was first established for no-fading channels, is omitted here for the sake of brevity.

11.3.4 Approximate MMSE Detection: Linear Serial Interference Cancellation

From the previous sections it is seen that computation of the MMSE receiver requires inversion of a matrix whose dimensions increase linearly with the product KP. Since the computational effort required to perform matrix inversion is generally proportional to the third power of the matrix dimension, in applications where computational resources are constrained it is of interest to devise alternative lower-complexity detection structures. To this end, consider, for instance, the MMSE detector that implements the decision rule in Equation 11.58, i.e.,

$$\widehat{\widetilde{\boldsymbol{b}}} = \operatorname{sgn}\left[\Re\left\{\underbrace{(\widetilde{\boldsymbol{C}}^H \widetilde{\boldsymbol{A}}^H \widetilde{\boldsymbol{R}}\widetilde{\boldsymbol{A}}\widetilde{\boldsymbol{C}} + 2\mathcal{N}_0 \boldsymbol{I}_{(2P+1)K})^{-1}\widetilde{\boldsymbol{x}}}_{f}\right\}\right] \tag{11.59}$$

Letting

$$\boldsymbol{J} = \widetilde{\boldsymbol{C}}^H \widetilde{\boldsymbol{A}}^H \widetilde{\boldsymbol{R}}\widetilde{\boldsymbol{A}}\widetilde{\boldsymbol{C}} + 2\mathcal{N}_0 \boldsymbol{I}_{(2P+1)K} \tag{11.60}$$

it is seen that the vector $\boldsymbol{f}$ to be plugged in the decision rule in Equation 11.59 can be obtained by solving the linear system

$$\boldsymbol{J}\boldsymbol{f} = \widetilde{\boldsymbol{x}} \tag{11.61}$$

Since, as already discussed, direct inversion of the matrix $\boldsymbol{J}$ may entail an intolerable computational burden, a suitable alternative is to resort to iterative techniques for solving linear systems, such as the Gauss–Seidel iterations [7]. In particular, given the system in Equation 11.61, the i-th entry of the solution at the m-th iteration, say $\boldsymbol{f}^{(m)}(i)$, of the Gauss–Seidel procedure is expressed as

$$\boldsymbol{f}^{(m)}(i) = \frac{1}{\boldsymbol{J}(i,i)}\left[\widetilde{\boldsymbol{x}}(i) - \sum_{j<i} \boldsymbol{J}(i,j)\boldsymbol{f}^{(m)}(j) - \sum_{j>i} \boldsymbol{J}(i,j)\boldsymbol{f}^{(m-1)}(j)\right] \tag{11.62}$$

By virtue of the Ostrowski–Reich theorem, since the matrix $\boldsymbol{J}$ is positive-definite, the above iterative procedure converges, for any starting point $\boldsymbol{f}^{(0)}$, to the solution of the system in Equation 11.61 [7]. The iteration in Equation 11.62 can be interpreted as a serial interference cancellation, since at each iteration, the already updated entries of the vector $\boldsymbol{f}$ are used to update its remaining entries; otherwise stated, the iterative Gauss–Seidel procedure can be regarded, under a multiuser detection perspective, as a serial interference cancellation receiver with soft symbol estimates. Note that in principle each iteration requires a computational burden that is quadratic in the product KP; however, this complexity is actually lower since $\boldsymbol{J}$ is a sparse matrix. It is also interesting to observe that the iteration in Equation 11.62 can be more compactly rewritten through matrix notation. Indeed, on decomposing

$$\boldsymbol{J} = \boldsymbol{J}^U + \boldsymbol{J}^D + \boldsymbol{J}^L \tag{11.63}$$

with $\boldsymbol{J}^U$, $\boldsymbol{J}^L$, and $\boldsymbol{J}^D$ denoting the upper triangular, lower triangular, and diagonal part of the matrix $\boldsymbol{J}$, respectively, it can be easily shown that the Gauss–Seidel iteration can be written as

$$\boldsymbol{f}^{(m)} = -(\boldsymbol{J}^D + \boldsymbol{J}^L)^{-1}\boldsymbol{J}^U\boldsymbol{f}^{(m-1)} + (\boldsymbol{J}^D + \boldsymbol{J}^L)^{-1}\widetilde{\boldsymbol{x}} \tag{11.64}$$

The same steps that lead to the approximate implementation of the MMSE detector (Equation 11.58) can also be used to obtain approximate lower-complexity implementations of the MMSE detection rules in Equation 11.52 and Equation 11.53. As an example, the latter rule is written as

$$\widehat{b}_k(m) = \text{sgn}\left[\Re\left\{ \boldsymbol{c}_k^H(m)\boldsymbol{S}_k(m)\Sigma_{\boldsymbol{C}}\widetilde{\boldsymbol{A}}^H\widetilde{\boldsymbol{R}}\underbrace{(\widetilde{\boldsymbol{R}}\widetilde{\boldsymbol{A}}\Sigma_{\boldsymbol{C}}\widetilde{\boldsymbol{A}}^H\widetilde{\boldsymbol{R}} + 2\mathcal{N}_0\widetilde{\boldsymbol{R}})^{-1}\widetilde{\boldsymbol{r}}}_{f} \right\}\right] \tag{11.65}$$

and upon letting

$$\boldsymbol{J} = \widetilde{\boldsymbol{R}}\widetilde{\boldsymbol{A}}\Sigma_{\boldsymbol{C}}\widetilde{\boldsymbol{A}}^H\widetilde{\boldsymbol{R}} + 2\mathcal{N}_0\widetilde{\boldsymbol{R}} \tag{11.66}$$

the iteration in Equation 11.62 with $\widetilde{\boldsymbol{r}}$ in place of $\widetilde{\boldsymbol{x}}$ can be used to approximate the vector in the decision rule in Equation 11.65.

11.3.5 Constrained ML Detection: Nonlinear Serial Interference Cancellation

Besides the use of linear multiuser detection schemes, another interesting approach to overcome the huge complexity of the optimum ML detector is to resort to constrained ML (CML) data detection, which has the potential of improving the performance of linear multiuser detectors. CML multiuser detection is based on the idea of maximizing the likelihood of the received signal by letting the unknown data symbols take value in a set that includes the values $+1$, -1. Indeed, it has been already discussed that if the symbols are modeled as real-valued quantities, the corresponding CML multiuser receiver is the decorrelating detector (Equation 11.45). In general, instead, any possible choice of the set in which the unknown symbols take values leads to a particular CML multiuser receiver. Hereafter, we assume that the data symbols are confined to lie within a hypercube (i.e., the set $[-1,+1]^{(2P+1)K})$) whose vertices are the $(2P+1)K$-tuples of unknown symbols. Otherwise stated, the CML detector selects as the transmitted sequence the solution to the following optimization problem:

$$\widetilde{\boldsymbol{b}}_{\text{CML}} = \arg \max_{\widetilde{\boldsymbol{b}} \in [-1,+1]^{(2P+1)K}} 2\Re[\widetilde{\boldsymbol{b}}^H\widetilde{\boldsymbol{x}}] - \widetilde{\boldsymbol{b}}^H\boldsymbol{H}\widetilde{\boldsymbol{b}} \tag{11.67}$$

Although the solution cannot be given in closed form, the above problem is a convex minimization over a convex set, whereby it has only one fixed point, which is the sought solution. Iterative methods can thus be employed to solve the problem in Equation 11.67; hereafter, we will apply a constrained gradient method that, given the fact that the constraint set is element-wise separable,[9] assumes a simplified formulation. In particular, we can use the nonlinear Gauss–Seidel algorithm, which amounts to sequentially minimizing the cost function in Equation 11.67 with respect to only one entry of the vector $\widetilde{\boldsymbol{b}}$ at a time. Setting to zero the gradient of the function in Equation 11.67 with respect to the i-th entry of $\widetilde{\boldsymbol{b}}$, say $\widetilde{\boldsymbol{b}}_i$, we have

$$2\widetilde{\boldsymbol{x}}(i) - 2\widetilde{\boldsymbol{b}}_i\boldsymbol{H}_{i,i} - 2\sum_{j\neq i}\boldsymbol{H}_{i,j}\widetilde{\boldsymbol{b}}_j = 0 \tag{11.68}$$

Solving with respect to $\widetilde{\boldsymbol{b}}_i$ yields

$$\widetilde{\boldsymbol{b}}_i = \frac{1}{\boldsymbol{H}_{i,i}}\left[-\sum_{j\neq i}\boldsymbol{H}_{i,j}\widetilde{\boldsymbol{b}}_j + \widetilde{\boldsymbol{x}}(i)\right] \tag{11.69}$$

[9] A convex set is said to be element-wise separable if it can be written as the Cartesian product of monodimensional intervals.

Finally, the iteration is completed by projecting x_i onto the interval $[-1, +1]$, i.e., by setting to zero its imaginary part and by clipping its real part. Interestingly, Equation 11.69 can be interpreted as an interference cancellation operation, thus implying that the CML problem (Equation 11.67) can be iteratively solved through the use of a nonlinear SIC receiver. This fact was first recognized in [17]. Since the constraint set is element-wise separable, if the matrix $\boldsymbol{H}$ has full column rank, so that the matrix $\boldsymbol{H}^H\boldsymbol{H}$ is positive-definite, the outlined nonlinear iterative algorithm always converges, for any starting point, to the fixed point of the cost function in Equation 11.67.

11.3.6 Performance Results

In the following, the performance of the multiuser receivers is illustrated through both analytical considerations and some sample plots illustrating numerical results. For the sake of simplicity, the theoretical analysis is restricted to the linear multiuser detectors, and numerical results are used to compare the remaining above illustrated receivers. However, for an analytical performance study of the optimum ML receiver we defer the reader to references [24, 25]. Unless otherwise stated, it is assumed in the following that the matrices $\widetilde{\boldsymbol{R}}$ and $\boldsymbol{H}$ are nonsingular.

11.3.6.1 Analysis

We start by deriving the error probability (or bit error rate (BER)) for the decorrelating detector in Equation 11.42. Note that

$$\boldsymbol{c}_k^H(m)\boldsymbol{Q}_k^{-1}(m)\boldsymbol{y}_k(m) = \boldsymbol{c}_k^H(m)\boldsymbol{Q}_k^{-1}(m)\boldsymbol{c}_k(m)A_k b_k(m) + \underbrace{\boldsymbol{c}_k^H(m)\boldsymbol{Q}_k^{-1}(m)\boldsymbol{S}_k(m)\widetilde{\boldsymbol{R}}^{-1}\widetilde{\boldsymbol{z}}}_{\nu_k(m)} \tag{11.70}$$

Since $\nu_k(m)$ is a zero-mean complex Gaussian random variate with variance $\boldsymbol{c}_k^H(m)\boldsymbol{Q}_k^{-1}(m)\boldsymbol{c}_k(m)$, it can be easily shown that, conditioned on the channel vector $\boldsymbol{c}_k(m)$, the probability of erroneous detection of the bit $b_k(m)$ for the decorrelating detector in Equation 11.42 is written as[10]

$$P_{k,m}(e|\boldsymbol{c}_k(m)) = \frac{1}{2}\text{erfc}\left(A_k\sqrt{\boldsymbol{c}_k^H(m)\boldsymbol{Q}_k^{-1}(m)\boldsymbol{c}_k(m)}\right) \tag{11.71}$$

Now, since for Rayleigh fading and uncorrelated scattering the vector $\boldsymbol{c}_k(m)$ is a complex Gaussian vector with diagonal covariance matrix $\boldsymbol{M_c}$, the quantity $\boldsymbol{c}_k^H(m)\boldsymbol{Q}_k^{-1}(m)\boldsymbol{c}_k(m)$ is a quadratic form of independent Gaussian random variates with characteristic function

$$\Psi(\omega) = \prod_{i=1}^{L}\frac{1}{1-j\omega\lambda_i} = \sum_{i=1}^{L}\frac{\beta_i}{1-j\omega\lambda_i} \tag{11.72}$$

with $\{\lambda_i\}_{i=1}^L$ the distinct eigenvalues[11] of $\boldsymbol{M_c}\,\boldsymbol{Q}_k^{-1}(m)$ and

$$\beta_i = \prod_{\ell\neq i}\frac{\lambda_i}{\lambda_i-\lambda_\ell} \tag{11.73}$$

the coefficients of the partial fraction expansion. The unconditional error probability is thus obtained by averaging with respect to the distribution of the quadratic form; letting $\gamma_{i,k} = A_k^2\lambda_i$, the result is [16]

$$P_{k,m}(e) = \sum_{i=1}^{L}\frac{\beta_i}{2}\left[1-\sqrt{\frac{\gamma_{i,k}}{1+\gamma_{i,k}}}\right] \tag{11.74}$$

---

[10] The complementary error function is defined as $\text{erfc}(x) = \frac{2}{\sqrt{\pi}}\int_x^\infty e^{-t^2}dt$.

[11] For ease of notation, we are omitting the dependence of the eigenvalues on the indices k and m.

It is interesting to note that the above expression depends on the user index k and the temporal index m. The dependence on k is due to the fact that in an asynchronous system subject to fading the received users' spreading codes are not equally correlated, while the dependence on the index m is due to the fact that we are considering a finite-length data packet and possibly time-varying fading. Moreover, it is seen from Equation 11.35 and Equation 11.39 that the matrix $\mathbf{Q}_k^{-1}(m)$ is proportional to $1/\mathcal{N}_0$, and so are the coefficients λ_i and $\gamma_{i,k}$ for any $i = 1, \ldots, L$. As a consequence, if we let

$$\gamma_{i,k}^n = \mathcal{N}_0 \gamma_{i,k}, \qquad \forall i = 1, \ldots, L$$

the error probability in Equation 11.74 can be also written as

$$P_{k,m}(e) = \sum_{i=1}^{L} \frac{\beta_i}{2} \left[1 - \sqrt{\frac{1}{1 + \frac{\mathcal{N}_0}{\gamma_{i,k}^n}}} \right] \tag{11.75}$$

Besides the error probability, another interesting performance measure is the near–far resistance, which gives information on the receiver behavior in the limiting situation of close-to-zero thermal noise PSD and increasingly large amplitudes of the interfering users. More precisely, the near–far resistance characterizes the performance degradation experienced by a particular receiver for the presence of the MAI, since it is related to the rate of decay of the error probability (for increasing useful signal-to-noise ratio) of the given receiver, compared to that of the optimum RAKE coherent receiver in a single-user channel [16]. Thus, if the near–far resistance equals unity, then the receiver is able to get rid of MAI without any performance impairment, and in fact, its BER has the same decay rate (for $\mathcal{N}_0 \to 0$) as the optimum receiver in a single-user transmission. If, conversely, the near–far resistance is zero, then the receiver suffers from an error probability floor and its performance is limited by the MAI level rather than by the thermal noise. The near–far resistance was first introduced in [11] for the case of AWGN channels, while in [24, 25] it is extended to the case of multipath channels. In order to derive its expression, it is useful to approximate the error probability in Equation 11.75 through the following McLaurin expansion:

$$P_{k,m}(e) = f_L \prod_{i=1}^{L} \frac{\mathcal{N}_0}{\gamma_{i,k}^n} + o\left(\mathcal{N}_0^L\right) = f_L \left(\frac{\mathcal{N}_0}{\gamma_e} \right)^L + o\left(\mathcal{N}_0^L\right) \tag{11.76}$$

with

$$f_L = \frac{(2L+1)(2L-1)\cdots 3 \cdot 1}{L! 2^{L+1}} \tag{11.77}$$

and

$$\gamma_e = \left(\prod_{i=1}^{L} \frac{1}{\gamma_{i,k}^n} \right)^{1/L} \tag{11.78}$$

On the other hand, if we consider the isolated transmission of the bit $b_k(m)$ in the absence of MAI and intersymbol interference, it can easily be shown that the corresponding error probability, say $P_{k,m}^I$, is written as

$$P_{k,m}^I = \sum_{i=1}^{L} \frac{\beta_i^I}{2} \left[1 - \sqrt{\frac{1}{1 + \frac{\mathcal{N}_0}{\gamma_{i,k}^I}}} \right] \tag{11.79}$$

with

$$\gamma_{i,k}^I = \frac{A_k^2 \lambda_i^I}{2}$$

$$\beta_i^I = \prod_{\ell \neq i} \frac{\lambda_i^I}{\lambda_i^I - \lambda_\ell^I} \tag{11.80}$$

and $\{\lambda_i^I\}_{i=1}^L$ the distinct eigenvalues of $\boldsymbol{M}_c\boldsymbol{S}_k(m)\widetilde{\boldsymbol{R}}\boldsymbol{S}_k^H(m)$. Similarly to Equation 11.75, the error probability in Equation 11.79 can be expanded, for small values of $\mathcal{N}_0$, as follows:

$$P_{k,m}^I(e) = f_L \prod_{i=1}^{L} \frac{\mathcal{N}_0}{\gamma_{i,k}^I} + o(\mathcal{N}_0^L) = f_L \left(\frac{\mathcal{N}_0}{\gamma_g}\right)^L + o(\mathcal{N}_0^L) \tag{11.81}$$

with

$$\gamma_g = \left(\prod_{i=1}^{L} \frac{1}{\gamma_{i,k}^I}\right)^{1/L} \tag{11.82}$$

It is thus seen from Equation 11.76 and Equation 11.81 that for a vanishingly small thermal noise level, the error probability expression for the decorrelating detector in Equation 11.42 and for an isolated binary phase shift keying (BPSK) transmission over a multipath channel depends on the geometric means, γ_e and γ_g, of the average received energy contrasts on each multipath component. It can thus be argued that the ability of the multiuser detector in Equation 11.42 to remove MAI may be described by the effective signal-to-noise ratio γ_e required by the optimum detector in an isolated transmission to achieve the same asymptotic error probability. Based on this observation, in [24] the *asymptotic multiuser efficiency* (AME) is first defined as

$$\text{AME} = \lim_{\mathcal{N}_0 \to 0} \frac{\gamma_e}{\gamma_g} \tag{11.83}$$

Finally, the near–far resistance is, as usual, defined as the infimum of the AME for arbitrarily varying other users' amplitudes. Based on Equation 11.78 and Equation 11.82, it is thus easy to show that the near–far resistance for the decorrelating receiver implementing the decision rule in Equation 11.42 and detecting the symbol $b_k(m)$ can be written as

$$\eta_{k,m} = 2\left[\frac{\det\left(\mathcal{N}_0 \boldsymbol{Q}_k^{-1}(m)\right)}{\det\left(\boldsymbol{S}_k(m)\widetilde{\boldsymbol{R}}\boldsymbol{S}_k^H(m)\right)}\right]^{1/L} \tag{11.84}$$

Let us now consider the decorrelating decision rule in Equation 11.43. We now have

$$\boldsymbol{c}_k^H(m)\boldsymbol{S}_k(m)\widetilde{\boldsymbol{R}}^{-1}\widetilde{\boldsymbol{r}} = \boldsymbol{c}_k^H(m)\boldsymbol{c}_k(m)A_k b_k(m) + \underbrace{\boldsymbol{c}_k^H(m)\boldsymbol{z}_{d,k}(m)}_{\nu_k'(m)} \tag{11.85}$$

Since $\nu_k'(m)$ is a complex zero-mean Gaussian random variate with variance $\boldsymbol{c}_k^H(m)\boldsymbol{Q}_k(m)\boldsymbol{c}_k(m)$, the conditional error probability can be easily shown to be expressed as

$$P_{k,m}(e|\boldsymbol{c}_k(m)) = \frac{1}{2}\operatorname{erfc}\left(\frac{A_k\|\boldsymbol{c}_k(m)\|^2}{\sqrt{\boldsymbol{c}_k^H(m)\boldsymbol{Q}_k(m)\boldsymbol{c}_k(m)}}\right) \tag{11.86}$$

Unfortunately, the above conditional error probability is not easily averaged over the channel statistics, and a numerical average has to be carried out in order to evaluate the unconditional performance. Alternatively, closed-form expressions for an upper and lower bound to Equation 11.86 can be also devised. Indeed, note that by denoting $\lambda_{\max}(\boldsymbol{A})$ and $\lambda_{\min}(\boldsymbol{A})$ as the largest and smallest eigenvalues of a square matrix $\boldsymbol{A}$, respectively, the following inequalities hold:

$$\max_{\boldsymbol{x}\neq 0} \frac{\boldsymbol{x}^H\boldsymbol{A}\boldsymbol{x}}{\boldsymbol{x}^H\boldsymbol{x}} = \lambda_{\max}(\boldsymbol{A}), \qquad \min_{\boldsymbol{x}\neq 0} \frac{\boldsymbol{x}^H\boldsymbol{A}\boldsymbol{x}}{\boldsymbol{x}^H\boldsymbol{x}} = \lambda_{\min}(\boldsymbol{A}) \tag{11.87}$$

which in turn imply

$$\min_{x\neq 0} \frac{\boldsymbol{x}^H \boldsymbol{x}}{\boldsymbol{x}^H \boldsymbol{A}\boldsymbol{x}} = \frac{1}{\lambda_{\max}(\boldsymbol{A})}, \qquad \max_{x\neq 0} \frac{\boldsymbol{x}^H \boldsymbol{x}}{\boldsymbol{x}^H \boldsymbol{A}\boldsymbol{x}} = \frac{1}{\lambda_{\min}(\boldsymbol{A})} \tag{11.88}$$

Based on Equation 11.88, and recalling that the complementary error function is monotonic and decreasing, the following bounds for the error probability in Equation 11.86 can easily be worked out:

$$\frac{1}{2}\text{erfc}\left(A_k\sqrt{\frac{\boldsymbol{c}_k^H(m)\boldsymbol{c}_k(m)}{\lambda_{\min}(\boldsymbol{Q}_k(m))}}\right) \leq P_{k,m}(e|\boldsymbol{c}_k(m)) \leq \frac{1}{2}\text{erfc}\left(A_k\sqrt{\frac{\boldsymbol{c}_k^H(m)\boldsymbol{c}_k(m)}{\lambda_{\max}(\boldsymbol{Q}_k(m))}}\right) \tag{11.89}$$

Given Equation 11.89, and denoting by $\{\lambda_i\}_{i=1}^{L}$ the distinct eigenvalues of the covariance matrix $\boldsymbol{M_C}$, the following bounds can be derived for the unconditional error probability of the decision rule in Equation 11.43:

$$\sum_{i=1}^{L}\frac{\beta_i}{2}\left[1-\sqrt{\frac{\gamma'_{i,k}}{1+\gamma'_{i,k}}}\right] \leq P_{k,m}(e) \leq \sum_{i=1}^{L}\frac{\beta_i}{2}\left[1-\sqrt{\frac{\gamma''_{i,k}}{1+\gamma''_{i,k}}}\right] \tag{11.90}$$

In the above equation,

$$\gamma'_{\ell,k} = \frac{A_k^2\lambda_\ell}{\lambda_{\min}(\boldsymbol{Q}_k(m))}, \qquad \gamma''_{\ell,k} = \frac{A_k^2\lambda_\ell}{\lambda_{\max}(\boldsymbol{Q}_k(m))}, \qquad \forall \ell = 1,\ldots,L$$

while the coefficients β_i are defined as in Equation 11.73. The bounds in Equation 11.90 can also be used to obtain closed-form expressions for upper and lower bounds to the system near–far resistance. In particular, following the same steps as those that led to Equation 11.84, it can be easily shown that

$$\frac{2\mathcal{N}_0\lambda_{\max}^{-1}(\boldsymbol{Q}_k(m))}{\left[\det\left(\boldsymbol{S}_k(m)\widetilde{\boldsymbol{R}}\boldsymbol{S}_k^H(m)\right)\right]^{1/L}} \leq \eta_{k,m} \leq \frac{2\mathcal{N}_0\lambda_{\min}^{-1}(\boldsymbol{Q}_k(m))}{\left[\det\left(\boldsymbol{S}_k(m)\widetilde{\boldsymbol{R}}\boldsymbol{S}_k^H(m)\right)\right]^{1/L}} \tag{11.91}$$

Finally, let us consider the decorrelating detector that operates after multipath combining and that implements the decision rule in Equation 11.45. Since

$$\boldsymbol{H}^{-1}\widetilde{\boldsymbol{x}} = \widetilde{\boldsymbol{b}} + \boldsymbol{H}^{-1}\widetilde{\boldsymbol{C}}^H\widetilde{\boldsymbol{A}}^H\widetilde{\boldsymbol{z}} \tag{11.92}$$

if we denote by $\boldsymbol{h}_k^T(m)$ the $(m+P)K+k+1$-th row of the matrix $\boldsymbol{H}^{-1}$, a decision on the bit $b_k(m)$ is taken according to the sign of the real part of the statistic

$$b_k(m) + \boldsymbol{h}_k^T(m)\widetilde{\boldsymbol{C}}^H\widetilde{\boldsymbol{A}}^H\widetilde{\boldsymbol{z}} \tag{11.93}$$

It is interesting to note that the test statistic is now dependent through the vector $\boldsymbol{h}_k^T(m)$ on all the channel coefficients and not only on the vector $\boldsymbol{c}_k(m)$. As a consequence, based on Equation 11.93, the probability of erroneous detection of the bit $b_k(m)$ for the detector in Equation 11.45, conditioned on $\widetilde{\boldsymbol{C}}$, is written as

$$P_{k,m}(e|\widetilde{\boldsymbol{C}}) = \frac{1}{2}\text{erfc}\left(\frac{1}{\sqrt{2\mathcal{N}_0\boldsymbol{h}_k(m)^T\widetilde{\boldsymbol{C}}^H\widetilde{\boldsymbol{A}}^H\widetilde{\boldsymbol{R}}\widetilde{\boldsymbol{A}}\widetilde{\boldsymbol{C}}\boldsymbol{h}_k^*(m)}}\right) = \frac{1}{2}\text{erfc}\left(\frac{1}{\sqrt{2\mathcal{N}_0[\boldsymbol{H}^{-1}]_{a,a}}}\right) \tag{11.94}$$

with $a = (m+P)K+k+1$. Unfortunately, averaging the above expression with respect to the channel matrix $\widetilde{\boldsymbol{C}}$ appears to be infeasible, whereby a semianalytical approach is to be pursued to obtain numerical values of the unconditional error probability. Lacking an expression for the unconditional error probability, it follows that no near–far resistance expression can be derived as well. However, since Equation 11.44

resembles the signal model of a synchronous DS/CDMA system subject to no fading [11, 22], a conditional near–far resistance expression can be derived in keeping with the approach in [11]. Assuming that the a-th column of the matrix $\boldsymbol{H}$ is not included in the range span of the remaining columns of $\boldsymbol{H}$,[12] we obtain the following expression:

$$\eta_{k,m|\widetilde{\boldsymbol{C}}} = \frac{([\boldsymbol{H}]_{a,a})^{+}}{[\boldsymbol{H}^{+}]_{a,a}} \tag{11.95}$$

with $(\cdot)^{+}$ denoting the Moore–Penrose generalized inverse. We maintain that Equation 11.95 represents the near–far resistance of the receiver in Equation 11.45 conditioned on the channel realization $\widetilde{\boldsymbol{C}}$, and that averaging Equation 11.95 with respect to the matrix $\widetilde{\boldsymbol{C}}$ does *not* yield the near–far resistance as defined in [11, 25, 24], where indeed the near–far resistance is derived from the *unconditional* error probability. Nevertheless, Equation 11.95 and its numerical average over the fading channel realizations turn out to be useful to assess the receiver sensitivity to the MAI strength, since nonzero values of Equation 11.95 and its numerical average ensure that the system is immune to interferers with arbitrarily large power.

Let us now consider the linear MMSE receivers. Based on the equivalence between the MMSE and the decorrelating receivers in the limit $\mathcal{N}_0 \to 0$, it is easily shown that these receivers have the same near–far resistance, and Equations 11.84, 11.91, and 11.95 represent the system near–far resistance also for the MMSE receivers in Equations 11.52, 11.53, and 11.58, respectively. As regards the system error probability, it is important to note that the MMSE receiver does not nullify the MAI, thus implying that the test statistic depends upon the interfering users' information symbols. Conditioned on the interfering symbols and channel coefficient realizations, error probability expressions can be easily derived following the same steps that have been taken for the analysis of the decorrelating detectors. However, these expressions must be numerically averaged, since obtaining closed-form formulas for the unconditional error probability appears to be hardly feasible. An alternative, yet common approach is to model the output MAI as a correlated Gaussian random process in order to avoid the numerical average over the random information symbols. As an example, consider the decision rule in Equation 11.54 with $\boldsymbol{D}$ given by Equation 11.57. Denoting by $\boldsymbol{d}_k(m)$ the $(m+P)K+k+1$-th column of the matrix $\boldsymbol{D}$, and by $\widetilde{\boldsymbol{b}}'$ the vector $\widetilde{\boldsymbol{b}}$ with the bit $b_k(m)$ set to zero, it can be easily shown that the test statistic for detecting the symbol $b_k(m)$ is expressed as

$$\boldsymbol{d}_k^H(m)\widetilde{\boldsymbol{r}} = \underbrace{\boldsymbol{d}_k^H(m)\widetilde{\boldsymbol{R}}\boldsymbol{S}_k^H(m)A_k\boldsymbol{c}_k(m)b_k(m)}_{\text{useful signal}} + \underbrace{\boldsymbol{d}_k^H(m)\widetilde{\boldsymbol{R}}\widetilde{\boldsymbol{A}}\widetilde{\boldsymbol{C}}\widetilde{\boldsymbol{b}}'}_{v_k(m)} + \boldsymbol{d}_k^H(m)\widetilde{\boldsymbol{z}} \tag{11.96}$$

Assuming that $v_k(m)$ is a zero-mean Gaussian random variate with variance

$$\sigma_k^2(m) = \boldsymbol{d}_k^H(m)\widetilde{\boldsymbol{R}}\widetilde{\boldsymbol{A}}\widetilde{\boldsymbol{C}}E\{\widetilde{\boldsymbol{b}}'\widetilde{\boldsymbol{b}}'^H\}\widetilde{\boldsymbol{C}}^H\widetilde{\boldsymbol{A}}^H\widetilde{\boldsymbol{R}}^H\boldsymbol{d}_k(m)$$

the conditional probability of erroneous detection for the MMSE receiver in Equation 11.58 can be approximated as

$$P_{k,m}(e|\widetilde{\boldsymbol{C}}) \approx \frac{1}{2}\operatorname{erfc}\left(\frac{\Re\left[\boldsymbol{d}_k^H(m)\widetilde{\boldsymbol{R}}\boldsymbol{S}_k^H(m)A_k\boldsymbol{c}_k(m)\right]}{\sqrt{\sigma_k^2(m) + 2\mathcal{N}_0\boldsymbol{d}_k^H(m)\widetilde{\boldsymbol{R}}\boldsymbol{d}_k(m)}}\right) \tag{11.97}$$

A semianalytic average of the above expression is then needed to obtain the unconditional error probability.

[12] Note that if this condition is met, the signature of interest lies in the interference subspace and the near–far resistance of any linear detector is zero.

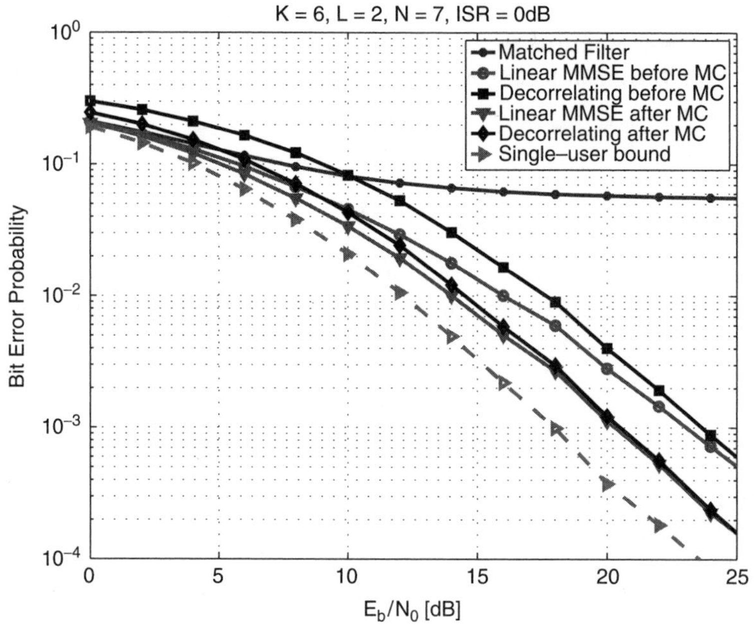

FIGURE 11.4 Error probability vs. $E_b/\mathcal{N}_0$ for several detectors. It is assumed that the receiver has perfect CSI.

11.3.7 Some Numerical Results

The results reported in Figure 11.4 to Figure 11.6 refer to a CDMA system with processing gain $N = 7$, subject to a two-path propagation channel with fast fading. The packet length is $2P+1=40$, the multipath delays are drawn from a uniform distribution in $[0, T_b]$, and the channel covariance matrix $\boldsymbol{M}_c = \text{Diag}(.7, .3)$. Figure 11.4 and Figure 11.5 report the bit error probability vs. the average received energy contrast $E_b/\mathcal{N}_0$ (E_b denotes the average received energy per transmitted bit) for a system with $K = 6$ users. In particular, the performance of the conventional matched filter receiver in Equation 11.24, of the decorrelating detectors in Equations 11.42 and 11.45, and of the MMSE detectors in Equations 11.52 and 11.58 are reported in Figure 11.4, while the performance of the iterative implementation of the CML receiver is reported in Figure 11.5. Moreover, in both plots a single-user bound, i.e., the system performance achievable in a single-user channel, is reported for comparison purposes. In these figures, and in the following, the term ISR (interference-to-signal ratio) is used to denote the ratio between the average energy of each interfering user and the average energy of the user of interest. Results clearly show that the conventional single-user receiver performance is affected by an error floor due to the MAI. On the contrary, the multiuser receivers do not exhibit this drawback and are capable of achieving satisfactory performance. In particular, the CML performs pretty close to the single-user bound; it is also seen that the linear multiuser detectors operating after multipath combining outperform the corresponding counterparts operating before multipath combining. Moreover, the linear MMSE receiver achieves a performance that is superior to that of the decorrelating detector. Finally, results not shown here for conciseness have demonstrated that the simplified detectors in Equations 11.43 and 11.53 perform a little worse than the receivers in Equations 11.42 and 11.52, respectively.

Figure 11.6 shows the near–far resistance of the linear detectors vs. the number of users K for a system with processing gain $N = 15$ and a two-path propagation channel. The experimental results confirm that the linear detectors are immune to the near–far effect and that the detectors operating after multipath combining are capable of offering, at the price of some complexity increase, improved performance with respect to the linear detectors operating before multipath combining.

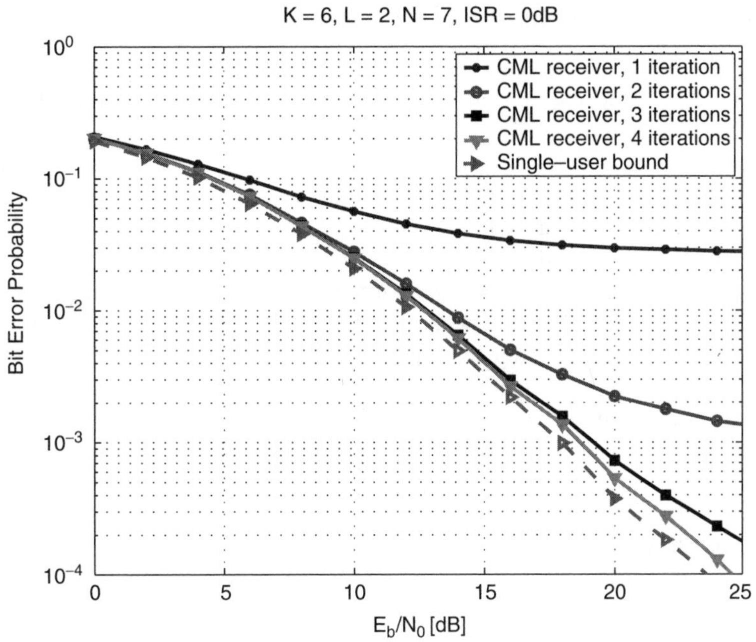

FIGURE 11.5 Error probability vs. $E_b/\mathcal{N}_0$ for the iterative implementation of the CML. For comparison purposes, the single-user bound is also reported.

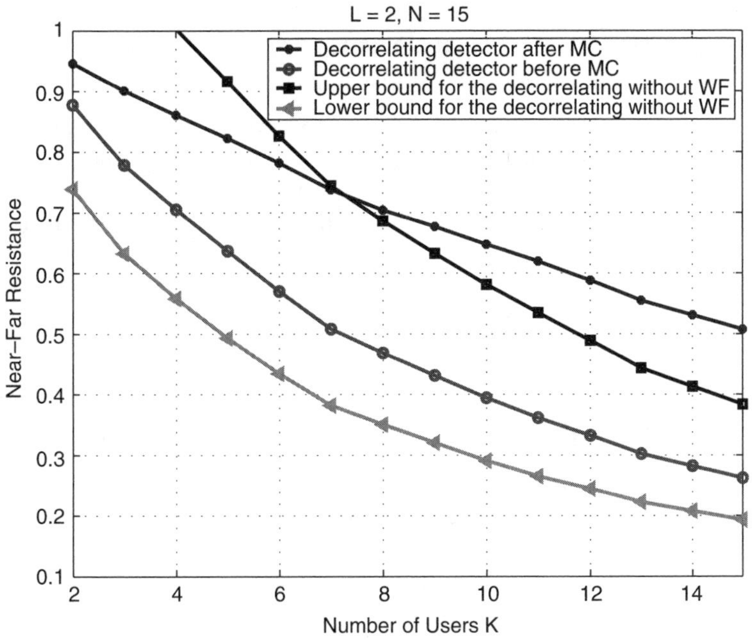

FIGURE 11.6 Near–far resistance vs. the number of users K for the decorrelating detectors.

11.3.8 Sliding-Window One-Shot Multiuser Receivers

The multiuser receivers so far developed perform a batch processing of the entire received data frame, which consists of $(2P+1)$ symbols per user. In some applications, however, complexity constraints and the need for small detection delays lead to consideration of sliding-window one-shot receivers. These receivers process windowed versions of the received signal; after having detected the information symbols transmitted in a given symbol interval, the processing window is shifted of T_b in order to enable detection of the information symbols transmitted in the following symbol interval.

The length of the processing window either may be as small as T_b (a reasonable choice if complexity is an issue and if $T_m \ll T_b$) or may comprise several bit intervals. As a general trend, the larger the processing window length, the better the system performance, even though this performance improvement comes at the expenses of increased computational complexity. In the following, we dwell on one-shot multiuser receivers, showing that the illustrated multiuser detection techniques can be extended to the case of one-shot multiuser detection with little effort. To fix the ideas, consider a processing window of length $2T_b$, denote by $\mathcal{I}_p = [pT_b, (p+2)T_b]$ the p-th processing window, and assume for simplicity that the received signal windowed to $\mathcal{I}_p$ contains contributions from the symbols $\boldsymbol{b}(p-1)$, $\boldsymbol{b}(p)$, and $\boldsymbol{b}(p+1)$, (Note, however, that in general, depending on how large T_m is and on the realization of the timing offsets $\{\tau_k\}_{k=0}^{K-1}$, the signal windowed to $\mathcal{I}_p$ may also contain the contribution from symbols preceding $\boldsymbol{b}(p-1)$.) In order to convert the received signal to discrete time a low-pass filter followed by a sampler at an integer multiple of the chip rate is usually adopted. For rectangular chip pulses, this operation amounts to considering the projections

$$r(n) = \sqrt{\frac{M}{T_c}} \int_{nT_c/M}^{(n+1)T_c/M} r(t)dt \tag{11.98}$$

with M denoting the oversampling factor, i.e., the number of samples per chip interval. Stacking the $2NM$ samples $r(MNp), \ldots, r(MN(p+2)-1)$ corresponding to the interval $\mathcal{I}_p$ in a vector, say $\boldsymbol{r}(p)$, we have

$$\boldsymbol{r}(p) = \sum_{k=0}^{K-1} A_k(b_k(p-1)\boldsymbol{S}_{k,-1}\boldsymbol{c}_k(p-1) + b_k(p)\boldsymbol{S}_{k,0}\boldsymbol{c}_k(p) + b_k(p+1)\boldsymbol{S}_{k,+1}\boldsymbol{c}_k(p+1)) + \boldsymbol{w}(p) \tag{11.99}$$

In the above equation, the $2NM \times L$-dimensional matrix $\boldsymbol{S}_{k,i}$ contains on its columns the discrete-time multipath replicas of the signature modulated by $b_k(p+i)$; i.e., the ℓ-th column of $\boldsymbol{S}_{k,i}$ is the discrete-time version of the signature $s_k(t - \tau_k - \tau_{k,\ell} - (p+i)T_b)$, windowed to $\mathcal{I}_p$. Moreover, $\boldsymbol{w}(p)$ is a complex zero-mean random Gaussian vector with covariance matrix $2\mathcal{N}_0\boldsymbol{I}_{2NM}$. Upon defining the $2NM \times 3KL$-dimensional matrix

$$\widetilde{\boldsymbol{S}} = [\boldsymbol{S}_{0,-1} \ \ldots \ \boldsymbol{S}_{K-1,-1} \ \boldsymbol{S}_{0,0} \ \ldots \ \boldsymbol{S}_{K-1,0} \ \boldsymbol{S}_{0,+1} \ \ldots \ \boldsymbol{S}_{K-1,+1}] \tag{11.100}$$

and

$$\widetilde{\boldsymbol{C}}(p) = \text{Diag}(\boldsymbol{C}(p-1), \boldsymbol{C}(p), \boldsymbol{C}(p+1)), \qquad 3KL \times 3K \tag{11.101}$$

$$\widetilde{\boldsymbol{b}}(p) = \begin{bmatrix} \boldsymbol{b}(p-1) \\ \boldsymbol{b}(p) \\ \boldsymbol{b}(p+1) \end{bmatrix}, \qquad 3K \times 1 \tag{11.102}$$

and

$$\widetilde{\boldsymbol{A}}_1 = \boldsymbol{I}_3 \otimes \boldsymbol{A}, \qquad 3KL \times 3KL \tag{11.103}$$

it can easily be shown that the vector $\boldsymbol{r}(p)$ in Equation 11.99 can be expressed as

$$\boldsymbol{r}(p) = \widetilde{\boldsymbol{S}}\widetilde{\boldsymbol{A}}_1\widetilde{\boldsymbol{C}}(p)\widetilde{\boldsymbol{b}}(p) + \boldsymbol{w}(p) \tag{11.104}$$

and, premultiplying the above expression by $\widetilde{\boldsymbol{S}}^H$, we have

$$\boldsymbol{y}(p) = \widetilde{\boldsymbol{S}}^H \boldsymbol{r}(p) = \underbrace{\widetilde{\boldsymbol{S}}^H \widetilde{\boldsymbol{S}}}_{\widetilde{\boldsymbol{R}}'} \widetilde{\boldsymbol{A}}_1\widetilde{\boldsymbol{C}}(p)\widetilde{\boldsymbol{b}}(p) + \underbrace{\widetilde{\boldsymbol{S}}^H \boldsymbol{w}(p)}_{\widetilde{\boldsymbol{z}}(p)} \tag{11.105}$$

Inspecting the above equation, it is seen that Equation 11.105 has exactly the same structure as $\widetilde{\boldsymbol{r}}$ in Equation 11.20; as a consequence, all the multiuser receivers so far illustrated can be easily coupled with the sliding-window approach by considering the observable Equation 11.105 in lieu of Equation 11.20. The only difference is that once the vector $\widetilde{\boldsymbol{b}}(p)$ has been detected, only the K central entries of this vector are to be considered, since the first K entries have already been detected based on the observation of $\boldsymbol{r}(p-1)$, while the last K entries of $\widetilde{\boldsymbol{b}}(p)$ will be detected by processing the vector $\boldsymbol{r}(p+1)$.

11.4 Multiuser Detection with Unknown CSI

The previous section has focused on the situation, somewhat ideal, that the CSI is either known or perfectly estimated at the receiver. In the following, we review multiuser detection techniques for the case that the channel impulse responses are not known at the receiver. Hereafter, we assume that the channel coherence time $(\Delta t)_c$ exceeds the frame duration $(2P+1)T_b$, so that the fading channel is time invariant over the received packet.

11.4.1 Signal Representation in Unknown CSI

In order to perform multiuser detection with no CSI, it is still convenient to adopt a sliding-window approach, i.e., to process, in each symbol interval, the signal as observed in a few signaling intervals and then let the processing window slide with steps of length T_b. A classical approach to multiuser detection with no CSI has thus been that of designing channel estimation and data detection algorithms based on the observation in Equation 11.104. On the other hand, considering the signal model in Equation 11.104 is equivalent to implicitly assuming that the delays τ_k and $\tau_{k,l}$, $\forall k, l$, are known to the receiver (note indeed that knowledge of the delays is necessary to ensure that the matrix $\widetilde{\boldsymbol{S}}$, which depends on the users' spreading codes and delays, is known). Here, we relax this hypothesis and consider the general case that both the channel delays and fading gain coefficients are unknown. Moreover, we consider a more general channel model; i.e., we assume that the k-th user channel can be deemed as a linear time-invariant filter whose impulse response $c_k(t)$ is supported on $[0, T_m]$. Note that this model, under the slow fading assumption, includes the L-path channel considered in the previous section. Based on the above assumptions, a different signal representation for the vector $\boldsymbol{r}(p)$ is needed. To begin with, we observe that the complex envelope of the received signal can now be written as

$$\begin{aligned} r(t) &= \sum_{k=0}^{K-1} \sum_{m=-P}^{P} A_k b_k(m) s_k(t - \tau_k - mT_b) * c_k(t) + w(t) \\ &= \sum_{k=0}^{K-1} \sum_{m=-P}^{P} b_k(m) h_k(t - mT_b) + w(t) \end{aligned} \tag{11.106}$$

In Equation 11.106, we have let

$$h_k(t) = A_k s_k(t-\tau_k) * c_k(t) = A_k \sum_{n=0}^{N-1} \beta_k(n) \psi_{T_c}(t - \tau_k - nT_c) * c_k(t) = \sum_{n=0}^{N-1} \beta_k(n) g_k(t - nT_c) \tag{11.107}$$

where all of the (unknown) channel characteristics have been placed in the unknown functions $g_k(t)$, which are defined as

$$g_k(t) = A_k \psi_{T_c}(t - \tau_k) * c_k(t) \tag{11.108}$$

Notice that if the multipath delay spread T_m is assumed to be equal for all of the channels, then the waveforms $h_k(t)$ have compact support in $[\tau_k, \tau_k + T_b + T_m] \subseteq [0, 2T_b]$, where the inclusion stems from the reasonable assumption that $\tau_k + T_m < T_b$, which we henceforth adopt.[13] Considering the projection operation in Equation 11.98, and stacking the $2NM$ samples corresponding to the interval $\mathcal{I}_p$ into the vector $\boldsymbol{r}(p)$, we have

$$\boldsymbol{r}(p) = [r(NMp), r(NMp+1), \ldots, r(NM(p+2)-1)]^T \tag{11.109}$$

It can be easily shown that

$$r(NMp+i) = \sum_{k=0}^{K-1} [b_k(p-1)h_k(i+NM) + b_k(p)h_k(i) + b_k(p+1)h_k(i-NM)] + w(NMp+i),$$
$$i = 0, \ldots, 2NM-1 \tag{11.110}$$

The term $h_k(i)$ can easily be shown to be given by

$$h_k(i) = \sqrt{\frac{M}{T_c}} \int_{iT_c/M}^{(i+1)T_c/M} \sum_{n=0}^{N-1} \beta_{n,k} g_k(t - nT_c)dt = \sum_{n=0}^{N-1} \beta_{n,k} g_k(i - nM) \tag{11.111}$$

where in turn we have

$$g_k(i) = \sqrt{\frac{M}{T_c}} \int_{iT_c/M}^{(i+1)T_c/M} g_k(t)dt \tag{11.112}$$

Since the waveform $g_k(t)$ is supported in the interval $[\tau_k, \tau_k + T_c + T_m] \subseteq [0, T_c + T_b]$, it follows that the samples $g_k(i)$ are equal to zero $\forall i \geq (N+1)M$. As a consequence, on defining the $(N+1)M$-dimensional vector

$$\boldsymbol{g}_k = [g_k(0), g_k(1), \ldots, g_k((N+1)M-1)]^T \tag{11.113}$$

it can be easily shown that the discrete-time observable $\boldsymbol{r}(p)$ in Equation 11.109 can be written as

$$\boldsymbol{r}(p) = \sum_{k=0}^{K-1} [b_k(p-1)\boldsymbol{h}_{k,-1} + \boldsymbol{h}_k b_k(p) + b_k(p+1)\boldsymbol{h}_{k,+1}] + \boldsymbol{w}(p) \tag{11.114}$$

[13]The following derivations also apply if this assumption is relaxed, but with minor modifications.

where $\boldsymbol{h}_k$, $\boldsymbol{h}_{k,-1}$, and $\boldsymbol{h}_{k,+1}$ are the discrete-time versions of the waveforms $h_k(t-pT_b)$, $h_k(t-(p-1)T_b)$, and $h_k(t-(p+1)T_b)$ windowed to the interval $\mathcal{I}_p$, respectively. Now define the following $2NM \times (N+1)M$ matrix:

$$\boldsymbol{C}_k = \begin{bmatrix} \beta_{0,k}\boldsymbol{I}_M & \boldsymbol{O}_M & \boldsymbol{O}_M & \dots & \boldsymbol{O}_M \\ \beta_{1,k}\boldsymbol{I}_M & \beta_{0,k}\boldsymbol{I}_M & \boldsymbol{O}_M & \dots & \boldsymbol{O}_M \\ v\vdots & \vdots & \dots & \dots & \vdots \\ \beta_{N-1,k}\boldsymbol{I}_M & \beta_{N-2,k}\boldsymbol{I}_M & \dots & \dots & \boldsymbol{O}_M \\ \boldsymbol{O}_M & \dots & \dots & \dots & \beta_{0,k}\boldsymbol{I}_M \\ \vdots & \vdots & \vdots & \vdots & \vdots \\ \boldsymbol{O}_M & \boldsymbol{O}_M & \dots & \boldsymbol{O}_M & \beta_{N-1,k}\boldsymbol{I}_M \end{bmatrix} \tag{11.115}$$

Denoting by $\boldsymbol{U}_k$ and $\boldsymbol{L}_k$ the $NM \times (N+1)M$-dimensional upper and lower semiblocks partitioning the matrix $\boldsymbol{C}_k$, i.e.,

$$\boldsymbol{C}_k = \begin{bmatrix} \boldsymbol{U}_k \\ \boldsymbol{L}_k \end{bmatrix}$$

we also define the following $2NM \times (N+1)M$-dimensional matrices:

$$\boldsymbol{C}_{k,-1} = \begin{bmatrix} \boldsymbol{L}_k \\ \boldsymbol{O}_{NM,(N+1)M} \end{bmatrix} \quad \text{and} \quad \boldsymbol{C}_{k,+1} = \begin{bmatrix} \boldsymbol{O}_{NM,(N+1)M} \\ \boldsymbol{U}_k \end{bmatrix} \tag{11.116}$$

Based on Equations 11.113, 11.115, and 11.116, it can be shown that

$$\boldsymbol{h}_{k,-1} = \boldsymbol{C}_{k,-1}\boldsymbol{g}_k, \qquad \boldsymbol{h}_k = \boldsymbol{C}_k\boldsymbol{g}_k, \qquad \text{and} \qquad \boldsymbol{h}_{k,+1} = \boldsymbol{C}_{k,+1}\boldsymbol{g}_k \tag{11.117}$$

thus implying that the vector $\boldsymbol{r}(p)$ in Equation 11.114 can also be expressed as

$$\boldsymbol{r}(p) = \sum_{k=0}^{K-1} [b_k(p-1)\boldsymbol{C}_{k,-1} + b_k(p)\boldsymbol{C}_k + b_k(p+1)\boldsymbol{C}_{k,+1}]\boldsymbol{g}_k + \boldsymbol{w}(p) \tag{11.118}$$

It is worth pointing out that Equation 11.117 is quite crucial for the design of any channel estimation and data detection algorithm. Indeed, it is seen from Equation 11.117 that in a system with no knowledge of the timing or channel, the discrete-time versions of the signatures can be deemed as the product of known matrices, containing shifted versions of the spreading codes, and an unknown vector, carrying information on all the unknown quantities. Notice also that the fact that the propagation delays are unknown is reflected by the redundancy needed in the signal representation. Otherwise stated, the receiver knows that the signature modulated by $b_k(p)$ is somewhere in the interval $[0, 2T_b]$, which is thus to be spanned entirely, whereas the availability of even incomplete information as to the channel state (i.e., a rough estimate of the users' delays) could allow spanning a reduced interval and would eventually lead to a vector $\boldsymbol{g}_k$ with a smaller number of entries. Finally, we note that upon partitioning the vector $\boldsymbol{r}(p)$ into two NM-dimensional vectors, i.e.,

$$\boldsymbol{r}(p) = \begin{bmatrix} \boldsymbol{r}'(p) \\ \boldsymbol{r}'(p+1) \end{bmatrix} \tag{11.119}$$

$\boldsymbol{r}'(p)$ is obviously the discrete-time observable windowed to $[pT_b, (p+1)T_b]$ and is expressed as

$$\boldsymbol{r}'(p) = \sum_{k=0}^{K-1} [b_k(p-1)\boldsymbol{L}_k + b_k(p)\boldsymbol{U}_k]\, \boldsymbol{g}_k + \boldsymbol{w}'(p) \tag{11.120}$$

11.4.2 Available Strategies to Cope with Missing CSI

Based on the signal model in either Equation 11.118 or Equation 11.120, the problem now arises to detect the information symbols with no knowledge of the channel vectors $\boldsymbol{g}_k$. The detection strategies that have been conceived to solve this problem mainly fall into two general categories. In the first category, the general approach consists of first estimating the unknown vectors $\boldsymbol{g}_k$ and then, once these vectors have been estimated, to apply conventional sliding-window multiuser receivers to detect the information symbols. The second category instead comprises algorithms that try to directly decode the information symbols without performing an explicit estimate of the channel (see, e.g., [10]). Additionally, the processing rules may be divided into trained procedures, which are based on the transmission of a known training sequence, and into blind procedures, which are able to perform data detection with no need for a known training sequence. Among these procedures, moreover, an important area is then represented by the so-called code-aided techniques, which try to estimate the information symbols of only one user of interest, based on the knowledge of the spreading code of that user only (see, e.g., [3, 15, 19]). Code-aided techniques, which can in turn be either trained or blind, are thus of interest in the downlink of cellular wireless networks, since they do not require knowledge of the spreading codes of the interfering users.

It is rather obvious that describing the vast bulk of estimation/detection strategies available in the literature would require much more space than this chapter allows. Accordingly, in the following we describe a few fundamental techniques that are representative of the illustrated categories. We start by first discussing a trained channel estimation algorithm, and then we focus on methods for direct data detection without channel estimation.

11.4.3 Channel Estimation Based on the Least Squares Criterion

Consider the received signal model in Equation 11.120. Letting

$$\overline{\boldsymbol{C}}_k = [\boldsymbol{L}_k\, \boldsymbol{U}_k], \qquad NM \times 2(N+1)M \tag{11.121}$$

$$\boldsymbol{B}_k(p) = \begin{bmatrix} b_k(p-1)\boldsymbol{I}_{(N+1)M} \\ b_k(p)\boldsymbol{I}_{(N+1)M} \end{bmatrix} \tag{11.122}$$

$$\overline{\boldsymbol{C}} = [\overline{\boldsymbol{C}}_0, \ldots, \overline{\boldsymbol{C}}_{K-1}], \qquad NM \times 2K(N+1)M \tag{11.123}$$

$$\boldsymbol{B}(p) = \text{Diag}\,(\boldsymbol{B}_0(p), \ldots, \boldsymbol{B}_{K-1}(p)), \qquad 2K(N+1)M \times K(N+1)M \tag{11.124}$$

and

$$\widetilde{\boldsymbol{g}} = \begin{bmatrix} \boldsymbol{g}_0 \\ \vdots \\ \boldsymbol{g}_{K-1} \end{bmatrix} \tag{11.125}$$

Equation 11.120 can also be written as

$$\boldsymbol{r}'(p) = \sum_{k=0}^{K-1} \overline{\boldsymbol{C}}_k \boldsymbol{B}_k(p)\boldsymbol{g}_k + \boldsymbol{w}'(p) \tag{11.126}$$

and as

$$\boldsymbol{r}'(p) = \overline{\boldsymbol{C}\boldsymbol{B}}(p)\widetilde{\boldsymbol{g}} + \boldsymbol{w}'(p) \tag{11.127}$$

Assume now that the receiver has knowledge of the information symbols $\boldsymbol{b}(0), \boldsymbol{b}(1), \ldots, \boldsymbol{b}(N_T)$, with N_T the length (expressed in symbol intervals) of the training sequence. Based on the observation of $\boldsymbol{r}'(1), \ldots, \boldsymbol{r}'(N_T)$, an estimate of the channel vector $\widetilde{\boldsymbol{g}}$, say $\widehat{\widetilde{\boldsymbol{g}}}$, can be obtained solving the following least squares problem:

$$\widehat{\widetilde{\boldsymbol{g}}} = \arg\min_{\widetilde{\boldsymbol{g}}} \sum_{p=1}^{N_T} \|\boldsymbol{r}'(p) - \overline{\boldsymbol{C}}\,\overline{\boldsymbol{B}}(p)\widetilde{\boldsymbol{g}}\|^2 \tag{11.128}$$

The solution to the above problem can be found through standard differentiation techniques and is expressed as

$$\widehat{\widetilde{\boldsymbol{g}}} = \left[\sum_{p=1}^{N_T} \overline{\boldsymbol{B}}^H(p)\overline{\boldsymbol{C}}^H\overline{\boldsymbol{C}\boldsymbol{B}}(p)\right]^{-1}\left[\sum_{p=1}^{N_T} \overline{\boldsymbol{B}}^H(p)\overline{\boldsymbol{C}}^H\boldsymbol{r}'(p)\right] \tag{11.129}$$

Note that, given the signal model in Equation 11.127, it can be shown that the least squares estimate (Equation 11.129) coincides with the ML estimate of the vector $\widetilde{\boldsymbol{g}}$; moreover, this estimate is also unbiased and asymptotically (for large N_T) consistent, and given the linear relationship between the noiseless data and the quantity to be estimated, it coincides with the minimum variance unbiased estimator (MVUE) of the vector $\widetilde{\boldsymbol{g}}$.

Equation 11.129 provides an estimation procedure for the vector $\widetilde{\boldsymbol{g}}$, which is a composite channel vector containing the channel vectors for all the active users. Moreover, Equation 11.129 exploits knowledge of training bits and spreading codes for all the active users. In view of this considerations, Equation 11.129 represents a *centralized* channel estimation procedure. On the other hand, it is also of interest to develop *decentralized* code-aided channel estimation procedures that can be implemented in mobile receivers on the downlink of a wireless network with no prior knowledge of the interfering signals. Let us thus assume that the h-th user is the one of interest and that the receiver has knowledge of the training symbols $\boldsymbol{b}_h(0), \boldsymbol{b}_h(1), \ldots, \boldsymbol{b}_h(N_T)$. Based on the observation of $\boldsymbol{r}'(1), \ldots, \boldsymbol{r}'(N_T)$, an estimate of the channel vector $\boldsymbol{g}_h$ of the h-th user, say $\widehat{\boldsymbol{g}}_h$, can be obtained by solving the following least squares problem:

$$\widehat{\boldsymbol{g}}_h = \arg\min_{\boldsymbol{g}_h} \sum_{p=1}^{N_T} \|\boldsymbol{r}'(p) - \overline{\boldsymbol{C}}_h\boldsymbol{B}_h(p)\boldsymbol{g}_h\|^2 \tag{11.130}$$

whose solution is expressed as

$$\widehat{\boldsymbol{g}}_h = \left[\sum_{p=1}^{N_T} \boldsymbol{B}_h^H(p)\overline{\boldsymbol{C}}_h^H\overline{\boldsymbol{C}}_h\boldsymbol{B}_h(p)\right]^{-1}\left[\sum_{p=1}^{N_T} \boldsymbol{B}_h^H(p)\overline{\boldsymbol{C}}_h^H\boldsymbol{r}'(p)\right] \tag{11.131}$$

Note that the estimation rule in Equation 11.131 neglects the MAI disturbance, whereby it is expected to perform worse than the centralized estimation rule in Equation 11.129. This behavior, on the other hand, can be also justified by noting that the decentralized estimation rule relies on minor prior information, compared with the centralized channel estimation procedure.

Figure 11.7 reports the normalized correlation coefficient between the actual channel vector and its estimated value vs. the bit energy contrast $E_b/\mathcal{N}_0$. A system with processing gain $N = 15$, oversampling

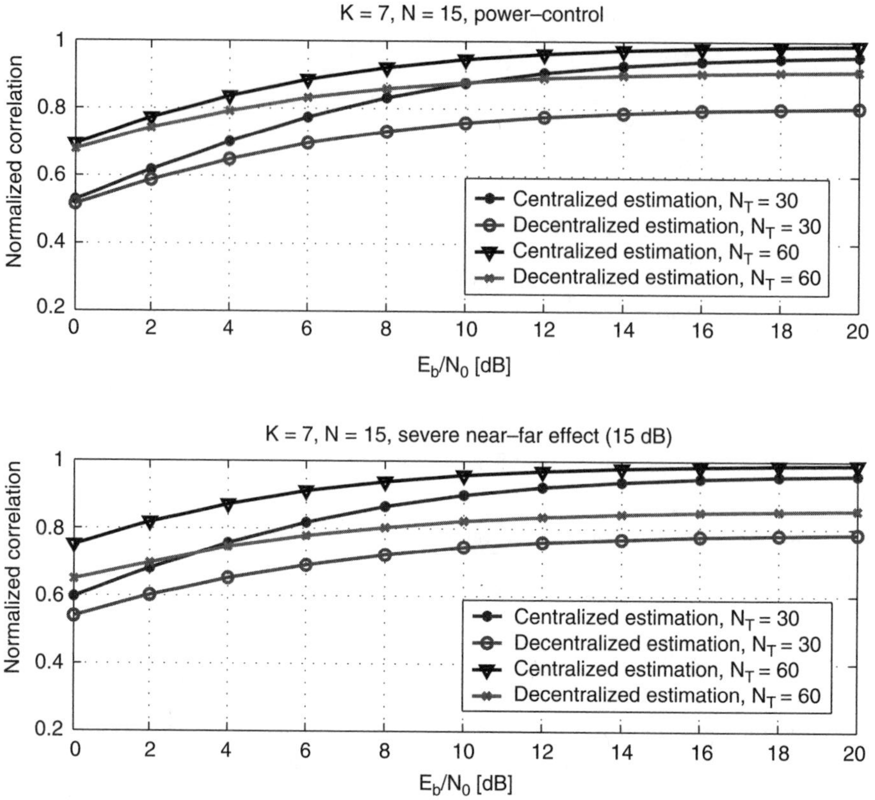

FIGURE 11.7 Normalized correlation coefficient vs. $E_b/\mathcal{N}_0$ for the trained least squares channel estimation procedures. The lower plot refers to the case that the CDMA signals may have powers unbalanced up to 15 dB.

factor $M = 2$, and $K = 7$ users has been considered. The results refer to a three-path propagation channel and are averaged over 2500 independent channel realizations. Both the situations of average power control and heavy near–far scenario with power disparities up to 15 dB are reported for two different values of the training length N_T. Results show that performance depends on the energy contrast and on the length of the training sequence. As expected, the centralized estimation procedure outperforms the decentralized one, and it is also seen that the lack of power control has a limited effect on the performance of the estimation procedures.

11.4.4 Trained Adaptive Code-Aided Symbol Detection

In the previous section channel estimation procedures were illustrated. As already discussed, once the channel vectors $\boldsymbol{g}_k$ have been obtained, they can be used in Equation 11.117 so that, through the signal model in Equation 11.114, the problem at hand can be conducted to that of sliding-window one-shot multiuser detection with known (i.e., estimated) channel. An alternative approach is to instead directly decode the information symbols with no explicit channel estimation. In this section we illustrate an adaptive code-aided symbol detection procedure that is based on the transmission of known training symbols. To this end, consider the signal model in Equation 11.118 and assume to be interested in detecting the bit $b_k(n)$ based on the linear processing rule

$$b_k(n) = \text{sgn}\left\{\Re\left[\boldsymbol{d}_k^H(n)\boldsymbol{r}(n)\right]\right\} \tag{11.132}$$

with $\boldsymbol{d}_k(n)$ a $2NM$-dimensional vector, to be designed according to a suitable optimization criterion. As an example, an effective strategy is to adopt an exponentially windowed time-averaged minimum square error criterion. More precisely, based on the observation of the data $\boldsymbol{r}(0), \ldots, \boldsymbol{r}(n)$, and on the knowledge of the training bits $b_k(0), \ldots, b_k(n)$, the vector $\boldsymbol{d}_k(n)$ can be sought as the solution to the problem

$$\boldsymbol{d}_k(n) = \arg\min_{\boldsymbol{d}_k(n)} \sum_{i=0}^{n} \lambda^{n-i} \left| \boldsymbol{d}_k^H(n)\boldsymbol{r}(i) - b_k(i) \right|^2 \tag{11.133}$$

In the above expression, λ is a real constant slightly smaller than unity and is termed the *forgetting factor*. Its purpose is to provide a tracking capability to the solution by ensuring that in the learning process that leads to the determination of $\boldsymbol{d}_k(n)$ the most recent observations are weighted more than past observations, so that the receiver can adapt to changes in the reference scenario due to, e.g., users leaving or entering the communication scene or variations in the signals' strength caused by the mobility of the users. Upon letting

$$\boldsymbol{R}(n) = \sum_{i=0}^{n} \lambda^{n-i} \boldsymbol{r}(i)\boldsymbol{r}^H(i) \tag{11.134}$$

and

$$\boldsymbol{q}_k(n) = \sum_{i=0}^{n} \lambda^{n-i} \boldsymbol{r}(i) b_k(i) \tag{11.135}$$

it is readily seen that the solution to Equation 11.133 is written as

$$\boldsymbol{d}_k(n) = \boldsymbol{R}^{-1}(n)\boldsymbol{q}_k(n) \tag{11.136}$$

It is interesting to note that based on the recursive least squares (RLS) algorithm, the batch solution (Equation 11.136) can be given a recursive formulation wherein the vector $\boldsymbol{d}_k(n)$ can be updated with a complexity that is quadratic[14] in the product $2NM$ and based on $\boldsymbol{d}_k(n-1)$ and $\boldsymbol{r}(n)$. Otherwise stated, as soon as new observations become available, it is possible to recursively update the vector $\boldsymbol{d}_k(\cdot)$ to be used in the decision rule in Equation 11.132. To elaborate, note that since

$$\boldsymbol{R}(n) = \lambda \boldsymbol{R}(n-1) + \boldsymbol{r}(n)\boldsymbol{r}^H(n) \tag{11.137}$$

a straightforward application of the matrix inversion lemma leads to

$$\boldsymbol{R}^{-1}(n) = \frac{1}{\lambda}\left[\boldsymbol{R}^{-1}(n-1) - \frac{\boldsymbol{R}^{-1}(n-1)\boldsymbol{r}(n)\boldsymbol{r}^H(n)\boldsymbol{R}^{-1}(n-1)}{\lambda + \boldsymbol{r}^H(n)\boldsymbol{R}^{-1}(n-1)\boldsymbol{r}(n)}\right] \tag{11.138}$$

Accordingly, following [6, Chapter 13], it can be shown that the vector $\boldsymbol{d}_k(n)$ can be obtained based on the following recursion:

$$\begin{aligned}
\boldsymbol{k}(n) &= \frac{\boldsymbol{R}^{-1}(n-1)\boldsymbol{r}(n)}{\lambda + \boldsymbol{r}^H(n)\boldsymbol{R}^{-1}(n-1)\boldsymbol{r}(n)} \\
\boldsymbol{R}^{-1}(n) &= \frac{1}{\lambda}[\boldsymbol{R}^{-1}(n-1) - \boldsymbol{k}(n)\boldsymbol{r}^H(n)\boldsymbol{R}^{-1}(n-1)] \\
\epsilon(n) &= b_k(n) - \boldsymbol{d}_k^H(n-1)\boldsymbol{r}(n) \\
\boldsymbol{d}_k(n) &= \boldsymbol{d}_k(n-1) + \epsilon^*(n)\boldsymbol{k}(n)
\end{aligned} \tag{11.139}$$

[14]Note that due to the batch matrix inversion, direct computation of Equation 11.136 entails a complexity that is cubic in the product $2NM$.

The last line in Equation 11.139 represents the update equation for the detection vector $\boldsymbol{d}_k(\cdot)$. Note that this equation, in turn, depends on the error vector $\epsilon(n)$, which can be built based on the knowledge of the training symbol $b_k(n)$. Typically, at the beginning of each data frame, training bits are placed so that adaptive algorithms like the one in Equation 11.139 can be run and lead to an estimate of the vector $\boldsymbol{d}_k(\cdot)$. Once the training phase is over, real data detection takes place and the adaptive algorithm switches to the so-called decision-directed mode, where (possibly erroneous) detected bits replace the known training symbols, and the error vector in the third line of Equation 11.139 is computed according to

$$\epsilon(n) = \text{sgn}\left\{\Re\left[\boldsymbol{d}_k^H(n-1)\boldsymbol{r}(n)\right]\right\} - \boldsymbol{d}_k^H(n-1)\boldsymbol{r}(n) \tag{11.140}$$

Obviously, in the decision-directed mode the adaptation algorithm is sensitive to errors in the symbol estimates. However, if the training phase has been sufficiently long and the signal-to-noise ratio is not very low, the switch to the decision-directed mode has a negligible effect on the system performance, unless the considered system is highly dynamic and the interference background is subject to rapid changes.

Figure 11.8 shows the mean square error $|\boldsymbol{d}_k^H(n)\boldsymbol{r}(n) - b_k(n)|^2$ of the RLS algorithm vs. the symbol interval number for a system with processing gain $N = 15$, oversampling factor $M = 2$, and $K = 7$ users. Two different values of $E_b/\mathcal{N}_0$ and of the ISR are considered. The reported results are an average over 1000 runs; at each run, the user delays and channel coefficients are independently updated. For the first 100 symbol intervals the algorithm is trained, and then it switches to decision-directed mode update (Equation 11.140). Figure 11.9 refers to the same scenario as Figure 11.8 and shows the system error probability vs. $E_b/\mathcal{N}_0$ at several symbol intervals, also in comparison with that achieved by the ideal MMSE receiver, which requires knowledge of the CSI, propagation delays, and signatures for all active users. It is seen from the figure that the RLS algorithm converges fast and that it also exhibits satisfactory

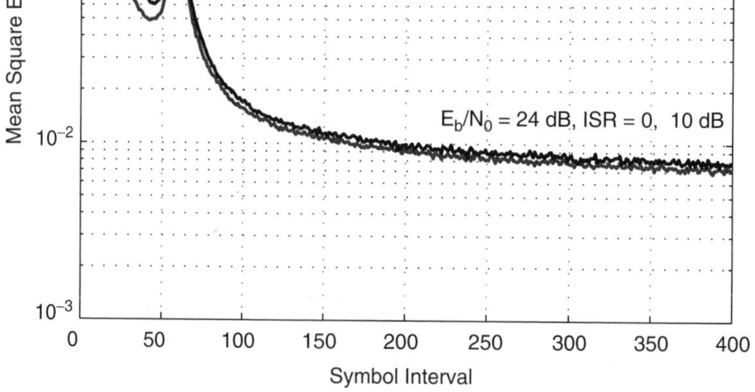

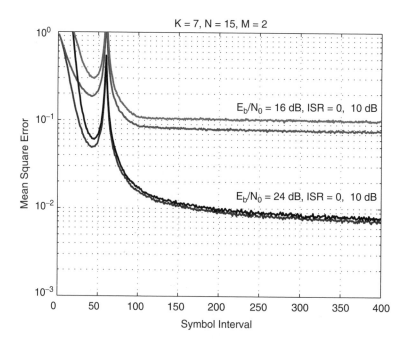

FIGURE 11.8 Mean square error vs. symbol interval for the trained code-aided RLS implementation of the MMSE receiver. After the first 100 symbol intervals the algorithm switches to the decision-directed mode and is driven by the symbol estimates. Note that the algorithm performance is affected very slightly by the near–far problem.

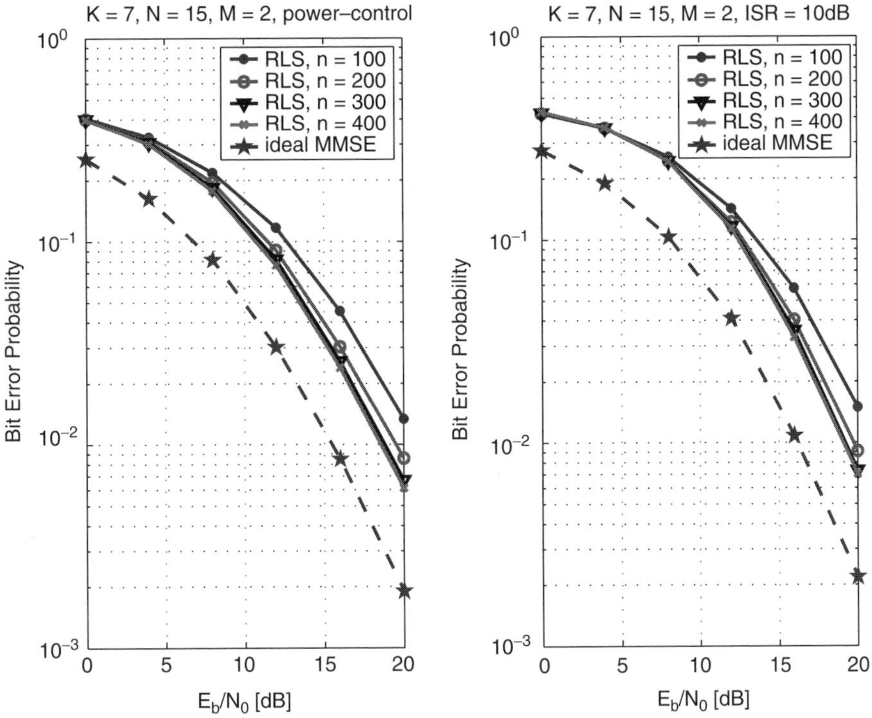

FIGURE 11.9 Error probability vs. $E_b/\mathcal{N}_0$ for the trained code-aided RLS implementation of the MMSE receiver. For comparison purposes, the performance of the ideal MMSE receiver, which assumes perfect knowledge of the CSI and the signatures for all active users, is also reported.

performance in the decision-directed mode. Moreover, the algorithm performance is slightly affected by the ISR, thus confirming that the RLS algorithm is not very sensitive to the near–far effect.

11.4.5 Code-Aided Joint Blind Multiuser Detection and Equalization

After having illustrated the trained RLS algorithm for code-aided data detection in systems with unknown CSI, the last part of this review is finally devoted to the discussion of some techniques for blind code-aided data detection in multiuser CDMA systems.

Given the fact that the CSI is unknown and no training sequence is available, the modulation format cannot be a coherent BPSK. We thus assume differential encoding and detection, implying that the information of the p-th signaling interval is contained in the bits $b_k(p)$ and $b_k(p-1)$. Thus, assuming that user 0 is the one of interest, the decision rule of a linear multiuser detector is

$$\widehat{d}_0(p) = \text{sgn}[\Re\{(\boldsymbol{m}^H\boldsymbol{r}(p))(\boldsymbol{m}^H\boldsymbol{r}(p-1))^*\}] \tag{11.141}$$

where $\boldsymbol{r}(p)$ is the $2NM$-dimensional data vector in Equation 11.118, $\widehat{d}_0(p)$ is the incremental phase between $b_0(p)$ and $b_0(p-1)$, and $\boldsymbol{m}$ is a $2NM$-dimensional vector to be suitably designed. In the following sections, we outline three different approaches to design this vector, based on the knowledge of the spreading code of the user of interest only.

11.4.6 Subspace-Based Blind MMSE Detection

To begin with, we note that given the channel vector realizations $g_0, \ldots, g_{K-1}$, the conditional covariance matrix of the observables in Equation 11.118 is expressed as[15]

$$R_{rr} = E\{r(p)r^H(p)\} = \sum_{k=0}^{K-1}[C_{k,-1} + C_k + C_{k,+1}]g_k g_k^H [C_{k,-1} + C_k + C_{k,+1}]^H + 2\mathcal{N}_0 I_{2NM} \quad (11.142)$$

and the linear MMSE detector for the zeroth user is written as

$$m = R_{rr}^{-1} h_0 = R_{rr}^{-1} C_0 g_0 \quad (11.143)$$

Given Equation 11.143, it is understood that the MMSE receiver can be implemented in a blind fashion if the receiver can estimate the matrix R_{rr} and the channel vector g_0. As to the former, the sample covariance matrix

$$\widehat{R}_{rr} = \frac{1}{2P+1} \sum_{p=-P}^{P} r(p)r^H(p) \quad (11.144)$$

can be adopted in Equation 11.143, since it can be shown that under mild conditions the estimate in Equation 11.144 converges almost surely to the conditional covariance matrix Equation 11.142. With regard to the vector g_0, we first note that if $3K < 2NM$ and the vectors in $\{h_{k,-1}, h_k, h_{k,+1}\}_{k=0}^{K-1}$ are linearly independent (this is a condition that is usually met for carefully chosen spreading codes and when the system is not heavily loaded), the matrix R_{rr} has $3K$ eigenvalues larger than $2\mathcal{N}_0$ and $2NM - 3K$ eigenvalues equal to $2\mathcal{N}_0$. The eigenvectors corresponding to the $3K$ largest eigenvalues define the signal subspace, and the eigenvectors associated with the remaining eigenvalues define the noise subspace. Accordingly, the following eigendecomposition can be shown to hold

$$R_{rr} = U_s \Lambda_s U_s^H + U_n \Lambda_n U_n^H \quad (11.145)$$

where Λ_s is a diagonal matrix containing the $3K$ dominant eigenvalues, Λ_n is a diagonal matrix containing the $2NM - 3K$ eigenvalues equal to $2\mathcal{N}_0$, and U_s and U_n contain on their columns the eigenvectors defining the signal and noise subspaces, respectively. Now, since the vector $h_0 = C_0 g_0$ is in the signal subspace, the norm of the vector $U_n^H C_0 g_0$ is zero, and this condition can be exploited to estimate the channel vector g_0. In practice, however, since the receiver has only an approximate knowledge of the noise subspace eigenvectors U_n (indeed, an estimate of U_n, say $\widehat{U}_n$, is obtained by collecting the $2NM - 3K$ smallest eigenvectors of the sample covariance $\widehat{R}_{rr}$), the vector g_0 can be found by solving the following constrained optimization problem[16] [23]:

$$\widehat{g}_0 = \arg\min_{g_0} \left\|\widehat{U}_n^H C_0 g_0\right\|^2 \qquad \text{subject to} \|\widehat{g}_0\|^2 = 1 \quad (11.146)$$

The above problem can be shown to admit a unique solution only if the dimensionality of the intersection between the column span of C_0 and the column span of U_s is one, i.e., if there is only one linear combination of the columns of C_0 that lies in the signal subspace. Applying standard Lagrangian techniques, it is shown that the solution to Equation 11.146 coincides with the eigenvector corresponding to the smallest eigenvalue of the matrix $C_0^H \widehat{U}_n \widehat{U}_n^H C_0^H$. Note that this solution is inherently affected by a phase ambiguity that is removed by the differential decoding rule in Equation 11.141.

[15] For ease of notation, we omit here and in what follows an explicit indication of the fact that the statistical expectation is conditioned on the channel vector realizations.

[16] The constraint is aimed at ruling out the trivial null solution.

11.4.7 Minimum Variance Blind Detection

Another effective design criterion for the vector $\boldsymbol{m}$ is to adopt a minimum variance approach [18]. The vector $\boldsymbol{m}$ to be adopted in the decision rule in Equation 11.141 is required to minimize the output variance subject to the constraint that the useful signal contribution is kept constant. Otherwise stated, the following optimization problem is considered:

$$\boldsymbol{m} = \arg\min_{\boldsymbol{x}}[\boldsymbol{x}^H \boldsymbol{R}_{rr} \boldsymbol{x}] \qquad \text{subject to } \boldsymbol{m}^H \boldsymbol{C}_0 = \boldsymbol{x}_0 \tag{11.147}$$

with $\boldsymbol{x}_0$ a vector that is specified below. Note that the constraint ensures that the useful signal contribution at the output of the filter $\boldsymbol{m}$ is $\boldsymbol{m}^H \boldsymbol{h}_0 = \boldsymbol{x}_0^H \boldsymbol{g}_0$, so that the minimization in Equation 11.147 does not affect the useful signal energy. The solution can be shown to be written as

$$\boldsymbol{m} = \boldsymbol{R}_{rr}^{-1} \boldsymbol{C}_0 \left(\boldsymbol{C}_0^H \boldsymbol{R}_{rr}^{-1} \boldsymbol{C}_0\right)^{-1} \boldsymbol{x}_0 \tag{11.148}$$

Note that, given Equation 11.148, the mean output energy (MOE) is expressed as

$$\text{MOE} = \boldsymbol{m}^H \boldsymbol{R}_{rr} \boldsymbol{m} = \boldsymbol{x}_0^H \left(\boldsymbol{C}_0^H \boldsymbol{R}_{rr}^{-1} \boldsymbol{C}_0\right)^{-1} \boldsymbol{x}_0 \tag{11.149}$$

Accordingly, given the fact that the channel vector $\boldsymbol{g}_0$ is unknown, a reasonable choice to optimize $\boldsymbol{x}_0$ is to maximize the output signal strength after the interference has been suppressed; otherwise stated, the vector $\boldsymbol{x}_0$ maximizes the MOE in Equation 11.149 subject to the constraint $\|\boldsymbol{x}_0\|^2 = 1$, i.e.,

$$\boldsymbol{x}_0 = \arg\max_{\boldsymbol{x}_0} \frac{\text{MOE}}{\boldsymbol{x}_0^H \boldsymbol{x}_0} \tag{11.150}$$

The above cost function is a Rayleigh quotient and the solution vector $\boldsymbol{x}_0$ to be finally substituted in Equation 11.148 is the eigenvector corresponding to the maximum eigenvalue of $(\boldsymbol{C}_0^H \boldsymbol{R}_{rr}^{-1} \boldsymbol{C}_0)^{-1}$. Note that in this case the vector $\boldsymbol{m}$ also has an ambiguous phase that is removed by the differential detection rule.

11.4.8 Two-Stage Blind Detection

Another possible strategy for designing the vector $\boldsymbol{m}$ in a blind fashion is to resort to a two-stage receiver, i.e., to deem the vector $\boldsymbol{m}$ as the following product: $\boldsymbol{m} = \boldsymbol{D}\boldsymbol{e}$, with $\boldsymbol{D}$ a rectangular matrix and $\boldsymbol{e}$ a column vector [4]. The first stage of the receiver is aimed at interference reduction, while the latter stage is aimed at error probability optimization. Since the useful signal signature at the output of the first stage is $\boldsymbol{D}^H \boldsymbol{C}_0 \boldsymbol{g}_0$, a sufficient condition to ensure that the useful signal is not nullified for any nonzero realization of the channel vector $\boldsymbol{g}_0$ is to require that the matrix $\boldsymbol{D}^H \boldsymbol{C}_0$ has full-column rank. Accordingly, the matrix $\boldsymbol{D}$ is $2NM \times (N+1)M$-dimensional, and it is sought as the solution to the following constrained optimization problem:

$$\boldsymbol{D} = \arg\min_{\boldsymbol{D}} E\{\|\boldsymbol{D}^H \boldsymbol{r}(p)\|^2\} \qquad \text{subject to } \det(\boldsymbol{D}^H \boldsymbol{C}_0) \neq 0 \tag{11.151}$$

and is expressed as [4]

$$\boldsymbol{D} = \left(\boldsymbol{R}_{rr} + \boldsymbol{C}_0 \boldsymbol{C}_0^H\right)^{-1} \boldsymbol{C}_0 \tag{11.152}$$

At the output of the filter $\boldsymbol{D}$, the interference contribution is strongly attenuated with respect to the useful signal, so that the output signal, $\boldsymbol{y}(p)$, can be written approximately as

$$\boldsymbol{y}(p) = b_0(p) \boldsymbol{D}^H \boldsymbol{C}_0 \boldsymbol{g}_0 + \boldsymbol{D}^H \boldsymbol{w}(p) \tag{11.153}$$

To design the second stage of the receiver, we choose $\boldsymbol{e}$ to maximize the signal-to-noise ratio, i.e., as the cascade of a noise-whitening filter and a filter matched to the resulting useful signature. As for the whitening transformation, it is easily determined by noticing that the covariance matrix of the noise is now $2\mathcal{N}_0 \boldsymbol{D}^H \boldsymbol{D}$. Since this matrix is positive-definite and Hermitian, the following Cholesky factorization applies:

$$\boldsymbol{D}^H \boldsymbol{D} = \boldsymbol{L}\boldsymbol{L}^H \tag{11.154}$$

with $\boldsymbol{L}$ a nonsingular lower-triangular matrix. The whitened observables are now given by

$$\boldsymbol{y}_w(p) = b_0(p)\boldsymbol{L}^{-1}\boldsymbol{D}^H \boldsymbol{C}_0 \boldsymbol{g}_0 + \boldsymbol{L}^{-1}\boldsymbol{D}^H \boldsymbol{w}(p) \tag{11.155}$$

Were $\boldsymbol{g}_0$ a known vector, the optimum processing at this point would be an ordinary matched filter. But since $\boldsymbol{g}_0$ is unknown to the receiver, a further step is necessary in the procedure in order to estimate the desired projection direction. To be more definite, consider the covariance matrix of the whitened observables:

$$\boldsymbol{R}_{\boldsymbol{y}_w \boldsymbol{y}_w} = E\left\{\boldsymbol{y}_w(p)\boldsymbol{y}_w^H(p)\right\} = \boldsymbol{L}^{-1}\boldsymbol{D}^H \boldsymbol{C}_0 \boldsymbol{g}_0 \boldsymbol{g}_0^H \boldsymbol{C}_0^H \boldsymbol{D}(\boldsymbol{L}^{-1})^H + 2\mathcal{N}_0 \boldsymbol{I}_{(N+1)M} \tag{11.156}$$

The covariance matrix in Equation 11.156 is thus the sum of a full-rank identity matrix and a unit-rank matrix, the latter admitting $\boldsymbol{L}^{-1}\boldsymbol{D}^H \boldsymbol{C}_0 \boldsymbol{g}_0$ as its unique dominant eigenvector. Consequently, $(N+1)M-1$ eigenvalues of this matrix are coincident, while the largest eigenvalue corresponds to an eigenvector that is parallel to $\boldsymbol{L}^{-1}\boldsymbol{D}^H \boldsymbol{C}_0 \boldsymbol{g}_0$. Thus the matched filter for the detection problem (Equation 11.155) is given by this eigenvector. Since the covariance matrix (Equation 11.156) is not actually known to the receiver, in practice it can be replaced by its sample estimate; i.e., we consider the matrix

$$\widehat{\boldsymbol{R}}_{\boldsymbol{y}_w \boldsymbol{y}_w} = \frac{1}{2P+1}\sum_{p=-P}^{P} \boldsymbol{y}_w(p)\boldsymbol{y}_w^H(p) \tag{11.157}$$

The estimated matched filter for the detection problem (Equation 11.155) is given by the principal eigenvector, which we denote by $\widehat{\boldsymbol{v}}$, of the matrix in Equation 11.157.

Summing up, the detector can be based upon the following procedure:

1. Observe the vector $\boldsymbol{r}(p)$.
2. Evaluate $\boldsymbol{y}(p) = \boldsymbol{D}^H \boldsymbol{r}(p)$.
3. Perform the Cholesky factorization $\boldsymbol{D}^H \boldsymbol{D} = \boldsymbol{L}\boldsymbol{L}^H$.
4. Transform the vector $\boldsymbol{y}(p)$ into the whitened vector $\boldsymbol{y}_w(p) = \boldsymbol{L}^{-1}\boldsymbol{y}(p)$.
5. Evaluate the sample covariance matrix $\widehat{\boldsymbol{R}}_{\boldsymbol{y}_w \boldsymbol{y}_w}$ of the vector $\boldsymbol{y}_w(p)$ through Equation 11.157.
6. Determine the eigenvector $\widehat{\boldsymbol{v}}$ corresponding to the maximum eigenvalue of the matrix $\widehat{\boldsymbol{R}}_{\boldsymbol{y}_w \boldsymbol{y}_w}$.
7. The vector $\boldsymbol{m}$ to be used in the differential decision rule in Equation 11.141 is given by

$$\boldsymbol{m} = \boldsymbol{D}(\boldsymbol{L}^{-1})^H \widehat{\boldsymbol{v}} \tag{11.158}$$

Note, finally, that in this case the solution is also affected by a phase ambiguity, which is removed by virtue of the differential symbol decoding.

11.4.9 Performance Results

Some performance results on the error probability of the three above code-aided blind detection schemes are reported in Figure 11.10 and Figure 11.11. A system with average power control (i.e., ISR = 0 dB), processing gain $N = 15$, and oversampling factor $M = 2$ is considered. Figure 11.10 shows the system error probability vs. $E_b/\mathcal{N}_0$ for $K = 5$, while Figure 11.11 shows the system error probability vs. the number of users at $E_b/\mathcal{N}_0 = 22$ dB and for $M = 1$. In Figure 11.10 the performance of the nonblind MMSE receiver is also reported for the sake of comparison. Results show that the subspace-based MMSE receiver achieves the best performance, followed by the two-stage receiver. Interestingly, both receivers

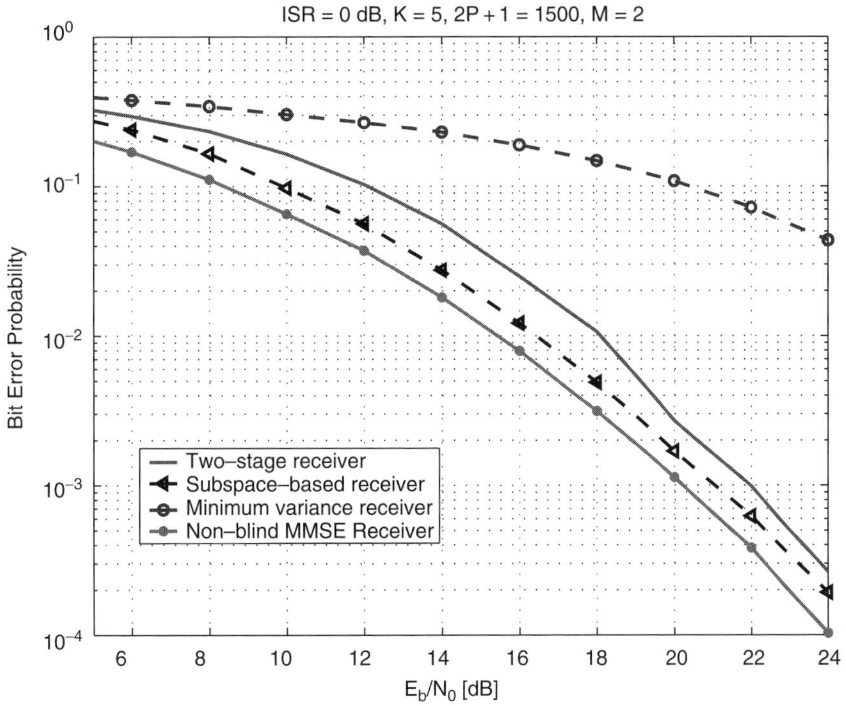

FIGURE 11.10 Error probability vs. $E_b/\mathcal{N}_0$ for the blind code-aided receivers and for the ideal nonblind MMSE receiver.

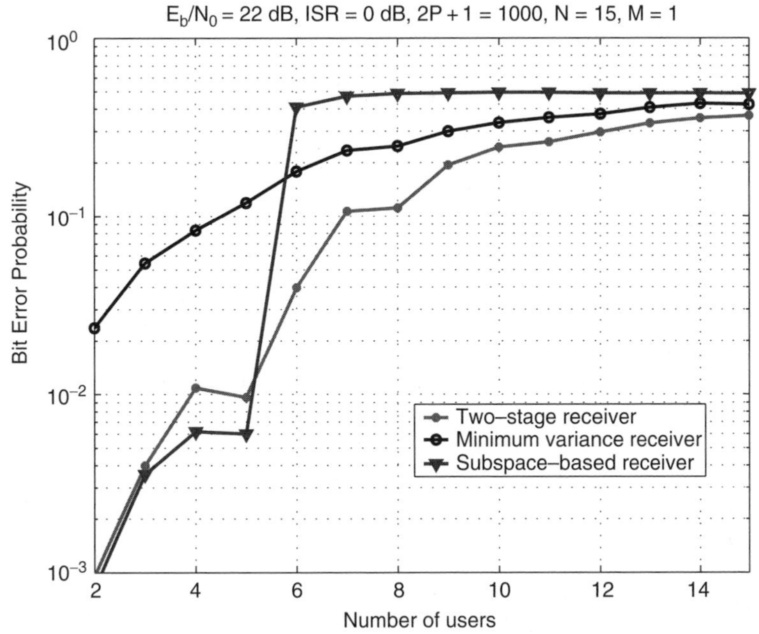

FIGURE 11.11 Error probability vs. the number of users for the blind code-aided receivers at $E_b/\mathcal{N}_0 = 22$ dB.

exhibit, at an error probability of 10^{-3}, a loss of about 1 to 2 dB with respect to the ideal MMSE receiver. On the contrary, Figure 11.11 shows that the performance of the blind detectors suddenly degrades for increasing K. A suitable means to contrast this effect is to increase the oversampling factor M.

11.5 Conclusions

An overview of multiuser detection techniques for mobile fading channels has been given in this chapter. First, the case that the receiver has full CSI has been considered. In particular, linear and nonlinear receivers have been illustrated, also with regard to the case of sliding-window one-shot data detection. Then the focus of the chapter shifted toward the more challenging situation that the CSI is not available at the receiver. With regard to the issue of multiuser channel state estimation, algorithms based on the least squares criterion have been discussed. With regard to the issue of direct data detection without explicit channel estimation, code-aided detection techniques have also been reviewed for situations where either a known training sequence is available at the receiver or the receiver is to perform blind data detection. It is important to point out that this review is by no means exhaustive of the vast body of work on multiuser detection for fading channels that is available in the open literature. As discussed in the Introduction, multiuser detection is a fascinating and intriguing research topic that has been attracting the interest of many researchers for decades. Rather, the author's aim in writing this chapter has been to provide the fundamental techniques for multiuser detection in fading channels, with a special emphasis on slow frequency-selective fading channels, which are of primary interest in high-data-rate mobile communications. It is, however, anticipated that future years will witness further remarkable innovations in this exciting field.

References

[1] P.A. Bello, "Characterization of randomly time-variant linear channels," *IEEE Trans. Commun. Syst.*, 11, 360–393, 1963.

[2] E. Biglieri, J.G. Proakis and S. Shamai (Shitz), "Fading channels: information-theoretic and communications aspects," *IEEE Trans. Inf. Theory*, 44, 2619–2692, 1998.

[3] S. Buzzi, M. Lops, and H.V. Poor, "Code-aided interference suppression for DS/CDMA overlay systems," *Proc. IEEE*, 90, 394–435, 2002.

[4] S. Buzzi, M. Lops and H.V. Poor, "Blind adaptive joint multiuser detection and equalization in dispersive differentially-encoded CDMA channels," *IEEE Trans. Signal Process.*, 51, 1880–1893, 2003.

[5] S. Hara and R. Prasad, "Overview of multicarrier CDMA," *IEEE Commun. Mag.*, 35, 126–133, 1997.

[6] S. Haykin, *Adaptive Filter Theory*, 3rd ed., Prentice-Hall, Upper Saddle River, NJ, 1996.

[7] D. Young, *Iterative Solution of Large Linear Systems*, Academic Press, New York, 1971.

[8] M. Latva-aho, "Advanced Receivers for Wideband CDMA Systems," Ph.D. thesis, University of Oulu, Finland, 1998.

[9] M. Latva-aho and M. Juntti, "LMMSE detection for DS-CDMA systems in fading channels," *IEEE Trans. Commun.*, 48, 194–199, 2000.

[10] X. Li and H.H. Fan, "Direct blind multiuser detection for CDMA in multipath without channel estimation," *IEEE Trans. Signal Process.*, 49, 63–73, 2001.

[11] R. Lupas and S. Verdú, "Linear multiuser detectors for synchronous code-division multiple access channels," *IEEE Trans. Inf. Theory*, 35, 123–136, 1989.

[12] Telecommunications Industry Association, *Mobile Station-Base Station Compatibility Standard for Dual-Mode Wideband Spread Spectrum Cellular System*, TIA/EIA/IS-95, Telecommunications Industry Association, July 1993.

[13] D. Parsons, *The Mobile Radio Propagation Channel*, Pentech Press, London, 1992.

[14] H.V. Poor and S. Verdú, "Probability of error in MMSE multiuser detection," *IEEE Trans. Inf. Theory*, 43, 858–871, 1997.

[15] H.V. Poor and X. Wang, "Code-aided interference suppression for DS/CDMA communications. I. Interference suppression capability" and "Code-aided interference suppression for DS/CDMA

communications. II. Parallel blind adaptive implementations," *IEEE Trans. Commun.*, 45, 1101–1122, 1997.

[16] J.G. Proakis, *Digital Communications*, 3rd ed., McGraw-Hill, Singapore, 1995.

[17] P.H. Tan, L.K. Rasmussen and T. J. Lim, "Constrained maximum-likelihood detection in CDMA," *IEEE Trans. Commun.*, 49, 142–153, 2001.

[18] M.K. Tsatsanis and Z.(D.) Xu, "Performance analysis of minimum variance CDMA receivers," *IEEE Trans. Signal Process.*, 46, 3014–3022, 1998.

[19] J.K. Tugnait and T. Li, "Blind asynchronous multiuser CDMA receivers for ISI channels using code-aided CMA," *IEEE J. Selected Areas Commun.*, 19, 1520–1530, 2001.

[20] S. Verdú, "Optimum sequence detection of asynchronous multiple-access Communications," in *Proceedings of the 1983 IEEE International Symposium Information Theory*, St. Jovite, Canada, September 1983, p. 80.

[21] S. Verdú, "Minimum probability of error for asynchronous Gaussian multiple access channels," *IEEE Trans. Inf. Theory*, 32, 85–96, 1986.

[22] S. Verdú, *Multiuser Detection*, Cambridge University Press, Cambridge, U.K., 1998.

[23] X. Wang and H.V. Poor, "Blind equalization and multiuser detection in dispersive CDMA channels," *IEEE Trans. Commun.*, 46, 91–103, 1998.

[24] Z. Zvonar, "Multiuser detection in asynchronous CDMA frequency-selective fading channels," *Wireless Personal Commun.*, 2, 373–392, 1996.

[25] Z. Zvonar and D. Brady, "Optimum detection in asynchronous multiple-access multipath Rayleigh fading channels," in *Proceedings of the 26th Annual Conference on Information, Sciences and Systems*, Princeton University, Princeton, NJ, March 1992, pp. 826–831.

[26] Z. Zvonar and D. Brady, "Multiuser detection in single-path fading channels," *IEEE Trans. Commun.*, 42, 1729–1739, 1994.

[27] Z. Zvonar and D. Brady, "Suboptimal multiuser detector for frequency-selective Rayleigh fading synchronous CDMA channels," *IEEE Trans. Commun.*, 43, 154–157, 1995.

[28] Z. Zvonar and D. Brady, "Differentially coherent multiuser detection in asynchronous CDMA flat Rayleigh fading channels," *IEEE Trans. Commun.*, 43, 1252–1255, 1995.

[29] Z. Zvonar and D. Brady, "Linear multipath-decorrelating receivers for CDMA frequency-selective fading channel," *IEEE Trans. Commun.*, 44, 650–653, 1996.

V

MIMO Systems

12 Principles of MIMO-OFDM Wireless Systems *Helmut Bölcskei* **12**-1
Introduction • The Broadband MIMO Fading Channel • Capacity of Broadband MIMO-OFDM Systems • Space–Frequency Coded MIMO-OFDM • Impact of Propagation Parameters on Space–Frequency Codes

13 Space–Time Coding and Signal Processing for Broadband Wireless Communications *Naofal Al-Dhahir* **13**-1
Introduction • Broadband Wireless Channel Model • Information-Theoretic Considerations • Signal Transmission Issues • Summary and Future Challenges

14 Linear Precoding for MIMO Systems *Anna Scaglione, Atul Salhotra, and Azadeh Vosoughi* **14**-1
Introduction • Optimum Precoding • Performance Analysis and Random Matrices • Channel Estimation for MIMO Systems Using Precoding Techniques • Conclusions

15 Performance Analysis of Multiple Antenna Systems *Ming Kang and Mohamed-Slim Alouini* **15**-1
Introduction • MIMO Systems without Co-Channel Interference • MIMO Systems in the Presence of Co-Channel Interference

12

Principles of MIMO-OFDM Wireless Systems

12.1 Introduction 12-1
12.2 The Broadband MIMO Fading Channel.............. 12-2
Basic Assumptions • Array Geometry • Fading Statistics • Ricean Component • Comments on the Channel Model
12.3 Capacity of Broadband MIMO-OFDM Systems 12-4
MIMO-OFDM • Capacity of MIMO-OFDM Spatial Multiplexing Systems • Impact of Propagation Parameters on Capacity • Numerical Results
12.4 Space–Frequency Coded MIMO-OFDM 12-9
Space–Frequency Coding • Error Rate Performance • Maximum Diversity Order and Coding Gain
12.5 Impact of Propagation Parameters on Space–Frequency Codes 12-14
Impact of Propagation Parameters • Simulation Results

Helmut Bölcskei
Swiss Federal Institute of Technology (ETH), Zurich

Abstract

The use of multiple antennas at both ends of a wireless link (multiple-input multiple-output (MIMO) technology) holds the potential to drastically improve the *spectral efficiency* and *link reliability* in future wireless communications systems. A particularly promising candidate for next-generation fixed and mobile wireless systems is the combination of MIMO technology with orthogonal frequency division multiplexing (OFDM). This chapter provides an overview of the basic principles of MIMO-OFDM.

12.1 Introduction

The major challenges in future wireless communications system design are *increased spectral efficiency* and *improved link reliability*. The wireless channel constitutes a hostile propagation medium, which suffers from *fading* (caused by destructive addition of multipath components) and *interference from other users*. Diversity provides the receiver with several (ideally independent) replicas of the transmitted signal and is therefore a powerful means to combat fading and interference resulting in improved link reliability. Common forms of diversity are time diversity (due to Doppler spread) and frequency diversity (due to delay spread). In recent years the use of spatial (or antenna) diversity has become very popular, which is mostly due to the fact that it can be provided without loss in spectral efficiency. Receive diversity, that is, the use of multiple antennas on the receive side of a wireless link, is a well-studied subject [1]. Driven by

0-8493-1657-X/$0.00+$1.50

mobile wireless applications, where it is difficult to deploy multiple antennas in the hand set, the use of multiple antennas on the transmit side combined with signal processing and coding has become known as *space–time coding* [2–4] and is currently an active area of research. The use of multiple antennas at both ends of a wireless link (multiple-input multiple-output (MIMO) technology) has recently been demonstrated to have the potential of achieving extraordinary data rates [5–9]. The corresponding technology is known as *spatial multiplexing* [5, 9] or BLAST [6, 10] and yields an impressive increase in spectral efficiency.

Most of the previous work in the area of MIMO wireless has been restricted to narrowband systems. Besides spatial diversity broadband MIMO channels, however, offer higher capacity and frequency diversity due to delay spread. *Orthogonal frequency division multiplexing* (OFDM) [11, 12] significantly reduces receiver complexity in wireless broadband systems. The use of MIMO technology in combination with OFDM, i.e., MIMO-OFDM [8, 9, 13], thus seems to be an attractive solution for future broadband wireless systems.

The purpose of this chapter is to provide a survey of the basic principles of MIMO-OFDM. The material summarized in this chapter appeared previously in more detail in [9, 13, 14].

Notation: $\mathcal{E}\{\cdot\}$ denotes the expectation operator, $\mathbf{I}_M$ is the $M \times M$ identity matrix, $\mathbf{0}_{M \times N}$ stands for the $M \times N$ all-zeros matrix, $\|\mathbf{A}\|_F^2 = \sum_{i,j} |[\mathbf{A}]_{i,j}|^2$ is the squared Frobenius norm of the matrix $\mathbf{A}$, $\mathbf{A} \otimes \mathbf{B}$ denotes the Kronecker product of the matrices $\mathbf{A}$ and $\mathbf{B}$, $\mathrm{Tr}(\mathbf{A})$ and $\det(\mathbf{A})$ stand for the trace and determinant of $\mathbf{A}$, respectively, $r(\mathbf{A})$ is the rank of $\mathbf{A}$, $\lambda_i(\mathbf{A})$ denotes the i-th eigenvalue of $\mathbf{A}$, $\sigma(\mathbf{A})$ stands for the eigenvalue spectrum of $\mathbf{A}$, $\mathbf{a}_i$ denotes the i-th column of $\mathbf{A}$, and $\mathbf{vec}(\mathbf{A}) = [\mathbf{a}_0^T\ \mathbf{a}_1^T\ \ldots\ \mathbf{a}_{N-1}^T]^T$. The superscripts T, $*$, and H denote transpose, element-wise conjugation, and conjugate transpose, respectively. A circularly symmetric complex Gaussian random variable is a random variable $z = (x + jy) \sim \mathcal{CN}(0, \sigma^2)$, where x and y are independent and identically distributed (i.i.d.) $\mathcal{N}(0, \sigma^2/2)$.

12.2 The Broadband MIMO Fading Channel

In this section, we shall introduce a model for MIMO broadband fading channels taking into account real-world propagation conditions.

12.2.1 Basic Assumptions

In the following, M_T and M_R denote the number of transmit and receive antennas, respectively. We assume that the discrete-time $M_R \times M_T$ matrix-valued channel has order $L-1$ with transfer function

$$\mathbf{H}(e^{j2\pi\theta}) = \sum_{l=0}^{L-1} \mathbf{H}_l e^{-j2\pi l\theta}, \quad 0 \leq \theta < 1 \tag{12.1}$$

where the $M_R \times M_T$ complex-valued random matrix $\mathbf{H}_l$ represents the l-th tap. One can think of each of the taps as corresponding to a significant scatterer cluster (Figure 12.1) with each of the paths emanating from within the same scatterer cluster experiencing the same delay. We write each of the taps as the sum of a fixed (possibly line-of-sight) component, $\bar{\mathbf{H}}_l = \mathcal{E}\{\mathbf{H}_l\}$, and a variable (or scattered) component $\widetilde{\mathbf{H}}_l$ as

$$\mathbf{H}_l = \bar{\mathbf{H}}_l + \widetilde{\mathbf{H}}_l, \quad l = 0, 1, \ldots, L-1.$$

The channel is said to be Rayleigh fading if $\bar{\mathbf{H}}_l = \mathbf{0}_{M_R \times M_T}$ for $l = 0, 1, \ldots, L-1$ and Ricean fading if $\bar{\mathbf{H}}_l \neq \mathbf{0}_{M_R \times M_T}$ for at least one $l \in [0, L-1]$. The elements of the matrices $\widetilde{\mathbf{H}}_l$ ($l = 0, 1, \ldots, L-1$) are (possibly correlated) circularly symmetric complex Gaussian random variables. Different scatterer clusters are assumed to induce uncorrelated fading (or equivalently different taps fade independently), i.e.,

$$\mathcal{E}\{\mathbf{vec}(\widetilde{\mathbf{H}}_l)(\mathbf{vec}(\widetilde{\mathbf{H}}_{l'}))^H\} = \mathbf{0}_{M_R M_T \times M_R M_T} \quad \text{for } l \neq l'. \tag{12.2}$$

Each scatterer cluster has a mean angle of departure from the transmit array and a mean angle of arrival at the receive array, denoted as $\bar{\theta}_{T,l}$ and $\bar{\theta}_{R,l}$ (Figure 12.1), respectively; a cluster angle spread as perceived by

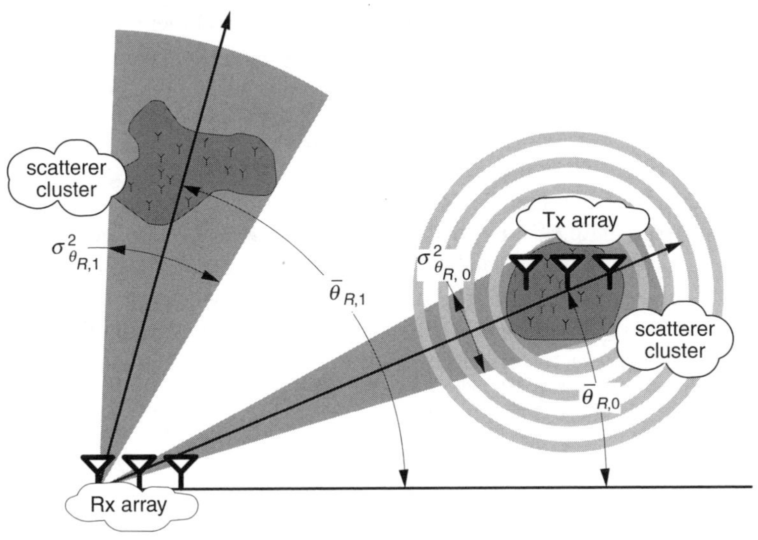

FIGURE 12.1 Schematic representation of MIMO broadband channel composed of multiple clustered paths. For simplicity, only the relevant angles for the receive array are shown — the transmit array situation is reciprocal. The two clusters correspond to different delay taps.

the transmitter $\sigma^2_{\theta_{T,l}}$ (proportional to the scattering radius of the cluster as observed by the transmitter); a cluster angle spread as perceived by the receiver $\sigma^2_{\theta_{R,l}}$ (proportional to the scattering radius of the cluster as observed by the receiver); and a path gain σ_l^2 (derived from the power delay profile of the channel). Finally, we define the total transmit and receive angle spreads for the l-th path cluster as the spreads of $\bar{\theta}_{T,l}$ and $\bar{\theta}_{R,l}$, respectively.

12.2.2 Array Geometry

We assume a uniform linear array at both the transmitter and receiver with identical unipolarized antenna elements. The relative transmit and receive antenna spacings are denoted as $\Delta_T = \frac{d_T}{\lambda}$ and $\Delta_R = \frac{d_R}{\lambda}$, respectively, where d_T and d_R stand for absolute antenna spacing and $\lambda = c/f_c$ is the wavelength of a narrowband signal with center frequency f_c.

12.2.3 Fading Statistics

Spatial fading correlation can occur at both the transmitter and receiver, the impact of which is modeled by decomposing the Rayleigh component of the l-th tap according to

$$\widetilde{\mathbf{H}}_l = \mathbf{R}_l^{1/2}\widetilde{\mathbf{H}}_{w,l}\left(\mathbf{S}_l^{1/2}\right)^T, \quad l = 0, 1, \ldots, L-1 \tag{12.3}$$

where $\mathbf{R}_l = \mathbf{R}_l^{1/2}\mathbf{R}_l^{1/2}$ and $\mathbf{S}_l = \mathbf{S}_l^{1/2}\mathbf{S}_l^{1/2}$ are the receive and transmit correlation matrices, respectively, and $\widetilde{\mathbf{H}}_{w,l}$ is an $M_R \times M_T$ matrix with i.i.d. $\mathcal{CN}(0, \sigma_l^2)$ entries. We note that the decomposition in Equation 12.3 does not incorporate the most general case of spatial fading correlation, but yields a reasonable compromise between analytical tractability and validity of the channel model.

Note that the power delay profile σ_l^2 has been incorporated into the matrices $\widetilde{\mathbf{H}}_{w,l}$. From Equation 12.2 we have $\mathcal{E}\{\mathbf{vec}(\widetilde{\mathbf{H}}_{w,l})(\mathbf{vec}(\widetilde{\mathbf{H}}_{w,l'}))^H\} = \mathbf{0}_{M_RM_T \times M_RM_T}$ for $l \neq l'$. In the following, we define $\rho(s\Delta, \bar{\theta}, \sigma_\theta)$ to be the fading correlation between two antenna elements spaced $s\Delta$ wavelengths apart. The correlation matrices $\mathbf{R}_l$ and $\mathbf{S}_l$ are consequently given by

$$[\mathbf{R}_l]_{m,n} = \rho((n-m)\Delta_R, \bar{\theta}_{R,l}, \sigma_{\theta_{R,l}})$$
$$[\mathbf{S}_l]_{m,n} = \rho((n-m)\Delta_T, \bar{\theta}_{T,l}, \sigma_{\theta_{T,l}}).$$

Let us next assume that the actual angle of departure for the l-th path cluster is given by $\theta_{T,l} = \bar{\theta}_{T,l} + \hat{\theta}_{T,l}$ with $\hat{\theta}_{T,l} \sim \mathcal{N}(0, \sigma^2_{\theta_{T,l}})$, and the actual angle of arrival is $\theta_{R,l} = \bar{\theta}_{R,l} + \hat{\theta}_{R,l}$ with $\hat{\theta}_{R,l} \sim \mathcal{N}(0, \sigma^2_{\theta_{R,l}})$. With these assumptions, we obtain [15]

$$\rho(s\Delta, \bar{\theta}, \sigma_\theta) \approx e^{-j2\pi s\Delta\cos(\bar{\theta})} e^{-\frac{1}{2}(2\pi s\Delta\sin(\bar{\theta})\sigma_\theta)^2} \tag{12.4}$$

which implies that the correlation function is essentially Gaussian with spread inversely proportional to the product of antenna spacing and cluster angle spread. Consequently, large antenna spacing and/or large cluster angle spread lead to small spatial fading correlation and vice versa. It must be stressed, however, that Equation 12.4 is an approximation that becomes inaccurate when the mean angle of arrival is close to zero or close to π, or when the cluster angle spread is large. We note that in the limiting case of zero receive angle spread, i.e., $\sigma_{\theta_{R,l}} = 0$, the receive correlation matrix has a rank of 1 with $\mathbf{R}_l = \mathbf{a}(\bar{\theta}_{R,l})\mathbf{a}^H(\bar{\theta}_{R,l})$, where

$$\mathbf{a}(\theta) = [1 \; e^{j2\pi\Delta\cos(\theta)} \; \ldots \; e^{j2\pi(M_R-1)\Delta\cos(\theta)}]^T. \tag{12.5}$$

Likewise, for $\sigma_{\theta_{T,l}} = 0$, we have $\mathbf{S}_l = \mathbf{a}(\bar{\theta}_{T,l})\mathbf{a}^H(\bar{\theta}_{T,l})$.

12.2.4 Ricean Component

The Ricean component of the l-th tap is modeled as

$$\bar{\mathbf{H}}_l = \sum_{i=0}^{P_l-1} \beta_{l,i}\, \mathbf{a}(\bar{\theta}_{R,l,i})\mathbf{a}^T(\bar{\theta}_{T,l,i}) \tag{12.6}$$

where $\bar{\theta}_{R,l,i}$ and $\bar{\theta}_{T,l,i}$ denote the angle of arrival and the angle of departure, respectively, of the i-th component of $\bar{\mathbf{H}}_l$, and $\beta_{l,i}$ is the corresponding complex-valued path amplitude. We can furthermore associate a Ricean K-factor with each of the taps by defining

$$K_l = \frac{\|\bar{\mathbf{H}}_l\|_F^2}{\mathcal{E}\{\|\widetilde{\mathbf{H}}_l\|_F^2\}}, \quad l = 0, 1, \ldots, L-1.$$

We note that a large cluster angle spread will in general result in a high-rank Ricean component.

12.2.5 Comments on the Channel Model

For the sake of simplicity of exposition, we assumed that different scatterer clusters can be resolved in time and hence correspond to different delays. In practice, this is not necessarily the case. We emphasize, however, that allowing different scatterer clusters to have the same delay does not in general yield significant new insights into the impact of the propagation conditions on the performance of MIMO-OFDM systems.

12.3 Capacity of Broadband MIMO-OFDM Systems

This section is devoted to the capacity of broadband MIMO-OFDM systems operating in spatial multiplexing mode. Spatial multiplexing [5, 9], also referred to as BLAST [6, 10], increases the capacity of wireless radio links drastically with no additional power or bandwidth consumption. The technology requires multiple antennas at both ends of the wireless link and realizes capacity gains (denoted as *spatial multiplexing gain*) by sending independent data streams from different antennas.

12.3.1 MIMO-OFDM

The main motivation for using OFDM in a MIMO channel is the fact that OFDM modulation turns a frequency-selective MIMO channel into a set of parallel frequency-flat MIMO channels. This renders

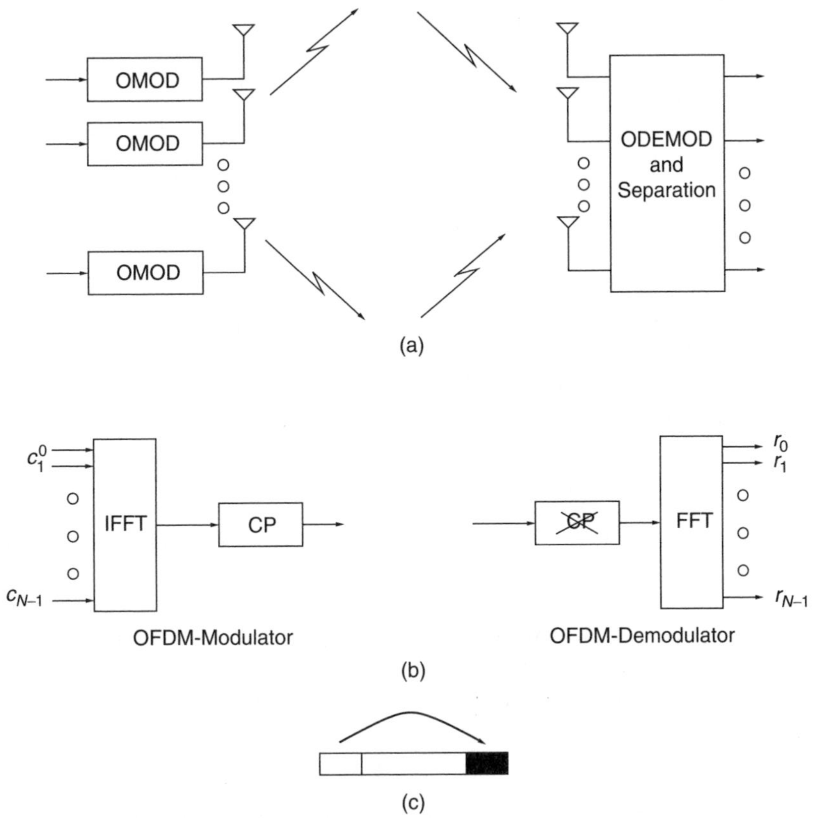

FIGURE 12.2 (a) Schematic of a MIMO-OFDM system (OMOD and ODEMOD denote an OFDM-modulator and demodulator, respectively). (b) Single-antenna OFDM modulator and demodulator. (c) Adding the cyclic prefix.

multichannel equalization particularly simple, since for each OFDM tone only a constant matrix has to be inverted [8, 9].

In a MIMO-OFDM system with N subcarriers (or tones) the individual data streams are first passed through OFDM modulators, which perform an *inverse fast Fourier transform* (IFFT) on blocks of length N followed by a parallel-to-serial conversion. A cyclic prefix (CP) of length $L_{cp} \geq L$ containing a copy of the last L_{cp} samples of the parallel-to-serial converted output of the N-point IFFT is then prepended. The resulting OFDM symbols of length $N + L_{cp}$ are launched simultaneously from the individual transmit antennas. The CP is essentially a guard interval that serves to eliminate interference between OFDM symbols and turns linear convolution into circular convolution such that the channel is diagonalized by the FFT. In the receiver the individual signals are passed through OFDM demodulators, which first discard the CP and then perform an N-point FFT. The outputs of the OFDM demodulators are finally separated and decoded. Figure 12.2 shows a schematic of a MIMO-OFDM system. For a more detailed discussion of the basic principles of OFDM, the interested reader is referred to [16]. The assumption of the length of the CP being greater than or equal to the length of the discrete-time baseband channel impulse response (i.e., $L_{cp} \geq L$) guarantees that the frequency-selective MIMO fading channel indeed decouples into a set of parallel frequency-flat MIMO fading channels [11]. Organizing the transmitted data symbols into frequency vectors $\mathbf{c}_k = [c_k^{(0)}\ c_k^{(1)}\ \ldots\ c_k^{(M_T-1)}]^T$ $(k = 0, 1, \ldots, N-1)$ with $c_k^{(i)}$ denoting the data symbol transmitted from the i-th antenna on the k-th tone, the reconstructed data vector for the k-th tone is given by [8, 9]

$$\mathbf{r}_k = \mathbf{H}(e^{j\frac{2\pi}{N}k})\,\mathbf{c}_k + \mathbf{n}_k, \quad k = 0, 1, \ldots, N-1 \tag{12.7}$$

where $\mathbf{n}_k$ is complex-valued circularly symmetric additive white Gaussian noise satisfying $\mathcal{E}\{\mathbf{n}_k\mathbf{n}_l^H\} = \sigma_n^2\mathbf{I}_{M_R}\delta[k-l]$.

12.3.2 Capacity of MIMO-OFDM Spatial Multiplexing Systems

In the following, we ignore the loss in spectral efficiency due to the presence of the CP (recall that the CP contains redundant information). We assume that the channel is purely Rayleigh fading, ergodic, remains constant over a block spanning at least one OFDM symbol, and changes in an independent fashion from block to block. For the sake of simplicity of exposition, we restrict our attention to the case of receive correlation only (i.e., the transmit antennas fade in an uncorrelated fashion).

In OFDM-based spatial multiplexing systems statistically independent data streams are transmitted from different antennas and different tones and the total available power is allocated uniformly across all space–frequency subchannels [9]. Assuming that coding and *interleaving* are performed across OFDM symbols and that the number of fading blocks spanned by a code word goes to infinity, whereas the block size (which equals the number of tones in the OFDM system multiplied by the number of OFDM symbols spanning one channel use, or equivalently one block) remains constant (and finite), an ergodic or Shannon capacity exists and is given by[1] [9]

$$C = \mathcal{E}\left\{\log\det\left(\mathbf{I}_{M_R} + \rho\,\boldsymbol{\Lambda}\widetilde{\mathbf{H}}_w\widetilde{\mathbf{H}}_w^H\right)\right\} \quad \text{bps/Hz} \tag{12.8}$$

where $\boldsymbol{\Lambda} = \text{diag}\{\lambda_i(\mathbf{R})\}_{i=0}^{M_R-1}$ with $\mathbf{R} = \sum_{l=0}^{L-1}\sigma_l^2\,\mathbf{R}_l$, $\rho = \frac{P}{M_T N\sigma_n^2}$ with P denoting the total available power, $\widetilde{\mathbf{H}}_w$ is an $M_R \times M_T$ i.i.d. random matrix with $\mathcal{CN}(0,1)$ entries, and the expectation is taken with respect to $\widetilde{\mathbf{H}}_w$. The operational meaning of C is as follows. At rates lower than C, the error probability (for a good code) decays exponentially with the transmission length. Capacity can be achieved in principle by transmitting a code word over a very large number of independently fading blocks.

It is instructive to study the case of fixed M_R with M_T large, where $\frac{1}{M_T}\widetilde{\mathbf{H}}_w\widetilde{\mathbf{H}}_w^H \to \mathbf{I}_{M_R}$ and consequently

$$C = \log\det(\mathbf{I}_{M_R} + \bar{\rho}\,\boldsymbol{\Lambda}) \tag{12.9}$$

with $\bar{\rho} = M_T\rho = \frac{P}{N\sigma_n^2}$. For small $\bar{\rho}$, it follows from Equation 12.9 that in the large M_T limit [9]

$$C \approx \log\left(1 + \bar{\rho}\,\text{Tr}(\mathbf{R})\right)$$

where all the higher-order terms in $\bar{\rho}$ have been neglected. Thus, in the low signal-to-noise ratio (SNR) regime the ergodic capacity is governed by $\text{Tr}(\mathbf{R})$. In the high SNR regime, we obtain a fundamentally different conclusion. Starting from Equation 12.9 we have

$$C = \sum_{i=0}^{r(\mathbf{R})-1} \log(1 + \bar{\rho}\,\lambda_i(\mathbf{R})) \tag{12.10}$$

with $\lambda_i(\mathbf{R})$ denoting the nonzero eigenvalues of $\mathbf{R}$. The rank and eigenvalue spread of the sum correlation matrix $\mathbf{R} = \sum_{l=0}^{L-1}\sigma_l^2\,\mathbf{R}_l$ therefore critically determine ergodic capacity in the high SNR regime. In fact, it follows directly from Equation 12.10 that the multiplexing gain in the large M_T limit is given by $r(\mathbf{R})$. Moreover, for a given $\text{Tr}(\mathbf{R})$, the right-hand side (RHS) in Equation 12.10 is maximized if $r(\mathbf{R}) = M_R$ and all the $\lambda_i(\mathbf{R})$ $(i = 0, 1, \ldots, M_R - 1)$ are equal [9]. A deviation of $\lambda_i(\mathbf{R})$ as a function of i from a constant function will therefore result in a loss in terms of ergodic capacity.

12.3.3 Impact of Propagation Parameters on Capacity

We shall next show how the propagation parameters impact the eigenvalues of the sum correlation matrix $\mathbf{R}$ and hence ergodic capacity. Since the individual correlation matrices $\mathbf{R}_l$ $(l = 0, 1, \ldots, L-1)$ are Toeplitz, the sum correlation matrix $\mathbf{R}$ is Toeplitz, as well. Using Equation 12.4 and applying Szegö's theorem [17]

[1]Throughout the chapter all logarithms are to the base 2.

to $\mathbf{R}$, we obtain the limiting ($M_R \to \infty$) distribution[2] of the eigenvalues of $\mathbf{R} = \sum_{l=0}^{L-1} \sigma_l^2 \mathbf{R}_l$ as

$$\lambda(\nu) = \sum_{l=0}^{L-1} \sigma_l^2 \underbrace{\vartheta_3\left(\pi(\nu - \Delta_R \cos(\bar{\theta}_{R,l})), e^{-\frac{1}{2}(2\pi \Delta_R \sin(\bar{\theta}_{R,l})\sigma_{\theta_{R,l}})^2}\right)}_{\lambda_l(\nu)}, \quad 0 \le \nu < 1 \tag{12.11}$$

with the third-order theta function given by $\vartheta_3(\nu, q) = \sum_{n=-\infty}^{\infty} q^{n^2} e^{2jn\nu}$. Although this expression yields the exact eigenvalue distribution only in the limiting case $M_R \to \infty$, good approximations of the eigenvalues for finite M_R can be obtained by sampling $\lambda(\nu)$ uniformly, which allows us to assume that the eigenvalue distribution in the finite M_R case follows the shape of $\lambda(\nu)$. This observation, combined with Equation 12.11, shall next be used to relate propagation and system parameters to the eigenvalues of $\mathbf{R}$ and hence ergodic capacity.

12.3.3.1 Impact of Cluster Angle Spread and Antenna Spacing

Let us start by investigating the influence of receive cluster angle spreads and receive antenna spacing on ergodic capacity. For the sake of simplicity, consider a single-tap channel (i.e., $L = 1$) with associated receive correlation matrix $\mathbf{R}_0$. The limiting eigenvalue distribution of $\mathbf{R} = \sigma_0^2 \mathbf{R}_0$ is given by

$$\lambda(\nu) = \sigma_0^2 \vartheta_3(\pi(\nu - \Delta_R \cos(\bar{\theta}_{R,0})), e^{-\frac{1}{2}(2\pi \Delta_R \sin(\bar{\theta}_{R,0})\sigma_{\theta_{R,0}})^2}).$$

Now, noting that the correlation function $\rho(s\Delta_R, \bar{\theta}_{R,0}, \sigma_{\theta_{R,0}})$ as a function of s is essentially a modulated Gaussian function with its spread decreasing for increasing antenna spacing and/or increasing cluster angle spread and vice versa, it follows that $\lambda(\nu)$ will be flatter in the case of large antenna spacing and/or large cluster angle spread (i.e., low spatial fading correlation). For small antenna spacing and/or small cluster angle spread $\lambda(\nu)$ will be peaky. Figure 12.3(a) and (b) show the limiting eigenvalue distribution of $\mathbf{R}_0$ for high and low spatial fading correlation, respectively. From our previous discussion, it thus follows that the ergodic capacity will decrease for increasing concentration of $\lambda(\nu)$ (or equivalently high spatial fading correlation) and vice versa.

12.3.3.2 Impact of Total Angle Spread

We shall next study the impact of total receive angle spread on ergodic capacity. Assume either that the individual scatterer cluster angle spreads are small or that antenna spacing at the receiver is small, or both. Hence, the individual $\lambda_l(\nu)$ will be peaky. Now, from Equation 12.11 we can see that the limiting distribution $\lambda(\nu)$ is obtained by adding the individual limiting distributions $\lambda_l(\nu)$ weighted by the σ_l^2. Note furthermore that $\lambda_l(\nu)$ is essentially Gaussian centered around $\Delta_R \cos(\bar{\theta}_{R,l})$. Now, if the total angle spread, i.e., the spread of the $\bar{\theta}_{R,l}$, is large, the sum limiting distribution $\lambda(\nu)$ can still be flat even though the individual $\lambda_l(\nu)$ are peaky. For given small cluster angle spreads, Figure 12.4(a) and (b) show example limiting eigenvalue distributions for a three-tap channel (assuming a uniform power delay profile) with a total angle spread of 22.5 and 90 degrees, respectively. We can clearly see the impact of total angle spread on the limiting eigenvalue distribution $\lambda(\nu)$ and hence on ergodic capacity. Large total angle spread renders $\lambda(\nu)$ flat and therefore increases ergodic capacity, whereas small total angle spread makes $\lambda(\nu)$ peaky and hence leads to reduced ergodic capacity.

12.3.3.3 Ergodic Capacity in the SISO and MIMO Cases

It is well known that in the single-input single-output (SISO) case delay spread channels do not offer an advantage over flat-fading channels in terms of ergodic capacity [18] (provided the receive SNR is kept

[2]Note that for $M_R \to \infty$ the eigenvalues of $\mathbf{R}$ are characterized by a periodic continuous function [17]. Thus, in the following, whenever we use the term *eigenvalue distribution*, we are actually referring to this function.

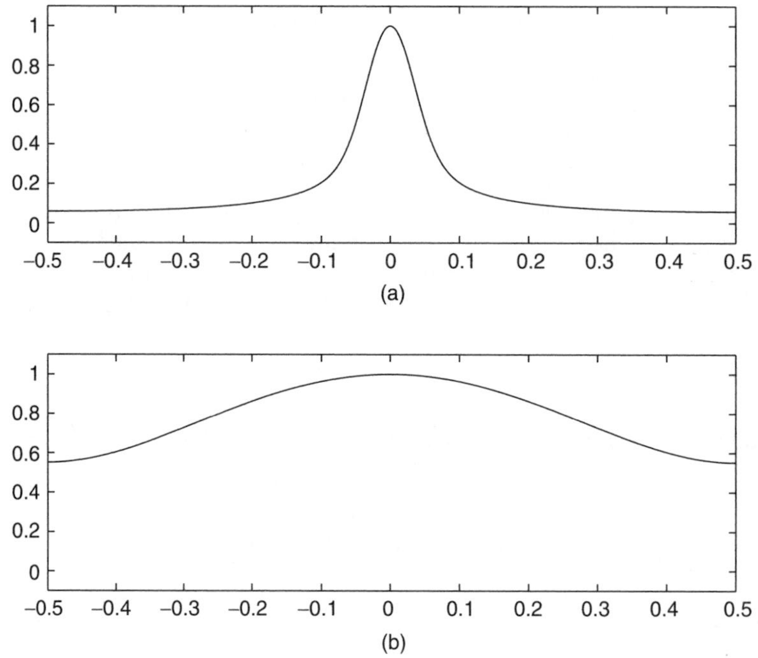

FIGURE 12.3 Limiting eigenvalue distribution of the correlation matrix $\mathbf{R}_0$ for the cases of (a) high spatial fading correlation and (b) low spatial fading correlation.

constant). This can easily be seen from Equation 12.8 by noting that in the SISO case $\mathbf{R} = \sum_{l=0}^{L-1} \sigma_l^2$, which implies that ergodic capacity is only a function of the total energy in the channel and does not depend on how this energy is distributed across taps. In the MIMO case the situation can be fundamentally different. Fix $\text{Tr}(\mathbf{R})$ and take a flat-fading scenario (i.e., $L = 1$) with small antenna spacing such that $\mathbf{R} = \sigma_0^2 \mathbf{R}_0$ has a

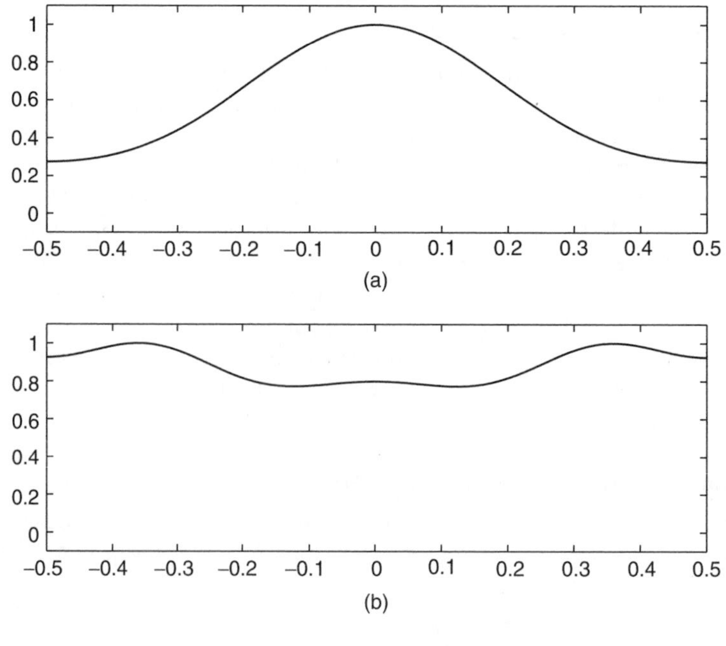

FIGURE 12.4 Limiting eigenvalue distribution of the sum correlation matrix $\mathbf{R} = \sum_{l=0}^{2} \sigma_l^2 \mathbf{R}_l$ for fixed cluster angle spread and for the cases of (a) small total angle spread and (b) large total angle spread.

rank of 1. In this case the matrix $\Lambda\widetilde{\mathbf{H}}_w\widetilde{\mathbf{H}}_w^H$ has a rank of 1 with probability 1 and hence only one spatial data pipe can be opened up, or equivalently, there is no spatial multiplexing gain. Now compare the flat-fading scenario to a frequency-selective fading scenario where $L \geq M_R$ and each of the $\mathbf{R}_l$ $(l = 0, 1, \ldots, L-1)$ has a rank of 1 but the sum correlation matrix $\mathbf{R} = \sum_{l=0}^{L-1} \sigma_l^2 \mathbf{R}_l$ has full rank. For this to happen a sufficiently large total angle spread is needed. Clearly, in this case $\min(M_T, M_R)$ spatial data pipes can be opened up and we will get a higher ergodic capacity because the rank of $\mathbf{R}$ is higher than in the flat-fading case. We note that in the case where all the correlation matrices satisfy $\mathbf{R}_l = \mathbf{I}_{M_R}$ $(l = 0, 1, \ldots, L-1)$ this effect does not occur. However, this scenario corresponds to fully uncorrelated spatial fading on all taps and is therefore unlikely. *We can conclude that in practice MIMO delay spread channels offer an advantage over MIMO flat-fading channels in terms of ergodic capacity*. However, we caution the reader that this conclusion is a result of the assumption that delayed paths increase the total angle spread. This assumption has been verified by measurement for outdoor MIMO broadband channels in the 2.5-GHz band [19].

12.3.4 Numerical Results

We conclude this section with a numerical result studying the impact of delay spread on ergodic capacity. In this example the power delay profile was taken to be exponential, tap spacing was uniform, the relative receive antenna spacing was set to $\Delta_R = \frac{1}{2}$, SNR was defined as $\text{SNR} = \bar{\rho} = \frac{P}{N\sigma_n^2}$, and the number of antennas was $M_T = M_R = 4$. In order to make the following comparisons fair, we normalize the energy in the channel by setting $\text{Tr}(\mathbf{R}) = 1$ for all cases. The cluster angle spreads were assumed to satisfy $\sigma_{\theta_{R,l}} = 0\ (l = 0, 1, \ldots, L-1)$. In the flat-fading case the mean angle of arrival was set to $\bar{\theta}_{R,0} = \pi/2$. In the frequency-selective case we assumed a total receive angle spread of 90 degrees. Figure 12.5(a) shows the ergodic capacity (in bps/Hz) (obtained through evaluation of Equation 12.8 using 1000 independent Monte Carlo runs) as a function of SNR for different values of L. We can see that ergodic capacity indeed increases for increasing L and hence increasing rank of $\mathbf{R}$. Moreover, we observe that increasing the number of resolvable (i.e., independently fading) taps beyond 4 does not further increase ergodic capacity. The reason for this is that the multiplexing gain $\min(r(\mathbf{R}), M_T)$ cannot exceed 4. Figure 12.5(b) shows the ergodic capacity for the same parameters as above except for the cluster angle spreads increased to $\sigma_{\theta_{R,l}} = 0.25\ (l = 0, 1, \ldots, L-1)$. In this case the rank of the individual correlation matrices $\mathbf{R}_l$ is higher than 1 and the improvement in terms of ergodic capacity resulting from the presence of multiple taps is less pronounced. We emphasize that the conclusions drawn in this simulation result are a consequence of the assumption that delayed paths increase the rank of the sum correlation matrix $\mathbf{R}$.

12.4 Space–Frequency Coded MIMO-OFDM

While spatial multiplexing realizes increased spectral efficiency, *space–frequency coding* [13, 14, 20] is used to improve link reliability through (spatial and frequency) *diversity gain*. In this section, we describe the basics of space–frequency coded MIMO-OFDM.

12.4.1 Space–Frequency Coding

Using the notation introduced in Section 12.3, we start from the input–output relation

$$\mathbf{r}_k = \sqrt{E_s}\,\mathbf{H}(e^{j\frac{2\pi}{N}k})\,\mathbf{c}_k + \mathbf{n}_k, \quad k = 0, 1, \ldots, N-1 \tag{12.12}$$

where the data symbols $c_k^{(i)}$ are taken from a finite complex alphabet and have an average energy of 1. The constant E_s is an energy normalization factor. Throughout the remainder of this chapter, we assume that the channel is constant over the duration of at least one OFDM symbol, the transmitter has no channel knowledge, and the receiver knows the channel perfectly. We furthermore assume that coding is performed only within one OFDM symbol such that one data burst consists of N vectors of size $M_T \times 1$, or equivalently, one spatial OFDM symbol.

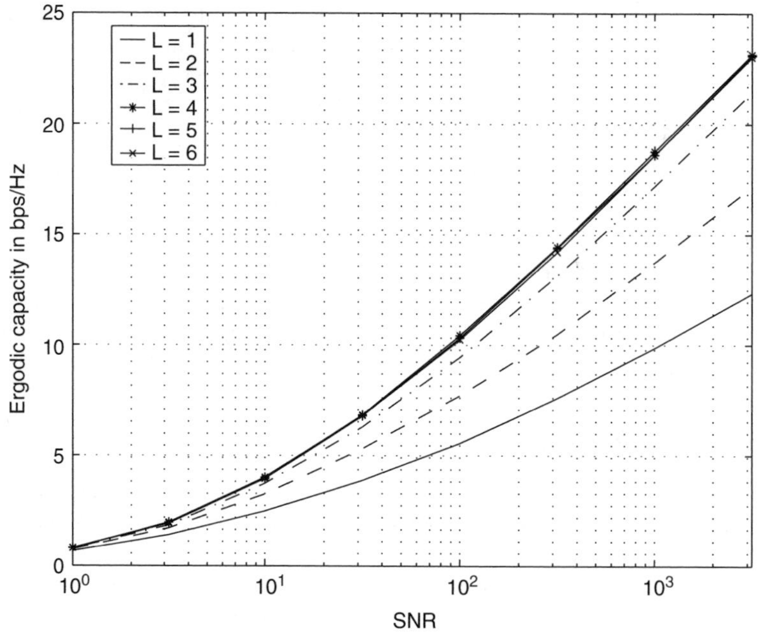

(a)

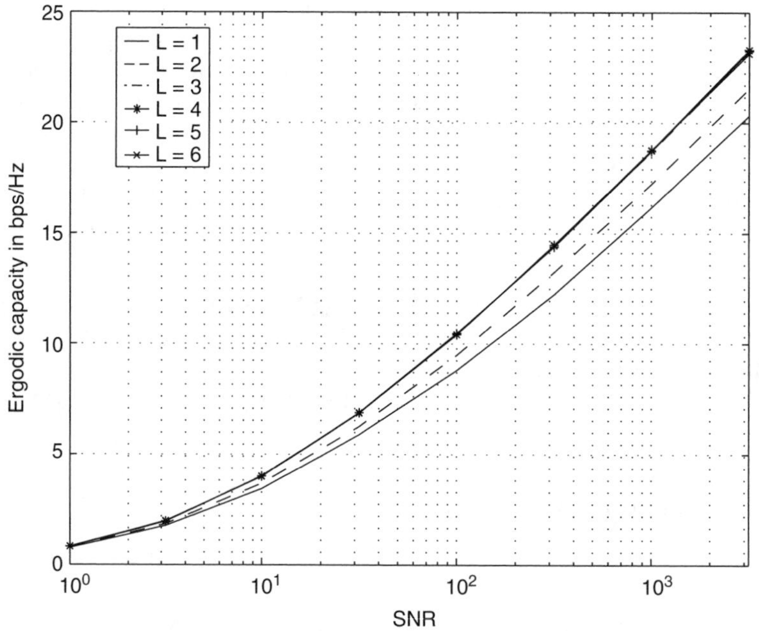

(b)

FIGURE 12.5 Ergodic capacity (in bps/Hz) as a function of SNR for various values of L and (a) small cluster angle spreads and (b) large cluster angle spreads.

The maximum likelihood (ML) decoder computes the vector sequence $\hat{\mathbf{c}}_k$ $(k = 0, 1, \ldots, N-1)$ according to

$$\widehat{\mathbf{C}} = \arg\min_{\mathbf{C}} \sum_{k=0}^{N-1} \|\mathbf{r}_k - \sqrt{E_s}\,\mathbf{H}(e^{j\frac{2\pi}{N}k})\,\mathbf{c}_k\|^2$$

where $\mathbf{C} = [\mathbf{c}_0\,\mathbf{c}_1\,\ldots\,\mathbf{c}_{N-1}]$ and $\widehat{\mathbf{C}} = [\hat{\mathbf{c}}_0\,\hat{\mathbf{c}}_1\,\ldots\,\hat{\mathbf{c}}_{N-1}]$. In the remainder of this section, we employ the channel model introduced in Section 12.2 in its full generality and assume that $N > M_T L$.

12.4.2 Error Rate Performance

Assume that $\mathbf{C} = [\mathbf{c}_0\,\mathbf{c}_1\,\ldots\,\mathbf{c}_{N-1}]$ and $\mathbf{E} = [\mathbf{e}_0\,\mathbf{e}_1\,\ldots\,\mathbf{e}_{N-1}]$ are two different space–frequency code words of size $M_T \times N$. The average (with respect to the random channel) probability that the receiver decides erroneously in favor of the signal $\mathbf{E}$, assuming that $\mathbf{C}$ was transmitted, can be upper-bounded by the pairwise error probability (PEP) as [14]

$$P(\mathbf{C} \rightarrow \mathbf{E}) \leq \exp\left(-\frac{E_s}{4\sigma_n^2}\|\bar{\mathbf{Y}}\|^2\right) \prod_{i=0}^{r(\mathbf{C}_Y)-1} \exp\left(\left(\frac{E_s}{4\sigma_n^2}\right)^2 \frac{|b_i|^2\lambda_i^2(\mathbf{C}_Y)}{1+\frac{E_s}{4\sigma_n^2}\lambda_i(\mathbf{C}_Y)}\right) \frac{1}{1+\frac{E_s}{4\sigma_n^2}\lambda_i(\mathbf{C}_Y)} \tag{12.13}$$

where

$$\mathbf{C}_Y = \sum_{l=0}^{L-1} \sigma_l^2 [\mathbf{D}^l(\mathbf{C}-\mathbf{E})^T \mathbf{S}_l (\mathbf{C}-\mathbf{E})^* \mathbf{D}^{-l}] \otimes \mathbf{R}_l \tag{12.14}$$

with $\mathbf{D} = \text{diag}\{e^{-j\frac{2\pi}{N}k}\}_{k=0}^{N-1}$; $\bar{\mathbf{Y}} = [\bar{\mathbf{y}}_0^T\,\bar{\mathbf{y}}_1^T\,\ldots\,\bar{\mathbf{y}}_{N-1}^T]^T$ with $\bar{\mathbf{y}}_k = \bar{\mathbf{H}}(e^{j\frac{2\pi}{N}k})(\mathbf{c}_k - \mathbf{e}_k)$ and $\bar{\mathbf{H}}(e^{j\frac{2\pi}{N}k}) = \sum_{l=0}^{L-1} \bar{\mathbf{H}}_l\, e^{-j\frac{2\pi}{N}kl}$; $\mathbf{C}_Y = \mathbf{U}\boldsymbol{\Phi}\mathbf{U}^H$ with $\boldsymbol{\Phi} = \text{diag}\{\lambda_0(\mathbf{C}_Y), \ldots, \lambda_{r(\mathbf{C}_Y)-1}(\mathbf{C}_Y), 0, \ldots, 0\}$; and $\mathbf{b} = [b_0\,b_1\,\ldots\,b_{M_R N-1}]^T = \text{diag}\{\frac{1}{\sqrt{\lambda_0(\mathbf{C}_Y)}}, \ldots, \frac{1}{\sqrt{\lambda_{r(\mathbf{C}_Y)-1}(\mathbf{C}_Y)}}, 0, \ldots, 0\}\mathbf{U}^H\bar{\mathbf{Y}}$.

12.4.3 Maximum Diversity Order and Coding Gain

Based on Equation 12.13, we are now ready to establish the maximum achievable diversity order and coding gain for various propagation scenarios. We focus on the high SNR regime, i.e., $\frac{E_s}{4\sigma_n^2} \gg 1$, and discuss the case of Rayleigh fading and Ricean fading separately.

12.4.3.1 Rayleigh Fading

Let us start by assuming that the channel is purely Rayleigh fading, i.e., $\bar{\mathbf{H}}_l = \mathbf{0}_{M_R \times M_T}$ for $l = 0, 1, \ldots, L-1$. In this case, Equation 12.13 simplifies to

$$\mathbf{P}(\mathbf{C} \rightarrow \mathbf{E}) \leq \left(\frac{E_s}{4\sigma_n^2}\right)^{-r(\mathbf{C}_Y)} \prod_{i=0}^{r(\mathbf{C}_Y)-1} \lambda_i^{-1}(\mathbf{C}_Y). \tag{12.15}$$

We recall the following definitions [2]. The *diversity order d* achieved by a space–frequency code is given by the minimum rank of $\mathbf{C}_Y$ over all code word pairs $\{\mathbf{C}, \mathbf{E}\}$. The *coding gain* is defined as the minimum of $(\prod_{i=0}^{d-1} \lambda_i(\mathbf{C}_Y))^{\frac{1}{d}}$ over all matrices $\mathbf{C}_Y$ with $r(\mathbf{C}_Y) = d$. In the case of i.i.d. channels, where $\mathbf{R}_l = \mathbf{I}_{M_R}$ and $\mathbf{S}_l = \mathbf{I}_{M_T}$ for $l = 0, 1, \ldots, L-1$, the maximum achievable diversity order is $d = M_T M_R |\mathcal{L}|$, where $\mathcal{L} = \{i : \sigma_i^2 > 0\}$ $(i \in [0, L-1])$, and $|\mathcal{L}|$ denotes the size of the set $\mathcal{L}$.

The propagation parameters determine the maximum achievable diversity order and coding gain through the rank and the eigenvalues of the matrix $\mathbf{C}_Y$. In the following, we shall therefore study the dependence of $\mathbf{C}_Y$ on the correlation matrices $\mathbf{R}_l$ and $\mathbf{S}_l$. According to the conclusions obtained, the discussion is organized into three different cases, namely, receive correlation only, transmit correlation only, and joint transmit and receive correlation.

12.4.3.1.1 *Receive Correlation Only*

In this case $\mathbf{S}_l = \mathbf{I}_{M_T}$ for $l = 0, 1, \ldots, L-1$, and Equation 12.14 specializes to

$$\mathbf{C}_Y = \sum_{l=0}^{L-1} \sigma_l^2 [\mathbf{D}^l(\mathbf{C}-\mathbf{E})^T(\mathbf{C}-\mathbf{E})^*\mathbf{D}^{-l}] \otimes \mathbf{R}_l.$$

We start with the assumption that all the correlation matrices $\mathbf{R}_l$ with $l \in \mathcal{L}$ have full rank. Defining

$$\mathbf{C}_u = \sum_{l=0}^{L-1} \sigma_l^2 [\mathbf{D}^l(\mathbf{C}-\mathbf{E})^T(\mathbf{C}-\mathbf{E})^*\mathbf{D}^{-l}] \tag{12.16}$$

we obtain [14]

$$P(\mathbf{C} \rightarrow \mathbf{E}) \leq \left(\frac{E_s}{4\sigma_n^2}\right)^{-r(\mathbf{C}_u)M_R} \prod_{i=0}^{r(\mathbf{C}_u)M_R-1} \theta_i^{-1} \prod_{i=0}^{r(\mathbf{C}_u)-1} \lambda_i^{-M_R}(\mathbf{C}_u) \tag{12.17}$$

where $0 < \lambda_{min}(\widehat{\mathbf{R}}) \leq \theta_i \leq \lambda_{max}(\widehat{\mathbf{R}})$ with $\widehat{\mathbf{R}} = \text{diag}\{\mathbf{R}_l\}_{l \in \mathcal{L}}$. Noting that the diversity order achieved by a space–frequency code in the i.i.d. case is given by $M_R \min_{\mathbf{C}_u} r(\mathbf{C}_u)$ with the minimum taken over all code word pairs $\{\mathbf{C}, \mathbf{E}\}$, it follows that the diversity order achieved by a space–frequency code in the presence of receive correlation only with all correlation matrices $\mathbf{R}_l$ ($l = 0, 1, \ldots, L-1$) full rank is equal to the diversity order achieved by the code in the absence of spatial fading correlation. The corresponding coding gain is given by the coding gain achieved in the uncorrelated case $(\prod_{i=0}^{r(\mathbf{C}_u)-1} \lambda_i(\mathbf{C}_u))^{\frac{1}{r(\mathbf{C}_u)}}$ multiplied by $(\prod_{i=0}^{r(\mathbf{C}_u)M_R-1} \theta_i)^{\frac{1}{r(\mathbf{C}_u)M_R}}$. In the special instance where the space–frequency code achieves full diversity gain in the i.i.d. case, i.e., the minimum rank of $\mathbf{C}_u$ over all pairs of code word matrices $\{\mathbf{C}, \mathbf{E}\}$ is $M_T|\mathcal{L}|$, we get

$$P(\mathbf{C} \rightarrow \mathbf{E}) \leq \left(\frac{E_s}{4\sigma_n^2}\right)^{-M_T M_R |\mathcal{L}|} (\det(\widehat{\mathbf{R}}))^{-M_T} \prod_{i=0}^{M_T|\mathcal{L}|-1} \lambda_i^{-M_R}(\mathbf{C}_u). \tag{12.18}$$

From the discussion above and Equation 12.18 it follows that the diversity order achieved in this case is $d = M_T M_R |\mathcal{L}|$, which is the maximum possible diversity order achievable in the i.i.d. case [13]. The coding gain is given by the coding gain achieved in the i.i.d. case multiplied by $(\det(\widehat{\mathbf{R}}))^{\frac{1}{M_R|\mathcal{L}|}}$. Using the normalization $\text{Tr}(\mathbf{R}) = M_R|\mathcal{L}|$, it is easily seen that $\det(\widehat{\mathbf{R}}) \leq 1$ where the upper bound is achieved if all the $\mathbf{R}_l$ with $l \in \mathcal{L}$ are unitary. Thus, we conclude that spatial fading correlation always leads to a loss in performance with the degradation being determined by the eigenvalue spread of $\widehat{\mathbf{R}}$.

Finally, if $\widehat{\mathbf{R}}$ is rank deficient (which is the case if at least one of the $\mathbf{R}_l$ is rank deficient) and the space–frequency code achieves full diversity gain in the i.i.d. case, the achievable diversity order is upper-bounded as [14]

$$d \leq M_T \sum_{l \in \mathcal{L}} r(\mathbf{R}_l) \tag{12.19}$$

12.4.3.1.2 *Transmit Correlation Only*

In the case of transmit correlation only, $\mathbf{R}_l = \mathbf{I}_{M_R}$ for $l = 0, 1, \ldots, L-1$. For nonsingular $\widehat{\mathbf{S}} = \text{diag}\{\mathbf{S}_l\}_{l \in \mathcal{L}}$ we have

$$P(\mathbf{C} \rightarrow \mathbf{E}) \leq \left(\frac{E_s}{4\sigma_n^2}\right)^{-M_R r(\mathbf{C}_u)} \prod_{i=0}^{r(\mathbf{C}_u)-1} \theta_i^{-M_R} \prod_{i=0}^{r(\mathbf{C}_u)-1} \lambda_i^{-M_R}(\mathbf{C}_u) \tag{12.20}$$

where $0 < \lambda_{min}(\widehat{\mathbf{S}}) \leq \theta_i \leq \lambda_{max}(\widehat{\mathbf{S}})$, which implies that the diversity order is given by $d = M_R \min_{\mathbf{C}_u} r(\mathbf{C}_u)$ with the minimum taken over all code word pairs $\{\mathbf{C}, \mathbf{E}\}$. Hence the achievable diversity order in the case of transmit correlation only with full-rank $\widehat{\mathbf{S}}$ is equal to the diversity order achieved by the space–frequency code in the i.i.d. case. For general $\widehat{\mathbf{S}}$, focusing on a minimum-rank error event $\mathbf{C}-\mathbf{E}$ (i.e., an error event that leads to minimum $r(\mathbf{C}_u)$), we get

$$M_R(r(\mathbf{C}_u)+r(\widehat{\mathbf{S}})-M_T|\mathcal{L}|)\leq d\leq M_R\min\{r(\mathbf{C}_u),r(\widehat{\mathbf{S}})\}. \tag{12.21}$$

We note that Equation 12.21 shows that if $\widehat{\mathbf{S}}$ is singular and the space–frequency code does not achieve full diversity gain in the i.i.d. case, the diversity order achieved in the correlated case can only be lower bounded by $M_R(r(\mathbf{C}_u)+r(\widehat{\mathbf{S}})-M_T|\mathcal{L}|)$. In this case, it is difficult to make statements on the exact diversity order achieved since d will be a function of the eigenspaces of the $\mathbf{S}_l$ and the eigenspace of $(\mathbf{C}-\mathbf{E})^*(\mathbf{C}-\mathbf{E})^T$. In general, as evidenced by Equation 12.21 in the presence of transmit correlation, it is crucial that the space–frequency code excites the range spaces of the $\mathbf{S}_l$ in order to obtain good performance in terms of error probability.

12.4.3.1.3 *Joint Transmit–Receive Correlation*

In the case of joint transmit–receive correlation, it follows from Equation 12.14 that the maximum achievable diversity order is given by $d=\sum_{l\in\mathcal{L}}r(\mathbf{S}_l)r(\mathbf{R}_l)$. Noting that the diversity order offered by the l-th tap is $r(\mathbf{R}_l)r(\mathbf{S}_l)$, this result says that the maximum achievable diversity order is simply the number of total degrees of freedom available in the channel.

12.4.3.2 Ricean Fading

So far, we have restricted our attention to Rayleigh fading. Let us next consider the Ricean case. Again assuming high SNR ($\frac{E_s}{4\sigma_n^2}\gg 1$), the PEP upper bound in Equation 12.13 simplifies to

$$\mathbf{P}(\mathbf{C}\to\mathbf{E})\leq\exp\left\{\frac{E_s}{4\sigma_n^2}\left(\sum_{i=0}^{r(\mathbf{C}_Y)-1}|b_i|^2\lambda_i(\mathbf{C}_Y)-\|\bar{\mathbf{Y}}\|^2\right)\right\}\left(\frac{E_s}{4\sigma_n^2}\right)^{-r(\mathbf{C}_Y)}\prod_{i=0}^{r(\mathbf{C}_Y)-1}\lambda_i^{-1}(\mathbf{C}_Y). \tag{12.22}$$

We can see that the PEP upper bound consists of an exponential term multiplied by the RHS of Equation 12.15, which is the PEP upper bound in the Rayleigh fading case. The behavior of the second and third terms of the RHS in Equation 12.22 was studied above in the section on Rayleigh fading. We shall therefore focus on the first term of the RHS of Equation 12.22, which represents the contribution due to the Ricean component, and start by noting that

$$\sum_{i=0}^{r(\mathbf{C}_Y)-1}|b_i|^2\lambda_i(\mathbf{C}_Y)-\|\bar{\mathbf{Y}}\|^2=\bar{\mathbf{Y}}^H\mathbf{U}\,\mathrm{diag}\{\underbrace{0,0,\ldots,0}_{r(\mathbf{C}_Y)},-1,-1,\ldots,-1\}\,\mathbf{U}^H\bar{\mathbf{Y}}$$

which after application of the Rayleigh–Ritz theorem [21] yields

$$e^{-\frac{E_s}{4\sigma_n^2}\|\bar{\mathbf{Y}}\|^2}\leq\exp\left\{\frac{E_s}{4\sigma_n^2}\left(\sum_{i=0}^{r(\mathbf{C}_Y)-1}|b_i|^2\lambda_i(\mathbf{C}_Y)-\|\bar{\mathbf{Y}}\|^2\right)\right\}\leq 1. \tag{12.23}$$

We can therefore conclude that the performance in the Ricean case depends strongly on the angle between $\bar{\mathbf{Y}}$ and the eigenspace of $\mathbf{C}_Y$. Assuming that $\bar{\mathbf{Y}}$ is such that the lower bound in Equation 12.23 is achieved, we get

$$\mathbf{P}(\mathbf{C}\to\mathbf{E})\leq e^{-\frac{E_s}{4\sigma_n^2}\|\bar{\mathbf{Y}}\|^2}\left(\frac{E_s}{4\sigma_n^2}\right)^{-r(\mathbf{C}_Y)}\prod_{i=0}^{r(\mathbf{C}_Y)-1}\lambda_i^{-1}(\mathbf{C}_Y).$$

In this case, the PEP upper bound is minimized if $\|\bar{\mathbf{Y}}\|^2$ is maximized, which implies that the error probability performance is determined by the properties of the matrices $\bar{\mathbf{H}}(e^{j\frac{2\pi}{N}k})$ $(k=0,1,\ldots,N-1)$, and hence the matrices $\bar{\mathbf{H}}_l$. In particular, we note that error events with code difference vectors lying in the null spaces of the $\bar{\mathbf{H}}(e^{j\frac{2\pi}{N}k})$ will yield small $\|\bar{\mathbf{Y}}\|^2$, and hence result in high error probability. For high-rate codes such as spatial multiplexing it is therefore crucial for the matrices $\bar{\mathbf{H}}(e^{j\frac{2\pi}{N}k})$ to have high rank in order to ensure good performance in terms of error probability. Let us first investigate the flat-fading case where $\bar{\mathbf{H}}(e^{j\frac{2\pi}{N}k}) = \bar{\mathbf{H}}_0$ for $k = 0, 1, \ldots, N-1$. Now, using Equation 12.6 it follows that $\bar{\mathbf{H}}_0$ has a high rank and is well conditioned if the transmit and receive angle spreads and the number of paths contributing to $\bar{\mathbf{H}}_0$ are sufficiently large. We can therefore expect that large P_0 and large transmit and receive angle spreads will lead to small error probability. In the frequency-selective case the matrices $\bar{\mathbf{H}}(e^{j\frac{2\pi}{N}k}) = \sum_{l=0}^{L-1} \bar{\mathbf{H}}_l\, e^{-j\frac{2\pi}{N}kl}$ can have full rank and be well conditioned even if the individual matrices $\bar{\mathbf{H}}_l$ are rank deficient (or equivalently, the individual cluster angle spreads are small), and the performance will depend on the total number of paths $\sum_{l=0}^{L-1} P_l$ and the angle spread measured over all the taps (i.e., the total angle spread). We can therefore conclude that in the MIMO Ricean case, the presence of frequency selectivity can be beneficial as it can help to restore the rank of the $\bar{\mathbf{H}}(e^{j\frac{2\pi}{N}k})$ matrices. The situation is fundamentally different in the SISO case, where frequency selectivity causes dips in the frequency response, and the performance degrades compared to the flat-fading case.

Low rate codes, such as space–frequency block codes, will be less sensitive to the rank and condition number of the matrices $\bar{\mathbf{H}}(e^{j\frac{2\pi}{N}k})$, as there will be fewer code difference vectors that tend to lie in the null spaces of the $\bar{\mathbf{H}}(e^{j\frac{2\pi}{N}k})$, and hence cause high error probability. More discussions and conclusions along these lines can be found in Simulation Example 4 in Section 12.5.

12.5 Impact of Propagation Parameters on Space–Frequency Codes

We have seen in the previous section that the rank and eigenvalues of the individual correlation matrices $\mathbf{S}_l$ and $\mathbf{R}_l$ determine the diversity gain and coding gain achieved by a space–frequency code. In this section, we shall first relate angle spread and antenna spacing to the eigenvalues of the correlation matrices, and then use these results to study the impact of the propagation environment on the performance of space–frequency codes.

12.5.1 Impact of Propagation Parameters

12.5.1.1 Impact of Cluster Angle Spread

Let us restrict our attention to the transmit correlation matrix $\mathbf{S}_l$ (the same analysis applies to the $\mathbf{R}_l$). Since $\mathbf{S}_l$ is Toeplitz, its limiting ($M_T \to \infty$) eigenvalue distribution can be obtained as $\lambda_l(\nu) = \vartheta_3(\pi(\nu - \Delta_T \cos(\bar{\theta}_{T,l})), e^{-\frac{1}{2}(2\pi\Delta_T \sin(\bar{\theta}_{T,l})\sigma_{\theta_{T,l}})^2})$ with the third-order theta function defined in Equation 12.11. Similar to the discussion in Section 12.3, it follows that $\lambda_l(\nu)$ will be flatter in the case of large antenna spacing and/or large cluster angle spread (i.e., low spatial fading correlation). For small antenna spacing and/or small cluster angle spread (i.e., high spatial fading correlation) $\lambda_l(\nu)$ will be peaky. For any M_T, using the fact that $\mathbf{S}_l = \mathbf{S}_l^H$, the eigenvalues of $\mathbf{S}_l$ can be lower- and upper-bounded by the infimum and the supremum of $\lambda_l(\nu)$. In particular, defining

$$m = \operatorname*{ess\,inf}_{\nu \in [0,1)} \lambda_l(\nu) \qquad M = \operatorname*{ess\,sup}_{\nu \in [0,1)} \lambda_l(\nu)$$

we have that [22]

$$m \le \lambda_i(\mathbf{S}_l) \le M, \qquad i = 0, 1, \ldots, M_T - 1.$$

This result can be used to provide an upper bound on the PEP in terms of $\lambda_l(\nu)$, and hence makes the dependence of the PEP on cluster angle spread and antenna spacing more explicit. For example, assuming flat fading (i.e., $L = 1$) and a nonsingular $\mathbf{S}_0$, we get from Equation 12.20 that

$$P(\mathbf{C} \to \mathbf{E}) \le \left(\frac{mE_s}{4\sigma_n^2}\right)^{-M_R\, r(\mathbf{C}_u)} \prod_{i=0}^{r(\mathbf{C}_u)-1} \lambda_i^{-M_R}(\mathbf{C}_u)$$

where $\mathbf{C}_u = \sigma_0^2(\mathbf{C}-\mathbf{E})^T(\mathbf{C}-\mathbf{E})^*$. Now, using the normalization $\mathrm{Tr}(\mathbf{S}_0) = M_T$, it follows that $m \leq 1$ with $m = 1$ if and only if $\mathbf{S}_0$ is unitary or, equivalently, spatial fading is uncorrelated. The presence of spatial fading correlation can therefore be interpreted as a reduction in the effective SNR by a factor of m.

12.5.1.2 Impact of Total Angle Spread

We shall next investigate the impact of total angle spread on the performance of space–frequency codes. Let us start with the case of receive correlation only and assume that the receive antenna spacing or the receive cluster angle spread is small so that the individual receive correlation matrices $\mathbf{R}_l$ are rank deficient. Consider the extreme scenario $\sigma_{\theta_{R,l}} = 0$ for $l = 0, 1, \ldots, L-1$, where $\mathbf{R}_l = \mathbf{a}(\bar{\theta}_{R,l})\mathbf{a}^H(\bar{\theta}_{R,l})$ $(l = 0, 1, \ldots, L-1)$. Next, assume that the $\mathbf{R}_l$ span mutually orthogonal subspaces; i.e., the $\mathbf{a}(\bar{\theta}_{R,l})$ are orthogonal to each other and hence $\mathbf{R}_l^{1/2}\mathbf{R}_{l'}^{1/2} = \mathbf{0}$ for $l \neq l'$. Of course, for this to hold we need $L \leq M_R$. Roughly speaking, mutual orthogonality of the $\mathbf{a}(\bar{\theta}_{R,l})$ requires that either the total receive angle spread be sufficiently large or M_R is large so that the receive array provides high spatial resolution. Now, if the correlation matrices $\mathbf{R}_l$ indeed span mutually orthogonal subspaces, it follows that $\sigma(\mathbf{C}_Y) = \{\sigma(\widetilde{\mathbf{C}}_Y), 0, \ldots, 0\}$ with [14]

$$\widetilde{\mathbf{C}}_Y = \mathrm{diag}\left\{\sigma_l^2[(\mathbf{C}-\mathbf{E})^*(\mathbf{C}-\mathbf{E})^T] \otimes \mathbf{R}_l\right\}_{l\in\mathcal{L}}. \tag{12.24}$$

The block diagonality of $\widetilde{\mathbf{C}}_Y$ implies several interesting properties. First, we note that Equation 12.24 immediately yields $d = d_f \sum_{l\in\mathcal{L}} r(\mathbf{R}_l)$, where d_f is the diversity order achieved by the space–frequency code in the flat-fading spatially uncorrelated case. Moreover, we can infer from Equation 12.24 that the space–frequency code design criteria reduce to the well-known space–time code design criteria first reported in [2]. Hence, in this case space–frequency code design for the broadband case reduces to classical space–time code design for the narrowband case. This result has a physically intuitive explanation, which is as follows. Start from the i.i.d. case (i.e., $\mathbf{R}_l = \mathbf{I}_{M_R}$ for $l = 0, 1, \ldots, L-1$), where the design criteria are the classical rank and determinant criteria [2] applied to the stacked matrix $[\sqrt{\sigma_0^2}\,(\mathbf{C}-\mathbf{E})^T \;\ldots\; \sqrt{\sigma_{L-1}^2}\,\mathbf{D}^{L-1}(\mathbf{C}-\mathbf{E})^T]$. In order to achieve good performance in terms of error probability we need the columns of $(\mathbf{C}-\mathbf{E})^T$ to be as orthogonal as possible to each other and the columns of $\mathbf{D}^l(\mathbf{C}-\mathbf{E})^T$ to be as orthogonal as possible to the columns of $\mathbf{D}^{l'}(\mathbf{C}-\mathbf{E})^T$ for $l \neq l'$. Next, write $(\mathbf{C}-\mathbf{E})^T = \mathbf{F}(\mathbf{C}_t-\mathbf{E}_t)^T$ where $\mathbf{F}$ with $[\mathbf{F}]_{m,n} = \frac{1}{\sqrt{N}}e^{-j\frac{2\pi}{N}mn}$ is the $N \times N$ FFT matrix and exploit $\mathbf{F}^{-1}\mathbf{D}^l\mathbf{F}(\mathbf{C}_t-\mathbf{E}_t)^T = (\mathbf{C}_{t-l}-\mathbf{E}_{t-l})^T$, where $(\mathbf{C}_{t-l}-\mathbf{E}_{t-l})$ denotes the matrix obtained by cyclically shifting the rows of $(\mathbf{C}_t-\mathbf{E}_t)$ by l positions to the right. The design criteria can now be rephrased (exploiting the unitarity of the FFT) as the columns of $(\mathbf{C}_t-\mathbf{E}_t)^T$ being as orthogonal as possible to each other and the columns of $(\mathbf{C}_{t-l}-\mathbf{E}_{t-l})^T$ being as orthogonal as possible to the columns of $(\mathbf{C}_{t-l'}-\mathbf{E}_{t-l'})^T$ for $l \neq l'$. This means that the cyclically shifted versions of $(\mathbf{C}_t-\mathbf{E}_t)^T$ should be as orthogonal as possible to each other, so that the receiver can separate them and exploit frequency diversity gain. This can be achieved by judicious space–frequency code design [23]. Now, in the case of receive correlation only, where the $\mathbf{R}_l$ span mutually orthogonal subspaces, the separation of the delayed replicas is provided by the channel as the different delayed versions excite mutually orthogonal subspaces. Roughly speaking, if the total receive angle spread is large or M_R is large, the spatial component of the channel orthogonalizes the delayed versions of the transmitted signal and frequency diversity gain is provided for free (assuming ML decoding).

We shall finally discuss the impact of total transmit angle spread on the performance of space–frequency codes in the case of transmit correlation only. As already mentioned, in the presence of transmit correlation, the performance of space–frequency codes depends very much on the geometry of the code relative to the geometry of the transmit correlation matrices. Let us start by considering the flat-fading case (i.e., $L = 1$), where $\mathbf{C}_Y = \sigma_0^2\,(\mathbf{C}-\mathbf{E})^T\mathbf{S}_0(\mathbf{C}-\mathbf{E})^*$. Now, assume that either the transmit cluster angle spread or the transmit antenna spacing is small so that $\mathbf{S}_0$ is rank deficient. In this case, the error events $(\mathbf{C}-\mathbf{E})^*$ lying in the null space of $\mathbf{S}_0$ will cause large PEP. Next, consider the case of multipath propagation where $L > 1$ and retain the assumption of small cluster angle spreads or small transmit antenna spacing. If the total transmit angle spread is large, the $\mathbf{S}_l$ will span different subspaces and hence have different null spaces. Consequently, if $(\mathbf{C}-\mathbf{E})^*$ happens to lie in the null space of one of the $\mathbf{S}_l$, it can still lie in the range of one of the other $\mathbf{S}_l$, and hence using Equation 12.14 it follows that the PEP performance may still be good. This effect will be studied further and quantified in Simulation Example 1 below.

Summarizing, we can conclude that increased total angle spread leads not only to an increased ergodic capacity, as observed in Section 12.3, but also to a generally improved performance from an error probability point of view.

12.5.2 Simulation Results

We conclude our discussion with simulation results that corroborate and quantify some of the analytical results provided in Sections 12.4 and 12.5. Unless specified otherwise, we simulated a space–frequency coded MIMO-OFDM system with $M_T = M_R = 2$, $L = 2$, $\Delta_T = \Delta_R = 1$, $\sigma_0^2 = \sigma_1^2 = \frac{1}{2}$, $N = 32$ tones, no *forward error* correction, and ML decoding. The SNR is defined as $\text{SNR} = 10\,\log_{10}(\frac{M_T E_s}{\sigma_n^2})$. In all simulations the channel was normalized to satisfy $\sum_{l \in \mathcal{L}} \mathcal{E}\{\|\mathbf{H}_l\|_F^2\} = M_T M_R$. We employed a space–frequency coding scheme where an arbitrary (inner) space–time code is used to transmit data in the first $\frac{N}{2}$ tones followed by a repetition of this block. This construction ensures that the additionally available frequency diversity is fully exploited [23].

12.5.2.1 Simulation Example 1

In this simulation example we study the impact of total transmit angle spread $\Delta\bar{\theta}_T = \bar{\theta}_{T,1} - \bar{\theta}_{T,0}$ in the presence of high transmit correlation (caused by small cluster angle spreads) on the performance of space–frequency codes. As inner codes we used quadrature phase shift keying (QPSK)-based spatial multiplexing [5, 6, 9], which amounts to transmitting independent data symbols on each tone and each antenna, and the Alamouti scheme [3] based on QPSK. The transmit cluster angle spreads were chosen as $\sigma_{\theta_{T,0}} = \sigma_{\theta_{T,1}} = 0$ so that $\mathbf{S}_l = \mathbf{a}(\bar{\theta}_{T,l})\mathbf{a}^H(\bar{\theta}_{T,l})$ $(l = 0, 1)$. The receive antennas were assumed to fade in an uncorrelated fashion (i.e., $\mathbf{R}_0 = \mathbf{R}_1 = \mathbf{I}_2$). The top diagram in Figure 12.6 shows the block error rate as a function of $\Delta\bar{\theta}_T$ with $\bar{\theta}_{T,0} = 0$ for spatial multiplexing at an SNR of 14 dB and for the Alamouti scheme at an SNR of 7 dB. We can clearly see that spatial multiplexing is very sensitive to total transmit angle spread and that the performance improves significantly for the case where the array response vectors $\mathbf{a}(\bar{\theta}_{T,0})$ and $\mathbf{a}(\bar{\theta}_{T,1})$ are close to orthogonal to each other. This is reflected by displaying the condition number of the 2×2 matrix $\mathbf{B} = [\mathbf{a}(\bar{\theta}_{T,0})\ \mathbf{a}(\bar{\theta}_{T,1})]$ in the bottom diagram of Figure 12.6. The lower-rate Alamouti scheme is virtually unaffected by the total transmit angle spread, which is due to the fact that it orthogonalizes the channel irrespectively of the channel realization, and hence its performance is independent of the rank properties of the channel. For spatial multiplexing the situation is different since the performance depends critically on the rank of the channel realizations. To be more specific, we note that each transmit correlation matrix $\mathbf{S}_l$ spans a one-dimensional subspace with the angle between these subspaces being a function of total angle spread (as evidenced by the bottom plot of Figure 12.6). If the total transmit angle spread is small, the two subspaces tend to be aligned so that transmit signal vectors that happen to lie in the orthogonal complement of this subspace will cause high error probability. If the total angle spread is large, the $\mathbf{S}_l$ tend to span different subspaces. Hence, if a transmit vector excites the null space of $\mathbf{S}_0$, it will lie in the range space of $\mathbf{S}_1$ and vice versa. This leads to good performance in terms of error rate because none of the transmit vectors will be attenuated by the channel (on average, where the average is with respect to the random channel).

12.5.2.2 Simulation Example 2

This example investigates the impact of transmit and receive antenna correlation on the performance of space–frequency codes. Figure 12.7 displays the block error rate for a system employing the two-antenna 16-state QPSK trellis code proposed in [2] as inner code. For $M_T = 2$ and $M_R = 3$, we show the cases of no correlation (i.i.d. channel), transmit correlation only ($\Delta_T = 0.1$, $\sigma_{\theta_{T,0}} = \sigma_{\theta_{T,1}} = 0.25$, $\bar{\theta}_{T,0} = \bar{\theta}_{T,1} = \pi/4$), and receive correlation only ($\Delta_R = 0.1$, $\sigma_{\theta_{R,0}} = \sigma_{\theta_{R,1}} = 0.25$, $\bar{\theta}_{R,0} = \bar{\theta}_{R,1} = \pi/4$), respectively. We note that this choice of channel parameters results in $|\rho(\Delta, \bar{\theta}_l, \sigma_{\theta_l})| = 0.994$ and $|\rho(2\Delta, \bar{\theta}_l, \sigma_{\theta_l})| = 0.976$ for $l = 0, 1$, which basically amounts to fully correlated spatial fading. Figure 12.7 shows that best performance is

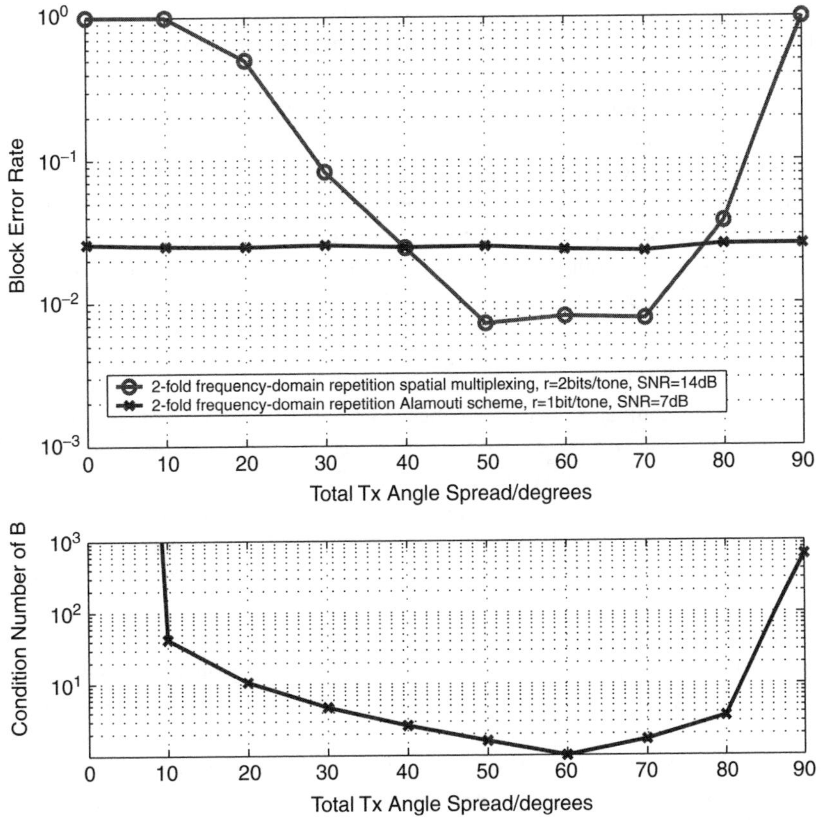

FIGURE 12.6 Impact of total transmit angle spread on performance of space-frequency codes.

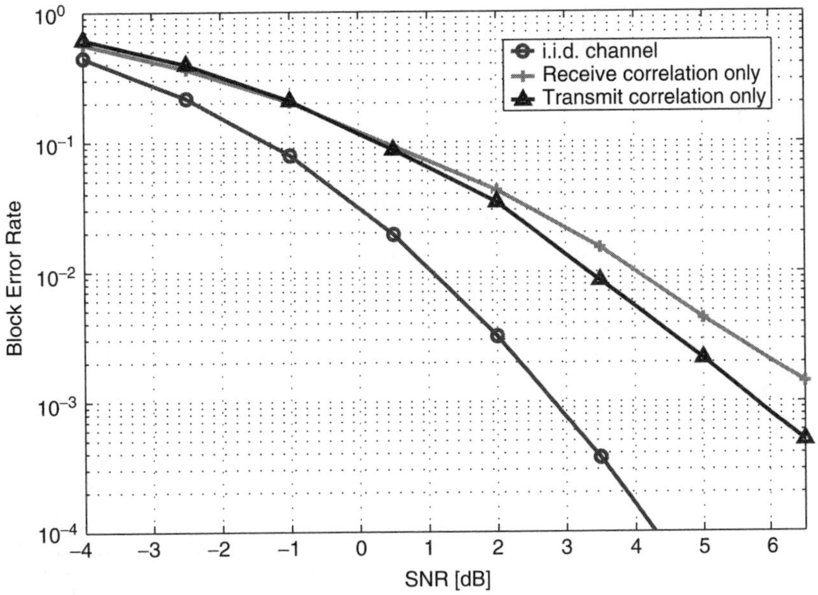

FIGURE 12.7 Impact of transmit and receive antenna correlation on the performance of space-frequency Trellis codes.

achieved in the i.i.d. case (as expected), performance degrades in the presence of transmit correlation only, and degrades even more in the presence of receive correlation only. This asymmetry between the transmit and receive correlation cases can be explained as follows. In the i.i.d. case the channel offers diversity order $d = M_T M_R L = 12$. For the correlated cases we make the simplifying assumption $|\rho(\Delta, \bar{\theta}_l, \sigma_{\theta_l})| = |\rho(2\Delta, \bar{\theta}_l, \sigma_{\theta_l})| = 1$ for $l = 0, 1$, which implies $r(\mathbf{S}_0) = r(\mathbf{S}_1) = r(\mathbf{R}_0) = r(\mathbf{R}_1) = 1$. In the case of transmit correlation only, the diversity order follows from Equation 12.21 as $d = M_R(r(\mathbf{S}_0) + r(\mathbf{S}_1)) = 6$, whereas in the case of receive correlation only Equation 12.19 yields $d \leq M_T(r(\mathbf{R}_0) + r(\mathbf{R}_1)) = 4$. These differences in the achievable diversity order are reflected in the error rate performance depicted in Figure 12.7. While not shown in the diagram, we emphasize that the performance gap between transmit correlation only and receive correlation only is even more pronounced for higher SNR.

12.5.2.3 Simulation Example 3

This simulation example compares the performance of orthogonal and nonorthogonal space–frequency codes in the presence of transmit correlation.[3] We have seen above that the impact of transmit antenna correlation is highly dependent on how the code excites the channel (i.e., the code geometry) relative to the channel geometry. Given that the transmitter does not know the channel, an orthogonal scheme such as the Alamouti scheme, which excites all spatial directions uniformly, should exhibit maximum robustness with respect to this kind of channel impairment. Nonorthogonal schemes such as spatial multiplexing do not excite all spatial directions uniformly and hence suffer from widely varying performance losses in the presence of transmit correlation. In order to demonstrate this effect, we compared a 16-QAM (quadrature amplitude modulation)-based Alamouti scheme as a simple orthogonal inner code to spatial multiplexing based on QPSK (nonorthogonal inner code). Note that this choice of symbol constellations ensures that the transmission rate is the same in both cases. The SNR was kept constant at 12 dB, while the absolute value of the transmit antenna correlation coefficient was varied from 0 to 1. The correlation matrix was chosen to be identical for both taps. Figure 12.8 shows the resulting bit error rates assuming Gray encoding. It is clearly seen that while with uncorrelated transmit antennas the Alamouti scheme performs inferior compared to spatial multiplexing, performance degradation with increasing correlation is much worse for spatial multiplexing, such that at full correlation the Alamouti scheme performs significantly better. These performance differences along with the results of Simulation Example 1 lead us to the conclusion that space–frequency block codes exhibit superior robustness with respect to varying propagation conditions when compared to (spatially) high-rate schemes such as spatial multiplexing.

12.5.2.4 Simulation Example 4

The last simulation example studies the impact of K-factor and geometry of the Ricean component on the performance of space–frequency codes. We compared a 2 bits/tone spatial multiplexing binary PSK (BPSK) code to a simple BPSK-based delay diversity scheme with delay 4 (achieved by premultiplying the signals with a linear phase factor in the frequency domain). In both cases, we used a rate 1/2, constraint length 4 convolutional code with generator polynomials 15 and 17, and random interleaving. Hence, the overall transmission rate for delay diversity was half the transmission rate for spatial multiplexing. We emphasize that the difference in overall transmission rate results from the fact that the spatial rate of delay diversity (one symbol per vector transmission) is half the spatial rate of spatial multiplexing (two symbols per vector transmission). This ensures that the performance differences between the two schemes are due to the different spatial signaling structures and rates. The receiver for spatial multiplexing first separates the spatial signals using a minimum mean square error (MMSE) front end and then performs soft Viterbi decoding, whereas in the case of delay diversity a soft Viterbi decoder only was used (spatial separation is not required because the two transmit antennas are effectively collapsed into one antenna). The matrices

[3] A space–frequency code is referred to as orthogonal or nonorthogonal if the inner space–time code is orthogonal or nonorthogonal, respectively.

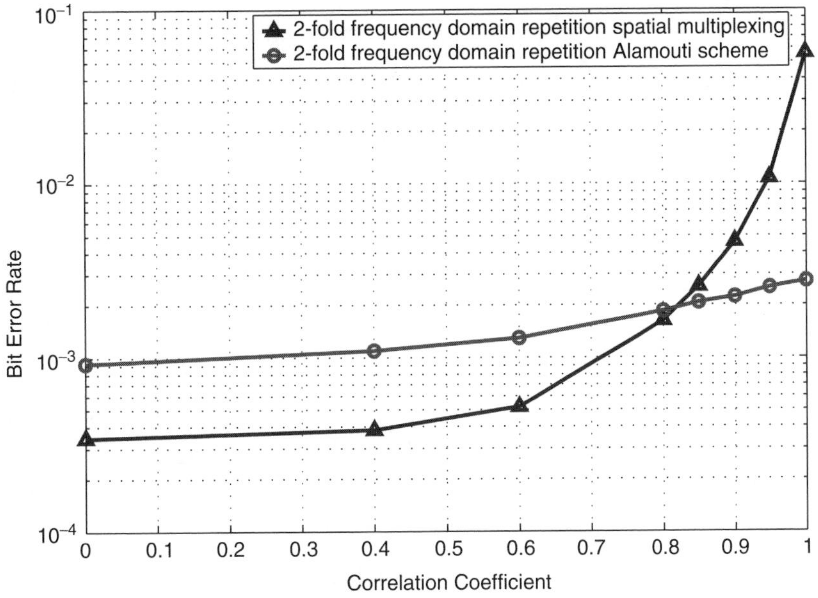

FIGURE 12.8 Impact of varying transmit antenna correlation on performance of space-frequency codes.

$\bar{\mathbf{H}}_l$ were generated using the parametric model described in Equation 12.6. The four cases depicted in Figure 12.9 and Figure 12.10, respectively, are:

- Scenario 1: Pure Rayleigh fading on both taps, i.e., $K_0 = K_1 = 0$.
- Scenario 2: In this and the remaining cases $K_0 = K_1 = 16$ dB. Only one significant path is present in each of the two taps, i.e., $P_0 = P_1 = 1$; therefore $r(\bar{\mathbf{H}}_0) = r(\bar{\mathbf{H}}_1) = 1$. The angle spread

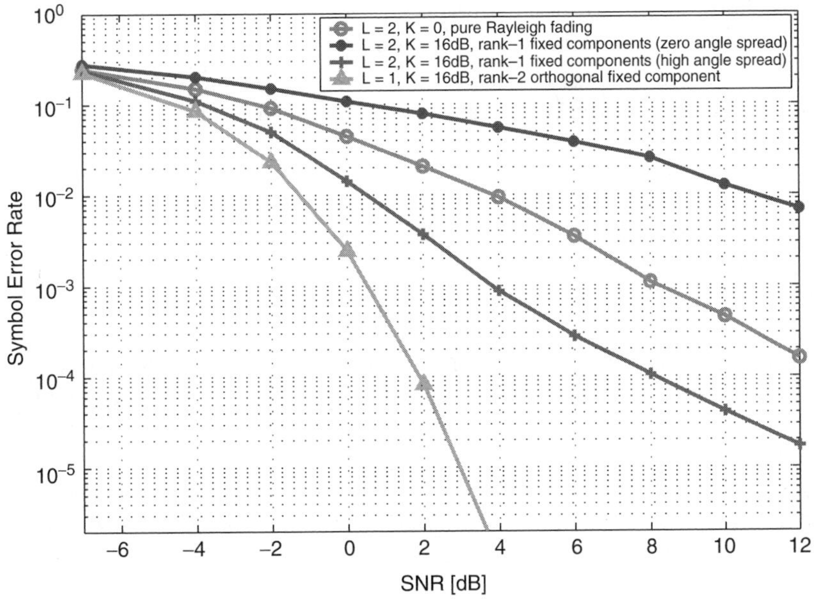

FIGURE 12.9 Impact of K-factor ($K = K_0 = K_1$) and geometry of Ricean component on performance of MIMO-OFDM spatial multiplexing.

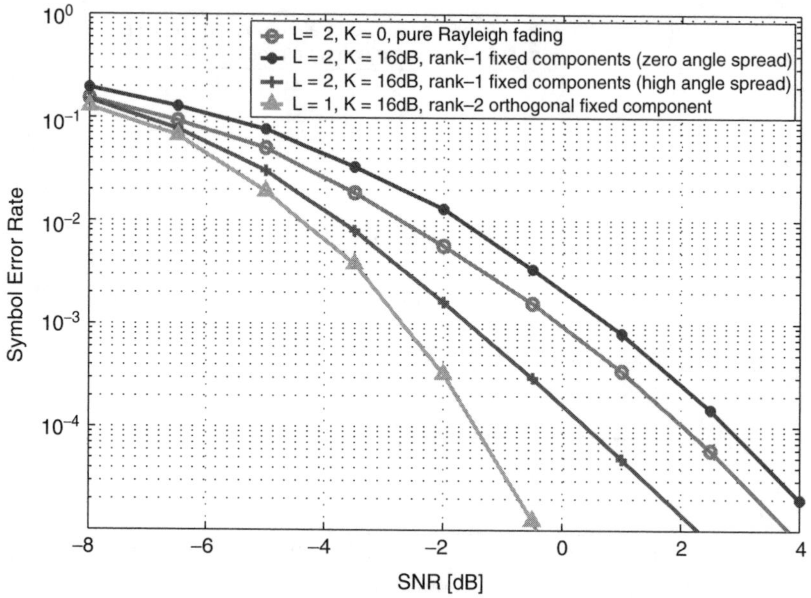

FIGURE 12.10 Impact of K-factor ($K = K_0 = K_1$) and geometry of Ricean component on performance of delay diversity combined with convolutional code.

across the two taps is zero so that $\bar{\mathbf{H}}_0 = \bar{\mathbf{H}}_1$. Due to the finite constellation used here, the specific realization of the Ricean channel component has a large impact on performance. In order to make our conclusions representative, we average over a uniform distribution of angles of arrival and departure (which determine the Ricean channel component) between 0 and 2π.

- Scenario 3: Again both taps have only one significant path, but there is an angle spread of $\frac{\pi}{2}$ across the taps, such that $\bar{\mathbf{H}}_0$ and $\bar{\mathbf{H}}_1$ are still rank 1, but not identical anymore. Also, we again average over a uniform angle of arrival and departure distribution, but the angle *difference* between the two taps is kept constant at $\frac{\pi}{2}$ for both the angle of arrival and departure.
- Scenario 4: Flat-fading scenario (i.e., $L = 1$), but two paths adding up at this single tap, i.e., $P_0 = 2$, resulting in a rank 2 matrix $\bar{\mathbf{H}}_0$. The angles of arrival and departure were chosen such that $\bar{\mathbf{H}}_0$ is orthogonal.

12.5.2.4.1 *Spatial Multiplexing*

From Figure 12.9 it is clearly seen that depending on the propagation environment a high K-factor can be either beneficial or detrimental. In Scenario 2, the matrices $\bar{\mathbf{H}}(e^{j\frac{2\pi}{N}k})$ have a rank of 1 for $k = 0, 1, \ldots, N-1$, thus severely impeding spatial multiplexing performance. Scenario 4 leads to good performance since $\bar{\mathbf{H}}(e^{j\frac{2\pi}{N}k}) = \bar{\mathbf{H}}_0$ is orthogonal for $k = 0, 1, \ldots, N-1$. Finally, we can see that in Scenario 3 we obtain significantly better performance than in Scenario 2, which is due to the fact that the high angle spread across the two taps results in full-rank $\bar{\mathbf{H}}(e^{j\frac{2\pi}{N}k})$, even though $\bar{\mathbf{H}}_0$ and $\bar{\mathbf{H}}_1$ each have a rank of 1. In Scenario 3 if only one tap were present the $\bar{\mathbf{H}}(e^{j\frac{2\pi}{N}k})$ $(k = 0, 1, \ldots, N-1)$ would be rank deficient, and hence the performance would be comparable to Scenario 2. It is therefore interesting to observe that the presence of frequency selectivity (provided that the delayed paths increase the total angle spread) can improve performance in the Ricean MIMO case. This is in contrast to the SISO Ricean case where frequency selectivity leads to performance degradation when compared to the frequency-flat case.

12.5.2.4.2 *Delay Diversity Combined with Convolutional Code*

The same trends are true for the lower-rate delay diversity scheme, but the differences are less pronounced. In particular, Scenario 2 exhibits a much smaller performance loss compared to the Rayleigh fading case.

This is due to the fact that since we are dealing with a scheme with half the spatial signaling rate, the rank properties and condition numbers of the $\bar{\mathbf{H}}(e^{j\frac{2\pi}{N}k})$ are less important than in the case of spatial multiplexing.

Glossary

Diversity gain — Improvement in link reliability obtained by transmitting the same data on independently fading branches.

Fading — Fluctuation in the signal level due to shadowing and multipath effects.

FFT — Fast Fourier transform.

Forward error correction (FEC) — A technique that inserts redundant bits during transmission to help detect and correct bit errors during reception.

IFFT — Inverse fast Fourier transform.

Interleaving — A form of data scrambling that spreads bursts of bit errors evenly over the received data allowing efficient forward error correction.

Link reliability — Measure for the quality of a wireless communication link.

Multiple-input multiple-output (MIMO) — Wireless communication system employing an antenna array at both ends of the link.

(Spatial) multiplexing gain — Capacity gain at no additional power or bandwidth consumption obtained through the use of multiple antennas at both ends of a wireless radio link.

Orthogonal frequency division multiplexing (OFDM) — Modulation scheme that divides the available frequency band into subcarriers (or tones) of smaller bandwidth and thereby drastically simplifies equalization.

Space–frequency coding — Coding technique that realizes spatial and frequency diversity gains without knowing the channel in the transmitter by spreading information across antennas (space) and frequency.

Space–time coding — Coding technique that realizes spatial diversity gain without knowing the channel in the transmitter by spreading information across antennas (space) and time.

Spatial multiplexing — Signaling technique to realize spatial multiplexing gain.

Further Information

The books [24–26] contain excellent introductions to and overviews of MIMO wireless and space–time coding. Four special issues have appeared on MIMO wireless in the past 2 years in the *IEEE Transactions on Signal Processing* (October 2002 and October 2003), *IEEE Journal on Selected Areas in Communications* (April and June 2003), and the *IEEE Transactions on Information Theory* (October 2003).

References

[1] W.C. Jakes, *Microwave Mobile Communications*, New York: Wiley, 1974.

[2] V. Tarokh, N. Seshadri, and A.R. Calderbank, "Space-time codes for high data rate wireless communication: Performance criterion and code construction," *IEEE Trans. Inf. Theory*, 44(2), 744–765, 1998.

[3] S.M. Alamouti, "A simple transmit diversity technique for wireless communications," *IEEE J. Selected Areas Commun.*, 16(8), 1451–1458, 1998.

[4] B.M. Hochwald and T.L. Marzetta, "Unitary space-time modulation for multiple-antenna communications in Rayleigh flat fading," *IEEE Trans. Inf. Theory*, 46(2), 543–564, 2000.

[5] A.J. Paulraj and T. Kailath, "Increasing capacity in wireless broadcast systems using distributed transmission/directional reception," U.S. Patent 5,345,599, 1994.

[6] G.J. Foschini, "Layered space-time architecture for wireless communication in a fading environment when using multi-element antennas," *Bell Labs Tech. J.*, 41–59, 1996.

[7] I.E. Telatar, "Capacity of multi-antenna Gaussian channels," *Eur. Trans. Telecommun.*, 10, 585–595, 1999.

[8] G.G. Raleigh and J.M. Cioffi, "Spatio-temporal coding for wireless communication," *IEEE Trans. Commun.*, 46(3), 357–366, 1998.

[9] H. Bölcskei, D. Gesbert, and A.J. Paulraj, "On the capacity of OFDM-based spatial multiplexing systems," *IEEE Trans. Commun.*, 50(2), 225–234, 2002.

[10] G.J. Foschini and M.J. Gans, "On limits of wireless communications in a fading environment when using multiple antennas," *Wireless Personal Commun.*, 6, 311–335, 1998.

[11] A. Peled and A. Ruiz, "Frequency domain data transmission using reduced computational complexity algorithms," in *Proceedings of the IEEE ICASSP-80*, Denver, CO, 1980, pp. 964–967.

[12] B. LeFloch, M. Alard, and C. Berrou, "Coded orthogonal frequency division multiplex," *Proc. IEEE*, 83(6), 982–996, 1995.

[13] H. Bölcskei and A.J. Paulraj, "Space-frequency coded broadband OFDM systems," in *Proceedings of IEEE WCNC-2000*, Chicago, IL, September 2000, pp. 1–6.

[14] H. Bölcskei, M. Borgmann, and A.J. Paulraj, "Impact of the propagation environment on the performance of space-frequency coded MIMO-OFDM," *IEEE J. Selected Areas Commun.*, 21(3), 427–439, 2003.

[15] D. Asztély, "On Antenna Arrays in Mobile Communication Systems: Fast Fading and GSM Base Station Receiver Algorithms," Technical Report IR-S3-SB-9611, Royal Institute of Technology, Stockholm, Sweden, March 1996.

[16] M. Sandell, "Design and Analysis of Estimators for Multicarrier Modulation and Ultrasonic Imaging," Ph.D. dissertation, Lulea University of Technology, Lulea, Sweden, 1996.

[17] U. Grenander and G. Szegö, *Toeplitz Forms and Their Applications*, New York: Chelsea Publishing Company, 1984.

[18] E. Biglieri, J. Proakis, and S. Shamai, "Fading channels: Information-theoretic and communications aspects," *IEEE Trans. Inf. Theory*, 44(6), 2619–2692, 1998.

[19] V. Erceg, Private communication, April 2001.

[20] B. Lu and X. Wang, "Space-time code design in OFDM systems," in *Proceedings of the IEEE GLOBECOM*, Vol. 2, San Francisco, CA, November 2000, pp. 1000–1004.

[21] R.A. Horn and C.R. Johnson, *Topics in Matrix Analysis*, New York: Cambridge Press, 1991.

[22] R.M. Gray, "Toeplitz and Circulant Matrices," Technical Report, Stanford University, ISL, August 2002. (Online: available at http://ee-www.stanford.edu/~gray/toeplitz.html.)

[23] H. Bölcskei, M. Borgmann, and A.J. Paulraj, "Space-frequency coded MIMO-OFDM with variable multiplexing-diversity tradeoff," in *Proceedings of the IEEE International Conference on Communications*, Anchorage, AK, May 2003, pp. 2837–2841.

[24] A.J. Paulraj, R.U. Nabar, and D.A. Gore, *Introduction to Space-Time Wireless Communications*, Cambridge, U.K.: Cambridge University Press, 2003.

[25] E.G. Larsson and P. Stoica, *Space-Time Block Coding for Wireless Communications*, Cambridge, U.K.: Cambridge University Press, 2003.

[26] G.B. Giannakis, Z. Liu, X. Ma, and S. Zhou, *Space-Time Coding for Broadband Wireless Communications*, New York: Wiley, 2004.

13

Space–Time Coding and Signal Processing for Broadband Wireless Communications

Naofal Al-Dhahir
The University of Texas at Dallas

13.1 Introduction 13-1
13.2 Broadband Wireless Channel Model 13-3
13.3 Information-Theoretic Considerations 13-5
Shannon Capacity of Fading ISI Channels • Diversity Order • Design Criteria for Space–Time Codes over Flat-Fading Channels
13.4 Signal Transmission Issues 13-7
Transmitter Techniques • Receiver Techniques
13.5 Summary and Future Challenges 13-23

Abstract

We present an overview of physical-layer techniques for space–time-coded signaling over broadband wireless channels. This includes a discussion of information-theoretic performance limits, transmitter techniques, and receiver signal processing techniques. The latter are classified into coherent or noncoherent techniques depending on whether channel knowledge at the receiver is required. Practical algorithms for multiple-antenna channel estimation under quasi-static and rapidly varying channel conditions are presented. Effective space–time algorithms for joint equalization/decoding and for interference cancellation are also described. We conclude with a discussion of future challenges. Throughout the chapter, our objective is to optimize the performance of space–time signal processing algorithms subject to practical complexity constraints.

13.1 Introduction

The salient feature of wireless transmission is the randomness of the communications channel, which leads to random fluctuations in the received signal commonly known as fading. This randomness can be exploited to enhance performance through *diversity*. We broadly define diversity as the method of

0-8493-1657-X/$0.00+$1.50

conveying information through multiple independent instantiations of these random fades. There are several forms of diversity; our focus in this chapter will be on *spatial* diversity through multiple independent transmit/receive antennas. Information theory has been used to show that multiple antennas have the potential to dramatically increase achievable bit rates [1], thus converting wireless channels from narrow to wide data pipes.

The earliest form of spatial transmit diversity is the delay diversity scheme proposed in [2, 3] where a signal is transmitted from one antenna, then delayed one time slot, and transmitted from the other antenna. Signal processing is used at the receiver to decode the superposition of the original and time-delayed signals. Viewing multiple-antenna diversity as independent conduits of information, more sophisticated transmission (coding) schemes can be designed to get closer to theoretical performance limits. Using this approach, we focus on space–time coding (STC) schemes defined by Tarokh et al. [4] and Alamouti [5], which introduce temporal and spatial correlation into the signals transmitted from different antennas without increasing the total transmitted power or the transmission bandwidth. There is, in fact, a diversity gain that results from multiple paths between the base station and user terminal, and a coding gain that results from how symbols are correlated across transmit antennas. Significant performance improvements are possible with only two antennas at the base station and one or two antennas at the user terminal, and with simple receiver structures. The second antenna at the user terminal can be used to further increase system capacity through interference suppression.

In only a few years, space–time codes have progressed from invention to adoption in the major wireless standards. For wideband code division multiple access (WCDMA) where short spreading sequences are used, transmit diversity provided by space–time codes represents the difference between data rates of 100 and 384 kb/s. Section 13.4 describes modern multiple-antenna coding and signal processing methods that take steps toward the theoretically predicted gains of Section 13.3. Our emphasis is on solutions that include channel estimation, joint decoding and equalization, and where the complexity of signal processing is practical. The new world of multiple transmit and receive antennas requires significant modification of techniques developed for single-transmit single-receive communication. Since receiver cost and complexity are important considerations, our treatment of innovation in signal processing is grounded in systems with one, two, or four transmit antennas and one or two receive antennas. For example, the interference cancellation techniques presented in Section 13.4 enable transmission of 1 Mb/s over a 200-kHz EDGE (enhanced data rate for GSM (Global System for Mobile) Communications Evolution) channel using four transmit and two receive antennas. Hence, our limitation on numbers of antennas does not significantly dampen user expectations.

Initial STC research efforts focused on narrowband flat-fading channels [4–6]. Successful implementation of STC over multiuser broadband frequency-selective channels requires the development of novel, practical, and high-performance signal processing algorithms for channel estimation, joint equalization/decoding, and interference suppression. This task is quite challenging due to the long delay spread of broadband channels, which increases the number of channel parameters to be estimated and the number of trellis states in joint equalization/decoding, especially with multiple transmit antennas. This, in turn, places significant additional computational and power consumption loads on user terminals. On the other hand, development and implementation of such advanced algorithms for broadband wireless channels promise even more significant performance gains than those reported for narrowband channels [4–6] due to availability of multipath (in addition to spatial) diversity gains that can be realized. By virtue of their design, space–time-coded signals enjoy rich structure that can (and should) be exploited to develop near-optimum reduced-complexity modem signal processing algorithms. The organization of this chapter is as follows. Section 13.2 lays out the broadband wireless channel model assumed and used as a basis for the information-theoretic and signal transmission developments in Sections 13.3 and 13.4, respectively. The former includes discussion of Shannon capacity, diversity order, and STC design criteria whereas the latter covers signal processing techniques for space–time transceivers. The chapter is concluded in Section 13.5 with a summary and discussion of several future challenges. Throughout this chapter, the notations $(\cdot)^*$ and $(\bar{\cdot})$ are used interchangeably to denote the complex conjugate transpose operation.

13.2 Broadband Wireless Channel Model

A typical outdoor wireless propagation environment is depicted in Figure 13.1 where the mobile wireless terminal is communicating with a wireless access point (base station). The signal transmitted from the mobile may reach the access point directly (line of sight) or through multiple reflections on local scatterers (buildings, mountains, etc.). As a result, the received signal is affected by multiple random attenuations and delays. Moreover, the mobility of either the nodes or the scattering environment may cause these random fluctuations to vary with time. Furthermore, a shared wireless environment may cause undesirable interference to the transmitted signal. The combined effect of these factors makes wireless a challenging communication environment.

For a transmitted signal $s(t)$, the continuous-time received signal $y_c(t)$ can be expressed as

$$y_c(t) = \int h_c(t;\tau)s(t-\tau)d\tau + z(t) \tag{13.1}$$

where $h_c(t;\tau)$ is the response of the time-varying channel[1] if an impulse is sent at time $t-\tau$, and $z(t)$ is the additive Gaussian noise. To collect discrete-time sufficient statistics we need to sample Equation 13.1 faster than the Nyquist rate. That is, we sample Equation 13.1 at a rate larger than $2(W_I + W_s)$, where W_I is the input bandwidth and W_s is the bandwidth of the channel time variation. In this chapter, we assume that this criterion is met and therefore we focus on the following discrete-time model:

$$y(k) = y_c(kT_s) = \sum_{l=0}^{\nu} h(k;l)x(k-l) + z(k) \tag{13.2}$$

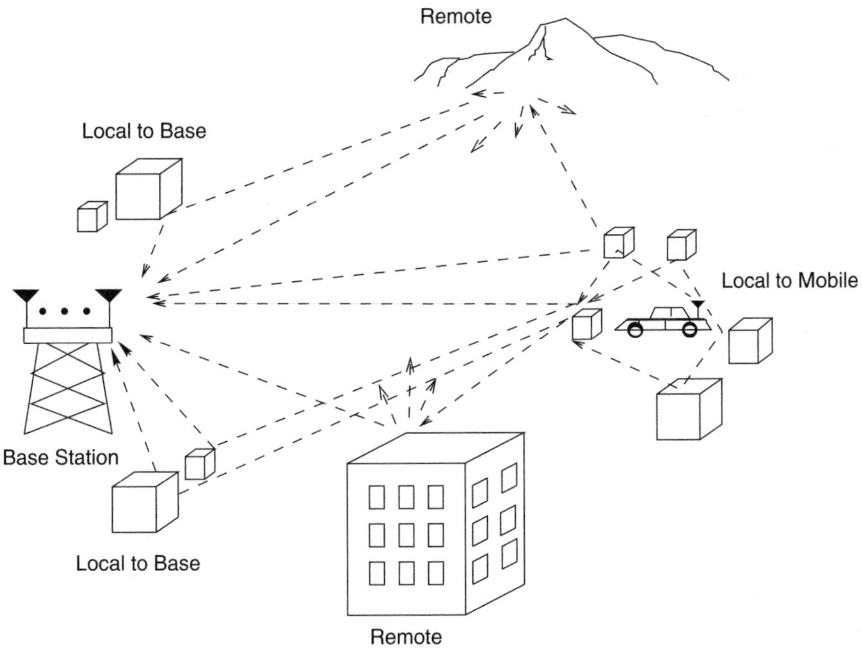

FIGURE 13.1 Radio propagation environment.

[1]Including the effects of transmit/receive filters.

where $y(k)$, $x(k)$, and $z(k)$ are the output, input, and noise samples at sampling instant k, respectively, and $h(k;l)$ represents the sampled time-varying channel impulse response (CIR) of finite memory ν. Any loss in modeling the channel as having a finite-duration impulse response can be made small by appropriately selecting ν.

Three key characteristics of broadband mobile wireless channels are time selectivity, frequency selectivity, and space selectivity. Time selectivity arises from mobility, frequency selectivity arises from broadband transmission, and space selectivity arises from the spatial interference patterns of the radio waves. The three respective key parameters in the characterization of mobile broadband wireless channels are *coherence time*, *coherence bandwidth*, and *coherence distance*. The coherence time is the time duration over which each CIR tap can be assumed constant. It is approximately equal to the inverse of the Doppler frequency.[2] The channel is said to be *time selective* if the symbol period is larger than the channel coherence time. The coherence bandwidth is the frequency duration over which the channel frequency response can be assumed flat. It is approximately equal to the inverse of the channel delay spread.[3] The channel is said to be *frequency selective* if the symbol period is smaller than the delay spread of the channel. Likewise, the coherence distance is the maximum spatial separation over which the channel response can be assumed constant. This can be related to the behavior of arrival directions of the reflected radio waves and is characterized by the *angular spread* of the multiple paths [7, 8]. The channel is said to be *space selective* between two antennas if their separation is larger than the coherence distance.

The channel memory causes interference among successive transmitted symbols that results in significant performance degradation unless corrective measures (known as equalization) are implemented. In this chapter, we shall use the terms *frequency-selective channel*, *broadband channel*, and *intersymbol interference (ISI) channel* interchangeably. The introduction of M_t transmit and M_r receive antennas leads to the following generalization of the basic channel model:

$$\mathbf{y}(k) = \sum_{l=0}^{\nu} \mathbf{H}(k;l)\mathbf{x}(k-l) + \mathbf{z}(k) \tag{13.3}$$

where the $M_r \times M_t$ complex matrix $\mathbf{H}(k;l)$ represents the lth tap of the channel matrix response with $\mathbf{x} \in \mathbb{C}^{M_t}$ as the input and $\mathbf{y} \in \mathbb{C}^{M_r}$ as the output. The input vector may have independent entries to achieve high throughput (e.g., through spatial multiplexing) or correlated entries through coding or filtering to achieve high reliability (better distance properties, higher diversity, spectral shaping, or desirable spatial profile). Throughout this article, the input is assumed to be zero mean and satisfy an average power constraint, i.e., $\mathbb{E}[||\mathbf{x}(k)||^2] \leq P$. The vector $\mathbf{z} \in \mathbb{C}^{M_r}$ models the effects of noise and interference.[4] It is assumed to be independent of the input and is modeled as a complex additive circularly symmetric Gaussian vector with $\mathbf{z} \sim \mathcal{CN}(0, \mathbf{R}_{zz})$, i.e., a complex Gaussian vector with mean $\mathbf{0}$ and covariance $\mathbf{R}_{zz}$. Finally, we modify the basic channel model to accommodate a block or frame of N consecutive symbols. Now Equation 13.3 can be expressed in matrix notation as

$$\mathbf{y} = \mathbf{H}\mathbf{x} + \mathbf{z} \tag{13.4}$$

where $\mathbf{y}, \mathbf{z} \in \mathbb{C}^{N.M_r}$, $\mathbf{x} \in \mathbb{C}^{M_t(N+\nu)}$, and $\mathbf{H} \in \mathbb{C}^{N.M_r \times M_t(N+\nu)}$. In each input block, we insert a guard sequence of length equal to the channel memory ν to eliminate interblock interference (IBI). In practice, the most common choices for the guard sequence are the all–zeros sequence (also known as *zero stuffing*) and the *cyclic prefix* (CP). When the channel is known at the transmitter, it is possible to increase throughput by optimizing the choice of the guard sequence.

---

[2]The Doppler frequency is a measure of the frequency spread experienced by a pure sinusoid transmitted over the channel. It is equal to the ratio of the mobile speed to the carrier wavelength.

[3]The channel delay spread is a measure of the time spread experienced by a pure impulse transmitted over the channel.

[4]Including co-channel interference, adjacent channel interference, and multiuser interference.

The channel model in Equation 13.4 includes several popular special cases. First, the quasi-static channel model follows by assuming the channel time invariant within the transmission block. In this case, using the cyclic prefix makes the channel matrix **H** *block circulant*, hence diagonalizable using the fast Fourier transform (FFT). Second, the flat-fading channel model follows by setting $\nu = 0$, which renders the channel matrix **H** a *block diagonal* matrix. Third, the channel model for single-antenna transmission, reception, or both follows directly by setting M_t, M_r, or both equal to 1, respectively.

13.3 Information-Theoretic Considerations

13.3.1 Shannon Capacity of Fading ISI Channels

Transmitter and receiver diversity for time-invariant ISI channels have been examined in [9–11] and references therein. In [9, 11], the *time-invariant* multiple-input multiple-output (MIMO) ISI channel is studied when *both* the transmitter and the receiver have perfect side information of the channel state. When dealing with *time-invariant* channels, the Fourier basis is the eigenbasis of the channel and one can easily develop a capacity argument using standard techniques [12]. Using such an argument, it is shown in [11] that the capacity scales linearly with the number of antennas ($\min\{M_t, M_r\}$) at high signal-to-noise ratio (SNR). Therefore, since the Fourier basis is optimal in this case, orthogonal frequency division multiplexing (OFDM) is a good choice as a modulation technique for the MIMO channel. In [13] only the receiver is assumed to know the fading channel state. However, a quasi-static assumption where the channel is assumed to be time invariant over the transmission block is used, making it suitable for slowly time-varying channels where transmission bandwidth $\gg$ Doppler spread. Here too, since the channel is block time invariant, the Fourier basis is the eigenbasis for the channel and therefore is the optimal transmission basis. The rate of reliable information transmission for the scalar ISI channel has been derived in [13] in terms of the expected mutual information. This argument can easily be generalized [14] to the multiple-antenna block-fading ISI channel as

$$\lim_{n\to\infty} R_n = (2\pi)^{-1}\mathbb{E}\left[\int_0^{2\pi} log(|\mathbf{I} + \mathbf{H}(f)\mathbf{S}(f)\mathbf{H}^H(f)/\sigma^2|)df\right] \tag{13.5}$$

where $\mathbf{H}(f)$ is the Fourier transform of the channel matrix and $\mathbf{S}(f)$ is the input power spectral density. In general, maximizing Equation 13.5 with respect to $\mathbf{S}(f)$ is a hard problem and further simplifying assumptions need to be made for solving this problem. However, schemes using a flat input spectrum in both time and space may be practical [1, 4, 15].

For time-varying channels, there is an inherent conflict between increasing the transmission block length (for coding arguments) and the block time invariance assumption. Therefore, as the block length increases, there could be time variations within a transmission block. In this case, the Fourier basis is no longer the optimal basis for transmission and would suffer loss of orthogonality since it would not be the eigenbasis for the time-varying channel. The general question of a characterization of capacity for such a channel is open. Some lower bounds on capacity have been developed [14, 16].

13.3.2 Diversity Order

The focus in Section 13.3.1 was on transmission rate. Another very practical performance criterion is probability of error. This is particularly important when we are coding over a small number of blocks (low delay) where the Shannon capacity is zero [13] and, therefore, we need to design for low error probability. By characterizing the error probability, we can also formulate design criteria for space–time codes in Section 13.3.3

Since we are allowed to transmit a coded sequence, we are interested in the probability that an erroneous code word **e** is mistaken for the transmitted code word **x**. This is called the *pairwise error probability* (PEP) and is then used to bound the error probability. This is analyzed under the condition that the receiver has perfect channel state information. However, a similar analysis can be performed when the receiver does not know the channel state information but has statistical knowledge of the channel.

For simplicity, we shall first present the result for a flat-fading Rayleigh channel (where $\nu = 0$). In the case when the receiver has perfect channel state information, we can bound the PEP between $\mathbf{x}$ and $\mathbf{e}$ (denoted by $P(\mathbf{x} \to \mathbf{e})$) as follows [4, 17]:

$$P(\mathbf{x} \to \mathbf{e}) \leq \left[\frac{1}{\prod_{n=1}^{M_t} \left(1 + \frac{E_s}{4N_0}\lambda_n\right)} \right]^{M_r} \tag{13.6}$$

where λ_n are the eigenvalues of the matrix $\mathbf{A}(\mathbf{x},\mathbf{e}) = \mathbf{B}^*(\mathbf{x},\mathbf{e})\mathbf{B}(\mathbf{x},\mathbf{e})$ and

$$\mathbf{B}(\mathbf{x},\mathbf{e}) = \begin{pmatrix} \mathbf{x}_1(1) - \mathbf{e}_1(1) & \cdots & \mathbf{x}_{M_t}(0) - \mathbf{e}_{M_t}(0) \\ \vdots & \vdots & \vdots \\ \mathbf{x}_1(N-1) - \mathbf{e}_1(N-1) & \cdots & \mathbf{x}_{M_t}(N-1) - \mathbf{e}_{M_t}(N-1) \end{pmatrix} \tag{13.7}$$

If q denotes the rank of $\mathbf{A}(\mathbf{x},\mathbf{e})$, (i.e., the number of nonzero eigenvalues) then we can rewrite Equation 13.6 as

$$P(\mathbf{x} \to \mathbf{e}) \leq \left[\prod_{n=1}^{q} \lambda_n \right]^{-M_r} \left(\frac{E_s}{4N_0} \right)^{-qM_r} \tag{13.8}$$

Thus, we define the notion of *diversity order* as follows: a scheme that has an average error probability $\bar{P}_e(SNR)$ as a function of SNR that behaves as

$$\lim_{SNR\to\infty} \frac{\log(\bar{P}_e(SNR))}{\log(SNR)} = -d \tag{13.9}$$

is said to have a diversity order of d.

In other words, a scheme with diversity order d has an error probability at high SNR behaving as $\bar{P}_e(SNR) \approx SNR^{-d}$. Given this definition, we can see that the diversity order in Equation 13.8 is at most qM_r. Moreover, in Equation 13.8 we obtain an additional coding gain of $(\prod_{n=1}^{q} \lambda_n)^{1/q}$.

Note that in order to obtain the average error probability, one can do a naive union bound using the pairwise error probability given in Equation 13.8. However, this bound may not be tight and a more careful upper bound for the error probability can be derived [18]. However, if we ensure that *every* pair of code words satisfies the diversity order in Equation 13.8, then clearly the average error probability satisfies it as well. Therefore, code design for diversity order through pairwise error probability is a sufficient condition, although more detailed criteria can be derived based on a more accurate expression for average error probability. The error probability analysis can easily be extended to the cases of quasi-static or time-varying ISI channels (see, e.g., [19]).

In order to design practical codes that achieve a performance target, we need to gain insights from the analysis to state design criteria. For example, in the flat-fading case of Equation 13.8 we can state the following rank and determinant design criteria [4, 17].

13.3.3 Design Criteria for Space–Time Codes over Flat-Fading Channels

- *Rank criterion:* To achieve maximum diversity M_tM_r, the matrix $\mathbf{B}(\mathbf{x},\mathbf{e})$ from Equation 13.7 has to be full rank for any code words $\mathbf{x},\mathbf{e}$. If the minimum rank of $\mathbf{B}(\mathbf{x},\mathbf{e})$ over all pairs of distinct codewords is q, then a diversity order of qM_r is achieved.
- *Determinant criterion:* For a given diversity order target of q, maximize $(\prod_{n=1}^{q} \lambda_n)^{1/q}$ over all pairs of distinct code words.

A similar set of design criteria can be stated for the quasi-static ISI fading channel [19].

Therefore, if we need to construct codes satisfying these design criteria, we can guarantee performance in terms of diversity order. The main problem in practice is to construct codes that do not have large decoding complexity. This sets up a familiar tension on the design in terms of satisfying the performance requirements and having low-complexity decoding. In Section 13.4, we outline constructions that explore this tension and are motivated by the theoretical considerations outlined above.

If coherent detection is difficult or too costly, one can do noncoherent detection for the multiple-antenna channel [20]. Though it is demonstrated in [20] that a training-based technique achieves the same capacity–SNR slope as the optimal, there might be a situation where inexpensive receivers are needed because channel estimation cannot be accommodated. In such a case, differential techniques that satisfy the diversity order might be desirable. There has been significant recent work on differential transmission with noncoherent detection (see, for example, [21] and references therein), and this is a topic we discuss in Section 13.4.2.2.

13.4 Signal Transmission Issues

This section deals with physical-layer techniques for MIMO broadband wireless channels under practical conditions. The main focus of this section is on signal processing algorithms in space–time transceivers. We start in Section 13.4.1 by examining transmitter techniques that improve the throughput or reliability of wireless communication over broadband MIMO channels. These techniques include spatial multiplexing (which improves throughput), transmit diversity (which improves reliability) including space–time trellis and block codes, and OFDM (which mitigates the channel's frequency selectivity). In Section 13.4.2, we describe in some detail two classes of receiver techniques for joint equalization and decoding of space–time-coded transmissions. The first class is coherent techniques that require channel estimation. We consider both quasi-static and rapidly varying channels for this class of techniques. We also consider adaptive techniques that do not explicitly estimate the channel but rather learn and track its characteristics using training symbols and previous decisions. The second class is noncoherent techniques where no channel estimation is required, which makes them especially attractive for rapidly varying channels where accurate channel estimation becomes very challenging.

13.4.1 Transmitter Techniques

In this section, we describe transmitter techniques that enhance the performance of multiple-transmit antenna systems either by increasing their bit rate through spatial multiplexing or by decreasing their average error rate through spatial diversity. We briefly discuss the fundamental trade-off in achieving these two performance objectives. Other forms of transmit diversity are also described and compared. Finally, an effective modulation technique known as OFDM, suitable for broadband channels, is described and its pros and cons delineated.

13.4.1.1 Spatial Multiplexing (BLAST)

By adding M_t transmit and M_r receive antennas, the achievable rate is further multiplied by $\min(M_t, M_r)$. A major challenge in realizing this significant additional throughput gain in practice is the development of cost-effective integrated multiple-chain radio frequency (RF) implementations and of low-complexity MIMO receiver architectures. An example of the latter is the Bell Labs Layered Space–Time (BLAST) architecture [22, 23] where the multiple transmitted data streams are separated and detected successively using a combination of array processing (nulling) and multiuser detection (interference cancellation) techniques.

In its most basic form known as Vertical (V)-BLAST, the receiver signal processing functions are similar to a decision feedback equalizer (DFE) operating in the spatial domain where the nulling operation is performed by the feedforward filter and the interference cancellation operation is performed by the feedback filter [24]. As in all feedback-based detection schemes, V-BLAST suffers from *error propagation* effects, which are further exacerbated by the fact that the first detected stream (typically chosen to be the one with the highest detection SNR) enjoys the smallest diversity order of $(M_r - M_t + 1)$ since $(M_t - 1)$

spatial degrees of freedom are used to null interference from the other $(M_t - 1)$ data streams yet to be detected.

Several techniques have been proposed to enhance the performance of V-BLAST such as using minimum mean square error (MMSE) interference cancellation instead of nulling, assigning variable data rates to the various streams where the first detected streams that pass through less reliable channels (smaller diversity order) are assigned a smaller data rate than streams passing through more reliable channels, performing maximum likelihood (ML) detection on the first few unreliable streams [24] or reduced-complexity ML detection for all streams using the sphere decoder algorithm [25],[5] and combining BLAST with soft-information-based iterative turbo detection algorithms [26, 27].

We conclude with the following two remarks. First, the BLAST architecture has been extended to the broadband channel scenario using a MIMO generalization of the classical DFE [28, 29]. Second, the presence of antenna correlation and the lack of scattering richness in the propagation environment reduce the achievable rates of spatial multiplexing techniques from their theoretical projections under ideal assumptions. Nevertheless, recent experimental results show that a substantial fraction of these theoretical rates is still achievable under practical propagation scenarios [30].

13.4.1.2 Transmit Diversity Techniques

Fading is a major performance-limiting impairment on wireless channels that arises mainly from destructive addition of multipaths in the propagation medium. Diversity techniques mitigate fading by transmitting multiple correlated replicas of the same information signal through independently fading channel realizations that are much less likely to fade simultaneously than each individually. Diversity techniques can be classified according to the domain in which they are created into three main categories: temporal, frequency, and spatial diversities (see, e.g., [4] for more discussion). Our focus in this article will be on spatial diversity where multiple antennas are used at the transmitter or receiver to provide multiple independently fading paths[6] for the transmitted signals that carry the same information (spatial redundancy). Spatial diversity techniques provide significant performance gains without sacrificing precious bandwidth or transmit power resources.

Spatial diversity techniques fall under two main categories: transmit diversity and receive diversity. Receive diversity combines multiple independently received signals (corresponding to the same transmitted signal), can utilize channel state information (CSI) available at the receiver, and is more suitable for the uplink (since it is more cost-effective to implement multiple antennas at the base station than at the terminal). Transmit diversity is more challenging to provision and realize because it involves the design of multiple correlated signals from a single information signal without utilizing CSI (typically not available accurately at the transmitter end). Furthermore, transmit diversity must be coupled with effective receiver signal processing techniques that can extract the desired information signal from the distorted and noisy received signal. Transmit diversity is more practical than receive diversity for enhancing the downlink (which is the bottleneck in broadband asymmetric applications such as Internet browsing and downloading) to preserve the small size and low-power consumption features of the user terminal. A common attribute of transmit and receive diversity is that both experience diminishing returns (i.e., diminishing SNR gains at a given error of probability) as the number of antennas increases [7], which makes them effective, from a performance–complexity trade-off point of view, for small numbers of antennas (typically less than four). This is in contrast with spatial multiplexing gains where the rate multiplexing gains continue to increase linearly with the number of antennas (assumed equal at both ends).

There are two main classes of multiple-antenna transmitter techniques: closed loop and open loop. The former uses a feedback channel to send CSI acquired at the receiver back to the transmitter to be used

---

[5] Using ML detection techniques also allows us to relax the assumption $M_t \leq M_r$ needed in the traditional BLAST algorithm [23].

[6] This can be ensured by placing the antennas sufficiently apart (more than the coherence distance).

in signal design, while the latter does not require CSI. Assuming availability of ideal (i.e., error-free and instantaneous) CSI at the transmitter, closed-loop techniques have an SNR advantage of $10\log_{10}(M_t)$ dB over open-loop techniques due to the array gain factor [5]. However, several practical factors degrade the performance of closed-loop techniques, including channel estimation errors at the receiver, errors in feedback link (due to noise, interference, and quantization effects), and feedback delay that causes a mismatch between available and actual CSI. All of these factors combined with the extra bandwidth and system complexity resources needed for the feedback link make open-loop techniques more attractive as a robust means for improving downlink performance for high-mobility applications, while closed-loop techniques (such as beam forming) become attractive under low-mobility conditions. Our focus in this article will be exclusively on open-loop spatial transmit diversity techniques due to their applicability to both scenarios.[7] Beam-forming techniques are discussed extensively in several tutorial papers such as [32, 33].

The simplest example of open-loop spatial transmit diversity techniques is *delay diversity* [2, 3] where the signal transmitted at sampling instant k from the ith antenna is $x_i(k) = x(k - l_i)$ for $2 \le i \le M_t$ and $x_1(k) = x(k)$ where l_i denotes the time delay (in symbol periods) on the ith transmit antenna. Assuming a single receive antenna, the D-transform[8] of the received signal is given by

$$y(D) = x(D)\left(h_1(D) + \sum_{i=2}^{M_t} D^{l_i} h_i(D)\right) + z(D) \tag{13.10}$$

It is clear from Equation 13.10 that delay diversity transforms spatial diversity into multipath diversity that can be realized through equalization [34]. For flat-fading channels, we can set $l_i = (i-1)$ and achieve full (i.e., order — M_t) spatial diversity using an ML equalizer with $(2^b)^{M_t-1}$ states where 2^b is the input alphabet size. However, for frequency-selective channels, a delay of at least $l_i = (i-1)(\nu+1)$ is needed to ensure that coefficients from the various spatial finite impulse response (FIR) channels do not interfere with each other causing a diversity loss. This, in turn, increases equalizer complexity to $(2^b)^{(M_t-1)(\nu+1)}$ states, which is prohibitive even for moderate b, M_t, and ν. In the next section, we describe another family of open-loop spatial transmit diversity techniques known as space–time block codes that achieve full spatial diversity with practical complexity even for frequency-selective channels with long delay spreads.

13.4.1.3 Space–Time Coding

Space–time coding has received considerable attention in academic and industrial circles [35, 36] due to its many advantages. First, it improves the downlink performance without the need for multiple receive antennas at the terminals. For example, for WCDMA, STC techniques where shown in [37] to result in substantial capacity gains due to the resulting "smoother" fading, which in turn makes power control more effective and reduces the transmitted power. Second, it can be elegantly combined with channel coding, as shown in [4], realizing a coding gain in addition to the spatial diversity gain. Third, it does not require CSI at the transmitter, i.e., operates in open-loop mode, thus eliminating the need for an expensive and, in case of rapid channel fading, unreliable reverse link. Finally, it has been shown to be robust against nonideal operating conditions such as antenna correlation, channel estimation errors, and Doppler effects [6, 38]. There has been extensive work on the design of space–time codes since its introduction in [4]. The combination of the turbo principle [39, 40] with space–time codes has been explored (see, for example, [41, 42] among several other references). In addition, the application of linear density parity check (LDPC) codes [43] to space–time coding has been explored (see, for example, [44] and references therein). We focus our discussion on the basic principles of space–time codes and describe two main flavors: trellis and block codes.

---

[7]It is also possible to combine closed-loop and open-loop techniques as shown recently in [31].

[8]The D-transform of a discrete-time sequence $\{x(k)\}_{k=0}^{N-1}$ is defined as $x(D) = \sum_{k=0}^{N-1} x(k)D^k$. It is derived from the Z-transform by replacing the unit delay Z^{-1} by D.

13.4.1.3.1 Space–Time Trellis Codes

The space–time trellis encoder maps the information bit stream into M_t streams of symbols (each belonging in a size-2^b signal constellation) that are transmitted simultaneously.[9] Space–time trellis code (STTC) design criteria are based on minimizing the PEP bound in Section 13.3.2.

As an example, we consider the 8-state 8-PSK (phase shift keying) STTC for two transmit antennas introduced in [4] whose trellis description is given in Figure 13.2, where the edge label c_1c_2 means that symbol c_1 is transmitted from the first antenna and symbol c_2 from the second antenna. The different symbol pairs in a given row label the transitions out of a given state, in order, from top to bottom. An equivalent and convenient (for reasons to become clear shortly) implementation of the 8-state 8-PSK STTC encoder is depicted in Figure 13.3. This equivalent implementation clearly shows that the 8-state 8-PSK STTC is identical to classical delay diversity transmission [34], *except* that the delayed symbol from the second antenna is multiplied by -1 if it is an odd symbol, i.e., $\in \{1, 3, 5, 7\}$. This slight modification

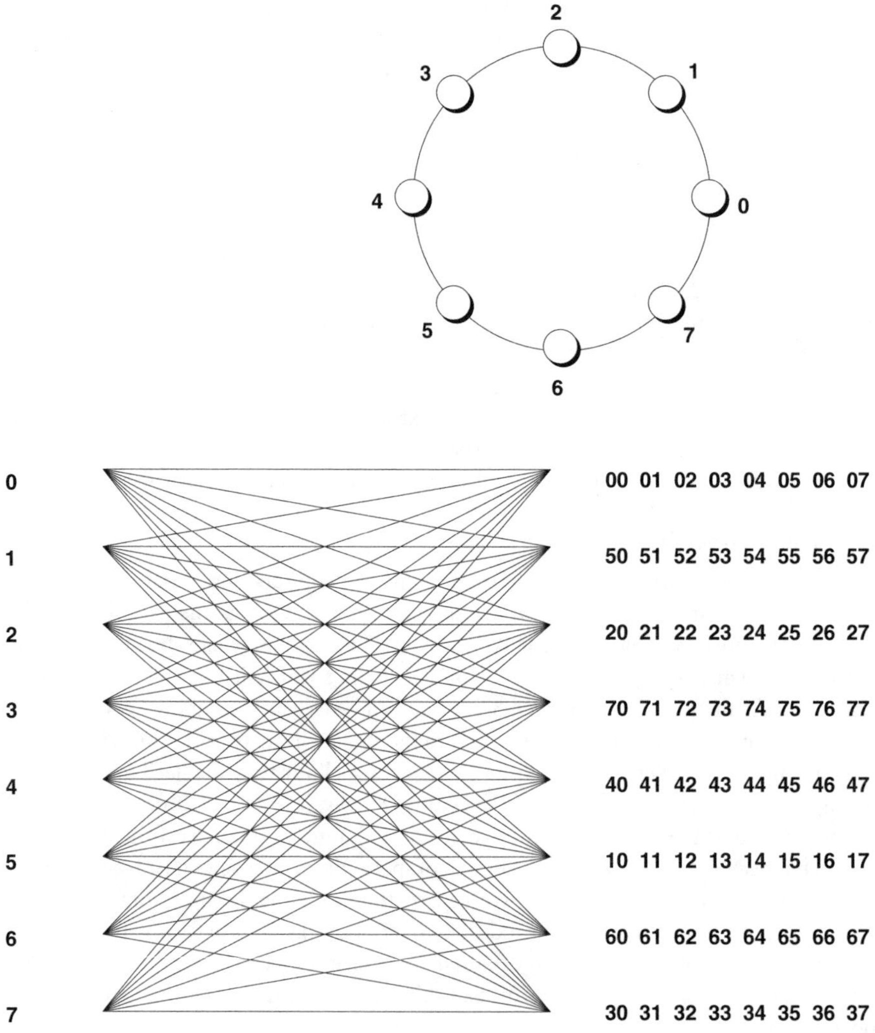

FIGURE 13.2 8-State 8-PSK space–time trellis code with two transmit antennas and a spectral efficiency of 3 bits/s/Hz.

[9]The total transmitted power is divided equally among the M_t transmit antennas.

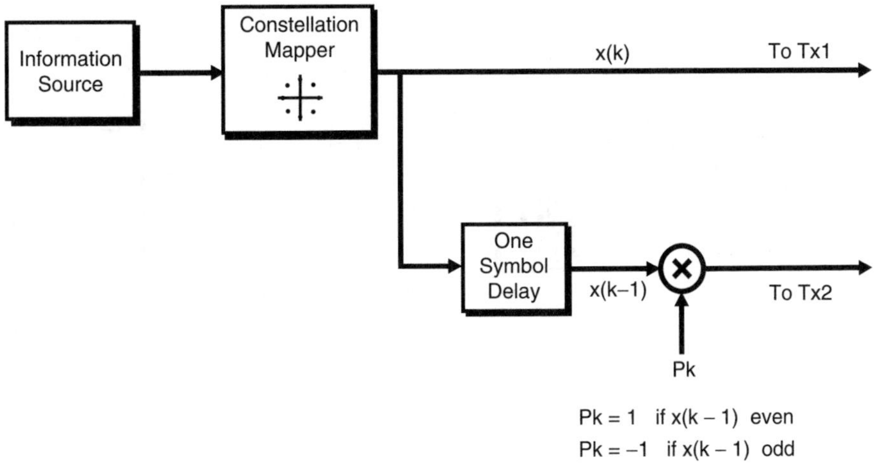

FIGURE 13.3 Equivalent encoder model for 8-state 8-PSK space–time trellis code with two transmit antennas.

results in additional coding gain over a flat-fading channel. We emphasize that this STTC does not achieve the maximum possible diversity gains (spatial and multipath) on frequency-selective channels; however, its performance is near optimum for practical ranges of SNR on wireless links[45].[10] Furthermore, when implementing the 8-state 8-PSK STTC described above over a frequency-selective channel, its structure can be exploited to reduce the complexity of joint equalization and decoding. This is achieved by embedding the space–time encoder in Figure 13.3 in the two channels $h_1(D)$ and $h_2(D)$, resulting in an equivalent single-input single-output (SISO) data-dependent CIR with memory $(\nu + 1)$ whose D-transform is given by

$$h_{eqv}^{STTC}(k, D) = h_1(D) + p_k D h_2(D) \tag{13.11}$$

where $p_k = \pm 1$ is data dependent. Therefore, trellis-based joint space–time equalization and decoding with $8^{\nu+1}$ states can be performed on this equivalent channel. Without exploiting the STTC structure, trellis equalization requires $8^{2\nu}$ states and STTC decoding requires 8 states.

The discussion in this section just illustrates one STTC example. Several other full-rate full-diversity STTCs for different signal constellations and different numbers of antennas were presented in [4].

13.4.1.3.2 Alamouti-Type Space–Time Block Codes

The decoding complexity of STTCs (measured by the number of trellis states at the decoder) increases *exponentially* as a function of the diversity level and transmission rate [4]. In addressing the issue of decoding complexity, Alamouti [5] discovered an ingenious space–time block coding scheme for transmission with two antennas. According to this scheme, input symbols are grouped in pairs where symbols x_k and x_{k+1} are transmitted at time k from the first and second antennas, respectively. Then, at time $k + 1$, symbol $-\bar{x}_{k+1}$ is transmitted from the first antenna and symbol $\bar{x}_k$ is transmitted from the second antenna, where $\bar{(.)}$ denotes complex conjugation (c.f., Figure 13.4). This imposes an orthogonal spatiotemporal structure on the transmitted symbols. Alamouti's space–time block codes (STBC) have been adopted in several wireless standards such as WCDMA [47] and CDMA2000 [48] due to the following attractive features. First, they achieve full diversity at full transmission rate for any (real or complex) signal constellation. Second, they do not require CSI at the transmitter (i.e., open loop). Third, maximum likelihood decoding involves only *linear* processing at the receiver (due to the orthogonal code structure).

[10]For examples of STTC designs for frequency-selective channels see, e.g., [46].

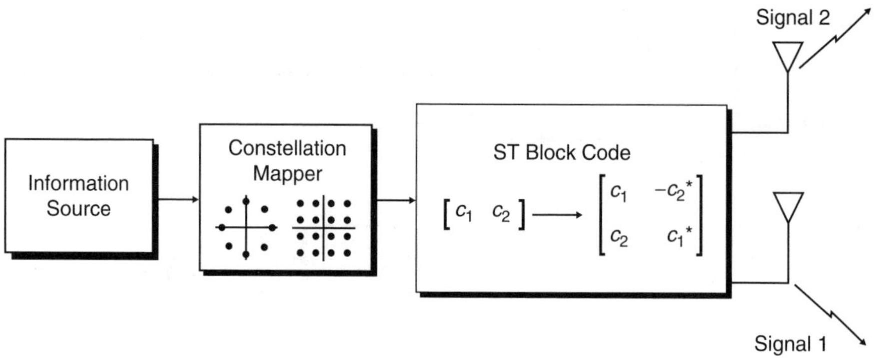

FIGURE 13.4 Spatial transmit diversity with Alamouti's space–time block code.

The Alamouti STBCs have been extended to the case of more than two transmit antennas [49] using the theory of orthogonal designs. There it was shown that, in general, full-rate orthogonal designs exist for all real constellations for two, four, and eight transmit antennas only, while for all complex constellations they exist only for two transmit antennas (the Alamouti scheme). However, for particular constellations, it might be possible to construct full-rate orthogonal designs for other numbers of transmit antennas. Moreover, if a rate loss is acceptable, full-rate orthogonal designs exist for an arbitrary number of transmit antennas [49].

Recently, STBCs have been extended to the frequency-selective channel case by implementing the Alamouti orthogonal signaling scheme at a *block* level instead of *symbol* level. Depending on whether the implementation is done in the time or frequency domain, three STBC structures for frequency-selective channels have been proposed: time reversal (TR)-STBC [50], OFDM-STBC [51], and frequency-domain equalized (FDE)-STBC [52]. As an illustration, we present next the space–time encoding scheme for FDE-STBC. Denote the nth symbol of the kth transmitted block from antenna i by $x_i^{(k)}$. At times $k = 0, 2, 4, \ldots$ pairs of length N blocks $\mathbf{x}_1^{(k)}(n)$ and $\mathbf{x}_2^{(k)}(n)$ (for $0 \leq n \leq N-1$) are generated by the mobile user. Inspired by Alamouti's STBCs, we encode the information symbols as follows [52]:

$$\mathbf{x}_1^{(k+1)}(n) = -\mathbf{x}_2^{*(k)}((-n)_N) \quad \text{and} \quad \mathbf{x}_x^{(k+1)}(n) = \mathbf{x}_1^{*(k)}((-n)_N)$$
$$\text{for } n = 0, 1, \ldots, N-1 \quad \text{and} \quad k = 0, 2, 4, \ldots \tag{13.12}$$

where $(\cdot)_N$ denotes the modulo-N operation. In addition, a cyclic prefix of length ν (the maximum order of the FIR wireless channel) is added to each transmitted block to eliminate IBI and make all channel matrices *circulant*. Refer to [53] for a detailed description and comparison of these schemes. The main point we would like to stress here is that these three STBC schemes can realize both spatial and multipath diversity gains at practical complexity levels. For channels with long delay spreads, the frequency-domain implementation using the FFT either in a single-carrier or multicarrier fashion becomes more advantageous from a complexity point of view.

13.4.1.4 Diversity vs. Throughput Trade-Off

In the previous sections, we have seen two types of gains from MIMO transmitter techniques: spatial multiplexing gains and diversity gains. As shown in [18], one can simultaneously achieve diversity gains and multiplexing gains. In this context, V-BLAST achieves maximum spatial multiplexing gain but offers no diversity gain, while space–time trellis and block codes achieve maximum diversity gains with no multiplexing gains. While there are many architectures that achieve both types of gains (an example is the 4-transmit 2-receive antenna architecture in [54]), the design of MIMO systems that achieve the entire optimal diversity-multiplexing trade-off curve in [18] remains a significant open research problem.

With a single receive antenna, the Alamouti scheme was shown in [18] to achieve the optimal rate-diversity trade-off curve. However, it becomes suboptimal with two (or more) receive antennas and when extended to the case of more than two transmit antennas using orthogonal designs [49]. This suboptimality arises from the orthogonality constraint imposed in this class of space–time codes to achieve maximum diversity gains (at a given rate) and to have a linear decoder complexity without signal constellation expansion.

13.4.1.5 Orthogonal Frequency Division Multiplexing

In OFDM, the high-rate input stream is demultiplexed and transmitted over N low-rate independent frequency subcarriers. This multicarrier transmission scheme is implemented digitally using the efficient FFT [55]. Since OFDM is a block transmission scheme, a guard sequence (of length at least equal to channel memory) is needed to eliminate IBI and ensure that individual subcarriers can be isolated at the receiver. In OFDM, the choice for guard sequence is a *cyclic prefix*, which makes the channel matrix *circulant*, and hence diagonalizable by FFT. If the FFT size is made large enough such that the width of each frequency bin is less than the *coherence bandwidth* of the channel, then no equalization is needed.[11] A large FFT size (compared to channel memory) also reduces the cyclic prefix guard sequence overhead at the expense of increased storage and processing requirements and increased delay, which might not be acceptable for delay-sensitive applications.

OFDM is very attractive as a modulation/equalization scheme for channels with long delay spreads (where trellis-based equalization is very complex). For time-selective channels, the subcarriers in an OFDM signal lose their orthogonality (or equivalently, the channel matrix is no longer circulant, and hence is no longer diagonalizable by the FFT), resulting in intercarrier interference (ICI). Successful application of OFDM to high-mobility scenarios is critically dependent on the implementation of effective channel estimation/tracking and ICI suppression schemes (see Section 13.4.2.1 and [56]).

OFDM offers great flexibility in that multiple signals with different rates and quality-of-service (QoS) requirements can be transmitted over the parallel frequency subchannels. Moreover, avoiding strong RF narrowband interference within the transmission bandwidth is easily accomplished by turning off the corresponding subchannels. OFDM was applied to MIMO broadband channels in [11, 57]. OFDM has two main drawbacks, namely, a high peak-to-average ratio (PAR), which results in larger back-off with nonlinear amplifiers [58], and high sensitivity to frequency errors and phase noise [59]. An alternative equalization scheme that overcomes these two drawbacks of OFDM while retaining its reduced implementation complexity advantage (but not its multirate capability) is the single-carrier (SC) frequency-domain equalizer [60].

13.4.2 Receiver Techniques

In this section, we present an overview of receiver signal processing algorithms needed for the detection of space–time-coded signals. These algorithms can be classified under two main categories: coherent and noncoherent. Coherent detection requires CSI either explicitly by estimating the channel matrix and feeding this estimate to joint equalization/decoding algorithms or implicitly where the optimum settings of the joint equalizer/decoder are computed adaptively using training symbols (discussed in Section 13.4.2.1). Noncoherent techniques (discussed in Section 13.4.2.2) do not require CSI and hence are more suitable for rapidly time-varying channels where it is more challenging to acquire accurate CSI or when cost and complexity constraints exclude the use of channel estimation modules.

To explain the main ideas without unduly complicating the presentation, we assume (unless otherwise stated) a single receive antenna. When multiple independent receive antennas are available, additional *receive diversity gains* can be realized by *combining* the outputs of these antennas using spatiotemporal processing (see, e.g., [61] and references therein).

[11]Except for a simple one-tap complex equalizer for each subchannel, assuming negligible intercarrier interference due to Doppler effects or frequency offset errors.

13.4.2.1 Coherent Techniques

We start by considering channel estimation techniques for both quasi-static (or slowly time-varying) channels and rapidly time-varying channels with appreciable channel variation within the block. This is followed by a discussion of joint equalization/decoding techniques for space–time trellis and block coded signals. Then we show how to improve the performance of OFDM receivers in fast time-varying channels. This is followed by a discussion of joint equalization and interference cancellation schemes in a multiuser environment. Finally, we conclude by describing an adaptive technique for joint space–time equalization and decoding.

13.4.2.1.1 Channel Estimation for Quasi-Static Channels

For quasi-static channels, CSI can be estimated at the receiver using a training sequence embedded in each transmission block. For single-transmit-antenna signaling, the training sequence is only required to have good (i.e., impulse-like) autocorrelation properties. However, for the M_t transmit antenna scenarios, the M_t training sequences should, in addition, have low (ideally zero) cross-correlation. In addition, it is desirable (in order to avoid amplifier nonlinear distortion) to use training sequences with constant amplitude. *Perfect root of unity sequences* (PRUSs) [62] have these ideal correlation and constant-amplitude properties. However, for a given training sequence length, PRUSs do not always belong to standard signal constellations such as PSK. Additional challenges in channel estimation for multiple-transmit-antenna systems over the single-transmit-antenna case are the increased number of channel parameters to be estimated and the reduced transmit power (by a factor of M_t) for each transmit antenna.

In [63], it was proposed to encode a single training sequence by a space-time encoder to generate the M_t training sequences.[12] Strictly speaking, this approach is suboptimum since the M_t transmitted training sequences are cross-correlated by the space–time encoder, which imposes a constraint on the possible generated training sequences. However, it turns out that, with proper design, the performance loss from optimal PRUS training is negligible [63]. Furthermore, this approach reduces the training sequence search space from $(2^b)^{M_t N_t}$ to $(2^b)^{N_t}$ (assuming equal input and output alphabet size 2^b and length — N_t training sequences), making exhaustive searches more practical and thus facilitating the identification of good training sequences from standard signal constellations such as PSK.

The search space can be further reduced by exploiting special characteristics of the particular STC. As an example, consider the 8-state 8-PSK STTCs for two transmit and one receive antenna of Section 13.4.1.3, whose equivalent CIR is given by Equation 13.11. For a given transmission block (over which the two channels $h_1(D)$ and $h_2(D)$ are constant), the input sequence determines the equivalent channel. By transmitting only even training symbols from the subconstellation $C_e = \{0, 2, 4, 6\}$, $p_k = +1$ and the equivalent channel is given by $h_e(D) = h_1(D) + Dh_2(D)$. On the other hand, transmitting only odd training symbols from the subconstellation $C_0 = \{1, 3, 5, 7\}$ results in $p_k = -1$ and the equivalent channel $h_o(D) = h_1(D) - Dh_2(D)$. After estimating $h_e(D)$ and $h_o(D)$, we can compute

$$h_1(D) = \frac{h_e(D) + h_o(D)}{2} \quad \text{and} \quad h_2(D) = \frac{h_e(D) - h_o(D)}{2D} \tag{13.13}$$

Consider a training sequence of the form $\mathbf{s} = [\mathbf{s}_e \ \mathbf{s}_o]$ where $\mathbf{s}_e$ has length $N_t/2$ and takes values in the C_e subconstellation and $\mathbf{s}_o$ has length $N_t/2$ and takes values in the C_o subconstellation. Note that if $\mathbf{s}_e$ is a good sequence in terms of MMSE for the estimation of $h_e(D)$, the sequence $\mathbf{s}_o$ created as $\mathbf{s}_o = a\ \mathbf{s}_e$ where $a = exp(\frac{i\pi k}{4})$ and any $k = 1, 3, 5, 7$ achieves the same MMSE for the estimation of $h_o(D)$. Thus, instead of searching over all possible 8^{N_t} sequences $\mathbf{s}$, we can further restrict the search space to the $4^{\frac{N_t}{2}}$ sequences $\mathbf{s}_e$. A reduced-size search can identify sequences $\mathbf{s}_e$ and $\mathbf{s}_o = a\ \mathbf{s}_e$ such that the channel estimation MMSE is achieved. We emphasize that similar reduced-complexity techniques can be developed for other STTCs by deriving their equivalent encoder models (as in Figure 13.3).

[12] We assume, for simplicity, the same space–time encoder for the training and information symbols. However, they could be different in general.

In summary, the special STC structure can be utilized to simplify training sequence design for multiple-antenna transmissions without sacrificing performance.

13.4.2.1.2 Channel Estimation and Tracking for Rapidly Time-Varying Channels

When the channel varies significantly within a transmission block, the channel matrix $\mathbf{H}$ loses its special structured (Toeplitz) form, which makes both the estimation and equalization of $\mathbf{H}$ more challenging. It is well known that multicarrier transmissions (such as OFDM) are more sensitive to time variations than single-carrier transmissions. In OFDM, rapid time variations of the underlying channels within a transmission block result in intercarrier interference. Depending on the Doppler frequency and block length chosen for transmission, ICI can potentially cause a severe deterioration of quality of service.

To simplify the presentation, consider the case $M_r = M_t = 1$. In a time-varying environment, estimation of $\mathbf{H}$ amounts to estimating N channels $\mathbf{h}_n := [h(n;0), \ldots, h(n;\nu-1)]^T, 0 \leq n \leq N-1$, that comprise the rows of $\mathbf{H}$. To reduce the number of parameters needed for channel estimation from $N\nu$ to less than N, [56] makes the assumption that some of the channels $\mathbf{h}_n$ can be obtained by linear interpolation. Such an assumption holds if there is not significant variation between channels $\mathbf{h}_n$ and $\mathbf{h}_{n+1}, 0 \leq n \leq (N-2)$. The matrix $\mathbf{H}$ is parametrized using a small number of its rows. Then the entire matrix is expressed as a function of these rows, therefore reducing the number of parameters to be estimated. Physically, this puts markers in time where the channel is estimated, and the estimates at other times are interpolated using these estimates.

The channel matrix $\tilde{\mathbf{H}}$ obtained by the interpolation of the channels $\mathbf{h}_m(1), \ldots, \mathbf{h}_m(M)$ (where $m(1), \ldots, m(M)$ are the channel rows used in the interpolation) is

$$\tilde{\mathbf{H}} = \sum_{1 \leq i \leq M} \mathbf{A}_{m(i)} H_C(\mathbf{h}_{m(i)}) \tag{13.14}$$

where $\mathbf{A}_{m(i)}$ is an $N \times N$ diagonal matrix with entries equal to the interpolation weights and $H_c(\mathbf{h})$ is the OFDM circulant channel matrix that corresponds to the FIR channel $\mathbf{h}$. Given the structured form of $\tilde{\mathbf{H}}$, its estimation amounts to estimating $M\nu$ parameters grouped in the $M\nu \times 1$ vector $\tilde{\mathbf{h}} := [\mathbf{h}^T_{m(1)}, \ldots, \mathbf{h}^T_{m(M)}]^T$, where $(\cdot)^T$ denotes the transpose. Using judiciously placed pilot tones, the latter can be obtained as the least squares solution $\hat{\tilde{h}} = \tilde{\mathbf{B}}^{\dagger}_{(P)}\mathbf{Y}_{(P)}$, where $(\cdot)^{\dagger}$ denotes the pseudo inverse and $\mathbf{Y}_{(P)}$ is the received vector when pilot tones are included in the transmission and $\tilde{\mathbf{B}}_{(P)}$ is a matrix with entries that are functions of the pilot tones and interpolation weights (see [56] for more details).

The quality of channel estimates depends critically on the placement of pilot tones on the FFT grid. In frequency-selective time-invariant channels, placing the pilot tones equi-spaced on the FFT grid is the optimal scheme [64]. On the other hand, work on PSAM (pilot symbol assisted modulation) [65] has suggested that, for Rayleigh flat-fading time-varying channels, pilot symbols should be placed periodically in the time domain to produce reliable channel estimates, which are then interpolated to allow for the coherent detection of the transmitted symbols. The periodic transmission of pilot symbols in the time domain suggests a grouping of pilot tones in the frequency domain. We have found that our channel estimation method produces the best results when the *pilot tones are partitioned into equi-spaced groups on the FFT grid* (see [56]).

Furthermore, under relatively mild Doppler, the quality of the channel estimates (and, consequently, the signal-to-ICI-plus noise ratio (SINR) gains of the ICI mitigating methods in [56]) can be improved by channel tracking. A simple tracking scheme is the following: an initial channel estimate $\hat{\mathbf{H}}_{(0)}$ can be obtained by transmitting a full training block; subsequent blocks contain pilot tones that are used to acquire new estimates $\mathbf{H}'_{(n)}$. The channel estimate for the nth OFDM block is obtained using a forgetting factor α as follows: $\hat{\mathbf{H}}_{(n)} = \alpha\hat{\mathbf{H}}_{(n-1)} + (1-\alpha)\mathbf{H}'_{(n)}$. Frequent retraining can further improve the quality of the channel estimates at the expense of the training symbols overhead.

13.4.2.1.3 Joint Equalization/Decoding of Space–Time Trellis Codes

For broadband transmissions, equalization is indispensable for mitigating *intersymbol interference.* STC makes equalization more challenging because it generates multiple *correlated* signals that are transmitted *simultaneously at equal power*. However, carefully designed joint equalization/decoding schemes can exploit this correlation to reduce implementation complexity while achieving significant performance gains over single-antenna transmissions (due to spatial and multipath diversity gains). In this section, we briefly describe examples of practical near-optimal joint equalization/decoding schemes for STTCs and STBCs. A detailed treatment of this subject is given in [53].

Both the STTCs and the CIRs are finite-state machines described by a trellis. Hence, optimum performance is achieved by joint STTC equalization/decoding on the combined trellis. The complexity of full joint trellis equalization and decoding increases *exponentially* with the channel memory,[13] signal constellation size, and number of transmit antennas. On the other hand, increasing these three parameters is an effective means to achieve high bit rates. This motivates the need for reduced-complexity joint equalization and decoding schemes that achieve a practical performance–complexity trade-off. An example of such a scheme is described next.

When implementing the 8-state 8-PSK STTC over frequency-selective channels, its rich structure can be exploited to reduce equalization/decoding complexity. This is achieved by performing trellis-based joint equalization and space–time decoding with $8^{\nu+1}$ states on the equivalent channel given in Equation 13.11. For channels with long memory, further complexity reductions (at some performance loss) are achieved by preceding the joint equalizer/decoder with a channel-shortening FIR prefilter. The objective of the prefilter is to shorten and shape the effective CIR seen by the equalizer to reduce its complexity. Prefilter design algorithms suitable for space–time-coded signals are described in [66, 67].

Several reduced-state equalization/decoding algorithms have been proposed in the literature, including delayed decision feedback sequence estimation (DDFSE) [68], reduced-state sequence estimation (RSSE) [69], T-BCJR [70], and M-BCJR [67, 70]. We have investigated and improved the M-BCJR algorithm and applied it to the problem of joint equalization and decoding of STTCs, as described briefly next. The findings of our study are detailed in [67].

The M-BCJR algorithm is a reduced-complexity version of the famed BCJR algorithm [71], where at each trellis step only the M active states associated with the highest metrics are retained. An improved version of the M-BCJR algorithm was proposed in [67] and applied to perform joint STTC equalization and decoding. More specifically, it was shown in [67] that preceding the M-BCJR equalizer/decoder with a channel-shortening prefilter improves its performance, especially for small values of M. Even better performance is achieved when a different prefilter is used for the forward and backward M-BCJR recursions. The value of M and the number of prefilter taps can be jointly optimized to achieve the best performance–complexity trade-offs. For channels with long delay spreads, FFT-based techniques such as OFDM emerge as attractive alternative candidates for STTC equalization/decoding [57].

13.4.2.1.4 Joint Equalization/Decoding of Space–Time Block Codes

Our focus will be on Alamouti-type STBCs with two transmit antennas. The treatment can be extended to more than two antennas using orthogonal designs [49] at the expense of some rate loss for complex signal constellations.

The main attractive feature of STBCs is the quaternionic[14] structure of the spatiotemporal channel matrix. This allows us to eliminate interantenna interference using a low-complexity linear combiner (which is a spatiotemporal matched filter and is also the maximum likelihood detector in this case). Then, joint equalization and decoding for each antenna stream proceeds using any of the well-known algorithms for the single-antenna case, which can be implemented in either the time or frequency domains. For illustration purposes, we describe next a joint equalization and decoding algorithm for the single-carrier frequency-domain equalizer (SC FDE)-STBC. A more detailed discussion and comparison is given in [53].

[13] For a given channel delay spread, the channel memory increases linearly with the transmission bandwidth.

[14] A 2×2 complex orthogonal matrix of the form $\begin{bmatrix} c_1 & c_2 \\ -\bar{c}_2 & \bar{c}_1 \end{bmatrix}$ is called a quaternion.

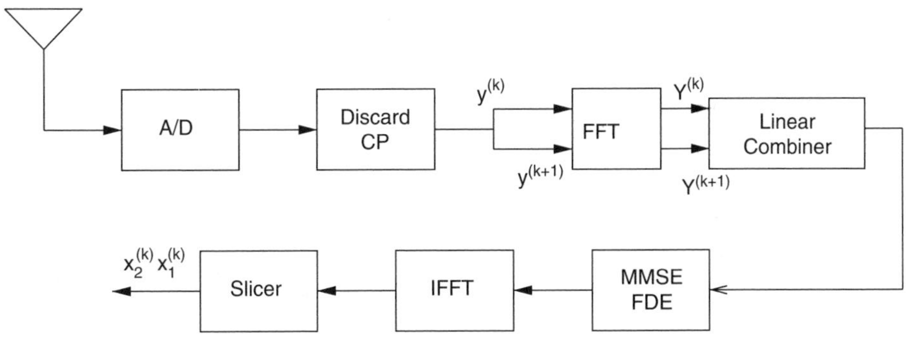

FIGURE 13.5 FDE-STBC receiver block diagram.

The SC FDE-STBC receiver block diagram is given in Figure 13.5. After analog-to-digital (A/D) conversion, the CP part of each received block is discarded. Mathematically, we can express the input–output relationship over the jth received block as follows:

$$\mathbf{y}^{(j)} = \mathbf{H}_1^{(j)}\mathbf{x}_1^{(j)} + \mathbf{H}_2^{(j)}\mathbf{x}_2^{(j)} + \mathbf{z}^{(j)} \tag{13.15}$$

where $\mathbf{H}_1^{(j)}$ and $\mathbf{H}_2^{(j)}$ are $N \times N$ circulant matrices whose first columns are equal to $\mathbf{h}_1^{(j)}$ and $\mathbf{h}_2^{(j)}$, respectively, appended by $(N - \nu - 1)$ zeros, and $\mathbf{z}^{(j)}$ is the noise vector. Since $\mathbf{H}_1^{(j)}$ and $\mathbf{H}_2^{(j)}$ are circulant matrices, they admit the eigendecompositions

$$\mathbf{H}_1^{(j)} = \mathbf{Q}^*\mathbf{\Lambda}_1^{(j)}\mathbf{Q}, \quad \mathbf{H}_2^{(j)} = \mathbf{Q}^*\mathbf{\Lambda}_2^{(j)}\mathbf{Q}$$

where $\mathbf{Q}$ is the orthonormal FFT matrix and $\mathbf{\Lambda}_1^{(j)}$ (respectively $\mathbf{\Lambda}_2^{(j)}$) is a diagonal matrix whose (n, n) entry is equal to the nth FFT coefficient of $\mathbf{h}_1^{(j)}$ (respectively $\mathbf{h}_2^{(j)}$). Therefore, applying the FFT to $\mathbf{y}^{(j)}$, we get (for $j = k, k + 1$)

$$\mathbf{Y}^{(j)} = \mathbf{Q}\mathbf{y}^{(j)} = \mathbf{\Lambda}_1^{(j)}\mathbf{X}_1^{(j)} + \mathbf{\Lambda}_2^{(j)}\mathbf{X}_2^{(j)} + \mathbf{Z}^{(j)}$$

The SC FDE-STBC encoding rule is given by [52]

$$\mathbf{X}_1^{(k+1)}(m) = \bar{\mathbf{X}}_2^{(k)}(m) \quad \text{and} \quad \mathbf{X}_2^{(k+1)}(m) = -\bar{\mathbf{X}}_1^{(k)}(m) \tag{13.16}$$

for $m = 0, 1, \ldots, N - 1$ and $k = 0, 2, 4, \ldots$. The length N blocks at the FFT output are then processed in pairs, resulting in the two blocks (we drop the time index from the channel matrices since they are assumed fixed over the two blocks under consideration)

$$\underbrace{\begin{bmatrix} \mathbf{Y}^{(k)} \\ \bar{\mathbf{Y}}^{(k+1)} \end{bmatrix}}_{\mathbf{Y}} = \underbrace{\begin{bmatrix} \mathbf{\Lambda}_1 & \mathbf{\Lambda}_2 \\ -\bar{\mathbf{\Lambda}}_2 & \bar{\mathbf{\Lambda}}_1 \end{bmatrix}}_{\mathbf{\Lambda}} \underbrace{\begin{bmatrix} \mathbf{X}_1^{(k)} \\ \mathbf{X}_2^{(k)} \end{bmatrix}}_{\mathbf{X}} + \underbrace{\begin{bmatrix} \mathbf{Z}^{(k)} \\ \bar{\mathbf{Z}}^{(k+1)} \end{bmatrix}}_{\mathbf{Z}} \tag{13.17}$$

where $\mathbf{X}_1^{(k)}$ and $\mathbf{X}_2^{(k)}$ are the FFTs of the information blocks $\mathbf{x}_1^{(k)}$ and $\mathbf{x}_2^{(k)}$, respectively, and $\mathbf{Z}$ is the noise vector. We used the encoding rule in Equation 13.16 to arrive at Equation 13.17. To eliminate *interantenna interference*, the linear combiner $\mathbf{\Lambda}^*$ is applied to $\mathbf{Y}$. Due to the quaternionic structure of $\mathbf{\Lambda}$, a second-order diversity gain is achieved. Then, the two decoupled blocks at the output of the linear combiner are equalized separately using the MMSE FDE [60], which consists of N complex taps per block that mitigate *intersymbol interference*. Finally, the MMSE FDE output is transformed back to the time domain using inverse FFT where decisions are made.

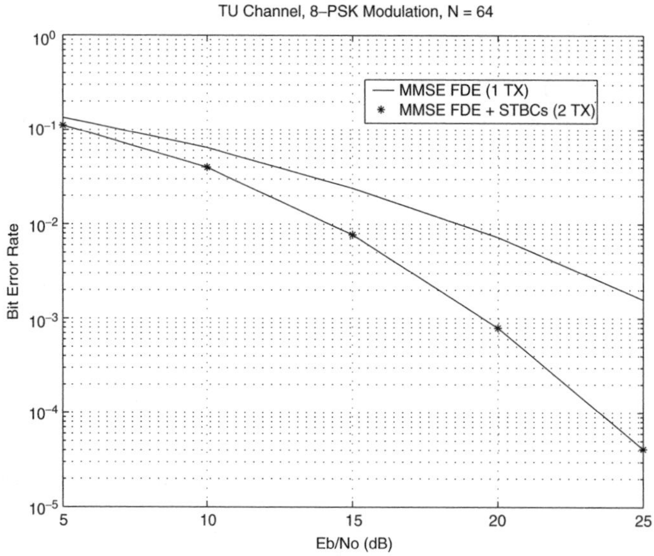

FIGURE 13.6 BER of SC MMSE FDE with and without STBCs for typical urban channel ($\nu = 3$) with 8-PSK modulation and $N = 64$.

Figure 13.6 shows the diversity advantage achieved in SC MMSE FDE-STBCs compared to single-antenna transmission. This figure assumes perfect channel knowledge at the receiver.[15] To investigate the effect of channel estimation errors on performance, we assumed that there were 26 training symbols per two blocks of information symbols (total length of 128 symbols). Two shifted PRUSs [62] were transmitted from antennas 1 and 2 during training. The two channel estimates were jointly computed using a standard least squares algorithm [74]. From Figure 13.7, it can be seen that the performance loss due to channel estimation errors is ≈1 to 1.5 dB.

13.4.2.1.5 OFDM with Fast Channel Variations

As discussed earlier, in the presence of channel variations within the transmission block, the subcarriers in OFDM transmission are no longer orthogonal and ICI can result in significant performance degradation. One approach, described in detail in [56], to mitigate ICI and restore (approximately) subcarrier orthogonality is to implement a time-domain MIMO prefilter **W** at the receiver front end that attempts to restore the *circulant* structure of the overall channel matrix and hence make it diagonalizable by FFT.

The matrix prefilter **W** can be designed (using channel knowledge acquired with the methods presented earlier for rapidly varying channels) to maximize SINR. It is shown in [56] that the design of **W** can be posed as a generalized eigenvalue problem, where **W** is calculated as a function of the estimated channel matrix **H**.

13.4.2.1.6 Joint Equalization and Interference Cancellation

We can double the number of STBC users (i.e., double system capacity) without additional radio spectrum resources by adding a second receive antenna at the base station and using interference cancellation techniques. These techniques exploit the rich STBC structure to reduce the number of receive antennas required (for effective joint equalization, decoding, and interference cancellation) compared to traditional antenna nulling techniques (see [54] for more discussion). With two receive antennas and two STBC users

[15] Performance can be further improved by adding a feedback section, as shown in [72, 73].

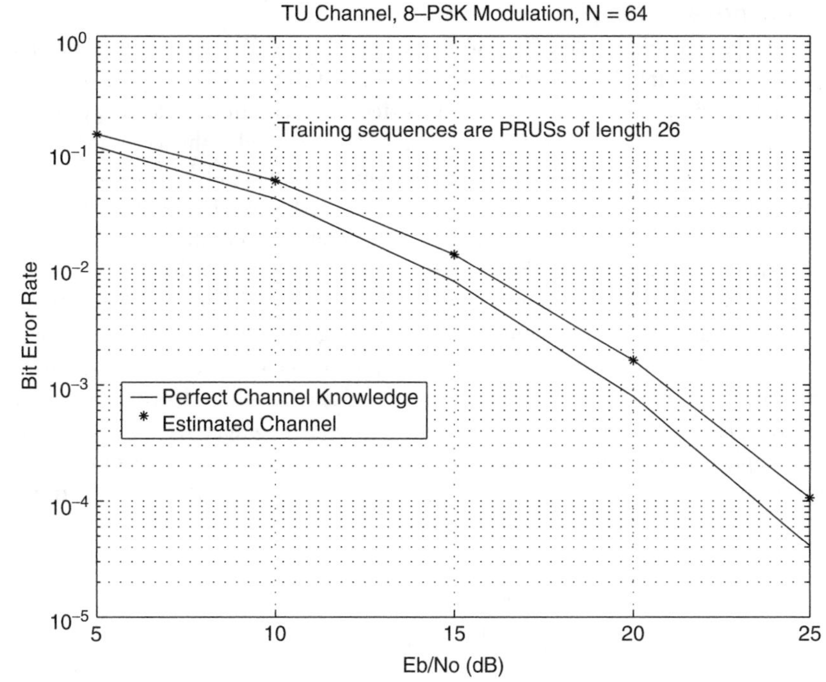

FIGURE 13.7 Effect of channel estimation on performance of SC FDE-STBCs.

(each equipped with two antennas), Equation 13.17 generalizes to

$$\begin{bmatrix} \mathbf{Y}_1 \\ \mathbf{Y}_2 \end{bmatrix} = \begin{bmatrix} \boldsymbol{\Lambda}_x & \boldsymbol{\Gamma}_x \\ \boldsymbol{\Lambda}_s & \boldsymbol{\Gamma}_s \end{bmatrix} \begin{bmatrix} \mathbf{X} \\ \mathbf{S} \end{bmatrix} + \begin{bmatrix} \mathbf{Z}_1 \\ \mathbf{Z}_2 \end{bmatrix} \tag{13.18}$$

where $\mathbf{Y}_1$ and $\mathbf{Y}_2$ are the processed signals from the first and second antennas while $\mathbf{Z}_1$ and $\mathbf{Z}_2$ are the corresponding noise vectors. The vector $\mathbf{S}$ consists of two subvectors representing the size — N FFTs of the two information blocks transmitted from the interfering user's first and second antennas. The two STBC users can be decoupled by applying the following linear zero-forcing[16] interference canceler:

$$\begin{bmatrix} \mathbf{R}_1 \\ \mathbf{R}_2 \end{bmatrix} \stackrel{def}{=} \begin{bmatrix} \mathbf{I} & -\boldsymbol{\Gamma}_x\boldsymbol{\Gamma}_s^{-1} \\ -\boldsymbol{\Lambda}_s\boldsymbol{\Lambda}_x^{-1} & \mathbf{I} \end{bmatrix} \begin{bmatrix} \mathbf{Y}_1 \\ \mathbf{Y}_2 \end{bmatrix} = \begin{bmatrix} \tilde{\boldsymbol{\Lambda}}_x & \mathbf{0} \\ \mathbf{0} & \tilde{\boldsymbol{\Gamma}}_s \end{bmatrix} \begin{bmatrix} \mathbf{X} \\ \mathbf{S} \end{bmatrix} + \begin{bmatrix} \tilde{\mathbf{Z}}_1 \\ \tilde{\mathbf{Z}}_2 \end{bmatrix}$$

where $(.)^{-1}$ denotes the inverse, $\tilde{\boldsymbol{\Lambda}}_x \stackrel{def}{=} \boldsymbol{\Lambda}_x - \boldsymbol{\Gamma}_x\boldsymbol{\Gamma}_s^{-1}\boldsymbol{\Lambda}_s$, and $\tilde{\boldsymbol{\Gamma}}_s \stackrel{def}{=} \boldsymbol{\Gamma}_s - \boldsymbol{\Lambda}_s\boldsymbol{\Lambda}_x^{-1}\boldsymbol{\Gamma}_x$. The critical observation to make here is that *both* $\tilde{\boldsymbol{\Lambda}}_x$ *and* $\tilde{\boldsymbol{\Gamma}}_s$ *are orthogonal Alamouti-like matrices.* Therefore, decoding proceeds as in the single-user case and the full diversity gain is guaranteed for both users.

Note that this technique can be easily extended to more than two co-channel users and other space–time-coded orthogonal designs [75]. The form of detection performed above is a simple decorrelating receiver [76] and hence ignores the noise, making it suitable only for high signal-to-noise ratios. As in multiuser detection, an MMSE detector can be constructed for the detection problem in Equation 13.18. Such an approach has been proposed for the flat-fading channel in [54]. In summary, all the sophisticated multiuser detection techniques can be applied in detecting the symbols in Equation 13.18 once we utilize the structure of the space–time-coded signal.

[16]Better performance is achieved by an MMSE canceler that avoids noise enhancement [54].

13.4.2.1.7 Adaptive Techniques

The coherent receiver techniques described so far require CSI, which is estimated and tracked using training sequences/pilot symbols inserted in each block and then used to compute the optimum joint equalizer/decoder settings. An alternative to this two-step channel estimate-based approach is *adaptive* space–time equalization/decoding where CSI is not explicitly estimated at the receiver. Adaptive receivers still require training overhead to converge to their optimum settings, which, in the presence of channel variations, are adapted using previous decisions to *track* these variations. The celebrated least mean squares (LMS) adaptive algorithm [77] is widely used in single-antenna communications systems today due to its low implementation complexity. However, it has been shown to exhibit slow convergence and suffer significant performance degradation (relative to performance achieved with the optimum settings) when applied to broadband MIMO channels due to the large number of parameters that need to be simultaneously adapted and the wide eigenvalue spread problems encountered on those channels. Faster convergence can be achieved by implementing a more sophisticated family of algorithms known as recursive least squares (RLS). However, their high computational complexity compared to LMS and their notorious behavior when implemented in finite precision limit their appeal in practice. It was shown in [78] that the orthogonal structure of STBCs can be exploited to develop fast-converging RLS-type adaptive FDE-STBCs at LMS-type complexity. A brief overview is given next.

From Figure 13.5,

$$\begin{bmatrix} \hat{\mathbf{X}}_1^{(k)} \\ \hat{\mathbf{X}}_2^{(k)} \end{bmatrix} = \begin{bmatrix} \mathbf{A}_1 & \mathbf{A}_2 \\ \bar{\mathbf{A}}_2 & -\bar{\mathbf{A}}_1 \end{bmatrix} \mathbf{Y} \tag{13.19}$$

where $\mathbf{Y}$ was defined in Equation 13.17 and the diagonal matrices $\mathbf{A}_1$ and $\mathbf{A}_2$ are given by

$$\mathbf{A}_1 = \mathbf{\Lambda}_1^*.diag\left\{\frac{1}{\tilde{\mathbf{\Lambda}}(i,i) + \frac{1}{SNR}}\right\}_{i=0}^{N-1}, \qquad \mathbf{A}_2 = \mathbf{\Lambda}_2^*.diag\left\{\frac{1}{\tilde{\mathbf{\Lambda}}(i,i) + \frac{1}{SNR}}\right\}_{i=0}^{N-1} \tag{13.20}$$

with $\tilde{\mathbf{\Lambda}}(i,i) = |\mathbf{\Lambda}_1(i,i)|^2 + |\mathbf{\Lambda}_2(i,i)|^2$. Alternatively, we can write

$$\begin{bmatrix} \hat{\mathbf{X}}_1^{(k)} \\ \hat{\mathbf{X}}_2^{(k)} \end{bmatrix} = \begin{bmatrix} diag(\mathbf{Y}^{(k)}) & -diag(\bar{\mathbf{Y}}^{(k+1)}) \\ diag(\mathbf{Y}^{(k+1)}) & diag(\bar{\mathbf{Y}}^{(k)}) \end{bmatrix} \begin{bmatrix} \bar{\mathbf{W}}_1 \\ \mathbf{W}_2 \end{bmatrix} = \mathbf{U}_k \mathcal{W} \tag{13.21}$$

where $\bar{\mathbf{W}}_1$ and $\mathbf{W}_2$ are vectors containing the diagonal elements of $\bar{\mathbf{A}}_1$ and $\mathbf{A}_2$, respectively, and $\mathcal{W}$ is a $2N \times 1$ vector containing the elements of $\bar{\mathbf{W}}_1$ and $\mathbf{W}_2$. The $2N \times 2N$ quaternionic matrix $\mathbf{U}_k$ contains the received symbols for blocks k and $k+1$. Equation 13.21 can be used to develop a frequency-domain block-adaptive RLS algorithm for $\mathbf{W}$ that, using the special quaternionic structure of the problem, can be simplified to the following LMS-type recursions (see [78] for details of the derivation):

$$\mathcal{W}_{k+2} = \mathcal{W}_k + \begin{bmatrix} \mathbf{P}_{k+2} & \mathbf{0} \\ \mathbf{0} & \mathbf{P}_{k+2} \end{bmatrix} \mathbf{U}_{k+2}(\mathbf{D}_{k+2} - \mathbf{U}_{k+2}\mathcal{W}_k) \tag{13.22}$$

where $\mathbf{D}_{k+2} = [\mathbf{X}_1^{(k+2)} \;\; \bar{\mathbf{X}}_2^{(k+2)}]^T$ for the training mode and $\mathbf{D}_{k+2} = [\hat{\mathbf{X}}_1^{(k+2)} \;\; \bar{\hat{\mathbf{X}}}_2^{(\mathbf{k+2})}]^T$ for the decision-directed mode. The $N \times N$ diagonal matrix $\mathbf{P}_{k+2}$ is computed by the recursion

$$\mathbf{P}_{k+2} = \lambda^{-1}(\mathbf{P}_k - \lambda^{-1}\mathbf{P}_k\mathbf{\Gamma}_{k+2}\mathbf{P}_k) \tag{13.23}$$

where the diagonal matrices $\mathbf{\Gamma}_{k+2}$ and $\mathbf{\Delta}_{k+2}$ are computed from the recursions

$$\mathbf{\Gamma}_{k+2} = diag(\mathbf{Y}^{(k)})\mathbf{\Delta}_{k+2}diag(\bar{\mathbf{Y}}^{(k)}) + diag(\mathbf{Y}^{(k+1)})\mathbf{\Delta}_{k+2}diag(\bar{\mathbf{Y}}^{(k+1)})$$

$$\mathbf{\Delta}_{k+2} = (\mathbf{I}_N + \lambda^{-1}(diag(\mathbf{Y}^{(k)})\mathbf{P}_k diag(\bar{\mathbf{Y}}^{(k)}) + diag(\mathbf{Y}^{(k+1)})\mathbf{P}_k diag(\bar{\mathbf{Y}}^{(k+1)})))^{-1}$$

The initial conditions are $\mathcal{W}_0 = \mathbf{0}$ and $\mathbf{P}_0 = \delta\mathbf{I}_N$ where δ is a large number, and the forgetting factor λ is chosen close to 1.

The block diagram of the adaptive FDE-STBCs is shown in Figure 13.9. Pairs of consecutive received blocks are transformed to the frequency domain using FFT, then the data matrix in Equation 13.21 is formed. The filter output (the product $\mathbf{U}_k \mathcal{W}_{k-2}$) is transformed back to the time domain using inverse IFFT and passed to a decision device to generate data estimates. The output of the adaptive equalizer is compared to the desired response to generate an error vector, which is in turn used to update the equalizer coefficients according to the RLS recursions. The equalizer operates in a training mode until it converges, then it switches to a decision-directed mode where previous decisions are used for tracking. When operating over fast time-varying channels, retraining blocks can be transmitted periodically to prevent equalizer divergence (see [78]).

13.4.2.2 Noncoherent Techniques

Noncoherent transmission schemes do not require channel estimation, hence eliminating the need for bandwidth-consuming training sequences and reducing terminal complexity. This becomes more significant for rapidly fading channels where frequent retraining is needed to track channel variations and for multiple-antenna broadband transmission scenarios where more channel parameters (several coefficients for each transmit–receive antenna pair) need to be estimated. One class of noncoherent techniques is blind identification and detection schemes. Here, the structure of the channel (finite impulse response), the input constellation (finite alphabet), and the output (cyclostationarity) are exploited to eliminate training symbols. Such techniques have a vast literature and we refer the interested reader to a good survey in [79]. Another class of noncoherent techniques is the generalized ML receiver in [80].

Several noncoherent space–time transmission schemes have been proposed for flat-fading channels, including differential STBC schemes with two [81] or more [82] transmit antennas and group differential STC schemes (see, e.g., [21] and references therein). Here, we describe a differential space–time transmission scheme for frequency-selective channels we recently proposed in [83] that achieves full diversity (spatial and multipath) at rate one[17] with two transmit antennas. This scheme is a differential form for the OFDM-STBC structure described in [51]. A time-domain differential space–time scheme with single-carrier transmission is presented in [83].

We consider two symbols, $X_1(m)$ and $X_2(m)$, drawn from a PSK constellation that in a conventional OFDM system would be transmitted over two consecutive OFDM blocks on the same subcarrier m. Following the Alamouti encoding scheme described in Section 13.4.1.3, the two source symbols are mapped as

$$\mathbf{X}^{(1)}(m) = [X_1(m), X_2(m)]^T, \quad \mathbf{X}^{(2)}(m) = [-\bar{X}_2(m), \bar{X}_1(m)]^T \tag{13.24}$$

where $\mathbf{X}^{(1)}$ represents the information-bearing vector for the first OFDM block and $\mathbf{X}^{(2)}$ corresponds to the second OFDM block.[18] Let N denote the FFT size, then $\mathbf{X}^{(1)}$ and $\mathbf{X}^{(2)}$ are length $2N$ vectors holding the symbols to be transmitted by the two transmit antennas. Consequently, after taking the FFT at the receiver, we have (at subcarrier m)

$$\begin{pmatrix} Y_1(m) & Y_2(m) \\ -\bar{Y}_2(m) & \bar{Y}_1(m) \end{pmatrix} = \begin{pmatrix} H_1(m) & H_2(m) \\ -\bar{H}_2(m) & \bar{H}_1(m) \end{pmatrix} \begin{pmatrix} X_1(m) & -\bar{X}_2(m) \\ X_2(m) & \bar{X}_1(m) \end{pmatrix} + \text{noise} \tag{13.25}$$

where $H_1(m)$ and $H_2(m)$ are the frequency responses of the two channels at subcarrier m.

For block k and subcarrier m, denote the source symbols as $\mathbf{u}_m^{(k)} = [u_{1,m}^{(k)} \quad u_{2,m}^{(k)}]^T$, the transmitted matrix as $\mathbf{X}_m^{(k)}$, and the received matrix as $\mathbf{Y}_m^{(k)}$. Then, in the absence of noise, Equation 13.25 is written

---

[17] This does not include the rate penalty incurred by concatenating OFDM-STBC with an outer code and interleaving across tones, which is common to all OFDM systems (see, e.g., [60] for more discussion).

[18] Intuitively, each OFDM subcarrier can be thought of as a flat-fading channel and the Alamouti code is applied to each of the OFDM subcarriers. As a result, the Alamouti code yields diversity gains at every subcarrier.

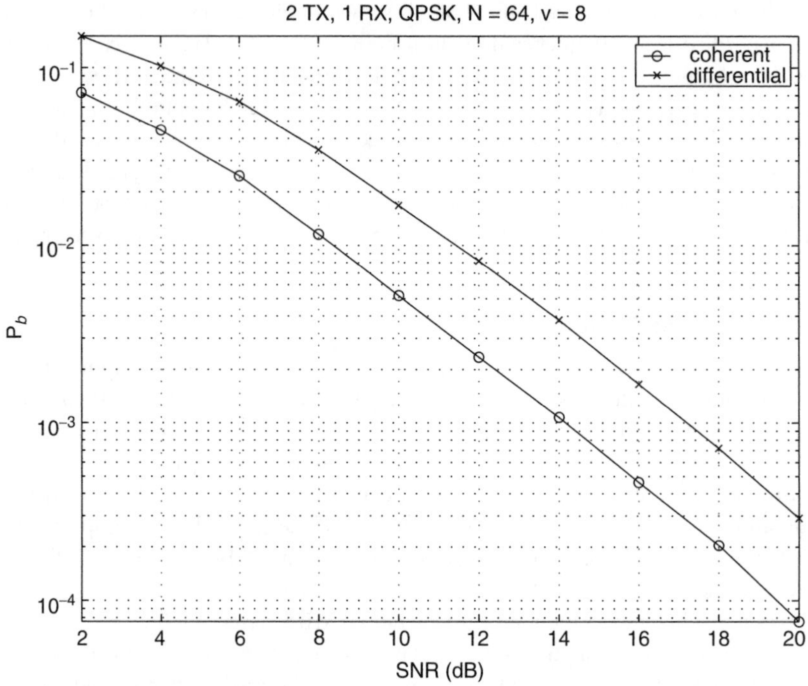

FIGURE 13.8 Performance comparison between coherent and differential OFDM-STBCs with 2 TX, 1 RX, QPSK modulation, FFT size of 64, $\nu = 8$.

as $\mathbf{Y}_m^{(k)} = \mathbf{H}_m \mathbf{X}_m^{(k)}$, where we assume that the channel is fixed over two consecutive blocks. Using the quaternionic structure of $\mathbf{H}_m$, it follows that

$$\bar{\mathbf{Y}}_m^{(k-1)}\,\mathbf{Y}_m^{(k)} = (|H_1(m)|^2 + |H_2(m)|^2)\,\bar{\mathbf{X}}_m^{(k-1)}\,\mathbf{X}_m^{(k)}$$

Since we would like to estimate the source symbols contained in $\mathbf{U}_m^{(k)} \overset{def}{=} \begin{pmatrix} u_{1,m}^{(k)} & -\bar{u}_{2,m}^{(k)} \\ u_{2,m}^{(k)} & \bar{u}_{1,m}^{(k)} \end{pmatrix}$, we define the differential transmission rule $\mathbf{X}_m^{(k)} = (\bar{\mathbf{X}}_m^{(k-1)})^{-1}\,\mathbf{U}_m^{(k)}$. Note that no inverse computation is needed in computing $(\bar{\mathbf{X}}_m^{(k-1)})^{-1}$ due to the quaternionic structure of $\bar{\mathbf{X}}_m^{(k-1)}$. Figure 13.8 illustrates the 3-dB SNR loss of differential OFDM-STBCs relative to their coherent counterpart (with perfect CSI assumed) for an indoor wireless environment.

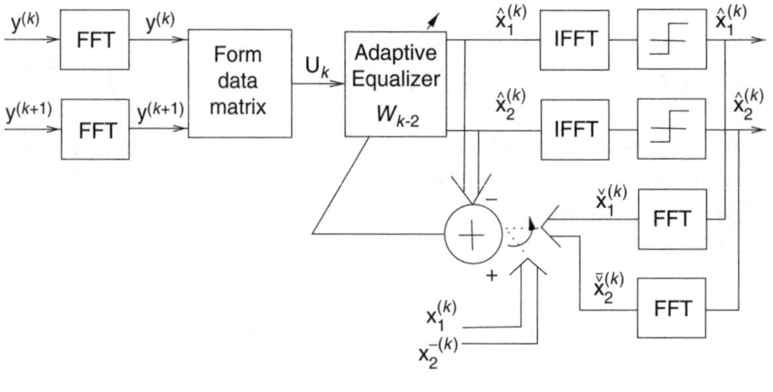

FIGURE 13.9 Block diagram of the adaptive space–time FDE-STBC equalizer/decoder.

13.5 Summary and Future Challenges

The main function of the physical-layer is delivery of the highest possible bit rates reliably over the wireless channel. To perform this function, several transmitter and receiver techniques are needed to mitigate various performance impairments. This includes the use of diversity techniques to mitigate fading effects by exploiting space selectivity of the channel, the use of MIMO antenna techniques to realize spatial rate multiplexing gains, and the use of OFDM (or other equalization techniques such as M-BCJR or SC FDE) to mitigate channel frequency selectivity. In addition, three main techniques can be used to mitigate Doppler effects due to channel time selectivity: channel estimation and tracking using pilot symbols/tones, adaptive filtering, and differential transmission/detection. The first technique is more complex but achieves superior performance for quasi-static or slowly fading channels.

When selecting the number of transmit/receive antennas, several practical considerations must be taken into account, as described next. Under strict delay constraints, achieving high diversity gains (i.e., high reliability) becomes critical in order to minimize the need for retransmissions. Since transmit/receive diversity gains experience diminishing returns as their numbers increase, complexity considerations dictate the use of small antenna arrays (typically no more than four antennas at each end). Current technology limitations favor using more antennas at the base station than at the user terminal.

For delay-tolerant applications (such as data file transfers), achieving high throughput takes precedence over achieving high diversity, and larger antenna arrays (of course, still limited by cost and space constraints) can be used to achieve high-spatial-rate multiplexing gains. Likewise, high-mobility channel conditions substantially impact the choice of system parameters such as the use of shorter blocks, lower carrier frequencies, and noncoherent or adaptive receiver techniques.

STTCs [4] use multiple transmit antennas to achieve diversity and coding gains. The first gain manifests itself as an increase in the slope of the bit error rate (BER) vs. SNR curve (on a log–log scale) at high SNR, while the latter gain manifests itself as a horizontal shift in that same curve. At low SNR, it becomes more important to maximize the coding gain, while at high SNR diversity gains dominate performance. For SNR ranges typically encountered on broadband wireless terrestrial links, it might be wise to sacrifice some diversity gain in exchange for more coding gain. For example, using only two transmit and one receive antenna for a channel with delay spread as high as 16 taps, the maximum (spatial and multipath) diversity gain possible is $16 \times 2 \times 1 = 32$. For typical SNR levels in the 10- to 25-dB range, it suffices to design STC that achieves a much smaller diversity level (e.g., up to 8) to limit the receiver complexity and to use the extra degrees of freedom in code design to achieve a higher coding gain. STC has also been shown to result in significant improvements in the networking throughput [84].

Many challenges still exist at the physical layer on the road to achieving high-rate and reliable wireless transmission. We conclude this chapter by enumerating some of these challenges

- **Code design:** A significant open problem is the design of practical space–time codes that achieve the optimal diversity-rate trade-off derived in [18] and have a practical decoding complexity.
- **Implementation issues:** These include the development of low-cost integrated multiple RF chains and of low-power parallelizable implementations of the STC receiver signal processing algorithms suitable for digital signal processing (DSP) and ASIC implementations at the high sampling rates used for broadband transmission/reception. While it is desirable (from an implementation point of view) to use a single receive antenna to preserve the desirable small form factor and low-power consumption of the user terminal, adding a second receive antenna multiplies the diversity gain by a factor of two (hence doubling the decay rate of average probability of error with SNR at high SNR) and enables interference cancellation. Implementation and manufacturing innovations are needed to make multiple-antenna user terminals commercially viable. Antennas capable of realizing polarization diversity gains [85] further enhance communication reliability by utilizing another source of diversity.
- **Receiver signal processing:** While effective and practical joint equalization and decoding schemes that exploit the multipath diversity available in frequency-selective channels have been developed,

the full exploitation of time diversity in fast time-varying channels remains elusive. The main challenge here is the development of practical adaptive algorithms that can track the rapid variations of the large number of taps in MIMO channels and equalizers. While some encouraging steps have been made in this direction [78, 86], the allowable Doppler rates (which depend on the mobile speed and carrier frequency) for high performance are still quite limited.

Acknowledgment

I thank the following individuals for many stimulating discussions and technical contributions to this chapter: Rob Calderbank, Suhas Diggavi, Christina Fragouli, Anastasios Stamoulis, and Waleed Younis.

References

[1] I. Telatar. Capacity of Multi-Antenna Gaussian Channels. *European Transactions on Telecommunications*, 585–595, 1999.

[2] J. Uddenfeldt and A. Raith. Cellular Digital Mobile Radio System and Method of Transmitting Information in a Digital Cellular Mobile Radio System. U.S. patent 5,088,108, February 1992.

[3] A. Wittneben. A New Bandwidth Efficient Transmit Antenna Modulation Diversity Scheme for Linear Digital Modulation. In *International Conference on Communications*, 1993, pp. 1630–1634.

[4] V. Tarokh, N. Seshadri, and A.R. Calderbank. Space-Time Codes for High Data Rate Wireless Communications: Performance Criterion and Code Construction. *IEEE Transactions on Information Theory*, 744–765, 1998.

[5] S. Alamouti. A Simple Transmit Diversity Technique for Wireless Communications. *IEEE Journal on Selected Areas in Communications*, 1451–1458, 1998.

[6] A. Naguib, V. Tarokh, N. Seshadri, and A.R. Calderbank. A Space-Time Coding Modem for High-Data-Rate Wireless Communications. *IEEE Journal on Selected Areas in Communications*, 1459–1477, 1998.

[7] W.C. Jakes. *Microwave Mobile Communications*. IEEE Press, Washington, DC, 1974.

[8] T. Rappaport. *Wireless Communications*. IEEE Press, Washington, DC, 1996.

[9] L.H. Brandenburg and A. Wyner. Capacity of the Gaussian Channel with Memory: The Multivariate Case. *Bell Systems Technical Journal*, 53, 745–779, 1974.

[10] P. Balaban and J. Salz. Optimum Diversity Combining and Equalization in Digital Data Transmission with Applications to Cellular Mobile Radio. *IEEE Transactions on Communications*, 40, 885–894, 1992.

[11] G. Raleigh and J. Cioffi. Spatio-Temporal Coding for Wireless Communication. *IEEE Transactions on Communications*, 357–366, 1998.

[12] R.G. Gallager. *Information Theory and Reliable Communications*. John Wiley & Sons, New York, 1968.

[13] L.H. Ozarow, S. Shamai, and A.D. Wyner. Information Theoretic Considerations for Cellular Mobile Radio. *IEEE Transactions on Vehicular Technology*, 43(2), 359–378, 1994.

[14] S.N. Diggavi. On Achievable Performance of Spatial Diversity Fading Channels. *IEEE Transactions on Information Theory*, 47(1), 308–325, 2001. (Also see *Proc. ISIT*, 396, 1998).

[15] G.J. Foschini and M.J. Gans. On Limits of Wireless Communication in a Fading Environment When Using Multiple Antennas. *Wireless Personal Communication*, 311–335, 1998.

[16] N. Al-Dhahir and S.N. Diggavi. On the Achievable Rates of Time-Varying Frequency-Selective Channels. In *Conference on Information Sciences and Systems*, Princeton, NJ, March 2002, pp. 860–865.

[17] J. Guey, M.P. Fitz, M. Bell, and W. Kuo. Signal Design for Transmitter Diversity Wireless Communication Systems over Rayleigh Fading Channels. *IEEE Transactions on Communications*, 47(4): 527–537, 1999.

[18] L. Zheng and D. Tse. Diversity and Multiplexing: A Fundamental Tradeoff in Multiple Antenna Channels. *IEEE Transactions on Information Theory*, 2002.

[19] S. Diggavi, N. Al-Dhahir, A. Stamoulis, and A.R. Calderbank. Great Expectations: The Value of Spatial Diversity in Wireless Networks. Invited Paper to Proceedings of the IEEE Special Issue on Gigabit Wireless. 92(2): 219–270, February 2004.

[20] L. Zheng and D. Tse. Communication on the Grassmann Manifold: A Geometric Approach to the Noncoherent Multiple-Antenna Channel. *IEEE Transactions on Information Theory*, 48(2): 359–383, 2002.

[21] B.L. Hughes. Differential Space-Time Modulation. *IEEE Transactions on Information Theory*, 46(7): 2567–2578, 2000.

[22] G. Foschini. Layered Space-Time Architecture for Wireless Communication in a Fading Environment When Using Multiple Antenna Elements. *Bell Labs Technical Journal*, 41–59, 1996.

[23] G.J. Foschini, G. Golden, R. Valenzuela, and P. Wolniansky. Simplified Processing for High Spectral Efficiency Wireless Communication Employing Multi-Element Arrays. *IEEE Journal on Selected Areas in Communications*, 17, 1841–1852, 1999.

[24] W. Choi, R. Negi, and J.M. Cioffi. Combined ML and DFE Decoding for the V-BLAST System. In *International Conference on Communications*, 2000, pp. 1243–1248.

[25] O. Damen, A. Chkeif, and J. Belfiore. Lattice Code Decoder for Space-Time Codes. *IEEE Communications Letters*, 161–163, 2000.

[26] S. Ariyavisitakul. Turbo Space-Time Processing to Improve Wireless Channel Capacity. *IEEE Transactions of Communications*, 1347–1359, 2000.

[27] M. Sellathurai and S. Haykin. TURBO-BLAST for Wireless Communications: Theory and Experiments. *IEEE Transactions on Signal Processing*, 2538–2546, 2002.

[28] N. Al-Dhahir and A.H.Sayed. The Finite-Length MIMO MMSE-DFE. *IEEE Transactions on Signal Processing*, 2921–2936, 2000.

[29] A. Lozano and C. Papadias. Layered Space-Time Receivers for Frequency-Selective Wireless Channels. *IEEE Transactions on Communications*, 65–73, 2002.

[30] M. Gans et al. Outdoor BLAST Measurement System at 2.44 GHz: Calibration and Initial Results. *IEEE Journal on Selected Areas in Communications*, 570–583, 2002.

[31] R. Soni, M. Buehrer, and R. Benning. Intelligent Antenna System for CDMA2000. *IEEE Signal Processing Magazine*, 54–67, 2002.

[32] L. Godara. Applications of Antenna Arrays to Mobile Communications. Part I. Performance Improvement, Feasibility, and System Considerations. *Proceedings of the IEEE*, 85, 1031–1060, 1997.

[33] L. Godara. Applications of Antenna Arrays to Mobile Communications. Part II. Beamforming and Direction-of-Arrival Considerations. *Proceedings of the IEEE*, 85, 1195–1245, 1997.

[34] N. Seshadri and J. Winters. Two Signaling Schemes for Improving the Error Performance of Frequency-Division-Duplex (FDD) Transmission Systems Using Transmitter Antenna Diversity. In *Vehicular Technology Conference (VTC)*, 1993, pp. 508–511.

[35] Special issue of *IEEE Transactions on Signal Processing* on space–time coding, October 2002.

[36] N. Al-Dhahir, C. Fragouli, A. Stamoulis, Y. Younis, and A.R. Calderbank. Space-Time Processing for Broadband Wireless Access. *IEEE Communications Magazine*, 136–142, 2002.

[37] S. Parkvall, M. Karlsson, M. Samuelsson, L. Hedlund and B. Goransson. Transmit Diversity in WCDMA: Link and System Level Results. In *Vehicular Technology Conference*, 2000, pp. 864–868.

[38] V. Tarokh, A. Naguib, N. Seshadri, and A.R. Calderbank. Space-Time Codes for High Data Rate Wireless Communication: Performance Criteria in the Presence of Channel Estimation Errors, Mobility, and Multiple Paths. *IEEE Transactions on Communications*, 199–207, 1999.

[39] C. Berrou and A. Glavieux. Near Optimum Error Correcting Coding and Decoding: Turbo Codes. *IEEE Transactions on Communications*, 44(10): 1261–1271, 1996. (Also see C. Berrou, A. Glavieux, and P. Thitimajshima, in *Proceedings of ICC'93*.)

[40] S. Benedetto and G. Montorsi. Unveiling Turbo Codes: Some Results on Parallel Concatenated Coding Schemes. *IEEE Transactions on Information Theory*, 42(2): 409–428, 1996.

[41] G. Bauch and N. Al-Dhahir. Reduced-Complexity Space-Time Turbo Equalization for Frequency-Selective MIMO Channels. *IEEE Transactions on Wireless Communications*, 819–828, 2002.

[42] Y. Liu, M.P. Fitz, and O.Y. Takeshita. Full Rate Space-Time Turbo Codes. *IEEE Journal on Selected Areas in Communications*, 19(5):969–980, 2001.

[43] R.G. Gallager. *Low Density Parity Check Codes*. MIT Press, Cambridge, MA, 1963. (Available at http://justice.mit.edu/people/gallager.html.)

[44] B. Lu, X. Wang, and K.R. Narayanan. LDPC-Based Space-Time Coded OFDM Systems over Correlated Fading Channels: Performance Analysis and Receiver Design. *IEEE Transactions on Communications*, 50(1):74–88, 2002.

[45] C. Fragouli, N. Al-Dhahir, and W. Turin. Effect of Spatio-Temporal Channel Correlation on the Performance of Space-Time Codes. In *International Conference on Communications*, Vol. 2, April 2002, pp. 826–830.

[46] Y. Liu, M. Fitz, and O. Takeshita. Space-Time Codes Performance Criteria and Design for Frequency Selective Fading Channels. In *International Conference on Communications*, Vol. 9, June 2001, pp. 2800–2804.

[47] Texas Instruments. Space-Time Block Coded Transmit Antenna Diversity for WCDMA, SMG2 Document 581/98. Submitted October 1998.

[48] TIA 45.5 Subcommittee. The CDMA 2000 Candidate Submission, Draft. June 1998.

[49] V. Tarokh, H. Jafarkhani, and A.R. Calderbank. Space-Time Block Codes from Orthogonal Designs. *IEEE Transactions on Information Theory*, 1456–1467, 1999.

[50] E. Lindskog and A. Paulraj. A Transmit Diversity Scheme for Delay Spread Channels. In *International Conference on Communications (ICC)*, June 2000, pp. 307–311.

[51] Z. Liu, G. Giannakis, A. Scaglione, and S. Barbarossa. Decoding and Equalization of Unknown Multipath Channels Based on Block Precoding and Transmit-Antenna Diversity. In *Asilomar Conference on Signals, Systems, and Computers*, 1999, pp. 1557–1561.

[52] N. Al-Dhahir. Single-Carrier Frequency-Domain Equalization for Space-Time Block-Coded Transmissions over Frequency-Selective Fading Channels. *IEEE Communications Letters*, 304–306, 2001.

[53] N. Al-Dhahir. Overview and Comparison of Equalization Schemes for Space-Time-Coded Signals with Application to EDGE. *IEEE Transactions on Signal Processing*, 2477–2488, 2002.

[54] A.F. Naguib, N. Seshadri, and A.R. Calderbank. Applications of Space-Time Block Codes and Interference Suppression for High Capacity and High Data Rate Wireless Systems. In *Thirty-Second Asilomar Conference on Signals, Systems and Computers*, 1998, pp. 1803–1810.

[55] S. Weinstein and P. Ebert. Data Transmission by Frequency-Division Multiplexing Using the Discrete Fourier Transform. *IEEE Transactions on Communications*, 19(5):628–634, 1971.

[56] A. Stamoulis, S. Diggavi, and N. Al-Dhahir. Intercarrier Interference in MIMO OFDM. *IEEE Transactions on Signal Processing*, 50(10):2451–2464, 2002.

[57] D. Agrawal, V. Tarokh, A. Naguib, and N. Seshadri. Space-Time Coded OFDM for High Data-Rate Wireless Communication over Wideband Channels. In *Vehicular Technology Conference*, May 1998, pp. 2232–2236.

[58] H. Sari, G. Karam, and I. Jeanclaud. An Analysis of Orthogonal Frequency Division Multiplexing for Mobile Radio Applications. In *Vehicular Technology Conference*, 1994, pp. 1635–1639.

[59] T. Pollet, M. Van Bladel, and M. Moeneclaey. BER Sensitivity of OFDM Systems to Carrier Frequency Offset and Wiener Phase Noise. *IEEE Transactions on Communications*, 191–193, 1995.

[60] H. Sari, G. Karam, and I. Jeanclaude. Transmission Techniques for Digital Terrestrial TV Broadcasting. *IEEE Communications Magazine*, 100–109, 1995.

[61] A. Paulraj and B. Ng. Space-Time Modems for Wireless Personal Communications. *IEEE Personal Communications*, 36–48, 1998.

[62] D. Chu. Polyphase Codes with Good Periodic Correlation Properties. *IEEE Transactions on Information Theory*, 18, 531–532, 1972.

[63] C. Fragouli, N. Al-Dhahir, and W. Turin. Reduced-Complexity Training Schemes for Multiple-Antenna Broadband Transmissions. In *Wireless Communications and Networking Conference*, Vol. 1, March 2002, pp. 78–83.

[64] R. Negi and J. Cioffi. Pilot Tone Selection for Channel Estimation in a Mobile OFDM System. *IEEE Transactions on Consumer Electronics*, 44(3):1122–1128, 1998.

[65] J.K. Cavers. An Analysis of Pilot Symbol Assisted Modulation for Rayleigh Fading Channels (Mobile Radio). *IEEE Transactions on Vehicular Technology*, 40, 686–693, 1991.

[66] W. Younis and N. Al-Dhahir. Joint Prefiltering and MLSE Equalization of Space-Time-Coded Transmissions over Frequency-Selective Channels. *IEEE Transactions on Vehicular Technology*, 144–154, 2002.

[67] C. Fragouli, N. Al-Dhahir, S. Diggavi, and W. Turin. Prefiltered Space-Time M-BCJR Equalizer for Frequency-Selective Channels. *IEEE Transactions on Communications*, 742–753, 2002.

[68] A. Duel-Hallen and C. Heegard. Delayed Decision-Feedback Sequence Estimation. *IEEE Transactions on Communications*, 428–436, 1989.

[69] M. Eyuboglu and S. Qureshi. Reduced-State Sequence Estimation for Coded Modulation of Intersymbol Interference Channels. *IEEE Journal on Selected Areas in Communications*, 989–999, 1999.

[70] V. Franz and J. Anderson. Concatenated Decoding with a Reduced-Search BCJR Algorithm. *IEEE Journal on Selected Areas in Communications*, 186–195, 1998.

[71] L. Bahl, J. Cocke, F. Jelinek, and J. Raviv. Optimal Decoding of Linear Codes for Minimizing Symbol Error Rate. *IEEE Transactions on Information Theory*, 20, 284–287, 1974.

[72] D. Falconer, S. Ariyavisitakul, A. Benyamin-Seeyar, and B. Eidson. Frequency Domain Equalization for Single-Carrier Broadband Wireless Systems. *IEEE Communications Magazine*, 40, 58–66, 2002.

[73] K. Berberidis and P. Karaivazoglou. An Efficient Block Adaptive Decision Feedback Equalizer Implemented in the Frequency Domain. *IEEE Transactions on Signal Processing*, 2273–2285, 2002.

[74] S. Crozier, D. Falconer, and S. Mahmoud. Least Sum of Squared Errors (LSSE) Channel Estimation. *IEE Proceedings — Part F*, 371–378, 1991.

[75] A. Stamoulis, N. Al-Dhahir, and A.R. Calderbank. Further Results on Interference Cancellation and Space-Time Block Codes. In *Thirty-Fifth Asilomar Conference on Signals, Systems and Computers*, Vol. 1, 2001, pp. 257–262.

[76] S. Verdu. *Multiuser Detection*. Cambridge University Press, Cambridge, U.K., 1998.

[77] S. Haykin. *Adaptive Filter Theory*, 2nd ed. Prentice Hall, New York, 1991.

[78] W. Younis, N. Al-Dhahir, and A.H. Sayed. Adaptive Frequency-Domain Equalization of Space-Time Block-Coded Transmissions. In *International Conference on Acoustics, Speech, and Signal Processing*, May 2002, pp. 2353–2356.

[79] L. Tong and S. Perreau. Multichannel Blind Identification: From Subspace to Maximum Likelihood Methods. *Proceedings of the IEEE*, 1951–1968, 1998.

[80] M. Uysal, N. Al-Dhahir, and C.N. Georghiades. A Space-Time Block-Coded OFDM Scheme for Unknown Frequency-Selective Fading Channels. *IEEE Communications Letters*, 5, 393–395, 2001.

[81] V. Tarokh and H. Jafarkhani. A Differential Detection Scheme for Transmit Diversity. *IEEE Journal on Selected Areas in Communications*, 18(7):1169–1174, 2000.

[82] H. Jafarkhani and V. Tarokh. Multiple Transmit Antenna Differential Detection from Generalized Orthogonal Designs. *IEEE Transactions on Information Theory*, 47(6):2626–2631, 2001.

[83] S. Diggavi, N. Al-Dhahir, A. Stamoulis, and A.R. Calderbank. Differential Space-Time Coding for Frequency-Selective Channels. *IEEE Communications Letters*, 253–255, 2002.

[84] A. Stamoulis and N. Al-Dhahir. Impact of Space-Time Block Codes on 802.11 Network Throughput, 2(5): 1029–1039, September 2003.

[85] L. Aydin, E. Esteves, and R. Padovani. Reverse Link Capacity and Coverage Improvement for CDMA Cellular Systems Using Polarization and Spatial Diversity. In *International Conference on Communications*, 2002, pp. 1887–1892.

[86] C. Komninakis, C. Fragouli, A. Sayed, and R. Wesel. Adaptive Multi-Input Multi-Output Fading Channel Equalization Using Kalman Estimation. In *International Conference on Communications*, 2000, pp. 1655–1659.

14

Linear Precoding for MIMO Systems

14.1 Introduction **14**-1
System Model

14.2 Optimum Precoding **14**-3
Jointly Optimum Design and Performance Bounds

14.3 Performance Analysis and Random Matrices **14**-10
Differential Forms and Random Matrix Techniques
• The Statistics of the Capacity

14.4 Channel Estimation for MIMO Systems Using Precoding Techniques **14**-19
Channel Estimation Algorithm • Cramer–Rao Lower Bound
• Numerical Results

14.5 Conclusions **14**-24

Anna Scaglione
Cornell University

Atul Salhotra
Cornell University

Azadeh Vosoughi
Cornell University

Abstract

Broadband transmission with antenna diversity at both the transmitter and receiver sides offers several degrees of freedom to design modems. They can be exploited to increase the resilience or increase the throughput of the communication link. The wireless channel effect can be mapped conveniently in a multi-input multi-output (MIMO) model, which allows the derivization and analysis of linear code designs using algebraic tools. We refer to these techniques as linear precoding methods. This chapter focuses on three aspects related to the design of such precoders: (1) the optimal design under average and peak power type of constraints; (2) the performance analysis in the presence of random fading; and (3) the performance of generalized training and semiblind techniques.

14.1 Introduction

In this chapter we consider a broadband power-limited source with K transmitters and R receivers. We assume that W is the bandwidth available and T is the desired duration for the transmission. The channel effect can be modeled as a time-varying linear distortion with memory between each antenna pair plus additive noise added at each receiver (Figure 14.1). The bandwidth and time limitations and the finite number of antennas constrain the continuous time-transmit waveform to be in a finite dimensional space, and the available dimensions over which communication takes place are generally referred to as degrees of freedom [10], [21], [42], [60]. The question that arises is how to design a modem that optimally utilizes the resources available [16]. To this end, in Section 14.1.1 we develop a discrete-time multi-input multi-output (MIMO) equivalent model that maps the effect of the link on the input signal into a linear

0-8493-1657-X/$0.00+$1.50

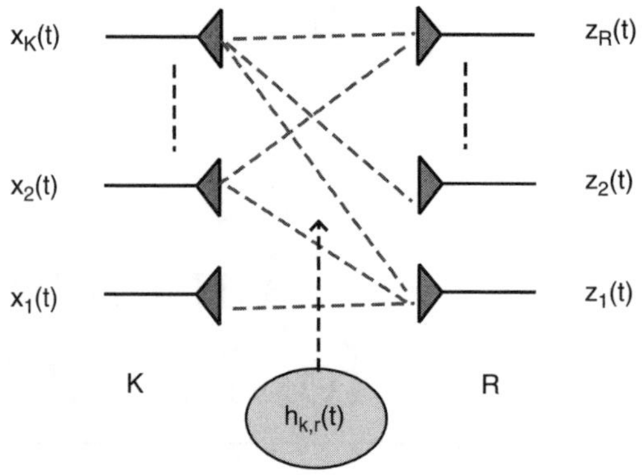

FIGURE 14.1 K transmitter and R receiver MIMO system.

operator $\boldsymbol{H}$ of finite dimension with an additive noise vector. The number of available dimensions of the space for our system are upper-bounded by $WT \min(K, R)$, as will be shown next in Section 14.2.1. For a more thorough discussion on trade-off between diversity and freedom in communications systems, refer to [18], [59]. After Section 14.1.1 we reduce the modem design into the design of the optimum linear input and output operators, with a norm constraint on the input mapping. We refer to this scheme as linear precoding. Several modulation and multiplexing schemes can be viewed as special cases of linear precoding, and therefore, the insight gained from optimal precoded transmission sheds light on the fundamental limits of communication over the MIMO channel affected by noise and linear distortion, such as wireless medium [43].

This chapter is divided into three parts: (1) the first part (Section 14.2) deals with the optimal design of precoders and decoders under average and peak power types of constraints; (2) the second part (Section 14.3) provides the basic tools necessary for the performance analysis in MIMO random fading channels; and (3) the third part (Section 14.4) analyzes the performance of training and precoding, trying to underline the trade-off between investing transmission resources on training vs. information symbols. Many references to the vast literature on MIMO systems are spread throughout the chapter.

Notation: Boldface upper- and lowercases denote matrices and column vectors, respectively. $tr(\boldsymbol{A})$ and $|\boldsymbol{A}|$ are the trace and determinant of $\boldsymbol{A}$, respectively. The column vector formed by stacking vertically the columns of $\boldsymbol{A}$ is $\boldsymbol{a} = vec(\boldsymbol{A})$. $\boldsymbol{I}_M$ is the identity matrix of size M. Complex conjugate, Hermitian, and transpose and pseudoinverse operations are represented by $(.)^*$, $(.)^H$, $(.)^T$, and $(.)^\dagger$, respectively. $vec^H(\boldsymbol{A})$ is the transpose conjugate of $vec(\boldsymbol{A})$. $\otimes$ is the Kronecker product. The (i, j) entry of matrix $\boldsymbol{A}$ is indicated with $[\boldsymbol{A}]_{ij}$. Continuous-time signal vectors are like $\mathbf{a}(t)$ and discrete-time vector sequences like $\mathbf{a}[n]$. Sequences of vectors obtained by stacking consecutive blocks, such as $\mathbf{a}_i = [\mathbf{a}[iM], \ldots, \mathbf{a}[iM + M - 1]]$, are characterized by the suffix i. To manipulate blocked matrices we introduce vectors of indices $\mathbf{k} = (k_1, \ldots, k_m)$ and the notation $\mathbf{A}[\mathbf{k}] \equiv (\mathbf{A}[k_1]^H, \ldots, \mathbf{A}[k_m]^H)^H$. We define a *tall matrix* as one that has more rows than columns. A matrix $\boldsymbol{A}_{m\times n}$ with rank less than full rank ($\min\{m, n\}$) is called a *rank-deficient* matrix. We use the acronym CSI to indicate the *channel status information.*

14.1.1 System Model

The system considered has K transmit and R receive antennas. The baseband equivalent transmitted signal is the vector $\boldsymbol{x}(t) := (x_1(t), \ldots, x_K(t))^T$ of complex envelopes emitted by the transmit antennas. We assume a digital link with linear modulation so that the vector $\boldsymbol{x}(t)$ is related to the (coded) symbol vector $\boldsymbol{x}[n]$ by

$$\boldsymbol{x}(t) = \sum_{n=-\infty}^{+\infty} \boldsymbol{x}[n] g_T(t - nT) \tag{14.1}$$

where $g_T(t)$ is the transmit pulse and $1/T$ is the rate with which the data $\boldsymbol{x}[n]$ are transmitted. Correspondingly, $\boldsymbol{z}(t) = \boldsymbol{y}(t) + \boldsymbol{n}(t)$ is the received $R \times 1$ vector, which contains the channel output $\boldsymbol{y}(t)$ and additive noise $\boldsymbol{n}(t)$. For a linear (generally time-varying) channel, the input–output (I/O) relationship can be cast in the form

$$\boldsymbol{y}(t) = \iint g_R(t-\theta)\mathbf{H}(\theta,\tau)\boldsymbol{x}(\theta-\tau)d\tau d\theta \tag{14.2}$$

where $g_R(t)$ is the impulse response of the low-pass receive filter (usually a square root raised cosine filter) matched to the transmit filter $g_T(t)$, and the (k,l)-th entry of matrix $\mathbf{H}(\theta,\tau)$ is the impulse response of the channel between the l-th transmit and the k-th receive antennas. Introducing the discrete-time time-varying impulse response

$$\mathbf{H}[k,n] \equiv \int_{-\infty}^{\infty}\int_{-\infty}^{\infty} \mathbf{H}(\theta,\tau)g_T(\theta-\tau-(k-n)T)g_R(kT-\theta)d\tau d\theta \tag{14.3}$$

we can write the vector of received samples $\boldsymbol{y}[k] := \boldsymbol{y}(kT)$ as

$$\boldsymbol{y}[k] = \sum_{n=-\infty}^{\infty} \mathbf{H}[k,k-n]\boldsymbol{x}[n] \tag{14.4}$$

If the channel discrete-time time-varying impulse response $\mathbf{H}[k,n]$ is causal and has finite memory L, we can write the I/O relationship (Equation 14.4) in block finite impulse response (FIR) form. Specifically, stacking $P = M + L$ transmit snapshots in a $PK \times 1$ vector $\boldsymbol{x}_i := vec([\boldsymbol{x}[iP],\ldots,\boldsymbol{x}[iP+P-1]])$ and M received snapshots in a $MR \times 1$ vector $\boldsymbol{y}_i := vec([\boldsymbol{y}[iP+L],\ldots,\boldsymbol{y}[iP+P-1]])$, where we eliminated the first L vectors to cancel the interblock interference (IBI), we have

$$\boldsymbol{y}_i = \boldsymbol{H}\boldsymbol{x}_i \tag{14.5}$$

where $\boldsymbol{H}$ is an $RM \times KP$ block-banded matrix.

If the channel is also time invariant (LTI), i.e., $\mathbf{H}[n,l] \equiv \mathbf{H}[l]$ where $\{\mathbf{H}[l]\}_{r,k}$ is the l-th sample of the impulse response characterizing the channel between the k-th transmit element and the r-th receive element, then $\boldsymbol{H}$ in Equation 14.5 becomes a block Toeplitz matrix:

$$\boldsymbol{H} = \begin{pmatrix} \mathbf{H}[L] & \cdots & \mathbf{H}[0] & 0 & \ldots & 0 \\ 0 & \mathbf{H}[L] & \cdots & \mathbf{H}[0] & \ddots & \vdots \\ \vdots & \ddots & \ddots & \ddots & \ddots & 0 \\ 0 & \cdots & 0 & \mathbf{H}[L] & \cdots & \mathbf{H}[0] \end{pmatrix}_{RM\times KP} \tag{14.6}$$

Although our designs are valid for any $\boldsymbol{H}$, in case of time-varying channels the assumption of knowledge of the CSI at the transmitter is not realistic, unless the channel can be considered time invariant for a sufficiently long interval.

14.2 Optimum Precoding

The problem of communicating reliably over Gaussian and fading channels has been studied extensively. The capacity, reliability function, error exponent, and optimal transmission technique (channel coding and decoding design) depend on the degree of transmitter and receiver knowledge of the channel parameters, i.e., of $\boldsymbol{H}$ and of the noise color. Refer to [3] and references therein for an excellent review of information-theoretic and communication aspects of fading channels. For instance, Tarokh [52] pioneered the design of space–time codes for transmission over MIMO channels subject to Rayleigh and Ricean fading, with CSI known at the receiver only. For the case when neither the transmitter nor the receiver knows the CSI, [22] shows that Cayley differential unitary space–time codes achieve the capacity of Rayleigh flat-fading channel. In this section, we investigate the design of transmitter space–time coding that we refer to as linear precoding [44], assuming the CSI is known at the receiver as well as at the transmitting end.[1] Linear precoding is a block transmission scheme that uses a pair of linear transformations $\boldsymbol{F}$ (precoder) and $\boldsymbol{G}$ (decoder) of the transmit symbols and receive samples, respectively, that operate *jointly* and *linearly* on the time and space dimensions. The framework is similar to the one used in e.g., [4], [13], [27], [57], [58]. The designs target different criteria of optimality and constraints, and our linear optimal solutions can appropriately take advantage of the CSI and best utilize resources while maintaining a reasonable complexity. Most of the results are presented in the forms of *lemmas* and *corollaries*. Refer to [44] for the proofs.

14.2.1 Jointly Optimum Design and Performance Bounds

We will precode $N \times 1$ vectors of symbols $\boldsymbol{s}_i$ using a precoder $\boldsymbol{F}$ as

$$\boldsymbol{x}_i = \boldsymbol{F}\boldsymbol{s}_i \tag{14.7}$$

where $N \leq \min(KP, RM)$. Note that this will require the knowledge of R at the transmitter; i.e., the design will depend on the number of receive antennas. Since N symbols will be embedded in $\boldsymbol{x}_i$ through the precoder $\boldsymbol{F}$, it will take $P = (M + L)$ snapshots to transmit N information symbols. If T is the time necessary to transmit one snapshot, the throughput is

$$\frac{N}{PT} \leq \frac{\min(K(M+L), RM)}{(M+L)T} \longrightarrow_{M \gg L} \frac{\min(K, R)}{T}$$

Assume that the $N \times 1$ vectors of symbols $\boldsymbol{s}_i$ transmitted every PT second satisfy the following:

1. The size N of the block $\boldsymbol{s}_i$ of encoded symbols satisfies $N \leq \text{rank}(\boldsymbol{H})$. This is necessary to guarantee symbol recovery, because it is otherwise impossible to invert the channel with a linear equalizer $\boldsymbol{G}$.
2. The transmit symbols are white, i.e., $\boldsymbol{R}_{ss} = \sigma_{ss}^2 \boldsymbol{I}$, the noise $\boldsymbol{n}_i$ is Gaussian with covariance $\boldsymbol{R}_{nn}$, the noise covariance matrix $\boldsymbol{R}_{nn}$ is positive-definite, and $\boldsymbol{n}_i$ and $\boldsymbol{s}_i$ are uncorrelated.

The receiver performs an appropriate inverse mapping $\boldsymbol{G}$ on the vector $\boldsymbol{z}_i = vec(\boldsymbol{z}[iP + L], \ldots, \boldsymbol{z}[iP + P - 1])$, estimating the symbols as $\hat{\boldsymbol{s}}_i = \boldsymbol{G}\boldsymbol{z}_i$. From Equation 14.5 we have

$$\hat{\boldsymbol{s}}_i = \boldsymbol{G}\boldsymbol{z}_i = \boldsymbol{G}\boldsymbol{H}\boldsymbol{F}\boldsymbol{s}_i + \boldsymbol{G}\boldsymbol{n}_i \tag{14.8}$$

A reasonable criterion to design a linear receiver $\boldsymbol{G}$, for given $\boldsymbol{F}$ and $\boldsymbol{H}$, is to minimize the mean square error (MSE) matrix that is given by

$$E\{(\hat{\boldsymbol{s}}_i - \boldsymbol{s}_i)(\hat{\boldsymbol{s}}_i - \boldsymbol{s}_i)^H\} = MSE(\boldsymbol{F}, \boldsymbol{G}) \tag{14.9}$$

where

$$MSE(\boldsymbol{F}, \boldsymbol{G}) := (\boldsymbol{G}\boldsymbol{H}\boldsymbol{F} - \boldsymbol{I})\boldsymbol{R}_{ss}(\boldsymbol{G}\boldsymbol{H}\boldsymbol{F} - \boldsymbol{I})^H + \boldsymbol{G}\boldsymbol{R}_{nn}\boldsymbol{G}^H \tag{14.10}$$

Because of number 1 above we can write

$$\boldsymbol{G}_{opt} = \boldsymbol{F}^H \boldsymbol{H}^H \left(\boldsymbol{H}\boldsymbol{F}\boldsymbol{F}^H \boldsymbol{H}^H + \boldsymbol{R}_{nn}\sigma_{ss}^{-2}\right)^{-1} \tag{14.11}$$

and

$$\overline{MSE}(\boldsymbol{F}) = \sigma_{ss}^2 \left(\boldsymbol{I} + \sigma_{ss}^2 \boldsymbol{F}^H \boldsymbol{H}^H \boldsymbol{R}_{nn}^{-1} \boldsymbol{H}\boldsymbol{F}\right)^{-1} \tag{14.12}$$

---

[1] CSI can be acquired at the transmitter either if a feedback channel is present or when the transmitter and receiver operate in a time division duplex so that the time-invariant MIMO channel transfer function is the same both ways.

Next, we determine $\boldsymbol{F}_{opt}$ based on different performance measures that depend upon $\overline{MSE}(\boldsymbol{F})$. A reasonable constraint is to bound the expected norm of the transmit vector $E\{\|\boldsymbol{x}_i\|^2\} = tr(\boldsymbol{F}\boldsymbol{F}^H)\sigma_{ss}^2$, which we refer to as the power constraint (PC):

$$tr(\boldsymbol{F}\boldsymbol{F}^H)\,\sigma_{ss}^2 = \mathcal{P}_0$$

An alternative is to constrain the maximum eigenvalue of the transmit vector covariance $\boldsymbol{F}\boldsymbol{F}^H\sigma_{ss}^2$, which also limits the power because $tr(\boldsymbol{F}\boldsymbol{F}^H)\,\sigma_{ss}^2 \leq \lambda_{max}(\boldsymbol{F}\boldsymbol{F}^H)\,N\sigma_{ss}^2$. This corresponds to

$$\lambda_{max}(\boldsymbol{F}\boldsymbol{F}^H)\,\sigma_{ss}^2 = \mathcal{L}_0 \tag{14.13}$$

Besides limiting the transmit power, the maximum eigenvalue constraint in Equation 14.13 imposes a limit on the peak power of the output. This is because

$$\max_{i,k}(|\{\boldsymbol{F}\boldsymbol{s}_i\}_k|^2) \leq \lambda_{max}(\boldsymbol{F}^H\boldsymbol{F})\max_i(\|\boldsymbol{s}_i\|^2) \tag{14.14}$$

where $\|\boldsymbol{s}_i\|^2$ is bounded since $\boldsymbol{s}_i$ is formed by symbols that are all bounded in amplitude. The advantage of this constraint is that it limits the signal peak, independent of the specific constellation used.

Finally, let us introduce the following eigenvalue decomposition (EVD)

$$\boldsymbol{H}^H\boldsymbol{R}_{nn}^{-1}\boldsymbol{H} = \bar{\boldsymbol{V}}\bar{\boldsymbol{\Lambda}}\bar{\boldsymbol{V}}^H \tag{14.15}$$

where $\bar{\boldsymbol{V}}$ may be tall if $\boldsymbol{H}^H\boldsymbol{R}_{nn}^{-1}\boldsymbol{H}$ is rank deficient and $\bar{\boldsymbol{\Lambda}}$ is a $Q \times Q$ diagonal matrix, where $Q := rank(\boldsymbol{H}^H\boldsymbol{R}_{nn}^{-1}\boldsymbol{H}) = rank(\boldsymbol{H})$. Note that in the rest of the section the matrices $\boldsymbol{V}$ and $\boldsymbol{\Lambda}$ are going to be defined according to the following definition, which is not a restriction:

The elements $\{\lambda_{qq}\}_{q=1}^{Q}$ in the diagonal of matrix $\bar{\boldsymbol{\Lambda}}$, which are the nonnull eigenvalues of $\boldsymbol{H}^H\boldsymbol{R}_{nn}^{-1}\boldsymbol{H}$, are arranged in decreasing order. Note that number 1 above requires $N \leq Q$. For convenience, we will denote by $\boldsymbol{\Lambda}$ the $N \times N$ diagonal matrix with diagonal entries $\{\lambda_{qq}\}_{q=1}^{N}$ ($\boldsymbol{\Lambda}$ is equal to the top left $N \times N$ block of $\bar{\boldsymbol{\Lambda}}$), and matrix $\boldsymbol{V}$ by the first N columns of $\bar{\boldsymbol{V}}$, which are the eigenvectors corresponding to the N largest eigenvalues $\{\lambda_{qq}\}_{q=1}^{N}$ of $\boldsymbol{H}^H\boldsymbol{R}_{nn}^{-1}\boldsymbol{H}$.

14.2.1.1 MMSE Criterion under Transmit Power and Maximum Eigenvalue Constraint

The minimum MSE (MMSE) design minimizes $tr(MSE(\boldsymbol{F},\boldsymbol{G}))$ jointly with respect to $\boldsymbol{F}$ and $\boldsymbol{G}$ under the transmit power constraint. The joint transmit and receive design that minimizes $tr(MSE(\boldsymbol{G},\boldsymbol{F}))$ can be obtained by minimizing $tr(\overline{MSE}(\boldsymbol{F}))$ (c.f. Equation 14.12) with respect to $\boldsymbol{F}$. The solution for $\boldsymbol{F}_{opt}$ is given in the following lemma.

Lemma 14.1 The solution of the optimization problem:

$$\min_{\boldsymbol{F}} tr(\overline{MSE}(\boldsymbol{F})) \quad \text{and} \quad tr\left(\boldsymbol{F}_{opt}\boldsymbol{F}_{opt}^H\right)\sigma_{ss}^2 = \mathcal{P}_0 \tag{14.16}$$

is given by $\boldsymbol{F}_{opt} = \boldsymbol{V}\boldsymbol{\Phi}$, where $\boldsymbol{\Phi}$ is an $N \times N$ diagonal matrix with the following (i,i) entry[2]:

$$|\phi_{ii}|^2 = \left(\frac{\mathcal{P}_0 + \sum_{n=1}^{\bar{N}}\lambda_{nn}^{-1}}{\sigma_{ss}^2\sum_{n=1}^{\bar{N}}\lambda_{nn}^{-1/2}}\lambda_{ii}^{-1/2} - \frac{1}{\lambda_{ii}\sigma_{ss}^2}\right)^+ \tag{14.17}$$

where $(x)^+ := \max(x,0)$ and $\bar{N} \leq N$ is such that $|\phi_{nn}|^2 > 0$ for $n \in [1,\bar{N}]$ and $|\phi_{nn}|^2 = 0$ for all other n.

Interestingly, the minimization of the determinant, in lieu of the trace, of the $\overline{MSE}(\boldsymbol{F})$ matrix with respect to $\boldsymbol{F}$ is equivalent to maximizing the information rate. The following lemma gives the result.

[2]Note that only the amplitude of ϕ_{ii} is fixed, while the phase is arbitrary; thus ϕ_{ii} can be a real number.

Lemma 14.2 The solution of the optimization problem

$$\min_F |\overline{MSE}(F)| \quad \text{and} \quad tr\left(F_{opt}F_{opt}^H\right)\sigma_{ss}^2 = \mathcal{P}_0 \tag{14.18}$$

is given by $F_{opt} = V\Phi$, where Φ is an $N \times N$ diagonal matrix with (i,i) entry

$$|\phi_{ii}|^2 = \left(\frac{\mathcal{P}_0 + \sum_{k=1}^{\bar{N}} \lambda_{kk}^{-1}}{\bar{N}\sigma_{ss}^2} - \frac{1}{\lambda_{ii}\sigma_{ss}^2}\right)^+ \tag{14.19}$$

and $\bar{N} \leq N$ is the number of positive $|\phi_{ii}|^2$.

The power loading on the eigenvectors V of $H^H R_{nn}^{-1} H$ of Lemma 14.2 is identical to the so-called water filling, obtained from the maximization of the mutual information on parallel Gaussian channels (see, e.g., [8], [14], [15]) and in the context of linear precoding, it leads exactly to the solution described in Lemma 2 [45]. In particular:

Corollary 14.1 For a Gaussian input s_i, if G has the following structure:

$$G = \tilde{\Gamma} F^H H^H R_{nn}^{-1} \tag{14.20}$$

where $\tilde{\Gamma}$ is an arbitrary $N \times N$ matrix, the mutual information $I(\hat{s}, x)$ per block *does not depend* on G (see also [45]), and is

$$I(\hat{s}, x) = \log\left|\sigma_{ss}^2 HFF^H H^H R_{nn}^{-1} + I\right| \tag{14.21}$$

The F_{opt} in Equation 14.19 maximizes the mutual information between transmit and receive data.

Lemma 14.3 The solution of the optimization problems

$$\min_F tr\left(\overline{MSE}(F)\right), \quad \min_F \quad |\overline{MSE}(F)| \text{ subject to } \lambda_{max}\left(F_{opt}F_{opt}^H\right)\sigma_{ss}^2 = \mathcal{L}_0$$

is given by $F_{opt} = \sqrt{\mathcal{L}_0/\sigma_{ss}^2}\, V$.

As with Lemma 14.2, it is worth noting that because of Equation 14.21, the solution in Lemma 14.3 also provides the maximum information rate under Equation 14.13.

14.2.1.2 Maximum $\lambda_{min}(SNR(F, G))$ under Power and Maximum Eigenvalue Constraints

It is difficult to deal with the designs that minimize the error probability because of the complexity involved in obtaining the results in closed form. The authors in [44] propose design criteria come close to the desired goal, though their optimization is *alphabet independent*. If $s_i(\mathcal{H}_k)$ denotes the symbol vector corresponding to hypothesis $\mathcal{H}_k$, and $\mathcal{D}_i$ the decision on the i-th symbol block, then the maximum likelihood (ML) decision rule is [25]

$$\mathcal{D}_i = \underset{\mathcal{H}_k}{\operatorname{argmin}} [\hat{s}_i - GHFs_i(\mathcal{H}_k)]^H (GR_{nn}G^H)^{-1} [\hat{s}_i - GHFs_i(\mathcal{H}_k)] \tag{14.22}$$

An indirect way of reducing the probability of error is to maximize the minimum distance between hypotheses. Consider the following SNR-like matrix as a sensible measure related to the probability of error:

$$SNR(F, G) := F^H H^H G^H (GR_{nn}G^H)^{-1} GHF\sigma_{ss}^2 \tag{14.23}$$

Properties of $SNR(F, G)$ may provide suboptimal, but more general design solutions that are not tied to a certain symbol alphabet. The criterion that is adopted is based on the observation that the minimum eigenvalue $\lambda_{min}(SNR(\boldsymbol{F},\boldsymbol{G}))$ provides a lower bound for the minimum distance:

$$\min_{h,k:h\neq k}[\boldsymbol{s}_i(\mathcal{H}_h)-\boldsymbol{s}_i(\mathcal{H}_k)]^H SNR(\boldsymbol{F},\boldsymbol{G})[\boldsymbol{s}_i(\mathcal{H}_h)-\boldsymbol{s}_i(\mathcal{H}_k)] \\ \geq \lambda_{min}(SNR(\boldsymbol{F},\boldsymbol{G}))\min_{h,k:h\neq k}\|\boldsymbol{s}_i(\mathcal{H}_h)-\boldsymbol{s}_i(\mathcal{H}_k)\|^2 \tag{14.24}$$

The corresponding solutions are given in the following lemma.

Lemma 14.4 The solution of the optimization problem

$$\max_{\boldsymbol{F},\boldsymbol{G}}\lambda_{min}(SNR(\boldsymbol{F},\boldsymbol{G})) \text{ subject to} \\ tr\left(\boldsymbol{F}_{opt}\boldsymbol{F}_{opt}^H\right)\sigma_{ss}^2=\mathcal{P}_0 \quad \text{and} \quad \lambda_{max}(\boldsymbol{F}\boldsymbol{F}^H)\sigma_{ss}^2=\mathcal{L}_0 \tag{14.25}$$

is given by $\boldsymbol{F}_{opt}=\boldsymbol{V\Phi}$ and $\boldsymbol{G}_{opt}=\tilde{\boldsymbol{\Gamma}}\boldsymbol{V}^H\boldsymbol{H}^H\boldsymbol{R}_{nn}^{-1}$ with the $N\times N$ matrix $\tilde{\boldsymbol{\Gamma}}$ being invertible [3] and $\boldsymbol{\Phi}$ being an $N\times N$ diagonal matrix having diagonal entries

$$|\phi_{ii}|^2=K'\lambda_{ii}^{-1} \tag{14.26}$$

where K' is given by $\frac{\mathcal{P}_0}{\sigma_{ss}^2\sum_k\lambda_{kk}^{-1}}$ for the power constraint and $\frac{\mathcal{L}_0\lambda_{NN}}{\sigma_{ss}^2}$ for the eigenvalue constraint.

Note that the solution under power constraint leads to

$$SNR(\boldsymbol{F}_{opt},\boldsymbol{G}_{opt})=\sigma_{ss}^2\boldsymbol{\Phi}^H\boldsymbol{\Lambda\Phi}\frac{\mathcal{P}_0}{\sum_k\lambda_{kk}^{-1}}\boldsymbol{I} \tag{14.27}$$

and the one under eigenvalue constraint gives $SNR(\boldsymbol{F}_{opt},\boldsymbol{G}_{opt})=\mathcal{L}_0\boldsymbol{I}$.

Interestingly, the solution of Lemma 14.4 coincides with the MMSE solution under the zero-forcing (ZF) constraint in [43, Theorem 14.3]. The ZF receiver in [43, Theorem 14.3] corresponds to selecting $\tilde{\boldsymbol{\Gamma}}=\boldsymbol{I}$, and in this case, the design of Lemma 14.4 leads to an ML detection scheme that performs separately a low-complexity quantization of the components of $\hat{\boldsymbol{s}}_i$. As a last remark, it is interesting to observe that for arbitrary $\boldsymbol{F}$ and $\boldsymbol{G}$ we can extend to the Gaussian MIMO case the capacity formula of the single-input single-output (SISO) additive white Gaussian noise (AWGN) channel as follows:

$$I(\hat{\boldsymbol{s}},\boldsymbol{x})=\log|\boldsymbol{I}+SNR(\boldsymbol{F},\boldsymbol{G})| \tag{14.28}$$

Next, we provide expressions for the performance measures such as the mutual information, the probability of error, and the mean square error achievable with the optimal precoding/decoding schemes discussed so far. As mentioned before, all optimal designs lead invariably to loading the power across the eigenvectors of $\boldsymbol{H}^H\boldsymbol{R}_{nn}^{-1}\boldsymbol{H}$, and the performance depends on the eigenvalues. Section 14.3 addresses the problem of calculating the closed-form statistics of the performance measures in random fading.

14.2.1.3 Equivalent Decomposition into Independent Subchannels

Lemma 14.5 All optimal designs we described so far have solutions of the following form:

$$\boldsymbol{F}_{opt}=\boldsymbol{V\Phi}, \qquad \boldsymbol{G}_{opt}=\boldsymbol{\Gamma\Lambda}^{-1}\boldsymbol{V}^H\boldsymbol{H}^H\boldsymbol{R}_{nn}^{-1} \tag{14.29}$$

where $\boldsymbol{\Phi}$ and $\boldsymbol{\Gamma}$ are diagonal matrices.

The matrices $\boldsymbol{F}_{opt}$ and $\boldsymbol{G}_{opt}$ in Equation 14.29, cascaded with the channel matrix $\boldsymbol{H}$ in between, are depicted in Figure 14.2. Matrix $\boldsymbol{V}$ tunes the transmit filters to the eigenstructure of the propagation channel, which depends on $\boldsymbol{H}$ and the additive Gaussian noise (AGN) covariance $\boldsymbol{R}_{nn}$.

[3] The receiver selection is not completely defined by the optimal design criterion. A similar observation was made in [45] in deriving the solution for the maximum information rate.

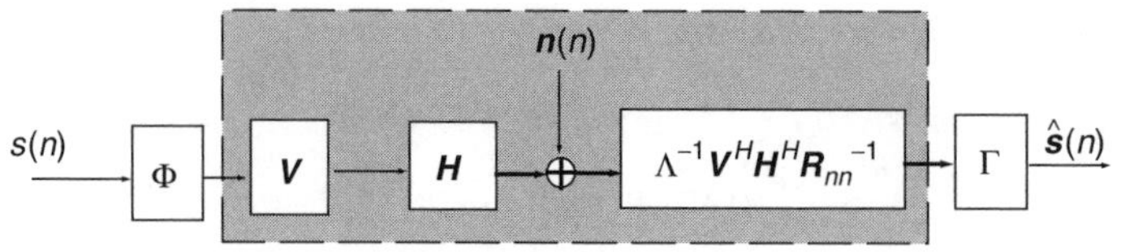

FIGURE 14.2 Optimal transceivers: matrix model.

The matrix equivalent of the cascade inside the box of Figure 14.2 is

$$\boldsymbol{\Lambda}^{-1}\boldsymbol{V}^H\boldsymbol{H}^H\boldsymbol{R}_{nn}^{-1}\boldsymbol{H}\boldsymbol{V}\boldsymbol{\Lambda}^{-1}\boldsymbol{V}^H\bar{\boldsymbol{V}}\bar{\boldsymbol{\Lambda}}\bar{\boldsymbol{V}}^H\boldsymbol{V} = \boldsymbol{I} \tag{14.30}$$

and the noise correlation at the output of the box is

$$\boldsymbol{\Lambda}^{-1}\boldsymbol{V}^H\boldsymbol{H}^H\boldsymbol{R}_{nn}^{-1}\boldsymbol{R}_{nn}\boldsymbol{R}_{nn}^{-1}\boldsymbol{H}\boldsymbol{V}\boldsymbol{\Lambda}^{-1} = \boldsymbol{\Lambda}^{-1} \tag{14.31}$$

Thus, the matrix channel is described by the diagonal transfer matrix $\boldsymbol{\Gamma\Phi}$ and additive noise with correlation matrix $\boldsymbol{\Gamma}^H\boldsymbol{\Gamma}\boldsymbol{\Lambda}^{-1}$. Hence, the N subchannels are decoupled and Figure 14.2 becomes equivalent to Figure 14.3, in which case the flat fading on each of the parallel subchannels corresponds to the diagonal elements of $\boldsymbol{\Gamma\Phi}$, and the noise components $\{\beta_i\}_k$, $k = 1, \ldots, N$ are uncorrelated with variance λ_{kk}^{-1}.

14.2.1.4 Performance Measures

The decomposition of the channel as shown above streamlines the performance analysis as seen from the following corollaries.

Corollary 14.2 With $\boldsymbol{\Phi}$ and $\boldsymbol{\Gamma}$ diagonal, the transceivers in Equation 14.29 render the MIMO linear AGN channel with memory equivalent to N parallel independent intersymbol interference (ISI)-free subchannels, each with flat fading gain $\phi_{kk}\gamma_{kk}$ and AGN $\{\beta_i\}_k$, with variance $1/\lambda_{kk}$ and $\{\beta_i\}_k$, $\{\beta_i\}_j$ uncorrelated for $k \neq j$; i.e.,

$$\{\hat{\boldsymbol{s}}_i\}_k = \phi_{kk}\,\gamma_{kk}\,\{\boldsymbol{s}_i\}_k + \gamma_{kk}\,\{\beta_i\}_k, \qquad k = 1, \ldots, N \tag{14.32}$$

The *SNR* at the output for the k-th subchannel is

$$SNR_k = \frac{\sigma_{ss}^2|\phi_{kk}|^2|\gamma_{kk}|^2}{\lambda_{kk}^{-1}|\gamma_{kk}|^2}\sigma_{ss}^2|\phi_{kk}|^2\lambda_{kk} \tag{14.33}$$

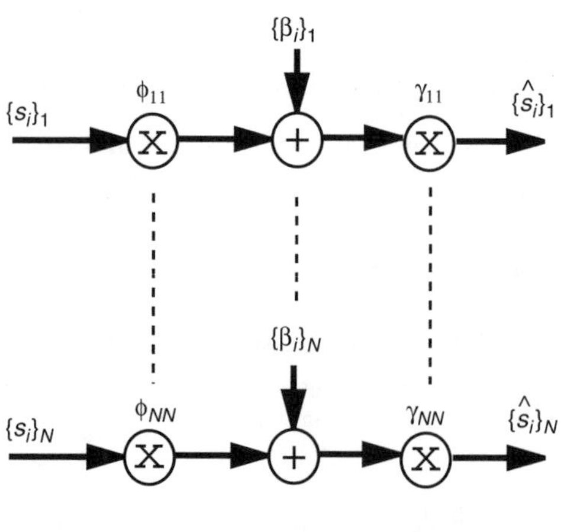

FIGURE 14.3 Equivalent subchannels.

The independence of the parallel subchannels implies the following.

Corollary 14.3 For the linear transceivers of Equation 14.29, the mutual information is given by

$$I(\hat{\boldsymbol{s}};\boldsymbol{x}) = \frac{1}{N}\sum_{i=1}^{N}\log_2(1+SNR_i) \tag{14.34}$$

Under constrained power $I(\hat{\boldsymbol{s}};\boldsymbol{x})$, Equation 14.34 achieves its maximum when $\{\phi_{ii}\}_{i=1}^{N}$ are given by Equation 14.19.

According to Equation 14.32, the set of equivalent parallel subchannels is also ISI-free. Thus, assuming that information is quantized and that $\{\boldsymbol{s}_i\}_k$ belong to a finite alphabet, the optimal decision scheme based on $\{\hat{\boldsymbol{s}}_i\}_k$ performs symbol by symbol detection.

Based on the model in Figure 14.3, we can also derive the following result.

Corollary 14.4 The MSE for the $\boldsymbol{F}$ and $\boldsymbol{G}$ in Equation 14.29 coincides with the cumulative MSE over the N independent subchannels in Figure 14.3, and is

$$MSE(\boldsymbol{\Gamma},\boldsymbol{\Phi}) = \sum_{i=1}^{N}\left[\frac{|\gamma_{ii}|^2}{\lambda_{ii}} + |\gamma_{ii}\phi_{ii}-1|^2\sigma_{ss}^2\right] \tag{14.35}$$

For the MMSE receiver $\boldsymbol{\Gamma} = \sigma_{ss}^2\boldsymbol{\Phi}^H(\boldsymbol{\Lambda}^{-1}+\sigma_{ss}^2\boldsymbol{\Phi}\boldsymbol{\Phi}^H)^{-1}$, the MSE is

$$\overline{MSE}(\boldsymbol{\Phi}) = \sum_{i=1}^{N}\frac{\sigma_{ss}^2}{1+|\phi_{ii}|^2\sigma_{ss}^2\lambda_{ii}} \quad \sum_{i=1}^{N}\frac{\sigma_{ss}^2}{1+SNR_i} \tag{14.36}$$

which reaches its minimum for a constrained power when ϕ_{ii} are given by Equation 14.17.

For the ZF receiver $\boldsymbol{\Gamma} = \boldsymbol{\Phi}^{\dagger}$ the MSE is

$$\overline{MSE}(\boldsymbol{\Phi}) = \sum_{i=1}^{N}\frac{1}{|\phi_{ii}|^2\lambda_{ii}} \quad \sum_{i=1}^{N}\frac{\sigma_{ss}^2}{SNR_i} \tag{14.37}$$

Corollaries 1 through 4 show that performance evaluation of these optimal designs requires only specifying the values of SNR_i. Table 14.1 gives the SNR_i characterizing the optimal designs presented in this section.

Comparing the SNR_i expressions, we observe the similarity between the MMSE and water-filling solutions of Lemmas 14.1 and 14.2: both of these solutions tend to exclude the most noisy subchannels (corresponding to the smallest λ_{ii} values) and the SNR_i grows as $\lambda_{ii}^{1/2}$ and λ_{ii}, respectively. The other designs do not include this control and thus waste power over these subchannels.

TABLE 14.1 Comparison of Optimal Loading for Different Coding Strategies

	Criteria	Constraint	$SNR_i = \|\phi_{ii}\|^2\sigma_{ss}^2\lambda_{ii}$
1	$\min(tr(MSE))$	CP	$(\mathcal{K}_{tr(MSE)}\,\lambda_{ii}^{1/2}-1)^+$
2	$\min(\|MSE\|)$	CP	$(\mathcal{K}_{\|MSE\|}\,\lambda_{ii}-1)^+$
3	$\min(\|MSE\|)$	$C\lambda_{max}$	$\mathcal{L}_0\,\lambda_{ii}$
4	$\min(tr(MSE))$	$C\lambda_{max}$	$\mathcal{L}_0\,\lambda_{ii}$
5	$\max(\lambda_{min}(SNR))$	CP	$\mathcal{K}_{SNR-CP}$
6	$\max(\lambda_{min}(SNR))$	$C\lambda_{max}$	$\mathcal{K}_{SNR-C\lambda_{max}}$

Note: CP = constraint on the average transmit power; $C\lambda_{max}$ = constraint on the maximum eigenvalue.

14.3 Performance Analysis and Random Matrices

We resort to random matrices in the context of MIMO because of the random fading between the transmit and receive elements of the antenna array. We have seen in Section 14.2 that the eigenvalues are the only channel matrix parameters that affect the performance of the optimal space–time precoders. This section is concerned mainly with the derivation of the statistics of MIMO frequency-selective channel capacity. In particular, we derive the characteristic function of the capacity of the Rayleigh faded channel for both low and high SNR scenarios. Deriving the performance of MIMO systems requires the nontrivial step of deriving the joint statistics of the eigenvalues of the random MIMO frequency response. The methodology used to derive the statistics of eigenvalues and eigenvectors is presented in Section 14.3.1, which is applied later in the derivations carried out in Section 14.3.2.

The focus of the classical random matrix theory is on the asymptotic probability distribution of the eigenvalues of a random matrix as the matrix dimensions tend to infinity [5], [17], [20], [24], [48]. The interplay between engineering insights and mathematical results has proven to be very fruitful: for example, in order to understand the scalability properties of the linear multiuser detectors, a random model for the spreading sequences is employed by Tse and others, who have used this model to compare the performances of the conventional matched filter receiver, the decorrelator, and the linear MMSE (LMMSE) receiver in the limit of the number of users and the processing gain [11], [54]. Li employed the same model and used the asymptotic eigenvalue moment results to analyze the capacity of multiuser code division multiple-access (CDMA) channel with frequency-selective fading [28], [29]. The same modeling paradigm of random spreading was used in [19], [26], [36], [41], [55] to analyze multiuser systems.

Our interest in this chapter is not in the CDMA systems but in the transmit and receive diversity schemes. We will now see how we can utilize the tools provided to us by the random matrix theory to derive the performance of MIMO systems.

14.3.1 Differential Forms and Random Matrix Techniques

In order to derive the statistics of the eigenvalues of these random matrices, the first step requires deriving the Jacobian of the change of variables from the original matrix to its factors. When the decomposition is unique, the number of independent variables in the matrix and in the corresponding factors is the same and the Jacobian matrix is square. This can be verified in the cases of eigenvalue decomposition (however, appropriate constraints need to be added for complex matrices), Cholesky decomposition for self-adjoint matrices, and QR or LU (lower–upper) decompositions, and for general matrices [9].

Exterior differential calculus, based on the seminal work of Élie Cartan [12], provides a useful tool to calculate the above Jacobians. The concept of *exterior product*, denoted by the symbol $\wedge$, was introduced by Hermann Günter Grassmann in 1844 and was utilized by Cartan in the study of differential forms. Ordinary vectors are 1-vectors; wedge products of p independent vectors generates the space of p-vectors. Given vectors α, β, γ, the basic axioms of Grassmann algebra are:

- Associativity: $(\alpha \wedge \beta) \wedge \gamma = \alpha \wedge (\beta \wedge \gamma)$
- Anticommutativity: $\alpha \wedge \beta = -\beta \wedge \alpha$
- Distributivity: $(a\alpha + b\beta) \wedge \gamma = a(\alpha \wedge \gamma) + b(\beta \wedge \gamma)$

The axioms are sufficient to establish that[4]

$$\alpha \wedge \alpha = 0 \quad \text{and} \quad (\boldsymbol{A}\alpha) \wedge \beta = |\boldsymbol{A}|(\alpha \wedge \beta) \tag{14.38}$$

[4]If $\boldsymbol{A}$ is $m \times n$ and $m > n$, or if it is rank deficient, $|\boldsymbol{A}|$ has to be replaced by 0. If $m \leq n$, $|\boldsymbol{A}|$ has to be replaced by the matrix compound $\wedge^m \boldsymbol{A}$, i.e., the matrix of all cofactors of order m, if $m \leq n$ [12].

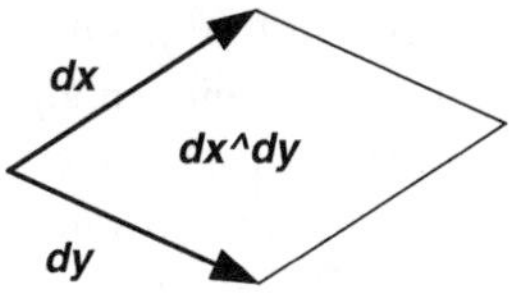

FIGURE 14.4 Illustration of wedge product.

Cartan's exterior differential calculus (a very clear book is [12]) is based on the observation that if we consider the sign in the Jacobian, products of differentials $dxdy$ behave as $dx \wedge dy$ (Figure 14.4): this can be easily observed by introducing a dummy transformation $x(u, v), y(u, v)$ and realizing that $dxdy = |\partial(x, y)/\partial(u, v)|dudv$ equals 0 for $u = v = x$ and equals $-dydx$ for $u = y$ and $v = x$. The rules of exterior differential calculus are derived by applying Grassman algebra to 1-forms such as dx or the gradient of a differentiable function ∇f. An r-form is

$$\alpha = \sum_{k_1<k_2\cdots<k_r} A(x_1, \ldots, x_n)dx_{k_1} \wedge \quad \ldots \quad \wedge dx_{k_r} \tag{14.39}$$

To complete the description of Cartan's differential forms, we give an axiomatic definition of the d operator:

- $d(r\text{-form}) = (r + 1)\text{-form}$.
- $d(\alpha + \beta) = d\alpha + d\beta$.
- If α is an r-form and β is an s-form, $d(\alpha \wedge \beta) = d\alpha \wedge \beta + (-1)^r \alpha \wedge d\beta$.
- $d(dx) = 0$ (Poincarè Lemma).

These rules are systematic and the results are simpler to grasp than the theory of manifolds. In addition, they provide a way of deriving the Jacobian of an arbitrary matrix factorization by applying the d operator first and then evaluating the $\wedge$ product of all the independent differentials. This last task is a bit involved since it requires the description of the group of matrices by means of their independent parameters [37]. The evaluation of this Jacobian is essential to derive the probability density function (pdf) of the factors from the pdf of the original matrix. We will borrow the notation from [9] and indicate by $d\boldsymbol{A}$ the matrix of differentials and by $(d\boldsymbol{A})$ the exterior product of the independent entries in $d\boldsymbol{A}$, for example:

- For an arbitrary $\boldsymbol{A}$, $(d\boldsymbol{A}) = \wedge_i \wedge_j da_{ij}$.
- If $\boldsymbol{A}$ is diagonal, $(d\boldsymbol{A}) = \wedge_i da_{ii}$.
- If $\boldsymbol{A} = \boldsymbol{A}^T$ or $\boldsymbol{A}$ is lower triangular, $(d\boldsymbol{A}) = \wedge_{1\le i\le j\le n} da_{ij}$.

When dealing with complex matrices we can apply the same rules remembering that any complex dz has an associated $(dz) = d\Re[z]d\Im[z]$ or, more precisely, $(dz) = d\Re[z] \wedge d\Im[z]$. Therefore, dz can be treated as a bidimensional vector. Since the multiplication of $z = x + jy$ by a complex number $\alpha = a + jb$ can be viewed as

$$(x, y)\begin{pmatrix} a & -b \\ b & a \end{pmatrix} \tag{14.40}$$

from Equation 14.38 it follows that $(d\alpha z) = |\alpha|^2 dxdy$. In general [17], see the following.

Lemma 14.6 If $\mathbf{w} = \mathbf{u} + j\mathbf{v}$ are analytical functions of $\mathbf{z} = \mathbf{x} + j\mathbf{y}$, then

$$det\left(\frac{\partial(\mathbf{u}, \mathbf{v})}{\partial(\mathbf{x}, \mathbf{y})}\right) = \left|det\left(\frac{\partial \mathbf{w}}{\partial \mathbf{z}}\right)\right|^2 \tag{14.41}$$

Other properties of the complex case are easily derived, for example, (1) $(dz) = -(dz^*)$; (2) $dz \wedge dz^* = 0$. Note that for $\boldsymbol{B} = \boldsymbol{X}\boldsymbol{A}$, $(d\boldsymbol{B}) = |\boldsymbol{X}|^n(d\boldsymbol{A})$ in $\mathbb{R}^n$. Because of Equation 14.38 and Lemma 14.6, orthogonal or unitary linear mappings of $\boldsymbol{A}$ do not change $(d\boldsymbol{A})$, i.e., if $\boldsymbol{Q}^H\boldsymbol{Q} = \boldsymbol{I}$ $(\boldsymbol{Q}^H d\boldsymbol{A}) \equiv (d\boldsymbol{A})$.

We need to evaluate $(d\boldsymbol{Q})$ to find the Jacobian for the QR, EVD, etc. A unitary $\boldsymbol{Q}$ can be described by n^2 smooth functions that can be integrated over nice enough intervals, which describe the so-called Stiefel manifold: clearly, the independent parameters of the Stiefel manifold are not the real and imaginary parts of the elements of $\boldsymbol{Q}$. In this case, n out of the n^2 parameters are redundant (in the sense that the decomposition is unique up to n parameters). This ambiguity could be removed by having the diagonal elements of $\boldsymbol{Q}$ set to be real. Since $\boldsymbol{Q}\boldsymbol{Q}^H = \boldsymbol{I} \rightarrow \boldsymbol{Q}d\boldsymbol{Q}^H = -d\boldsymbol{Q}\boldsymbol{Q}^H$, $\boldsymbol{Q}d\boldsymbol{Q}^H$ is antisymmetric and the diagonal elements of $\boldsymbol{Q}d\boldsymbol{Q}^H$ are purely imaginary. Note also that when $\boldsymbol{Q}$ is $m \times n$ and semiunitary with $n \leq m$, we have $2mn - n(n+1)$ real parameters (the roles are reversed if $n > m$) and we can always define an $m \times m$ matrix $\bar{\boldsymbol{Q}} = (\boldsymbol{Q}, \boldsymbol{Q}^{\perp})$ such that $\bar{\boldsymbol{Q}}^H\boldsymbol{Q} = \boldsymbol{I}_{m,n}$, so that $(d\boldsymbol{Q}) = (\bar{\boldsymbol{Q}}^H d\boldsymbol{Q})$.

The uniform pdf in the Stiefel group of orthogonal or unitary matrices is called *Haar distribution* [17, Chapter 1]. The element of volume of Stiefel manifold can be found by extending the ideas given by Edelman[5] [9]. For the case of a unitary group, the volume element is

$$(\bar{\boldsymbol{Q}}^H d\boldsymbol{Q}) = \bigwedge_{i \geq j} \boldsymbol{q}_i^H d\boldsymbol{q}_j \tag{14.42}$$

where $\bar{\boldsymbol{Q}} = [\boldsymbol{q}_1 \cdots \boldsymbol{q}_m]$ is the same as before and $\boldsymbol{q}_i$ is a complex unit vector (for details, refer to [46]). The volume of $(\bar{\boldsymbol{Q}}^H d\boldsymbol{Q})$ integrated over $\boldsymbol{Q}^H\boldsymbol{Q} = \boldsymbol{I}$, for $\boldsymbol{Q}$ unitary, is

$$Vol(\boldsymbol{Q}_{m,n}) = \frac{2^n(\pi)^{mn-n(n-1)/2}}{\prod_{i=0}^{n-1}\Gamma(m-i)} \tag{14.43}$$

when the n constraints are added to $\boldsymbol{Q}_{m,n}$ (for example, the diagonal elements are constrained to be real):

$$\overline{Vol}(\boldsymbol{Q}_{m,n}) = \frac{(\pi)^{(m-1)n-n(n-1)/2}}{\prod_{i=0}^{n-1}\Gamma(m-i)} \tag{14.44}$$

Now we mention some of the important matrix distributions, computed using the above-mentioned tools, that will be key to arriving at the characteristic function expression for the capacity of the MIMO channel. Consider the following examples.

Example 14.1

Consider first the matrix of the form $\boldsymbol{A} = \boldsymbol{B}^H\boldsymbol{B}$, where $\boldsymbol{B}$ is a random $m \times n$ matrix with continuous pdf and we will assume that $m \geq n$, in which case $\boldsymbol{A}$ is full rank with a probability of 1.[6] We are interested in deriving the pdf of $\boldsymbol{A}$ given the pdf of $\boldsymbol{B}$. Let us denote by $p_{\boldsymbol{A}}(\boldsymbol{A})$ and $p_{\boldsymbol{B}}(\boldsymbol{B})$ the pdfs of the random matrices $\boldsymbol{A}$ and $\boldsymbol{B}$, respectively: the pdf of $\boldsymbol{A}$ is called generalized Wishart distribution. When calculating the density of the square envelope of a complex Gaussian random variable, say b, it is common to calculate the density of phase and amplitude of b, say $(\rho, \phi) \rightarrow b = \rho e^{j\phi}$, and then calculate the density of $a = \rho^2$. Similarly, in the multivariate case, we can resort to a QR decomposition for $\boldsymbol{B}$ and express its relationship with $\boldsymbol{A}$ through the fact that the Cholesky decompositions of $\boldsymbol{A}$ are unique and such that

$$\boldsymbol{B} = \boldsymbol{Q}\boldsymbol{R} \Leftrightarrow \boldsymbol{A} = \boldsymbol{R}^H\boldsymbol{R} \tag{14.45}$$

At this point we are left with a problem analogous to the scalar case, which is that of finding the density of $\boldsymbol{R}$ and derive from it the density of $\boldsymbol{R}^H\boldsymbol{R}$. To derive the density of $\boldsymbol{R}$ we can use differential calculus. With

---

[5] In [9] the derivation of the volume element of the orthogonal matrix group is found to be $(\bar{\boldsymbol{Q}}^T d\boldsymbol{Q}) = \bigwedge_{i>j} \boldsymbol{q}_i^T d\boldsymbol{q}_j$. Note that the elements $\boldsymbol{q}_i^T d\boldsymbol{q}_i = 0$ for the real case.

[6] In case $m < n$, $\boldsymbol{A}$ has $n - m$ zero eigenvalues. Because the nonnull eigenvalues of $\boldsymbol{B}^H\boldsymbol{B}$ and $\boldsymbol{B}\boldsymbol{B}^H$ coincide, the case $m \geq n$ is general enough to provide the distribution of the nonzero eigenvalues for any choice of n, m.

$(d\boldsymbol{R}) = \wedge_{i<j}(dr_{ij})$, using the rules of differential calculus [46],

$$(d\boldsymbol{A}) = 2^n \prod_{i=1}^{n} (|r_{ii}|)^{2(n-i)+1} (d\boldsymbol{R})$$

$$\Rightarrow p_{\boldsymbol{A}}(\boldsymbol{A})(d\boldsymbol{A}) = p_{\boldsymbol{A}}(\boldsymbol{R}^H\boldsymbol{R}) \prod_{i=1}^{n} 2^n \, (|r_{ii}|)^{2(n-i)+1} \, (d\boldsymbol{R}) \tag{14.46}$$

For QR factorization to be unique, we constrain the diagonal elements of $\boldsymbol{R}$ to be real. Now,

$$(d\boldsymbol{B}) = (\bar{\boldsymbol{Q}}^H d\boldsymbol{B}) = \prod_{i=1}^{n} (|r_{ii}|)^{2(m-i)+1} (d\boldsymbol{R})(d\boldsymbol{Q}) \tag{14.47}$$

where $(d\boldsymbol{Q}) = (\bar{\boldsymbol{Q}}^H d\boldsymbol{Q})$ is the element of volume of the Stiefel manifold. Hence,

$$p_{\boldsymbol{B}}(\boldsymbol{B})(d\boldsymbol{B}) = p_{\boldsymbol{B}}(\boldsymbol{QR}) \prod_{i=1}^{n} (|r_{ii}|)^{2(m-i)+1} (d\boldsymbol{R})(d\boldsymbol{Q}) \tag{14.48}$$

and, with $\sqrt{\boldsymbol{A}} \triangleq \boldsymbol{R}$, from Equation 14.46 and Equation 14.48 and $|\boldsymbol{A}| = \prod_{i=1}^{n} |r_{ii}|^2$ it follows that

$$p_{\boldsymbol{A}}(\boldsymbol{A}) = 2^{-n} |\boldsymbol{A}|^{m-n} \int p_{\boldsymbol{B}}(\boldsymbol{Q}\sqrt{\boldsymbol{A}})(\bar{\boldsymbol{Q}}^H d\boldsymbol{Q}) \tag{14.49}$$

which is the form of the so-called *generalized Wishart density* [17]. Generalizing the results in [17] to the complex case in Equation 14.46 implies the following lemma.

Lemma 14.7 When the pdf $p_{\boldsymbol{B}}(\boldsymbol{B}) = p(\boldsymbol{B}^H\boldsymbol{B})$, then:
1. $\boldsymbol{Q}$ and $\boldsymbol{R}$ in the QR decomposition $\boldsymbol{B} = \boldsymbol{QR}$ are independent. The pdf of $\boldsymbol{Q}$ is uniform over the unit $\boldsymbol{QQ}^H = \boldsymbol{I}$ (*Haar distribution*) and $\boldsymbol{R}$ is

$$p_{\boldsymbol{R}}(\boldsymbol{R}) = \prod_{i=1}^{n} \left(|r_{ii}|^2\right)^{m-n} p(\boldsymbol{R}^H\boldsymbol{R}) Vol(\boldsymbol{Q}_{m,n}) \tag{14.50}$$

2. The pdf of $\boldsymbol{A}$ is (c.f. Equation 14.43)

$$p_{\boldsymbol{A}}(\boldsymbol{A}) = 2^{-n} |\boldsymbol{A}|^{m-n} p(\boldsymbol{A}) Vol(\boldsymbol{Q}_{m,n}) \tag{14.51}$$

Example 14.2

In this example, we briefly go over the steps to derive the density of the eigenvalues of $\boldsymbol{A}$ using EVD. The Jacobian of the EVD of $\boldsymbol{A} = \boldsymbol{U}\Lambda\boldsymbol{U}^{\mathbf{H}}$ can be obtained by constraining the diagonal elements of $\boldsymbol{U}$ to be real so that the EVD is unique. Noting that $(\boldsymbol{U}^H d\boldsymbol{A}\boldsymbol{U}) = |\boldsymbol{U}^H\boldsymbol{U}|(d\boldsymbol{A}) = (d\boldsymbol{A})$ and that $d\boldsymbol{U}^H\boldsymbol{U} = -\boldsymbol{U}^H d\boldsymbol{U}$,

$$\begin{aligned}(d\boldsymbol{A}) &\equiv (\boldsymbol{U}^H d\boldsymbol{A}\boldsymbol{U}) = (\boldsymbol{U}^H d\boldsymbol{U}\boldsymbol{\Lambda}\boldsymbol{U}^H\boldsymbol{U} + \boldsymbol{U}^H\boldsymbol{U}d\boldsymbol{\Lambda}\boldsymbol{U}^H\boldsymbol{U} + \boldsymbol{U}^H\boldsymbol{U}\boldsymbol{\Lambda}d\boldsymbol{U}^H\boldsymbol{U}) \\ &= (\boldsymbol{U}^H d\boldsymbol{U}\boldsymbol{\Lambda} - \boldsymbol{\Lambda}\boldsymbol{U}^H d\boldsymbol{U} + d\boldsymbol{\Lambda}) \\ &= \prod_{1\le i<k\le n} (\lambda_k - \lambda_i)^2 (d\boldsymbol{\Lambda})(\boldsymbol{U}^H d\boldsymbol{U})\end{aligned} \tag{14.52}$$

Equations 14.49 and 14.52 can be used to address the general case of $\boldsymbol{A} = \boldsymbol{B}^H\boldsymbol{B}$:

$$p_{\Lambda}(\boldsymbol{\Lambda}) = 2^{-n} \prod_{1\le i<k\le n} (\lambda_k - \lambda_i)^2 \left(\prod_{i=1}^{n} \lambda_i\right)^{m-n} \Psi(\lambda_1, \ldots, \lambda_n) \tag{14.53}$$

$$\Psi(\lambda_1, \ldots, \lambda_n) \triangleq \int p_{\boldsymbol{B}}(\boldsymbol{Q}\sqrt{\boldsymbol{\Lambda}}\boldsymbol{U}^H)(\bar{\boldsymbol{Q}}^H d\boldsymbol{Q})(\boldsymbol{U}^H d\boldsymbol{U}) \tag{14.54}$$

When in Lemma 14.7 $p(\boldsymbol{A}) \equiv p(\Lambda)$, the density of the eigenvalues is simple to derive: for example, in the multivariate Gaussian case $\{\boldsymbol{B}\}_{i,j} \sim \mathcal{N}(0, \sigma^2)$, $p(\boldsymbol{A}) = (\pi\sigma^2)^{-mn} \exp(-\frac{tr(\boldsymbol{A})}{\sigma^2})$ (c.f. Equation 14.51), and for $\lambda_i > 0$,

$$p_\Lambda(\boldsymbol{\Lambda}) = \chi_1 \prod_{1\leq i<k\leq n} (\lambda_k - \lambda_i)^2 e^{-\frac{\sum_i \lambda_i}{\sigma^2}} \left(\prod_{i=1}^{n} \lambda_i\right)^{m-n} \tag{14.55}$$

where $\chi_1 = 2^{-n}(\pi\sigma^2)^{-mn} Vol(\boldsymbol{Q}_{m,n})\overline{Vol}(\boldsymbol{U}_{n,n})$.

Using Wigner's approach, the density function obtained by averaging over all permutations $p_\Lambda(\Lambda)$ is $\frac{1}{n!} p_\Lambda(\Lambda)$, thus [20] the following lemma.

Lemma 14.8 For $m \geq n$ and any continuous real $f(\boldsymbol{A}) = \sum_{i=1}^{n} f(\lambda_i(\boldsymbol{A}))$,

$$E\{f(\boldsymbol{A})\} = \int_0^\infty f(x)\mu_n^{m-n}(x)dx \tag{14.56}$$

$$\mu_n^{m-n}(x) \triangleq \frac{1}{n!}\int_0^\infty \cdots \int_0^\infty p_\Lambda(x, \lambda_2, \ldots, \lambda_n) d\lambda_2 \ldots d\lambda_n \tag{14.57}$$

Note that for $f(\boldsymbol{A}) = \sum_{i=1}^{n} \delta(x - \lambda_i(\boldsymbol{A}))$, $E\{f(\boldsymbol{A})\}$ in Equation 14.56 is the *empirical distribution* of the eigenvalues or, in other words, the average histogram of the eigenvalues of random matrix samples.

When $p_\Lambda(\Lambda)$ is as in Equation 14.55 [5], with $\alpha = m - n$,

$$\mu_n^\alpha(x) = \frac{1}{n}\sum_{k=0}^{n-1} \phi_k^\alpha(x)^2 \tag{14.58}$$

where, denoting by $L_k^\alpha(x)$ the Laguerre polynomials of order α,

$$\phi_k^\alpha(x) = \left[\frac{k!}{\Gamma(k+\alpha+1)} x^\alpha e^{-x}\right]^{1/2} L_k^\alpha(x) \tag{14.59}$$

14.3.2 The Statistics of the Capacity

With all the tools that the random matrix theory has endowed us with, we are now ready to take a step forward and calculate the statistics of the channel capacity. Deriving other performance measures (see Section 14.2) poses similar challenges. In this section we will use random matrix techniques to derive the statistics of the channel capacity for a frequency-selective MIMO channel that is spatially and temporally correlated [46], [47]. This analysis summarizes many of the results obtained in the literature on this topic under more restrictive assumptions [7], [14], [49], [53], [56] and is particularly useful in the context of broadband space–time communications. Specifically, the assumptions made are:

- The noise is AWGN with variance $\sigma_n^2 = 1$.
- $\{\mathbf{H}[l]\}_{r,t}^*$ are spatially and temporally uncorrelated circularly symmetric zero-mean complex Gaussian random variables (Rayleigh fading) with $\boldsymbol{R}_H[l_1, l_2, r_1, r_2, t_1, t_2] \triangleq E\{\{\mathbf{H}[l_1]\}_{r_1,t_1}^*, \{\mathbf{H}[l_2]\}_{r_2,t_2}\} = \delta(t_1 - t_2)\,\delta(r_1 - r_2)\boldsymbol{R}_H(l_2, l_1)$.
- The number of frequency bins is an integer multiple of the channel duration, i.e., $K = Q(L + 1)$, where L is the length of the channel.
- $\boldsymbol{R}_H(l_1, l_2) = \boldsymbol{R}_H(l_2 - l_1)$. In general, this condition rarely applies because the paths are likely not to have the same average power. However, this assumption describes a worst-case scenario in terms of the frequency selectivity of the channel and helps simplify the derivations considerably.

Consider the MIMO channel discussed in the Section 14.1.1. Since the additive Gaussian noise is spatially and temporally white, space–time OFDM will convert our frequency-selective MIMO system into a set of

K parallel independent MIMO systems. If the channel matrix $\boldsymbol{H}$ is sandwiched between the two matrices

$$\mathbf{E}_T \equiv (\bar{\mathbf{W}}_K \otimes \mathbf{I}_{N_T \times N_T}), \quad \mathbf{E}_R \equiv \left(\mathbf{W}_K^H \otimes \mathbf{I}_{N_R \times N_R}\right)$$

where $\bar{\mathbf{W}}_{K+L,K}$ is an extended $(K+L) \times K$ inverse fast Fourier transform (IFFT) matrix with a proper phase shift that creates the so-called cyclic prefix and $\mathbf{W}_K$ is the $K \times K$ IFFT matrix, the equivalent channel is

$$\tilde{\boldsymbol{H}} \equiv \mathbf{E}_R \boldsymbol{H} \mathbf{E}_T diag(\tilde{\mathbf{H}}[\boldsymbol{k}]), \quad \boldsymbol{k} \equiv (0, \ldots, K-1)$$

where $\tilde{\mathbf{H}}[k]$ is the MIMO transfer function at the k-th frequency bin:

$$\tilde{\mathbf{H}}[k] = \sum_{l=0}^{L} \mathbf{H}[l] e^{-j2\pi \frac{kl}{K}} \tag{14.60}$$

Let us denote

$$n \triangleq \min(N_T, N_R), \quad m \triangleq \max(N_T, N_R) \tag{14.61}$$

where N_T and N_R represent the number of transmit and receive antennas,[7] respectively. Denoting by γ the signal-to-noise ratio dictated by the large-scale fading and receiver noise power, the conditional channel capacity is

$$C = \log |\boldsymbol{I} + \gamma \tilde{\boldsymbol{H}}^H \tilde{\boldsymbol{H}}|$$

Therefore, the average capacity is

$$E\{C\} = \sum_{k=0}^{K-1} \sum_{l=1}^{n} E\{\log(1 + \gamma \lambda_l[k])\}$$

and the characteristic function of C is

$$\Phi_C(s) = E\{e^{sC}\} = E\left\{\prod_{k=0}^{K-1} |\boldsymbol{I} + \gamma \tilde{\mathbf{H}}^H[k]\tilde{\mathbf{H}}[k]|^s\right\}$$

The interesting and challenging aspect of the MIMO case is that the performance is expressed in terms of the eigenvalues of the matrix $\tilde{\boldsymbol{H}}^H \tilde{\boldsymbol{H}}$, and thus the results for the scalar case are not generalized in a straightforward manner to MIMO systems. The derivation of $\Phi_C(s)$ is, in general, more complicated since it requires averaging over the joint density of the eigenvalues of all $\tilde{\mathbf{H}}[k]^H \tilde{\mathbf{H}}[k]$, $k = 0, \ldots, K-1$, and the matrices are dependent. However, it is worth noticing that an approximate result for $\gamma \ll 1$ can be obtained quite easily. The case when $\gamma \gg 1$ will be discussed later.

Proposition 14.1 For $\gamma \ll 1$,

$$\Phi_C(s) \approx E\{|\boldsymbol{I} + \gamma K \mathbf{H}^H[\boldsymbol{d}]\mathbf{H}[\boldsymbol{d}]|^s\} \tag{14.62}$$

The eigenvalues of the product $\mathbf{H}^H[\boldsymbol{d}]\mathbf{H}[\boldsymbol{d}]$ can be calculated as described in Section 14.3.1. The interesting consequence of Equation 14.62 is that at low SNR (in the so-called *low-power regime* [30]), the statistics of the capacity of the frequency-selective channel are approximately equivalent to the ones of a MIMO flat-fading channel with $N_R(L+1)$ antennas rather than N_R antennas.

[7] Here $N_T = K$ and $N_R = R$ in Section 14.1.1.

Calculating the dimensions m and n in Equation 14.61 with $N_R(L+1)$ in place of N_R, the approximate characteristic function in Equation 14.62 can be expressed in a rather complex closed form [7], [56], which is the exact solution for the Rayleigh flat-fading case:

$$\Phi_C(s) \approx \chi_3 |\mathbf{G}|$$

where $\mathbf{G}$ is $n \times n$ and

$$\{\mathbf{G}\}_{i,j} = G(i+j-2), \qquad i,j = 1,\ldots,m$$

with

$$G(k) = \frac{1}{\Gamma(-s/\ln 2)} \gamma^{m-n-k-1} \Gamma(1+k+m-n) \Gamma\left(-1-k-m+n-\frac{s}{\ln 2}\right)$$
$$ {}_1F_1\left(1+k+m-n, 2+k+m-n+\frac{s}{\ln 2}, \frac{1}{\gamma}\right)$$
$$+\gamma^{\frac{s}{\ln 2}} \Gamma\left(1+k+m-n+\frac{s}{\ln 2}\right) {}_1F_1\left(-\frac{s}{\ln 2}, -k-m+n-\frac{s}{\ln 2}, \frac{1}{\gamma}\right)$$

The expression in Equation 14.62 for really small γ can be written as

$$\Phi_C(s) \approx E\{|1+\gamma K tr(\mathbf{H}^H[\boldsymbol{d}]\mathbf{H}[\boldsymbol{d}])|^s\}$$
$$= E\{e^{s\ln(1+\gamma K vec(\mathbf{H}[\boldsymbol{d}])^H vec(\mathbf{H}[\boldsymbol{d}]))}\}$$

where $\mathbf{H}[\boldsymbol{d}] \sim \mathcal{N}(0, \boldsymbol{R}_r \otimes \boldsymbol{R}_t \otimes \boldsymbol{R}_H)$,[8] where $\boldsymbol{R}_r$, $\boldsymbol{R}_t$, and $\boldsymbol{R}_H$ describe the correlation among the receive elements, transmit elements, and paths, respectively. Approximating $\ln(1+x)$ as x for small x and using the multivariate Gaussian density for $\mathbf{H}[\boldsymbol{d}]$, we obtain the following:

$$\Phi_C(s) \approx \frac{1}{|\boldsymbol{I} - \gamma s K \overline{\boldsymbol{R}}|}, \quad \overline{\boldsymbol{R}} = \boldsymbol{R}_r \otimes \boldsymbol{R}_t \otimes \boldsymbol{R}_H \tag{14.63}$$

The capacity distribution is approximately a standard χ*-square* distribution. If the number of degrees of freedom of the χ*-square* distribution is large, it can be further approximated by a Gaussian distribution.

To address the opposite case of high γ, $K \geq L$ is considered; i.e., the number of subcarriers is larger than the channel order. We can decompose $p_{\tilde{H}}(\tilde{\mathbf{H}}[\boldsymbol{k}])$ as follows:

$$p_{\tilde{H}}(\tilde{\mathbf{H}}[\boldsymbol{k}]) p(\tilde{\mathbf{H}}[\bar{\mathbf{p}}] \mid \tilde{\mathbf{H}}[\mathbf{p}]) p(\tilde{\mathbf{H}}[\mathbf{p}])$$

where $\boldsymbol{k} = (0,\ldots,K-1)$; $\mathbf{p} = (k_0,\ldots,k_L)$ is a vector that has, as elements, $L+1$ distinct but otherwise arbitrary indices extracted from $\boldsymbol{k}$; and $\bar{\mathbf{p}}$ is the vector of the complementary indices. The blocks of $\tilde{\mathbf{H}}[\mathbf{p}](\tilde{\mathbf{H}}^H[k_0],\ldots,\tilde{\mathbf{H}}^H[k_L])^H$ are in a one-to-one mapping with the blocks of $\mathbf{H}[\boldsymbol{d}] = (\mathbf{H}^H[0],\ldots,\mathbf{H}^H[L])^H$; in fact, Equation 14.60 for each antenna pair represents a system of linear equations, each corresponding to a different index $k_i \in \mathbf{p}$, with coefficients forming a full-rank Vandermonde matrix $\boldsymbol{W}_{L+1}(\mathbf{p})$:

$$\{\boldsymbol{W}_{L+1}(\mathbf{p})\}_{li} = e^{-j\frac{2\pi}{K}lk_i}, \quad l \in [0,L],\ k_i \in \mathbf{p},\ i \in [0,L]$$

and we can write

$$\tilde{\mathbf{H}}[\mathbf{p}] = (\boldsymbol{W}_{L+1}(\mathbf{p}) \otimes \boldsymbol{I})\mathbf{H}[\boldsymbol{d}], \quad \mathbf{p} = (k_0,\ldots,k_L)^T, \quad \boldsymbol{d} = (0,\ldots,L)^T \tag{14.64}$$

[8]It is assumed that the correlation does not change with time and the following separation model applies: $E\{\{\mathbf{H}[l]\}^*_{r_1,t_1}\{\mathbf{H}[l]\}_{r_2,t_2}\}\{\boldsymbol{R}_r\}_{r_1,r_2}\{\boldsymbol{R}_t\}_{t_1,t_2}$.

Thus, for any h_j we can write

$$\tilde{\mathbf{H}}[h_j] = \sum_{l=0}^{L} \mathbf{H}[l] e^{-j2\pi h_j l/K} = \sum_{k_i \in \mathbf{p}} \tilde{\mathbf{H}}[k_i] P_{k_i}(e^{-j2\pi h_j/K}) \tag{14.65}$$

where

$$P_{k_i}(z) \triangleq \prod_{j \neq i, 0 \leq j \leq L} \frac{z - e^{-j2\pi k_j/K}}{e^{-j2\pi k_i/K} - e^{-j2\pi k_j/K}} \tag{14.66}$$

with $(k_i, k_j) \in \mathbf{p}$.

From Equations 14.65 and 14.66, it follows that $p(\tilde{\mathbf{H}}[\bar{\mathbf{p}}] \mid \tilde{\mathbf{H}}[\mathbf{p}])$ is a product of Dirac deltas. With

$$\boldsymbol{P}[\mathbf{p}, h_j] \triangleq ([P_{k_0}(e^{-j2\pi h_j/K}), \ldots, P_{k_L}(e^{-j2\pi h_j/K})] \otimes \boldsymbol{I})$$

we have

$$p(\tilde{\mathbf{H}}[\bar{\mathbf{p}}] \mid \tilde{\mathbf{H}}[\mathbf{p}]) = \prod_{h_j \in \bar{\mathbf{p}}} \delta\left(\tilde{\mathbf{H}}[h_j] - \boldsymbol{P}[\mathbf{p}, h_j]\tilde{\mathbf{H}}[\mathbf{p}]\right)$$

$$p_{\tilde{H}}(\tilde{\mathbf{H}}[\mathbf{p}]) = |\boldsymbol{W}_{L+1}(\mathbf{p})|^{-N_R N_T} p_H((\boldsymbol{W}_{L+1}(\mathbf{p}) \otimes \boldsymbol{I})^{-1}\tilde{\mathbf{H}}[\mathbf{p}])$$

Gathering these results we can state the following lemma (valid for any γ).

Lemma 14.9 For an FIR N_T-input N_R-output MIMO frequency-selective channel having a probability density function of the MIMO impulse response $p_H(\mathbf{H}(\boldsymbol{d})), \boldsymbol{d} = (0, \ldots, L), \mathbf{H}(\boldsymbol{d}) = (\mathbf{H}^H(0), \ldots, \mathbf{H}^H(L))^H$, the characteristic function of the mutual information is equal to

$$\Phi_c(s) = \chi_2 \int \prod_{h=0}^{K-1} \left| \boldsymbol{I} + \gamma \tilde{\mathbf{H}}^H[\mathbf{p}] \boldsymbol{P}^H[\mathbf{p}, h] \boldsymbol{P}[\mathbf{p}, h] \tilde{\mathbf{H}}[\mathbf{p}] \right|^s$$
$$\times\, p_H\big((\boldsymbol{W}_{L+1}^{-1}(\mathbf{p}) \otimes \boldsymbol{I})\tilde{\mathbf{H}}[\mathbf{p}]\big)(d\tilde{\mathbf{H}}[\mathbf{p}])$$

where $\boldsymbol{W}_{L+1}(\mathbf{p})$ is defined in Equation 14.64, $\tilde{\mathbf{H}}[\mathbf{p}]$ is defined in Equation 14.64, $\boldsymbol{W}_{L+1}^{-1}(\mathbf{p})$ can be expressed in terms of the coefficients of the Lagrange polynomials in Equation 14.66, and $\chi_2 = |\boldsymbol{W}_{L+1}(\mathbf{p})|^{-N_R N_T}$.

Under the assumption that $K = Q(L+1)$, we can arrive at a closed-form simpler expression for $\Phi_c(s)$. Choosing $\mathbf{p} = (0, Q, \ldots, QL)$, since $e^{-j\frac{2\pi}{Q(L+1)} lQd} = e^{-j\frac{2\pi}{(L+1)} ld}$, $\boldsymbol{W}_{L+1}(\mathbf{p})$ is unitary. It should be noted that the coefficients of the linear combination in Equation 14.65 that corresponds to $\tilde{\mathbf{H}}[lQ + q]$ are for the most part highly concentrated around $l - n = 0$, which suggests the approximation

$$\tilde{\mathbf{H}}[lQ + q] \approx P_{lQ}(e^{-j2\pi(lQ+q)/K})\tilde{\mathbf{H}}[lQ] = \alpha(q)\tilde{\mathbf{H}}[lQ]$$

where

$$\alpha(q) \triangleq \frac{e^{j2\pi\left(\frac{qL}{Q(L+1)}\right)}}{L+1} \frac{\sin\left(\frac{q}{Q}\right)}{\sin\left(\frac{\pi q}{Q(L+1)}\right)}$$

If $\mathbf{p}$ is selected to have uniformly spaced frequency indices, in force of the Szëgo theorem for $L \gg 1$, the elements of $\tilde{\mathbf{H}}[\mathbf{p}]$ will be approximately uncorrelated not only in space but also across the frequency bins. Since the correlation matrix of $\mathbf{H}(\boldsymbol{d})$ is $(\boldsymbol{I} \otimes \boldsymbol{R}_H)$, using the central limit theorem the pdf of $\tilde{\mathbf{H}}[\mathbf{p}]$ is approximately $\backsim \mathcal{N}(\mathbf{0}, (\boldsymbol{W}_{L+1}^H(\mathbf{p})\boldsymbol{R}_H \boldsymbol{W}_{L+1}(\mathbf{p}) \otimes \boldsymbol{I}))$, where $(\boldsymbol{W}_{L+1}^H \boldsymbol{R}_H \boldsymbol{W}_{L+1}) \approx diag(\sigma_h^2[\mathbf{p}])$, where $\sigma_h^2[\mathbf{p}] = (\sigma_h^2[0], \ldots, \sigma_h^2[LQ])$ and $\sigma_h^2[lQ] \sum_n \boldsymbol{R}_H[n] e^{-j2\pi nl/(L+1)}$. Thus we have the following proposition.

Proposition 14.2 Under the assumptions stated in the beginning of the section for $L \gg 1$ and assuming $E\{\tilde{\mathbf{H}}[\mathbf{p}]\} = \mathbf{0}$, the $\tilde{\mathbf{H}}^H[lQ]$ are approximately Gaussian and independent and

$$\Phi_c(s) \approx \chi_3 \prod_{l=0}^{L} \int \prod_{q=0}^{Q-1} \left| I + \gamma\alpha(q)\tilde{\mathbf{H}}^H[lQ]\tilde{\mathbf{H}}[lQ] \right|^s e^{-\frac{tr(\tilde{H}^H[lQ]\tilde{H}[lQ])}{\sigma_h^2[l]}} (d\tilde{\mathbf{H}}[lQ]) \tag{14.67}$$

where $\chi_3 = \prod_{l=0}^{L}(\pi\sigma_h^2[l])^{N_T N_R} = \pi^{N_T N_R(L+1)}|\boldsymbol{R}_H|^{N_T N_R}$. The integral on the right side of Equation 14.67 is analogous to Equation 14.63 [7], [56].

We proceed in our approximation and exploit the fact that $\gamma \gg 1$. Also, we can include the spatial correlation for the separable model, as was done for the low γ case, with the covariance matrix of $\mathbf{H}[\boldsymbol{d}]$ being $\boldsymbol{R}_r \otimes \boldsymbol{R}_t \otimes \boldsymbol{R}_H$. Since the discrete Fourier transform (DFT) operates in time, the spatial correlation of the MIMO frequency response $\tilde{\mathbf{H}}[\mathbf{p}]$ for $L \gg 1$ tends to be $\boldsymbol{R}_r \otimes \boldsymbol{R}_t \otimes diag(\sigma_h^2[\mathbf{p}])$, or in other words,

$$E\left[vec(\tilde{\mathbf{H}}[lQ])vec(\tilde{\mathbf{H}}[lQ])^H\right] = \sigma^2[lQ](\boldsymbol{R}_t \otimes \boldsymbol{R}_r)$$

With this in mind, we can state the following proposition.

Proposition 14.3 Under the same assumptions and using the same approximations that led to Proposition 2, if $\gamma \gg 1$,

$$\Phi_C(s) \approx e^{sn(L+1)\log\prod_{q=0}^{Q-1}\left(\alpha(q)\gamma|R_H|^{\frac{1}{L+1}}|R_t|^{\frac{1}{n}}|R_r|^{\frac{1}{n}}\right)} \prod_{i=1}^{n}\left[\frac{\Gamma(Qs+m-n+i)}{\Gamma(m-n+i)}\right]^{L+1} \tag{14.68}$$

Refer to the [46] for details.

After having obtained the expressions for the characteristic function of the mutual information, we claim that the capacity can be approximated by a Gaussian random variable. We now proceed to find the parameters of the Gaussian distribution for the extreme cases of γ considered. We first take up the high γ case for which Equation 14.68 shows that the channel gain takes the form of the geometric mean of the eigenvalues of the channel covariance matrix. The form of the characteristic function in Equation 14.68 motivates the idea of approximating the pdf of capacity with a Gaussian pdf whose parameters can be easily computed. The first factor in Equation 14.68 implies that the capacity pdf $p_C(C)$ is a shifted version of the inverse Laplace transform of the term

$$\left(\prod_{i=1}^{n}\frac{\Gamma(Qs+m-n+i)}{\Gamma(m-n+i)}\right)^{L+1} \tag{14.69}$$

The factor in Equation 14.69 is the L-th power of a product of functions. Since in our approximations $L \gg 1$, we can infer that the inverse Laplace transform of Equation 14.69 will be very close to a Gaussian pdf, due to the central limit theorem.

From the first- and second-order derivatives of $\frac{\Gamma(Qs+m-n+i)}{\Gamma(m-n+i)}$ in $s = 0$, one can easily obtain the first-order moment $\mu_i^{(1)}$ and the variance σ_i^2 of its inverse Laplace transform, which are

$$\mu_i = Q\psi^{(0)}(m-n+i), \qquad \sigma_i^2 = Q^2\psi^{(2)}(m-n+i) \tag{14.70}$$

where $\psi^n(x)$ is the n-th derivative of the polygamma function, also known as the digamma function [1]. Therefore, we have

$$p_C(C) \approx \frac{e^{-\frac{\left(C-\left(\prod_{q=0}^{Q-1}\alpha(q)\gamma|R_H|^{\frac{1}{L+1}}|R_t|^{1/n}|R_r|^{1/n}\right)-K\sum_{i=1}^{n}\psi^{(0)}(m-n+i)\right)^2}{2KQ\sum_{i=1}^{n}\psi^{(2)}(m-n+i)}}}{\sqrt{2\pi KQ\sum_{i=1}^{n}\psi^{(2)}(m-n+i)}}$$

It should be noted that the variance of the capacity in Equation 14.70 obtained in the high SNR regime is independent of γ, and the same conclusion was reached in [32], [35], [49], [56]. This is in direct contrast to the low γ scenario, which will be considered later. Here we also notice that even if under many approximations, the effect of correlation and SNR is only to shift the mean of the distribution, but the same parameters have no impact on the capacity variance, which is only a function of m, n. The plots were obtained in [46] to show that the Gaussian approximation was valid.

As pointed out earlier in this section, the capacity for the case of $\gamma \ll 1$ takes the following form:

$$C \approx \gamma K tr(\mathbf{H}^H[\boldsymbol{d}]\mathbf{H}[\boldsymbol{d}])$$

This is a χ-*square* distribution, which, under the limit of a large number of antennas and paths, will closely approximate the Gaussian density. However, even for the reasonable values of N_R and N_T, we show numerically that the Gaussian fit is accurate even when γ is really small [46]. To do so, we first find parameters of the corresponding Gaussian random variable using Equation 14.63. First- and second-order derivatives of $\Phi_C(s)$ evaluated at $s = 0$ yield the mean and variance:[9]

$$\mu = \gamma K tr(\boldsymbol{R}_r)tr(\boldsymbol{R}_t)tr(\boldsymbol{R}_H), \qquad \sigma^2 = (\gamma K)^2 tr\left(\boldsymbol{R}_r^2\right)tr\left(\boldsymbol{R}_t^2\right)tr\left(\boldsymbol{R}_H^2\right)$$

Hence, in this case we have the following:

$$p_C(C) \approx \frac{e^{-\frac{(C-\gamma K tr(\overline{\boldsymbol{R}}))^2}{2(\gamma K)^2 tr(\overline{\boldsymbol{R}}^2)}}}{\sqrt{2\pi(\gamma K)^2 tr(\overline{\boldsymbol{R}}^2)}}$$

It is interesting to note that mean and variance are proportional to the signal-to-noise ratio (γ). Also, while there is no explicit dependence on m and n, the dependence on the correlation matrices of paths and the array elements is through their trace.

So far, we have looked at the design aspects of linear precoders and decoders for the MIMO channels. In doing so, the channel was assumed to be known at both the transmitter and receiver. However, this may not always be the case. Hence, the task of estimating the channel for such precoded MIMO systems should be dealt with, and this is the focus of the next section.

14.4 Channel Estimation for MIMO Systems Using Precoding Techniques

Earlier work [44] has shown that the redundancy introduced by the precoder can be exploited to blindly estimate deterministic frequency-selective channels up to a scale. To acquire the channel status information (CSI) without ambiguity, training is required. Training symbols can be inserted in the data stream (the pilot symbol assisted modulation (PSAM) technique). Inserted training symbols are separated from the information symbols either in frequency [38] or in time [33]. Alternatively, one could use the superimposed pilot sequence technique [23], in which a known pilot sequence is linearly added to the unknown data sequence.

Manton et al. [31] introduced the affine precoding scheme as a general framework in which PSAMs with pilot tones or superimposed training sequences can be treated as special cases. In this scheme, the transmitted and received data blocks are presented as

$$\boldsymbol{x}_i = \boldsymbol{F}\boldsymbol{s}_i + \boldsymbol{t} \qquad \boldsymbol{y}_i = \boldsymbol{H}\boldsymbol{x}_i + \boldsymbol{n}_i$$

For a frequency-selective SISO channel, the optimal design of the affine precoders for joint channel estimation and symbol recovery is derived in [31]. In [39], [40] the problems of channel estimation and symbol recovery are separated and only the training symbols are used to estimate $\boldsymbol{H}$. Under the constraint that $\boldsymbol{B}^{\dagger}\boldsymbol{H}\boldsymbol{F} = 0$ (where $\boldsymbol{B}$ is a column-wise circulant matrix with first column $\boldsymbol{t}$), the authors derive the optimal pair $(\boldsymbol{F}, \boldsymbol{t})$, which reduces the least squares (LS) channel estimation error. In [6] the authors

[9] We use the identity $\frac{\partial}{\partial\alpha}\{\ln|\boldsymbol{A}|\} = tr\{\boldsymbol{A}^{-1}\frac{\partial \boldsymbol{A}}{\partial\alpha}\}$.

investigate the problem of channel estimation with superimposed pilot symbols for a quasi-statistic flat-fading MIMO system that uses the space–time orthogonal block codes. It is shown that under certain training schemes, superimposed training increases the Cramer–Rao bound (CRB).

We consider a block flat-fading memoryless MIMO channel in additive white Gaussian noise in which the channel matrix in Equation 14.6 reduces to $\boldsymbol{H} = \boldsymbol{I}_P \otimes \mathbf{H}[0]$. The frequency-selective case can be treated similarly, with some further complication in the notation that we avoid here. We provide a channel estimator and CRB expression that can be used as a benchmark to evaluate the performance of any unbiased estimator. In contrast to most previous works in which the symbols are considered as deterministic unknown parameters, we assume that the symbols are Gaussian. It is known that for an AWGN channel, the Gaussian distribution of input maximizes the mutual information between the input and output of the channel [8]. Furthermore, the assumption of Gaussian symbols allows the derivation of a new CRB and a new estimator technique that utilize not only training, but also a subspace method that capitalizes the redundant structure of $\boldsymbol{F}$. This model highlights two important aspects: (1) the interesting trade-off between transmitting at full rate and achieving coherent reception, and (2) the different way in which the components associated with the symbols (blind) contribute to the channel estimator compared to the training components (nonblind). We make the following assumptions:

1. The channel is deterministic and unknown.
2. The transmit symbols $\boldsymbol{s}_i$ and the noise $\boldsymbol{n}_i$ are white, i.e., $\boldsymbol{R}_{ss} = \sigma_{ss}^2 \boldsymbol{I}$ and $\boldsymbol{R}_{nn} = \sigma_{nn}^2 \boldsymbol{I}$, and $\boldsymbol{n}_i$ and $\boldsymbol{s}_i$ are uncorrelated.

We have $\boldsymbol{y}_i \sim \mathcal{CN}(\mu, \boldsymbol{R})$ where

$$\mu = \boldsymbol{H}\boldsymbol{t}, \qquad \boldsymbol{R} = \sigma_{ss}^2 \boldsymbol{H}\boldsymbol{F}\boldsymbol{F}^H \boldsymbol{H}^H + \sigma_{nn}^2 \boldsymbol{I} \tag{14.71}$$

in which $\boldsymbol{h}$ is an $RK \times 1$ vector containing the parameters to be estimated. To obtain a meaningful estimate of $\boldsymbol{h}$, we need at least as many observation samples as unknown parameters, which implies $RP > RK$. This model is general enough to cover the frequency-selective MIMO channel presented in Section 14.2.

14.4.1 Channel Estimation Algorithm

The ML channel estimate for a general structure channel matrix $\boldsymbol{H}$ is $\hat{\boldsymbol{H}} = \arg\max_H \ln p(\boldsymbol{y}_i|\boldsymbol{H})$, i.e., the solution of the system of equations

$$\frac{\partial \ln p(\boldsymbol{y}_i|\boldsymbol{H})}{\partial \boldsymbol{H}^H} = \mathbf{0} \tag{14.72}$$

The log-likelihood function can be written as $\ln p(\boldsymbol{y}_i|\boldsymbol{H}) Const. - \ln|\boldsymbol{R}| - \boldsymbol{z}_i^H \boldsymbol{R}^{-1}\boldsymbol{z}_i$, where $\boldsymbol{z}_i := \boldsymbol{y}_i - \boldsymbol{H}\boldsymbol{t}$ and $\boldsymbol{R}$ is given by Equation 14.71. Therefore,

$$\frac{\partial \ln p(\boldsymbol{y}_i|\boldsymbol{H})}{\partial \boldsymbol{H}_{kl}^*} - \frac{\partial \ln|\boldsymbol{R}|}{\partial \boldsymbol{H}_{kl}^*} - \frac{\partial \boldsymbol{z}_i^H \boldsymbol{R}^{-1}\boldsymbol{z}_i}{\partial \boldsymbol{H}_{kl}^*} \tag{14.73}$$

Since [25]

$$\frac{\partial \ln|\boldsymbol{R}|}{\partial \boldsymbol{H}_{kl}^*} tr\left(\boldsymbol{R}^{-1}\frac{\partial \boldsymbol{R}}{\partial \boldsymbol{H}_{kl}^*}\right) \tag{14.74}$$

and

$$\frac{\partial \boldsymbol{z}_i^H \boldsymbol{R}^{-1}\boldsymbol{z}_i}{\partial \boldsymbol{H}_{kl}^*} = tr\left(\boldsymbol{R}^{-1}\frac{\partial \boldsymbol{z}_i\boldsymbol{z}_i^H}{\partial \boldsymbol{H}_{kl}^*}\right) + tr\left(\frac{\partial \boldsymbol{R}^{-1}}{\partial \boldsymbol{H}_{kl}^*}\boldsymbol{z}_i\boldsymbol{z}_i^H\right) \tag{14.75}$$

in which

$$\frac{\partial \boldsymbol{R}^{-1}}{\partial \boldsymbol{H}_{kl}^*} = -\boldsymbol{R}^{-1}\frac{\partial \boldsymbol{R}}{\partial \boldsymbol{H}_{kl}^*}\boldsymbol{R}^{-1} \tag{14.76}$$

it is not difficult to show that

$$\frac{\partial \ln p(\boldsymbol{y}_i|\boldsymbol{H})}{\partial \boldsymbol{H}^H} = -\sigma_{ss}^2 \boldsymbol{R}^{-1}\boldsymbol{H}\boldsymbol{F}\boldsymbol{F}^H + \boldsymbol{R}^{-1}\boldsymbol{z}_i\boldsymbol{t}^H + \sigma_{ss}^2 \boldsymbol{R}^{-1}\boldsymbol{z}_i\boldsymbol{z}_i^H\boldsymbol{R}^{-1}\boldsymbol{H}\boldsymbol{F}\boldsymbol{F}^H \tag{14.77}$$

where the regularity condition can easily be verified, i.e., $E\{\frac{\partial \ln p(\boldsymbol{y}_i|\boldsymbol{H})}{\partial \boldsymbol{H}^H}\} = 0$. Since $\boldsymbol{R}^{-1}$ is a common positive definite factor, it cannot be separately set to zero. Therefore, a possible solution for Equation 14.77 is

$$\boldsymbol{z}_i\boldsymbol{t}^H + \sigma_{ss}^2\left(\boldsymbol{z}_i\boldsymbol{z}_i^H\boldsymbol{R}^{-1} - \boldsymbol{I}\right)\boldsymbol{H}\boldsymbol{F}\boldsymbol{F}^H = \boldsymbol{0}_{RP\times N} \tag{14.78}$$

If $\boldsymbol{\Pi}_F^{\perp}$ and $\boldsymbol{\Pi}_t^{\perp}$[10] span together the entire space, we can derive two equivalent sets of equations from Equation 14.78: (1) $\boldsymbol{z}_i\boldsymbol{t}^H\boldsymbol{\Pi}_F^{\perp} = 0$ and (2) $(\sigma_{ss}^2\boldsymbol{z}_i\boldsymbol{z}_i^H\boldsymbol{R}^{-1} - \sigma_{ss}^2\boldsymbol{I})\boldsymbol{H}\boldsymbol{F}\boldsymbol{F}^H\boldsymbol{\Pi}_t^{\perp} = 0$. These equations separate the channel estimator performed using training from the channel estimator done through the symbols (blind). However, the second equation is a nonlinear system of equations that does not lead to a simple estimator. We propose an estimation algorithm that is a combination of subspace and the least squares method.

Let $\boldsymbol{h} = vec(\mathbf{H}[0]^T)$ be the $RK \times 1$ vector of parameters to be estimated. We exploit the block diagonal structure of $\boldsymbol{H} = \boldsymbol{I} \otimes \mathbf{H}[0]$ to provide alternative representations for $\boldsymbol{Ht}$ and $\boldsymbol{HF}$. Assume $\boldsymbol{a} \in \mathcal{C}^{KP\times 1}$ and let $\boldsymbol{a} = [\boldsymbol{a}_1^T \cdots \boldsymbol{a}_P^T]^T$ where $\boldsymbol{a}_k$ is a $K \times 1$ subvector of $\boldsymbol{a}$. We introduce a mapping Φ from $\boldsymbol{a}$ to a matrix $\boldsymbol{A}$, $\Phi : \mathcal{C}^{KP\times 1} \to \mathcal{C}^{RP\times RK}$ such that $\boldsymbol{A} = \Phi(\boldsymbol{a}) = [\boldsymbol{I}_R \otimes \boldsymbol{a}_1, \ldots, \boldsymbol{I}_R \otimes \boldsymbol{a}_P]^T$. We rewrite $\boldsymbol{Ht}$ as $\boldsymbol{Th}$ where $\boldsymbol{T} = \Phi(\boldsymbol{t})$. Similarly, we write $\boldsymbol{HF}$ as $\boldsymbol{HF}\mathcal{F}(\boldsymbol{I}_N \otimes \boldsymbol{h})$, where $\mathcal{F} := [\Phi(\boldsymbol{f}_1)\Phi(\boldsymbol{f}_2)\cdots\Phi(\boldsymbol{f}_N)]$ and $\boldsymbol{f}_k$ is the k-th column of $\boldsymbol{F}$.[11] Equivalently, the observation vector $\boldsymbol{y}_i$ and the covariance matrix $\boldsymbol{R}$ can be represented as

$$\begin{aligned}\boldsymbol{y}_i &= \mathcal{F}(\boldsymbol{I}_N \otimes \boldsymbol{h})\boldsymbol{s}_i + \boldsymbol{Th} + \boldsymbol{n}_i \\ \boldsymbol{R} &= \sigma_{ss}^2\mathcal{F}(\boldsymbol{I}_N \otimes \boldsymbol{h}\boldsymbol{h}^H)\mathcal{F}^H + \sigma_{nn}^2\boldsymbol{I}\end{aligned} \tag{14.79}$$

Right multiplying $\boldsymbol{y}_i$ by $\boldsymbol{M} := (\boldsymbol{\Pi}_T^{\perp}\mathcal{F})^{\dagger}\boldsymbol{\Pi}_T^{\perp}$, we obtain $\boldsymbol{M}\boldsymbol{y}_i = (\boldsymbol{I}_N \otimes \boldsymbol{h})\boldsymbol{s}_i + \boldsymbol{M}\boldsymbol{n}_i$. To guarantee the existence of $\boldsymbol{M}$ we require $P > NK$. Consider the partitions of $\boldsymbol{M} = [\boldsymbol{M}_1^H \cdots \boldsymbol{M}_N^H]^H$. We have

$$\boldsymbol{v}_k := \boldsymbol{M}_k\boldsymbol{y}_i = \boldsymbol{h}\boldsymbol{s}_i(k) + \boldsymbol{M}_k\boldsymbol{n}_i, \quad k = 1, 2, \ldots, N \tag{14.80}$$

where $\boldsymbol{v}_k \sim \mathcal{CN}(0, \sigma_{ss}^2\boldsymbol{h}\boldsymbol{h}^H + \sigma_{nn}^2\boldsymbol{M}_k\boldsymbol{M}_k^H)$. For general precoder $\boldsymbol{F}$ and training $\boldsymbol{t}$, $\boldsymbol{v}_k$s are dependent and have different distributions. We then form the matrix

$$\frac{1}{N}\sum_{k=1}^{N}\boldsymbol{v}_k\boldsymbol{v}_k^H = \frac{1}{N}\boldsymbol{h}\boldsymbol{h}^H\sum_{k=1}^{N}|\boldsymbol{s}_i(k)|^2 + \frac{1}{N}\sum_{k=1}^{N}\sum_{l=1}^{N}\boldsymbol{M}_k\boldsymbol{n}_i\boldsymbol{n}_i^H\boldsymbol{M}_l^H \tag{14.81}$$

Let $\lambda_1 \geq \lambda_2 \geq \cdots \geq \lambda_{RK}$ denote the eigenvalues of the right-hand-side matrix in Equation 14.81 and $\tilde{\boldsymbol{g}}_1, \ldots, \tilde{\boldsymbol{g}}_{RK}$ be the unit-norm eigenvectors associated with $\lambda_1, \lambda_2, \ldots, \lambda_{RK}$. Define $\tilde{\boldsymbol{G}} := [\tilde{\boldsymbol{g}}_2, \ldots, \tilde{\boldsymbol{g}}_{RK}]$. From Equation 14.81 we see that at high SNR, the eigenvector corresponding to the maximum eigenvalue determines the direction of vector $\boldsymbol{h}$, i.e., $\tilde{\boldsymbol{g}}_1 = \frac{\boldsymbol{h}}{\|\boldsymbol{h}\|}$. We view the second term in the right-hand side of Equation 14.81 as a small perturbation. Thus we can write

$$\tilde{\boldsymbol{G}}^H\boldsymbol{h} \cong \boldsymbol{0}_{RK\times 1} \tag{14.82}$$

On the other hand, right multiplying $\boldsymbol{y}_i$ by $\boldsymbol{\Pi}_{\mathcal{F}}^{\perp}$ we obtain $\boldsymbol{\Pi}_{\mathcal{F}}^{\perp}\boldsymbol{y}_i = \boldsymbol{\Pi}_{\mathcal{F}}^{\perp}\boldsymbol{Th} + \boldsymbol{\Pi}_{\mathcal{F}}^{\perp}\boldsymbol{n}_i$. At high SNR, an estimate of the channel $\boldsymbol{h}$ is obtained by solving

$$\boldsymbol{\Pi}_{\mathcal{F}}^{\perp}\boldsymbol{y}_i \cong \boldsymbol{\Pi}_{\mathcal{F}}^{\perp}\boldsymbol{Th} \tag{14.83}$$

[10]Given $\boldsymbol{A}$, $\boldsymbol{\Pi}_A := \boldsymbol{A}(\boldsymbol{A}^H\boldsymbol{A})^{-1}\boldsymbol{A}^H$ is the orthogonal projection matrix onto $\boldsymbol{A}$, such that $\boldsymbol{\Pi}_A\boldsymbol{A} = \boldsymbol{A}$ and $\boldsymbol{\Pi}_A^{\perp} := \boldsymbol{I} - \boldsymbol{\Pi}_A$ is the complement orthogonal projection matrix onto $\boldsymbol{A}$ such that $\boldsymbol{\Pi}_A^{\perp}\boldsymbol{A} = 0$.

[11]The mapping $\Phi(.)$ can be generalized to treat the case of $\boldsymbol{h} = vec(\mathbf{H}[0]^T, \ldots, \mathbf{H}[L]^T)$ for frequency-selective channels.

Combining Equations 14.82 and 14.83, we obtain the channel estimate as

$$h = \begin{bmatrix} \Pi_{\mathcal{F}}^{\perp} T \\ \tilde{G}^H \end{bmatrix}^{\dagger} \begin{bmatrix} \Pi_{\mathcal{F}}^{\perp} y_i \\ 0 \end{bmatrix}$$

Under certain constraints on (F, t) the additive noise term in Equation 14.80 is whitened and the estimator resembles the well-known MUSIC estimator [50]. Let (F, t) be such that $\mathcal{F}^H \mathcal{F} = I_{NRK}$, $T^H T = I_{RK}$, and $\mathcal{F}^H T = 0_{NRK \times RK}$. Then, M becomes $\mathcal{F}^H$ and the v_k values are independent and identically distributed (i.i.d), i.e., $v_k \sim \mathcal{CN}(0, \sigma_{ss}^2 hh^H + \sigma_{nn}^2 I_{RK})$. Let $C := [F, t]$ and $c_i = [c_{i1}^T \cdots c_{iP}^T]^T$ where c_{ik} is a $K \times 1$ subvector of c_i, the i-th column of C. The necessary and sufficient condition on (F, t) such that they admit the above constraints is

$$\sum_{m=1}^{P} c_{im} c_{jm}^H = \delta(i-j) I_k$$

For the special case of $K = 1$, C is semiunitary, i.e., $C^H C = I_{N+1}$.

14.4.2 Cramer–Rao Lower Bound

The CRB is a lower bound on the covariance matrix of any unbiased estimate of the deterministic parameters we desire to estimate. It provides us a benchmark against which we can compare the performance of the derived estimator. For a general structure matrix H, let $\Delta = vec(\frac{\partial \ln p(y_i|H)}{\partial H^*})$. We show that the unconstrained Fisher information matrix (FIM) $\mathcal{J}$ has the following form:

$$\mathcal{J} = E\{\Delta\Delta^H\} = \left(tt^H + \sigma_{ss}^4 FF^H H^H R^{-1} H FF^H\right)^T \otimes R^{-1}$$

Since the number of unknown parameters ($P^2 KR$) is more than the number of known parameters (KP), $\mathcal{J}$ is singular. To retrieve the invertibility, we use the *a priori* knowledge of block diagonal structure of H. The general closed form of constrained CRB is derived in [51]. Let $\theta \in R^{n \times 1}$ be the vector of deterministic parameters to be estimated from the observation vector y_i. Assume the estimator of θ, denoted by $\hat{\theta}$, is unbiased. We establish $k < n$ equality constraints on the elements of θ such that $g(\hat{\theta}) = 0$. The gradient matrix $G \in R^{k \times n}$ is defined by

$$G(\theta) = \frac{\partial g(\theta)}{\partial \theta^T}$$

G is assumed to be full row rank for any θ satisfying the constraints. Consider $U \in R^{k \times (n-k)}$ whose columns are the orthogonal basis for the null space of G so that $GU = 0$, where $U^T U = I$. Let $\mathcal{J}$ be the unconstrained FIM, which is singular. If $U^T \mathcal{J} U$ is invertible, the constrained CRB is

$$E\{(\hat{\theta} - \theta)(\hat{\theta} - \theta)^T\} \geq U(U^T \mathcal{J} U)^{-1} U^T$$

For the specific case of H block diagonal, it is actually more convenient to use the structure of H directly. For real parameter estimation, the general CRB expression of Gaussian observation is derived in [25]. Extending the result to the case of complex parameter estimation, we find the FIM as follows:

$$\mathcal{J}_{kl}^c = E\left\{\frac{\partial \ln p(y_i|h)}{\partial h_k^*} \frac{\partial \ln p(y_i|h)}{\partial h_l}\right\} = [T^H R^{-1} T]_{kl} + \underbrace{tr\left(R^{-1} \frac{\partial R}{\partial h_k^*} R^{-1} \frac{\partial R}{\partial h_l}\right)}_{\Sigma_{kl}}$$

Let B_k be $B_k := \frac{\partial hh^H}{\partial h_k^*} = [0_{RK \times (k-1)}\ h\ 0_{RK \times (RK-k)}]$. We express Σ_{kl} as

$$\Sigma_{kl} = \sigma_{ss}^4 tr\left(R^{-1} \mathcal{F}(I_N \otimes B_k) \mathcal{F}^H R^{-1} \mathcal{F}(I_N \otimes B_l^H) \mathcal{F}^H\right) \tag{14.84}$$

Let $D := \mathcal{F}^H R^{-1} \mathcal{F}$. Using two trace properties $tr(A^H B) = vec^H(A^H)vec(B)$ and $vec(ACB) = (B^T \otimes A)vec(C)$, we can simplify Equation 14.84 as

$$\Sigma_{kl} = \sigma_{ss}^4 vec^H\left(I_N \otimes B_k^H\right)(D^* \otimes D)vec\left(I_N \otimes B_l^H\right)$$

Define $\mathcal{B} := [vec(I \otimes B_1^H)vec(I \otimes B_2^H) \cdots vec(I \otimes B_{RK}^H)]$. We can write Σ as $\Sigma = \sigma_s^4 \mathcal{B}^H(D^* \otimes D)\mathcal{B}$ and obtain the FIM expression

$$\mathcal{J}^c = T^H R^{-1} T + \sigma_s^4 \mathcal{B}^H (D^* \otimes D)\mathcal{B}$$

14.4.3 Numerical Results

We investigate the trade-off between the power allocated for training and for the information symbols by comparing the average channel CRB and the MSE of the proposed estimator and the mutual information between input and output of the channel $I(x_i, y_i)$. We assume that the total transmitted power $\mathcal{P}_0 = \sigma_s^2 tr(FF^H) + \|t\|^2$ is constant and the fraction of power assigned to training is $\zeta = \frac{\|t\|^2}{\mathcal{P}_0}$. We set $K = 2$, $R = 2$, $P = 8$, $N = 3$ and $\sigma_{ss}^2 = 1$. The simulation results are averaged over 500 sets of independent Rayleigh fading channels. The training vector is $t^H = [1\ 0\ 0\ 1\ zeros(1, 12)]$ and the Hermitian of the precoder F^H is a 3×16 matrix whose first row is $[zeros(1, 4)\ 1\ 0\ 0\ 1\ zeros(1, 8)]$ with the subsequent rows being the right circulant shift of their preceding row by 4 places such that $F^H t = 0$. Without loss of generality, we assume that $\mathcal{P}_0 = 1$ and therefore the signal-to-noise ratio (SNR) is $SNR = -10log_{10}\sigma_n^2$. For each SNR, t and F are scaled such that the power constraint is satisfied.

Figure 14.5(a) and (b) illustrate the difference between MSE of the subspace and LS channel estimate and compare their performance against CRB as training power changes. It is not surprising to observe that increasing ζ improves the estimation accuracy and therefore decreases the CRB. However, this improvement comes at the expense of decreasing the maximum transmission rate. This is illustrated in Figure 14.5(d)

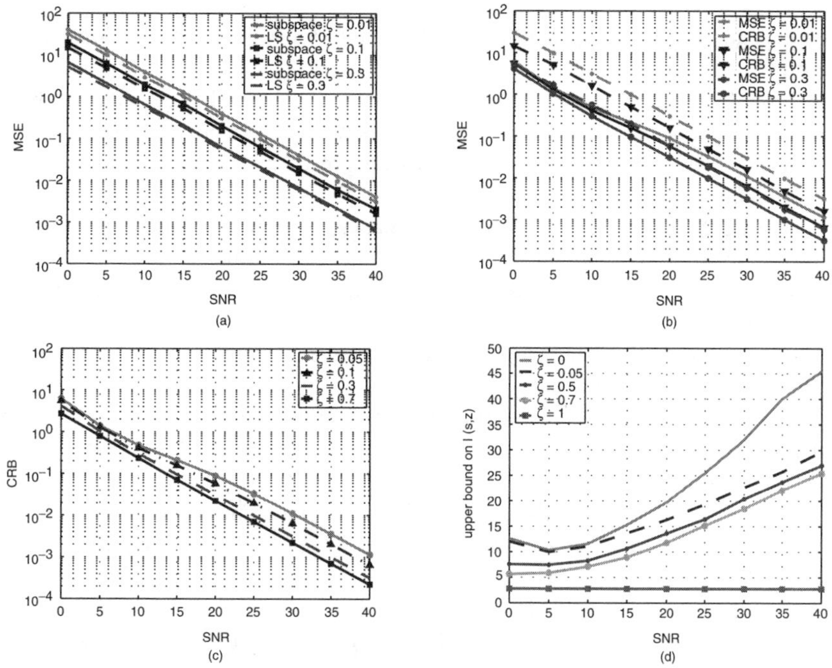

FIGURE 14.5 (a) comparison between Subspace and Least Square estimator as function of ζ; (b) comparison between MSE and CRB for $\zeta = 0.1$ and $\zeta = 0.7$; (c) CRB as a function of ζ; (d) upper bound on $I(x_i, y_i)$.

where we show the upper bound on $I(\boldsymbol{x}_i, \boldsymbol{y}_i)$ derived in [34] as a function of covariance of channel measurement error, assuming the error covariance is the CRB matrix derived in Section 14.4.2.

14.5 Conclusions

In this chapter we have looked at three principal aspects related to the design of linear precoders. We first dealt with the optimal design of precoder and decoder under average and peak power type of constraints. Simple closed-form solutions were obtained that are scalable with respect to the number of antennas, size of the coding block, and transmit average/peak power. The performance of our MIMO system was analyzed in the presence of random fading, and random matrix theory emerged as a useful tool for analysis. We also analyzed the performance of precoding with superimposed training. We studied the trade-off between the power allocated for training and for information symbols.

References

[1] E. Artin, *The Gamma Function*, Holt, Rinehart and Winston, New York, 1964.

[2] S. Benedetto and E. Biglieri, *Principles of Digital Transmission with Wireless Applications*, Kluwer Academic, Dordrecht, Netherlands, 1999.

[3] E. Biglieri, J. Proakis, and S. Shamai, "Fading Channels: Information-Theoretic and Communications Aspects," *IEEE Trans. Inf. Theory*, 44(6), 2619–2692, 1998.

[4] L.H. Brandenburg and A.D. Wyner, "Capacity of the Gaussian Channel with Memory: The Multivariate Case," *Bell Syst. Tech. J.*, 53, 745–778, 1974.

[5] B.V. Bronk, "Exponential Ensembles for Random Matrices," *J. Math. Phys.*, 6, 228–237, 1965.

[6] C. Budianu and L. Tong, "Channel Estimation for Space-Time Orthogonal Block Codes," *IEEE Trans. Signal Process.*, 50, 2515–2528, 2002.

[7] M. Chiani, "Evaluating the Capacity Distribution of MIMO Rayleigh Fading Channels," in *Book of Extended Abstracts for IEEE International Symposium on Advances in Wireless Communications*, ISCW'02, Canada, September 2002.

[8] T.M. Cover and J.A. Thomas, *Elements of Information Theory*, John Wiley & Sons, Inc., New York, 1991.

[9] A. Edelman, "Random Eigenvalues," Notes for Math 273 University of California–Berkeley, February 27, 1999, available at http://www.math.berkeley.edu/edelman/math273.html.

[10] R. Etkin and D. Tse, "Degrees of Freedom in Underspread MIMO Fading Channels," submitted to *IEEE Trans. Inf. Theory*.

[11] J. Evans and D.N.C. Tse, "Large System Performance of Linear Multiuser Receivers in Multipath Fading Channels," *IEEE Trans. Inf. Theory*, 46(6), 2059–2077, 2000.

[12] H. Flanders, *Differential Forms with Applications to the Physical Sciences*, 2nd ed., Dover Publications Inc., New York, 1989.

[13] G.D. Forney, Jr., and M.V. Eyuboğlu, "Combined Equalization and Coding Using Precoding," *IEEE Commun. Mag.*, 25–34, 1991.

[14] G.J. Foschini, "Layered Space-Time Architecture for Wireless Communication in a Fading Environment When Using Multielement Antennas," *Bell Labs. Tech. J.*, 1(2), 41–59, 1996.

[15] R. Gallager, *Elements of Information Theory*, Section 8, Wiley, New York, 1968.

[16] A. Ganesan and A.M. Sayeed, "Bandwidth-Efficient Exploitation of the Degrees of Freedom in a Multipath-Fading Channel," in *Proceedings of IEEE ISIT*, Sorrento, Italy, July 2000, 161.

[17] V. Girko, *Theory of Random Determinant*, Kluwer Publishers, Dordrecht, Netherlands, 1990.

[18] M. Godavarti and A.O. Hero III, "Diversity and Degrees of Freedom in Wireless Communications," preprint, available at http://www.eecs.umich.edu/hero/Preprints/it_div_vs_dof1.pdf.

[19] A.J. Grant and P.D. Alexander, "Random Sequence Multisets for Synchronous Code-Division Multiple-Access Channels," *IEEE Trans. Inf. Theory*, 44, 2832–2836, 1998.

[20] Haagerup and S. Thorbjrnsen, "Random Matrices with Complex Gaussian Entries," preprint (49 pp.), Odense, 1998, available at http://www.imada.ou.dk/haagerup/2000-.html.

[21] L.W. Hanlen and M. Fu, "Wireless Communication Systems with Spatial Diversity: A Volumetric Approach," in International Conference on Communications, 2003.

[22] B. Hassibi and B.M. Hochwald, "Cayley Differential Unitary Space-Time Codes," *IEEE Trans. Inf. Theory*, 48, 1485–1503, 2002.

[23] P. Hoeher and F. Tufvesson, "Channel Estimation with Superimposed Pilot Sequence," *Proceedings of GLOBECOM Conference*, Rio de Janeiro, Brazil, 1999, pp. 2162–2166.

[24] A.T. James, "Distributions of Matrix Variates and Latent Roots Derived from Normal Samples," *Ann. Math. Stat.*, 35, 475–501, 1964.

[25] S.M. Kay, *Fundamentals of Statistical Signal Processing, Estimation/Detection Theory*, Vols. 1 and 2, Prentice Hall, New York, 1993.

[26] Kiran and D.N.C. Tse, "Effective Bandwidths and Effective Interference for Linear Multiuser Receivers in Asynchronous Systems," *IEEE Trans. Inf. Theory*, 46, 1426–1447, 2000.

[27] J.W. Lechleider, "The Optimum Combination of Block Codes and Receivers for Arbitrary Channels," *IEEE Trans. Commun.*, 615–621, 1990.

[28] L. Li, A.M. Tulino, and S. Verdú, "Analysis and Design of MMSE Multiuser Detectors Using Random Matrix Methods," *IEEE Trans. Inf. Theory*, 2002.

[29] L. Li, A.M. Tulino, and S. Verdú, "Asymptotic Eigenvalue Moments for Linear Multiuser Detection," invited paper, in Conference on Information Sciences and Systems, Vol. 1, September 2001, pp. 273–304.

[30] A. Lozano, A. Tulino, and S. Verdù, "Multi-Antenna Capacity in the Low-Power Regime," in *DIMACS Workshop on Signal Processing for Wireless Transmission*, Rutgers, October 2002.

[31] J.H. Manton, I.Y. Mareels, and Y. Hua, "Affine Precoders for Reliable Communications," in *Proceedings of ICASSP*, Vol. 5, Istanbul, Turkey, June 2000, pp. 2741–2744.

[32] T.L. Marzetta, B.M. Hochwald, and V. Tarokh, "Multi-Antenna Channel-Hardening and Its Implications for Rate Feedback and Scheduling," *IEEE Trans. Inf. Theory*, 2002 (preprint available at http://mars.bell-labs.com/cm/ms/what/mars/papers/ratefeedback/).

[33] T.L. Marzetta, "BLAST Training: Estimating Channel Characteristics for High Capacity Space-Time Wireless," in *Proceedings of 37th Allerton Conference*, Monticello, IL, September 1999.

[34] M. Medard, "The Effect upon Channel Capacity in Wireless Communications of Perfect and Imperfect Knowledge of the Channel," *IEEE Trans. Inf. Theory*, 46, 933–945, 2000.

[35] A.L. Moustakas, S.H. Simon, and A.M. Sengupta, "MIMO Capacity through Correlated Channels in the Presence of Correlated Interferers and Noise: A (Not So) Large N Analysis," preprint available at http://mars.bell-labs.com/cm/ms/what/mars/papers/physics/index1.html.

[36] R.R. Müller, P. Scramm, and J.B. Huber, "Spectral Efficency of CDMA Systems with Lnear Interference Suppression," in *Proceedings Workshop Kommunikationstechnik*, Ulm, Germany, January 1997, pp. 93–97.

[37] F.D. Murnaghan, *The Theory of Group Representations*, Johns Hopkins Press, Baltimore, 1938.

[38] R. Negi and J. Cioffi, "Pilot Tone Selection for Channel Estimation in a Mobile OFDM System," *IEEE Trans. Consumer Electron.*, 44, 1122–1128, 1998.

[39] S. Ohno and G.B. Giannakis, "Optimal Training and Redundant Precoders for Block Transmissions with Application to Wireless OFDM," *Proc. ICASSP*, 4, 2389–2392, 2001.

[40] S. Ohno and G.B. Giannakis, "Superimposed Training on Redundant Precoding for Low-Complexity Recovery of Block Transmission," *IEEE ICC*, 5, 1501–1505, 2001.

[41] P.B. Rapajic, M.L. Honig, and G. Woodward, "Multiuser Decision Feedback Detection: Performance Bounds and Adaptive Algorithms," in *Proceedings of ISIT*, Cambridge, MA, August 1998, p. 34.

[42] A.M. Sayeed and V.V. Veeravalli, "The Essential Degrees of Freedom in Space-Time Fading Channels," in *PIMRC'02*, Lisbon, Portugal, September 2002, pp. 1512–1516.

[43] A. Scaglione, G.B. Giannakis, and S. Barbarossa, "Redundant Filterbank Precoders and Equalizers. Part I. Unification and Optimal Designs," *IEEE Trans. Signal Process.*, 47(7), 1988–2006, 1999.

[44] A. Scaglione, P. Stoica, S. Barbarossa, G.B. Giannakis, and H. Sampath, "Optimal Designs for Space-Time Linear Precoders and Decoders," *IEEE Trans. Signal Process.*, 50(5), 1051–1064, 2002.

[45] A. Scaglione, S. Barbarossa, and G.B. Giannakis, "Filterbank Transceivers Optimizing Information Rate in Block Transmissions over Dispersive Channels," *IEEE Trans. Inf. Theory*, 45(3), 1019–1032, 1999.

[46] A. Scaglione and A. Salhotra, "The Statistics of the MIMO Frequency Selective Fading AWGN Channel Capacity," *IEEE Trans. Inf. Theory*, 2002.

[47] A. Scaglione, "Statistical Analysis of the Capacity of MIMO Frequency Selective Rayleigh Fading Channels with Arbitrary Number of Inputs and Outputs," in *Proceedings of IEEE ISIT*, 2000, pp. 278–278.

[48] J.W. Silverstein, "Strong Convergence of the Empirical Distribution of Eigenvalues of Large Dimensional Random Matrices," *J. Multivariate Analy.*, 55(2), 331–339, 1995.

[49] P.J. Smith and M. Shafi, "On a Gaussian Approximation to the Capacity of Wireless MIMO Systems," in *International Conference on Communications*, New York, April–May 2002.

[50] P. Stoica and A. Nehori, "MUSIC, Maximum Likelihood, and Cramer-Rao Bound," *IEEE Trans. ASSP*, 37(5), 720–741, 1989.

[51] P. Stoica and B.N. Ng, "On the Carmer-Rao Bound under Parametric Constriants," *IEEE Signal Process. Lett.*, 5(7), 177–179, 1998.

[52] V. Tarokh, N. Seshadri, and A.R. Calderbank, "Space-Time Codes for High Data Rate Wireless Communication: Performance Criterion and Code Construction," *IEEE Trans. Inf. Theory*, 44(2), 744–765, 1998.

[53] I.E. Telatar, "Capacity of Multi-Antenna Gaussian Channels," *Eur. Trans. Telecommun.*, 10, 585–595, 1999.

[54] D.N.C. Tse and S.V. Hanly, "Linear Multiuser Receivers: Effective Interference, Effective Bandwidth and User Capacity," *IEEE Trans. Inf. Thoery*, 45(2), 641–657, 1999.

[55] S. Verdú and S. Shamai, "Spectral Efficiency of CDMA with Random Spreading," *IEEE Trans. Inf. Theory*, 45, 622–640, 1999.

[56] Z. Wang and G.B. Giannakis, "Outage Mutual Information of Space-Time MIMO Channels," in *Proceedings of 40th Allerton Conference*, October 2002.

[57] J. Yang and S. Roy, "Joint Transmitter and Receiver Optimization for Multiple-Input-Multiple-Output (MIMO) with Decision Feedback," *IEEE Trans. Inf. Theory*, 42, 3221–3231, 1994.

[58] J. Yang and S. Roy, "On Joint Transmitter and Receiver Optimization for Multiple-Input-Multiple-Output (MIMO) Transmission Systems," *IEEE Trans. Commun.*, 42, 3221–3231, 1994.

[59] L. Zheng and D. Tse, "Diversity and Multiplexing: A Fundamental Tradeoff in Multiple Antenna Channels," *IEEE Trans. Inf. Theory*, 49, 2003.

[60] L. Zheng and D. Tse, "Communicating on the Grassmann Manifold: A Geometric Approach to the Non-Coherent Multiple Antenna Channel," *IEEE Trans. Inf. Theory*, 48, 359–383, 2002.

15

Performance Analysis of Multiple Antenna Systems*

15.1 Introduction **15**-2

15.2 MIMO Systems without Co-Channel Interference **15**-3
System Model and Problem Statement • MIMO Channel Capacity without Channel State Information at the Transmitter • Capacity/Outage Probability of MIMO MRC (Beam-Forming) Systems with Perfect Channel State Information at the Transmitter • Water-Filling Capacity and Beam Forming Performance of Correlated MIMO Rayleigh Channels with Covariance Feedback

15.3 MIMO Systems in the Presence of Co-Channel Interference **15**-19
Problem Statement • Capacity CCDF of MIMO Optimum Combining Scheme with Perfect CSI at Transmitter • Statistics of the MIMO Capacity with Co-Channel Interference and without Feedback • Rician/Rayleigh Fading Scenarios in an Interference-Limited Environment

Ming Kang
University of Minnesota

Mohamed-Slim Alouini
University of Minnesota

Abstract

This chapter presents exact results on the performance of wireless communications systems employing multiple antennas at both the transmitter and receiver (known also as multiple-input multiple-output (MIMO) systems), in terms of the channel capacity, capacity complementary cumulative distribution function, outage probability, and average bit error rate. We study effects of various important factors, such as the co-channel interference (CCI), the line-of-sight path (Rician fading), fading correlations, and different levels of availability of channel state information (CSI) at the transmitter (perfect CSI, partial CSI, and no CSI). Numerical examples are also provided and discussed to illustrate the mathematical formalism and to show the impact of these factors.

*This work was supported in part by the National Science Foundation Grant CCR-9983462.

0-8493-1657-X/$0.00+$1.50

15.1 Introduction

Wireless communications systems equipped with multiple antennas at both the transmitter and the receiver have been widely regarded as an important technology to improve performance. On the one hand, these so-called multiple-input multiple-output (MIMO) systems can increase diversity to combat the fading phenomenon inherent in wireless channels. The basic idea is that when a signal is transmitted through a MIMO channel, multiple independently faded replicas of the signal can be obtained and combined at the receiver to reduce the error probability and improve the reliability of the communication. On the other hand, MIMO systems can accommodate a much higher data rate due to the increased degrees of freedom for communications offered by the randomness of fading channels [1]. It was shown in [2] that in a $T \times R$ (T antennas at the transmitter and R antennas at the receiver) independent and identically distributed (i.i.d.) Rayleigh fading channel, the channel capacity increases linearly with $\min(T, R)$. This capacity growing potential is remarkable since each 3-dB increase in signal-to-noise ratio (SNR) results in a roughly $\min(T, R)$ bps/Hz capacity gain when the SNR is high, compared to 1 bps/Hz capacity gain for traditional single-antenna systems. This gain is mainly due to the spatial multiplexing effect; that is, multiple independent data streams can be simultaneously transmitted over parallel subchannels created by the randomly faded path gains. One of the schemes to exploit this benefit is known as the vertical Bell Labs space–time architecture, or V-BLAST [3].

The Shannon capacity of a channel defines its theoretical upper bound for the maximum rate of data transmission at an arbitrarily small bit error probability, without any delay or complexity constraints. Therefore, this capacity represents not only an optimistic upper bound but also a benchmark against which to compare the spectral efficiency of all practical communications schemes. Performance limits of MIMO systems, including the channel capacity, have been studied mainly in an i.i.d. Rayleigh fading environment [1][4]. In particular, for the case when the receiver had perfect channel state information (CSI) but the transmitter had neither instantaneous nor statistical CSI, the capacity of MIMO fading channels was treated as a random variable and the capacity complementary cumulative distribution function (CCDF) was studied by simulations [1]. The exact capacity mean was derived in [4]. An approximate capacity characteristic function was obtained in [5]. In [6], it was observed that the capacity of MIMO channels is approximately Gaussian; thus only the capacity mean and the variance are needed to obtain an accurate approximation to the capacity CCDF. They then derived the capacity variance and Gaussian approximated capacity CCDF for the i.i.d. Rayleigh fading scenario [6].

In a practical wireless communications system, fading conditions are usually more complicated. For example, when there exists a line-of-sight path between the transmitter and the receiver, the channel is often modeled as Rician faded. Furthermore, signals transmitted by or received at different antenna elements are likely to be correlated in practice when, for example, antenna elements are not sufficiently separated [7]. In the first part of this chapter, we study the effects of the Rician fading and fading correlations on the system performance. We assume that the receiver has the perfect CSI and consider three levels of availability of CSI at the transmitter, namely, no CSI, partial CSI through covariance feedback, and perfect CSI.

In the second part of this chapter, we study the effect of co-channel interference (CCI) on the system performance, including the capacity. It is well known that wireless communications systems often adopt the frequency reuse concept to improve the spectral efficiency of the cellular network. This introduces CCI that ultimately limits the system performance. The effect of CCI has been studied for wireless communications systems with multiple antennas only at the receiver (see, for example, [8]). Asymptotic studies have been conducted in the context of code-division-multiple-access (CDMA) systems (see, for example, [9][10]), and similar techniques were used to carry out the asymptotic study on the capacity of MIMO Rayleigh fading channels in the presence of CCI [11]. In this chapter, we will present some exact results on the capacity of MIMO fading channels in the presence of CCI. We assume that the receiver has perfect CSI of both desired user and CCI, while the transmitter either has no CSI or has perfect CSI.

Due to the space limitation, most results are given directly without proof. Interested readers should refer to [12] where, as an example of the calculations involved to obtain these results, the proof of Theorem 7 in Section 15.2.3 is given in detail.

15.2 MIMO Systems without Co-Channel Interference

15.2.1 System Model and Problem Statement

In this section, we consider a single-user Gaussian channel with T antenna elements at the transmitter and R antenna elements at the receiver. The discrete equivalent $R \times 1$ received vector at the receiver can be modeled as

$$y = Hx + n \tag{15.1}$$

where n is the complex additive white Gaussian noise (AWGN) vector with zero mean and covariance matrix $\sigma_n^2 I_R$, where I_R is the $R \times R$ identity matrix. In Equation 15.1, H denotes the $R \times T$ channel gain matrix with entries $\{H\}_{i,j}$ being the complex channel gain from the jth transmitter antenna element to the ith receiver antenna element. x is the transmitted vector with the power constraint $E(x^H x) \leq \Omega$, where $(\cdot)^H$ denotes the conjugate transpose operator and $E(\cdot)$ is the expected value. Conditioned on the channel realization H, the MIMO channel capacity described above is well known to be given by [1]

$$C = \max_{K, \mathrm{tr}(K) \leq \Omega} \log_2 \left(\det \left(I_R + \frac{1}{\sigma_n^2} H K H^H \right) \right) \text{ bps/Hz} \tag{15.2}$$

where $K = E(xx^H)$, $\det(\cdot)$ denotes the determinant operator, and $\mathrm{tr}(\cdot)$ is the trace operator. The optimum choice of K depends on how much CSI is available at the transmitter.

15.2.2 MIMO Channel Capacity without Channel State Information at the Transmitter

If perfect CSI is assumed at the receiver while the transmitter has neither instantaneous nor statistical channel information, the optimum choice of K to maximize the mutual information is to split the total power equally among the transmitter antenna elements [1][4], i.e., $K = \frac{\Omega}{T} I_T$.

15.2.2.1 Rician Fading

When there exists a line-of-sight path between the transmitter and the receiver, the channel fading can be accurately described by the Rician model. When the channel experiences independent Rician type of fading, the columns of H are independent complex Gaussian vectors with covariance matrix $\Sigma = \sigma^2 I_R$ and mean matrix $E(H) = M$. The moment-generating function (MGF) of capacity is given by the following theorem [13].

Theorem 15.1 *Let $E(H) = M$, $\Sigma = \sigma^2 I_R$, $s = \min(T, R)$, $t = \max(T, R)$, and $0 < \lambda_1 < \lambda_2 < \cdots < \lambda_s < \infty$ be s nonzero distinct eigenvalues of $M^H \Sigma^{-1} M$, then the MGF of the capacity C, $\mathcal{M}_C(\tau) = E(e^{\tau C})$, is given by*

$$\mathcal{M}_C(\tau) = E(e^{\tau C}) = \frac{e^{-\mathrm{tr}(\Lambda)}}{(\Gamma(t - s + 1))^s \det(V)} \det(\Psi(\tau)) \tag{15.3}$$

where $\Gamma(\cdot)$ is the gamma function [14, Equation 8.31.1], $\Lambda = \mathrm{diag}(\lambda_1, \ldots, \lambda_s)$, V is an $s \times s$ matrix whose determinant is a Vandermonde determinant given by

$$\det(V) = \det\left(\lambda_i^{s-j}\right) = \det \begin{pmatrix} \lambda_1^{s-1} & \lambda_1^{s-2} & \cdots & 1 \\ \lambda_2^{s-1} & \lambda_2^{s-2} & \cdots & 1 \\ \vdots & \vdots & \ddots & \vdots \\ \lambda_s^{s-1} & \lambda_s^{s-2} & \cdots & 1 \end{pmatrix} = \prod_{1 \leq i < j \leq s} (\lambda_i - \lambda_j) \tag{15.4}$$

and $\Psi(\tau)$ is an $s \times s$ matrix function of τ whose entries are given by

$$\{\Psi(\tau)\}_{i,j} = \int_0^{\infty} y^{t-i}(1+\rho y)^{\tau/\ln(2)} e^{-y} {}_0F_1(t-s+1, y\lambda_j)dy,$$
$$i,\ j = 1,\ldots,s \tag{15.5}$$

where ${}_0F_1(\cdot,\cdot)$ is the generalized hypergeometric function ${}_pF_q(a_1,\ldots,a_p,b_1,\ldots,b_q,z)$ defined in [14, Equation 9.14.1] with the parameters $p=0$ and $q=1$, $\ln(\cdot)$ is the natural logarithm function, and $\rho = \frac{\Omega\sigma^2}{\sigma_n^2 T}$ is the normalized transmitting power per transmitter branch.

With the MGF in hand, the nth moment of C, $E(C^n)$, can now be obtained based on $E(C^n) = \frac{d^n(\mathcal{M}_C(\tau))}{d\tau^n}\Big|_{\tau=0}$, where the derivatives are taken by the product rule of the differentiation.

Theorem 15.2 *The mean capacity $E(C)$ for the non-i.i.d. Rician case is given by*

$$E(C) = \frac{e^{-\mathrm{tr}(\boldsymbol{\Lambda})}}{\ln(2)(\Gamma(t-s+1))^s \det(\boldsymbol{V})} \sum_{k=1}^{s} \det(\boldsymbol{\Theta}(k)) \tag{15.6}$$

where $\boldsymbol{\Theta}(k), k=1,\ldots,s$, are $s \times s$ matrices with entries

$$\{\boldsymbol{\Theta}(k)\}_{i,j} = \begin{cases} \int_0^{\infty} y^{t-i}\ln(1+\rho y)e^{-y}{}_0F_1(t-s+1,y\lambda_j)dy, & j=k \\ \Gamma(t-i+1){}_1F_1(t-i+1,t-s+1,\lambda_j), & j\neq k \end{cases} \tag{15.7}$$

where ${}_1F_1(\cdot,\cdot,\cdot)$ is the confluent hypergeometric function [14, Equation 9.210.1].

With the mean capacity in hand, to find the variance of the capacity $\mathrm{Var}(C) = E(C^2) - (E(C))^2$, we just need to find the second moment of the capacity $E(C^2)$, which we give in what follows.

Theorem 15.3 *The second moment of the capacity for the non-i.i.d. Rician case is given by*

$$E(C^2) = \frac{e^{-\mathrm{tr}(\boldsymbol{\Lambda})}}{\ln^2(2)(\Gamma(t-s+1))^s \det(\boldsymbol{V})} \sum_{k=1}^{s}\sum_{l=1}^{s} \det(\boldsymbol{\Delta}(k,l)) \tag{15.8}$$

where $\boldsymbol{\Delta}(k,l), k,\ l=1,\ldots,s$, are $s \times s$ matrices with entries

- *$k=l$:*

$$\{\boldsymbol{\Delta}(k,l)\}_{i,j} = \begin{cases} \int_0^{\infty} y^{t-i}\ln^2(1+\rho y)e^{-y}{}_0F_1(t-s+1,y\lambda_j)dy, & j=k=l \\ \Gamma(t-i+1){}_1F_1(t-i+1,t-s+1,\lambda_j), & \text{otherwise} \end{cases} \tag{15.9}$$

- *$k \neq l$:*

$$\{\boldsymbol{\Delta}(k,l)\}_{i,j} = \begin{cases} \int_0^{\infty} y^{t-i}\ln(1+\rho y)e^{-y}{}_0F_1(t-s+1,y\lambda_j)dy, & j=k \text{ or } j=l \\ \Gamma(t-i+1){}_1F_1(t-i+1,t-s+1,\lambda_j), & \text{otherwise} \end{cases} \tag{15.10}$$

15.2.2.2 I.i.d. Rician Fading and i.i.d. Rayleigh Fading Channels

Note that in Theorem 1 non-i.i.d. Rician actually means that the noncentrality matrix $\boldsymbol{M}^H\boldsymbol{\Sigma}^{-1}\boldsymbol{M}$ has s nonzero distinct eigenvalues, although in many cases even when $\boldsymbol{H}$ has non-i.i.d. entries, the noncentrality matrix may have some identical nonzero eigenvalues or may have less than s nonzero eigenvalues. However, as we shall see later, all the cases when some of the eigenvalues of the noncentrality matrix are identical or zeroes are simply the limiting cases of Theorem 1. We will explicitly derive the important special case when $\boldsymbol{H}$ has i.i.d. entries, in which case $\boldsymbol{M}^H\boldsymbol{\Sigma}^{-1}\boldsymbol{M}$ has only one nonzero eigenvalue, say λ_1, in order to give an example to illustrate the techniques involved. The result is summarized in the following corollary [13].

Corollary 15.1 When $\boldsymbol{M}$ has η at each entry and $\boldsymbol{\Sigma} = \sigma^2 \boldsymbol{I}_R$ (i.i.d. Rician case), then $\boldsymbol{M}^H \boldsymbol{\Sigma}^{-1} \boldsymbol{M}$ has only one nonzero eigenvalue $\lambda_1 = \frac{st|\eta|^2}{\sigma^2}$, where $|\cdot|$ denotes the modulus of a complex number. In this case, the MGF of the capacity reduces to

$$\mathcal{M}_C(\tau) = E(e^{\tau C}) = \frac{e^{-\lambda_1}}{\Gamma(t-s+1)\lambda_1^{s-1}} \frac{\det(\boldsymbol{\Psi}_{\text{iid}}(\tau))}{\prod_{m=1}^{s-1} \Gamma(t-m)\Gamma(s-m)} \tag{15.11}$$

where $\boldsymbol{\Psi}_{\text{iid}}(\tau)$ is an $s \times s$ matrix whose entries in the first column are the same as those of $\boldsymbol{\Psi}(\tau)$ defined by Equation 15.5, i.e., $\{\boldsymbol{\Psi}_{\text{iid}}(\tau)\}_{i,1} = \{\boldsymbol{\Psi}(\tau)\}_{i,1}, i = 1, \ldots, s$, and the entries from the second column to the sth column are given by

$$\begin{aligned}\{\boldsymbol{\Psi}_{\text{iid}}(\tau)\}_{i,j} &= \int_0^{\infty} (1+\rho y)^{\tau/\ln(2)} y^{t+s-i-j} e^{-y} dy \\ &= \rho^{-(t+s-i-j+1)} \Gamma(t+s-i-j+1)\, \mathrm{U}\left(t+s-i-j+1, t+s-i-j+2+\frac{\tau}{\ln(2)}, \frac{1}{\rho}\right), \\ & i = 1, \ldots, s,\ j = 2, \ldots, s,\end{aligned} \tag{15.12}$$

where $\mathrm{U}(\cdot,\cdot,\cdot)$ is the confluent hypergeometric function of the second kind [14, Equation. 9.210.2].

An important special case of Corollary 15.1 is when $\boldsymbol{M} = \boldsymbol{0}$ (i.e., Rayleigh fading). In such case the MGF can be shown to reduce to the result summarized in the following corollary [13].

Corollary 15.2 When $E(\boldsymbol{H}) = \boldsymbol{0}$ (i.i.d. Rayleigh fading case), the MGF of the capacity reduces to

$$\mathcal{M}_C(\tau) = E(e^{\tau C}) = \frac{\det(\boldsymbol{\Psi}_{\text{c}}(\tau))}{\prod_{m=1}^{s} \Gamma(t-m+1)\Gamma(s-m+1)} \tag{15.13}$$

where $\boldsymbol{\Psi}_{\text{c}}(\tau)$ is an $s \times s$ Hankel matrix function of τ with entries given by

$$\begin{aligned}\{\boldsymbol{\Psi}_{\text{c}}(\tau)\}_{i,j} &= \rho^{-(t+s-i-j+1)} \Gamma(t+s-i-j+1) \\ &\quad \times \mathrm{U}\left(t+s-i-j+1, t+s-i-j+2+\frac{\tau}{\ln(2)}, \frac{1}{\rho}\right), \\ &\quad i, j = 1, \ldots, s\end{aligned} \tag{15.14}$$

Note that the capacity MGF for i.i.d. Rayleigh channels was also independently and simultaneously obtained in [15] and [16] using a different approach from [13].

Figure 15.1 plots the mean capacity of MIMO channels vs. the scaled transmitting signal-to-noise ratio (SNR) $\frac{\Omega\sigma^2}{\sigma_n^2}$ in decibel with $T = R = 3$ and $\sigma^2 = \sigma_n^2 = 1$. It indicates that more antenna elements will increase the channel capacity. Figure 15.2 plots the mean capacity of MIMO channels vs. the transmitting SNR $\frac{\Omega}{\sigma_n^2}$ in decibel with $T = R = 3$ when $\sigma_n^2 = 1$. We fix $\eta^2 + \sigma^2 = 1$ to investigate the effect of Rician fading. In this formulation, existence of a line-of-sight path, or the Rician fading, can be viewed as a fading condition in between the Rayleigh fading ($\eta = 0$ and $\sigma^2 = 1$) and the deterministic channel ($\eta = 1$ and $\sigma^2 = 0$). It can be seen that under this formulation, a stronger line-of-sight path decreases the channel capacity; i.e., MIMO systems can benefit from a richly scattering environment.

15.2.2.3 Correlated Rayleigh Fading Channels

If we assume that there exists correlated Rayleigh type of fading, then the channel matrix $\boldsymbol{H}$ is a complex Gaussian random matrix with zero mean and covariance matrix $\boldsymbol{\Sigma} \otimes \boldsymbol{B}$, where $\otimes$ stands for the Kronecker product [17] and $\boldsymbol{B}$: $T \times T$, $\boldsymbol{\Sigma}$: $R \times R$ are Hermitian positive-definite matrices. $\boldsymbol{B}$ can be viewed as the

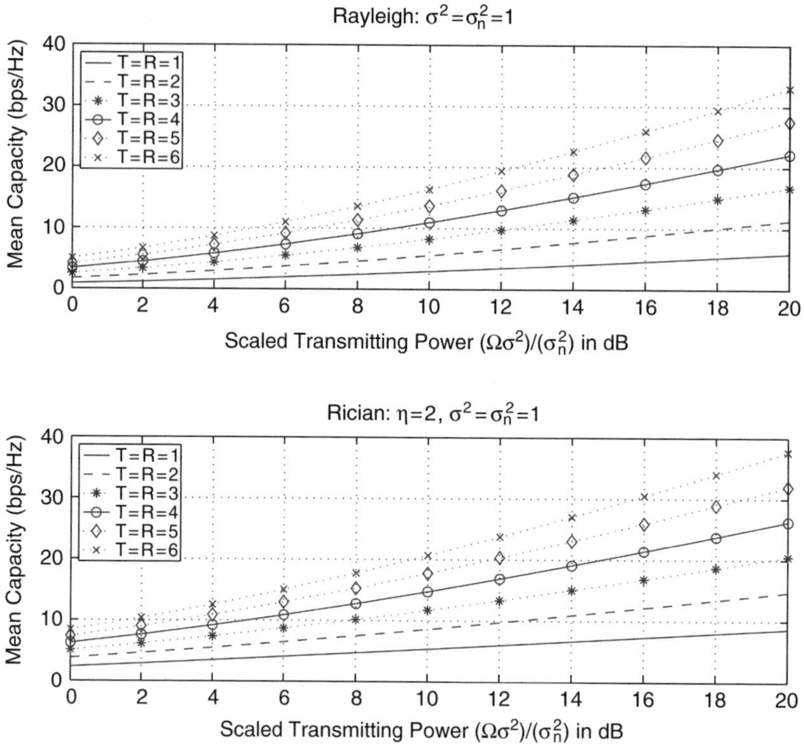

FIGURE 15.1 Mean capacity of MIMO channels vs. the scaled transmitting power $\frac{\Omega\sigma^2}{\sigma_n^2}$ in decibel with different numbers of transmitter/receiver antennas $T = R$ for (1) i.i.d. Rayleigh fading with $\sigma^2 = \sigma_n^2 = 1$ and (2) i.i.d. Rician fading with $\eta = 2$ and $\sigma^2 = \sigma_n^2 = 1$.

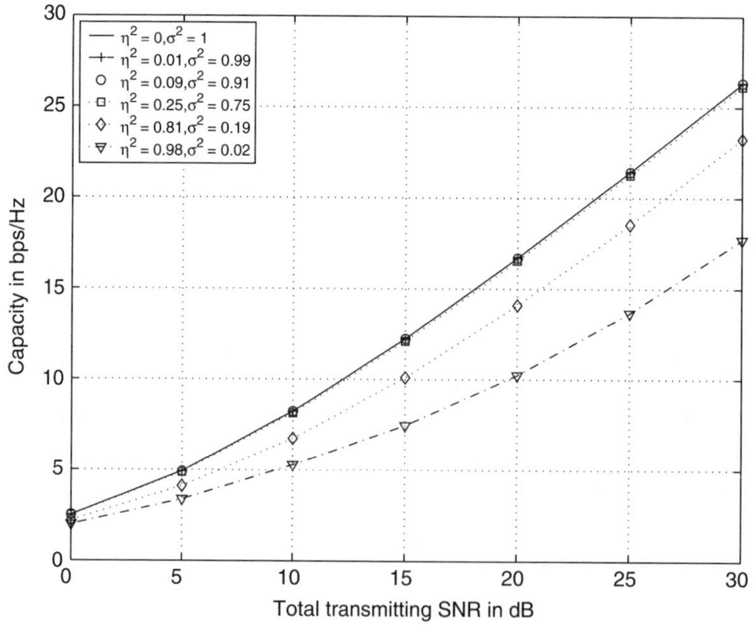

FIGURE 15.2 Mean capacity of MIMO channels vs. the transmitting SNR $\frac{\Omega}{\sigma_n^2}$ in decibel with $T = R = 3$ when $\sigma_n^2 = 1$ and $\eta^2 + \sigma^2 = 1$.

covariance matrix at the transmitter and $\boldsymbol{\Sigma}$ can be viewed as the covariance matrix at the receiver. The probability density function (PDF) of $\boldsymbol{H}$ is well known to be given by [18, Equation 56]

$$\pi^{-TR} \det(\boldsymbol{\Sigma})^{-T} \det(\boldsymbol{B})^{-R} \exp(-\mathrm{tr}(\boldsymbol{\Sigma}^{-1}\boldsymbol{H}\boldsymbol{B}^{-1}\boldsymbol{H}^{H})) \tag{15.15}$$

where $(\cdot)^{-1}$ denotes the matrix inverse. In order to get easy-to-compute expressions and gain further insight, we assume that antenna correlations exist only at either the transmitter or receiver but not both. This assumption has its justification considering the fact that in a practical mobile system, the antenna elements at the base station can be implemented far apart from each other to make the spatial correlations among them small enough to be considered as uncorrelated, while the antenna elements at the mobile unit are usually closer to each other due to the space limitation; thus, correlations among them are often nonnegligible. Therefore, this assumption corresponds to the uplink or downlink communication between the base station and mobile unit. We further assume for the moment that the transmitter has no knowledge about the channel while perfect CSI is assumed at the receiver; thus, $\boldsymbol{K} = \frac{\Omega}{T}\boldsymbol{I}_T$. The correlations among antenna elements exist at either the transmitter or receiver but not both; i.e., either $\boldsymbol{B}$ or $\boldsymbol{\Sigma}$ is an identity matrix while the other one is any Hermitian positive-definite matrix. We denote the covariance matrix at the correlated side as $\boldsymbol{\Phi}$. We give here the general case when the eigenvalues of the covariance matrix at the correlated side are distinct. (We call this case the general case since those cases when some of the eigenvalues are equal can be obtained by taking the limit in the distinct eigenvalues case.) In [19], we also explicitly derived the results for certain special correlation models of interest.

- MGF of capacity [19]

Theorem 15.4 *Let $s = \min(T, R)$, $t = \max(T, R)$, and $\boldsymbol{\Phi}$ the Hermitian positive-definite covariance matrix for the correlated side, while the covariance matrix of the other side is assumed to be the identity matrix. Then $\boldsymbol{\Phi}$ is either a $t \times t$ matrix or an $s \times s$ matrix depending on whether the correlated side has more or less antenna elements, and these two cases will result in slightly different results.*

1. *If $\boldsymbol{\Phi}$ is an $s \times s$ Hermitian positive-definite matrix with distinct eigenvalues $0 < \phi_1 < \cdots < \phi_s$, then the MGF of the channel capacity defined in Equation 15.2, $\mathcal{M}_C(\tau) = E(e^{\tau C})$, is given by*

$$\mathcal{M}_C(\tau) = \frac{\det(\boldsymbol{\Psi}_1)}{\det(\boldsymbol{V}_1)\prod_{i=1}^{s}\Gamma(t-i+1)} \tag{15.16}$$

where $P = \frac{\Omega}{\sigma_n^2 T}$ is the transmitting signal-to-noise ratio per transmitter antenna branch, $\boldsymbol{\Psi}_1$ is an $s \times s$ matrix with entries given by

$$\{\boldsymbol{\Psi}_1\}_{i,j} = \int_0^\infty (1+Py)^{\tau/\ln(2)} y^{t-i} e^{-\frac{y}{\phi_j}}\, dy, \quad i, j = 1, \ldots, s \tag{15.17}$$

which can be evaluated in terms of hypergeometric functions using Equation 15.127, and $\boldsymbol{V}_1$ is an $s \times s$ matrix with the determinant to be given by

$$\det(\boldsymbol{V}_1) = \det\left((-1)^{s-j}\phi_i^{t-s+j}\right) = \left(\prod_{i=1}^{s}\phi_i^t\right)\prod_{1\le l<k\le s}\left(\frac{1}{\phi_k} - \frac{1}{\phi_l}\right) \tag{15.18}$$

2. *If $\boldsymbol{\Phi}$ is a $t \times t$ Hermitian positive-definite matrix with distinct eigenvalues $0 < \phi_1 < \cdots < \phi_t$, then the MGF of the channel capacity defined in Equation 15.2, $\mathcal{M}_C(\tau) = E(e^{\tau C})$, is given by*

$$\mathcal{M}_C(\tau) = \frac{(-1)^{s(t-s)}\det(\boldsymbol{\Psi}_2)}{\det(\boldsymbol{V}_2)\prod_{i=1}^{s}\Gamma(s-i+1)} \tag{15.19}$$

where $\boldsymbol{\Psi}_2$ *is a* $t \times t$ *matrix given by*

$$\boldsymbol{\Psi}_2 = \begin{pmatrix} \boldsymbol{\Psi}_{2A} \\ \boldsymbol{\Psi}_{2B} \end{pmatrix} \tag{15.20}$$

where $\boldsymbol{\Psi}_{2A}$ *is a* $(t-s) \times t$ *matrix with entries given by*

$$\{\boldsymbol{\Psi}_{2A}\}_{i,j} = \left(\frac{-1}{\phi_j}\right)^{t-s-i}, \quad i = 1, \ldots, t-s,\ j = 1, \ldots, t \tag{15.21}$$

$\boldsymbol{\Psi}_{2B}$ *is an* $s \times t$ *matrix with entries given by*

$$\{\boldsymbol{\Psi}_{2B}\}_{i,j} = \int_0^{\infty} (1+Py)^{\tau/\ln(2)} y^{s-i} e^{-\frac{y}{\phi_j}}\, dy, \quad i = 1, \ldots, s,\ j = 1, \ldots, t \tag{15.22}$$

which can be evaluated in terms of hypergeometric functions with the help of Equation 15.127, and $\boldsymbol{V}_2$ *is a* $t \times t$ *matrix defined as*

$$\det(\boldsymbol{V}_2) = \det\left((-1)^{t-j}\phi_i^{s-t+j}\right) = \left(\prod_{i=1}^{t} \phi_i^s\right) \prod_{1 \le l < k \le t} \left(\frac{1}{\phi_k} - \frac{1}{\phi_l}\right), \quad i, j = 1, \ldots, t \tag{15.23}$$

Note that the case when $t = s$ *is included in case 1.*

- Mean capacity

Theorem 15.5 *The mean capacity with the MGF obtained in Theorem 15.4 is given by:*

1. *If* $\boldsymbol{\Phi}$ *is an* $s \times s$ *Hermitian positive-definite matrix with distinct eigenvalues* $0 < \phi_1 < \cdots < \phi_s$, *then the mean capacity* $E(C)$ *is given by*

$$E(C) = \frac{\sum_{k=1}^{s} \det(\boldsymbol{\Psi}_1(k))}{\ln(2)\det(\boldsymbol{V}_1)\prod_{i=1}^{s}\Gamma(t-i+1)} \tag{15.24}$$

where $\boldsymbol{V}_1$ *is defined in Theorem 15.4 and* $\boldsymbol{\Psi}_1(k), k = 1, \ldots, s$, *are* $s \times s$ *matrices with entries given by*

$$\{\boldsymbol{\Psi}_1(k)\}_{i,j} = \begin{cases} \int_0^{\infty} \ln(1+Py) y^{t-i} e^{-\frac{y}{\phi_j}}\, dy, & \text{if } i = k \\ \phi_j^{t-i+1}\Gamma(t-i+1), & \text{if } i \neq k \end{cases} \tag{15.25}$$

where the integral in Equation 15.25 can be evaluated in terms of the complementary incomplete gamma function with the help of Equation 15.128.

2. *If* $\boldsymbol{\Phi}$ *is a* $t \times t$ *Hermitian positive-definite matrix with distinct eigenvalues* $0 < \phi_1 < \cdots < \phi_t$, *then the mean capacity* $E(C)$ *is given by*

$$E(C) = \frac{(-1)^{s(t-s)}\sum_{k=1}^{s} \det(\boldsymbol{\Psi}_2(k))}{\ln(2)\det(\boldsymbol{V}_2)\prod_{i=1}^{s}\Gamma(s-i+1)} \tag{15.26}$$

where $\boldsymbol{\Psi}_2(k), k = 1, \ldots, s$, *are* $t \times t$ *matrices with entries given by*

$$\boldsymbol{\Psi}_2(k) = \begin{pmatrix} \boldsymbol{\Psi}_{2A} \\ \boldsymbol{\Psi}_{2B}(k) \end{pmatrix} \tag{15.27}$$

where $\mathbf{\Psi}_{2A}$ *and* $\mathbf{V}_2$ *are defined in Theorem 15.4 and* $\mathbf{\Psi}_{2B}(k), k = 1, \ldots, s$*, are* $s \times t$ *matrices with entries given by*

$$\{\mathbf{\Psi}_{2B}(k)\}_{i,j} = \begin{cases} \int_0^\infty \ln(1 + Py) y^{s-i} e^{-\frac{y}{\phi_j}} dy, & \text{if } i = k \\ \phi_j^{s-i+1}\Gamma(s - i + 1), & \text{if } i \neq k \end{cases} \tag{15.28}$$

where the integral in Equation 15.28 can be evaluated in terms of the complementary incomplete gamma function with the help of Equation 15.128.

Note that the case when $t = s$ *is included in case 1.*

- Capacity variance. With the mean capacity in hand, to find the variance of the capacity $\text{Var}(C) = E(C^2) - (E(C))^2$, we just need to find the second moments of the capacity $E(C^2)$, which we give as the following theorem.

Theorem 15.6

1. *If* $\mathbf{\Phi}$ *is an* $s \times s$ *Hermitian positive-definite matrix with distinct eigenvalues* $0 < \phi_1 < \cdots < \phi_s$*, then the second moment of the capacity* $E(C^2)$ *is given by*

$$E(C^2) = \frac{\sum_{k=1}^{s} \sum_{l=1}^{s} \det(\mathbf{\Psi}_1(k,l))}{\ln^2(2) \det(\mathbf{V}_1) \prod_{i=1}^{s} \Gamma(t - i + 1)} \tag{15.29}$$

where $\mathbf{\Psi}_1(k,l), k, l = 1, \ldots, s$*, are* $s \times s$ *matrices with entries given by*

$$\{\mathbf{\Psi}_1(k,l)\}_{i,j} = \begin{cases} \int_0^\infty \ln^2(1 + Py) y^{t-i} e^{-\frac{y}{\phi_j}} dy, & i = k = l \\ \int_0^\infty \ln(1 + Py) y^{t-i} e^{-\frac{y}{\phi_j}} dy, & i = k \text{ or } i = l, k \neq l \\ \phi_j^{t-i+1}\Gamma(t - i + 1), & i \neq k, i \neq l \end{cases} \tag{15.30}$$

where the integrals in Equation 15.30 can be evaluated in terms of the Meijer's G function or the complementary incomplete gamma function with the help of Equation 15.128 or Equation 15.129.

2. *If* $\mathbf{\Phi}$ *is a* $t \times t$ *Hermitian positive-definite matrix with distinct eigenvalues* $0 < \phi_1 < \cdots < \phi_t$*, then the second moment of the capacity* $E(C^2)$ *is given by*

$$E(C^2) = \frac{(-1)^{s(t-s)} \sum_{k=1}^{s} \sum_{l=1}^{s} \det(\mathbf{\Psi}_2(k,l))}{\ln^2(2) \det(\mathbf{V}_2) \prod_{i=1}^{s} \Gamma(s - i + 1)} \tag{15.31}$$

where $\mathbf{\Psi}_2(k,l), k, l = 1, \ldots, s$*, are* $t \times t$ *matrices with entries given by*

$$\mathbf{\Psi}_2(k,l) = \begin{pmatrix} \mathbf{\Psi}_{2A} \\ \mathbf{\Psi}_{2B}(k,l) \end{pmatrix} \tag{15.32}$$

where $\mathbf{\Psi}_{2A}$ *is defined in Theorem 15.4 and* $\mathbf{\Psi}_{2B}(k,l), k, l = 1, \ldots, s$*, are* $s \times t$ *matrices with entries given by*

$$\{\mathbf{\Psi}_{2B}(k,l)\}_{i,j} = \begin{cases} \int_0^\infty \ln^2(1 + Py) y^{s-i} e^{-\frac{y}{\phi_j}} dy, & i = k = l \\ \int_0^\infty \ln(1 + Py) y^{s-i} e^{-\frac{y}{\phi_j}} dy, & i = k \text{ or } i = l, k \neq l \\ \phi_j^{s-i+1}\Gamma(s - i + 1), & i \neq k, i \neq l \end{cases} \tag{15.33}$$

where the integrals in Equation 15.33 can be evaluated in terms of the Meijer's G function or the complementary incomplete gamma function with the help of Equation 15.128 or Equation 15.129.

Note that the case when $t = s$ *is included in case 1.*

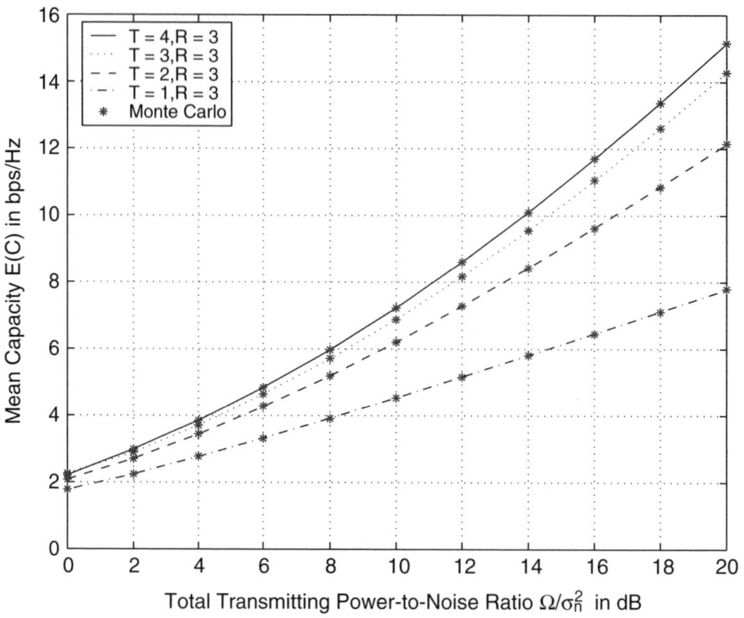

FIGURE 15.3 Mean capacity $E(C)$ in bits per second/herzt vs. the total transmitting SNR $\frac{\Omega}{\sigma_n^2}$ in decibel with the number of transmit antenna elements T as a parameter when the number of receive antenna elements $R = 3$ and the receiver correlation matrix $\mathbf{\Phi}$ has eigenvalues 0.1, 0.5, and 2.4, compared with Monte Carlo simulations.

Figure 15.3 shows the mean capacity $E(C)$ in bits per second/herzt vs. the total transmitting SNR $\frac{\Omega}{\sigma_n^2}$ in decibel with the number of transmit antenna elements T as a parameter when the number of receive antenna elements $R = 3$ and the receiver correlation matrix $\mathbf{\Phi}$ has eigenvalues 0.1, 0.5, and 2.4 (arbitrarily picked with the constraint that they add up to 3), compared with Monte Carlo simulations. Figure 15.4 shows the mean capacity $E(C)$ in bits per second/herzt vs. the total transmitting power–noise ratio $\frac{\Omega}{\sigma_n^2}$ in decibel with the intraclass[1] correlation ξ as a parameter when the number of transmit/receive antenna elements $T = R = 3$. The capacity results for the intraclass model a explicitly given in [19].

15.2.2.4 Capacity CCDF

In [6], Smith and Shafi showed that the MIMO channel capacity can be accurately approximated by a Gaussian distributed random variable. Therefore, with the mean and variance in hand, the capacity CCDF C_{ccdf}, defined as the probability that the capacity exceeds an acceptable threshold C_{th}, can be obtained for all above cases using

$$C_{\text{ccdf}} = \Pr(C \geq C_{\text{th}}) = \frac{1}{2}\text{erfc}\left(\frac{C_{\text{th}} - E(C)}{\sqrt{2\,\text{Var}(C)}}\right) \tag{15.34}$$

where erfc$(\cdot)$ is the complementary error function defined by

$$\text{erfc}(x) = \frac{2}{\sqrt{\pi}}\int_x^{\infty} \exp(-t^2)dt \tag{15.35}$$

This function is a standard built-in function in many popular scientific computation softwares such as Matlab and Mathematica. With the help of this Gaussian approximation method, the capacity CCDF of

[1]The intraclass correlation model is defined as $E(x_i x_j^H) = 1$ if $i = j$, and $0 < E(x_i x_j^H) = \xi < 1$ if $i \neq j$.

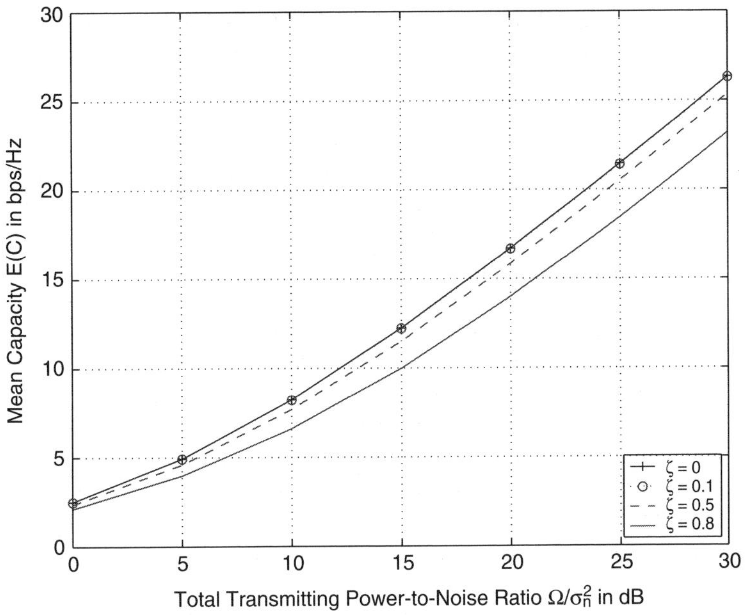

FIGURE 15.4 Mean capacity $E(C)$ in bits per second/herzt vs. the total transmitting power–noise ratio $\frac{\Omega}{\sigma_n^2}$ in decibel with the intraclass correlation ξ as a parameter when the number of transmit/receive antenna elements $T = R = 3$.

MIMO channels can be obtained for (1) non-i.i.d. Rician, (2) i.i.d. Rician, and (3) correlated Rayleigh scenarios based on the means and variances obtained in this section. Figure 15.5 plots the Gaussian approximated capacity CCDF vs. the capacity threshold C_{th} with different configurations of transmitter/receiver antennas $T = R$ for i.i.d. Rician fading when $\eta = 2$, $\sigma^2 = \sigma_n^2 = 1$, and $\Omega = 12\text{dB}$. It can be seen that the Gaussian approximation works quite well even for a small number of antenna elements.

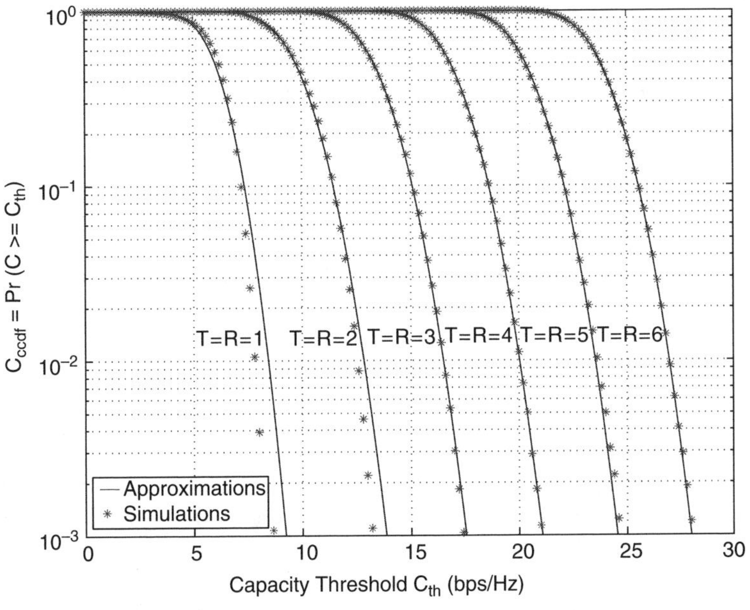

FIGURE 15.5 C_{ccdf} of of MIMO channels vs. the capacity threshold C_{th} with different configurations of transmitter/receiver antennas $T = R$ for i.i.d. Rician fading when $\eta = 2$, $\sigma^2 = \sigma_n^2 = 1$, and $\Omega = 12$ dB.

15.2.3 Capacity/Outage Probability of MIMO MRC (Beam-Forming) Systems with Perfect Channel State Information at the Transmitter

15.2.3.1 System Models and Problem Statement

If both the transmitter and receiver have perfect CSI, the well-known optimum transmitting strategy to achieve MIMO channel capacity is to transmit independent Gaussian inputs along the eigenvectors of $\boldsymbol{H}^H\boldsymbol{H}$ and allocate power to each data stream using the water-filling algorithm with respect to the eigenvalues of $\boldsymbol{H}^H\boldsymbol{H}$ [1][20]. In this way the stronger modes will get more power while the weaker modes will be allocated less or no power. However, similar to the case where the transmitter has no CSI, the capacity is achieved with vector codes, and this increases significantly the decoding complexity at the receiver [21]. For this reason, instead of maximizing the capacity, one can maximize the output signal-to-noise ratio using the so-called MIMO MRC scheme [12][22][23][24][25]. The transmitting strategy of such systems is to allocate all power along the eigenvector corresponding to the largest eigenvalue of $\boldsymbol{H}^H\boldsymbol{H}$, and thus this MIMO MRC scheme is also called optimum beam forming. This MIMO MRC or beam-forming scheme is interesting since it (1) provides the maximum output SNR at the receiver, (2) it only involves scalar codes, which greatly reduces the complexity of both preprocessing and the detection, and (3) in some cases, the capacity achieved by MRC or beam forming is very close to the water-filling capacity discussed in [21]. The performance of such systems in terms of the system outage probability (the CDF of the output SNR) was studied by Dighe et al. in an i.i.d. Rayleigh fading environment [24] and later was extended to the independent but not necessarily identically distributed Rician fading scenario by the authors in [12][25]. The CDF of the output SNR for the MIMO MRC systems over correlated Rayleigh channels was obtained in [19]. First, we briefly describe the system model of the MIMO MRC or beam-forming scheme. The received $R \times 1$ vector is given by

$$\boldsymbol{y} = \boldsymbol{H}\boldsymbol{w}_t s_D + \boldsymbol{n} \tag{15.36}$$

where s_D is the transmitted signal and $\boldsymbol{n}$ is the complex Gaussian noise vector with zero mean and covariance matrix $\sigma_n^2 \boldsymbol{I}_R$. Without loss of generality, we assume that s_D has unit average power. In Equation 15.36, $\boldsymbol{w}_t$ represents the weight vector at the transmitter with $\|\boldsymbol{w}_t\|^2 = \Omega_D$ (i.e, the power of the vector $\boldsymbol{w}_t$ is restricted to be Ω_D, and again, $\boldsymbol{H}$ is the $R \times T$ channel gain matrix for the desired user, where $\{\boldsymbol{H}\}_{i,j}$ is the complex channel gain from the jth transmitter antenna element to the ith receiver antenna element. The solution to maximize the output SNR is the so-called MRC scheme [22][23]. This scheme jointly chooses transmit combining weight vector $\boldsymbol{w}_t$ and the receive combining weight vector $\boldsymbol{w}_r$ so that the combiner output SNR is maximized . The jointly optimum weights are given by [23]

$$\begin{aligned} \boldsymbol{w}_t &= \sqrt{\Omega}\boldsymbol{U}_{\max} \\ \boldsymbol{w}_r &= \boldsymbol{H}^H\boldsymbol{U}_{\max} \end{aligned} \tag{15.37}$$

where $\boldsymbol{U}_{\max}$ is the normalized eigenvector of $\boldsymbol{H}^H\boldsymbol{H}$ corresponding to its largest eigenvalue $\lambda_{\max}$. The resulting maximum SNR μ at the MRC combiner output is given by

$$\mu = \frac{\Omega}{\sigma_n^2}\lambda_{\max} \tag{15.38}$$

The outage probability of MIMO MRC systems is an important statistical measure to assess the quality of service provided by the system. It is defined as the probability of failing to achieve a specified SNR value μ_{th} sufficient for satisfactory reception. Therefore, the outage probability is simply the CDF of the output SNR evaluated at μ_{th}. Specifically, the outage probability is defined by

$$\text{P}_{\text{out}} = \Pr(\mu < \mu_{\text{th}}) = \Pr\left(\lambda_{\max} < \frac{\sigma_n^2}{\Omega}\mu_{\text{th}}\right) \tag{15.39}$$

The channel capacity achieved by Gaussian input s_D with the MRC or beam-forming scheme is given by

$$C = \log_2(1+\mu) \quad \text{bps/Hz} \tag{15.40}$$

where μ is the combiner output SNR given by Equation 15.38. Thus the capacity CCDF C_{ccdf} is given by

$$C_{\text{ccdf}} = \Pr(C \geq C_{\text{th}}) = 1 - \Pr(\mu < 2^{C_{\text{th}}} - 1) \tag{15.41}$$

i.e., the C_{ccdf} is simply the CCDF of the output SNR evaluated at $2^{C_{\text{th}}} - 1$.

15.2.3.2 MIMO MRC Systems Outage Probability

- Non-i.i.d. Rician case

Theorem 15.7 *[12][25]. If columns of $\boldsymbol{H}$ are independent R-variate complex Gaussian vectors with covariance matrix $\boldsymbol{\Sigma} = \sigma^2 \boldsymbol{I}_R$ and $E(\boldsymbol{H}) = \boldsymbol{M}$ (independent but not necessarily identically distributed, or non-i.i.d., Rician), let $s = \min(T, R)$ and $t = \max(T, R)$. If the noncentrality matrix $\boldsymbol{M}^H \boldsymbol{\Sigma}^{-1} \boldsymbol{M}$ has s distinct nonzero eigenvalues $0 < \lambda_1 < \lambda_2 < \cdots < \lambda_s$, then the CDF of the output SNR μ defined by Equation 15.38 is given by*

$$\Pr(\mu \leq \mu_{\text{th}}) = \frac{e^{-\text{tr}(\boldsymbol{\Lambda})}}{\det(\boldsymbol{V})(\Gamma(t-s+1))^s} \det(\boldsymbol{\Psi}(x)) \tag{15.42}$$

where $\boldsymbol{\Lambda} = \text{diag}(\lambda_1, \ldots, \lambda_s)$, $x = \frac{\sigma_n^2}{\Omega \sigma^2} \mu_{\text{th}}$,

$$\det(\boldsymbol{V}) = \det\left(\lambda_i^{s-j}\right) = \prod_{i<j}^{s} (\lambda_i - \lambda_j) \tag{15.43}$$

and $\boldsymbol{\Psi}(x)$ is an $s \times s$ matrix function of $x \in (0, \infty)$ whose entries are given by

$$\begin{aligned}
\{\boldsymbol{\Psi}(x)\}_{i,j} &= \int_0^x y^{t-i} e^{-y} {}_0F_1(t-s+1; y\lambda_j) dy \\
&= \Gamma(t-i+1) {}_1F_1(t-i+1; t-s+1; \lambda_j) \\
&\quad - e^{\lambda_j} \Gamma(t-s+1) \sum_{l=1}^{s-i+1} \frac{(s-i)!}{(l-1)!} \binom{t-i}{t-s+l-1} \\
&\quad \times \lambda_j^{l-1} Q_{t-s+l}\left(\sqrt{2\lambda_j}, \sqrt{2x}\right) - e^{-x} \Gamma(t-s+1) \\
&\quad \times \sum_{l=1}^{s-i} \sum_{k=0}^{s-i-l} 2^{-l-k} \frac{(s-i-1-k)!}{(l-1)!} \binom{t-i}{t-s+l+k} \\
&\quad \times \left(\sqrt{2\lambda_j}\right)^{s+l-t-1} \left(\sqrt{2x}\right)^{t-s+2k+l+1} I_{t-s+l-1}\left(2\sqrt{\lambda_j x}\right), \\
&\quad i, j = 1, \ldots, s
\end{aligned} \tag{15.44}$$

where $Q_p(\cdot, \cdot)$ is the pth-order generalized Marcum Q function [26] and $I_p(\cdot)$ denotes the pth-order modified Bessel function of the first kind [27, Equation 9.6.20].

- I.i.d. Rician and i.i.d. Rayleigh fading scenarios. The i.i.d. Rician fading and i.i.d. Rayleigh fading scenarios can be treated as special cases of the above non-i.i.d. Rician scenario. The results were explicitly given in [12][25] but are omitted here due to space limitations.
- One-sided correlated Rayleigh fading

Theorem 15.8 *[19]. Let $\boldsymbol{H}$ be the same as in Theorem 15.4, $s = \min(T, R)$, $t = \max(T, R)$, $x = \frac{\sigma_n^2}{\Omega}\mu_{\text{th}}$, and $\boldsymbol{\Phi}$ be the Hermitian positive-definite covariance matrix for the correlated side while the covariance of the other side is assumed to be the identity matrix. Then:*

a. *If $\boldsymbol{\Phi}$ is an $s \times s$ Hermitian positive-definite matrix with distinct eigenvalues $0 < \phi_1 < \cdots < \phi_s$, then the CDF of the output SNR μ defined in Equation 15.38 is given by*

$$\Pr(\mu < \mu_{\text{th}}) = \frac{\det(\boldsymbol{\Delta}_1(x))}{\det(\boldsymbol{V}_1)\prod_{i=1}^{s}\Gamma(t-i+1)} \tag{15.45}$$

where $\boldsymbol{V}_1$ is defined in Theorem 15.4 and $\boldsymbol{\Delta}_1(x)$ is an $s \times s$ matrix with entries given by

$$\{\boldsymbol{\Delta}_1(x)\}_{i,j} = \phi_j^{t-i+1}\gamma(t-i+1, x/\phi_j),\quad i,j = 1,\ldots,s \tag{15.46}$$

where $\gamma(\cdot,\cdot)$ is the incomplete gamma function defined by [14, Equation 8.350.1].

b. *If $\boldsymbol{\Phi}$ is a $t \times t$ Hermitian positive-definite matrix with distinct eigenvalues $0 < \phi_1 < \cdots < \phi_t$, then the CDF of the output SNR μ defined in Equation 15.38 is given by*

$$\Pr(\mu < \mu_{\text{th}}) = \frac{(-1)^{s(t-s)}\det(\boldsymbol{\Delta}_2(x))}{\det(\boldsymbol{V}_2)\prod_{i=1}^{s}\Gamma(s-i+1)} \tag{15.47}$$

where $\boldsymbol{\Delta}_2(x)$ is a $t \times t$ matrix given by

$$\boldsymbol{\Delta}_2(x) = \begin{pmatrix} \boldsymbol{\Psi}_{2A} \\ \boldsymbol{\Delta}_{2B}(x) \end{pmatrix} \tag{15.48}$$

where $\boldsymbol{V}_2$ and $\boldsymbol{\Psi}_{2A}$ are defined in Theorem 15.4 and $\boldsymbol{\Delta}_{2B}(x)$ is an $s \times t$ matrix with entries given by

$$\{\boldsymbol{\Delta}_{2B}(x)\}_{i,j} = \phi_j^{s-i+1}\gamma(s-i+1, x/\phi_j),\quad i = 1,\ldots,s;\ j = 1,\ldots,t \tag{15.49}$$

Note that the case when $t = s$ is included in case 1.

15.2.3.3 Capacity CCDF of MIMO MRC Systems

With the outage probability of MIMO MRC systems in hand, the capacity CCDF of such systems can be easily deduced from Equation 15.41 for (1) non-i.i.d. Rician, (2) i.i.d. Rician, and (3) correlated Rayleigh with an arbitrary one-sided covariance matrix. Figure 15.6 plots the capacity CCDF of a MIMO MRC system when $\frac{\Omega\sigma^2}{\sigma_n^2} = 3$ dB, $\{\boldsymbol{H}\}_{i,j}$ are i.i.d. complex Gaussian random variables with mean $1 + \sqrt{-1}$ and $\sigma^2 = 2$, and $\sigma_n^2 = 2$.

15.2.4 Water-Filling Capacity and Beam Forming Performance of Correlated MIMO Rayleigh Channels with Covariance Feedback

15.2.4.1 Water-Filling Capacity

Finding the optimum transmitting strategy is equivalent to the optimization problem given by

$$C = \max_{\boldsymbol{K}:\text{tr}(\boldsymbol{k})\le\Omega} \log_2\det\left(\boldsymbol{I}_R + \frac{1}{\sigma_n^2}\boldsymbol{HKH}^H\right) \tag{15.50}$$

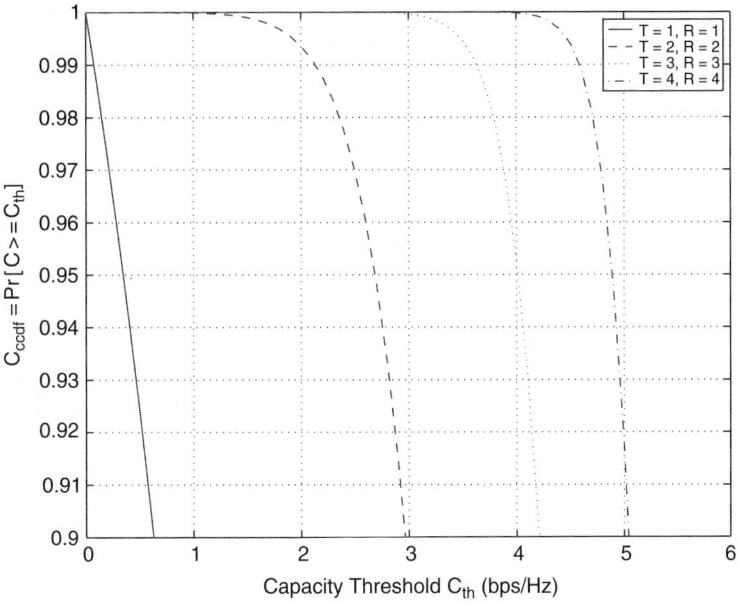

FIGURE 15.6 Capacity CCDF of MIMO MRC system vs. capacity when $\frac{\Omega\sigma^2}{\sigma_n^2} = 3$ dB, $\{\boldsymbol{H}\}_{i,j}$ are i.i.d. complex Gaussian random variables with mean $1 + \sqrt{-1}$ and $\sigma^2 = 2$, and $\sigma_n^2 = 2$.

We assume that the transmit antennas are correlated with covariance $\boldsymbol{B}$ and the receive antennas are uncorrelated. If we further assume that $\boldsymbol{B}$ is known at the transmitter, the optimum choice of $\boldsymbol{K}$ is $\boldsymbol{K} = \boldsymbol{U}_B \boldsymbol{\Lambda}_K \boldsymbol{U}_B{}^H$ [21], where $\boldsymbol{U}_B$ is the $T \times T$ unitary matrix composed of eigenvectors of $\boldsymbol{B}$, i.e., $\boldsymbol{\Lambda}_B = \boldsymbol{U}_B{}^H \boldsymbol{B} \boldsymbol{U}_B$, where $\boldsymbol{\Lambda}_B = \text{diag}(\lambda_{B,1}, \lambda_{B,2}, \ldots, \lambda_{B,T})$ is the diagonal matrix composed of eigenvalues of $\boldsymbol{B}$, $\lambda_{B,1} \geq \lambda_{B,2} \geq \cdots \geq \lambda_{B,T} \geq 0$, and $\boldsymbol{\Lambda}_K = \text{diag}(p_1, p_2, \ldots, p_T)$ is the diagonal power allocation matrix given by the water-filling-like numerical searching algorithm.[2] Note that $\boldsymbol{\Lambda}_K$ may not be of full rank. If we denote the rank of $\boldsymbol{\Lambda}_K$ as L, then $L \leq T$ and thus the resulting $\boldsymbol{\Lambda}_K = \text{diag}(p_1, \ldots, p_L, 0 \ldots, 0)$. The resulting instantaneous water-filling capacity is given by

$$C = \log_2 \det \left(\boldsymbol{I}_R + \frac{1}{\sigma_n^2} \boldsymbol{H} \boldsymbol{U}_B \boldsymbol{\Lambda}_K \boldsymbol{U}_B{}^H \boldsymbol{H}^H \right) \text{ bps/Hz} \tag{15.51}$$

Since $\boldsymbol{H} = \boldsymbol{Z}\boldsymbol{B}^{\frac{1}{2}}$, where $\boldsymbol{Z}$ is an $R \times T$ matrix whose entries are i.i.d. complex Gaussian random variables with zero mean and variance 1, Equation 15.51 can be rewritten as

$$\begin{aligned} C &= \log_2 \det \left(\boldsymbol{I}_R + \frac{1}{\sigma_n^2} \boldsymbol{Z}\boldsymbol{B}^{\frac{1}{2}} \boldsymbol{U}_B \boldsymbol{\Lambda}_K \boldsymbol{U}_B{}^H \boldsymbol{B}^{\frac{1}{2}} \boldsymbol{Z}^H \right) \\ &= \log_2 \det \left(\boldsymbol{I}_R + \frac{1}{\sigma_n^2} \boldsymbol{Z}\boldsymbol{U}_B \boldsymbol{\Lambda}_B{}^{\frac{1}{2}} \boldsymbol{U}_B{}^H \boldsymbol{U}_B \boldsymbol{\Lambda}_K \boldsymbol{U}_B{}^H \boldsymbol{U}_B \boldsymbol{\Lambda}_B{}^{\frac{1}{2}} \boldsymbol{U}_B{}^H \boldsymbol{Z}^H \right) \\ &= \log_2 \det \left(\boldsymbol{I}_R + \frac{1}{\sigma_n^2} \boldsymbol{Z}\boldsymbol{U}_B \boldsymbol{\Lambda}_B{}^{\frac{1}{2}} \boldsymbol{\Lambda}_K \boldsymbol{\Lambda}_B{}^{\frac{1}{2}} \boldsymbol{U}_B{}^H \boldsymbol{Z}^H \right) \end{aligned} \tag{15.52}$$

[2]The solution to the optimization problem in the partial CSI case cannot be explicitly given and has to be obtained by a numerical search algorithm [21]. But the solution will also allocate more power to the stronger eigenmodes, which is in agreement with the water-filling principle [21]. Therefore, we refer to this numerical search algorithm as "water-filling-like" and refer to the capacity achieved by this optimum power allocation scheme as "water-filling capacity."

Note that $\boldsymbol{Z}$ is invariant under unitary transformation; i.e., $\boldsymbol{Z}\boldsymbol{U}_{\boldsymbol{B}}$ has the same distribution as $\boldsymbol{Z}$. Therefore, the capacity given by Equation 15.52 is statistically equivalent to

$$C = \log_2 \det\left(\boldsymbol{I}_R + \frac{1}{\sigma_n^2}\boldsymbol{Z}\boldsymbol{\Lambda}_{\boldsymbol{B}}^{\frac{1}{2}}\boldsymbol{\Lambda}_{\boldsymbol{K}}\boldsymbol{\Lambda}_{\boldsymbol{B}}^{\frac{1}{2}}\boldsymbol{Z}^H\right) \tag{15.53}$$

Now it is easy to see that finding the statistics of water-filling capacity given by Equation 15.53 is exactly the same problem as finding the statistics of the capacity of MIMO one-sided correlated Rayleigh channels without CSI at the transmitter and which we studied in the previous section. Now the MGF, mean, and second moment of water-filling capacity with covariance feedback can be deduced as the following theorem [28].

Theorem 15.9 *The MGF, mean, variance, and Gaussian-approximated CCDF of the water-filling capacity with covariance feedback are given by Theorems 15.4, 15.5, and 15.6, respectively, with parameters* $T = L$, $P = \frac{1}{\sigma_n^2}$, *and* $\boldsymbol{\Phi} = \boldsymbol{\Lambda}_{\boldsymbol{B}}^{\frac{1}{2}}\boldsymbol{\Lambda}_{\boldsymbol{K}}\boldsymbol{\Lambda}_{\boldsymbol{B}}^{\frac{1}{2}}$.

15.2.4.2 Beam-Forming Performance

The beam-forming transmitting strategy with the covariance feedback is to transmit along the eigenvector of $\boldsymbol{B}$ corresponding to the largest eigenvalue of $\boldsymbol{B}$. We denote the largest eigenvalue of $\boldsymbol{B}$ as $\lambda_{B,\max}$ and the associated eigenvector as $\boldsymbol{U}_{B,\max}$. Now $\boldsymbol{\Lambda}_{\boldsymbol{K}} = \mathrm{diag}(\Omega, 0, \ldots, 0)$, i.e., $\boldsymbol{K}$ is of rank 1. Substituting these into Equation 15.53, we get the achieved capacity (with complex Gaussian input) of beam forming as

$$C_{bf} = \log_2 \det\left(\boldsymbol{I}_{\boldsymbol{R}} + \frac{\lambda_{B,\max}\Omega}{\sigma_n^2}\boldsymbol{z}\boldsymbol{z}^H\right) \text{ bps/Hz} \tag{15.54}$$

where $\boldsymbol{z}$ is an $R \times 1$ vector whose entries are i.i.d. complex Gaussian random variables with zero mean and variance 1. Since $\det(\boldsymbol{I}_n + \boldsymbol{X}\boldsymbol{Y}) = \det(\boldsymbol{I}_m + \boldsymbol{Y}\boldsymbol{X})$ for $\boldsymbol{X}$: $n \times m$ and $\boldsymbol{Y}$: $m \times n$, Equation 15.54 can be written as

$$C_{bf} = \log_2 \det\left(1 + \frac{\lambda_{B,\max}\Omega}{\sigma_n^2}\boldsymbol{z}^H\boldsymbol{z}\right) = \log_2\left(1 + \frac{\lambda_{B,\max}\Omega}{2\sigma_n^2}\chi_{2R}^2\right) \tag{15.55}$$

where χ_{2R}^2 stands for χ^2 random variable with $2R$ degrees of freedom. Note that Equation 15.55 is equivalent to the capacity of a $1 \times R$ i.i.d. Rayleigh channel with transmitting power constraint $\Omega\lambda_{B,\max}$. Therefore, the statistics of C_{bf} can now be easily obtained. For example, the CDF of the capacity is given by

$$\Pr(C_{bf} < C_{\text{th}}) = \Pr\left(\chi_{2R}^2 < \frac{2\sigma_n^2(2^{C_{\text{th}}} - 1)}{\Omega\lambda_{B,\max}} \stackrel{\Delta}{=} x\right) = \frac{\gamma(R, x/2)}{\Gamma(R)} \tag{15.56}$$

The PDF of capacity, $f_C(C)$, is given by

$$f_C(C) = \frac{\sigma_n^2 2^C \ln(2)}{\Omega\lambda_{B,\max}\Gamma(R)}\left(\frac{\sigma_n^2(2^C - 1)}{\Omega\lambda_{B,\max}}\right)^{R-1}\exp\left(-\frac{\sigma_n^2(2^C - 1)}{\Omega\lambda_{B,\max}}\right), C \geq 0 \tag{15.57}$$

The capacity mean can also be obtained as

$$E(C) = \frac{e^{\alpha}}{\ln(2)}\sum_{k=1}^{R}\Gamma(-R + k, \alpha)\alpha^{R-k} \tag{15.58}$$

where

$$\alpha = \frac{\sigma_n^2}{\Omega\lambda_{B,\max}} \tag{15.59}$$

On the other hand, to obtain the statistics of the output SNR, we may model the beam-forming scheme as

$$y = Hw_t s_D + n \tag{15.60}$$

where $w_t = \sqrt{\Omega} U_{B,\max}$ for beam forming with respect to B. Therefore, Equation 15.60 becomes

$$y = \sqrt{\Omega} s_D H U_{B,\max} + n \tag{15.61}$$

At the receiver side, since the receiver has perfect CSI and thus knows H and B, the receiving weight vector w_r is the matched filter to $HU_{B,\max}$. Therefore, the output of the receiver combiner output is

$$U_{B,\max}^H H^H y = \sqrt{\Omega} U_{B,\max}^H H^H s_D H U_B + U_{B,\max}^H H^H n \tag{15.62}$$

Hence, the instantaneous combiner output SNR μ_{bf} is given by

$$\mu_{bf} = \frac{\Omega \left\| U_{B,\max}^H H^H H U_{B,\max} \right\|^2}{\sigma_n^2 \left\| U_{B,\max}^H H^H \right\|^2} = \frac{\Omega}{\sigma_n^2} U_{B,\max}^H H^H H U_{B,\max} \tag{15.63}$$

Noting that $H = ZB^{\frac{1}{2}}$ and that $U_{B,\max}$ is the eigenvector of B corresponding to $\lambda_{B,\max}$, we can write (in the sense of statistical equivalence) μ_{bf} as

$$\mu_{bf} = \frac{\Omega}{\sigma_n^2} \lambda_{B,\max} U_{B,\max}^H Z^H Z U_{B,\max} \tag{15.64}$$

Since $ZU_{B,\max}$ is an $R \times 1$ vector whose entries are i.i.d. complex Gaussian random variables with zero mean and variance 1, Equation 15.64 becomes

$$\mu_{bf} = \frac{\Omega \lambda_{B,\max}}{2\sigma_n^2} \chi_{2R}^2 \tag{15.65}$$

Therefore, all statistical properties of the beam-forming output SNR can be obtained from those of χ^2 random variables. For example, the CDF of μ_{bf} is given by

$$\Pr(\mu_{bf} < \mu_{\text{th}}) = \Pr\left(\chi_{2R}^2 < \frac{2\sigma_n^2 \mu_{\text{th}}}{\Omega \lambda_{B,\max}} \right) = \frac{\gamma\left(R, \frac{\sigma_n^2 \mu_{\text{th}}}{\Omega \lambda_{B,\max}}\right)}{\Gamma(R)} \tag{15.66}$$

and the MGF of μ_{bf} is given by

$$\mathcal{M}_{\mu_{bf}}(\tau) = E(e^{\tau \mu_{bf}}) = \left(\frac{1}{1 - \frac{\Omega \lambda_{B,\max}}{\sigma_n^2} \tau} \right)^R \tag{15.67}$$

The average symbol error rate (SER) of M-ary phase-shift-keying (M-PSK) signals is given by [29, Section 5.4.1, Equation 5.67] as

$$P_s(E) = \frac{1}{\pi} \int_0^{(M-1)\pi/M} \mathcal{M}_{\mu_{bf}}\left(-\frac{g_{\text{psk}}}{\sin^2 \phi} \right) d\phi \tag{15.68}$$

which can be evaluated as

$$P_s(E) = \mathrm{F}_1 \left(R + \frac{1}{2}, R, \frac{1}{2}, R + \frac{3}{2}, \sin^2\left(\frac{(M-1)\pi}{M} \right), \frac{-\sin^2\left(\frac{(M-1)\pi}{M}\right)\alpha}{g_{\text{psk}}} \right) \times \frac{\alpha^R \sin^{2R+1}((M-1)\pi/M)}{\pi(2R+1)(g_{\text{psk}})^R} \tag{15.69}$$

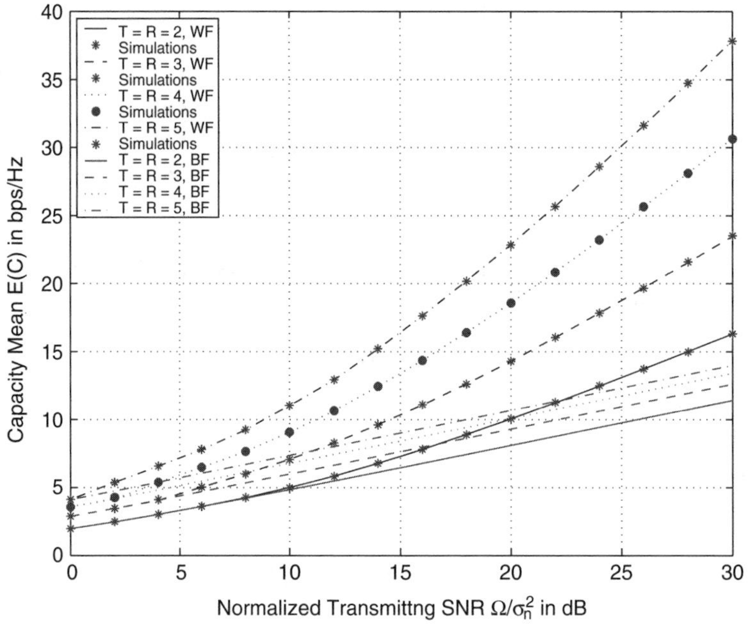

FIGURE 15.7 Capacity mean $E(C)$ in bits per second/herzt vs. the normalized transmitting SNR $\frac{\Omega}{\sigma_n^2}$ in decibel for (1) the WF scheme and (2) the BF scheme.

where $g_{\text{psk}} = \sin^2(\pi/M)$, α is defined in Equation 15.59, and $\text{F}_1(\cdot,\cdot,\cdot,\cdot,\cdot,\cdot)$ is the Appell hypergeometric function of two variables defined by [14, Equation 9.180.1]. The average bit error rate (BER) of binary signals ($M =$ (BFSK)) are special cases of Equation 15.69 with $g_{\text{psk}} = 1$ for binary PSK (BPSK), $g_{\text{psk}} = 0.5$ for orthogonal binary FSK (BFSK), and $g_{\text{psk}} = 0.715$ for BFSK with minimum correlation. Moreover,

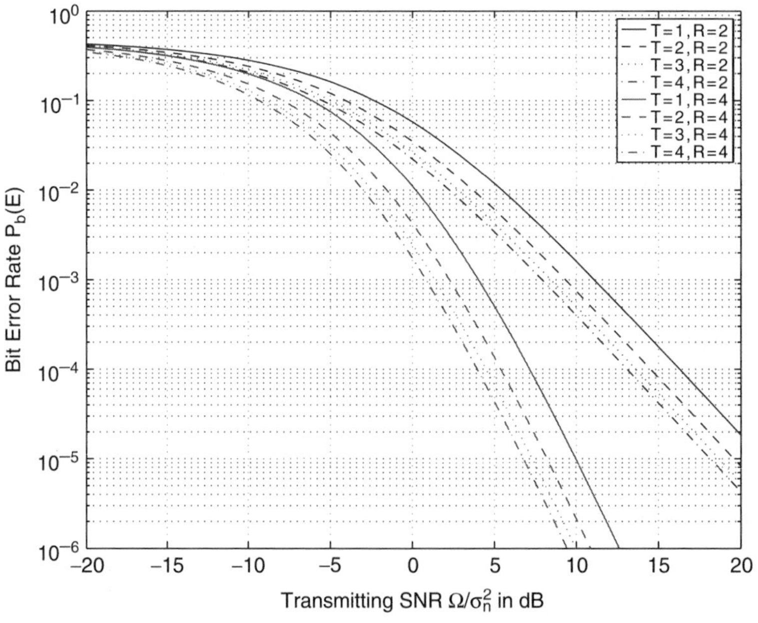

FIGURE 15.8 Average BER vs. the transmitting SNR $\frac{\Omega}{\sigma_n^2}$ in decibel for BPSK signals and when (1) $R = 2$ and (2) $R = 4$.

when $M = 2$, the average BER of binary signals reduces to

$$P_b(E) = \left(\frac{\alpha}{g_{\text{psk}}}\right)^R \frac{\Gamma(R+1.5)}{(2R+1)\Gamma(R+1)\sqrt{\pi}}\, {}_2F_1\left(R+0.5, R; R+1; \frac{-\alpha}{g_{\text{psk}}}\right) \tag{15.70}$$

where ${}_2F_1(\cdot, \cdot; \cdot; \cdot)$ is the Gaussian hypergeometric function [14, Equation 9.100]. Figure 15.7 shows the mean capacity $E(C)$ in bits per second/herzt vs. the transmitting SNR $\frac{\Omega}{\sigma_n^2}$ in decibel for (1) the water-filling-like (WF) scheme and (2) the beam-forming (BF) scheme with covariance feedback. It shows that more capacity gain can be achieved with the optimum power allocation when the transmitting SNR is high. Figure 15.8 quantifies the average BER vs. the transmitting SNR $\frac{\Omega}{\sigma_n^2}$ in decibel for BPSK signals when (1) $R = 2$ and (2) $R = 4$.

15.3 MIMO Systems in the Presence of Co-Channel Interference

15.3.1 Problem Statement

We consider a wireless link equipped with T antenna elements at the transmitter and R antenna elements at the receiver. It is assumed that there are L interfering users each equipped with T_i antenna elements, $i = 1, \ldots, L$. The received $R \times 1$ vector at the desired user's receiver can thus be modeled as

$$\boldsymbol{y} = \boldsymbol{H}_D \boldsymbol{x}_D + \sum_{i=1}^{L} \sqrt{\Omega_i}\boldsymbol{H}_i \boldsymbol{x}_i + \boldsymbol{n} \tag{15.71}$$

where $\boldsymbol{H}_D$: $R \times T$ and $\boldsymbol{x}_D$: $T \times 1$ are the normalized channel matrix and the transmitted data vector for the desired user, respectively. Similarly, Ω_i, $\boldsymbol{H}_i$: $R \times T_i$, and $\boldsymbol{x}_i$: $T_i \times 1$ are the short-term average power, the normalized channel matrix, and the transmitted complex Gaussian vectors for the ith co-channel interferer. The additive noise $\boldsymbol{n}$: $R \times 1$ is assumed to be a complex Gaussian vector with zero mean and covariance matrix $\sigma_n^2 \boldsymbol{I}_R$. The interference-plus-noise $\sum_{i=1}^{L} \sqrt{\Omega_i}\boldsymbol{H}_i \boldsymbol{x}_i + \boldsymbol{n}$, conditioned on $\boldsymbol{H}_i$, $i = 1, \ldots, L$, is complex Gaussian with covariance matrix

$$\boldsymbol{B}_I = \sum_{i=1}^{L} \Omega_i \boldsymbol{H}_i \boldsymbol{K}_i \boldsymbol{H}_i^H + \sigma_n^2 \boldsymbol{I}_R \tag{15.72}$$

where $\boldsymbol{K}_i = E(\boldsymbol{x}_i \boldsymbol{x}_i^H)$. We further assume that the desired user has total transmitting power constraint of Ω_D, i.e., $\boldsymbol{K}_D = E(\boldsymbol{x}_D \boldsymbol{x}_D^H)$ and $\text{tr}(\boldsymbol{K}_D) \leq \Omega_D$. Now we take a closer look at $\boldsymbol{B}_I$ defined in Equation 15.72.

1. Let us assume that $\boldsymbol{H}_i, i = 1, \ldots, L$, are zero-mean complex Gaussian matrices with i.i.d. rows, but the columns are possibly correlated with covariance matrix $\boldsymbol{A}_i$: $T_i \times T_i$ (semicorrelated Rayleigh fading interferers), then $\boldsymbol{H}_i$ can be represented as $\boldsymbol{Z}_i \boldsymbol{A}_i^{\frac{1}{2}}$, where $\boldsymbol{Z}_i$: $R \times T_i$ denotes the complex Gaussian matrix whose entries are i.i.d. complex Gaussian random variables with zero mean and covariance 1. In this case, the term $\boldsymbol{H}_i \boldsymbol{K}_i \boldsymbol{H}_i^H$ in $\boldsymbol{B}_I$ can be written as $\boldsymbol{Z}_i(\boldsymbol{A}_i^{\frac{1}{2}} \boldsymbol{K}_i \boldsymbol{A}_i^{\frac{1}{2}})\boldsymbol{Z}_i^H$; i.e., this interferer can be viewed to be subject to i.i.d. Rayleigh fading but with the 'effective' signal covariance $(\boldsymbol{A}_i^{\frac{1}{2}} \boldsymbol{K}_i \boldsymbol{A}_i^{\frac{1}{2}})$. This means that the correlations among the transmitter antenna elements of each interferer do not change the problem if we properly adjust the $\boldsymbol{K}_i$. The same arguments apply to the correlations among different interferers. Therefore, it is sufficient to study the case when $\boldsymbol{H}_i, i = 1, \ldots, L$, are uncorrelated and each has i.i.d. complex Gaussian entries with zero mean and covariance 1, even if there exist possible correlations among transmitter antenna elements of each interferer or among different interferers.
2. Since the $\boldsymbol{K}_i, i = 1, \ldots, L$, are positive-semidefinite, we can write $\boldsymbol{K}_i = \boldsymbol{U}_i \boldsymbol{\Lambda}_i \boldsymbol{U}_i^H$, where $\boldsymbol{U}_i$ is the unitary matrix consisting of the eigenvectors of $\boldsymbol{K}_i$, and $\boldsymbol{\Lambda}_i$ the diagonal matrix composed of

the corresponding eigenvalues (real and nonnegative). Note that $\mathbf{\Lambda}_i$ is a $T_i \times T_i$ matrix but may have rank $n_i \le T_i$ depending on the number of independent streams transmitted each time. Because $\boldsymbol{H}_i$ is invariant under the unitary transformation $\boldsymbol{H}_i \boldsymbol{U}_i$ (i.e., they have the same distribution), it is clear that assuming $\boldsymbol{K}_i$ to be a diagonal matrix does not make us lose any generality. As such, from now on we will assume that $\boldsymbol{K}_i$ is diagonal.

3. Let this diagonal $\boldsymbol{K}_i = \text{diag}(K_{i,1}, \ldots, K_{i,n_i}, 0, \ldots)$, then $\Omega_i \boldsymbol{H}_i \boldsymbol{K}_i \boldsymbol{H}_i^H$ can be written as $\Omega_i \boldsymbol{H}_i \boldsymbol{K}_i \boldsymbol{H}_i^H = \sum_{l=1}^{n_i} \Omega_i K_{i,l} \boldsymbol{H}_{i,l} \boldsymbol{H}_{i,l}^H$, where $\boldsymbol{H}_{i,l}$ is the lth column of $\boldsymbol{H}_i$. Now we get $\boldsymbol{B}_I = \sum_{i=1}^{L} \sum_{l=1}^{n_i} \Omega_i K_{i,l} \boldsymbol{H}_{i,l} \boldsymbol{H}_{i,l}^H + \sigma_n^2 \boldsymbol{I}_R$. If we let $\sum_{i=1}^{L} n_i = N_I$ and $P_m, m = 1, \ldots, N_I$, be the mth smallest $\Omega_i K_{i,l}$ for $i = 1, \ldots, L$ and $l = 1, \ldots, n_i$, and $\boldsymbol{H}_I$ be the $R \times N_I$ matrix whose mth column is the channel vector corresponding to P_m, we can write Equation 15.71 the matrix form

$$\boldsymbol{y} = \boldsymbol{H}_D \boldsymbol{x}_D + \boldsymbol{H}_I \boldsymbol{P}_I^{\frac{1}{2}} \boldsymbol{x}_I + \boldsymbol{n} \tag{15.73}$$

where $\boldsymbol{P}_I = \text{diag}(P_1, P_2, \ldots, P_{N_I})$ and $\boldsymbol{B}_I$ becomes

$$\boldsymbol{B}_I = \boldsymbol{H}_I \boldsymbol{P}_I \boldsymbol{H}_I^H + \sigma_n^2 \boldsymbol{I}_R \tag{15.74}$$

Now that we obtained an equivalent model to Equation 15.71, this equivalent model Equation 15.73 can be viewed as a system with N_I independent interferers; each has one transmitter antenna element, P_k $(k = 1, \ldots, N_I)$ short-term average power, and transmits a normalized signal $\boldsymbol{x}_{I,k}$ over channel vector $\boldsymbol{H}_{I,k}$. But we should keep in mind that based on the previous arguments, we are in fact including in our analysis the more general case where there are correlations among interferers' transmitter antenna elements, among different interferers, and among the interfering signals, as well as different numbers of transmit antenna elements and transmitting strategies of interferers. We assume that a centralized mechanism does not exist; thus, the joint optimization of the total system capacity for both the desired user and the interferers is not possible. This is particularly true when some or all of the co-channel interferers are from neighboring cells, in which case the transmitter has no control over the transmitting strategies of interferers. Therefore, the desired user simply optimizes the capacity of its own link.

15.3.2 Capacity CCDF of MIMO Optimum Combining Scheme with Perfect CSI at Transmitter

When the transmitter has perfect knowledge of the channels of the desired user and CCI, the optimum transmitting strategy to maximize the single-link capacity is to transmit independent complex Gaussian inputs along the eigenvectors of $\boldsymbol{H}_D^H \boldsymbol{B}_I^{-1} \boldsymbol{H}_D$ and allocate power by performing the water-filling algorithm on $\boldsymbol{H}_D^H \boldsymbol{B}_I^{-1} \boldsymbol{H}_D$. On the other hand, we can use the optimum combining (beam-forming) scheme, which maximizes the output signal-to-interference-plus-noise ratio (SINR). This optimum combining scheme transmits one stream each time and thus involves only scalar codes, which greatly decreases the detection complexity at the receiver. For such optimum combining systems, we can rewrite the $R \times 1$ received vector at the receiver as

$$\boldsymbol{y} = \boldsymbol{H}_D \boldsymbol{w}_t s_D + \boldsymbol{H}_I \boldsymbol{P}_I^{\frac{1}{2}} \boldsymbol{x}_I + \boldsymbol{n} \tag{15.75}$$

where $\boldsymbol{w}_t$ is the transmitting weight vector with $\boldsymbol{w}_t^H \boldsymbol{w}_t = \Omega_D$ and s_D is the scalar information symbol of the desired user with power 1. The joint optimum transmit/receive weight vectors to maximize the output SINR were derived in [30] as

$$\begin{aligned} \boldsymbol{w}_t &= \sqrt{\Omega_D} \boldsymbol{U}_{\max} \\ \boldsymbol{w}_r &= \boldsymbol{B}_I^{-1} \boldsymbol{H}_D \boldsymbol{w}_t \end{aligned} \tag{15.76}$$

where $\boldsymbol{U}_{\max}$ $(\|\boldsymbol{U}_{\max}\| = 1)$ denotes the eigenvector corresponding to the largest eigenvalue of the quadratic form

$$\boldsymbol{F} = \boldsymbol{H}_D^H \boldsymbol{B}_I^{-1} \boldsymbol{H}_D = \boldsymbol{H}_D^H \left(\boldsymbol{H}_I \boldsymbol{P}_I \boldsymbol{H}_I^H + \sigma_n^2 \boldsymbol{I}_R \right)^{-1} \boldsymbol{H}_D \tag{15.77}$$

and the maximum SINR is given by

$$\mu = \Omega_D \lambda_{\max} \tag{15.78}$$

where $\lambda_{\max}$ is the largest eigenvalue of the matrix $\boldsymbol{F}$ defined in Equation 15.77. The outage probability is defined as the CDF of the output SINR evaluated at μ_{th}, i.e.,

$$\text{P}_{\text{out}} = \Pr(\mu < \mu_{\text{th}}) = \Pr\left(\lambda_{\max} < \frac{\mu_{\text{th}}}{\Omega_D} \stackrel{\Delta}{=} x\right) \tag{15.79}$$

First, we study the problems concerning the distributions of eigenvalues of the quadratic form $\boldsymbol{X}(\boldsymbol{YTY}+a\boldsymbol{I})^{-1}\boldsymbol{X}$, where $\boldsymbol{X}$ and $\boldsymbol{Y}$ are complex Gaussian matrices, $\boldsymbol{T}$ a Hermitian positive-semidefinite matrix, and a a scalar. Specifically, we assume that the desired signal is subject to i.i.d. Rayleigh fading; i.e., the entries of $\boldsymbol{H}_D$ are all i.i.d. complex Gaussian random variables with zero mean and variance 1. $\boldsymbol{H}_I$ is an $R \times N_I$ complex Gaussian matrix with zero mean. The outage probability of SINR μ can be obtained as follows [31].

15.3.2.1 Outage Probability

1. When $R \leq T$ and $N_I > R$,

$$\text{P}_{\text{out}} = \frac{\det(\boldsymbol{C}_1(x))}{\prod_{k=1}^{R} \Gamma(T-k+1)\Gamma(R-k+1)\det\left(P_i^{j-1}\right)} \tag{15.80}$$

where $\Gamma(\cdot)$ is the gamma function, $\det(P_i^{j-1})$ is the Vandermonde determinant defined by

$$\det\left(P_i^{j-1}\right) = \prod_{1 \leq l < k \leq N_I} (P_k - P_l) \tag{15.81}$$

and the $N_I \times N_I$ matrix $\boldsymbol{C}_1(x)$ is defined by

$$\boldsymbol{C}_1 = \begin{pmatrix} \boldsymbol{C}_{1A} \\ \boldsymbol{C}_{1B}(x) \end{pmatrix} \tag{15.82}$$

where the $(N_I - R) \times N_I$ block $\boldsymbol{C}_{1A}$ is defined by

$$\{\boldsymbol{C}_{1A}\}_{i,j} = P_j^{i-1}, i = 1, \ldots, N_I - R;\ j = 1, \ldots, N_I \tag{15.83}$$

and the $R \times N_I$ block $\boldsymbol{C}_{1B}(x)$ is defined by

$$\begin{aligned} \{\boldsymbol{C}_{1B}(x)\}_{i,j} = {} & P_j^{N_I-R-1} e^{\frac{\sigma_n^2}{P_j}} \Gamma(T-i+1) \left[P_j^i\, \Gamma\left(i, \frac{\sigma_n^2}{P_j}\right) \right. \\ & \left. - \sum_{k=0}^{T-i} \frac{x^k}{k!} \left(\frac{P_j}{1+xP_j}\right)^{k+i} \Gamma\left(k+i, (x+1/P_j)\sigma_n^2\right) \right], \\ & i = 1, \ldots, R;\ j = 1, \ldots, N_I \end{aligned} \tag{15.84}$$

where $\Gamma(\cdot,\cdot)$ is the complementary incomplete gamma function defined by [14, Equation 8.350.2].

2. When $R \leq T$ and $N_I \leq R$,

$$\text{P}_{\text{out}} = \frac{\left(\prod_{n=1}^{N_I} P_n^{-(R-N_I+1)}\right) \left(\sigma_n^2\right)^{T(R-N_I)} \det(\boldsymbol{C}_2(x))}{\left(\prod_{k=1}^{R} \Gamma(T-k+1)\Gamma(R-k+1)\right) \det\left(P_i^{j-1}\right)} \tag{15.85}$$

where the $R \times R$ matrix $\mathbf{C}_2(x)$ is defined by

$$\mathbf{C}_2(x) = \begin{pmatrix} \mathbf{C}_{2A}(x) \\ \mathbf{C}_{2B}(x) \end{pmatrix} \tag{15.86}$$

where the $(R - N_I) \times R$ block $\mathbf{C}_{2A}(x)$ is given by

$$\{\mathbf{C}_{2A}(x)\}_{i,j} = (-1)^{i-1}\left(\sigma_n^2\right)^{-(T+i-j)}\gamma\left(T+i-j, \sigma_n^2 x\right),$$
$$i = 1, \ldots, R - N_I, \quad j = 1, \ldots, R \tag{15.87}$$

where the $N_I \times R$ block $\mathbf{C}_{2B}(x)$ is defined by

$$\{\mathbf{C}_{2B}(x)\}_{i,j} = e^{\frac{\sigma_n^2}{P_i}}\Gamma(T-j+1)\left[P_i^j\,\Gamma\left(j, \frac{\sigma_n^2}{P_i}\right)\right.$$
$$\left. - \sum_{k=0}^{T-j}\frac{x^k}{k!}\left(\frac{P_i}{1+xP_i}\right)^{k+j}\Gamma\left(k+j, (x+1/P_i)\sigma_n^2\right)\right],$$
$$i = 1, \ldots, N_I, \quad j = 1, \ldots, R \tag{15.88}$$

3. When $R > T$ and $N_I \geq R$,

$$\mathrm{P_{out}} = \frac{(-1)^{T(R-T)}\det(\mathbf{C}_3(x))}{\left(\prod_{k=1}^{R}\Gamma(R-k+1)\right)\left(\prod_{k=1}^{T}\Gamma(T-k+1)\right)\det\left(P_i^{j-1}\right)}, \tag{15.89}$$

where $\mathbf{C}_3(x)$ is an $N_I \times N_I$ matrix given by

$$\mathbf{C}_3(x) = \begin{pmatrix} \mathbf{C}_{3A} \\ \mathbf{C}_{3B} \\ \mathbf{C}_{3B}(x) \end{pmatrix} \tag{15.90}$$

where the $(N_I - R) \times N_I$ block $\mathbf{C}_{3A}$ is defined by

$$\{\mathbf{C}_{3A}\}_{i,j} = P_j^{i-1}, \; i = 1, \ldots, N_I - R, \; j = 1, \ldots, N_I \tag{15.91}$$

the $(R - T) \times N_I$ block $\mathbf{C}_{3B}$ is defined by

$$\{\mathbf{C}_{3B}\}_{i,j} = P_j^{N_I-R+T+i-1}e^{\frac{\sigma_n^2}{P_j}}\Gamma\left(T+i, \frac{\sigma_n^2}{P_j}\right),$$
$$i = 1, \ldots, R-T, \quad j = 1, \ldots, N_I \tag{15.92}$$

and the $T \times N_I$ block $\mathbf{C}_{3B}(x)$ is given by

$$\{\mathbf{C}_{3B}(x)\}_{i,j} = P_j^{N_I-R-1}e^{\frac{\sigma_n^2}{P_j}}\Gamma(T-i+1)\left[P_j^i\,\Gamma\left(i, \frac{\sigma_n^2}{P_j}\right)\right.$$
$$\left. - \sum_{k=0}^{T-i}\frac{x^k}{k!}\left(\frac{P_j}{1+xP_j}\right)^{k+i}\Gamma\left(k+i, (x+1/P_j)\sigma_n^2\right)\right],$$
$$i = 1, \ldots, T, \quad j = 1, \ldots, N_I \tag{15.93}$$

4. When $R \geq N_I > T$,

$$\mathrm{P}_{\text{out}} = \frac{(-1)^{T(R-T)}\left(\sigma_n^2\right)^{T(R-N_I)}}{\left(\prod_{k=1}^{T}\Gamma(T-k+1)\right)} \frac{\det(\boldsymbol{C}_4(x))}{\left(\prod_{k=1}^{N_I}\Gamma(R-k+1)P_k^{R-N_I+1}\right)\det\left(P_i^{j-1}\right)} \tag{15.94}$$

where $\boldsymbol{C}_4(x)$ is an $N_I \times N_I$ matrix given by

$$\boldsymbol{C}_4(x) = \begin{pmatrix} \boldsymbol{C}_{4A} \\ \boldsymbol{C}_{4B}(x) \end{pmatrix} \tag{15.95}$$

where the $(N_I - T) \times N_I$ block $\boldsymbol{C}_{4A}$ is defined by

$$\begin{aligned} \{\boldsymbol{C}_{4A}\}_{i,j} &= \sum_{k=0}^{T}\binom{T}{k}\left(\sigma_n^2\right)^{T-k} P_j^{R-N_I+i+k}\Gamma(R-N_I+i+k), \\ & i = 1,\ldots,N_I - T, \ j = 1,\ldots,N_I \end{aligned} \tag{15.96}$$

and the $T \times N_I$ block $\boldsymbol{C}_{4B}$ is defined by

$$\begin{aligned} \{\boldsymbol{C}_{4B}(x)\}_{i,j} &= e^{\frac{\sigma_n^2}{P_j}}\Gamma(T-i+1)\left[P_j^i\,\Gamma\left(i,\frac{\sigma_n^2}{P_j}\right)\right. \\ &\quad \left. - \sum_{k=0}^{T-i}\frac{x^k}{k!}\left(\frac{P_j}{1+xP_j}\right)^{k+i}\gamma\left(k+i,(x+1/P_j)\sigma_n^2\right)\right] \\ &\quad - \sum_{k=0}^{R-N_I-1}\frac{(-1)^k}{k!\left(\sigma_n^2\right)^{T-i+k+1}}\Gamma\left(T-i+k+1,\sigma_n^2 x\right) \\ &\quad \times \sum_{m=0}^{T}\binom{T}{m}\left(\sigma_n^2\right)^{T-m}P_j^{k+m+1}\Gamma(k+m+1), \\ & i = 1,\ldots,T, \ j = 1,\ldots,N_I \end{aligned} \tag{15.97}$$

5. When $R \geq T$ and $N_I < T$,

$$\begin{aligned} \mathrm{P}_{\text{out}} &= \frac{(-1)^{T(R-T)}\left(\sigma_n^2\right)^{T(R-N_I)}}{\left(\prod_{k=1}^{T}\Gamma(T-k+1)\right)\left(\prod_{k=1}^{T}\Gamma(R-k+1)\right)} \\ &\quad \times \frac{\det(\boldsymbol{C}_5(x))}{\left(\prod_{k=1}^{N_I}P_k^{R-N_I+1}\right)\det\left(P_i^{j-1}\right)} \end{aligned} \tag{15.98}$$

where $\boldsymbol{C}_5(x)$ is a $T \times T$ matrix given by

$$\boldsymbol{C}_5(x) = \left(\boldsymbol{C}_{5A}(x) \quad \boldsymbol{C}_{4B}(x)\right) \tag{15.99}$$

where the $T \times (T - N_I)$ block $\boldsymbol{C}_{5A}(x)$ is defined by

$$\begin{aligned} \{\boldsymbol{C}_{5A}(x)\}_{i,j} &= \frac{(-1)^{R-T+j-1}}{\left(\sigma_n^2\right)^{R-i+j}}\gamma\left(R-i+j,\sigma_n^2 x\right), \\ & i = 1,\ldots,T, \ j = 1,\ldots,T-N_I \end{aligned} \tag{15.100}$$

and the $T \times N_I$ block $\boldsymbol{C}_{4B}(\boldsymbol{x})$ is defined in Equation 15.97.

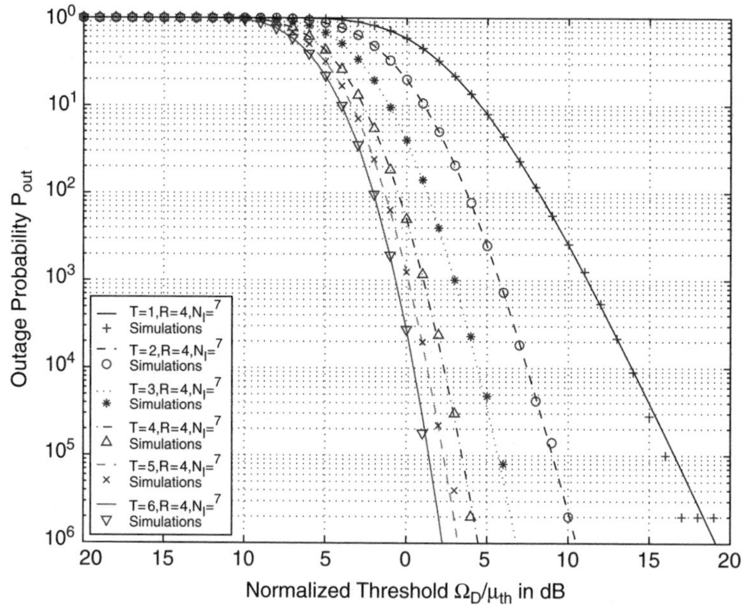

FIGURE 15.9 Outage probability of MIMO optimum combining systems vs. the normalized threshold $\frac{\Omega_D}{\mu_{\text{th}}}$ in decibel with T as a parameter when $R = 4$, $N_I = 7$, $\sigma_n^2 = 0.5$, and the average interfering powers are i.i.d. uniform $(0, 1)$ and normalized so that the total interfering power is N_I.

15.3.2.2 Capacity of Optimum Combining

The channel capacity achieved by the complex Gaussian input s_D with the optimum combining or beamforming scheme is given by $C = \log_2(1 + \mu)$ bps/Hz, where μ is the combiner output SINR given by Equation 15.78. Thus the capacity CCDF is $C_{\text{ccdf}} = \Pr(C \geq C_{\text{th}}) = 1 - \Pr(\mu < 2^{C_{\text{th}}} - 1)$; i.e., the C_{ccdf} is simply the complementary CDF of the output SINR evaluated at $2^{C_{\text{th}}} - 1$. Figure 15.9 plots the outage probability of MIMO optimum combining systems vs. the normalized threshold $\frac{\Omega_D}{\mu_{\text{th}}}$ in decibel with T as a parameter when $R = 4$, $N_I = 7$, $\sigma_n^2 = 0.5$, and the average interfering powers are i.i.d. uniform $(0, 1)$ and normalized so that the total interfering power is N_I. Note also the good match between the analytical curves and the Monte Carlo simulations.

15.3.3 Statistics of the MIMO Capacity with Co-Channel Interference and without Feedback

15.3.3.1 Problem Statement

In this section, we will consider the impact of co-channel interference on the capacity of MIMO channels modeled as Equation 15.73 (which is the equivalent simplified model of Equation 15.71). Similar to the previous subsection, we assume that a centralized mechanism does not exist; thus, the joint optimization of the total system capacity for both the desired user and the interferers is not possible. The instantaneous mutual information (MI) between the input and output of the desired user, after whitening $\boldsymbol{y}$ by $\boldsymbol{B}_I^{-\frac{1}{2}}$, can be expressed as [32]

$$I(\boldsymbol{x}_D; (\boldsymbol{y}, \boldsymbol{K}_I)) = \log_2\left(\det\left(\boldsymbol{I}_R + \boldsymbol{H}_D\boldsymbol{K}_D\boldsymbol{H}_D^H\boldsymbol{B}_I^{-1}\right)\right) \tag{15.101}$$

The optimization problem is to choose $\boldsymbol{K}_D$ to maximize the mutual information given to $\boldsymbol{B}_I$ subject to the total transmitting power constraint Ω_D, i.e.,

$$C = \max_{\text{tr}(\boldsymbol{K}_D)\leq\Omega_D} \log_2\left(\det\left(\boldsymbol{I}_R + \boldsymbol{H}_D\boldsymbol{K}_D\boldsymbol{H}_D^H\boldsymbol{B}_I^{-1}\right)\right) \tag{15.102}$$

When the receiver has perfect CSI of both the desired user and CCI while the transmitter has no CSI about $\boldsymbol{H}_D$ and $\boldsymbol{H}_I$, the optimum choice of $\boldsymbol{K}_D = \frac{\Omega_D}{T}\boldsymbol{I}_T$, i.e., transmitting T independent complex Gaussian inputs with equal powers. The resulting capacity is given by [32]

$$C = \log_2\left(\det\left(\boldsymbol{I}_R + \frac{\Omega_D}{T}\boldsymbol{H}_D\boldsymbol{H}_D^H\boldsymbol{B}_I^{-1}\right)\right)$$
$$= \sum_{k=1}^{\min(T,R)} \log_2\left(1 + \frac{\Omega_D}{T}\phi_k\right) \quad (15.103)$$

where $0 < \phi_1 < \phi_2 < \cdots < \phi_{\min(T,R)}$ are the nonzero eigenvalues of $(\boldsymbol{B}_I^{-\frac{1}{2}}\boldsymbol{H}_D)(\boldsymbol{B}_I^{-\frac{1}{2}}\boldsymbol{H}_D)^H$. Note that $(\boldsymbol{B}_I^{-\frac{1}{2}}\boldsymbol{H}_D)(\boldsymbol{B}_I^{-\frac{1}{2}}\boldsymbol{H}_D)^H$ has the same nonzero eigenvalues as those of

$$\boldsymbol{F} = \boldsymbol{H}_D^H\boldsymbol{B}_I^{-1}\boldsymbol{H}_D = \boldsymbol{H}_D^H\left(\boldsymbol{H}_I\boldsymbol{P}_I\boldsymbol{H}_I^H + \sigma_n^2\boldsymbol{I}_R\right)^{-1}\boldsymbol{H}_D \quad (15.104)$$

When the transmitter has perfect knowledge of the channels of both the desired user and the interferers, the optimum capacity-achieving strategy is to transmit independent complex Gaussian inputs along the eigenvectors of $\boldsymbol{F}$ and allocate powers based on the water-filling principle, i.e., $p_k = (\psi - \frac{1}{\phi_k})^+$ with $(\cdot)^+$ defined as $(y)^+ = \max(y, 0)$, where ψ is a constant depending on the total power constraint Ω_D, i.e., $\sum p_k = \Omega_D$. From the water-filling power allocation, we expect that when Ω_D is low, the water-filling algorithm tends to allocate all power to a single data stream along the eigenmode corresponding to the largest eigenvalue of $\boldsymbol{F}$, which is equivalent to the optimum combining scheme studied in the previous section. However, when Ω_D increases, the power allocation will allow more than one data stream to be transmitted each time, and when the Ω_D is high enough, all subchannels will transmit and the power difference among different data streams becomes negligible and thus very close to the equal-power allocation when the CSI is not available at the transmitter when $T \leq R$ (there is a capacity loss when $T > R$ without CSI feedback, since in this case the total power is divided among T antenna elements but there are only R subchannels). Since the water-filling principle is the optimum strategy, the above analysis tells us that when the transmitting SNR is very low, the CSI becomes critical and the capacity achieved by the optimum combining scheme is very close to the optimum water-filling capacity, which uses vector codes. On the other hand, when the SNR is high enough, the CSI feedback does not provide significant capacity gain over the equal-power allocation scheme where no CSI at the transmitter is needed when $T \leq R$, and the capacity of the optimum combining scheme will be significantly lower than the equal-power allocation scheme. We now focus on the case when the CSI is not available at the transmitter and further assume that the desired user is subject to i.i.d. Rayleigh fading; thus, $\boldsymbol{H}_D$ has i.i.d. entries of complex Gaussian random variables with zero mean and variance 1. The channel matrix of CCI $\boldsymbol{H}_I$, as justified before, is also an i.i.d. zero-mean complex Gaussian matrix with i.i.d. entries with variance 1. The MGF of the capacity can be obtained as follows [31].

15.3.3.2 Capacity MGF $E(e^{\tau C})$

1. When $R \leq T$ and $N_I > R$,

$$E(e^{\tau C}) = \frac{\det(\boldsymbol{G}_1)}{\prod_{k=1}^{R}\Gamma(T-k+1)\Gamma(R-k+1)\det\left(P_i^{j-1}\right)} \quad (15.105)$$

where the $N_I \times N_I$ matrix $\boldsymbol{G}_1$ is defined by

$$\boldsymbol{G}_1 = \begin{pmatrix} \boldsymbol{C}_{1A} \\ \boldsymbol{G}_{1B} \end{pmatrix} \quad (15.106)$$

where the $(N_I - R) \times N_I$ block $\mathbf{C_{1A}}$ is defined by Equation 15.83 and the $R \times N_I$ block $\mathbf{G_{1B}}$ is defined by

$$\{\mathbf{G_{1B}}\}_{i,j} = P_j^{N_I-R-1} e^{\sigma_n^2/P_j} \int_0^{\infty} \frac{(1+\rho y)^{\tau/\ln(2)} y^{T-i}}{(1/P_j + y)^{T+1}} \times \Gamma\left(T+1, \sigma_n^2(1/P_j + y)\right) dy, \quad i = 1, \ldots, R, \; j = 1, \ldots, N_I \tag{15.107}$$

where $\rho = \frac{\Omega_D}{T}$

2. When $R \le T$ and $N_I \le R$,

$$E(e^{\tau C}) = \frac{\left(\prod_{n=1}^{N_I} P_n^{-(R-N_I+1)}\right) \left(\sigma_n^2\right)^{T(R-N_I)} \det(\mathbf{G_2})}{\left(\prod_{k=1}^{R} \Gamma(T-k+1)\Gamma(R-k+1)\right) \det\left(P_i^{j-1}\right)} \tag{15.108}$$

where the $R \times R$ matrix $\mathbf{G_2}(x)$ is defined by

$$\mathbf{G_2} = \begin{pmatrix} \mathbf{G_{2A}} \\ \mathbf{G_{2B}} \end{pmatrix} \tag{15.109}$$

where the $(R - N_I) \times R$ block $\mathbf{G_{2A}}$ is given by

$$\{\mathbf{G_{2A}}\}_{i,j} = (-1)^{i-1} \frac{\Gamma(T+i-j)}{\rho^{T+i-j}} U\left(T+i-j, T+i-j+1+\tau/\ln(2), \frac{\sigma_n^2}{\rho}\right), \quad i = 1, \ldots, R-N_I, \; j = 1, \ldots, R \tag{15.110}$$

where U $(\cdot,\cdot,\cdot)$ is the confluent hypergeometric function of the second kind [14, Equation 9.210.2] and the $N_I \times R$ block $\mathbf{G_{2B}}(x)$ is defined by

$$\{\mathbf{G_{2B}}\}_{i,j} = e^{\sigma_n^2/P_i} \int_0^{\infty} \frac{(1+\rho y)^{\tau/\ln(2)} y^{T-j}}{(1/P_i + y)^{T+1}} \Gamma\left(T+1, \sigma_n^2(1/P_i + y)\right) dy, \quad i = 1, \ldots, N_I, \; j = 1, \ldots, R \tag{15.111}$$

3. When $R > T$ and $N_I \ge R$,

$$E(e^{\tau C}) = \frac{(-1)^{T(R-T)}}{\left(\prod_{k=1}^{R} \Gamma(R-k+1)\right)} \frac{\det(\mathbf{G_3})}{\left(\prod_{k=1}^{T} \Gamma(T-k+1)\right) \det\left(P_i^{j-1}\right)} \tag{15.112}$$

where $\mathbf{G_3}$ is an $N_I \times N_I$ matrix given by

$$\mathbf{G_3} = \begin{pmatrix} \mathbf{C_{3A}} \\ \mathbf{C_{3B}} \\ \mathbf{G_{3C}} \end{pmatrix} \tag{15.113}$$

where the $(N_I - R) \times N_I$ block $\mathbf{C_{3A}}$ is defined by Equation 15.91, the $(R - T) \times N_I$ block $\mathbf{C_{3B}}$ is defined by Equation 15.92, and the $T \times N_I$ block $\mathbf{G_{3C}}$ is given by

$$\{\mathbf{G_{3B}}\}_{i,j} = P_j^{N_I-R-1} e^{\sigma_n^2/P_j} \int_0^{\infty} \frac{(1+\rho y)^{\tau/\ln(2)} y^{T-i}}{(1/P_j + y)^{T+1}} \times \Gamma\left(T+1, \sigma_n^2(1/P_j + y)\right) dy, \quad i = 1, \ldots, T, \; j = 1, \ldots, N_I \tag{15.114}$$

4. When $R \geq N_I > T$,

$$E(e^{\tau C}) = \frac{(-1)^{T(R-T)}\left(\sigma_n^2\right)^{T(R-N_I)}}{\left(\prod_{k=1}^{T}\Gamma(T-k+1)\right)} \frac{\det(\mathbf{G_4})}{\left(\prod_{k=1}^{N_I}\Gamma(R-k+1)P_k^{R-N_I+1}\right)\det\left(P_i^{j-1}\right)} \tag{15.115}$$

where $\mathbf{G_4}$ is an $N_I \times N_I$ matrix given by

$$\mathbf{G_4} = \begin{pmatrix} \mathbf{C_{4A}} \\ \mathbf{G_{4B}} \end{pmatrix} \tag{15.116}$$

where the $(N_I - T) \times N_I$ block $\mathbf{C_{4A}}$ is defined by Equation 15.96 and the $T \times N_I$ block $\mathbf{G_{4B}}$ is defined by

$$\begin{aligned} \{\mathbf{G_{4B}}\}_{i,j} &= e^{\sigma_n^2/P_j}\int_0^\infty \frac{(1+\rho y)^{\tau/\ln(2)}y^{T-i}}{(1/P_j+y)^{T+1}}\Gamma\left(T+1,\sigma_n^2(1/P_j+y)\right)dy \\ &- \sum_{k=0}^{R-N_I-1}\frac{(-1)^k}{k!}\frac{\Gamma(T+i+k+1)}{\rho^{T+i+k+1}}\sum_{m=0}^{T}\binom{T}{m}\left(\sigma_n^2\right)^{T-m} \\ &\times P_j^{k+m+1}\Gamma(k+m+1)U\left(T+i+k+1, T+i+k+2+\tau/\ln(2), \frac{\sigma_n^2}{\rho}\right) \\ &i = 1,\ldots,T, \; j = 1,\ldots,N_I \end{aligned} \tag{15.117}$$

5. When $R \geq T$ and $N_I < T$,

$$\begin{aligned} E(e^{\tau C}) &= \frac{(-1)^{T(R-T)}\left(\sigma_n^2\right)^{T(R-N_I)}}{\left(\prod_{k=1}^{T}\Gamma(T-k+1)\right)\left(\prod_{k=1}^{T}\Gamma(R-k+1)\right)} \\ &\times \frac{\det(\mathbf{G_5})}{\left(\prod_{k=1}^{N_I}P_k^{R-N_I+1}\right)\det\left(P_i^{j-1}\right)} \end{aligned} \tag{15.118}$$

where $\mathbf{G_5}$ is a $T \times T$ matrix given by

$$\mathbf{G_5} = \begin{pmatrix} \mathbf{G_{5A}} & \mathbf{G_{4B}} \end{pmatrix} \tag{15.119}$$

where the $T \times (T - N_I)$ block $\mathbf{G_{5A}}$ is defined by

$$\begin{aligned} \{\mathbf{G_{5A}}\}_{i,j} &= (-1)^{R-T+j-1}\frac{\Gamma(R-i+j)}{\rho^{R-i+j}}U\left(R-i+j, R-i+j+1+\tau/\ln(2), \frac{\sigma_n^2}{\rho}\right), \\ &i = 1,\ldots,T, \; j = 1,\ldots,T-N_I \end{aligned} \tag{15.120}$$

and the $T \times N_I$ block $\mathbf{G_{4B}}$ is defined in Equation 15.117.

The expressions of the mean and variance of the capacity are given explicitly in the journal version of [31] but are omitted here due to space limitations. An approximated capacity CCDF can also be obtained. Figure 15.10 plots the mean capacity of MIMO Rayleigh/Rayleigh channels without CSI at the transmitter vs. the total transmitting power Ω_D in decibel with the number of interferers N_I as a parameter when $T = R = 3$, $\sigma_n^2 = 1$, and interfering powers are the N_I largest numbers from the set $\{1.5, 1.6, 1.8, 2.0\}$.

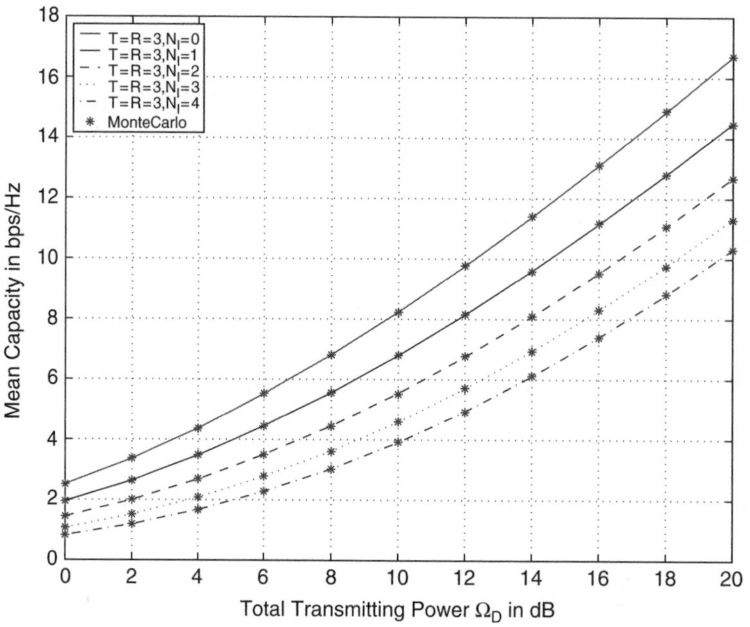

FIGURE 15.10 Mean capacity of MIMO Rayleigh/Rayleigh channels vs. the total transmitting power Ω_D in decibel with the number of interferers N_I as a parameter when $T = R = 3, \sigma_n^2 = 1$, and interfering powers are the N_I largest numbers from the set $\{1.5, 1.6, 1.8, 2.0\}$.

15.3.4 Rician/Rayleigh Fading Scenarios in an Interference-Limited Environment

In microcell mobile communications systems, a typical scenario consists of a direct line-of-sight path for the desired user accompanied by multiple diffuse interferers. This can be appropriately modeled by a desired user subject to Rician type of fading and the co-channel interferers subject to Rayleigh fading, which we refer to as the Rician/Rayleigh fading scenario. In this case, $\boldsymbol{H}_D$: $R \times T$ and $\boldsymbol{H}_I$: $R \times N_I$ are independent matrices whose columns are i.i.d. complex multivariate Gaussian vectors with covariance matrix $\boldsymbol{\Sigma}$, $E(\boldsymbol{H}_D) = \boldsymbol{M}$, and $E(\boldsymbol{H}_I) = \boldsymbol{0}$. To make the problem mathematically tractable, we assume that the system is interference limited; thus, the thermal noise can be neglected. This assumption (which is going to be validated later by numerical results) is more valid when the signal-to-noise ratio and the interference-to-noise ratio are high or when the system operates in an interference-rich (N_I is reasonably large) environment. We further assume that the interfering signals come from the same distance from the receiver and the shadowing effects are small, and thus $\Omega_{I,1} = \cdots = \Omega_{I,N_I} \stackrel{\Delta}{=} \Omega_I$. Now $\boldsymbol{B}_I$ given by Equation 15.74 reduces to $\boldsymbol{B}_I = \boldsymbol{H}_I \boldsymbol{H}_I^H$ and is positive-definite with probability 1 when $N_I \geq R$ (Therefore $\boldsymbol{B}_I$ is invertible with probability 1), and $\boldsymbol{F}$ defined by Equation 15.77 reduces to $\boldsymbol{F}_1 = \frac{1}{\Omega_I} \boldsymbol{H}_D^H (\boldsymbol{H}_I \boldsymbol{H}_I)^{-1} \boldsymbol{H}_D$, which is a Hermitian nonnegative-definite matrix. The MGFs of the capacity without feedback can be obtained for these cases, and due to space limitations, the results are just summarized without proof in what follows [33]. Note that the CDF of the signal-to-interference ratio (and therefore the capacity CCDF) of the MIMO optimum combining scheme with perfect CSI at the transmitter was also obtained in [34][35][36] under the same assumptions.

1. Distinct eigenvalues case: Let $\boldsymbol{H}_D$: $R \times T$ and $\boldsymbol{H}_I$: $R \times N_I$, $N_I \geq R$, be independent matrices whose columns are independent R-variate complex normal vectors with covariance matrix $\boldsymbol{\Sigma}$, $E(\boldsymbol{H}_D) = \boldsymbol{M}$, and $E(\boldsymbol{H}_I) = \boldsymbol{0}$. Let $s = \min(T, R)$, $t = \max(T, R)$, and $r = \min(N_I, N_I + T - R)$. If $\boldsymbol{M}^H \boldsymbol{\Sigma}^{-1} \boldsymbol{M}$ has s nonzero distinct eigenvalues $0 < \lambda_1 < \lambda_2 < \cdots < \lambda_s$, then the MGF of the

capacity $E(e^{\tau C})$ is given by

$$E(e^{\tau C}) = c_1 \det(\boldsymbol{\Psi}_{nc}) \tag{15.121}$$

where $c_1 = \frac{e^{-\mathrm{tr}(\boldsymbol{\Lambda})}}{\det(\boldsymbol{V})} \prod_{l=1}^{s} \frac{\Gamma(r+t-s+1)}{\Gamma(t-s+1)\Gamma(r-l+1)}$, $\boldsymbol{\Lambda} = \mathrm{diag}(\lambda_1, \ldots, \lambda_s)$ and $\det(\boldsymbol{V})$ is the Vandermonde determinant consisting of $\lambda_1, \ldots, \lambda_s$, i.e., $\{\boldsymbol{V}\}_{i,j} = \lambda_i^{s-j}$, with $\det(\boldsymbol{V}) = \prod_{i<j}^{s}(\lambda_i - \lambda_j)$, and $\boldsymbol{\Psi}_{nc}$ is an $s \times s$ matrix whose entry at the ith row and jth column is given by

$$\begin{aligned}\{\boldsymbol{\Psi}_{nc}\}_{i,j} &= \int_0^{\infty} (1+\rho_2 y)^{\tau/\ln(2)} \frac{y^{t-i}}{(1+y)^{t+r-s+1}} {}_1F_1\left(t+r-s+1; t-s+1; \frac{y\lambda_j}{1+y}\right) dy \\ &= \sum_{m=0}^{r} \frac{(-r)_m(-\lambda_j)^m}{(t-s+1)_m m!} B\left(t-i+m+1,\ r-s+i+\frac{\tau}{\ln(2)}\right) \\ &\quad \times \Phi_1\left(t-i+m+1, -\frac{\tau}{\ln(2)}, r+t-s+m+\frac{\tau}{\ln(2)}, (1-\rho_2), \lambda_j\right)\end{aligned} \tag{15.122}$$

where $\rho_2 = \frac{\Omega_D}{\Omega_I T}$, $(a)_n = a(a+1)\cdots(a+n-1)$ with $(a)_0 = 1$, $B(\cdot,\cdot)$ the beta function defined in [14, Equation 8.380.1], and $\Phi_1(\cdot,\cdot,\cdot,\cdot,\cdot)$ is the confluent hypergeometric function of two variables defined in [14, Equation 9.261.1].

2. Central case: When $E(\boldsymbol{H}_D) = \boldsymbol{0}$ and $E(\boldsymbol{H}_I) = \boldsymbol{0}$, the capacity MGF $E(e^{\tau C})$ is given by

$$E(e^{\tau C}) = c_2 \det(\boldsymbol{\Psi}_c) \tag{15.123}$$

where $c_2 = \prod_{n=1}^{s} \frac{\Gamma(t+r-n+1)}{\Gamma(t-n+1)\Gamma(r-n+1)\Gamma(s-n+1)}$ and $\boldsymbol{\Psi}_c$ is an $s \times s$ matrix with entries given by

$$\begin{aligned}\{\boldsymbol{\Psi}_c\}_{i,j} &= B(t+s-i-j+1, r-s+i-\tau/\ln(2)) \\ &\quad \times {}_2F_1\left(-\frac{\tau}{\ln(2)}, t+s-i-j+1; t+r-j+1-\frac{\tau}{\ln(2)}; 1-\rho_2\right), \\ &\quad i,j = 1,\ldots,s\end{aligned} \tag{15.124}$$

Note that Equation 15.123 is applicable to any positive-definite covariance matrix $\boldsymbol{\Sigma}$ due to the fact that in the central case, $\boldsymbol{F}_1$ is invariant under simultaneous transformations of $\boldsymbol{\Sigma}^{-\frac{1}{2}}\boldsymbol{H}_D$ and $\boldsymbol{\Sigma}^{-\frac{1}{2}}\boldsymbol{H}_I$.

3. I.i.d. Rician/Rayleigh case: When $\{\boldsymbol{H}_D\}_{i,j}$ are i.i.d. complex Gaussian random variables with mean η and variance σ^2, i.e., $\{\boldsymbol{H}_D\}_{i,j} \sim \mathcal{CN}(\eta, \sigma^2)$, and $\{\boldsymbol{H}_I\}_{i,j}$ are i.i.d. with $\{\boldsymbol{H}_I\}_{i,j} \sim \mathcal{CN}(0, \sigma^2)$, then $\boldsymbol{M}^H\boldsymbol{\Sigma}^{-1}\boldsymbol{M}$ has only one nonzero eigenvalue $\lambda_1 = \frac{st|\eta|^2}{\sigma^2}$, where $|\cdot|$ denotes the modulus of a complex number. In these conditions, the capacity MGF reduces to

$$E(e^{\tau C}) = c_3 \det(\boldsymbol{\Psi}_{\mathrm{iid}}) \tag{15.125}$$

where

$$c_3 = \frac{e^{-\lambda_1}}{\lambda_1^{s-1}} \frac{\Gamma(r+t-s+1)\Gamma(s)\Gamma(t)}{\Gamma(t-s+1)\Gamma(t+r)} \prod_{n=1}^{s} \frac{\Gamma(t+r-n+1)}{\Gamma(t-n+1)\Gamma(r-n+1)\Gamma(s-n+1)} \tag{15.126}$$

where the first column ($j = 1$) of $\boldsymbol{\Psi}_{\mathrm{iid}}$ is the same as that of $\boldsymbol{\Psi}_{\mathrm{nc}}$, and the second column to the sth column (i.e., $j = 2, \ldots, s$) of $\boldsymbol{\Psi}_{\mathrm{iid}}$ is the same as that of $\boldsymbol{\Psi}_{\mathrm{c}}$.

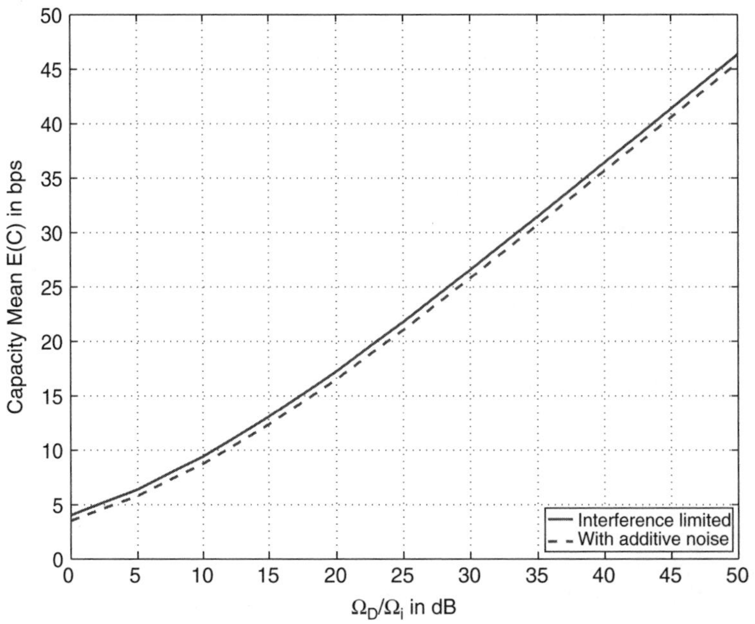

FIGURE 15.11 Capacity mean vs. the normalized transmitting power $\frac{\Omega_D}{\Omega_I}$ in decibel with a Rician factor $K_D = \frac{|\eta|^2}{\sigma_n^2} = 4$ when $T = R = 3$ and $N_I = 4$ for (1) analytical curves without noise and (2) simulation curves with additive noise when the interference-to-noise ratio is 6 dB.

In Figure 15.11, we compare the capacity mean obtained under the assumption that the system is interference limited with the simulations of a more realistic system where the noise is nonnegligible (we use an interference-to-noise ratio of 6 dB). We can see that our analytical results tightly upper-bound the simulation results and can therefore be used as a good estimate of the capacity of MIMO systems in the presence of CCI.

Appendix A: Some Integral Identities

1. In the MGF of capacity formulas, one can use [14, Equation 3.383.5]:

$$\int_0^\infty (1 + ax)^{-\nu} x^{q-1} e^{-px} dx = a^{-q}\Gamma(q)U\left(q, q + 1 - \nu, \frac{p}{a}\right) \tag{15.127}$$

   where Re $q > 0$, Re $p > 0$, Re $a > 0$, and ν is a complex number. In Equation 15.127, $\Gamma(\cdot)$ is the gamma function [14, Equation 8.310.1] and $U(\cdot,\cdot,\cdot)$ is the confluent hypergeometric function of the second kind [14, Equation 9.210.2].

2. In the mean capacity formulas, one can use [37, Equation 78]:

$$\int_0^\infty \ln(1 + ay) y^{n-1} e^{-cy} dy = \Gamma(n) e^{c/a} \sum_{k=1}^{n} \frac{\Gamma(-n + k, c/a)}{c^k a^{n-k}} \tag{15.128}$$

   where $\Gamma(\cdot,\cdot)$ is the complementary incomplete gamma function [14, Equation 8.350.2].

3. In the capacity variance formulas, one can use

$$\int_0^{\infty} \ln^2(1+ay)y^{n-1}e^{-cy}dy = \frac{2e^{c/a}}{a^n}\sum_{m=0}^{n-1}\binom{n-1}{m} \times (-1)^m G_{3\,4}^{4\,0}\left(\frac{c}{a}\middle|_{0,\,m-n,\,m-n,\,m-n}^{m-(n-1),\,m-(n-1),\,m-(n-1)}\right) \quad (15.129)$$

where $G_{p\,q}^{m\,n}(x|_{b_1,\dots,b_q}^{a_1,\dots,a_p})$ is the Meijer's G function defined by [14, Equation 9.301].

References

[1] G.J. Foschini and M.J. Gans, "On limits of wireless communication in a fading environment when using multiple antennas," *Wireless Personal Commun.*, 6, 311–335, 1998.

[2] G. Foschini, "Layered space-time architecture for wireless communication in a fading environment when using multi-element antennas," *Bell Labs Tech. Report*, Autumn 1996.

[3] G. Foschini, G.D. Golden, P.W. Wolniansky, and R.A. Valenzuela, "Simplified processing for wireless communication at high spectral efficiency," *IEEE J. Selected Areas Commun.*, 17, 1841–1852, 1999.

[4] I.E. Telatar, "Capacity of multi-antenna Gaussian channels," *Eur. Trans. Telecomm.*, 10(6), pp. 585–596, 1999.

[5] A. Scaglione, "Statistical analysis of the capacity of MIMO frequency selective Rayleigh fading channels with arbitrary number of inputs and outputs," in *Proceedings of the IEEE International Symposium Information Theory (ISIT 2002)*, Lausanne, Switzerland, June 2002, p. 278.

[6] P.J. Smith and M. Shafi, "On a Gaussian approximation to the capacity of wireless MIMO systems," in *Proceedings of the IEEE International Conference Communication (ICC'2002)*, New York, April 2002, pp. 406–410.

[7] C. Chuah, J. Kahn, D. Tse, and R. Valenzuela, "Capacity scaling in MIMO wireless systems under correlated fading," *IEEE Trans. Inf. Theory*, 48(3), 637–650, 2002.

[8] A. Shah and A.M. Haimovich, "Performance analysis of optimum combining in wireless communications with Rayleigh fading and cochannel interference," *IEEE Trans. Commun.*, 46(4), 473–479, 1998.

[9] S. Verdu and S. Shamai (Shitz), "Spectral efficiency of CDMA with random spreading," *IEEE Trans. Inf. Theory*, 45, 622–640, 1999.

[10] D.N.C. Tse and S.V. Hanly, "Linear multiuser receivers: effective interference, effective bandwidth and user capacity," *IEEE Trans. Inf. Theory*, 45, 641–657, 1999.

[11] A. Lozano and A.M. Tulino, "Capacity of multiple-transmit multiple-receive antenna architectures," *IEEE Trans. Inf. Theory*, 48, 3117–3128, 2002.

[12] M. Kang and M.-S. Alouini, "Largest eigenvalue of complex Wishart matrices and performance of MIMO MRC systems," *IEEE J. Selected Areas Commun.*, 21, 418–426, 2003.

[13] M. Kang and M.-S. Alouini, "On the capacity of MIMO Rician channels," in *Proceedings of the 40th Annual Allerton Conference on Communication, Control, and Computing (Allerton 2002)*, Monticello, IL, October 2002.

[14] I.S. Gradshteyn and I.M. Ryzhik, *Table of Integrals, Series, and Products*, 5th ed., Orlando, FL: Academic Press, 1994.

[15] M. Chiani, "Evaluating the capacity distribution of MIMO Rayleigh channels," in *Proceedings of the IEEE International Symposium Advance Wireless Communication (ISWC 2002)*, Victoria, Canada, September 2002, pp. 3–4.

[16] Z. Wang and G.B. Giannakis, "Outage mutual information of space-time MIMO channels," in *Proceedings of the 40th Annual Allerton Conference on Communication, Control, and Computing (Allerton 2002)*, Monticello, IL, October 2002.

[17] R.A. Horn and C.R. Johnson, *Matrix Analysis*, Cambridge, U.K.: Cambridge University Press, 1985.

[18] C.G. Khatri, "On certain distribution problems based on positive definite quadratic functions in normal vectors," *Ann. Math. Stat.*, 37(2), 468–470, 1966.

[19] M. Kang and M.-S. Alouini, "Impact of correlation on the capacity of MIMO channels," in *Proceedings of the IEEE International Conference Communication (ICC 2003)*, Anchorage, AK, May 2003.

[20] T.M. Cover and J.A. Thomas, *Elements of Information Theory*, New York: John Wiley, 1991.

[21] S.A. Jafar, S. Vishwanath, and A.J. Goldsmith, "Channel capacity and beamforming for multiple transmit and receive antennas with covariance feedback," in *Proceedings of the IEEE International Conference Communication (ICC 2001)*, Helsinki, Finland, April 2001, pp. 2266–2270.

[22] T.K.Y. Lo, "Maximum ratio transmission," *IEEE Trans. Commun.*, 47, 1458–1461, 1999.

[23] C.-H. Tse, K.-W. Yip, and T.-S. Ng, "Performance tradeoffs between maximum ratio transmission and switched-transmit diversity," in *Proceedings of the 11th IEEE International Symposium on Personal, Indoor and Mobile Radio Communication (PIMRC 2000)*, London, September 2000, pp. 1485–1489.

[24] P.A. Dighe, R.K. Mallik, and S.R. Jamuar, "Analysis of transmit-receive diversity in Rayleigh fading," in *Proceedings of the IEEE Global Telecommunication Conference (Globecom 2001)*, San Antonio, TX, November 2001, pp. 1132–1136.

[25] M. Kang and M.-S. Alouini, "Performance analysis of MIMO MRC systems over Rician fading channels," in *Proceedings of the IEEE Vehicular Technology Conference (VTC 2002—Fall)*, Vancouver, Canada, September 2002, pp. 869–873.

[26] A.H. Nuttall, "Some Integrals Involving the *Q* Function," Technology Report 4297, Naval Underwater Systems Center, New London Lab, New London, 1972.

[27] M. Abramowitz and I.A. Stegun, *Handbook of Mathematical Functions with Formulas, Graphs, and Mathematical Tables*, 9th ed., New York: Dover Publications, 1970.

[28] M. Kang and M.-S. Alouini, "Water-filling capacity and beamforming performance of MIMO channels with covariance feedback," in *Proceedings of the IEEE Workshop on Signal Processing Advance on Wireless Communication (SPAWC 2003)*, Rome, Italy, June 2003.

[29] M.K. Simon and M.-S. Alouini, *Digital Communications over Generalized Fading Channels: A Unified Approach to Performance Analysis*, New York: John Wiley & Sons, 2000.

[30] K.-K. Wong, R.S.K. Cheng, K.B. Letaief, and R.D. Murch, "Adaptive antennas at the mobile and base stations in an OFDM/TDMA system," *IEEE Trans. Commun.*, 49(1), 195–206, 2001.

[31] M. Kang, L. Yang, and M.-S. Alouini, "Performance analysis of MIMO channels in presence of co-channel interference and additive Gaussian noise," in *Proceedings of the 35th Annual Conference on Information Sciences and Systems (CISS 2003)*, Johns Hopkins University, Baltimore, MD, March 2003.

[32] R.S. Blum, J.H. Winters, and N.R. Sollenberger, "On the capacity of cellular systems with MIMO," *IEEE Commun. Lett.*, 242–244, 2002.

[33] M. Kang, L. Yang, and M.-S. Alouini, "Capacity of MIMO Rician channels with multiple correlated Rayleigh co-channel interferers," in *IEEE Global Telecommunication Conference (Globecom 2003)*, San Francisco, CA, December 2003.

[34] M. Kang and M.-S. Alouini, "Performance analysis of MIMO systems with co-channel interference over Rayleigh fading channels," in *Proceedings of the IEEE International Conference Communication (ICC 2002)*, New York, April 2002, pp. 391–395.

[35] M. Kang and M.-S. Alouini, "Quadratic forms in complex Gaussian matrices and performance analysis of MIMO systems with co-channel interference," in *Proceedings of the IEEE International Symposium Information Theory (ISIT 2002)*, Lausanne, Switzerland, June 2002, pp. 160–160.

[36] M. Kang and M.-S. Alouini, "Quadratic forms in complex Gaussian matrices and performance analysis of MIMO systems with co-channel interference," *IEEE Trans. Wireless Commun.*, 3(2), 418–431, March 2004.

[37] M.-S. Alouini and A. Goldsmith, "Capacity of Rayleigh fading channels under different adaptive transmission and diversity techniques," *IEEE Trans. Veh. Technol.*, VT-48(4), 1165–1181, 1999.

VI

Equalization and Receiver Design

16 Equalization Techniques for Fading Channels *Geert Leus and Marc Moonen* **16-1**
Introduction • Wireless Channel Model • System Model • Block Equalization • Serial Linear Equalization • Serial Decision Feedback Equalization • Frequency-Domain Equalization for TIV Channels • Existence of Zero-Forcing Solution • Complexity • Channel Estimation and Direct Equalizer Design • Performance Results • Summary

17 Low-Complexity Diversity Combining Schemes for Mobile Communications
Hong-Chuan Yang and Mohamed-Slim Alouini **17-1**
Introduction • System and Channel Models • Dual-Branch Switch and Stay Combining • Multibranch Switched Diversity • Generalized Switch and Examine Combining (GSEC) • Further Remarks

18 Overview of Equalization Techniques for MIMO Fading Channels
Noura Sellami, Inbar Fijalkow, and Mohamed Siala **18-1**
Introduction • Frequency-Selective MIMO Channel Model • Block Linear and Decision-Feedback Equalizers • List-Type Equalizers • The Multidimensional Whitened Matched Filter • The Block Equalizers vs. the Prefiltered List-Type MAP • Conclusion

19 Neural Networks for Transmission over Nonlinear Channels
Mohamed Ibnkahla, Yuan Jun, and Rober Boutros **19-1**
Introduction • Identification of Memoryless Nonlinear Amplifiers • Modeling and Identification of Nonlinear Channels with Memory • Channel Equalization • Conclusion

16

Equalization Techniques for Fading Channels

16.1 Introduction 16-1
16.2 Wireless Channel Model 16-2
TIV Channels • TV Channels
16.3 System Model 16-6
TIV Channels • TV Channels
16.4 Block Equalization 16-7
Block Linear Equalization • Block Decision Feedback Equalization
16.5 Serial Linear Equalization 16-10
TIV Channels • TV Channels • Equalizer Design
16.6 Serial Decision Feedback Equalization 16-14
TIV Channels • TV Channels • Equalizer Design
16.7 Frequency-Domain Equalization for TIV Channels .. 16-17
FD Linear Equalization • FD Decision Feedback Equalization
16.8 Existence of Zero-Forcing Solution 16-21
Linear Equalizers • Decision Feedback Equalizers
16.9 Complexity 16-22
Design Complexity • Implementation Complexity
16.10 Channel Estimation and Direct Equalizer Design ... 16-23
16.11 Performance Results 16-24
TIV Channels • TV Channels
16.12 Summary 16-27

Geert Leus
Delft University of Technology

Marc Moonen
Katholieke Universiteit Leuven

16.1 Introduction

Due to the distortive character of the propagation environment, transmitted data symbols will spread out in time and will interfere with each other, a phenomenon called intersymbol interference (ISI). The degree of ISI depends on the data rate: the higher the data rate, the more ISI is introduced. On the other hand, changes in the propagation environment, e.g., due to mobility in wireless communications, introduce channel time variation, which could be very harmful. Mitigating these fading channel effects, also referred to as *channel equalization*, constitutes a major challenge in current and future communication systems.

In order to design a good channel equalizer, a practical channel model has to be derived. First, we can write the overall system as a symbol rate single-input multiple-output (SIMO) system, where the

0-8493-1657-X/$0.00+$1.50

multiple outputs are obtained by multiple receive antennas and fractional sampling. Then, looking at a fixed time window, we can distinguish between *time-invariant* (TIV) and *time-variant* (TV) channels. For TIV channels, we will model the channel by a TIV finite impulse response (FIR) channel, whereas for TV channels, it will be convenient to model the channel time variation by means of a basis expansion model (BEM), leading to a BEM FIR channel [13], [32], [39].

For TIV channels, channel equalizers have been extensively studied in literature (see, for instance, [11], [14, Chapter 5], [18, Chapter 10], [29, Chapter 10] and references therein). For TV channels, on the other hand, they have only been introduced recently. Instead of focusing on complex maximum likelihood (ML) or maximum *a posteriori* (MAP) equalizers, we will discuss more practical *finite-length* linear and decision feedback equalizers. We derive minimum mean square error (MMSE) solutions, which strike an optimal balance between ISI removal and noise enhancement. By setting the signal power to infinity, these MMSE solutions can easily be transformed into zero-forcing (ZF) solutions that completely remove the ISI. We mainly focus on equalizer design based on channel knowledge, and briefly mention channel estimation algorithms and direct equalizer design algorithms, which do not require channel knowledge.

In this chapter, we distinguish between block equalizers and serial equalizers (as already mentioned, only practical finite-length versions will be considered). Block equalizers treat both TIV and TV channels in a similar fashion, and will therefore be described in a unified way. *Block linear equalizers* (BLEs) [15], [33], as well as *block decision feedback equalizers* (BDFEs) [15], [36], will be discussed.

What the serial equalizers are concerned, we will focus on both *serial linear equalizers* (SLEs) and *serial decision feedback equalizers* (SDFEs). It will be convenient to use the same models for the serial equalizer and for the channel. Hence, for a TIV FIR channel, we will use a TIV FIR serial equalizer [1], [41], whereas for a BEM FIR channel, we will use a BEM FIR serial equalizer [2], [3], [19]. For an SDFE, this means that both the feedforward and the feedback filters are modeled this way. Note that in the past a TIV FIR serial equalizer has been employed to equalize a BEM FIR channel, but this requires a symbol rate SIMO channel with many outputs for the linear ZF solution to exist [23]. However, when a BEM FIR serial equalizer is used to equalize a BEM FIR channel, only a symbol rate SIMO channel with two outputs is required for the linear ZF solution to exist [3], [19]. We will discuss serial equalization for TIV and TV channels in parallel, in order to show the similarities between the two approaches.

Finally, for TIV channels, it is also possible to adopt *frequency-domain* (FD) *equalization*, which can be viewed as a structured block equalization. We will discuss *FD linear equalizers* (FDLEs) [6], [10], [31] as well as *FD decision feedback equalizers* (FDDFEs) [4], [10].

Note that throughout this chapter, we will mainly focus our attention on wireless communications. However, most of the proposed techniques can also be adopted for other types of communications, i.e., wireline communications, optical communications, underwater communications, etc.

Notation: We use upper- and lowercase bold-faced letters to denote matrices and column vectors, respectively. Superscripts $*$, T, and H represent complex conjugate, transpose, and Hermitian, respectively. We denote the Kronecker delta by $\delta[n]$ and the Kronecker product by $\otimes$. The convolution operation is represented by $\star$. We denote the $N \times N$ identity matrix as $\mathbf{I}_N$ and the $M \times N$ all-zero matrix as $\mathbf{0}_{M\times N}$. For a column vector $\mathbf{x}$, diag$\{\mathbf{x}\}$ denotes the diagonal matrix with $\mathbf{x}$ on the diagonal, whereas for a square matrix $\mathbf{X}$, diag$\{\mathbf{X}\}$ denotes the diagonal matrix with the diagonal of $\mathbf{X}$ on the diagonal. Next, $[\mathbf{x}]_{i_1:i_2}$ denotes the subvector of $\mathbf{x}$ containing entries i_1 to i_2 (if $i_1 : i_2$ is replaced by i, only the ith entry is considered), and $[\mathbf{X}]_{r_1:r_2,c_1:c_2}$ denotes the submatrix of $\mathbf{X}$ on the intersection of rows r_1 to r_2 and columns c_1 to c_2 (if $r_1 : r_2$ ($c_1 : c_2$) is replaced by r (c), only the rth row (cth column) is considered; if r_1 and r_2 (c_1 and c_2) are omitted, all rows (columns) are considered). Finally, Q$\{\cdot\}$ represents a decision device that optimally maps soft symbol estimates into hard symbol estimates.

16.2 Wireless Channel Model

In this section, we discuss the channel model, which is, of course, a crucial ingredient when deriving means to mitigate fading channel effects. As already mentioned, we will focus our attention on wireless channels, but note that the proposed channel model also holds for many other applications. More specifically, we consider a baseband description of a wireless system with one transmit and M receive antennas. For the mth receive antenna, the symbol sequence $x[n]$ is filtered by the transmit filter $g_{\text{tr}}(t)$, distorted by the physical channel $g_{\text{ch}}^{(m)}(t;\tau)$, corrupted by additive noise $v^{(m)}(t)$, and finally filtered by the receive filter $g_{\text{rec}}(t)$. The received signal at the mth receive antenna $y^{(m)}(t)$ can then be written as

$$y^{(m)}(t) = \sum_{n=-\infty}^{\infty} g^{(m)}(t; t - nT)x[n] + w^{(m)}(t)$$

where T is the symbol period, $w^{(m)}(t) := g_{\text{rec}}(t) \star v^{(m)}(t)$ is the additive noise signal at the mth receive antenna, and [12, Chapter 1]

$$g^{(m)}(t;\tau) := \int_{-\infty}^{\infty}\int_{-\infty}^{\infty} g_{\text{rec}}(s)g_{\text{tr}}(\tau - \theta - s)g_{\text{ch}}^{(m)}(t - s;\theta)dsd\theta \tag{16.1}$$

is the composite channel for the mth receive antenna. Note that the larger the number of receive antennas M, the smaller the probability that at some time all M channels are in a deep fade. As a result, the larger the number of receive antennas M, the better the performance. This phenomenon is known as *receive antenna diversity* [28].

Symbol rate sampling, i.e., sampling the M receive antennas at rate $1/T$, is one option, but when the channel bandwidth is larger than $1/(2T)$, the rate $1/T$ is lower than the Nyquist rate. This causes aliasing, which could deteriorate the performance. *Fractional sampling*, i.e., sampling the M receive antennas at rate P/T with $P > 1$, can solve this problem [38], [42]. However, note that since the channel bandwidth is never significantly larger than $1/T$, the performance will not increase much when increasing P beyond $P = 2$.

Focusing on the general case, where the M receive antennas are sampled at rate P/T with $P \geq 1$, each rate P/T received sequence can be split into P rate $1/T$ received sequences. The pth rate $1/T$ received sequence at the mth receive antenna $y^{(mP+p)}[n] := y^{(m)}((nP + p)T/P)$ can be written as

$$y^{(mP+p)}[n] := \sum_{\nu=-\infty}^{\infty} g^{(mP+p)}[n;\nu]x[n-\nu] + w^{(mP+p)}[n] \tag{16.2}$$

where $w^{(mP+p)}[n] := w^{(m)}((nP+p)T/P)$ and $g^{(mP+p)}[n;\nu] := g^{(m)}((nP+p)T/P; (\nu P + p)T/P)$. Hence, we obtain a symbol rate single-input multiple-output system with $A = MP$ outputs, which are obtained by multiple receive antennas and fractional sampling.

Most wireless links experience multipath propagation, where clusters of reflected or scattered rays arrive at the receiver. All the rays within the same cluster experience the same delay, but each of them is characterized by its own complex gain and frequency offset. Hence, we can express the physical channel $g_{\text{ch}}^{(m)}(t;\tau)$ as [5, Chapter 3], [9], [12, Chapter 1], [16, Chapter 1]

$$g_{\text{ch}}^{(m)}(t;\tau) = \sum_c \delta\left(\tau - \tau_c^{(m)}\right) \sum_r G_{c,r}^{(m)} e^{j2\pi f_{c,r}^{(m)} t} \tag{16.3}$$

where $\tau_c^{(m)}$ is the delay of the cth cluster related to the mth receive antenna, and $G_{c,r}^{(m)}$ and $f_{c,r}^{(m)}$ are the respective complex gain and frequency offset of the rth ray of the cth cluster related to the mth receive antenna. Assuming the time variation of the physical channel $g_{\text{ch}}^{(m)}(t;\tau)$ over the span of the receive filter $g_{\text{rec}}(t)$ is negligible, we can replace $g_{\text{ch}}^{(m)}(t - s;\theta)$ by $g_{\text{ch}}^{(m)}(t;\theta)$ in Equation 16.1, leading to

$$\begin{aligned}
g^{(m)}(t;\tau) &= \int_{-\infty}^{\infty}\left(\int_{-\infty}^{\infty} g_{\text{rec}}(s)g_{\text{tr}}(\tau - \theta - s)ds\right) g_{\text{ch}}^{(m)}(t;\theta)d\theta \\
&= \int_{-\infty}^{\infty} \psi(\tau - \theta)g_{\text{ch}}^{(m)}(t;\theta)d\theta \\
&= \sum_c \psi\left(\tau - \tau_c^{(m)}\right) \sum_r G_{c,r}^{(m)} e^{j2\pi f_{c,r}^{(m)} t}
\end{aligned}$$

where $\psi(t) := g_{\mathrm{rec}}(t) \star g_{\mathrm{tr}}(t)$. This means that the channel $g^{(mP+p)}[n;\nu]$ can be expressed as

$$g^{(mP+p)}[n;\nu] = g^{(m)}((nP+p)T/P;(\nu P+p)T/P)$$
$$= \sum_{c} \psi\left((\nu P+p)T/P - \tau_c^{(m)}\right) \sum_{r} G_{c,r}^{(m)} e^{j2\pi f_{c,r}^{(m)}(nP+p)T/P} \quad (16.4)$$

The above channel model has a rather complex structure, which complicates, if not prevents, the development of a low-complexity equalization structure that blends well with the channel structure. Moreover, the above channel model contains a large number of parameters, which causes a major problem when trying to estimate the channel. Hence, we will have to look for other channel models that are well structured and contain a smaller number of parameters. Therefore, we will look at a limited time window $t \in (0, NT)$, which corresponds to $n \in \{0, 1, \ldots, N-1\}$. Depending on the ratio of NT over $1/f_{\max}$, where $f_{\max}$ is the overall *Doppler spread* of all M channels,

$$f_{\max} := \max_{m,c,r} \left\{ \left| f_{c,r}^{(m)} \right| \right\}$$

we call the channel related to the ath output $g^{(a)}[n;\nu]$ time-invariant or time-variant (note that $a \in \{0, 1, \ldots, A-1\}$).

16.2.1 TIV Channels

We refer to a channel as TIV if the channel time variation over NT is negligible, i.e., if NT is much smaller than $1/f_{\max}$. Assuming that each composite channel satisfies $g^{(m)}(t;\tau) = 0$ for $\tau \notin [0, (L+1)T]$, each TIV channel $g^{(a)}[n;\nu]$ can be modeled for $n \in \{0, 1, \ldots, N-1\}$ by a so-called TIV FIR channel:

$$h^{(a)}[n;\nu] = \sum_{l=0}^{L} \delta[\nu - l] h_l^{(a)} \quad (16.5)$$

The above TIV FIR channel is well structured and contains a small number of parameters. Hence, we have obtained a practical channel model.

From Equation 16.2, the TIV FIR input–output relation for $n \in \{0, 1, \ldots, N-1\}$ can be written as (see also Figure 16.1)

$$y^{(a)}[n] = \sum_{l=0}^{L} h_l^{(a)} x[n-l] + w^{(a)}[n] \quad (16.6)$$

16.2.2 TV Channels

We refer to a channel as TV if the channel time variation over NT is not negligible, i.e., if NT is not much smaller than $1/f_{\max}$. Assuming that each composite channel satisfies $g^{(m)}(t;\tau) = 0$ for $\tau \notin [0, (L+1)T]$,

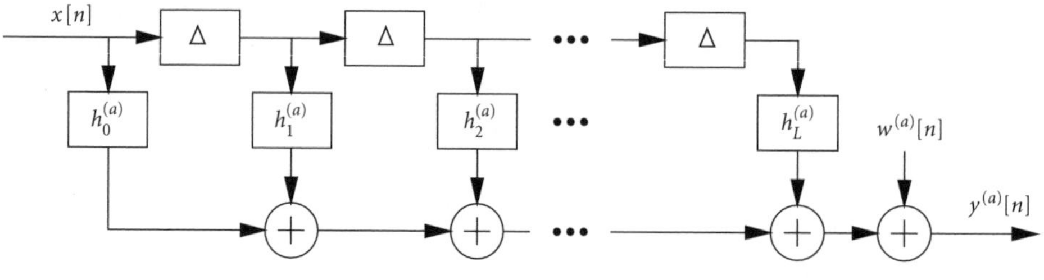

FIGURE 16.1 TIV FIR input–output relation (Δ represents a unit delay).

each TV channel $g^{(a)}[n;\nu]$ can be modeled for $n \in \{0,1,\ldots,N-1\}$ by a so-called TV FIR channel:

$$h^{(a)}[n;\nu] = \sum_{l=0}^{L} \delta[\nu - l] h_l^{(a)}[n] \tag{16.7}$$

Like the TIV FIR channel, the TV FIR channel is well structured, but in contrast to the TIV FIR channel, the TV FIR channel contains a large number of parameters, which is objectionable. The key in finding a TV channel model that is well structured *and* contains a small number of parameters is to model the channel time variation using a so-called basis expansion model [13], [32], [39].

Assuming each composite channel satisfies $g^{(m)}(t;\tau) = 0$ for $\tau \notin [0,(L+1)T]$, each TV channel $g^{(a)}[n;\nu]$ can be modeled for $n \in \{0,1,\ldots,N-1\}$ by a so-called BEM FIR channel:

$$h^{(a)}[n;\nu] = \sum_{l=0}^{L} \delta[\nu - l] \sum_{q=-Q/2}^{Q/2} h_{q,l}^{(a)} e^{j2\pi qn/K} \tag{16.8}$$

where the parameters Q and K should be selected such that $Q/(2KT) \approx f_{\max}$. Note that in general Q can be kept very small as long as NT is smaller than $1/(2f_{\max})$, as illustrated in the next example. Hence, we have again obtained a practical channel model that is well structured and contains a small number of parameters.

From Equation 16.2, the BEM FIR input–output relation for $n \in \{0,1,\ldots,N-1\}$ can be written as (see also Figure 16.2)

$$y^{(a)}[n] = \sum_{l=0}^{L} \sum_{q=-Q/2}^{Q/2} h_{q,l}^{(a)} e^{j2\pi qn/K} x[n-l] + w^{(a)}[n] \tag{16.9}$$

Example: Consider a channel $g_{\text{ch}}^{(0)}(t)$ consisting of five clusters of 100 reflected or scattered rays. The delay of the cth cluster is given by $\tau_c = cT/2$ $(c \in \{0,1,2,3,4\})$. Assuming that $g_{\text{tr}}(t)$ and $g_{\text{rec}}(t)$ are rectangular functions over $(0, T)$ with height $1/T$, and thus $\psi(t) = g_{\text{rec}}(t) \star g_{\text{tr}}(t)$ is a triangular function over $(0, 2T)$ with height 1, we can assume that $L = 3$. The complex gain and frequency offset of the rth ray of the cth cluster are given by $G_{c,r}^{(0)} = e^{j\theta_{c,r}^{(0)}}/\sqrt{100}$ and $f_{c,r}^{(0)} = \cos(\phi_{c,r}^{(0)}) f_{\max}$, where $\theta_{c,r}^{(0)}$ and $\phi_{c,r}^{(0)}$ are uniformly distributed over $(0, 2\pi)$. Assuming that $f_{\max} = 1/(400T)$, we now show that the BEM FIR channel is very accurate when Q and K are selected such that $Q/(2KT) \approx f_{\max} = 1/(400T)$. To illustrate that Q can be kept very small as long as NT is smaller than $1/(2f_{\max})$, we consider the extreme case of $Q = 2$ and $NT = 1/(2f_{\max}) = 200T$. To satisfy $Q/(2KT) \approx f_{\max} = 1/(400T)$, we then take $K = 400$. Assuming fractional sampling with a factor $P = 2$, Figure 16.3 shows the modulus of the eight TV channel taps

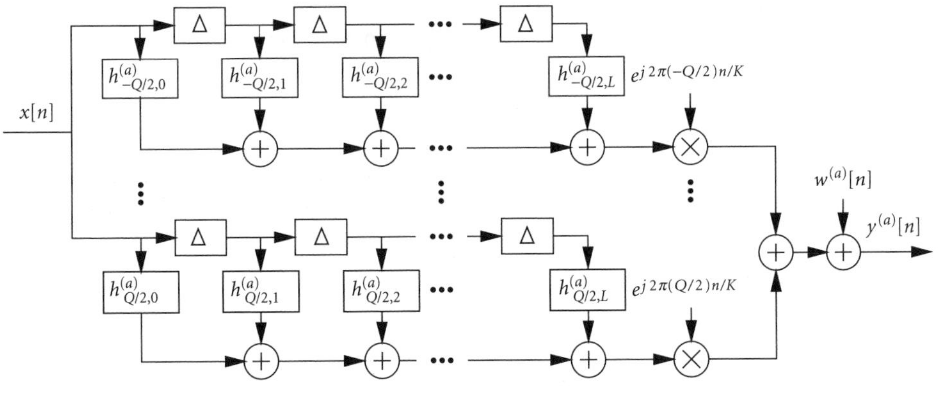

FIGURE 16.2 BEM FIR input–output relation (Δ represents a unit delay).

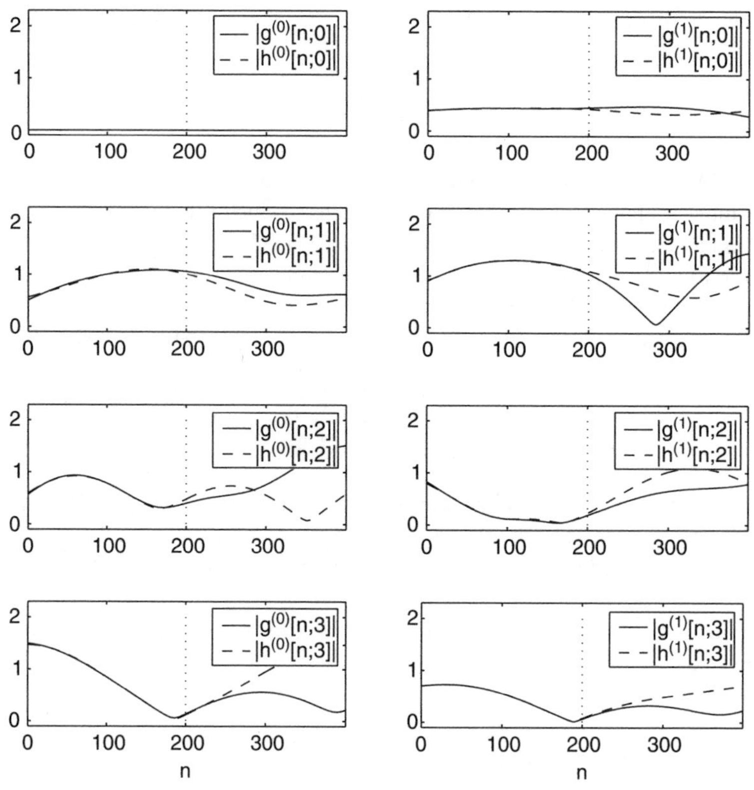

FIGURE 16.3 Illustration of the tight fit of the BEM FIR channel.

$\{\{g^{(a)}[n;l]\}_{a=0}^{1}\}_{l=0}^{3}$ and the modulus of the eight BEM FIR channel taps $\{\{h^{(a)}[n;l]\}_{a=0}^{1}\}_{l=0}^{3}$ obtained by least squares fitting over $n \in \{0, 1, \ldots, 199\}$. Clearly the approximation for $n \in \{0, 1, \ldots, 199\}$ is very good.

16.3 System Model

From now on we will adopt the TIV FIR channel of Equation 16.5 for TIV channels and the BEM FIR channel of Equation 16.8 for TV channels. We restrict our attention to outputs $y^{(a)}[n]$ for $n \in \{0, 1, \ldots, N-1\}$.

Defining the $(N+L) \times 1$ data symbol block $\mathbf{x} := [x[-L], \ldots, x[N-1]]^T$, the $N \times 1$ received sample block at the ath output $\mathbf{y}^{(a)} := [y^{(a)}[0], \ldots, y^{(a)}[N-1]]^T$ can be written as

$$\mathbf{y}^{(a)} = \mathbf{H}^{(a)}\mathbf{x} + \mathbf{w}^{(a)} \tag{16.10}$$

where $\mathbf{w}^{(a)}$ is similarly defined as $\mathbf{y}^{(a)}$ and $\mathbf{H}^{(a)}$ is an $N \times (N+L)$ channel matrix. The definition of the latter depends on whether we are dealing with TIV or TV channels. In either case, defining $\mathbf{y} := [\mathbf{y}^{(0)T}, \ldots, \mathbf{y}^{(A-1)T}]^T$, we obtain

$$\mathbf{y} = \mathbf{H}\mathbf{x} + \mathbf{w}$$

where $\mathbf{w}$ is similarly defined as $\mathbf{y}$ and $\mathbf{H} := [\mathbf{H}^{(0)T}, \ldots, \mathbf{H}^{(A-1)T}]^T$. Note that throughout this chapter we will assume perfect knowledge of $\mathbf{H}$. In Section 16.10, we will give a few hints on how to estimate $\mathbf{H}$ in practice.

16.3.1 TIV Channels

In case of TIV channels, the $N \times (N + L)$ channel matrix $\mathbf{H}^{(a)}$ is given by

$$\mathbf{H}^{(a)} = \sum_{l=0}^{L} h_l^{(a)} \mathbf{Z}_l \tag{16.11}$$

where $\mathbf{Z}_l := [\mathbf{0}_{N\times(L-l)}, \mathbf{I}_N, \mathbf{0}_{N\times l}]$. Substituting Equation 16.11 into Equation 16.10, the $N \times 1$ received sample block at the ath output can then be written as

$$\mathbf{y}^{(a)} = \sum_{l=0}^{L} h_l^{(a)} \mathbf{Z}_l \mathbf{x} + \mathbf{w}^{(a)} \tag{16.12}$$

16.3.2 TV Channels

In case of TV channels, the $N \times (N + L)$ channel matrix $\mathbf{H}^{(a)}$ is given by

$$\mathbf{H}^{(a)} = \sum_{l=0}^{L} \sum_{q=-Q/2}^{Q/2} h_{q,l}^{(a)} \mathbf{D}_q \mathbf{Z}_l \tag{16.13}$$

where $\mathbf{D}_q := \text{diag}\{[1, e^{j2\pi q/K}, \ldots, e^{j2\pi q(N-1)/K}]^T\}$ and $\mathbf{Z}_l := [\mathbf{0}_{N\times(L-l)}, \mathbf{I}_N, \mathbf{0}_{N\times l}]$. Substituting Equation 16.13 into Equation 16.10, the $N \times 1$ received sample block at the ath output can then be written as

$$\mathbf{y}^{(a)} = \sum_{l=0}^{L} \sum_{q=-Q/2}^{Q/2} h_{q,l}^{(a)} \mathbf{D}_q \mathbf{Z}_l \mathbf{x} + \mathbf{w}^{(a)} \tag{16.14}$$

16.4 Block Equalization

For block equalization, we will assume that $\mathbf{x} = [\mathbf{0}_{1\times L}, \mathbf{s}^T, \mathbf{0}_{1\times L}]^T$, where $\mathbf{s}$ is an $(N - L) \times 1$ data symbol block. This corresponds to zero padding-based block transmission where L zeros are padded after each data symbol block of length $N - L$. The received sample block at the ath output can then be written as

$$\mathbf{y}^{(a)} = \bar{\mathbf{H}}^{(a)} \mathbf{s} + \mathbf{w}^{(a)} \tag{16.15}$$

where $\bar{\mathbf{H}}^{(a)} := [\mathbf{H}^{(a)}]_{:,L+1:N}$. We further obtain

$$\mathbf{y} = \bar{\mathbf{H}} \mathbf{s} + \mathbf{w}$$

where $\bar{\mathbf{H}} := [\bar{\mathbf{H}}^{(0)T}, \ldots, \bar{\mathbf{H}}^{(A-1)T}]^T$

Zero padding can be viewed as a special case of known symbol padding, where the same L known symbols are padded after each data symbol block of length $N - L$. When not all zero, these known symbols can aid synchronization and channel estimation (for TIV channels this has been discussed in [7], [20], [30]). However, for the sake of simplicity, we will stick to zero padding. All results presented for zero padding can easily be modified for the more general known symbol padding case.

Zero padding can also be viewed as a special case of linear precoding [33], [34]. Actually, for TIV channels, zero padding turns out to be the best type of linear precoding in terms of performance at high SNR [33], [34]. For TV channels, on the other hand, this is not the case, and other linear precoding strategies with a better performance have been suggested [25] (see also [22] for a multiuser scenario). However, since we do not want to focus on linear precoder design, we will stick to zero padding. All results presented for zero padding can easily be modified for the more general linear precoding case.

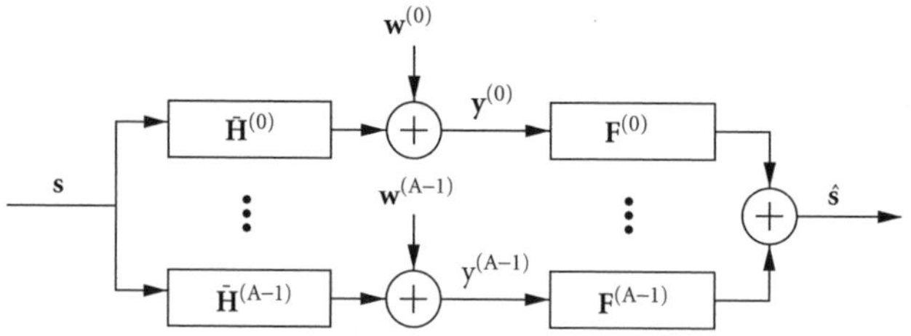

FIGURE 16.4 Block linear equalization.

We could also adopt cyclic prefix-based block transmission [31]. However, as we will show later on, zero padding-based block transmission is closely related to cyclic prefix-based block transmission. Hence, we will not discuss it in this chapter. A comprehensive overview of different types of block transmission is presented in [45].

16.4.1 Block Linear Equalization

In this section, we discuss block linear equalization [15], [33] (see also [17], [24] for a similar approach in the CDMA context). We consider zero padding-based block transmission (see Equation 16.15). As illustrated in Figure 16.4, we adopt a block linear equalizer consisting of a block filter $\mathbf{F}^{(a)}$ for the ath output, in order to find an estimate of $\mathbf{s}$:

$$\hat{\mathbf{s}} = \sum_{a=0}^{A-1} \mathbf{F}^{(a)}\mathbf{y}^{(a)} = \left(\sum_{a=0}^{A-1} \mathbf{F}^{(a)}\bar{\mathbf{H}}^{(a)}\right)\mathbf{s} + \sum_{a=0}^{A-1} \mathbf{F}^{(a)}\mathbf{w}^{(a)}$$

Defining $\mathbf{F} := [\mathbf{F}^{(0)}, \ldots, \mathbf{F}^{(A-1)}]$, we then obtain

$$\hat{\mathbf{s}} = \mathbf{F}\mathbf{y} = \mathbf{F}\bar{\mathbf{H}}\mathbf{s} + \mathbf{F}\mathbf{w}$$

Let us focus on the MMSE BLE, which minimizes the MSE $\mathcal{J} = \mathcal{E}\{\|\mathbf{s} - \hat{\mathbf{s}}\|^2\}$. Defining the data and noise covariance matrices as $\mathbf{R}_s := \mathcal{E}\{\mathbf{s}\mathbf{s}^H\}$ and $\mathbf{R}_w := \mathcal{E}\{\mathbf{w}\mathbf{w}^H\}$, respectively, the MSE can be expressed as

$$\mathcal{J} = \text{tr}\{\mathbf{F}(\bar{\mathbf{H}}\mathbf{R}_s\bar{\mathbf{H}}^H + \mathbf{R}_w)\mathbf{F}^H - 2\Re\{\mathbf{R}_s\bar{\mathbf{H}}^H\mathbf{F}^H\} + \mathbf{R}_s\}$$

Solving $\partial\mathcal{J}/\partial\mathbf{F} = \mathbf{0}$, we obtain

$$\begin{aligned}\mathbf{F}_{MMSE} &= \mathbf{R}_s\bar{\mathbf{H}}^H(\bar{\mathbf{H}}\mathbf{R}_s\bar{\mathbf{H}}^H + \mathbf{R}_w)^{-1} \\ &= \left(\bar{\mathbf{H}}^H\mathbf{R}_w^{-1}\bar{\mathbf{H}} + \mathbf{R}_s^{-1}\right)^{-1}\bar{\mathbf{H}}^H\mathbf{R}_w^{-1}\end{aligned}$$

where the second equality is obtained by using the matrix inversion lemma. Assuming that $\bar{\mathbf{H}}$ has full column rank, the corresponding ZF BLE can be obtained by setting the signal power to infinity ($\mathbf{R}_s^{-1} = \mathbf{0}$):

$$\mathbf{F}_{ZF} = \left(\bar{\mathbf{H}}^H\mathbf{R}_w^{-1}\bar{\mathbf{H}}\right)^{-1}\bar{\mathbf{H}}^H\mathbf{R}_w^{-1}$$

Assuming the data sequence and additive noises are mutually uncorrelated and white with variance σ_s^2 and σ_v^2, respectively, the data and noise covariance matrices can be computed in closed form:

$$\mathbf{R}_s = \sigma_s^2\mathbf{I}_{N-L}$$

$$\mathbf{R}_w = \sigma_v^2\mathbf{I}_M \otimes \begin{bmatrix} \mathbf{\Phi}_{N,0} & \cdots & \mathbf{\Phi}_{N,P-1} \\ \vdots & & \vdots \\ \mathbf{\Phi}_{N,-P+1} & \cdots & \mathbf{\Phi}_{N,0} \end{bmatrix}$$

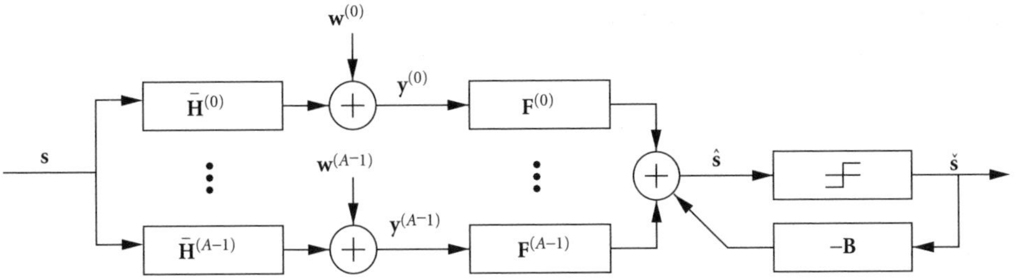

FIGURE 16.5 Block decision feedback equalization.

where $\mathbf{\Phi}_{I,p}$ is the $I \times I$ matrix defined as

$$[\mathbf{\Phi}_{I,p}]_{i,i'} := \int_{-\infty}^{\infty} g_{\text{rec}}(\tau) g_{\text{rec}}(\tau + (i' - i)T + pT/P) d\tau$$

16.4.2 Block Decision Feedback Equalization

In this section, we discuss block decision feedback equalization [15], [36] (see also [8], [17] for a similar approach in the CDMA context). We again consider zero padding-based block transmission (see Equation 16.15). As illustrated in Figure 16.5, we adopt a block decision feedback equalizer consisting of a block feedforward filter $\mathbf{F}^{(a)}$ for the ath output and a block feedback filter $\mathbf{B}$, in order to find an estimate of $\mathbf{s}$:

$$\hat{\mathbf{s}} = \sum_{a=0}^{A-1} \mathbf{F}^{(a)} \mathbf{y}^{(a)} - \mathbf{B}\check{\mathbf{s}} = \left(\sum_{a=0}^{A-1} \mathbf{F}^{(a)} \bar{\mathbf{H}}^{(a)} \right) \mathbf{s} - \mathbf{B}\check{\mathbf{s}} + \sum_{a=0}^{A-1} \mathbf{F}^{(a)} \mathbf{w}^{(a)} \tag{16.16}$$

where $\check{\mathbf{s}} := \mathrm{Q}\{\hat{\mathbf{s}}\}$. In order to feedback decisions in a causal way, we require $\mathbf{B}$ to be a zero diagonal upper triangular matrix. Defining $\mathbf{F} := [\mathbf{F}^{(0)}, \ldots, \mathbf{F}^{(A-1)}]$, and assuming past decisions are correct (a common assumption in DFE design), i.e., $\check{\mathbf{s}} = \mathbf{s}$, we obtain

$$\hat{\mathbf{s}} = \mathbf{F}\mathbf{y} - \mathbf{B}\mathbf{s} = \mathbf{F}\bar{\mathbf{H}}\mathbf{s} - \mathbf{B}\mathbf{s} + \mathbf{F}\mathbf{w}$$

Let us focus on the MMSE BDFE, which minimizes the MSE $\mathcal{J} = \mathcal{E}\{\|\mathbf{s} - \hat{\mathbf{s}}\|^2\}$. In a fashion similar to that for the BLE, the MSE can be expressed as

$$\mathcal{J} = \mathrm{tr}\{\mathbf{F}(\bar{\mathbf{H}}\mathbf{R}_s\bar{\mathbf{H}}^H + \mathbf{R}_w)\mathbf{F}^H - 2\Re\{(\mathbf{B} + \mathbf{I}_{N-L})\mathbf{R}_s\bar{\mathbf{H}}^H\mathbf{F}^H\} + (\mathbf{B} + \mathbf{I}_{N-L})\mathbf{R}_s(\mathbf{B} + \mathbf{I}_{N-L})^H\} \tag{16.17}$$

Solving $\partial\mathcal{J}/\partial\mathbf{F} = \mathbf{0}$, we obtain

$$\mathbf{F}_{MMSE} = (\mathbf{B} + \mathbf{I}_{N-L})\mathbf{R}_s\bar{\mathbf{H}}^H(\bar{\mathbf{H}}\mathbf{R}_s\bar{\mathbf{H}}^H + \mathbf{R}_w)^{-1} \tag{16.18}$$

$$= (\mathbf{B} + \mathbf{I}_{N-L})\left(\bar{\mathbf{H}}^H\mathbf{R}_w^{-1}\bar{\mathbf{H}} + \mathbf{R}_s^{-1}\right)^{-1}\bar{\mathbf{H}}^H\mathbf{R}_w^{-1} \tag{16.19}$$

where the second equality is again obtained by using the matrix inversion lemma. Next, substituting Equation 16.18 into Equation 16.17 results, after some calculation, in

$$\mathcal{J} = \mathrm{tr}\{(\mathbf{B} + \mathbf{I}_{N-L})\mathbf{R}_{MMSE}(\mathbf{B} + \mathbf{I}_{N-L})^H\}$$

where $\mathbf{R}_{MMSE} = (\bar{\mathbf{H}}^H\mathbf{R}_w^{-1}\bar{\mathbf{H}} + \mathbf{R}_s^{-1})^{-1}$. Solving $\partial\mathcal{J}/\partial\mathbf{B} = \mathbf{0}$ under the constraint that $\mathbf{B}$ is a zero diagonal upper triangular matrix, we finally obtain

$$\mathbf{B}_{MMSE} = \mathrm{diag}\{\mathrm{chol}\{\mathbf{R}_{MMSE}\}\}^{-1}\mathrm{chol}\{\mathbf{R}_{MMSE}\} - \mathbf{I}_{N-L}$$

where chol{**A**} represents the upper triangular matrix that satisfies the following Cholesky decomposition: $\mathbf{A} = \text{chol}\{\mathbf{A}\}^H \text{chol}\{\mathbf{A}\}$. To summarize, the MMSE BDFE is given by

$$\mathbf{F}_{MMSE} = (\mathbf{B}_{MMSE} + \mathbf{I}_{N-L})\left(\bar{\mathbf{H}}^H \mathbf{R}_w^{-1} \bar{\mathbf{H}} + \mathbf{R}_s^{-1}\right)^{-1} \bar{\mathbf{H}}^H \mathbf{R}_w^{-1}$$
$$\mathbf{R}_{MMSE} = \left(\bar{\mathbf{H}}^H \mathbf{R}_w^{-1} \bar{\mathbf{H}} + \mathbf{R}_s^{-1}\right)^{-1}$$
$$\mathbf{B}_{MMSE} = \text{diag}\{\text{chol}\{\mathbf{R}_{MMSE}\}\}^{-1} \text{chol}\{\mathbf{R}_{MMSE}\} - \mathbf{I}_{N-L}$$

Assuming $\bar{\mathbf{H}}$ has full column rank, the corresponding ZF BDFE can again be obtained by setting the signal power to infinity ($\mathbf{R}_s^{-1} = \mathbf{0}$):

$$\mathbf{F}_{ZF} = (\mathbf{B}_{ZF} + \mathbf{I}_{N-L})\left(\bar{\mathbf{H}}^H \mathbf{R}_w^{-1} \bar{\mathbf{H}}\right)^{-1} \bar{\mathbf{H}}^H \mathbf{R}_w^{-1}$$
$$\mathbf{R}_{ZF} = \left(\bar{\mathbf{H}}^H \mathbf{R}_w^{-1} \bar{\mathbf{H}}\right)^{-1}$$
$$\mathbf{B}_{ZF} = \text{diag}\{\text{chol}\{\mathbf{R}_{ZF}\}\}^{-1} \text{chol}\{\mathbf{R}_{ZF}\} - \mathbf{I}_{N-L}$$

16.5 Serial Linear Equalization

In this section, we discuss serial linear equalization. We do not focus on zero padding-based block transmission, but rather on the serial transmission model (see Equations 16.12 and 16.14). Hence, we assume that all entries of **x** contain data symbols. Note, however, that we will not estimate the edges of **x** and only estimate the middle part of **x** (denoted as $\mathbf{x}_\star$). The edges are either estimated in a previous step (top entries of **x**) or will be estimated in a next step (bottom entries of **x**).

We adopt a serial linear equalizer consisting of a serial filter $f^{(a)}[n;\nu]$ for the ath output, in order to find an estimate of $x[n-d]$ (see Figure 16.6):

$$\hat{x}[n-d] = \sum_{a=0}^{A-1} \sum_{\nu=-\infty}^{\infty} f^{(a)}[n;\nu] y^{(a)}[n-\nu] \tag{16.20}$$

where d represents the synchronization delay. To discuss the structure of this SLE in more detail, we distinguish between TIV and TV channels. Both cases will give rise to a related data model, which allows us to treat the equalizer design in a joint fashion.

16.5.1 TIV Channels

Since for a TIV channel, the TIV FIR channel of Equation 16.5 was applied, it is also convenient to use a TIV FIR serial filter $f^{(a)}[n;\nu]$ [41]. In other words, we design each serial equalizer $f^{(a)}[n;\nu]$ to have

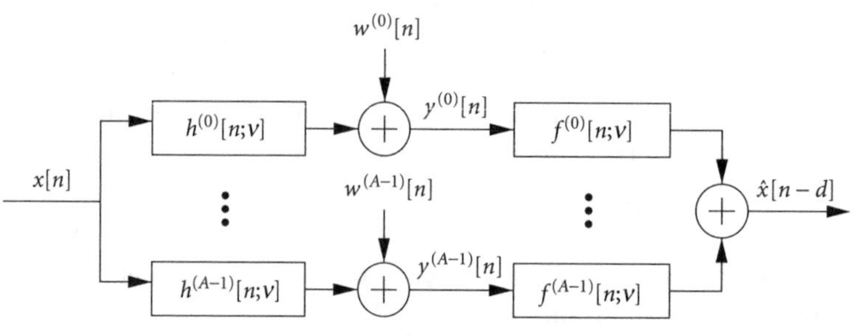

FIGURE 16.6 Serial linear equalization.

$L' + 1$ TIV taps:

$$f^{(a)}[n; \nu] = \sum_{l'=0}^{L'} \delta[\nu - l'] f_{l'}^{(a)}$$

An estimate of $x[n-d]$ is then computed as

$$\hat{x}[n-d] = \sum_{a=0}^{A-1} \sum_{l'=0}^{L'} f_{l'}^{(a)} y^{(a)}[n-l'] \tag{16.21}$$

Defining the l'th time-shifted received sequence related to the ath output as

$$\mathbf{y}_{l'}^{(a)} := \bar{\mathbf{Z}}_{l'} \mathbf{y}^{(a)}$$

where $\bar{\mathbf{Z}}_{l'} := [\mathbf{0}_{(N-L')\times(L'-l')}, \mathbf{I}_{N-L'}, \mathbf{0}_{(N-L')\times l'}]$, and introducing

$$\mathbf{x}_\star := [x[L'-d], \ldots, x[N-d-1]]^T$$

an estimate of $\mathbf{x}_\star$ is obtained as

$$\hat{\mathbf{x}}_\star^T = \sum_{a=0}^{A-1} \mathbf{f}^{(a)T} \mathbf{Y}^{(a)}$$

where $\mathbf{f}^{(a)}$ is the $(L'+1) \times 1$ vector given by $\mathbf{f}^{(a)} := [f_{L'}^{(a)}, \ldots, f_0^{(a)}]^T$ and $\mathbf{Y}^{(a)}$ is the $(L'+1) \times (N-L')$ matrix given by $\mathbf{Y}^{(a)} := [\mathbf{y}_{L'}^{(a)}, \ldots, \mathbf{y}_0^{(a)}]^T$.

Let us now rewrite $\mathbf{Y}^{(a)}$ as a function of the TIV FIR channel parameters and data symbols. The l'th time-shifted received sequence related to the ath output can be written as

$$\begin{aligned}
\mathbf{y}_{l'}^{(a)} &:= \bar{\mathbf{Z}}_{l'} \mathbf{y}^{(a)} \\
&= \sum_{l=0}^{L} h_l^{(a)} \bar{\mathbf{Z}}_{l'} \mathbf{Z}_l \mathbf{x} + \mathbf{w}_{l'}^{(a)} \\
&= \sum_{l=0}^{L} h_l^{(a)} \tilde{\mathbf{Z}}_{l+l'} \mathbf{x} + \mathbf{w}_{l'}^{(a)}
\end{aligned}$$

where $\mathbf{w}_{l'}^{(a)}$ is defined similarly as $\mathbf{y}_{l'}^{(a)}$ and $\tilde{\mathbf{Z}}_k := [\mathbf{0}_{(N-L')\times(L+L'-k)}, \mathbf{I}_{N-L'}, \mathbf{0}_{(N-L')\times k}]$. Introducing $k := l + l'$ and defining $\mathbf{x}_k := \tilde{\mathbf{Z}}_k \mathbf{x}$ (note that $\mathbf{x}_\star = \mathbf{x}_d$), we can also write this as

$$\mathbf{y}_{l'}^{(a)} = \sum_{k=0}^{L+L'} h_{k-l'}^{(a)} \mathbf{x}_k + \mathbf{w}_{l'}^{(a)}$$

Defining $\mathbf{X} := [\mathbf{x}_{L+L'}, \ldots, \mathbf{x}_0]^T$, $\mathbf{Y}^{(a)}$ can then be expressed as

$$\mathbf{Y}^{(a)} = \mathcal{H}^{(a)} \mathbf{X} + \mathbf{W}^{(a)}$$

where $\mathbf{W}^{(a)}$ is defined similarly as $\mathbf{Y}^{(a)}$ and $\mathcal{H}^{(a)}$ is the $(L'+1) \times (L+L'+1)$ Toeplitz matrix given by

$$\mathcal{H}^{(a)} := \begin{bmatrix} h_L^{(a)} & \ldots & h_0^{(a)} & & 0 \\ & \ddots & & \ddots & \\ 0 & & h_L^{(a)} & \ldots & h_0^{(a)} \end{bmatrix}$$

Defining $\mathbf{Y} := [\mathbf{Y}^{(0)T}, \ldots, \mathbf{Y}^{(A-1)T}]^T$, we then obtain

$$\mathbf{Y} = \mathcal{H}\mathbf{X} + \mathbf{W} \tag{16.22}$$

where $\mathbf{W}$ is defined similarly as $\mathbf{Y}$ and $\mathcal{H} := [\mathcal{H}^{(0)T}, \ldots, \mathcal{H}^{(A-1)T}]^T$. Hence, we obtain

$$\hat{\mathbf{x}}_\star^T = \sum_{a=0}^{A-1} \mathbf{f}^{(a)T}\mathbf{Y}^{(a)} = \mathbf{f}^T\mathbf{Y} = \mathbf{f}^T\mathcal{H}\mathbf{X} + \mathbf{f}^T\mathbf{W} \tag{16.23}$$

where $\mathbf{f} := [\mathbf{f}^{(0)T}, \ldots, \mathbf{f}^{(A-1)T}]^T$

16.5.2 TV Channels

Since for a TV channel the BEM FIR channel of Equation 16.8 is applied, it is also convenient to use a BEM FIR serial filter $f^{(a)}[n;\nu]$ [3], [19]. In other words, we design each serial filter $f^{(a)}[n;\nu]$ to have $L'+1$ TV taps, where the time variation of each tap is modeled by $Q'+1$ complex exponentials:

$$f^{(a)}[n;\nu] = \sum_{l'=0}^{L'} \delta[\nu - l'] \sum_{q'=-Q'/2}^{Q'/2} e^{j2\pi q' n/K} f_{q',l'}^{(a)}$$

An estimate of $x[n-d]$ is then computed as

$$\hat{x}[n-d] = \sum_{a=0}^{A-1}\sum_{l'=0}^{L'}\sum_{q'=-Q'/2}^{Q'/2} e^{j2\pi q' n/K} f_{q',l'}^{(a)} y^{(a)}[n-l']$$

Defining the q'th frequency-shifted and l'th time-shifted received sequence related to the ath output as

$$\mathbf{y}_{q',l'}^{(a)} := \bar{\mathbf{D}}_{q'}\bar{\mathbf{Z}}_{l'}\mathbf{y}^{(a)}$$

where $\bar{\mathbf{D}}_{q'} := \mathrm{diag}\{[1, e^{j2\pi q'/K}, \ldots, e^{j2\pi q'(N-L'-1)/K}]^T\}$ and $\bar{\mathbf{Z}}_{l'} := [\mathbf{0}_{(N-L')\times(L'-l')}, \mathbf{I}_{N-L'}, \mathbf{0}_{(N-L')\times l'}]$, and introducing

$$\mathbf{x}_\star := [x[L'-d], \ldots, x[N-d-1]]^T$$

an estimate of $\mathbf{x}_\star$ is obtained as

$$\hat{\mathbf{x}}_\star^T = \sum_{a=0}^{A-1} \mathbf{f}^{(a)T}\mathbf{Y}^{(a)}$$

where $\mathbf{f}^{(a)}$ is the $(L'+1)(Q'+1)\times 1$ vector given by $\mathbf{f}^{(a)} := [f_{Q'/2,L'}^{(a)}, \ldots, f_{Q'/2,0}^{(a)}, \ldots, f_{-Q'/2,0}^{(a)}]^T$ and $\mathbf{Y}^{(a)}$ is the $(L'+1)(Q'+1)\times(N-L')$ matrix given by $\mathbf{Y}^{(a)} := [\mathbf{y}_{Q'/2,L'}^{(a)}, \ldots, \mathbf{y}_{Q'/2,0}^{(a)}, \ldots, \mathbf{y}_{-Q'/2,0}^{(a)}]^T$.

Let us now rewrite $\mathbf{Y}^{(a)}$ as a function of the BEM FIR channel parameters and data symbols. Using the property $\bar{\mathbf{Z}}_{l'}\mathbf{D}_q = e^{j2\pi q(L'-l')/K}\bar{\mathbf{D}}_q\bar{\mathbf{Z}}_{l'}$, the q'th frequency-shifted and l'th time-shifted received sequence related to the ath output can be written as

$$\begin{aligned}
\mathbf{y}_{q',l'}^{(a)} &:= \bar{\mathbf{D}}_{q'}\bar{\mathbf{Z}}_{l'}\mathbf{y}^{(a)} \\
&= \sum_{l=0}^{L}\sum_{q=-Q/2}^{Q/2} h_{q,l}^{(a)} e^{j2\pi q(L'-l')/K}\bar{\mathbf{D}}_{q'}\bar{\mathbf{D}}_q\bar{\mathbf{Z}}_{l'}\mathbf{Z}_l\mathbf{x} + \mathbf{w}_{q',l'}^{(a)} \\
&= \sum_{l=0}^{L}\sum_{q=-Q/2}^{Q/2} e^{j2\pi q(L'-l')/K} h_{q,l}^{(a)}\bar{\mathbf{D}}_{q+q'}\tilde{\mathbf{Z}}_{l+l'}\mathbf{x} + \mathbf{w}_{q',l'}^{(a)}
\end{aligned}$$

where $\mathbf{w}^{(a)}_{q',l'}$ is defined similarly as $\mathbf{y}^{(a)}_{q',l'}$ and $\tilde{\mathbf{Z}}_k := [\mathbf{0}_{(N-L')\times(L+L'-k)}, \mathbf{I}_{N-L'}, \mathbf{0}_{(N-L')\times k}]$. Introducing $k := l + l'$ and $p := q + q'$, and defining $\mathbf{x}_{p,k} := \bar{\mathbf{D}}_p \tilde{\mathbf{Z}}_k \mathbf{x}$ (note that $\mathbf{x}_\star = \mathbf{x}_{0,d}$), we can also write this as

$$\mathbf{y}^{(a)}_{q',l'} = \sum_{k=0}^{L+L'} \sum_{p=-(Q+Q')/2}^{(Q+Q')/2} e^{j2\pi(p-q')(L'-l')/K} h^{(a)}_{p-q',k-l'} \mathbf{x}_{p,k} + \mathbf{w}^{(a)}_{q',l'}$$

Defining $\mathbf{X} := [\mathbf{x}_{Q/2+Q'/2,L+L'}, \ldots, \mathbf{x}_{Q/2+Q'/2,0}, \ldots, \mathbf{x}_{-Q/2-Q'/2,0}]^T$, $\mathbf{Y}^{(a)}$ can then be expressed as

$$\mathbf{Y}^{(a)} = \mathcal{H}^{(a)}\mathbf{X} + \mathbf{W}^{(a)}$$

where $\mathbf{W}^{(a)}$ is defined similarly as $\mathbf{Y}^{(a)}$ and $\mathcal{H}^{(a)}$ is the $(Q' + 1)(L' + 1) \times (Q + Q' + 1)(L + L' + 1)$ matrix given by

$$\mathcal{H}^{(a)} := \begin{bmatrix} \boldsymbol{\Omega}^{Q/2}\mathcal{H}^{(a)}_{Q/2} & \cdots & \boldsymbol{\Omega}^{-Q/2}\mathcal{H}^{(a)}_{-Q/2} & & \mathbf{0} \\ & \ddots & & \ddots & \\ \mathbf{0} & & \boldsymbol{\Omega}^{Q/2}\mathcal{H}^{(a)}_{Q/2} & \cdots & \boldsymbol{\Omega}^{-Q/2}\mathcal{H}^{(a)}_{-Q/2} \end{bmatrix}$$

with $\mathcal{H}^{(a)}_q$ the $(L' + 1) \times (L + L' + 1)$ Toeplitz matrix given by

$$\mathcal{H}^{(a)}_q := \begin{bmatrix} h^{(a)}_{q,L} & \cdots & h^{(a)}_{q,0} & & 0 \\ & \ddots & & \ddots & \\ 0 & & h^{(a)}_{q,L} & \cdots & h^{(a)}_{q,0} \end{bmatrix}$$

and $\boldsymbol{\Omega} := \operatorname{diag}\{[1, e^{j2\pi/K}, \ldots, e^{j2\pi L'/K}]^T\}$. Defining $\mathbf{Y} := [\mathbf{Y}^{(0)T}, \ldots, \mathbf{Y}^{(A-1)T}]^T$, we then obtain

$$\mathbf{Y} = \mathcal{H}\mathbf{X} + \mathbf{W} \tag{16.24}$$

where $\mathbf{W}$ is defined similarly as $\mathbf{Y}$ and $\mathcal{H} := [\mathcal{H}^{(0)T}, \ldots, \mathcal{H}^{(A-1)T}]^T$. Hence, we obtain

$$\hat{\mathbf{x}}^T_\star = \sum_{a=0}^{A-1} \mathbf{f}^{(a)T}\mathbf{Y}^{(a)} = \mathbf{f}^T\mathbf{Y} = \mathbf{f}^T\mathcal{H}\mathbf{X} + \mathbf{f}^T\mathbf{W} \tag{16.25}$$

where $\mathbf{f} := [\mathbf{f}^{(0)T}, \ldots, \mathbf{f}^{(A-1)T}]^T$

16.5.3 Equalizer Design

Noticing the equivalence between Equation 16.23 and Equation 16.25 (although with different matrix/vector definitions), we can now proceed with the SLE design for TIV and TV channels in a joint fashion.

Let us focus on the MMSE SLE, which minimizes the MSE $\mathcal{J} = \mathrm{E}\{\|\mathbf{x}_\star - \hat{\mathbf{x}}_\star\|^2\}$. Defining the data and noise covariance matrices as $\mathbf{R}_X := \mathrm{E}\{\mathbf{X}\mathbf{X}^H\}$ and $\mathbf{R}_W = \mathrm{E}\{\mathbf{W}\mathbf{W}^H\}$, respectively, the MSE can be expressed as

$$\mathcal{J} = \mathbf{f}^T(\mathcal{H}\mathbf{R}_X\mathcal{H}^H + \mathbf{R}_W)\mathbf{f}^* - 2\Re\{\mathbf{e}^T\mathbf{R}_X\mathcal{H}^H\mathbf{f}^*\} + \mathbf{e}^T\mathbf{R}_X\mathbf{e}^*$$

For TIV channels, $\mathbf{e}$ is the $(L + L' + 1) \times 1$ unit vector with a 1 in position $d + 1$. For TV channels, $\mathbf{e}$ is the $(Q + Q' + 1)(L + L' + 1) \times 1$ unit vector with a 1 in position $(Q + Q')(L + L' + 1)/2 + d + 1$. Solving $\partial\mathcal{J}/\partial\mathbf{f} = \mathbf{0}$, we obtain

$$\begin{aligned} \mathbf{f}^T_{MMSE} &= \mathbf{e}^T\mathbf{R}_X\mathcal{H}^H(\mathcal{H}\mathbf{R}_X\mathcal{H}^H + \mathbf{R}_W)^{-1} \\ &= \mathbf{e}^T\left(\mathcal{H}^H\mathbf{R}_W^{-1}\mathcal{H} + \mathbf{R}_X^{-1}\right)^{-1}\mathcal{H}^H\mathbf{R}_W^{-1} \end{aligned} \tag{16.26}$$

where the second equality is obtained by using the matrix inversion lemma. Assuming that $\mathcal{H}$ has full column rank, the corresponding ZF SLE can be obtained by setting the signal power to infinity ($\mathbf{R}_X^{-1} = \mathbf{0}$):

$$\mathbf{f}_{ZF}^T = \mathbf{e}^T \left(\mathcal{H}^H \mathbf{R}_W^{-1} \mathcal{H}\right)^{-1} \mathcal{H}^H \mathbf{R}_W^{-1} \tag{16.27}$$

Assuming the data sequence and the additive noises are mutually uncorrelated and white with variances σ_x^2 and σ_v^2, respectively, the data and noise covariance matrices can be computed in closed form. For TIV channels, the data and noise covariance matrices are given by

$$\mathbf{R}_X = \sigma_x^2 \mathbf{I}_{L+L'+1}$$

$$\mathbf{R}_W = \sigma_v^2 \mathbf{I}_M \otimes \begin{bmatrix} \mathbf{\Phi}_{L'+1,0} & \cdots & \mathbf{\Phi}_{L'+1,P-1} \\ \vdots & & \vdots \\ \mathbf{\Phi}_{L'+1,-P+1} & \cdots & \mathbf{\Phi}_{L'+1,0} \end{bmatrix}$$

For TV channels, the data and noise covariance matrices are given by

$$\mathbf{R}_X = \sigma_x^2 \mathbf{J}_{Q+Q'+1} \otimes \mathbf{I}_{L+L'+1}$$

$$\mathbf{R}_W = \sigma_v^2 \mathbf{I}_M \otimes \begin{bmatrix} \mathbf{J}_{Q'+1} \otimes \mathbf{\Phi}_{L'+1,0} & \cdots & \mathbf{J}_{Q'+1} \otimes \mathbf{\Phi}_{L'+1,P-1} \\ \vdots & & \vdots \\ \mathbf{J}_{Q'+1} \otimes \mathbf{\Phi}_{L'+1,-P+1} & \cdots & \mathbf{J}_{Q'+1} \otimes \mathbf{\Phi}_{L'+1,0} \end{bmatrix}$$

where $\mathbf{J}_I$ is the $I \times I$ matrix defined as

$$[\mathbf{J}_I]_{i,i'} = \sum_{n=0}^{N-L'-1} e^{j2\pi(i-i')n/K}$$

16.6 Serial Decision Feedback Equalization

In this section, we discuss serial decision feedback equalization. As before, we do not focus on zero padding-based block transmission, but rather on the serial transmission model (see Equations 16.12 and 16.14).

We adopt a serial decision feedback equalizer consisting of a serial feedforward filter $f^{(a)}[n;\nu]$ for the ath output and a serial feedback filter $b[n;\nu]$, in order to find an estimate of $x[n-d]$ (see Figure 16.7):

$$\hat{x}[n-d] = \sum_{a=0}^{A-1} \sum_{\nu=-\infty}^{\infty} f^{(a)}[n;\nu] y^{(a)}[n-\nu] - \sum_{\nu=-\infty}^{\infty} b[n;\nu] \check{x}[n-d-\nu]$$

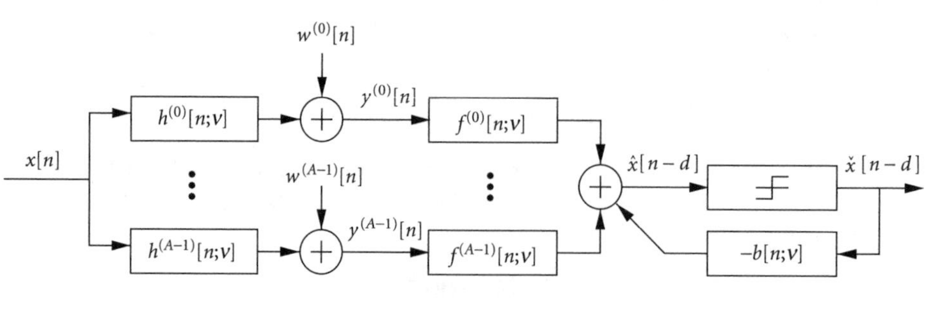

FIGURE 16.7 Serial decision feedback equalization.

where d again represents the synchronization delay and $\check{x}[n] := Q\{\hat{x}[n]\}$. To discuss the structure of this SDFE in more detail, we again distinguish between TIV and TV channels. Both cases will again give rise to a related data model, which allows us to treat the equalizer design in a joint fashion.

16.6.1 TIV Channels

Since for a TIV channel, the TIV FIR channel of Equation 16.5 is applied, it is also convenient to use a TIV FIR serial feedforward filter $f^{(a)}[n;\nu]$ and a TIV FIR serial feedback filter $b[n;\nu]$ [1]. In other words, we design each serial feedforward filter $f^{(a)}[n;\nu]$ to have $L'+1$ TIV taps:

$$f^{(a)}[n;\nu] = \sum_{l'=0}^{L'} \delta[\nu - l'] f_{l'}^{(a)} \tag{16.28}$$

and the serial feedback filter $b[n;\nu]$ to have $L''+1$ TIV taps:

$$b[n;\nu] = \sum_{l''=0}^{L''} \delta[\nu - l''] b_{l''} \tag{16.29}$$

where in order to feedback decisions in a causal way, we require $b_0 = 0$. An estimate of $x[n-d]$ is then computed as

$$\hat{x}[n-d] = \sum_{a=0}^{A-1} \sum_{l'=0}^{L'} f_{l'}^{(a)} y^{(a)}[n-l'] - \sum_{l''=1}^{L''} b_{l''} \check{x}[n-d-l''] \tag{16.30}$$

Using the notation introduced in Section 16.5.1, and assuming that past decisions are correct, we can write this as

$$\hat{\mathbf{x}}_{\star}^T = \sum_{a=0}^{A-1} \mathbf{f}^{(a)T}\mathbf{Y}^{(a)} - \mathbf{b}^T\mathbf{P}\mathbf{X} = \mathbf{f}^T\mathbf{Y} - \mathbf{b}^T\mathbf{P}\mathbf{X}$$

$$= \mathbf{f}^T\mathcal{H}\mathbf{X} - \mathbf{b}^T\mathbf{P}\mathbf{X} + \mathbf{f}^T\mathbf{W} \tag{16.31}$$

where $\mathbf{b}$ is the $(L''+1)\times 1$ vector given by $\mathbf{b} = [b_{L''},\ldots,b_1,0]^T$ and $\mathbf{P}$ is the $(L''+1)\times(L+L'+1)$ selection matrix given by

$$\mathbf{P} := [\mathbf{0}_{(L''+1)\times(L+L'-L''-d)}, \mathbf{I}_{L''+1}, \mathbf{0}_{(L''+1)\times d}]$$

16.6.2 TV Channels

Since for a TV channel, the BEM FIR channel of Equation 16.8 is applied, it is also convenient to use a BEM FIR serial feedforward filter $f^{(a)}[n;\nu]$ and a BEM FIR serial feedback filter $b[n;\nu]$ [2]. In other words, we design each serial feedforward filter $f^{(a)}[n;\nu]$ to have $L'+1$ TV taps, where the time variation of each tap is modeled by $Q'+1$ complex exponentials:

$$f^{(a)}[n;\nu] = \sum_{l'=0}^{L'} \delta[\nu - l'] \sum_{q'=-Q'/2}^{Q'/2} e^{j2\pi q' n/K} f_{q',l'}^{(a)} \tag{16.32}$$

and the serial feedback filter $b[n;\nu]$ to have $L''+1$ TV taps, where the time variation of each tap is modeled by $Q''+1$ complex exponentials:

$$b[n;\nu] = \sum_{l''=0}^{L''} \delta[\nu - l''] \sum_{q''=-Q''/2}^{Q''/2} e^{j2\pi q'' n/K} b_{q'',l''} \tag{16.33}$$

where in order to feedback decisions in a causal way, we require $b_{-Q''/2,0} = \cdots = b_{Q''/2,0} = 0$. An estimate of $x[n-d]$ is then computed as

$$\hat{x}[n-d] = \sum_{a=0}^{A-1}\sum_{l'=0}^{L'}\sum_{q'=-Q'/2}^{Q'/2} e^{j2\pi q'n/K} f_{q',l'}^{(a)} y^{(a)}[n-l'] - \sum_{l''=1}^{L''}\sum_{q''=-Q''/2}^{Q''/2} e^{j2\pi q''n/K} b_{q'',l''}\check{x}[n-d-l''] \quad (16.34)$$

Using the notation introduced in Section 16.5.2, and assuming that past decisions are correct, we can write this as

$$\hat{\mathbf{x}}_\star^T = \sum_{a=0}^{A-1} \mathbf{f}^{(a)T}\mathbf{Y}^{(a)} - \mathbf{b}^T\mathbf{P}\mathbf{X} = \mathbf{f}^T\mathbf{Y} - \mathbf{b}^T\mathbf{P}\mathbf{X}$$

$$= \mathbf{f}^T\boldsymbol{\mathcal{H}}\mathbf{X} - \mathbf{b}^T\mathbf{P}\mathbf{X} + \mathbf{f}^T\mathbf{W} \quad (16.35)$$

where $\mathbf{b}$ is the $((Q''+1)L''+1)\times 1$ vector given by $\mathbf{b} = [b_{Q''/2,L''},\ldots,b_{Q''/2,1},\ldots,b_{1,1},b_{0,L''},\ldots,b_{0,1},0,$ $b_{-1,L''},\ldots,b_{-1,1},\ldots,b_{-Q''/2,1}]^T$ and $\mathbf{P}$ is the $((Q''+1)L''+1)\times(Q+Q'+1)(L+L'+1)$ selection matrix given by

$$\mathbf{P} := \begin{bmatrix} & \mathbf{I}_{Q''/2}\otimes\mathbf{P}_1 & & & \\ \mathbf{0}_{\alpha\times\beta} & & \mathbf{P}_2 & & \mathbf{0}_{\alpha\times\beta} \\ & & & \mathbf{I}_{Q''/2}\otimes\mathbf{P}_1 & \end{bmatrix}$$

with $\alpha := (Q''+1)L''+1$, $\beta := (Q+Q'-Q'')(L+L'+1)/2$, $\mathbf{P}_1 := [\mathbf{0}_{L''\times(L+L'-L''-d)}, \mathbf{I}_{L''}, \mathbf{0}_{L''\times(d+1)}]$, and $\mathbf{P}_2 := [\mathbf{0}_{(L''+1)\times(L+L'-L''-d)}, \mathbf{I}_{L''+1}, \mathbf{0}_{(L''+1)\times d}]$.

16.6.3 Equalizer Design

As in Section 16.5.3, noticing the equivalence between Equation 16.31 and Equation 16.35, we can proceed with the SDFE design for TIV and TV channels in a joint fashion.

Let us focus on the MMSE SDFE, which minimizes the MSE $\mathcal{J} = \mathrm{E}\{\|\mathbf{x}_\star - \hat{\mathbf{x}}_\star\|^2\}$. In a fashion similar to that for the SLE, the MSE can be expressed as

$$\mathcal{J} = \mathbf{f}^T(\boldsymbol{\mathcal{H}}\mathbf{R}_X\boldsymbol{\mathcal{H}}^H + \mathbf{R}_W)\mathbf{f}^* - 2\Re\{(\mathbf{b}+\mathbf{e})^T\mathbf{P}\mathbf{R}_X\boldsymbol{\mathcal{H}}^H\mathbf{f}^*\} + (\mathbf{b}+\mathbf{e})^T\mathbf{P}\mathbf{R}_X\mathbf{P}^H(\mathbf{b}+\mathbf{e})^* \quad (16.36)$$

For TIV channels, $\mathbf{e}$ is the $(L''+1)\times 1$ unit vector with a 1 in position $L''+1$. For TV channels, $\mathbf{e}$ is the $((Q''+1)L''+1)\times 1$ unit vector with a 1 in position $Q''L''/2+L''+1$. Solving $\partial\mathcal{J}/\partial\mathbf{f} = \mathbf{0}$, we obtain

$$\mathbf{f}_{MMSE}^T = (\mathbf{b}+\mathbf{e})^T\mathbf{P}\mathbf{R}_X\boldsymbol{\mathcal{H}}^H\left(\boldsymbol{\mathcal{H}}\mathbf{R}_X\boldsymbol{\mathcal{H}}^H + \mathbf{R}_W\right)^{-1} \quad (16.37)$$

$$= (\mathbf{b}+\mathbf{e})^T\mathbf{P}\left(\boldsymbol{\mathcal{H}}^H\mathbf{R}_W^{-1}\boldsymbol{\mathcal{H}} + \mathbf{R}_X^{-1}\right)^{-1}\boldsymbol{\mathcal{H}}^H\mathbf{R}_W^{-1} \quad (16.38)$$

where the second equality is again obtained by using the matrix inversion lemma. Next, substituting Equation 16.37 into Equation 16.36 results, after some calculation, in

$$\mathcal{J} = (\mathbf{b}+\mathbf{e})^T\mathbf{R}_{MMSE}(\mathbf{b}+\mathbf{e})^*$$

where $\mathbf{R}_{MMSE} = \mathbf{P}(\boldsymbol{\mathcal{H}}^H\mathbf{R}_W^{-1}\boldsymbol{\mathcal{H}} + \mathbf{R}_X^{-1})^{-1}\mathbf{P}^H$. Solving $\partial\mathcal{J}/\partial\mathbf{b} = \mathbf{0}$ under the constraint that $\mathbf{e}^T\mathbf{b} = 0$, we finally obtain

$$\mathbf{b}_{MMSE}^T = \frac{\mathbf{e}^T\mathbf{R}_{MMSE}^{-1}}{\mathbf{e}^T\mathbf{R}_{MMSE}^{-1}\mathbf{e}} - \mathbf{e}^T$$

To summarize, the MMSE SDFE is given by

$$\mathbf{f}_{MMSE}^{T} = (\mathbf{b}_{MMSE} + \mathbf{e})^{T}\mathbf{P}\left(\boldsymbol{\mathcal{H}}^{H}\mathbf{R}_{W}^{-1}\boldsymbol{\mathcal{H}} + \mathbf{R}_{X}^{-1}\right)^{-1}\boldsymbol{\mathcal{H}}^{H}\mathbf{R}_{W}^{-1}$$
$$\mathbf{R}_{MMSE} = \mathbf{P}\left(\boldsymbol{\mathcal{H}}^{H}\mathbf{R}_{W}^{-1}\boldsymbol{\mathcal{H}} + \mathbf{R}_{X}^{-1}\right)^{-1}\mathbf{P}^{H}$$
$$\mathbf{b}_{MMSE}^{T} = \frac{\mathbf{e}^{T}\mathbf{R}_{MMSE}^{-1}}{\mathbf{e}^{T}\mathbf{R}_{MMSE}^{-1}\mathbf{e}} - \mathbf{e}^{T}$$

Assuming $\boldsymbol{\mathcal{H}}$ has full column rank, the corresponding ZF SDFE can again be obtained by setting the signal power to infinity ($\mathbf{R}_{X}^{-1} = \mathbf{0}$):

$$\mathbf{f}_{ZF}^{T} = (\mathbf{b}_{ZF} + \mathbf{e})^{T}\mathbf{P}\left(\boldsymbol{\mathcal{H}}^{H}\mathbf{R}_{W}^{-1}\boldsymbol{\mathcal{H}}\right)^{-1}\boldsymbol{\mathcal{H}}^{H}\mathbf{R}_{W}^{-1}$$
$$\mathbf{R}_{ZF} = \mathbf{P}\left(\boldsymbol{\mathcal{H}}^{H}\mathbf{R}_{W}^{-1}\boldsymbol{\mathcal{H}}\right)^{-1}\mathbf{P}^{H}$$
$$\mathbf{b}_{ZF}^{T} = \frac{\mathbf{e}^{T}\mathbf{R}_{ZF}^{-1}}{\mathbf{e}^{T}\mathbf{R}_{ZF}^{-1}\mathbf{e}} - \mathbf{e}^{T}$$

16.7 Frequency-Domain Equalization for TIV Channels

For TIV channels, a popular method to reduce the implementation complexity of block equalization is based on frequency-domain processing. To explain this FD equalization, we resort again to the zero padding-based block transmission applied for block equalization.

Rewriting the data model for zero padding Equation 16.15 as

$$\mathbf{y}^{(a)} = \mathbf{H}_{c}^{(a)}\mathbf{u} + \mathbf{w}^{(a)} \tag{16.39}$$

where $\mathbf{u} := [\mathbf{s}^{T}, \mathbf{0}_{1\times L}]^{T}$ and $\mathbf{H}_{c}^{(a)} := [[\mathbf{H}^{(a)}]_{:,L+1:N}, [\mathbf{H}^{(a)}]_{:,1:L} + [\mathbf{H}^{(a)}]_{:,N+1:N+L}]$, we observe a similarity with the data model for cyclic prefix-based block transmission [31], with the exception that the symbols in the cyclic prefix are now zero. Hence, for TIV channels, where $\mathbf{H}_{c}^{(a)}$ is circulant, we can simplify Equation 16.39 using fast Fourier transform (FFT) operations, as for cyclic prefix-based block transmission [45].

Defining the N-point normalized FFT of $\mathbf{u}$ as $\tilde{\mathbf{u}} := \mathbf{G}\mathbf{u}$ and the N-point normalized FFT of $\mathbf{y}^{(a)}$ as $\tilde{\mathbf{y}}^{(a)} := \mathbf{G}\mathbf{y}^{(a)}$, we obtain

$$\begin{aligned}\tilde{\mathbf{y}}^{(a)} &= \mathbf{G}\mathbf{H}_{c}^{(a)}\mathbf{G}^{H}\tilde{\mathbf{u}} + \tilde{\mathbf{w}}^{(a)} \\ &= \tilde{\mathbf{H}}^{(a)}\tilde{\mathbf{u}} + \tilde{\mathbf{w}}^{(a)}\end{aligned} \tag{16.40}$$

where $\tilde{\mathbf{w}}^{(a)}$ is defined similarly as $\tilde{\mathbf{y}}^{(a)}$ and $\tilde{\mathbf{H}}^{(a)} := \mathrm{diag}\{\sqrt{N}\mathbf{G}[h_{0}^{(a)}, \ldots, h_{L}^{(a)}, \ldots, 0]^{T}\}$. Defining $\tilde{\mathbf{y}} := [\tilde{\mathbf{y}}^{(0)T}, \ldots, \tilde{\mathbf{y}}^{(A-1)T}]^{T}$, we then obtain

$$\tilde{\mathbf{y}} = \tilde{\mathbf{H}}\tilde{\mathbf{u}} + \tilde{\mathbf{w}} \tag{16.41}$$

where $\tilde{\mathbf{w}}$ is defined similarly as $\tilde{\mathbf{y}}$ and $\tilde{\mathbf{H}} := [\tilde{\mathbf{H}}^{(0)T}, \ldots, \tilde{\mathbf{H}}^{(A-1)T}]^{T}$. This data model will allow us to use simplified FD processing, as illustrated next. Once an estimate $\hat{\tilde{\mathbf{u}}}$ of $\tilde{\mathbf{u}}$ is obtained, an estimate of $\mathbf{s}$ can be computed as

$$\hat{\mathbf{s}} = [\mathbf{I}_{N-L}, \mathbf{0}_{(N-L)\times L}]\mathbf{G}^{H}\hat{\tilde{\mathbf{u}}}$$

Note that $\hat{\mathbf{s}}$ also implies an estimate $\hat{x}[n]$ of $x[n]$.

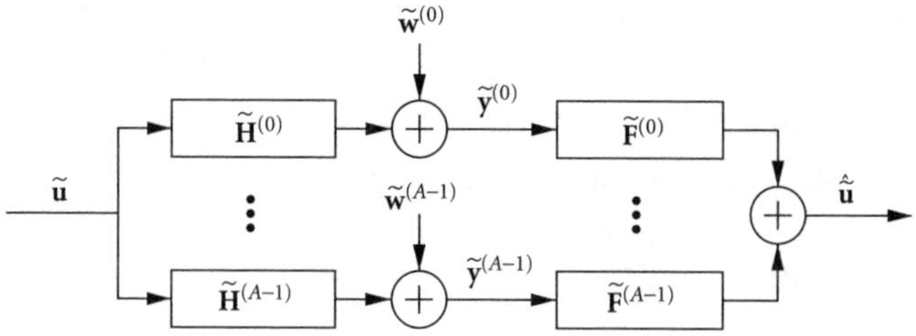

FIGURE 16.8 Frequency-domain linear equalization.

16.7.1 FD Linear Equalization

As illustrated in Figure 16.8, an FD linear equalizer computes an estimate of $\tilde{\mathbf{u}}$ using an FD filter $\tilde{\mathbf{F}}^{(a)}$ for the ath output [6], [10], [31]:

$$\hat{\tilde{\mathbf{u}}} = \sum_{a=0}^{A-1} \tilde{\mathbf{F}}^{(a)}\tilde{\mathbf{y}}^{(a)} = \left(\sum_{a=0}^{A-1} \tilde{\mathbf{F}}^{(a)}\tilde{\mathbf{H}}^{(a)}\right)\tilde{\mathbf{u}} + \sum_{a=0}^{A-1} \tilde{\mathbf{F}}^{(a)}\tilde{\mathbf{w}}^{(a)}$$

Defining $\tilde{\mathbf{F}} := [\tilde{\mathbf{F}}^{(0)}, \ldots, \tilde{\mathbf{F}}^{(A-1)}]$, we then obtain

$$\hat{\tilde{\mathbf{u}}} = \tilde{\mathbf{F}}\tilde{\mathbf{y}} = \tilde{\mathbf{F}}\tilde{\mathbf{H}}\tilde{\mathbf{u}} + \tilde{\mathbf{F}}\tilde{\mathbf{w}}$$

Let us focus on the MMSE FDLE, which minimizes the MSE $\mathcal{J} = \mathcal{E}\{\|\tilde{\mathbf{u}} - \hat{\tilde{\mathbf{u}}}\|^2\}$. Defining the FD data and noise covariance matrices as $\mathbf{R}_{\tilde{u}} := \mathcal{E}\{\tilde{\mathbf{u}}\tilde{\mathbf{u}}^H\}$ and $\mathbf{R}_{\tilde{w}} := \mathcal{E}\{\tilde{\mathbf{w}}\tilde{\mathbf{w}}^H\}$, respectively, the MSE can be expressed as

$$\mathcal{J} = \mathrm{tr}\{\tilde{\mathbf{F}}(\tilde{\mathbf{H}}\mathrm{diag}\{\mathbf{R}_{\tilde{u}}\}\tilde{\mathbf{H}}^H + \mathrm{diag}\{\mathbf{R}_{\tilde{w}}\})\tilde{\mathbf{F}}^H - 2\Re\{\mathrm{diag}\{\mathbf{R}_{\tilde{u}}\}\tilde{\mathbf{H}}^H\tilde{\mathbf{F}}^H\} + \mathrm{diag}\{\mathbf{R}_{\tilde{u}}\}\}$$

Solving $\partial\mathcal{J}/\partial\tilde{\mathbf{F}} = \mathbf{0}$, we obtain

$$\begin{aligned}\tilde{\mathbf{F}}_{MMSE} &= \mathrm{diag}\{\mathbf{R}_{\tilde{u}}\}\tilde{\mathbf{H}}^H(\tilde{\mathbf{H}}\mathrm{diag}\{\mathbf{R}_{\tilde{u}}\}\tilde{\mathbf{H}}^H + \mathrm{diag}\{\mathbf{R}_{\tilde{w}}\})^{-1} \\ &= (\tilde{\mathbf{H}}^H\mathrm{diag}\{\mathbf{R}_{\tilde{w}}\}^{-1}\tilde{\mathbf{H}} + \mathrm{diag}\{\mathbf{R}_{\tilde{u}}\}^{-1})^{-1}\tilde{\mathbf{H}}^H\mathrm{diag}\{\mathbf{R}_{\tilde{w}}\}^{-1}\end{aligned}$$

where the second equality is obtained by using the matrix inversion lemma. Assuming $\tilde{\mathbf{H}}$ has full column rank, the corresponding ZF FDLE can be obtained by setting the signal power to infinity ($\mathbf{R}_{\tilde{u}}^{-1} = \mathbf{0}$):

$$\tilde{\mathbf{F}}_{ZF} = (\tilde{\mathbf{H}}^H\mathrm{diag}\{\mathbf{R}_{\tilde{w}}\}^{-1}\tilde{\mathbf{H}})^{-1}\tilde{\mathbf{H}}^H\mathrm{diag}\{\mathbf{R}_{\tilde{w}}\}^{-1}$$

Assuming the data sequence and additive noises are mutually uncorrelated and white with variances σ_s^2 and σ_v^2, respectively, the FD data and noise covariance matrices can be computed in closed form:

$$\mathbf{R}_{\tilde{u}} = \sigma_s^2\mathbf{G}\begin{bmatrix}\mathbf{I}_{N-L} & \mathbf{0}_{(N-L)\times L} \\ \mathbf{0}_{L\times(N-L)} & \mathbf{0}_{L\times L}\end{bmatrix}\mathbf{G}^H$$

$$\mathbf{R}_{\tilde{w}} = \sigma_v^2\mathbf{I}_M \otimes \begin{bmatrix}\mathbf{G}\boldsymbol{\Phi}_{N,0}\mathbf{G}^H & \cdots & \mathbf{G}\boldsymbol{\Phi}_{N,P-1}\mathbf{G}^H \\ \vdots & & \vdots \\ \mathbf{G}\boldsymbol{\Phi}_{N,-P+1}\mathbf{G}^H & \cdots & \mathbf{G}\boldsymbol{\Phi}_{N,0}\mathbf{G}^H\end{bmatrix}$$

16.7.2 FD Decision Feedback Equalization

Due to the inherent delay of FD processing, the feedback part of any FD decision feedback equalization approach has to be implemented in the time domain, e.g., by means of a serial filter. Therefore, as illustrated in Figure 16.9, an FD decision feedback equalizer computes an estimate of $\tilde{\mathbf{u}}$ using an FD feedforward filter

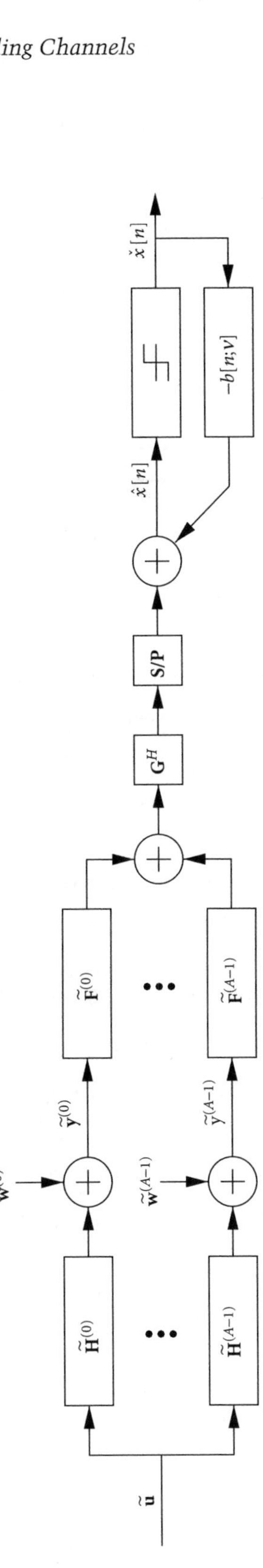

FIGURE 16.9 Frequency-domain decision feedback equalization.

$\tilde{\mathbf{F}}^{(a)}$ for the ath output and a serial feedback filter $b[n;\nu]$ [4], [10]:

$$\hat{\mathbf{u}} = \sum_{a=0}^{A-1} \tilde{\mathbf{F}}^{(a)}\tilde{\mathbf{y}}^{(a)} - \sum_{n=0}^{N-1} [\mathbf{G}]_{:,n+1} \sum_{\nu=-\infty}^{\infty} b[n;\nu]\check{x}[n-\nu]$$

$$= \left(\sum_{a=0}^{A-1} \tilde{\mathbf{F}}^{(a)}\tilde{\mathbf{H}}^{(a)}\right)\tilde{\mathbf{u}} - \sum_{n=0}^{N-1} [\mathbf{G}]_{:,n+1} \sum_{\nu=-\infty}^{\infty} b[n;\nu]\check{x}[n-\nu] + \sum_{a=0}^{A-1} \tilde{\mathbf{F}}^{(a)}\tilde{\mathbf{w}}^{(a)}$$

where $\check{x}[n] = \mathrm{Q}\{\hat{x}[n]\}$. In order to avoid overlap with the previous symbol block and allow for a simple computation, we design the serial feedback filter $b[n;\nu]$ as a TIV FIR filter with $L+1$ taps:

$$b[n;\nu] = \sum_{l=0}^{L} \delta[\nu - l] b_l$$

where in order to feedback decisions in a causal way, we require $b_0 = 0$. Hence, we can write

$$\hat{\mathbf{u}} = \left(\sum_{a=0}^{A-1} \tilde{\mathbf{F}}^{(a)}\tilde{\mathbf{H}}^{(a)}\right)\tilde{\mathbf{u}} - \sum_{n=0}^{N-1} [\mathbf{G}]_{:,n+1} \sum_{l=1}^{L} b_l\check{x}[n-l] + \sum_{a=0}^{A-1} \tilde{\mathbf{F}}^{(a)}\tilde{\mathbf{w}}^{(a)}$$

Defining $\tilde{\mathbf{F}} := [\tilde{\mathbf{F}}^{(0)}, \ldots, \tilde{\mathbf{F}}^{(A-1)}]$, and assuming past decisions are correct, i.e., $\check{x}[n] = x[n]$, we then obtain

$$\hat{\mathbf{u}} = \tilde{\mathbf{F}}\tilde{\mathbf{y}} - \mathbf{G}\mathbf{B}_c\mathbf{u} = \tilde{\mathbf{F}}\tilde{\mathbf{H}}\tilde{\mathbf{u}} - \mathbf{G}\mathbf{B}_c\mathbf{G}^H\tilde{\mathbf{u}} + \tilde{\mathbf{F}}\tilde{\mathbf{w}}$$

where $\mathbf{B}_c$ is a circulant matrix with first column $[\mathbf{b}^T, \mathbf{0}_{1\times(N-L-1)}]^T$, where $\mathbf{b} := [0, b_1, \ldots, b_L]^T$. Let us focus on the MMSE FDDFE, which minimizes the MSE $\mathcal{J} = \mathcal{E}\{\|\tilde{\mathbf{u}} - \hat{\mathbf{u}}\|^2\}$. In a fashion similar to that for the FDLE, the MSE can be expressed as

$$\begin{aligned}\mathcal{J} = \mathrm{tr}\{&\tilde{\mathbf{F}}(\tilde{\mathbf{H}}\mathrm{diag}\{\mathbf{R}_{\tilde{u}}\}\tilde{\mathbf{H}}^H + \mathrm{diag}\{\mathbf{R}_{\tilde{w}}\})\tilde{\mathbf{F}}^H - 2\Re\{\mathbf{G}(\mathbf{B}_c + \mathbf{I}_N)\mathbf{G}^H\mathrm{diag}\{\mathbf{R}_{\tilde{u}}\}\tilde{\mathbf{H}}^H\tilde{\mathbf{F}}^H\} \\ &+ \mathbf{G}(\mathbf{B}_c + \mathbf{I}_N)\mathbf{G}^H\mathrm{diag}\{\mathbf{R}_{\tilde{u}}\}\mathbf{G}(\mathbf{B}_c + \mathbf{I}_N)^H\mathbf{G}^H\}\end{aligned} \tag{16.42}$$

where $\mathbf{B}_c + \mathbf{I}_N$ is a circulant matrix with first column $[(\mathbf{b}+\mathbf{e})^T, \mathbf{0}_{1\times(N-L-1)}]^T$, where $\mathbf{e}$ is the $(L+1)\times 1$ unit vector with a 1 in the first position. Solving $\partial\mathcal{J}/\partial\tilde{\mathbf{F}} = \mathbf{0}$, we obtain

$$\tilde{\mathbf{F}}_{MMSE} = \mathbf{G}(\mathbf{B}_c + \mathbf{I}_N)\mathbf{G}^H\mathrm{diag}\{\mathbf{R}_{\tilde{u}}\}\tilde{\mathbf{H}}^H(\tilde{\mathbf{H}}\mathrm{diag}\{\mathbf{R}_{\tilde{u}}\}\tilde{\mathbf{H}}^H + \mathrm{diag}\{\mathbf{R}_{\tilde{w}}\})^{-1} \tag{16.43}$$

$$= \mathbf{G}(\mathbf{B}_c + \mathbf{I}_N)\mathbf{G}^H(\tilde{\mathbf{H}}^H\mathrm{diag}\{\mathbf{R}_{\tilde{w}}\}^{-1}\tilde{\mathbf{H}} + \mathrm{diag}\{\mathbf{R}_{\tilde{u}}\}^{-1})^{-1}\tilde{\mathbf{H}}^H\mathrm{diag}\{\mathbf{R}_{\tilde{w}}\}^{-1} \tag{16.44}$$

where the second equality is again obtained by using the matrix inversion lemma. Next, substituting Equation 16.43 into Equation 16.42 results, after some calculation, in

$$\begin{aligned}\mathcal{J} &= \mathrm{tr}\{\mathbf{G}(\mathbf{B}_c + \mathbf{I}_N)\mathbf{G}^H(\tilde{\mathbf{H}}^H\mathrm{diag}\{\mathbf{R}_{\tilde{w}}\}^{-1}\tilde{\mathbf{H}} + \mathrm{diag}\{\mathbf{R}_{\tilde{u}}\}^{-1})^{-1}\mathbf{G}(\mathbf{B}_c + \mathbf{I}_N)^H\mathbf{G}^H\} \\ &= (\mathbf{b}+\mathbf{e})^T\mathbf{R}_{MMSE}(\mathbf{b}+\mathbf{e})^*\end{aligned}$$

where $\mathbf{R}_{MMSE} = N[\mathbf{G}^T(\tilde{\mathbf{H}}^H\mathrm{diag}\{\mathbf{R}_{\tilde{w}}\}^{-1}\tilde{\mathbf{H}} + \mathrm{diag}\{\mathbf{R}_{\tilde{u}}\}^{-1})^{-1}\mathbf{G}^*]_{1:L+1,1:L+1}$. Solving $\partial\mathcal{J}/\partial\mathbf{b} = \mathbf{0}$ under the constraint $\mathbf{e}^T\mathbf{b} = 0$, we finally obtain

$$\mathbf{b}_{MMSE}^T = \frac{\mathbf{e}^T\mathbf{R}_{MMSE}^{-1}}{\mathbf{e}^T\mathbf{R}_{MMSE}^{-1}\mathbf{e}} - \mathbf{e}^T$$

To summarize, the MMSE FDDFE is given by

$$\tilde{\mathbf{F}}_{MMSE} = \mathbf{G}(\mathbf{B}_{c,MMSE} + \mathbf{I}_N)\mathbf{G}^H(\tilde{\mathbf{H}}^H\mathrm{diag}\{\mathbf{R}_{\tilde{w}}\}^{-1}\tilde{\mathbf{H}} + \mathrm{diag}\{\mathbf{R}_{\tilde{u}}\}^{-1})^{-1}\tilde{\mathbf{H}}^H\mathrm{diag}\{\mathbf{R}_{\tilde{w}}\}^{-1}$$

$$\mathbf{R}_{MMSE} = N[\mathbf{G}^T(\tilde{\mathbf{H}}^H\mathrm{diag}\{\mathbf{R}_{\tilde{w}}\}^{-1}\tilde{\mathbf{H}} + \mathrm{diag}\{\mathbf{R}_{\tilde{u}}\}^{-1})^{-1}\mathbf{G}^*]_{1:L+1,1:L+1}$$

$$\mathbf{b}_{MMSE}^T = \frac{\mathbf{e}^T\mathbf{R}_{MMSE}^{-1}}{\mathbf{e}^T\mathbf{R}_{MMSE}^{-1}\mathbf{e}} - \mathbf{e}^T$$

Assuming $\tilde{\mathbf{H}}$ has full column rank, the corresponding ZF FDDFE can again be obtained by setting the signal power to infinity ($\mathbf{R}_{\tilde{u}}^{-1} = \mathbf{0}$):

$$\tilde{\mathbf{F}}_{ZF} = \mathbf{G}(\mathbf{B}_{c,ZF} + \mathbf{I}_N)\mathbf{G}^H(\tilde{\mathbf{H}}^H \text{diag}\{\mathbf{R}_{\tilde{w}}\}^{-1}\tilde{\mathbf{H}})^{-1}\tilde{\mathbf{H}}^H \text{diag}\{\mathbf{R}_{\tilde{w}}\}^{-1}$$

$$\mathbf{R}_{ZF} = N[\mathbf{G}^T(\tilde{\mathbf{H}}^H \text{diag}\{\mathbf{R}_{\tilde{w}}\}^{-1}\tilde{\mathbf{H}})^{-1}\mathbf{G}^*]_{1:L+1,1:L+1}$$

$$\mathbf{b}_{ZF}^T = \frac{\mathbf{e}^T\mathbf{R}_{ZF}^{-1}}{\mathbf{e}^T\mathbf{R}_{ZF}^{-1}\mathbf{e}} - \mathbf{e}^T$$

16.8 Existence of Zero-Forcing Solution

Comparing the MMSE with the ZF solution, the MMSE solution always leads to a better performance than the ZF solution. However, the *existence* of the ZF solution generally gives a good indication of the performance at high SNR. For instance, when the ZF solution does not exist with probability 1, e.g., when it never exists because certain dimensionality conditions are not satisfied, the performance will saturate at high SNR. When the ZF solution exists with probability 1, the performance will always increase with increasing SNR. The smaller the region for which one comes close to a channel realization for which the ZF solution does not exist, the steeper the slope of the performance curve (or the higher the collected *diversity*). We will now briefly discuss the existence of the ZF solution for the different equalizers introduced previously.

16.8.1 Linear Equalizers

Let us first focus on the LEs. The ZF BLE exists if and only if the channel matrix $\bar{\mathbf{H}}$ has full column rank, which requires that $\bar{\mathbf{H}}$ has at least as many rows as columns. Since $\bar{\mathbf{H}}$ has AN rows and $N - L$ columns, this is always the case. However, this does not mean that $\bar{\mathbf{H}}$ always has full column rank. The latter is only true for TIV channels. On the other hand, judging from Equation 16.27, one would think that the ZF SLE exists if and only if the channel matrix $\mathcal{H}$ has full column rank. However, this is only true if $\mathcal{H}$ is column reduced (see [35] for TIV channels and [2], [19] for TV channels). Assuming this is the case, the existence of the ZF SLE is equivalent with $\mathcal{H}$ having full column rank, which requires that $\mathcal{H}$ has at least as many rows as columns. For TIV channels, this happens when $A(L'+1) \geq L + L' + 1$, whereas for TV channels, this happens when $A(Q'+1)(L'+1) \geq (Q+Q'+1)(L+L'+1)$. Clearly, these inequalities can only be satisfied if $A \geq 2$ with a sufficiently large L' for TIV channels and a sufficiently large Q' and L' for TV channels. Hence, we need at least two outputs, which can, for instance, be achieved by sampling $M=2$ receive antennas at rate $1/T$ ($P=1$) or sampling $M=1$ receive antenna at rate $2/T$ ($P=2$). More detailed sufficient conditions for the ZF SLE to exist can be found in [35] for TIV channels and [2], [19] for TV channels. As already discussed, for TIV channels we can also adopt FD processing. The ZF FDLE exists if and only if $\tilde{\mathbf{H}}$ has full column rank, which requires that $\tilde{\mathbf{H}}$ has at least as many rows as columns. Since $\tilde{\mathbf{H}}$ has AN rows and N columns, this is again always the case. However, in contrast to $\bar{\mathbf{H}}$, $\tilde{\mathbf{H}}$ does not always have full column rank for TIV channels. It becomes singular when all A channels have a common zero on the N-point FFT grid, i.e., when $\tilde{\mathbf{H}}$ has a zero column.

16.8.2 Decision Feedback Equalizers

As far as DFEs are concerned, note that the MMSE and ZF DFEs that we have proposed earlier assume that past decisions are correct, which basically makes the DFEs look linear. Only in this context do the statements we made at the beginning of this section hold. Judging from the equations we presented for the different ZF DFEs, we would tend to think that a ZF DFE exists if and only if the corresponding ZF LE exists. However, other (more complicated) equations could be derived from which we could see that a ZF DFE can also exist when the corresponding ZF LE does not exist. Suffice it to illustrate this for the SDFE. Defining $\mathbf{P}^{\perp}$ as the orthogonal complement of $\mathbf{P}$, i.e., $\mathbf{P}^{\perp T}\mathbf{P} = \mathbf{0}$, it is clear from Equation

16.31 and Equation 16.35 that ISI is completely removed and thus a ZF SDFE is obtained if $\mathbf{f}^T\mathcal{H}\mathbf{P}^{\perp T}=\mathbf{0}$, $\mathbf{f}^T\mathcal{H}\mathbf{e}=1$, and $(\mathbf{b}+\mathbf{e})^T=\mathbf{f}^T\mathcal{H}\mathbf{P}$. A sufficient condition for this to be satisfied is that $[\mathcal{H}\mathbf{P}^{\perp T},\mathcal{H}\mathbf{e}]$ has full column rank, which requires that $[\mathcal{H}\mathbf{P}^{\perp T},\mathcal{H}\mathbf{e}]$ has at least as many rows as columns. For TIV channels, this happens when $A(L'+1)\geq(L+L'+1)-L''$, whereas for TV channels, this happens when $A(Q'+1)(L'+1)\geq(Q+Q'+1)(L+L'+1)-(Q''+1)L''$. Again, these inequalities can be satisfied if $A\geq 2$, but this time they can also be satisfied if $A=1$, i.e., when the ZF SLE does not exist.

16.9 Complexity

In this section, we discuss some complexity issues of the above equalization structures. We can distinguish between design complexity and implementation complexity. The design complexity is the computational cost to design the equalizer, whereas the implementation complexity is the computational cost to equalize the channel once the equalizer has been designed. The block size N will play an important role in these complexities. In the following, we always assume that N is chosen large enough such that blind channel estimation becomes feasible or the overhead of the training symbols for training-based channel estimation does not decrease the data transmission rate too much (this basically boils down to choosing $N \gg L+1$ for TIV channels and $N \gg (Q+1)(L+1)$ for TV channels).

16.9.1 Design Complexity

Although many equalizer design procedures are possible (see Section 16.10), we will consider equalizer design based on channel knowledge. For the sake of simplicity, we will not exploit the band structure of $\bar{\mathbf{H}}^{(a)}$ or the special structure of $\mathcal{H}^{(a)}$ in the design complexity calculations. Let us first take a look at the linear equalization approaches. To design a BLE, we have to compute the inverse of an $(N-L)\times(N-L)$ matrix, which requires $\mathcal{O}((N-L)^3)$ flops. On the other hand, to compute an SLE, we need the inverse of a $D\times D$ matrix, where $D=L+L'+1$ for TIV channels and $D=(Q+Q'+1)(L+L'+1)$ for TV channels, which requires $\mathcal{O}(D^3)$ flops. As will be illustrated in Section 16.11, with a D that is much smaller than $N-L$, the performance of the SLE can approach the performance of the BLE for $A>1$. As a result, the SLE can have a much smaller design complexity than the BLE, without a significant loss in performance for $A>1$. For $A=1$, there is a loss in performance at high SNR, since in contrast to the performance of the BLE, the performance of the SLE saturates at high SNR.

Let us now focus on decision feedback equalization approaches. To design the feedback part of a BDFE, we have to compute the Cholesky decomposition of an $(N-L)\times(N-L)$ matrix. Hence, next to the $\mathcal{O}((N-L)^3)$ flops to design the feedforward part (similar to the complexity to design a BLE), we have an extra cost of $\mathcal{O}((N-L)^3)$ flops to design the feedback part. On the other hand, to compute the feedback part of an SDFE, we need the inverse of a $D''\times D''$ matrix, where $D''=L''+1$ for TIV channels and $D''=(Q''+1)L''+1$ for TV channels. Hence, next to the $\mathcal{O}(D^3)$ flops to design the feedforward part (similar to the complexity to design a SLE), we have an extra cost of $\mathcal{O}(D''^3)$ flops to design the feedback part. As will be illustrated in Section 16.11, with a D that is much smaller than $N-L$ and a D'' that is about half the size of D (and thus also much smaller than $N-L$), the performance of the SDFE can approach the performance of the BDFE. As a result, the SDFE can have a much smaller design complexity than the BDFE, without a significant loss in performance. This even holds for $A=1$, since like the performance of the BDFE, the performance of the SDFE does not saturate at high SNR.

Adopting an FDLE (FDDFE) for TIV channels, the FFT processing is computationally the most expensive and results in a complexity of $\mathcal{O}(N\log_2 N)$ flops. Hence, the FDLE (FDDFE) for TIV channels has a much smaller design complexity than the BLE (BDFE), while their performances are comparable, as will be illustrated in Section 16.11. The comparison with the design complexity of the SLE (SDFE) for TIV channels depends on the specific scenario.

16.9.2 Implementation Complexity

The implementation complexity will be defined here as the number of multiply–add (MA) operations required to estimate the transmitted data symbols. For the BLE, estimating the transmitted data symbols requires $N(N-L)$ MA operations per output, with an extra $(N-L)(N-L-1)/2$ MA operations for the BDFE. On the other hand, for the SLE, estimating the transmitted data symbols requires $(N-L')D'$ MA operations per output, with an extra $(N-L')(D''-1)$ MA operations for the SDFE, where $D' = L'+1$ for TIV channels and $D' = (Q'+1)(L'+1)$ for TV channels, and D'' is defined as before. Previously, we mentioned that with a D (and thus also a D') that is much smaller than $N-L$, the performance of the SLE can approach the performance of the BLE for $A > 1$. Hence, the SLE can also have a much smaller implementation complexity than the BLE, without a significant loss in performance for $A > 1$. Above, we also mentioned that with a D (and thus also a D') that is much smaller than $N-L$ and a D'' that is about half the size of D (and thus much smaller than $(N-L-1)/2$), the performance of the SDFE can approach the performance of the BDFE. Hence, the SDFE can also have a much smaller implementation complexity than the BDFE, without a significant loss in performance.

Like the design complexity, the implementation complexity of the FDLE (FDDFE) for TIV channels is completely determined by the complexity of the FFT processing, which is again much smaller than the implementation complexity of the BLE (BDFE). As before, the comparison with the implementation complexity of the SLE (SDFE) for TIV channels depends on the specific scenario. However, we should keep in mind that FD processing is only useful for TIV channels.

16.10 Channel Estimation and Direct Equalizer Design

Until now we have focused on equalizer design based on channel knowledge. In practice, however, the channel has to be estimated. There are basically two ways to estimate the channel: training-based or blind (intermediate so-called semiblind approaches are also possible). When training-based channel estimation is adopted, training symbols are inserted in $\mathbf{x}$ for serial transmission or in $\mathbf{s}$ for zero padding-based block transmission. We refer the interested reader to [44] for TIV channels and [26] for TV channels. When blind channel estimation is adopted, no training symbols are inserted. For serial transmission over TIV channels, many blind channel estimation algorithms have been proposed based on the data model in Equation 16.22 [11], [27], [41]. For serial transmission over TV channels, most of these algorithms can be extended by observing the similarity between the data models in Equations 16.22 and 16.24 [21]. Instead of only working with time-shifted versions of the received sequences, one then has to make use of time- and frequency-shifted versions of the received sequences. For zero padding-based block transmission over TIV channels, an interesting blind channel estimation algorithm has been developed in [34]. For zero padding-based block transmission over TV channels, this algorithm can be extended by employing a special type of linear precoding as described in [37].

Next to equalizer design based on channel knowledge, there also exist direct equalizer design algorithms, which do not require channel knowledge. They have mainly been developed for SLEs and can easily be extended to SDFEs. Again, one can distinguish between training-based and blind approaches. Training-based approaches are fairly easy to develop. Looking at Equation 16.23 and Equation 16.25, training-based direct equalizer design algorithms basically try to estimate $\mathbf{f}$ based on knowledge of $\mathbf{Y}$ and partial knowledge of $\mathbf{x}_\star$ via least squares fitting, for instance. Blind approaches are more difficult to derive. For TIV channels, many blind direct equalizer design algorithms have been proposed based on the data model in Equation 16.22 [11], [40], [41], [43], [46]. For TV channels, most of these algorithms can again be extended by observing the similarity between the data models in Equations 16.22 and 16.24. As before, instead of only working with time-shifted versions of the received sequences, one then has to make use of time- and frequency-shifted versions of the received sequences. Such direct equalizer design algorithms for TV channels are currently under investigation.

16.11 Performance Results

In this section, we compare the performances of the different equalizers discussed in this chapter. We only focus on the MMSE equalizers, which have a better performance than the ZF equalizers. We generate M ($M = 1, 2$) channels $g_{\text{ch}}^{(m)}(t)$ consisting of five clusters of 100 reflected or scattered rays. The delay of the cth cluster is given by $\tau_c = cT/2$ ($c \in \{0, 1, 2, 3, 4\}$). Assuming that $g_{\text{tr}}(t)$ and $g_{\text{rec}}(t)$ are rectangular functions over $(0, T)$ with height $1/T$, and thus $\psi(t) = g_{\text{rec}}(t) \star g_{\text{tr}}(t)$ is a triangular function over $(0, 2T)$ with height 1, we can thus assume that $L = 3$. The complex gain and frequency offset of the rth ray of the cth cluster are given by $G_{c,r}^{(m)} = e^{j\theta_{c,r}^{(m)}}/\sqrt{100}$ and $f_{c,r}^{(m)} = \cos(\phi_{c,r}^{(m)}) f_{\max}$, where $\theta_{c,r}^{(m)}$ and $\phi_{c,r}^{(m)}$ are uniformly distributed over $(0, 2\pi)$. We further consider fractional sampling with a factor of $P = 1, 2$. The modulation we use is quadrature phase shift keying (QPSK) with unit modulus. We assume the data sequence and the additive noises are mutually uncorrelated and white. The SNR is defined as $SNR = 5/\sigma_v^2$, where σ_v^2 is the variance of the additive noise. The factor 5 is due to the fact that we consider 5 clusters. For TIV channels, we fit the TIV FIR channel of Equation 16.5 to the true channel for $n \in \{0, 1, \ldots, N-1\}$, whereas for TV channels, we fit the BEM FIR channel of Equation 16.8 to the true channel for $n \in \{0, 1, \ldots, N-1\}$. In both cases, we use the obtained channel model parameters to design our equalizer. In practice, we have to estimate the channel model parameters of Equation 16.5 or Equation 16.8 using some channel estimation method. This can be a training-based method or a blind method (see Section 16.10). Although we make abstraction of this channel estimation procedure in this chapter, it will determine the block size N that we adopt in the simulations.

16.11.1 TIV Channels

For the TIV channels case, we consider $f_{\max} = 0$, and use the TIV FIR channel of Equation 16.5 with $L = 3$ to design our equalizers. Since $f_{\max} = 0$, the TIV FIR assumption will hold for any block size N. We consider a block size $N = 64$. This block size is large enough such that blind channel estimation becomes feasible or the overhead of the training symbols for training-based channel estimation does not decrease the data transmission rate too much. Moreover, it is also the block size that has been adopted for the IEEE 802.11a and HIPERLAN/2 WLAN (wireless local area network) standards (in the context of OFDM (orthogonal frequency division multiplexing)). Let us first compare the block equalizers with the frequency-domain equalizers for TIV channels. Figure 16.10 shows the performance of the BLE, BDFE, FDLE, and FDDFE in TIV channels for $M = 1, 2$ and $P = 1, 2$. We observe that the performance of the FDLE (FDDFE) approaches the performance of the BLE (BDFE) for all cases. However, we see that when fractional sampling is employed ($P = 2$), the FDLE (FDDFE) is not capable of improving the performance as much as the BLE (BDFE) does, but the difference between the two approaches is still rather small. Let us next compare the block equalizers with the serial equalizers for TIV channels. For the serial equalizers, we take $L' = 7$, $d = (L + L')/2 = 5$, and $L'' = L + L' - d = 5$. Figure 16.11 shows the performance of the BLE, BDFE, SLE, and SDFE in TIV channels for $M = 1, 2$ and $P = 1, 2$. We observe that the performance of the SLE approaches the performance of the BLE, except for the case $M = P = 1$ at high SNR, and the performance of the SDFE approaches the performance of the BDFE for all cases.

As mentioned before, the design complexity of the BLE is $\mathcal{O}\{(N-L)^3\}$ flops, with an extra $\mathcal{O}\{(N-L)^3\}$ flops for the BDFE, where $(N - L)^3 \approx 227{,}000$. On the other hand, the design complexity of the SLE is $\mathcal{O}\{D^3\}$ flops, with an extra $\mathcal{O}\{D''^3\}$ flops for the SDFE, where $D^3 = (L + L' + 1)^3 \approx 1300$ and $D''^3 = (L'' + 1)^3 \approx 200$. Hence, the design complexity of the SLE (SDFE) is clearly much smaller than the design complexity of the BLE (BDFE). A similar observation holds for the implementation complexity. The BLE requires $N(N - L) = 3904$ MA operations per output, with an extra $(N - L)(N - L - 1)/2 = 1830$ MA operations for the BDFE, whereas the SLE requires $(N - L')D' = (N - L')(L' + 1) = 456$ MA operations per output, with an extra $(N - L')(D'' - 1) = (N - L')L'' = 285$ MA operations for the SDFE. The major computational cost of designing or implementing the FDLE (FDDFE) is the FFT processing, which results in $\mathcal{O}\{N \log_2 N\}$ flops, where $N \log_2 N = 384$. Hence, compared to the BLE (BDFE), the

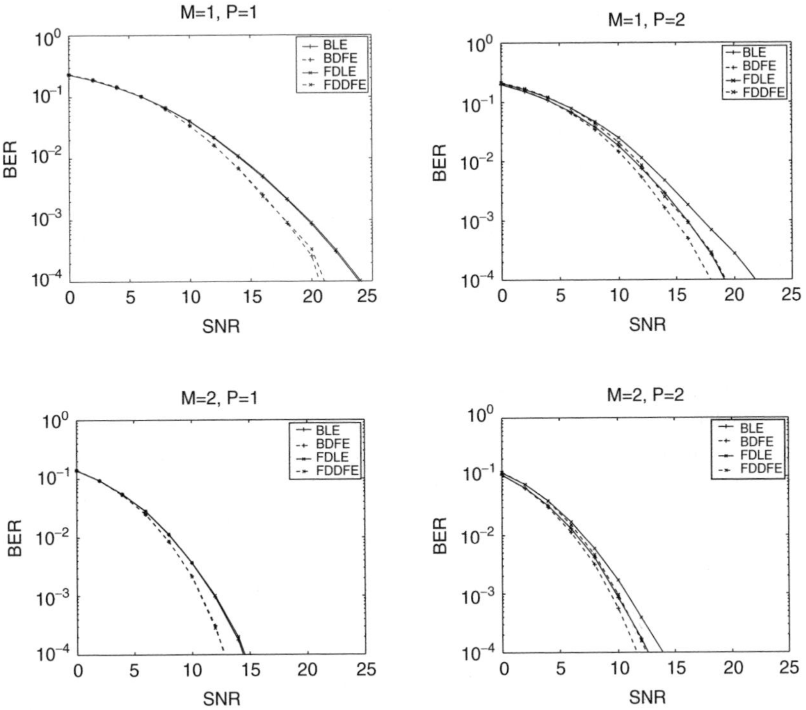

FIGURE 16.10 Comparison of different block and frequency-domain equalizers in TIV channels for $M = 1, 2$ and $P = 1, 2$.

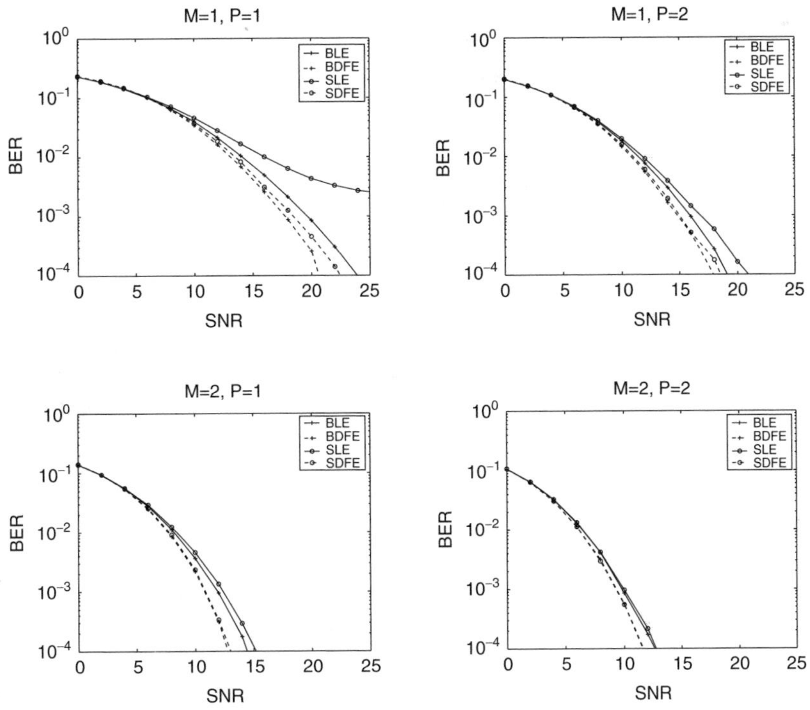

FIGURE 16.11 Comparison of different block and serial equalizers in TIV channels for $M = 1, 2$ and $P = 1, 2$.

design and implementation complexities of the FDLE (FDDFE) are much smaller. Compared to the SLE (SDFE), they are comparable for the chosen block size N. However, they would be larger for a larger block size N.

16.11.2 TV Channels

For the TV channels case, we consider $f_{\max} = 1/(400T)$, and use the BEM FIR channel of Equation 16.8 with $L = 3$ to design our equalizers. We know that in the best-case scenario we can take $Q = 2$, and then the number of channel model parameters that would have to be estimated in practice is three times as large as in the TIV channels case. Hence, it would then make sense to take N about three times as large as in the TIV channels case. Let us, for instance, take $N = 200$. In that case, $NT = 200T \leq 1/(2f_{\max}) = 200T$, which means that Q can be kept small to obtain a good BEM FIR approximation, as mentioned in Section 16.2.2. Therefore, we can indeed take $Q = 2$ as assumed above. To satisfy $Q/(2KT) \approx f_{\max} = 1/(400T)$, we then take $K = 400$. As illustrated in Example 1, these parameters lead to a tight fit of the BEM FIR channel to the true channel. Since frequency-domain equalizers are not useful for TV channels, we only compare the block equalizers with the serial equalizers for TV channels. For the serial equalizers, we take $L' = 7$, $d = (L + L')/2 = 5$, $L'' = L + L' - d = 5$, $Q' = 6$, and $Q'' = (Q + Q')/2 = 4$. Figure 16.12 shows the performance of the BLE, BDFE, SLE, and SDFE in TV channels for $M = 1, 2$ and $P = 1, 2$. As for the TIV channels, we observe that the performance of the SLE approaches the performance of the BLE, except for the case $M = P = 1$ at high SNR, and the performance of the SDFE approaches the performance of the BDFE for all cases.

Let us again take a look at the design and implementation complexities of the different methods. For the BLE and BDFE, N has changed compared to the TIV channels case. In other words, the design complexity now depends on $(N - L)^3 \approx 7,645,400$. For the SLE and SDFE, D and D'' have changed compared to the TIV channels case. More specifically, we now have $D^3 = ((L + L' + 1)(Q + Q' + 1))^3 \approx 970,300$ and

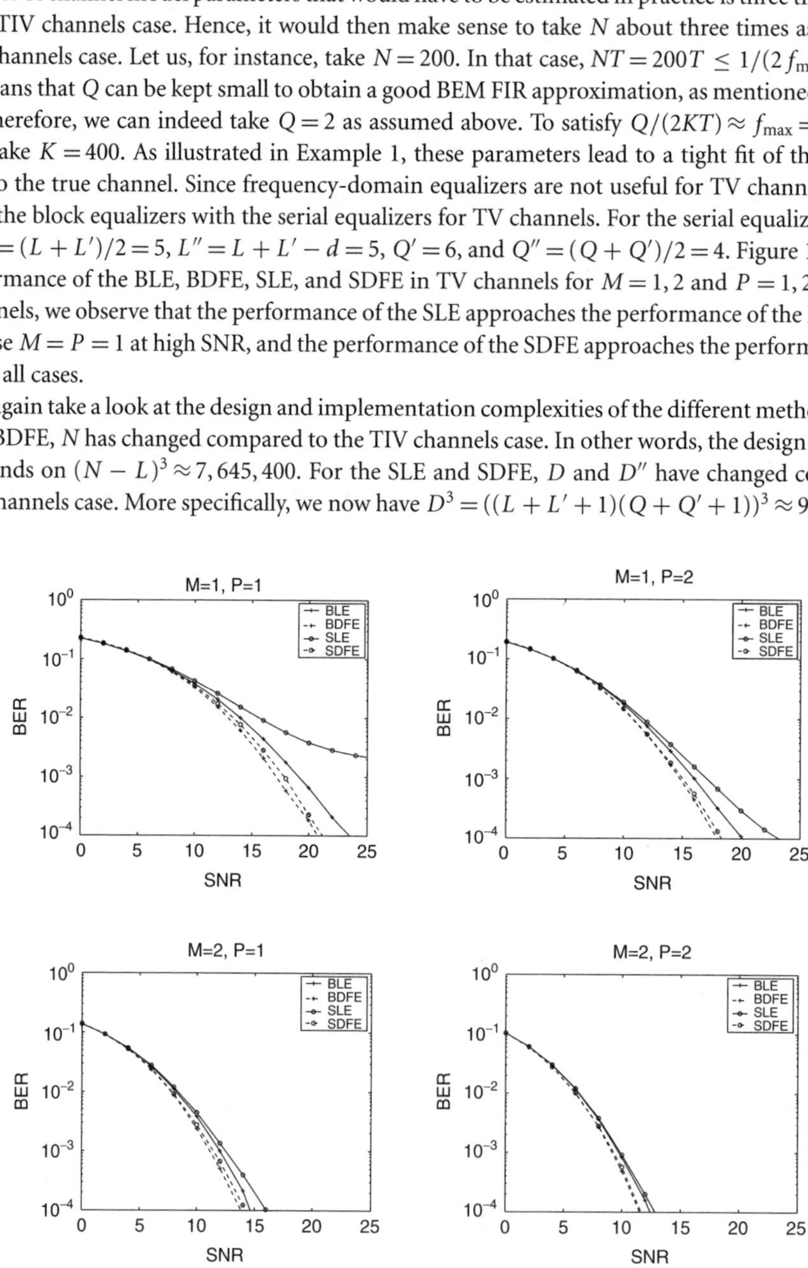

FIGURE 16.12 Comparison of different block and serial equalizers in TV channels for $M = 1, 2$ and $P = 1, 2$.

$D''^3 = ((Q'' + 1)L' + 1)^3 \approx 17,600$. As for the TIV channels case, we observe that the design complexity of the SLE (SDFE) is much smaller than the design complexity of the BLE (BDFE). A similar observation holds for the implementation complexity. The BLE requires $N(N - L) = 39,400$ MA operations per output, with an extra $(N - L)(N - L - 1)/2 = 19,306$ MA operations for the BDFE, whereas the SLE requires $(N - L')D' = (N - L')(Q' + 1)(L' + 1) = 10,808$ MA operations per output, with an extra $(N - L')(D'' - 1) = (N - L')(Q'' + 1)L'' = 4825$ MA operations for the SDFE. Note, however, that the relative difference between the design or implementation complexity of the SLE (SDFE) and the BLE (BDFE) is smaller than for the TIV channels case. One could argue that for a smaller block size N, the difference would even disappear, but then the block size would not be large enough such that blind channel estimation becomes feasible or the overhead of the training symbols for training-based channel estimation does not decrease the data transmission rate too much.

16.12 Summary

In this chapter, we have presented different practical finite-length equalization structures for TIV and TV channels. We have investigated linear and decision feedback block equalizers, linear and decision feedback serial equalizers, and linear and decision feedback frequency-domain equalizers (the latter only apply to TIV channels).

All these channel equalizers are based on a practical channel model. Writing the overall system as a symbol rate SIMO system, where the multiple outputs are obtained by multiple receive antennas and fractional sampling, we can distinguish between TIV and TV channels by looking at the channel time variation over a fixed time window. For TIV channels, we have modeled the channel by a TIV FIR channel, whereas for TV channels, it has been convenient to model the channel time variation by means of a basis expansion model, leading to a BEM FIR channel. Note that for the serial equalizers, we have used the same model for the equalizer as for the channel. Hence, for a TIV FIR channel, we have adopted a TIV FIR serial equalizer, whereas for a BEM FIR channel, we have adopted a BEM FIR serial equalizer. Note that in contrast with equalizing a BEM FIR channel with a TIV FIR serial equalizer, which requires a symbol rate SIMO channel with many outputs for the linear ZF solution to exist, equalizing a BEM FIR channel with a BEM FIR serial equalizer only requires a symbol rate SIMO channel with two outputs for the linear ZF solution to exist.

A complexity analysis and some illustrative simulation results have indicated that the SLE can have a much smaller design and implementation complexity than the BLE, without a significant loss in performance for a system with multiple outputs. For a system with a single output, there is a loss in performance at high SNR, since in contrast to the performance of the BLE, the performance of the SLE saturates at high SNR. Similarly, the SDFE can have a much smaller design and implementation complexity than the BDFE, without a significant loss in performance. This even holds for a system with a single output, since like the performance of the BDFE, the performance of the SDFE does not saturate at high SNR. Finally, the FDLE (FDDFE) for TIV channels has a much smaller design and implementation complexity than the BLE (BDFE), while their performances are comparable. The comparison with the design and implementation complexities of the SLE (SDFE) for TIV channels depends on the specific scenario. However, note that FD processing can only be applied for TIV channels.

Acknowledgments

This research work was supported by the Belgian Programme on Interuniversity Attraction Poles, initiated by the Belgian Federal Science Policy Office, IUAP P5/22 (Dynamical Systems Systems and Control: Computation, Identification and Modeling) and P5/11 (Mobile Multimedia Communication Systems and Networks), by the Concerted Research Action GOA-MEFISTO-666 (Mathematical Engineering for Information and Communication Systems Technology) of the Flemish Government, by Research Project

FWO nr.G.0196.02 (Design of Efficient Communication Techniques for Wireless Time-Dispersive Multi-user MIMO Systems), and by NWO-STW under the VICI Programme (DTC.5893).

References

[1] N. Al-Dhahir and J.M. Cioffi. MMSE Decision-Feedback Equalizers: Finite-Length Results. *IEEE Trans. Inf. Theory*, 41(4):961–975, 1995.

[2] I. Barhumi, G. Leus, and M. Moonen. Time-Varying FIR Decision Feedback Equalization of Doubly-Selective Channels. In *Proceedings of IEEE Global Communications Conf. (GLOBECOM '03)*, San Francisco, CA, December 2003.

[3] I. Barhumi, G. Leus, and M. Moonen. Time-Varying FIR Equalization of Doubly-Selective Channels. In *Proceedings of IEEE International Conference on Communications (ICC '03)*, Anchorage, AK, May 2003.

[4] N. Benvenuto and S. Tomasin. On the Comparison between OFDM and Single Carrier Modulation with a DFE Using a Frequency-Domain Feedforward Filter. *IEEE Trans. Commun.*, 50(6):947–955, 2002.

[5] J.K. Cavers. *Mobile Channel Characteristics*. Kluwer, Dordrecht, Netherlands, 2000.

[6] A. Czylwik. Comparison between Adaptive OFDM and Single Carrier Modulation with Frequency Domain Equalization. In *Proceedings of IEEE Vehicular Technology Conference (VTC)*, Phoenix, AZ, May 1997, pp. 865–869.

[7] L. Deneire, B. Gyselinckx, and M. Engels. Training Sequence vs. Cyclic Prefix: A New Look on Single Carrier Communication. In *Proceedings of GLOBECOM*, San Fransisco, CA, November/December 2000.

[8] A. Duel-Hallen. A Family of Multiuser Decision-Feedback Detectors for Asynchronous Code-Division Multiple-Access Channels. *IEEE Trans. Commun.*, 43(2/3/4):421–434, 1995.

[9] A. Duel-Hallen, S. Hu, and H. Hallen. Long-Range Prediction of Fading Channels. *IEEE Signal Process. Mag.*, 17(3), 62–75, 2000.

[10] D. Falconer, S.L. Ariyavisitakul, A. Benyamin-Seeyar, and B. Eidson. Frequency Domain Equalization for Single-Carrier Broadband Wireless Systems. *IEEE Commun. Mag.*, 40(4):58–66, 2002.

[11] G.B. Giannakis, Y. Hua, P. Stoica, and L. Tong, Eds. *Signal Processing Advances in Wireless and Mobile Communications: Trends in Channel Estimation and Equalization*, Vol. 1. Prentice Hall, New York, 2000.

[12] G.B. Giannakis, Y. Hua, P. Stoica, and L. Tong, Eds. *Signal Processing Advances in Wireless* and *Mobile Communications: Trends in Single- and Multi-User Systems*, Vol. 2. Prentice Hall, New York, 2000.

[13] G.B. Giannakis and C. Tepedelenlioğlu. Basis Expansion Models and Diversity Techniques for Blind Equalization of Time-Varying Channels. *Proc. IEEE*, 86(10), 1969–1986, 1998.

[14] J.D. Gibson, Ed. *The Mobile Communications Handbook*, 2nd ed. CRC Press/IEEE Press, Boca Ratom, FL, 1999.

[15] A. Ginesi, G.M. Vitetta, and D.D. Falconer. Block Channel Equalization in the Presence of a Cochannel Interferent Signal. *IEEE J. Selected Areas Commun.*, 17(11):1853–1862, 1999.

[16] W.C. Jakes, Ed. *Microwave Mobile Communications*. Wiley, New York, 1974.

[17] A. Klein, G. Kawas Kaleh, and P.W. Baier. Zero Forcing and Minimum Mean-Square-Error Equalization for Multiple-Access Channels. *IEEE Trans. Vehicular Technol.*, 45(2):276–287, 1996.

[18] E.A. Lee and D.G. Messerschmitt. *Digital Communications*, 2nd ed. Kluwer, Dordrecht, Netherlands, 1994.

[19] G. Leus, I. Barhumi, and M. Moonen. MMSE Time-Varying FIR Equalization of Doubly-Selective Channels. In *Proceedings of IEEE International Conference on Acoustics, Speech, and Signal Processing (ICASSP '03)*, Hong Kong, April 2003.

[20] G. Leus and M. Moonen. Semi-Blind Channel Estimation for Block Transmission with Non-Zero Padding. In *Proceedings of Asilomar Conference on Signals, Systems and Computers*, Pacific Grove, CA, November 2001, pp. 762–766.

[21] G. Leus and M. Moonen. Deterministic Subspace Based Blind Channel Estimation for Doubly-Selective Channels. In *Proceedings of the IEEE Workshop on Signal Processing Advances in Wireless Communications (SPAWC '03)*, Rome, June 2003.

[22] G. Leus, S. Zhou, and G.B. Giannakis. Orthogonal Multiple Access over Time- and Frequency-Selectvie Channels. *IEEE Trans. Inf. Theory*, 49(8):1942–1950, 2003.

[23] H. Liu and G.B. Giannakis. Deterministic Approaches for Blind Equalization of Time-Varying Channels with Antenna Arrays. *IEEE Trans. Signal Process.*, 46(11):3003–3013, 1998.

[24] R. Lupas and S. Verdú. Linear Multiuser Detectors for Synchronous Code-Division Multiple-Access Channels. *IEEE Trans. Inf. Theory*, 35(1):123–136, 1989.

[25] X. Ma and G.B. Giannakis. Maximum-Diversity Transmissions over Doubly-Selective Wireless Channels. *IEEE Trans. Inf. Theory*, 49(7):1832–1840, 2003.

[26] X. Ma, G.B. Giannakis, and S. Ohno. Optimal Training for Block Transmissions over Doubly-Selective Wireless Fading Channels. *IEEE Trans. Signal Process.*, 51(5):1351–1366, 2003.

[27] E. Moulines, P. Duhamel, J.-F. Cardoso, and S. Mayrargue. Subspace Methods for the Blind Identification of Multichannel FIR Filters. *IEEE Trans. Signal Process.*, 43(2):516–525, 1995.

[28] A.J. Paulraj and C.B. Papadias. Space-Time Processing for Wireless Communications. *IEEE Signal Process. Mag.*, 14(6):49–83, 1997.

[29] J.G. Proakis. *Digital Communications*, 4th ed. McGraw-Hill, New York, 2001.

[30] O. Rousseaux, G. Leus, and M. Moonen. Training Based Maximum Likelihood Channel Identification. In *Proceedings of IEEE Workshop on Signal Processing Advances for Wireless Communications (SPAWC)*, Rome, June 2003.

[31] H. Sari, G. Karam, and I. Jeanclaude. Transmission Techniques for Digital Terrestrial TV Broadcasting. *IEEE Commun. Mag.*, 33(2):100–109, 1995.

[32] A.M. Sayeed and B. Aazhang. Joint Multipath-Doppler Diversity in Mobile Wireless Communications. *IEEE Trans. Commun.*, 47(1):123–132, 1999.

[33] A. Scaglione and G.B. Giannakis. Redundant Filterbank Precoders and Equalizers. Part I. Unification and Optimal Designs. *IEEE Trans. Signal Process.*, 47(7):1988–2006, 1999.

[34] A. Scaglione and G.B. Giannakis. Redundant Filterbank Precoders and Equalizers. Part II. Blind Channel Estimation, Synchronization, and Direct Equalization. *IEEE Trans. Signal Process.*, 47(7):2007–2022, 1999.

[35] D.T.M. Slock. Blind Fractionally-Spaced Equalization, Perfect-Reconstruction Filter Banks and Multichannel Linear Prediction. In *Proceedings of IEEE International Conference on Acoustics, Speech, and Signal Processing (ICASSP '94)*, Adelaide, Australia, April 1994, pp. IV/585–IV/588.

[36] A. Stamoulis, G.B. Giannakis, and A. Scaglione. Block for Decision-Feedback Equalizers for Filterbank Precoded Transmissions with Blind Channel Estimation Capabilities. *IEEE Trans. Commun.*, 49(1):69–83, 2001.

[37] C. Tepedelenlioğlu and G.B. Giannakis. Transmitter Redundancy for Blind Estimation and Equalization of Time- and Frequency-Selective Channels. *IEEE Trans. Signal Process.*, 48(7):2029–2043, 2000.

[38] J.R. Treichler, I. Fijalkow, and C.R. Johnson, Jr. Fractionally Spaced Equalizers: How Long Should They Really Be? *IEEE Signal Process. Mag.*, 13(3), 65–81, 1996.

[39] M.K. Tsatsanis and G.B. Giannakis. Modeling and Equalization of Rapidly Fading Channels. *Int. J. Adaptive Control Signal Process.*, 10(2/3):159–176, 1996.

[40] M.K. Tsatsanis and Z. Xu. Performance Analysis of Minimum Variance Receivers. *IEEE Trans. Signal Process.*, 46(11):3014–3022, 1998.

[41] J.K. Tugnait, L. Tong, and Z. Ding. Single-User Channel Estimation and Equalization. *IEEE Signal Process. Mag.*, 17(3):17–28, 2000.

[42] G. Ungerboeck. Fractional Tap-Spacing Equalizer and Consequences for Clock Recovery in Data Modems. *IEEE Trans. Commun.*, 24(8):856–864, 1976.

[43] A.-J. van der Veen, S. Talwar, and A. Paulraj. A Subspace Approach to Blind Space-Time Signal Processing for Wireless Communication Systems. *IEEE Trans. Signal Process.*, 45(1):173–190, 1997.

[44] H. Vikalo, B. Hassibi, B. Hochwald, and T. Kailath. Optimal Training for Frequency-Selective Channels. In *Proceedings of ICASSP*, Salt Lake City, UT, May 2001.

[45] Z. Wang and G.B. Giannakis. Wireless Multicarrier Communications: Where Fourier Meets Shannon. *IEEE Signal Process. Mag.*, 17(3), 29–48, 2000.

[46] D. Gesbert, P. Duhamel, and S. Mayrargue. On-Line Blind Multichannel Equalization Based on Mutually Referenced Filters. *IEEE Trans. Signal Process.*, 45(9):2307–2317, 1997.

17

Low-Complexity Diversity Combining Schemes for Mobile Communications*

17.1 Introduction 17-2
17.2 System and Channel Models 17-3
System Model • Channel Model
17.3 Dual-Branch Switch and Stay Combining 17-4
Dual-Branch SSC Schemes • Markov Chain-Based Analysis • Statistics of Overall Combiner Output • Application to Performance Analysis and Comparisons
17.4 Multibranch Switched Diversity 17-14
L-Branch SSC • L-Branch SEC
17.5 Generalized Switch and Examine Combining (GSEC) 17-20
Mode of Operation of GSEC • MGF of Combiner Output • Application to Performance Analysis • Complexity Savings over GSC
17.6 Further Remarks 17-28

Hong-Chuan Yang
University of Victoria

Mohamed-Slim Alouini
University of Minnesota

Abstract

Switched diversity offers one of the lowest-complexity solutions to mitigate the effect of fading in wireless communication systems. In this chapter, starting from traditional dual-branch switch and stay combining (SSC) schemes, we develop and analyze several switching-based diversity combining schemes. More specifically, we first present a Markov chain-based analytical framework for the performance analysis of various switching strategies used in conjunction with dual-branch SSC systems. This analysis leads to a thorough comparison and trade-off study between different SSC strategies. Then we generalize switched diversity to a multibranch scenario. We show that while SSC does not benefit from additional diversity paths, the performance of switch and examine combining (SEC) improves with additional branches. Finally, noting the deficiency of pure switched diversity schemes in taking advantage of available diversity paths, we

*This work was supported in part by National Science Foundation Grants CCR-9983462 and ECS-9979443. The work of H.-C. Yang was also supported in part by the Doctoral Dissertation Fellowship from the University of Minnesota and start-up funds from the University of Victoria.

0-8493-1657-X/$0.00+$1.50

propose generalized switch and examine combining (GSEC) as a good candidate combining scheme for diversity-rich environments. A description of the GSEC mode of operation shows that this scheme conserves a fixed complexity as the number of diversity paths increases, and offers a considerable complexity savings compared to traditional maximum ratio combining (MRC) and even other low-complexity schemes such as generalized selection combining (GSC). For all of these schemes, we present closed-form expressions for the statistics of the combiner output, including the moment-generating function, cumulative distribution function, and probability density function. The resulting expressions allow for accurate performance analysis of these diversity techniques over most fading models of interest. As an illustration of the mathematical formalism, selected numerical examples are provided together with some related discussions and interpretations. It is observed that compared with the more popular selection-based combining schemes, switching-based combining schemes lead to simpler receiver structure, less signal processing, and lower power consumption at the cost of certain performance loss.

17.1 Introduction

Multipath fading has a deleterious impact on the performance of wireless communications systems. To support high-quality multimedia communication services, future wireless systems must efficiently mitigate the fading effect. Diversity combining techniques represent one of the most effective solutions to fading mitigation [1]. They can significantly improve the performance of wireless communications systems by means of multiple transmission and reception and appropriate combining of the differently faded replicas of the same information-bearing signal.

Among different combining schemes, there is a trade-off between performance and complexity [2, Chapter 12]. From a performance perspective, the optimal combining solution is the well-known maximum ratio combining (MRC). However, the implementation of MRC is complex, and this complexity grows as the number of diversity paths increases. Indeed, MRC not only requires the same number of radio frequency (RF) chains as the number of available diversity paths, but also mandates the complete and simultaneous knowledge of the channel condition for each diversity path. On the other hand, selection combining (SC) [3, Section 6.2] has much lower complexity. It uses only the best diversity paths of all available ones. This nevertheless requires the estimation and comparison of the quality indicators of all diversity paths.

Switch and stay combining (SSC) is another low-complexity combining scheme [1], [4]–[10]. It further reduces the complexity by eliminating the need for checking the path quality of all diversity paths. In particular, with SSC, the receiver uses the current branch as long as it is acceptable (e.g., the path quality is above a fixed threshold). Whenever the current branch becomes unacceptable, the receiver switches and stays on another branch regardless of its quality. As such, the receiver with SSC needs only to monitor the quality of the currently used branch. In addition, SSC only requires comparisons between a channel estimate and a fixed threshold where the receiver with SC needs to compare quality estimates of different diversity paths. Therefore, SSC also mandates simpler comparison circuitry than SC.

Many switching strategies have been proposed and analyzed for dual-branch SSC systems over the past four decades. For example, Blanco and Zdunek first introduced a specific discrete-time strategy of SSC [4,5]. Subsequently, Abu-Dayya and Beaulieu proposed a simplified SSC strategy [6,7]. While considerable work has been carried out on the performance analysis of these two SSC schemes, to the best knowledge of the authors, the difference and trade-off between them have not been addressed so far. In this chapter, we present a Markov chain-based analytical framework for the performance analysis of various switching strategies used in conjunction with dual-branch SSC systems [11]. Aside from putting under the same umbrella all of the previously known performance results regarding SSC systems, it also allows solving for the outage probability and average error rate performance, as well as switching rate of different strategies in very general fading scenarios. As a result, we are able to conduct a thorough comparison and highlight the main differences and trade-offs involved between these various SSC strategies.

In general, the more the available diversity paths, the larger the potential diversity benefit. Consequently, emerging wireless communications systems tend to employ physical-layer solutions operating in a diversity-rich environment, i.e., with a large number of diversity paths. Examples include ultrawideband (UWB) systems, wideband code division multiple-access (CDMA) systems [12,13], millimeter-wave systems [14], and multiple-input-multiple-output (MIMO) systems [15]. However, previous works on SSC have focused on the dual-branch case. In this chapter, we generalize switched diversity to the multibranch scenario and investigate its operation and performance [16]. In particular, we present closed-form expressions for the statistics of the combiner output, including the cumulative distribution function (CDF), probability density function (PDF), and moment-generating function (MGF). These expressions allow for accurate performance analysis of multibranch switched combining schemes over most fading models of interest. We show that while SSC does not benefit from additional diversity paths, the performance of switch and examine combining (SEC) improves with additional branches.

The complexity savings of switched combining schemes over the optimal MRC scheme come at the cost of considerable performance loss, especially in a diversity-rich environment. Similar to SC, the performance of SEC is limited by the fact that only one antenna branch is used for data reception. Recently, to bridge the performance gap between MRC and SC, the so-called generalized selection combining (GSC) scheme, also known as hybrid selection/maximum ratio combining (H-S/MRC), was proposed and extensively studied (see, for example, [12], [14], [17]– [21] and references therein). With GSC, the receiver applies MRC to the L_c strongest (with highest received signal power) paths among the L available ones, where L_c is usually much smaller than L. Since only L_c, instead of L, paths need to be processed in the fashion of MRC, GSC provides considerable complexity savings over conventional MRC. However, its operation requires the estimation and full ranking of all L diversity paths, which entails a certain complexity, especially when L is large. In this chapter, by following the hybrid combining approach, we propose generalized switch and examine combining (GSEC) as a good candidate combining scheme for diversity-rich environments [22]. A description of the GSEC mode of operation shows that this scheme conserves a fixed complexity as the number of diversity paths increases and offers a considerable complexity savings over traditional MRC and even the competing low-complexity GSC scheme. We study the performance of GSEC over independent and identically distributed (i.i.d.) diversity paths. We also provide a thorough trade-off study of performance vs. complexity between GSEC and GSC.

The rest of this chapter is organized as follows. Section 17.2 contains the general description of the system and channel model under consideration. Section 17.3 addresses the analysis and comparison of different SSC schemes, while Section 17.4 studies the operation and performance of multibranch switched combining schemes. The new GSEC scheme is presented and investigated in Section 17.5. Finally, Section 17.6 provides some concluding remarks. In all sections, as an illustration of the mathematical formalism, selected numerical examples are provided together with some related discussions and interpretations.

17.2 System and Channel Models

17.2.1 System Model

In this work, we consider receiver space diversity systems, as shown in Figure 17.1. In particular, multiple antennas are used at the receiver to create different faded replicas of the transmitted signal. The schemes presented in this work can be easily adapted to be applicable to transmit diversity systems, MIMO systems, CDMA systems, and UWB systems. For example, in CDMA and UWB systems, the antenna branches may correspond to different resolvable multipaths. We assume that the diversity combining schemes are

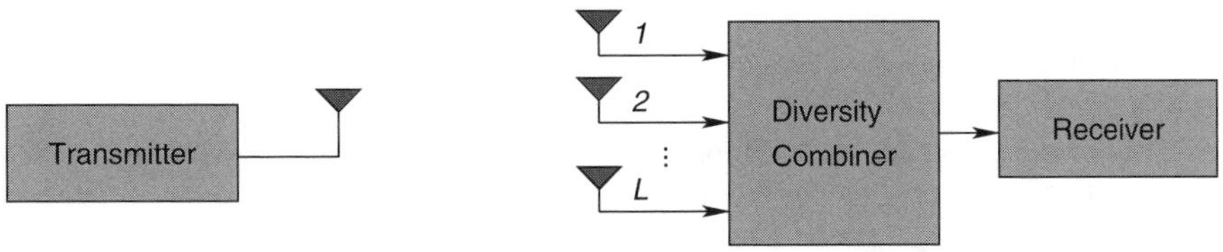

FIGURE 17.1 Block diagram of receiver space diversity systems.

TABLE 17.1 Statistics of the Faded Signal Power s for the Three Fading Models under Consideration

Model	Rayleigh	Rice	Nakagami-m
Parameter	·	$K = n^2 \geq 0$	$m \geq \frac{1}{2}$
PDF ($p(x)$)	$\frac{1}{\Omega} e^{-\frac{x}{\Omega}}$	$\frac{(1+n^2)\, e^{-n^2}}{\Omega} e^{-\frac{1+n^2}{\Omega} x} I_0\left(2n\sqrt{\frac{1+n^2}{\Omega} x}\right)$	$\left(\frac{m}{\Omega}\right)^m \frac{x^{m-1}}{\Gamma(m)} e^{-\frac{mx}{\Omega}}$
CDF ($P(x)$)	$1 - e^{-\frac{x}{\Omega}}$	$1 - Q_1\left(n\sqrt{2}, \sqrt{\frac{2(1+n^2)}{\Omega} x}\right)$	$1 - \frac{\Gamma\left(m, \frac{m}{\Omega} x\right)}{\Gamma(m)}$
MGF ($M(t)$)	$(1 - t\Omega)^{-1}$	$\frac{1+n^2}{1+n^2-t\Omega} \exp\left(\frac{t\Omega n^2}{1+n^2-t\Omega}\right)$	$\left(1 - \frac{t\Omega}{m}\right)^{-m}$

implemented in a discrete-time fashion. More specifically, short guard periods are periodically inserted into the transmitted signal. During these guard periods, the receiver performs a series of operations, including channel estimations and necessary comparisons, to reach the appropriate diversity combining decision.

17.2.2 Channel Model

We adopt a block fading channel model. More specifically, the data burst is assumed to experience roughly the same fading as its preceding guard period. The fading conditions during different guard periods and data burst pairs are assumed to be independent. Let $s_i(n)$ be the received signal power of the ith branch during the nth guard period. We denote by $S(n)$ the overall received signal power after switched diversity is applied. Therefore, if the ith branch is used for the nth data burst, we have $S(n) = s_i(n)$.

We assume that the received signal on each antenna branch follows either the Rayleigh, Rice, or Nakagami-m fading model. In Table 17.1, we summarize the PDF, $p(x)$; CDF, $P(x)$; and MGF, $M(t) = \int_0^\infty e^{tx} p_x(x)dx$ of the faded signal power s for the three fading models under consideration. The signal-to-noise ratio (SNR) γ is proportional to the faded signal power. Therefore, Table 17.1 also gives the statistics of the SNR for single branch case. In Table 17.1, Ω is the average fading power, $\Gamma(\cdot)$ is the Gamma function [23, Section 8.31], $I_0(\cdot)$ is the modified Bessel function of the first kind with zero order [23, Section 8.43], $\Gamma(\cdot,\cdot)$ is the incomplete Gamma function [23, Section 8.35], and $Q_1(\cdot,\cdot)$ is the first-order Marcum Q function [24].

17.3 Dual-Branch Switch and Stay Combining

In this section, we focus on low-complexity combining schemes for dual-branch diversity systems, i.e., $L = 2$ in Figure 17.1. Both SC and SSC can be used with dual-branch systems. With SC, the receiver always uses the better branch, usually the one with the highest received signal power. Therefore, during each guard period, the receiver using SC needs to estimate the received signal power of both diversity paths and compare the two power estimates. With SSC, however, the receiver uses the current branch until it becomes unacceptable, and then it switches and stays on the other branch regardless of its quality. The branch acceptance is determined by comparing the current power estimate with a preselected fixed threshold, which can be implemented with a threshold detector. Note that the major complexity savings of SSC over SC is that SSC needs only the power estimate of the current path and simpler comparison to reach a combining decision.

Many switching strategies have been proposed and analyzed for SSC over the past four decades [4]–[7]. In this section, we first provide detailed descriptions for different dual-branch SSC schemes. We then present a Markov chain-based analytical framework for the performance analysis and comparison of various SSC schemes. Based on this framework, a thorough comparison and trade-off study between different SSC systems is carried out [11].

TABLE 17.2 Complexity Comparison of SC and Three SSC Schemes in Terms of Operations Needed for Combining Decision

	No. of Power Estimations	Type of Comparison	No. of Comparisons
SC	2	Two estimates	1
SSC-A	1	An estimate and a fixed value	1
SSC-B	1 or 2	An estimate and a fixed value	1 or 2
SSC-C	1	An estimate and a fixed value	1

17.3.1 Dual-Branch SSC Schemes

In what follows, we describe the mode of operation of various SSC strategies and briefly discuss their hardware implementation complexity. The comparison result of hardware complexity of SC and different SSC strategies is also summarized in Table 17.2. In particular, we used two criteria in the complexity comparison of different schemes: the number of power estimation and the type and number of comparisons needed during the guard period for each scheme.

1. *Type 1 SSC (SSC-A) [6]:* The most popular SSC strategy, denoted in what follows by SSC-A, was originally proposed and studied by Abu-Dayya and Beaulieu in [6]. With SSC-A, the receiver estimates the channel and switches to the other branch if and only if the estimate is less than the switching threshold. With SSC-A, the combined signal power sequence $S(n)$ is governed by the following mathematical relationship:

$$\text{if } S(n-1) = s_i(n-1)$$
$$\text{then } S(n) = \begin{cases} s_i(n), & \text{if } s_i(n) \geq s_{T_i} \\ s_{\bar{i}}(n), & \text{if } s_i(n) < s_{T_i} \end{cases} \tag{17.1}$$

where the overbar on the i subscript means that if $i = 1$, then $\bar{i} = 2$ and vice versa, and s_{T_i} is the switching threshold of the ith branch $(i = 1, 2)$. Note that unlike previous works on SSC, we now allow that different antenna branches have different switching thresholds. This may require additional complexity. However, if the antenna branches are not identically faded, to achieve the best performance in terms of minimizing the average error rate, we need to use different switching thresholds for different antenna branches [11]. Using a common switching threshold results into a certain amount of performance degradation. Note that we can accommodate different thresholds by using two different threshold detectors. In any guard period, only one channel is estimated and only one comparison between the channel estimate and a constant for the currently used branch is necessary.

2. *Type 2 SSC (SSC-B) [4]:* Blanco and Zdunek analyzed an alternative discrete-time realization of SSC in [4]. We denote it by SSC-B in what follows. In SSC-B, the receiver makes its switching decision based on the channel estimations of the currently used branch in two consecutive guard periods, the current and the one immediately previous to it. A switch is initiated whenever a downward crossing of the switching threshold occurs. Mathematically speaking, this strategy can be described by

$$\text{if } S(n-1) = s_i(n-1)$$
$$\text{then } S(n) = \begin{cases} s_i(n), & \text{if } s_i(n-1) < s_{T_i} \\ & \text{or } s_i(n-1) \geq s_{T_i} \text{ and } s_i(n) \geq s_{T_i} \\ s_{\bar{i}}(n), & \text{if } s_i(n-1) \geq s_{T_i} \text{ and } s_i(n) < s_{T_i} \end{cases} \tag{17.2}$$

Implicitly, this strategy requires the estimation of the switch-to branch in the same guard period so that the receiver always has enough information for the switching decision in the next guard period. So, besides requiring one bit of memory for the comparison result in the previous guard period, this strategy may also need to perform one (when branch switching does not occur) or two (when branch switching does occur) estimations and comparisons between an estimated quantity and a constant in each guard period.

3. *Type 3 SSC (SSC-C):* If we relax the implicit requirement of SSC-B dealing with the channel estimate in a previous guard period and if we do not allow branch switching when the receiver does not have

enough information, a simpler strategy results, denoted by SSC-C. Again, a switch is initiated whenever a downward crossing of the switching threshold occurs. This strategy can be mathematically described by

$$\text{if } S(n-1) = s_i(n-1)$$
$$\text{then } S(n) = \begin{cases} s_i(n), & \text{if } s_i(n-1) < s_{T_i} \\ & \text{or } s_i(n-1) \geq s_{T_i} \text{ and } s_i(n) \geq s_{T_i} \\ & \text{or } s_i(n-1) \text{ unknown} \\ s_{\bar{i}}(n), & \text{if } s_i(n-1) \geq s_{T_i} \text{ and } s_i(n) < s_{T_i} \end{cases} \tag{17.3}$$

It is easy to see that this strategy requires only one channel estimation and one comparison between an estimated quantity and a constant in each guard period, in addition to one bit of memory.

17.3.2 Markov Chain-Based Analysis

With the assumption of block fading channel, all SSC strategies satisfy the Markovian property. More specifically, future switches depend only on the currently used antenna branch and its fading condition. A Markov chain can be built based on the switching strategy of the chosen scheme. In particular, with the appropriate definition of the state space, the transition probability matrix P [25, Chapter 4] of the Markov chain can be obtained in terms of the statistics of each branch.

1. *SSC-A:* In [10], a two-state Markov chain has been used for SSC-A to determined the probability of usage of each branch at any data burst. In particular, State i corresponds to "Antenna i is used when the estimation is performed" [10, Section II], $i = 1, 2$. Here, we use a different state space definition. More specifically, the state space is defined in the following manner:

State 1 $\Leftrightarrow S(n) = s_1(n)$ and $s_1(n) < s_{T_1}$ State 2 $\Leftrightarrow S(n) = s_1(n)$ and $s_1(n) \geq s_{T_1}$

State 3 $\Leftrightarrow S(n) = s_2(n)$ and $s_2(n) < s_{T_2}$ State 4 $\Leftrightarrow S(n) = s_2(n)$ and $s_2(n) \geq s_{T_2}$

Accordingly, based on the switching strategy of SSC-A, the transition probability matrix, P^a, for this Markov chain is given by

$$P^a = \begin{bmatrix} 0 & 0 & q_2 & 1-q_2 \\ q_1 & 1-q_1 & 0 & 0 \\ q_1 & 1-q_1 & 0 & 0 \\ 0 & 0 & q_2 & 1-q_2 \end{bmatrix}_{4\times 4} \tag{17.4}$$

where

$$q_i = \Pr[s_i(n) < s_{T_i}] = P_i(s_{T_i}), \quad i = 1, 2 \tag{17.5}$$

and $P_i(\cdot)$ is the CDF of the received signal power for the ith antenna, as given in Table 17.1 for some popular fading channel models. As will be seen in subsequent sections, this definition of the Markov chain leads to the most general statistical description of SSC-A output statistics. Its generalization will render similar general results for SSC-B and SSC-C possible for various fading scenarios.

2. *SSC-B:* For SSC-B, since the switching decision is based on two consecutive comparison results, we define the state space of the Markov chain using two comparison results. As a result, we end up with a six-state Markov chain whose state space is defined as

State 1 $\Leftrightarrow S(n) = s_1(n)$ and $s_1(n-1) \geq s_{T_1}$ and $s_1(n) < s_{T_1}$

State 2 $\Leftrightarrow S(n) = s_1(n)$ and $s_1(n-1) < s_{T_1}$

State 3 $\Leftrightarrow S(n) = s_1(n)$ and $s_1(n-1) \geq s_{T_1}$ and $s_1(n) \geq s_{T_1}$

State 4 $\Leftrightarrow S(n) = s_2(n)$ and $s_2(n-1) \geq s_{T_2}$ and $s_2(n) < s_{T_2}$

State 5 $\Leftrightarrow S(n) = s_2(n)$ and $s_2(n-1) < s_{T_2}$

State 6 $\Leftrightarrow S(n) = s_2(n)$ and $s_2(n-1) \geq s_{T_2}$ and $s_2(n) \geq s_{T_2}$

Correspondingly, based on the switching strategy of SSC-B, the transition probability matrix, P^b, of this Markov chain is given by

$$P^b = \begin{bmatrix} 0 & 0 & 0 & \left(1-\frac{q_{12}}{q_1}\right)q_2 & \frac{q_{12}}{q_1} & \left(1-\frac{q_{12}}{q_1}\right)(1-q_2) \\ (1-q_1)q_1 & q_1 & (1-q_1)^2 & 0 & 0 & 0 \\ q_1 & 0 & 1-q_1 & 0 & 0 & 0 \\ \left(1-\frac{q_{12}}{q_2}\right)q_1 & \frac{q_{12}}{q_2} & \left(1-\frac{q_{12}}{q_2}\right)(1-q_1) & 0 & 0 & 0 \\ 0 & 0 & 0 & (1-q_2)q_2 & q_2 & (1-q_2)^2 \\ 0 & 0 & 0 & q_2 & 0 & 1-q_2 \end{bmatrix}_{6\times 6}$$

where

$$q_{12} = \Pr\left[s_1(n) < s_{T_1} \text{ and } s_2(n) < s_{T_2}\right] = P_{1,2}(s_{T_1}, s_{T_2}) \tag{17.6}$$

where $P_{1,2}(\cdot,\cdot)$ is the joint CDF of the received signal powers of the two diversity paths, which is available in closed form in terms of first-order Marcum Q function for the Rayleigh fading environment [26, Appendix A] and in terms of an infinite series for the Nakagami fading environment [27]. When the fading is independent over two diversity paths, $q_{12} = P_{1,2}(s_{T_1}, s_{T_2}) = q_1 q_2$, where q_i is defined in Equation 17.5 for $i = 1, 2$.

3. *SSC-C:* As expected, the state space definition for SSC-C is very similar to that for SSC-B. However, we need to include an additional pair of states, which corresponds to the situation that a branch has just been switched to and hence not enough information for a switching decision is available. In this case, the receiver has to wait for another guard period to decide whether to switch. We define an eight-state Markov chain for SSC-C as

$$\begin{aligned}
&\text{State 1} \Leftrightarrow S(n) = s_1(n) \text{ and } s_1(n-1) \geq s_{T_1} \text{ and } s_1(n) < s_{T_1} \\
&\text{State 2} \Leftrightarrow S(n) = s_1(n) \text{ and } s_1(n-1) < s_{T_1} \\
&\text{State 3} \Leftrightarrow S(n) = s_1(n) \text{ and } s_1(n-1) \geq s_{T_1} \text{ and } s_1(n) \geq s_{T_1} \\
&\text{State 4} \Leftrightarrow S(n) = s_1(n) \text{ and } s_1(n-1) \text{ unknown} \\
&\text{State 5} \Leftrightarrow S(n) = s_2(n) \text{ and } s_2(n-1) \geq s_{T_2} \text{ and } s_2(n) < s_{T_2} \\
&\text{State 6} \Leftrightarrow S(n) = s_2(n) \text{ and } s_2(n-1) < s_{T_2} \\
&\text{State 7} \Leftrightarrow S(n) = s_2(n) \text{ and } s_2(n-1) \geq s_{T_2} \text{ and } s_2(n) \geq s_{T_2} \\
&\text{State 8} \Leftrightarrow S(n) = s_2(n) \text{ and } s_2(n-1) \text{ unknown}
\end{aligned}$$

Accordingly, based on the switching strategy of SSC-C, the transition probability matrix, P^c, of this Markov chain is given by

$$P^c = \begin{bmatrix} 0 & 0 & 0 & 0 & 0 & 0 & 0 & 1 \\ (1-q_1)q_1 & q_1 & (1-q_1)^2 & 0 & 0 & 0 & 0 & 0 \\ q_1 & 0 & 1-q_1 & 0 & 0 & 0 & 0 & 0 \\ (1-q_1)q_1 & q_1 & (1-q_1)^2 & 0 & 0 & 0 & 0 & 0 \\ 0 & 0 & 0 & 1 & 0 & 0 & 0 & 0 \\ 0 & 0 & 0 & 0 & (1-q_2)q_2 & q_2 & (1-q_2)^2 & 0 \\ 0 & 0 & 0 & 0 & q_2 & 0 & 1-q_2 & 0 \\ 0 & 0 & 0 & 0 & (1-q_2)q_2 & q_2 & (1-q_2)^2 & 0 \end{bmatrix}_{8\times 8}$$

where again q_i is defined in Equation 17.5 for $i = 1, 2$.

17.3.3 Statistics of Overall Combiner Output

Let M denote the size of the state space of the Markov chains (i.e., $M = 4$ for SSC-A, $M = 6$ for SSC-B, and $M = 8$ for SSC-C). Conditioning on the current state of the receiver, the overall CDF of the combiner output for all three SSC schemes, $P_{\text{ssc}}(\cdot)$, can be written as

$$P_{\text{ssc}}(x) = \sum_{i=1}^{M} \pi_i \, P^{(i)}(x) \tag{17.7}$$

where $P^{(i)}(\cdot)$ are the state-dependent CDFs of the combiner output and π_i are the stationary probabilities of the ith state.

Based on Markov chain theory, we can obtain π_i in closed form by solving the following system of linear equations:

$$\begin{cases} \vec{\pi} P = \vec{\pi}; \\ \sum_{i=1}^{M} \pi_i = 1 \end{cases} \tag{17.8}$$

where P is the transition probability matrix of the Markov chains and $\vec{\pi}$ is the row vector defined as $\vec{\pi} = [\pi_1, \pi_2, \ldots, \pi_M]$. For example, after substituting Equation 17.4 into Equation 17.8 and solving Equation 17.8 for $\vec{\pi}$, we obtain the stationary probability vector for SSC-A, $\vec{\pi}^a$, as

$$\vec{\pi}^a = \left[\frac{q_1 q_2}{q_1 + q_2}, \frac{(1 - q_1) q_2}{q_1 + q_2}, \frac{q_1 q_2}{q_1 + q_2}, \frac{(1 - q_2) q_1}{q_1 + q_2} \right] \tag{17.9}$$

where q_i is defined as in Equation 17.5.

It can be shown that all the state-dependent CDFs $P^{(i)}(\cdot)$ for the three SSC schemes take three simple general forms. In particular, for States 2 and 5 of SSC-B and States 2, 4, 6, and 8 of SSC-C, $P^{(i)}(\cdot)$ takes the form of a single-branch CDF:

$$P(x) = P_i(x) \tag{17.10}$$

For States 2 and 4 of SSC-A, States 3 and 6 of SSC-B, and States 3 and 6 of switch and stay combining-B, $P^{(i)}(\cdot)$ takes the form of a truncated single-branch CDF:

$$P(x) = \frac{P_i(x) - P_i(s_{T_i})}{1 - P_i(s_{T_i})}, \quad x \geq s_{T_i} \tag{17.11}$$

Finally, for States 1 and 3 of SSC-A, States 1 and 4 of switch and stay combining-B, and States 1 and 5 of SSC-B, $P^{(i)}(\cdot)$ takes the form of a conditional CDF:

$$P(x) = \frac{P_{i,j}(s_{T_i}, x)}{P_i(s_{T_i})} \Longrightarrow P_j(x) \text{ if independent paths} \tag{17.12}$$

As an example, after appropriate substitutions of Equations 17.9 to 17.12 into Equation 17.7 and some manipulations, the CDF of the combiner output for SSC-A is obtained, in the most general case, as

$$\begin{aligned} P^a_{\text{ssc}}(x) &= \sum_{i=1}^{4} \pi_i^a P^{(i)}(x) \\ &= \begin{cases} A, & x < s_{T_2} \\ A + \dfrac{(1 - q_2) q_1}{q_1 + q_2} \dfrac{P_2(x) - q_2}{1 - q_2}, & s_{T_2} \leq x < s_{T_1} \\ A + \dfrac{(1 - q_1) q_2}{q_1 + q_2} \dfrac{P_1(x) - q_1}{1 - q_1} + \dfrac{(1 - q_2) q_1}{q_1 + q_2} \dfrac{P_2(x) - q_2}{1 - q_2}, & s_{T_1} \leq x \end{cases} \end{aligned} \tag{17.13}$$

where

$$A = \frac{q_1 q_2}{q_1 + q_2}\left(\frac{P_{1,2}(s_{T_1}, x)}{q_1} + \frac{P_{1,2}(x, s_{T_2})}{q_2}\right) \tag{17.14}$$

Note that this formula generalizes previous results by allowing unequal switching thresholds on the two branches. For instance, if $s_{T_1} = s_{T_2}$, Equation 17.13 combined with Equation 17.14 reduces to [10, Equation 4]. If $s_{T_1} = s_{T_2} = s_T$ and the branches are correlated but identically distributed, Equation 17.13 combined with Equation 17.14 reduces to [7, Equation 6]. If $s_{T_1} = s_{T_2} = s_T$ and the branches are independent but not identically distributed, Equation 17.13 combined with Equation 17.14 reduces to [9, Equation 62].

With the CDF of the combiner output available, the overall PDF and MGF of the combiner output of all three SSC schemes, $p_{\mathrm{ssc}}(\cdot)$ and $M_{\mathrm{ssc}}(\cdot)$, can be routinely calculated as

$$p_{\mathrm{ssc}}(x) = \sum_{i=1}^{M} \pi_i \frac{dP^{(i)}(x)}{dx} \tag{17.15}$$

and

$$M_{\mathrm{ssc}}(t) = \sum_{i=1}^{M} \pi_i \int_0^{\infty} \frac{dP^{(i)}(x)}{dx} e^{xt}\, dx \tag{17.16}$$

Since all state-dependent CDFs for different SSC strategies take three general simple forms, the state-dependent PDFs and MGFs can be easily obtained. For example, the PDF and MGF of the SSC-B combiner output over generically correlated and unbalanced Rayleigh paths can be obtained as

$$\begin{aligned} p_{\mathrm{ssc}}^{b}(x) &= \sum_{i=1}^{6} \pi_i^b p^{(i)}(x) \\ &= \begin{cases} B, & x < s_{T_2} \\ B + \dfrac{\pi_6^b}{1-q_2}\dfrac{1}{\Omega_2}\exp\left(-\dfrac{x}{\Omega_2}\right), & s_{T_2} \le x < s_{T_1} \\ B + \dfrac{\pi_3^b}{1-q_1}\dfrac{1}{\Omega_1}\exp\left(-\dfrac{x}{\Omega_1}\right) + \dfrac{\pi_6^b}{1-q_2}\dfrac{1}{\Omega_2}\exp\left(-\dfrac{x}{\Omega_2}\right), & s_{T_1} \le x \end{cases} \end{aligned} \tag{17.17}$$

where

$$\begin{aligned} B &= \frac{\pi_2^b}{\Omega_1}\exp\left(-\frac{x}{\Omega_1}\right) + \frac{\pi_5^b}{\Omega_2}\exp\left(-\frac{x}{\Omega_2}\right) \\ &\quad + \frac{\pi_1^b}{q_1\Omega_1}\exp\left(-\frac{x}{\Omega_1}\right)\left[1 - Q_1\left(\sqrt{\frac{2\rho x}{(1-\rho)\Omega_1}}, \sqrt{\frac{2 s_{T_2}}{(1-\rho)\Omega_2}}\right)\right] \\ &\quad + \frac{\pi_4^b}{q_2\Omega_2}\exp\left(-\frac{x}{\Omega_2}\right)\left[1 - Q_1\left(\sqrt{\frac{2\rho x}{(1-\rho)\Omega_2}}, \sqrt{\frac{2 s_{T_1}}{(1-\rho)\Omega_1}}\right)\right] \end{aligned} \tag{17.18}$$

and

$$\begin{aligned} M_{\mathrm{ssc}}^{b}(t) &= \sum_{i=1}^{6} \pi_i^b M^{(i)}(t) \\ &= (1-\Omega_1 t)^{-1}\left[\pi_2^b + \frac{\pi_3^b}{1-q_1}\exp\left(-(1-\Omega_1 t)\frac{s_{T_1}}{\Omega_1}\right) + \frac{\pi_4^b}{q_1}\exp\left(-\frac{s_{T_1}}{\Omega_1}\frac{1-\Omega_2 t}{1-(1-\rho)\Omega_2 t}\right)\right] \\ &\quad + (1-\Omega_2 t)^{-1}\left[\pi_5^b + \frac{\pi_6^b}{1-q_2}\exp\left(-(1-\Omega_2 t)\frac{s_{T_2}}{\Omega_2}\right) + \frac{\pi_1^b}{q_2}\exp\left(-\frac{s_{T_2}}{\Omega_2}\frac{1-\Omega_1 t}{1-(1-\rho)\,\Omega_1 t}\right)\right] \end{aligned} \tag{17.19}$$

respectively. $\pi_i^b, i = 1, \ldots, 6$, are the stationary probabilities for SSC-B, given by

$$\vec{\pi}^b = \left[\frac{(1-q_1)q_1(1-q_2)q_2}{(q_1+q_2)(1+q_1q_2+q_{12})-(q_1+q_2)^2-2q_1q_2q_{12}}, \frac{q_1q_{12}(1-q_2)}{(q_1+q_2)(1+q_1q_2+q_{12})-(q_1+q_2)^2-2q_1q_2q_{12}}, \frac{(1-q_1)^2(1-q_2)q_2}{(q_1+q_2)(1+q_1q_2+q_{12})-(q_1+q_2)^2-2q_1q_2q_{12}}, \frac{(1-q_1)q_1(1-q_2)q_2}{(q_1+q_2)(1+q_1q_2+q_{12})-(q_1+q_2)^2-2q_1q_2q_{12}}, \frac{(1-q_1)q_{12}q_2}{(q_1+q_2)(1+q_1q_2+q_{12})-(q_1+q_2)^2-2q_1q_2q_{12}}, \frac{(1-q_1)q_1(1-q_2)q_2}{(q_1+q_2)(1+q_1q_2+q_{12})-(q_1+q_2)^2-2q_1q_2q_{12}}\right] \tag{17.20}$$

If the branches are i.i.d., Equation 17.17 combined with Equation 17.18 reduces to [4, Equation 2.15].

17.3.4 Application to Performance Analysis and Comparisons

Based on the Markov chain-based analytical framework, we now conduct a thorough comparison between various SSC strategies. In particular, the trade-off of performance and switching rate among different SSC schemes is accurately quantified and illustrated.

1. *Performance comparison:* With the generic formulas for the statistics of the combiner output with the three switch-combining strategies in hand, the performance analysis of these strategies can now be carried out. Note that since performance measures such as bit error rate (BER) and SNR are usually averaged quantities over combiner output, they can also be expressed as weighted sums of their corresponding state-dependent quantities. Thus, the performance analysis of different SSC schemes is also simplified and unified.

In particular, with the MGF of the combiner output obtained in many situations, we can evaluate the average bit and symbol error rate of different modulation schemes over various fading channel scenarios using the MGF-based approach [28]. For example, for binary differential phase shift keying (BDPSK) and binary frequency shift keying (BFSK) modulations, the average BER is given by $P_b(E) = M_{\mathrm{ssc}}(-g)/2$, where $g = 1$ for BDPSK and $g = 1/2$ for BFSK. Also, for most of the other modulation schemes, a single finite-interval integration suffices for direct computation of the desired average error rate [28]. For example, the average symbol error probability, $P_S(E)$, of M-ary phase shift keying (PSK) modulation is given by

$$P_S(E) = \frac{1}{\pi}\int_0^{\frac{(M-1)\pi}{M}} M_{\mathrm{ssc}}\left(-\frac{g_{\mathrm{PSK}}}{\sin^2\phi}\right) d\phi \tag{17.21}$$

where $g_{\mathrm{PSK}} = \sin^2(\frac{\pi}{M})$, while average symbol error probability of M-ary quadrature amplitude modulation (QAM) is given by

$$P_{\mathrm{QAM}}(E) = \left(1 - \frac{1}{\sqrt{M}}\right)\frac{4}{\pi}\int_0^{\pi/2} M_{\mathrm{ssc}}\left(\frac{g_{\mathrm{QAM}}}{\sin^2\theta}\right) d\theta - \left(1 - \frac{1}{\sqrt{M}}\right)^2 \frac{4}{\pi}\int_0^{\pi/4} M_{\mathrm{ssc}}\left(\frac{g_{\mathrm{QAM}}}{\sin^2\theta}\right) d\theta \tag{17.22}$$

where $g_{\mathrm{QAM}} = \frac{3\log_2(M)}{2(M-1)}$.

Figure 17.2 shows the BER performance of the three SSC schemes for BPSK as a function of the common switching threshold γ_T over i.i.d. Rayleigh branches. It is clear that while the BER of these three SSC schemes

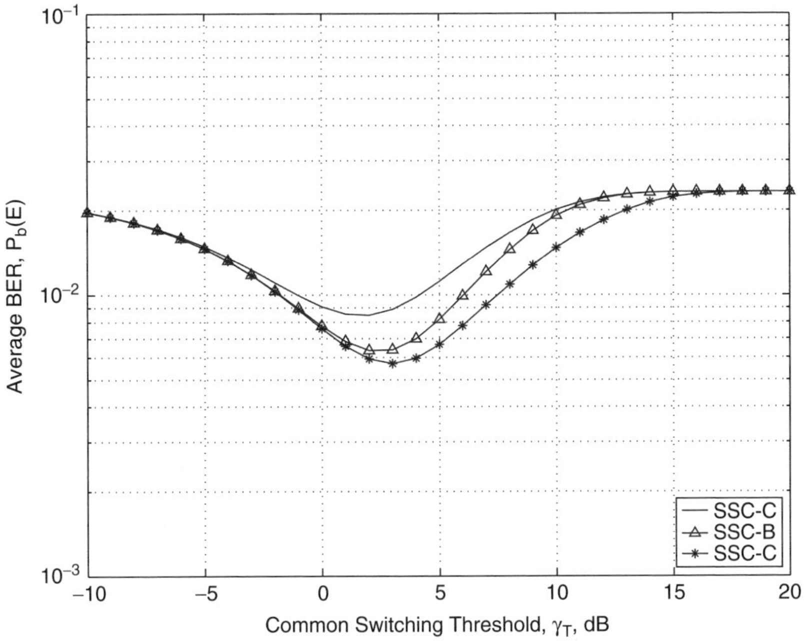

FIGURE 17.2 Average BER of BPSK for three SSC schemes over i.i.d. Rayleigh branches as a function of common switching threshold, γ_T, with $\bar{\gamma}_1 = \bar{\gamma}_2 = 10$ dB.

has a similar behavior as switching threshold varies, SSC-A always outperforms the other two schemes and SSC-C leads to the worst performance. Note also that the optimal choice of switching thresholds exists for all three SSC schemes, but if the switching threshold becomes too small or too large, all three SSC schemes give exactly the same performance.

Intuitively, if the switching threshold is too small, the currently used branch is always acceptable. Therefore, there is no branch switching and thus no diversity. If the switching threshold is too high, for SSC-A, the receiver will always be switching branches, while for SSC-B and SSC-C, there will be no switching since it is difficult to have downward crossing. Both of these situations lead to the no-diversity case. Since the average BER is a continuous function of the switching threshold γ_T, there exists an optimal choice of γ_T in terms of minimizing the average BER. This optimal value γ_T^* is a solution of the equation

$$\frac{dP_b(E)}{d\gamma_T}|_{\gamma_T=\gamma_T^*} = 0 \tag{17.23}$$

Because the average BER is usually a complicated function of γ_T, it is difficult to obtain a compact expression for γ_T^* by solving Equation 17.23, except for certain simple special cases [28, Section 9.8.1.4]. In the remainder of this chapter, unless otherwise noted, the optimal switching thresholds are obtained through numerically minimizing the average BER expression, as given in Equation 17.21 while using an appropriate MGF, for a given average SNR value.

Figure 17.3 compares the BER performance of three SSC schemes for BPSK in an i.i.d. Rayleigh fading environment with optimal switching threshold. For comparison purposes, the error performances of no-diversity and SC cases are also plotted. It can be seen that SSC-A and SSC-B have approximately the same performance and both of them outperform SSC-C by approximately 1 dB. Note also that while the three SSC schemes do not provide as good a performance as SC, they still offer a considerable diversity benefit over the no-diversity case.

Outage probability is an important performance measure for wireless communications systems. It is typically defined as the probability that the combined SNR γ fails to meet a required SNR threshold γ_{th}.

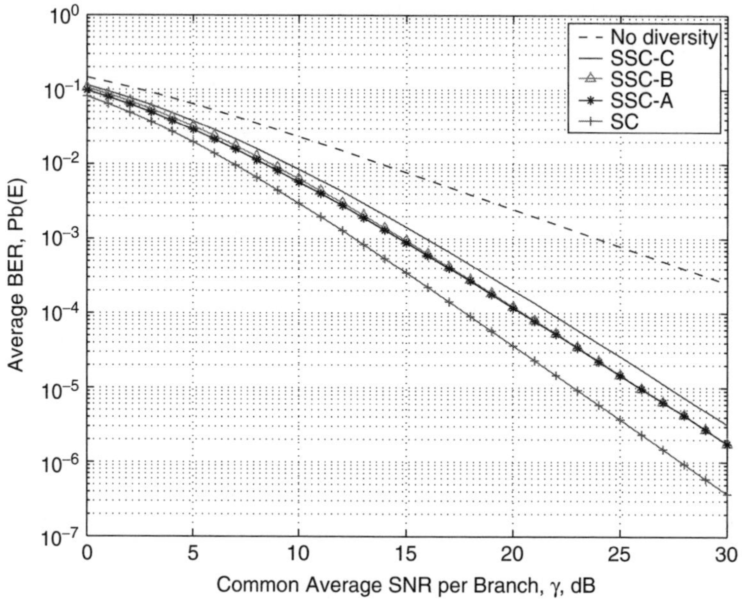

FIGURE 17.3 Average BER of BPSK for three SSC schemes over i.i.d. Rayleigh branches as a function of common average SNR, $\bar{\gamma}$, with optimal switching thresholds.

Hence, it corresponds to the CDF of the combined SNR evaluated at γ_{th}. Noting that SNR is proportional to signal power, with the CDF of the combiner output signal power obtained, we can easily calculate the outage probability for the three SSC schemes by using the appropriate outage threshold. Figure 17.4 compares the outage probability of three SSC schemes over an independent Nakagami fading environment.

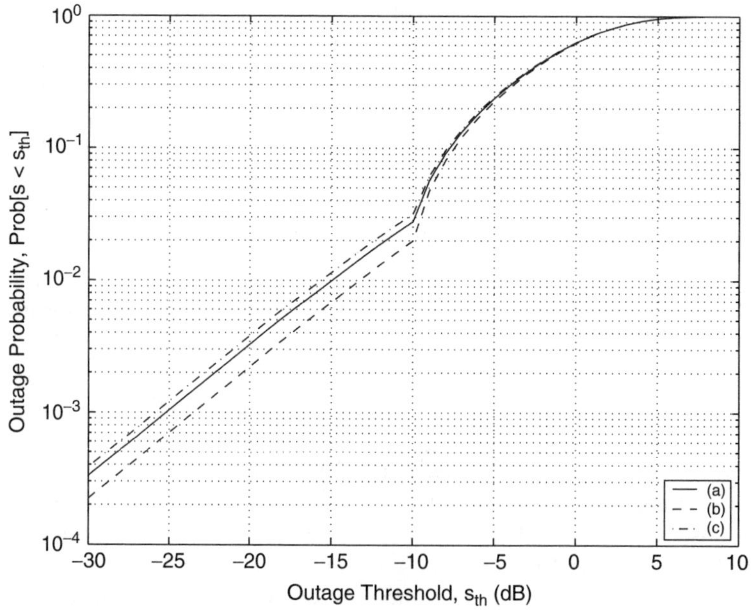

FIGURE 17.4 Outage probability (or equivalently the CDF) for three SSC schemes over independent Nakagami branches as a function of outage threshold, γ_{th}, with $\bar{\gamma}_1 = 8$ dB, $\bar{\gamma}_2 = 12$ dB, $\gamma_{T_1} = 7$ dB, and $\gamma_{T_2} = 11$ dB.

It can be seen that similar to BER performance, SSC-A always outperforms the other two schemes and SSC-C leads to the worst performance. Note also that as the Nakagami parameter m increases, the outage performance improves for all three schemes and at the same time the performance of SSC-B approaches that of SSC-A from that of SSC-C.

2. *Switching rate comparison:* The execution of switching branches will not only consume power, but also reduce the data throughput in a transmit switched diversity configuration [29]. As such, reducing the rate of switches and eliminating unnecessary switches is of great importance, especially for mobile terminals with limited battery lifetime. The trade-off of switching rate and performance between three SSC strategies is fundamental in the design process. Therefore, we now accurately quantify the switching rate for the three SSC strategies.

With the Markov chain built for three SSC strategies, we can easily compute their average switching rates. More specifically, since for SSC-A branch switching occurs with a probability of 1 if the receiver is in States 1 and 3, the probability of switching at any guard period, P_s^a, is given by

$$P_s^a = \frac{2q_1q_2}{q_1 + q_2} \tag{17.24}$$

where q_i is defined in Equation 17.5 for $i = 1, 2$. Similarly, the switching probabilities for SSC-B, P_s^b, and for SSC-C, P_s^c, are given by

$$P_s^b = \frac{2q_1(1 - q_1)\,q_2(1 - q_2)}{(q_1 + q_2)(1 + q_1q_2 + q_{12}) - (q_1 + q_2)^2 - 2q_1q_2q_{12}} \tag{17.25}$$

and

$$P_s^c = \frac{2q_1(1 - q_1)q_2(1 - q_2)}{q_1(1 - q_1) + q_2(1 - q_2)} \tag{17.26}$$

respectively. Correspondingly, given the frequency of guard period $1/T$, the switching rate can be easily obtained as P_s/T.

Figure 17.5 compares the switching rates of the three combining strategies over independent Nakagami branches. It can be seen that as switching threshold γ_T increases, the switching rate of SSC-A monotonely

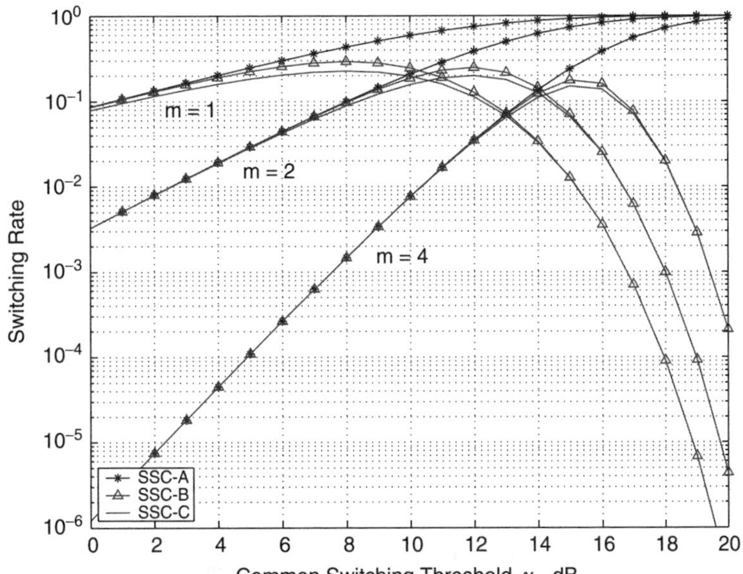

FIGURE 17.5 Switching rate for three SSC schemes over independent Nakagami branches as a function of common switching threshold, γ_T, with $\bar{\gamma}_1 = 8$ dB and $\bar{\gamma}_2 = 12$ dB.

TABLE 17.3 Average BER and Switching Rate of Three SSC Schemes in Rayleigh Fading While Using the Optimal Switching Threshold ($\bar{\gamma}_1 = 8$ and $\bar{\gamma}_2 = 12$ dB)

	SSC-A	SSC-B	SSC-C
Optimal switching threshold, in dB	2.65	2.07	1.28
Average BER	0.0058	0.0062	0.0083
Switching rate	0.1530	0.1317	0.1007

increases, whereas those of SSC-B and SSC-C will decrease after reaching their corresponding maximum values. Note that SSC-C always gives the smallest switching rate among all three SSC schemes, as expected. Note also that as the Nakagami parameter m increases, the switching rate for all schemes decreases for the low and moderate choices of switching thresholds.

It is also of interest to see how much the switching rates of three SSC schemes differ when the optimal switching thresholds in terms of minimizing the average BER are used. For this purpose, in Table 17.3, we present the BERs and switching rates of three SSC schemes while using the optimal switching threshold over Rayleigh fading branches. As we can see, the performance advantage of SSC-A comes at the cost of higher power consumption even with the optimal switching threshold.

17.4 Multibranch Switched Diversity

In this section, we consider low-complexity combining schemes for L-branch diversity systems. The generalization of SC to the L-branch scenario is straightforward. The receiver now needs to select the best branch among all L available ones. As a result, although only one branch will be used for data reception, the receiver with SC needs to estimate all L diversity paths and perform $L - 1$ comparisons in each guard period before reaching the combining decision.

Noting that the complexity savings of dual SSC schemes over SC are less channel estimation and fewer and simpler comparisons, we consider the operation and performance of switching-based combining schemes in an L-branch scenario in this section [16]. After proving that SSC does not benefit from additional diversity paths, we focus on the performance analysis of switch and examine combining. In particular, the closed-form expressions for the CDF, PDF, and MGF of the SEC combiner output are derived and applied to its performance analysis over various fading models of interest.

17.4.1 L-Branch SSC

While all three SSC schemes can be used in an L-branch scenario, we only consider the operation and performance of SSC-A in the following presentation. As we will see, the observations obtained in this section apply to all three SSC schemes.

We assume that the receiver can cyclically switch between the L antenna branches. Branch switching occurs only when the received signal power of the currently used branch is below the switching threshold s_T,[1] and as such, this branch becomes unacceptable. With SSC, the receiver settles on the next antenna branch no matter what the channel condition corresponding to that branch is. The mode of operation of the SSC scheme can be described mathematically by the sequence $S(n)$ given by

$$S(n) = s_i(n) \quad \text{iff} \quad \begin{cases} S(n-1) = s_i(n-1) \text{ and } s_i(n) \geq s_T \\ \text{or} \\ S(n-1) = s_{((i-1))_L}(n-1) \text{ and } s_{((i-1))_L}(n) < s_T \end{cases} \tag{17.27}$$

[1]For simplicity, we now use the same switching threshold for all diversity paths.

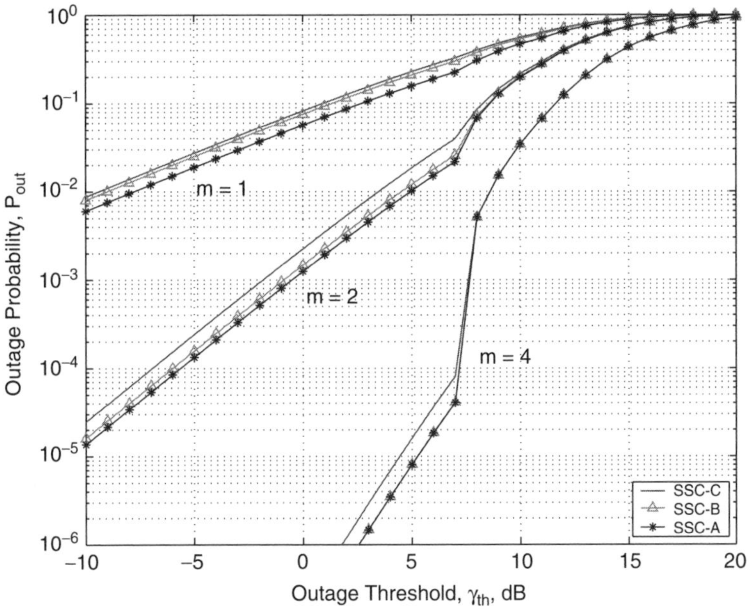

FIGURE 17.6 Outage probability (or equivalently the CDF) ofSSC for three scenarios with identical Rayleigh branches ($\Omega_i = 0$ dB, $i = 0, \ldots, L-1$): (a) $L = 3$, $\rho_{0,1} = 0.795$, $\rho_{1,2} = 0.795$, $\rho_{2,0} = 0.605$; (b) $L = 2$, $\rho_{0,1} = \rho_{1,0} = 0.605$; (c) $L = 2$, $\rho_{0,1} = \rho_{1,0} = 0.795$.

where $i = 0, \ldots, L-1$ and $((i-1))_L$ denotes the $i-1$ modulo Lth branch, e.g., if $L = 4$, then $((2-3))_L = 3$ and $((3+1))_L = 0$. As we can see, since during each guard period, at most two diversity paths are involved in the diversity combining decision, all SSC schemes cannot benefit from additional diversity paths. In particular, if the diversity branches are equi-correlated and identically distributed, L-branch SSC has the exact same performance as dual-branch SSC. These observations can be further illustrated with the following numerical examples.

Figure 17.6 shows the outage probability (or equivalently the CDF) of the signal power at an SSC output for three different scenarios over identical Rayleigh faded branches. The first scenario corresponds to the three-branch case with cross-correlation between successive branches equal to 0.795, 0.795, and 0.605, which corresponds to the three-element linear antenna array studied in [30]. The second scenario corresponds to the dual-branch case with a branch correlation of0.605, while the third scenario corresponds to the dual-branch case with a branch correlation of 0.795. As can be seen by comparing these three scenarios, SSC with three correlated branches is worse than SSC with two slightly correlated branches but better than SSC with two highly correlated branches.

Figure 17.7 shows the outage probability (or equivalently the CDF) of the signal power at SSC output for three different scenarios over independent Rayleigh fading. The first scenario corresponds to the three-branch case with local means of −3, 0, and 3 dB. The second scenario corresponds to the dual-branch case with local means of 0 and 3 dB, while the third scenario corresponds to the dual-branch case with local means of −3 and 0 dB. It is observed that SSC with three branches is worse than SSC with the two best branches of the three but better than SSC with the two worst branches of the three.

17.4.2 L-Branch SEC

It is because the receiver simply stays on the switch-to branch irrespective of its quality that SSC cannot benefit from additional diversity paths. In this section, we consider another switching-based combining scheme, namely, switch and examine combining, and investigate its operation and performance over the generalized fading channel. We show that unlike SSC, SEC can indeed benefit from additional diversity paths.

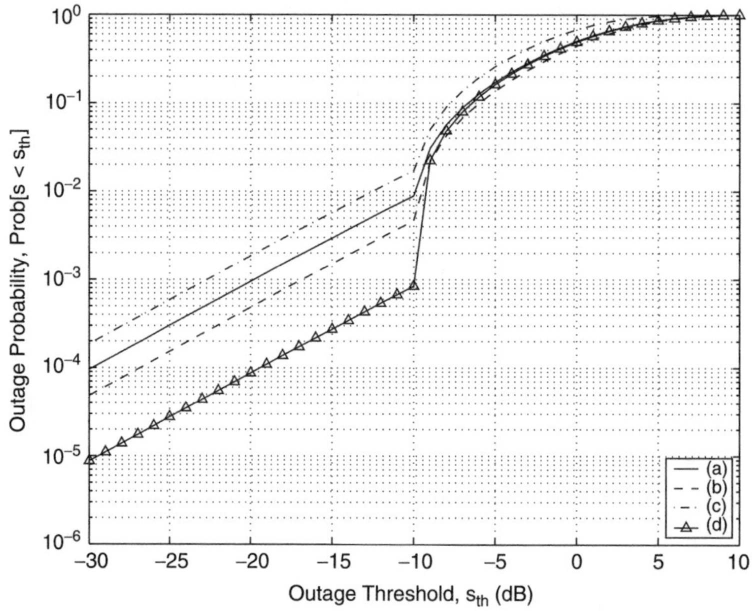

FIGURE 17.7 Outage probability (or equivalently the CDF) of SSC and SEC with two and three independent Rayleigh branches: (a) SSC with $L = 3$, $\Omega_0 = -3$ dB, $\Omega_1 = 0$ dB, $\Omega_2 = 3$ dB; (b) SSC/SEC with $L = 2$, $\Omega_0 = 0$ dB, $\Omega_1 = 3$ dB; (c) SSC/SEC with $L = 2$, $\Omega_0 = 0$ dB, $\Omega_1 = -3$ dB; (d) SEC with $L = 3$, $\Omega_0 = -3$ dB, $\Omega_1 = 0$ dB, $\Omega_2 = 3$ dB.

1. *Mode of operation:* Similar to the SSC case, the receiver with SEC cyclically switches between the L antenna branches. Branch switching occurs only when the received signal power of the currently used branch is below the threshold s_T, and as such, this branch becomes unacceptable. Unlike SSC, the receiver examines the received signal power of the switch-to branch and switches again if it is not acceptable. The receiver will repeat this process until either it finds an acceptable branch or all L antenna branches have been examined. In the latter case, it uses the last examined antenna branch for the data reception.

Mathematically, the SEC scheme mode of operation can be described by the sequence $S(n)$ given by

$$S(n) = s_i(n) \text{iff} \begin{cases} S(n-1) = s_i(n-1) \text{ and } s_i(n) \geq s_T \\ \text{or} \\ S(n-1) = s_i(n-1) \text{ and } s_k(n) < s_T,\ k = 0, \ldots, L-1 \\ \text{or} \\ S(n-1) = s_{((i-1))_L}(n-1) \text{ and} \\ \quad s_{((i-1))_L}(n) < s_T \text{ and } s_i(n) \geq s_T \\ \vdots \\ \text{or} \\ S(n-1) = s_{((i-j))_L}(n-1) \text{ and} \\ \quad s_{((i-j+k))_L}(n) < s_T,\ k = 0, \ldots, j-1, \text{ and } s_i(n) \geq s_T \\ \vdots \\ \text{or} \\ S(n-1) = s_{((i+1))_L}(n-1) \text{ and} \\ \quad s_{((i+1+k))_L}(n) < s_T,\ k = 0, \ldots, L-2, \text{ and } s_i(n) \geq s_T \end{cases} \tag{17.28}$$

where $i = 0, \ldots, L-1$. As we can see, SEC is actually a generalization of SSC-A. However, the same generalization cannot be applied to SSC-B and SSC-C due to the causality problem.

2. *Statistics of combiner output:* We now assume that the L diversity paths are independent but not necessarily identically faded. Since the events that antenna i is used for transmission during time interval n, i.e., $S(n) = s_i(n), i = 0, \ldots, L-1$, are mutually exclusive, the CDF of the received signal power after L-branch SEC diversity reception, $P_S(\cdot)$, can be written as

$$\begin{aligned} P_S(x) &= \Pr[S(n) \leq x] \\ &= \sum_{i=0}^{L-1} \Pr[S(n) = s_i(n) \text{ and } s_i(n) \leq x] \end{aligned} \tag{17.29}$$

Applying Equation 17.28, the summand in Equation 17.29 can be shown to be given by

$$\Pr[S(n) = s_i(n) \text{ and } s_i(n) \leq x] = \\ \begin{cases} \Pr[S(n-1) = s_i(n-1)] \Pr[s_{((i-1-k))_L}(n) < s_T, \\ \quad k = 0, \ldots, L-2 \text{ and } s_i(n) \leq x], & x < s_T \\ \Pr[S(n-1) = s_i(n-1)] \Pr[s_T \leq s_i(n) \leq x] \\ \quad + \Pr[S(n-1) = s_i(n-1)] \Pr[s_k(n) < s_T,\ k = 0, \ldots, L-1] \\ \quad + \sum_{j=1}^{L-1} \Pr[S(n-1) = s_{((i-j))_L}(n-1)] \Pr[s_{((i-j+k))_L}(n) < s_T, \\ \quad k = 0, \ldots, j-1 \text{ and } s_T \leq s_i(n) \leq x], & x \geq s_T \end{cases} \tag{17.30}$$

After substituting Equation 17.30 into Equation 17.29 and some manipulations with the help of independent branch assumption, we obtain the CDF of SEC combiner output as

$$P_S(x) = \begin{cases} \sum_{i=0}^{L-1} \pi_i P_i(x) \prod_{k=0,k\neq i}^{L-1} P_k(s_T), & x < s_T \\ \sum_{i=0}^{L-1} \Big\{ \pi_i \prod_{k=1}^{L} P_k(s_T) \\ \quad + \sum_{j=0}^{L-1} \pi_{((i-j))_L} [P_i(x) - P_i(s_T)] \prod_{k=0}^{j-1} P_{((i-j+k))_L}(s_T) \Big\}, & x \geq s_T \end{cases} \tag{17.31}$$

where π_i is the probability that the receiver uses the ith antenna. It can be obtained as the stationary distribution of an L-state Markov chain whose transition probability matrix entries are

$$p^{i,j} = \begin{cases} 1 - P_i(s_T) + \prod_{k=0}^{L-1} P_k(s_T), & i = j \\ [1 - P_j(s_T)] \prod_{k=i}^{j-1} P_k(s_T), & i < j \\ [1 - P_j(s_T)] \prod_{k=i}^{L-1} P_k(s_T) \prod_{k=0}^{j} P_k(s_T), & j < i \end{cases} \tag{17.32}$$

where $i, j = 0, \ldots, L-1$. It can be shown, by solving Equation 17.8 while using Equation 17.32 for P, that the π_i values are given by

$$\pi_i = \left[\sum_{j=0}^{L-1} \left(\frac{P_{L-1}(s_T)(1 - P_j(s_T))}{P_j(s_T)(1 - P_{L-1}(s_T))} \right) \right]^{-1} \frac{P_{L-1}(s_T)(1 - P_i(s_T))}{P_i(s_T)(1 - P_{L-1}(s_T))} \tag{17.33}$$

Correspondingly, the PDF and MGF of the combiner output of SEC can be obtained as

$$p_S(x) = \begin{cases} \sum_{i=0}^{L-1} \pi_i \left[\prod_{k=0,k\neq i}^{L-1} P_k(s_T) \right] p_i(x), \ x < s_T \\ \sum_{i=0}^{L-1} \Big[\sum_{j=0}^{L-1} \pi_{((i-j))_L} \\ \quad \times \prod_{k=0}^{j-1} P_{((i-j+k))_L}(s_T) \Big] p_i(x), \ x \geq s_T \end{cases} \tag{17.34}$$

and

$$M_S(t) = \sum_{i=1}^{L} \left[\pi_i \left(\prod_{k=0,k\neq i}^{L-1} P_k(s_T) \right) M_i(t) + \sum_{j=0}^{L-2} \pi_{((i-j))_L} \prod_{k=0}^{j-1} P_{((i-j+k))_L}(s_T) \times \int_{s_T}^{\infty} e^{tx} p_i(x)dx \right] \tag{17.35}$$

respectively, where $p_i(\cdot)$ and $M_i(\cdot)$ are the PDF and MGF of the ith branch, respectively, as given in Table 17.1 for various fading scenarios.

Figure 17.7 compares the outage probability of SSC and SEC over independent but not identically distributed Rayleigh branches. Comparing curves (a) and (d), we can see that with three branches, SEC clearly outperforms SSC, especially for the low values of the outage threshold. Since dual-branch SSC and dual-branch SEC have exactly the same performance by intuition, curves (b) and (c), also show the performance of dual-branch SEC. Consequently, comparing curves (b), (c), and (d), we can conclude that SEC will always benefit from increasing the number of available branches, and this performance gain depends essentially on the fading statistics of the additional branches.

3. *Error performance of SEC:* In this section, we illustrate the usefulness of the results derived in previous sections by specializing and applying them to the error performance evaluation of SEC. As noted in the previous section, with the MGF of the combiner output obtained for SEC, we can now evaluate the average bit and symbol error rates of different modulation schemes with SEC over various fading scenarios using the MGF-based approach [28]. For most of the cases, a single finite-interval integration suffices for direct computation of the desired average error rate.

In Figure 17.8, we plot the average BER of BPSK with SEC over i.i.d. Rayleigh fading paths vs. the switching threshold. As we can see, increasing L improves the performance of SEC for moderate choice of switching threshold. We also notice that there exist optimal choices of the switching threshold for different values of L. Intuitively, if the switching threshold is very small compared to the average SNR, the current branch will

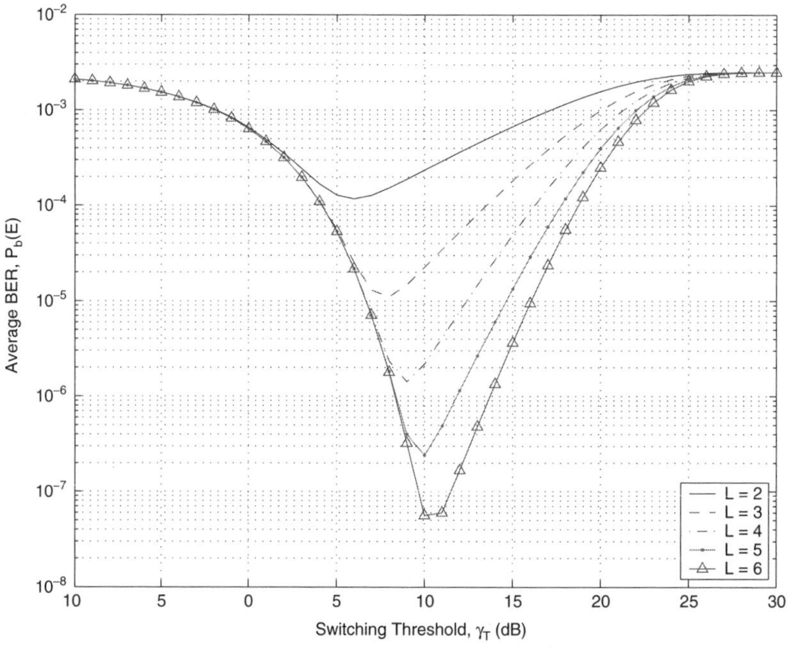

FIGURE 17.8 Average BER of BPSK with SEC over L i.i.d. Rayleigh branches as a function of switching threshold γ_T with $\bar{\gamma} = 20$ dB.

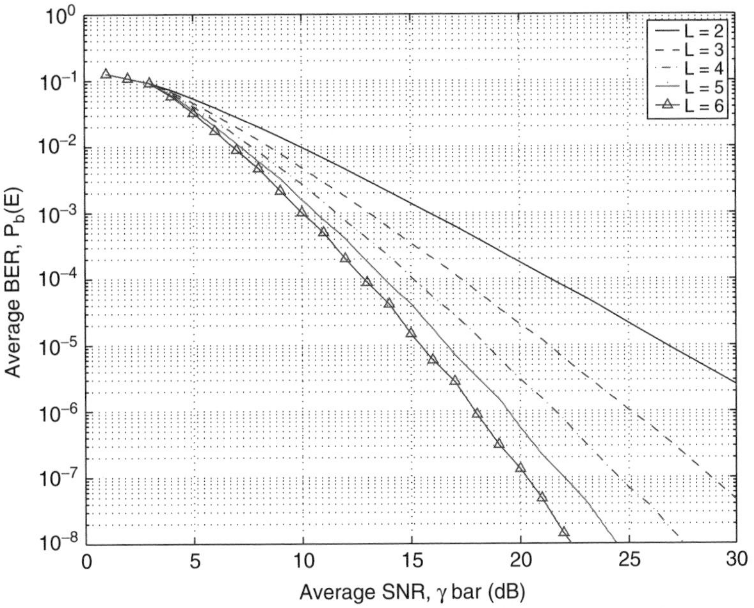

FIGURE 17.9 Average BER of BPSK with SEC over L i.i.d. Rayleigh branches as a function of average SNR $\bar{\gamma}$ with optimal γ_T.

always be acceptable. Thus, the SEC combiner uses one branch most of the time. As a result, the additional branches cannot contribute. If the switching is very high in comparison with the average SNR, all branches will be unacceptable. Thus, the additional branches will not be able to provide improvements either.

Figure 17.9 shows the average BER of BPSK with SEC over i.i.d. Rayleigh fading with optimal switching threshold. The optimal switching thresholds are obtained through numerically minimizing Equation 17.21

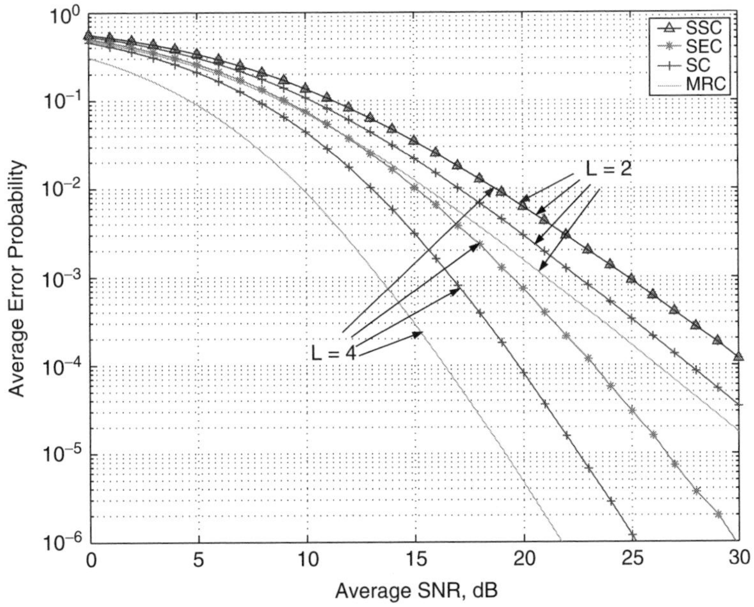

FIGURE 17.10 Comparison of the error performance for 8-PSK of SSC, SEC, SC, and MRC over i.i.d. Rayleigh fading branches with $L = 2$ and $L = 4$.

while using Equation 17.35 for a given average SNR value. It is clearly observed that, with optimal choice of switching threshold γ_T, the BER performance will benefit from increasing L over all regions of average SNR. As an additional numerical example, Figure 17.10 compares the error performances for 8-PSK of SSC, SEC, SC, and MRC over i.i.d. Rayleigh fading branches with $L = 2$ and $L = 4$. Note that as L increases, the error performance of SEC improves while that of SSC remains the same, which again shows that the error performance of SSC will not benefit from additional branches. It can also be seen that both SSC and SEC lead to higher error probability than MRC and SC, as expected, and that the performance gap between SEC and MRC becomes larger as L increases.

17.5 Generalized Switch and Examine Combining (GSEC)

In this section, based on our results from the previous sections, we propose and study a new low-complexity combining scheme for diversity-rich environments [22]. This new scheme, termed *generalized switch and examine combining* (GSEC), extends the notion of SEC studied in previous section to multi-input-multi-output scenarios and then cascades it with a traditional MRC combiner. In particular, we study the operation and performance of GSEC over i.i.d. Rayleigh fading paths. We also provide a thorough trade-off study of performance and complexity between GSEC and GSC.

17.5.1 Mode of Operation of GSEC

Figure 17.11 shows the block diagram of the GSEC combiner. The signal replicas received over L diversity paths with SNR $\gamma_i, i = 1, \ldots, L$, are fed into an L-input-L_c-output SEC (L_c/L SEC) combiner. During each guard period, the L_c/L SEC combiner selects L_c different paths out of the total L ones as per the rule discussed next. The L_c output signals of the L_c/L SEC combiner, whose SNRs are denoted by s_j, $j = 1, \ldots, L_c$, are then combined in the fashion of traditional MRC and applied to an appropriate demodulator/detector. Therefore, the overall combiner output SNR, Γ, is given by

$$\Gamma = \sum_{j=1}^{L_c} s_j \tag{17.36}$$

When $L_c = L$, we have

$$\Gamma = \sum_{j=1}^{L} s_j = \sum_{j=1}^{L} \gamma_j \tag{17.37}$$

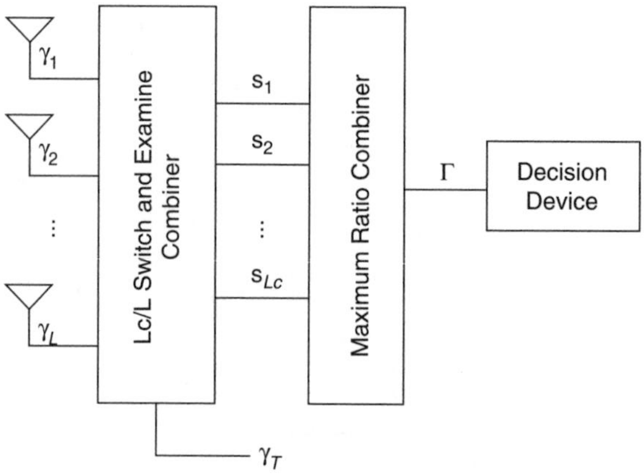

FIGURE 17.11 Structure of a GSEC combiner.

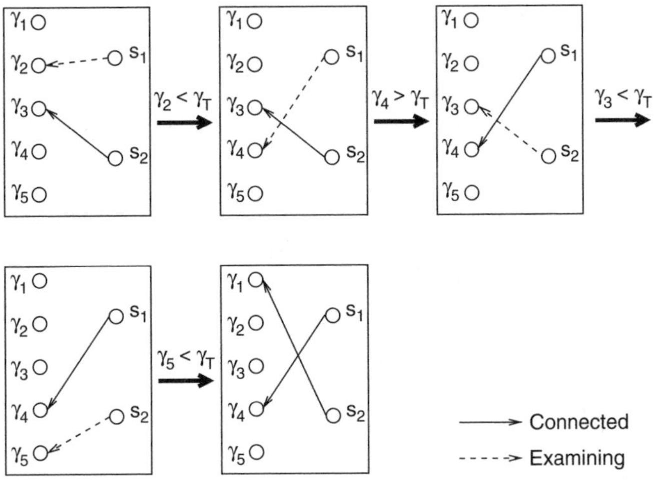

FIGURE 17.12 Sample operation of the L_c/L SEC combiner.

and thus the overall combiner is equivalent to a traditional MRC combiner, whereas when $L_c = 1$, it becomes the SEC combiner, which was studied in the previous section.

During each guard period, the L_c/L SEC combiner tries to connect acceptable diversity paths to the MRC inputs by examining as many paths as necessary and possible. Since the diversity paths are assumed to be identically faded, they can be examined in any order without preference. To reduce unnecessary path switches and save power, it is better to examine the quality of those diversity paths to which the MRC inputs are currently connected. For each unacceptable path, i.e., whose instantaneous SNR is below a preselected fixed threshold, denoted by γ_T, the combiner tries to replace it with an acceptable path by switching and examining the other $L - L_c$ diversity paths. If an acceptable one is found, the combiner connects the corresponding MRC input to this path and the examining for the next unacceptable path starts from the next available path in succession. If no acceptable path can be found, the combiner uses the last available diversity path for the current MRC input and stops examining the quality of the other MRC inputs.

As an example, Figure 17.12 shows a sample path update process of an SEC combiner with $L = 5$ and $L_c = 2$. Before the process starts, we have $s_1 = \gamma_2$ and $s_2 = \gamma_3$. We assume that the instantaneous SNRs of every diversity path satisfy $\gamma_1 < \gamma_T, \gamma_2 < \gamma_T, \gamma_3 < \gamma_T, \gamma_4 > \gamma_T$, and $\gamma_5 < \gamma_T$. The combiner will first examine γ_2. Since $\gamma_2 < \gamma_T$, the combiner will then check γ_4. Now that diversity path 4 is acceptable, the combiner sets $s_1 = \gamma_4$. The combiner then examines γ_3. Since $\gamma_3 < \gamma_T$, the combiner then checks γ_5. After finding that path 5 is not acceptable, the combiner uses the last available diversity path for current MRC input, i.e., it sets $s_2 = \gamma_1$.

17.5.2 MGF of Combiner Output

Based on the mode of operation of L_c/L SEC, we note that the number of s_j in the summation in Equation 17.36 that are greater than or equal to the threshold γ_T take only values from 0 to L_c. Since they are mutually exclusive and disjoint events, we can apply the total probability theorem and write the MGF of combiner output Γ in the following weighted sum form:

$$M_\Gamma(t) = \sum_{i=0}^{L_c} \pi_i \, M_\Gamma^i(t) \tag{17.38}$$

where $M_\Gamma^i(\cdot)$ is the MGF of Γ corresponding to the event that exactly i out of L_c s_j values in the summation of Equation 17.36 are greater than or equal to γ_T, and π_i is the probability that there are exactly is_j values in the summation of Equation 17.36 that are greater than or equal to γ_T.

Mathematically speaking, π_i is given by

$$\pi_i = \Pr[s_j \geq \gamma_T, \quad \text{for exact } i \ j\text{'s} \in \{1, 2, \ldots, L_c\}] \tag{17.39}$$

Equivalently, based on the mode of operation of L_c/L SEC, Equation 17.39 can be rewritten in terms of the SNR of L diversity paths, γ, as

$$\pi_i = \begin{cases} \Pr[\gamma_j \geq \gamma_T, & \text{for exact } i \ j\text{'s} \in \{1, 2, \ldots, L\}], \quad i = 0, \ldots, L_c - 1 \\ \Pr[\gamma_j \geq \gamma_T, & \text{for more than } i - 1 \ j\text{'s} \in \{1, 2, \ldots, L\}], \quad i = L_c \end{cases} \tag{17.40}$$

Noting that the diversity paths are assumed to be i.i.d. faded, it is not difficult to show that the π_i values are given by

$$\pi_i = \begin{cases} \binom{L}{i} [P_\gamma(\gamma_T)]^{L-i}[1 - P_\gamma(\gamma_T)]^i, & i = 0, \cdots, L_c - 1 \\ \sum_{j=L_c}^{L} \binom{L}{j} [P_\gamma(\gamma_T)]^{L-j}[1 - P_\gamma(\gamma_T)]^j, & i = L_c \end{cases} \tag{17.41}$$

where $P_\gamma(\cdot)$ is the CDF of the received SNR, which is common to all diversity paths and given in Table 17.1 for the three fading models under consideration.

To derive the expression of $M_\Gamma^i(\cdot)$, let us first consider the conditional PDF, $p_s^+(\cdot)$, of s_i given that s_i is greater than or equal to γ_T. This conditional PDF can be shown to be given by

$$p_s^+(x) = \frac{p_\gamma(x)}{1 - P_\gamma(\gamma_T)}, \quad x \geq \gamma_T \tag{17.42}$$

where $p_\gamma(\cdot)$ is the PDF of the received SNR, which is again common to all diversity paths and given in Table 17.1. Correspondingly, the conditional MGF, $M_s^+(\cdot)$, of s_i given that s_i is greater than or equal to γ_T is given by

$$\begin{aligned} M_s^+(t) &= \int_{-\infty}^{+\infty} p_s^+(x)\, e^{tx}\, dx \\ &= \frac{1}{1 - P_\gamma(\gamma_T)} \int_{\gamma_T}^{+\infty} p_\gamma(x)\, e^{tx} dx \end{aligned} \tag{17.43}$$

Similarly, it can be shown that the conditional MGF, $M_s^-(\cdot)$, of s_i given that s_i is less than γ_T is given by

$$M_s^-(t) = \frac{1}{P_\gamma(\gamma_T)} \int_0^{\gamma_T} p_\gamma(x)\, e^{tx}\, dx \tag{17.44}$$

Since the diversity paths are assumed to be independently faded, we can obtain $M_\Gamma^i(\cdot)$ as

$$M_\Gamma^i(t) = [M_s^+(t)]^i [M_s^-(t)]^{L_c - i} \tag{17.45}$$

Finally, after substituting Equation 17.41 and Equation 17.45 into Equation 17.38, the generic expression for the MGF of the overall combiner output S is given by

$$\begin{aligned} M_\Gamma(t) &= \sum_{i=0}^{L_c-1} \binom{L}{i} [P_\gamma(\gamma_T)]^{L-i}[1 - P_\gamma(\gamma_T)]^i [M_s^+(t)]^i [M_s^-(t)]^{L_c-i} \\ &\quad + \sum_{j=L_c}^{L} \binom{L}{j} [P_\gamma(\gamma_T)]^{L-j}[1 - P_\gamma(\gamma_T)]^j [M_s^+(t)]^{L_c} \end{aligned} \tag{17.46}$$

Both Equation 17.43 and Equation 17.44 are available in closed form for the fading model under consideration. Their expressions are summarized in Table 17.4 for convenience. Thus, we have obtained closed-form

TABLE 17.4 Conditional MGF $M_s^+(t)$ and $M_s^-(t)$ for the Three Fading Models under Consideration

Model	Rayleigh	Rice	Nakagami-m
$M_s^+(t)$	$\frac{1}{1-t\bar{\gamma}}e^{t\gamma_T}$	$\frac{1+K}{1+K-t\bar{\gamma}}e^{\frac{t\bar{\gamma}K}{1+K-t\bar{\gamma}}}\frac{Q_1\left(\sqrt{\frac{2K(1+K)}{1+K-t\bar{\gamma}}},\sqrt{2(1+K-t\bar{\gamma})\frac{\gamma_T}{\bar{\gamma}}}\right)}{Q_1\left(\sqrt{2K},\sqrt{2(1+K)\frac{\gamma_T}{\bar{\gamma}}}\right)}$	$\left(1-\frac{t\bar{\gamma}}{m}\right)^{-m}\frac{\Gamma\left(m,\frac{m\gamma_T}{\bar{\gamma}}-t\gamma_T\right)}{\Gamma\left(m,\frac{m\gamma_T}{\bar{\gamma}}\right)}$
$M_s^-(t)$	$\frac{1}{1-t\bar{\gamma}}\frac{1-e^{t\gamma_T-\frac{\gamma_T}{\bar{\gamma}}}}{1-e^{-\frac{\gamma_T}{\bar{\gamma}}}}$	$\frac{1+K}{1+K-t\bar{\gamma}}e^{\frac{t\bar{\gamma}K}{1+K-t\bar{\gamma}}}\frac{1-Q_1\left(\sqrt{\frac{2K(1+K)}{1+K-t\bar{\gamma}}},\sqrt{2(1+K-t\bar{\gamma})\frac{\gamma_T}{\bar{\gamma}}}\right)}{1-Q_1\left(\sqrt{2K},\sqrt{2(1+K)\frac{\gamma_T}{\bar{\gamma}}}\right)}$	$\left(1-\frac{t\bar{\gamma}}{m}\right)^{-m}\frac{1-\Gamma\left(m,\frac{m\gamma_T}{\bar{\gamma}}-t\gamma_T\right)}{1-\Gamma\left(m,\frac{m\gamma_T}{\bar{\gamma}}\right)}$

expressions of the MGF of combined SNR Γ with GSEC over generalized fading channels. When $L_c = L$, it can be shown, with the help of the following relationship,

$$[1 - P_\gamma(\gamma_T)]M_s^+(t) + P_\gamma(\gamma_T)M_s^-(t) = M_\gamma(t) \tag{17.47}$$

where $M_\gamma(\cdot)$ is the common MGF of the received SNR, that Equation 17.46 reduces to $[M_\gamma(t)]^L$, i.e., the GSEC combiner is equivalent to an L-branch MRC combiner, as one might expect. It can also be shown that when $L_c = 1$, Equation 17.46 simplifies to the MGF of the output SNR of a traditional L-branch SEC combiner [16, Equation 35].

17.5.3 Application to Performance Analysis

In this section, capitalizing on the MGF of the overall combiner output Γ derived in the previous section, we analyze the performance of GSEC over fading channels in terms of the outage probability and average error probability.

1. *Outage probability:* For GSEC, outage probability, P_{out}, can be computed by evaluating the CDF of Γ at γ_{th}. While it is very difficult, if not impossible, to obtain the CDF of Γ in an elegant or tractable closed form for Nakagami and Rician fading cases, we obtain the outage probability of GSEC over i.i.d. Rayleigh fading as

$$\begin{aligned}
P_{\text{out}} = & \sum_{i=0}^{L_c-1}\binom{L}{i}\left[1-\exp\left(-\frac{\gamma_T}{\bar{\gamma}}\right)\right]^{L-L_c} \\
& \times \sum_{j=0}^{\min[L_c,\lfloor\frac{\gamma_{\text{th}}}{\gamma_T}\rfloor]-i}\binom{L_c-i}{j}(-1)^j\exp\left(-\frac{(i+j)\gamma_T}{\bar{\gamma}}\right) \\
& \times\left[1-\exp\left(-\frac{\gamma_{\text{th}}-(i+j)\,\gamma_T}{\bar{\gamma}}\right)\sum_{k=0}^{L_c-1}\frac{1}{k!}\left(\frac{\gamma_{\text{th}}-(i+j)\,\gamma_T}{\bar{\gamma}}\right)^k\right] \\
& + I_{L_c\gamma_T}(\gamma_{\text{th}})\sum_{j=L_c}^{L}\binom{L}{j}\left[1-\exp\left(-\frac{\gamma_T}{\bar{\gamma}}\right)\right]^{L-j}\exp\left(-\frac{j\gamma_T}{\bar{\gamma}}\right) \\
& \times\left[1-\exp\left(-\frac{\gamma_{\text{th}}-L_c\,\gamma_T}{\bar{\gamma}}\right)\sum_{k=0}^{L_c-1}\frac{1}{k!}\left(\frac{\gamma_{\text{th}}-L_c\,\gamma_T}{\bar{\gamma}}\right)^k\right]
\end{aligned} \tag{17.48}$$

where $I_{L_c\gamma_T}(x)$ is an indicator function, which is equal to 1 if $x \geq L_c\gamma_T$ and zero otherwise.

For the case of $\gamma_T = 0$, noting that $1 - \exp(-\frac{\gamma_T}{\bar{\gamma}}) = 0$ and $I_{L_c\gamma_T}(\gamma_{\text{th}}) = 1$, it can be easily shown that Equation 17.48 simplifies to

$$P_{\text{out}} = 1 - \exp\left(-\frac{\gamma_{\text{th}}}{\bar{\gamma}}\right)\sum_{k=0}^{L_c-1}\frac{1}{k!}\left(\frac{\gamma_{\text{th}}}{\bar{\gamma}}\right)^k \tag{17.49}$$

which is the outage probability of MRC with L_c i.i.d. Rayleigh fading paths [31, Equation 6.69].

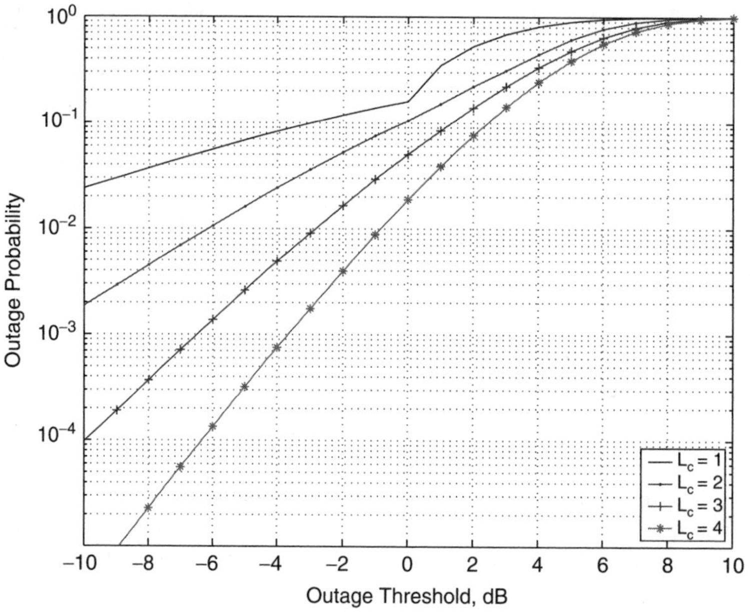

FIGURE 17.13 Outage probability of GSEC with $L = 4$ for different L_c as a function of the normalized outage threshold $\gamma_{\text{th}}/\bar{\gamma}$ ($\gamma_T/\bar{\gamma} = 0$ dB).

It is also of interest to consider the case of $\gamma_{\text{th}} < \gamma_T$. Note that in this case $\lfloor \frac{\gamma_{\text{th}}}{\gamma_T} \rfloor = 0$ and $I_{L_c\gamma_T}(\gamma_{\text{th}}) = 0$. As such, only the term corresponding to $i = j = 0$ of the double summation in Equation 17.48 will survive. The outage probability expression now becomes

$$P_{\text{out}} = \left[1 - \exp\left(-\frac{\gamma_T}{\bar{\gamma}}\right)\right]^{L-L_c} \times \left[1 - \exp\left(-\frac{\gamma_{\text{th}}}{\bar{\gamma}}\right) \sum_{k=0}^{L_c-1} \frac{1}{k!}\left(\frac{\gamma_{\text{th}}}{\bar{\gamma}}\right)^k\right] \tag{17.50}$$

If, in addition, $\gamma_T = +\infty$, Equation 17.50 reduces to Equation 17.49, giving the outage probability of an L_c-branch MRC, as expected.

Figure 17.13 shows the outage performance of GSEC over i.i.d. Rayleigh fading. It can be observed that with L fixed to 4 and L_c increasing from 1 to 4, the outage performance improves but with diminishing gains. Actually, the outage probability of GSEC in this case decreases from the outage probability for a four-branch SEC [16] when $L_c = 1$ to the outage probability for a four-branch MRC when $L_c = L = 4$, as expected.

2. *Average error probability:* Using the closed-form expression for the MGF of the combiner output SNR obtained in the previous section, we can evaluate the average bit and symbol error rates of different modulation schemes with GSEC over various fading scenarios by following the MGF-based approach [28].

Figure 17.14 plots the average bit error probability of BPSK with GSEC as a function of the switching threshold for $L = 4$ and different values of L_c. It can be observed that there exists an optimal choice of the switching threshold in the minimum average error probability sense except for the $L_c = 4$ case, which corresponds to a traditional four-branch MRC. The reason for this is that when the switching threshold is too high or too low, there is no diversity benefit from the SEC combiner. The combiner becomes equivalent to an L_c-branch GSC. Note also that the performance advantage of optimal threshold diminishes as L_c

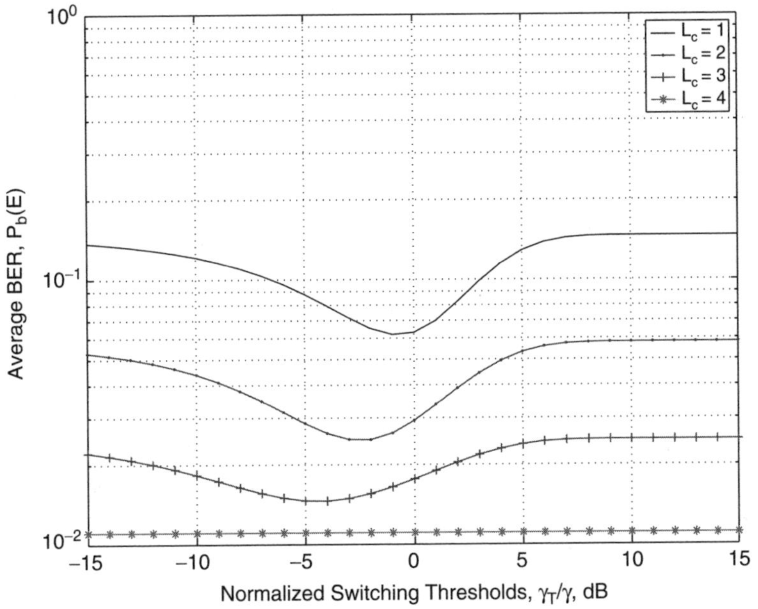

FIGURE 17.14 Average BER of BPSK with GSEC over $L = 4$ i.i.d. Rayleigh fading paths as a function of the common switching threshold γ_T ($\bar{\gamma} = 3$ dB).

increases. Figure 17.15 compares the error performance of GSEC with optimal switching threshold and GSC. The optimal switching thresholds are obtained through numerically minimizing Equation 17.21 while using Equation 17.46 for a given average SNR value. It can be easily observed that the simplicity of GSEC over GSC comes at the cost of a certain performance loss. As an additional numerical example,

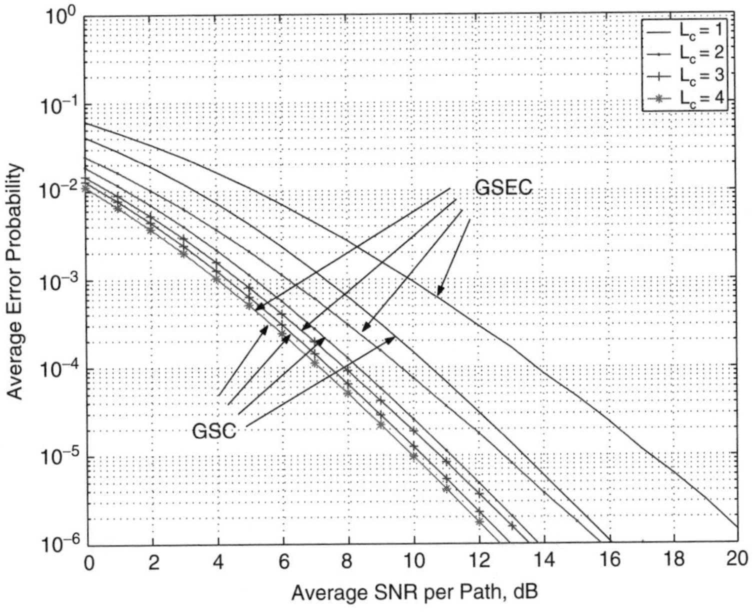

FIGURE 17.15 Comparison of the error performance of GSEC with optimal switching threshold and GSC over i.i.d. Rayleigh fading with $L = 4$.

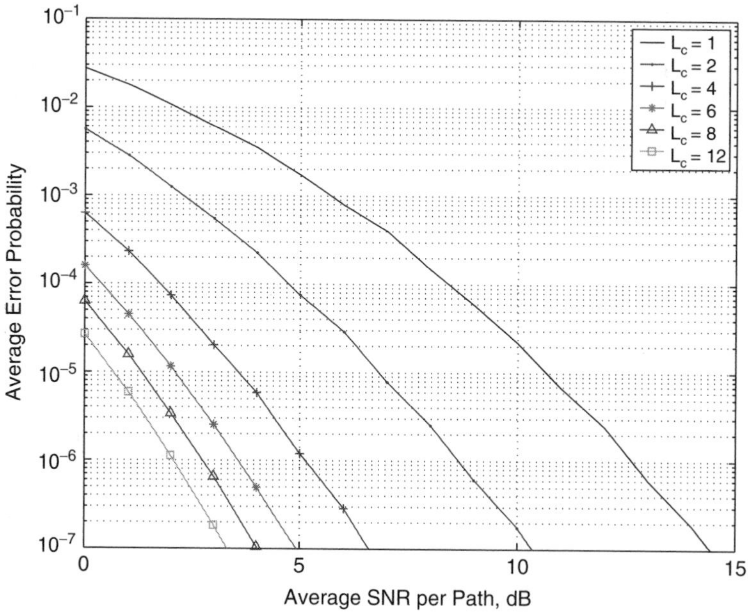

FIGURE 17.16 Average BER of BPSK with GSEC with optimal switching threshold over 12 i.i.d. Rayleigh fading paths as a function of the average SNR $\bar{\gamma}$.

Figure 17.16 shows the error performance of GSEC in a diversity-rich environment with $L = 12$ available diversity paths. It can be seen that beyond $L_c = 6$, the performance gain is not that significant. This implies that GSEC can take advantage of the large number of diversity paths available by approaching the performance of traditional MRC while remaining with a relative lower complexity compared to competing combining schemes such as GSC.

17.5.4 Complexity Savings over GSC

In comparison with GSC, GSEC not only comes with simpler comparison circuitry and no need for full ranking of all diversity paths, but also requires fewer comparisons and less channel estimates on average. In particular, based on the operation of GSEC, we can observe that the receiver using GSEC needs to perform at most L comparisons between the channel estimates and the threshold during each guard period, whereas for GSC, to select the best L_c paths, the receiver needs to perform approximately $L_c \times L$ comparisons, all of which are between the channel estimates. Note also that unlike GSC where all paths are estimated in one shot, the path examination of GSEC is performed only if necessary. If the first L_c examined path is acceptable, the receiver will not estimate the other diversity paths. In this section, we calculate the average number of channel estimates of GSEC, N, during a guard period. This study allows a thorough comparison between GSEC and GSC.

Note that because of MRC combining, the receiver needs to estimate at least L_c diversity paths. In the case that there are less than L_c acceptable paths among the total L ones, based on the mode of operation of GSEC, whenever the receiver encounters $L - L_c$ unacceptable paths, it will apply MRC to the remaining paths, which necessitates the estimation of all available diversity paths. As a result, in this case, the receiver needs to estimate all L available diversity paths. It is not difficult to show that the probability that there are less than L_c acceptable paths is given by

$$P_B = \sum_{L-L_c+1}^{L} \binom{L}{k} [P_\gamma(\gamma_T)]^{L-k}[1 - P_\gamma(\gamma_T)]^k \tag{17.51}$$

On the other hand, if there are at least L_c acceptable diversity paths, based on the mode of operation of GSEC, the combiner will stop examining paths whenever it has found L_c acceptable ones. In this case, the number of channel estimates during a guard period takes values from L_c to L. Note that the probability that k channel estimates are performed in a guard period, denoted by $P_A^{(k)}$, is equal to the probability that the L_cth acceptable path is the kth path examined, or equivalently, exactly $k - L_c$ ones of the first $k - 1$ examined paths are unacceptable, whereas the kth examined path is acceptable. It can be shown, with the assumption of i.i.d. diversity paths, that $P_A^{(k)}$ is mathematically given by

$$P_A^{(k)} = \binom{k-1}{k-L_c} [1 - P_\gamma(\gamma_T)]^{L_c} [P_\gamma(\gamma_T)]^{k-L_c} \tag{17.52}$$

Finally, by combining these two cases, we obtain the expression for the overall average number of channel estimates needed by GSEC during a guard period:

$$\begin{aligned} N &= \sum_{L_c}^{L} k\, P_A^{(k)} + L\, P_B \\ &= \sum_{k=L_c}^{L} k\, [1 - P_\gamma(\gamma_T)]^{L_c} \binom{k-1}{k-L_c} [P_\gamma(\gamma_T)]^{k-L_c} \\ &\quad + L \sum_{L-L_c+1}^{L} \binom{L}{k} [P_\gamma(\gamma_T)]^{L-k} [1 - P_\gamma(\gamma_T)]^{k} \end{aligned} \tag{17.53}$$

In the case of $\gamma_T = 0$, since $P_\gamma(\gamma_T) = 0$, it can be shown that $N = L_c$, as expected. On the other hand, when $\gamma_T \to \infty$ and thus $P_\gamma(\gamma_T) = 1$, it can be similarly shown that $N = L$, also as expected.

The result in Equation 17.53 is very general and applies to different fading scenarios of interest by substituting the appropriate CDF given in Table 17.1. Figure 17.17 plots the average number of path estimates of GSEC with $L = 4$ as a function of the switching threshold for the Rayleigh fading case. As we can see, as the switching threshold γ_T increases, the number of path estimates N increases from L_c

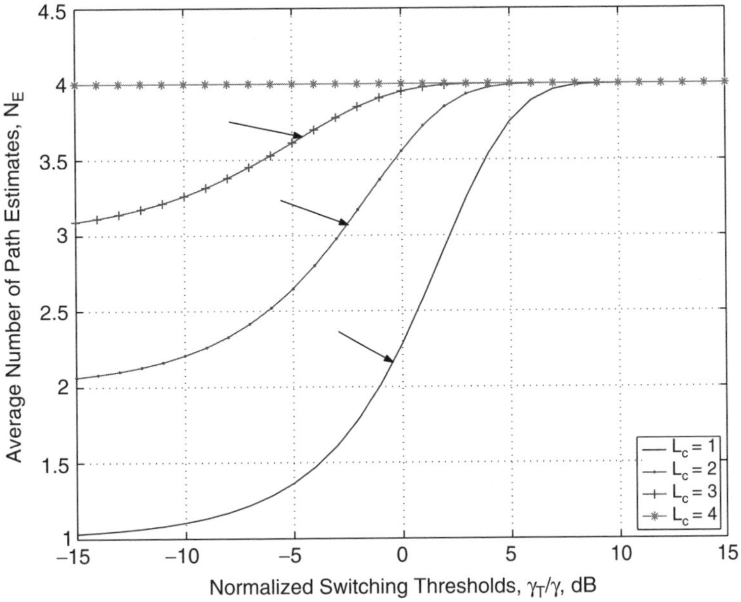

FIGURE 17.17 Average number of channel estimates of GSEC with $L = 6$ as a function of the common switching threshold γ_T ($\bar{\gamma} = 3$ dB).

to L. Intuitively, that is because if the threshold is large, it becomes more difficult for the receiver to find acceptable paths, and as such, more paths need to be estimated. We also marked the optimal operating points for GSEC in terms of minimizing the average error rate, which were read from Figure 17.14. Note that even for the best performance, GSEC does not need to estimate all the diversity paths, whereas for GSC, all L diversity paths need to be estimated in each guard period.

17.6 Further Remarks

Switched diversity offers one of the lowest-complexity solutions to mitigating the fading effect in wireless communications systems. It is observed that compared to the more popular selection-based combining schemes, switching-based combining schemes lead to a simpler receiver structure, less signal processing, and lower power consumption at the cost of a certain performance loss. Related works on switched diversity also show that it is applicable to both the transmitter and receiver sides [16] and that it can be implemented in both predetection and postdetection fashions [32]. Switched diversity schemes can also provide considerable performance improvement in the presence of co-channel interference [33].

References

[1] W.C. Jakes, *Microwave Mobile Communication*, 2nd ed., Piscataway, NJ: IEEE Press, 1994.

[2] R. Prasad, *Universal Wireless Personal Communications*, Boston, MA: Artech House, 1998.

[3] G.L. Stüber, *Principles of Mobile Communications*, 2nd ed., Norwell, MA: Kluwer Academic Publishers, 2000.

[4] M.A. Blanco and K.J. Zdunek, "Performance and optimization of switched diversity systems for the detection of signals with Rayleigh fading," *IEEE Trans. Commun.*, COM-27(12), 1887–1895, 1979.

[5] M.A. Blanco, "Diversity receiver performance in Nakagami fading," in *Proceedings of IEEE Southeastern Conference*, Orlando, FL, April 1983, pp. 529–532.

[6] A.A. Abu-Dayya and N.C. Beaulieu, "Analysis of switched diversity systems on generalized-fading channels," *IEEE Trans. Commun.*, COM-42(11), 2959–2966, 1994.

[7] A.A. Abu-Dayya and N.C. Beaulieu, "Switched diversity on microcellular Ricean channels," *IEEE Trans. Veh. Technol.*, VT-43(4), 970–976, 1994.

[8] G. Femenias and I. Furió, "Analysis of switched diversity TCM-MPSK systems on Nakagami fading channels," *IEEE Trans. Veh. Technol.*, VT-46(1), 102–107, 1997.

[9] Y.-C. Ko, M.-S. Alouini, and M.K. Simon, "Analysis and optimization of switched diversity systems," *IEEE Trans. Veh. Technol.*, VT-49(5), 1569–1574, 2000.

[10] C. Tellambura, A. Annamalai, and V.K. Bhargava, "Unified analysis of switched diversity systems in independent and correlated fading channel," *IEEE Trans. Commun.*, COM-49(11), 1955–1965, 2001.

[11] H.-C. Yang and M.-S. Alouini, "Analysis and comparison of various switched diversity strategies," in *Proceedings of the IEEE Semiannual Veh. Technol. Conference (VTC Fall 2002)*, Vancouver, British Columbia, Canada, September 2002, pp. 1948–1952 (full paper to appear in *IEEE Trans. Commun.*).

[12] N. Kong, T. Eng, and L. B. Milstein, "A selection combining scheme for RAKE receivers," in *Proceedings of IEEE International Conference of University Personal Communication ICUPC' 95*, Tokyo, Japan, November 1995, pp. 426–429.

[13] M.Z. Win and Z.A. Kostić, "Virtual path analysis of selective Rake receiver in dense multipath channels," *IEEE Commun. Lett.*, 3(11), 308–310, 1999.

[14] Y. Roy, J.-Y. Chouinard, and S.A. Mahmoud, "Selection diversity combining with multiple antennas for MM-wave indoor wireless channels," *IEEE J. Selected Areas Commun.*, SAC-14(4), 674–682, 1998.

[15] A.F. Molisch, M.Z. Win, and J.H. Winters, "Capacity of MIMO systems with antenna selection," in *Proceedings of IEEE International Conference on Commununications, (ICC '01)*, Vol. 2, Helsinki, Finland, June 2001, pp. 570–574.

[16] H.-C. Yang and M.-S. Alouini, "Performance analysis of multibranch switched diversity systems," *IEEE Trans. Commun.*, COM-51(5), 782–794, 2003.

[17] N. Kong and L.B. Milstein, "Average SNR of a generalized diversity selection combining scheme," *IEEE Commun. Lett.*, 3(3), 5759, 1999 (see also *Proceedings of IEEE International Conference on Communications ICC '98*, Atlanta, Georgia, June 1998, pp. 1556–1560).

[18] M.Z. Win and J.H. Winters, "Analysis of hybrid selection/maximal-ratio combining in Rayleigh fading," *IEEE Trans. Commun.*, COM-47(12), 1773–1776, 1999 (see also *Proceedings of IEEE International Conference on Communications (ICC '99)*, Vancouver, British Columbia, Canada, June 1999, pp. 6–10).

[19] M.-S. Alouini and M.K. Simon, "An MGF-based performance analysis of generalized selective combining over Rayleigh fading channels," *IEEE Trans. Commun.*, COM-48(3), 401–415, 2000.

[20] Y. Ma and C. Chai, "Unified error probability analysis for generalized selection combining in Nakagami fading channels," *IEEE J. Selected Areas Commun.*, SAC-18(11), 2198–2210, 2000.

[21] A. Annamalai and C. Tellambura, "Analysis of hybrid selection/maximal-ratio diversity combiner with Gaussian errors," *IEEE Trans. Wireless Commun.*, TWC-1(3), 498–512, 2002.

[22] H.-C. Yang and M.-S. Alouini, "Performance analysis of generalized switch and examine combining," in *Proceedings of IEEE International Conference on Communications (ICC 2003)*, Anchorage, Alaska, May 2003, pp. 2218–2222 (full paper to appear in *IEEE Trans. Commun.*).

[23] I.S. Gradshteyn and I.M. Ryzhik, *Table of Integrals, Series, and Products*, 5th ed., San Diego, CA: Academic Press, 1994.

[24] A.H. Nuttall, "Some integrals involving the Q_M function," *IEEE Trans. Inf. Theory*, 1, 95–96, 1975.

[25] S. Ross, *An Introduction to Probability Models*, 6th ed., San Diego, CA: Academic Press, 1997.

[26] M. Schwartz, W.R. Bennett, and S. Stein, *Communication Systems and Techniques*, New York: McGraw-Hill, 1966.

[27] C.C. Tan and N.C. Beaulieu, "Infinite series representation of the bivariate Rayleigh and Nakagami-*m* distributions," *IEEE Trans. Commun.*, COM-45(10), 1159–1161, 1997.

[28] M.K. Simon and M.-S. Alouini, *Digital Communications over Generalized Fading Channels: A Unified Approach to Performance Analysis*, New York, John Wiley & Sons, 2000.

[29] J.H. Winters, "Switched diversity with feedback for DPSK mobile radio systems," *IEEE Trans. Veh. Technol.*, VT-32(2), 134–150, 1983.

[30] Q.T. Zhang, "Exact analysis of postdetection combining for DPSK and NFSK systems over arbitrarily correlated Nakagami channels," *IEEE Trans. Commun.*, TCOM-46(11), 1459–1467, 1998.

[31] T.S. Rappaport, *Wireless Communications: Principles and Practice*, Upper Saddle River, NJ: PTR Prentice Hall, 1996.

[32] H.-C. Yang, M.-S. Alouini, and M.K. Simon, "Average error rate of NCFSK with multiple-branch post-detection switched diversity," in *Proceedings of the IEEE Semiannual Vehicular Technology Conference (VTC Fall 2002)*, Vancouver, British Columbia, Canada, September 2002, pp. 366–370 (full paper to appear in *Wiley J. Wireless Commun. Mobile Comput.*).

[33] H.-C. Yang and M.-S. Alouini, "Outage probability of dual-branch diversity systems in presence of co-channel interference," *IEEE Trans. Wireless Commun.*, WCOM-2(2), 310–319, 2003.

18

Overview of Equalization Techniques for MIMO Fading Channels

Noura Sellami
Université de Cergy Pontoise

Inbar Fijalkow
Université de Cergy Pontoise

Mohamed Siala
Sup'Com

18.1 Introduction **18**-1
18.2 Frequency-Selective MIMO Channel Model **18**-2
General Framework • Simulation Framework
18.3 Block Linear and Decision-Feedback Equalizers **18**-4
Block Linear Equalizers • Block Decision-Feedback Equalizers
18.4 List-Type Equalizers **18**-8
The Single Antenna Case • Generalization to the MIMO Case • Simulation Results
18.5 The Multidimensional Whitened Matched Filter **18**-12
Whitened Matched Filter • Prefiltered List-Type MAP Equalizer • WMF Implementation Using Linear Prediction • Energy Concentration • Simulation Results
18.6 The Block Equalizers vs. the Prefiltered List-Type MAP **18**-17
Complexity Comparison • Performance Comparison
18.7 Conclusion **18**-19

18.1 Introduction

Future wireless communications promise to offer a variety of multimedia services that require reliable transmission at high data rates. Foschini and Gans [12] have demonstrated the enormous capacity potential gain of wireless communication systems with antenna arrays at both the transmitter and receiver. In order to achieve the promised high data rates over frequency selective multiple-input multiple-output (MIMO) channels, such as mobile radio channels, the receiver has to combat both intersymbol interference (ISI) and spatial interference (between antennas) by using an equalizer.

It is well known that the best performance in terms of error rate can be achieved through trellis equalization based on maximum likelihood sequence estimation (MLSE) or symbol-by-symbol maximum *a posteriori* probability (MAP) estimation [1]. However, the complexity of these methods is proportional to the number of states of the trellis that grow exponentially with the product of the channel memory and the number of transmit antennas [3]. When the channel memory becomes large and high-order constellations are used, the algorithms become impractical. Therefore, it is interesting to develop reduced-complexity algorithms without significant degradation in performance.

0-8493-1657-X/$0.00+$1.50

In this chapter, we will present two families of suboptimum equalizers achieving a good complexity/performance trade-off:

- **Block linear and decision-feedback equalizers** with a zero-forcing (ZF) or minimum mean square error (MMSE) optimization criterion [15]. These equalizers were first proposed for multiuser code-division multiple-access (CDMA) block transmission systems to jointly detect the different users [14]. In view of the analogy between the CDMA systems and multiple-antenna systems, these equalizers can be easily adapted to the multiple-antenna case.
- **List-type equalizers**, which are suboptimum versions of the Viterbi equalizer (generalized Viterbi algorithm [13]) or the MAP equalizer (list-type MAP equalizer [20]) based on **state reduction** and **per survivor processing** (PSP). The trellis has a reduced number of states taking into account a reduced number of taps of the channel. The remaining intersymbol and co-channel interference is cancelled by internal per survivor processing, as done in a classical decision-feedback equalizer (DFE) [10][13][17]. In order to avoid error propagation, more than one survivor is retained per state (a list of S survivors, $S > 1$) [13]. However, these equalizers still suffer from error propagation, especially when the channel is not minimum phase. To induce the performance of these equalizers, it is desirable to use a receiver filter that concentrates the channel energy on its first taps. In the scalar case, this is easily achieved by using a minimum phase factorization of the channel. This approach is extended in [23] to the MIMO case using a whitened matched filter (WMF), which makes the channel 'minimum phase' and keeps the noise white. Simulation results show that the use of the WMF yields significant improvement, particularly over severe channels.

We will give simulation results corresponding to these two families of equalizers and compare their performances and complexities of implementation.

Content: This chapter is organized as follows. Section 18.2 describes the frequency-selective MIMO channel model. Section 18.3 presents the block linear equalizers (BLEs) and block decision-feedback equalizers (BDFEs). Section 18.4 deals with the list-type equalizers. Section 18.5 describes the multidimensional WMF and gives simulation results for the list-type MAP equalizer prefiltered by the WMF. And Section 18.6 draws a comparison between the block equalizers and the prefiltered list-type MAP equalizer.

Notations: Throughout this chapter, scalars and matrices are lower and uppercase, respectively, and vectors are underlined lowercase. The symbols $(.)^T$, $(.)^H$, and $(.)^{-1}$ denote transposition, transconjugation, and inversion, respectively. Moreover, Trace(A) denotes the trace of matrix A and $E(.)$ denotes the expected value operator. The $n \times n$ identity matrix is denoted by I_n. Finally, $\lfloor x \rfloor$ denotes the highest integer not bigger than x.

18.2 Frequency-Selective MIMO Channel Model

18.2.1 General Framework

Due to multipath propagation and time dispersion, the wireless MIMO channel we consider is a frequency-selective fading channel from each transmit antenna to each receive antenna. The number of transmit antennas is denoted by N and the number of receive antennas is denoted by M. The antennas are separated far enough apart to provide independent channels from each transmit to each receive antenna. Each channel is modeled as Rayleigh fading with a memory of L symbols (see below). The memory L is considered to be the same for all channels. Actually, if the transmit antennas are located at the same base station, the assumption can be justified by the fact that the number of multipath components with different delays is dictated by large structures and reflecting objects [2]. Foschini and Gans [12] have demonstrated the enormous capacity potential gain of wireless MIMO systems. Many approaches have been studied in order to exploit this capacity [5]. The information theory shows that with N transmit antennas and M receive antennas ($N \leq M$), N independent data streams have to be transmitted simultaneously in order to achieve the channel capacity. This is the principle of **spatial multiplexing**. A system following this approach is the

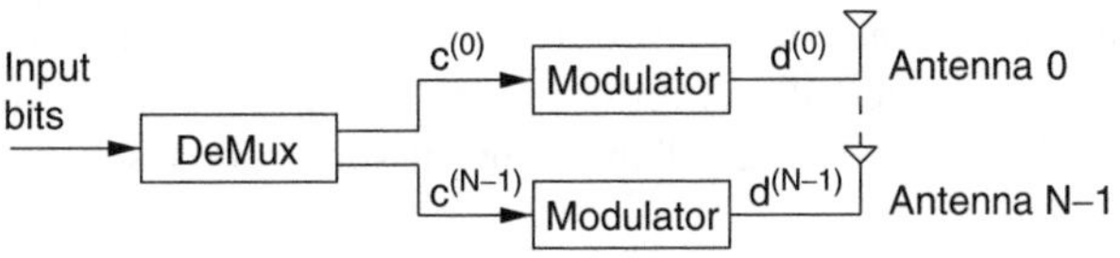

FIGURE 18.1 Transmitter structure.

Bell Labs layered space–time (BLAST) architecture proposed in [11]. Another approach consists in using **transmit diversity**. Instead of transmitting independent data streams, the same information is transmitted in a judicious way simultaneously by the transmit antennas. This is the idea of the space–time codes proposed in [25][26] by Tarokh et al. In this chapter, we consider a system based on spatial multiplexing. As shown in Figure 18.1, the input information bit sequence is first demultiplexed into N streams that are mapped to PSK/QAM (phase shift keying/quadrature amplitude modulation) symbols and transmitted simultaneously by the N transmitting antennas. The same modulation constellation with size Q is used for each stream. Thus, $\log_2(Q)$ information bits are mapped into one Q-*ary* symbol. We assume that transmissions are organized into bursts of T symbols. The channel is supposed to be invariant during a burst and to change independently from burst to burst. This assumption is reasonable for a mobile system like the Global System for Mobile Communications (GSM), which employs frequency hopping. Furthermore, we assume in this chapter that the channel is unknown to the transmitter and perfectly known at the receiver. Channel knowledge at the receiver can be obtained by sending training sequences and estimating the channel.

The received baseband signal sampled at the symbol rate at antenna j at time k is a linear combination of the N transmitted signals perturbed by noise:

$$r_k^{(j)} = \sum_{i=0}^{N-1}\sum_{l=0}^{L-1} h_{i,j}(l)d_{k-l}^{(i)} + n_k^{(j)}, \quad 0 \le j \le M-1 \tag{18.1}$$

In this expression, $n_k^{(j)}$ are modeled as independent samples of a zero-mean white complex Gaussian noise with variance $\sigma^2 = N_0$ and $h_{i,j}(l)$ is the lth tap gain from transmit antenna i to receive antenna j. The tap gains $h_{i,j}(l)$ are modeled as independent complex Gaussian random variables with zero mean and variance $\sigma_h^2(l)$. We assume that $\sum_{l=0}^{L-1}\sigma_h^2(l) = 1$.

Let $\underline{d}_k = (d_k^{(0)}, \ldots, d_k^{(N-1)})^T$ be the N-long vector of modulated symbols transmitted from the N transmitting antennas at time k and $\underline{n}_k = (n_k^{(0)}, \ldots, n_k^{(M-1)})^T$ be the M-long noise vector at the M receiving antennas.

The output of the channel is the M-long vector $\underline{r}_k = (r_k^{(0)}, \ldots, r_k^{(M-1)})^T$ with Z-transform:

$$\underline{r}(z) = H(z)\underline{d}(z) + \underline{n}(z) \tag{18.2}$$

where $H(z) = \sum_{l=0}^{L-1} H(l)z^{-l}$ and $(H(l))_{j,i} = h_{i,j}(l)$.

The problem we address is then to recover the information bits from the noisy observation $\underline{r}(z)$.

18.2.2 Simulation Framework

In our simulations, we focus on a MIMO system with two transmit antennas and two receive antennas. We use a frequency-selective fading channel with memory $L = 5$. The channel is considered to be time invariant during the transmission of a burst of $T = 512$ information bits and changes independently from burst to burst. We assume that the channel is perfectly known at the receiver. We consider three types of frequency-selective channels described by the standard deviation of the Rayleigh distribution of their taps:

Channel 1: $\underline{\sigma}_h = (0.227; 0.460; 0.688; 0.460; 0.227)$
Channel 2: $\underline{\sigma}_h = (1/\sqrt{5}; 1/\sqrt{5}; 1/\sqrt{5}; 1/\sqrt{5}; 1/\sqrt{5})$
Channel 3: $\underline{\sigma}_h = (0.716; 0.501; 0.429; 0.214; 0.071)$

where $\underline{\sigma}_h = (\sigma_h(0), \ldots, \sigma_h(L-1))$.

These channels were chosen because they have different energy profiles. The modulation used is binary (BPSK). We plot the bit error rate (BER) with respect to averaged E_b/N_0 per receive antenna. These simulation conditions will be used to obtain all the following simulation results.

18.3 Block Linear and Decision-Feedback Equalizers

In this section, we consider filter-based equalizers for block transmission systems that are derived from the single antenna case [15]. In [15], linear and decision-feedback structures are introduced with zero-forcing and minimum mean square error block optimization criteria. It has been shown that they perform better than conventional ZF and MMSE equalizers. In [14], these equalizers have been considered for the uplink of a CDMA system with multiple antennas at the receiver. The MIMO system presented in Section 18.2 is a particular case of the system studied in [14] when no spreading is used. In the following, we will adapt the equalizers presented in [14][15] to our system. These equalizers are:

- Block linear equalizers: ZF-BLE (ZF block linear equalizer) and MMSE-BLE (MMSE block linear equalizer)
- Block decision-feedback equalizers: ZF-BDFE (ZF block decision-feedback equalizer) and MMSE-BDFE (MMSE block decision-feedback equalizer)

We give here the expressions of these four equalizers. The transmitted symbols are supposed to be uncorrelated.

18.3.1 Block Linear Equalizers

Let $\underline{d} = (\underline{d}_0^T, \ldots, \underline{d}_{T-1}^T)^T$ be the $NT \times 1$ vector of modulated symbols transmitted by the N transmit antennas during the transmission of a block, $\underline{r} = (\underline{r}_0^T, \ldots, \underline{r}_{T+L-2}^T)^T$ the $M(T+L-1) \times 1$ whole received vector at the M receiving antennas, and $\underline{n} = (\underline{n}_0^T, \ldots, \underline{n}_{T+L-2}^T)^T$ the noise vector. The received vector $\underline{r}$ can be written as

$$\underline{r} = G\underline{d} + \underline{n}. \tag{18.3}$$

In this expression, G is the $M(T+L-1) \times NT$ matrix defined as

$$G = \begin{pmatrix} G_0 & 0 & . & . & . & 0 \\ G_1 & G_0 & 0 & . & . & . \\ . & G_1 & G_0 & . & . & . \\ . & . & G_1 & . & 0 & . \\ G_{L-1} & . & . & . & G_0 & 0 \\ 0 & G_{L-1} & . & . & G_1 & G_0 \\ . & 0 & G_{L-1} & . & . & G_1 \\ . & . & 0 & . & . & . \\ . & . & . & . & G_{L-1} & . \\ 0 & . & . & . & 0 & G_{L-1} \end{pmatrix} \tag{18.4}$$

where a block G_l, for $0 \leq l \leq L$, is an $M \times N$ matrix with entries $(G_l)_{j,i} = h_{i,j}(l)$, for $0 \leq i \leq N-1$, $0 \leq j \leq M-1$.

We give in the following the expressions of the block linear equalizers:

- ZF-BLE: The ZF-BLE equalizer, the output of which is denoted by $\hat{\underline{d}}_{ZF-BLE}$, minimizes the quadratic form

$$(\underline{r} - G\hat{\underline{d}}_{ZF-BLE})^H(\underline{r} - G\hat{\underline{d}}_{ZF-BLE}) \tag{18.5}$$

which leads to the unbiased estimate

$$\begin{aligned}\hat{\underline{d}}_{ZF-BLE} &= (G^H G)^{-1} G^H \underline{r} \\ &= \underline{d} + (G^H G)^{-1} G^H \underline{n}\end{aligned} \tag{18.6}$$

This estimate contains no ISI and no spatial interference perturbation term but only the desired symbols perturbed with filtered noise. This equalizer is called zero forcing since it totally eliminates ISI and spatial interference. It is given by the Moore–Penrose pseudoinverse of G, the matrix $G^H G$ being supposed to be full rank.

We use the signal-to-noise ratio (SNR) in the decision variable on the symbol transmitted as a measure of performance. It is defined as the variance of the signal term in the decision variable divided by the variance of the disturbance terms (interference and noise). According to Equation 18.6, this SNR for the decision variable on the symbol transmitted by antenna i at time k is given by

$$\gamma_{ZF-BLE}(i,k) = \frac{1}{\sigma^2((G^H G)^{-1})_{p,p}} \tag{18.7}$$

where $p = k + Ti$, $0 \leq k \leq T-1$ and $0 \leq i \leq N-1$.

- MMSE-BLE: The MMSE-BLE equalizer, the output of which is denoted by $\hat{\underline{d}}_{MMSE-BLE}$, minimizes the quantity

$$E((\hat{\underline{d}}_{MMSE-BLE} - \underline{d})^H (\hat{\underline{d}}_{MMSE-BLE} - \underline{d})) \tag{18.8}$$

which leads to

$$\begin{aligned}\hat{\underline{d}}_{MMSE-BLE} &= (\sigma^2 I_{TN} + G^H G)^{-1} G^H \underline{r} \\ &= (I_{TN} + \sigma^2 (G^H G)^{-1})^{-1} \hat{\underline{d}}_{ZF-BLE}\end{aligned} \tag{18.9}$$

This estimate can be seen as the output of the ZF-BLE followed by a Wiener filter $W = (I_{TN} + \sigma^2(G^H G)^{-1})^{-1}$. The Wiener filter reduces the performance degradation due to the fact that in the ZF-BLE the decisions do not take into account the noise correlations existing in the decision variables (see [15] for more details).

Since $\underline{r} = G\underline{d} + \underline{n}$, we can write

$$\hat{\underline{d}}_{MMSE-BLE} = diag(W)\underline{d} + (W - diag(W))\underline{d} + W(G^H G)^{-1} G^H \underline{n} \tag{18.10}$$

where $diag(W)$ represents a diagonal matrix containing the diagonal elements of the matrix W. According to Equation 18.10, the estimate $\hat{\underline{d}}_{MMSE-BLE}$ contains the desired symbols, an interference term (ISI and spatial interference), and a noise term. The SNR at the output of the MMSE-BLE per symbol transmitted by antenna i at time k is given by

$$\gamma_{MMSE-BLE}(i,k) = \frac{((I_{TN} + \sigma^2(G^H G)^{-1})^{-1})_{p,p}}{1 - ((I_{TN} + \sigma^2(G^H G)^{-1})^{-1})_{p,p}} \tag{18.11}$$

where $p = k + Ti$, $0 \leq k \leq T-1$, and $0 \leq i \leq N-1$. This quantity is in general larger than $\gamma_{ZF-BLE}(i,k)$ given in Equation 18.7 [15].

18.3.2 Block Decision-Feedback Equalizers

The ZF-BLE and MMSE-BLE were modified in [15] to produce the ZF-BDFE and MMSE-BDFE, respectively. Figure 18.2 shows the block diagram of a block decision-feedback equalizer. Nonlinearity is introduced into the system by feeding back previous estimates of data symbols in order to remove the interference. The received vector $\underline{r}$ is filtered by a matrix filter S, depending on the chosen criterion, leading

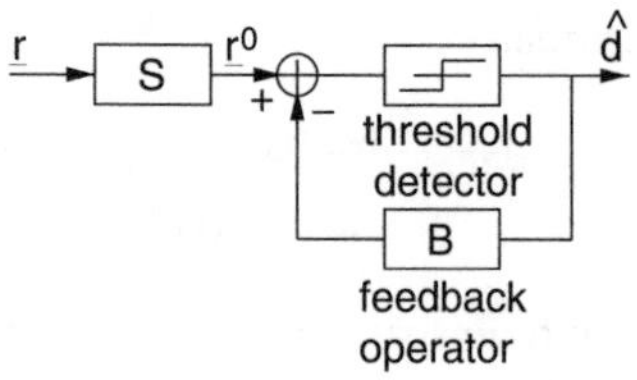

FIGURE 18.2 The structure of a block decision-feedback equalizer.

to the vector $\underline{r}'$. Previous data estimates are processed through a feedback filter B and subtracted from $\underline{r}'$. The resultant signal is fed into a threshold detector for data estimation. We give in the following the expressions of the ZF-BDFE and MMSE-BDFE:

- ZF-BDFE: In order to calculate this estimator, we consider the Cholesky decomposition of $G^H G$:

$$G^H G = (\Sigma\Psi)^H(\Sigma\Psi) \tag{18.12}$$

where Ψ is an upper triangular matrix with ones along the diagonal and Σ a diagonal matrix with real entries.

We consider the received modified signal:

$$\begin{aligned} \underline{r}' &= \Psi(G^H G)^{-1}G^H\underline{r} \\ &= \Sigma^{-1}((\Sigma\Psi)^H)^{-1}G^H\underline{r} \\ &= \underline{d} + (\Psi - I_{NT})\underline{d} + \Sigma^{-1}\Sigma^{-1}(\Psi^H)^{-1}G^H\underline{n} \end{aligned} \tag{18.13}$$

In this expression, the noise is white with covariance matrix $\Sigma^{-1}\Sigma^{-1}$. Since Ψ is upper triangular, the decision on a symbol of the vector $\underline{d}$ can be obtained by using the past decisions on the previous symbols. Thus, by taking the decisions in reverse order of the index of the components of $\underline{d}$, the term $(\Psi - I_{NT})$ in Equation 18.13 depends only on the symbols already estimated. Decisions are made according to the recursive formula

$$\hat{d}_{ZF-BDFE,TN-1} = D\left\{r'_{TN-1}\right\}$$

$$\hat{d}_{ZF-BDFE,TN-1-j'} = D\left\{r'_{TN-1-j'} - \sum_{j"=1}^{j'}(\Psi - I_{NT})_{NT-1-j',NT-1-j'+j"}\,\hat{d}_{ZF-BDFE,NT-1-j'+j"}\right\} \tag{18.14}$$

where $D\{.\}$ is a decision operation performed in a threshold detector with respect to the used constellation and $1 \leq j' \leq TN$. In the BPSK case, $D\{.\}$ is the sign function. For other constellations it takes the value of the nearest constellation point. The matrix S here (see Figure 18.2) is equal to $\Psi(G^H G)^{-1}G^H$ and B is equal to $(\Psi - I_{NT})$. If all the past decisions are correct, the SNR at the output of the ZF-BDFE per symbol transmitted by antenna i at time k is given by

$$\gamma_{ZF-BDFE}(i,k) = (\Sigma)^2_{p,p} \tag{18.15}$$

where $p = k + Ti$, $0 \leq k \leq T-1$ and $0 \leq i \leq N-1$. This quantity is in general larger than $\gamma_{ZF-BLE}(i,k)$ given in Equation 18.7 [15].

When some of the past decisions are erroneous, error propagation can occur. To fight against this phenomenon, we can sort the components of $\underline{d}$ in such a way that the most reliable symbols, corresponding to the best channels, are detected first [14][15].

- MMSE-BDFE: In order to calculate this estimator, we consider the Cholesky decomposition of $(I_{NT} + \frac{1}{\sigma^2} G^H G)$:

$$I_{NT} + \frac{1}{\sigma^2} G^H G = (\Sigma' \Psi')^H (\Sigma' \Psi') \tag{18.16}$$

where Ψ' is an upper triangular matrix with ones along the diagonal and Σ' a diagonal matrix with real entries.

The structure of the MMSE-BDFE is the same as that of the ZF-BDFE when Ψ is replaced by Ψ', Σ by Σ', and $\hat{d}_{ZF-BDFE}$ by $\hat{d}_{MMSE-BDFE}$. Here, the received modified signal writes

$$\underline{r}' = \Sigma'^{-1} ((\Sigma' \Psi')^H)^{-1} G^H \underline{r} \tag{18.17}$$

If all the past decisions are correct, the SNR at the output of the MMSE-BDFE per symbol transmitted by antenna i at time k is given by

$$\gamma_{MMSE-BDFE}(i, k) = \left(\Sigma'\right)^2_{p,p} - 1 \tag{18.18}$$

where $p = k + Ti$, $0 \leq k \leq T - 1$, and $0 \leq i \leq N - 1$. This quantity is in general larger than $\gamma_{ZF-BDFE}(i, k)$ given in Equation 18.15 [15].

18.3.2.1 Simulation Results

Figure 18.3 shows the performance of the four equalizers for channel 1 presented in Section 18.2.2. The simulation conditions are given in Section 18.2.2. Simulations show that the performance is almost the same for the three channels. It is worth noting that this result is specific to the considered channels and cannot be generalized. As expected, the performance of the MMSE equalizers is better than that of the corresponding ZF equalizers. Moreover, the performance of the equalizers with decision feedback is

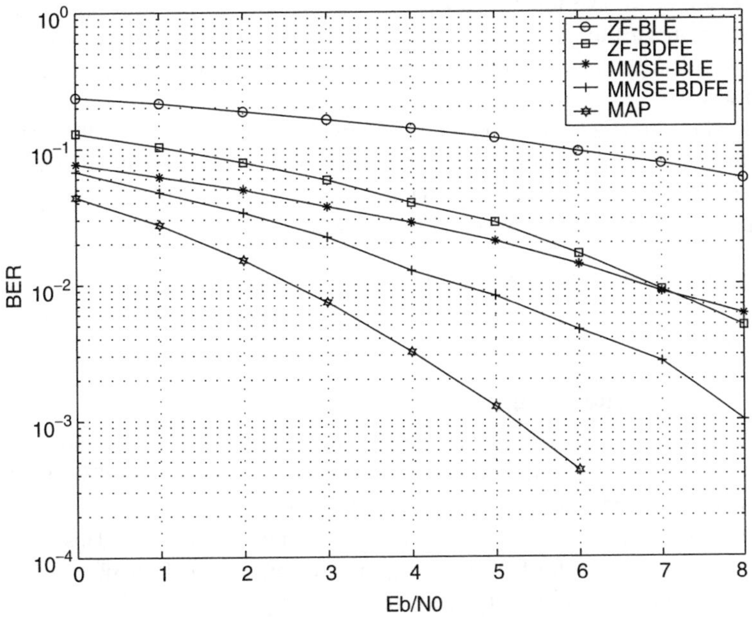

FIGURE 18.3 BER performance of the ZF-BLE, ZF-BDFE, MMSE-BLE, MMSE-BDFE, and MAP equalizer for the three channels presented in Section 18.2.2.

better than that of the corresponding equalizers without decision feedback. The performance of the MAP equalizer is also depicted in Figure 18.3. For a BER $= 10^{-2}$, the performance of the MMSE-BDFE is within 1.6 dB of that achieved by the MAP equalizer. We can conclude that the performance of these equalizers is relatively distant from the optimal one. Thus, it is interesting to study other types of equalizers that would achieve a better complexity/performance trade-off. In the following, we will present a second family of suboptimum equalizers called the **list-type equalizers.**

18.4 List-Type Equalizers

Trellis equalization performed by a Viterbi [27] or MAP [1] equalizer achieves good performance but is generally too complex to be implemented. In order to reduce the complexity of these equalizers, List-type equalizers, which are state reduction algorithms of the Viterbi or MAP equalizer using per survivor processing, have been proposed in [13][20].

In this section, we first present these algorithms, referred to as the generalized Viterbi algorithm and the list-type MAP algorithm, in the single-antenna case. Then, in view of the importance of soft equalization, we will focus on the list-type MAP equalizer and generalize it to the MIMO case.

18.4.1 The Single Antenna Case

18.4.1.1 Viterbi Algorithm

Before introducing the generalized Viterbi algorithm, we first review the Viterbi algorithm. A frequency-selective single-input single-output (SISO) channel with L taps can be modeled as a discrete-time finite-state machine, whose state $\mu(n)$ at time n is defined by the $L-1$ most recent channel inputs $\mu(n) = (d_{n-1}, d_{n-2}, \ldots, d_{n-L+2}, d_{n-L+1})$, where d_{n-1} is the channel input at time $n-1$. Thus, the channel can be described by a trellis with Q^{L-1}states, Q being the constellation size.

The Viterbi algorithm can be described recursively as follows [27]:

1. Initial condition: At time n, the algorithm retains for each state $\mu(n)$ of the trellis a path $u(n) = (\mu(0), d_0, d_1, \ldots, d_{n-1})$ leading to $\mu(n)$.
2. Path extension: At time $n+1$, the retained paths are extended by one branch (or transition) $\xi(n) = (\mu(n), d_n)$ yielding new candidates that will be stored in $Q^{(L-1)}$ lists (Q candidates converge at each state).
3. Path selection: From each list, the best candidate with the largest metric is selected for the next step. The path metric is the cumulative sum of the branch metrics along the path. We recall that the metric of a branch $\xi(n) = (\mu(n), d_n)$ is equal to the squared euclidean distance $|r_n - \sum_{l=0}^{L-1} h(l)d_{n-l}|^2$, r_n being the output of the SISO channel at time n and $h(l)$ its lth tap gain.

Starting from the first level, the Viterbi algorithm repeats these procedures until states at the last level are reached. In the following, the paths retained at each state are called the survivors and the extensions of the survivors are called the candidates.

18.4.1.2 Generalized Viterbi Algorithm

In order to reduce the complexity of the Viterbi algorithm while keeping good performance, a **reduced-state** version referred to as delayed decision-feedback sequence estimation (DDFSE) has been proposed by Duel-Hallen and Heegard in [10]. As for the Viterbi algorithm, at each step, the trellis states correspond to all the possible values taken by a finite number J of previous inputs. This memory J is taken equal to the channel memory L for the Viterbi algorithm, while for the DDFSE J is less than L. Actually, for the DDFSE, the channel memory is divided into two parts: one is used to construct the trellis on which the Viterbi algorithm is based, and the other is cancelled by considering the path converging to each state (**per survivor processing**). Thus, the DDFSE is a hybrid of the DFE and the Viterbi algorithm.

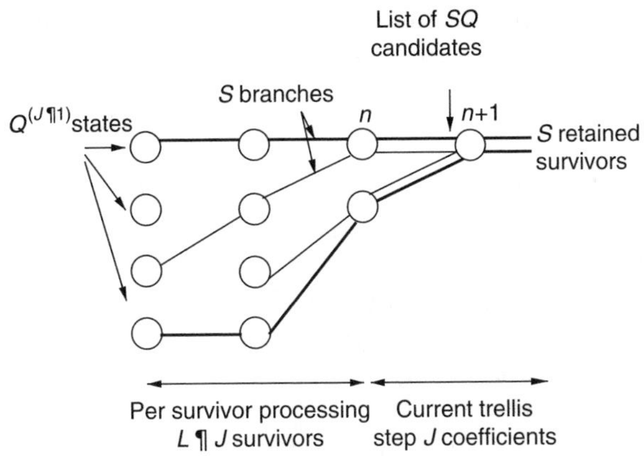

FIGURE 18.4 Illustration of the generalized Viterbi algorithm.

The number of states in the Viterbi algorithm increases exponentially with the channel memory. However, all states do not have the same importance, since the majority of them have low metrics compared to the best one. Thus, one may think that it is reasonable to neglect the survivors having low metrics without discarding the best one [13]. This is the principle of the M-algorithm; at each step all the candidates are stored in a single list and the M best survivors are selected. According to this terminology, the Viterbi algorithm is a multiple-list single-survivor algorithm while the M-algorithm is a single-list multiple-survivor one. In order to improve the performance of these suboptimum versions of the Viterbi algorithm (DDFSE and M-algorithm), Hashimoto proposed in [13] a multiple-list multiple-survivor algorithm referred to as the **generalized Viterbi algorithm (GVA)**. This algorithm combines state reduction and per survivor processing, with many survivors retained at each state (in each list).

In the GVA, as in the DDFSE, the trellis considers a reduced channel memory $J < L$. As shown in Figure 18.4, the generalized Viterbi algorithm can be described as follows:

1. Initial condition: At time n, the algorithm retains for each state S survivors.
2. Path extension: At time $n+1$, the retained paths are extended by one branch yielding new candidates (SQ candidates converge at each state), which will be stored in $Q^{(J-1)}$ lists.
3. Path selection: From each list, the S best candidates are selected as the survivors for the next step.

Hence, the GVA gives a twofold generalization of the Viterbi algorithm:

1. The channel memory considered to construct the trellis is reduced ($J < L$). Thus the trellis has $Q^{(J-1)}$ states instead of $Q^{(L-1)}$ states. The branch metric calculation takes into account the intersymbol and co-channel interference due to the past symbols through the use of per survivor processing.
2. More than one survivor (S survivors) is retained at each state in order to fight against error propagation.

18.4.1.3 List-Type MAP Algorithm

Recently, the iterative processing principle used in turbo coding [4] has gained a lot of attention and has been successfully applied to a wide variety of joint detection and decoding problems [7]. The basic idea behind iterative processing is to exchange extrinsic (soft) information among the receiver modules in order to achieve successively refined performance. Thus, in view of the importance of soft equalization, we are interested in soft output versions of the GVA. Soft equalization can be performed by the MAP equalizer [1], which produces soft decisions on the transmitted symbols by computing their *a posteriori* probabilities (APPs) given the received sequence. A soft output version of the GVA following the idea of the MAP algorithm, called **list-type MAP** algorithm, has been proposed by Penther et al. [20] for SISO channels. In the following, we present a generalization of this soft input/soft output algorithm to MIMO channels [23].

18.4.2 Generalization to the MIMO Case

As explained in [13][17], the trellis has $Q^{(J-1)N}$ states, where J is the reduced memory of the channel ($J < L$), N is the number of transmit antennas, and Q is the constellation size. The first J taps of the channel are processed by the trellis transitions of the list-type equalizer and the remaining taps are processed by per survivor processing with S survivors at each state.

Our objective is to compute the *a posteriori* probabilities $P(d_n^{(i)}|\underline{r})$ for $0 \leq i \leq N-1$ and $0 \leq n \leq T-1$, given the received vector during a burst.

Let $\mu(n) = (\underline{d}_{n-1}^T, \underline{d}_{n-2}^T, \ldots, \underline{d}_{n-J+2}^T, \underline{d}_{n-J+1}^T)$ be the state at time n. As described in [1], the APPs are obtained after calculation of the forward probability $\alpha_n(\mu)$ and the backward probability $\beta_n(\mu)$. Here, the forward probability $\alpha_n^{S_i}(\mu)$ will also depend on the survivors S_i $(i = 1, \ldots, S)$. At time n, a state μ is associated with a survivor list $S_{i,n}$ and a forward probability list $\alpha_n^{S_i}(\mu)$. At time $n+1$, SQ^N branches converge to the following state μ' and their path metrics are defined as $\alpha_n^{S_i}(\mu)\gamma_n^{S_i}(\mu,\mu')$, where $\gamma_n^{S_i}(\mu,\mu')$ is the branch transition probability between states μ and μ' for the survivor S_i. This transition can be expressed as the product of an *a priori* probability and the channel transition probability:

$$\gamma_n^{S_i}(\mu,\mu') = p(\underline{r}_n|\mu,\mu',S_i)P(\mu'|\mu) \tag{18.19}$$

Since the channel noise is white and Gaussian, the channel transition probability is given by

$$p(\underline{r}_n|\mu,\mu',S_i) = \frac{1}{(\pi\sigma^2)^M}\exp\left(-\frac{\left\|\underline{r}_n - \sum_{l=0}^{J-1} H(l)\underline{\tilde{d}}_{n-l} - \sum_{l=J}^{L-1} H(l)\underline{\hat{d}}_{n-l}\right\|^2}{\sigma^2}\right) \tag{18.20}$$

where $\underline{\tilde{d}}_{n-l} = (\tilde{d}_{n-l}^{(0)}, \ldots, \tilde{d}_{n-l}^{(N-1)})^T$ is the vector of elements of the trellis transition and $\underline{\hat{d}}_{n-l} = (\hat{d}_{n-l}^{(0)}, \ldots, \hat{d}_{n-l}^{(N-1)})^T$ is the vector of the symbols estimated during the per survivor processing for the survivor S_i of interest. All symbols are assumed equally likely, so $P(\mu'|\mu) = 1/Q^N$.

The SQ^N path metrics are sorted by order and the S highest will define, for the state μ' at time $n+1$, the survivor list $S_{i',n+1}$ and the forward probabilities $\alpha_{n+1}^{S_{i'}}(\mu')$. Each forward probability $\alpha_{n+1}^{S_{i'}}(\mu')$ is calculated as the sum of all the path metrics leading to μ', which are inferior or equal to the path metric associated with $S_{i'}$ [20]. The backward probability $\beta_n(\mu)$ is calculated recursively using the survivors obtained in the preceding step as

$$\beta_n(\mu) = \sum_{\mu'}\sum_{i=1}^{S}\gamma_n^{S_i}(\mu,\mu')\beta_{n+1}(\mu') \tag{18.21}$$

Let $(m_1, \ldots, m_Q)$ be the constellation points. The equalizer calculates the APP $P(d_n^{(i)} = m_q|\underline{r})$ for each possible symbol $d_n^{(i)} = m_q, q = 1, \ldots, Q$ as

$$P\left(d_n^{(i)} = m_q|\underline{r}\right) = \sum_{f=1}^{Q^{N-1}}\sum_{\mu=1}^{Q^{(J-1)N}}\sum_{j=1}^{S}\alpha_n^{S_j}(\mu)\gamma_n^{S_j}\left(\mu, S_f^{m_q}(\mu)\right)\beta_{n+1}\left(S_f^{m_q}(\mu)\right) \tag{18.22}$$

where $S_f^{m_q}(\mu)$, for $1 \leq f \leq Q^{N-1}$, are the states following the state μ if the input symbol corresponding to the ith transmitting antenna is m_q.

The probability that an information bit $c_k^{(i)}$ is equal to zero is then calculated as

$$P\left(c_k^{(i)} = 0\middle|\underline{r}\right) = \sum_{d_n^{(i)} \in d^0} P\left(d_n^{(i)}\middle|\underline{r}\right) \tag{18.23}$$

where $n = \lfloor \frac{k}{\log_2(Q)} \rfloor$ and d^0 is the set of symbols $d_n^{(i)}$ corresponding to $c_k^{(i)} = 0$.

The probability $P(c_k^{(i)} = 1|\underline{r})$ is obtained as

$$P\left(c_k^{(i)} = 1\middle|\underline{r}\right) = 1 - P\left(c_k^{(i)} = 0\middle|\underline{r}\right) \tag{18.24}$$

The list-type MAP equalizer encompasses the MAP equalizer obtained by letting $J = L$ and $S = 1$, and the soft DDFSE obtained by letting $J < L$ and $S = 1$.

The algorithm can be simplified without substantial degradation in performance by omitting the backward probabilities in Equation 18.22 [20]. In this case, a delay equal to the reduced channel memory J is introduced in the computation of the APP

$$P\left(d_{n-J+1}^{(i)} = m_q\middle|\underline{r}\right) = \sum_{b=1}^{Q^{N-1}} \sum_{\mu=1}^{Q^{(J-1)N}} \sum_{j=1}^{S} \alpha_n^{S_j}\left(S_b^{m_q}(\mu)\right) \gamma_n^{S_j}\left(S_b^{m_q}(\mu), \mu\right) \tag{18.25}$$

where $S_b^{m_q}(\mu)$, for $1 \le b \le Q^{N-1}$, are the states leading to the state μ if the input symbol transmitted by the ith transmit antenna at time $n - J + 1$ is m_q.

18.4.3 Simulation Results

Figure 18.5 and Figure 18.6 depict the bit error rate of the list-type MAP equalizer vs. E_b/N_0 for channels 1, 2, and 3, for $S \in \{1, 2, 8\}$, respectively, when $J = 2$ (the trellis has 4 states) and when $J = 3$ (the trellis has 16 states). Simulation results show that the performance depends on the channel energy profile. Actually, for channels 1 and 2, the performance is bad when $J = 2$ and $J = 3$. The increase of the number of

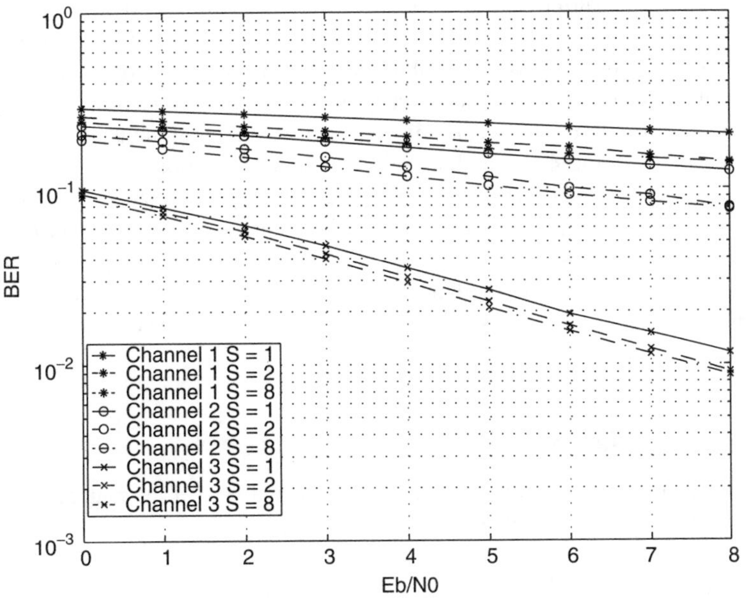

FIGURE 18.5 BER performance of the list-type MAP equalizer for a reduced channel memory $J = 2$ and $S \in \{1, 2, 8\}$.

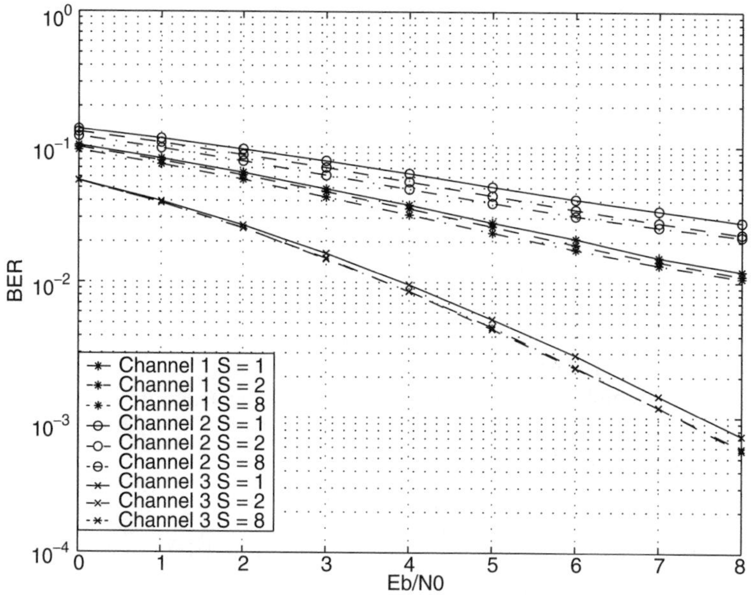

FIGURE 18.6 BER performance of the list-type MAP equalizer for a reduced channel memory $J = 3$ and $S \in \{1, 2, 8\}$.

survivors does not lead to a noticeable improvement of performance. Better performance can be achieved by increasing J, but this would yield an exponential increase of the algorithm complexity.

It is worth mentioning that the performance obtained by using the GVA is almost the same as that obtained by using the list-type MAP equalizer.

18.5 The Multidimensional Whitened Matched Filter

Since the list-type MAP equalizer considers a reduced number of taps of the channel (the first J taps) to construct the trellis, it is desirable to concentrate the channel energy on these taps to reduce error propagation. This can be done by using a multidimensional **whitened matched filter** as a prefilter for the equalizer, analogously to the scalar case. Actually, in the case of a SISO frequency-selective channel, a WMF is used to transform the received signal into a sequence with minimum phase channel response and additive white noise. This procedure is a first step in the implementation of some equalizers, including decision-feedback detectors and delayed decision-feedback sequence estimators [10]. This is achieved by factoring the channel spectrum into a product of a minimum phase filter and its time inverse.

In the following, we will present a solution to design the multidimensional whitened matched filter based on prediction theory results [6][8][9]. It will be shown that the prefiltered MIMO channels have the property of energy concentration, as for single-antenna channels [23]. The WMF is then used as a prefilter for the list-type MAP equalizer.

18.5.1 Whitened Matched Filter

As in the SISO case, the first step is to feed the received signal $\underline{r}(z)$ to the matched filter $H^H(z^{-1}) = \sum_{l=0}^{L-1} H^H(l)z^l$. The output signal is then

$$\underline{x}(z) = H^H(z^{-1})\underline{r}(z) = S(z)\underline{d}(z) + \underline{b}(z) \tag{18.26}$$

where $S(z) = H^H(z^{-1})H(z)$ and $\underline{b}(z) = H^H(z^{-1})\underline{n}(z)$ is a colored noise with spectrum $N_0 S(z)$.

For the multidimensional case, some results have been derived using the linear prediction theory for vector wide-sense stationary (WSS) processes [6][8][9]. An interesting result is stated below.

Theorem: Multidimensional spectral factorization [9]

Given an N-dimensional WSS process $\underline{v}(z)$ with spectrum $S_v(z)$, there exists a factorization

$$S_v(z) = B_-^H(z^{-1})B_-(z) \tag{18.27}$$

such that $B_-(z)$ is causal and stable and has a causal inverse. The filter $B_-^{-1}(z)$ is stable if $S_v(z)$ is nonsingular on the unit circle. This factorization is called the **minimum phase factorization** of $S_v(z)$.

According to this theorem, there exists an $N \times N$ causal and stable matrix filter $B_-(z)$ with causal and stable inverse, which verifies $B_-^H(z^{-1})B_-(z) = H^H(z^{-1})H(z)$ as long as $M \geq N$ (more receivers than transmitters). The filter $(B_-^H(z^{-1}))^{-1}$ is a whitening filter for a process with spectrum $S(z)$. Assuming that $(B_-^H(z^{-1}))^{-1}$ can be perfectly known, and passing $\underline{x}(z)$ through this filter, we have the following as the output of the whitened matched filter:

$$\underline{y}(z) = (B_-^H(z^{-1}))^{-1}\underline{x}(z) = B_-(z)\underline{d}(z) + \underline{n}_1(z) \tag{18.28}$$

where $\underline{n}_1(z)$ is a white Gaussian noise with spectrum $N_0 I_N$, I_N being the $(N \times N)$ identity matrix.

18.5.2 Prefiltered List-Type MAP Equalizer

In [23], we proposed to use the WMF as a prefilter for the list-type MAP equalizer. The list-type MAP then has at its input the vector $\underline{y}$. Thus, it will compute the APP $P(d_n^{(i)}|\underline{y})$ for $0 \leq i \leq N-1$ and $0 \leq n \leq T-1$, knowing the vector $\underline{y}$.

Since the noise at the input of the equalizer is white and Gaussian, the equations derived in Section 18.4.2 can be used here to compute the APP while replacing $\underline{r}$ by $\underline{y}$, H by B_-, and M by N.

18.5.3 WMF Implementation Using Linear Prediction

Several algorithms have been presented in the literature to determine the spectral factors $B_-(z)$ and $B_-^H(z^{-1})$ [18]. In the following, based on prediction theory results [6], we briefly explain how to simply compute an approximation of $(B_-^H(z^{-1}))^{-1}$.

As we will see, the algorithm based on the prediction theory provides the factorization $S(z) = B_-(z)B_-^H(z^{-1})$. Since we want to obtain the spectral factorization given in Equation 18.27, we begin by factoring $S^H(z) = B_1(z)B_1^H(z^{-1})$, then we take $B_-(z) = B_1^H(z)$.

Let $\underline{s}(z)$ denote an N-dimensional WSS process with spectrum $S^H(z) = H^H(z)H(z^{-1})$, $\underline{i}(z)$ its innovations and $L(z) = \sum_{k=0}^{\infty} L(k)z^{-k}$ its innovations filter ($L(z)$ is an $N \times N$ polynomial matrix). At time n, $\underline{s}_n$ can be written as

$$\underline{s}_n = \sum_{k=0}^{\infty} L(k)\underline{i}_{n-k} \tag{18.29}$$

This is called the **Wold decomposition** [6].

Consider the linear predictor $A(z) = \sum_{k=1}^{\infty} A(k)z^{-k}$ of $\underline{s}_n$ ($A(z)$ is an $N \times N$ polynomial matrix). The estimation $\hat{\underline{s}}_n$ of $\underline{s}_n$ in terms of its entire past is then

$$\begin{aligned}\hat{\underline{s}}_n &= \sum_{k=1}^{\infty} A(k)\underline{s}_{n-k} \\ &= \sum_{k=1}^{\infty} L(k)\underline{i}_{n-k}\end{aligned} \tag{18.30}$$

$\hat{\underline{s}}_n$ can be approximated by the estimation of $\underline{s}_n$ in terms of its D most recent past values:

$$\hat{\underline{s}}_n \simeq \sum_{k=1}^{D} A(k)\underline{s}_{n-k} \tag{18.31}$$

The estimation error is given by

$$\underline{e}_n = \underline{s}_n - \hat{\underline{s}}_n = L(0)\underline{i}_n \tag{18.32}$$

The mean square value of the estimation error is minimum if the error $\underline{e}_n$ is orthogonal to $\underline{s}_{n-k}$, for $1 \leq k \leq s D$. This yields the Yule–Walker equations [19]:

$$\begin{cases} E\left(\underline{e}_n \underline{s}_{n-k}^H\right) = 0, \quad 1 \leq k \leq D \\ E\left(\underline{e}_n \underline{s}_n^H\right) = L(0)L(0)^H \end{cases} \tag{18.33}$$

we obtain,

$$\begin{cases} A = (A(1), \ldots, A(D)) = RR_D^{-1} \\ R(0) - \sum_{k=1}^{D} A(k)R(k)^H = L(0)L(0)^H \end{cases} \tag{18.34}$$

In these expressions, R_D is the $DN \times DN$ covariance matrix of the random vector $\underline{s}_n^D = (\underline{s}_n^T, \ldots, \underline{s}_{n-D+1}^T)^T$, $R(i) = E(\underline{s}_k \underline{s}_{k-i}^H)$ and $R = (R(1), \ldots, R(D))$. Equation 18.32 writes, by considering the Z-transform,

$$\underline{s}(z) \simeq A_1(z)^{-1}L(0)\underline{i}(z) \tag{18.35}$$

where $A_1(z) = (I_N - \sum_{k=1}^{D} A(k)z^{-k})$. By comparing Equations 18.29 and 18.35, we deduce that the matrix $A_1(z)^{-1}L(0)$ is an approximation of the filter $L(z)$ given by the Wold decomposition.

From Equation 18.35 and since the spectrum of $\underline{i}(z)$ is the identity matrix, we can write

$$H^H(z)H(z^{-1}) \simeq A_1(z)^{-1}L(0)L(0)^H \left(A_1^H(z^{-1})\right)^{-1} \tag{18.36}$$

By taking the transconjugate of this expression, we obtain

$$H^H(z^{-1})H(z) \simeq A_1(z^{-1})^{-1}L(0)L(0)^H \left(A_1^H(z)\right)^{-1} \tag{18.37}$$

Thus, an approximation of the matrix $(\underline{B}^H(z^{-1}))^{-1}$ is given by $L(0)^{-1}A_1(z^{-1})$. The implementation of $(\underline{B}^H(z^{-1}))^{-1}$ is given by solving Equation 18.34. The autocorrelation matrices $R(i)$ are obtained by identifying the terms in z^{-i} in the equality

$$H^H(z)H(z^{-1}) = \sum_{i=-\infty}^{+\infty} R(i)z^{-i} \tag{18.38}$$

We choose arbitrarily $L(0)$ as a lower triangular matrix verifying Equation 18.34, since we do not care which minimum phase factor is considered.

18.5.4 Energy Concentration

Now we want to verify that the prefiltered MIMO channels have the property of energy concentration, as for SISO channels. We start by recalling the property for the SISO case [19].

Lemma: energy concentration for SISO channels [19]

If B_- is an impulse response of a minimum phase filter, and H a filter having the same spectrum, then for any n_0,

$$\sum_{i=0}^{n_0} |B_-(i)|^2 \geq \sum_{i=0}^{n_0} |H(i)|^2 \tag{18.39}$$

It has been shown in [6] that the property of energy concentration stated above holds for minimum phase MIMO channels obtained by spectral factorization for $n_0 = 0$.

Theorem [6]

Let $\underline{v}$ be an N-dimensional stochastic process, $B_-(z)$ an $N \times N$ filter matrix corresponding to the Wold decomposition of $\underline{v}$, and $H(z)$ a spectral factor of the matrix spectrum of $\underline{v}$ such as $B_-^H(z^{-1})B_-(z) = H^H(z^{-1})H(z)$, then

$$B_-(0)^H B_-(0) \geq H(0)^H H(0) \tag{18.40}$$

To our best knowledge, there is no generalization of this theorem when $n_0 \geq 1$. In the sequel, however, we will verify through simulations that this property still holds for orders $n_0 \geq 1$. In order to measure the gain in terms of energy on the first taps that can be obtained by using the WMF, we consider the following quantities:

$$gain_0(H) = \frac{Trace(B_-(0)^H B_-(0) - H(0)^H H(0))}{Trace\left(\sum_{i=0}^{L-1} H(i)^H H(i)\right)} \tag{18.41}$$

$$gain_{n_0}(H) = \frac{Trace(\sum_{i=0}^{n_0}(B_-(i)^H B_-(i) - H(i)^H H(i)))}{Trace\left(\sum_{i=0}^{L-1} H(i)^H H(i)\right)} \tag{18.42}$$

The measure $gain_{n_0}(H)$ represents the difference between the total energy of the n_0 first taps of $B_-(z)$ and $H(z)$. It can be easily shown from Equations 18.34 and 18.38 that the matrix $B_-(z)=(B_-^H(z^{-1}))^{-1}H^H(z^{-1})H(z)$ verifies $B_-(z) = \sum_{l=0}^{L-1} B_-(l)z^{-l}$ and $B_-(l) = 0$, for $l > L-1$. Thus, $gain_{n_0}(H) = 0$, for $n_0 \geq L-1$. Table 18.1 shows the mean and standard deviation of the measure $gain_0$ over 1000 realizations for the three frequency-selective channels described in Section 18.2.2.

Channel 1 has the highest gain since the powers of its delayed paths are larger than those of its direct path. Second in terms of gain, channel 2 is quite severe because each path has the same averaged power. Third, channel 3 is close to the minimum phase.

Table 18.2 shows the mean and standard deviation of $gain_{n_0}(H)$ when $n_0 = 1$ and $n_0 = 2$ for the same channels.

For the three channels, Table 18.2 shows that the energy concentration property is verified when $n_0 = 1$ and $n_0 = 2$. We notice that for channel 3 the mean and standard deviation of the different gains are very close. Hence, the gains can be very small for a few realizations. In order to achieve a good trade-off between complexity and performance, a test can be performed on the calculated gains for each realization of the

TABLE 18.1 The Mean and Standard Deviation of $gain_0(H)$ for the Three Channels Presented in Section 18.2.2

	Channel 1	Channel 2	Channel 3
Mean of $gain_0$	0.540	0.369	0.159
Standard deviation of $gain_0$	0.114	0.107	0.121

TABLE 18.2 The Mean and Standard Deviation of $gain_{n_0}$ (H) When $n_0 = 1$ and $n_0 = 2$ for the Three Channels Presented in Section 18.2.2

	Channel 1	Channel 2	Channel 3
Mean of $gain_1$	0.542	0.312	0.089
Standard deviation of $gain_1$	0.138	0.118	0.081
Mean of $gain_2$	0.219	0.238	0.0213
Standard deviation of $gain_2$	0.094	0.104	0.0196

channel. If $gain_{n_0}(H)$ is less than a determined threshold, the received signals are not prefiltered by the WMF before the equalization, since the improvement will be very little. Otherwise, the WMF is used.

18.5.5 Simulation Results

In this section, we give the simulation results for the list-type MAP equalizer prefiltered by the WMF. The calculation of the WMF is performed by using a predictor of degree $D = 10$. Simulations show that this value of D is sufficient to obtain an accurate spectral factorization.

Figure 18.7 and Figure 18.8 show the performance of the list-type MAP equalizer for the three channels of interest with and without prefiltering. Figure 18.7 shows the performance when $J = 2$ (the trellis has four states) and $S = 2$. As expected from Section 18.5.4, the channel gaining most improvement by the use of the WMF is channel 1 followed by channel 2 and then channel 3. Figure 18.8 shows the equalizer performance for $J = 3$ (16 states) and $S = 2$. In this case, the prefiltering yields a gain even if the reduced memory is higher ($J = 3$). This gain is less important than that obtained when $J = 2$, but remains significant. We also plot the BER performance of the MAP equalizer (256 states). For channel 1, when the WMF is used, for BER $= 10^{-2}$, the performance is within 2.5 dB of that of the MAP equalizer when $J = 2$ and 1 dB when $J = 3$. The equalizer presented here allows approach of the MAP equalizer performance while using a reduced-complexity algorithm.

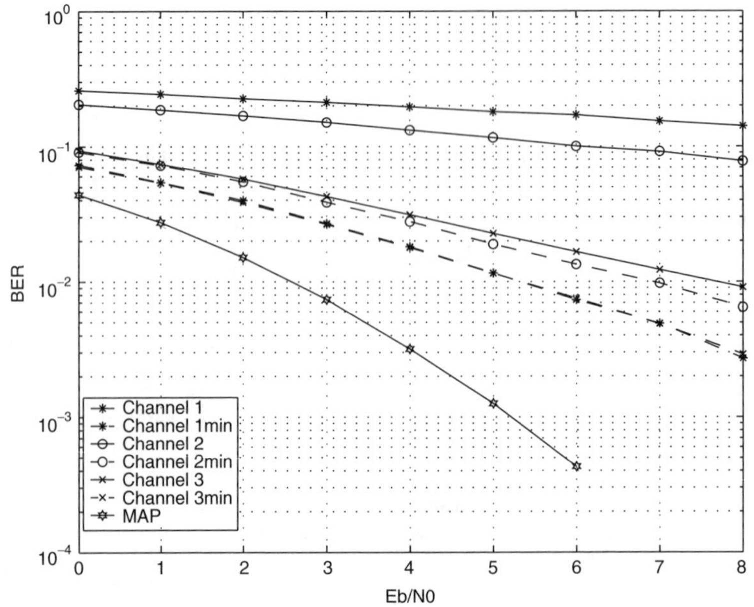

FIGURE 18.7 BER performance of the list-type MAP equalizer with and without the prefilter for a reduced channel memory $J = 2$ and a number of survivors $S = 2$. The dotted curves correspond to a receiver with a prefilter.

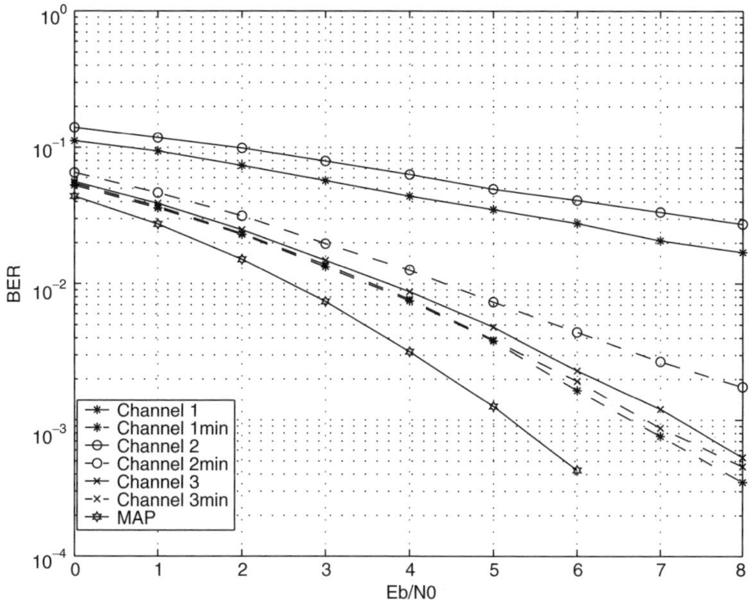

FIGURE 18.8 BER performance of the list-type MAP equalizer with and without the prefilter for a reduced channel memory $J = 3$ and a number of survivors $S = 2$. The dotted curves correspond to a receiver with a prefilter.

Remark The performance of the block equalizers presented in Section 18.3 depends on the components of the matrix $G^H G$ (see Equations 18.7 and 18.11). The components of this matrix do not change for the minimum phase channel obtained by using the WMF since the spectrum of the channel is equal to the spectrum of its minimum phase version. Thus, there is no point in applying the WMF before these equalizers.

18.6 The Block Equalizers vs. the Prefiltered List-Type MAP

Now we propose to draw a comparison between the block equalizers presented in Section 18.3 and the prefiltered list-type MAP equalizer based on their complexities and performances. We assume here that all of the parameters of the transmission are set except the size T of a block of symbols.

In [16], the computation of the complexities of the four block equalizers (ZF-BLE, MMSE-BLE, ZF-BDFE, and MMSE-BDFE) has been provided. It has been shown that these equalizers have similar complexities. Since MMSE-BDFE achieves the best performance, we will be interested in the following in this equalizer, and we will compare it to the list-type MAP equalizer.

18.6.1 Complexity Comparison

In order to evaluate the complexities, we consider the number of complex multiplications made by each algorithm. Table 18.3 gives the number of multiplications required for the MMSE-BDFE. The different steps of the algorithm are quite simple due to the structure of the channel matrix G given in Equation 18.4, which is block Toeplitz and contains many zero blocks [21].

Table 18.4 gives the number of multiplications required for the list-type MAP equalizer. We consider here the simplified version of the algorithm obtained by omitting the backward probabilities in Equation 18.22 (see Section 18.4.2).

TABLE 18.3 Number of Multiplications Required for MMSE-BDFE

Step	Number of Multiplications
$G^H \underline{r}$	$NTLM$
$J = G^H G + I_{NT}\sigma^2$	$\frac{L(L-1)}{2} N^2 M$
Cholesky decomposition of J	$\frac{TL^2 N^3}{4}$
$\underline{r}' = \Sigma'^{-1}((\Sigma'\Psi')^H)^{-1} G^H \underline{r}$	$(LN-1)N(T-L) + 6NT$
$\underline{r}' - (\Psi' - I_{NT})\hat{d}_{MMSE-BDFE}$	$N^2 LT$

TABLE 18.4 Number of Multiplications Required for the List-Type MAP Equalizer

Step	Complexity
Computation of γ	$2TQ^{JN}N^2(S(L-J)+J)$
α−recursion	$TQ^{JN}S$

Figure 18.9 Number of multiplications required for MMSE-BDFE and the List-type MAP equalizer for different values of T. We set the following parameters: $L=5$, $J=3$, $S=2$, $N=M=2$, and $Q=2$ (BPSK). We notice that the list-type MAP equalizer is much more complex to implement than the MMSE-BDFE.

18.6.2 Performance Comparison

Figure 18.10 compares the performance of MMSE-BDFE with that of the prefiltered list-type MAP for $J = 3$ and $S = 2$ for the three channels described in Section 18.2.2. For channels 1 and 3, the performance

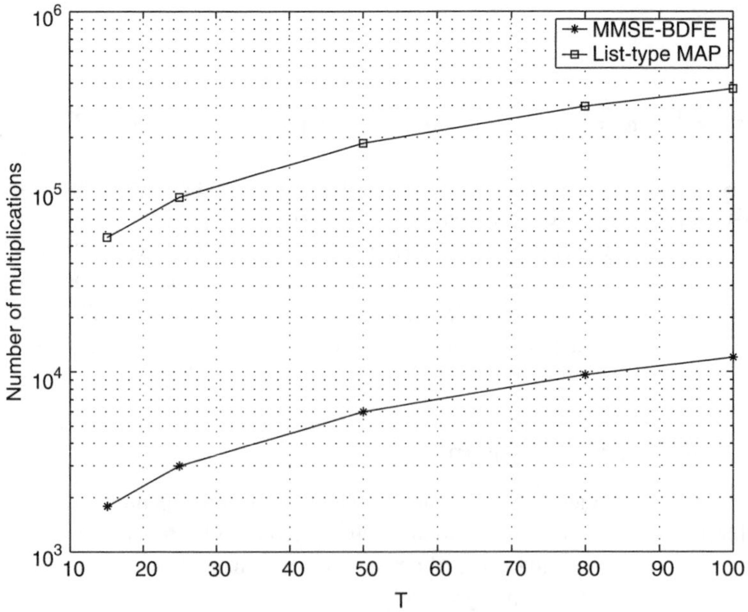

FIGURE 18.9 Number of multiplications required for MMSE-BDFE and the prefiltered list-type MAP equalizer for different values of the size T of a block of symbols transmitted by one transmit antenna.

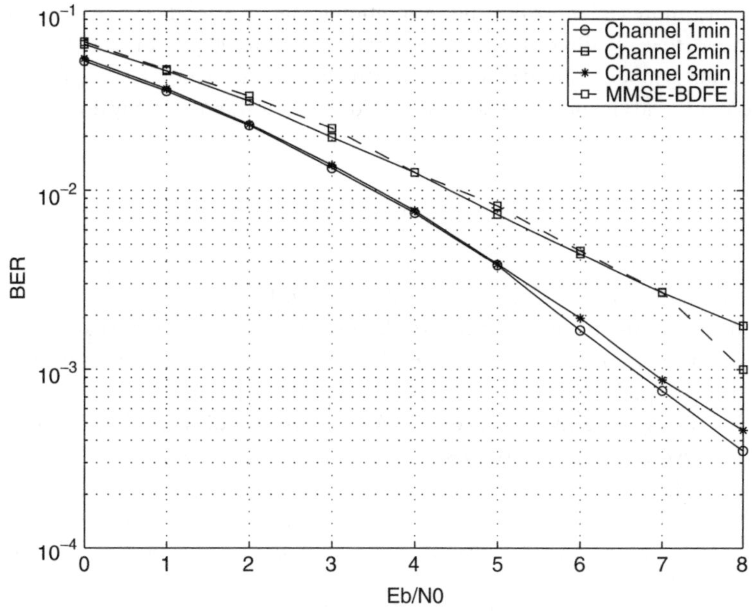

FIGURE 18.10 Comparison between the performance of the MMSE-BDFE equalizer and that of the prefiltered list-type MAP equalizer for the three channels of interest.

of the prefiltered list-type MAP is better than that of MMSE-BDFE. For BER= 10^{-2}, the gain is 1 dB. For channel 2, the performances of both equalizers are almost identical. We can conclude that in most cases, the prefiltered list-type MAP equalizer achieves better performance than MMSE-BDFE.

18.7 Conclusion

In this chapter, we considered the problem of equalization of a frequency-selective MIMO channel. The receiver has to combat both intersymbol interference and spatial interference (between antennas). Since optimum equalizers using MAP (forward–backward) or the Viterbi algorithm are too complex to be implemented, we considered reduced-complexity solutions. We presented two possible suboptimum solutions:

1. Block linear and decision-feedback equalizers based on ZF or MMSE criteria proposed first for CDMA block transmission systems [14] [15].
2. List-type equalizers, which are suboptimum versions of MAP or the Viterbi, algorithm, based on state reduction, to reduce the complexity, and per survivor processing, retaining many survivors at each state to fight against error propagation. In order to improve the performance of this equalizer, a whitened matched filter can be used to concentrate the channel energy on its first taps.

Then we drew a comparison between the complexities and performances of these two types of equalizers. This comparison showed that the prefiltered list-type MAP equalizer is more complex but achieves better performance than MMSE-BDFE. Moreover, it is more suitable for iterative processing where detection and channel decoding are iterated several times to improve performance [7]. Actually, as the MAP equalizer, the list-type MAP equalizer can easily use the APP provided by the channel decoder as *a priori*, contrary to MMSE-BDFE. This is an important feature since iterative turbo processing is a good solution to recover the performance loss due to the use of a suboptimum equalizer [2][22][23].

In this chapter, we assumed that the channel is perfectly known to the receiver. This is obviously not the case in practice. In wireless communications systems, the receiver considers a channel estimation based on training sequences. In general, as the number of subchannels in the MIMO channel increases, channel estimation becomes more and more difficult. A solution to obtain a good channel estimate is to use long training sequences. However, this causes a decrease in the data throughput. In order to reduce the amount of required training symbols, joint data detection and channel estimation techniques can be used [24]. In this case, decoding and channel estimation are performed jointly and already decoded symbols help to improve the channel estimation.

References

[1] L.R. Bahl, J. Cocke, F. Jelinek, and J. Raviv, "Optimal decoding of linear codes for minimizing symbol error rate," *IEEE Trans. Inf. Theory*, IT-32, 284–287, 1974.

[2] G. Bauch and N. Al-Dhahir, "Iterative equalization and decoding with channel shortening filters for space-time coded modulation," in *IEEE Vehicular Technology Conference (VTC)*, Boston, MA, September 2000, pp. 1575–1582.

[3] G. Bauch, A.F. Naguib, and N. Seshadri, "MAP equalization of space-time coded signals over frequency selective channels," in *Wireless Communications and Networking Conference*, September 1999.

[4] C. Berrou, A. Glavieux, and P. Thitimajshima, "Near Shannon limit error-correcting coding and decoding: turbo-codes," in *IEEE International Conference Communications*, Geneva, Switzerland, May 1993, pp. 1064–1070.

[5] H. Bölcskei and A.J. Paulraj, "Multiple-input multiple-output (MIMO) wireless systems," chapter in *The Communications Handbook*, 2nd ed., J. Gibson, Ed., CRC Press, Boca Raton, FL, 2002, pp. 90.1–90.14.

[6] P.E. Caines, *Linear Stochastic Systems*, Wiley, New York, 1988.

[7] C. Douillard, M. Jézéquel, C. Berrou, A. Picart, P. Didier, and A. Glavieux, "Iterative correction of intersymbol interference: turbo-equalization," *Eur. Trans. Telecommun.*, 6(5), 507–511, 1995.

[8] A. Duel-Hallen, "Equalizers for multiple input/multiple output channels and PAM systems with cyclostationary input sequences," *IEEE J. Selected Areas Commun.*, 10(3), 630–639, 1992.

[9] A. Duel-Hallen, "A family of multiuser decision-feedback detectors for asynchronous code-division multiple-access channels," *IEEE Trans. Commun.*, 43(2/3/4), 421–433, 1995.

[10] A. Duel-Hallen and C. Heegard, "Delayed decision-feedback sequence estimation," *IEEE Trans. Commun.*, 37(5), 428–436, 1989.

[11] G.J. Foschini, "Layered space-time architecture for wireless communication in a fading environment when using multiple antennas," *Bell Labs Tech. J.*, 1(2), 41–59, 1996.

[12] G.J. Foschini and M.J. Gans, "On limits of wireless communications in a fading environment when using multiple antennas," *Wireless Personal Commun.*, 6(3), 311–335, 1998.

[13] T. Hashimoto, "A list-type reduced-constraint generalization of the Viterbi algorithm," *IEEE Trans. Inf. Theory*, IT-33, 866–876, 1987.

[14] P. Jung and J. Blanz, "Joint detection with coherent receiver antenna diversity in CDMA mobile radio systems," *IEEE Trans. Veh. Technol.*, 44(1), 1995.

[15] G. Kawas Kaleh, "Channel equalization for block transmission systems," *IEEE J. Selected Areas Commun.*, 13(1), 110–121, 1995.

[16] E.L. Kuan and L. Hanzo, "Joint detection CDMA techniques for third-generation transceivers," in *Proceedings of ACTS Mobile Communication Summit 98*, Rhodes, Greece, June 1998.

[17] H. Kubo, K. Murakami, and T. Fujino, "A list-output Viterbi equalizer with two kinds of metric criteria," *IEEE International Conference on Universal Personal Communications*, October 1998, pp. 1209–1213.

[18] V. Kucera, "Factorization of rational spectral matrices: a survey of methods," in *International Conference on Control*, Vol. 2, 1991, pp. 1074–1078.

[19] A. Papoulis, *Signal Analysis*, McGraw-Hill, New York, 1977.

[20] B. Penther, D. Castelain, and H. Kubo, "A modified turbo-detector for long delay spread channels," in *International Symposium on Turbo-Codes*, Brest, France, 2000.

[21] Y. Pigeonnat, "Joint detection for UMTS: complexity and alternative solutions," in *IEEE Vehicular Technology Conference (VTC)*, Vol. 1, July 1999, pp. 546–550.

[22] A. Roumy, I. Fijalkow, and D. Pirez, "Turbo multiuser detection for coded asynchronous DS-CDMA over frequency selective channels," *J. on Commun. Networks*, 3(3), 202–210, 2001.

[23] N. Sellami, I. Fijalkow, and M. Siala, "Low-complexity iterative receiver for space-time coded signals over frequency selective channels, " *EURASIP J. Appl. Signal Process.*, 517–524, 2002.

[24] N. Sellami, S. Lasaulce, and I. Fijalkow, "Turbo channel estimation for coded DS-CDMA systems over frequency selective channels," *Signal Processing Advances in Wireless Communications (SPAWC '03)*, Rome, Italy, June 2003.

[25] V. Tarokh, N. Seshadri, and A.R. Calderbank, "Space-time codes for high data rate wireless communication: performance criterion and code construction," *IEEE Trans. Inf. Theory*, 44(2), 744–765, 1998.

[26] V. Tarokh, H. Jafarkhani, and A.R. Calderbank, "Space-time block codes from orthogonal designs," *IEEE Trans. Inf. Theory*, 45(5), 1456–1467, 1999.

[27] A.J. Viterbi and J.K. Omura, *Principles of Digital Communication and Coding*, McGraw-Hill, New York, 1979.

19

Neural Networks for Transmission over Nonlinear Channels

19.1 Introduction 19-1
19.2 Identification of Memoryless Nonlinear Amplifiers ... 19-4
Natural Gradient Learning • Influence of the HPA Modeling Error on the Symbol-Error-Rate Performance • Application to Adaptive Predistortion
19.3 Modeling and Identification of Nonlinear Channels with Memory 19-15
Neural Network Channel Identification • Learning Algorithm • Application to MLSE Receiver Design • Simulation Examples
19.4 Channel Equalization 19-21
Neural Network Structure • BP Algorithm • NG Algorithm • Simulation Examples
19.5 Conclusion 19-24

Mohamed Ibnkahla
Queen's University

Jun Yuan
Queen's University

Rober Boutros
Queen's University

Abstract

This chapter applies neural networks (NNs) to transmission over nonlinear channels. We present applications to satellite communication channels composed of a nonlinear high-power amplifier (HPA) followed by a linear filter. The applications include adaptive predistortion, HPA modeling, maximum likelihood sequence estimation (MLSE) receiver design, and adaptive equalization. We show, in particular, that the natural gradient (NG) descent method used for NN training outperforms the classical ordinary gradient descent-based backpropagation (BP) algorithm in terms of convergence speed, mean square error (MSE), and bit error rate (BER) performance.

19.1 Introduction

Large-scale deployment of broadband multimedia communications is expected to allow users, regardless of their geographic location, to connect with the information, people, and services they need, to better compete in the international marketplace and to have equal access to leading-edge health care, education, government services, interactive multimedia, Internet applications, etc. Exploring ways to facilitate delivery of high-speed applications and services via satellites becomes an urgent and real challenge. In this regard, satellites will play an important role in future fourth-generation (4-G) mobile communications systems.

0-8493-1657-X/$0.00+$1.50

Significant research efforts are being made worldwide in the field of satellite mobile communications in order to meet the 4-G quality-of-service (QoS) requirements. These requirements include high-data-rate transmissions and very low bit error rates (BERs).

Satellite channels, that can be considered as nonlinear dynamical systems, face two major challenges: nonlinear distortions caused by the use of onboard nonlinear amplifiers (such as traveling wave tube (TWT) amplifiers and solid-state power amplifiers (SSPAs)), and multipath fading caused by the downlink propagation channel [5]. These problems have to be mitigated in order to achieve high transmission rates and acceptable BER performance. This chapter shows how the use of neural networks can be an important step in achieving this goal. Here we focus on channel identification and equalization issues.

Channel identification [26] allows estimating and characterizing of the communication channel. The resulting channel model may be used for several purposes [5], e.g., receiver design, performance evaluation of the communication channel, power control, adaptive modulation, etc. Several techniques have been proposed in the literature for nonlinear channel identification. Most of these techniques are based on parameterized nonlinear models such as Wiener–Hammerstein models, Volterra series, wavelet networks, neural networks, etc. The parameter estimation can be performed using nonadaptive techniques such as least squares methods and higher-order statistics-based methods, and adaptive techniques such as adaptive gradient learning [5, 12, 14, 15, 26, 31, 34, 37, 38].

In the area of nonlinear system identification, neural networks have been widely used as powerful modeling tools for nonlinear dynamical systems [14, 20, 28]. This has been motivated by the universal approximation property of neural networks [13, 16]. Indeed, several feedforward neural network (NN) structures have been shown to be universal approximators; i.e., they can approximate any continuous real-valued function on a bounded subset of R^n (where R is the set of real numbers) [4, 13, 14, 16]. These results deal with static or memoryless mappings. Most work on dynamical system identification has been empirical in nature, and the choice of the NN structure depends closely on the specificities of the application [20, 32]. For example, Narenda and Parthasarathy [28, 29] have used several multilayer perceptron (MLP) structures (trained with static and dynamic back propagation (BP) methods) and nonlinear autoregressive moving average (NARMA) representations for identification and control of dynamical systems with applications to robotics, autonomous navigation, aeronautics, etc.

A general structure capable of approximating input–output mappings of nonlinear discrete-time systems was introduced in [38]. The structure is composed of two stages: a dynamical stage, followed by a memoryless nonlinear stage. A necessary and sufficient condition was given for a large set of structures of this form to be capable of approximating a wide class of nonlinear discrete-time systems. A review of the uniform approximation property of dynamic networks created by a cascade of linear filters and static nonlinearities is presented in [32]. To further understand the concept of optimal signal processing, the author in [32] integrates concepts from function approximation, linear and nonlinear regression, dynamic modeling, and delay operators.

The BP algorithm [14, 35], which is used to train MLP networks, has two major drawbacks: First, its convergence is slow due to the existence of plateau regions in its cost function, which can be inadequate for on-line training. Second, the NN parameters may be trapped in a nonoptimal local minimum, leading to suboptimal approximation of the system [14, 15]. Natural gradient (NG) descent learning [2, 3], on the other hand, has been shown to have better convergence capabilities than the classical BP algorithm because it takes into account the geometry of the manifold in which the NN weights evolve. Therefore, NG learning can better avoid the plateau phenomena, which characterize the BP learning curves [21]. This chapter will make a comparative study between the two approaches.

Adaptive equalizers are used to reduce the channel disturbances such as intersymbol interferences (ISIs), fading, noise, nonlinear distortions, co-channel and adjacent channel interferences, etc. In future 4-G communications, the challenge is more difficult as high data rates and low BERs will be required, and consequently, the design of efficient and fast methods for communication channel equalization will become an absolute necessity.

Neural networks have been proposed in the literature as powerful nonlinear equalizers [14]. Various structures, such as multilayer perceptrons [14, 17], radial basis function (RBF) networks [4, 9], and self-organizing maps (SOMs) [24] have been employed. For communication channels using two-dimensional signaling schemes, such as quadrature amplitude modulation (QAM) and phase shift keying (PSK) modulation, complex-valued NN equalizers have been proposed, which allows direct adaptive processing of complex-valued signals [14]. For highly nonlinear channel equalization, recurrent NNs have been employed because they can exhibit a rich and complex highly nonlinear dynamical behavior [30].

MLP structures are very popular since they can perform highly nonlinear mappings with memory. Several MLP structures have been proposed in the literature. For example, the MLP can be fully connected or it can be composed of separable blocks, e.g., a linear block with memory, followed by a memoryless nonlinearity [18, 19]. The choice of the structure depends on the particular equalization problem (i.e., channel memory, type of nonlinearity, whether the nonlinear and linear memory parts are separable, etc.) [1, 19, 20, 32].

Neural networks have the capability to be combined with other techniques such as fuzzy systems, genetic algorithms, and linear adaptive filtering [14, 20, 32]. This is due to their flexible structure and capability of self-organization and learning. In the area of channel equalization, NNs can be combined with other (classical) equalizers by performing partial equalization tasks, such as decision process, inverting a nonlinearity, tracking variable parameters in a channel, etc. For example, in this chapter a memoryless MLP structure is combined with the classical LMS equalizer. The task of the LMS equalizer is to mitigate ISI, while the MLP structure aims to reduce the channel nonlinear distortions.

Another typical example concerns self-organizing maps, which can be easily combined with conventional equalizers such as the LMS algorithm or the decision feedback equalizer (DFE) [8, 20, 25]. The SOM is particularly useful when two-dimensional signaling systems with high-order constellations (such as M-QAM) are employed. This can be intuitively understood when we know, for example, that a 16-QAM constellation can be represented by a 4×4 SOM. The capability of SOMs to self-organize allows them to be excellent candidates for blind, semiblind, or supervised equalization of time-varying wireless channels [24, 25].

DFEs have emerged in the last two decades as an important class of equalizers, and have shown excellent results for the equalization of channels with high ISI [33]. To further enhance the performance of DFEs, combined MLP-DFE structures have been proposed in the literature, which outperformed the LMS-DFE structure [40]. A lattice filter structure used as a whitening process has been employed for nonlinear channel equalization [41]. This structure results in a substantial improvement in terms of convergence rate, steady-state MSE, and BER, compared to the classical MLP-DFE.

Maximum likelihood sequence estimation (MLSE) receivers have been extensively studied in the literature [5]. This chapter proposes a new MLSE receiver based on NN channel estimation. The chapter compares the proposed NN MLSE to NN equalizers trained with the NG and BP algorithms, and to the ideal MLSE receiver (that assumes perfect channel knowledge).

The study of the convergence properties of neural network algorithms is very important, as it allows understanding the behavior of NNs and therefore gives ideas into how to improve their performances. Several NN structures and algorithms have been statistically analyzed. For example, Bershad et al. analyze in [6] gradient descent identification of a nonlinear system with memory composed of a filter followed by a nonlinearity. The adaptive system is composed of a filter followed with a memoryless nonlinearity, assuming the knowledge of the system nonlinearity to within a parameter factor. In particular, the study shows that the adaptive filter converges to a scaled version of the unknown system filter. In [7], the same authors study stochastic gradient tracking of time-varying polynomial Wiener systems. They show how the adaptive variables vary through time, as functions of the variations of the unknown system. In [15] the behavior of several linear adaptive algorithms is analyzed. A statistical analysis of neural network identification of a nonlinear dynamical system (composed of a filter followed by a memoryless nonlinearity) is presented in [18]. The neural network is composed of a linear adaptive filter followed by a two-layer memoryless neural network, and the BP algorithm is used for training. The study confirms that the adaptive filter converges to a scaled version of the unknown system filter.

The chapter is organized as follows. The following section is devoted to NN memoryless channel identification with applications to M-QAM transmission over nonlinear memoryless channels and predistortion of nonlinear amplifiers. Section 19.3 investigates NN modeling of nonlinear channels with memory, with application to MLSE receiver design. Section 19.4 applies neural networks to channel equalization. Finally, comparison between different NN equalizers and MLSE receivers are given in Section 19.5.

19.2 Identification of Memoryless Nonlinear Amplifiers

High-power amplifiers (HPAs) used in satellite communications behave, in general, as nonlinear memoryless channels as their output depends on the current input [5, 36, 39].

Let the HPA input signal be expressed as $x(n) = r(n)e^{j\phi_0(n)}$; then its output can be expressed as

$$y(n) = f(x(n)) = A(r(n))e^{j[\phi(r(n))+\phi_0(n)]} \tag{19.1}$$

where $A(r)$ and $\phi(r)$ are called amplitude-to-amplitude (AM/AM) and amplitude-to-phase (AM/PM) conversions, respectively.

The amplifier gain is defined as

$$G(r) = \frac{A(r)}{r}$$

The well-known Saleh [36] parameterized analytical formulas are examples of classical HPA models, in which the AM/AM and AM/PM conversions are modeled, respectively, as

$$A(r) = \frac{\alpha_a r}{1+\beta_a r^2} \qquad \phi(r) = \frac{\alpha_P r^2}{1+\beta_P r^2}$$

Parameters α_a, β_a, α_P, and β_P are chosen in order to fit with the amplifier input–output measured data. For example, they can take the following values for a specific TWT amplifier: $\alpha_a = 2$, $\beta_a = 1$, $\alpha_P = 4$, and $\beta_P = 9$.

The back-off is defined as the ratio between the amplifier input saturation power (P_{sat}) and the input signal power (P_{in}):

$$BO(dB) = 10\log\left(\frac{P_{sat}}{P_{in}}\right) \tag{19.2}$$

Efficient analytical HPA models are needed for several reasons, such as implementing digital communication channel simulators, analyzing and evaluating the performance of satellite communication links, and studying the effect of nonlinearity on system behavior (e.g., spectral regrowth, co-channel interference, ISI, etc.) [5]. Therefore, it is important to obtain efficient analytical models to accurately approximate the HPA physical behavior [5, 36].

This section employs multilayer neural networks to model HPA amplifiers using NG descent for the learning process. The main advantage of NN models [20] over classical models is that they offer better MSE approximation performance than classical HPA models. They are adaptive, which makes them appropriate for on-line modeling. Classical models are based on off-line optimization procedures, which do not allow on-line modeling. Moreover, NNs can model a variety of HPAs using the same parametrized structure (e.g., a structure with five neurons); only the weight values change from one HPA model to another. On the contrary, most classical models have been designed for specific families of HPAs (e.g., HPAs having a given asymptotic behavior). Therefore, they may not be appropriate for new generations of HPAs used in modern communications systems.

In this section we present the neural network structure and NG algorithm. We then analyze the influence of the HPA modeling error on the symbol error rate (SER) performance. Finally, we present an application on satellite channel predistortion.

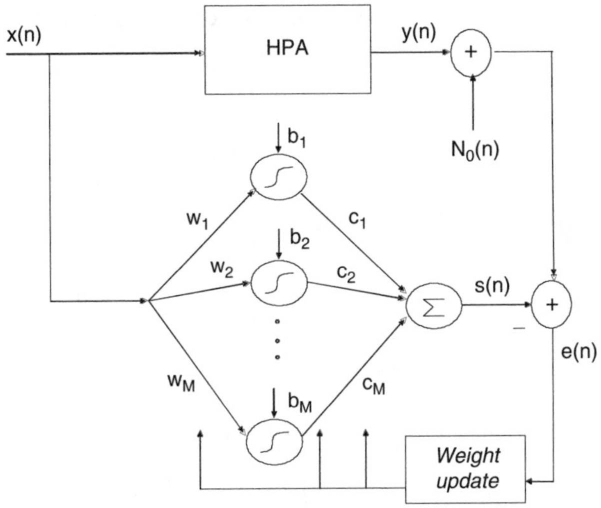

FIGURE 19.1 NN adaptive modeling of a nonlinear HPA.

19.2.1 Natural Gradient Learning

Figure 19.1 shows the modeling scheme of an HPA characteristic using a two-layer neural network. The NN has a scalar input, M neurons in the first layer, and a scalar output. A set of input–output data (collected from the physical HPA measurements, or from the satellite channel input–output signals) is presented to the NN. The network output at time n is expressed as

$$s(n) = \sum_{k=1}^{M} c_k g(w_k x(n) + b_k) \tag{19.3}$$

where $x(n)$ is the input sample at time n and g is the activation function (which is taken in this chapter as the hyperbolic tangent function). $\{w_k\}$, $\{b_k\}$, and $\{c_k\}$, $k = 1, \ldots, M$, are the NN weights. We denote by θ the vector containing the network weights: $\theta = [w_1, \ldots, w_M, b_1, \ldots, b_M, c_1, \ldots, c_M]^t$, where $()^t$ denotes the transpose.

To perform the modeling task, the NN weights are updated using a learning algorithm that minimizes the squared error between the HPA output and the NN output. The cost function is expressed as

$$J(n) = \frac{1}{2}e^2(n)$$

where

$$e(n) = y(n) + N_0(n) - s(n)$$

$y(n)$ is the HPA output and $N_0(n)$ is a zero-mean white Gaussian noise with variance σ^2.

The learning algorithm minimizes the cost function by following the steepest descent direction. If the space of parameters is an orthonormal coordinate system, then the steepest descent corresponds to the ordinary gradient. It has been shown that in the case of multilayer NNs, the steepest descent direction (or the NG) of the loss function is actually given by [2, 3]

$$-\tilde{\nabla}_\theta J(n) = -G^{-1}\nabla_\theta J(n)$$

where ∇_θ is the (ordinary) gradient with respect to θ. In our case, the gradient is expressed as

$$\nabla_\theta(l(n)) = -e(n)\cdot\nabla_\theta s(n) = -e(n)\cdot\begin{pmatrix} x(n)c_1 g'(w_1x(n)+b_1) \\ \vdots \\ x(n)c_M g'(w_Mx(n)+b_M) \\ g(w_1x(n)+b_1) \\ \vdots \\ g(w_Mx(n)+b_M) \\ c_1g'(w_1x(n)+b_1) \\ \vdots \\ c_Mg'(w_Mx(n)+b_M) \end{pmatrix} \tag{19.4}$$

and G^{-1} is the inverse of the Fisher information matrix (FIM) [2]:

$$G = [g_{i,j}(\theta)] = \left[E\left(\frac{\partial J(n)}{\partial\theta_i}\frac{\partial J(n)}{\partial\theta_j}\right)\right] \tag{19.5}$$

Thus, the natural gradient learning algorithm adjusts the neural network weights as

$$\theta(n+1) = \theta(n) - \mu G^{-1}\nabla_\theta(J(n)) \tag{19.6}$$

where μ is a small positive constant.

The calculation of the expectations in the FIM requires the knowledge of the probability distribution functions (PDFs) of x and s, which are not always available. Moreover, the calculation of the inverse of the FIM is computationally very costly. A Kalman filter technique has been proposed in [3] for an on-line estimation of the FIM inverse:

$$\hat{G}^{-1}(n+1) = \frac{1}{1-\varepsilon_n}\hat{G}^{-1}(n) - \frac{\varepsilon_n}{1-\varepsilon_n}\frac{\hat{G}^{-1}(n)\nabla_\theta s(n)(\nabla_\theta s(n))^t\hat{G}^{-1}(n)}{(1-\varepsilon_n)+\varepsilon_n(\nabla_\theta s(n))^t\hat{G}^{-1}(n)\nabla_\theta s(n)} \tag{19.7}$$

where $\nabla_\theta s(n)$ is the ordinary gradient of s (see Equation 19.4).

This equation involves an updating rate ε_n. When ε_n is small (which will be assumed in this chapter), we may approximate Equation 19.7 by a simpler one [3]:

$$\hat{G}^{-1}(n+1) = (1+\varepsilon_n)\,\hat{G}^{-1}(n) - \varepsilon_n\hat{G}^{-1}(n)\nabla_\theta s(n)(\nabla_\theta s(n))^t\hat{G}^{-1}(n) \tag{19.8}$$

When the updating rate $\varepsilon_n = c_\varepsilon/n$ (where c_ε is a small constant), then Kalman filtering is equivalent to an on-line calculation of the arithmetic mean. When ε_n is a constant, the convergence is faster than the previous case, but the algorithm may be less stable. In order to reach a good trade-off between convergence speed and stability, we will use a search-and-converge schedule in which ε_n is defined by

$$\varepsilon_n = \frac{\varepsilon_0 + \frac{c_\varepsilon n}{\tau}}{1 + \frac{c_\varepsilon n}{\tau\varepsilon_0} + \frac{n^2}{\tau}} \tag{19.9}$$

where τ is a positive time constant and ε_0 is a small positive constant.

Here, small n corresponds to a search phase (ε_n is close to ε_0) and large n corresponds to a converge phase ($\varepsilon_n$is equivalent to c_ε/n for large n).

Simulation Example

A neural network composed of $M = 5$ neurons is used to model the AM/AM characteristic of a TWT amplifier. The NG and BP algorithms are compared. The simulations show that the value of $\mu = 0.005$ represents a good trade-off (for each algorithm) between convergence speed and stability. Figure 19.2a shows the learning curves for both algorithms; it can be seen that the NG descent has a much faster convergence. The resulting mean square error (MSE) approximation performances are 9 10^{-6} and

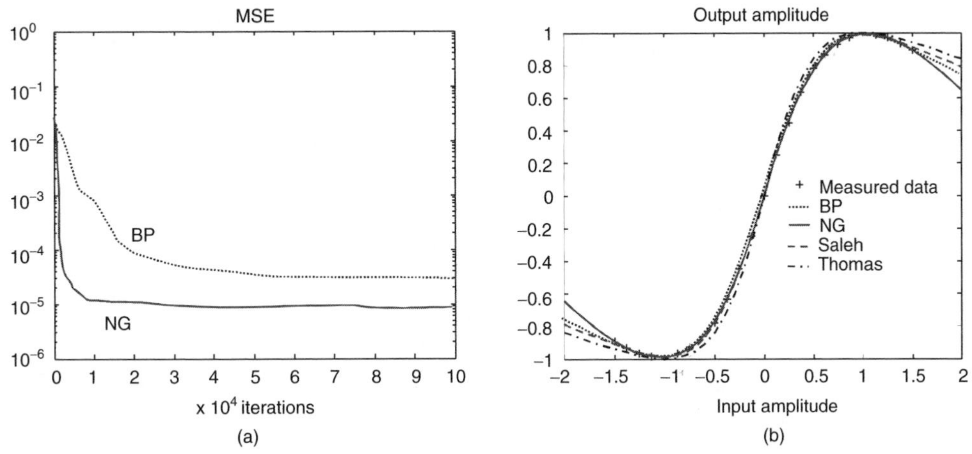

FIGURE 19.2 Modeling the AM/AM conversion of Intelsat IV TWT amplifier: learning curves of the BP and NG algorithm (a); and comparison between the difference models (b).

3.0 10^{-5} for the NG and BP, respectively. Figure 19.2b illustrates the resulting AM/AM models and the original TWT input–output data that has been used for learning. Note that for this amplifier, classical Saleh [36] and Thomas et al. [39] models give 1.4 10^{-4} and 2.0 10^{-4} MSE errors, respectively. Therefore, the NG approach outperforms classical models and the BP algorithm.

This significant MSE improvement performed by the NG approach can have an important impact on the satellite system performance. To illustrate this, the following section studies the effect of the modeling error on the communication channel symbol-error-rate performance.

19.2.2 Influence of the HPA Modeling Error on the Symbol-Error-Rate Performance

As we have mentioned above, HPA analytical models are mainly used to analyze and design nonlinear communication channels. Therefore, it is important to see how the modeling error will affect the channel performance. We consider here a simplified satellite channel composed of an HPA; this will allow us to focus on the effect of the HPA modeling error.

The transmitted signal $x(n)$ is M-QAM modulated ($x(n) \in \{x_i(n), i = 1, \ldots, M\}$). The received baseband signal can be written as

$$y(n) = f(x(n)) + N_0(n)$$

where $N_0(n)$ is an additive white Gaussian noise and $f(.)$ represents the exact nonlinear HPA transfer function:

$$f(x(n)) = A(|x(n)|)e^{j[\phi(|x(n)|)+\phi_0(n)]} \tag{19.10}$$

The receiver is taken as a maximum likelihood detector [5], which detects symbol x_j if it satisfies

$$x_j = Arg(\underset{i}{Min}\,|y(n) - f_{CM}(x_i)|)$$

where $f_{CM}(.)$is the channel model (estimated at the receiver).

As we have seen in Section 19.2.1, the channel model is subject to a modeling error (ME). Therefore, we can write

$$f_{CM}(x_i(n)) = (1 + \Delta_i)A(|x_i(n)|)e^{j[\phi(|x_i(n)|)(1+\delta_i)+\phi_0(n)]} \tag{19.11}$$

where Δ_i and δ_i represent the modeling errors on the amplitude and phase, respectively (Figure 19.3).

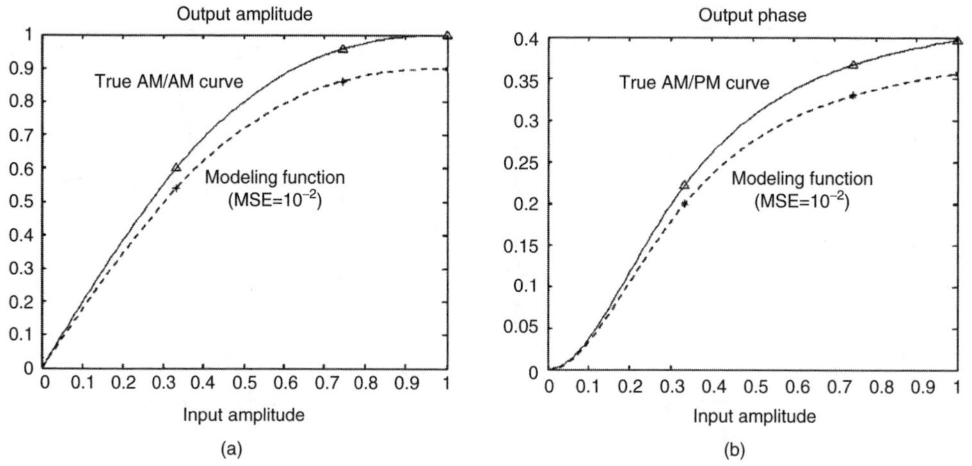

FIGURE 19.3 Illustration of the modeling error on the AM/AM and AM/PM curves (positive axis), the points represent the effect on the 3 different amplitudes of a 16-QAM constellation.

Figure 19.3 illustrates the effect of the ME on the AM/AM and AM/PM conversions for a 16-QAM input signal (note that this constellation has three different amplitudes). Figure 19.4 illustrates the effect on the decision boundaries of a 16-QAM constellation. When there is an error on the amplitude, the decision boundaries get contracted or expanded with respect to the original constellation (Figure 19.4a), depending on the error sign. When the error affects the phase, the decision boundaries are rotated (Figure 19.4b). Finally, when the error affects both amplitude and phase, the decision boundaries are contracted (or expanded) and rotated (Figure 19.4c).

For simplicity, we will assume the same constant error Δ (respectively δ) that affects the amplitudes (respectively phases) of the 16-QAM symbols (the same approach can easily be derived for different Δ_i and δ_i values). We will use Craig's method [10] to evaluate the exact symbol error probability. To do so, we need to integrate over all subregions of every possible signal point in the constellation (see Figure 19.5

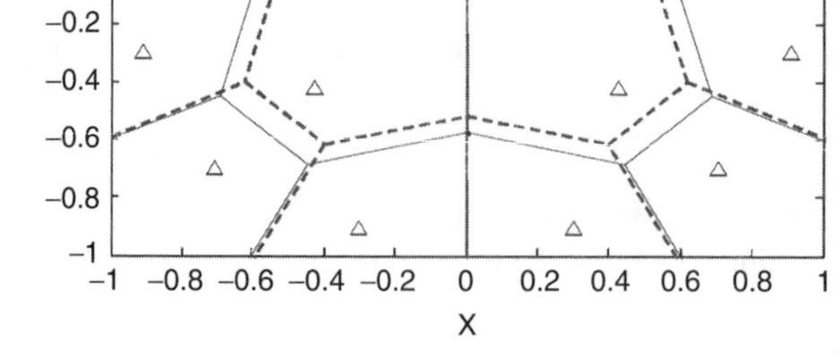

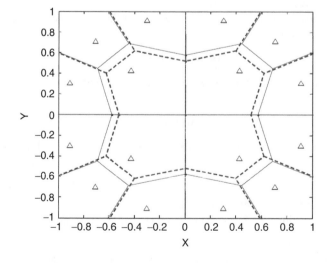

FIGURE 19.4a 16-QAM receiver decision boundaries for the true function (solid line), and the model with an error on the amplitude (dashed line). (Only the exact symbols were plotted.)

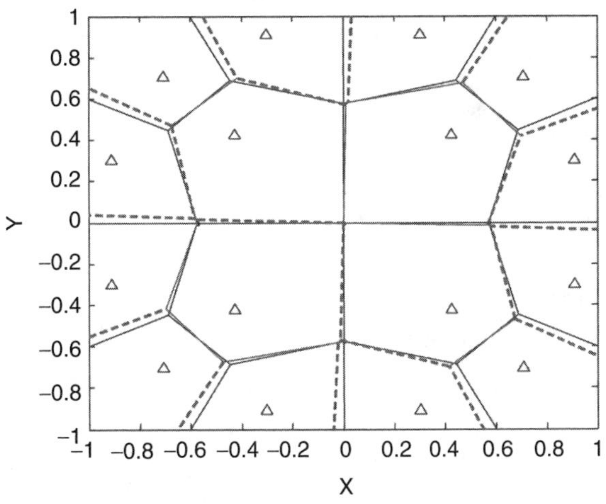

FIGURE 19.4b 16-QAM receiver decision boundaries for the true function (solid line), and the model with an error on the phase (dashed line).

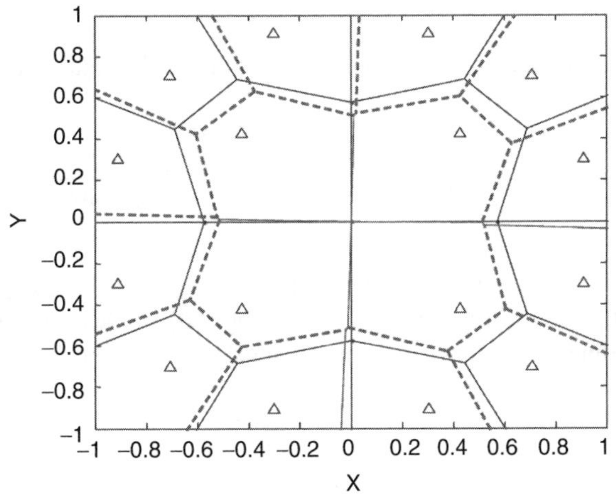

FIGURE 19.4c 16-QAM receiver decision boundaries for the true function (solid line) and the model with errors on both amplitude and phase (dashed line).

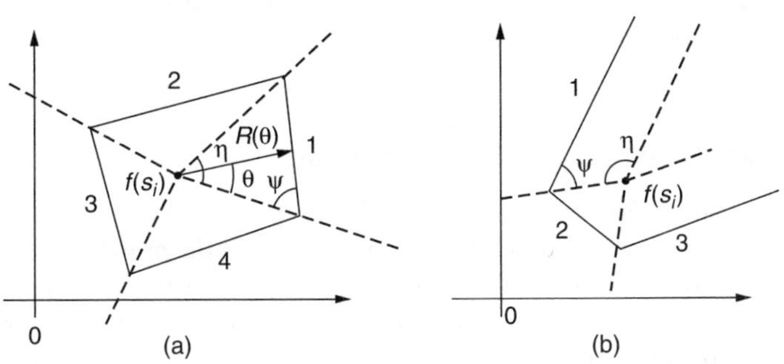

FIGURE 19.5 Two types of decision regions: closed regions (a) and open regions (b).

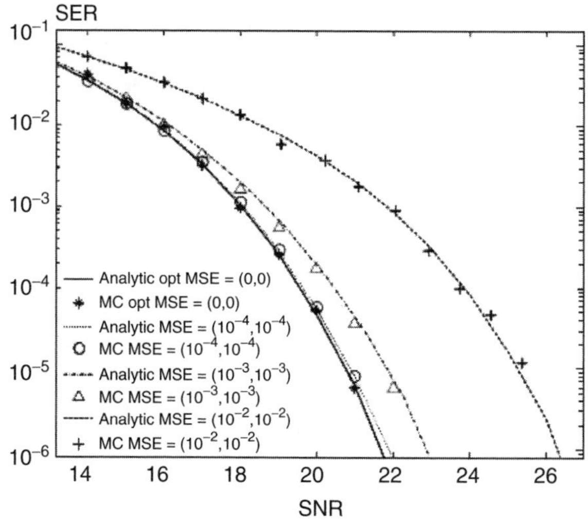

FIGURE 19.6 Craig's approximation and MC simulations for ME in AM&PM (MSE=(a,b), a for the amplitude modeling, b for the phase modeling).

for the cases of open and closed regions). The error probability can be evaluated as

$$P(e|\Delta,\delta) = \sum_{i=1}^{M} P(e|x_i,\Delta,\delta)P(x_i) = \sum_{i=1}^{M}\sum_{j=1}^{G_i} \frac{P(x_i)}{2\pi} \int_{0}^{\eta_{i,j}} \exp\left\{-\frac{b_{i,j}\gamma_s \sin^2\psi_{i,j}}{\sin^2(\theta+\psi_{i,j})}\right\} d\theta \tag{19.12}$$

where $P(x_i)$ is the prior probability of symbol x_i; $b_{i,j}$, $\psi_{i,j}$, and $\eta_{i,j}$ are the geometrical parameters corresponding to symbol x_i and its corresponding subregion j (which are determined by the input constellation and modeling errors Δ and δ); γ_s is the signal-to-noise ratio (SNR) per symbol; and G_i is the total number of subregions corresponding to symbol x_i.

Simulation Example

Figure 19.6 compares the analytical SER given by Equation 19.3 to Monte Carlo (MC) estimations for different values of Δ and δ. The figure shows a good fit between analytical and MC results. The curves show that a modeling error can highly affect the channel SER performance. For example, for an SER of 10^{-4}, an MSE of 10^{-3} introduces a 1 dB loss in the SNR (with respect to the ideal zero-error modeling case). This loss increases as the MSE increases. These results were expected, since the error, as we have seen above, introduces significant changes to the optimal decision boundaries.

19.2.3 Application to Adaptive Predistortion

In this section we consider a nonlinear memoryless channel composed of a TWT amplifier. Predistortion [11, 22] allows the linearization of nonlinear channels so that the overall transfer function (of the predistorter followed by the HPA) is linear (Figure 19.7). A popular method used in predistortion is the Volterra series, where the HPA response is usually approximated by a truncated power series and then a predistortion circuit is built to cancel the effect of nonlinearity up to the order of the power series. The coefficients are usually complex [22, 23]. Lookup table predistortion is starting to receive great attention, and this is mainly because of the fast development of digital signal processors [27].

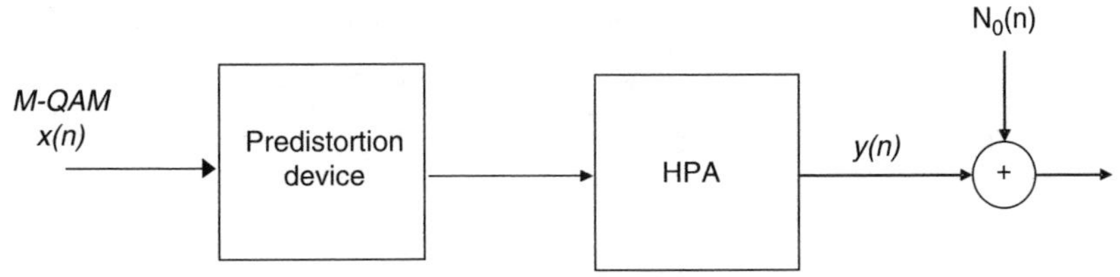

FIGURE 19.7 Principle of predistortion.

In this chapter, we use adaptive neural networks to build our predistorter. The predistorter is designed in three steps:

1. Two neural networks are used to adaptively model the AM/AM and AM/PM conversions of the TWT (as explained in Section 19.1).
2. Another NN is used to adaptively approximate the inverse of the AM/AM conversion (Figure 19.8). This can be done by directly inverting the TWT AM/AM characteristic or by inverting the AM/AM model (Figure 19.8a).
3. Finally, the NN phase model (obtained in step 1) together with the NN AM/AM inverse (obtained in step 2) constitute the predistorter, as depicted in Figure 19.8b. We compensate for the phase by subtracting the value of the phase given by the NN AM/PM model.

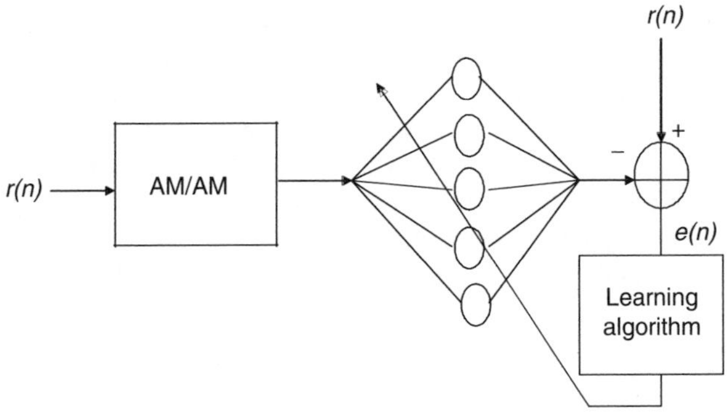

FIGURE 19.8a Inverting the AM/AM conversion.

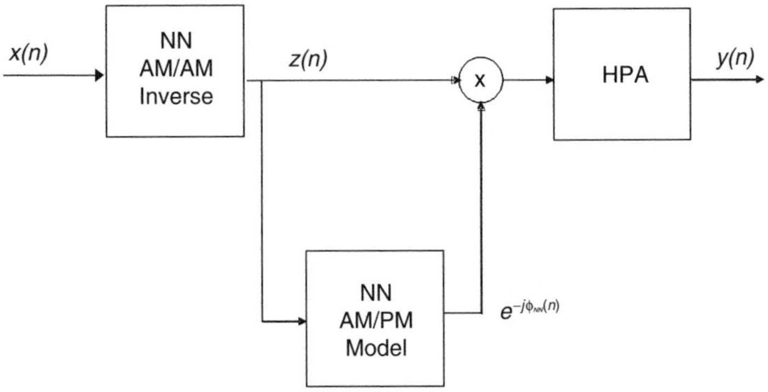

FIGURE 19.8b Overall predistortion scheme.

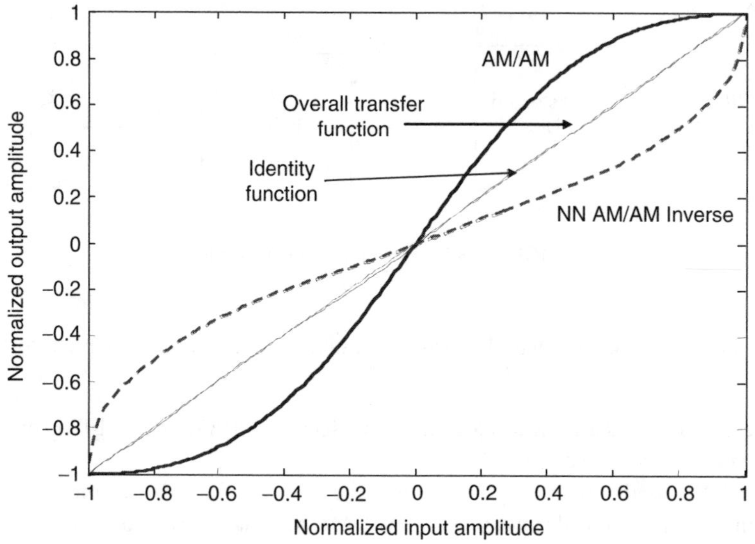

FIGURE 19.9 NN AM/AM inverse, overall transfer function, and comparison with the ideal identity function.

Simulation Example

The following simulations used a known 256-QAM sequence to train the different NNs. During the learning process, the HPA was operated at its maximal efficiency (i.e., the back-off was chosen such that the highest amplitude in the constellation corresponded to the HPA saturation point). The NG descent was used to train the different NN structures, which were composed of five neurons for AM/AM and AM/PM conversion modeling and eight neurons for the AM/AM inverse approximation. A total of 150,000 iterations were needed by the NG algorithm to achieve an MSE of 3.6 10^{-5} with respect to the ideal inverse. Note that the BP algorithm achieves an MSE of 10^{-3} after 750,000 iterations. Because of this poor performance, the BP will not be included in the BER comparisons. Figure 19.9 shows the HPA AM/AM conversion, the NG NN AM/AM inverse, and the overall predistorter + channel amplitude conversion. The latter is very close to the identity function.

Figure 19.10 shows the 256-QAM transmitted constellation (Figure 19.10a), output constellation without predistortion, which is distorted especially at the edges (Figure 19.10b), and predistorter + channel

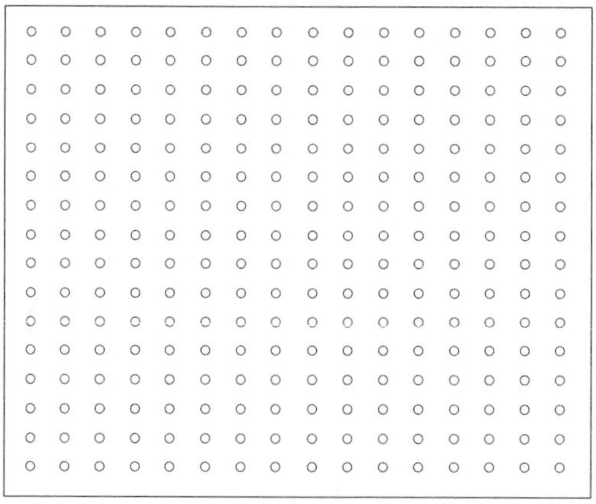

(a)

FIGURE 19.10a 256-QAM input constellation.

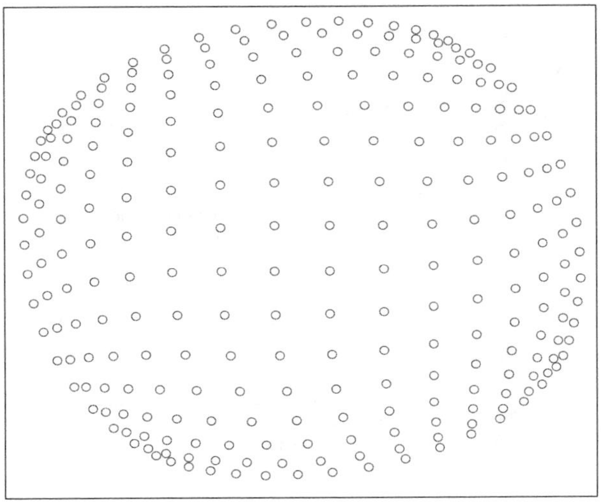

FIGURE 19.10b HPA output constellation (without predistortion).

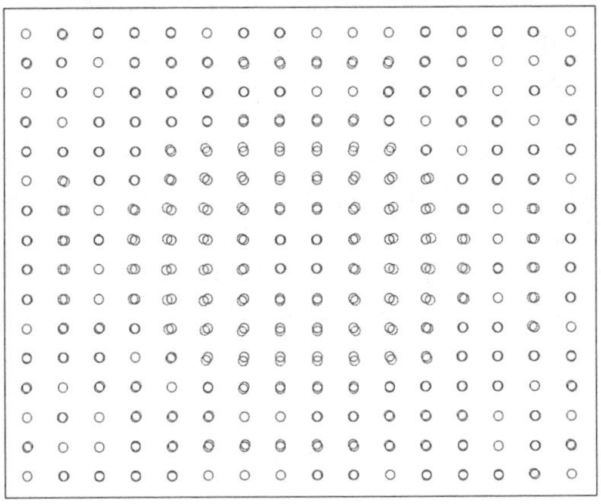

FIGURE 19.10c Superposition of the 256-QAM ideal transmitted constellation and that obtained after predistortion: The difference between the two is minor.

constellation output (Figure 19.10c), which is very close to the transmitted constellation. This means that both amplitude and phase nonlinear distortions have been almost totally mitigated.

The predistorter + channel SER probability for 256-QAM modulation is given in Figure 19.11a, the HPA being used at its maximal efficiency. The SER of the channel without predistortion (using an optimal maximum likelihood detector with respect to the distorted constellation) is also given for comparison. The performance of the NG NN nonlinear channel + predistorter is very close to that of the linear channel, which is confirmed by Figure 19.10c. Note that the Volterra approach performs very poorly.

The total degradation (TD) is another way to evaluate predistorter performance. It is defined by

$$TD(dB) = BO(dB) + \Delta(dB) \tag{19.13}$$

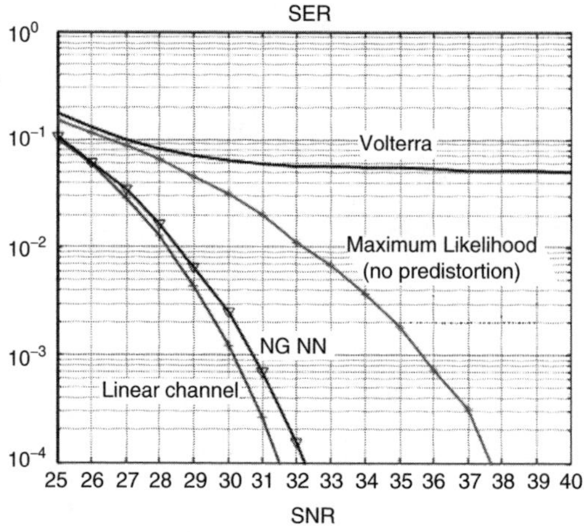

FIGURE 19.11a Symbol error rate vs. SNR for 256-QAM ($BO = 4.22$ dB).

where the total degradation, *TD*, is defined as the sum of the amplifier input back-off (BO) and the increment Δ in the SNR required to maintain a given SER (which will be taken here equal to 10^{-4}) with respect to the linear channel. Since increasing the back-off will reduce the channel nonlinear behavior and at the same time will decrease the power efficiency of the amplifier, then for each predistortion technique we can obtain the minimal $TD_{\min}$. The lower $TD_{\min}$ is, the more efficient a particular predistortion technique is. Figure 19.11b compares the TD performance of the NG algorithm to that of the Volterra series. The value of $TD_{\min}$ for the NG predistorter is 0.85 dB obtained at the lowest possible BO performed by the 256 constellation (here it is 4.22 dB). For the Volterra approach $TD_{\min} = 3.83$ dB obtained at a 7.32-dB backoff. The NG approach allows a gain of 3.85 dB compared to Volterra. Note that when the BO is greater than 6.6 dB, Volterra gives a lower TD than the NG approach. This is expected because the NG learning process is performed when the amplifier is operated at its maximal efficiency

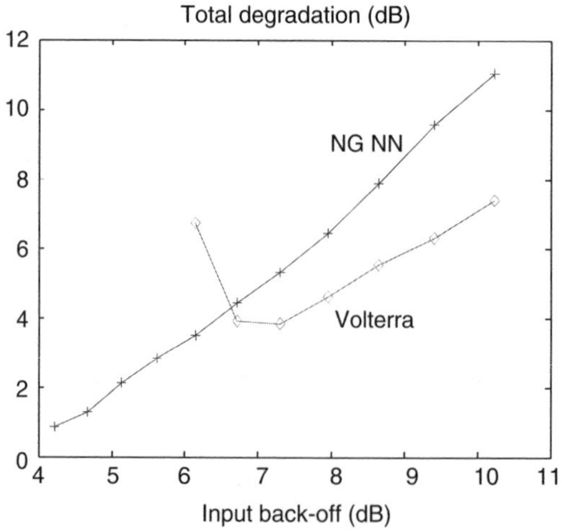

FIGURE 19.11b Total degradation: comparison between Volterra approach and NG NN approach.

(BO = 4.22 dB); therefore, the inverse function is better approximated near the saturation region than in the linear region.

19.3 Modeling and Identification of Nonlinear Channels with Memory

The channel to be considered in the remainder of the chapter is composed of a memoryless nonlinearity followed by a finite impulse response (FIR) filter H. This corresponds to a downlink (satellite or terrestrial) wireless channel, where the nonlinearity represents the HPA and the filter represents the multipath propagation channel (Figure 19.12).

The channel output can be expressed as

$$y(n) = \sum_{k=0}^{N_H-1} h_k f(x(n-k)) + N_0(n) \tag{19.14}$$

where $f(.)$ is the nonlinear function of the HPA, $H(z) = \sum_{k=0}^{N_H-1} h_k z^{-k}$ is the linear filter, and N_0 is an additive white Gaussian noise.

In this section, we will present a neural network approach to identify both nonlinearity and propagation channel. Then, in order to mitigate the nonlinear distortions and ISI effects, we propose an MLSE receiver based on the NN channel model.

19.3.1 Neural Network Channel Identification

The neural network channel identification scheme is composed of a memoryless NN followed by an adaptive linear filter Q (Figure 19.12). The NN aims to identify the HPA transfer function, whereas the adaptive filter Q aims to identify the unknown propagation channel (i.e., filter H).

The memoryless NN consists of two subnetworks called NNG and NNP (Figure 19.12), each of them has M (real-valued) neurons in the first layer and a scalar output. NNG and NNP are expected to model and identify the amplifier gain and phase conversions, respectively.

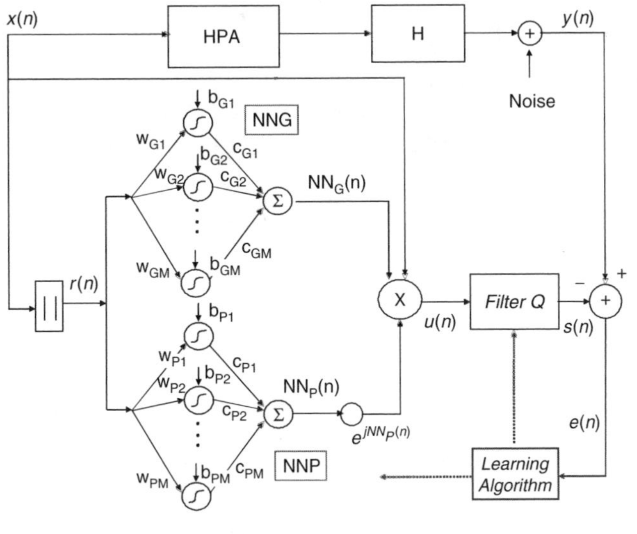

FIGURE 19.12 Identification scheme of a nonlinear channel with memory.

The filter-memoryless neural network structure has been shown to outperform fully connected complex-valued multilayer neural networks with memory when applied to satellite channel identification (see, e.g., [19, 20]).

The two subnetworks have the same input, which is the amplitude of the transmitted symbol ($r(n) = |x(n)|$) in the case of supervised learning, i.e., training sequence (TS) mode.

The output of the neural network is expressed as

$$u(n) = x(n)NN_G(r(n))e^{jNN_P(r(n))} \tag{19.15}$$

where

$$NN_G(r(n)) = \sum_{i=1}^{M} c_{G_i} g(w_{G_i} r(n) + b_{G_i}) \quad \text{(NNG output)} \tag{19.16}$$

$$NN_P(r(n)) = \sum_{i=1}^{M} c_{P_i} g(w_{P_i} r(n) + b_{P_i}) \quad \text{(NNP output)} \tag{19.17}$$

$\{w_{G_i}, c_{G_i}, b_{G_i}\}$ and $\{w_{P_i}, c_{P_i}, b_{P_i}\}$ are the weights of subnetworks NNG and NNP, respectively.

Finally, Q is an adaptive FIR filter $Q(z) = \sum_{k=0}^{N_Q-1} q_k z^{-k}$, its output is given by

$$s(n) = \sum_{k=0}^{N_Q-1} q_k u(n-k) \tag{19.18}$$

The system parameter vector will be denoted by θ, which includes all parameters to be updated, i.e., the weights of subnetworks NNG and NNP and filter Q:

$$\theta = [w_{G1}, \ldots, w_{GM}, b_{G1}, \ldots, b_{GM}, c_{G1}, \ldots, c_{GM}, w_{P1}, \ldots, w_{PM}, b_{P1}, \ldots, b_{PM}, \\ c_{P1}, \ldots, c_{PM}, q_0, \ldots, q_{N_Q-1}]^t$$

19.3.2 Learning Algorithm

The neural network is used to identify the channel by supervised learning. At each iteration, a pair of channel input–output signals is presented to the NN. The NN parameters are then updated in order to minimize the squared error $J(n)$ between the channel output and the NN output:

$$J(n) = \frac{1}{2}\|e(n)\|^2 = \frac{1}{2}\left[e_R^2(n) + e_I^2(n)\right]$$

where

$$e(n) = y(n) - s(n) = e_R(n) + je_I(n)$$

Indices R and I refer to the real and imaginary parts, respectively.

As discussed in Section 19.2, the steepest descent is given by the natural gradient:

$$-\tilde{\nabla}_{\theta(n)} J(n) = -G^{-1}\nabla_{\theta(n)} J(n) \tag{19.19}$$

where G^{-1} is the inverse of the Fisher information matrix.

The neural network weights are updated as follows:

$$\theta(n+1) = \theta(n) - \mu \tilde{\nabla}_{\theta(n)} J(n) \tag{19.20}$$

$$\tilde{\nabla}_{\theta(n)} J(n) = G^{-1}(n) \nabla_{\theta(n)} J(n) \tag{19.21}$$

G^{-1}can be approximated on-line using Kalman processing (as in Section 19.2):

$$\hat{G}^{-1}(n+1) = (1+\varepsilon_n)\hat{G}^{-1}(n) - \varepsilon_n \hat{G}^{-1}(n) \nabla_\theta s(n) (\nabla_\theta s(n))^t \hat{G}^{-1}(n) \tag{19.22}$$

Since we deal here with a complex-valued output, the ordinary gradient is written as

$$\nabla_{\theta(n)} J(n) = e_R(n) \nabla_{\theta(n)} e_R(n) + e_I(n) \nabla_{\theta(n)} e_I(n) \tag{19.23}$$

where $\nabla_{\theta(n)} J(n)$ represents the ordinary gradient of $J(n)$ with respect to θ (see the Appendix).

19.3.3 Application to MLSE Receiver Design

The above identification structure is used at the receiver side as a neural network channel estimator (NNCE), as illustrated in Figure 19.13. A Viterbi decoder uses the channel model provided by the NNCE to estimate the transmitted sequence. It is expected that the better the accuracy of the NNCE, the better the performance of the Viterbi decoder. There are two training modes for the NNCE. The first is the training sequence (TS) mode, which uses a known input sequence to train the NNCE. The second is the decision-directed (DD) mode, which uses the estimated sequence (after detection) for training when a TS is not available.

19.3.4 Simulation Examples

This section presents computer simulations to illustrate the performance of the NN channel identification scheme. The modulation scheme is 16-QAM and the amplifier is used at its maximal efficiency.

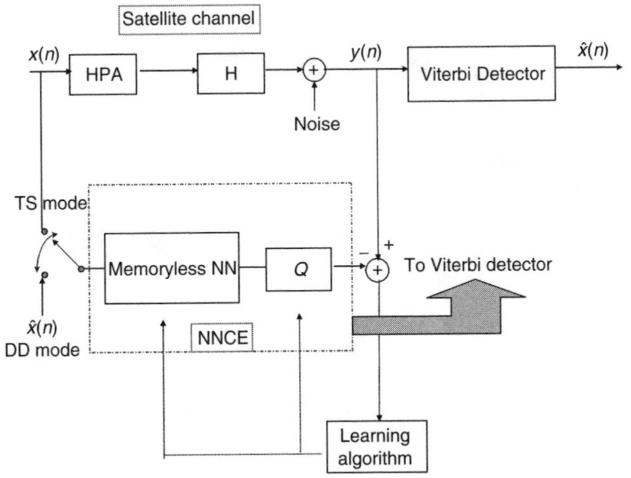

FIGURE 19.13 Neural network MLSE receiver.

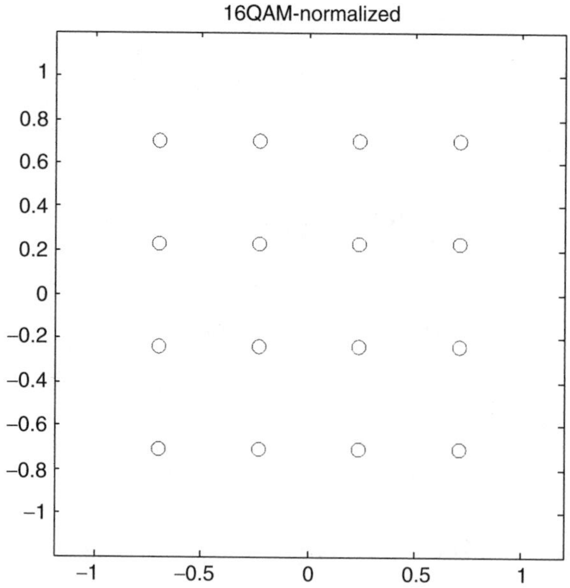

FIGURE 19.14a Transmitted 16-QAM constellation.

Figure 19.14 illustrates the effect of the satellite channel on rectangular 16-QAM constellation. The transmitted constellation is illustrated in Figure 19.14a. Figure 19.14b shows the output constellation when $H(Z) = 1$; i.e., the signal is affected only by the HAP nonlinearity and additive noise. It can be seen that the constellation is rotated because of the phase conversion, and symbols are closer to each other because of the amplitude nonlinearity. The effects of ISIs caused by the delayed path, in addition to the amplifier nonlinear distortions and additive noise, are illustrated in Figure 19.14c.

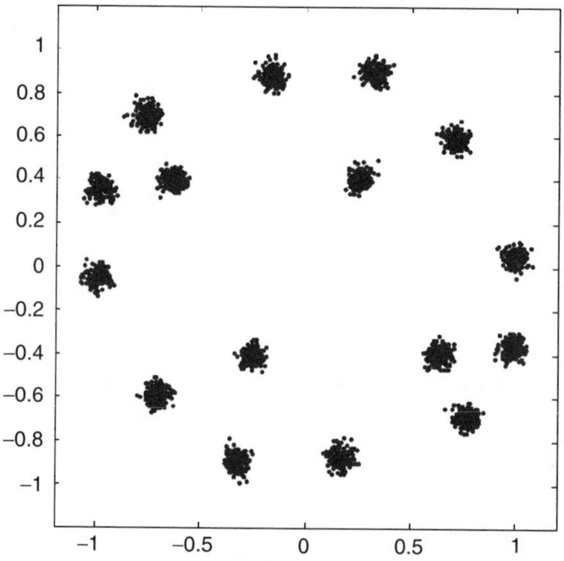

FIGURE 19.14b Signal constellation at the channel output when no memory is present ($H = 1$, $BO = 2.55$ dB).

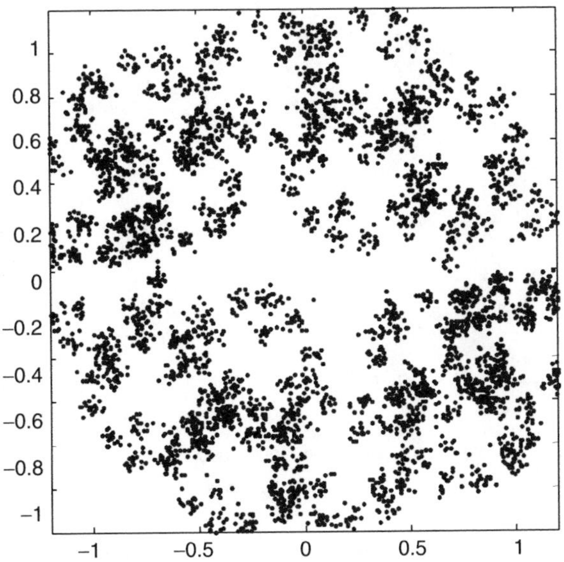

FIGURE 19.14c Signal constellation at the channel output ($H(z) = 1 + 0.3\,z^{-1}$, $BO = 2.55$ dB).

In what follows, the channel filter is taken as $H(z) = 1 + 0.3\,z^{-1}$. The following parameters are taken for the NG NNCE algorithm: $\varepsilon_0 = 0.005$, $c_\varepsilon = 1$, and $\tau = 70{,}000$. Each subnetwork is composed of $M = 5$ neurons. The Viterbi decoding block contains $N_1 = 1$ training symbol and $N_2 = 9$ information symbols. The receiver used a training sequence of 3000 transmitted symbols, after which the decision-directed mode is activated.

Figure 19.15 compares the learning curve (i.e., MSE vs. time) of the NG NNCE to that of the classical BP NNCE for different values of μ (the same initial weight values have been taken for the two

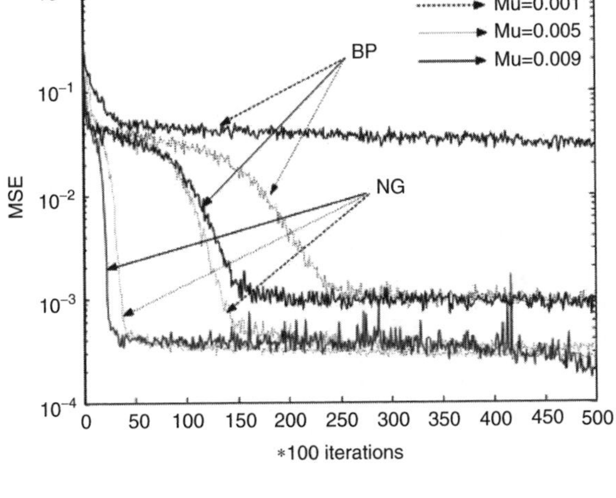

FIGURE 19.15 BP and NG identification of the nonlinear dynamical channel: influence of the learning rate μ on the convergence speed and MSE.

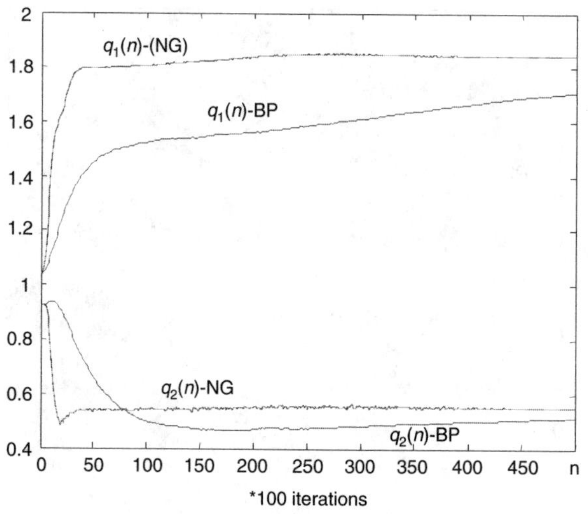

FIGURE 19.16 Evolution of adaptive filter Q weights (comparison between BP and NG, $\mu = 0.005$).

algorithms). It can be seen that the NG has faster convergence speed and yields smaller MSEs. Both NG and BP NNCEs have successfully identified the different parts of the unknown channel: the propagation channel H (Figure 19.16), the HPA AM/AM conversion (Figure 19.17a), and the TWT AM/PM conversion (Figure 19.17b). Note that the adaptive filter Q has converged to a scaled version of H (i.e., $Q(\infty) \approx \alpha\ H$). Figure 19.16 shows the evolution of the weights during the learning process. The scale factor α is equal to 1.84 for the NG algorithm and 1.71 for the BP algorithm. The value of the scale factor is not important

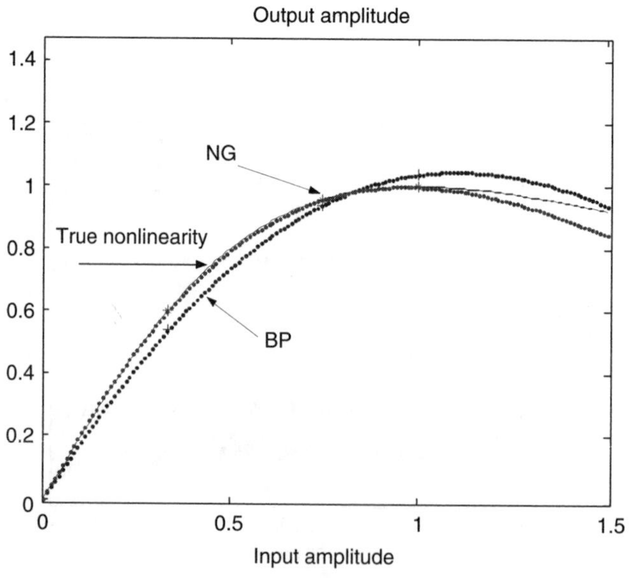

FIGURE 19.17a TWT AM/AM characteristic: True curve and normalized neural network models, (+) and (*) represent the three 16-QAM amplitudes and their corresponding outputs for BP and NG, respectively.

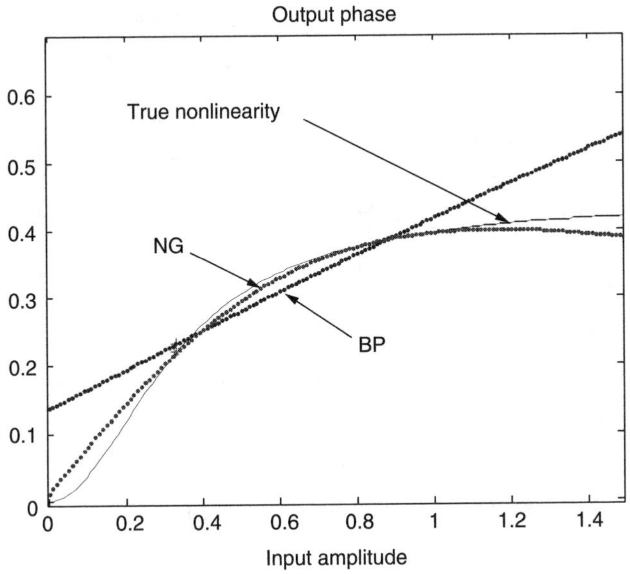

FIGURE 19.17b TWT AM/PM characteristic: True curve and normalized neural network models, (+) and (*) represent the three 16-QAM amplitudes and their corresponding outputs for BP and NG, respectively.

since subnetwork NNG, which controls the gain, makes the necessary scale compensation. Interesting studies on the convergence of the ordinary gradient descent identification of similar systems can be found in [6, 7, 18]. Several structures have been studied in these references. It is shown, in particular, how the scale factor is divided over the different parts of the adaptive system.

Figure 19.17 shows that the NG algorithm yields better AM/AM and AM/PM approximation than the BP algorithm. Since we used 16-QAM modulation, it is worth noting that the nonlinearity characteristic is expected to be better approximated around the three possible amplitudes of the 16-QAM constellation, as illustrated in Figure 19.17.

19.4 Channel Equalization

This section proposes a linear filter followed by a memoryless neural network for channel equalization.

19.4.1 Neural Network Structure

The channel model studied in this section is composed of a memoryless HPA followed by an FIR filter H (the channel presented in Section 19.3). Channel equalization is used in order to adaptively reduce ISI and nonlinear distortions. The structure we have chosen for the equalizer is composed of a linear FIR filter Q' followed by a two-layer memoryless complex-valued NN (Figure 19.18). The adaptive filter Q' aims to invert the linear part of the system (adaptive deconvolution), and therefore it is expected to mitigate ISI. The memoryless NN aims to invert the memoryless nonlinearity, and therefore it is expected to mitigate nonlinear distortions. Moreover, the channel is assumed to be unknown *a priori*; therefore, the equalizer needs to be adaptive. Finally, we consider here M-QAM modulation (which justifies the use of a complex-valued equalizer).

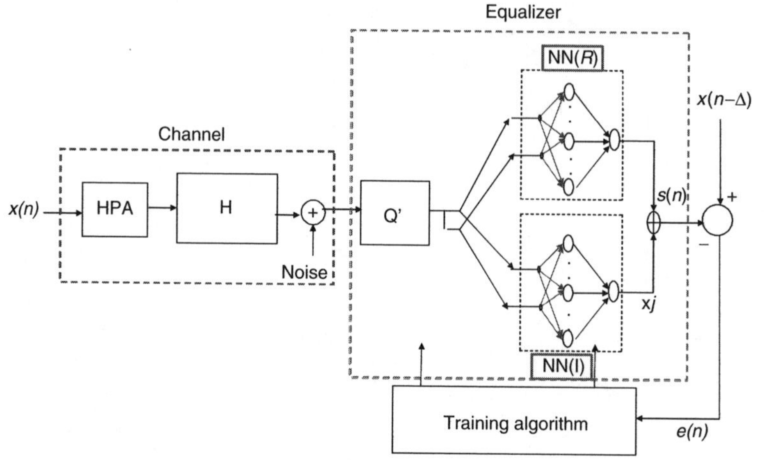

FIGURE 19.18 Adaptive equalization of a nonlinear channel with memory.

The filter output can be written as

$$z(n) = \sum_{k=0}^{N_{q'}-1} q'_k y(n-k) = z^R(n) + jz^I(n) \tag{19.24}$$

where $Q'(z) = \sum_{k=0}^{N_{q'}-1} q'_k z^{-k}$.

The neural network is composed of two subnetworks, one for the real part (R), the other for the imaginary part (I). Each subnetwork has two scalar inputs ($z^R(n)$ and $z^I(n)$), M neurons in the first layer, and a scalar output. The two outputs are then combined to form the complex output $s(n)$:

$$s(n) = s^R(n) + js^I(n) \tag{19.25}$$

where $s^R(n)$ and $s^I(n)$ are the outputs of the subnetworks R and I, respectively:

$$s^R(n) = \sum_{k=1}^{M} c_k^R g\left(w_k^{a,R} z^R(n) + w_k^{b,R} z^I(n) + b_k^R\right) \tag{19.26}$$

$$s^I(n) = \sum_{k=1}^{M} c_k^I g\left(w_k^{a,I} z^R(n) + w_k^{b,I} z^I(n) + b_k^I\right) \tag{19.27}$$

The equalizer output is expressed as

$$s(n) = s^R(n) + js^I(n)$$

$\{w_k^{a,R}, w_k^{b,R}, b_k^R, c_k^R\}$ and $\{w_k^{a,I}, w_k^{b,I}, b_k^I, c_k^I\}$, $k = 1, \ldots, M$, are the weights of subnetworks R and I, respectively.

The NN parameters are updated to minimize the loss function $J(n)$ between the desired output (which is a known delayed transmitted sequence, $x(n-\Delta)$) and the equalizer output $s(n)$:

$$J(n) = \tfrac{1}{2}e(n)^2 = \tfrac{1}{2}((e^R(n))^2 + (e^I(n))^2) = \tfrac{1}{2}\|x(n-\Delta) - s(n)\|^2 \tag{19.28}$$

Let the parameter set of subnetworks R and I be denoted, respectively, as

$$\theta^R = \left[w_1^{a,R}, \ldots, w_M^{a,R}, w_1^{b,R}, \ldots, w_M^{b,R}, c_1^R, \ldots, c_M^R, b_1^R, \ldots, b_M^R\right]^t$$

and

$$\theta^I = \left[w_1^{a,I}, \ldots, w_M^{a,I}, w_1^{b,I}, \ldots, w_M^{b,I}, c_1^I, \ldots, c_M^I, b_1^I, \ldots, b_M^I\right]^t$$

19.4.2 BP Algorithm

For subnetwork R, the NG algorithm updates the weights as follows:

$$\theta^R(n+1) = \theta^R(n) + \mu e^R(n)\nabla_{\theta^R} s^R(n) \tag{19.29}$$

where ∇_{θ^R} represents the ordinary gradient with respect to vector θ^R.

The same approach is performed for the imaginary part:

$$\theta^I(n+1) = \theta^I(n) + \mu e^I(n)\nabla_{\theta^I} s^I(n) \tag{19.30}$$

19.4.3 NG Algorithm

For subnetwork R, the NG algorithm updates the weights as follows:

$$\theta^R(n+1) = \theta^R(n) + \mu e^R(n)\hat{G}^{R-1}(n)\nabla_{\theta^R} s^R(n) \tag{19.31}$$

where $\hat{G}^R$ is obtained using a Kalman filter technique:

$$\hat{G}^{R-1}(n+1) = (1+\varepsilon_n)\hat{G}^{R-1}(n) - \varepsilon_n \hat{G}^{R-1}(n)\nabla_{\theta^R} s^R(n)(\nabla_{\theta^R} s^R(n))^t \hat{G}^{R-1}(n) \tag{19.32}$$

The same approach is performed for the imaginary part:

$$\theta^I(n+1) = \theta^I(n) + \mu e^I(n)\hat{G}^{I-1}(n)\nabla_{\theta^I} s^I(n) \tag{19.33}$$

$$\hat{G}^{I-1}(n+1) = (1+\varepsilon_n)\hat{G}^{I-1}(n) - \varepsilon_n \hat{G}^{I-1}(n)\nabla_{\theta^I} s^I(n)(\nabla_{\theta^I} s^I(n))^t \hat{G}^{I-1}(n) \tag{19.34}$$

Finally, for the sake of simplicity, filter Q' is updated using the classical LMS algorithm (ordinary gradient descent):

$$Q'(n+1) = Q'(n) + \mu(e^R(n)\nabla_Q s^R(n) + e^I(n)\nabla_Q s^I(n)) \tag{19.35}$$

Actually, simulation results showed that for both identification (of a filter followed by a nonlinearity) and equalization problems, applying the NG descent on the adaptive filter does not introduce significant improvement of the performance of the system, as long as the NG is applied to the nonlinear memoryless neural network.

Refer to [21] for a detailed discussion on a linear filter-memoryless nonlinearity identification problem. It is shown that if the NG is applied to the nonlinear memoryless part of the network, then applying the NG to the linear part instead of the ordinary gradient does not result in a significant improvement in terms of convergence speed and MSE performance. A discussion is also presented on whether to consider the space of parameters as a whole or to apply the NG on separated subspaces, i.e., by neglecting the coupling terms between the different subspaces (w, c, and b) in the FIM. In particular, it is shown that neglecting the coupling terms between the linear and nonlinear parts does not significantly affect the performance of the system. See also [42] for NG descent applied to linear systems.

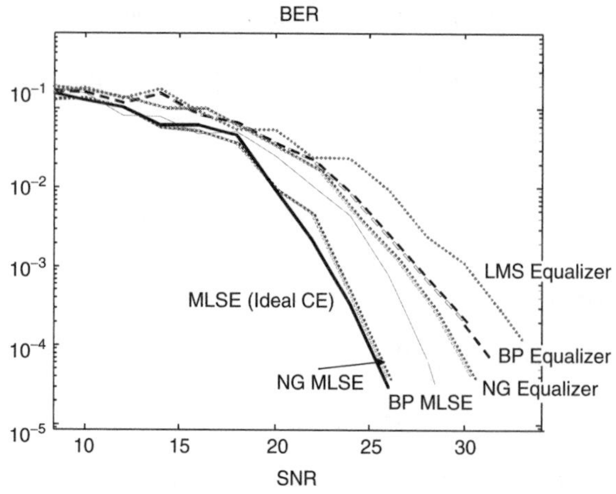

FIGURE 19.19 BER vs. SNR: Comparison between different receivers, $H(z) = 1 + 0.3\,z^{-1}$, $BO = 2.55\,dB$.

TABLE 19.1 Performance Comparison between MLSE Receivers and Equalizers

	Ideal MLSE	NG MLSE	BP MLSE	NG NN Equalizer	BP NN Equalizer	LMS Equalizer
SNR required to achieve 10^{-4} BER	25 dB	25.05 dB	27.75 dB	29.5 dB	31.5 dB	34
Gain with respect to LMS equalizer	9 dB	8.95 dB	6.25 dB	4.5 dB	2.5 dB	

19.4.4 Simulation Examples

The performances of two multilayer NN equalizers (trained with the BP and NG algorithms, respectively) are compared here to the NG and BP MLSE receivers, the classical LMS equalizer [15], and the ideal MLSE receiver in which the channel is assumed perfectly known.

The LMS filter was composed of 10 weights. Each NN equalizer was composed of a linear filter Q' (with 10 weights) followed by a two-layer memoryless neural network (Figure 19.18), with five neurons in the real (R) part and five neurons in the imaginary (I) part, and a complex-valued scalar output. The NNCE has the same structure as the one described in Section 19.3.4.

Figure 19.19 shows the BER curves of the different algorithms. Table 19.1 illustrates the SNR gains performed by each algorithm (the LMS equalizer is taken as reference). The NG MLSE outperforms all other algorithms. Its BER performance is very close to that of the ideal MLSE receiver. This is justified since the NG NNCE allows an excellent identification of the different parts of the channel, in particular at the 16-QAM constellation points (Figure 19.17). Note, finally, that the NG NN equalizer outperforms the BP NN equalizer, which suggests that the NG equalizer allows better inversion of the linear and nonlinear parts of the channel.

19.5 Conclusion

This chapter proposes different NN algorithms and structures for nonlinear channel identification and equalization. The first application concerns modeling memoryless nonlinear HPAs. Neural networks trained with the NG descent outperform classical HPA models. A memoryless NN followed by an adaptive filter is proposed for the identification of a nonlinear dynamical channel that is composed of an HPA followed by a linear filter. This structure allows characterizing the HPA nonlinear transfer function as well as the linear function. This structure is applied to MLSE receiver design for satellite communications. The NN MLSE receiver is shown to perform similarly to the ideal MLSE receiver in which the channel is assumed perfectly known. An adaptive equalizer composed of a linear filter followed by a memoryless nonlinearity has been proposed to mitigate both channel ISI and nonlinear distortions.

Acknowledgments

This work has been supported in part by the Natural Sciences and Engineering Research Council of Canada (NSERC), the Canadian Space Agency/Canadian Institute for Telecommunications Research (CSA/CITR), the Ontario Premier's Research Excellence Award (PREA), and Communications and Information Technology Ontario (CITO).

References

[1] T. Adali, X. Liu, and M.K. Sonmez, "Conditional learning neural networks and its application to channel equalization," *IEEE Trans. Signal Process.*, 45(4), 1997.

[2] S.I. Amari, "Natural gradient works efficiently in learning," *Neural Comput.*, 10, 251–276, 1998.

[3] S.I. Amari, H. Park, and K. Fukumizu, "Adaptive method for realizing natural gradient learning for multi-layer perceptrons," *Neural Comput.*, 12, 1399–1409, 2000.

[4] A. Back, "Radial basis functions," in *Handbook of Neural Network Signal Processing*, Y.-H. Hu and J.-N. Hwang, Eds., CRC Press, Boca Raton, FL, 2002, pp. 3.1–3.23.

[5] S. Benedetto and E. Biglieri, *Digital Transmission with Wireless Applications*, Kluwer Academic Publishers, Dordrecht, Netherland, 1999.

[6] N. Bershad, P. Celka, and J.M. Vesin, "Stochastic analysis of gradient adaptive identification of nonlinear systems with memory for Gaussian data and noisy input and output measurements," *IEEE Trans. Signal Process.*, 47(3), 675–689, 1999.

[7] N. Bershad, P. Celka, and J.M. Vesin, "Analysis of stochastic gradient tracking of time-varying polynomial Wiener systems," *IEEE Trans. Signal Process.*, 48(6), 1676–1686, 2000.

[8] S. Bouchired, M. Ibnkahla, D. Roviras, and F. Castanie, "Equalization of satellite UMTS channels using neural networks," in *Proceedings of IEEE ICASSP*, May 1999.

[9] S. Chen, S. McLaughlin, and B. Mulgrew, "Complex-valued radial basis function networks. Part II. Application to digital communication channel equalization," *Signal Process.*, 36(2), 175–188, 1994.

[10] J.W. Craig, "A new simple and exact result for calculating the probability of error for two-dimensional signal constellations," in *Proceedings of IEEE Military Communications Conference MILCOM'91*, Boston, MA, 1991.

[11] A. D'Andrea, V. Lottici, and R. Reggiannini, "RF power amplifier linearization through amplitude and phase predistortion," *IEEE Trans. Commun.*, 44(11), 1996.

[12] M. Ghogho, S. Meddeb, and J. Bakkoury, "Identification of time-varying nonlinear channels using polynomial filters," in *Proceedings of IEEE Workshop on Non-Linear Signal and Image Processing*, Michigan, 1997.

[13] F. Girosi and T. Poggio, "Networks and the best approximation property," *Biol. Cybernet.*, 63, 169–176, 1990.

[14] S. Haykin, *Neural Networks: A Comprehensive Foundation*, IEEE Press, Washington, DC, 1997.

[15] S. Haykin, *Adaptive Filter Theory*, Prentice Hall, New York, 1996.

[16] K. Hornik, "Multilayer feedforward neural networks are universal approximators," *Neural Networks*, 2, 359–366, 1989.

[17] J. Hwang, S. Kung, M. Niranjan, and J. Principe, "The past, present and future of neural networks for signal processing," *IEEE Signal Process. Mag.*, 28–48, 1997.

[18] M. Ibnkahla, "Statistical analysis of neural network modeling and identification of nonlinear channels with memory," *IEEE Trans. Signal Process.*, 1508–1517, 2002.

[19] M. Ibnkahla, N.J. Bershad, J. Sombrin, and F. Castanié, "Neural network modeling and identification of non-linear channels with memory: algorithms, applications and analytic models," *IEEE Trans. Signal Process.*, 46(5), 1998.

[20] M. Ibnkahla, "Applications of neural networks to digital communications: a survey," in *EURASIP Signal Processing*, Elsevier, Amsterdam, 2000, pp. 1185–11215.

[21] M. Ibnkahla, "Nonlinear system identification using neural networks trained with natural gradient descent," *EURASIP J. Appl. Signal Process.*, 1229–1237, Dec. 2003.

[22] G. Karam and H. Sari, "Analysis of predistortion, equalization and ISI cancellation in digital radio systems with nonlinear transmit amplifiers," *IEEE Trans. Commun.*, 37(12), 1989.

[23] G. Karam and H. Sari, "A data predistortion technique with memory for QAM radio systems," *IEEE Trans. Commun.*, 39(2), 336–344, 1991.

[24] T. Kohonen, *Self-Organizing Maps*, Springer-Verlag, Heidelberg, 1995.

[25] T. Kohonen, E. Oja, O. Simula, A. Visa, and J. Kangas, "Engineering applications of the self-organizing map," *IEEE Proc.*, 84, 1358–1389, 1996.

[26] L. Ljung, *System Identification: Theory for the User*, Prentice Hall, New York, 1999.

[27] K. Muhonen, M. Kavehrad, and R. Krishnamoorthy, "Look-up table technique for adaptive digital predistortion: a development and comparison," *IEEE Trans. Vehicular Technol.*, 49, 1995–2001, 2000.

[28] K.S. Narenda, "Neural networks for control," *IEEE Proc.*, 84(10), 1385–1406, 1996.

[29] K.S. Narendra and K. Parthasarathy, "Identification and control of dynamical systems using neural networks," *IEEE Trans. Neural Networks*, 1, 4–27, 1990.

[30] R. Parisi, E. Di Claudio, G. Orlandi, and B. Rao, "Fast adaptive digital equalization by recurrent neural networks," *IEEE Trans. Signal Process.*, 45(11), 2731–2739, 1997.

[31] S. Prakriya and D. Hatzinakos, "Blind identification of LTI-ZMNL-LTI nonlinear channel models," *IEEE Trans. Signal Process.*, 43, 3007–3013, 1995.

[32] J. Principe, "Dynamic neural networks and optimal signal processing," in *Handbook of Neural Network Signal Processing*, Y.-H. Hu and J.-N. Hwang, Eds., CRC Press, Boca Raton, FL, 2002, pp. 6.1–6.28.

[33] S. Qureshi, "Adaptive equalization," *IEEE Proc.*, 73(9), 1349–1387, 1985.

[34] J. Ralston, A. Zoubir, and B. Bouashash, "Identification of a class of nonlinear systems under stationary non-Gaussian excitation," *IEEE Trans. Signal Process.*, 45, 719–735, 1997.

[35] D. Rumelhart, G. Hinton, and R. Williams, "Learning internal representations by error propagation," in *Parallel Distributed Processing*, D. Rumelhart and J. McClelland Eds., MIT Press, Cambridge, MA, 1986, 318–362.

[36] A. Saleh, "Frequency-independent and frequency-dependent nonlinear models of TWT amplifiers," *IEEE Trans. Commun.*, 29(11), 1981.

[37] J. Sjoberg et al., "Nonlinear black box modeling in system identification: a unified overview," *Automatica*, 31, 1691–1724, 1995.

[38] B. Stiles, I. Sandberg, and J. Ghosh, "Complete memory structures for approximating nonlinear discrete-time mappings," *IEEE Trans. Neural Networks*, 8(6), 1397–1409, 1997.

[39] M. Thomas, M. Weidner, and S. Durrani, "Digital amplitude-phase keying with M-ary alphabets," *IEEE Trans. Commun.*, 22, 168–180, 1974.

[40] A. Zerguine, A. Shafi, and M. Bettayeb, "Multi-layer perceptron-based DFE with lattrice structure," *IEEE Trans. Neural Networks*, 12(3), 532–545, 2001.

[41] A. Zerguine and A. Shafi, "Performance of the multilayer perceptron-based decision feedback equalizer with lattrice structure in nonlinear channels", in *Soft Computing in Communications*, L. Wang, Ed., Springer-Verlag, Heidelberg, 2003, pp. 31–53.

[42] L.-Q. Zhang, S. Amari, and A. Cichocki, "Semiparametric model and superefficiency in blind deconvolution," *Signal Process.*, 81, 2535–2553, 2001.

Appendix

Computation of the gradients:

$$\nabla_\theta e_R(n) = \begin{bmatrix} \sum_{k=0}^{N_Q-1} q_k r^2(n-k)\cos(NN_P(r(n-k))+\phi(n))c_{G1}g'(w_{G1}r(n-k)+b_{G1}) \\ \cdot \\ \cdot \\ \sum_{k=0}^{N_Q-1} q_k r^2(n-k)\cos(NN_P(r(n-k))+\phi(n))c_{GM}g'(w_{GM}r(n-k)+b_{GM}) \\ \sum_{k=0}^{N_Q-1} q_k r(n-k)\cos(NN_P(r(n-k))+\phi(n))c_{G1}g'(w_{G1}r(n-k)+b_{G1}) \\ \cdot \\ \cdot \\ \sum_{k=0}^{N_Q-1} q_k r(n-k)\cos(NN_P(r(n-k))+\phi(n))c_{GM}g'(w_{GM}r(n-k)+b_{GM}) \\ \sum_{k=0}^{N_Q-1} q_k r(n-k)\cos(NN_P(r(n-k))+\phi(n))g(w_{G1}r(n-k)+b_{G1}) \\ \cdot \\ \cdot \\ \sum_{k=0}^{N_Q-1} q_k r(n-k)\cos(NN_P(r(n-k))+\phi(n))g(w_{GM}r(n-k)+b_{GM}) \\ -\sum_{k=0}^{N_Q-1} q_k r^2(n-k)\sin(NN_P(r(n-k))+\phi(n))c_{P1}g'(w_{P1}r(n-k)+b_{P1}) \\ \cdot \\ \cdot \\ -\sum_{k=0}^{N_Q-1} q_k r^2(n-k)\sin(NN_P(r(n-k))+\phi(n))c_{PM}g'(w_{PM}r(n-k)+b_{PM}) \\ -\sum_{k=0}^{N_Q-1} q_k r(n-k)\sin(NN_P(r(n-k))+\phi(n))c_{P1}g'(w_{11}r(n-k)+b_{P1}) \\ \cdot \\ \cdot \\ -\sum_{k=0}^{N_Q-1} q_k r(n-k)\sin(NN_P(r(n-k))+\phi(n))c_{PM}g'(w_{PM}r(n-k)+b_{PM}) \\ -\sum_{k=0}^{N_Q-1} q_k r(n-k)\sin(NN_P(r(n-k))+\phi(n))g(w_{P1}r(n-k)+b_{P1}) \\ \cdot \\ \cdot \\ -\sum_{k=0}^{N_Q-1} q_k r(n-k)\sin(NN_P(r(n-k))+\phi(n))g(w_{PM}r(n-k)+b_{PM}) \\ u_R(n) \\ \cdot \\ \cdot \\ u_R(n-N_Q+1) \end{bmatrix}$$

$$\nabla_{\theta} e_I(n) = \begin{bmatrix} \sum_{k=0}^{N_Q-1} q_k r^2(n-k)\sin(NN_P(r(n-k))+\phi(n)) c_{G1} g'(w_{G1} r(n-k)+b_{G1}) \\ \cdot \\ \cdot \\ \sum_{k=0}^{N_Q-1} q_k r^2(n-k)\sin(NN_P(r(n-k))+\phi(n)) c_{GM} g'(w_{GM} r(n-k)+b_{GM}) \\ \sum_{k=0}^{N_Q-1} q_k r(n-k)\sin(NN_P(r(n-k))+\phi(n)) c_{G1} g'(w_{G1} r(n-k)+b_{G1}) \\ \cdot \\ \cdot \\ \sum_{k=0}^{N_Q-1} q_k r(n-k)\sin(NN_P(r(n-k))+\phi(n)) c_{GM} g'(w_{GM} r(n-k)+b_{GM}) \\ \sum_{k=0}^{N_Q-1} q_k r(n-k)\sin(NN_P(r(n-k))+\phi(n)) g(w_{G1} r(n-k)+b_{G1}) \\ \cdot \\ \cdot \\ \sum_{k=0}^{N_Q-1} q_k r(n-k)\sin(NN_P(r(n-k))+\phi(n)) g(w_{GM} r(n-k)+b_{GM}) \\ -\sum_{k=0}^{N_Q-1} q_k r^2(n-k)\cos(NN_P(r(n-k))+\phi(n)) c_{P1} g'(w_{P1} r(n-k)+b_{P1}) \\ \cdot \\ \cdot \\ -\sum_{k=0}^{N_Q-1} q_k r^2(n-k)\cos(NN_P(r(n-k))+\phi(n)) c_{PM} g'(w_{PM} r(n-k)+b_{PM}) \\ -\sum_{k=0}^{N_Q-1} q_k r(n-k)\cos(NN_P(r(n-k))+\phi(n)) c_{P1} g'(w_{P1} r(n-k)+b_{P1}) \\ \cdot \\ \cdot \\ -\sum_{k=0}^{N_Q-1} q_k r(n-k)\cos(NN_P(r(n-k))+\phi(n)) c_{PM} g'(w_{PM} r(n-k)+b_{PM}) \\ -\sum_{k=0}^{N_Q-1} q_k r(n-k)\cos(NN_P(r(n-k))+\phi(n)) g(w_{P1} r(n-k)+b_{P1}) \\ \cdot \\ \cdot \\ -\sum_{k=0}^{N_Q-1} q_k r(n-k)\cos(NN_P(r(n-k))+\phi(n)) g(w_{PM} r(n-k)+b_{PM}) \\ u_I(n) \\ \cdot \\ \cdot \\ u_I(n-N_Q+1) \end{bmatrix}$$

VII

Voice over IP

20 Voice over IP and Wireless: Principles and Challenges
Fernando Díaz-de-María, Ascensión Gallardo-Antolín, and Carmen Peláez-Moreno **20**-1
Introduction • Speech Coding for IP and Wireless: Principles • Voice over IP • Voice over Wireless • Voice over IP over Wireless • Voice-Enabled Services over IP and Wireless • Conclusions and Challenges

20

Voice over IP and Wireless: Principles and Challenges

Fernando Díaz-de-María
Universidad Carlos III de Madrid

Ascensión Gallardo-Antolín
Universidad Carlos III de Madrid

Carmen Peláez-Moreno
Universidad Carlos III de Madrid

20.1 Introduction 20-2
20.2 Speech Coding for IP and Wireless: Principles 20-2
Some Preliminary Notions • Speech Coding Basics • Speech Coding for IP and Wireless
20.3 Voice over IP 20-7
An Overview • Technological Barriers • Quality of Service • Standard Speech Coders for VoIP • Packet loss Recovery Techniques • Other Packet-Based Alternatives for Voice Transport: Trends
20.4 Voice over Wireless 20-11
Wireless Voice Communications Systems • Fundamental Issues in Speech Coding for Wireless • Standard Speech Coders for Wireless • Some Trends
20.5 Voice over IP over Wireless 20-15
20.6 Voice-Enabled Services over IP and Wireless 20-16
Introduction • Technologies for Voice-Enabled Interfaces • Architectures for Automatic Speech Recognition over Communication Networks • ASR over IP • ASR over Wireless
20.7 Conclusions and Challenges 20-21

Abstract

Now that wireless, Internet Protocol (IP)-based and combinations of both communication systems have become pervasive networking standards, the transmission of voice over IP and wireless becomes a topic of high relevance. This chapter briefly describes the fundamental issues involved in speech coding for those environments including descriptions of the most suitable speech coding standards, the specific troubles that this transmission encounters, the technological barriers together with their current solutions and the most active lines of research associated. A short introduction to the challenges that the transmission of voice over IP over wireless encompasses is provided as well. Human-to-machine scenarios are also considered, as they are likely to be very common in the new applications that make use of these modern networks. Special emphasis will be placed on how network-related distortions affect the performance of automatic speech recognizers. The chapter is completed with the conclusions and some of the envisaged challenges.

0-8493-1657-X/$0.00+$1.50

20.1 Introduction

The voice is the vehicle for the most important type of communication between human beings: speech communication. Consequently, voice transmission has played, since the origins of telecommunications, a distinguished role in the conception, development, and deployment of communications systems. At present, wireless and Internet protocol (IP)-based (and combinations of both) communications systems are the most outstanding paradigms, therefore voice over IP and wireless has become a relevant topic.

Traditionally, compression rate, speech quality under several conditions, and coding delay or complexity were the most relevant issues to be taken into account when designing a voice communications system. Nowadays, other network-dependent factors have to be considered. Robustness against transmission errors, frame erasures or packet losses, performance of tandem configurations, coding of nonspeech signals (e.g., background noises, music, etc.), and scalability and adaptation to a quality of service (QoS) are some illustrative examples. Furthermore, wideband speech coding has attracted much attention in recent years. Advances in speech coding — which have significantly reduced the bit rate for this high-quality speech — together with expected increased capacity of wireless systems and IP-based transport networks have boosted interest in this field.

The purpose of this chapter is to provide a brief description of the fundamental issues regarding speech coding, the specific problems of voice over IP and wireless, current solutions, and the most active lines of research. The reader will be directed to appropriate references for details. The chapter is organized as follows. Section 20.2 describes the principles of speech coding and the specific techniques used in IP and wireless environments. Section 20.3 is devoted to voice over IP. Specifically, technological barriers and current solutions are described and standard speech coders are concisely presented. Section 20.4 is dedicated to voice over wireless. Fundamental topics are addressed, standard speech coders are revised, and some relevant trends are highlighted. Section 20.5 provides a short introduction to the voice over IP over the wireless paradigm. Section 20.6 describes how network-related distortions affect the performance of human-to-machine communications, with emphasis on automatic speech recognition over IP and wireless. Finally, some conclusions are drawn and envisaged challenges are outlined in Section 20.7.

20.2 Speech Coding for IP and Wireless: Principles

20.2.1 Some Preliminary Notions

Before focusing on speech coding itself, the fundamental properties of the speech signal should be revised, since the coding procedures are specifically tailored for exploiting such properties:

- Speech is a nonstationary signal; i.e., its statistical properties change over time.
- It exhibits a high dynamic range: there are large amplitude differences between the highest and lowest energy segments.
- It is sometimes almost periodic (vocal cords vibrating), while other times it is noise-like (no vibration). Those speech sounds made when the vocal cords vibrate are known as voiced, while those that are noise-like are called unvoiced. Furthermore, mixed (simultaneously voiced and unvoiced) sounds very commonly appear.
- The speech signal is very redundant. Two types of redundancies can be distinguished: those involving consecutive samples (short-term) and those involving one-period-apart samples (characteristic of voiced sounds and known as long-term).

A simplified model of the human speech production mechanism (known as a source–filter model), which integrates to a limited extent the speech characteristics described above, has been extensively employed in the field of speech coding. In this model, illustrated in Figure 20.1, speech is seen as the response of a (typically, all-pole) filter to a source signal (also referred to as excitation). The filter intends to model the articulation by capturing the main vocal tract resonances, while the excitation is expected to account for

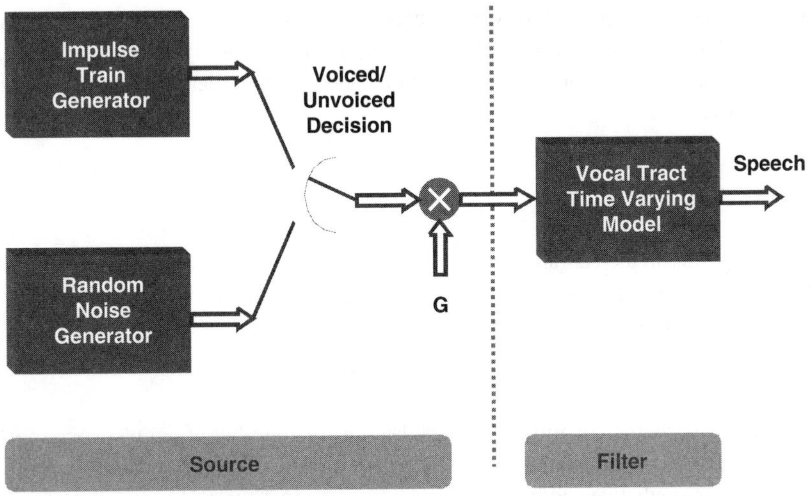

FIGURE 20.1 Source-filter model of the human speech production mechanism. Speech is seen as the response of a filter to a source signal. The filter tries to model the articulation by capturing the main vocal tract resonances, while the excitation is expected to account for either the noise-like (random noise generator) or periodic nature (impulse train generator) of unvoiced or voiced segments, respectively. The energy of the speech is controlled by the parameter G, a multiplicative factor applied to the excitation.

either the noise-like (random noise generator in Figure 20.1) or periodic nature (impulse train generator) of unvoiced or voiced segments, respectively. The energy of the speech is controlled by the parameter G, a multiplicative factor applied to the excitation (see Figure 20.1).

Speech coding is inherently a lossy process. Since the nature of the speech signal is analog, an analog-to-digital conversion is necessary. This conversion conceptually entails two different processes: sampling (continuous-time to discrete-time) and quantization (real-valued to discrete-valued). The bit rate (BR), number of bits per second, is obtained as the product of the sampling rate and the number of bits per sample. Since the sampling rate is given (bandwidth of the signal), speech coding algorithms aim to reduce the number of bits per sample.

The bandwidth considered is another relevant subject in speech coding (it determines the sampling rate). Either narrowband (300 to 3400 Hz) or wideband (50 to 7000 Hz) speech signals are contemplated. Narrowband speech offers a limited quality, appropriate for telephone applications. Wideband speech provides increased naturalness and intelligibility, eases speaker recognition, and is intended for higher-quality-demanding applications.

20.2.2 Speech Coding Basics

20.2.2.1 Differential Coding

Speech coding mainly consists of removing redundancy and encoding the remaining nonredundant part of speech in a perceptually satisfactory manner. Redundancy is removed by time-variant or adaptive methods to match the nonstationary nature of the speech signal. Differential (predictive) coding is the most usual paradigm for redundancy removal. For every sample of the speech signal, a prediction is made based on previous ones, and it is the difference between this sample and its prediction that is actually quantized and transmitted. The better the prediction is, the lower the dynamic range of the difference signal and, consequently, the more efficient the quantization becomes. Furthermore, if exactly the same prediction is made at the encoder and decoder ends, the difference signal quantization error is the only distortion in the whole differential coding system.

According to the two types of redundancies present in the speech signal, two types of predictions are usually implemented. Short-term prediction involves a number of adjacent previous samples, while long-term prediction — appropriated for voiced speech segments — uses samples of the previous period to estimate those of the current one. Both predictors (short- and long-term) are linear, i.e., each sample is estimated as a linear combination of previous ones. Thus, a P-th-order short-term prediction can be formulated as follows:

$$\tilde{x}[n] = \sum_{k=1}^{P} a_k x[n-k] \tag{20.1}$$

where $\tilde{x}[n]$ denotes the predicted value, a_k, $1 \leq k \leq P$, are the weighting coefficients, and $x[n-k]$, $1 \leq k \leq P$, are the previous speech samples. For the long-term case, considering the simplest first-order case, the prediction would be

$$\tilde{x}[n] = b\, x[n-M] \tag{20.2}$$

where b is the weighting coefficient and M the pitch lag (i.e., the number of samples corresponding to the fundamental period of the speech signal).

Furthermore, linear predictive coding (LPC) has been the most successful coding technique over the years. There are several reasons for this, and here we highlight a few:

- Linear prediction (LP) has proven to be a very efficient technique for redundancy removal.
- The power spectrum of the all-pole filter derived from the short-term LP coefficients,

$$|H(e^{j\omega})|^2 = \frac{1}{\left|1 - \sum_{k=1}^{P} a_k\, e^{-jk\omega}\right|^2} \tag{20.3}$$

  is an approximation of the power spectrum of the original speech signal. For typical values of P (8 to 16), the LP spectrum becomes a good approximation of the speech signal spectral envelope.
- It is very convenient for hardware implementations.

On the other hand, LP coefficients are not directly quantized. Some alternative sets of coefficients with preferable quantization properties are used instead. Line spectral pairs (LSPs) [45] are the most popular choice. They behave well when interpolated (typically, short-term LP coefficients are estimated once every 10 to 30 ms, and interpolated versions are obtained more frequently to provide smooth transitions) and are very efficiently vector quantized.

20.2.2.2 Adaptive Quantization

The quantization of a nonstationary and high-dynamic-range signal like speech leads inherently to a signal-dependent signal-to-noise ratio (SNR). In other words, the quality of the quantization is time varying. Logarithmic quantization is the simplest solution. The original dynamic range is log-compressed and the SNR becomes approximately constant. Adaptive quantization is the second alternative. It consists of periodically updating the quantization step to follow the properties of the incoming signal. Adaptive quantization is clearly superior to logarithmic quantization when few bits per sample are available.

20.2.2.3 Noise Masking

The noise-masking ability of the human auditory system is also a relevant issue in speech coding. Since the final receptor of the speech is usually a human being — sometimes, and even more frequently as time goes on, an automatic speech recognizer could be found instead — speech coders can take advantage of known properties of our auditory system. Basically, a human being does not perceive a low-intensity signal (in this case, quantization error) due to a simultaneously occurring stronger signal (speech), as long as both are close enough to each other in frequency and the latter is sufficiently strong with respect to the former. Typically, the way to exploit this property in the field of speech coding is conforming the spectrum of the quantization error following (parallel to) that of the actual speech by means of what is called a perceptual weighting filter. Thus, more error is concentrated in the higher-energy frequency bands of the speech, where the error could likely be masked.

20.2.2.4 Quality Measures

Speech coder quality is assessed through subjective listening tests. The most widely used is known as mean opinion score (MOS). Listeners are asked to assign one of five possible numerical scores on a scale from 1 (unsatisfactory) to 5 (excellent) for a set of reference utterances. Finally, an average value is obtained.

20.2.3 Speech Coding for IP and Wireless

In this section, the fundamentals of the three traditional speech coder types, namely, waveform, parametric, and hybrid coders, are presented. Afterwards, a brief discussion about the more relevant coding attributes for IP and wireless is included.

20.2.3.1 Waveform Coders

As its name denotes, waveform coders attempt to preserve the waveform of the original signal. They operate on a sample-by-sample basis, and consequently do not incur in any time delay (to be more precise, the algorithmic delay is one sampling period), and typically operate at high bit rates (32 kb/s and above). Since they aim to approximate the waveform, some objective and simple quality measures like SNR can give us an approximate indication of their quality.

20.2.3.2 Parametric Coders

Parametric coders assume a simplified model of the speech signal. Thus, this model, irrespective of the waveform, keeps the perceptually relevant characteristics of the speech signal. These coders work on a frame-by-frame basis. For every frame, the model parameters are estimated, quantized and transmitted. They generally operate at low bit rates (below 4 kb/s) and must be subjectively evaluated. Within this group of coders, the LPC vocoder[1] has been the most outstanding representative. This vocoder is based on the simplified speech production mechanism of Figure 20.1, the time-varying vocal tract being modeled through an all-pole LP filter and the excitation through either a noise or an impulse generator. Nowadays, we would draw attention to two of them: mixed excitation linear prediction (MELP) and waveform interpolation (WI) vocoders:

- MELP is the new generation of LPC vocoders. Very likely, the hard voiced/unvoiced decision turned out to be the major limitation of the traditional LPC vocoder. MELP coders consider a weighted combination of both types of excitations as the way to overcome that limitation.
- WI coders are based on interpreting the voiced periods of speech as a sequence of pitch-cyclic waveforms. Thus, the slow rate of variation of these waveforms facilitates prediction and interpolation. They provide outstanding voice quality at low bit rates, the computational complexity being their main drawback.

20.2.3.3 Hybrid Coders

Hybrid coders fill the gap between waveform and parametric coders. On the one hand, they assume a speech model; on the other, they try to preserve the (perceptually weighted) original waveform. Hybrid coders work on a frame-by-frame basis and typically operate at medium bit rates (between 4 and 16 kb/s). These coders are also known as analysis-by-synthesis (AbS) coders. This alternative denomination tells us more about their fundamentals. AbS coders are based on the previously described source–filter model. The distinguishing characteristic of AbS coders is that the excitation (source) to the filter is chosen by synthesis. The excitation is determined over blocks that are shorter than the frame and that are known as

[1]The term *vocoder* is a combination of two words: *vo*ice *coder*.

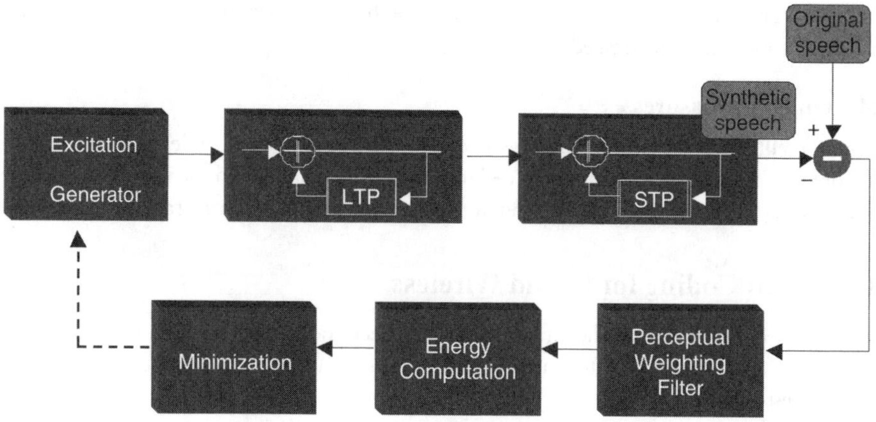

FIGURE 20.2 Block diagram of the hybrid coder. Synthetic speech is generated by passing an excitation through the cascade of LPT and STP inverse filters. The difference between the original and the synthetic speech is computed and perceptually weighted. Next, the energy of this weighted difference is calculated and used as the score to select the optimum excitation.

subframes. Typically, one frame is divided into two to four subframes. Several candidate excitations are tested, and the one that results in the smallest (perceptually weighted) difference between the synthetic and the original speech is chosen as the quantized excitation. During the synthesis stage, the excitation coming from the excitation generator is fed into the cascade made up of the inverse of the long-term and short-term prediction (LPT and STP, respectively) filters. Figure 20.2 illustrates this procedure.

The code excited linear prediction (CELP) coder is the preponderant hybrid algorithm. Almost every standard coder operating at 4 to 16 kb/s is CELP type. In this case, all the possible excitations are stored in a codebook. Only the index of the selected excitation sequence should be transmitted to the decoder end (which stores a copy of the same codebook). On the one hand, this is a very efficient and flexible (as very different kinds of sequences populate the codebook) way to represent the excitation. On the other hand, its computational cost is conceptually quite high in comparison to other types of coders. Nevertheless, since 1984 (the proposal of the CELP algorithm) the original cost has been dramatically reduced. Furthermore, the success of this architecture stimulated the implementation of the long-term predictor as an adaptive codebook. As previously described, the long-term predictor estimates the current sample of the excitation to the all-pole synthesis filter as a weighted version of a one-pitch-lag-behind sample (see Equation 20.2). Therefore, the long-term prediction can be accomplished by means of a search in a codebook that stores (a concatenation of) the last excitation sequences.

It is also worthwhile to mention another type of coder that, though we would classify as parametric, involves some CELP concepts. Harmonic vector excitation (HVXC) coders, adopted as a part of the MPEG-4 standard, perform a vector quantization of the excitation for voiced frames and employ a CELP algorithm for unvoiced ones.

20.2.3.4 Relevant Coder Attributes

Several, often conflicting issues should be taken into account when designing a speech coder. In the context of this chapter, we have selected the following: bit rate, quality, complexity, delay and robustness.

20.2.3.4.1 Bit Rate

Obviously, the higher the bit rate, the better the quality. Speech coders for IP are either waveform (16 to 64 kb/s) or hybrid (5.3 to 16 kb/s), while for wireless the selected coders are either hybrid (3.45 to 13 kb/s) or parametric (1.2 or 2.4 kb/s, generally). Traditionally, as required by communication networks, speech coding algorithms have been designed to provide a fixed bit rate (FBR), i.e., a constant number of bits per second. However, in the last several years, two types of non-FBRs have shown an extraordinary development: variable bit rate (VBR) and adaptive multirate (AMR).

- VBR or source-controlled BR: On the one hand, the speech signal inherently demands VBR. Voiced segments require higher bit rates than unvoiced segments for the same perceptual quality. On the other hand, this VBR can be naturally conveyed through modern access mechanisms (code-division multiple-access, CDMA) or transport networks (which are packet based).
- AMR or network-controlled BR: Modern wireless voice communications systems dynamically assign bit rate for source and channel coding. Thus, speech coders are designed to operate at several BRs, which can be selected depending on network conditions.

20.2.3.4.2 Complexity

Complexity is another topic to be taken into account. Maintaining the quality, the bit rate can be reduced by increasing the complexity. In the IP environment, complexity is not critical, since current PCs are powerful enough to run any standard coder in real time. In wireless, it is more relevant because complexity means power consumption and, consequently, reduction of battery life. Nevertheless, from the power consumption point of view the coder itself is not as relevant as the radio frequency stage. Thus, the voice activity detector (VAD) — See Section 20.4.2.4 — which allows reduction of the transmitted power during silent periods, becomes a key subsystem.

20.2.3.4.3 Delay

Delay is also a relevant issue in both IP and wireless. Perhaps it is more critical for voice over IP (VoIP), since there are several subsystems that add significant delays (packaging, PC audio cards, routers, etc.). The one-way coding delay should be kept below 150 ms. This figure could be relaxed to 300 ms for lowly interactive conversations. Above 300 ms, the impairment becomes very unpleasant.

The one-way coding delay for hybrid or parametric coders is mainly determined by the algorithmic delay (roughly 3.5 times this last figure), which is the time interval between two consecutive frames plus some additional look-ahead. (The analysis window used to obtain the LP coefficients of the present frame also comprises the next subframe of that which follows to allow the interpolation of filter coefficients without incurring an additional algorithmic delay. Thus, the interpolation process relies only on current and previous frame coefficients, though the coefficients corresponding to the current one take into account the first subframe of the next.) The algorithmic delay typically takes values between 15 and 35 ms. The one-way coding delay is consequently between 52.5 and 122.5 ms.

20.2.3.4.4 Robustness

Robustness becomes a major topic in speech coding for IP and wireless. First of all, VoIP can be affected by packet loss. Thus, speech coders should be designed to gracefully degrade when one or even several consecutive entire frames are lost (one packet could contain one or more frames, and packet loss typically occurs in bursts). Moreover, wireless communications systems are prone to transmission errors (and again the occurrence pattern is bursty). Though some parts of the systems try to minimize the impact of transmission errors (interleaver and channel coder), the speech coder itself should be as robust as possible.

20.3 Voice over IP

20.3.1 An Overview

During the final years of the 20th century, IP-based networks became omnipresent. Their rapid spread and potential general-purpose orientation have attracted so much attention that many of the still-existing specific-purpose networks, like the Public Switched Telephone Network (PSTN), have been somehow left behind. Among the most significant advantages of IP networks is their ability to transport any kind of digital media (voice, text, video, audio, etc.), allowing not only the substitution of several specific-purpose networks by a single one, but also the combination and manipulation of various media for creating new applications and services.

However, this claimed general-purpose orientation is not exempt from technical challenges and problems concerning the specific nature of the individual medium being transported. In Section 20.3.2, the reasons for these difficulties in the particular case of voice transport will be examined, followed by an outline of the most significant of the proposed methods for providing QoS in such situations (Section 20.3.3).

The bandwidth restrictions imposed by these types of networks (as well as by wireless networks) require the use of low-to medium-bit-rate speech coders. Thus, Section 20.3.4 will be devoted to the description of the most often employed standard speech coders for VoIP. In Section 20.3.5 we will describe the basics of packet loss recovery techniques.

We will finish (Section 20.3.6) with a brief exposition of incipient developments of packet-based alternatives to IP networks and their behavior in voice transport.

20.3.2 Technological Barriers

IP networks were not originally conceived to transport voice and therefore cannot guarantee some of the fundamental requirements of this kind of *inelastic* traffic due to its real-time constant-bit-rate delivery demand, as opposed to *elastic* traffic, like data, whose principal needs are integrity and reliability.

The following is a description of the issues that make IP networks unsuitable (*a priori*) for voice transport.

20.3.2.1 End-to-End Delay and Jitter

As previously mentioned, for a conversation to take place fluently, a maximum end-to-end delay must be observed. VoIP systems produce a delay that is typically higher than that of other voice transmission systems due to the large number of subsystems involved: audio capture system, speech coders (see coding delay in Section 20.2.3.4), packaging, protocol syntax, routers, gateways, etc.

Furthermore, the variability of this delay, called jitter, is another important barrier. It is due to the fact that the IP does not establish a dedicated communication channel, but proceeds on a packet-by-packet basis. This means that for every packet carrying, in some situations, not more than 10 ms of speech, a new path to its destination must be found, producing different delays for each packet. The impossibility of predicting this delay makes it difficult to properly design the dimensions of the speech buffers required to manage speech decoding at the destination. Besides, it makes voice frames appear in an order different from that in which they were created.

20.3.2.2 Packet Loss

Transport control protocol (TCP) is the transport protocol usually associated with IP. Because it was originally designed to carry data information, it employs a mechanism for flow control based on discarding those packets that cannot be delivered due to node congestions. This triggers the resending of the lost packet by the emitting source once a timeout expires without having received a proper acknowledgment. This way the source detects the network congestion and moderates its packet emission rate.

This kind of behavior is not advisable in the case of voice traffic. If a voice packet is not delivered at the time it should be reproduced, resending it is of no use. Therefore, to minimize these effects in most VoIP applications, the transport protocol employed is user datagram protocol (UDP), which does not guarantee the delivery of every packet but does not load the network unnecessarily.

In addition to network congestion, there is still another cause of packet losses: unaffordable delays. When a speech frame reaches its destination too late to be reproduced, it is considered a missing frame.

20.3.2.3 Throughput

The throughput is the nominal bandwidth available for effective transmission over the network. An individual user has at his disposal only a portion of the total throughput and needs the use of medium-bit-rate speech coders.

20.3.2.4 Internet Availability and Reliability

The availability of a network is measured as the time rate that the network is not functioning for any reason, and its reliability as the maximum amount of time that passes in case of malfunctioning before the service is resumed.

Because the Internet is composed of many different networks, its availability and reliability cannot be guaranteed. On the contrary, PSTN is considered by many national governments to be an essential and basic service, thus requiring strict limits for these two indicators. If IP networks are to substitute basic telephony service, they must enhance these two aspects.

20.3.2.5 Security and Confidentiality

Some other important aspects of telephony service are security and confidentiality. Though important research efforts are addressing this topic, secure user identification and validation (for service charge purposes, for example) are not as easy in IP networks as in the PSTN.

20.3.3 Quality of Service

As has become apparent from the previous section, several problems must be addressed to provide voice services with at least the same quality now offered by circuit-switched networks. QoS is a very broad and dynamic subject. For our purposes, we do not intend to offer a comprehensive review of this field but only to outline some of its more well-established tools.

Protocol H.323 [31] was the one initially chosen to implement the first packet-based voice services due to its immediate availability. It was originally designed to manage packet-based multimedia communications over local area networks and was originally developed as an adaptation of H.320. It utilizes the real-time transport protocol/real-time transport control protocol (RTP/RTCP) from the Internet Engineering Task Force (IETF), along with internationally standardized coders, and it is also being used for video and other communications over the Internet. It specifies call control, including signaling, registration, and admissions mechanisms, as well as a means of providing supplementary services like those now implemented in the PTSN.

Session initiation protocol (SIP) [41] is a certified standard of the IETF designed to provide multimedia call control and enhanced telephony services. It is transport-layer independent because it can be used with any packet-based protocol, and its most significant qualities — if compared with H.323 — are its reduced complexity and enhanced scalability. As we will see in Section 20.5, SIP was the selected call control protocol in the Internet multimedia subsystem (IMS) specified for providing wireless VoIP services in the Universal Mobile Telecommunications System (UMTS).

Concerning the provision of reliable and constant-bit-rate transport of voice packets (or any other inelastic kind of traffic), the IETF has developed the aforementioned RTP/RTCP [43] set of protocols. The former provides time signaling, loss packet detection, security, and content identification, and the latter informs the sender and destination about the performance of the network, measuring delay, jitter, and packet loss rate.

However, these protocols cannot guarantee the availability of resources in the network and therefore the QoS. For that purpose, IETF specifies the resource reservation protocol (RSVP) [23], which allows users to make reservations on the networks' nodes of an established session. However, this protocol has severe scalability problems due to the amount of computation needed in the nodes. Therefore, proposals based on differentiated services (diffserv), where the traffic is classified in the boundary nodes, seem to be more appropriate for core networks.

20.3.4 Standard Speech Coders for VoIP

Speech coders play a key role in the quality of the speech transmitted over IP networks. The principal parameters of the most employed coders in VoIP applications are summarized in Table 20.1. It is worth mentioning that the quality shown, measured through the mean opinion score scale, corresponds to clean

TABLE 20.1 VoIP Standard Speech Coders

Speech Coder	Technology	Frame Rate (ms)	Look-Ahead (ms)	Bit Rate (Kb/s)	MOS
ITU-T G.711	PCM	0.125	—	64	4.3
ITU-T G.726	ADPCM	0.125	—	16	2
				24	3.2
				32	4
				40	4.2
ITU-T G.727	Embedded ADPCM	0.125	—	16	2.8
				24	3.7
				32	3.9
				40	4.1
ITU-T G.729A	CS-ACELP	10	5	8	3.8
ITU-T G.723.1	MP-MLQ	30	7.5	6.3	3.9
	ACELP			5.3	3.8

operation conditions. Readers must be warned about the different origins of these MOS figures for different coders, which allows only an approximated comparison between coder quality (due to the subjective nature of the measure).

ITU-T G.711 [28] is a waveform speech coder that uses the classical pulse code modulation (PCM) technique where the quantization steps follow the μ-law (U.S., Japan) or the A-law (Europe). This is one of the simplest procedures by which to code a speech signal, and since the signal is coded on a sample-by-sample basis, it produces the smallest algorithmic delay possible. However, the bandwidth required for its application makes its use unaffordable in many situations.

ITU-T G.726 [27] uses the adaptive differential pulse code modulation (ADPCM) technique and provides several operation modes using different bandwidths (40, 32, 24, and 16 Kb/s), the most commonly employed being 32 Kb/s, denominated primary mode.

As can be inferred from Table 20.1, the lowest-bandwidth operation modes are very inefficient compared with the analysis-by-synthesis coders that will be described next. However, their advantages are their simplicity and low algorithmic delay.

ITU-T G.727 is an embedded ADPCM where the number of bits employed for the differential quantization of each sample accounts for the different possible operation modes or bit rates, in such a way that every operation mode can be supplemented with the additional bits necessary to jump to another operation mode. This way a 40 Kb/s coded speech can be decoded using any of the operation modes, decoding the appropriate subset of parameters [44].

ITU-T G.723.1 [26] is an analysis-by-synthesis linear predictive coder and provides a dual coding rate of 5.3 and 6.3 Kb/s. It is possible to switch between both rates at a frame level, and also, an option for variable-rate operation is available using a VAD, which compresses the silent portions.

The excitation signal is composed of periodic and nonperiodic components. The construction of the periodic component involves the estimation of a pitch lag and a fifth-order predictor for modeling the long-term correlations.

The nonperiodic component is computed using different techniques depending on the coding rate used. For the higher rate, 6.3 Kb/s, the encoder uses a multipulse maximum likelihood quantization (MP-MLQ), while at 5.3 Kb/s, it employs an algebraic code excited linear prediction (ACELP) scheme.

Finally, ITU-T G.729 [29] specifies an 8 Kb/s speech coder using a conjugate-structure algebraic code excited linear predictive coder (CS-ACELP). ITU-T G.729A [30] is a reduced-complexity version of the aforementioned speech coder that produces a slightly inferior quality.

20.3.5 Packet loss Recovery Techniques

Packet loss recovery techniques can be classified into two groups: those relying on sender actions and those based on the receiver (for a more thorough review, see, for example, [38] or [42]).

Sender-based techniques can be further divided into those implying a retransmission of the lost packet at the request of the receiver and those based on interleaving and forward error correction (FEC). The former are not feasible in real-time conversational applications due to the long delay they imply. Interleaving techniques are normally used in wireless applications where the most common problem is bit errors occurring in bursts. However, in the case of IP networks, where the problem of packet losses also tends to happen in bursts, interleaving techniques at a packet level produce an inadmissible delay. Finally, many applications rely on FEC techniques that, nevertheless, increase the employed bandwidth.

Receiver-based techniques, on the other hand, only require actions at the receiver end, therefore producing no extra delay and bandwidth occupation. Most of the speech coders define specific means of recovering from packet losses (maybe affecting several frames). Thus, for example, the G.723.1 speech coder has been designed to be robust against frame erasures: when the decoder is in concealment mode, it uses the previous LSP vector to predict the current one and generates a synthetic voiced or unvoiced excitation signal based upon a decision made about the last good frame. The decoded speech is attenuated if bad frames continue to arrive, until it is completely muted after three consecutive losses.

G.729 also implements a similar procedure to recover from frame erasures. However, as the LP parameters are differentially encoded, the effects of packet losses last longer.

20.3.6 Other Packet-Based Alternatives for Voice Transport: Trends

Despite the fact that the most common packet-based protocol is the combination of TCP/IP, there are other packet-based alternatives that are being implemented by major carriers and, many times, though these other technologies can be employed to carry voice themselves, are subsequently employed to carry TCP/IP.

Some of the most commonly implemented alternatives are asynchronous transfer mode (ATM) and frame relay and, recently, multiprotocol label switching (MPLS). The latter is gaining support and is being implemented in many backbone networks. MPLS is able to interwork with IP diffserv (see Section 20.3.3) and can also be implemented over ATM, thus exploiting the already deployed ATM networks. Moreover, it implements a comprehensive QoS on aggregated traffic that can result in an important advantage for voice transport [50].

20.4 Voice over Wireless

20.4.1 Wireless Voice Communications Systems

In this section devoted to voice over wireless, we will refer to the North American IS-54 and IS-136 time-division multiple-access (TDMA) and IS-95 code-division multiple-access (CDMA) systems, the European Global System for Mobile (GSM) TDMA, second-generation wireless systems, and third-generation wide-band code-division multiple-access (WCDMA) systems (UMTS and CDMA2000).

Compared to other voice communications systems, the most relevant characteristic of wireless voice communications systems is the unavoidable presence of transmission errors (mostly occurring in bursts). Consequently, one of the main objectives of wireless voice communications systems should be to provide high intelligibility and high-quality speech in a variety of operational channel conditions. Speech coding is only one of the processes involved. Interleaving and channel coding also play relevant roles.

The speech coder itself should be robust. The effect of residual errors (those still present after the error correction process) on decoded speech has to be limited as much as possible. Interleaving is the most effective technique to deal with the bursty patterns of transmission errors. Channel coding is an essential subsystem in wireless voice communications. Otherwise, voice over wireless would not be feasible. Although source and channel coding have traditionally been separately designed, joint source–channel coding (see Section 20.4.4.1) is expected to make significant advances in this field.

Voice over wireless communications systems also include other series of subsystems aimed at reducing average power consumption and, consequently, increasing the system capacity, like voice activity detection (VAD), the discontinuous transmission (DTX) mechanism, and the multiplexing access mechanism. These subsystems will receive more attention in the next section.

20.4.2 Fundamental Issues in Speech Coding for Wireless

The previous general considerations regarding wireless voice communications systems have significant implications on the design of speech coders for wireless communications systems. Some of them are discussed below.

20.4.2.1 Channel Quality and Adaptive Operation

Wireless channel quality is typically characterized through the carrier-to-interference ratio (C/I). Speech coders for wireless should provide high speech quality in typical operation conditions (C/I > 10 dB). Furthermore, satisfactory speech quality is desirable even in poor channel conditions (C/I < 7 dB). In order to manage such requirements, recent standard coders are able to adapt their bit rates according to the varying channel quality or network load. Thus, channel capacity is dynamically balanced between source and channel coding, producing high-quality speech under good channel conditions and at least highly intelligible speech under poor conditions.

20.4.2.2 Background Noise

The good reproduction of background noise by speech coders significantly contributes to the quality perception of the user. If the background noise sounds unnatural, the user will negatively judge the quality even though the speech quality is good. Coding of background noise becomes an important issue in wireless speech coding because as the bit rate decreases, the coder becomes more model dependent (speech dependent), which makes it less appropriate for nonspeech signals.

20.4.2.3 Tandeming

Successive encoding and decoding processes involving the same or different coders, called tandeming, are very common in practice. In fact, during the selection and characterization phases of any standard coder, the candidates are always tested under such conditions. Although once the speech signal has been encoded there is apparently no reason to decode it until it reaches the end user (assuming that the entire communication path is digital), this is not the habitual situation in real operation. Actually, when the signal goes through international links, it is usually decoded and reencoded, using typical waveform standard speech coders to undergo the international segment, to be decoded and reencoded again, now using one of the corresponding wireless standards, when it enters the mobile network. On the other hand, even when the signal does not transverse borders, it occasionally suffers the same tandeming process when the near- and far-end telephone operators are different. Finally, for networking reasons, and more frequently than one may suspect, the speech signal goes through two or more encoding–decoding stages.

This fact should be taken into account when a speech coder is designed. For example, most of the current speech coders incorporate a postfiltering stage to attenuate the (audible) quantization noise in the perceptually less relevant frequency bands for the speech signal (coarsely speaking, both noise and speech are suppressed in those bands perceptually less relevant for the speech). In case of tandeming, the postfiltering stages corresponding to each coder are cascaded. Therefore, postfiltering should be designed for a given number of tandeming stages.

20.4.2.4 Voice Activity Detection

VAD has become an essential subsystem in wireless speech communications systems. When voice activity is not detected, the encoder is able to switch to a lower-bit-rate mode typically used for coding the background noise. In some wireless systems, a DTX mechanism, governed by the VAD subsystem, makes it possible to stop transmission during speech pauses, in this way reducing the average transmitted power. Also during these pauses, some characteristics of the background noise are transmitted in order to reproduce at the receiver end a synthetic comfort noise similar to the original one.

20.4.2.5 Unequal Error Protection (UEP)

Channel encoders (which are explicitly designed for each standard speech coder) classify the source bits into several categories depending on their relative perceptual impact. In this way, not only are some

TABLE 20.2 Wireless Network Standard Speech Coders

Speech Coder	Technology	Frame Rate (ms)	Look-Ahead (ms)	Bit Rate (Kb/s)	MOS
GSM FR	RPE-LTP	20	—	13	3.7
GSM EFR	ACELP	20	—	12.2	4.1
GSM HR	VSELP	20	—	5.6	3.5
GSM/UMTS AMR-NB	ACELP	20	5	12.2	4.1
				10.2	4.0
				7.95	3.9
				7.40	3.8
				6.70	3.8
				5.9	3.7
				5.15	3.5
				4.75	3.5
ITU-T GSM/UMTS AMR-WB	ACELP	20	—	6.6	4.0
				8.85	4.5
				12.65	4.6
				14.25	4.6
				15.85	4.6
				18.25	4.6
				19.85	4.7
				23.05	4.6
				23.85	4.6
NA-CDMA IS-96	QCELP	20	5	0.8–8.5	3.3
NA-TDMA IS-641	ACELP	20	—	7.4	3.8
NA-TDMA IS-54	VSELP	20	5	7.95	3.5
NA-CDMA IS-127	RCELP	20	—	1.2–9.6	3.8

errors detected and even corrected, but their influence on the (decoded) speech perceptual quality is also minimized.

20.4.2.6 Frame Erasures

Furthermore, when the channel decoder considers that a speech frame is seriously damaged (because the most critical parameters are unreliable due to errors), the frame is labeled (by the channel decoder) as "bad" and discarded (not conventionally decoded). Instead, the frame is substituted by an attenuated version of the last reliably received frame. If various consecutive frames are seriously damaged, the attenuation increases, and when the number of replaced frames is more than five (100 ms), the decoder mutes the output.

20.4.3 Standard Speech Coders for Wireless

Table 20.2 summarizes the main parameters defining the coders employed in wireless systems. As in Table 20.1, the MOS quality figures come from different sources, and consequently, the comparisons between coders qualities can only be approximated. They are always obtained under clean conditions (without any channel distortion or tandeming).

20.4.3.1 European GSM and UMTS Standards

In this section, the speech coders employed in the European GSM and UMTS wireless systems will be outlined. The European Telecommunications Standards Institute (ETSI) has standardized two types of speech traffic channels: full-rate (22.8 Kb/s) and half-rate (11.4 Kb/s). However, due to the unreliability of the radio transmission channel, a considerable portion of the bit rate capacity is devoted to channel coding.

Phase 1 of GSM standardization initially described the full-rate (FR) speech coder [8], which made use of 13 Kb/s, leaving the remaining 9.8 Kb/s to the channel coder. Phase 2 introduced two new coders: the enhanced full-rate (EFR) [10] for the full-rate speech channel and the half-rate (HR) [9] for the half-rate speech channel. The former employs 12.2 Kb/s and the latter 5.6 Kb/s. Finally, an adaptive multirate (AMR)

[11] set of coders has been defined to obtain the best speech quality, dynamically balancing the allocation of bits between the speech and channel coder. Thus, the AMR system adapts to channel conditions by selecting the appropriate mode (full- or half-rate) and the speech coding rate that allows sufficient error protection for the actual error level of the channel. These last coders are also being employed in UMTS.

The GSM FR speech coder operates at 13 Kb/s. It is a regular pulse excited linear predictive coder with a long-term predictor (RPE-LTP) [48]. For the channel coding, GSM uses convolutional encoding and block interleaving.

The HR coder uses the vector sum excited linear prediction (VSELP) paradigm [19]. Four voicing modes can be selected for the periodic component of the excitation, and a weighted sum of a reduced number of basis vectors, which define the stochastic codebook, produces the nonperiodic component of the excitation.

The GSM EFR uses the algebraic CELP paradigm (ACELP) [10] and encodes speech at 12.2 Kb/s. The excitation is composed of an adaptive codebook contribution (using a long-term predictor with a noninteger delay precision) and an algebraic fixed codebook.

This coder employs the FR speech channel just like the GSM FR coder and therefore uses the same channel coder, but leaves 0.8 Kb/s more than the GSM FR for protection. These bits are employed to provide extra protection to some of the most important bits in the bit stream.

The adaptive multirate–narrowband (AMR-NB) speech coder [11] provides eight modes corresponding to the bit rates shown in Table 20.2, being the uppermost the same as GSM EFR. Complementary channel encoders are used to complete either the FR speech channel bandwidth or the HR (when possible).

The algorithm for speech encoding is roughly the same as GSM EFR. However, for each mode, a different number of tracks and pulses per track are chosen for the fixed algebraic codebook to attain the desired number of bits.

The adaptive multirate–wideband (AMR-WB) [4] uses an extended audio bandwidth, from 50 Hz to 7 KHz, giving superior speech quality, and using (like its narrowband counterpart) ACELP technology. It provides several modes of operation that can be selected for different levels of background noise, leaving the rest of the available bandwidth for channel protection.

20.4.3.2 North American Cellular Systems

The Interim Standard (IS)-54 speech coder was standardized for use in the TDMA digital cellular system in North America. It is a VSELP coder that processes speech every 20 ms using a look-ahead of 5 ms and a total bit rate of 7.95 Kb/s.

IS-641 is very similar to the GSM EFR coder. It uses a 5.6 Kb/s source coder and a channel coder up to 13 Kb/s. The excitation signal is also computed as the sum of an adaptive and algebraic fixed contribution, structured in the same manner as the excitation in GSM EFR.

IS-96 is Qualcomm's proprietary version of a CELP speech coder, known as QCELP, designed for use in the CDMA digital cellular system in North America. It runs at 8.5 Kb/s when speech is being produced and drops to 0.8 Kb/s in silent intervals.

IS-127 is an enhanced variable-rate coder (EVRC) designed to replace IS-96 in the IS-95 CDMA cellular system. It is a relaxation code excited linear predictive (RCELP) speech coder that supports three bit rates (9.6, 4.8, and 1.2 Kb/s). The actual rate is determined by the contents of the current speech and the history of the characteristics of previous frames.

20.4.4 Some Trends

In this section, the most relevant research activities in the context of speech coding for wireless communications are briefly presented.

20.4.4.1 Joint Channel–Source Coding

Traditionally, source and channel coders have been separately designed. This approach, called tandem source–channel coding, would be asymptotically optimal if no residual redundancies remained after source coding. However, this does not occur in practice because of the constraints of speech coding algorithms in complexity and delay. As a consequence, the global system performance degrades at low SNRs.

To overcome these limitations, several techniques for jointly designing source and channel coding have been recently developed. This approach is known as joint source–channel coding and basically consists of exploiting the residual redundancies of the speech signal in order to guarantee high speech quality and intelligibility, even under very bad channel conditions.

One technique based on this idea is source-controlled channel coding (SCCD), in which correlations in source coder parameters are used for minimizing the bit rate allocated for the channel coding stage [21].

Another promising approach is soft bit source decoding (SBSD), which proposes estimating the speech parameters during the source decoding stage, taking into account both the reliability of the information provided by the channel decoder and the knowledge of previous frames of the signal [14].

Both approaches can be concatenated for improved results, as shown in [22].

20.4.4.2 Robustness Issues for Low-Bit-Rate Coders

Nowadays, and likely due to the development of the future NATO Narrow Band Voice Coder (STANAG 4591), two main issues have concentrated the attention of speech coding researchers for wireless communications: robustness of speech coders when operating under hard acoustic conditions and development of coders at very low bit rates with reasonable speech quality and high intelligibility.

The NATO Narrow Band Voice Coder is being developed not only for military applications, but also for professional and commercial use. Its most relevant requirements are [46]:

- Very low bit rates (1200 and 2400 bps) with small delay and reasonable complexity
- Robustness in a wide range of acoustic environments: background noise, tandem conditions (see Section 20.4.2.3), and transmission errors
- Robustness for a wide range of users, taking into account not only interspeaker variability (e.g., language independence, even nonnative speakers), but also intraspeaker variability (e.g., whispered voice)

Among the different proposals for this standard, it is worth mentioning the mixed excitation linear prediction coder [49] at 1200 bps and the harmonic and stochastic excitation (HSX), which is a dual coder at 1200 and 2400 bps [20]. HSX is an LP-based coder with a mixed excitation model. In particular, the periodic impulse train covers the lower frequencies of the excitation and the noise covers the upper frequencies. However, its main novelty is the explicit incorporation of two noise reduction techniques into the speech coder (a combination of Wiener and Lim filters) for achieving environmental robustness at low SNRs.

20.4.4.3 Selectable Mode Vocoder (SMV [18])

This new variant of CELP coders has been selected by the Telecommunication Industry Association (TIA) and the Third Generation Partnership Project 2 (3GPP2) for CDMA applications and also forms the basis for the newly emerging ITU-T 4 Kb/s speech coding standard.

The distinguishing characteristic of SMV is its ability to select the best bit rate as a function of the type of input speech frame (silence, noise-like, unvoiced, onset, nonstationary voiced, and stationary voiced).

20.5 Voice over IP over Wireless

The mobility offered by wireless networks and the ubiquity of IP, together with their increasing ability to transport multimedia content, make the combination of both an appealing all-in-one solution. This way, any type of multimedia content could be transmitted over IP regardless of the physical layout of the network.

However, even considering the enhanced spectral efficiency of the third generation of wireless systems, the fact that IP was originally designed for wirelines makes it an inefficient protocol for wireless transmissions. Issues such as bandwidth efficiency, security, and confidentiality are of vital importance for the development of IP-based multimedia services over these networks:

- The bandwidth efficiency of IPs is very poor when packets are filled with speech frames due to their short durations in comparison with the amount of headings necessary to send them. On the other hand, if many speech frames are accommodated in a single IP packet to make better use of the available bandwidth, the delay due to the buffering of those frames becomes unpleasant in conversational applications. Compressed heading solutions have been proposed to improve performance [5].
- The security and privacy characteristics of wireless channels are issues that IPs must face. This topic has been evaluated within the IEEE 802.11i standard for wireless local area networks (WLANs) and proposals based on Ipsec protocols are also being used in other types of wireless networks [25].

Besides, again, the elastic traffic orientation of IP networks makes them unsuitable for transporting real-time voice. Many efforts are being made to enhance the behavior of wireless networks to enable them to provide QoS, extending and adapting some of the solutions described in Section 20.3.3.

Thus, an Internet multimedia subsystem has been defined in UMTS to provide an end-to-end QoS guarantee by using policy-based control principles. SIP (see Section 20.3.3) has been chosen as the call control standard. The latest revision of the SIP specification includes enhancements to optimize SIP for wireless networks. One of the enhancements, for example, allows SIP messages to be compressed so that they can be efficiently transported over the wireless interface.

The QoS of WLAN is considered in the IEEE 802.11e, which defines means of establishing priorities of medium access within the packets, so that real-time inelastic traffic could more easily meet the requirements of delay and jitter.

20.6 Voice-Enabled Services over IP and Wireless

20.6.1 Introduction

The rapid growth of the Internet and wireless communications is enabling a wide variety of applications in which the integration of different multimedia data (text, audio, image, video) is desirable. In this context, the ability to provide the client with on-line, cost-effective, friendly interfaces will acutely influence the success of an application. Because speech is the most natural form of human communication, spoken interfaces are crucial for these applications because they enable a user to interact with a computer as if it were a conversational partner.

The most common application of speech technologies over IP and wireless networks is access to remote information about flight or train schedules, hotels, and other services. Furthermore, other emerging applications can be envisaged: e-mail processing (reading and writing) on cellular phones, dictation over the phone to personal assistants, etc.

20.6.2 Technologies for Voice-Enabled Interfaces

Interactive spoken interfaces integrate several technologies based on speech (speech recognition and synthesis modules) and natural language processing (discourse analysis, sentence interpretation, and automatic response generation modules) to enable natural human-to-machine interaction. In the context of this section, we focus on speech processing technologies and, specifically, on automatic speech recognition (ASR), though a brief presentation and discussion concerning text-to-speech conversion (TTS) systems and speaker recognition is also presented.

- Speech Synthesis (Text-to-Speech Conversion)

  TTS systems are computer-based modules that are capable of reading or automatically pronouncing any text or sentence aloud. In conversational interfaces, TTS systems are useful for accessing textual information and, in combination with a language generation technique, for verbalizing dialogues generated by the system.

- Speech Recognition

  Automatic speech recognition deals with the automatic transcription of words or sentences uttered by a user. Its automation with other linguistic processing techniques such as discourse analysis and sentence interpretation, allows the correction of unrecognized words, improving the performance of the overall system.

- Speaker Identification and Verification

  Speaker recognition aims to automatically recognize the identity of a speaker using the specific characteristics of his or her speech. Two different tasks can be distinguished: speaker identification (which consists of identifying the person who has pronounced a given utterance) and speaker verification (in this case, the system accepts or rejects the identity claimed by the speaker). This technique can be used for enabling control of access to various services by voice, for example, banking over IP networks, booking services, or remote access to computers.

The rest of this section is devoted to ASR, which is one of the key technologies for enabling natural conversational interactions between humans and machines. Speaker recognition systems are affected in a similar way as ASR systems by the specific distortions of wireless and IP environments. Interested readers should refer to [3] and [39] for a more detailed discussion. With respect to the text-to-speech systems, information on how IP and wireless networks affect the perceived quality has been very limited. Although there are no objective studies about the suitability of certain speech coders for TTS tasks, Muthusamy et al. [35] report that subjective listening evaluations show that speech coders slightly degrade the perceptual quality of synthetic speech (especially the GSM FR coder).

20.6.3 Architectures for Automatic Speech Recognition over Communication Networks

Most of the ASR systems consist of two main modules: the parameterizer or feature extractor and the classifier or recognizer. The first extracts a compact representation of the speech signal suitable for ASR. The second compares the utterance to its acoustic database, which contains the models for the basic acoustic units. In addition, the ASR system is assisted by a dictionary (a set of words in the lexicon of the task) and a language model (which contains information about the allowable combination of words and phrases). The classifier module is usually based on the hidden Markov model (HMM) paradigm. It is worth noting for posterior discussions that classification is much more computationally demanding than parameterization.

The integration of ASR systems in voice-enabled services over communication networks such as GSM, UMTS, and IP can be implemented according to three possible architectures ([7], [15]), depending on how the speech recognition processing (basically, feature extractor and classifier) is distributed between the terminal and the machine running the ASR application, i.e., between the client and server sides. These architectures are (1) terminal-based speech recognition (TSR) or local speech recognition (client-only processing), (2) network-based speech recognition (NSR) or remote speech recognition (server-only processing), and (3) distributed speech recognition (DSR) (client–server processing).

20.6.3.1 Local or Terminal-Based Speech Recognition

The best way to avoid both coding distortion and transmission errors is to perform the speech recognition at the user local terminal. Nevertheless, this approach has three disadvantages. First, the application must reside at the local terminal, and therefore the interface design will be tightly constrained by the hardware (memory and computational load) resources of the terminal. Second, application updating is much more difficult at the terminal than at the server. And third, it is not possible to reproduce the speech signal at the remote end.

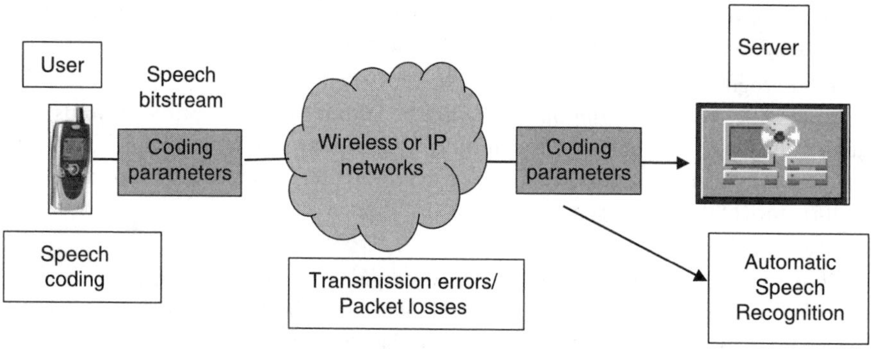

FIGURE 20.3 Remote or network-based speech recognition (NSR).

20.6.3.2 Remote or Network-Based Speech Recognition

The best alternative for reducing the computational load at the local terminal is to let the server perform the entire recognition process, as shown in Figure 20.3. This way, the terminal is totally independent of the kind and complexity of the application running at the remote end, the only requirement being the use of the same standard coder in both the local terminal and the remote server. This fact becomes relevant for the design of applications that integrate voice, data, or any other kind of media, since it allows universal access from almost any terminal.

In this case, voice should be compressed and transmitted over the network using a specific speech coder. Two different approaches to ASR have been reported. In the first approach (decoded speech-based recognition), the voice is decoded at the receiver end (as usual), and subsequently, the recognition is performed. In the second approach (bit-stream-based or digital speech-based recognition), the recognition feature vectors are extracted directly from the encoded speech (i.e., the bit stream), instead of decoding it and subsequently extracting the feature vectors [16], [33], [36]. Though this approach requires specific processing for each coder, it avoids the coding distortion and it makes the ASR system more robust against transmission errors.

20.6.3.3 Distributed Speech Recognition

A compromise between the two previous solutions consists of performing part of the recognition process at the client end (namely, the parameterization) and the remainder at the server end (see, for example, [12] or [40]). In fact, the terminal only extracts the feature vectors (suitable for speech recognition) and their transmission can be performed over a data channel, as illustrated in Figure 20.4.

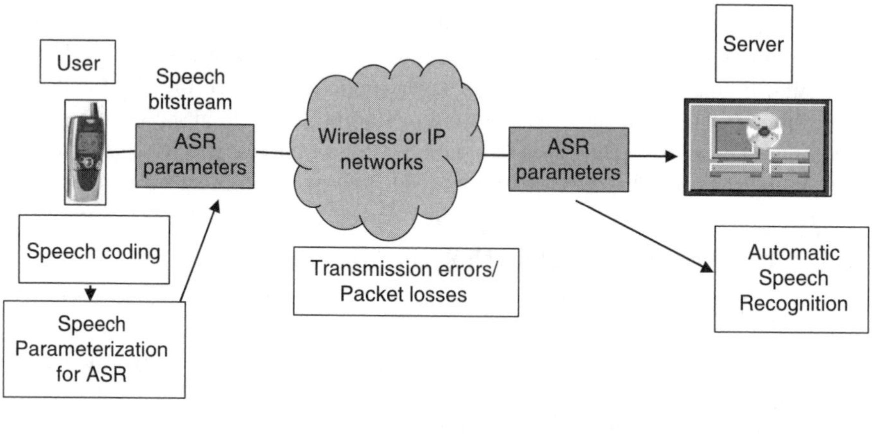

FIGURE 20.4 Distributed speech recognition (DSR).

With respect to its shortcomings, first, there is the impossibility of reproducing the speech signal at the remote end, and second, there is the necessity of a standardized front end so that every client terminal computes and transmits the same parameters. In this context, ETSI has proposed the AURORA standard [12] for feature extraction and compression.

The advantages of DSR rely on the fact that the bandwidth required to transmit the recognition parameters is very small (4.4 Kb/s in the AURORA standard), while the computational effort needed for the parameter extraction is not so high. Besides, distortion due to the speech coding is completely avoided. Also, DSR allows the application of channel coding (or error concealment techniques) designed *ad hoc* for the protection and correction of errors introduced by the communication channel.

Many efforts are currently being made to explore which of these two choices, NSR or DSR, is more suitable for enabling ASR tasks over wireless and IP communications, as discussed in the following sections.

20.6.4 ASR over IP

In ASR systems over IP networks, two major difficulties are to be considered: coding distortion and packet loss. In this section, we describe the more relevant studies on the influence of these factors on ASR systems and some of the existing solutions.

20.6.4.1 Coding Distortion

The voice coders G.723.1 and G.729, included in the H.323 protocol suite [31], are the most commonly used. These coders introduce some distortions in the speech signal, which cause corresponding performance degradation in ASR systems, as shown in [24] and [36]. In fact, even when working with matched conditions (i.e., training the system using decoded speech), the speech recognition performance substantially decreases when coders operating at bit rates under 16 kb/s are used (which is the case for G.723.1 and G.729).

However, coding distortion only affects ASR performance when NSR architecture is used (in particular, in the decoded speech-based recognition approach).

20.6.4.2 Packet Loss

As previously mentioned, the inadequacy of IP networks for real-time traffic such as voice appears in the form of packet loss, either because the packets are actually lost or because they arrive too late. Depending on the implementation, one packet can contain one or more speech frames. In addition, packet losses usually happen in bursts, producing a significant number of missing speech frames.

Obviously, packet loss deteriorates the quality of the decoded speech and then can severely affect the recognition performance of the ASR systems. Error concealment, i.e., a receiver-based strategy (see Section 20.3.5), is the most common alternative. Error concealment techniques were originally proposed in the speech coding context. They aim to reconstruct a perceptually acceptable voice operating only at the decoder end by means of techniques specifically designed for every standard coder (see Section 20.3.5).

Nevertheless, error concealment techniques in speech coding are tightly constrained by strict delay and computational requirements. These requirements are notably more relaxed in the ASR context, which has stimulated the research of *ad hoc* methods for error concealment in ASR, for instance, dropping the corrupted packet from the received sequence, replacing the lost packet with some averaged training data, and linear or nonlinear interpolation of neighboring frames [37]. Moreover, they can combine with FEC techniques (see Section 20.3.5) with very promising results, as shown in [6].

20.6.5 ASR over Wireless

Wireless environments entail three main problems for the ASR systems: noisy scenarios, source coding distortion, and transmission errors. Besides, other subsystems of the wireless communications system also affect the performance of an ASR system, for example, the presence of the VAD subsystem or the insertion of comfort noise (see Section 20.4.2.4). The VAD, though conservatively designed, occasionally causes some clipping of the speech signal, while the insertion of comfort noise inevitably (though slightly) disturbs the estimation, at the remote server, of the background noise characteristics.

20.6.5.1 Noisy Scenarios

The inherent mobility entailed by wireless communications systems allows the user to employ phone almost anywhere: public places, within a car, the roadside, etc. As a consequence, reliable ASR systems should be robust to any kind of background noise as well as to speaker's reactions (speech changes) to these noises.

Current research addressing the noise problem, i.e., trying to reduce the mismatches between training and testing conditions, advances along three separate directions[2]: (1) speech enhancement, (2) more robust parameterization techniques impervious to noise effects, and (3) techniques for adapting clean-speech hidden Markov models to noisy speech conditions (for a complete review, see, for example, [32]). These techniques were initially developed for PC or (fixed) telephone-based ASR applications, but they can be applied to GSM and UMTS environments. For example, Fingscheidt et al. [15] and Kim et al. [33] have proposed the application of different techniques of speech enhancement for DSR and NSR, respectively. Muthusamy et al. [35] have analyzed the behavior of model compensation on noisy coded speech using different standards (GSM FR, GSM EFR, G.729, and G.726). In the same context, an enhanced version of the ETSI AURORA standard [13] introduces several modules for noise reduction.

20.6.5.2 The Influence of Coding Distortion on Speech Recognition

As in IP communications, coding distortion only affects ASR performance in the NSR architecture (in the decoded speech-based recognition approach).

In the context of the GSM environment, Hirsch [24] showed that speech coded by using GSM FR, GSM HR, and GSM EFR standards leads to a deterioration of the recognition performance. The lowest accuracy is achieved by the HR scheme (the lowest bit rate), while the best performance is obtained when using the GSM EFR (the more modern of the two FR standards).

For UMTS, the influence of the AMR coder is also noticeable: all the AMR modes (see Table 20.2) cause significant degradation of the performance of the ASR system, especially the minimum allowed (4.7 Kb/s) [15], [24], [35].

Finally, it is necessary to take into account the effect of speech coder tandeming, which is very common in practice (see Section 20.4.2.3). In [17] and [34] an analysis of these situations can be found, showing that tandeming dramatically reduces the recognition rate of the system.

20.6.5.3 Transmission Errors and Lost Frames

Despite the presence of channel coding, transmission errors and frame erasures inevitably appear in wireless communications environments (see Sections 20.4.2.5 and 20.4.2.6).

Several studies (for example, [16] for NSR architectures and [2] for DSR) have been conducted to analyze the effects of these two types of channel errors in ASR performance. The main conclusion is that both types of distortion drastically degrade the ASR system performance. Moreover, in [2] it has been shown that speech recognition is more sensitive to transmission errors than to (moderate) frame erasures.

In order to improve the robustness of ASR systems to channel errors, there are several recent works [2], [47] that aim to design a channel coding specially tailored to ASR features. This approach is useful in DSR architectures in which ASR parameters are quantized and transmitted to the remote server, where recognition actually takes place.

Following this idea, in [1] it is shown that no major degradations in ASR performance are observed for medium or good channel conditions. However, for bad channel conditions, a dramatic decrease of the recognition rate is found. This observation has motivated the design of more sophisticated channel coders for ASR feature's transmission, based on the idea of joint channel–source coding (see Section 20.4.4.1).

---

[2]Some authors divide them in only two classes, merging the first two.

For example, in [2], a novel method has been proposed to obtain a confidence measure on the information received at the channel decoding state to use it to conceal the missing frames at the recognition stage (joint channel decoding–speech recognition).

For the NSR architecture, the bit-stream-based recognition approach (see Section 20.6.3.2) has proven to be very effective for alleviating the effect of transmission errors [16], [17].

20.7 Conclusions and Challenges

In this chapter we have reviewed the principles and outlined the challenges of the transmission of voice over IP and wireless networks. These are currently the dominating networks, and they will eventually converge to a seamless wireless and wireline network running IP.

However, the transmission of voice over these networks has to overcome a series of problems. In this chapter we have described the most relevant difficulties for IP and wireless networks separately.

In particular, bandwidth limitations make speech coders gain a prominent role in the development of current solutions. However, they produce a coding distortion that unavoidably increases when the bandwidth restrictions become important.

Transmission errors also represent a handicap to obtain good-quality voice transmission. We can find two types of errors: packet losses in IP networks and bit errors in wireless systems. Furthermore, they usually occur in bursts, which makes them especially damaging.

Delays and variations of this delay (jitter) can also represent a serious problem in conversational applications, in particular in IP networks, as well as in tandeming stages along the communication path. Moreover, the complexity of the coding algorithms and its implications on battery life become important issues in small-size wireless devices.

Several techniques have been devised to enhance the performance of these communications systems, like better exploitation of the bandwidth using VAD or efficient content-adapted protection against errors (UEP). New protocols are being developed to account for the peculiarities of voice traffic. Furthermore, new speech coding algorithms capable of adapting to variable network conditions (AMR) and concealment techniques ready to cope with the persisting errors are being currently proposed.

Nevertheless, there are unresolved challenges or techniques that are still not fully developed, like the provision of QoS guarantees, joint channel–source coding, robustness against environmental noises and hostile channels, and the achievement of toll quality at 4 Kb/s.

Special attention must be paid to the transmission of voice over IP over wireless channels. All the aspects that were reviewed for both VoIP and voice over wireless must be taken into account, making this means of transmission an extremely challenging problem.

Lastly, the challenges that face the applications that are developed using the voice transmitted as such for human-to-machine communications need to be carefully considered. Automatic speech recognition is a key speech technology whose use could contribute to the naturalness of the application interfaces and that becomes especially influenced by the network transporting the voice.

Acknowledgments

This work has been partially supported by Spanish CICYT Grant TIC2002-02025 and Spanish Regional Grant CAM-07T-0031-2003.

List of Acronyms and Abbreviations

3GPP2: Third Generation Partnership Project 2
AbS: analysis-by-synthesis
ACELP: algebraic code excited linear prediction
ADPCM: adaptive differential pulse code modulation
AMR: adaptive multirate

AMR-NB: adaptive multirate–narrowband
AMR-WB: adaptive multirate–wideband
ASR: automatic speech recognition
ATM: asynchronous transfer mode
BR: bit rate
CDMA: code-division multiple-access
CELP: code-excited linear prediction
CS-ACELP: conjugate-structure algebraic code excited linear prediction
C/I: carrier-to-interference ratio
Diffserv: differentiated services
DSR: distributed speech recognition
DTX: discontinuous transmission
EFR: enhanced full-rate speech coder
ETSI: European Telecommunications Standards Institute
FBR: fixed bit rate
FEC: forward error correction
FR: full-rate speech coder
GSM: Global System for Mobile Communications
HMM: hidden Markov model
HR: half-rate speech coder
HSX: harmonic and stochastic excitation
HVXC: harmonic vector excitation
IETF: Internet Engineering Task Force
IMS: Internet multimedia subsystem
IS: Interim Standard
LP: linear prediction
LPC: linear predictive coding
LSP: line spectral pair
MELP: mixed excitation linear prediction
MOS: mean opinion score
MPLS: multiprotocol label switching
MP-MLQ: multipulse maximum likelihood quantization
NSR: network-based speech recognition
PCM: pulse code modulation
PSTN: Public Switched Telephone Network
QCELP: Qualcomm's propietary version of CELP
QoS: quality of service
RCELP: relaxation code excited liner prediction
RPE-LTP: regular pulse excited linear predictive coder with a long-term predictor
RSVP: resource reservation protocol
RTCP: real-time transport control protocol
RTP: real-time transport protocol
SBSD: soft bit source decoding
SCCD: source-controlled channel coding
SIP: session initiation protocol
SNR: signal-to-noise ratio
SMV: selectable mode vocoder
TCP: transport control protocol
TDMA: time-division multiple-access
TIA: Telecommunication Industry Association
TSR: terminal-based speech recognition

TTS: text-to-speech system
UDP: user datagram protocol
UEP: unequal error protection
UMTS: Universal Mobile Telecommunications System
VAD: voice activity detector/detection
VBR: variable bit rate
VoIP: voice over IP
VSELP: vector sum excited linear prediction
WCDMA: wideband code-division multiple-access
WI: waveform interpolation
WLAN: wireless local area network

References

[1] Aurora, "Recognition with WI007 Compression and Transmission over GSM Channel," Document AU/266/00, Ericsson, December 2000.
[2] Bernard, A. and Alwan, A., "Low-Bitrate Distributed Speech Recognition for Packet-Based and Wireless Communications," *IEEE Transactions on Speech and Audio Processing*, 10(8), 570–579, 2002.
[3] Besacier, L., Grassi, S., Dufaux, A., Ansorage, M. and Pellandini, F., "GSM Speech Coding and Speaker Recognition," in *Proceedings of ICASSP*, Istanbul, Turkey, 2000, pp. 1085–1088.
[4] Bessette, et al., "The Adaptive Multirate Wideband Speech Coder (AMR-WB)," *IEEE Transactions on Speech and Audio Processing*, 10(8), 2002.
[5] Bormann, C., "Robust Header Compression (ROHC): Framework and Four Profiles: RTP, UDP, ESP, and Uncompressed," IETF RFC 3095, July 2001.
[6] Boulis, C., Ostendorf, M., Riskin, E.A. and Otterson, S., "Graceful Degradations of Speech Recognition Performance over Packet-Erasure Networks," *IEEE Transactions on Speech and Audio Processing*, 10(8), 580–590. 2002.
[7] Digalakis, V.V., Neumeyer, L.G. and Perakakis, M., "Quantization of Cepstral Parameters for Speech Recognition over the World Wide Web," *IEEE Journal on Selected Areas in Communications*, 17(1), 82–90, 1999.
[8] ETSI, "Digital Cellular Telecommunications System; Full Rate Speech Transcoding," Recommendation GSM 6.10, February 1992.
[9] ETSI, "Digital Cellular Telecommunications Systems; Half Rate Speech. Part 2. Half Rate Speech Transcoding," Recommendation GSM 6.20, December 1995.
[10] ETSI, "Enhance Full Rate (EFR) Speech Transcoding," Recommendation GSM 6.60, March 1997.
[11] ETSI, "Digital Cellular Telecommunication Systems: Adaptive Multi-Rate (AMR) Speech Transcoding," Recommendation GSM 06.90, April 2000.
[12] ETSI, "Speech Processing, Transmission and Quality Aspects (STQ); Distributed Speech Recognition; Front-End Feature Extraction Algorithm; Compression Algorithms," ES 201 108, April 2000.
[13] ETSI, "Speech Processing, Transmission and Quality Aspects (STQ); Distributed Speech Recognition; Advanced Front-end Feature Extraction Algorithm; Compression Algorithms," ES 202 050, July 2002.
[14] Fingscheidt, T. and Vary, P., "Softbit Speech Decoding: A New Approach to Error Concealment," *IEEE Transactions on Speech and Audio Processing*, 240–251, 2001.
[15] Fingscheidt, T., Aalburg, S., Stan, S. and Beaugeant, C., "Network-Based vs. Distributed Speech Recognition in Adaptive Multi-Rate Wireless Systems," in *Proceedings of ICSLP '02*, Denver, CO, 2002, pp. 2209–2212.
[16] Gallardo-Antolín, A., Díaz-de-María, F. and Valverde, F., "Avoiding Distortions Due to Speech Coding and Transmission Errors in GSM ASR Tasks," in *Proceedings of ICASSP*, Vol. 1, Phoenix, AZ, 1999, pp. 277–280.

[17] Gallardo-Antolín, A., Peláez-Moreno, C., and Díaz-de-María, F., "A Robust Front-End for ASR over IP and GSM Networks: An Integrated Scenario," in *Proceedings of Eurospeech*, Vol. 2, Aalborg, Dinamarca, September 2001, pp. 1103–1106.
[18] Gao, Y. et al., "The SMV Algorithm Selected by TIA and 3GPP2 for CDMA Applications," in *Proceedings of ICASSP*, Salt Lake City, UT, 2001, pp. 709–712.
[19] Gerson, I.A. and Jasiuk, M.A., "Vector Sum Excited Linear Prediction (VSELP)," in *Advances in Speech Coding*, Kluwer, Norwell, MA, 1991, pp. 69–79.
[20] Guilmin, G., Gournay, P. and Chartier, F., "Description of the French NATO Candidate," in *IEEE Workshop on Speech Coding*, Tsukuba, Ibaraki, Japan, 2002.
[21] Hagenauer, J., "Source-Controlled Channel Decoding," *IEEE Transactions on Communications*, 43, 2449–2457, 1995.
[22] Heinen, S. and Vary, P., "Joint Source-Channel MMSE-Decoding of Speech Parameters," in *Proceedings of ICASSP*, Istanbul, Turkey, 2000, pp. 1507–1510.
[23] Herzog, S., "RSVP Extensions for Policy Control," IETF RFC 2750, January 2000.
[24] Hirsch, H.G., "The Influence of Speech Coding on Recognition Performance in Telecommunication Networks," in *Proceedings of ICSLP 2002*, Denver, CO, 2002, pp. 1877–1880.
[25] Ipsec IETF Work Group, http://www.ietf.org/html.charters/ipsec-charter.html.
[26] ITU-T, "Dual Rate Speech Coder for Multimedia Communications Transmitting at 5.3 and 6.3 kbit/s," Recommendation G.723.1, ITU-T, 1996.
[27] ITU-T, "40, 32, 24, 16 kbps Adaptive Differential Pulse Code Modulation (ADPCM)," Recommendation G.726, ITU-T, 1990.
[28] ITU-T, "Pulse Code Modulation (PCM) of Voice Frequencies," Recommendation G.711, ITU-T, 1988.
[29] ITU-T, "Coding of Speech at 8 kbit/s Using Conjugate-Structure Algebraic-Code-Excited Linear-Prediction (CS-ACELP)," Recommendation G.729, ITU-T, 1996.
[30] ITU-T, "Reduced Complexity 8 kbit/s CS-ACELP Speech Coder," Recommendation G.729, Annex A, ITU-T, 1996.
[31] ITU-T, "Packet-Based Multimedia Communications Systems," H.323 Standard Recommendation, ITU-T, 2000.
[32] Junqua, J.C., Ed., "Robustness in Language and Speech Technology," Kluwer Academic, Dordrecht, Netherlands, 2001.
[33] Kim, H.K., Cox, R.V. and Rose, R.C., "Performance Improvement of a Bitstream-Based Front-End for Wireless Speech Recognition in Adverse Environments," *IEEE Transactions on Speech and Audio Processing*, 10(8), 591–604, 2002.
[34] Lilly, B.T. and Paliwal, K.K., "Effect of Speech Coders on Speech Recognition Performance," in *Proceedings of ICSLP-96*, Vol. 4, Philadelphia, PA, 1996, pp. 2344–2347.
[35] Muthusamy, Y., Gong, Y. and Gupta, R., "The Effects of Speech Compression on Speech Recognition and Text-to-Speech Analysis," in *Proceedings of ICSLP-2002*, Denver, CO, 2002, pp. 2229–2232.
[36] Peláez-Moreno, C., Gallardo-Antolín, A. and Díaz-de.-María, F., "Recognizing Voice Over IP. A Robust Front-End for Speech Recognition on the WWW," *IEEE Transactions on Multimedia*, 3, 209, 2001.
[37] Peláez-Moreno, C., Gallardo-Antolín, A., Parrado-Hernández, E. and Díaz-de-María, F., "SVM-Based Lost Packet Concealment for ASR Applications over IP," in *Proceedings of the European Signal Processing Conference (EUSIPCO)*, Vol. 3, Toulouse, France, September 2002, pp. 529–532.
[38] Perkins, C., Hodson, O. and Hardman, V., "A Survey of Packet-Loss Recovery Techniques for Streaming Audio," *IEEE Network*, 12(5), 40–48, 1998.
[39] Quatieri, T.F., Singer, E., Dunn, R.B., Reynolds, D.A. and Campbel, J.P., "Speaker and Language Recognition Using Speech Coder Parameters," in *Proceedings of Eurospeech*, Vol. 2, Budapest, Hungary, 1999, pp. 787–790.

[40] Ramaswamy, G.N. and Gopalakrishnan, P.S., "Compression of Acoustic Features for Speech Recognition in Network Environments," in *Proceedings of ICASSP*-98, Vol. II, Seattle, WA, 1998, pp. 977–80.
[41] Rosenberg, J., et al., "SIP: Session Initiation Protocol," IETF RFC 3261, June 2002.
[42] Sanneck, H., "Packet Loss Recovery and Control for Voice Transmission over the Internet," GMD Research Series 8, 2000.
[43] Schulzrinne, H., Casner, S., Frederick, R. and Jacobson, V., "RTP: A Transport Protocol to Real-Time Applications" (work in progress), IETF, 2000.
[44] Sherif, M.H. et al., "Overview and Performance of CCITT/ANSI Embedded ADPCM Algorithms," *IEEE Transactions on Communications*, 41(2), 1993.
[45] Soong, F.K. and Huang, B.-H., "Line Spectum Pair (LSP) and Speech Data Compression," in *Proceedings of ICASSP-84*, San Diego, CA, 1984, pp. 1.10.1–1.10.4.
[46] Street, M., "Host Laboratory Role in the Selection of the Future NATO Narrow Band Voice Coder," in *IEEE Workshop on Speech Coding*, Tsukuba, Ibaraki, Japan, 2002.
[47] Tan, Z.-H. and Dalsgaard, P., "Channel Error Protection Scheme for Distributed Speech Recognition," in *Proceedings of ICSLP 2002*, Denver, CO, 2002, pp. 2225–2228.
[48] Vary, P., Hoffman, R., Hellwig, K. and Sluyter, R., "A Regular-Pulse Excited Linear Predictive Coder," *Speech Communication*, 7(2), 209–215, 1988.
[49] Wang, T., Koishida, K., Cuperman, V., Gersho, A. and Collura, J.S., "A 1200/2400 BPS Coding Suite Based on MELP," in *IEEE Workshop on Speech Coding*, Tsukuba, Ibaraki, Japan, 2002.
[50] Wright, D., "Voice over MPLS Compared to Voice over Other Packet Transport Technologies," *IEEE Communications Magazine*, 40(11), 124–132, 2002.

VIII

Wireless Geolocation Techniques

21 **Geolocation Techniques for Mobile Radio Systems** *James J. Caffery, Jr. and Saipradeep Venkatraman* **21**-1
Introduction • Geolocation Methods • Geolocation Algorithms • Location Parameter Estimation • Impairments to Accuracy • Provisions in the Standards • Summary

22 **Adaptive Arrays for GPS Receivers** *Moeness Amin, Wei Sun, and Alan Lindsey*... **22**-1
Introduction • GPS Signal Model • Interference Suppression Techniques in GPS • Multipath Mitigation in GPS • Conclusions

21

Geolocation Techniques for Mobile Radio Systems

James J. Caffery, Jr.
University of Cincinnati

Saipradeep Venkatraman
University of Cincinnati

21.1 Introduction **21**-1
FCC Regulations for E-911 • Location-Based Services (LBS)
21.2 Geolocation Methods **21**-3
21.3 Geolocation Algorithms **21**-5
Geometric Solutions • Least Squares Estimation • Other Location Algorithms • Hybrid Location Methods • Performance
21.4 Location Parameter Estimation **21**-12
AOA Estimation • Range Estimation • Range Difference Estimation • Joint Parameter Estimation
21.5 Impairments to Accuracy **21**-14
Multipath Propagation • Hearability • Non-Line-of-Sight Propagation
21.6 Provisions in the Standards **21**-17
3G Location Solutions • Locating Legacy Terminals
21.7 Summary **21**-18

21.1 Introduction

With the emergence of enhanced services in third-generation (3G) wireless communications systems comes the necessity of developing technologies that will enable and support those services. In many applications, wireless position location, or geolocation, is a key component. For wireless communication networks, an inherently suitable approach for wireless geolocation is known as *radiolocation*, in which the parameters that are used for location estimation are obtained from radio signal measurements between a target and one or more fixed stations. In the realm of cellular and PCS systems, the mobile subscribers, or mobile stations (MSs), serve as the target, while the base stations (BSs) serve the role of the fixed stations. Geolocation, then, can be categorized in two ways: handset-based location and network-based location. For handset-based location, parameters that are used to compute an MS's position are measured at the MS and either used to calculate position in the MS or transferred to a central processing facility for position estimation. In network-based solutions, the parameters used for computing location are measured at the BSs and transferred to the central facility for location determination. Both approaches have their advantages and drawbacks. Network-based solutions do not require modifications to the handset but may be limited in

0-8493-1657-X/$0.00+$1.50

their utility due to the "hearability" problem (see Section 21.5.2). Handset-based solutions, on the other hand, require handset modification but have the potential of greater location accuracy when modified with a Global Positioning System (GPS) capable receiver. Both approaches take advantage of the existing wireless communications infrastructure without the need for supplementary technology such as dead reckoning.

GPS, a satellite-based location system, is a popular solution for providing location in terrestrial wireless communication networks. GPS is a proven technology and provides high accuracy when a line-of-sight (LOS) path exists between the receiver and at least four satellites. However, GPS is often inoperable in areas where satellites are blocked, such as in buildings and built-up urban areas. Further, the time-to-first-fix (TTFF) for a conventional GPS receiver from a "cold" start can take several minutes. Additionally, adding GPS functionality to a handset can be costly, bulky, and drain battery power at an unacceptable rate [1]. A proposed alternative to conventional GPS, assisted GPS (AGPS), shows promise of overcoming some of these limitations (see Section 21.6). But regardless of whether GPS or AGPS is chosen as the location solution by service providers, GPS solutions only apply to new handsets and cannot be used with legacy terminals. Consequently, radiolocation systems based on the wireless communications infrastructure are still an important area of research.

Of particular interest to wireless subscribers, industry, and government is the provision of emergency services to wireless callers. Over the past decade, the percentage of emergency calls made from wireless phones has drastically increased with the increasing reliance on wireless phones in the general public. Without accurate location information for the increasing volume of wireless calls, the emergency operator must determine the caller's position by verbally querying the caller. This can take some time and is prone to error if the caller is under duress or is in an unfamiliar area. In addition to the potential loss of life and property, the time required to process each call in this fashion places an increasing strain on the emergency system resources as the percentage of wireless emergency calls increases.

21.1.1 FCC Regulations for E-911

Much of the activity over the past several years in the area of wireless geolocation has been driven by the 1996 Report and Order of the Federal Communications Commission (FCC) [2]. The mandate established guidelines for wireless service providers to provide both basic and enhanced 911 (E-911) capabilities. Implementation of location services was to be carried out in two phases. In Phase I, wireless carriers were required to provide public safety answering points (PSAPs) with a call-back number and the location of the cell site or sector that was receiving the 911 call. Such capabilities allowed the PSAP to call back if the call was disconnected. Phase II, which was to be completed by October 2001, required the wireless carriers to provide the location of all 911 calls to the PSAP. At first, it was anticipated that the wireless carriers would deploy network-based location technologies to meet the Phase II requirements of 100-m accuracy for 67% of 911 calls and 300-m accuracy for 95% of the calls. By 1999, technological advances allowed for the implementation of handset-based technologies. As a result, in October 1999, the FCC revised its rules and imposed stricter accuracy requirements of 50 m for 67% of 911 calls and 150 m for 95% of the calls for handset-based solutions. The network-based requirements remained unchanged.

21.1.2 Location-Based Services (LBS)

The FCC requirements may provide the driving force behind geolocation technological development and deployment, particularly in the U.S., but several new and enhanced services can also benefit from the technology. Location information can be used to support commercial services such as navigation or route guidance, electronic yellow pages, and location-sensitive billing [3, 4]. The so-called 'value-added services' provide wireless subscribers with location-based Internet services, local broadcasting, and local traffic information [5]. Security issues such as improved fraud detection also benefit with the ability to locate fraudulent users more quickly and efficiently. Additionally, the emerging personal safety services will rely heavily on accurate location [6].

In terms of system enhancement, location technologies provide wireless carriers with the means of improving wireless communications system design and performance. Location has been used to improve routing for *ad hoc* networks [7] and for network planning where radio resources are dynamically redistributed to improve coverage and capacity in areas regularly visited by users [8]. Combinations of location with geographic information services (GIS) also provide the potential for greatly improving the efficiency of resource allocation in a network [9]. Additionally, location information can play an important role in assisting handoffs between cells and in designing hard and soft handoff schemes for wireless networks [10, 11].

An application that has recently received considerable attention in the U.S. is the area of Homeland Security. Position reporting is essential in providing a defense against emerging national threats as well as handling conventional emergencies. Accurate location information is essential in allowing law enforcement to respond quickly to reports of suspicious activity. In the event of an attack, a large number of emergency calls may be originated and accurate location information would allow screening of calls that are placed in response to the same event.

21.2 Geolocation Methods

The parameters that are often measured and used for location include angles of arrival (AOAs), signal strength, times of arrival (TOAs), and time differences of arrival (TDOAs). The fundamental methods of radiolocation that use these parameters can be grouped into three categories: direction finding, ranging, and range differencing.

Direction-finding, or AOA, methods measure the angle, θ, at which a signal is incident at a BS whose position is known. Given the known position and bearing of the incoming signal, a linear line of position (LOP) can be formed on which the MS lies. For location in two dimensions, two such measurements are required at two BSs to form two linear LOPs whose intersection is the estimate of the MS position, as illustrated in Figure 21.1. The AOA information can be obtained through the use of antenna arrays or directional antennas. For more available measurements (i.e., an overdetermined system), other approaches for determining location can be formed as described in Section 21.3.2.

Ranging methods are based on measuring the distance between the MS and multiple BSs. Since a circle is a curve of constant distance from a point, LOPs for ranging are circular. Because of this, range measurements between the MS and three BSs are required in order to resolve the ambiguity that only two measurements produce in two-dimensional location. However, if additional information is available, such as the cell sector in which an MS resides in a cellular network, one of the two possible solutions can be eliminated, and thus, two-dimensional location may be performed with only two BSs. The location of the MS lies at the intersection of the circular LOPs, as illustrated in Figure 21.2. Range estimates can be obtained based on signal strength or TOA measurements.

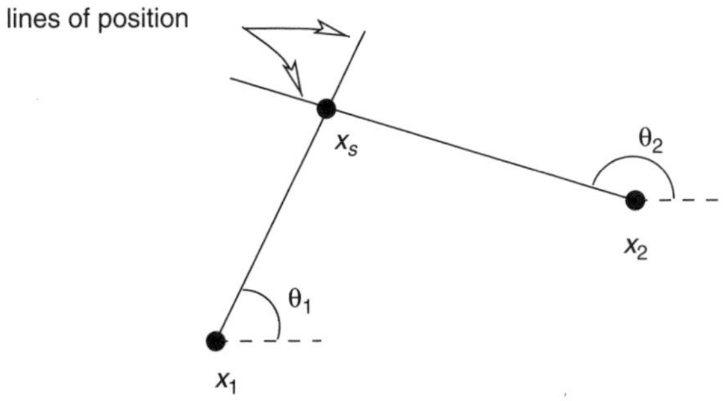

FIGURE 21.1 The AOA method of location estimation.

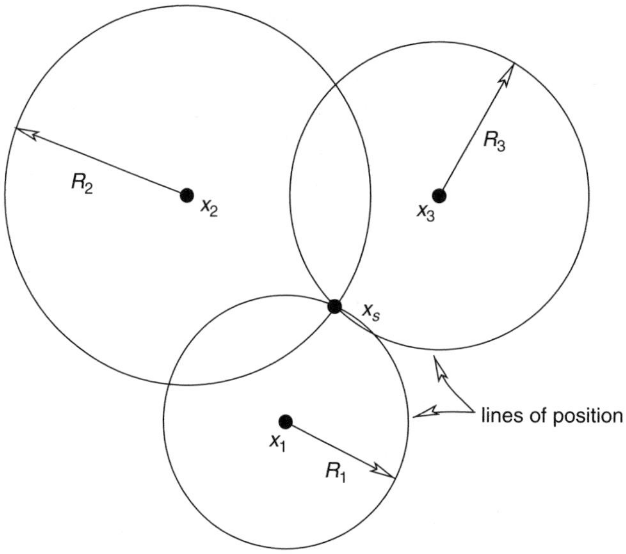

FIGURE 21.2 Location estimation based on range measurements.

Range-differencing methods determine the MS position by measuring the differences in distances between two BSs and an MS. Since a hyperbola is a curve of constant distance difference between a point on the hyperbola and its two foci, the LOPs for range differencing are hyperbolic. Since two BSs are required to form a single LOP, three BSs are required to generate two LOPs whose intersection provide the location of the MS, as illustrated in Figure 21.3. Note that each set of foci (i.e., BSs) produce two possible hyperbolic curves. The appropriate curve can be chosen by noting the sign of the measured range difference with respect to the two BSs. Range differences can be obtained by differencing measured ranges or by direct computation.

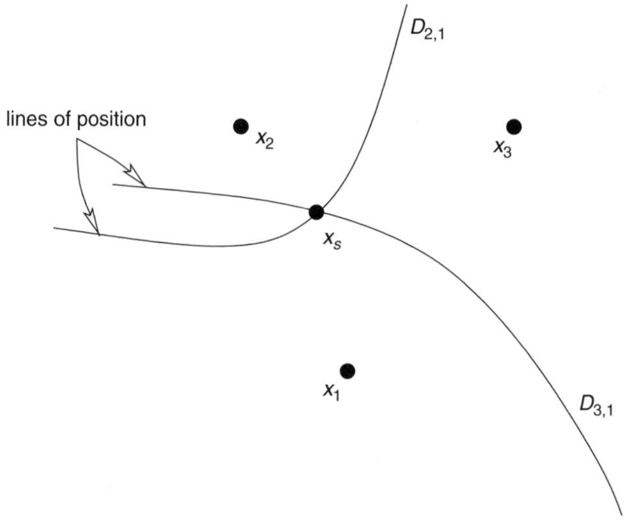

FIGURE 21.3 Range differencing used for location estimation.

21.3 Geolocation Algorithms

21.3.1 Geometric Solutions

A straightforward approach for estimating the location of the MS for direction finding, ranging, and range differencing is to compute the point of intersection of the LOPs for each of those methods. Solutions formed in this manner take advantage of the underlying geometric formulation of the location problem. We will assume in this section that the number of BSs used is the minimum required for two-dimensional location. The MS position in an overdetermined system, in which the number of AOA, range, or range-difference measurements is greater than the minimum required, is better estimated with algorithms discussed in Section 21.3.2.

Suppose that the MS is located at position $\mathbf{x}_s = [x_s, y_s]^T$, while the BSs are located at $\mathbf{x}_i = [x_i, y_i]^T$, $i = 1, \ldots, N_B$. For AOA location estimation, the measured angle at the ith BS, $i = 1, 2$, is given by

$$\theta_i = \tan^{-1}\left(\frac{y_i - y_s}{x_i - x_s}\right) \tag{21.1}$$

where it is assumed that each AOA is measured with respect to a common baseline, for instance, the x-axis. The linear LOP for the ith BS that passes through point $\mathbf{x}_i$ can then be represented as

$$y_s = x_s \tan\theta_i + (y_i - x_i \tan\theta_i) \tag{21.2}$$

Equating the LOPs for $i = 1, 2$ produces the estimate of x_s:

$$x_s = \frac{y_2 - y_1 - x_2 \tan\theta_2 + x_1 \tan\theta_1}{\tan\theta_1 - \tan\theta_2} \tag{21.3}$$

which can be substituted into Equation 21.2 for $i = 1$ or $i = 2$ to form the estimate of y_s.

For ranging, the distance between the MS and ith BS can be represented as

$$R_i = \|\mathbf{x}_i - \mathbf{x}_s\| = \sqrt{(x_i - x_s)^2 + (y_i - y_s)^2} \tag{21.4}$$

for $N_B = 3$. We note that with measurement errors, the three circular LOPs may not overlap at a single point. Regardless, we may still form a location estimate by recognizing that we can transform the circular LOPs to linear LOPs by using the lines connecting the points of intersection for each pair of range circles [12, 13], as illustrated in Figure 21.4. The two LOPs can be expressed as

$$(x_2 - x_1)x_s + (y_2 - y_1)y_s = \frac{1}{2}\left(\|\mathbf{x}_2\|^2 - \|\mathbf{x}_1\|^2 + R_1^2 - R_2^2\right)$$

$$(x_3 - x_2)x_s + (y_3 - y_2)y_s = \frac{1}{2}\left(\|\mathbf{x}_3\|^2 - \|\mathbf{x}_2\|^2 + R_2^2 - R_3^2\right)$$

It is then straightforward to show that

$$x_s = \frac{(y_2 - y_1)C_3 - (y_3 - y_2)C_1}{[(x_3 - x_2)(y_2 - y_1) - (x_2 - x_1)(y_3 - y_2)]} \tag{21.5}$$

$$y_s = \frac{(x_2 - x_1)C_3 - (x_3 - x_2)C_1}{[(y_3 - y_2)(x_2 - x_1) - (y_2 - y_1)(x_3 - x_2)]} \tag{21.6}$$

where

$$C_1 = \frac{1}{2}\left(\|\mathbf{x}_2\|^2 - \|\mathbf{x}_1\|^2 + R_1^2 - R_2^2\right)$$

$$C_3 = \frac{1}{2}\left(\|\mathbf{x}_3\|^2 - \|\mathbf{x}_2\|^2 + R_2^2 - R_3^2\right)$$

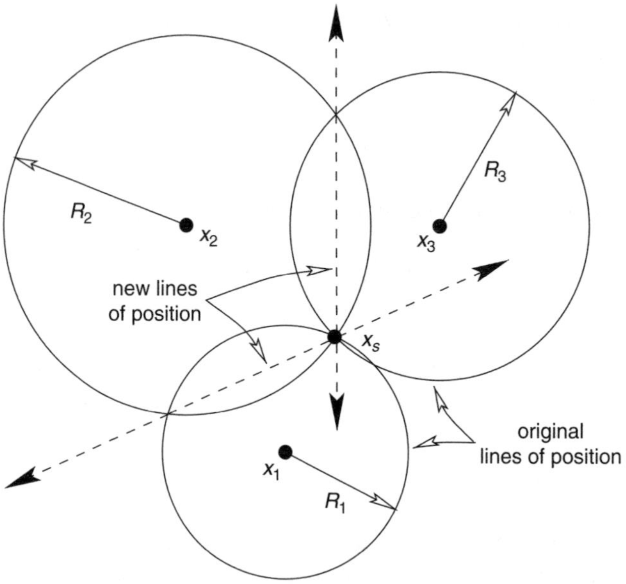

FIGURE 21.4 Generation of linear LOPs from circular LOPs for ranging location estimation.

For range differencing, we define the difference in distances to the MS from the ith and jth BSs as

$$D_{i,j} = \|\mathbf{x}_i - \mathbf{x}_s\| - \|\mathbf{x}_j - \mathbf{x}_s\| \tag{21.7}$$

which is a hyperbolic equation with foci at the two BSs. Note that for BSs i, j, and k, $D_{k,i} = D_{k,j} - D_{i,j}$. It can be shown that the x- and y-coordinates of the MS location estimate are related by [13]

$$y_s = mx_s + b \tag{21.8}$$

where

$$m = \frac{D_{3,1}(x_2 - x_1) - D_{2,1}(x_3 - x_1)}{D_{2,1}(y_3 - y_1) - D_{3,1}(y_2 - y_1)}$$

$$b = \frac{D_{2,1}\|\mathbf{x}_3\|^2 - D_{3,1}\|\mathbf{x}_2\|^2 + D_{3,2}\|\mathbf{x}_1\|^2 + c^2 D_{3,1} D_{2,1} D_{3,2}}{2\left[D_{2,1}(y_3 - y_1) - D_{3,1}(y_2 - y_1)\right]}$$

By squaring and differencing Equation 21.7 for $D_{2,1}$ and $D_{3,1}$, and then substituting for y_s in Equation 21.8, we obtain the quadratic in x_s:

$$Ax_s^2 + Bx_s + C = 0 \tag{21.9}$$

where

$$A = 4c^2 D_{2,1}^2 (1 + m)^2 - u^2$$
$$B = -4c^2 D_{2,1}^2 [2x_1 - 2m(y_1 - b)] + 2u$$
$$C = 4c^2 D_{2,1}^2 \left(\|\mathbf{x}_1\|^2 - 2y_1 b + b^2\right) - v^2$$

and

$$u = 2[x_1 - x_2 + m(y_1 - y_2)]$$
$$v = \|\mathbf{x}_2\|^2 - \|\mathbf{x}_1\|^2 + 2b(y_1 - y_2) - c^2 D_{2,1}^2$$

which can then be used to compute y_s in Equation 21.8. The solution to Equation 21.9 yields two solutions for x_s leading to an ambiguity. The two solutions arise from the fact that there are two possible branches to the hyperbola. The ambiguity can be resolved by noting the sign of one of the range differences, for instance, $D_{2,1}$. If $D_{2,1} > 0$, then the MS lies closer to BS 1 than BS 2. Unfortunately, a consequence of this approach is that there are frequently two or more possible solutions and there is no way to distinguish them without more information [14]. Transformation of the hyperbolic LOPs to linear LOPs using geometric arguments for range difference estimation with four BSs in two dimensions is given in [15]. In this approach, the MS location is at the intersection of the linear LOPs, which are obtained from the major axes of two conics on which the BSs lie. For range-differencing location in three dimensions, an extension of the approach used above is developed in [16] using four BSs. For the overdetermined case, in which there are more equations than unknowns, a method is developed in [17] for direct computation of a location estimate.

21.3.2 Least Squares Estimation

When there are errors in the measured AOAs, ranges, or range differences, the solutions above can be utilized with direct substitution of the measured quantities. However, statistical solutions are more justifiable in the presence of measurement errors and are more suitable for overdetermined systems, unlike the solutions in the previous section. To facilitate a statistical approach to the problem of MS location estimation, we define a noisy vector of measurements at the N_B BSs as

$$\mathbf{m} = \mathbf{F}(\mathbf{x}_s) + \mathbf{n} \tag{21.10}$$

where $\mathbf{n}$ is the measurement noise vector with zero-mean and covariance matrix $\mathbf{Q}$. The vector $\mathbf{F}(\mathbf{x}_s)$ contains the AOAs, ranges, and range differences defined in Equations 21.1, 21.4, and 21.7, respectively. The goal is to form a location estimate given the measurements in Equation 21.10.

A common approach for determining an estimate from a noisy set of measurements is the method of least squares (LS) estimation, which forms a location estimate, $\hat{\mathbf{x}}_s$, by minimizing

$$\mathcal{E}(\hat{\mathbf{x}}_s) = [\mathbf{m} - \mathbf{F}(\hat{\mathbf{x}}_s)]^T\mathbf{W}[\mathbf{m} - \mathbf{F}(\hat{\mathbf{x}}_s)] \tag{21.11}$$

Even when the noise cannot be assumed to be Gaussian, this is a reasonable approach for location estimation. For simplicity, we will assume that $\mathbf{W} = \mathbf{I}$.

Since the measured parameters are nonlinear functions of the MS position, the LS problem is a nonlinear one. A straightforward approach is to minimize Equation 21.11 using a gradient descent method [18]. With this approach, an initial guess is made of the MS location and successive estimates are updated according to

$$\hat{\mathbf{x}}_s^{(k+1)} = \hat{\mathbf{x}}_s^{(k)} - \boldsymbol{\mu}\,\nabla\mathcal{E}\left(\hat{\mathbf{x}}_s^{(k)}\right) \tag{21.12}$$

where the matrix $\boldsymbol{\mu} = \text{diag}[\mu_x, \mu_y]$ is the step size, $\hat{\mathbf{x}}_s^{(k)}$ is the estimate at iteration k, and $\nabla = \partial/\partial\mathbf{x}$ denotes the gradient vector with respect to $\mathbf{x}$. Because the system of equations is nonlinear, the error surface is multimodal and the algorithm could converge to one of the local minima and not the global minimum. Initialization of the algorithm, $\mathbf{x}_s^{(0)}$, is also important.

In order to mold the problem into a linear LS problem, the nonlinear function $\mathbf{F}(\mathbf{x}_s)$ can be linearized using a Taylor series expansion [19, 20, 21] about some reference point $\mathbf{x}_o$ so that

$$\mathbf{F}(\mathbf{x}_s) \approx \mathbf{F}(\mathbf{x}_o) + \mathbf{H}(\mathbf{x}_s - \mathbf{x}_o) \tag{21.13}$$

where $\mathbf{H}$ is the Jacobian matrix of $\mathbf{F}(\mathbf{x}_s)$ evaluated at $\mathbf{x}_o$. Then, the LS solution can be formed as

$$\hat{\mathbf{x}}_s = \mathbf{x}_o + (\mathbf{H}^T\mathbf{H})^{-1}\mathbf{H}^T[\mathbf{m} - \mathbf{F}(\mathbf{x}_o)] \tag{21.14}$$

This approach can be performed iteratively, with each successive estimate being closer to the final estimate [19, 21]. Alternatively, if we ignore the linearization errors, a direct LS solution can be obtained by

$$\hat{\mathbf{x}}_s = (\mathbf{H}^T\mathbf{H})^{-1}\mathbf{H}^T\mathbf{m} \tag{21.15}$$

For AOA location in the overdetermined case, a system of linear equations in the unknown MS position, $\mathbf{x}_s$, can be generated without linearization by utilizing the relationship

$$\mathbf{x}_s = \mathbf{x}_i + R_i \begin{bmatrix} \cos\theta_i \\ \sin\theta_i \end{bmatrix} \tag{21.16}$$

Solving these two equations for R_i and equating the results produces the system, $i = 1, \ldots, N_B$, of linear relationships

$$\mathbf{A}\mathbf{x}_s = \mathbf{b} \tag{21.17}$$

where

$$\mathbf{A} = \begin{bmatrix}\boldsymbol{\omega}_1 \cdots \boldsymbol{\omega}_{N_B}\end{bmatrix}^T \quad \text{and} \quad \mathbf{b} = \begin{bmatrix} \boldsymbol{\omega}_1^T\mathbf{x}_1 \\ \vdots \\ \boldsymbol{\omega}_{N_B}^T\mathbf{x}_{N_B} \end{bmatrix} \tag{21.18}$$

with $\boldsymbol{\omega}_i = [\sin\theta_i \;\; -\cos\theta_i]^T$ [22]. The LS estimate is then obtained as

$$\hat{\mathbf{x}}_s = (\mathbf{A}^T\mathbf{A})^{-1}\mathbf{A}^T\mathbf{b} \tag{21.19}$$

For ranging (specifically, TOA), an LS approach that avoids linearization of the nonlinear functions $\mathbf{F}(\mathbf{x}_s)$ uses the linear LOPs given in Equation 21.2 for the overdetermined case. For $N_B > 3$ BSs, $N_B - 1$ nonredundant linear LOPs can be formed and represented in matrix notation as in Equation 21.17 where [12]

$$\mathbf{A} = \begin{bmatrix} (\mathbf{x}_2 - \mathbf{x}_1)^T \\ (\mathbf{x}_3 - \mathbf{x}_2) \\ \vdots \\ \mathbf{x}_{N_B} - \mathbf{x}_{N_B-1} \end{bmatrix} \quad \text{and} \quad \mathbf{b} = \frac{1}{2}\begin{bmatrix} \|\mathbf{x}_2\|^2 - \|\mathbf{x}_1\|^2 + R_1^2 - R_2^2 \\ \|\mathbf{x}_3\|^2 - \|\mathbf{x}_2\|^2 + R_2^2 - R_3^2 \\ \vdots \\ \|\mathbf{x}_{N_B}\|^2 - \|\mathbf{x}_{N_B-1}\|^2 + R_{N_B-1}^2 - R_{N_B}^2 \end{bmatrix} \tag{21.20}$$

so that the LS solution is as given in Equation 21.19.

For range differencing, or TDOA, several other algorithms have been developed that exploit the hyperbolic equations. One LS-based location algorithm, called the spherical interpolation (SI) method, is developed in [23]. By mapping the spatial origin to the first BS and through some simple algebraic manipulation, an error expression can be obtained as

$$\mathbf{e} = \boldsymbol{\delta} - 2R_s\mathbf{m} - 2\mathbf{S}\mathbf{x}_s \tag{21.21}$$

where $R_s = \|\mathbf{x}_s\|$, $\mathbf{m}$ is the measured TDOA vector,

$$\boldsymbol{\delta} = \begin{bmatrix} R_s^2 - D_{2,1}^2 \\ \vdots \\ R_s^2 - D_{N_B,1}^2 \end{bmatrix} \quad \text{and} \quad \mathbf{S} = \begin{bmatrix} \mathbf{x}_2^T \\ \vdots \\ \mathbf{x}_{N_B}^T \end{bmatrix} \tag{21.22}$$

Note that Equation 21.21 is linear in $\mathbf{x}_s$ given R_s, and it is also linear in R_s given $\mathbf{x}_s$. Consequently, a two-step LS solution is developed that first provides the linear LS solution of $\mathbf{x}_s$ given R_s, which is then substituted back into Equation 21.21 to form the linear LS solution of R_s given $\mathbf{x}_s$. The result of the method is the LS estimate

$$\hat{\mathbf{x}}_s = \frac{1}{2}[\mathbf{S}^T\mathbf{S}]^{-1}\mathbf{S}^T[\boldsymbol{\delta} - 2\hat{R}_s\mathbf{m}] \tag{21.23}$$

where $\hat{R}_s$ is the LS solution for R_s. A variation of this algorithm is called the spherical intersection (SX) method, which inserts the linear LS estimate of $\mathbf{x}_s$ given R_s into $R_s = \mathbf{x}_s^T\mathbf{x}_s$ [24]. The result is a quadratic equation in R_s, the positive root of which is used to compute the location estimate.

Another TDOA LS solution that avoids linearization uses a transformation of the TDOA measurements [25]. The simple transformation results in the new 'measurements' given by

$$\boldsymbol{\varphi} = \boldsymbol{\Delta}\mathbf{x}_s + R_1\mathbf{m} \tag{21.24}$$

where

$$\boldsymbol{\varphi} = \frac{1}{2}\begin{bmatrix} \|\mathbf{x}_2\|^2 - \|\mathbf{x}_1\|^2 - D_{2,1}^2 \\ \vdots \\ \|\mathbf{x}_N\|^2 - \|\mathbf{x}_1\|^2 - D_{N,1}^2 \end{bmatrix} \quad \text{and} \quad \boldsymbol{\Delta} = \begin{bmatrix} (\mathbf{x}_2 - \mathbf{x}_1)^T \\ \vdots \\ (\mathbf{x}_N - \mathbf{x}_1)^T \end{bmatrix} \tag{21.25}$$

The term R_1 is nonlinear in $\mathbf{x}_s$ and can be removed by using a projection matrix, $\mathbf{P}$, that has $\mathbf{m}$ in its null space. Projecting Equation 21.24 with $\mathbf{P}$ results in

$$\mathbf{P}\boldsymbol{\varphi} = \mathbf{P}\boldsymbol{\Delta}\mathbf{x}_s \tag{21.26}$$

which leads to the linear LS estimate of the MS location:

$$\hat{\mathbf{x}}_s = (\boldsymbol{\Delta}^T\mathbf{P}^T\mathbf{P}\boldsymbol{\Delta})^{-1}\boldsymbol{\Delta}^T\mathbf{P}^T\mathbf{P}\boldsymbol{\varphi} \tag{21.27}$$

For location in n dimensions, this method requires $n + 2$ BSs. It is shown in [25] that this result is equivalent to the SI method. A more general form for this LS solution is developed in [26]. Another two-step minimization method is developed in [27] that is optimal for small TDOA estimation errors.

21.3.3 Other Location Algorithms

While the LS approach to MS location is the predominant solution used in the literature, other approaches have been developed as well in order to generate improved location estimators. A Bayesian approach that partitions the location area into grids and uses signal strength measurements is presented in [28]. A maximum likelihood (ML) approach that applies to signal strength, TOA, and TDOA location systems is presented in [29] with a modification for the inclusion of sector cell information in [30]. Another ML technique is developed in [31] for signal strength measurements using robust functions, while an approximate ML and approximate minimum mean square error method are developed in [32] using nonparametric estimates of the density functions for TDOA location. ML and other parametric approaches are developed in [33] for AOA, TOA, and TDOA. Finally, a signal strength ML method is presented in [34], while a method of weighting cell-ID location estimates (see Section 21.6) using signal strength information is developed in [35].

21.3.4 Hybrid Location Methods

In many instances, several different types of measurements are available that can be used to improve an MS location estimate. For example, the linearized LS result in Equation 21.15 for AOA only could be modified to account for TOA and TDOA by augmenting the vector $\mathbf{F}(\mathbf{x}_s)$ in Equation 21.10, and hence $\mathbf{H}$ to include the appropriate measurements [36].

Hybrid TOA/TDOA methods are discussed in [14] and [37]. The combination is shown in [14] to improve the limitations of both methods when applied individually, assuming the TOAs and TDOAs are independent. The hybrid solution discussed in [37] uses a constrained weighted LS approach where the

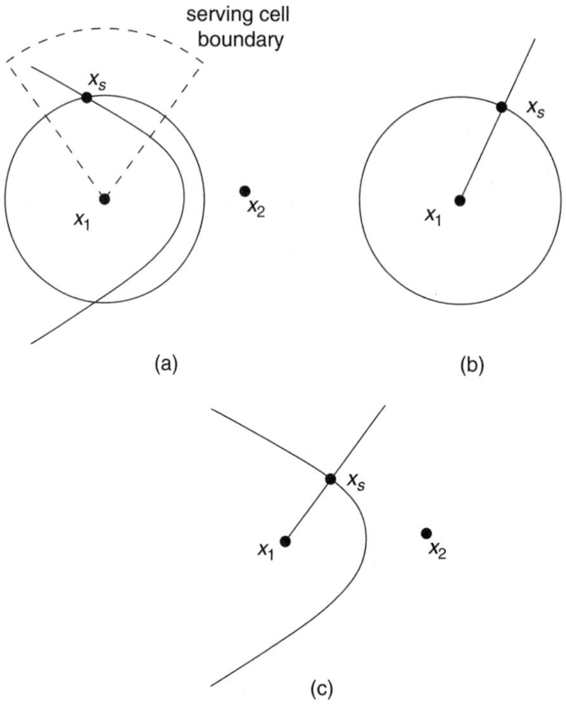

FIGURE 21.5 Examples of the geometry of (a) hybrid TOA/TDOA, (b) hybrid AOA/TOA, and (c) hybrid AOA/TDOA location techniques.

solution is constrained to be in the MS's serving cell/sector and within the range of the TOA measurement, as illustrated in Figure 21.5(a). A benefit of the approach is its ability to estimate the location of an MS with only two BSs.

Hybrid AOA/TOA methods are investigated in [36] and [38]. The simplest form of AOA/TOA location is classical radar, where a direction and range from a known point provide a location estimate. With a measured range R_i and direction θ_i from a BS located at $\mathbf{x}_i$, the MS position estimate is computed from Equation 21.16, which defines the intersection of the linear and circular LOPs from the AOA and TOA measurements, respectively, as illustrated in Figure 21.5(b). This approach is particularly useful near the BS where line-of-sight (LOS) propagation is more likely and the effect of angular error is reduced. It has the additional benefit that location estimation is possible with only a single BS. Reference [38] uses this method in order to find a starting point for the nonlinear LS algorithm in [18].

Hybrid AOA/TDOA methods can also be implemented in which the MS position lies at the intersection of the linear and hyperbolic LOPs from the AOA and TDOA measurements, respectively [39]. Thus, the location of the MS can be estimated with only two BSs, as illustrated in Figure 21.5(c). Other methods are developed in [39] and [40] that can accommodate more than two BSs. In [40], Kalman filtering with appropriate models for the measured TDOAs and a single AOA measured at the serving BS is employed to track an MS in motion. In [39], an LS location estimate based on a Taylor series linearization, as in [36], and a two-step LS approach, which extends [27] to include AOA measurements, are developed.

21.3.5 Performance

To illustrate the relative performance of several of the algorithms discussed above, consider a network of BSs where the cells have an ideal hexagonal shape and assume that the BS serving the target MS is in the center of the surrounding BSs. Further, assume that the cell radii are each 2 km and TOA/TDOA measurements

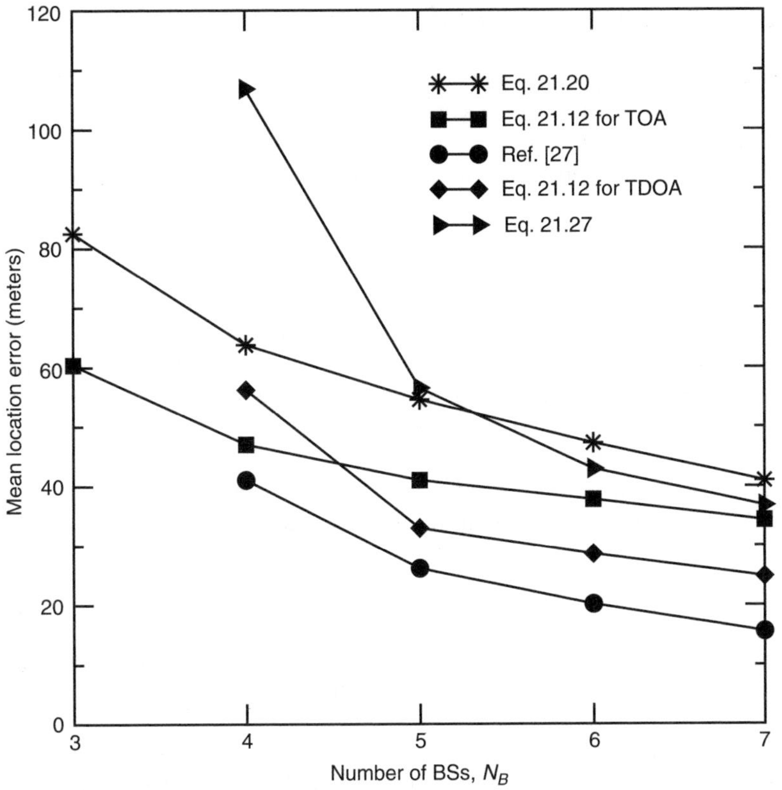

FIGURE 21.6 TOA and TDOA performance comparison.

have a standard deviation of 50 m. Figure 21.6 shows the mean location error of the TOA algorithms in Equations 21.12 and 21.20 and the TDOA algorithms in Equations 21.12, and 21.27 and reference [27] vs. the number of BSs used for location. As the figure illustrates, higher accuracy is achieved when using more BSs. In addition, the TDOA algorithms perform somewhat better than the TOA algorithms for the scenario considered.

A key factor that affects performance is the geometry of the BSs used for location. A common measure of the performance is the geometric dilution of precision (GDOP), which is defined as

$$\text{GDOP} = \frac{\sqrt{\text{E}\left[(\hat{\mathbf{x}}_s - \hat{\boldsymbol{\mu}})^T(\hat{\mathbf{x}}_s - \hat{\boldsymbol{\mu}})\right]}}{\sigma_r} \tag{21.28}$$

where σ_r denotes the fundamental TOA/TDOA error and $\hat{\boldsymbol{\mu}} = \text{E}[\hat{\mathbf{x}}_s]$ is the mean location estimate. For AOA, σ_r^2 is the average variance of the distance between each BS and a reference point near the true position of the MS. The GDOP is an indicator of the extent to which the fundamental ranging error is magnified by the geometric relation between the MS and BSs. It serves as a useful criterion for selecting the set of BSs from a large set to produce the minimum location error. In addition, it may aid cell site planning for cellular networks that plan to provide location services to their users.

In the simulation results of Figure 21.6, the GDOP is relatively small (i.e., near unity). However, if any of the BSs are shifted closer to a nearby BS, the resulting GDOP might increase dramatically, thereby causing increased location error. See [41] for more information regarding the effects of geometry on GDOP and its importance for location accuracy.

21.4 Location Parameter Estimation

In the preceding sections, the AOAs, ranges (based on signal strength or TOA), and range differences (TDOAs) were assumed to be known. In practice, these parameters must be estimated before the location of the MS can be computed, and the accuracy with which this is accomplished affects the overall location accuracy. In the following, we summarize several methods for providing parameters for direction finding, ranging, and range differencing.

21.4.1 AOA Estimation

The AOAs are typically estimated through the use of antenna arrays at the BSs. An antenna array is composed of a number of antenna elements whose signals are often combined to produce a beam in a desired direction. In mobile radio systems, the antenna arrays are typically located only at the BSs, since it is difficult to place an array in an MS handset due to its physical size. Thus, AOA estimation is generally used for network-based solutions.

Consider an antenna array in which samples from L source signals are sampled at M antenna array elements. The $M \times 1$ received signal at the array can be expressed as

$$\mathbf{v}(t) = \mathbf{A}(\boldsymbol{\Theta})\mathbf{s}(t) + \mathbf{n}(t) \tag{21.29}$$

where $\mathbf{v}(t) = [v_1(t), \ldots, v_M(t)]^T$, $\mathbf{A}(\boldsymbol{\Theta}) = [\mathbf{a}(\theta_1), \ldots, \mathbf{a}(\theta_L)]$ is the matrix of steering vectors, and $\mathbf{s}(t) = [s_1(t), \ldots, s_L(t)]^T$ is the vector of source signals. With this formulation, there are several approaches for estimating $\boldsymbol{\Theta}$ for the L source signals.

An ML-based AOA estimator is presented in [42] for $L < M$ and linearly independent steering vectors. Due to the high complexity of the ML solution, other methods have been developed to reduce the complexity. One such technique, the *alternating projection* method, uses an iterative algorithm to transform the multidimensional, nonlinear problem to a sequence of simpler one-dimensional problems [42]. Details of other ML methods can be found in [43, 44].

Another approach for AOA estimation forms estimates by minimizing the output power of the antenna array beam former while a distortionless response is maintained along the direction of a signal of interest. Hence, the approach is known as the minimum variance distortionless response (MVDR) method [45]. This constrained minimization problem can be solved via Lagrangian multipliers [45] to produce the estimated AOAs.

An LS approach is developed in [46] in which the AOAs are computed from the minimization of

$$J = \mathrm{E}[|\mathbf{v}(n) - \mathbf{A}(\boldsymbol{\Theta})\mathbf{s}(n)|^2] \tag{21.30}$$

which is defined from Equation 21.29. From the estimate of $\mathbf{A}(\boldsymbol{\Theta})$, the AOAs can be determined.

A form of AOA estimation is the signal subspace approach made popular by Schmidt [47] with the MUSIC algorithm. Subspace algorithms operate by separating a signal subspace from a noise subspace and exploiting the statistical properties of each. MUSIC, in particular, exploits the eigenstructure of the input covariance matrix $\mathbf{R}_{vv} = \mathrm{E}[\mathbf{v}\mathbf{v}^H]$. The eigenvectors of the L larger eigenvalues are used to form the signal subspace, while the $M - L$ remaining eigenvectors are used to form the noise subspace. This approach utilizes the fact that the eigenvectors of the signal subspace (in which $\mathbf{a}(\theta)$ lies) and noise subspace are orthogonal.

Variants of the MUSIC algorithm have been developed to improve its resolution and decrease its computational complexity, including Root-MUSIC [48] and Cyclic MUSIC [49]. Other improved subspace-based AOA estimation techniques include the ESPRIT algorithm [50] and its variants, and a minimum-norm approach [51]. Further details of subspace-based estimation techniques can be found in [43, 52, 53].

Classic approaches for AOA estimation include Capon's ML method [54] and Burg's maximum entropy (ME) method [55]. Closed-form solutions for AOA estimation, as opposed to maximizing spectra, as in

MUSIC, are thoroughly described in [56]. Methods of location specific to mobile radio channels are given in [57, 58, 59].

21.4.2 Range Estimation

Estimates of the range between an MS and BS can be obtained by measuring signal strength or the TOA at either the MS or BS. Given the transmit power, the range between the MS and BS can be estimated by measuring the received signal power and applying an empirical model describing signal attenuation with distance [60, 61].

Signal strength is estimated by approximating the local mean μ_p of the received squared envelope $r^2(l)$ at position l according to [62]

$$\hat{\mu}_p = \frac{1}{2L}\int_{x-L}^{x+L} r^2(l)dl \tag{21.31}$$

which averages the received signal strength over a window length of $2L$ samples. If $2L$ is too large, $\hat{\mu}_p$ will not describe the lognormal fading, and if too small, it will not average out the multipath fading [62]. Unfortunately, range estimates produced by Equation 21.31 may still have large variances due to the variations of the mobile radio channel. Consequently, a fuzzy logic approach was developed in [63] that uses membership functions to provide a measure of reliability of the estimated range. Alternatively, using premeasured signal strength contours centered at the BSs [64, 65] can reduce the effect of the channel variability.

Range can also be estimated by measuring the time required for a signal to propagate between the MS and BS, i.e., TOA estimation. The measured time is converted into a distance by multiplying the TOA by the propagation speed, which is usually the speed of light for mobile radio channels. For code division multiple-access (CDMA)-based signaling, conventional coarse acquisition, such as the sliding correlator [66] or matched filter [67], or fine acquisition, such as the delay-lock loop (DLL) [68], can be used assuming the code offset between the MS and BS is known.

Recently, there has been a plethora of activity in TOA estimation for location. ML-based TOA estimators are presented in [69] and [70], while a high-resolution estimator based on the minimum variance estimates and normalized minimum variance of the power delay profile is presented in [71]. Similarly, a generalized likelihood ratio test (GLRT) is used in [72] to detect the first arriving path from estimated channel observations. Successive cancellation for TOA estimation in multipath channels is used with either a correlation or MUSIC approach at each iteration in [73]. Signal subspace-based techniques have also been developed for TOA estimation, particularly for CDMA systems, in [74, 75, 76] for static channels.

21.4.3 Range Difference Estimation

A straightforward approach for computing range differences, i.e., TDOAs, is to form differences of pairs of measured TOAs. However, more direct methods exist. A classic approach for TDOA estimation is based on generalized cross-correlation (GCC), which maximizes the cross-correlation between two signals at a pair of BSs [77, 78, 79]:

$$\hat{\tau} = \max_{\tau}\left\{\frac{1}{T}\int_0^T r_1(t)r_2(t+\tau)dt\right\} \tag{21.32}$$

where $r_i(t)$ is the received signal at the ith BS. Improvement in the accuracy of the delay estimate, $\hat{\tau}$, can be obtained by filtering the two received signals prior to cross-correlation [77, 80, 81, 82]. Other methods of TDOA estimation such as cyclic cross-correlation (CYC-COR), spectral coherence alignment (SPECCOA), and band-limited spectral correlation ratio (BL-SPECCOR) use methods that exploit the cyclostationarity of signals [83].

21.4.4 Joint Parameter Estimation

Recently, estimators that jointly estimate more than one type of location parameter (e.g., joint AOA/TOA) simultaneously have been developed. These are useful for hybrid location estimation schemes, such as those discussed in Section 21.3.4, as well as for receiver design in space–time communications systems. Most joint estimators are based on ML techniques and signal subspace approaches, such as MUSIC or ESPRIT, and are developed for joint AOA/TOA estimation of a single user's multipath signal components at a receiver. The methods provide the added benefit that no additional procedures are necessary to pair or match the jointly estimated parameters.

The ML approach in [84] for joint AOA/TOA estimation in static channels presents an iterative scheme that transforms a multidimensional ML criterion into two sets of one-dimensional problems. Both a deterministic and a stochastic ML algorithm are developed in [85] for joint AOA/TOA estimation in time-varying channels. A closed-form subspace method in [86] uses a two-dimensional ESPRIT-like shift invariance technique to jointly estimate the TOAs and AOAs. Another ESPRIT-based estimator in [87] uses separate one-dimensional spatial and temporal ESPRIT algorithms (S-ESPRIT and T-ESPRIT, respectively) in a tree structure for jointly estimating the TOAs and AOAs of multipath components. Subspace approaches for joint TOA/AOA estimation of multipath components based on MUSIC can be found in [85] for time-varying channels and in [88] for fast fading channels, which exploit the stationarity of the channel.

21.5 Impairments to Accuracy

Most of the algorithms described in Section 21.2 generally assume that an LOS path exists between the MS and each BS and that the AOA, range, and range difference measurements are only corrupted by additive white Gaussian noise. In mobile radio channels, however, other impairments exist that must be mitigated. These include the propagation phenomena of multipath fading and non-line-of-sight (NLOS) propagation. Both of these cause errors, sometimes significant, in the measured parameters and, consequently, affect the accuracy of the location estimates. In addition, it may be difficult to obtain the required parameter measurements for a given algorithm due to the inability of a BS to "hear" an MS (or vice versa). These issues are briefly discussed below. Due to the drastic impact of NLOS propagation, it is treated in more detail.

21.5.1 Multipath Propagation

Multipath propagation is described by the reception of multiple signals at an MS or BS. These signals can combine constructively (larger signal) or destructively (smaller signal) and result in the phenomenon known as *fading*. The multiple signals that are received make it difficult to accurately determine the signal strength, AOAs, TOAs, or TDOAs in wireless systems. The high variability of the received signal strength in mobile radio channels limits its use for accurate location. Although it is possible to estimate the multiple AOAs of multipath signals at an antenna array using appropriate processing techniques, the AOA corresponding to the true AOA must be determined. Methods for improving AOA estimates in mobile radio channels can be found in [57, 58, 59].

For TOA and TDOA location systems, multipath propagation causes errors in the timing estimates even when there is an LOS path between the MS and BS. TOA and TDOA estimators that are based on correlation techniques are influenced by the presence of multipath, especially when the multipath components cannot be resolved. Improvements in the conventional correlation-based code tracking loops are given in [89, 90]. Subspace-based estimation methods are given in [91, 92] for estimating TOAs in multipath channels. Further, TOA estimators for multipath channels are given in [93] using a block LS method, in [94] using an adaptive filtering operation, and in [95] using the independent fading of the paths and the pulse shape symmetry.

21.5.2 Hearability

Hearability is a major limiting factor for both network-based and handset-based methods and adversely affects the deployment of a location scheme. In order to provide an unambiguous position fix, purely time-based location techniques require a minimum of three BSs to simultaneously transmit to or receive from the MS to be located. With two BSs or less, it becomes necessary to consider AOA-only schemes or the hybrid techniques in Section 21.3.4.

A challenge common to all standards is the effect of the environment type on hearability and the choice of location scheme to deploy. In rural environments where the separation between cell sites is large, it is often difficult for the MS to hear from or be heard at BSs other than its serving BS, unless the MS is located in a handoff region. Measurements performed using a Grayson Wireless CellScope^{TM} in an AMPS (see Section 21.6) system showed that the probability of the number of BSs that could be heard by a receiver varied with the environment type. For instance, the likelihood of finding three BSs with a received signal strength indicator (RSSI) stronger than −100 dB was only 35% in rural areas, whereas it was approximately 84% in urban areas [96]. It is clear, then, that range-based methods that require three or more BSs are not suitable for most rural areas.

Hearability also depends on the standards and technology that are employed. For wireless standards that incorporate CDMA, the transmission powers of the MS are controlled to reduce the interference to other users and to conserve the MS battery power. Consequently, for network-based location, the carrier-to-interference ratio is received below the detection threshold at the neighboring BSs. A proposed solution to improve hearability in the IS-95 system utilizes a power-up function (PUF) to provide high-powered bursts from the MS that could be heard at the adjacent cells [97].

Alternatively, for handset-based schemes in the 3G wideband CDMA (WCDMA) system, it has been proposed that idle periods in the downlink (IPDLs) be added during position location, where the serving BS ceases transmission for a very short period, thus enabling the MS to make measurements from distant BSs [98]. The length of the idle period has to be short enough that the impact on system capacity is minimal and long enough to make accurate timing measurements.

21.5.3 Non-Line-of-Sight Propagation

For the time-based location systems, NLOS propagation introduces a large positive error in the TOA and TDOA measurements even when high-resolution timing techniques are employed. The typical error introduced by NLOS propagation in time-based ranging systems has been measured in the Global System for Mobile Communications (GSM), which indicates that the NLOS error can average 400 to 700 m [99]. Results in [13, 18] show that the NLOS error can lead to location estimates that are several hundred meters in error, depending on the characteristics of the scattering about the MS. Expressions have been derived in [13] for the location estimate bias when NLOS-corrupted TOA or TDOA measurements are used in linearlized LS solutions and when applying the algorithm in [25] to TDOA measurements. Although it was shown that the linearized LS method was more accurate than that in [25], the FCC requirements were achieved less than 20% of the time. The increase in the mean square location error in the presence of measurement biases for the linearized LS algorithm is derived in [100].

For an AOA system, the absence of an LOS path greatly diminishes its use for providing location estimates. Unfortunately, in macrocellular systems, an LOS path rarely exists. In microcells, it is often assumed that the cells are small enough that the MS has a direct path to its *serving* BS, although it is unlikely that an LOS path exists to other BSs (depending on the deployment).

Until recently, few researchers addressed the NLOS problem in wireless location systems. The first studies to address NLOS as a key impairment for wireless location systems can be found in [99, 101]. Since then, several techniques have been developed and these are briefly outlined below.

The method in [102] attempts to reconstruct LOS TOA measurements from a history of LOS and NLOS TOA measurements, which can span tens of seconds. In this approach, a time series of range measurements is recorded as an MS moves throughout a cell. A pth-order polynomial is fit to the range measurements vs. time to produce a smoothed version of the range measurement curve. The smoothed curve is then compared to the actual range measurements. Since the measurements were recorded as the MS moved in the cell, it is likely that both LOS and NLOS range measurements were made at a BS. For LOS range measurements, we would expect to see a random variation that is defined by the statistics of standard measurement noise. This noise is usually assumed to be Gaussian with zero mean and a relatively small variance. For NLOS BS, measurements in [99] describe range measurements with a large, nonzero mean and much larger variance. From the time series of range measurements, the variance can be determined and compared to known values (i.e., the statistics of both the LOS and NLOS noise/error are known *a priori*) to determine if a BS is LOS or NLOS.

The implementation of this method requires a few key assumptions. First, the statistics of the NLOS bias are known. Second, the MS must be in motion *throughout* the cell in order to obtain the statistics necessary for the LOS reconstruction. Finally, several of the measurements in the measurement history must be made to an LOS BS. Without the recognition of standard measurement noise, which can only occur with an LOS BS, the method fails.

The method in [103] expands on the ideas in [102] and utilizes a time history of TOA measurements to determine whether a BS is LOS or NLOS. The decision is based on the standard deviation of the measurements, with the assumption that the standard deviation is higher for NLOS BSs. While this may be true for many scenarios, it is not necessarily so for all of them, or even the majority. For a low-mobility, or stationary, MS, using standard deviation to identify NLOS BSs may not work since TOA estimates with similar large errors may be made consistently, thus producing a small standard deviation.

Other extensions to the work in [102] generalized the NLOS identification problem to a decision theoretical framework [104]. Again, using the assumption that NLOS and LOS BSs generate different standard deviations of TOA measurements, a binary hypothesis test was used to identify NLOS BSs. The approach used familiar ideas from estimation theory, such as the GLRT and uniformly most powerful (UMP) tests. A further approach for identifying NLOS BSs in [105] used statistical methods such as skewness and kurtosis tests and tests for normality, such as the Shapiro–Wilk and Anderson–Darling tests.

In [106], a simple method is presented to distinguish between LOS and NLOS range measurements in a mixed measurement set when the LOS path from the MS to the serving BS is intermittently blocked. The approach exploits the knowledge of the measurement noise statistics to determine which measurements fall below the true range and then reconstructs the LOS range.

Other methods at mitigating NLOS effects for TOA location attempt to selectively remove or weight NLOS-corrupted measurements by examining the standard deviation of the measurements. In [107], an initial estimate of the MS location is determined from all range measurements from which a standard deviation is formed, using the computed position as the mean value. If the difference between the measured TOA and the TOA for the computed position is greater than a multiple of the computed standard deviation, then that TOA measurement is weighted proportionately to minimize its effect [107]. A similar algorithm using measured AOAs is presented in [108]. In that paper, the author ignored those AOA measurements that differed from the original calculated position by more than some multiple of the standard deviation of the AOA measurements. A TDOA-based algorithm is considered in [109], where it was assumed that only one BS had an NLOS path to the MS. The TDOA measurement between this BS and the reference LOS serving BS is identified as an NLOS measurement by comparing the magnitudes of the TDOA residuals in a manner similar to that in [107]. Then the MS location and the NLOS TDOA error are computed using LS methods. These approaches only perform well when there are a large number of BSs available, with the majority being LOS with the MS.

A range-based constrained optimization procedure for tracking an MS in an NLOS environment is presented in [110]; it utilizes constraints derived from the fact that the measured ranges are greater than the true ranges and that the incremental changes in the MS position are bounded by some finite number if the maximum velocity is known. This algorithm also requires that some of the communicating BSs have LOS links to the MS, and the accuracy is expected to be low if unfavorable NLOS conditions persist for several measurement instants.

Some statistical algorithms have been designed to work in NLOS environments that are described by channel scattering models such as the ring/disk of scatterers and the Gaussian scattering models [111, 112].

These algorithms aim to estimate the true ranges from the MS to a minimum of three NLOS BSs and then employ traditional location algorithms in [13, 19] to determine the MS location. In [112], the probability density functions (PDFs) of the TOAs are derived using the scattering models and then ML expectation–maximization (EM) and Bayesian estimators are applied to the multipath TOA measurements at each BS to estimate the true LOS ranges. In [111], the statistics of the measured TOAs of several multipaths are matched with those produced by the scattering models to estimate the LOS ranges. The merits of these methods are that they require the availability of only three BSs for location and can be applied for the case of a stationary user. However, it becomes important to characterize the scattering environment in a particular area before the algorithm can be applied.

Methods of scaling NLOS-corrupted TOA and AOA measurements to near their true LOS values are presented in [105] and [113]. In [105], the scale factors are determined through a constrained optimization process in which the constraints are provided by physical constraints on the NLOS error. This approach is suitable for situations where all BSs are NLOS. Likewise, the method in [113] computes scale factors through knowledge of multipath AOAs. In this approach, LOS propagation is assumed to the serving BS, but not to neighboring BSs.

All of the solutions described above have advantages and disadvantages. However, it is clear that the NLOS problem must be properly addressed in order to obtain accurate location estimates in wireless communication networks.

21.6 Provisions in the Standards

In this section, we provide a summary of the standards and specifications established by various regulatory bodies to achieve worldwide consensus on the location technologies suitable for each generation of mobile wireless systems.

21.6.1 3G Location Solutions

First-generation (1G) systems such as the analog Advanced Mobile Phone System (AMPS) were aimed at providing voice-only communications. Second-generation (2G) systems led the evolution to digital wireless communications, offering voice and low-rate data services, and include the GSM, Digital AMPS (DAMPS)/IS-136, and CDMA/IS-95B. The personal wireless systems currently deployed around the world are 1G and 2G systems. 3G systems are aimed at providing enhanced services with multimedia capabilities to 2G platforms at high data rates. The main 3G proposals include the WCDMA system, a Euro/Japanese initiative; cdma2000, proposed by the U.S. Telecommunications Industry Association (TIA); and the EDGE system, where EDGE stands for enhanced data rates for GSM evolution. In the push for high-speed and value-added services, analog 1G systems are fast becoming obsolete and the roll-out of 3G systems was expected to begin in late 2003. In the following, we will look at the standards set forth for the future 3G systems and currently deployed 2G systems.

The standardization of positioning for the 3G mobile communications systems has been handled by the 3rd Generation Partnership Project (3GPP) and 3GPP2. The focus of 3GPP has been on WCDMA systems and GSM, while 3GPP2 has been working on standards for cdma2000 and cdmaOne systems [114].

For the Universal Terrestrial Radio Access Network (UTRAN) that employs WCDMA and operates in either time division duplexing (TDD) or frequency division duplexing (FDD) mode, 3GPP has specified cell-ID, observed TDOA (OTDOA), and AGPS as the location technologies. The cell-ID method approximates the MS location at its serving BS and the accuracy of positioning depends on the coverage area of the BS. By augmenting cell-ID with additional information such as the round-trip time (RTT) in FDD mode and the received (RX) timing deviation in TDD mode, accuracy can be improved by limiting the location of the MS to within range circles derived from these time measurements. Although cell-ID is not a mandatory method, it can be used if the more accurate OTDOA or AGPS fails to locate the MS.

AGPS can be implemented as an MS-assisted solution or an MS-based solution. In the MS-assisted solution, the network conveys the visible satellite list, satellite signal Doppler and code phase, approximate handset position, and ephemeris to the MS to assist with reducing the GPS acquisition time. The MS calculates the pseudoranges to the satellites and returns them to the network server where the MS position is determined. The MS-based solution maintains a fully functional GPS receiver in the handset that computes the positions of the satellites and ultimately its position. AGPS allows for MS positioning indoors or in severely faded conditions because the receiver in the MS can detect and demodulate the satellite navigation message sequences that are an order of magnitude weaker than those required by conventional GPS receivers. Only a partial GPS receiver is required in the MS to achieve this functionality, but legacy terminals cannot be used and new handsets are required to deploy the technology.

OTDOA is based on TDOA and locates the MS by means of trilateration. Similar to AGPS, it can be implemented as either MS-assisted OTDOA or MS-based OTDOA. The MS determines the geometric time differences between signals arriving from each pair of BSs, and the final computation for the MS location can be accomplished at the network (MS-assisted OTDOA) or at the MS (MS-based OTDOA). The computation also requires the relative time difference (RTD) if the BSs are unsynchronized in FDD mode. To address one of the shortcomings of TDOA-based techniques, namely, hearability, IPDL has been specified as an option. Since the IPDL method is based on downlink transmission, the location service can be provided efficiently to a large number of MSs simultaneously.

For the GSM EDGE Radio Access Network (GERAN), cell-ID, enhanced observed time difference (E-OTD), and AGPS have been standardized. E-OTD is a TDOA positioning method similar to OTDOA, but it operates in TDMA-based networks. The location technologies specified by 3GPP2 for the time-synchronized systems, cdma2000 and cdmaOne, are advanced forward link trilateration (AFLT) and AGPS. AFLT is similar to OTDOA and involves measuring the time difference between CDMA pilot signal pairs.

For the 2G systems, E-OTD has been finalized by the GSM standard committees (T1P1.5 and ETSI) in LCS Release 98 and Release 99. Other technologies that have been specified for GSM include uplink TOA and AGPS. Uplink TOA is based on measuring, at three or more location-measuring units (LMUs) in the infrastructure, the TOA of the access burst generated by having the handset perform an asynchronous handover. The MS's position is then determined by the TDOA principle of hyperbolic trilateration. For the 2G CDMA (IS-95) system, AFLT and AGPS have been standardized by the CDMA standard committee (TR45.5).

21.6.2 Locating Legacy Terminals

In the U.S., the FCC mandate on E-911 requires that any location technology that is deployed have the ability to locate all existing wireless terminals. It is only recently that handsets modified to accommodate location services have been introduced in the market. The vast majority of wireless phones in use do not have the necessary software or hardware to support handset-based location, and hence, it is essential to adopt a network-based technology to locate these legacy terminals.

Although cell-ID and enhanced cell-ID offer the most convenient solution to locate all terminals, accuracy depends heavily on the base station density and is generally too low to meet the FCC accuracy requirement. Among uplink methods that do not require handset modification, uplink TDOA (UTDOA) is being considered as a solution that is compatible with all of the major air interfaces and capable of locating existing and future phones on the operator's network. It can also be combined with the lower-accuracy cell-ID to offer uninterrupted location services in areas with poor hearability. As the market penetration of modified handsets grows, UTDOA can be complemented with other standardized handset-based solutions such as E-OTD and AGPS.

21.7 Summary

Geolocation will play an important role in many future services in wireless communication networks. Many technologies currently exist, from GPS to terrestrial-based solutions using the wireless network infrastructure, for locating MSs. However, the harsh nature of the mobile radio channel severely limits the accuracy with which a location estimate can be made. Many attempts have been made to improve performance when the radio signals are corrupted by multipath fading and NLOS error. The keys to highly accurate location lie in developing technologies and other solutions that can mitigate these impairments.

References

[1] B. Richton, G. Vannucci, and S. Wilkus, "Assisted GPS for wireless phone location," in *Next Generation Wireless Networks*, S. Tekinay, Ed., Elsevier Science Publishers, Amsterdam, 1993, chap. 4.

[2] FCC, "Revision of the commissions rules to ensure compatibility with enhanced 911 emergency calling systems, RM-8143," Docket 94-102, July 1996.

[3] L. Stilp, "Carrier and end-user applications for wireless location systems," in *Proceedings of the SPIE*, 1996, pp. 119–126.

[4] M. Sunay and I. Tekin, "Mobile location tracking in ds cdma networks using forward link time difference of arrival and its application to zone-based billing," in *GLOBECOM*, 1999, pp. 143–147.

[5] S. Wang, M. Green, and M. Malkawi, "Mobile positioning technologies and location services," in *IEEE Radio and Wireless Conference*, 2002, pp. 9–12.

[6] A. Mihovska and J. Pereira, "Location-based VAS: killer applications for the nextgeneration mobile internet," in *Personal, Indoor and Mobile Radio Conference*, 2001, pp. B50–B54.

[7] Y.-B. Ko and N. Vaidya, "Location-aided routing in mobile ad hoc networks," *Wireless Networks*, 6(4), 307–321, 2000.

[8] S. Sharma and A. Nix, "Dynamic W-CDMA network planning using mobile location," in *IEEE Vehicular Technology Conference*, 2002, pp. 1182–1186.

[9] P. Krishnamurthy, H. Kariimi, and P. Liangsutthisakon, "Telegeoinformatics: a novel approach towards efficient resource allocation and application and protocol development for wireless communications," in *International Conference on Parallel and Distributed Processing Techniques and Applications*, 2000, pp. 2065–2071.

[10] S. Wang, S. Sridhar, and M. Green, "Adaptive soft handoff method using mobile location information," in *IEEE Vehicular Technology Conference — Spring*, 2002, pp. 1936–1940.

[11] S. Wang, M. Green, and I. Tekin, "Adaptive handoff method using mobile location information," in *IEEE Emerging Technologies Symposium on BroadBand Communications for the Internet Era*, 2001, pp. 97–101.

[12] J. Caffery, Jr., "A new approach to the geometry of TOA location," in *IEEE Vehicular Technology Conference*, Vol. 4, September 2000, pp. 1943–1949.

[13] J. Caffery, Jr., *Wireless Location in CDMA Cellular Radio Systems*, Kluwer Academic Publishers, Dordrecht, Netherlands, 1999.

[14] G. Yost and S. Panchapakesan, "Automatic location identification using a hybrid technique," in *IEEE Vehicular Technology Conference*, 1998, pp. 276–267.

[15] R. Schmidt, "A new approach to geometry of range difference location," *IEEE Transactions on Aerospace and Electronic Systems*, AES-8, 821–835, 1972.

[16] B. Fang, "Simple solutions for hyperbolic and related position fixes," *IEEE Transactions on Aerospace and Electronic Systems*, 26, 784, 1990.

[17] K. Ho and Y. Chan, "Solution and performance analysis of geolocation by TDOA," *IEEE Transactions on Aerospace and Electronic Systems*, 29, 1311–1322, 1993.

[18] J. Caffery and G. Stüber, "Subscriber location in CDMA cellular networks," *IEEE Transactions on Vehicular Technology*, 47, 406–416, 1998.

[19] W. Foy, "Position-location solutions by Taylor-series estimation," *IEEE Transactions on Aerospace and Electronic Systems*, AES-12, 187–193, 1976.

[20] D. Torrieri, "Statistical theory of passive location systems," *IEEE Transactions on Aerospace and Electronic Systems*, AES-20, 183–197, 1984.

[21] G. Turin, W. Jewell, and T. Johnston, "Simulation of urban vehicle-monitoring systems," *IEEE Transactions on Vehicular Technology*, VT-21, 9–16, 1972.

[22] A. Pagés-Zamora, J. Vidal, and D. Brooks, "Closed-form solution for positioning based on angle of arrival measurements," in *IEEE International Conference on Personal, Indoor and Mobile Radio Communications*, 2002, pp. 1522–1526.

[23] J. Smith and J. Abel, "The spherical interpolation method of source localization," *IEEE Journal of Oceanic Engineering*, OE-12, 246–252, 1987.

[24] J. Smith and J. Abel, "Closed-form least squares source location estimation from range-difference measurements," *IEEE Transactions on Acoustics, Speech, and Signal Processing*, ASSP-35, 1661–1669, 1987.

[25] B. Friedlander, "A passive location algorithm and its accuracy analysis," *IEEE Journal of Oceanic Engineering*, OE-12, 234–244, 1987.

[26] Y. Rasshcheplyayev and V. Shcherbachev, "Estimation of coordinates in range-difference radar systems on the basis of projective transformations of observation equations," *Journal of Communications Technology and Electronics*, 39, 1627–1636, 1995.

[27] Y. Chan and K. Ho, "A simple and efficient estimator for hyperbolic location," *IEEE Transactions on Signal Processing*, 42, 1905–1915, 1994.

[28] K. Budka, D. Calin, B. Chen, and D. Jeske, "A Bayesian method to improve mobile geolocation accuracy," in *IEEE VTC Spring 2002*, Vol. 2, 2002, pp. 1021–1025.

[29] M. Aso, M. Kawabata, and T. Hattori, "A new location estimation method based on maximum likelihood function in cellular systems," in *IEEE VTS Fall VTC 2001*, Vol. 1, 2001, pp. 106–110.

[30] M. Aso, T. Saikawa, and T. Hattori, "Mobile station location estimation using the maximum likelihood method in sector cell systems," in *IEEE Vehicular Technology Conference*, Vol. 2, 2002, pp. 1192–1196.

[31] M. McGuire, K. Plataniotis, and A. Venetsanopoulos, "Robust estimation of mobile terminal position," *IEE Electronic Letters*, 36, 1426–1428, 2000.

[32] M. McGuire, K. Plataniotis, and A. Venetsanopoulos, "Location of mobile terminals using time measurements and survey points," in *IEEE Pacific Rim Conference*, Vol. 2, 2002, pp. 635–638.

[33] R. Estrada, D. Munoz-Rodriquez, C. Molina, and K. Basu, "Cellular position location techniques: a parameter detection approach," in *IEEE Vehicular Technology Conference*, Vol. 2, 1999, pp. 1166–1171.

[34] R. Yamamoto, H. Matsutani, H. Matsuki, T. Oono, and H. Ohtsuka, "Position location technologies using signal strength in cellular systems," in *IEEE Vehicular Technology Conference — Spring*, 2001, pp. 2570–2574.

[35] M. Berg, "Performance of mobile station location methods in a Manhattan microcellular environment," in *IEEE International Conference on Third Generation Wireless and Beyond*, 2001.

[36] A. Pagés-Zamora and J. Vidal, "Evaluation of the improvement in the position estimate accuracy of umts mobiles with hybrid positioning techniques," in *IEEE Vehicular Technology Conference Spring*, Vol. 4, 2002, pp. 1631–1635.

[37] M. Spirito, "Mobile station location with heterogenous data," in *IEEE VTC 2001 (Fall)*, 2000, pp. 1583–1589.

[38] P. Deng and P. Fan, "An AOA assisted TOA positioning system," in *International Conference on Communication Technology Proceedings*, Vol. 2, 2000, pp. 1501–1504.

[39] L. Cong and W. Zhuang, "Hybrid TDOA/AOA mobile user location for wideband CDMA cellular systems," *IEEE Transactions on Wireless Communications*, 1, 439–447, 2002.

[40] N. Thomas, D. Cruickshank, and D. Laurenson, "Performance of a TDOA-AOA hybrid mobile location system," in *Second International Conference on 3G Mobile Communication Technologies*, 2001, pp. 216–220.

[41] P. Massatt and K. Rudnick, "Geometric formulas for dilution of precision calculations," *Journal of the Institute of Navigation*, 37, 379–391, 1991.

[42] I. Ziskind and M. Wax, "Maximum likelihood localization of multiple sources by alternating projection," *IEEE Transactions on Acoustics, Speech, and Signal Processing*, 36, 1553–1560, 1988.

[43] P. Gething, *Radio Direction Finding and Superresolution*, Peter Peregrinus Ltd., England, 1991.

[44] B. Agee and R. Calabretta, "ARMA-like and ML-like copy/DF approaches for signal specific emitter location," in *Proceedings Fifth Workshop on Spectrum Estimation and Modeling*, 1990, pp. 134–138.
[45] S. Haykin, *Adaptive Filter Theory*, Prentice Hall, New York, 1991.
[46] A. Swindlehurst, "Synchronization and spatial signature estimation for multiple known co-channel signals," in *Asilomar Conference on Signals, Systems and Computers*, 1995.
[47] R. Schmidt, "A Signal Subspace Approach to Multiple Emitter Location and Spectral Estimation," Ph.D. thesis, Stanford University, California, November 1981.
[48] A. Barabell, "Improving the resolution performance of eigenstructure-based direction finding algorithms," in *International Conference on Acoustics Speech and Signal Processing*, 1983, pp. 336–339.
[49] S. Schell, Calabretta, W. Gardner, and B. Agee, "Cyclic MUSIC algorithms for signal selective DOA estimation," in *International Conference on Acoustics Speech and Signal Processing*, 1988, pp. 2278–2281.
[50] A. Paulraj, R. Roy, and T. Kailath, "Estimation of signal parameters via rotational invariance techniques — esprit," in *Proceedings of Asilomar Conference on Circuits and Systems*, 1986, pp. 83–89.
[51] R. Kumaresan and D. Tufts, "Estimating the angles of arrival of multiple plane waves," *IEEE Transactions on Aerospace and Electronic Systems*, AES-19, 134–139, 1983.
[52] J. Cadzow, *The Signal Subspace Directions-of-Arrival Algorithm*, Elsevier Science Publishers, Amsterdam, 1993, chap. 5.
[53] Paulraj, et al., *Subspace Methods for Directions-of-Arrival Estimation*, Elsevier Science Publishers, Amsterdam, 1993, chap. 6.
[54] J. Capon, "High resolution frequency-wave number spectrum analysis," *Proceedings of IEEE*, 57, 1408–1418, 1969.
[55] J. Burg, "Maximum entropy spectral analysis," in *37th Annual Meeting of the Society of Exploration Geophysicists*, 1967.
[56] C. Rao and B. Zhou, *Closed Form Solutions to the Estimates of Directions of Arrival Using Data from an Array of Sensors*, Elsevier Science Publishers, Amsterdam, 1993, chap. 17.
[57] T. Biedka, J. Reed, and B. Woerner, "Direction finding methods for CDMA systems," in *Asilomar Conference on Signals, Systems and Computers*, 1997, pp. 637–641.
[58] R. Klukas and M. Fattouche, "Line-of-sight angle of arrival estimation in the outdoor multipath environment," *IEEE Transactions on Vehicular Technology*, 47, 342–351, 1998.
[59] Y. Jong and M. Herben, "High-resolution angle-of-arrival measurement of the mobile radio channel," *IEEE Transactions on Antennas and Propagation*, 47, 1677–1687, 1999.
[60] M. Hata, "Empirical formula for propagation loss in land mobile radio services," *IEEE Transactions on Vehicular Technology*, VT-29, 317–325, 1980.
[61] G. Ott, "Vehicle location in cellular mobile radio systems," *IEEE Transactions on Vehicular Technology*, VT-26, 43–46, 1977.
[62] W. Lee, "Estimate of local average power of a mobile radio signal," *IEEE Transactions on Vehicular Technology*, VT-34, 22–27, 1985.
[63] H.-L. Song, "Automatic vehicle location in cellular communications systems," *IEEE Transactions on Vehicular Technology*, 43, 902–908, 1994.
[64] W. Figel, N. Shepherd, and W. Trammell, "Vehicle location by a signal attenuation method," *IEEE Transactions on Vehicular Technology*, VT-18, 105–110, 1969.
[65] W. Smith, Jr., "Passive location of mobile cellular telephone terminals," in *IEEE International Carnahan Conference on Security Technology*, 1991, pp. 221–225.
[66] J. Holmes, *Coherent Spread Spectrum Systems*, John Wiley & Sons, Inc., New York, 1982.
[67] Simon, Omura, Scholtz, and Levitt, *Spread Spectrum Communications*, Vol. III, Computer Science Press, Rockville, MD, 1985.
[68] J. Spilker, "Delay-lock tracking of binary signals," *IEEE Transactions on Space Electronics and Telemetry*, SET-9, 1–8, 1963.

[69] N. Yousef and A. Sayed, "A new adaptive estimation algorithm for wireless location finding systems," in *Asilomar Conference on Signals, Systems and Computers*, 1999, pp. 491–495.

[70] N. Yousef and A. Sayed, "Robust time-delay and amplitude estimation for cdma location finding," in *IEEE Vehicular Technology Conference — Fall*, 1999, pp. 2163–2167.

[71] J. Vidal and M. Nájar, "High resolution time-of-arrival detection for wireless positioning systems," in *IEEE Vehicular Technology Conference — Fall*, 2002, pp. 2283–2287.

[72] R. Játiva and J. Vidal, "First arrival detection for positioning in mobile channels," in *IEEE Personal, Indoor and Mobile Radio Conference*, 2002, pp. 1540–1544.

[73] L. Krasny and H. Koorapaty, "Enhanced time of arrival estimation with successive cancellation," in *IEEE Vehicular Technology Conference — Spring*, 2002, pp. 851–855.

[74] S. Bensley and B. Aazhang, "Subspace-based channel estimation for code division multiple access communication systems," *IEEE Transactions on Communications*, 44, 1009–1020, 1996.

[75] E. Ström, S. Parkvall, S. Miller, and B. Ottersten, "Propagation delay estimation in asynchronous direct-sequence code-division multiple access systems," *IEEE Transactions on Communications*, 44, 84–93, 1996.

[76] J. Joutsensalo, "Algorithms for delay estimation and tracking in CDMA," in *IEEE Vehicular Technology Conference*, 1997, pp. 366–370.

[77] C. Knapp and G. Carter, "The generalized correlation method for estimation of time delay," *IEEE Transactions on Acoustics, Speech, and Signal Processing*, ASSP-24, 320–327, 1976.

[78] W. Gardner and C. Chen, "Signal-selective time-difference-of-arrival estimation for passive location of man-made signal sources in highly corruptive environments. Part I. Theory and method," *IEEE Transactions on Signal Processing*, 40, 1168–1184, 1992.

[79] W. Gardner and C. Chen, "Signal-selective time-difference-of-arrival estimation for passive location of man-made signal sources in highly corruptive environments. Part II. Algorithms and performance," *IEEE Transactions on Signal Processing*, 40, 1185–1197, 1992.

[80] G. Carter, A. Nuttall, and P. Cable, "The smoothed coherence transform," *Proceedings of the IEEE*, 61, 1479–1498, 1973.

[81] W. Hahn and S. Tretter, "Optimum processing for delay-vector estimation in passive signal arrays," *IEEE Transactions on Information Theory*, IT-19, 608–614, 1973.

[82] P. Roth, "Effective measurements using digital signal analysis," *IEEE Spectrum*, 8, 62–70, 1971.

[83] W. Gardner, *Cyclostationarity in Communications and Signal Processing*, IEEE Press, Washington, DC, 1994.

[84] M. Wax and A. Leshem, "Joint estimation of time delays and directions of arrival of multiple reflections of a known signal," *IEEE Transactions on Signal Processing*, 45, 2477–2484, 1997.

[85] G. Raleigh and T. Boros, "Joint space-time parameter estimation for wireless communication channels," *IEEE Transactions on Signal Processing*, 46, 1333–1343, 1998.

[86] A.-J. van der Veen, M. Vanderveen, and A. Paulraj, "Joint angle and delay estimation using shift-invariance techniques," *IEEE Transactions on Signal Processing*, 46, 406–418, 1998.

[87] Y.Y.-Wang, J.-T. Chen, and W.-H. Fang, "Joint estimation of the doa and delay based on the tst-esprit in a wireless channel," in *IEEE Workshop on Signal Processing Advances in Wireless Communications*, 2001, pp. 302–305.

[88] Z. Gu, E. Gunawan, and Z. Yu, "Joint spatiotemporal parameter estimation for ds-cdma system in fast fading multipath channel," in *IEEE Vehicular Technology Conference*, 2001, pp. 28–32.

[89] W. Sheen and G. Stüber, "A new tracking loop for direct sequence spread spectrum systems on frequency-selective fading channels," *IEEE Transactions on Communications*, 43, 3063–3072, 1995.

[90] D. van Nee, "The multipath estimating delay lock loop: approaching theoretical accuracy limits," in *IEEE Position, Location and Navigation Symposium*, 1994, pp. 246–251.

[91] S. Bensley and B. Aazhang, "Subspace-based estimation in multipath channel parameters for CDMA communication systems," in *IEEE Global Telecommunication Conference, Communication Theory Mini-Conference*, 1994, pp. 154–157.

[92] E. Ström, S. Parkvall, S. Miller, and B. Ottersten, "DS-CDMA synchronization in time-varying fading channels," *IEEE Journal on Selected Areas in Communications*, 1636–1642, 1996.

[93] N. Yousef and A. Sayed, "Robust multipath resolving in fading conditions for mobile-positioning systems," in *National Radio Science Conference*, 2000, pp. C19/1–8.

[94] N. Yousef and A. Sayed, "Adaptive multipath resolving for wireless location systems," in *Asilomar Conference on Signals, Systems and Computers*, 2001, pp. 1507–1511.

[95] N. Yousef and A. Sayed, "Detection of fading overlapping multipath components for mobile positioning systems," in *IEEE Conference on Communications*, 2001, pp. 3102–3106.

[96] J. Reed, K. Krizman, B. Woerner, and T. Rappaport, "An overview of the challenges and progress in meeting the e-911 requirement for location service," *IEEE Communications Magazine*, 36, 30–37, 1998.

[97] A. Ghosh and R. Love, "Mobile station location in DS-CDMA system," in *IEEE Vehicular Technology Conference*, 1998, pp. 254–258.

[98] Y. Zhao, "Standardization of mobile phone positioning for 3g systems," *IEEE Communications Magazine*, 40, 108–114, 2002.

[99] M. Silventoinen and T. Rantalainen, "Mobile station emergency locating in GSM," in *IEEE International Conference on Personal Wireless Communications*, 1996, pp. 232–238.

[100] H. Koorapaty, H. Grubeck, and M. Cedervall, "Effect of biased measurement errors on accuracy of position location methods," in *IEEE GLOBECOM*, 1998, pp. 1497–1502.

[101] J. Caffery, Jr., and G. Stüber, "Overview of radiolocation in CDMA cellular systems," *IEEE Communications Magazine*, 36, 38–45, 1998.

[102] M. Wylie and J. Holtzmann, "The non-line of sight problem in mobile location estimation," in *IEEE International Conference on Universal Personal Communications*, 1996, pp. 827–831.

[103] S.-S. Woo, H.-R. You, and J.-S. Koh, "The NLOS mitigation technique for position location using IS-95 CDMA networks," in *IEEE Vehicular Technology Conference — Fall*, Vol. 4, September 2000, pp. 2556–2560.

[104] J. Borras, P. Hatrack, and N. Mandayam, "Decision theoretic framework for nlos identification," in *Vehicular Technology Conference*, Vol. 2, 1998, pp. 1583–1587.

[105] S. Venkatraman and J. Caffery, Jr., "A statistical approach to non-line-of-sight BS identification," in *International Symposium on Wireless Personal Multimedia Communications*, 2002, pp. 296–300.

[106] M. Wylie-Green and S. Wang, "Robust range estimation in the presence of non-line-of-sight error," in *IEEE Vehicular Technology Conference — Fall*, 2001, pp. 101–105.

[107] P.-C. Chen, "A non-line-of-sight mitigation algorithm in location estimation," in *Wireless Communications and Networking Conference*, 1999, pp. 316–320.

[108] L. Xiong, "A selective model to suppress NLOS signals in angle-of-arrival (AOA) location estimation," in *9th IEEE International Symposium Personal, Indoor and Mobile Radio Communication*, Vol. 1, 1998, pp. 461–465.

[109] L. Cong and W. Zhuang, "Non-line-of-sight error mitigation in TDOA mobile location," in *Globecom 2001*, 2001, pp. 680–684.

[110] Y. Qi and H. Kobayashi, "Mitigation of NLOS effects in TOA positioning," in *Conference on Information Sciences and Systems at the Johns Hopkins University*, March 2001.

[111] S. Al-Jazzar, J. Caffery, Jr., and H.-R. You, "A scattering model based approach to nlos mitigation in toa location systems," in *IEEE Vehicular Technology Conference — Spring*, 2002, pp. 861–865.

[112] S. Al-Jazzar and J. Caffery, Jr., "Ml and Bayesian toa location estimators for nlos environments," in *IEEE Vehicular Technology Conference — Fall*, 2002, pp. 1178–1181.

[113] S. Venkatraman and J. Caffery, Jr., "Multipath-aided location estimatoin using angles-of-arrival," in *AEROSENSE: Location Services and Navigation Technologies*, 2003.

[114] 3GPP, "Report on Location Services, TS RAN R2.03 V.1.0," Technology Report, 2000.

22

Adaptive Arrays for GPS Receivers*

Moeness Amin
Villanova University

Wei Sun
Villanova University

Alan Lindsey
Air Force Research Lab

22.1 Introduction 22-1
22.2 GPS Signal Model 22-5
22.3 Interference Suppression Techniques in GPS 22-6
Broadband Interference Suppression Using Space–Time Array • Narrowband FM Interference Suppression Using Time–Frequency Method • A Self-Coherence Antijamming GPS Receiver
22.4 Multipath Mitigation in GPS 22-19
Bias Due to Signal Multipath • Single-Antenna Multipath Mitigation Techniques • Time-Delay and Carrier-Phase Estimation Using Antenna Array
22.5 Conclusions 22-24

Abstract

This chapter discusses the application of adaptive array processing techniques in the Global Positioning System (GPS). In GPS, the interference can be combated in the time, space, or frequency domain, or in a domain of joint variables. Space–time adaptive arrays, which utilize both spatial and temporal degrees of freedom, have been adopted in GPS for narrowband and broadband interference suppression, as well as multipath mitigation. In this chapter, we discuss commonly used array processing techniques for interference suppression and multipath mitigation in GPS. A novel self-coherent GPS antijamming receiver is also presented with computer simulations.

22.1 Introduction

Global Positioning System (GPS) is a satellite-based all-weather positioning, navigation, and timing system developed by the U.S. Department of Defense (DoD) in the late 1960s and early 1970s [1]. Today, there is hardly any doubt that, beyond its original military purpose, GPS has proven to be a great asset in a variety of civilian applications, due to its global coverage, precision in navigation, and low-cost receivers [2], [3]. While providing precise position, velocity, and time is the ultimate goal of GPS, it also provides signals for land surveying and mapping, geographic analyses, and tracking of people, vehicles, and other objects.

*This work is supported by the Air Force Research Lab, grant No. F30602-00-1-0514.

0-8493-1657-X/$0.00+$1.50

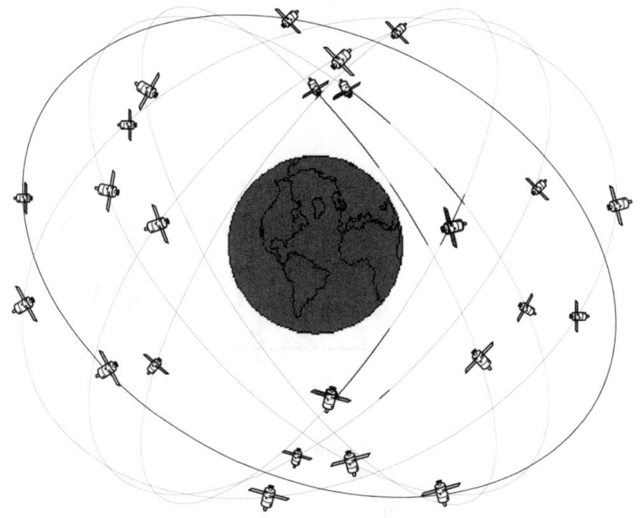

FIGURE 22.1 GPS constellation.

Time and frequency dissemination, based on the precise clocks on board the satellites, is yet another use for GPS.

GPS is composed of three components: a constellation of satellites, ground stations, and receivers [4]. The nominal GPS constellation, also called *space segment*, consists of 24 satellites with well-known positions in six earth-centered orbital plans with four equally spaced (60° apart) satellites in each plan (Figure 22.1). The ground stations, including a master control station, form the *control segment*, which tracks each satellite and periodically uploads to the satellite its prediction of future satellite positions and satellite clock time corrections. The GPS receivers, which are referred to as *GPS user segments*, track the ranging signals of selected satellites and convert them into position, velocity, and time estimates used for navigation, positioning, time dissemination, etc.

GPS utilizes the concept of one-way time-of-arrival (TOA) ranging to determine user position [4] by measuring the direct-path signal travel time from a satellite at the known location to a user's receiving device. Due to the lack of time synchronization between the satellite onboard clock and the receiver local clock, the difference between the satellite clock time and user clock time is termed *pseudorange*, which is the range from the satellite plus the clock offset. The pseudorange is measured in a GPS receiver by evaluating the GPS signal time delay from the satellite [5], [6]. By measuring the propagation time of signals broadcast from multiple satellites, the geolocation of the receiver is where the pseudoranges from a set of satellites intersect, as displayed in Figure 22.2.

GPS is a code-division multiple-access (CDMA) system that employs direct-sequence spread-spectrum (DS-SS) signaling [7]. The GPS satellites transmit signals at two L-band frequencies: $L1 = 1.57542$ GHz and $L2 = 1.2276$ GHz [4] (Figure 22.3). For each satellite, two different pseudorandom noise (PRN) codes, a coarse/acquisition (C/A) code, and a precision (P) code (encrypted into the Y code in the antispoofing mode) are used to modulate the navigation symbols. The navigation symbols, which contain the satellite clock and orbital parameters used in the calculation of the user position, are binary phase shift key (BPSK) symbols transmitted at a data rate of 50 bps [4]. The C/A code is a Gold code [8] with a chip rate of 1.023 Mchips/sec (or code period $P = 1023$) and repeats every millisecond, while the P code has a chip rate of 10.23 Mchips/sec and repeats about every week. The $L1$ carrier is modulated by both codes, while the $L2$ carrier is modulated by either the C/A or the P code, but not both at the same time.

A GPS receiver tracks a satellite by first reproducing the PRN code transmitted by the target satellite, then shifting the phase of the replica code until it correlates with the satellite's PRN code. Maximum correlation occurs when the locally generated PRN code matches the code of the incoming satellite signal,

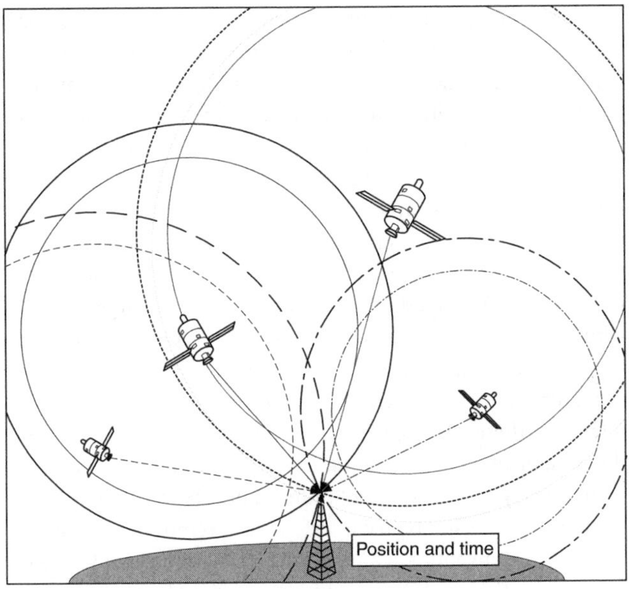

FIGURE 22.2 Intersection of GPS range spheres.

and consequently, the satellite clock can be derived [5]. A conventional receiver's tracking device consists of the delay lock loop (DLL) and the phase lock loop (PLL) [9].

Despite the ever increasing civilian applications, the main drawback of GPS remains its high sensibility to multipath and interference [1], [2], [10], [11], [12], [13], which are the two main sources of errors in range and position estimations. The effect of interference on the GPS receiver is to reduce the signal-to-noise ratio (SNR) of the GPS signal such that the receiver is unable to obtain measurements from the GPS satellite [4]. This causes the receiver to lose its ability to navigation. The spread-spectrum (SS) scheme, which underlies the GPS signal structure, provides a certain degree of protection against interference. However, when the interferer power becomes much stronger than the signal power, the spreading gain alone is insufficient to yield any meaningful information. For the C/A signal, the GPS receiver is vulnerable

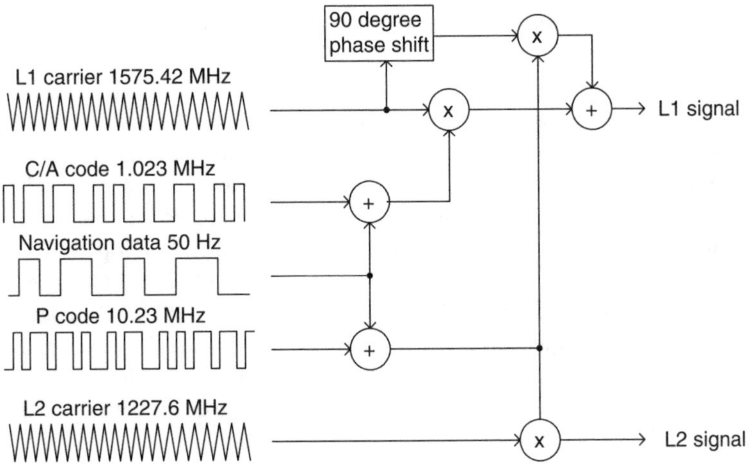

FIGURE 22.3 GPS signal structure.

to strong interferers whose power exceeds the approximately 30-dB gain ($10 \log 1023 \approx 30$ dB) offered via the spreading/despreading process. It is thus desirable that GPS receivers operate efficiently in the presence of rich multipath and strong interference, whether it is intentional or unintentional.

Multipath occurs when a signal emitted from a GPS satellite reaches the receiver following multiple propagation paths, due to reflections from nearby objects. In communications systems, multipath components can be used as sources of diversity to improve the system performance. In GPS, on the other hand, only the direct path signal carries the pertained geolocation information and the signal multipath results in respective erroneous pseudorange measurements. Multipath can cause the GPS receiver's tracking loops to lose lock of the satellite. This occurs when the receiver attempts to correlate with both the direct signal and its reflections [5], [14], [15]. Multipath components with delays larger than one chip, however, will not cause severe pseudorange measurement error due to the low autocorrelation values of the Gold code.

Interference suppression and multipath mitigation in SS communications systems have been active research topics for a long time, and many techniques have been reported in the literature (see, e.g., [10], [16], [17], [18] and references therein). In GPS, interference and multipath can be combated in the time, space, or frequency domain, or in a domain of joint variables, e.g., time–frequency [19], [20] or space–time [21], [22]. Time–frequency signal representations equip the receiver with the ability to detect the time–frequency signature of the nonstationary interferer and remove it through synthesis and subspace projection methods [23], [24]. Conventional antenna arrays, which are only based on spatial processing, are among the simplest, and yet effective, techniques for narrowband interference suppression. Space–time processing relies on antenna arrays to provide the receiver with spatial and temporal selectivity. The spatial selectivity allows discrimination between the GPS signals and interference based on their respective angles of arrival (AOA) [25], [26], [27]. The temporal dimension can be used for broadband interference and multipath cancellation. Generally, the criteria for determining the optimal array weights include maximum signal-to-interference-plus-noise ratio (MSINR), minimum mean square error (MMSE), and minimum output power (MOP) [22]. Array signal processing techniques are applied in [28], [29] for pseudorange and carrier phase measurements, where the interference and signal multipath are modeled as a circularly symmetric and spatially correlated Gaussian random process with zero-mean to reflect the fact that in GPS, only the direct-path signal carries information concerning the pseudorange calculation. In [22], adaptive space–time array processing (STAP) techniques are used to suppress strong interferers and signal multipath, both components responsible for broadening and biasing the cross-correlation function of the data and the receiver codes, and thereby degrading performance. A multistage nested Wiener filter (MSNWF) is developed in [30] based on the MOP principle to combat both narrowband and broadband interference. In [27], estimation of TOA is performed conditioned to a common group delay experienced by all the beams formed for the satellites in the field of view (FOV) (at least four satellites are required to acquire the three-dimensional position plus time [4]). It is noted, however, that the existing techniques for GPS interference cancellations do not fully utilize the GPS signal structure, namely, the repetitive property of the GPS C/A code.

Similar to interference, multipath distorts the C/A code and P code modulations, as well as the carrier phase observations. The errors produced by multipath have been studied in [6], [10], [13], [31], [32], [33], [34], [35]. One of the widely used multipath mitigation techniques is narrow correlator [36], [37], which is one of the first low-cost receiving techniques that employs narrower early–late spacing than the conventional early–late correlator for improved code tracking accuracy in multipath environments. Another major category of multipath mitigation techniques is based on the multipath elimination technology (MET) [38] or multipath elimination delay lock loop (MEDLL) [39], [40]. Other techniques include strobe and edge correlators [41], [42] and high-resolution correlators (HRCs) [43]. The above-mentioned techniques are all single-antenna techniques. The ability to discriminate the direct-path signal from its reflections in the single-antenna receiver is limited because of the time resolution. The spatial diversity introduced by a multiantenna GPS receiver can be used to resolve the direct-path signal from interference and multipath [28] in the spatial domain.

In this chapter, we concentrate on interference suppression and multipath mitigation in GPS receivers equipped with an adaptive antenna array. We first review some widely used adaptive array processing techniques for both broadband and narrowband interference suppression in GPS. We then present a novel GPS processor that is based on the inherent self-coherent feature of the GPS signal to remove strong interference from the received signal. Simulation results are provided to demonstrate the performance of the self-coherence-based GPS receiver. Finally, we present effective multipath mitigation techniques for single- and multi-antenna receivers.

22.2 GPS Signal Model

The GPS BPSK-modulated DS-SS signal, in the discrete-time form at the chip rate, is given by

$$s(t) = A \sum_{n=-\infty}^{\infty} b(n)p(t - nT_c) \tag{22.1}$$

where $b(n)$ is the satellite navigation symbol with amplitude A and $p(t)$ denotes the C/A code (Gold code) with chip duration T_c and code period P. In GPS, the C/A code repeats itself $D = 20$ times within one symbol. The signal reaching the GPS receiver is the aggregate of the GPS signals of satellites currently in the FOV, their multipaths, additive white Gaussian noise (AWGN), and broadband/narrowband interference. Thus, the signal received at the GPS receiver, after the frequency synchronization with the carrier, can be expressed as

$$x(n) = \sum_{q=0}^{Q} s_q(nT_s - \tau_q)e^{j\phi_q} + \sum_{k=1}^{K} B_k u_k(n) + v(n) \tag{22.2}$$

where T_s is the Nyquist sampling interval, Q is the number of multipath components, with subscript 0 designated to the direct-path signal, $s_q(n)$, τ_q, and ϕ_q are the signal sample, time delay, and phase shift of the qth multipath component, respectively, K is the number of interferers, $u_k(n)$ is the waveform of the kth interferer with amplitude B_k, and $v(n)$ is the AWGN with zero-mean and variance σ_v^2.

If an M-element linear array is employed at the receiver, then the baseband received vector comprising data across the array at time n is given by [6]

$$\begin{aligned}\mathbf{x}(n) &= \sum_{q=0}^{Q} s_q(nT_s - \tau_q)\mathbf{a}_q e^{j\phi_q} + \sum_{k=1}^{K} B_k u_k(n)\mathbf{d}_k + \mathbf{v}(n) \\ &= s_0(nT_s - \tau_0)\mathbf{a}_0 e^{j\phi_0} + \sum_{q=1}^{Q} s_q(nT_s - \tau_q)\mathbf{a}_q e^{j\phi_q} + \sum_{k=1}^{K} B_k u_k(n)\mathbf{d}_k + \mathbf{v}(n)\end{aligned} \tag{22.3}$$

where $\mathbf{a}_q$ and $\mathbf{d}_k$ are, respectively, $M \times 1$ spatial signatures of the qth satellite multipath and the kth interferer, and $\mathbf{v}(n)$ consists of noise samples. Let $\mathbf{s}(n) \triangleq s_0(nT_s - \tau_0)\mathbf{a}_0 e^{j\phi_0}$ denote the data vector across the array due to the direct-path signal. Then, Equation 22.3 can be rewritten as

$$\mathbf{x}(n) = \mathbf{s}(n) + \tilde{\mathbf{s}}(n) + \mathbf{u}(n) + \mathbf{v}(n) \tag{22.4}$$

where $\tilde{\mathbf{s}}(n) \triangleq \sum_{q=1}^{Q} s_q(nT_s - \tau_q)\mathbf{a}_q e^{j\phi_q}$ and $\mathbf{u}(n) \triangleq \sum_{k=1}^{K} B_k u_k(n)\mathbf{d}_k$. Assuming a direct path to the satellite located at the direction θ, we can express the spatial signature $\mathbf{a}_0$ as

$$\mathbf{a}_0 = \mathbf{a}(\theta) \triangleq [1, e^{j2\pi f_c \Delta_t}, \ldots, e^{j2\pi f_c (M-1)\Delta_t}]^T \tag{22.5}$$

where $(\cdot)^T$ is the matrix/vector transpose, f_c is the carrier frequency, $\Delta_t = D_M/c \sin\theta$ is the interelement path delay of the source in the direction of θ, c is the propagation speed of the waveform, and D_M is the sensor spacing.

Unlike other SS communication systems such as CDMA, where the so-called near–far phenomena may cause severe impairment at the receiver, in GPS, interference from other satellites is negligible because

(1) signals from all satellites are received at approximately the same power levels (no near–far problem), and (2) the weak cross-correlations of the Gold codes and inherent high processing gain are sufficient to attenuate interference from other satellites. For this reason, only one satellite is considered in Equation 22.2, though multiple satellites are needed for navigation.

22.3 Interference Suppression Techniques in GPS

It is well known that interferers can severely impair the acquisition and tracking performance of the GPS receiver. While narrowband interference in GPS can be effectively mitigated by temporal or frequency-domain filtering [44], broadband interference must be combated using adaptive antenna array techniques that utilize the spatial dimension of the array.

A variety of adaptive algorithms have been developed to suppress interference in GPS, each with different optimization strategies and constraints. In general, all adaptive beam-forming algorithms attempt to reduce the undesired interference while minimally affecting the desired GPS signals. The removal of interference is performed by weighting and combining the signals received at each sensor in a manner such that attenuation is applied in the direction of the interference sources and gain is provided in the direction of the desired sources.

22.3.1 Broadband Interference Suppression Using Space–Time Array

To combat broadband interference, space–time adaptive array is required. We consider an M-sensor space–time array at the receiver, with an Lth-order FIR (finite impulse response) filter attached to each sensor [22], as shown in Figure 22.4. The STAP is able to null interference and signal multipath as long as the interelement time-delay T satisfies $T < 1/B$, where B is the array operating bandwidth and $(L-1)T$ of each FIR filter is sufficient to encompass the differential multipath delays.

Let $\mathbf{x}(n-l)$ denote the received signal at the lth tap of the FIR filter across the array, $l = 0, \ldots, L-1$. Define $\mathbf{w}_l \triangleq [w_{1l} \cdots w_{Ml}]^T$ as the coefficient vector at the lth tap of the FIR filter across the array. The array weight matrix can be defined as

$$\mathbf{W} \triangleq [\mathbf{w}_0 \cdots \mathbf{w}_{L-1}] \tag{22.6}$$

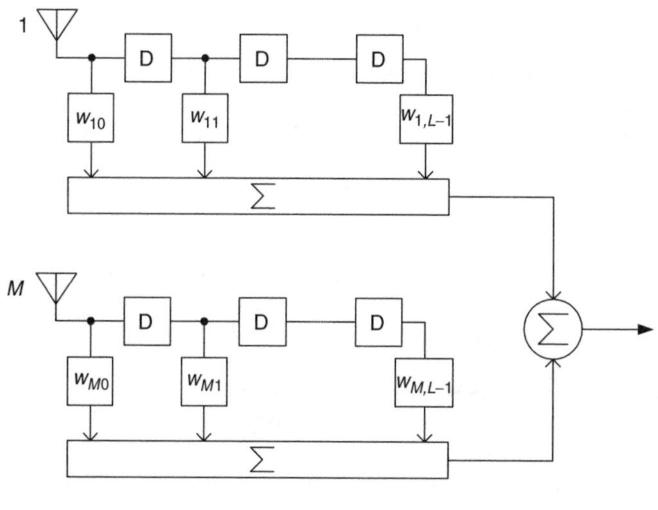

FIGURE 22.4 Adaptive space–time array processor.

The output of the array is then given by

$$z(n) = \mathbf{w}^H \mathbf{x}_{ML}(n) \tag{22.7}$$

where $\mathbf{w} \triangleq [\mathbf{w}_0^T \cdots \mathbf{w}_{L-1}^T]^T$, $\mathbf{x}_{ML}(n) \triangleq [\mathbf{x}^T(n) \cdots \mathbf{x}^T(n-L+1)]^T$, and $(\cdot)^H$ denotes the complex conjugate transpose. The average output power is calculated as

$$E\{|z(n)|^2\} = \mathbf{w}^H E\{\mathbf{x}_{ML}(n)\mathbf{x}_{ML}^H(n)\}\mathbf{w} = \mathbf{w}^H \mathbf{R}_s \mathbf{w} + \mathbf{w}^H \mathbf{R}_u \mathbf{w} + \mathbf{w}^H \mathbf{R}_v \mathbf{w} \tag{22.8}$$

where $E\{\cdot\}$ denotes the statistical expectation. It is assumed that the GPS signal, interference, and noise are uncorrelated. The covariance matrices of GPS signal, interference, and noise, respectively, are $\mathbf{R}_s \triangleq E\{[\mathbf{s}_{ML}(n) + \tilde{\mathbf{s}}_{ML}(n)][\mathbf{s}_{ML}(n) + \tilde{\mathbf{s}}_{ML}(n)]^H\}$, $\mathbf{R}_u \triangleq E\{\mathbf{u}_{ML}(n)\mathbf{u}_{ML}^H(n)\}$, and $\mathbf{R}_v \triangleq E\{\mathbf{v}_{ML}(n)\mathbf{v}_{ML}^H(n)\}$, where $\mathbf{s}_{ML}(n)$, $\tilde{\mathbf{s}}_{ML}(n)$, $\mathbf{u}_{ML}(n)$, and $\mathbf{v}_{ML}(n)$ are similarly defined as $\mathbf{x}_{ML}(n)$. The optimum weight vector $\mathbf{w}$ can be obtained via a number of methods, namely, the MSINR, MMSE, and MOP methods.

22.3.1.1 MSINR

The MSINR method seeks the array weights by maximizing the array output SINR, which is defined as [45]

$$\text{SINR} \triangleq \frac{\mathbf{w}^H \mathbf{R}_s \mathbf{w}}{\mathbf{w}^H(\mathbf{R}_u + \mathbf{R}_v)\mathbf{w}} \tag{22.9}$$

The desired weight vector $\mathbf{w}$ is the eigenvector corresponding to the maximum eigenvalue $\lambda_{\max}$ of the generalized eigenvalue problem:

$$\mathbf{R}_s \mathbf{w} = \lambda_{\max}(\mathbf{R}_u + \mathbf{R}_v)\mathbf{w} \tag{22.10}$$

In this case, the MSINR is given by $\lambda_{\max}$. It is clear from Equation 22.9 that the SINR method requires the knowledge of the transmitted signal in order to calculate the signal power at the receiver. The SINR in Equation 22.9 is obtained by taking both the direct-path signal and the signal multipath into account. It is important to note, however, that if only the direct-path signal is considered in solving Equation 22.9, the respective SINR will only decrease a few tenths of a decibel compared to the case when multipath is included [22].

22.3.1.2 MMSE

The MMSE method obtains the array weight vector by minimizing the error between the array output and a known reference signal. Let $s_d(n)$ denote the desired GPS signal, which is known at the receiver. The error between the array output and the desired signal is defined as $e(t) \triangleq (zn) - s_d(n)$ and the mean square error (MSE) is given by

$$\begin{aligned}\varepsilon &\triangleq E\{|e(n)|^2\} = E\{|\mathbf{w}^H \mathbf{x}_{ML}(n) - s_d(n)|^2\} \\ &= \mathbf{w}^H E\left\{\mathbf{x}_{ML}(n)\mathbf{x}_{ML}^H(n)\right\}\mathbf{w} - \mathbf{w}^H E\left\{\mathbf{x}_{ML}(n)s_d^*(n)\right\} - \mathbf{w}^T E\left\{\mathbf{x}_{ML}^*(n)s_d(n)\right\} + E\left\{s_d(n)s_d^*(n)\right\}\end{aligned} \tag{22.11}$$

Minimizing the above MSE yields the solution

$$\mathbf{w}_{\text{MMSE}} = (\mathbf{R}_s + \mathbf{R}_u + \mathbf{R}_v)^{-1}\mathbf{r}_{xs} \tag{22.12}$$

where $\mathbf{r}_{xs} \triangleq E\{\mathbf{x}_{ML}(n)s_d^*(n)\}$.

The vector $\mathbf{r}_{xs}$ in the MMSE method, as well as the covariance matrix $\mathbf{R}_s$ in the MSINR method, can be calculated as follows. Assume that the satellite is located at the angle (θ, ψ). For GPS, the temporal autocorrelation of the transmitted signal, which is essentially the autocorrelation function of the Gold code, denoted as $R_p(\tau)$, is known. For an adaptive array with M antennas and L taps per antenna, define the $L \times 1$ vector as

$$\mathbf{r} \triangleq [R_p(0) \cdots R_p(L-1)]^T \tag{22.13}$$

The $ML \times ML$ matrix $\mathbf{R}_s$ is calculated in the absence of interference and noise. Let $\eta(m)$ denote the phase shift for the GPS satellite at the angle (θ, ψ) to the mth antenna. Then the first column of matrix $\mathbf{R}_s$, which is $\mathbf{r}_{xs}$ in the MMSE method, is given by $[\mathbf{r}^T e^{j\eta(1)} \cdots \mathbf{r}^T e^{j\eta(M)}]^T$. The rest of matrix $\mathbf{R}_s$ can be similarly calculated.

22.3.1.3 MOP

The minimum output power is premised on the fact that the GPS signals are well below the noise floor [21]. By minimizing the total array output power, the interference/noise power level can be effectively reduced.

The output power of the space–time array is given by

$$P_{\text{out}} = E\{|z(n)|^2\} = \mathbf{w}^H \mathbf{R}_x \mathbf{w} \tag{22.14}$$

where $\mathbf{R}_x \triangleq E\{\mathbf{x}_{ML}(n)\mathbf{x}_{ML}^H(n)\}$. If the array weights are obtained by simply minimizing P_{out}, an obvious shortcoming is that no attempt is made to minimize the distortion on the GPS signal. With the GPS satellite positions assumed to be known, the constrained MOP method can be formulated as

$$\mathbf{w}_{\text{MOP}} = \min_{\mathbf{w}} \mathbf{w}^H \mathbf{R}_x \mathbf{w}, \quad \text{subject to } \mathbf{A}^H(\theta)\mathbf{w} = \mathbf{1}_L \tag{22.15}$$

where $\mathbf{A}(\theta) \triangleq \mathbf{I}_L \otimes \mathbf{a}(\theta)$, $\mathbf{I}_L$ is an $L \times L$ identity matrix, $\mathbf{1}_L$ is an $L \times 1$ unit vector, and $\otimes$ denotes the Kronecker product. The constraint in Equation 22.15 suggests that for each time tap, we have $\mathbf{a}^H(\theta)\mathbf{w}_l = 1$, $l = 0, \ldots, L-1$. Using Lagrange multipliers, the solution to Equation 22.15 is given by

$$\mathbf{w}_{\text{MOP}} = \mathbf{R}_x^{-1}\mathbf{A}(\theta)\left[\mathbf{A}^H(\theta)\mathbf{R}_x^{-1}\mathbf{A}(\theta)\right]^{-1}\mathbf{1}_L \tag{22.16}$$

With multiple satellites in the FOV, the above processes are repeated for each GPS satellite to obtain the optimum solution.

Any beam formed by a STAP processor will have an underlying temporal filtering characteristic toward the signal of interest [27]. With the STAP array shown in Figure 22.4, the effective FIR filter coefficients for the direct-path signal can be calculated as

$$\mathbf{h} = \mathbf{W}^H \mathbf{a}(\theta) \tag{22.17}$$

with $\mathbf{W}$ given by Equation 22.6. In GPS navigation, usually at least four satellites are required to calculate the receiver's three-dimensional position and time [4]. Assuming that there are F satellites in the FOV, we denote $\mathbf{h}_f$, $f = 1, \ldots, F$, as the corresponding effective FIR filter of the fth satellite. If no further constraints are imposed, it is possible that the group delay for each filter could be different and thus bias the GPS solution [27]. Two methods have been proposed in [27] to deal with the time-bias problem. The first approach constrains each STAP beam such that a fixed group delay common to all the beams is imposed; i.e., the vector $\mathbf{h}_f$, $f = 1, \ldots, F$, can be chosen such that the group delay of the effective filter is equal to a constant for the GPS waveform. This constraint forces $\mathbf{h}_f$ to be orthogonal to the vector $[R_s'(0) \cdots R_s'((L-1)\Delta_T)]^T$, where $R_s'(\tau)$ is the derivation of the autocorrelation function of the GPS signal and Δ_T is the fixed delay.

The second approach uses equalizers at the beam outputs such that the differential group delay seen from beam to beam is zero.

22.3.1.4 Signal Distortion Introduced by Processor

The main goal of the signal processing techniques in the GPS receivers discussed above is to minimize the effect of interference while preserving the GPS signal. However, since the processor does not have a uniform frequency response across the operating band, it is possible that it will introduce a distortion of the desired GPS signal. The distortion introduced by the GPS processor is investigated in [22] in terms of the cross-correlation between the array output and a desired signal.

Consider the space–time array shown in Figure 22.4. Without loss of generality, we assume that the satellite is located at the polar angle (θ, ψ) and the satellite carrier frequency is f_c. At the receiver, to take the possible steering vector mismatch into account, a different position $(\hat{\theta}, \hat{\psi})$ is used in the calculation of the cross-correlation function instead of the exact one. Correspondingly, the array weight becomes $w_{ml}(\hat{\theta}, \hat{\psi})$. Let $H(f)$ denote the frequency response of the GPS processor. Using $(\hat{\theta}, \hat{\psi})$, it is straightforward that the response of the FIR filter to the satellite is [22]

$$H(f, \hat{\theta}, \hat{\psi}) = \sum_{m=1}^{M} \sum_{l=0}^{L-1} \hat{w}_{ml} e^{j2\pi f(t_m + lT)} \tag{22.18}$$

where $t_m \triangleq d_m/c$, $d_m \triangleq x_m \sin\theta \cos\psi + y_m \sin\theta \sin\psi + z_m \cos\theta$, (x_m, y_m, z_m) is the three-dimensional position of the mth antenna, and $\hat{w}_{ml} = w_{ml}(\hat{\theta}, \hat{\psi}) e^{j2\pi f_c t_m}$.

Denote $S(f)$ as the Fourier transform of the GPS signal $s(t)$. The array output in the frequency domain is then $Z(f) = H(f)S(f)$, which, when expressed in the time domain, is given by

$$z(t) = \int_{-\infty}^{\infty} Z(f) e^{j2\pi f t} df = \int_{-\infty}^{\infty} H(f) S(f) e^{j2\pi f t} df \tag{22.19}$$

Let $s_d(t)$ be the reference signal known at the receiver. The cross-correlation between the array output $z(t)$ and $s_d(t)$ is calculated as

$$R_{zs}(\tau) = E\left\{z(t) s_d^*(t+\tau)\right\} = \int_{-\infty}^{\infty} P_s(f) H(f) e^{-j2\pi f \tau} df \tag{22.20}$$

where $P_s(f)$ is the power spectrum of the signal.

Since $P_s(f)$ is a positive symmetric function of f, the correlation peak will occur at $\tau = 0$ if $H(f)$ is absent. However, the presence of $H(f)$ can broaden the correlation function and the phase of $H(f)$ may bias the correlation peak. The shift introduced by $H(f)$ can be corrected as follows. If $H(f)$ is known, we can always construct another filter $G(f) = H^*(f)$ and sequentially attach it to the adaptive processor. Then the combined processor has a response $G(f)H(f) = |H(f)|^2$ and the cross-correlation becomes

$$\tilde{R}_{zs}(\tau) = \int_{-\infty}^{\infty} |H(f)|^2 P(f) e^{-j2\pi f \tau} df \tag{22.21}$$

It is obvious that the peak of $\tilde{R}_{zs}(\tau)$ now lies at $\tau = 0$. Another option is to equalize only the phase using an all-pass filter.

Though $G(f)$ helps in correcting the bias caused by the phase of $H(f)$, it offers no cure for multipath. In the presence of multipath, the Fourier transform of the GPS signal is replaced by $S(f)[1+\delta(f)]$, where $\delta(f) = \sum_{q=1}^{Q} s_q e^{-j2\pi f \tau_q}$, each multipath component with amplitude s_q and time-delay $\tau_q, q = 1, \ldots, Q$. The corresponding cross-correlation function is

$$\tilde{R}_{zs}\mathrm{m}(\tau) = \int_{-\infty}^{\infty} P_s(f) |H(f)|^2 [1 + \delta(f)] e^{-j2\pi f \tau} df \tag{22.22}$$

from which it is evident that the signal multipath can introduce a broadening of the correlation peak.

22.3.2 Narrowband FM Interference Suppression Using Time–Frequency Method

A subspace projection-based array processing technique is proposed in [46] for suppressing interference with a rapidly time-varying spectrum. The subspace projection techniques based on the time–frequency distributions of the FM interferer have been proposed in [23], [47]. The concept of subspace projection for instantaneously narrowband jammer suppression is to estimate the instantaneous frequency of the jammer using time-frequency signal representation techniques, and the result is then used to construct the interference subspace. The jammer components are removed by the projection of the received data onto the subspace that is orthogonal to the jammer subspace [46].

Consider an M-element linear space-only array. Assume that the interference waveform $u_k(n)$ in Equation 22.3 represents an instantaneously narrowband FM signal with constant amplitude $u_k(n) = e^{j\varphi_k(n)}$, $k = 1, \ldots, K$, and $\mathbf{v}(n)$ is the noise vector, which is assumed to be temporally and spatially white with zero-mean and $E\{\mathbf{v}(n)\mathbf{v}^H(l)\} = \sigma_v^2 \delta_{n,l} \mathbf{I}_M$.

The received signal is chip rate sampled. By collecting P consecutive samples at the receiver, we define the $MP \times 1$ vector,

$$\mathbf{x} \triangleq [\mathbf{x}^T(n) \cdots \mathbf{x}^T(n-(P-1))]^T = \mathbf{s} + \mathbf{u} + \mathbf{v} \tag{22.23}$$

as the received spatial–temporal data block, where $\mathbf{s}$ only contains samples of the direct-path GPS signal. Signal multipath is ignored from the received signal. If it is assumed that the navigation symbol "1" is transmitted and the received signal is aligned with the locally generated PRN code replica, we can express $\mathbf{s}$ in Equation 22.23 as

$$\mathbf{s} = [p(n) \cdots p(n-(P-1))]^T \otimes \mathbf{a}_0 \triangleq \mathbf{p} \otimes \mathbf{a}_0 \tag{22.24}$$

Similarly, define

$$\mathbf{u}_k \triangleq [u_k(n) \cdots u_k(n-(P-1))]^T \tag{22.25}$$

where $u_k(n)$ is the normalized jammer waveform, provided by using instantaneous frequency estimations, including time–frequency distributions. The vector $\mathbf{u}$ in Equation 22.23 is then given by $\mathbf{u} = \sum_{k=1}^{K} B_k \tilde{\mathbf{u}}_k$, where $\tilde{\mathbf{u}}_k \triangleq \mathbf{u}_k \otimes \mathbf{d}_k$. The overall interference subspace is spanned by $\tilde{\mathbf{U}} \triangleq [\tilde{\mathbf{u}}_1 \cdots \tilde{\mathbf{u}}_K]$. The interference is suppressed by projecting $\mathbf{x}$ onto the orthogonal subspace of $\tilde{\mathbf{U}}$. The orthogonal subspace projection matrix of $\tilde{\mathbf{U}}$ is given by [23]

$$\tilde{\mathbf{U}}_\perp = \mathbf{I}_{MP} - \tilde{\mathbf{U}}(\tilde{\mathbf{U}}^H\tilde{\mathbf{U}})^{-1}\tilde{\mathbf{U}}^H = \mathbf{I}_{MP} - \frac{1}{MP}\sum_{k=1}^{K} \tilde{\mathbf{u}}_k \tilde{\mathbf{u}}_k^H \tag{22.26}$$

Applying $\tilde{\mathbf{U}}_\perp$ on $\mathbf{x}$ yields the projection of the received signal vector onto the orthogonal subspace:

$$\mathbf{x}_\perp \triangleq \tilde{\mathbf{U}}_\perp \mathbf{x} = \tilde{\mathbf{U}}_\perp \mathbf{s} + \tilde{\mathbf{U}}_\perp \mathbf{v} \tag{22.27}$$

which shows that the jammers are completely removed from the received signal [46].

The performance of this subspace projection-based receiver is evaluated in terms of SINR after the interference suppression and despreading. It is noted that $\mathbf{s}$ represents the spatial–temporal signature of the GPS signal. Therefore, we can use $\mathbf{s}$ to perform the despreading after the interference suppression,

$$z = \mathbf{s}^H \mathbf{x}_\perp = \mathbf{s}^H \tilde{\mathbf{U}}_\perp \mathbf{s} + \mathbf{s}^H \tilde{\mathbf{U}}_\perp \mathbf{v} \tag{22.28}$$

Let $z_g \triangleq \mathbf{s}^H \tilde{\mathbf{U}}_\perp \mathbf{s}$ represent the term in Equation 22.28 due to the GPS signal after the despreading. Then, after some straightforward calculations, we have

$$\begin{aligned} z_g &= MP - \frac{1}{MP}\sum_{k=1}^{K} \mathbf{s}^H \tilde{\mathbf{u}}_k \tilde{\mathbf{u}}_k^H \mathbf{s} \\ &= MP - \frac{1}{MP}\sum_{k=1}^{K} (\mathbf{p}^H \mathbf{u}_k)(\mathbf{a}_0^H \mathbf{d}_k)(\mathbf{d}_k^H \mathbf{a}_0)(\mathbf{u}_k^H \mathbf{p}) \\ &= MP\left(1 - \sum_{k=1}^{K} |\alpha_k|^2 |\beta_k|^2\right) \end{aligned} \tag{22.29}$$

where $\alpha_k \triangleq P\mathbf{p}^H\mathbf{u}_k$ and $\beta_k \triangleq M\mathbf{a}_0^H\mathbf{d}_k$, and

$$E\{z\} = z_g = MP\left(1 - \sum_{k=1}^{K} |\alpha_k|^2|\beta_k|^2\right) \tag{22.30}$$

$$\text{var}\{z\} = E\{|\mathbf{s}^H\tilde{\mathbf{U}}_\perp\mathbf{v}|^2\} = \sigma_v^2 z_g \tag{22.31}$$

where we have invoked the white Gaussian assumption of the noise and var$\{\cdot\}$ denotes the variance of the argument.

The array output z in Equation 22.28 is obtained by considering only one block of P samples within the GPS symbol. At the symbol level, the array output after despreading is the summation of $D = 20$ blocks, i.e., $z = \sum_{d=1}^{D} z_d$, where z_d is given by Equation 22.28. It has been shown that the receiver SINR after the subspace projection and despreading is given by [46]

$$\text{SINR} = \frac{E^2\{z\}}{\text{var}\{z\}} = 2\frac{MP}{\sigma_v^2}\left[D - \sum_{k=1}^{K}\sum_{d=1}^{D} |\alpha_{kd}|^2|\beta_{kd}|^2\right] \tag{22.32}$$

The temporal and spatial coefficients appear as multiplicative products in the SINR expression in Equation 22.32. This suggests that the spatial and temporal signatures play equivalent roles in the receiver performance. In the absence of jammers, no excision is necessary, and the SINR of the receiver output will become $2MPD/\sigma_v^2$, which represents the upper bound of the interference suppression performance. Indeed, the term $2MP\sum_{k=1}^{K}\sum_{d=1}^{D} |\alpha_{kd}|^2|\beta_{kd}|^2$ is the reduction in the receiver performance caused by the proposed interference suppression technique. It reflects the energy of the signal component that is in the jammer subspace. We note that if the jammers and the DS-SS signal are orthogonal, either in the spatial domain ($\alpha = 0$) or in the temporal domain ($\beta = 0$), the interference excision is achieved with no loss in performance. In the general case, β, for FM interference, takes a small value, which is much smaller than α. Therefore, the difference in the temporal signatures of the incoming signals allows the proposed projection technique to excise FM jammers effectively with only insignificant signal loss.

Interference suppression using arrays is improved in several ways. First, the employment of an antenna array can lead to an accurate instantaneous frequency estimation of the jammers [48]. Second, in comparison to the single-sensor case when $M = 1$ and $\alpha_{kq} = 1$, and the SINR becomes

$$\text{SINR} = 2\frac{P}{\sigma_v^2}\left[D - \sum_{k=1}^{K}\sum_{d=1}^{D} |\beta_{kd}|^2\right] \tag{22.33}$$

which indicates that multisensor receivers, at minimum, improve SINR by the array gain. This is true, independent of the underlying fading channels and scattering environment. Finally, spatial selectivity, highlighted by the role of α, is used to discriminate against the jammer signal.

The above analysis assumes the perfect knowledge of the jammer waveform. In practice, instantaneous frequency, or phase, values of the jammer have to be estimated and, as such, are subject to errors. In the simulations, the effect of phase errors on the receiver performance is studied. Consider a periodic chirp jammer whose period is equal to one block length of the GPS C/A symbol. The angle of arrival of the satellite signal is 50°. An $M = 2$ array is considered with half-wavelength spacing. The signal-to-noise ratio and signal-to-interference-plus-noise ratio are defined, respectively, as $\text{SNR} = 10\log_{10} 1/\sigma_v^2$ and $\text{SINR} = 10\log_{10} 1/(\sigma_i^2 + \sigma_v^2)$, all in decibel, where unit signal power is assumed for simplicity, σ_v^2 is the noise power, and σ_i^2 is the jammer power. The jammer-to-signal ratio is defined as $\text{JSR} = 10\log_{10}\sigma_i^2$, also in decibel. The JSR is set to 50 dB and SNR is equal to -20 dB. Figure 22.5 depicts the simulated values of the receiver SINR vs. the phase error variance, varying from 0 to 0.01 for all blocks. The AOAs of the jammer signals are set to 50°, 35°, and 65°, respectively. It is clear from the figure that as the error variance increases, the output SINR decreases. The SINR of the single-sensor case is also plotted for comparison. Unlike the result of exact instantaneous frequency estimation, where antenna arrays bring a constant 3-dB array gain, the receiver SINR in the presence of those errors is dependent on the spatial signatures of the signal and

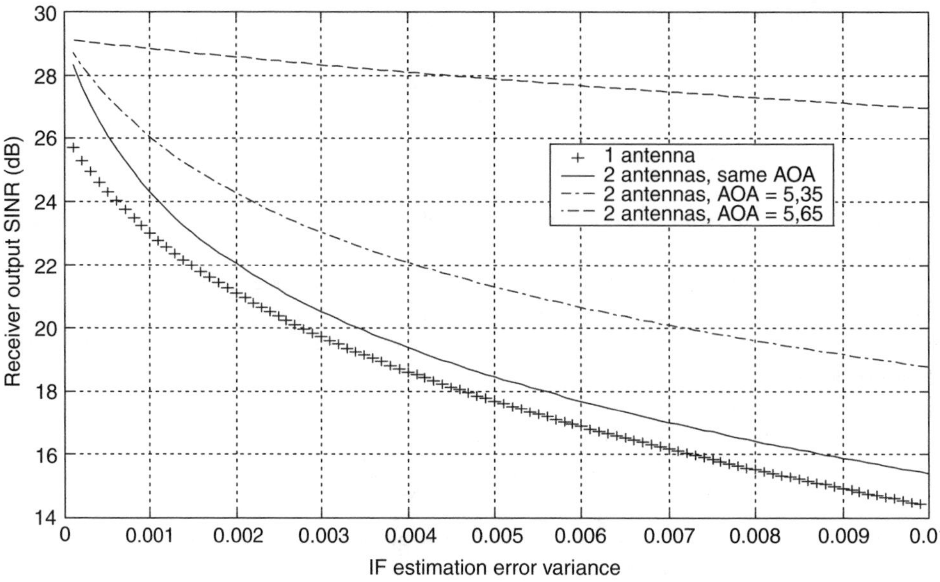

FIGURE 22.5 Receiver SINR vs. error variance.

jammer. For small spatial cross-correlation coefficients, the use of antenna array allows the receiver to be more robust to the instantaneous frequency estimation errors. The relation between the receiver SINR and the jammer AOA is shown in Figure 22.6, whereas the spatial cross-correlation coefficient vs. the jammer AOA is depicted in Figure 22.7. In this case, phase error variance was kept constant at 0.01. It is important to observe that the peak and the null in Figure 22.6 correspond, respectively, to the lowest and highest values of the spatial correlation parameter between the GPS signal and the jammer.

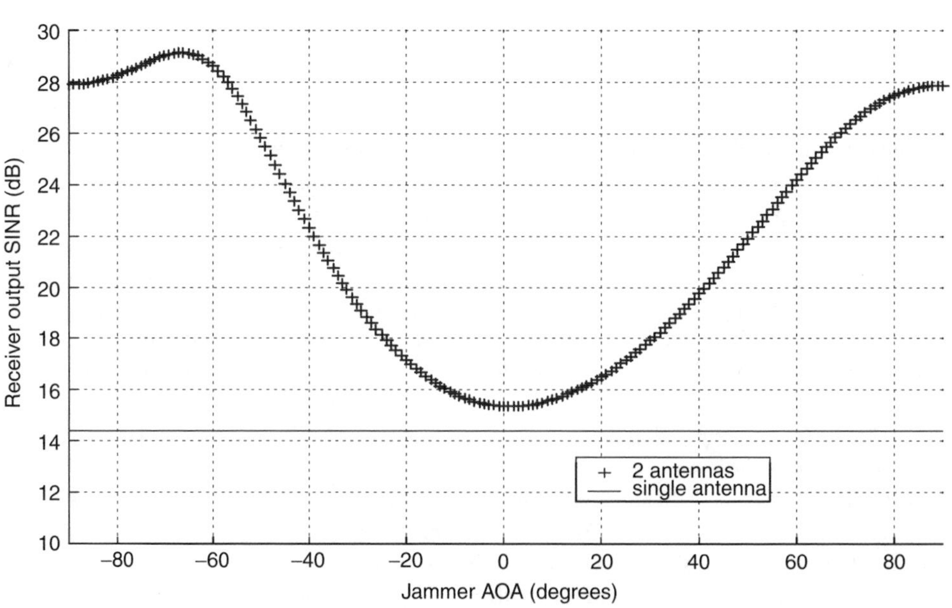

FIGURE 22.6 Receiver SINR vs. jammer AOA.

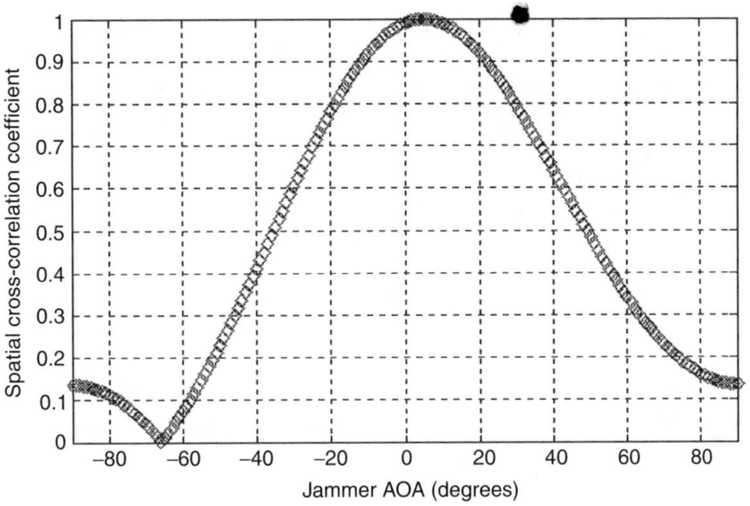

FIGURE 22.7 Spatial cross-correlation vs. jammer AOA.

22.3.3 A Self-Coherence Antijamming GPS Receiver

The interference suppression techniques discussed in Section 22.3.1 require some known information about the satellite at the receiver. For example, the MSINR and MMSE methods need the transmitted signal information in order to calculate the signal power and obtain the reference signal, while the MOP approach uses the knowledge of the satellite direction to preserve the GPS signal. When the GPS receiver is operating in a persistent jamming environment, however, this information may not be available, especially during the initial signal searching and acquisition stage. Thus, it is desirable to develop an interference suppression technique that does not rely on the *a priori* knowledge of the transmitted GPS signal.

It is known that the C/A code is repeated 20 times within the navigation symbols. As a result, the GPS signal exhibits strong self-coherence between chip samples that are separated by integer multiples of the spreading gain, and this unique structure can be utilized in interference suppression. Indeed, under the assumption that the interferers do not have the same structure as that of the GPS signal, a self-coherence antijamming GPS receiver can be constructed based on the SCORE (spectral self-coherence restoral) algorithm [49]. The receiver excises broadband interference by exploiting the structure of the GPS signal without knowing either the transmitted signal or the direction of the satellite.

A block diagram of the self-coherence GPS receiver equipped with an M-element linear uniform array is shown in Figure 22.8, while Figure 22.9 illustrates the structure of the received noise-free GPS signal.

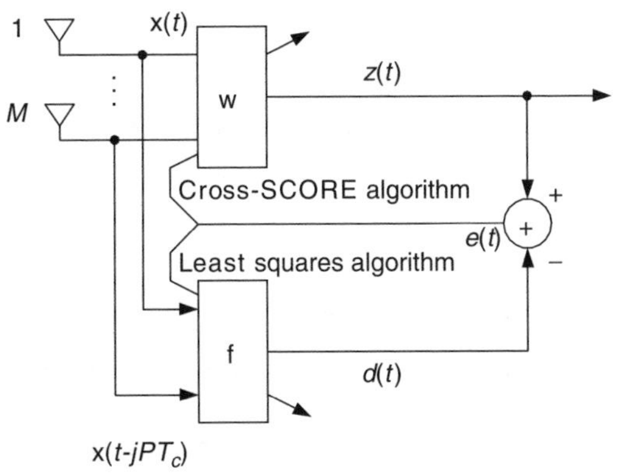

FIGURE 22.8 Block diagram of the self-coherence GPS receiver.

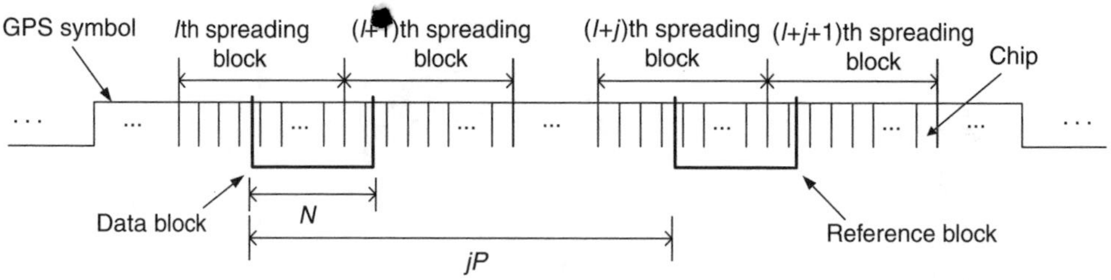

FIGURE 22.9 Noise-free GPS signal structure and data and reference blocks' formation.

Specifically, two blocks of data are formed at the receiver: a *data block*, which spans N consecutive samples, and a *reference block* with the same number of samples as the data block. The distance between the respective samples in the data and reference blocks is set equal to jP chips, where $1 \leq j < 20$. The repetition of the spreading code suggests that the nth sample in the reference block has the same value as its counterpart in the data block, providing that the two samples belong to the same symbol. The data in the data block are processed by a beam-former $\mathbf{w}$. In addition, an auxiliary beam-former $\mathbf{f}$ is also installed to provide a reference signal by processing samples in the reference block, as shown in Figure 22.8. The weight vector $\mathbf{w}$ is adaptively updated according to the cross-SCORE algorithm (refer to [50] for details on the SCORE algorithm), while $\mathbf{f}$ is renewed using the least squares algorithm.

The signal model for the self-coherence GPS receiver is derived from Equation 22.3, which is repeated as follows:

$$\mathbf{x}(n) = \sum_{q=0}^{Q} s_q(nT_s - \tau_q)\mathbf{a}_q e^{j\phi_q} + \sum_{k=1}^{K} B_k u_k(n)\mathbf{d}_k + \mathbf{v}(n) \tag{22.34}$$

According to the formulation of the data and reference blocks, the counterpart of $\mathbf{x}(n)$ in the reference block within the same symbol can be written as

$$\begin{aligned}\mathbf{x}(n - jP) &= \sum_{q=0}^{Q} s_q(nT_s - \tau_q - jP)\mathbf{a}_q e^{j\phi_q} + \sum_{k=1}^{K} B_k u_k(n - jP)\mathbf{d}_k + \mathbf{v}(n - jP) \\ &= \sum_{q=0}^{Q} s_q(nT_s - \tau_q)\mathbf{a}_q e^{j\phi_q} + \sum_{k=1}^{K} B_k u_k(n - jP)\mathbf{d}_k + \mathbf{v}(n - jP) \\ &= \mathbf{s}(n) + \tilde{\mathbf{s}}(n) + \mathbf{u}(n - jP) + \mathbf{v}(n - jP)\end{aligned} \tag{22.35}$$

where we have assumed that, when considered within the same symbol, $s_q(nT_s - \tau_q) = s_q(nT_s - \tau_q - jP)$, $q = 0, \ldots, Q$.

From Figure 22.8, the beam-former output and the reference signal, respectively, are given by $z(n) \triangleq \mathbf{w}^H\mathbf{x}(n)$ and $d(n) \triangleq \mathbf{f}^H\mathbf{x}(n - jP)$. Accordingly, we can define the following covariances:

$$R_{zd} \triangleq E\{z(n)d^H(n)\} = \mathbf{w}^H E\{\mathbf{x}(n)\mathbf{x}^H(n - jP)\}\mathbf{f} \tag{22.36}$$

$$R_{zz} \triangleq E\{z(n)z^H(n)\} = \mathbf{w}^H E\{\mathbf{x}(n)\mathbf{x}^H(n)\}\mathbf{w} \tag{22.37}$$

$$R_{dd} \triangleq E\{d(n)d^H(n)\} = \mathbf{f}^H E\{\mathbf{x}(n - jP)\mathbf{x}^H(n - jP)\}\mathbf{f} \tag{22.38}$$

Under the assumption that the GPS signal, interference, and noise are independent, then

$$\mathbf{R}_{xx} \triangleq E\{\mathbf{x}(n)\mathbf{x}^H(n)\} = E\{\mathbf{x}(n - jP)\mathbf{x}^H(n - jP)\} = \mathbf{R}_s + \mathbf{R}_u + \mathbf{R}_v \tag{22.39}$$

The three terms in Equation 22.39, denote the covariance matrices of the GPS, interference, and noise, respectively. Providing that only the GPS signals are correlated when delayed jP samples, the cross-correlation matrix between the corresponding data vectors in the data and reference blocks becomes

$$\mathbf{R}_{xx}^{(P)} \triangleq E\{\mathbf{x}(n)\mathbf{x}^H(n-jP)\} = \mathbf{R}_s \tag{22.40}$$

To determine the weight vector $\mathbf{w}$, we define the following cost function [50]:

$$C(\mathbf{w},\mathbf{f}) \triangleq \frac{|R_{zd}|^2}{R_{zz}R_{dd}} = \frac{|\mathbf{w}^H\mathbf{R}_{xx}^{(P)}\mathbf{f}|^2}{[\mathbf{w}^H\mathbf{R}_{xx}\mathbf{w}][\mathbf{f}^H\mathbf{R}_{xx}\mathbf{f}]} = \frac{|\mathbf{w}^H\mathbf{R}_s\mathbf{f}|^2}{[\mathbf{w}^H\mathbf{R}_{xx}\mathbf{w}][\mathbf{f}^H\mathbf{R}_{xx}\mathbf{f}]} \tag{22.41}$$

Then $\mathbf{w}$ is obtained by maximizing $C(\mathbf{w},\mathbf{f})$, i.e., the cross-correlation between $z(n)$ and $d(n)$.

In practice, the covariance matrices $\mathbf{R}_s$ and $\mathbf{R}_{xx}$ in Equation 22.41 are unknown and have to be replaced by their sample estimates. Define the $M \times N$ data and reference matrices as $\mathbf{X}_N \triangleq [\mathbf{x}(n)\cdots\mathbf{x}(n-(N-1))]$ and $\mathbf{X}_{N\text{ref}} \triangleq [\mathbf{x}(n-jP)\cdots\mathbf{x}(n-(N-1)-jP)]$, where N is the block length and $N \leq P$. Thus, the sample covariance matrices are given by

$$\hat{\mathbf{R}}_{xx} \triangleq \frac{1}{N}\mathbf{X}_N\mathbf{X}_N^H \tag{22.42}$$

$$\hat{\mathbf{R}}_{xx}^{(P)} \triangleq \frac{1}{N}\mathbf{X}_N\mathbf{X}_{N\text{ref}}^H \tag{22.43}$$

Using the covariance matrix estimates, Equations 22.36 to 22.38 become, respectively, $\hat{R}_{zd} = \mathbf{w}^H\hat{\mathbf{R}}_{xx}^{(P)}\mathbf{f}$, $\hat{R}_{zz} = \mathbf{w}^H\hat{\mathbf{R}}_{xx}\mathbf{w}$, and $\hat{R}_{dd} = \mathbf{f}^H\hat{\mathbf{R}}_{xx}\mathbf{f}$.

Let $e(n) \triangleq d(n) - z(n)$ be the error between the output of the beam former and the reference signal. For a fixed beam-former $\mathbf{w}$, the error $e(n)$ is minimized in the least squares sense when $\mathbf{f}$ is given by $\mathbf{f}_{\text{LS}} = \hat{\mathbf{R}}_{xx}^{-1}\hat{\mathbf{r}}_{xz}$, where

$$\hat{\mathbf{r}}_{xz} \triangleq \frac{1}{N}\sum_{i=0}^{N-1}\mathbf{x}(n-i-jP)z^H(n-i) = \hat{\mathbf{R}}_{xx}^{(P)H}\mathbf{w} \tag{22.44}$$

Substituting $\mathbf{f}$ in Equation 22.41 by $\mathbf{f}_{\text{LS}}$, the cost function becomes

$$C(\mathbf{w},\mathbf{f}_{\text{LS}}) = \frac{\mathbf{w}^H\hat{\mathbf{R}}_{xx}^{(P)}\hat{\mathbf{R}}_{xx}^{-1}\hat{\mathbf{R}}_{xx}^{(P)H}\mathbf{w}}{\mathbf{w}^H\hat{\mathbf{R}}_{xx}\mathbf{w}} \tag{22.45}$$

The weight vector $\mathbf{w}$ that maximizes $C(\mathbf{w},\mathbf{f}_{\text{LS}})$ is readily shown to be the eigenvector corresponding to the largest eigenvalue of the following generalized eigenvalue problem:

$$\hat{\mathbf{R}}_{xx}\mathbf{w} = \lambda_{\max}\hat{\mathbf{R}}_{xx}^{(P)}\hat{\mathbf{R}}_{xx}^{-1}\hat{\mathbf{R}}_{xx}^{(P)H}\mathbf{w} \tag{22.46}$$

where $\lambda_{\max}$ is the largest eigenvalue.

To demonstrate the performance of the self-coherence receiver, we use a linear uniform array consisting of $M = 7$ sensors with half-wavelength spacing. GPS navigation symbols are in the BPSK format and spread by the C/A code with a processing gain of $P = 1023$. At the receiver, chip-rate sampling is performed and $N = 800$ samples are collected in both the data and reference blocks.

Figure 22.10(a) shows the beam pattern when interference is absent and the data and reference blocks are within the same symbol. The SNR is −30 dB. It is clear that the antenna pattern is formed toward the satellite located at $\theta = 30°$. Figure 22.10(b) displays the antenna gain of the receiver when two jammers located at 40 and 60° enter the system. The JSR of each jammer is 30 dB and SINR = −33 dB. The satellite's location is at 20°. Also shown in Figure 22.10(b) is the antenna gain obtained using the MMSE method, as discussed in Section 22.3.1.2. Unlike the self-coherence receiver, which requires neither the knowledge of the position of the satellite nor the transmitted symbols, the MMSE-based receiver determines the weight vector by minimizing the mean square difference between the array output and the desired signal assumed known at the receiver. Figure 22.10(b) clearly shows that for both schemes nulls are placed at the jammer locations while high gains are obtained toward the direction of the satellite.

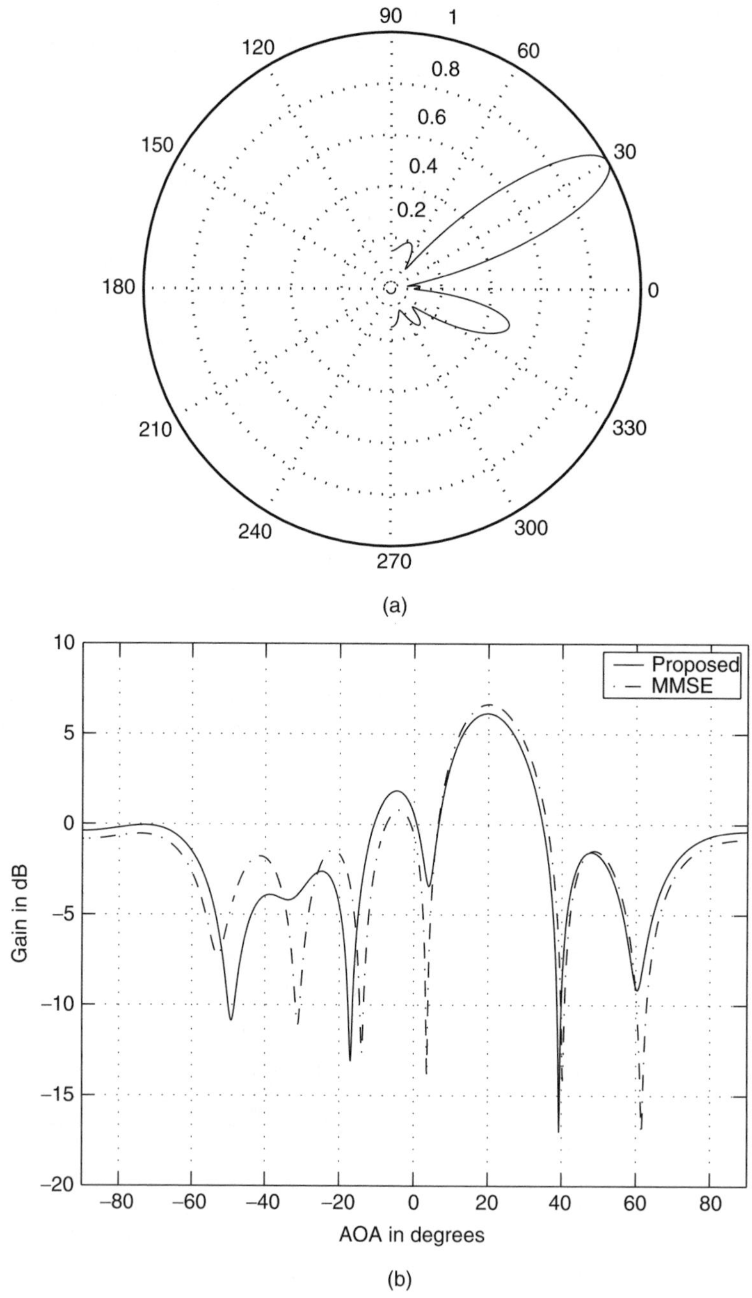

FIGURE 22.10 (a) Beam pattern with SNR = −30 dB when the data and reference blocks are within the same symbol. (b) Antenna gains of the proposed scheme and the MMSE scheme with SINR = −33 dB and JSR = 30 dB.

When the signal multipath is present at the receiver, the GPS receiver will treat the multipath components just like a direct signal since the signal multipath possesses exactly the same structure as the direct signal. However, the signal multipath widens the correlation function. For illustration, we add four multipath components to the received signal and calculate the correlation function at the output of the receiver according to [22]. For the direct GPS signal, we set SNR = −30 dB. The time delays are $0.1T_c, 0.2T_c, 0.3T_c$, and $0.4T_c$, and the carrier phases are random variables uniformly distributed over $(0, 2\pi)$. The powers of the multipath signals are set to be one tenth of that of the direct signal. The correlation functions with

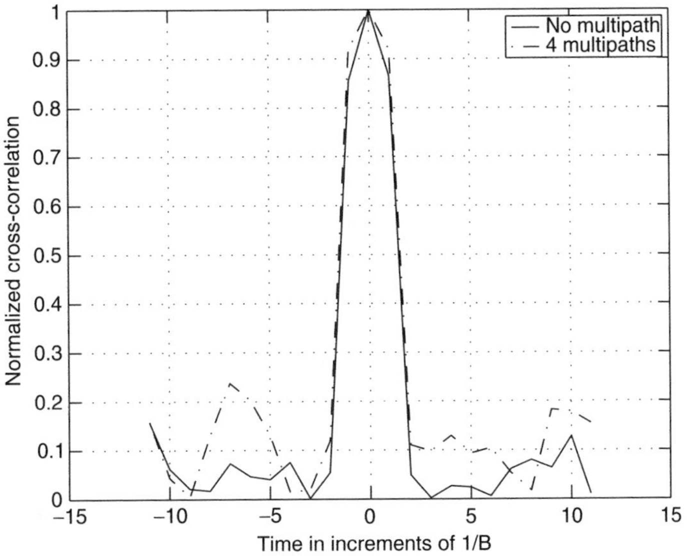

FIGURE 22.11 Correlations with and without multipath with SNR = −30 dB.

and without multipath are shown in Figure 22.11. It is observed from Figure 22.11 that multipath widens the autocorrelation function as compared to the one obtained without multipath. Since the self-coherence receiver is a spatial array without temporal filtering, it will not introduce additional delays and the peak of the autocorrelation function occurs at $t = 0$.

With multiple satellites in the FOV, the proposed scheme is expected to pass signals from all satellites with high gains since all satellite signals share the same periodic structure. For demonstration, four satellites located at $\theta_1 = 10°$, $\theta_2 = 30°$, $\theta_3 = 50°$, and $\theta_4 = 70°$ are selected from the constellation. In this case, $M = 9$ sensors are used and SNR $= -30$ dB. Figure 22.12 shows that as expected, four beams are generated toward the four satellites.

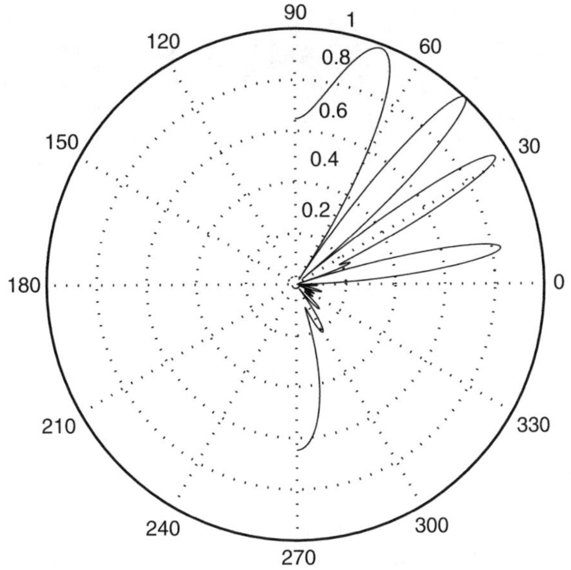

FIGURE 22.12 Beam pattern with four satellites and SNR = −30 dB.

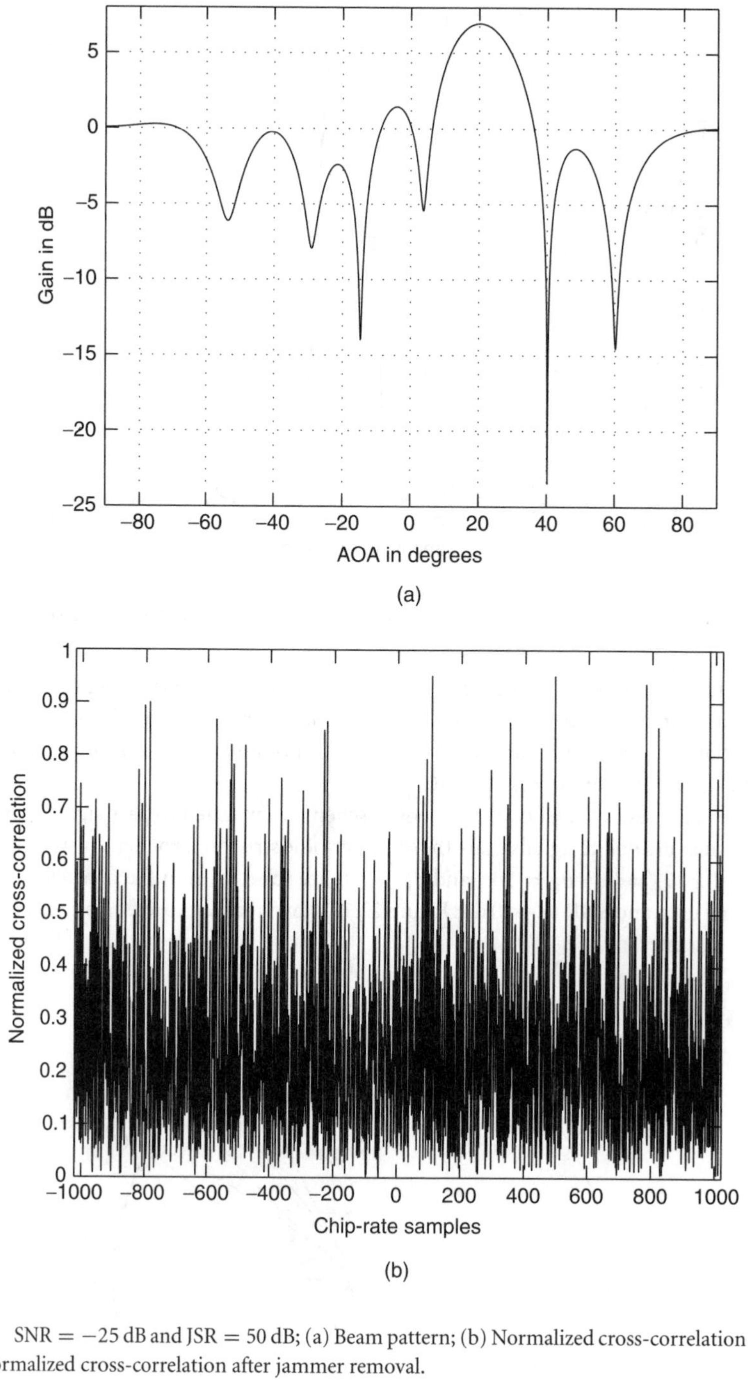

FIGURE 22.13 SNR = −25 dB and JSR = 50 dB; (a) Beam pattern; (b) Normalized cross-correlation before jammer removal; (c) Normalized cross-correlation after jammer removal.

In GPS, synchronization between the satellite and the receiver is achieved by cross-correlating the beamformer outputs with a locally generated Gold sequence [9]. When the phase of the GPS receiver replica code matches the phase of code sequence emitted from the satellite, there is maximum correlation [4]. In the simulation, the satellite is located at 20° and the two jammers are placed at 40° and 60°. The normalized cross-correlation together with the antenna beam pattern are shown in Figure 22.13 for SNR = −25 dB. Also shown in the figure is the normalized cross-correlations obtained before the jammers are removed. Figure 22.13 shows that the proposed GPS receiver can effectively cancel directional jammers and achieve

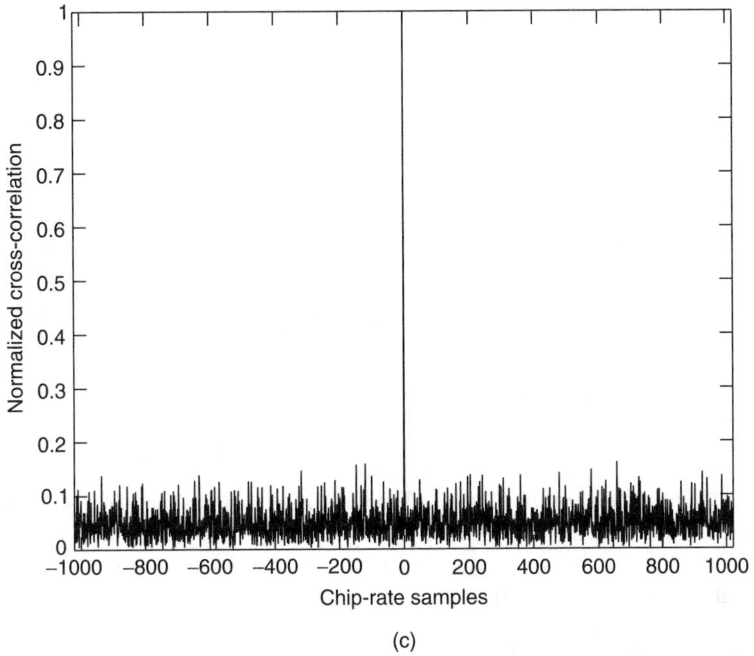

FIGURE 22.13 (*Continued.*)

synchronization even when the JSR is as high as 50 dB (Figure 22.13(c)). Without interference suppression, however, synchronization fails, as shown in Figure 22.13(b).

22.4 Multipath Mitigation in GPS

Multipath is another dominant error source in high-accuracy GPS applications. In the presence of multipath, most GPS positioning methods suffer a degradation in accuracy and an increase in processing time. The degradation of the pseudoranges is caused by the distortion of the correlation peak by the presence of the indirect signal. Distortion of the correlation peak causes the zero crossing of the early–late curve to be shifted, and thus the receiver will determine an erroneous pseudorange.

Unlike other error sources, multipath is normally uncorrelated between antenna locations. Hence, differencing techniques will not cancel the multipath incurred errors. Also, modeling multipath for each antenna location is difficult and impractical [40]. Various techniques have been proposed to eliminate/reduce the multipath induced bias. Narrow correlator is one of the most widely used approaches that improves the C/A code tracking performance by reducing the space between the early and late correlators in the multipath environments. Before discussing some common techniques in multipath mitigations, we first show the effect of signal multipath in biasing the correlation peak of the early–late receiver.

22.4.1 Bias Due to Signal Multipath

Figure 22.14 shows the configuration of the early–late correlator with E, P, and L representing the early, punctual, and late correlators, respectively. $R(\tau)$ is the noise-free cross-correlation function between the incoming signal and the locally generated reference code.

If the interference is ignored, the baseband received signal can be expressed in continuous-time format as (cf. Equation 22.2)

$$x(t) = \sum_{q=0}^{Q} s_q(t-\tau_q)e^{j\phi_q} + v(t) \tag{22.47}$$

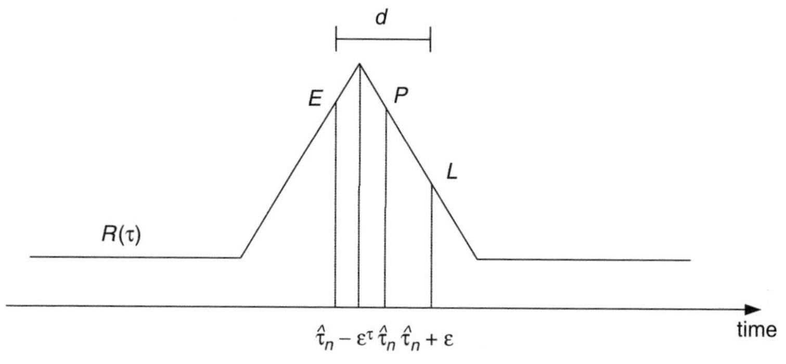

FIGURE 22.14 Early–late correlator configuration.

where $\tau \triangleq \tau_0$ is the time delay of the direct-path signal to be estimated and $\phi \triangleq \phi_0$ is the phase shift. The multipath components in Equation 22.47 can be written as a time-shifted version of the direct-path signal [6], i.e., $s_q(t - \tau_q) = s_0(t - \tau - \tilde{\tau}_q)$, where $\tilde{\tau}_q = \tau_q - \tau$ is the differential time delay between the direct-path signal and the qth multipath component.

22.4.1.1 Early–Late Discrimination Functions

Signals enter the noncoherent early–late correlator at IF [9] and are filtered by an IF filter with bandwidth B. The early–late DLL discriminators are based on in-phase (I) and quadra-phase (Q) samples of the filtered baseband signal cross-correlated with the reference PRN code. The cross-correlation process can be implemented by correlating the incoming I and Q samples with either the early (E) and late (L) reference codes (called the *early-minus-late mode*) or the punctual (P) and early-minus-late ($E - L$) reference codes (called the *dot-product mode*) [37]. The corresponding noncoherent discriminators are given as:

- Normalized early-minus-late discriminator:

$$d_\tau = \frac{\mathrm{I}_E^2 + \mathrm{Q}_E^2 - \mathrm{I}_L^2 - \mathrm{Q}_L^2}{\mathrm{I}_E^2 + \mathrm{Q}_E^2 + \mathrm{I}_L^2 + \mathrm{Q}_L^2} \tag{22.48}$$

- Normalized dot-product discriminator:

$$d_\tau = \frac{\mathrm{I}_{E-L}\mathrm{I}_P + \mathrm{Q}_{E-L}\mathrm{Q}_P}{\mathrm{I}_P^2 + \mathrm{Q}_P^2} \tag{22.49}$$

The I and Q components of the incoming signal at time t can be expressed as

$$\begin{aligned} \mathrm{I}(t) &= \sqrt{2\sigma}\, R_h(\tau)\cos\phi + v_I(t) \\ \mathrm{Q}(t) &= \sqrt{2\sigma}\, R_h(\tau)\sin\phi + v_Q(t) \end{aligned} \tag{22.50}$$

where σ is the SNR in a predetection bandwidth (usually 50 Hz) [37], and $v_I(t)$ and $v_Q(t)$ are the I and Q components of the noise, respectively. In Equation 22.50, $R_h(\tau)$ is the cross-correlation between the incoming filtered signal PRN code and the locally generated reference code:

$$R_h(\tau) = \int_0^\infty R_p(\xi)h(\tau - \xi)d\xi \tag{22.51}$$

where $h(t)$ is the impulse response of the receiver IF filter, whose frequency response is $H(f)$, and $R_p(\xi)$ is the autocorrelation function of the reference code $p(t)$, given by [6]

$$R_p(\xi) \triangleq \int_{T_P} p(t)p(t-\xi)dt = \begin{cases} 1 - \frac{P+1}{P}|\frac{\xi}{T_c}| \simeq 1 - |\frac{\xi}{T_c}|, & \text{for } |\xi| \leq T_c \\ -\frac{1}{P} \simeq 0, & \text{for } |\xi| > T_c \end{cases} \tag{22.52}$$

where T_c is the chip interval, $P = 1023$, and $T_p = PT_c$.

With an infinite precorrelation bandwidth and $|\tau| \leq d/2$, where d is the time interval between the early and late correlators, the expected value of d_τ is [37]

$$E\{d_\tau\} = 4\sigma(2-d)\tau, \quad \text{for early-minus-late power} \tag{22.53}$$

$$E\{d_\tau\} = 4\sigma\tau(1-|\tau|), \quad \text{for dot-product} \tag{22.54}$$

22.4.1.2 Time-Delay Estimation

For a noncoherent DLL, the time-delay τ can be estimated from the cross-correlation between the received signal and the direct-path signal. Let $\hat{\tau}$ denote an estimate of τ and assume that the early and late correlators are placed quasi-distance from $\hat{\tau}$. Then $\hat{\tau}$ can be obtained by minimizing the error function [51]

$$e(\hat{\tau}) \triangleq \left| R_{xs}\left(\hat{\tau} + \frac{d}{2}\right)\right|^2 - \left| R_{xs}\left(\hat{\tau} - \frac{d}{2}\right)\right|^2 \tag{22.55}$$

where $R_{xs}(\tau)$ is the cross-correlation between the received signal $x(t)$ and the direct-path signal $s_0(t)$:

$$R_{xs}(\tau) \triangleq \frac{1}{T_d}\int_{T_d} x^*(t)s_0(t-\tau)dt \tag{22.56}$$

where T_d is the integration period. Let $S(f)$ denote the Fourier transform of $s_0(t)$. Then, $s_0(t)$ can be expressed in $S(f)$ as

$$s_0(t) = \int_{-B/2}^{B/2} S(f)e^{j2\pi ft}df \tag{22.57}$$

Substituting $x(t)$ and $s_0(t)$ in Equation 22.56 with Equations 22.47 and 22.57, $R_{xs}(\tau)$ becomes [51]

$$R_{xs}\left(\hat{\tau} \pm \frac{d}{2}\right) = P_c \sum_{q=0}^{Q} \int_{B/2}^{B/2} \cos\left[2\pi f\left(\varepsilon \mp \frac{d}{2} + \tilde{\tau}_q\right)\right] P_s(f)df + R_{vs}\left(\pm\frac{d}{2}\right) \tag{22.58}$$

where P_c is the signal carrier power, $P_s(f) \triangleq \frac{|S(f)|^2}{T_d P_c}$ is the normalized power spectrum of the signal such that $\int_{-\infty}^{\infty} P_s(f)df = 1$, $R_{vs}(\tau)$ is the cross-correlation between the noise and the GPS signal, and $\varepsilon \triangleq \tau - \hat{\tau}$ is the estimation error. If the filter bandwidth and the estimation error satisfy $B\varepsilon < 0.25$, it is shown in [51] that Equation 22.58 can be simplified and the error function defined in Equation 22.55 can be written in a compact form as

$$e(\hat{\tau}) = \Phi\varepsilon + \gamma + \bar{v} \tag{22.59}$$

where $\bar{v}$ represents the noise, and Φ and γ are defined as

$$\Phi \triangleq 4\pi P_c^2 \sum_{g=0}^{Q}\sum_{q=0}^{Q}[\Xi_g \Upsilon_q + \Upsilon_g \Xi_q - \Psi_g \Omega_q - \Omega_g \Psi_q] \tag{22.60}$$

$$\gamma \triangleq -2P_c^2 \sum_{g=0}^{Q}\sum_{q=0}^{Q}[\Xi_g \Psi_q + \Psi_g \Omega_q] \tag{22.61}$$

where $\Xi_q = \int_{-B/2}^{B/2} \cos(2\pi f \tau_q)\cos(\pi f d) P_s(f) df$, $\Upsilon_q = \int_{-B/2}^{B/2} \cos(2\pi f \tau_q)\sin(\pi f d) P_s(f) df$, $\Psi_q = -\int_{-B/2}^{B/2} \sin(2\pi f \tau_q)\sin(\pi f d) P_s(f) df$, and $\Omega_q = -\int_{-B/2}^{B/2} \sin(2\pi f \tau_q)\cos(\pi f \delta) P_s(f) df$. If the noise term in Equation 22.59 is ignored and the error is set to zero, then $\varepsilon = -\gamma/\Phi$, from which we know that

$$\hat{\tau} = \tau + \frac{\gamma}{\Phi} \tag{22.62}$$

The analysis provided above clearly indicates that multipath causes bias in the time delay estimation and, subsequently, in the pseudorange measurement.

22.4.2 Single-Antenna Multipath Mitigation Techniques

22.4.2.1 Narrow Correlator

Conventional DLL discriminators use 1.0 chip correlator spacing, a concept usually found in P code tracking in early GPS receivers. However, for P code discriminator, Doppler and other disturbances may cause the receiver to lose code lock if narrower spacing is used [37]. This is because the P code chip is already relatively short. On the other hand, for C/A code tracking, since the C/A code has a much longer chip duration than the P code (10 times), narrow correlator can be implemented together with carrier aiding to minimize the Doppler effect.

Variable spacings have been implemented in the NovAtel GPSCard™ for C/A code tracking [36]. The receiver has a variable correlator spacing capability, with the spacing of the early and late reference code changing from 0.05 to 1.0 C/A code chip. The C/A code tracking performance of this narrow correlator receiver is better than 10 cm at nominal SNRs, which is essentially equivalent to the P code tracking. The narrow spacing also reduces the effects of multipath. This is because distortion of the cross-correlation function near its peak due to multipath is less severe than that at regions away from the peak. Thus, if the receiver could track near the peak, the effects of multipath would be reduced [37]. It has been shown that in the presence of multipath, the narrow spacing C/A code tracking provides superior performance, approaching that obtainable using a P code receiver [37].

22.4.2.2 Multipath Elimination Delay Lock Loop

MEDLL is the modified version of the conventional DLL, which estimates the multipath components in the received signal [39]. In order to achieve high accuracy, it is necessary to eliminate multipath errors in both the DLL and PLL discriminator and track direct signals transmitted from GPS satellites. The MEDLL approach decouples the correlation function into its direct and multipath components.

According to the maximum likelihood (ML) principle, the MEDLL method estimates the amplitude, time delay, and phase shift of each multipath component and subtracts these components from the measured correlation function [40]. Eventually, only the estimate of the direct-path signal correlation function is left and parameters of interest can be derived by using a standard early–late DLL. Simulation results in [40] show that the MEDLL performs as good as or even better than the P code DLL in the presence of multipath, while significantly outperforming the narrow correlator. The MEDLL receiver can virtually eliminate all multipath biases for delays greater than 0.1 chips [39].

22.4.3 Time-Delay and Carrier-Phase Estimation Using Antenna Array

Maximum likelihood-based approaches have been used extensively in parameter estimation [5], [6], [32], [34], [40], [52], [53], [54], [55], [56], [57]. In GPS, however, only the direct-path signal is the information-bearing signal; i.e., the pseudorange is derived from the direct-path signal. Therefore, the time-delay estimation in GPS needs only to focus on the direct-path signal. This is observed in [28] and utilized to develop an ML-based time delay and carrier phase estimation using antenna array.

Assuming an M-element linear array, a simplified signal model is adopted in [28]:

$$\mathbf{x}(n) = \tilde{A}_0 s_0(nT_s - \tau_0)\mathbf{a}_0 + \mathbf{v}(n) \tag{22.63}$$

where $\tilde{A}_0$ is a complex variable representing both the unknown amplitude and the carrier phase of the direct-path signal to be estimated, and $\mathbf{v}(n)$ is the vector that contains contributions from the interference, signal multipath, and noise. $\mathbf{v}(n)$ is modeled as a complex Gaussian process which satisfies $E\{\mathbf{v}(n)\} = 0$ and $E\{\mathbf{v}(n)\mathbf{v}^H(l)\} = \delta_{n,l}\Psi$, where Ψ is an unknown matrix to be estimated. In Equation 22.63, the steering vector of the direct-path signal $\mathbf{a}_0$ as well as the waveform of $s_0(t)$ are assumed to be known [28].

By collecting N data samples within one navigation symbol at the receiver, we define

$$\mathbf{X} \stackrel{\triangle}{=} [\mathbf{x}(1)\cdots\mathbf{x}(N)] = \tilde{A}_0\mathbf{a}_0\mathbf{s}^T(\tau_0) + \mathbf{V} \tag{22.64}$$

where $\mathbf{s}(\tau_0) \stackrel{\triangle}{=} [s(T_s - \tau_0)\cdots s(NT_s - \tau_0)]^T$ and $\mathbf{V} \stackrel{\triangle}{=} [\mathbf{v}(1)\cdots\mathbf{v}(N)]$. The unknown parameters, $(\tilde{A}_0, \tau_0, \Psi)$, are estimated using the maximum likelihood method. Let

$$\hat{\mathbf{R}}_v = \frac{1}{N}\sum_{n=1}^{N}[\mathbf{x}(n) - \tilde{A}_0 s_0(nT_s - \tau_0)\mathbf{a}_0][\mathbf{x}(n) - \tilde{A}_0 s_0(nT_s - \tau_0)\mathbf{a}_0]^H \tag{22.65}$$

Then, by ignoring terms irrelevant to the parameters of interest, it is straightforward to obtain the log-likelihood function of $\{\mathbf{x}(n)\}_{n=1}^N$ as

$$\mathcal{C}(\tilde{A}_0, \tau_0, \Psi) = -\ln|\Psi| - \operatorname{tr}\{\Psi^{-1}\hat{\mathbf{R}}_v\} \tag{22.66}$$

where $|\cdot|$ and $\operatorname{tr}\{\cdot\}$ stand for the determinant and trace of a matrix, respectively. Minimizing Equation 22.66 with respect to Ψ leads to the ML estimate of the unknown covariance matrix $\hat{\Psi} = \hat{\mathbf{R}}_v$. Further, define the following sample estimations: $\hat{\mathbf{R}}_{xx} \stackrel{\triangle}{=} \mathbf{X}\mathbf{X}^H/N$, $\hat{\mathbf{r}}_{xs}(\tau_0) \stackrel{\triangle}{=} \mathbf{X}\mathbf{s}^*(\tau_0)/N$, $\hat{P}_s \stackrel{\triangle}{=} \mathbf{s}^H(\tau_0)\mathbf{s}(\tau_0)/N$, and $\hat{\mathbf{W}}(\tau_0) = \hat{\mathbf{R}}_{xx} - \hat{\mathbf{r}}_{xs}(\tau_0)\hat{\mathbf{r}}_{xs}^H(\tau_0)/\hat{P}_s$. Then it has been shown in [28] that the ML estimates of τ_0 and A_0 can be obtained as

$$\hat{\tau}_{0\mathrm{ML}} = \arg\max_{\tau_0} \frac{\left|\mathbf{a}_0^H\hat{\mathbf{R}}_{xx}^{-1}\hat{\mathbf{r}}_{xs}(\tau_0)\right|^2}{\hat{P}_s - \hat{\mathbf{r}}_{xs}^H(\tau_0)\hat{\mathbf{R}}_{xx}^{-1}\hat{\mathbf{r}}_{xs}(\tau_0)} \tag{22.67}$$

$$\hat{\tilde{A}}_{0\mathrm{ML}} = \left.\frac{\mathbf{a}_0^H\hat{\mathbf{W}}^{-1}(\tau_0)\hat{\mathbf{r}}_{xs}(\tau_0)}{\hat{P}_s\mathbf{a}_0^H\hat{\mathbf{W}}^{-1}(\tau_0)\mathbf{a}_0}\right|_{\tau=\hat{\tau}_{0\mathrm{ML}}} \tag{22.68}$$

which shows that the estimation of τ_0 involves the search of the one-dimensional parameter space τ_0.

A hybrid beam former is constructed in [28] to implement the above ML estimator. Let $\mathbf{w}$ denote the beam-former weight vector. Then, the output of the beam former is $\mathbf{z} = \mathbf{w}^H\mathbf{X}$ and the MSE between $\mathbf{z}$ and the reference signal $\tilde{A}_0\mathbf{s}^T(\tau)$ is given by

$$\mathcal{E}(\mathbf{w}, \tilde{A}_0, \tau_0) = \frac{1}{N}\|\mathbf{z} - \tilde{A}_0\mathbf{s}^T(\tau)\|^2 \tag{22.69}$$

Then, by using the assumption that the steering vector $\mathbf{a}_0$ is known, the weight vector $\mathbf{w}$, $\tilde{A}_0$, and τ_0 can be obtained by minimizing $\mathcal{E}(\mathbf{w}, \tilde{A}_0, \tau_0)$ under the spatial constraint that $\mathbf{w}^H\mathbf{a}_0 = 1$, i.e.,

$$\begin{gathered}\left(\mathbf{w}_{\mathrm{MSE}}, \tilde{A}_{0\mathrm{MSE}}, \tau_{0\mathrm{MSE}}\right) = \arg\min_{\mathbf{w},\tilde{A}_0,\tau_0} \mathcal{E}(\mathbf{w}, \tilde{A}_0, \tau_0),\\ \text{subject to } \mathbf{w}^H\mathbf{a}_0 = 1\end{gathered} \tag{22.70}$$

where the resulting $\mathbf{w}_{\mathrm{MSE}}$ is a hybrid beam-former since it is obtained using both spatial and temporal references. $\mathbf{w}_{\mathrm{MSE}}$ can be obtained in two steps. First, $\mathbf{w}_{\mathrm{MSE}}$ is calculated for fixed $\tilde{A}_0$ and τ_0 as

$$\mathbf{w}_{\mathrm{MSE}}(\tilde{A}_0, \tau_0) = \tilde{A}_0^*\hat{\mathbf{R}}_{xx}^{-1}\hat{\mathbf{r}}_{xs}(\tau_0) + \frac{1 - \tilde{A}_0^*\mathbf{a}_0^H\hat{\mathbf{R}}_{xx}^{-1}\hat{\mathbf{r}}_{xs}(\tau_0)}{\mathbf{a}_0^H\hat{\mathbf{R}}_{xx}^{-1}\mathbf{a}_0}\hat{\mathbf{R}}_{xx}^{-1}\mathbf{a}_0 \tag{22.71}$$

which indicates that the beam-former is a weighted linear combination of the beam-former with temporal reference given by $\hat{\mathbf{R}}_{xx}^{-1}\hat{\mathbf{r}}_{xs}(\tau_0)$ and the beam-former with spatial reference given by $\hat{\mathbf{R}}_{xx}^{-1}\mathbf{a}_0$. In the second step, $\tilde{A}_0$ and τ_0 are estimated from the output of the beam-former. Simulation results in [28] show that the estimation error decreases with the number of antennas. This justifies the use of antenna array for improving the time delay and carrier phase estimation.

22.5 Conclusions

In this chapter, we discussed the applications of the adaptive antenna array in GPS interference suppression, multipath mitigation, and time delay and carrier phase estimations. The ultimate goal of GPS is to provide accurate navigation information. Therefore, it is desirable that GPS receivers operate efficiently in the strong jamming and rich multipath environments. Space–time adaptive arrays equip the GPS receivers with the ability to excise both narrowband and broadband interferers and to mitigate multipath. Some widely used techniques utilizing antenna arrays have been presented. It is concluded from the analysis and simulations that multiantenna GPS receivers are needed for effective geolocation in the presence of jamming and local scatters.

References

[1] J. McNeff, "The global positioning system," *IEEE Transactions on Microwave Theory and Techniques*, 50(3), 645–652, 2002.

[2] M.S. Braasch and A.J. van Dierendonck, "GPS receiver architectures and measurements," *Proceedings of the IEEE*, 87(1), 48–64, 1999.

[3] R. Blazquez, J.M. Blas, and J.I. Alonso, "Implementation of adaptive algorithm for array processing in real time applications to low cost GPS receivers," in *Proceedings of the IEEE 49th Vehicular Technology Conference*, 1999, pp. 463–467.

[4] E.D. Kaplan, Eds., *Understanding GPS: Principles and Applications*, Norwood, MA: Artech House, 1996.

[5] W. Zhang and J. Tranquilla, "Modeling and analysis for the GPS pseudo-range observable," *IEEE Transactions on Aerospace and Electronic Systems*, 31(2), 739–751, 1995.

[6] J. Soubielle, I. Fijalkow, P. Duvaut, and A. Bibaut, "GPS positioning in a multipath environment," *IEEE Transactions on Signal Processing*, 50(1), 141–150, 2002.

[7] R. Ziemer and R. Peterson, *Digital Communication and Spread Spectrum Systems*, New York: Macmillan, 1985.

[8] R. Gold, "Optimal binary sequences for spread spectrum multiplexing," *IEEE Transactions on Information Theory*, IT-13, 619–621, 1967.

[9] B.W. Parkinson and J.J. Spilker, Jr., Eds., *Global Positioning System: Theory and Applications*, Vol. I, Washington, DC: American Institute of Aeronautics, Inc., Progress in Astronautics and Aeronautics, 1996.

[10] R.D. Van Nee, "Spread-spectrum code and carrier synchronization errors caused by multipath and interference," *IEEE Transactions on Aerospace and Electronic Systems*, 29, 1359–1365, 1993.

[11] P.W. Ward, "GPS receiver interference monitoring, mitigation and analysis techniques," *Journal of the Institute of Navigation*, 41(4), 367–391, 1995.

[12] D. Moelker, E. van der Pol, and Y. Bar-Ness, "Multiple antennas for advanced GNSS multipath mitigation and multipath direction finding," in *Proceedings of ION GPS 97*, 1997, pp. 541–550.

[13] M. Braasch and M. DiBenedetto, "Spread spectrum ranging multipath model and validation," *IEEE Transactions on Aerospace and Electronic Systems*, 37, 298–303, 2001.

[14] R. De Gaudenzi, "Direct-sequence spread-spectrum chip tracking in the presence of unresolvable multipath components," *IEEE Transactions on Vehicular Technology*, 48(5), 1573–1583, 1999.

[15] B.R.C. Macabiau and A. Benhallam, "Tracking performance of GPS receivers with more than one multipath," in *Proceedings of the 3rd European Symposium on Global Navigation Satellite Systems*, 1999, pp. 1186–1189.

[16] R. Iltis and L. Milstein, "Performance analysis of narrowband interference rejection techniques in DS spread-spectrum systems," *IEEE Transactions on Communications*, 32(11), 1169–1177, 1984.

[17] L. Rusch and H. Poor, "Narrowband interference suppression in CDMA spread spectrum communications," *IEEE Transactions on Communications*, 42(2), 1969–1979, 1994.

[18] L. Rusch and H. Poor, "Multiuser detection techniques for narrowband interference suppression in spread spectrum communications," *IEEE Transactions on Communications*, 43(2), 1725–1737, 1995.

[19] Y. Zhang, M.G. Amin, and A.R. Lindsey, "Anti-jamming GPS receivers based on bilinear signal distributions," in *Proceedings of the 2001 Military Communications Conference (MILCOM 2001)*, Vol. 2, 2001, pp. 1070–1074.

[20] M.G. Amin and Y. Zhang, "Spatial time-frequency distributions and their applications," in *Proceedings of the 6th International Symposium on Signal Processing and Its Applications*, Vol. 1, 2001, 254–255.

[21] M.D. Zoltowski and A.S. Gecan, "Advanced adaptive null steering concepts for GPS," in *Proceedings of the 1995 Military Communications Conference (MILCOM 1995)*, Vol. 3, 1995, pp. 1214–1218.

[22] R.L. Fante and J.J. Vaccaro, "Wideband cancellation of interference in a GPS receive array," *IEEE Transactions on Aerospace and Electronic Systems*, 36(2), 549–564, 2000.

[23] Y. Zhang and M. Amin, "Array processing for nonstationary interference suppression in DS/SS communications using subspace projection techniques," *IEEE Transactions on Signal Processing*, 49(12), 3005–3014, 2001.

[24] Y. Zhang, A. Lindsey, and M. Amin, "Combined synthesis and projection techniques for jammer suppression in DS/SS communications," in *Proceedings of the IEEE International Conference on Acoustics, Speech, and Signal Processing, 2002 (ICASSP '02)*, Vol. 3, 2002, pp. 2757–2760.

[25] R.L. Fante and J.J. Vacarro, "Cancellation of jammers and jammer multipath in a GPS receiver," *IEEE Aerospace and Electronic System Magazine*, 13(11), 25–28, 1998.

[26] D. Moelker, E. van der Pol, and Y. Bar-Ness, "Adaptive antenna arrays for interference cancellation in GPS and GLONASS receivers," in *Proceedings of the IEEE 1996 Position Location and Navigation Symposium*, 1996, pp. 191–198.

[27] G. Hatke, "Adaptive array processing for wideband nulling in GPS systems," in *Proceedings of the 32nd Asilomar Conference on Signals, Systems and Computers*, Vol. 2, 1998, pp. 1332–1336.

[28] G. Seco and J. Fernandez-Rubio, "Array signal processing techniques for pseudorange and carrier phase measurement," in *Proceedings of the 2nd European Symposium on Global Navigation Satellite Systems (GNSS98)*, Vol. IX-P-10, 1998, pp. 20–23.

[29] G. Seco, "Antenna arrays for multipath and interference mitigation in GNSS receivers," Ph.D. dissertation, Polytechnic University of Catalonia, Barcelona, December 2000, downloadable from `http://gps-tsc.upc.es/comm/gonzalo/gonzalo.html`.

[30] W. Myrick, M. Zoltowski, and J. Goldstein, "GPS jammer suppression with low-sample support using reduced-rank power minimization," in *Proceedings of the 10th IEEE Workshop on Statistical Signal and Array Processing*, 2000, pp. 514–518.

[31] W. Zhuang, "Performance analysis of GPS carrier phase observable," *IEEE Transactions on Aerospace and Electronic Systems*, 32(2), 754–767, 1996.

[32] P. Axelrad, C.J. Comp, and P.F. MacDoran, "SNR-based multipath error correction for GPS differential phase," *IEEE Transactions on Aerospace and Electronic Systems*, 32(2), 650–660, 1996.

[33] C. Comp and P. Axelrad, "Adaptive SNR-based carrier phase multipath mitigation technique," *IEEE Transactions on Aerospace and Electronic Systems*, 34(1), 264–276, 1998.

[34] J. Ray, M. Cannon, and P. Fenton, "GPS code and carrier multipath mitigation using a multiantenna system," *IEEE Transactions on Aerospace and Electronic Systems*, 37(1), 183–195, 2001.

[35] J. Kelly and M. Braasch, "Validation of theoretical GPS multipath bias characteristics," in *Proceedings of the IEEE Aerospace Conference*, Vol. 3, 2001, pp. 1317–1325.

[36] P. Fenton, B. Falkenberg, T. Ford, K. Ng, and A. van Dierendonck, "NovAtels GPS receive: the high performance OEM sensor of the future," in *Proceedings of ION GPS 91*, 1991, pp. 49–58.

[37] A. van Dierendonck, P. Fenton, and T. Ford, "Theory and performance of narrow correlator spacing in a GPS receiver," *Journal of the Institute of Navigation*, 39(2), 265–283, 1992.

[38] B. Townsend and P. Fenton, "A practical approach to the reduction of pseudorange multipath errors in a L1 GPS receiver," in *Proceedings of the 1994 ION GPS Conference*, 1994, pp. 20–23.

[39] R.J. van Nee, "The multipath estimation delay lock loop: Approaching theoretical limits," in *Proceedings of the 2nd IEEE Symposium on Spread Spectrum Techniques and Applications*, 1992, pp. 39–42.

[40] B. Townsend and P. Fenton, "Performance evaluation of the multipath estimating delay lock loop," in *Proceedings of the 1995 ION National Technical Meeting*, 1995.

[41] L. Garin, F. van Diggelen, and J. Rousseau, "Strobe and edge correlator multipath mitigation for code," in *Proceedings of ION GPS 96*, 1996, pp. 657–664.

[42] L. Garin and J. Rousseau, "Enhanced strobe correlator multipath rejection for code and carrier," in *Proceedings of ION GPS 97*, 1997, pp. 559–568.

[43] G. McGraw and M. Braasch, "GNSS multipath mitigation using gated and high resolution correlator concepts," in *Proceedings of ION National Technical Meeting*, 1999, pp. 333–342.

[44] R. Rifkin and J. Vaccaro, "Comparison of narrowband adaptive filter technologies for GPS," in *Proceedings of the IEEE 2000 Position Location and Navigation Symposium*, 2000, pp. 125–131.

[45] D. Moelker and Y. Bar-Ness, "An optimal array processor for GPS interference cancellation," in *Proceedings of the 15th AIAA/IEEE Digital Avionics Systems Conference*, 1996, pp. 285–290.

[46] M. Amin, L. Zhao, and A. Lindsey, "Subspace array processing for the suppression of FM jamming in GPS receivers," *IEEE Transactions on Aerospace and Electronic Systems*, to appear.

[47] M. Amin and A. Akansu, "Time-frequency for interference excision in spread-spectrum communications," *IEEE Signal Processing Magazine: Highlights of Signal Processing for Communications: Celebrating a Half Century of Signal Processing*, 16(2), 1999.

[48] P.M.R. Landry, Jr., and D. Lekaim, "Interference mitigation in spread spectrum systems by wavelet coefficients thresholding," *European Transactions on Telecommunications*, 9, 191–202, 1998.

[49] W. Sun and M. Amin, "A self-coherence based GPS anti-jamming receiver," in *Proceedings of the 2003 IEEE Workshop on Statistical Signal Processing*, 2003, pp. 53–56.

[50] B. Agee, S. Schell, and W. Gardner, "Spectral self-coherence restoral: A new approach to blind adaptive signal extraction using antenna array," *Proceedings of the IEEE*, 78(4), 753–767, 1990.

[51] R. Fante and J. Vaccaro, "Evaluation and reduction of multipath-induced bias on GPS time-of-arrival," 2001, *IEEE Trans. on Aerospace and Electric Systems*, 39(3), 911–920, 2003.

[52] W.J. Hurd, J. Statman, and V. Vilnrotter, "High dynamic GPS receiver using maximum likelihood estimation and frequency tracking," *IEEE Transactions on Aerospace and Electronic Systems*, 23(4), 425–437, 1987.

[53] A.J. van der Veen, M.C. Vanderveen, and A. Paulraj, "Joint angle and delay estimation using shift-invariance techniques," *IEEE Transactions on Signal Processing*, 46(2), 405–418, 1998.

[54] M. Wax and A. Leshem, "Joint estimation of time delays and directions of arrival of multiple reflections of a known signal," *IEEE Transactions on Signal Processing*, 45(10), 2477–2484, 1997.

[55] J. Li and R.T. Compton, Jr., "Maximum likelihood angle estimation for signals with known waveforms," *IEEE Transactions on Signal Processing*, 41(9), 2850–2862, 1993.

[56] M. Cedervall and R.L. Moses, "Efficient maximum likelihood DOA estimation for signals with known waveforms in the presence of multipath," *IEEE Transactions on Signal Processing*, 45(3), 808–811, 1997.

[57] Z.-S. Liu, J. Li, and S.L. Miller, "An efficient code-timing estimator for receiver diversity DS-CDMA systems," *IEEE Transactions on Communications*, 46(6), 826–835, 1998.

IX

Power Control and Wireless Networking

23 Transmitter Power Control in Wireless Networking: Basic Principles and Core Algorithms *Nicholas Bambos and Savvas Gitzenis* **23**-1
Introduction • Power Control for Streamed Continuous Traffic • Power Control for Packetized Data Traffic • Final Remarks

24 Signal Processing for Multiaccess Communication Networks *Qing Zhao and Lang Tong* **24**-1
Introduction • MPR at the Physical Layer • The Interface between the Physical Layer and the MAC Layer: Resolvability • Impact of MPR on the Performance of Existing MAC Protocols • MAC Layer Design for Networks with MPR • Approach High Performance from Both Physical and MAC Layers • Conclusion

23

Transmitter Power Control in Wireless Networking: Basic Principles and Core Algorithms

Nicholas Bambos
Stanford University

Savvas Gitzenis
Stanford University

23.1 Introduction 23-1
23.2 Power Control for Streamed Continuous Traffic 23-3
The Target SIR Formulation of the Power Control Problem • Autonomous Distributed Power Control • DPC with Active Link Protection • Admission Control under DPC/ALP • Noninvasive Channel Probing and Selection
23.3 Power Control for Packetized Data Traffic 23-15
Power-Controlled Multiple Access: The Basic Model • Optimally Emptying the Transmitter Buffer • The Case of Per-Slot-Independent Interference • Structural Properties: Backlog Pressure and Phased Back-Off • Design of PCMA Algorithms: Responsive Interference
23.4 Final Remarks 23-20

Abstract

Transmitter power control is key in wireless networking for maintaining the quality of each wireless link at a desirable level via adaptation to channel interference and variations, while mitigating the interference the link generates on others and resulting in increased network capacity via higher spatial spectrum reuse.

In this chapter, we present the fundamentals of power control focusing on basic principles and core algorithms. For methodological and exposition purposes, we first investigate the case of wireless links supporting delay-sensitive continuous traffic (voice, video) where the link quality is largely characterized by the signal-to-interference ratio and bit error rate. Then we consider the case of packetized delay-tolerant traffic and address the fundamental power vs. delay trade-off. Several key power control algorithms are discussed.

23.1 Introduction

The advent and rapid expansion of wireless networking in the 1990s have prompted intense investigation of transmitter power control fundamentals, algorithms, and architectures for over a decade. Power control

0-8493-1657-X/$0.00+$1.50

is a key element of wireless networking for various reasons. Adjusting its transmitter power to adapt to interference, mobility variations, and channel impairments, a communication link can maintain a desirable quality of service (QoS). By doing this carefully and systematically, it will generate the minimum possible interference on other links sharing the channel, resulting in higher network capacity via better sharing of the radio bandwidth and denser spectrum reuse. Additionally, the transmitter will minimize the power drain and maximize the battery lifespan between recharges, a key operational limitation in modern mobile devices. QoS support, interference management, battery life, and various other factors make power control a very important element of current and projected wireless network designs.

Early cellular wireless networks circumvented the need for full power control by spatially separating transmissions using the same radio resource (time slot and frequency band in TDMA/FDMA (time division/frequency division multiple-access) systems, spreading code in CDMA (code division multiple-access) systems, etc.), resulting in low spatial densities of spectrum reuse. Power control was simply used in CDMA systems primarily to mitigate the near–far effect. Due to the scarcity of the radio spectrum and the exponentially increasing user demand for wireless mobile information services, the pressure to pack more radio transmissions into the same fixed radio spectrum is rising fast. To achieve this objective, there are basically two approaches:

1. Use transmitter power control to better manage and mitigate interference in order to increase the spatial spectrum reuse. This increases the network capacity for a given network infrastructure (access point) density.
2. Increase the infrastructure density by deploying more access points (base stations, wireless local area networks (LANs), etc.) and getting those access (and, perhaps, relay) points closer to the wireless terminals. This makes wireless terminal-to-access links smaller, hence needing less power for a given QoS level. In this way, denser spatial packing of links (and higher spectrum reuse) can be achieved. Besides the higher infrastructure cost, there is also more control and signalling needed in denser infrastructure environments to manage mobility, as more access point handoffs occur per unit of time for a given level of mobility.

Moving forward onto next-generation networks, both approaches are advancing in unison. Therefore, the importance of power control is critical. It is essential to better understand its fundamentals, as well as various important architectural and implementation aspects. In this chapter, we mostly focus on the fundamentals of power control, which reveal some important design insights. Particular implementation architectures are not discussed in the limited space of this chapter, but the core algorithms presented are specifically chosen to be distributed and autonomous, on-line adaptive (agile), robust and scalable, with an eye toward high-performance design and practical implementation.

It is interesting to consider the logic of power control from an individual link's point of view in order to understand the global nature of power control from the network's point of view. Intuitively speaking, the communication link should choose its transmission power to be large enough to adequately compensate for radio propagation losses and interference/noise effects, yet low enough to generate minimal interference on other network links sharing the same channel and disturb those minimally. According to the above logic, all links are coupled to each other, in the sense that one's benefit (communication) is everyone else's cost (interference). However, interference has a positive aspect too. Indeed, it provides collective feedback information from the channel/network that each link observes in order to assess the stress level of its channel/network environment. This is valuable information that the link can use to control its power, as well as autonomously perform various network control functions like admission, congestion, and handoff control in a distributed scalable manner. It should be mentioned that power control may be considered to operate at the media access control (MAC) layer of wireless networking; however, it touches on and actually couples with many other functions. Those range from signal processing, modulation, transmission, etc., on one side, to admission and handoff control on the other (as seen later).

Early work on power control [5, 8, 9, 10, 15, 20, 21] brought attention to the importance of the problem and provided some significant insights and results. A decade of further investigation [6, 16, 17, 18, 22, 23,

24, 26, 27, 29, 31] has revealed several interesting key facts and has crystallized the big picture of power control. We aim to sketch out this picture in this chapter, from an admittedly particular — yet central and insightful — point of view.

For methodological and presentation purposes, we first consider in Section 23.2 the power control problem for continuous streamed delay-sensitive traffic (voice, video) where the main quality-of-service metric is the bit error rate, expressed as a function of the signal-to-interference (plus noise) ratio (SIR). We then consider in Section 23.3 the case of packetized delay-tolerant data traffic. In this case, during interference highs the dilemma arises whether data transmission should be halted and delay incurred, expecting that the interference might soon subside and power benefits might be realized by transmitting lots of data at a low power cost. This core *delay vs. power* trade-off is captured below in a simple model and analyzed, yielding some interesting insights about the nature of the power control problem. The overall emphasis is on presenting the fundamental principles of power control, leading to basic core algorithms and protocols, without expanding into various operational and implementation details, some of which can be found in the background references. To achieve this we utilize simple — yet canonical — models of wireless networking, whose analysis provides significant insight as well as several graphs that visually demonstrate important aspects of power control dynamics.

23.2 Power Control for Streamed Continuous Traffic

In this section, we study the case of power control for streamed continuous delay-sensitive traffic (voice, video). In particular, voice service was the main driver in first-generation cellular wireless networks, while streamed multimedia services are planned for next-generation networks. The SIR of the wireless link is the basic QoS metric managed via control of the transmitter power, mapping to control of the bit error rate on the link.

We consider the wireless network as a collection of radio links, where each link is a single-hop radio transmission from a transmitter to an intended receiver. Chains of consecutive links may correspond to multihop communication paths, but they are also treated as sets of individual links for power control purposes. If there are several communication channels, we assume that the interference between links operating in different ones is negligible. That is, channels are orthogonal, and so only co-channel interference need be considered. We can therefore reduce the network picture to that of a *collection of interfering links* in a *single channel*, rendering the notions of *network admission* and *channel access* equivalent.

In the cellular communication network paradigm, links correspond to upstream or downstream transmissions between mobiles and base stations. In the *ad hoc* networking paradigm, links may correspond to single-hop transmissions between mobile terminals and fixed access points or other mobile terminals. Each channel is basically a *communication resource* shared by the various interfering links using it. For example, the channels can be nonoverlapping frequency bands in FDMA systems, nonoverlapping time slots in TDMA systems, distinct spreading codes in CDMA systems, or combinations of the above in hybrid systems.

23.2.1 The Target SIR Formulation of the Power Control Problem

A key performance metric of the radio link is its bit error rate, which to first approximation is a decreasing function of the SIR observed at its receiver node. We can therefore consider the QoS of a wireless link as an increasing function of its SIR and use the latter in our problem formulation. Given N interfering links in a channel, we denote the SIR of the ith link by

$$R_i = \frac{G_{ii} P_i}{\eta_i + \sum_{j \neq i} G_{ij} P_j}, \quad i, j \in \{1, 2, \ldots, N\} \tag{23.1}$$

where P_i is the power of the transmitter of link i, η_i is the thermal noise power at the receiver of link i, and G_{ij} is the power gain (actually loss) from the transmitter of the jth link to the receiver of the ith one. The power gain G_{ij} may incorporate free space loss, multipath fading, shadowing, and other radio-wave

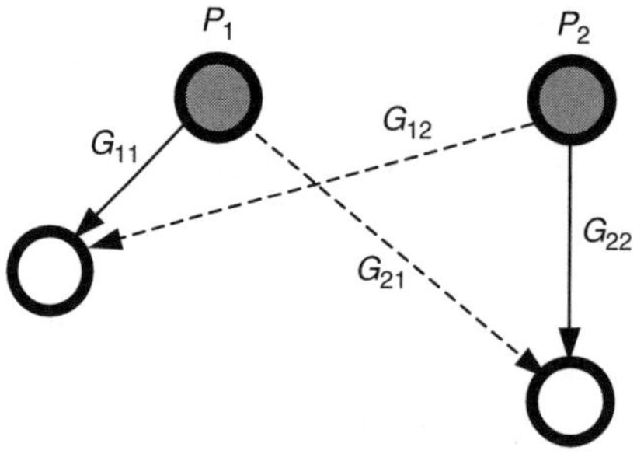

FIGURE 23.1 The simplest possible wireless network consisting of two interfering links in a channel. The power gains G_{ij} from the transmitters (dark circles) to the receivers (light circles) are depicted, as well as the transmitted powers.

propagation effects, as well as modulation effects (processing gain, etc.) To keep the presentation simple, in this section we assume that the G_{ij} are deterministic (fluctuations have been averaged out) and stay constant over time (no mobility).

For each link i we consider a target SIR $\gamma_i > 0$ that the link must maintain (or perhaps exceed) throughout the communication in order to operate properly and sustain acceptable quality of service. Recalling that the QoS is an increasing function of the SIR, we see that we need to have

$$R_i(\mathbf{P}) \geq \gamma_i, \quad \text{for all } i \in \{1, 2, \ldots, N\} \tag{23.2}$$

in order for the links to coexist smoothly in the channel. According to Equation 23.1, each SIR is a function of the whole power vector[1] $\mathbf{P} = (P_1, P_2, \ldots, P_N)^T$. Given a constant power gain matrix $\mathbf{G} = [G_{ij}]$ and noise vector $\boldsymbol{\eta} = (\eta_1, \eta_2, \ldots, \eta_N)$, the issue is to find the optimal power vector $\mathbf{P}$ that satisfies Equation 23.2, if any such vector exists.

In particular, the *feasibility* problem is to determine whether Equation 23.2 has a solution $\mathbf{P}$ satisfying Equation 23.1 and $P_i > 0$ for all $i = \{1, \ldots, N\}$. If that is feasible, then the *optimal power allocation* problem is to select the best power vector among potentially several feasible ones. Before proceeding with the general case, let us first consider a two-link network ($N = 2$), which will provide us with some useful insights.

23.2.1.1 Example of the Simple Two-Link Network

Consider the simplest possible network of Figure 23.1, which is comprised of only two links interfering in a channel. For exposition purposes, we analyze it in two steps, addressing the power feasibility and allocation problems. First, let us ignore the thermal noise, setting $\eta_1 = \eta_2 = 0$. The SIR Equations 23.1 and 23.2 yield $R_1 = \frac{G_{11}}{G_{12}}\frac{P_1}{P_2} \geq \gamma_1$ and $R_2 = \frac{G_{22}}{G_{21}}\frac{P_2}{P_1} \geq \gamma_2$, which can be put into linear inequality form:

$$G_{11}P_1 - \gamma_1 G_{12} P_2 \geq 0 \tag{23.3}$$

$$G_{22}P_2 - \gamma_2 G_{21} P_1 \geq 0 \tag{23.4}$$

Figure 23.2 shows the power feasibility region $\mathbf{D}(\mathbf{G}, \gamma)$ in the (P_1, P_2) power space. Any power vector $\mathbf{P} = (P_1, P_2)$ in $\mathbf{D}$ satisfies both inequalities above. The cone $\mathbf{D}$ is a function of the gain matrix $\mathbf{G} = [G_{ij}]$, the target SIR vector $\gamma = (\gamma_1, \gamma_2)$, and the noise vector $\boldsymbol{\eta}$, which is set to zero in the current discussion. Note that ε_1 and ε_2 are the lines where Equations 23.3 and 23.4 become equalities, respectively. The slopes of ε_1 and ε_2 are $\frac{1}{\gamma_1}\frac{G_{11}}{G_{12}}$ and $\frac{\gamma_2 G_{21}}{G_{22}}$, respectively, and according to Figure 23.2 we have:

1. When $\frac{1}{\gamma_1}\frac{G_{11}}{G_{12}} > \frac{\gamma_2 G_{21}}{G_{22}}$, the feasibility region $\mathbf{D}$ is a full cone and any power vector in it satisfies Equations 23.1 and 23.2.

[1]The symbol $(\cdot)^T$ denotes the transpose column vector.

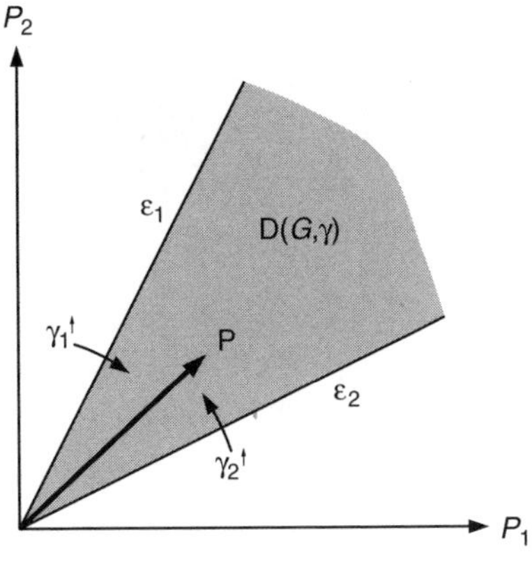

FIGURE 23.2 The power feasibility region $\mathbf{D}(\mathbf{G}, \gamma)$ (shaded region) of the network of Figure 23.1 corresponds to the set of power vectors $\mathbf{P} = (P_1, P_2)^T$ satisfying Equations 23.3 and 23.4. It is a cone with boundary lines $\varepsilon_1, \varepsilon_2$ where Equations 23.3 and 23.4, respectively, hold true with equality. Increasing γ_1 tilts ε_1 toward ε_2 (and conversely for γ_2), resulting in the shrinking of **D**. The noise vector $(\eta_1, \eta_2) = (0, 0)$ is artificially set to zero in this case (it is considered again in Figure 23.3).

2. When $\frac{G_{11}}{\gamma_1 G_{12}} = \frac{\gamma_2 G_{21}}{G_{22}}$, the two lines $\varepsilon_1, \varepsilon_2$ coincide and **D** reduces to a line.
3. When $\frac{G_{11}}{\gamma_1 G_{12}} < \frac{\gamma_2 G_{21}}{G_{22}}$, then $\mathbf{D} = \emptyset$ is empty; that is, no power vector $\mathbf{P} = (P_1, P_2)^T$ can satisfy Equation 23.2 of both links, so the system is *infeasible*.

The third case is due to the fact that the SIR requirements γ_1 and γ_2 are too high for the system to accommodate, given the power gains **G**. In general, observe that for any $i = 1, 2$ and $j \neq i$, (1) an increase in any γ_i, (2) an increase in any interlink power gain G_{ij}, or (3) a decrease in any intralink power gain G_{ii} results in line ε_i tilting toward line ε_j (until crossing it) and the feasibility region **D** shrinking. The reverse actions make **D** expand.

Taking now into consideration the thermal noise $(\eta_1, \eta_2) \neq (0, 0)$, the inequalities in Equation 23.2 take the following form in the two-link network of Figure 23.1:

$$G_{11} P_1 - \gamma_1 G_{12} P_2 \geq \gamma_1 \eta_1 \tag{23.5}$$

$$G_{22} P_2 - \gamma_2 G_{21} P_1 \geq \gamma_2 \eta_2 \tag{23.6}$$

Figure 23.3 shows the new feasibility region $\mathbf{D}(\mathbf{G}, \gamma, \boldsymbol{\eta})$, which is now a function of the gain matrix **G**, the target SIR vector $\boldsymbol{\gamma}$, and the noise vector $\boldsymbol{\eta}$. The tip of the cone $\mathbf{P}^*$ is the *Pareto-optimal* power vector, in the sense that any other feasible power vector **P** requires at least as much power from every transmitter [29], that is, $\mathbf{P} \geq \mathbf{P}^*$ component-wise. The inclusion of the noise (η_1, η_2) has the effect that cone boundary lines ε_1' and ε_2' of **D** — where Equations 23.5 and 23.6 respectively, hold true with equality — are shifted right and up by $\frac{\gamma_1 \eta_1}{G_{11}}$ and $\frac{\gamma_2 \eta_2}{G_{22}}$ relative to the original ε_1 and ε_2, respectively. Therefore, an increase in the target SIRs γ_i has the effect that not only does ε_i' tilt toward ε_j', but also the distance of ε_i' from ε_i increases. This results in $\mathbf{P}^*$ moving away from **0**, and hence all transmitters increasing their powers to satisfy Equation 23.2 for the increased γ_i.

The above discussed behavior and the gleaned geometric intuition extend directly to the general case of $N > 2$ links. The feasibility region **D** is then an N-dimensional polyhedral cone with boundary hyperplanes where Equation 23.2 is satisfied with equality. The tip of the feasibility cone is the Pareto-optimal power vector $\mathbf{P}^*$, at which the links should operate, in order to achieve their target SIRs at minimum possible power. Increasing any γ_i or any interlink gain G_{ij} and decreasing any intralink gain G_{ii} result in **D** shrinking and $\mathbf{P}^*$ moving away from **0**; eventually **D** "evaporates" and $\mathbf{P}^*$ disappears at infinity, as the system becomes infeasible. Adding more links in the channel results in introducing more "cutting" hyperplanes that define the feasibility cone, which again shrinks and sees its tip move away from **0**.

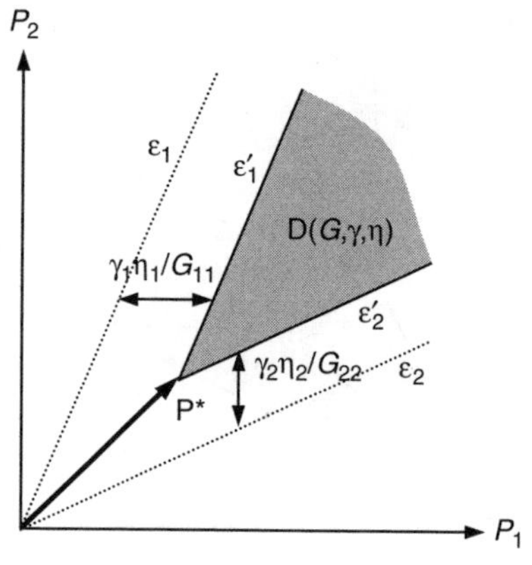

FIGURE 23.3 The feasibility region $\mathbf{D}(\mathbf{G}, \gamma, \eta)$ (shaded area) of the two-link network of Figure 23.1 where each power vector satisfies both Equations 23.5 and 23.6. It is a cone with boundary lines $\varepsilon'_1, \varepsilon'_2$ where Equations 23.5 and 23.6 hold true with equality, respectively. The tip of the cone is the Pareto-optimal power vector $\mathbf{P}^*$ where the target SIRs (γ_1, γ_2) are achieved at minimum power.

23.2.1.2 The Optimal Power Vector P*

The inequalities in Equation 23.2 can be put into a linear form $G_{ii} P_i - \gamma_i \sum_{i \neq j} G_{ij} P_j \geq \gamma_i \eta_i$ and further into a matrix form

$$(\mathbf{I} - \mathbf{F})\mathbf{P} \geq \mathbf{u} \quad \text{and} \quad \mathbf{P} > \mathbf{0} \tag{23.7}$$

where

$$\mathbf{u} = \left(\frac{\gamma_1 \eta_1}{G_{11}}, \frac{\gamma_2 \eta_2}{G_{22}}, \ldots, \frac{\gamma_N \eta_N}{G_{NN}} \right)^T \tag{23.8}$$

is the column vector of noise powers, normalized by the target SIR requirements over the intralink power gains, and $\mathbf{F}$ is the matrix with entries

$$\mathbf{F}_{ij} = \begin{cases} 0 & \text{if } i = j \\ \frac{\gamma_i G_{ij}}{G_{ii}} & \text{if } i \neq j \end{cases} \tag{23.9}$$

where $i, j \in \{1, 2, \ldots, N\}$. Note that the latter is the matrix of interlink power gains only, normalized by target SIRs over intralink gains.

The Perron–Frobenious theory of nonnegative matrices (having positive or zero entries only, like $\mathbf{F}$) [11, 12] addresses the existence of feasible power vectors (Equation 23.7) as follows. It is reasonable to assume that the nonnegative matrix $\mathbf{F}$ is irreducible; that is, there are no totally isolated groups of links that do not interact with each other. Then the *maximum modulus eigenvalue* $\rho_{\mathbf{F}}$ of $\mathbf{F}$ is real, positive, and simple, while the corresponding eigenvector is positive component-wise. The following statements are equivalent:

1. There exists a power vector $\mathbf{P} > 0$ such that $(\mathbf{I} - \mathbf{F})\mathbf{P} \geq \mathbf{u}$.
2. $\rho_{\mathbf{F}} < 1$.
3. $(\mathbf{I} - \mathbf{F})^{-1} = \sum_{k=0}^{\infty} \mathbf{F}^k$ exists and is positive component-wise.

If Equation 23.7 has a solution, by the third fact mentioned above we have that

$$\mathbf{P}^* = (\mathbf{I} - \mathbf{F})^{-1}\mathbf{u} \tag{23.10}$$

is the Pareto-optimal solution of Equation 23.7, in the sense that any other power vector $\mathbf{P}$ satisfying the equation would require as much power from every transmitter [29], or $\mathbf{P} \geq \mathbf{P}^*$ component-wise.

Therefore, if it is possible to satisfy the SIR requirements for all links simultaneously, the optimal power allocation strategy is to set the transmitter power to $\mathbf{P}^*$ and minimize the overall transmission power throughout the network.

23.2.2 Autonomous Distributed Power Control

Can links achieve the globally optimal power vector $\mathbf{P}^*$ without any interlink communication in a totally distributed and autonomous manner, even without any prior knowledge of the power gains matrix $\mathbf{G}$ and noise vector η? Fortunately, the answer is yes, as was explicitly observed in the early 1990s by Foschini and Miljanic [8] and others. To see this, consider first the following iterative algorithm over subsequent time slots $k = \{1, 2, 3, \ldots\}$

$$\mathbf{P}(k+1) = \mathbf{F}\mathbf{P}(k) + \mathbf{u} \tag{23.11}$$

and note that by recursively substituting Equation 23.11 into itself we get $\mathbf{P}(k) = \mathbf{F}^k\mathbf{P}(0) + \sum_{i=0}^{k-1} \mathbf{F}^i\mathbf{u}$. Therefore, the iteration converges to the the Pareto-optimal power vector

$$\lim_{k\to\infty} \mathbf{P}(k) = \lim_{k\to\infty} \mathbf{F}^k\mathbf{P}(0) + \lim_{k\to\infty} \sum_{l=0}^{k-1} \mathbf{F}^l\mathbf{u} = 0 + \left[\sum_{l=0}^{\infty} \mathbf{F}^l\right]\mathbf{u} = (\mathbf{I} - \mathbf{F})^{-1}\mathbf{u} = \mathbf{P}^* \tag{23.12}$$

when $\mathbf{P}^*$ exists or, equivalently, $\rho_\mathbf{F} < 1$. This can be easily seen by the basic facts of the Perron–Frobenious theory mentioned in the previous section, and recalling that $\rho_\mathbf{F} < 1$ implies $\lim_{k\to\infty} \mathbf{F}^k\mathbf{P}(0) = 0$.

Observe now that the distributed power control (DPC) algorithm (Equation 23.11) can be equivalently written for each link i as the iteration

$$P_i(k+1) = \frac{\gamma_i}{G_{ii}}\left[\sum_{j\neq i} G_{ij}P_j(k)\right] + \frac{\gamma_i}{G_{ii}}\eta_i = \frac{\gamma_i}{G_{ii}}\left[\sum_{j\neq i} G_{ij}P_j(k) + \eta_i\right] = \frac{\gamma_i}{R_i(k)}P_i(k) \tag{23.13}$$

since from the expression of the SIR (Equation 23.1) we have $\sum_{j\neq i} G_{ij}P_j(k) + \eta_i = \frac{G_{ii}P_i(k)}{R_i(k)}$. Therefore,

$$P_i(k+1) = \frac{\gamma_i}{R_i(k)}P_i(k) \tag{23.14}$$

for each $i = 1, 2, \ldots, N$ and $k = 1, 2, 3, \ldots$. Observe that under Equation 23.14 each link independently and autonomously measures its own SIR $R_i(k)$ and adjusts its transmission power to $P_i(k+1)$ according to its own SIR target γ_i, without explicit knowledge of any other transmitter power P_j, any power gain G_{ij}, or any noise level η_j. No interlink communication is required, but only intralink communication of the SIR $R_i(k)$ observed locally at the receiver of each link i back to its transmitter, where the power $P_i(k+1)$ is computed and set. The SIR information communicated is minimal (a single number per power update), so it can be carried on either a low-rate reverse virtual link or a global control channel, which may also carry other control information (acknowledgments, etc.). In "symmetric" networks — like cellular ones — with uplink/downlink communication, it may also be piggy-backed on the reverse link. The lack of any interlink communication justifies calling Equation 23.14 distributed, down to the level of the individual link, which is the functional primitive of the wireless network model.

An interesting observation regarding Equation 23.14 is that each link i autonomously updates its power trying to hit its SIR target γ_i — as if the interference induced from the other links remained constant because those would not update their powers. If the latter were indeed the case, link i would hit γ_i in a single step. In reality, all other links adjust their powers too, so $R_i(k)$ will asymptotically converge to γ_i (if feasible) as $k \to \infty$. It is worth noting that the convergence $R_i(k) \to \gamma_i$ is geometrically fast — as easily seen by the equivalent algorithm in Equation 23.11 — and $\rho_\mathbf{F} < 1$ is the convergence rate. If $\rho_\mathbf{F} > 1$, the

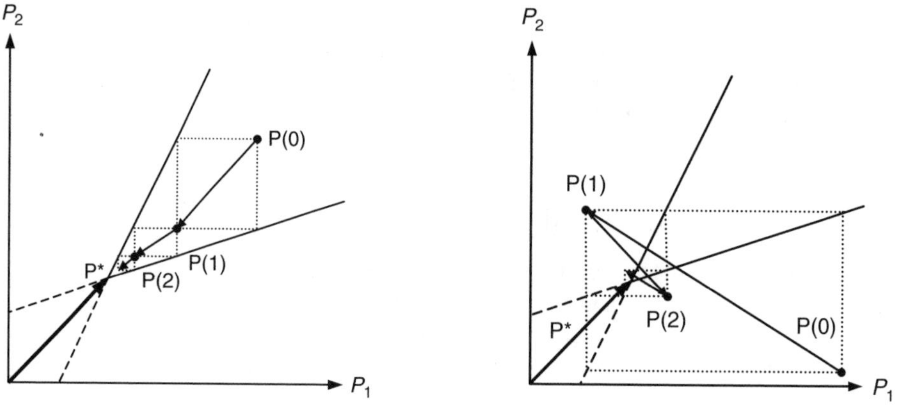

FIGURE 23.4 A graphical view of the convergence of the DPC algorithm (Equation 23.14) for the two-link network of Figure 23.1. The power vector evolution at consecutive iterations is shown. In the left graph, the initial power vector **P**(0) is within the feasibility region/cone, while in the right one it is outside. At each iteration, each transmitter autonomously changes its power to hit the cone boundary, which corresponds to hitting its target SIR if the other transmitters do not change their powers.

DPC algorithm obviously diverges and the link powers explode to infinity. Figure 23.4 graphically shows the power vector dynamics of a simple two-link network under DPC, while the evolutions of both power and SIR are plotted for a three-link network in Figure 23.5.

23.2.3 DPC with Active Link Protection

The power and SIR evolution dynamics of the raw DPC algorithm (Equations 23.14 and 23.11, respectively) are problematic because when new links power up in the channel, existing ones may see their SIRs dive

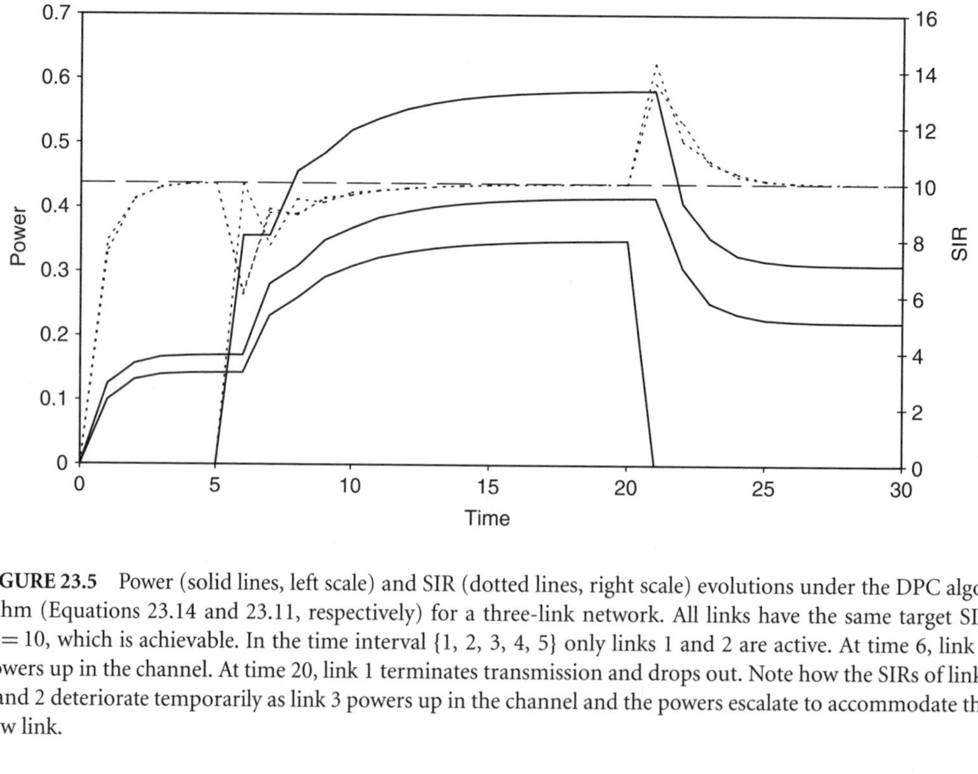

FIGURE 23.5 Power (solid lines, left scale) and SIR (dotted lines, right scale) evolutions under the DPC algorithm (Equations 23.14 and 23.11, respectively) for a three-link network. All links have the same target SIR $\gamma_i = 10$, which is achievable. In the time interval {1, 2, 3, 4, 5} only links 1 and 2 are active. At time 6, link 3 powers up in the channel. At time 20, link 1 terminates transmission and drops out. Note how the SIRs of links 1 and 2 deteriorate temporarily as link 3 powers up in the channel and the powers escalate to accommodate the new link.

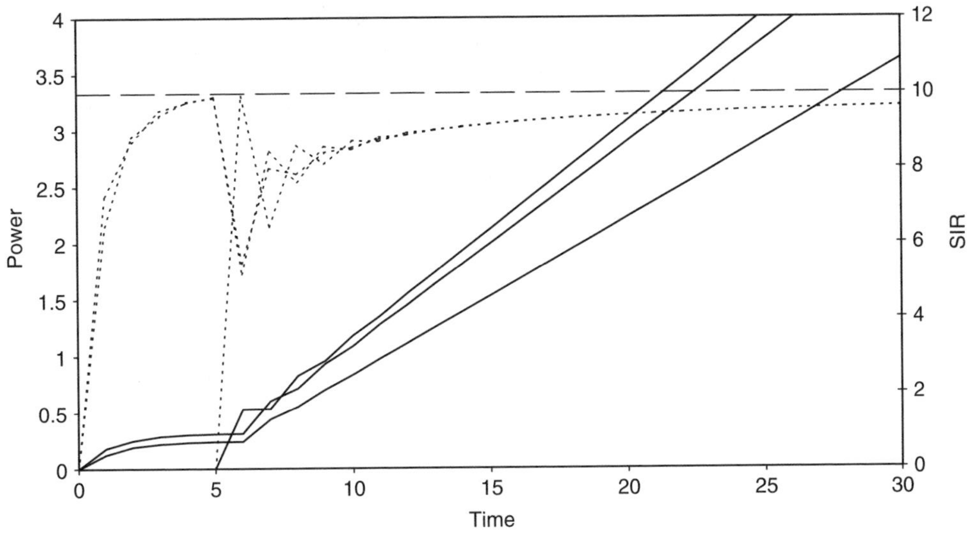

FIGURE 23.6 Power (solid lines, left scale) and SIR (dotted lines, right scale) evolutions under the DPC algorithm (Equations 23.14 and 23.11, respectively) for a three-link network with a common target SIR $\gamma_i = 10$, which is *infeasible*. In the time interval {1, 2, 3, 4, 5} only links 1 and 2 are active and actually admissible at $\gamma_i = 10$. At time 6, link 3 powers up in the channel. Since all three links are not admissible in the channel at $\gamma_i = 10$, their powers explode to infinity while their SIRs saturate below the target $\gamma_i = 10$.

significantly below their target SIRs γ for several iterations. This effect may cause established links to actually disconnect, which is unacceptable from a service support point of view. This is clearly shown in Figure 23.5, while in Figure 23.6 the case of new links powering up and driving even established ones unstable is demonstrated.

The underlying problem is that the raw DPC algorithm (Equation 23.14) does not support admission and congestion control in the channel, which are fundamental networking functions [22, 27, 29]. A novel idea/concept is needed to allow for such support, and this turns out to be the active link protection (ALP), discussed in this section. Indeed, since new links are expected to arrive and power up often in a wireless network, it is essential to protect already active (operational) ones from having their SIRs degrade below their target γ and disconnect. Even more importantly, we need to exercise admission control in the channel, rejecting any new link that makes the network infeasible. Can this be done in a distributed, autonomous, scalable fashion, without requiring any interlink communication? Fortunately, the answer is yes [22, 27, 29], as discussed below.

Let $\mathcal{L}$ be the set of all interfering links in the channel. At each power update step $k = 1, 2, \ldots$, let us partition $\mathcal{L}$ into the subset $\mathcal{A}_k$ of *active* or *operational* links or the subset $\mathcal{B}_k$ of *inactive* or *new* links according to the criterion of whether the links' SIRs are currently above or below their target γ, that is:

$$\mathcal{A}_k = \{i \in \mathcal{L} : R_i(k) \geq \gamma_i\} \tag{23.15}$$

$$\mathcal{B}_k = \{i \in \mathcal{L} : R_i(k) < \gamma_i\} \tag{23.16}$$

Let also

$$\delta = 1 + \epsilon > 1 \tag{23.17}$$

be a control parameter ($\epsilon > 0$), whose purpose and significance will become clear below. Then, the **DPC/ALP algorithm** works as follows:

$$P_i(k+1) = \begin{cases} \frac{\delta \gamma_i}{R_i(k)} P_i(k), & i \in \mathcal{A}_k \\ \delta P_i(k), & i \in \mathcal{B}_k \end{cases} \tag{23.18}$$

Note that under DPC/ALP, each *active* link updates its power similarly to the standard DPC algorithm (Equation 23.14), but shoots for a *boosted SIR target* $\delta\gamma_i$ for reasons explained below; each *new/inactive* link, however, powers up gradually at geometric rate δ. The intuition behind DPC/ALP is the following. Each active link shooting for a slightly higher (than needed) SIR target $\delta\gamma_i$ results in an *SIR protection margin* $\epsilon\gamma_i = (\delta - 1)\gamma_i$ acting as a cushion. It protects the link's SIR from dropping below its true target γ_i, cushioning the jolts induced by the new links powering up. On the other hand, since new links power up gradually, their degrading effect on active ones is bounded and has already been anticipated by raising the SIR target to $\delta\gamma_i$ and introducing the protection/cushioning zone of $(\delta - 1)\gamma_i$ thickness.

Several key properties of the DPC/ALP dynamics can be rigorously established and are leveraged later. First, note that there is a **bounded power overshoot** property:

$$P_i(k+1) \leq \delta P_i(k) \tag{23.19}$$

for each link i at each step $k = 1, 2, \ldots$, when the links operate under the DPC/ALP algorithm with any fixed $\delta > 1$. Indeed, this follows immediately from the definition of $\mathcal{A}_k$ and Equation 23.18. This property guarantees that active link powers may increase smoothly — up to a factor of δ on each update — while inactive link ones increase by exactly a factor δ in each update.

The second property is key **active link protection.** That is, when the links operate under DPC/ALP with any fixed $\delta > 1$, we have

$$R_i(k) \geq \gamma_i \Rightarrow R_i(k+1) \geq \gamma_i \tag{23.20}$$

for each link i at each time $k = 1, 2, \ldots$. That means that if a link i is active ($R_i(k) \geq \gamma_i$ and so $i \in \mathcal{A}_k$) in the kth time slot, it will also be active ($R_i(k+1) \geq \gamma_i$ and so $i \in \mathcal{A}_{k+1}$) in the $(k+1)^{st}$ time slot. Thus, *an active link will remain active forever* (until completing communication), while a new link may become active at some point and remain so forever after; that is, $i \in \mathcal{A}_k \Rightarrow i \in \mathcal{A}_{k+1}$ and $\mathcal{A}_k \subseteq \mathcal{A}_{k+1}$, but $\mathcal{B}_{k+1} \subseteq \mathcal{B}_k$ at each time k. This property in Equation 23.20 can be easily seen as follows. Observe that the interference $I_i(k+1) = \sum_{j\neq i} G_{ij} P_j(k+1) + \eta_i \leq \delta\{\sum_{j\neq i} G_{ij} P_j(k) + \eta_i\} = \delta I_i(k)$ or $I_i(k+1) \leq \delta I_i(k)$ at every time k, because of the property in Equation 23.19 and $\delta > 1$. Using this fact and Equation 23.18, we get for any active link $i \in \mathcal{A}_k$ that $R_i(k+1) = \frac{G_{ii} P_i(k+1)}{I_i(k+1)} = \frac{G_{ii}}{I_i(k+1)} \frac{\delta\gamma_i}{R_i(k)} P_i(k) = \delta\gamma_i \frac{I_i(k)}{I_i(k+1)} \geq \delta\gamma_i \frac{1}{\delta} = \gamma_i$, hence Equation 23.20.

Finally, there is a property of **new link improvement** at every time step. That is, when the links operate under DPC/ALP with any fixed $\delta > 1$, we have

$$R_i(k+1) \geq R_i(k), \text{ for any new link } i \in \mathcal{B}_k \tag{23.21}$$

at each time $k = 1, 2, 3 \ldots$. This is easily seen since $I_i(k+1) \leq \delta I_i(k)$ as argued above, and $P_i(k+1) = \delta P_i(k)$ for $i \in \mathcal{B}_k$. Thus, each of the new links increases its SIR at each step, so it may eventually rise above its target SIR γ and become active, staying such forever after. Note that this introduces a natural concept of channel admission, which is exploited later.

Note that the geometric convergence of the plain DPC algorithm, coupled with the geometric power-up of the inactive links, guarantees a geometric convergence of the DPC/ALP. For small δ (close to 1), the ALP component dominates the convergence rate; for large δ, the inherent speed (dominated by ρ_F) of the plain DPC becomes prevalent in the DPC/ALP.

23.2.4 Admission Control under DPC/ALP

Let us now study how the DPC/ALP algorithm (Equation 23.18) supports the concept of admission control of new links in the channel. Recall that network admission is equivalent to the link becoming active in the channel by raising its SIR above its target γ. According to the property in Equation 23.20, once a link becomes active it is guaranteed to remain so until voluntarily completing transmission. This property is the basis of the network admission concept. A new link arriving and seeking admission to the channel starts powering up from a low initial power and, according to Equation 23.21, sees its SIR improve at

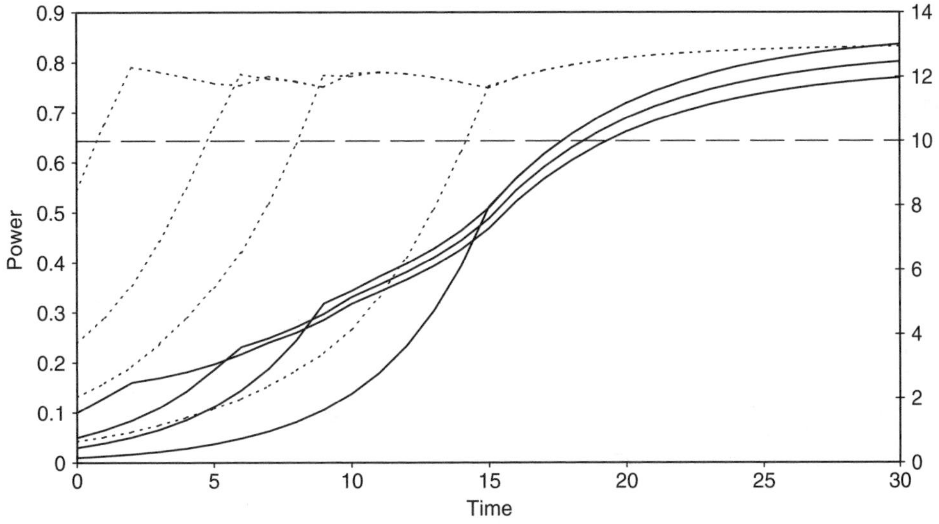

FIGURE 23.7 Power (solid lines, left scale) and SIR (dotted lines, right scale) evolutions under the DPC/ALP algorithm (Equation 23.18) of a four-link network with common SIR target $\gamma_i = 10$ and $\delta = 1.3$. All links are admissible at SIR level $\delta\gamma$; that is, they are δ-compatible. Each link starts inactive and when its SIR rises above γ_i the link turns active and eventually its SIR converges to $\delta\gamma_i$.

every step. Hence, it will either rise above its SIR target γ and become active (and stay so forever after) or saturate below its target if it is not admissible in the channel.

Let us examine some key cases of provable [22, 27, 29] link behavior and make some interesting observations. Suppose that the network starts at time 0 with a set $\mathcal{A} = \mathcal{A}_0$ of active links and a set $\mathcal{B} = \mathcal{B}_0$ of new (inactive) ones, and let us consider their evolution under DPC/ALP.

First consider the case where all links $i \in \mathcal{A} \cup \mathcal{B}$ can be admitted to the channel at SIR levels $\delta\gamma_i$ — call them δ-compatible. That means that there is a Pareto-optimal finite power vector $\mathbf{P}_\delta^* = (\mathbf{I} - \delta\mathbf{F})^{-1}\mathbf{u} < \infty$ that the system should operate at. Under DPC/ALP each new link will become active in finite time and be admitted to the channel. The power of each link will eventually converge $\lim_{k\to\infty} P_i(k) = [\mathbf{P}_\delta^*]_i$ to the optimal power value and its SIR to its boosted target value $\lim_{k\to\infty} R_i(k) = \delta\gamma_i$, which is targeted by DPC/ALP. Throughout convergence, no active link will ever drop below its actual target value γ. This is demonstrated in Figure 23.7.

What happens if all links $i \in \mathcal{A} \cup \mathcal{B}$ are *not* admissible at SIR levels $\delta\gamma_i$ (hence, $(\mathbf{I} - \delta\mathbf{F})^{-1}\mathbf{u} = \infty$) but are all admissible at SIR levels γ_i (hence, $(\mathbf{I} - \mathbf{F})^{-1}\mathbf{u} < \infty$)? In that case, all new links will actually become active in finite time. However, as active links will keep shooting for $\delta\gamma_i$ under DPC/ALP, all powers will eventually explode to infinity $\lim_{k\to\infty} P_i(k) = \infty$, while all SIRs will converge $\gamma_i < \lim_{k\to\infty} R_i(k) < \delta\gamma_i$ below the target $\delta\gamma_i$. This problematic power explosion situation is demonstrated in Figure 23.8.

By DPC/ALP shooting for the boosted SIR target $\delta\gamma_i$ rather than the required γ_i, some network capacity is basically traded for enhanced performance $(\delta - 1)\gamma_i$ per link. However, since δ can be chosen to be arbitrarily small, the capacity loss can be made arbitrarily small as well. In any case, any small loss of capacity is overcompensated by the active link protection and the other benefits in network performance discussed before. How should δ be chosen? When the channel is not congested (has low $\mathbf{P}^*$, $\rho_\mathbf{F}$ significantly lower than 1, etc.), we can choose a δ significantly higher than 1, so that new links increase their powers fast and gain admission to the channel. On the contrary, as the channel becomes congested ($\mathbf{P}^*$ gets high, $\rho_\mathbf{F}$ gets close to 1, etc.), the network becomes more sensitive to link admissions and we should choose a δ closer to 1 in order not to lose network capacity and also gracefully admit new links if possible.

In general, during evolution under DPC/ALP each new (inactive) link sees its SIR increase in every step because of the property in Equation 23.21; hence, either it will reach its target SIR γ and become active after a few steps or it will saturate below γ and stick there — this is a clear indication that it is not

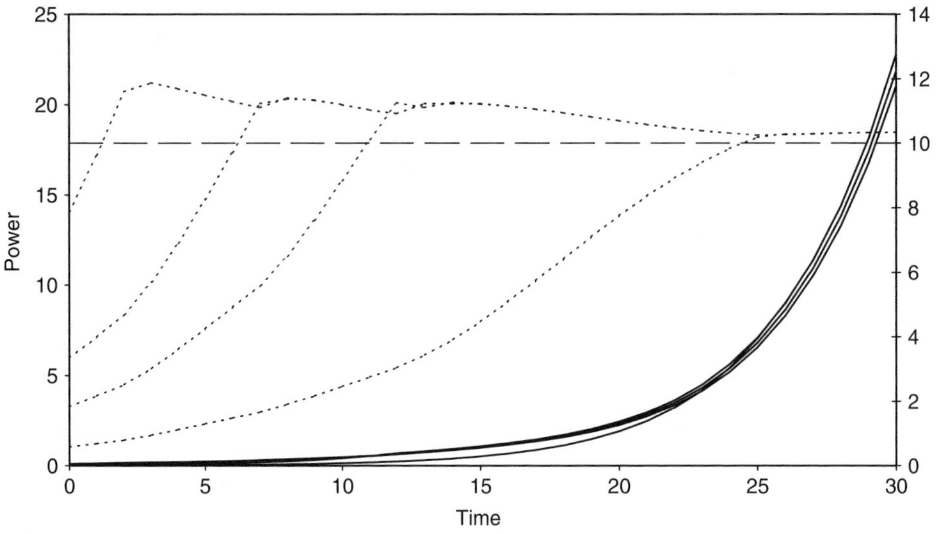

FIGURE 23.8 Power (solid lines, left scale) and SIR (dotted lines, right scale) evolutions under the DPC/ALP algorithm (Equation 23.18) of a four-link network with common SIR target $\gamma_i = 10$ and $\delta = 1.3$. All links are admissible at SIR target γ_i but not at $\delta\gamma_i$. Since DPC/ALP shoots for $\delta\gamma_i$, all powers eventually explode to infinity and the SIRs saturate below the DPC/ALP boosted target $\delta\gamma_i$, but still above the absolute target γ_i.

admissible to the channel. Therefore, some of the new links gain admission, while others are kept out by saturating below their target SIR; it may actually be beneficial to the network if some new links back off (set their power to zero) and try again later for admission. By backing off, they reduce interference on other new links, giving them a better chance of gaining admission. Figure 23.9 demonstrates such a case. The network consists of four links, but initially only one of them gains admission. The remaining three

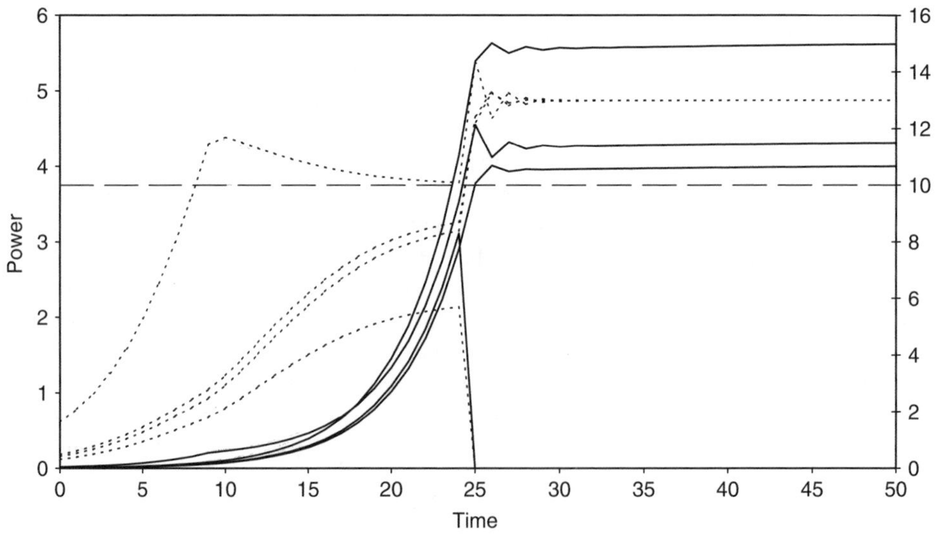

FIGURE 23.9 Power (solid lines, left scale) and SIR (dotted lines, right scale) evolutions under the DPC/ALP algorithm (Equation 23.18) of a four-link network with common SIR target $\gamma_i = 10$ and $\delta = 1.3$. One link gains admission soon, but the rest saturate below γ. When the link with the lowest SIR voluntarily drops out at time 25 (sets its power to zero), the other two soon gain admission.

links saturate below their SIR target γ_i. However, when a link drops out (the one with the lowest SIR in the plot) by setting its power to zero, the other two links soon get admitted. Had no link dropped out, the power levels of all the links would have exploded to infinity.

This example shows that we should augment the DPC/ALP algorithm so that new links voluntarily drop out when they sense that the channel is congested; hence, it is unlikely that they will become admitted. These links may either try admission to a different channel immediately (if available) or back off for some time before trying again at the same (or a different) channel. During their back-off, they should remain dormant (set the transmitter power level equal to zero) and then restart their power-up from the beginning. Depending on the criterion used for dropping out, we can design different algorithms for **voluntary dropout (VDO)** of new links as follows.

23.2.4.1 Voluntary Dropout

The first algorithm is based on the idea of a *time-out* when a new link tries for a long time to gain admission without success. Specifically, if a link powers up but remains inactive longer than time T_i (time-out), it computes a dropout time horizon D_i as a decreasing function of the distance between its target γ_i and its current SIR $R_i(T_i)$. The link continues to compete for admission for another D_i iterations. If at time $T_i + D_i$ it has not become active, it voluntarily drops out, setting its transmission power to zero. A new admission attempt may be reinitiated later in the same or another channel. The elimination of the link from the channel reduces the interference on other links seeking admission and increases their chances of success. T_i and D_i can be either fixed quantities or chosen at random (e.g., with a geometric distribution). Nonetheless, they are design parameters that should be optimized by testing. Indeed, too short a time-out period will cause the link to drop out possibly unnecessarily, whereas too long a time-out will result in the link heavily congesting the channel. Similarly, too short a back-off period will result in frequent readmission trials, which will increase the congestion of the channel and possibly lower the admission chances by having too many inactive links seeking admission; on the other hand, too long a back-off period will result in large delays before a link gets admitted.

Taking a different approach, we present an alternative algorithm for dropping out. The basic idea is that the link should keep trying to get admission as long as it observes some adequate SIR improvement over a time window of M_i steps. If persistently for more than M_i steps no such improvement has occurred, the link decides whether to drop out or to continue by flipping a coin. Coin flips in subsequent steps are independent, and the dropout probability is a decreasing function of the difference between the target and the current SIRs. As we have already seen, the idea behind this algorithm lies in the fact that if a link is inadmissible, then it will saturate below its target. Hence, consecutive power update steps will not bring about any significant SIR improvement, so the link will start the randomized dropout process. If, however, some other link drops out sooner, the link under consideration may experience some significant SIR increase, which will abort the dropout process and give it more time to try for admission. More details can be found in [29].

23.2.4.2 Forced Dropout

As we have already seen, admission of new links to the channel leads to an increase of the Pareto-optimal power vector $\mathbf{P}^*$, as the Perron–Frobenious eigenvalue ρ_F moves closer to 1. In a real system, however, the transmitters would be limited by some maximum transmission power $P_i^{\max}$. What if the transmitter has to exceed $P_i^{\max}$ in order to keep the link active? Clearly, we need a **forced dropout (FDO)** mechanism [29] that will force the inactive links to dropout when the active links are close to their maximum transmission powers. This mechanism should be as follows: When an active link senses that it is about to exceed its power limit, it should transmit a *distress* signal (which can be a special tone in some control channel/slot) at a certain power level. This signal is received by the links in its vicinity and instructs any inactive link that hears it above a certain power level to drop out immediately in order to de-congest the neighborhood of the active link, permitting it to relax its transmission power level. This distress signal should be transmitted by any active link $i \in A_k$ whenever its transmission power enters the region $(P_i^{\max}/\delta, P_i^{\max})$—as it follows from Equation 23.19—in order to avoid the forbidden region of $(P_i^{\max}, \infty)$.

Note that this property is not true for the plain DPC, as the power overshoots are not bounded in the DPC.

Forced dropout is also useful in the case of 1-compatible, but not δ-compatible, links [29]. Although not probable in a dynamic environment where multiple links seek admission, when it occurs, active link powers diverge to infinity, triggering a distress signal from the link that first reaches its power limit. Thus, newly admitted links should monitor for the distress signal for some short time horizon after their admission and drop out if they hear any. After the time horizon has passed, the links should consider themselves permanently admitted.

23.2.4.3 Initial Power Level of New Links

When the new links initiate their admission process, they start transmitting at some power P_o, which may cause an active link to drop its SIR below its target γ_i [29]. To avoid this link being considered inactive, we should modify the rule of distinguishing between active and inactive links to include a "once active, always active" clause. Thus, even if the active link mildly dips below γ_i for a short time, it will still be considered an active link and rapidly recover due to the fast geometric convergence of the DPC algorithm. However, by choosing P_o to be small enough, we can limit the disturbance to active links. Indeed, through a simple calculation [29] we see that if $P_o \leq \frac{(\delta-1)\eta_{\min}}{G_{\max}}$, no active link will drop below γ_i upon the appearance of the new one, given that all the active links have attained their enhanced targets $\delta\gamma_i$. $G_{\max}$ is the maximum interlink power gain, and $\eta_{\min}$ is the minimum thermal noise among the active links.

23.2.5 Noninvasive Channel Probing and Selection

Another key idea is that of channel probing [23, 30], which can alternatively be employed for admission control, or even in conjunction with the ideas discussed before. Channel probing can be used by a new link to quickly decide whether it is admissible in a channel, by collecting quantitative information about its congestion state in a fully autonomous manner. Hence, when there are several channels available, probing can be used to identify the best channel to seek admission into. The implementation of autonomous channel probing is based on the following provable facts.

Consider a set of initially active links $\mathcal{A} = \mathcal{A}_0$ and a set of initially inactive (new) ones $\mathcal{B} = \mathcal{B}_0$ interfering in a channel. Suppose that:

1. All links follow the DPC/ALP power update algorithm (Equation 23.18) with some fixed $\delta > 1$.
2. No new link in $\mathcal{B}$ becomes active before iteration K.
3. All active links $i \in \mathcal{A}$ are such that $\gamma_i \leq R_i(0) \leq \delta\gamma_i$ (are initially stable).

Then it can be shown [23, 30] that the SIR of each new link $i \in \mathcal{B}$ evolves within a certain envelope while the link powers up in the channel, as follows:

$$\frac{1}{\frac{X_i(\delta)}{\delta^k} + Y_i(\delta)} \leq R_i(k) \leq \frac{1}{\frac{X_i(1)}{\delta^k} + Y_i(1)}, \quad \text{for each } i \in \mathcal{B} \tag{23.22}$$

for all $k \leq K$, where $X_i(\delta)$ and $Y_i(\delta)$ are functions of $\mathbf{G}, \gamma, \boldsymbol{\eta}$, and $\delta > 1$. Actually, as $\delta \downarrow 1$, we have $X_i(\delta) \downarrow X_i(1) = X_i$ and $Y_i(\delta) \downarrow Y_i(1) = Y_i$; hence, the lower bound converges to the upper bound of the SIR evolution envelope. This means that for $\delta \approx 1$ the envelope shrinks to a trajectory

$$R_i(k) \approx \frac{1}{\frac{X_i}{\delta^k} + Y_i}, \quad \text{when } \delta \approx 1 \tag{23.23}$$

for $k = 0, 1, 2, \ldots, K$. Based on this, each new link $i \in \mathcal{B}$ can *autonomously* perform the following probing functions and make the corresponding decisions. First, by observing its own SIR for a few initial steps, it estimates the parameters X_i and Y_i via curve fitting. Based on these parameter estimates, the link predicts its SIR evolution in the future power-up steps. In principle, two consecutive SIR measurements (at low powers $P_i(0)$ and $P_i(0)\delta$) would be enough to estimate

$$X_i \approx \frac{\delta}{\delta - 1}\left\{\frac{1}{R_i(0)} - \frac{1}{R_i(1)}\right\} \quad \text{and} \quad Y_i \approx \frac{1}{\delta - 1}\left\{\frac{\delta}{R_i(1)} - \frac{1}{R_i(0)}\right\} \tag{23.24}$$

In practice, however, a link may need a few more steps to account for randomness and measurement errors. In any case, the transmitted power during these few initial steps is very low; hence, the probing process is fast and noninvasive, generating minimal interference in the channel.

If the new link i were to continue powering up in the channel without ever becoming active, then from Equation 23.23 we see that $R_i(k) \to \frac{1}{Y_i} < \gamma_i$, as $k \to \infty$ (since $\delta > 1$). Indeed, the SIR will keep increasing and eventually saturate below the target γ_i. Having estimated Y_i by probing the channel, the link may immediately decide to drop out when $\frac{1}{Y_i} < \gamma_i$, without stressing the channel any further.

If, however, the estimated parameter Y_l is such that $\frac{1}{Y_i} > \gamma_i$, then the link is admissible in the channel and should proceed to power up in it. Indeed, the link will gain admission and become active at time k_i^* approximately, when its SIR reaches the required threshold $R_i(k_i^*) \approx \gamma_i$, that is,

$$\frac{1}{\frac{X_i}{\delta^{k_i^*}} + Y_i} \approx \gamma_i \tag{23.25}$$

The link can then autonomously estimate its admission delay time $k_i^* \approx \frac{1}{\log(\delta)} \log(\frac{X_i}{\frac{1}{\gamma_i} - Y_i})$ and the power level at which it will gain admission $P_i(k_i^*) \approx \delta^{k_i^*} P_i(0) \approx (\frac{X_i}{\frac{1}{\gamma_i} - Y_i}) P_i(0)$, using the estimates X_i and Y_i obtained in the probing phase.

To conclude, the link can make a decision to proceed powering up and enter (become active in) the channel, if it finds the admission delay and entry power acceptable. If not, it can drop out from this channel and probe another one. Moreover, if multiple channels are available, the link can probe several of them and decide which one to join according to some criteria, for example, admission power and delay. We do not elaborate any further on these issues here, but complete details can be found in [23, 30].

23.3 Power Control for Packetized Data Traffic

The power control problem formulation employed in the previous section was geared toward wireless links carrying continuous delay-sensitive traffic (primarily voice or video), where power control is used to maintain a desired SIR level. This formulation, however, is not appropriate for packetized data traffic, which may be bursty and delay tolerant. In particular, it does not allow us to capture the fundamental *power vs. delay trade-off* of data traffic, which entails the following. During high interference periods, when high power is required to reliably transmit data, it may be better to buffer traffic and incur a delay cost, in anticipation that interference will soon subside and power benefits will be realized by reliably transmitting high data volumes at low power. What is needed is an enhanced power control model/formulation [24, 25, 26, 28, 31], capturing the above trade-off. We provide such a model and discuss it in this section, leading to identification of a new class of algorithms for power-controlled multiple access (PCMA).

23.3.1 Power-Controlled Multiple Access: The Basic Model

We take the local perspective of a single link, sharing a wireless channel with many others, which creates a fluctuating interference environment. Time is slotted. Let $I(k)$ be the interference that the link experiences at each receiver throughout time slot $k = 1, 2, 3, \ldots$; the interference remains constant within a slot but may change across different slots. The link's transmitter is equipped with a buffer, where packets may be stored waiting to be transmitted. Let $B(k)$ be the packet backlog in the buffer, that is, the number of packets in the transmitter buffer awaiting transmission. At the beginning of each time slot k, the link checks the interference $I(k)$ and the transmitter backlog $B(k)$ and decides what power $P(k)$ to use to transmit the head packet (if any) of the transmitter buffer (FIFO). If a packet is transmitted at power P across the channel, when the interference is I, then it is successfully received and decoded with probability $S(P, I)$ and so is removed from the buffer. Successful packet reception events are statistically independent in different time slots. The function $S(P, I)$ should obviously be increasing in P and decreasing in I; two typical functional forms in communications systems are the following:

$$S_1(P, I) = \frac{P}{\alpha P + \beta I} = \frac{\frac{P}{I}}{\alpha \frac{P}{I} + \beta} \tag{23.26}$$

where $\alpha \geq 1$ and $\beta > 0$, and

$$S_2(P, I) = 1 - e^{-\delta \frac{P}{I}} \tag{23.27}$$

where $\delta > 0$. Note that $\frac{P}{I}$ is the SIR in the time slot. If the packet is not successfully received at the receiver because it was corrupted during transmission (with probability $\{1 - S(P, I)\}$), then it will have to be transmitted again and again until it is successfully received. We assume that the transmitter is immediately notified at the end of each slot (perhaps over a separate control channel), whether or not the packet transmitted in that slot was successfully received, through some highly reliable ACK/NACK process that takes negligible time. Assume that the transmitter first checks the interference at the beginning of each time slot k, then selects a power $P(k)$ at which to transmit the head packet, and — if successful — removes the packet from the queue at the end of the time slot.

To spotlight and capture the fundamental power vs. delay trade-off, we make the assumption that the interference $I(k)$ induced on the link is *nonresponsive* to the powers $P(k)$ transmitted by it, as if the interference were determined by an extraneous (to the system) agent that is insensitive to the transmitter power of the link under consideration. This is a substantial simplification, since we know that all channel links are entangled to each other through mutual interference. Increasing the power of the link would increase the interference it generates on other channel links; those in turn would increase their powers, resulting in higher interference on the former link. The nonresponsive interference assumption, however, does have high value in the following sense. It allows for obtaining a tractable model whose optimal solution provides substantial insight into the general model of responsive interference. This insight obtained on the reduced model leads to the design of efficient PCMA algorithms, which perform surprisingly well when used in the full system with responsive interference [24, 25, 31]. This is further discussed later. Based on that, we proceed with the nonresponsive interference assumption, used as a conceptual device and a methodological step in our study. We assume that the interference $I(k)$ evolves according to an irreducible Markov chain with transition probabilities $P[I(k+1) = J \,|\, I(k) = I] = Q_{IJ}$ and steady-state π_I, where I, J are interference states from a finite set of all possible ones.

The power control problem can now be expressed as follows. First, let us define the cost structure of the system. In each time slot k, the link incurs:

1. A power cost $P(k)$, that is, the power transmitted in the channel in the time slot
2. A delay cost $D(B(k))$, which is an increasing function of the packet backlog $B(k)$ in the time slot

Thus the system incurs per time slot a cost equal to the sum of the above two costs. Then, the overall cost is equal to the cost incurred per slot summed over all time slots throughout the evolution of the system. Observing the backlog $B(k)$ and interference $I(k)$ in each slot k, the **power control problem** is to decide what power $P(k)$ to use for transmitting the head packet of the buffer, so as to minimize the overall average cost throughout the evolution of the system.

23.3.2 Optimally Emptying the Transmitter Buffer

In the simplest form of the problem, we consider the case of starting with B packets in the buffer and interference state I, and optimally controlling the transmitter power $P(1), P(2), P(3), \ldots$ in order to empty the buffer incurring the minimal overall cost. This problem can be solved using the framework of dynamic programming [13], as follows.

Let $V(B, I)$ be the *cost-to-go*, that is, the minimal expected cost that will be incurred under optimal power control until the buffer empties, given that the system starts with backlog B and interference I. The standard dynamic programming recursion (for $B > 0$) is simply

$$V(B,I) = \inf_{P \geq 0} \left\{ P + D(B) + S(P,I) \left[\sum_J Q_{IJ} V(B-1,J) \right] + [1 - S(P,I)] \sum_J Q_{IJ} V(B,J) \right\} \quad (23.28)$$

with initial value $V(0, I) = 0$. Q_{IJ} is the probability of the interference state switching from I to J, and the summation above is performed over all possible states J. The first two terms $P + D(B)$ of Equation 23.28 correspond to the cost incurred in the current time slot, while the other two terms express the collective future cost conditioned on successful and unsuccessful reception of the packet correspondingly. Rearranging the terms, for $B > 0$, we get

$$V(B,I) = \inf_{P \geq 0} \{P - S(P,I)X(B,I) + Y(B,I)\} \quad (23.29)$$

where

$$X(B,I) = \sum_J Q_{IJ}[V(B,J) - V(B-1,J)] \quad (23.30)$$

$$Y(B,I) = D(B) + \sum_J Q_{IJ} V(B,J) \quad (23.31)$$

Note that $V(B, J) - V(B-1, J) \geq 0$ for every $B > 0, I$, since the cost to empty the buffer increases when there are more packets. Therefore, $X(B, I) \geq 0$. Recursively solving the above equations, we obtain the optimal power control policy $P^* = P^*(B, I)$ as a function of the system state (B, I). Hence, the optimal transmitter power in slot k is $P^*(k) = P^*(B(k), I(k))$.

In order to solve the equations, we need to have a good knowledge (empirical or analytic) of the function $S(P, I)$ and a reliable statistical profile of interference behavior $\{Q_{IJ}\}$. It is then possible to solve the dynamic programming recursion off-line and store the optimal power values $P^*(B, I)$ in a lookup table accessible by the transmitter on-line.

23.3.3 The Case of Per-Slot-Independent Interference

Dynamic programming recursions can rarely be solved in a closed form. Fortunately, a nice semianalytical solution can be obtained for our model, in the special case when the interference is *independent and identically distributed* (i.i.d.) on different time slots. It is worth studying because it provides significant insight into the structure of optimal power control in the general case.

Under i.i.d. interference, the transition probabilities Q_{IJ} become equal to the stationary distribution probabilities π_J, and thus X and Y in Equations 23.30 and 23.31, respectively, lose their dependence on I, greatly reducing the complexity of the problem. Indeed, $X = X(B) = \sum_J \pi_J [V(B, J) - V(B-1, J)]$ and $Y = Y(B) = \sum_J \pi_J [D(B) + V(B, J)]$ under i.i.d. interference with distribution π_J. Revisiting Equation 23.29, if $S(P, I)$ is convex in P, then we can see that the solution P^* of the equation

$$\frac{\partial S(P,I)}{\partial P} = \frac{1}{X(B)} \quad (23.32)$$

if positive, minimizes the right-hand side of Equation 23.29. To better study the nature of the solution, let us consider two typical functional forms for $S(P, I)$ as follows. When $S_1(P, I) = \frac{P}{\alpha P + \beta I}$ with $\alpha \geq 1$ and

$\beta > 0$, from Equation 23.32 we get

$$P_1^*(B,I) = \begin{cases} \frac{1}{\alpha}\left[\sqrt{\beta X(B) I} - \beta I\right], & I \leq \frac{X(B)}{\beta} \\ 0, & I > \frac{X(B)}{\beta} \end{cases} \tag{23.33}$$

while when $S_2(P,I) = 1 - e^{-\delta\frac{P}{I}}$ with $\delta > 0$, Equation 23.32 results in

$$P_2^*(B,I) = \begin{cases} \frac{I}{\delta}\ln\frac{\delta X(B)}{I}, & I \leq \delta X(B) \\ 0, & I > \delta X(B) \end{cases} \tag{23.34}$$

as the optimal power control function. More details can be found in [31].

23.3.4 Structural Properties: Backlog Pressure and Phased Back-Off

Figure 23.10 shows plots of the optimal powers P_1^* (Equation 23.33) and P_2^* (Equation 23.34) against the interference I, for various $X = X(B)$ values. The explicit dependence on B has been suppressed in the graphs and has been incorporated in $X(B)$, which is an increasing function of B reflecting the backlog pressure (observe that the cost-to-go difference $V(B,I) - V(B-1,I)$ will be increasing in B, as $D(B)$ is increasing in B).

Note that although $S_1(P,I)$ and $S_2(P,I)$ have quite different analytic forms, the plots of the corresponding optimal power levels P_1^* and P_2^* are very similar. Indeed, observe that for a fixed backlog B (and thus a fixed backlog pressure $X(B)$), the optimal power P^* goes through three different phases in response to interference: *aggressive*, *soft back-off*, and *hard back-off*. In the first phase (low interference zone), the transmitter tries to aggressively match the interference and increases its power so as to maintain the needed success probability $S(P,I)$ and alleviate the delay cost. This effort gradually relaxes and eventually at some interference level the power P^* attains a maximum and the second soft back-off phase begins. In this phase, matching the interference is no more cost-efficient, and hence the transmitter starts to withdraw from the channel. The transmitter *softly backs off* of the channel by gradually decreasing its power until it hits zero. In the third phase, the transmitter completely refrains from any transmission (hard back-off) because setting $P^* = 0$ as the interference is too high to even make a transmission attempt. Intuitively, the two back-off phases can be explained as being preferable to incur some delay cost instead of spending enormous amounts of power (energy) for successful transmission when the interference is high.

As already mentioned, the effect of the backlog B on P^* is indirectly considered through $X(B)$, which is increasing in B. Observe that the boundaries of the three phases of P^* depend on the value of $X(B)$; an increase in $X(B)$ expands the length of the aggressive phase and the power levels used. In essence, $X(B)$ corresponds to the **backlog pressure**, that is, the pressure coming from the delayed packets, which forces the transmitter to be more aggressive at any fixed interference level I.

The above observations regarding the dependence of P^* on I and B, although originating from two particular functional forms of $S(P,I)$, are actually ubiquitous for any regular $S(P,I)$, and thus can be used in designing efficient PCMA algorithms. The properties and behaviors discussed are largely structural in nature and very robust with respect to particular choices of parameters and functions $S(P,I)$.

Incorporating a finite ceiling $P_{\max}$ for the transmitter power, that is, $P \in [0, P_{\max}]$, the solution changes slightly by having to take into account the new constraint in the minimization. Specifically, $P_1^*(B,I) = \min\{\frac{1}{\alpha}[\sqrt{\beta X(B) I} - \beta I], P_{\max}\}\mathbf{1}_{\{I \leq \frac{X(B)}{\beta}\}}$, where $\mathbf{1}_{\{\cdot\}}$ is the standard indicator[2] function. Moreover, $P_2^*(B,I) = \min\{\frac{I}{\delta}\ln\frac{\delta X(B)}{I}, P_{\max}\}\mathbf{1}_{\{I \leq \delta X(B)\}}$.

Finally, we can easily extend the model [24, 25, 26, 28, 31] to include packet arrivals, a finite buffer and corresponding overflow cost, and even packet deadlines and expiration cost. We can similarly use

[2]The indicator function $\mathbf{1}_{\{x\}}$ is equal to x when $x > 0$ and 0 otherwise.

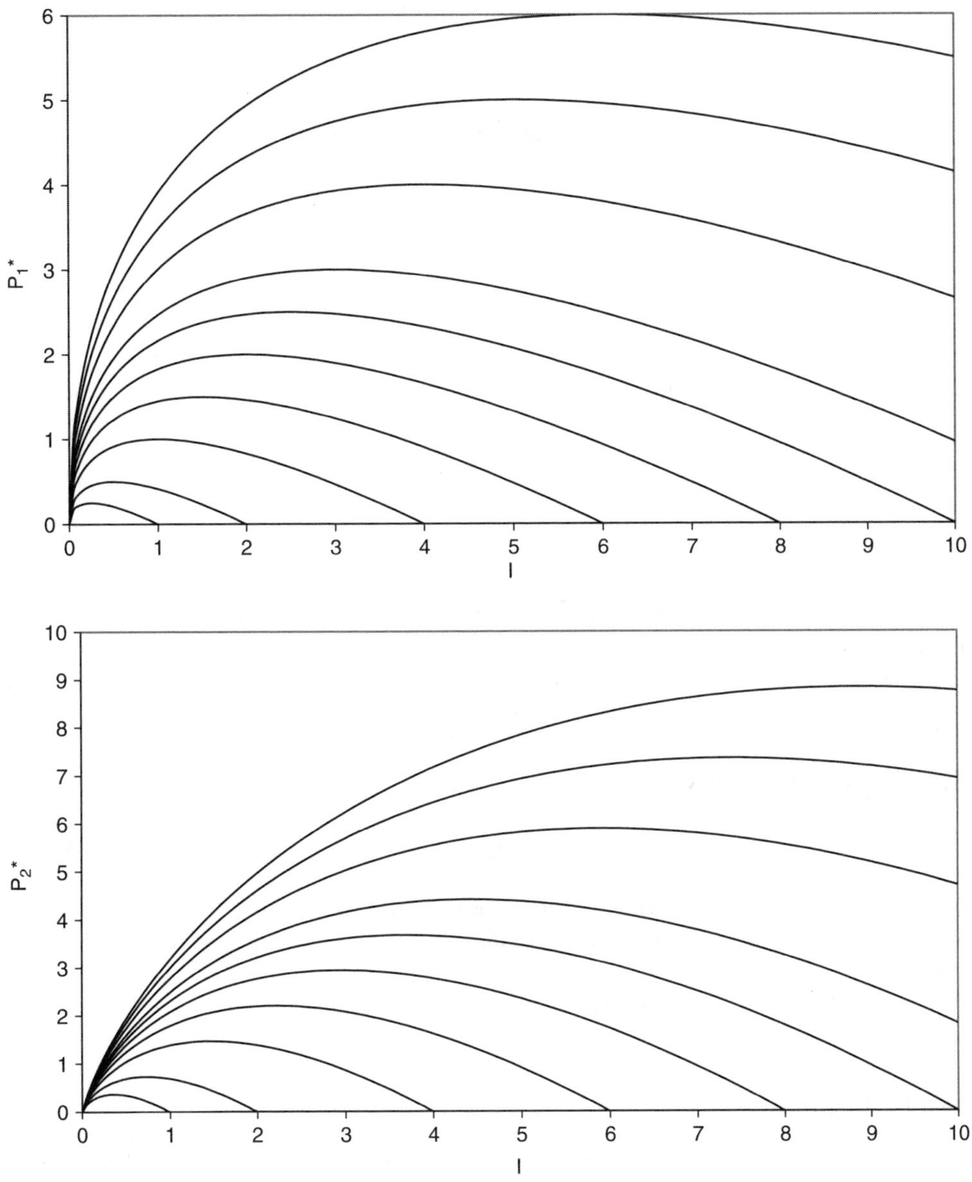

FIGURE 23.10 Plots of optimal power vs. interference function for various fixed back-log pressures $X(B) = 1, 2, 4, 6, 8, 10, 12, 16, 20, 24$. On the left, Equation 23.33 is plotted for $\alpha = \beta = 1$, while on the right Equation 23.34 is plotted for $\delta = 1$. In both plots, the family of curves are ordered, the lowest being for $X(B) = 1$ and the highest for $X(B) = 24$.

dynamic programming in all these cases, and the resulting optimal power control solutions have very similar structural properties to those discussed in the simple model analyzed above.

23.3.5 Design of PCMA Algorithms: Responsive Interference

Note that the DPC algorithm (Equation 23.14) previously introduced for continuous traffic can be written as $P_i(k+1) = \frac{\gamma_i}{R_i(k)} P_i(k) = \frac{\gamma_i}{G_{ii}} I_i(k)$ for each link i. At every step, it tries to match the interference and keep the SIR constant. Based on that observation, we can naturally extend it to the case of packetized traffic as

$$P_i(k+1) = \frac{\gamma_i}{G_{ii}} I_i(k) \mathbf{1}_{\{B_i(k)>0\}} \tag{23.35}$$

that is, each link applies DPC in each slot, when its backlog is nonzero.

Now, based on the optimal power control solutions in Equations 23.33 and 23.34, especially their versions including a power ceiling P_{max}, we can introduce the following two PCMA algorithms. Each link i observes its own backlog $B_i(k)$ and interference $I_i(k)$ at its receiver at the current time slot and sets its power in the next time slot as follows:

$$P_i(k+1) = \min\left\{\frac{1}{\alpha}\left[\sqrt{\beta X(B_i(k)) I_i(k)} - \beta I_i(k)\right], P_{\max}\right\} \mathbf{1}_{\{I_i(k) \le \frac{X(B_i(k))}{\beta}\}} \tag{23.36}$$

with $\alpha \geq 1$ and $\beta > 0$. Another PCMA algorithm is to set the power in the next time slot according to

$$P_i(k+1) = \min\left\{\frac{I_i(k)}{\delta} \ln \frac{I_i(k)}{\delta X(B_i(k))}, P_{\max}\right\} \mathbf{1}_{\{I_i(k) \le \delta X(B_i(k))\}} \tag{23.37}$$

with $\delta > 0$. The backlog pressure $X(B)$ is taken to be an increasing function of the backlog B, for example, $X(B) = \Phi B + \Psi$, with $\Phi, \Psi > 0$. These algorithms are designed to operate in the realistic responsive interference environment, where all links are entangled to each other via interference.

It is interesting to note that simulations experiments [24, 31] show that both PCMA algorithms perform very well, and certainly outperform by a large margin the DPC algorithm in Equation 23.35 in almost all operational scenarios. We do not elaborate further within the limited scope of this chapter, but only mention that the topic of optimal power control of wireless data traffic is vast and deserves further investigation.

23.4 Final Remarks

We have discussed fundamental aspects and core algorithms for power control of both continuous delay-sensitive traffic and packetized data traffic. The emphasis has been on understanding basic principles and simple — yet robust and scalable — algorithms with high practical potential. Technology-specific implementations of the algorithms have not been discussed — the limited space of this chapter would do them injustice.

The topic of transmitter power control is still under intense development and much further research should be expected in the near future, as wireless networking technology advances.

Acknowledgments

We thank the referees for thoroughly reading this chapter and making important suggestions regarding improvements and clarifications.

References

[1] J. C-I. Chuang and N. R. Sollenberger. Performance of autonomous dynamic channel assignment and power control for TDMA/FDMA wireless access. *IEEE Journal on Selected Areas in Communications*, 12(8):1314–1323, 1994.

[2] L. Kleinrock and J. Silvester. Spatial reuse in multihop packet radio networks. *Proc. IEEE*, 71(1), 1987.

[3] H. Alavi and R. W. Nettleton. Downstream power control for a spread spectrum cellular mobile radio system. *Proc. IEEE Globecom*, 1982.

[4] J. Zander. Performance of optimum transmitter power control in cellular radio systems. *IEEE Transactions on Vehicular Technology*, 41(1):57–62, 1992.

[5] J. Zander. Distributed cochannel interference control in cellular radio systems. *IEEE Transactions on Vehicular Technology*, 41(3):305–311, 1992.

[6] M. Andersin, Z. Rosberg and J. Zander. Soft and safe admission control in cellular networks. *IEEE/ACM Transactions on Networking*, 5(2):255–265, 1997.

[7] J. Zander. Radio resource management in future wireless networks; requirements and limitations. *IEEE Communications*, pp. 30–36, 1997.

[8] G. J. Foschini and Z. Miljanic. A simple distributed autonomous power control algorithm and its convergence. *IEEE Transactions on Vehicular Technology*, 42(4):641–646, 1993.

[9] G. J. Foschini and Z. Miljanic. Distributed autonomous wireless channel assignement with power control. *IEEE Transactions on Vehicular Technology*, 44(3):420–429, 1995.

[10] D. Mitra. An asynchronous distributed algorithm for power control in cellular radio systems. Proc. WINLAB '93 Workshop, 1993.

[11] E. Seneta. *Non-Negative Matrices*. George Allen and Unwin, London, 1973.

[12] F. R. Gantmacher. *The Theory of Matrices*. Chelsea, NY, 1960.

[13] D. Bertsekas. *Dynamic Programming*. Prentice Hall, New York, 1987.

[14] S. Grandhi, R. Vijayan, D. Goodman, and J. Zander. Centralized power control in cellular radio systems. *IEEE Transactions on Vehicular Technology*, 42(4):466–468, 1993.

[15] S. Grandhi, R. Vijayan, and D. Goodman. Distributed power control in cellular radio systems. *IEEE Transactions in Communications*, 42:226–228, 1994.

[16] S. Grandhi, R. Yates, and D. Goodman. Resource allocation for cellular radio systems. *IEEE Transactions on Vehicular Technology*, 46:581–587, 1997.

[17] S. Hanly. Capacity and power control in spread spectrum macrodiversity radio networks. *IEEE Transactions on Communications*, 44(8):247–256, 1996.

[18] R. Yates. A framework for uplink power control in cellular radio systems. *IEEE Journal of Selected Areas in Communications*, 13(7):1341–1347, 1995.

[19] R.Yates and C. Huang. Integrated power control and base station assignment. *IEEE Transactions on Vehicular Technology*, 44:638–644, 1995.

[20] N. Bambos and G. J. Pottie. On power control in high capacity cellular radio networks. *Proc. GLOBECOM '92*, 2:863–867, 1992.

[21] S. Chen, N. Bambos, and G. Pottie. Admission control schemes for wireless communication networks with adjustable transmitter powers. *Proc. IEEE INFOCOM '94*, Toronto, 1994.

[22] N. Bambos, S. C. Chen, and G. Pottie. Radio link admission algorithms for wireless networks with power control and active link quality protection. *Proc. IEEE INFOCOM '95*, Boston, MA, 1995.

[23] N. Bambos, S. Chen, and D. Mitra. Channel probing for distributed access control in wireless communication networks. *Proc. GLOBECOM '95*, pp. 322–325, 1995.

[24] N. Bambos and S. Kandukuri. Power controlled multiple access (PCMA) in wireless communication networks. *Proc. of IEEE INFOCOM 2000*, pp. 386–395, 2000.

[25] S. Kandukuri and N. Bambos. Multimodal dynamic multiple access (MDMA) in wireless packet networks. *Proc. of IEEE INFOCOM 2001*, pp. 199–208, 2001.

[26] J. Rulnick and N. Bambos. Mobile power management for wireless communication networks. *Wireless Networks*, 3:314, 1997.

[27] N. Bambos. Toward power-sensitive network architectures in wireless communications: concepts, issues, and design aspects. *IEEE Personal Communications*, 50–59, 1998.

[28] J. Rulnick and N. Bambos. Power-induced time division on asynchronous channels. *Wireless Networks*, 5:71–80, 1999.

[29] N. Bambos, S. C. Chen, and G. Pottie. Channel access algorithms with active link protection for wireless communication networks with power control. *IEEE/ACM Transactions on Networking*, 8(5):583–597, 2000.

[30] N. Bambos, J-W. Kim, S. Chen, and D. Mitra. Non-invasive channel probing for distributed admission control and channel allocation in wireless networks. Technical Report NetLab-2002-01/02, Engineering Library, Stanford University, Stanford, CA, January 2002.

[31] N. Bambos and S. Kandukuri. Power-controlled multiple access schemes for next-generation wireless packet networks. *IEEE Wireless Communications*, 58–64, 2002.

24

Signal Processing for Multiaccess Communication Networks*

24.1 Introduction 24-1
24.2 MPR at the Physical Layer 24-2
The Model • Assumptions and Properties • The Training-Based Zero-Forcing Receiver • The Semiblind Least Squares Smoothing Receiver • Further Remarks
24.3 The Interface between the Physical Layer and the MAC Layer: Resolvability 24-8
Channel Reception Matrix • Resolvability • From Resolvability Function to Reception Matrix • Further Remarks
24.4 Impact of MPR on the Performance of Existing MAC Protocols 24-11
Resolvability Comparison • Network Performance Comparison • Further Remarks
24.5 MAC Layer Design for Networks with MPR 24-13
The Network Model • The Multiqueue Service Room Protocol • The Dynamic Queue Protocol • Further Remarks
24.6 Approach High Performance from Both Physical and MAC Layers 24-16
24.7 Conclusion 24-17

Qing Zhao
Cornell University

Lang Tong
Cornell University

24.1 Introduction

In multiaccess wireless networks where a common channel is shared by a population of users, both the reception capability of the common wireless channel and the efficiency of the medium access control (MAC) protocol affect the network performance. Traditionally, MAC protocols are designed based on a collision channel model where any concurrent transmissions result in the destruction of all transmitted packets.

*This work was supported in part by the Multidisciplinary University Research Initiative (MURI) under the Office of Naval Research Contract N00014-00-1-0564, the Army Research Office under Grant ARO-DAAB19-00-1-0507, and the Army Research Laboratory CTA on Communication and Networks under Grant DAAD19-01-2-0011.

0-8493-1657-X/$0.00+$1.50

Numerous protocols, such as ALOHA, the tree algorithm, the first-come first-serve (FCFS) algorithm [5], and a class of adaptive schemes [6, 14, 15, 17], have been proposed to coordinate the transmissions of all users for the efficient utilization of the limited channel reception capability.

The advent of sophisticated signal processing techniques has changed the underlying assumptions made by conventional MAC protocols. In networks with space-division multiple access or code-division multiple access (CDMA), it is possible to receive some or all simultaneously transmitted packets that, in the conventional channel, would result in a collision and require retransmissions. In addition to the direct throughput improvement resulting from the recovery of colliding packets, the traffic load caused by retransmissions is reduced, which further decreases the frequency of collision.

While promising improvement in the overall performance of the network, this multipacket reception (MPR) capability presents new challenges to MAC designers. First, the MPR capability enriches the channel outcome, which makes it more difficult to infer the state of a user from the feedback information. Second, the MPR capability opens new options for collision resolution that, in the conventional collision channel, can be achieved only by splitting of users. To fully exploit the MPR capability, new MAC protocols need to be designed.

In this chapter, we present signal processing techniques that enable multipacket reception at the physical layer and examine their impact on the performance and design of MAC protocols. We give a summary of the results presented in [34–37]. More details can be found in [33].

24.2 MPR at the Physical Layer

We consider here a packetized system where data are generated in equal-sized packets. Transmission time is slotted, and each packet requires one time slot to transmit. MPR in such a network can be modeled as a signal separation problem in a multi-input multi-output (MIMO) system, illustrated in Figure 24.1, where $s^{(1)}(n), \ldots, s^{(K)}(n)$ are K simultaneously transmitted packets and $y_1(n), \ldots, y_P(n)$ are observations at the receiver. Here the multiple receivers can be antenna array elements or virtual receivers from temporal processing. The channel transfer function $H(z)$ depends on the form of modulation, the transmission protocol, and the configuration of transceiver antenna arrays. In a wireless mobile network, it is unrealistic to assume that $H(z)$ is known to the receiver in advance. The problem of MPR is then to design an estimator $F(z)$ without the knowledge of $H(z)$ to extract some or all transmitted packets.

In this section, we present the basic ideas of two signal processing techniques that enable MPR at the physical layer. Specifically, we consider the training-based zero-forcing receiver and the semiblind least squares smoothing (LSS) receiver proposed in [36].

24.2.1 The Model

We give in Figure 24.2 a more detailed description of the MIMO channel $H(z)$. Suppose that each packet contains N_0 symbols $s^{(k)}(0), \ldots, s^{(k)}(N_0 - 1)$ and $s^{(k)}(n) = 0$ for $n < 0$. The input–output relation of

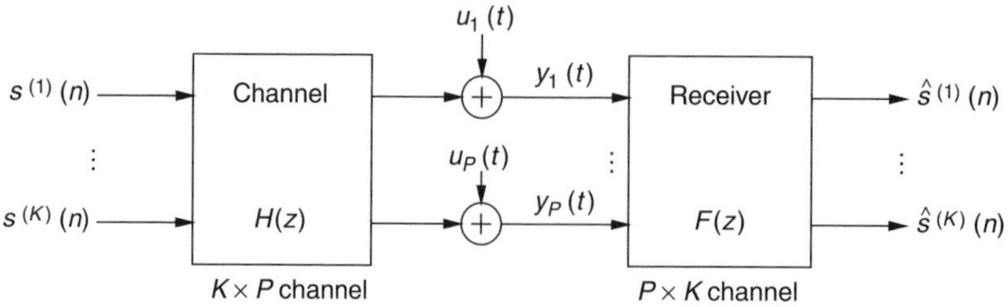

FIGURE 24.1 A general model for MPR.

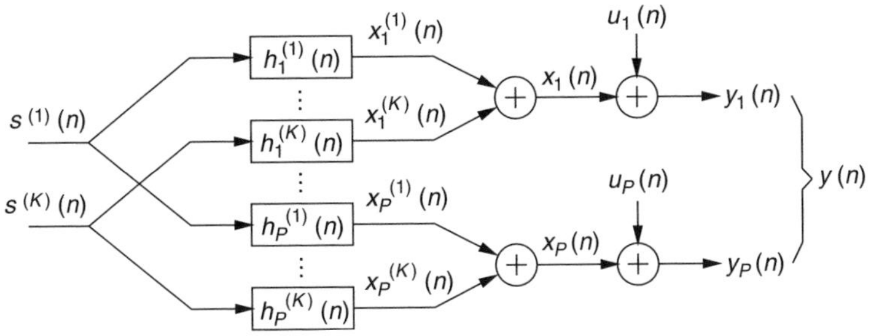

FIGURE 24.2 Packet collision: a MIMO system.

this MIMO system is given by

$$\mathbf{x}(n) \triangleq [x_1(n), \ldots, x_P(n)]^t = \sum_{k=1}^{K} \sum_{i=0}^{L_k} \mathbf{h}^{(k)}(i) s^{(k)}(n-i) \tag{24.1}$$

$$\mathbf{y}(n) = \mathbf{x}(n) + \mathbf{v}(n), \quad n = 0, \ldots, N_0 - 1 \tag{24.2}$$

where $\mathbf{y}(n) \in \mathcal{C}^{P\times 1}$ is the received signal and $\mathbf{v}(n) \in \mathcal{C}^{P\times 1}$ is the additive noise. We have assumed that the composite channel response $\{\mathbf{h}^{(k)}(i)\}$ of the kth colliding user has a duration of L_k symbol intervals.[1] Our goal here is to estimate $s^{(k)}(n)$ from $\mathbf{y}(n)$ $(n = 0, \ldots, N_0 - 1)$ without knowing $\mathbf{h}^{(k)}$ $(k = 1, \ldots, K)$, where $\mathbf{h}^{(k)}$ is defined as

$$\mathbf{h}^{(k)} \triangleq \begin{pmatrix} \mathbf{h}^{(k)}(0) \\ \vdots \\ \mathbf{h}^{(k)}(L_k) \end{pmatrix} \in \mathcal{C}^{(L_k+1)P\times 1} \tag{24.3}$$

For $N < N_0$, define the output block row vector $\mathbf{x}_n$ and the input row vector $\mathbf{s}_n^{(k)}$ as

$$\mathbf{x}_n \triangleq [\mathbf{x}(n), \ldots, \mathbf{x}(n+N-1)] \in \mathcal{C}^{P\times N}$$
$$\mathbf{s}_n^{(k)} \triangleq [s^{(k)}(n), \ldots, s^{(k)}(n+N-1)] \in \mathcal{C}^{1\times N}$$

From Equation 24.1 and Equation 24.2, we have

$$\mathbf{x}_n = \sum_{k=1}^{K} \sum_{i=0}^{L_k} \mathbf{h}^{(k)}(i) \mathbf{s}_{n-i}^{(k)}, \quad \mathbf{y}_n = \mathbf{x}_n + \mathbf{v}_n \tag{24.4}$$

where $\mathbf{y}_n \in \mathcal{C}^{P\times N}$ and $\mathbf{v}_n \in \mathcal{C}^{P\times N}$ are defined similarly as $\mathbf{x}_n \in \mathcal{C}^{P\times N}$. Consider w consecutive output block row vectors; we define

$$\mathbf{X}_w(n) \triangleq \begin{pmatrix} \mathbf{x}_n \\ \vdots \\ \mathbf{x}_{n+w-1} \end{pmatrix} \in \mathcal{C}^{wP\times N} \tag{24.5}$$

We thus have, from Equation 24.4,

$$\mathbf{X}_w(n) = \sum_{k=1}^{K} \mathbf{H}_w^{(k)} \mathbf{S}_w^{(k)}(n) = \mathbf{H}_w \mathbf{S}_w(n) \tag{24.6}$$
$$\mathbf{Y}_w(n) = \mathbf{X}_w(n) + \mathbf{V}_w(n)$$

[1] L_k is the maximum of the channel durations of the P subchannels shown in Figure 24.2.

where

$$\mathbf{H}_w \triangleq \left[\mathbf{H}_w^{(1)}, \ldots, \mathbf{H}_w^{(K)}\right] \in \mathcal{C}^{wP\times(Kw+\mathcal{L})} \tag{24.7}$$

$$\mathbf{H}_w^{(k)} \triangleq \begin{pmatrix} \mathbf{h}^{(k)}(L_k) & \cdots & \mathbf{h}^{(k)}(0) & & \\ & \ddots & & \ddots & \\ & & \mathbf{h}^{(k)}(L_k) & \cdots & \mathbf{h}^{(k)}(0) \end{pmatrix} \in \mathcal{C}^{wP\times(w+L_k)}$$

$$\mathbf{S}_w(n) \triangleq \begin{pmatrix} \mathbf{S}_w^{(1)}(n) \\ \vdots \\ \mathbf{S}_w^{(K)}(n) \end{pmatrix} \in \mathcal{C}^{(Kw+\mathcal{L})\times N}, \quad \mathbf{S}_w^{(k)}(n) \triangleq \begin{pmatrix} \mathbf{s}_{n-L_k}^{(k)} \\ \vdots \\ \mathbf{s}_n^{(k)} \\ \vdots \\ \mathbf{s}_{n+w-1}^{(k)} \end{pmatrix} \in \mathcal{C}^{(w+L_k)\times N} \tag{24.8}$$

$$\mathcal{L} \triangleq \sum_{i=1}^{K} L_i$$

24.2.2 Assumptions and Properties

We pose the following two assumptions for the training-based methods presented in Section 24.2.3 and the semiblind technique discussed in Section 24.2.4:

A1. There exists a w such that $\mathbf{H}_w$ (as defined in Equation 24.7) has full column rank.
A2. For the w specified in A1, $\mathbf{S}_w(n)$ (as defined in Equation 24.8) has full row rank.

Properties of A1 are summarized below:

P1. The row space $\mathcal{R}\{\mathbf{X}_w(n)\}$ of $\mathbf{X}_w(n)$ is isomorphic to the row space $\mathcal{R}\{\mathbf{S}_w(n)\}$ of $\mathbf{S}_w(n)$, i.e.,

$$\mathcal{R}\{\mathbf{X}_w(n)\} = \mathcal{R}\{\mathbf{S}_w(n)\} \tag{24.9}$$

P2. w has a lower bound given by

$$w \geq w_0 \triangleq \begin{cases} \lceil \frac{\mathcal{L}}{P-K} \rceil & \text{if } \mathcal{L} \neq 0 \\ 1 & \text{if } \mathcal{L} = 0 \end{cases} \tag{24.10}$$

Property P1 indicates that without knowing the input sequences, the row span of the input matrix $\mathbf{S}_w(n)$ can be obtained from the output $\mathbf{X}_w(n)$. Property P2 results from a necessary condition of A1. In order to make $\mathbf{H}_w$ full column rank, w should be sufficiently large so that $\mathbf{H}_w$ has no more columns than rows.

24.2.3 The Training-Based Zero-Forcing Receiver

In this section, we present the training-based zero-forcing (ZF) receiver. Consider an example where we have three colliding packets with channel order 1, 1, and 2, respectively. Suppose that $\mathbf{H}_w$ has full column rank and $\mathbf{S}_w(t)$ has full row rank for $w = 2$. From Equation 24.6 we have

$$\mathbf{X}_2(t) \triangleq \begin{pmatrix} \mathbf{x}_t \\ \mathbf{x}_{t+1} \end{pmatrix} = \mathbf{H}_2\mathbf{S}_2(t) \tag{24.11}$$

From the isomorphism between $\mathcal{R}\{\mathbf{X}_2(t)\}$ and $\mathcal{R}\{\mathbf{S}_2(t)\}$ as given in Equation 24.9, we have

$$\mathbf{s}_t^{(k)} \in \mathcal{R}\{\mathbf{X}_2(t)\}, \quad k = 1, 2, 3 \tag{24.12}$$

which implies that all colliding packets can be obtained as linear combinations in the row space of $\mathbf{X}_2(t)$. Under the assumption that $\mathbf{S}_2(t)$ has full row rank, we can show, from the isomorphic relation, that $\mathcal{R}\{\mathbf{X}_2(t)\}$ has dimension 10. Let $\{\mathbf{u}_1, \ldots, \mathbf{u}_{10}\}$ be a basis of $\mathcal{R}\{\mathbf{X}_2(t)\}$. We have, from Equation 24.12,

$$\mathbf{s}_t^{(k)} = \left[a_1^{(k)}, \ldots, a_{10}^{(k)}\right] \begin{pmatrix} \mathbf{u}_1 \\ \vdots \\ \mathbf{u}_{10} \end{pmatrix} \triangleq \mathbf{a}^{(k)}\mathbf{U}, \quad k = 1, 2, 3 \tag{24.13}$$

For the ZF receiver, the receiver coefficient vector $\mathbf{a}^{(k)}$ can be obtained by imposing the least squares criterion on the known symbols embedded in $\mathbf{s}_t^{(k)}$. Specifically, let α_k denote the vector containing the positions of the known symbols in $\mathbf{s}_t^{(k)}$. From Equation 24.13 we have

$$\mathbf{s}_t^{(k)}(\alpha_k) = \mathbf{a}^{(k)}\mathbf{U}(\alpha_k) \tag{24.14}$$

where $\mathbf{A}(\alpha_k)$ denotes the matrix that consists of the columns in $\mathbf{A}$ whose indexes are in α_k. If $\mathbf{U}(\alpha_k)$ is of full row rank, the optimal ZF receiver for $\mathbf{s}_t^{(k)}$ is given by

$$\mathbf{a}^{(k)} = \mathbf{s}_t^{(k)}(\alpha_k)\mathbf{U}'(\alpha_k)(\mathbf{U}(\alpha_k)\mathbf{U}'(\alpha_k))^{-1} \tag{24.15}$$

The full row rank condition on $\mathbf{U}(\alpha_k)$ requires that $\mathbf{U}(\alpha_k)$ has at least as many columns as rows. This implies that the minimum number $\mathcal{N}_{ZF}$ of known symbols required by the ZF receiver to obtain the colliding packets from $\mathbf{X}_w(n)$ is determined by the dimension of $\mathcal{R}\{\mathbf{X}_w(n)\}$, i.e.,

$$\mathcal{N}_{ZF} = Kw + \mathcal{L} \geq \begin{cases} K\lceil \frac{\mathcal{L}}{P-K} \rceil + \mathcal{L} & \text{if } \mathcal{L} \neq 0 \\ K & \text{if } \mathcal{L} = 0 \end{cases} \tag{24.16}$$

where the lower bound on $\mathcal{N}_{ZF}$ comes from the lower bound on w given in Property P2.

24.2.4 The Semiblind Least Squares Smoothing Receiver

Equation 24.16 shows that the minimum number of known symbols required by the ZF receiver is a monotone increasing function of the number of colliding packets (K) and the summation of the channel orders ($\mathcal{L}$). This observation suggests that in order to reduce the number of required known symbols, we should reduce multiaccess interference (decrease K) and intersymbol interference (decrease $\mathcal{L}$) in the received signal. The basic idea of the semiblind LSS receiver is to generate from the received signal an innovation sequence that contains less multiple-access interference (MAI) and intersymbol interference (ISI). The colliding packets can then be obtained from this innovation sequence with fewer known symbols.

We illustrate the basic idea of the semiblind LSS receiver with the same example given in Section 24.2.3. Consider first the detection of the two packets $\mathbf{s}_t^{(1)}$ and $\mathbf{s}_t^{(2)}$ with the smallest channel order from $\mathbf{X}_2(t)$. To reduce MAI and ISI, we introduce a smoothing operation on $\mathbf{X}_2(t)$ to obtain the innovation with respect to the future and past data. Specifically, consider $w = 2$ consecutive future and past data vectors $\mathbf{X}_2(t+2)$ and $\mathbf{X}_2(t-2)$. Under A1, we have the following two isomorphic relations:

$$\mathcal{R}\{\mathbf{X}_2(t+2)\} = \mathcal{R}\{\mathbf{S}_2(t+2)\}, \quad \mathcal{R}\{\mathbf{X}_2(t-2)\} = \mathcal{R}\{\mathbf{S}_2(t-2)\} \tag{24.17}$$

The overall data matrix $\mathbf{X}_6(t-2)$ that contains the future, current, and past data is given by

$$\mathbf{X}_6(t-2) \triangleq \begin{pmatrix} \mathbf{X}_2(t-2) \\ \mathbf{X}_2(t) \\ \mathbf{X}_2(t+2) \end{pmatrix} = \mathbf{H}_6\mathbf{S}_6(t-2) \tag{24.18}$$

Under the assumption that $\mathbf{H}_6$ has full column rank and $\mathbf{S}_6(t-2)$ has full row rank, $\mathbf{X}_6(t-2)$ spans a 22-dimensional row space that is isomorphic to $\mathcal{R}\{\mathbf{S}_6(t-2)\}$. This isomorphism is illustrated in Figure 24.3

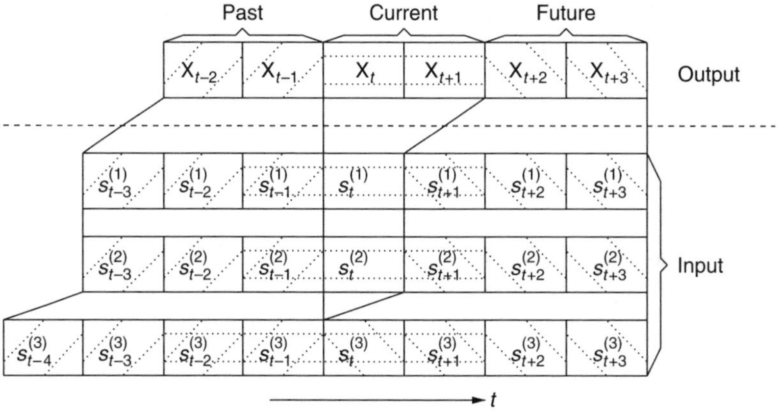

FIGURE 24.3 Isomorphism between input and output subspaces.

where the output block row vectors in $\mathbf{X}_6(t-2)$ and the input row vectors in $\mathbf{S}_6(t-2)$ are plotted. The other two pairs of isomorphic spaces specified in Equation 24.17 are also illustrated with right and left slashes, respectively. The input row vectors involved in the current data $\mathbf{X}_2(t)$ are shaded with horizontal lines. Figure 24.3 shows that among all input vectors contained in $\mathbf{X}_2(t)$, only $\mathbf{s}_t^{(1)}$ and $\mathbf{s}_t^{(2)}$ are outside the space spanned by the future and past data. All the other input vectors in $\mathbf{X}_2(t)$ are contained in either $\mathcal{R}\{\mathbf{X}_2(t+2)\}$ or $\mathcal{R}\{\mathbf{X}_2(t-2)\}$. For uncorrelated and zero-mean input signals, $\mathbf{s}_t^{(1)}$ and $\mathbf{s}_t^{(2)}$ are, asymptotically, the innovations of $\mathbf{X}_2(t)$ with respect to $\mathbf{X}_2(t+2)$ and $\mathbf{X}_2(t-2)$. It then follows that the asymptotic smoothing error $\mathbf{E}_2(t)$ of $\mathbf{X}_2(t)$ by the future data $\mathbf{X}_2(t+2)$ and the past data $\mathbf{X}_2(t-2)$ has the following form:

$$\begin{aligned}\mathbf{E}_2(t) &\triangleq \mathcal{P}^{\perp}_{[\mathbf{X}_2'(t+2),\mathbf{X}_2'(t-2)]'}\{\mathbf{X}_2(t)\}\\ &= \mathbf{h}_1\mathbf{s}_t^{(1)} + \mathbf{h}_2\mathbf{s}_t^{(2)}\end{aligned} \tag{24.19}$$

where $\mathcal{P}^{\perp}_{\mathbf{A}}\{\mathbf{X}\}$ denotes the projection error of $\mathbf{X}$ into the row space of $\mathbf{A}$ and $\mathbf{h}_i \triangleq [\mathbf{h}_i'(0),\ldots,\mathbf{h}_i'(L_i)]'$ is the vector of channel coefficients for the ith user.

From Equation 24.19 we note that the smoothing error $\mathbf{E}_2(t)$ contains only two multiaccess interferers; the MAI from the third user and the ISI of the first two users are completely removed from $\mathbf{X}_2(t)$ by the smoothing operation. The original system with $K=3$, $L=4$ has been converted to a system with $K=2$ and $L=0$. With $\mathbf{h}_1$ and $\mathbf{h}_2$ linearly independent, $\mathbf{E}_2(t)$ spans a two-dimensional row space from which $\mathbf{s}_t^{(1)}$ and $\mathbf{s}_t^{(2)}$ can be obtained with two known symbols from each.

After recovering the first two users' signals, we can also obtain their channel coefficients from $\mathbf{E}_2(t)$. Hence, signals from the first two users can be subtracted from the output, and we then have a system with $K=1$ and $L=2$. With the output after the subtraction denoted as $\bar{\mathbf{x}}_t$, we now apply the same process to the third user. Consider the smoothing error of $\bar{\mathbf{X}}_3(t)$ by $\bar{\mathbf{X}}_2(t+3)$ and $\bar{\mathbf{X}}_2(t-2)$. With similar analysis, we have the asymptotic smoothing error $\mathbf{E}_3(t)$ as

$$\begin{aligned}\mathbf{E}_3(t) &\triangleq \mathcal{P}^{\perp}_{[\bar{\mathbf{X}}_2'(t+3),\bar{\mathbf{X}}_2'(t-2)]'}\{\bar{\mathbf{X}}_3(t)\}\\ &= \mathbf{h}_3\mathbf{s}_t^{(3)}\end{aligned}$$

With one known symbol from the third user to remove the scalar ambiguity, both $\mathbf{s}_t^{(3)}$ and $\mathbf{h}_3$ can be obtained from $\mathbf{E}_3(t)$.

We now consider the general case where we have K packets involved in a collision. Suppose $L_k \in \{l_1,\ldots,l_J\}$ with $l_1 < l_2 < \cdots < l_J$. The number of colliding packets that come from channels with order l_i is k_i ($\sum_{i=1}^{J} k_i = K$). Without loss of generality, we assume that the packets are arranged according to their channel orders and the packets from channels with order l_1 are the first k_1 packets. Then we have the following theorem that characterizes the smoothing error:

Theorem 24.1 *Suppose that A1 holds for w and* $2w + l_1 + 1$. *Then for uncorrelated and zero-mean input signals, the asymptotic smoothing error* $\mathbf{E}_{l_1+1}(t)$ *is given by*

$$\mathbf{E}_{l_1+1}(t) \triangleq \mathcal{P}^{\perp}_{[\mathbf{X}'_w(t+l_1+1),\mathbf{X}'_w(t-w)]'}\{\mathbf{X}_{l_1+1}(t)\}$$
$$= \mathbf{h}_1\mathbf{s}_t^{(1)} + \cdots + \mathbf{h}_{k_1}\mathbf{s}_t^{(k_1)} \tag{24.20}$$

Theorem 24.1 summarizes the key result upon which the semiblind LSS approach is based. From Equation 24.20 we note that the smoothing error contains $k_1 \leq K$ multiaccess interferers and no ISI. With k_1 known symbols from the first k_1 users, their packets can be obtained as linear combinations in the k_1-dimensional row space spanned by the smoothing error. After recovering $\mathbf{s}_t^{(1)}, \ldots, \mathbf{s}_t^{(k_1)}$, we can also obtain $\mathbf{h}_1, \ldots, \mathbf{h}_{k_1}$ from $\mathbf{E}_{l_1+1}(t)$. Consequently, interference from the first k_1 users can be subtracted from the received signal and the same process can be applied to users with channel order l_2, which now is the smallest channel order.

A schematic diagram of this approach is shown in Figure 24.4, where we assume, without loss of generality, that $J = 3$. Figure 24.4 shows that the semiblind LSS receiver consists of two operations: a blind smoothing and a nonblind linear separation. The minimum number of known symbols required by each linear separator is also marked in Figure 24.4.

Unlike the ZF receiver, the minimum number of known symbols required by the semiblind LSS receiver to recover all colliding packets does not depend on the specific values of L_k $(1 \leq k \leq K)$. Instead, it is determined by the diversity of the channel orders as specified by the following equation:

$$\mathcal{N}_{LSS} = \max\{k_1, \ldots, k_J\} \leq K \tag{24.21}$$

As demonstrated by Theorem 24.1, the semiblind LSS receiver obtains packets with the smallest channel order first and then subtracts them from the observation. This successive demodulation makes this receiver particularly attractive in near–far scenarios where users with smaller channel orders are usually closer to the receiver, and hence have higher SNRs. The subtraction of stronger users from the observation facilitates the detection of weaker ones.

24.2.5 Further Remarks

There has been considerable research in blind and semiblind signal separation in recent years (see a survey on this topic in [19]). Without a sufficient number of training symbols, the key to signal separation is to utilize the structure of the channel and the characteristics of the input sources. The semiblind LSS

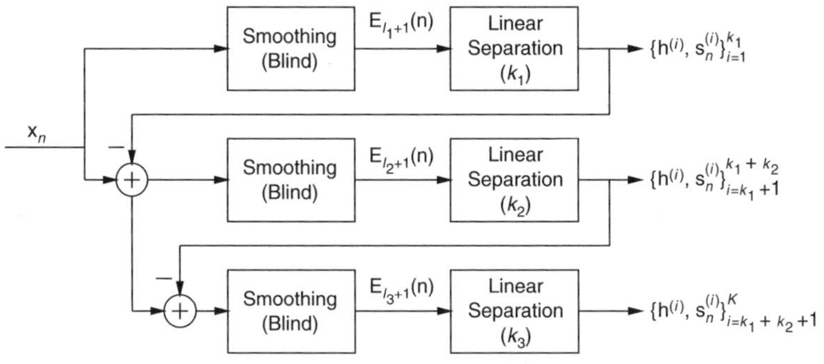

FIGURE 24.4 A schematic diagram of the semiblind LSS approach.

receiver presented in this section exploits the diversity of the propagation channel from each transmitting node to the receiver. The constant modulus property of communication signals can also be exploited to separate multiple sources by minimizing the signal dispersion using the constant modulus algorithm (CMA) [26, 27]. The finite alphabet property of communication signals provides yet another possibility for packet separation [18, 28]. In a transmitter-oriented CDMA system, discrimination among spreading codes also enables source separation [31]. For related topics in this growing field of research, refer to [13].

24.3 The Interface between the Physical Layer and the MAC Layer: Resolvability

In order to study the impact of MPR at the physical layer on the performance and design of the MAC layer, the MPR capability needs to be quantitatively characterized. In this section, we present a mathematical model for channels with MPR capability. We then introduce the resolvability function as a performance measure for receivers with MPR capability and analyze the training-based ZF receiver and semiblind LSS receiver.

24.3.1 Channel Reception Matrix

A general model for channels with multipacket reception capability has been developed in [11, 12], where the MPR capability is characterized by the probability $C_{n,k}$:

$$C_{n,k} \triangleq \mathrm{P}[k \text{ packets are correctly received} \mid n \text{ are transmitted}] \; (1 \leq n \leq M,\ 0 \leq k \leq n)$$

The multipacket reception matrix of the channel in a network with M users is then defined as

$$\mathbf{C} = \begin{pmatrix} C_{1,0} & C_{1,1} & & & \\ C_{2,0} & C_{2,1} & C_{2,2} & & \\ \vdots & \vdots & \vdots & & \\ C_{M,0} & C_{M,1} & C_{M,2} & \cdots & C_{M,M} \end{pmatrix} \tag{24.22}$$

Let

$$C_n \triangleq \sum_{k=1}^{n} kC_{n,k} \tag{24.23}$$

denote the expected number of correctly received packets when total n packets are transmitted. We then define the capacity of an MPR channel and the optimal number n_0 of transmissions as

$$\eta \triangleq \max_{n=1,\ldots,M} C_n, \quad n_0 \triangleq \min\{\arg \max_{n=1,\ldots,M} C_n\} \tag{24.24}$$

We can see that to achieve the channel capacity η, n_0 packets should be transmitted simultaneously.

This general model for MPR channels applies to the conventional collision channel and channels with capture as special examples. The reception matrices of the conventional noiseless collision channel and channels with capture are given by

$$\begin{pmatrix} 0 & 1 & 0 & \cdots & 0 \\ 1 & 0 & 0 & \cdots & 0 \\ \vdots & \vdots & \vdots & & \vdots \\ 1 & 0 & 0 & \cdots & 0 \end{pmatrix}, \quad \begin{pmatrix} 1-p_1 & p_1 & 0 & \cdots & 0 \\ 1-p_2 & p_2 & 0 & \cdots & 0 \\ \vdots & \vdots & \vdots & & \vdots \\ 1-p_M & p_M & 0 & \cdots & 0 \end{pmatrix} \tag{24.25}$$

where p_i is the probability of capture given i simultaneous transmissions.

24.3.2 Resolvability

In this section, we introduce the resolvability function as a performance measure for receivers with MPR capability. We show later that the resolvability function is closely related to the channel reception matrix **C** defined in Equation 24.22.

Definition 24.1 Suppose that there are K colliding packets, each coming (independently) from a channel with channel order $L_k \in \{l_1, \ldots, l_J\}$ where $l_1 < l_2 < \cdots < l_J$. The probability that L_k is l_j is assumed to be p_j. The collision resolvability function $\mathcal{R}_K(k, m)$ of a receiver is the probability that, given m known symbols per packet, at least $1 \leq k \leq K$ colliding packets can be resolved using the received data from the slot when collision occurs.

In general, the resolvability function is an attribute of a packet separation algorithm. It may be a function of channel characteristics, signal-to-noise ratio, and the performance of error control codes. To simplify the analysis, we consider here the uncoded system with no noise. The resolvability functions of the ZF receiver and the LSS receiver are given by the following theorems. Proofs of Theorems 24.2 to 24.4 can be found in [34].

Theorem 24.2 *Let* $\mathbf{l} = [l_1, \ldots, l_J]^t$ *be the vector of channel orders, and* $\mathbf{p} = [p_1, \ldots, p_J]^t$ *the vector of probabilities that a packet comes from channels with order* l_i. *Define* $\mathbf{k} = [k_1, \ldots, k_J]$ *as the random vector whose* i*th entry is the number of colliding packets that come from channels with order* l_i. *Then*

$$\mathcal{R}_K^{ZF}(k, m) = \begin{cases} 0 & \text{if } m < K \\ \sum_{\mathcal{L} \neq 0, K\lceil \frac{\mathcal{L}}{P-K} \rceil + \mathcal{L} \leq m} \sum_{\mathbf{k}, \mathbf{l}^t \mathbf{k} = \mathcal{L}} f(\mathbf{k}, \mathbf{p}) & \text{if } m \geq K \text{ and } l_1 \neq 0 \\ p_1^K + \sum_{\mathcal{L} \neq 0, K\lceil \frac{\mathcal{L}}{P-K} \rceil + \mathcal{L} \leq m} \sum_{\mathbf{k}, \mathbf{l}^t \mathbf{k} = \mathcal{L}} f(\mathbf{k}, \mathbf{p}) & \text{others} \end{cases} \tag{24.26}$$

where $f(\mathbf{k}, \mathbf{p})$ *is the multinomial mass function*

$$f(\mathbf{k}, \mathbf{p}) \triangleq \frac{K!}{k_1! \cdots k_J!} p_1^{k_1} \cdots p_J^{k_J}, \quad \sum_i k_i = K \tag{24.27}$$

Theorem 24.3 *Under the same definition in Theorem 24.2, the resolvability function for the semiblind least squares smoothing algorithm with successive demodulation is given by*

$$\mathcal{R}_K^{LSS}(k, m) = \sum_{\mathbf{k} \in \mathcal{K}(k)} f(\mathbf{k}, \mathbf{p}) \tag{24.28}$$

where $f(\mathbf{k}, \mathbf{p})$ *is the multinomial function defined in Equation 24.27, and*

$$\mathcal{K}(k) \triangleq \left\{ \mathbf{k} : k_i \leq m, i = 1, \ldots, I_k, \sum_{i=1}^{I_k - 1} k_i < k, \sum_{i=1}^{I_k} k_i \geq k \right\} \tag{24.29}$$

We now draw comparison between the ZF and semiblind LSS receivers in terms of resolvability. We are particularly interested in two boundary points in the (k, m) plane of $\mathcal{R}_K(k, m)$. One point, denoted as $m_0(k)$, is the "takeoff" point, which indicates the minimum number of known symbols for a receiver to resolve at least k packets from K colliding ones with nonzero probability. The other one, denoted as $m_1(k)$, indicates the minimum number of known symbols for a receiver to guarantee the recovery of at least k colliding packets.

Theorem 24.4 *Define*

$$m_0(k) \triangleq \arg\min_m \mathcal{R}_K(k,m) \neq 0 \tag{24.30}$$

$$m_1(k) \triangleq \arg\min_m \mathcal{R}_K(k,m) = 1 \tag{24.31}$$

We then have:

P1.

$$m_0^{LSS}(k) = \min\left\{\left\lceil \frac{k}{J-1}\right\rceil, \left\lceil \frac{K}{J}\right\rceil\right\} \tag{24.32}$$

$$m_0^{ZF}(k) = \begin{cases} K\left\lceil \frac{Pl_1}{P-K}\right\rceil & if\, l_1 \neq 0 \\ K & if\, l_1 = 0 \end{cases} \tag{24.33}$$

P2.

$$m_1^{LSS}(k) = K \tag{24.34}$$

$$m_1^{ZF}(k) = \begin{cases} K\left\lceil \frac{Pl_J}{P-K}\right\rceil & if\, l_J \neq 0 \\ K & if\, l_J = 0 \end{cases} \tag{24.35}$$

P3. $R_K^{LSS}(k,m) \geq R_K^{ZF}(k,m)$ *with equality if and only if* $m \geq m_1^{ZF}(k)$ *(when* $R_K^{LSS}(k,m) = R_K^{ZF}(k,m) = 1$*) or* $m < m_0^{LSS}(k)$ *(when* $R_K^{LSS}(k,m) = R_K^{ZF}(k,m) = 0$*).*

A proof of Theorem 24.4 can be found in [34].

24.3.3 From Resolvability Function to Reception Matrix

With the resolvability function $\mathcal{R}_K^{LSS}(k,m)$, the channel reception matrix for an M-user multiaccess network that employs the semiblind LSS receiver and has m known symbols embedded in each packet can be obtained as follows:

$$C_{n,k} = \mathcal{R}_n^{LSS}(k,m) - \mathcal{R}_n^{LSS}(k+1,m) \tag{24.36}$$

where $1 \leq n \leq M, 1 \leq k < n$. For $k = n$, we have

$$C_{n,n} = \mathcal{R}_n^{LSS}(n,m) \tag{24.37}$$

For $k = 0$, we easily have

$$C_{n,0} = 1 - \sum_{k=1}^{n} C_{n,k} \tag{24.38}$$

The channel reception matrix for a multiaccess network employing the training-based ZF receiver can be similarly obtained from $\mathcal{R}_K^{ZF}(k,m)$.

24.3.4 Further Remarks

The reception matrix **C** defined in Equation 24.22 provides a simple yet powerful interface between the physical and MAC layers. It allows tractable analysis on the interaction between these two layers. Though applicable to many scenarios, this model of the MPR physical layer does not characterize asymmetry among users that may arise in systems with antenna arrays. In [2, 21, 22], a more general MPR model is proposed where packets from different nodes can have different reception probabilities, depending on the identities of all transmitting nodes.

24.4 Impact of MPR on the Performance of Existing MAC Protocols

The training-based ZF receiver and the semiblind LSS receiver can be employed by both *ad hoc* and centrally controlled networks to provide MPR capability. In this section, we consider a centrally controlled network with a finite population of users. Numerical studies on the resolvability of the ZF receiver and the LSS receiver and the effects of their MPR capabilities on the network performance are presented. We consider an example where we have $P = 12, \mathbf{l} = [2, 3, 4]^t$, and $\mathbf{p} = [\frac{1}{3}, \frac{1}{3}, \frac{1}{3}]^t$ (definitions of $\mathbf{l}$ and $\mathbf{p}$ are given in Theorem 24.3).

24.4.1 Resolvability Comparison

Based on Theorem 24.2 and Theorem 24.3, the resolvability functions of the ZF receiver and the semiblind LSS receiver can be calculated. We illustrate the results with $K = 4$ and $k = 1$ in Figure 24.5, from which we observe

$$m_0^{LSS}(1) = 1, \quad m_1^{LSS}(1) = 4 \tag{24.39}$$

$$m_0^{ZF}(1) = 12, \quad m_1^{ZF}(1) = 24 \tag{24.40}$$

These results confirm Equation 24.32 to 24.35 and demonstrate the improvement in resolvability provided by the semiblind LSS receiver over the training-based ZF receiver.

With the resolvability function, the channel reception matrix can be obtained as given in Section 24.3.3.

24.4.2 Network Performance Comparison

Here we study the effect of the MPR capability of the ZF receiver and the semiblind LSS receiver on the throughput performance of a multiaccess network using slotted ALOHA with delayed first transmission.

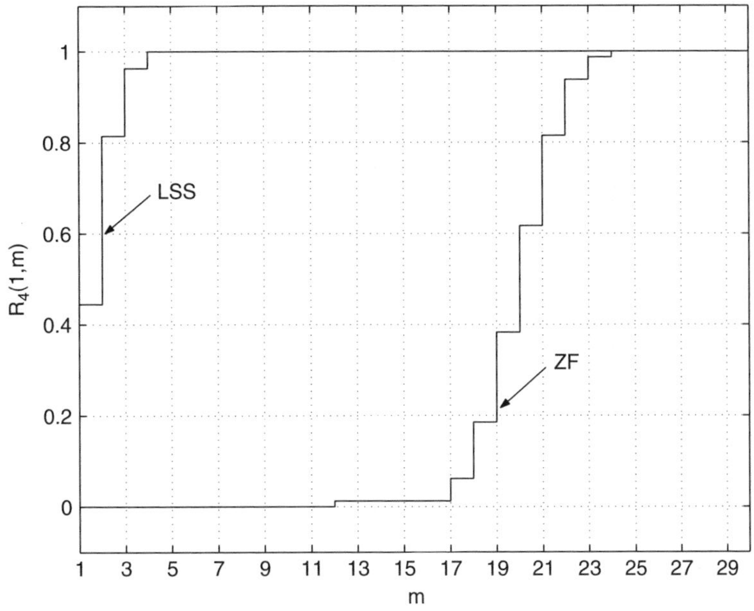

FIGURE 24.5 Resolvability functions ($K = 4, k = 1$).

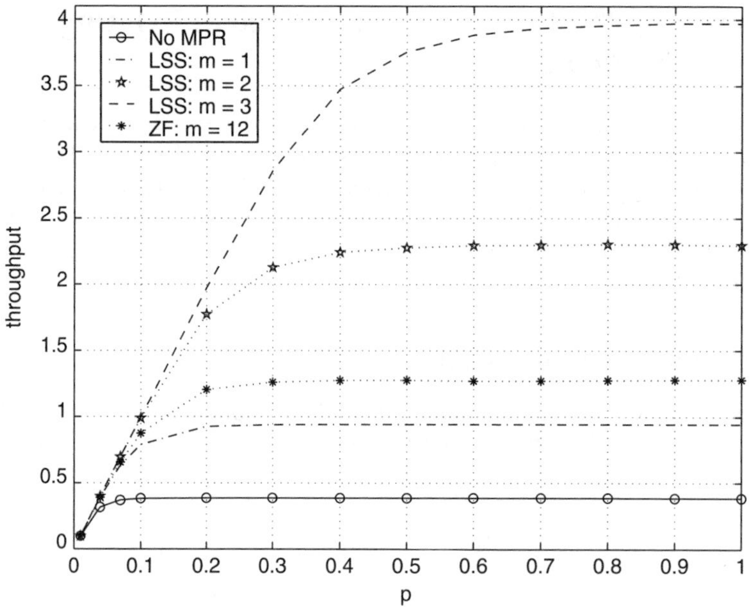

FIGURE 24.6 The impact of MPR on the throughput performance of slotted ALOHA.

We assume that there are $M = 10$ users in the network. Each user has a single buffer. With probability p, a user independently generates a packet within each slot. At the beginning of each slot, a packet, newly generated or backlogged, is transmitted with probability p_r, which is chosen to maximize the throughput. Given the channel reception matrix **C**, the throughput of such a system can be analyzed [5]. The results (as a function of p) are shown in Figure 24.6 where we consider the semiblind LSS receiver at $m = 1, 2, 3$ and the ZF receiver at $m = 12$. We can see that with one known symbol to remove a scalar ambiguity, the LSS receiver provides significant improvement in the network throughput, compared with that of networks without MPR capability. The throughput gain achieved by the semiblind LSS receiver with 2 known symbols is almost twice as large as that achieved by the training-based ZF receiver with 12 known symbols.

24.4.3 Further Remarks

There is a growing body of research on the impact of MPR capability on the performance of existing MAC protocols. Being the first random access protocol, the application of ALOHA to networks with MPR capability has been thoroughly studied. In [1, 8, 29, 38] and references therein, slotted ALOHA is applied to networks with capture effect. In [11, 12], a general model for channels with MPR capability is developed and the performance of slotted ALOHA analyzed for the infinite population case. In [4], the impact of MPR on the performance of slotted ALOHA is compared between two types of CDMA wireless local area networks (LANs) — centrally controlled and *ad hoc* networks — based on a finite population model. Based on a more general MPR model [21, 22] study the impact of MPR on the stability of ALOHA. Other random access protocols such as the FCFS algorithm and the window protocol [23] have also been extended to networks with capture effect and their performance evaluated [3, 20, 25, 32]. The application of contention free-scheme time-division multiple access (TDMA) to networks with MPR capability is another interesting research topic. In [7, 16], the authors address the use of TDMA in fully connected half-duplex *ad hoc* networks with MPR provided by multiple independent collision channels. In [24], dynamic time slot allocation is introduced for cellular systems with antenna arrays. Given a set of active users (users with packets to transmit), the proposed dynamic slot allocation scheme assigns an appropriate number of active users to each time slot to utilize the MPR capability provided by the antenna array.

24.5 MAC Layer Design for Networks with MPR

MPR at the physical layer undermines assumptions upon which many existing MAC protocols are built. To fully exploit the MPR capability at the physical layer, new MAC protocols need to be designed. In this section, we address the issue of MAC layer design for the optimal utilization of the MPR capability. Two protocols, the multiqueue service room (MQSR) and the dynamic queue protocol, are presented. The MQSR protocol is optimal in terms of per-slot throughput, while the dynamic queue protocol achieves a comparable performance with a much simpler implementation.

24.5.1 The Network Model

We consider a communication network with M users who transmit data to a central controller through a common channel. Each user generates data in the form of equal-sized packets. Transmission time is slotted, and each packet requires one time slot to transmit. Each user has a single buffer. With probability p, a user independently generates a packet within each slot, but only accepts this packet if its buffer is currently empty. A successfully transmitted packet leaves its buffer. Packets generated by a user with a full buffer are assumed lost.

The slotted MPR channel is characterized by the reception matrix given in Equation 24.22. We assume that the central controller can distinguish without error between empty and nonempty slots. Furthermore, if some packets are successfully demodulated at the end of a slot, the central controller can identify the source of these packets and inform their sources so that their buffers can be released. However, if at least one packet is successfully demodulated at the end of a slot, the central controller does not assume the knowledge whether there are other packets transmitted but not successfully received in this slot.

24.5.2 The Multiqueue Service Room Protocol

The multiqueue service room protocol [37] is designed explicitly for general MPR channels. The protocol is designed to accommodate groups of users with different delay requirements. Here we consider the case when there is only one group of users with the same delay requirement. The basic structure of the MQSR protocol is illustrated in Figure 24.7. When the network starts, all M users in the network are waiting in a queue to enter the service room for channel access. Users enter the service room in turn and stay ordered inside the service room. The service room consists of an access set and a waiting set. Users and only users in the access set transmit, in the current slot, packets generated before entering the service room. Packets generated by a user when it is inside the service room are held in the user's buffer (if the buffer is empty) and cannot be transmitted until next time this user enters the service room. After entering the service room, a user stays there until either the central controller detects that its packet generated before entering the service room has been successfully transmitted or it enters the service room with an empty buffer. At this time, we say this user is processed. A processed user leaves the service room and goes to the end of the queue.

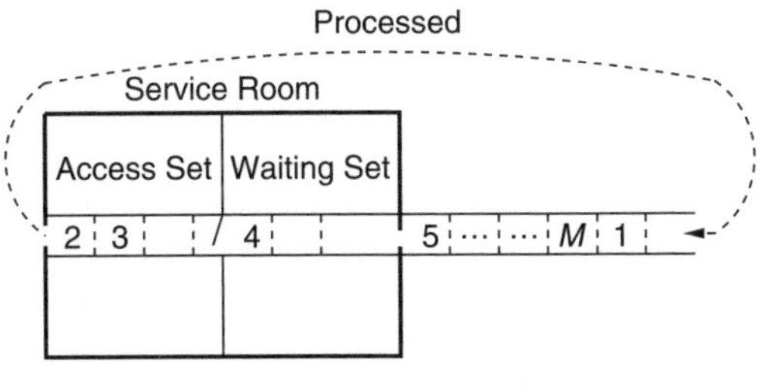

FIGURE 24.7 The basic structure of the MQSR protocol.

At the end of slot $t-1$, the central controller first removes, from the access set, users who are processed in slot $t-1$. It then chooses the access set for slot t by specifying the size $K(t)$ $(1 \leq K(t+1) \leq M)$ of the access set. If $K(t)$ is smaller than the number of users currently inside the access set, the last users in the access set are pushed back to the waiting set. If, on the other hand, $K(t)$ is larger than the number of users currently inside the access set, users in the waiting set, and users outside the service room if necessary, will join the access set to access the channel in slot t.

With this structure, the only parameter to be designed is $K(t)$, the size of the access set for slot t. The criterion here is to choose $K(t)$ so that the throughput of slot t is maximized. This problem can be formulated as follows. Let $X_i(t)$ denote the number of packets held by the ith user that may be transmitted in slot t. Let $S(t)$ be the number of successfully transmitted packets in slot t. We then have

$$K(t) = \arg \max_{k=1,\ldots,M} E[S(t) \mid K(t) = k, I_{[1,t-1]}] \tag{24.41}$$

$$= \arg \max_{k=1,\ldots,M} \sum_{n=1}^{k} C_n P\left[\sum_{i=1}^{k} X_i(t) = n \mid I_{[1,t-1]}\right] \tag{24.42}$$

where $I_{[1,t-1]}$ denotes the available information at the beginning of slot t, which includes the initial condition of the network in the form of the distribution of $X_i(1)$ $(i = 1, \ldots, M)$, the access sets for slot 1 to $t-1$, and the channel outcome for slot 1 to $t-1$.

In order to compute $P[\sum_{i=1}^{k} X_i(t) = n \mid I_{[1,t-1]}]$ for all possible k, we need the joint distribution of $X_i(t)$ $(i = 1, \ldots, M)$, which is updated at the beginning of slot t based on the joint distribution of $X_i(t-1)$ $(i = 1, \ldots, M)$ by incorporating the information we obtained from the outcome of slot $t-1$. Detailed derivation of the MQSR protocol can be found in [37].

24.5.3 The Dynamic Queue Protocol

The MQSR protocol optimally exploits all available information. It maximizes per-slot throughput under a set of heterogeneous delay constraints. The difficulty of the MQSR protocol, however, lies in its computational complexity, which grows exponentially with the number of users in the network. In this section, we present the dynamic queue protocol that achieves a performance comparable to that of the MQSR protocol with an on-line implementation as simple as that of slotted ALOHA.

Observe that the computational complexity of the MQSR protocol mainly comes from the update of joint distribution of all users' states at the beginning of each slot. In the dynamic queue protocol, the knowledge on the states of users is only updated periodically to reduce complexity. To prevent severe performance loss, the dynamic queue protocol ensures that the state of every user remains unchanged between two consecutive updates.

The basic idea of the dynamic queue protocol is to divide the time axis into transmission periods (TPs) where the ith TP is dedicated to the transmission of the packets generated in the $(i-1)$th TP. We assume that besides the packet waiting for transmission in the current TP, each user can hold at most one packet newly generated in the current TP and to be transmitted in the next one. Thus, in each TP, each user has at most one packet to transmit. Let q_i denote the probability that a user has a packet to transmit in the ith $(i \geq 1)$ TP. Recall that p denotes the probability that a user generates a packet within one time slot. We have

$$q_i = 1 - (1-p)^{L_{i-1}} \tag{24.43}$$

where L_{i-1} $(i \geq 2)$ denotes the length of the $(i-1)$th TP defined as the number of slots it contains. Thus, q_i carries our knowledge on the state of each user at the beginning of the ith TP.

At the beginning of the ith TP, all M users are waiting in a queue for the transmission of their packets. Based on q_i and the channel MPR capability, the size N_i of the access set that contains users who can access the channel simultaneously in the ith TP is chosen. Then the first N_i users in the queue are enabled to access the channel in the first slot of the ith TP. At the end of this slot, the central controller detects whether

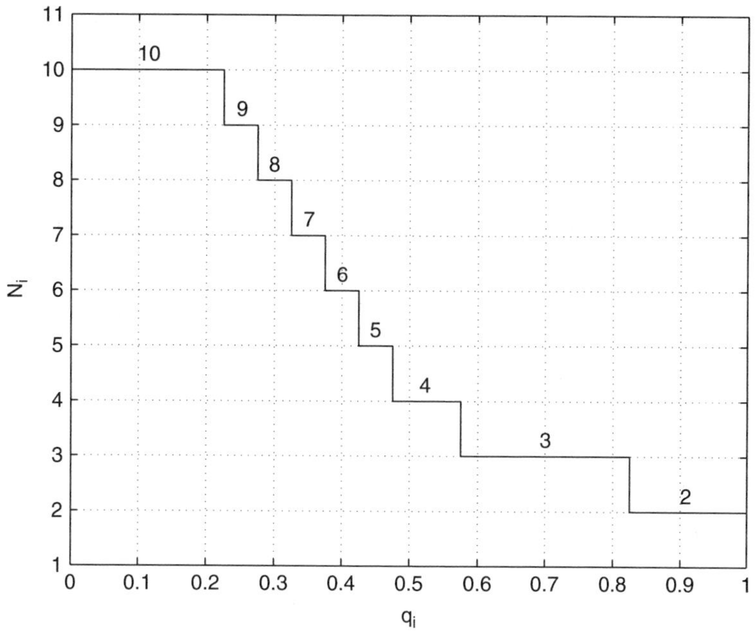

FIGURE 24.8 The optimal size of access set for the dynamic queue protocol.

this slot is empty. If it is empty, all these N_i users are processed and the next N_i users in the queue are enabled in the next slot. On the other hand, if this slot is not empty and k ($k \geq 0$) packets are successfully received, the sources of these k packets are processed; the remaining $N_i - k$ users along with the next k users in the queue are enabled to access the channel in the next slot. This procedure continues until all M users are processed.

With this structure, the only parameter to be designed is N_i, the size of the access set for the ith ($i \geq 1$) TP. The criterion of choosing N_i is to minimize the expected length of the ith TP. Specifically,

$$N_i = \arg \min_{N=1,\ldots,M} E[L_i \mid N] \tag{24.44}$$

where $E[L_i \mid N]$ is the expected length of the ith TP when the size of the access set is chosen to be N. It is shown [35] that $E[L_i \mid N]$ can be computed as the absorbing time of a finite-state discrete-time Markov chain where the state at the beginning of a slot is defined as the number of unprocessed users along with the number of packets that will be transmitted in this slot. The size N_i of the access set is then chosen as the one that minimizes $E[L_i \mid N]$. Furthermore, the optimal size of the access set can be computed off-line. By varying q_i from 0 to 1, we can construct a table that specifies the interval of q_i on which a size N ($N = 1, \ldots, M$) of the access set is optimal (a typical lookup table is illustrated in Figure 24.8). Thus, when the network starts, the optimal size of the access set for each TP can be obtained by looking up this table; little on-line computation is required to implement the dynamic queue protocol.

A steady-state performance analysis of the dynamic queue protocol can be found in [35].

24.5.4 Further Remarks

MAC protocol design for the conventional collision channel has a long history. Excellent surveys on this topic can be found in [9, 10]. Surprisingly, the problem of designing MAC protocols explicitly for MPR channels has not been attacked until recently. Besides the two protocols presented in this section, the opportunistic ALOHA proposed in [2, 30] is another MAC scheme designed for an MPR physical layer. The novelty of opportunistic ALOHA lies in the use of channel state information characterized by the propagation channel gain. It is shown in [2, 30] that if each node chooses optimally its transmission probability based on the quality of the channel, an asymptotic throughput no smaller than the spreading gain of the system can be achieved with arbitrarily small power from each node. This makes the protocol particularly attractive for networks with stringent energy constraint, for example, wireless sensor networks.

24.6 Approach High Performance from Both Physical and MAC Layers

In this section, we demonstrate the network performance gain resulting from a joint design of the physical and MAC layers. At the physical layer, we employ the semiblind LSS receiver that provides MPR capability. At the MAC layer, we use the MQSR and dynamic queue protocols that fully exploit the MPR capability at the physical layer.

We consider a multiaccess network with 10 users transmitting to a central controller. The channel durations of transmitted packets take, independently, three possible values with equal probability. The MPR capability provided by the LSS receiver with one known symbol can be obtained from Theorem 24.3. The capacity of this MPR channel is 4/3, which can be achieved by transmitting $n_0 = 2$ packets in each slot.

We first construct the lookup table for the dynamic queue protocol that specifies the q_i intervals in which a possible size (from 1 to 10) of access set is optimal. As shown in Figure 24.8, the result demonstrates clearly the trend that the heavier the traffic is (larger q_i), the smaller the access set should be, as intuition suggests. Note that the optimal size of access set equals n_0, which is greater than 1 at the heaviest traffic load ($q_i = 1$), indicating that contention is preferable at any traffic load for this MPR channel.

Shown in Figure 24.9 is the throughput performance at various packet generation rates p. Compare the performance of the MQSR protocol with that of the slotted ALOHA with optimal retransmission probability in the noiseless collision channel (without MPR). The throughput gain from 0.4 to 1 is achieved by the optimal protocol at the MAC layer. Compare the performance of the MQSR protocol in the noiseless collision channel with that in the MPR channel using the semiblind LSS receiver. The throughput gain

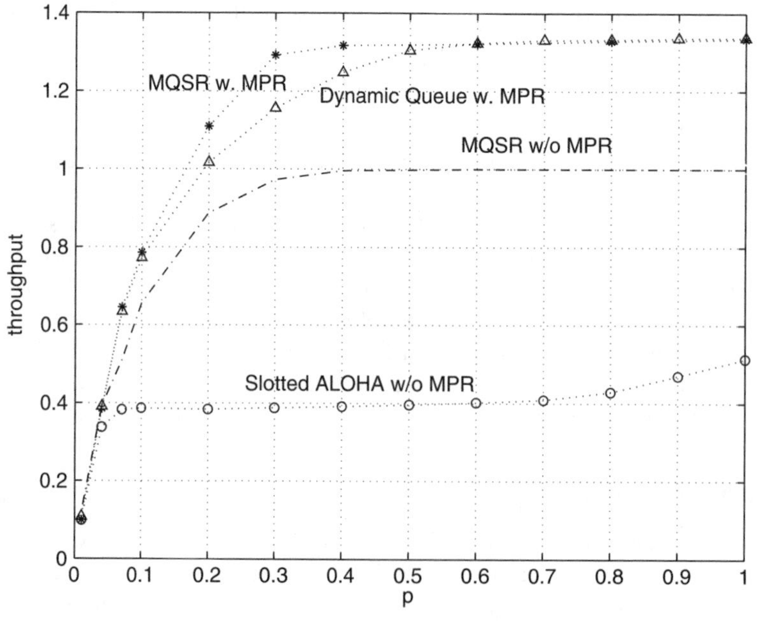

FIGURE 24.9 High-performance approaches from both physical and MAC layers.

from 1 to 4/3 is provided by the signal processing technique at the physical layer. By introducing MPR at the physical layer and employing optimal protocols at the MAC layer, a more than 200% gain is achieved, compared to slotted ALOHA in the conventional collision channel.

From Figure 24.9 we see that the MQSR protocol achieves the channel capacity at heavy traffic load, as proved in [37]. Figure 24.9 also demonstrates that the dynamic queue protocol, with an on-line implementation as simple as that of ALOHA, achieves a performance comparable to that of the optimal MQSR protocol.

24.7 Conclusion

The objective of this chapter is to explore the idea of cross-layer design of wireless networks, which is motivated by the need to provide a greater level of adaptivity to variations of wireless channels. We have examined one aspect of the interaction between the physical and MAC layers. In particular, we presented signal processing techniques that provide MPR capability at the physical layer and studied their impact on the performance and design of the MAC layer. Significant performance gain can be achieved by a joint design of these two layers.

References

[1] N. Abramson. The Throughput of Packet Broadcasting Channels. *IEEE Trans. Commun.*, 25(1): 117–128, January 1977.

[2] S. Adireddy and L. Tong. Exploiting Decentralized Channel State Information for Random Access. *Submitted to IEEE Trans. Info. Theory*, November 2002. http://acsp.ece.cornell.edu/pubJ.html.

[3] D.E. Ayyildiz and H. Delic. Adaptive Random Access Algorithm with Improved Delay Performance. *Int. J. Commun. Syst.*, 14:531–539, 2001.

[4] Q. Bao and L. Tong. A Performance Comparison between Ad Hoc and Centrally Controlled CDMA Wireless LANs. *IEEE Trans. Wireless Commun.*, 1(1), Jan 2001.

[5] D.P. Bertsekas and R. Gallager. *Data Networks*. Prentice Hall, New York, 1992.

[6] J.I. Capetanakis. Generalized TDMA: The Multi-Accessing Tree Protocol. *IEEE Trans. Commun.*, 27(10):1476–1484, Oct. 1979.

[7] I. Chlamtac and A. Farago. An Optimal Channel Access Protocol with Multiple Reception Capacity. *IEEE Trans. Computers*, 43(4):480–484, April 1994.

[8] G. del Angel and T.L. Fine. Randomized Power Control Strategies for Optimization of Multiple Access Radio Systems. In *Proc. 38th Allerton Conference on Communication, Control and Computing*, October 2000.

[9] Anthony Ephremides and Bruce Hajek. Information Theory and Communication Networks: An Unconsummated Union. *IEEE Trans. Inform. Theory*, 44(6):2416–2434, October 1998.

[10] R.G. Gallager. A Perspective on Multiaccess Channels. *IEEE Trans. Information Theory*, IT-31(2): 124–142, March 1985.

[11] S. Ghez, S. Verdú, and S. Schwartz. Stability Properties of Slotted Aloha with Multipacket Reception Capability. *IEEE Trans. Automatic Control*, 33(7):640–649, July 1988.

[12] S. Ghez, S. Verdú, and S. Schwartz. Optimal Decentralized Control in the Random Access Multipacket Channel. *IEEE Trans. Automatic Control*, 34(11):1153–1163, Nov. 1989.

[13] G. Giannakis, Y. Hua, P. Stoica, and L. Tong. *Signal Processing Advances in Wireless Communications: Vol. I, Trends in Channel Estimation and Equalization, Vol. II, Trends in Single- and Multi-User Systems*. PTR Prentice-Hall, Englewood Cliffs, NJ, 2001.

[14] M.G. Hluchyj. "Multiple Access Window Protocol: Analysis for Large Finite Populations." In *Proc. IEEE Conf. on Decision and Control*, New York, 1982, pp. 589–595.

[15] M.G. Hluchyj and R.G. Gallager. "Multiaccess of A Slotted Channel by Finitely Many Users." In *Proc. Nat. Telecomm. Conf.*, New Orleans, LA., Aug. 1981, pp. D4.2.1–D4.2.7.

[16] S. Kim and J. Yeo. Optimal Scheduling in CDMA Packet Radio Networks. *Computers and Operations Research*, 25:219–227, March 1998.

[17] L. Kleinrock and Y. Yemini. An Optimal Adaptive Scheme for Multiple Access Broadcast Communication. In *Proc. International Conference on Communications*, pp. 7.2.1–7.2.5, June 1978.

[18] H. Liu and G. Xu. Closed-Form Blind Symbol Estimation in Digital Communications. *IEEE Trans. on Signal Processing*, 43(11):2714–2723, Nov. 1995.

[19] R. Liu and L. Tong Ed. Special Issue on Blind System Identification and Estimation. *IEEE Proceedings*, 86(10), October 1998.

[20] D.F. Lyons and P. Papantoni-Kazakos. A Window Random Access Algorithm for Environments with Capture. *IEEE Trans. on Communications*, 37(7):766–770, July 1989.

[21] V. Naware and L. Tong. Smart Antennas, Dumb Scheduling for Medium Access Control. In *Proc. CISS 2003*, Baltimore, MD, March 2003.

[22] V. Naware and L. Tong. Stability of Slotted ALOHA with Spatial Diversity. In *Proc. of Int. Conf. on Comm. 2003*, Anchorage, AL, May 2003.

[23] M. Paterakis and P. Papantoni-Kazakos. A Simple Window Random Access Algorithm with Advantageous Properties. *IEEE Trans. Information Theory*, 35(5):1124–1130, September 1989.

[24] F. Shad, T. D. Todd, V. Kezys, and J. Litva. Dynamic Slot Allocation (DSA) in Indoor SDMA/TDMA Using a Smart antenna base station. *IEEE Trans. Networking*, 9:69–81, 2001.

[25] M. Sidi and I. Cidon. Splitting Protocols in Presence of Capture. *IEEE Trans. Info. Theory*, IT-31: 295–301, March 1985.

[26] J.R. Treichler and B.G. Agee. A New Approach to Multipath Correction of Constant Modulus Signals. *IEEE Trans. ASSP.*, ASSP-31(2):459–472, April 1983.

[27] A. J. van der Veen. Algebraic Methods for Deterministic Blind Beamforming. *IEEE Proc.*, 86(10): 1987–2008, Oct 1998.

[28] A. J. van der Veen, S. Talwar, and A. Paulraj. A Subspace Approach to Blind Space-Time Signal Processing for Wireless Communication Systems. *IEEE Trans. Signal Processing*, SP-45(1):173–190, Jan 1997.

[29] C. Vanderplas and J.P.M. Linnartz. Stability of Mobile Slotted ALOHA Network with Rayleigh Fading, Shadowing, and Near-far Effect. *IEEE Trans. Veh. Tech.*, 39:359–366, November 1990.

[30] P. Venkitasubramaniam, S. Adireddy, and L.Tong. Sensor Networks with Mobile Agents: Optimal Random Access and Coding. *submitted to IEEE Journal on Sel. Areas in Comm.: Special Issue on Sensor Networks*, July 2003.

[31] S. Verdú. *Multiuser Detection*. Cambridge University Press, Cambridge, UK, 1998.

[32] B. Yucel and H. Delic. Mobile Radio Window Random Access Algorithm with Diversity. *IEEE Trans. Vehicular Technology*, 49(6):2060–2070, November 2000.

[33] Q. Zhao. Multipacket Reception in Wireless Communication Networks. Ph.D. dissertation, Cornell University, Ithaca, NY, 2001.

[34] Q. Zhao, J. Bao, and L. Tong. Signal processing based collision resolution in slotted Aloha wireless ad hoc networks. *Signal Processing Advances in Wireless Communications, Vol. II*, Giannakis et. al., Eds., Prentice Hall, New York, 2001, pp. 315–354.

[35] Q. Zhao and L. Tong. A Dynamic Queue Protocol for Multiaccess Wireless Networks with Multipacket Reception. *IEEE Transactions on Wireless Communications*, 2003, http://people.ece.cornell.edu/ltong/.

[36] Q. Zhao and L. Tong. Semi-Blind Collision Resolution in Random Access Ad Hoc Wireless Networks. *IEEE Trans. Signal Processing*, 48(9), Sept. 2000.

[37] Q. Zhao and L. Tong. A Multiqueue Service Room MAC Protocol for Wireless Networks with Multipacket Reception. *IEEE/ACM Trans. on Networking*, 11(1), Feb. 2003.

[38] M. Zorzi. Mobile Radio Slotted ALOHA with Capture and Diversity. *Wireless Networks*, 1:227–239, May 1995.

X

Emerging Techniques and Applications

25 Time–Frequency Signal Processing for Wireless Communications
Boualem Boashash, A. Belouchrani, Karim Abed-Meraim, and Nguyen Linh-Trung **25-1**
Introduction • Time–Frequency Signal Processing Tools • Spread-Spectrum Communications Systems Using TFSP • Time–Frequency Array Signal Processing • Other TFSP Applications in Wireless Communications • Conclusion

26 Monte Carlo Signal Processing for Digital Communications: Principles and Applications *Xiaodong Wang* **26-1**
Introduction • MCMC Methods • SMC Methods • Concluding Remarks

27 Principles of Chaos Communications *Andreas Abel and Wolfgang Schwarz* **27-1**
What Is Chaos? • Communication: Requirements and Resources • Chaos in Communications • Communication Using Broadband Chaotic Carriers • Chaos for Spreading Code Generation • Chaotic vs. Classical Communications

28 Adaptation Techniques and Enabling Parameter Estimation Algorithms for Wireless Communications Systems *Hüseyin Arslan* **28-1**
Introduction • Overview of Adaptation Schemes • Parameter Measurements • Applications of Adaptive Algorithms: Case Studies • Future Research for Adaptation • Conclusion

25

Time–Frequency Signal Processing for Wireless Communications

Boualem Boashash
Queensland University of Technology

A. Belouchrani
Ecole Nationale Polytechnique

Karim Abed-Meraim
Telecom

Nguyen Linh-Trung
Queensland University of Technology

25.1 Introduction 25-2
25.2 Time–Frequency Signal Processing Tools 25-3
Limitations of Traditional Signal Representations • Joint Time–Frequency Representations • Quadratic Time–Frequency Distributions
25.3 Spread-Spectrum Communications Systems Using TFSP 25-13
Channel Modeling and Identification • Interference Mitigation
25.4 Time–Frequency Array Signal Processing 25-22
The Spatial Time–Frequency Distributions • STFD Structure in Wireless Communications • Advantages of STFDs over Covariance Matrix • Selection of Autoterms and Cross-Terms in the Time–Frequency Domain • Time–Frequency Direction-of-Arrival Estimation • Time–Frequency Source Separation
25.5 Other TFSP Applications in Wireless Communications 25-32
Precoding for LTV Channels • Signaling Using Chirp Modulation • Detection of FM Signals in Rayleigh Fading • Mobile Velocity/Doppler Estimation Using Time–Frequency Processing
25.6 Conclusion 25-35

Abstract

This chapter is intended to relate recent advances in the field of time–frequency signal processing (TFSP) to the need for further capacity of wireless communications systems. It first presents, in a brief and heuristic approach, the fundamentals of TFSP. It then describes the TFSP-based methodologies that are used in wireless communications with special emphasis on spread-spectrum techniques and time–frequency array processing. Topics discussed include channel modeling and identification, estimation of scattering function, interference mitigation, direction of arrival estimation, time–frequency MUSIC, and time–frequency source separation. Finally, other emerging applications of TFSP to wireless communications are discussed.

0-8493-1657-X/$0.00+$1.50

25.1 Introduction

Wireless communications is growing at an explosive rate, stimulated by a host of important emerging applications ranging from third-generation mobile telephony, wireless personal communications, and wireless subscriber loops, to radio and infrared indoor communications, nomadic computing, and wireless tactical military communications. Signal processing has played a key role in providing solutions to key problems encountered in communications in general, and in wireless communications in particular [1]. An important branch of signal processing called time–frequency signal processing (TFSP) has emerged over the past decades [2, 3]. It provides effective tools for analyzing nonstationary signals where the frequency content of signals varies in time, as well as for analyzing linear time-varying systems. The purpose of this chapter is to review the methodologies of TFSP applied to wireless communications.

Fundamental issues in wireless communications include the problems of *interference mitigation in CDMA* (code-division multiple access) or multicarrier CDMA (MC-CDMA), and *array processing* for source localization and signal separation.

Along with channel fading, there exist many types (inherent and noninherent) of interference causing degradation in the system performance, hence reducing the system capacity. In addition to the high capacity resulting from the smart way of providing multiple access, CDMA has shown advantages over many other multiple-access schemes in terms of reducing the effects of interference and multipathfading [13]. CDMA has in fact been selected to be the basic building block for third-generation wireless communications (wideband CDMA) [14], and the promising candidate for the fourth generation (generalized multicarrier CDMA) [15]. However, to achieve the best system performance, it is still very crucial to minimize the effects of various types of interference in CDMA systems, namely, narrowband interference (NBI) [16], multiple-access interference (MAI) [17], and, for high-data-rate applications, intersymbol interference (ISI).

When multiple sensors are available at the receiver side, array processing techniques can be used to achieve source (mobile) localization or source separation. Source localization is of great importance in radar/sonar applications but also in wireless mobile communications. Mobile localization can add a number of important services to the capabilities of cellular systems, including help for mobile navigation, emergency services (also known as E-911 problem), location-sensitive billing, fraud protection, and person/asset tracking [74].

The problem of source separation or blind source separation (BSS) arises when considering MIMO (multi-input multi-output) systems where BSS techniques are used to solve the MAI problem and to extract the desired information for each user.

On the other hand, received signals are generally nonstationary due to source signal's nonstationaries or to the time-varying nature of propagation channels. Indeed, the wireless communication environment exhibits a multipath propagation phenomenon with Doppler effect, where the received signal is not only affected by additive Gaussian noise but also by a sum of attenuated, delayed, and Doppler-shifted versions of the transmitted signal [5, 6]. As a result, the received signals are affected in strength and shape, depending on different environments (indoor, outdoor, urban, suburban), speed of mobile agent or surrounding movements, and signaling (bandwidth, data rate, modulation, and carrier frequency). The transmitted signals undergo serious fading through the propagation channel. Especially in wideband wireless communications, the underlying channel exhibits a random linear time-varying (LTV) characteristic and is most commonly assumed to be a wide-sense stationary uncorrelated scattering (WSSUS) process [9–12].

All these wireless communication problems involve a time-varying context, and therefore the use of TFSP techniques should lead to improvements in system performance.

The focus of this chapter is to review such TFSP techniques and clarify the methodologies of TFSP that are most relevant for use in wireless communications. The structure of the chapter is as follows: Section 25.2 briefly introduces the basics of TFSP in order to provide a better insight for the review in the parts that follow. Section 25.3 reviews some applications of TFSP techniques in spread-spectrum communication systems. Section 25.4 describes array signal processing techniques using time–frequency distributions (TFDs). And finally, Section 25.5 briefly reviews some other issues encountered in wireless communications where the TFSP techniques play a central role.

25.2 Time–Frequency Signal Processing Tools

TFSP is a relatively new field comprised of signal processing methods, techniques, and algorithms in which the two natural variables time, t, and frequency, f, are used *concurrently* in contrast with the traditional signal processing methods where time and frequency variables are used exclusively and independently. The observation of natural phenomena indicates that these two variables, t and f, are usually simultaneously present in signals (e.g., natural frequency-modulated (FM) signals such as the song of some birds). Such signals are called nonstationary because their spectral characteristics vary with time [21]. TFSP is designed to deal effectively with such signals by allowing their detailed and precise analysis and processing. It also enables the design and synthesis of signals and systems with specific time and frequency characteristics, suitable to applications such as wireless communications.

25.2.1 Limitations of Traditional Signal Representations

The spectrum of a signal (deterministic or random) gives no indication as to how the frequency content of a signal changes with time, information that is important when one deals with a large class of nonstationary signals such as FM signals. This frequency variation often contains critical information about the signal and the process inherent to real-life applications.[1]

The limitation of 'classical' spectral representations is better illustrated by the fact that we can find totally different signals (related to different physical phenomena), $s_a(t)$ and $s_b(t)$, which yet have the same spectra (that is, magnitude spectra) — see Figure 25.2. Representing signals in a way that is useful for precise characterization and identification serves as a part of the motivation for devising a more sophisticated and practical nonstationary signal analysis tool, which preserves all the information of the signal, and therefore discriminates signals in a better way, using one single complete representation instead of attempting to interpret magnitude and phase spectra separately.

25.2.2 Joint Time–Frequency Representations

25.2.2.1 Finding Hidden Information Using Time–Frequency Representations

Revealing the time and frequency dependence of a signal, such as a linear FM (LFM) signal, is achieved by using a joint time–frequency representation such as the one shown in Figure 25.1.[2]

In this representation, the start and stop time instants are clearly identifiable, as is the time variation of the frequency content of the signal described by the linear pattern of peaks. (This information cannot be retrieved solely from either the instantaneous power $|s(t)|^2$ or the spectrum representations $|S(f)|^2$. It is lost when the Fourier transform is squared in modulus and the phase of the spectrum is thereby discarded.) The spectrum phase contains the actual information about the internal organization of the signal, such as details of time instants at which the signal has energy above or below a particular threshold, and the order of appearance in time of the different frequencies present in the signal. The difficulty of interpreting and analyzing a phase spectrum makes the concept of a joint time–frequency signal representation attractive.

Example 25.1

Figure 25.2 illustrates another example of two slightly different signals with the same spectrum that could not be properly analyzed without a joint time–frequency representation. Both signals contain three LFMs whose start and stop times are different. The differences are not shown easily in the t or f domain, but

---

[1]This information is encoded in the phase spectrum. However, it generally is not used, as it is difficult to interpret and analyze a phase spectrum.

[2]In this example the Wigner–Ville distribution (WVD) is used, as it provides the optimal joint time–frequency distribution (TFD) for an LFM signal [21].

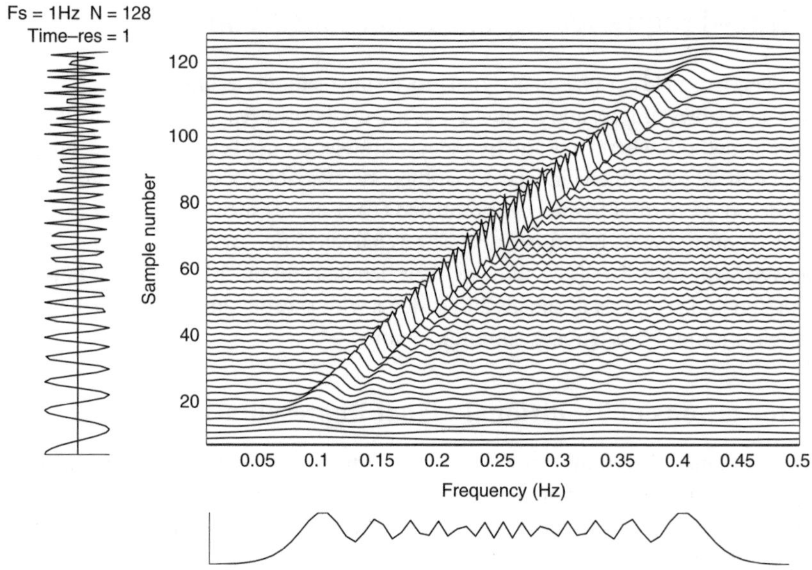

FIGURE 25.1 Time–frequency representation of a linear FM signal: the signal's time domain representation appears on the left, and its spectrum on the bottom.

appear clearly in the $t - f$ representation, which allows precise and simultaneous measurements of actual frequencies and their epochs. The B distribution (BD) [18] is used to represent two signals in the $t - f$ domain.

Figure 25.2 also indicates another significant reason to use joint time–frequency representations of signals: it reveals whether the signal is monocomponent or multicomponent, a fact that cannot be revealed by conventional spectral analysis, especially when individual components are also time varying, such as the six chirps in the figure.

25.2.2.2 What Is a Time–Frequency Representation?

TFSP is a natural extension of both time domain and frequency domain processing that involves representing signals in a complete space that can display all the signal information in a more accessible way [2]. Such a representation is intended to provide a *distribution* of signal energy, $E(t, f)$, vs. both time and frequency simultaneously. For this reason, the representation is commonly called a TFD.

A concept intimately related to joint time–frequency representation is that of instantaneous frequency (IF) and time delay (TD). The IF corresponds to the frequency of a sine wave that locally (at a given time) fits the signal under analysis. The TD is a measure of the order of arrival of the frequencies.

The TFD is expected to visually exhibit in the $t - f$ domain the time–frequency law of each signal component, thereby making the estimation of the IF, $f_i(t)$, and TD, $\tau_d(f)$, easier, and could also provide additional information about relative component amplitudes, and the spectral spread of the component around the IF (the spread is known as the instantaneous bandwidth, $B_i(t)$).

The TFD is often expected to satisfy a certain number of properties that are intuitively desirable for a practical analysis. Let us denote by $\rho_z(t, f)$ a TFD that is a time–frequency representation of signal $z(t)$. $\rho_z(t, f)$ is expected to satisfy the following properties:

- P1a: The TFD should be real and its integration over the whole time–frequency domain results in the total signal energy E_z:

$$\int_{-\infty}^{\infty}\int_{-\infty}^{\infty} \rho_z(t, f)dt\, df = E_z \tag{25.1}$$

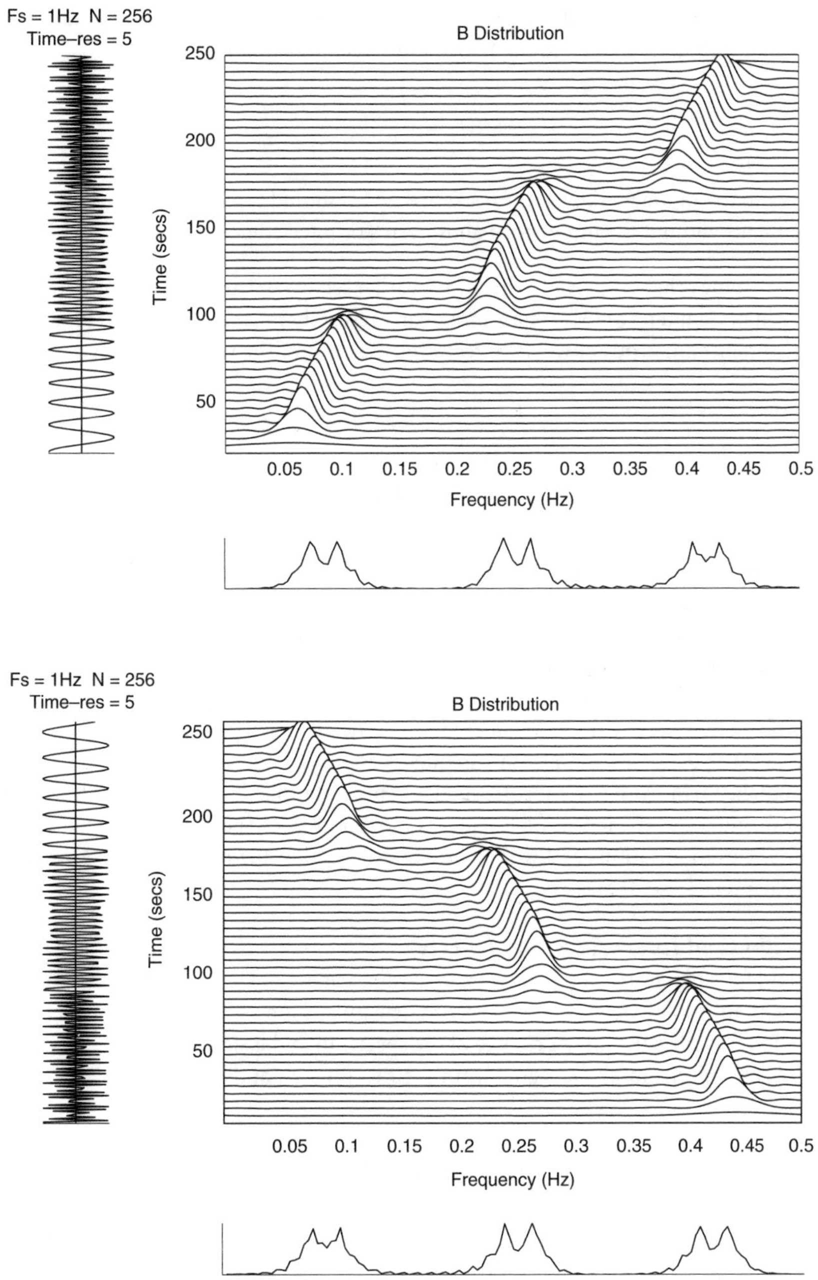

FIGURE 25.2 Time–frequency representation of two different three-component signals (using B distribution).

- P1b: It would also be desirable that the energy in a certain region R in the $t-f$ plane, E_{z_R}, be expressible as in Equation 25.1, but with limits of integration restricted to the boundaries $(\Delta t, \Delta f)$ of the region R:

$$E_{z_R} = \int_{\Delta t} \int_{\Delta f} \rho_z(t, f) dt\, df \tag{25.2}$$

which is a portion of signal energy in the band Δf and time interval Δt.

- P2: The peak of the TFD and the first moment of the time–frequency representation with respect to frequency should be equal to the IF of a monocomponent signal:

$$f_i(t) = \frac{\int_{-\infty}^{\infty} f\rho_z(t, f)df}{\int_{-\infty}^{\infty} \rho_z(t, f)df} \tag{25.3}$$

- P3: For multicomponent signals, the peaks of the TFD should exhibit the various IF laws of individual components without the nuisance of ghost terms or interferences.

There are also some other properties, which were earlier seen as strictly needed, but were found later not to be, as detailed in Chapter 3 of [3]. For example, early researchers indicated that a TFD should reduce to the spectrum and instantaneous power by integrating over one of the variables, so that

$$\int_{-\infty}^{\infty} \rho_z(t, f)dt = |S_z(f)|^2 \tag{25.4}$$

$$\int_{-\infty}^{\infty} \rho_z(t, f)df = |z(t)|^2 \tag{25.5}$$

These two conditions are often called marginal conditions. However, it was shown that some high resolution TFDs could be generated that did not meet the marginals [3].

Another property that was originally seen as desirable is positivity, but it was shown to be incompatible with P2 [21], which is more important in practice.

A number of questions arise from the above: Can we design a TFD that meets the specifications listed above? If yes, how can we do it? What are the significant signal characteristics and parameters that will impact the construction of a joint time–frequency representation? How do these relate to the TFD? How do we obtain them from the time–frequency representation? The answers to these questions are important in formulating efficient time–frequency methodologies specifically adapted for applications such as wireless communications and will be briefly discussed in the next sections. More details can be found in [3, Chapters 2 and 3] and [19].

25.2.2.2.1 Physical Interpretation of TFDs

Most TFDs used for a practical time–frequency signal analysis are not necessarily positive-definite, so they do not represent an instantaneous energy spectral density at time t and frequency f. For example, the Page distribution describes the rate of change of the spectrum and is defined as the time derivative of the running spectrum (spectrum of the signal from $-\infty$ to time t), and hence can take both positive and negative values:

$$\rho_{z_P}(t, f) = \frac{\partial}{\partial t}\left|\int_{-\infty}^{t} z(u)e^{-j2\pi f u}du\right|^2$$

To relate $\rho_z(t, f)$ to the physical quantities used in practical experimentation, we can interpret $\rho_z(t, f)$ as a measure of energy flow through the spectral window $(f - \Delta f/2, f + \Delta f/2)$ during the time interval $(t - \Delta t/2, t + \Delta t/2)$. The signal energy localized in this time–frequency domain, $(\Delta t, \Delta f)$, is then given by

$$E_{\Delta t,\Delta f} = \int_{t-\Delta t/2}^{t+\Delta t/2} \int_{f-\Delta f/2}^{f+\Delta f/2} \rho_z(t, f)dt\, df \tag{25.6}$$

The larger the window, the more likely $E_{\Delta t,\Delta f}$ will correspond to a true measure of physical energy. The window should be chosen large enough so that $\Delta t\Delta f$ satisfies Heisenberg's uncertainty relation [2]: $\Delta t\Delta f \geq 1/(4\pi)$.

25.2.2.2.2 Instantaneous Frequency and Group Delay

The IF, $f_i(t)$, of a monocomponent signal is a measure of the localization in time of the individual frequency components of the signal [2].

The IF, $f_i(t)$, of a monocomponent *analytic* signal[3] $z(t) = a(t)e^{j\phi(t)}$ is given by

$$f_i(t) = \frac{1}{2\pi}\frac{d\phi(t)}{dt} \tag{25.7}$$

The IF of a monocomponent *real* signal $s(t) = a(t)\cos\phi(t)$ is defined as the IF of the analytic signal $z(t)$ corresponding to $s(t)$. The expressions of the IF given above do not apply directly to multicomponent signals. For such signals, the expression of the IF needs to be applied to its individual components to have a meaningful physical interpretation.

The twin of the IF concept in the frequency domain is called the time delay, $\tau_d(f)$.

The TD of a monocomponent *analytical* signal $z(t)$ is defined as

$$\tau_d(f) = -\frac{1}{2\pi}\frac{d\theta(f)}{df} \tag{25.8}$$

where

$$\mathcal{F}\{z(t)\} = Z(f) = A(f)e^{j\theta(f)} \tag{25.9}$$

$\mathcal{F}\{\cdot\}$ stands for Fourier transform (FT). Equations 25.7 and 25.8 are similar except for the minus sign, in the same way that the FT and inverse FT (IFT) are similar except for a minus sign in the exponent of the basis function $e^{j2\pi ft}$.

The TD of a monocomponent *real* signal $s(t)$ is defined as the TD of the analytic signal $z(t)$ corresponding to $s(t)$.

The order of appearance of each time-varying frequency component is the TD. The global order of appearance of the frequencies is called the group delay (a mean value of individual TDs).

The IF and TD describe the internal organization of the signal. Neglecting this information would result in a lack of precision of the information characterizing the signal, as illustrated in [3, p. 7, Figure 1.1.2].

25.2.3 Quadratic Time–Frequency Distributions

25.2.3.1 Time-Varying Spectrum and the Wigner–Ville Distribution

To determine why time information appears to be lost when we take the power spectral distribution (PSD) and how it can be recovered, let us consider a complex random process $Z(t)$ with realizations $z(t,\epsilon)$, where ϵ represents the ensemble index identifying each realization. To improve clarity, we simply replace $z(t,\epsilon)$ by $z(t)$. (ϵ is implicit when we say that $z(t)$ is random.)

The autocorrelation function of $z(t)$ may be defined in symmetric form as

$$\begin{aligned} R_z(t,\tau) &= E\{z(t+\tau/2)z^*(t-\tau/2)\} \\ &= E\{K_z(t,\tau)\} \end{aligned} \tag{25.10}$$

where $K_z(t,\tau) = z(t+\tau/2)z^*(t-\tau/2)$ is the signal kernel and $E\{\cdot\}$ denotes the expected value operator.

The Wiener–Khintchine theorem states that for a stationary signal, the signal power spectral density equals the FT of its autocorrelation function.

[3]The analytic signal $z(t)$ associated with the real signal $x(t)$ is defined as

$$z(t) = x(t) + j\mathcal{H}\{x(t)\}$$

where $\mathcal{H}\{\ \}$ represents the Hilbert transform [21].

By extension to TD random signals, the time-varying PSD, $S_z(t, f)$, is defined as the FT of the time-varying autocorrelation function, $R_z(t, \tau)$, i.e.,

$$\begin{aligned} S_z(t, f) &= \underset{\tau \to f}{\mathcal{F}} \quad [R(t, \tau)] \\ &= \underset{\tau \to f}{\mathcal{F}} \quad E\{K_z(t, \tau)\} \end{aligned} \tag{25.11}$$

$$\begin{aligned} &= E\left\{ \underset{\tau \to f}{\mathcal{F}} \quad [K_z(t, \tau)] \right\} \\ &= E\{W_z(t, f)\} \end{aligned} \tag{25.12}$$

where the interchange of E and FT is made under the assumptions verified by the class of bounded amplitude finite-duration signals [3], and $W_z(t, f)$ is referred to as the Wigner–Ville distribution (WVD), expressed as

$$\begin{aligned} W_z(t, f) &= \underset{\tau \to f}{\mathcal{F}} [K_z(t, \tau)] \\ &= \underset{\tau \to f}{\mathcal{F}} [z(t + \tau/2) z^*(t - \tau/2)] \end{aligned} \tag{25.13}$$

The problem of estimating the time-varying PSD of a random process $z(t)$ is thus one of averaging the WVD of the process over ϵ. If only one realization of the signal is available, assuming the signal is locally ergodic[4] over a window, an estimate of the time-varying PSD is obtained by smoothing the WVD over the window of local ergodicity [21].

25.2.3.2 Time-Varying Spectrum Estimates and Quadratic TFDs

If $z(t)$ is deterministic, from Equation 25.12, $S_z(t, f)$ reduces to $W_z(t, f)$, i.e.,

$$S_z(t, f) = E\{W_z(t, f)\} = W_z(t, f) \tag{25.14}$$

- The signals we consider have a finite-duration T. This fact can be expressed by introducing a finite-duration time window $g_1(t)$, hence convolving $S_z(t, f)$ in the frequency direction with $G_1(f) = \mathcal{F}[g_1(t)]$.[5]
- In practice, signals also have finite bandwidth restrictions. This introduces a frequency window $G_2(f)$, convolving $S_z(t, f)$ in the time direction with $g_2(t)$ (the IFT of $G_2(f)$: $g_2(t) = \mathcal{F}^{-1}\{G_2(f)\}$).
- By combining the separable windowing effects of $g_1(t)$ in time and $G_2(f)$ in frequency, the above leads to

$$\hat{S}_z(t, f) = W_z(t, f) \star_f G_1(f) \star_t g_2(t)$$

  where $\star_t$ and $\star_f$ indicate convolution in time and frequency, respectively, and $G_1(f)$ and $g_2(t)$ are even functions (such as those traditionally used in spectral analysis and digital filter design).
- The above formulation was introduced step by step to illustrate the two-dimensional convolution that is inherent to signals that have nearly finite duration and bandwidth (most real-life signals). This formulation is a special case where the two-dimensional windowing in t and f is separable.
- In general, we need to introduce an even function[6] $\gamma(t, f)$ that may or may not be separable and that reflects the overall duration–bandwidth limitation in both time and frequency. This leads to

[4] In essence, a random process is ergodic if its ensemble averages equal its time averages.

[5] For simplicity, $\mathcal{F}$ instead of $\underset{t \to f}{\overset{\mathcal{F}}{\to}}$ where no ambiguity occurs.

[6] A real and even signal has a real and even FT.

the following general formulation of time–frequency representations:

$$\hat{S}_z(t, f) \equiv \rho_z(t, f) = W_z(t, f) \star_t \star_f \gamma(t, f) \tag{25.15}$$

where the double star indicates convolution in both time and frequency, $\hat{S}_z(t, f)$ is the time-varying estimate of the PSD, $W_z(t, f)$ is the WVD, and $\gamma(t, f)$ is an even function which defines the TFD and its properties.

By expanding $W_z(t, f)$ and the double convolution in Equation 25.15, we obtain the following general quadratic form:

$$\rho_z(t, f) = \int_{-\infty}^{\infty}\int_{-\infty}^{\infty}\int_{-\infty}^{\infty} e^{j2\pi\nu(u-t)} g(\nu, \tau) z\left(u + \frac{\tau}{2}\right) z^*\left(u - \frac{\tau}{2}\right) e^{-j2\pi f\tau}\, d\nu\, du\, d\tau \tag{25.16}$$

where $g(\nu, \tau)$ is the two-dimensional FT of $\gamma(t, f)$. The function $g(\nu, \tau)$ determines how the signal energy is distributed in time and frequency. It is analogous to the windows used in the classical spectral analysis. By appropriately choosing $g(\nu, \tau)$, we can obtain most of the popular time–frequency representations of signals [21].

25.2.3.3 Time, Lag, Frequency, and Doppler Domains and the Ambiguity Function

We have defined in Equation 25.13 the WVD of signal $z(t)$ as the FT of the signal kernel $K_z(t, \tau)$, i.e.,

$$W_z(t, f) = \underset{\tau \to f}{\mathcal{F}} \{K_z(t, \tau)\}$$

The time-average autocorrelation function of a deterministic signal $z(t)$ may be defined as the inverse FT of its energy spectrum $|Z(f)|^2$, i.e.,

$$R_z(\tau) = \mathcal{F}^{-1}\{|Z(f)|^2\}$$

$R_z(\tau) = \int_{-\infty}^{\infty} z(t + \frac{\tau}{2}) z^*(t - \frac{\tau}{2}) dt$ is a measure of the similarity between the signal and its time-delayed copies and is obtained by taking the integral over time of the signal kernel $K_z(t, \tau)$.

Another quantity that makes use of $K_z(t, \tau)$ is the symmetric ambiguity function (AF) [21]:

$$\begin{aligned} A_z(\nu, \tau) &= \int_{-\infty}^{\infty} K_z(t, \tau) e^{-j2\pi\nu t} dt \\ &= \int_{-\infty}^{\infty} z\left(t + \frac{\tau}{2}\right) z^*\left(t - \frac{\tau}{2}\right) e^{-j2\pi\nu t} dt \end{aligned} \tag{25.17}$$

Equation 25.17 represents the basic radar equation obtained by correlating a signal with the same signal delayed in time by lag τ and shifted in frequency by Doppler ν. The ambiguity domain also represents directly the effects of multipath-delay spread and Doppler-shift spread, which characterize the time-varying nature of wireless communication channels. It can therefore be used to analyze the behavior of signals in wireless communications and control the effects of spreading.

The key point here is the role played by $K_z(t, \tau)$ in the formulation of many important signal processing concepts. We showed that time–frequency, time-lag, and Doppler-lag representations (as well as Doppler-frequency representations) can be obtained from the signal kernel by means of FT and IFT. This is illustrated in Figure 25.3.

Figure 25.3 indicates that the two-dimensional FT of the WVD of signal $z(t)$ equals the symmetric AF $A_z(\nu, \tau)$ of $z(t)$ (as indicated by the two vertical arrows):

$$W_z(t, f) \, {}^t_f \rightleftarrows {}^\nu_\tau \, A_z(\nu, \tau)$$

This property is important in TFSP. It indicates that the twin of the time–frequency domain is the lag-Doppler domain (also called the ambiguity domain). The (t, f) domain represents the signal as a function of actual time and actual frequency, while the (ν, τ) domain represents the signal as a function

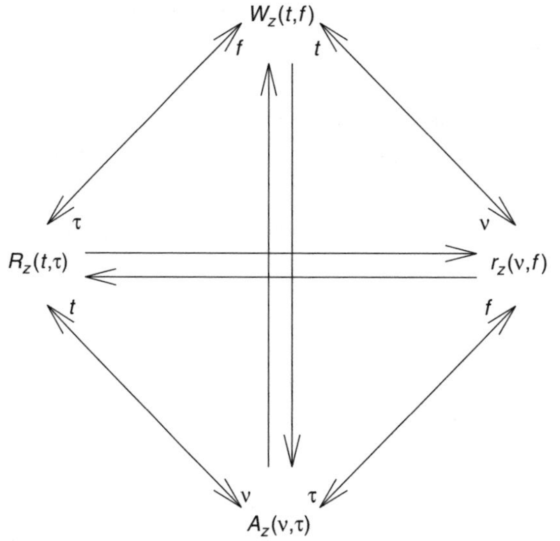

FIGURE 25.3 Various representations that can be obtained from the signal kernel $K_z(t,\tau)$.

of time shifts and frequency shifts. Hence, a double convolution in (t, f) results (by two-dimensional FT) in a multiplication in the ambiguity domain. This leads to the interpretation of TFD design, defined by Equation 25.15, as a two-dimensional filtering procedure in the ambiguity domain. Equation 25.16 can therefore be rewritten as the two-dimensional FT of the symmetric AF $A_z(\nu,\tau)$ filtered by the kernel filter $g(\nu,\tau)$:

$$\rho_z(t, f)\,{}^{t}_{f} \rightleftharpoons {}^{\nu}_{\tau}\, g(\nu,\tau)A_z(\nu,\tau) \tag{25.18}$$

Choosing the kernel filter $g(\nu,\tau)$ most relevant to an application results in a specific TFD. For an all-pass filter, $g(\nu,\tau) = 1$, $\rho_z(t, f)$ reduces to the WVD, which may be expressed as

$$W_z(t, f) = \underset{\tau\to f}{\mathcal{F}}[K_z(t,\tau)] = \underset{\tau\to f}{\mathcal{F}}[z(t+\tau/2)z^*(t-\tau/2)] \tag{25.19}$$

For $g(\nu,\tau)$ chosen to be the AF of the time analysis window $h(t)$, it reduces to the spectrogram, which may be expressed as

$$\rho_{spec}(t, f) = |\mathcal{S}(t, f)|^2 = \left|\int_{-\infty}^{\infty} s(\tau)h(\tau - t)e^{-j2\pi f\tau}d\tau\right|^2 \tag{25.20}$$

where $\mathcal{S}(t, f)$ is the short-time Fourier transform (STFT).

Using a symmetrical rectangular window of width T, $h(t) = rect(t/T)$, the above equation reduces to

$$\rho_{spec}(t, f) = \left|\int_{t-T/2}^{t+T/2} s(\tau)e^{-j2\pi f\tau}d\tau\right|^2 \tag{25.21}$$

This is obtained from the general form (Equation 25.20) by selecting the kernel filter to be

$$g(\nu,\tau) = rect(\tau/2T)\frac{\sin\pi(T-|\tau|)\nu}{\pi\nu} \tag{25.22}$$

The next section presents some important considerations relevant to the selection of suitable kernel filters.

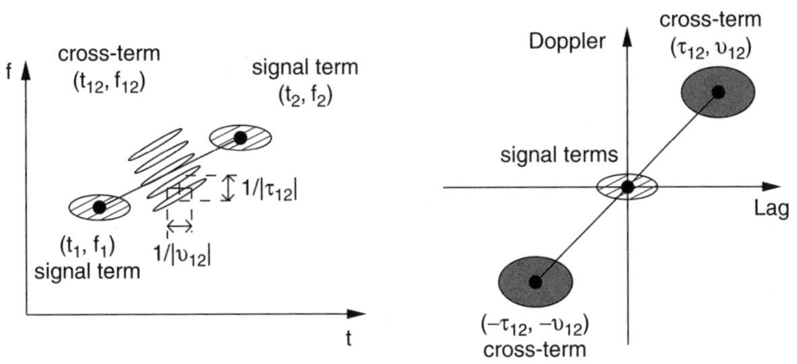

FIGURE 25.4 Position of signal terms and cross-terms in both time–frequency and Doppler-lag domains.

25.2.3.4 Quadratic TFDs, Multicomponent Signals, Cross-Terms Reduction, and Criteria for the Design of Quadratic TFDs

Equation 25.16 defines TFDs that are quadratic (or bilinear) in the signal $z(t)$. This implies that if the signal $z(t)$ includes two components $z_1(t)$ and $z_2(t)$, then its quadratic formulation will include not only these two components but also additional components corresponding to their cross-product $z_1(t)z_2(t)$. These additional components are often called cross-terms and are considered "artifacts" or "ghosts" appearing unexpectedly in the $t - f$ representation (see Figure 25.4). A similar effect occurs when we take the spectrum of $z_1(t) + z_2(t)$ and obtain cross-spectral components that are zero only when $z_1(t)$ and $z_2(t)$ do not overlap in frequency (see Figure 25.5) [3].

Thus, the introduction of either noise or some other deterministic components introduces significant cross-terms into the representation. In some applications these cross-terms may be useful as they provide

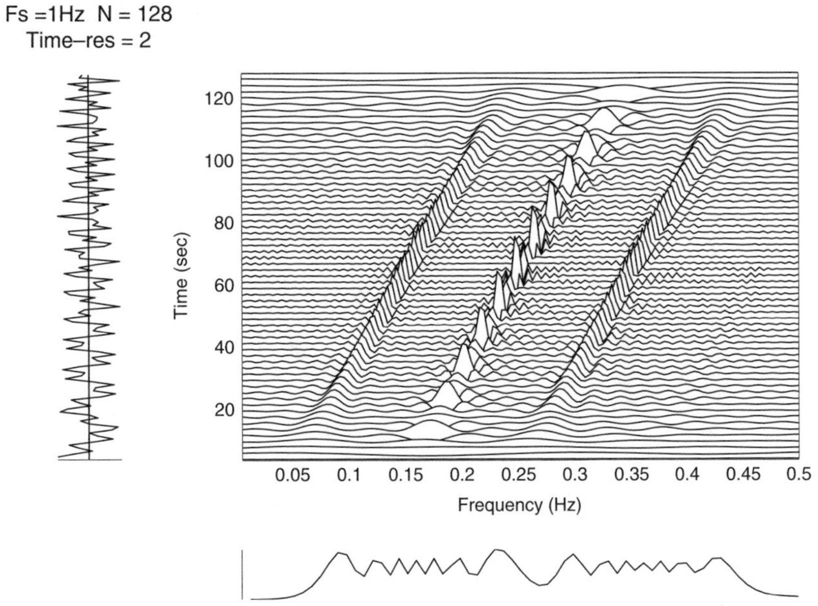

FIGURE 25.5 WVD of a signal composed of two linear frequency modulations exhibiting large positive and negative amplitudes commonly known as cross-terms.

additional features that can be used for signal identification and recognition. However, in most cases, they are considered undesirable interference terms that distort the reading of the representation, and we want to design TFDs that suppress them best.

25.2.3.4.1 Reduced Interference Distributions

Several TFDs have been designed for this purpose. One of the best known is the Gaussian distribution (also called Choi–Williams distribution (CWD)) whose kernel filter $g(\nu, \tau)$ is a two-dimensional Gaussian function centered around the origin in the ambiguity plane and whose spread is controlled by a parameter σ (hence controlling the amount of cross-terms reduction and autoterms resolution). Another recently introduced TFD is the BD [18], whose time-lag kernel filter is defined as

$$G(t, \tau) = \left(\frac{|\tau|}{\cosh^2(t)} \right)^{\beta} \tag{25.23}$$

where $0 < \beta \leq 1$ is an application-dependent parameter that controls the sharpness of the cutoff of the two-dimensional filter in the ambiguity domain, resulting in a trade-off between time–frequency resolution and cross-terms elimination. Cross-terms can be reduced by making β small. An improved version, the modified BD (MBD) was defined in [3, Article 5.7].

Resolution performance of various TFDs, when used to represent multicomponent signals, can be measured using an objective measure that takes into account the key attributes of TFDs (such as the amplitudes of autoterms and cross-terms, autoterms' bandwidth and sidelobes' amplitudes) [18].

25.2.3.4.2 Comparison of Quadratic TFDs

Using the above mentioned-objective measure, the BD and its modified version (MBD) [3] were found to be among the closest to the ideal distribution; the BD is essentially cross-term-free and has high resolution in the time–frequency plane (see Figure 25.6). This TFD outperforms the spectrogram and other existing reduced interference distributions in the analysis of multicomponent signals, and is practically 'equivalent' to the WVD in the analysis and estimation of a monocomponent linear FM signal (see [3, 18, 19] for more details). For this reason, we will often refer to it in this chapter in conjunction with other methods like the WVD, spectrogram, and Gaussian distribution.

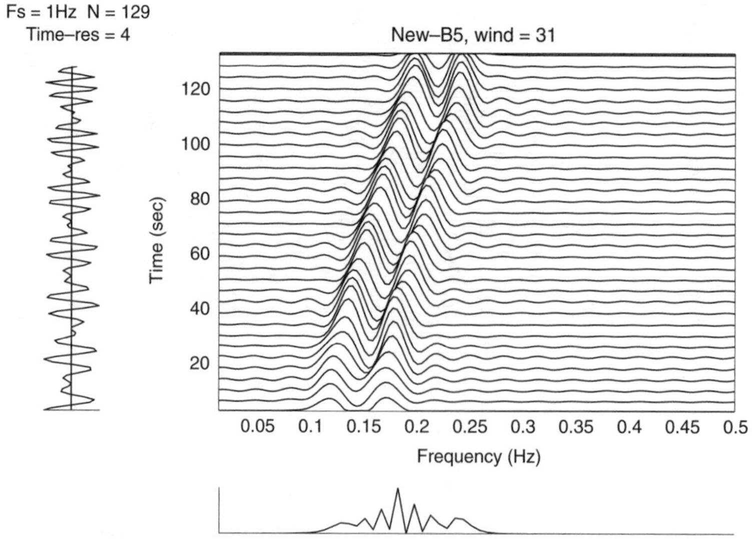

FIGURE 25.6 The B distribution of two closely spaced linear FM signals.

25.2.3.5 Time–Frequency Signal Synthesis

Whereby TF signal analysis algorithms are used to analyze the time-varying frequency behavior of signals, TF signal synthesis algorithms are used to synthesize (or estimate) signals whose TFDs exhibit some given time-varying frequency characteristics. This problem can be formulated as follows: if $z(t)$ is a signal of interest with $\rho_z(t, f)$ being its TFD in the general quadratic class, the synthesis problem is to find the analytic signal $\hat{z}(t)$ whose TFD, $\rho_{\hat{z}}(t, f)$, best approximates $\rho_z(t, f)$ according to some defined criteria. One of the earliest algorithms for TF signal synthesis was based on the WVD in [20]. Some improvements and extensions were made in [21–25]. Several other methods can be found in [3].

The basis of the WVD-based algorithm is the inversion property of the WVD [2]:

$$z(t) = \frac{1}{z^*(0)} \int W_z(t/2, f)\, e^{j2\pi f t}\, df \tag{25.24}$$

implying that the signal may be reconstructed to within a complex exponential constant $e^{j\alpha} = z^*(0)/|z(0)|$ given $|z(0)| \neq 0$.

25.2.3.6 Other Properties

Another important property of quadratic TFDs is their compatibility with filtering. This property expresses the fact that if a signal $y(t)$ is the convolution of $x(t)$ and $h(t)$ (i.e., $y(t)$ is the output of a linear time-invariant filter with impulse response $h(t)$ whose input is $x(t)$), the TFD of $y(t)$ is the time convolution between the TFD of $x(t)$ and the WVD of $h(t)$. Mathematically, if

$$y(t) = x(t) * h(t)$$

then

$$\rho_y(t, f) = \rho_x(t, f) \underset{t}{*} W_h(t, f) \tag{25.25}$$

The properties of quadratic TFDs discussed above are shown in the next sections to be relevant to providing efficient solutions to several problems arising in wireless communications, allowing for improved system performance. For space reasons, techniques such as IF estimation [53] and cross-WVD (XWVD) [3, Chapter 3] are not discussed here. Interested readers can refer to the references provided.

25.3 Spread-Spectrum Communications Systems Using TFSP

Spread-spectrum communications use signals whose bandwidth is much wider than the information bandwidth. This is achieved by the direct-sequence (DS) technique in which the transmitted signal is spread over a wide bandwidth by means of a code independent of the data. The availability of this code at the receiver enables the despreading and recovery of data, while spreading and mitigating the interference (see Figure 25.7). Spread-spectrum techniques offer a number of important advantages, such as code-division multiple access, low probability of intercept, communications over multipath propagation channels, and resistance to intentional jamming. The processing gain of a DS spread-spectrum system, generally defined as the ratio between the transmission and the data bandwidths, provides the system with a high degree of interference mitigation. However, in some cases, the interference might be much stronger than the useful signal, e.g., when the useful signal is affected by fading and the gain due to the coding might be insufficient to decode the useful signal reliably. Therefore, time–frequency (TF) signal processing techniques have been used in conjunction with the signal spreading to augment the processing gain, permitting greater interference protection without an increase in bandwidth. In this section, we review some of these TF signal processing techniques for channel identification and interference mitigation.

25.3.1 Channel Modeling and Identification

In this section we show how time–frequency representations of a linear time-varying propagation channel can be exploited for channel estimation either by direct use of the observation signal TFD as in [7], by

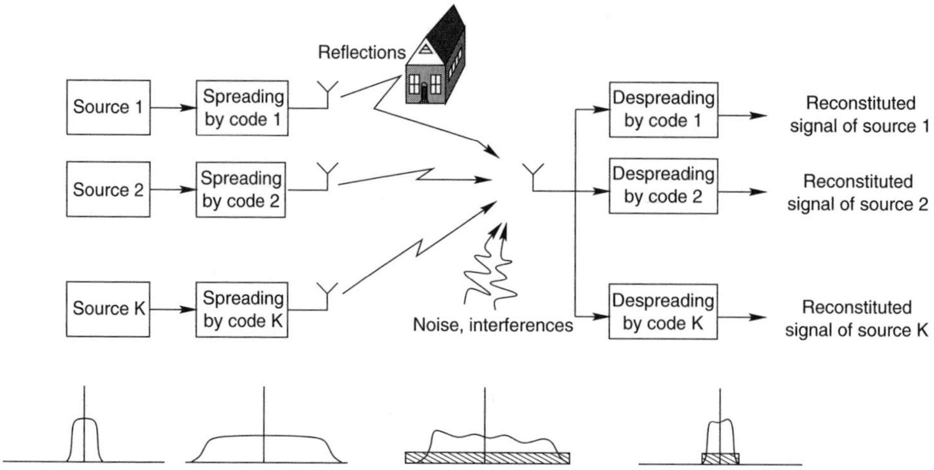

FIGURE 25.7 DS-CDMA communications system.

time–frequency polynomial modeling as in [8], or by using time–frequency canonical channel modeling, which we describe in Sections 25.3.1.1 and 25.3.1.2 since it is the most commonly used LTV channel estimation method.

25.3.1.1 Wireless Communication LTV Channel Model

A complex baseband received signal, $r(t)$, through a wireless mobile communication channel can be modeled[7] as follows [9]:

$$\begin{aligned} r(t) &= \int h(t,\tau)s(t-\tau)\,d\tau + \varepsilon(t) \\ &= \int\int \mathcal{U}(\nu,\tau)s(t-\tau)\,e^{j2\pi\nu t}\,d\tau\,d\nu + \varepsilon(t) \\ &= x(t) + \varepsilon(t) \end{aligned} \tag{25.26}$$

where $h(t,\tau)$ is the channel impulse response representing the LTV behavior; $s(t)$ is the complex baseband transmitted signal; $\varepsilon(t)$ is the additive white Gaussian noise with zero mean and variance σ_ε^2; τ and ν denote the delay and Doppler-shift variables, respectively; and $\mathcal{U}(\nu,\tau)$, the Fourier transform of $h(t,\tau)$ from t to ν, is called the delay-Doppler spread function of the LTV channel. By applying the Fourier transform among the variables t, f, τ, and ν, we can define several system functions [9, 12], with their relationships shown in Figure 25.8, which resembles the dual relationships of time–frequency representations in TFSP, as illustrated by Figure 25.3.

As previously mentioned, the delay–Doppler spread function is often modeled as a wide-sense stationary Gaussian process with uncorrelated scattering [9] whose second-order statistics can be represented by

$$E\{\mathcal{U}(\nu',\tau')\cdot\mathcal{U}^*(\nu,\tau)\} = P_{\mathcal{U}}(\nu,\tau)\cdot\delta(\nu'-\nu)\delta(\tau'-\tau) \tag{25.27}$$

where $P_{\mathcal{U}}(\nu,\tau)$ is the *scattering function* (SF) of the channel. It follows that the WSSUS channel may be represented as a collection of nonscintillating uncorrelated scatterers that cause both multipath delays and Doppler shifts.

[7] In practice, the double integral is bounded by the ranges of multipath delays and Doppler shifts; however, without loss of generality, we use the full range $(-\infty,\infty)$ and drop them for short notation.

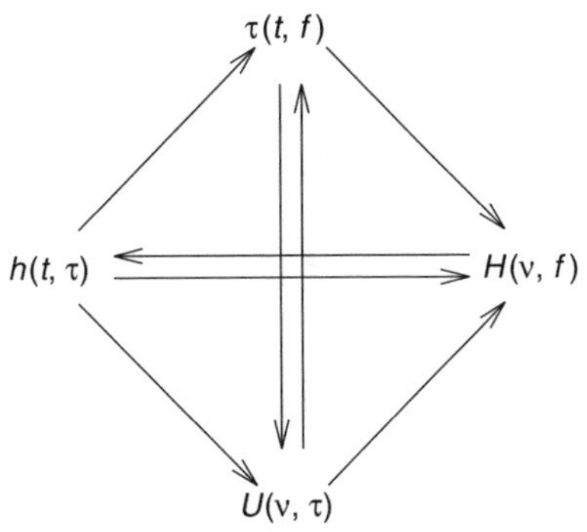

FIGURE 25.8 Relationship among Bello [9] system functions. $\mathcal{T}(t, f)$ and $\mathcal{H}(\nu, f)$ are the time-varying transfer function and output Doppler spread function, respectively.

Restrictions of practical channels on, for instance, time duration, bandwidth, fading rate, delay spread, and Doppler spread, allow a simplified representation of LTV channels in terms of canonical elements. Such channel representations, called *canonical channel* models, can simplify the analysis of the performance of communications systems. A common class of canonical channel representations is the class of sampling models that are applied when a system function vanishes for values of an independent variable (i.e., t, f, τ, or ν) outside some interval or when the input or output time function is time limited or band limited. Various sampling models can be found in [9]. However, we consider here the situation wherein the input signal $s(t)$ is band limited in $[f_o - B_o/2, f_o + B_o/2]$ and the output signal $x(t)$ (noise-free) is time limited in $[t_o - T_o/2, t_o + T_o/2]$.

Applying the sampling theorem on both t and f according to the above time and frequency constraints, the received signal may be expressed as [9]

$$x(t) = \sum_m \sum_n \mathcal{U}_{mn}\, s\left(t - \frac{m}{B_o}\right) e^{j2\pi(n/T_o)(t-m/B_o)} \tag{25.28}$$

where

$$\mathcal{U}_{mn} = \frac{1}{T_o B_o} \tilde{\mathcal{U}}\left(\frac{m}{B_o}, \frac{n}{T_o}\right) \tag{25.29}$$

with $\tilde{\mathcal{U}}(\nu, \tau)$ being some smoothed version[8] of $\mathcal{U}(\nu, \tau)$.

When the channel is random, the coefficients $\mathcal{U}_{mn}$ become random variables. Under the WSSUS assumption, the correlation of $\mathcal{U}_{mn}$ can be expressed in terms of the SF, $P_{\mathcal{U}}(\nu, \tau)$. Additionally, if the SF varies very little for changes in τ of the order $1/B_o$ or ν of the order $1/T_o$, this correlation is approximated as

$$E\left\{\mathcal{U}_{mn}\mathcal{U}^*_{rs}\right\} = \begin{cases} \frac{1}{T_o B_o} P_{\mathcal{U}}\left(\frac{m}{B_o}, \frac{n}{T_o}\right) & \text{for } m = r,\ n = s \\ 0 & \text{otherwise} \end{cases} \tag{25.30}$$

which means that the delay–Doppler coefficients $\mathcal{U}_{mn}$ are uncorrelated at different values of multipath delays and Doppler shifts. In practice, the multipath-delay τ vanishes outside $[0, T_m]$ and Doppler shift

[8]The expression of $\tilde{\mathcal{U}}(\nu, \tau)$ is given by

$$\tilde{\mathcal{U}}(\nu, \tau) = \int\int e^{j2\pi f_0(\tau-\eta)} e^{-j2\pi t_o(\nu-\mu)} \operatorname{sinc}[B_o(\tau - \eta)] \operatorname{sinc}[T_o(\nu - \mu)]\, \mathcal{U}(\mu, \eta)\, d\mu\, d\eta$$

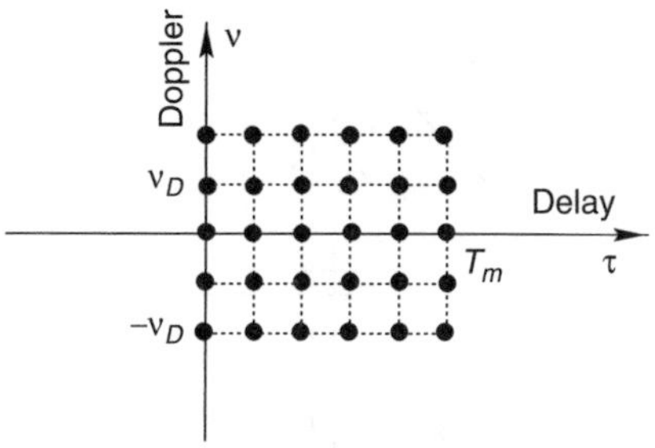

FIGURE 25.9 Second-order statistics of frequency-selective fast fading channel with WSSUS assumption. (T_m and ν_D are the maximum multipath delay and Doppler shift for a particular mobile environment.)

vanishes outside $[-\nu_D, \nu_D]$, where T_m and ν_D are the maximum multipath delay and maximum Doppler shift, respectively. Consequently, the statistics of the WSSUS channel through the above correlation may be sketched on the delay–Doppler plane as shown in Figure 25.9 [26].

We will shortly see how this canonical model can be used to approach the problem of channel estimation. Other issues relating to this model are not reviewed in this chapter but may be of interest, such as multiuser detection in [26] and blind multiuser equalization in [27].

25.3.1.2 Estimation of LTV channels

We review here the problem of pilot-based coherent estimation of LTV channels for CDMA systems using the minimum mean square error (MMSE) approach, which is based on the use of a canonical LTV channel [29]. It is known that RAKE receivers are commonly used for channel equalization as well as signal detection in frequency-selective fading environments such as in CDMA systems. However, in TF-selective fading scenarios, the performance of RAKE receivers is degraded. An extension of the canonical channel model given in Equation 25.28 was proposed in [28] to overcome this problem.

Consider a TF-selective fading environment for which T_m and ν_D denote the maximum multipath delay and Doppler shift, respectively. The transmitted signal $s(t)$ in a DS-CDMA scheme has the duration T and bandwidth B ($B = 1/T_c$, where T_c is the chip duration). The complex baseband signal admits the following canonical representation in terms of fixed multipath delays and Doppler shifts [29]:

$$r(t) \approx \sum_{n=0}^{N} \sum_{k=-K}^{K} \mathcal{U}_{kn}\, \xi_{kn}(t) + \varepsilon(t), \quad 0 < t \leq T \tag{25.31}$$

where $N = \lceil T_m B \rceil$ and $K = \lceil T \nu_D \rceil$ denote the numbers of resolvable multipath-delay and Doppler-shift components, respectively, and

$$\xi_{kn}(t) = s(t - n/B) \exp\{j2\pi(k/T)t\}$$

It is seen in Equation 25.31 that $x(t)$ is a linear combination of the set of quasi-orthogonal basis functions[9] $\{\xi_{kn}(t)\}$. In other words, thanks to the canonical representation, the TF-selective fading channel has been decomposed into a bank of $(N + 1)(2K + 1)$ TF-non-selective (flat) fading subchannels. A common approach, then, to the problem of channel estimation is to use matched filtering at the outputs of these subchannels.

Given a frame of $(2I + 1)$ pilot symbols, it is required that we estimate the set of channel coefficients $\{\mathcal{U}_{kn}\}$ at the symbol centered in the considered frame (i.e., at symbol 0). The received signal for the entire

[9]The spreading function $s(t)$ is usually designed so that its nonzero lag correlation coefficients are close to zero.

frame becomes

$$r(t) \approx \sum_{i=-I}^{I} \sum_{n=0}^{N} \sum_{k=-K}^{K} \mathcal{U}_{kn}(i)\, \xi_{kn}(t - iT) + \varepsilon(t) \tag{25.32}$$

The matched-filtered outputs for the entire frame, i.e., $\mathbf{z} = [\mathbf{z}^T(-I), \ldots, \mathbf{z}^T(I)]^T$ with output for each symbol $\mathbf{z}(i) = \langle r(t), \xi(t - iT)\rangle \triangleq \int_{-\infty}^{+\infty} r(t)\xi^*(t - iT)\, dt$, can be expressed as

$$\mathbf{z} = \mathbf{Q}\mathbf{u} + \varepsilon$$

where the channel coefficient vector $\mathbf{u} = [\mathbf{u}(-I)^T \cdots \mathbf{u}(0)^T \cdots \mathbf{u}(I)^T]^T$ with $\mathbf{u}(i) = [\mathcal{U}_{1,1}(i), \ldots, \mathcal{U}_{(2K+1),(N+1)}(i)]^T$; the correlation matrix $\mathbf{Q} = \text{diag}\,(\mathbf{P}, \ldots, \mathbf{P})$ with $\mathbf{P} = \int \xi^*(t - iT)\xi^T(t - iT)\, dt = \int_0^T \xi^*(t)\xi^T(t)\, dt$; $\xi(t)$ being the temporal waveform vector: $\xi(t) = [\xi_{1,1}(t), \ldots, \xi_{(2K+1),(N+1)}(t)]^T$; and the colored Gaussian noise vector $\varepsilon = [\varepsilon(-I)^T \cdots \varepsilon(I)^T]^T$ with each colored Gaussian noise vector $\varepsilon(i) = \langle \varepsilon(t), \xi(t - iT)\rangle$ having the covariance matrix $\mathcal{N}_0\mathbf{P}$.

As mentioned earlier, we need to estimate $\mathbf{u}(0)$. The linear MMSE estimation problem is formulated as

$$\mathbf{L}_{opt} = \underset{\mathbf{L}}{\text{argmin}}\ E\{\|\mathbf{u}(0) - \mathbf{L}^H\mathbf{z}\|^2\} \tag{25.33}$$

The solution of the minimization problem in Equation 25.33 is the Wiener filter given by

$$\mathbf{L}_{opt} = (\mathbf{Q}E\{\mathbf{u}\mathbf{u}^H\} + \mathcal{N}_0\mathbf{I})^{-1}\{\mathbf{u}\mathbf{u}^H(0)\} \tag{25.34}$$

The MMSE channel estimate $\hat{\mathbf{u}}(0) = \mathbf{L}_{opt}^H\mathbf{z}$ is then used for symbol detection using RAKE (or maximum ratio combiner) receiver. For the binary phase shift keying (BPSK) signaling scheme and under the assumption of negligible ISI, the estimated symbols are given by

$$\hat{b}(i) = \text{sign}(\text{real}(\hat{\mathbf{u}}(i)^H\mathbf{z}(i)))$$

More analysis on the estimation performance and practical implementation of the MMSE channel estimate in Equation 25.34 under the assumption of uncorrelated $\mathcal{U}_{kn}$ components in a particular symbol of interest[10] or the assumption of quasi-orthogonality of the basis functions $\{\xi_{kn}(t)\}$ can be found in [29, 30].

25.3.1.3 Estimation of Scattering Function

In certain wireless communications systems, e.g., radar or acoustic communications systems, one is interested in estimating the scattering function that reveals the TF-selective behavior of the fading channel under the WSSUS assumption [31–38]. In the problem of SF estimation, a common approach is to use the input–output relationship, described through the general TFDs [39] as

$$E\{\rho_x(t, f)\} = \rho_s(t, f) \underset{t\ f}{\star\star} P_{\mathcal{U}}(t, f) \tag{25.35}$$

where $\rho_s(t, f)$ is a TFD of the input $s(t)$ and $E\{\rho_x(t, f)\}$ is the expected value of a TFD of the output $x(t)$. In the AF domain, the previous relation becomes

$$E\{\mathcal{A}_x(\nu, \tau)\} = \mathcal{A}_s(\nu, \tau) \cdot R_{\mathrm{T}}(\nu, \tau) \tag{25.36}$$

where $R_{\mathrm{T}}(\nu, \tau)$ is the double Fourier transform of the channel SF $P_{\mathcal{U}}(t, f)$. Since the general AF (GAF), $\mathcal{A}_s(\nu, \tau)$, or general TFD, $\rho_s(t, f)$, includes the expression of the kernel, $g(\nu, \tau)$ (see Figure 25.3), and

[10]Note that the channel coefficients $\mathcal{U}_{kn}$ may be correlated in time across the symbols.

this kernel can be made arbitrary, two general classes of SF estimators were then proposed based on *deconvolution* and *direct implementation*, respectively.

The class of deconvolution estimators is defined based on the division of Equation 25.36 by the GAF of the input signal

$$\hat{P}_{\mathcal{U}}(t, f) \triangleq \mathcal{F}^{-1}_{\nu \to f}\{\mathcal{F}_{\tau \to f}\{\mathcal{A}_x(\nu, \tau)\}/\mathcal{A}_s(\nu, \tau)\} \tag{25.37}$$

Similar to the approach in [35], a zero-division problem in Equation 25.37 is encountered. A classical solution is to threshold the symmetric AF $A_s(\nu, \tau)$ at the points equal to zero (see [35] for more details).

On the other hand, one can choose the kernel $g(\nu, \tau)$ so that the TFD $\rho_s(t, f)$ in Equation 25.35 is impulse-like; the left-hand side of Equation 25.35, then approximates to $P_{\mathcal{U}}(t, f)$. Thus, the other class of SF estimators, namely, direct implementation, can be defined as

$$\hat{P}_{\mathcal{U}}(t, f) \triangleq E\{\rho_x(t, f)\} \tag{25.38}$$

We must note here that an impulse representation in the time–frequency plane does not exist due to the constraint of minimum time–frequency bandwidth according to Heisenberg's uncertainty principle. Therefore, we opt to choose an approximation in the sense of good localization in the time–frequency plane. For example, the TFD kernel may be approximated by the Hermite functions [40], which are defined as

$$g_n(x) = (-1)^n e^{x^2/2} \frac{d^n}{dx^n} e^{-x^2/2} \tag{25.39}$$

25.3.2 Interference Mitigation

25.3.2.1 TV-NBI Suppression in DS-CDMA

Spread-spectrum communication, based on which DS-CDMA is implemented, is known to have the capability of suppressing NBI. However, this NBI suppression becomes ineffective when the interfering signal is too powerful. In some of these cases, the interference immunity can be improved significantly by using signal processing techniques that complement the spread-spectrum modulation [16]. These active suppression techniques not only improve error rate performance, but also lead to an increase in capacity of CDMA cellular systems [41]. There have been several models proposed for narrowband interferers existing in communication channels, such as a deterministic sinusoidal signal [16], an autoregressive-modeled signal [42, 43], and a narrowband digital communications signal [44, 45]. A tremendous amount of research on NBI suppression can be seen in [16, 43, 44, 46–52]. None of these methods, however, is capable of suppressing NBI with time-varying spectral characteristics such as a linear FM signal (chirp signal).

To suppress this type of time-varying NBI (TV-NBI), the IF of the TV-NBI is first estimated using some TFDs; then a time-varying zero filter is used to suppress the interference. The effectiveness of TFDs in providing accurate estimates of the IF has extensively been shown in the literature [2, 4, 53, 54]. Using this approach, the problem of TV-NBI suppression has been carried out in [55–59].

Consider the transmission of a spread-spectrum signal $s(t)$ through an AWGN channel characterized by zero-mean Gaussian random process $\epsilon(t)$ and being interfered with by K different linear FM interferences $\zeta_k(t)$. The received signal model may be expressed in the form

$$r(t) = s(t) + \sum_{k=1}^{K} \zeta_k(t) + \epsilon(t) \tag{25.40}$$

Each interference signal belongs to the class of linear FM signals, with power P_{ζ_k}, that can be expressed as

$$\zeta_k(t) = \sqrt{P_{\zeta_k}}\, e^{j\phi(t,\theta_k)} = \sqrt{P_{\zeta_k}}\, e^{j(2\pi f_k t + \pi g_k t^2)}$$

where $\boldsymbol{\theta}_k = (f_k, g_k)$. The IF of each monocomponent linear FM signal is defined as

$$f_i(t, \boldsymbol{\theta}_k) \triangleq \frac{1}{2\pi} \frac{d\phi(t, \boldsymbol{\theta}_k)}{dt} = f_k + g_k t \tag{25.41}$$

For monocomponent TV-NBI, i.e., $K = 1$, the Born–Jordan distribution and the cone-shape ZAM distribution were used to estimate the IF in [55], and the spectrogram was used in [60, 61]. A problem related to this method is that if the signal-to-interference ratio is high, the estimation of the interference parameters might fail, and the suppression filter could track the useful signal, instead of the interference.

This problem can be improved by using the Wigner–Hough transform [57], defined as [62]

$$WH_r(\boldsymbol{\theta}_k) \triangleq \int_{-\infty}^{\infty} W_r(t, f_i(t; \boldsymbol{\theta}_k))\, dt \tag{25.42}$$

where $W_r(t, f)$ denotes the WVD of $r(t)$ and $f_i(t; \boldsymbol{\theta}_k)$ are the individual IFs of the interferences as given in Equation 25.41. This method also has the ability to deal with multicomponent TV-NBI. The integration in Equation 25.42 over all possible lines of the WVD, which is obtainable by applying a Hough transform, or equivalently, a Radon transform of the WVD, gives rise to peaks in the final parameter space; each peak corresponds to one linear FM signal, whose modulation parameters, (f_k, g_k), are the coordinates of the peaks.

We know that the WVD is optimal in representing monocomponent linear FM signals. The estimation of the IF using the peak of the XWVD achieves the best performance [63]. This technique, therefore, can be effectively used to estimate monocomponent linear FM TV-NBI. For multicomponent TV-NBI, the WVD is not optimal anymore. An alternative solution in this case is to use the BD, in Equation 25.23, since it outperforms some other reduced interference distributions when comparing the capacity of reducing the cross-terms [18].

25.3.2.2 Signal Modulation Design for ISI Mitigation

Apart from the DS-CDMA system mentioned previously, many other CDMA system concepts have been proposed, among which MC-CDMA is a promising system compared to DS-CDMA [15, 64]. MC transmission is a method to design a bandwidth-efficient communication system in the presence of channel distortion (ISI, especially for high-data-rate communications) by dividing the available channel bandwidth into a number of subchannels such that each channel is nearly ideal. The idea of using MC transmission comes from the advantage of this system in overcoming the effect of signal fading on time-varying channels [13]. The typical MC transmission system is the orthogonal frequency division multiplexing (OFDM) system. The combination of OFDM and CDMA allows for optimal detection performance, use of the available spectrum in an efficient way, retention of many advantages of a CDMA system, and exploitation of frequency diversity [65–68].

In MC transmission, the modulation scheme is done based on a set of basis functions (one is a time–frequency-shifted version of another) constructed by Gaussian pulse [69] or Nyquist pulse [70]. These pulses are required to have two important characteristics: (1) orthogonality to avoid ISI and interchannel interference (ICI), and (2) good localization to avoid symbol energy smearing out over the channel and perturbing neighboring symbols. For wireless mobile communication channels with TF-dispersive characteristics, the above two conditions become critical since the localization of the pulses is dispersed. There is a need to design better localized basis functions under such TF-dispersive conditions of the wireless mobile channels, in other words, to design a new set of basis functions so that the effects of ISI and ICI are minimized. The design of different pulses can be carried out using TF analysis in the ambiguity domain, where Hermite pulses have been proved to have better TF localization than Gaussian or Nyquist under TF-dispersive channels [40].

Consider an MC modulation scheme used in such a TF-dispersive mobile channel. Let I be the number of channels in the scheme in which $f_{c_i} = i\Delta f_c$ $(i = 0, \ldots, I-1)$ is the set of carrier frequencies and Δf_c is the spacing between adjacent carriers (usually $\Delta f_c \leq W$, where W is the bandwidth of the signal).

The transmitted signal is given by

$$x(t) = \sum_{i=0}^{I-1} \sum_{n=-\infty}^{\infty} c_{n,i}\, \xi_{n,i}(t) \tag{25.43}$$

where $c_{n,i}$ are the information-bearing symbols and $\xi_{n,i}$ are the basis functions, defined as

$$\xi_{n,i}(t) = \lambda(t - nT)\, e^{j2\pi f_i t} \tag{25.44}$$

The envelope function $\lambda(t)$ is called an elementary pulse (T being the symbol duration). Demodulation is performed by a projection of the received signal on the complex conjugates of the basis functions

$$r_{p,q} = \int r(t)\, \xi^*_{p,q}(t)\, dt \tag{25.45}$$

The orthogonality condition requires

$$\int \xi_{n,i}(t)\, \xi^*_{p,q}(t)\, dt = \int \lambda(t)\lambda^*(t + (n-p)T)\, e^{-j2\pi(i-q)\Delta f_c t}\, dt =$$

$$= \delta_{n,p}\delta_{i,q} = \begin{cases} 1 & n = p \text{ and } i = q \\ 0 & \text{otherwise} \end{cases} \tag{25.46}$$

By noting that the integral in Equation 25.46 can be considered a sampling of the AF, $\mathcal{A}_\lambda(\nu, \tau)$ (a shifted version of the symmetrical AF defined in Equation 25.17 of the envelope function $\lambda(t)$), Equation 25.46 can be expressed as

$$\int \xi_{n,i}(t)\, \xi^*_{p,q}(t)\, dt = \mathcal{A}_\lambda((i-q)\Delta f_c, (n-p)T)$$

The second required condition is good localization in the sense of a minimum energy spread in order to avoid the symbol energy smearing out over the dispersive channel and perturbing neighboring symbols. If ΔT and ΔW represent the time dispersion and frequency dispersion,[11] respectively, the following conditions need to be verified so the channel can be considered as frequency-non-selective and slow fading for each carrier:

$$\nu_D \ll \Delta f_c \qquad \Delta W \ll B_{coh}$$
$$T_m \ll T \qquad \Delta T \ll T_{coh}$$

where $B_{coh} \approx 1/T_m$ and $T_{coh} \approx 1/\nu_D$ are the coherence bandwidth and coherence time, respectively.

Several drawbacks of the Nyquist or Gaussian pulse, being the elementary pulse due to the lack of orthogonality and localization in TF-dispersive channels, lead to the design of a new pulse [40] based on the Hermite function given by (Equation 25.39).

Orthogonality is optimized by evaluating the AF of $\lambda(t)$, where $\lambda(t)$ is a linear combination of $D_{4n}(t) = f_{4n}(\sqrt{2\pi}t)$, as given by

$$\lambda(t) = \sum_{k=0}^{N_F-1} F_{4k} D_{4k}(t)$$

where N_F is the number of coefficients F_{4k} that can be found when imposing the condition for orthogonality.

[11] ΔT and ΔW are defined as $(\Delta T)^2 = \int t^2\, |\lambda(t)|^2\, dt$ and $(\Delta W)^2 = \int f^2\, |\Lambda(f)|^2\, df$ ($\Lambda(f)$ being the FT of $\lambda(t)$).

Performance evaluation shows that the Hermite pulse is better than the Nyquist pulse with cosine roll-off for a channel with characteristics of $T_m \leq 0.01$ and $\nu_D \leq 0.01$. The Hermite pulse loses its advantages at around $\nu_D = 0.1$. However, better performance of the Hermite pulse in the presence of time dispersion is encouraging for its application in multiuser multicarrier systems where different users are transmitting on neighboring carriers.

25.3.2.3 Multiple-Access Interference in MC-CDMA

Also based on the use of the ambiguity domain, a design of new signature waveforms for the MC-CDMA system was proposed in [71]. In MC-CDMA, the signature waveforms are normally designed in the frequency domain, whereas in DS-CDMA, they are designed in the time domain. There exist two major problems for multiuser access in an MC system: (1) the high effort in signal processing due to the need of a complex-valued RAKE or multiuser detector [73], and (2) the design of optimized code sets with reduced dynamic range and proper autocorrelation and cross-correlation properties. A possibility to the orthogonalized codes by shifting the signal in frequency has been shown in [72]. Consequently, a K-users-1-code MC-CDMA system was proposed in [71]. To each user, the same code is assigned with different frequency shifts.

More precisely, consider a multicarrier spread-spectrum signal defined as

$$s(t) = b(t) \cdot \sum_{q=0}^{Q-1} c(q)\, e^{j2\pi qt/T} = b(t) \cdot c_0(t)$$

where $b(t)$ is the data signal, $c(q)$ are the complex code coefficients in the frequency domain, and Q is the spreading factor. The transmitted signal of the kth user is shifted from that of the first user as

$$s_k(t) = b_k(t) \cdot \sum_{q=0}^{Q-1} c(q)\, e^{j2\pi(q+k)t/T} = b_k(t) \cdot c_0(t)e^{j2\pi kt/T} = b_k(t) \cdot c_k(t)$$

Due to the multipath fading, the received signal for the up-link of K users is given by

$$r(t) = \sum_{k=0}^{K-1} \sum_{p=0}^{P-1} \alpha_{k,p}(t)\, b_k(t - \tau_{k,p})\, c_k(t - \tau_{k,p})\, e^{j2\pi \nu_k t/T} + \epsilon(t)$$

where P is the number of propagation paths (assuming the same for all users) and $\alpha_{k,p}(t)$, $\tau_{k,p}$, and ν_k are the attenuation, multipath delay, and Doppler shift of the kth user's signal, respectively.

Using a time domain RAKE receiver with ideal path synchronization, we get as an expression for the detected symbol $\hat{b}_{k,p}$ of user k in path p

$$\hat{b}_{k,p} = \frac{1}{T} \int_0^T r(t) c_k^*(t - \tau_{k,p})\, dt$$

The detection problem can be simplified to analyzing the correlation functions $\mathcal{A}_{k_1,k_2}(\nu, \tau)$, which depend on the data signal $b(t)$ and the code $c(t)$. The whole interference could be evaluated by summing these terms, $\mathcal{A}_{k_1,k_2}(\nu, \tau)$, for all users and all paths with the correct path weight. Without loss of generality, choosing $k_1 = 1$ and $k_2 = k$, we have

$$\mathcal{A}_{1,k}(\nu_k, \tau_k) = \frac{1}{T} \int_0^T s_k(t) s_1^*(t - \tau_k)\, e^{j2\pi \nu_k t/T}\, dt \tag{25.47}$$

which is a representation of the signal in the ambiguity domain. By not considering the influence of the data signal, Equation 25.47 becomes

$$\mathcal{A}_{0,k}(\nu_k, \tau_k) = \sum_{n_1=0}^{Q-1}\sum_{n_2=0}^{Q-1} c(n_1)c^*(n_2)\, e^{j2\pi[n_2(\tau_k/T-1)+n_1+k+\nu_k]} \times \operatorname{sinc}[\pi(n_1 - n_2 + k + \nu_k)]$$

In the ambiguity domain, the values outside $\mathcal{A}_{1,k}(0,0)$ could be considered interference values of a user with delay $\tau = \tau_k$ and frequency shift $\nu = (k + \nu_k)/T$ to user 1. In other words, we have imperfect time and frequency synchronization.

Assuming that the delay τ and frequency deviation ν are uniformly distributed over $(0, T)$ and $(-\nu_k, \nu_k)$, respectively, the mean interference value is given by

$$\langle|\mathcal{A}_{1,k}(\tau_k, \nu_k)|^2\rangle = \frac{1}{2\nu_k}\int_0^T \int_{(k-\nu_k)/T}^{(k+\nu_k)/T} |\mathcal{A}(\nu,\tau) + \mathcal{A}^*(\nu,\tau - T)|^2\, d\tau\, d\nu \tag{25.48}$$

Numerical evaluation shows that codes with more concentrated interference power have a better mean interference and a better performance in the whole system.

In the above section, TFSP has been considered for one-dimensional signals. In the next section, TFDs are applied to multidimensional signals provided by multiantennae in order to solve relevant problems that arise in wireless communications.

25.4 Time–Frequency Array Signal Processing

Conventional array signal processing algorithms assume stationary signals and mainly employ the covariance matrix of the data array. When the frequency content of the measured signals is time varying (i.e., nonstationary signals), this class of approaches can still be applied. However, the achievable performances in this case are reduced with respect to those that would be achieved in a stationary environment. In the last decades, the stationarity hypothesis was motivated by the crucial need in practice of estimating sample statistics by resorting to temporal averaging under the additional assumption of ergodic signals. Instead of considering the nonstationarity as a shortcoming and trying to design algorithms robust with respect to nonstationarity, it would be better to take advantage of the nonstationarity by considering it as a source of information. The latter can then be exploited in the design of efficient algorithms in such nonstationary environments.

The question now is, How can we exploit the nonstationarity in array processing? This can be done by resorting to the spatial time–frequency distributions (STFDs), which are a generalization of the TFDs to a vector of multisensor signals (see Figure 25.10). Under a linear model, the STFDs and the commonly known covariance matrix exhibit the same eigenstructure. In wireless communications involving multiantennae, the aforementioned structure is often exploited to estimate some signal parameters through subspace-based techniques.

Algorithms based on STFDs properly use the time–frequency information to significantly improve performance. This improvement comes essentially from the fact that the effects of spreading the noise power while localizing the source energy in the time–frequency domain increase the signal-to-noise ratio (SNR).

STFD-based algorithms exploit the time–frequency representation of the signals together with the spatial diversity provided by the multiantennae.

The concept of the STFD was introduced for the first time in 1996 [75]. It was used successfully in solving the problem of the blind separation of nonstationary signals [75–78]. This concept was then applied to solve the problem of direction-of-arrival (DOA) estimation [79]. Since then, several works were conducted in this area using the new concept of STFD [80–89].

The following notations are used throughout the rest of this chapter. For a given matrix $\mathbf{A}$, the symbols $\mathbf{A}^T$, $\mathbf{A}^*$, $\mathbf{A}^H$, $\mathbf{A}^\#$, trace($\mathbf{A}$), and norm($\mathbf{A}$) respectively denote the transpose, conjugate, conjugate transpose, Moore–Penrose pseudoinverse, trace, and (Euclidean) norm of $\mathbf{A}$.

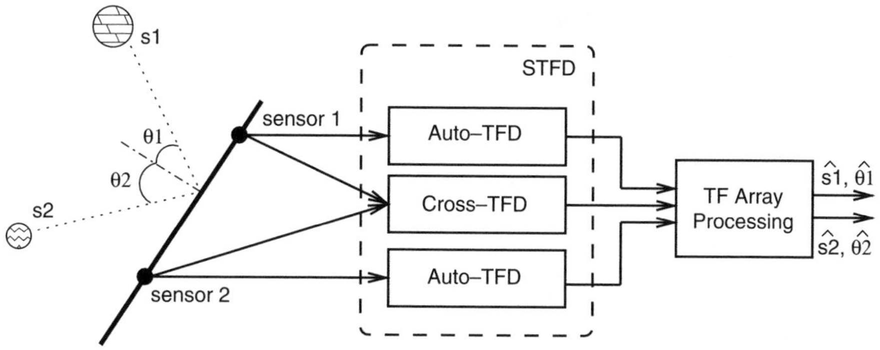

FIGURE 25.10 Time–frequency array signal processing.

25.4.1 The Spatial Time–Frequency Distributions

Given an analytic vector signal $\mathbf{z}(t)$, the spatial instantaneous autocorrelation function is defined as

$$\mathbf{K}_{\mathbf{zz}}(t,\tau) = \mathbf{z}\left(t+\frac{\tau}{2}\right)\mathbf{z}^H\left(t-\frac{\tau}{2}\right) \tag{25.49}$$

Define also the smoothed spatial instantaneous autocorrelation function as

$$\mathbf{S}_{\mathbf{zz}}(t,\tau) = G(t,\tau) \underset{t}{\star} \mathbf{K}_{\mathbf{zz}}(t,\tau) \tag{25.50}$$

where $G(t,\tau)$ is some time-lag kernel. The time convolution operator $\underset{t}{\star}$ is applied to each entry of the matrix $\mathbf{K}_{\mathbf{zz}}(t,\tau)$. The class of quadratic STFDs then defined as

$$\mathbf{D}_{\mathbf{zz}}(t,f) = \mathcal{F}_{\tau\to f}\{\mathbf{S}_{\mathbf{zz}}(t,\tau)\} \tag{25.51}$$

where the Fourier transform $\mathcal{F}$ is applied to each entry of the matrix $\mathbf{S}_{\mathbf{zz}}(t,\tau)$.

The discrete-time definition equivalent to Equations 25.50 and 25.51 leads to the simple implementation of STFD and is expressed as

$$\mathbf{D}_{\mathbf{zz}}(n,k) = \mathcal{DF}_{m\to k}\{G(n,m) \underset{n}{\star} \mathbf{K}_{\mathbf{zz}}(n,m)\} \tag{25.52}$$

which can also be expressed as

$$\mathbf{D}_{\mathbf{zz}}(n,k) = \sum_{m=-M}^{M}\sum_{p=-M}^{M} G(p-n,m)\mathbf{z}(p+m)\mathbf{z}^H(p-m)e^{-j4\pi\frac{mk}{N}} \tag{25.53}$$

where the discrete Fourier transform $\mathcal{DF}$ and the discrete-time convolution operator $\overset{\star}{n}$ are applied to each entry of the matrix $G(n,m) \overset{\star}{n} \mathbf{K}_{\mathbf{zz}}(n,m)$ and matrix $\mathbf{K}_{\mathbf{zz}}(n,m)$, respectively.

Note that the STFD of a vector signal is a matrix whose diagonal entries are the classical auto-TFDs of the vector components, and the off-diagonal entries are the cross-TFDs.

A more general definition of the STFD can be given as

$$\mathbf{D}_{\mathbf{zz}}(n,k) = \sum_{m=-M}^{M}\sum_{p=-M}^{M} \mathbf{G}(p-n,m)\odot(\mathbf{z}(p+m)\mathbf{z}^H(p-m)e^{-j4\pi\frac{mk}{N}}) \tag{25.54}$$

where $\odot$ designates the Hadamard product and $[\mathbf{G}(n,m)]_{ij} = G_{ij}(n,m)$ is the time-lag kernel associated with the pair of sensor signals $z_i(n)$ and $z_j(n)$ (kernels G_{ij} might be chosen according to the nature of considered signals $z_i(n)$ and $z_j(n)$ if such *a priori* information is available).

25.4.1.1 Structure under linear model

Consider the following linear model of the vector signal $\mathbf{z}(n)$:

$$\mathbf{z}(n) = \mathbf{A}\mathbf{s}(n) \tag{25.55}$$

where $\mathbf{A}$ is a $K \times L$ matrix $(K \geq L)$ and $\mathbf{s}(n)$ is a $L \times 1$ vector, which is referred to as the source signal vector.

Under this linear model the STFDs take the following structure:

$$\mathbf{D}_{\mathbf{zz}}(n,k) = \mathbf{A}\mathbf{D}_{\mathbf{ss}}(n,k)\mathbf{A}^H \tag{25.56}$$

where $\mathbf{D}_{\mathbf{ss}}(n,k)$ is the source STFD of vector $\mathbf{s}(n)$ whose entries are the auto- and cross-TFDs of the source signals.

The auto-STFD denoted by $\mathbf{D}^a_{\mathbf{zz}}(n,k)$ is the STFD, $\mathbf{D}_{\mathbf{zz}}(n,k)$, evaluated at autoterm points only. Correspondingly, the cross STFD $\mathbf{D}^c_{\mathbf{zz}}(n,k)$ is the STFD, $\mathbf{D}_{\mathbf{zz}}(n,k)$, evaluated at cross-term points.

Note that the diagonal (off-diagonal) elements of $\mathbf{D}_{\mathbf{ss}}(n,k)$ are autoterms (cross-terms). Thus, the auto- (cross-) STFD $\mathbf{D}^a_{\mathbf{ss}}(n,k)$ ($\mathbf{D}^c_{\mathbf{ss}}(n,k)$) is diagonal (off-diagonal[12]) for each time–frequency point that corresponds to a source autoterm (cross-term), provided the window effect is neglected.

25.4.1.2 Structure under Unitary Model

Denote by $\mathbf{W}$ a $L \times K$ whitening matrix such that

$$(\mathbf{W}\mathbf{A})(\mathbf{W}\mathbf{A})^H = \mathbf{U}\mathbf{U}^H = \mathbf{I} \tag{25.57}$$

Pre- and postmultiplying the STFD $\mathbf{D}_{\mathbf{zz}}(n,k)$ by $\mathbf{W}$ leads to the whitened STFD, defined as

$$\underline{\mathbf{D}}_{\mathbf{zz}}(n,k) = \mathbf{W}\mathbf{D}_{\mathbf{zz}}(n,k)\mathbf{W}^H = \mathbf{U}\mathbf{D}_{\mathbf{ss}}(n,k)\mathbf{U}^H \tag{25.58}$$

where the second equality stems from the definition of $\mathbf{W}$ and Equation 25.56. This above whitening leads to a linear model with a unitary mixing matrix.

Note that the whitening matrix can be computed in different ways. It can be obtained, for example, as an inverse square root of the observation covariance matrix [78] or computed from the STFD matrices as shown in [90].

At autoterm points, the whitened auto-STFD has the following structure:

$$\underline{\mathbf{D}}_{\mathbf{zz}}(n,k) = \mathbf{U}\mathbf{D}^a_{\mathbf{ss}}(n,k)\mathbf{U}^H \tag{25.59}$$

where $\mathbf{D}^a_{\mathbf{ss}}(n,k)$ is diagonal. However, at cross-term points, the whitened cross STFD exhibits the following structure:

$$\underline{\mathbf{D}}_{\mathbf{zz}}(n,k) = \mathbf{U}\mathbf{D}^c_{\mathbf{ss}}(n,k)\mathbf{U}^H \tag{25.60}$$

where $\mathbf{D}^c_{\mathbf{ss}}(n,k)$ is off-diagonal.

The above-defined STFDs permit the application of subspace techniques to solve a large class of channel estimation and equalization, blind source separation, and high-resolution DOA estimation problems. For the blind source separation problem, the STFDs allow the separation of Gaussian sources with identical spectral shape but with different time–frequency signatures [78]. In the area of DOA finding, the estimation of the signal and noise subspaces from the STFDs highly improves the angular resolution performance.

[12] A matrix is off-diagonal if its diagonal entries are zeros.

25.4.2 STFD Structure in Wireless Communications

In wireless communications, when L user signals arrive at a K-element antenna, the linear data model

$$\mathbf{z}(n) = \mathbf{A}\mathbf{s}(n) + \mathbf{n}(n) \tag{25.61}$$

is commonly assumed, where we recall that $\mathbf{z}(n)$ is the $K \times 1$ data vector received at the antennae and $\mathbf{s}(n)$ is the $L \times 1$ user data vector, the spatial matrix $\mathbf{A} = [\mathbf{a}_1 \cdots \mathbf{a}_n]$ represents the propagation matrix,[13] $\mathbf{a}_i$ is the steering vector corresponding to the ith user, and $\mathbf{n}(n)$ is an additive noise vector whose entries are modeled as stationary, temporally and spatially white, zero-mean random processes, and independent of the user-emitted signals.

Under the above assumptions, the expectation of the TFD matrix between the user signal and the noise vectors vanishes, i.e.,

$$E\{\mathbf{D}_{\mathbf{sn}}(n,k)\} = 0 \tag{25.62}$$

and it follows that

$$\tilde{\mathbf{D}}_{\mathbf{zz}}(n,k) = \mathbf{A}\tilde{\mathbf{D}}_{\mathbf{ss}}(n,k)\mathbf{A}^H + \sigma^2\mathbf{I} \tag{25.63}$$

with

$$\tilde{\mathbf{D}}_{\mathbf{zz}}(n,k) = E\{\mathbf{D}_{\mathbf{zz}}(n,k)\} \tag{25.64}$$

$$\tilde{\mathbf{D}}_{\mathbf{ss}}(n,k) = E\{\mathbf{D}_{\mathbf{ss}}(n,k)\} \tag{25.65}$$

where σ^2 is the noise power and $\mathbf{I}$ is the identity matrix. Under the same assumptions, the data covariance matrix, which is commonly used in array signal processing, is given by

$$\mathbf{R}_{\mathbf{zz}} = \mathbf{A}\mathbf{R}_{\mathbf{ss}}\mathbf{A}^H + \sigma^2\mathbf{I} \tag{25.66}$$

where

$$\mathbf{R}_{\mathbf{zz}} = E\{\mathbf{z}(n)\mathbf{z}(n)^H\} \tag{25.67}$$

$$\mathbf{R}_{\mathbf{ss}} = E\{\mathbf{s}(n)\mathbf{s}(n)^H\} \tag{25.68}$$

From Equations 25.63 and 25.66, it becomes clear that the STFDs and the covariance matrix exhibit the same eigenstructure. This structure is often exploited to estimate some signal parameters through subspace-based techniques.

25.4.3 Advantages of STFDs over Covariance Matrix

The STFDs allow the processing of the received data in both the spatial domain and the two-dimensional time–frequency domain simultaneously. In time–frequency array signal processing, the STFDs are eigen-decomposed, instead of the traditional covariance matrix $\mathbf{R}_{\mathbf{zz}}$, to separate the signal subspace and noise subspace. Thanks to the availability of time-varying filtering in the time–frequency domain, the STFD-based approaches can handle signals corrupted by interference occupying the same frequency band or the same time slot, but with different time–frequency signatures; thus, signal selectivity is increased with respect to covariance matrix-based methods. In addition, the effect of spreading noise power while localizing the user energy in the time–frequency plane amounts to increased robustness of the STFD-based approaches with respect to noise. In other words, the eigenvectors of the signal subspace obtained from an STFD matrix that is made up of signal autoterms are more robust to noise than those obtained from the

[13]This matrix is also known as the mixing matrix.

covariance matrix. Hence, the performance of the STFD-based approaches can be significantly improved, particularly when the input SNR is low[14] (typically, an SNR of 0 dB or lower). Moreover, in [86] it is proved that the traditional covariance-based subspace methods are low-dimensional cases of the STFD subspace methods.

If one selects the kernel $G(n, m)$ in Equation 25.53 so that the corresponding TFD satisfies the marginal condition

$$\sum_k \tilde{D}_{z_i z_j}(n, k) = E\{z_i(n) z_j(n)^*\} \tag{25.69}$$

then

$$\sum_k \tilde{\mathbf{D}}_{\mathbf{zz}}(n, k) = E\{\mathbf{z}(n)\mathbf{z}(n)^H\} = \mathbf{R}_{\mathbf{zz}} \tag{25.70}$$

The above equation shows that the projection of the STFD over the time domain is nothing more than the traditional covariance matrix. Hence, the space spanned by $\mathbf{R}_{\mathbf{zz}}$ is the projection of the space spanned by $\tilde{\mathbf{D}}_{\mathbf{zz}}(n, k)$ over the space that is orthogonal to the frequency dimension. This means that the space spanned by $\tilde{\mathbf{D}}_{\mathbf{zz}}(n, k)$ is the extension of the space spanned by $\mathbf{R}_{\mathbf{zz}}$ toward a higher-dimension space. Therefore, the $\mathbf{R}_{\mathbf{zz}}$-based techniques can be seen as a low-dimension special case of the $\tilde{\mathbf{D}}_{\mathbf{zz}}(n, k)$-based ones. This is quite straightforward since the STFD-based methods are multidimensional (spatial–time–frequency) processing methods. Obviously, the details and signatures of the signal will be described more accurately and finely in higher-dimension space. In fact, this is the reason that the STFD-based methods have better performance, such as signal selectivity, interference suppression, and high resolution, over the conventional covariance matrix-based approaches.

25.4.4 Selection of Autoterms and Cross-Terms in the Time–Frequency Domain

STFD-based methods require computation of STFDs at different time–frequency points. At autoterm points, where the diagonal structure of the source STFD is enforced, the data STFDs are either incorporated into a joint diagonalization (JD) technique or eigendecomposed after simple averaging over the source signatures of interest to estimate the mixing, or the array manifold matrix. At cross-term points, where this time the off-diagonal structure of the source STFD is enforced, the data STFDs are incorporated in an off-diagonalization technique to achieve the task of the mixing/propagation matrix identification.

An intuitive procedure to select the autoterms is to consider the time–frequency points corresponding to the maximum energy in the time–frequency plane [75]. The above intuitive procedure has shown some limitations in practical situations. A projection-based selection procedure of cross-terms and autoterms has been proposed in [83]. The latter exploits the off-diagonal structure of the cross-source STFDs and proceeds on whitened data STFDs. More precisely, for a cross-source STFD, we have

$$\text{Trace}(\underline{\mathbf{D}}^c_{\mathbf{zz}}(n, k)) = \text{Trace}(\mathbf{U}\mathbf{D}^c_{\mathbf{ss}}(n, k)\mathbf{U}^H) = \text{Trace}(\mathbf{D}^c_{\mathbf{ss}}(n, k)) \approx 0 \tag{25.71}$$

Based on this observation, the following testing procedure applies:

$$\text{if } \frac{\text{Trace}\{\hat{\underline{\mathbf{D}}}_{\mathbf{zz}}(n, k)\}}{\text{norm }\{\hat{\underline{\mathbf{D}}}_{\mathbf{zz}}(n, k)\}} < \epsilon \rightarrow \text{decide that } (n, k) \text{ is a cross-term point}$$

where the threshold ϵ is a positive scalar (typically $\epsilon = 0.9$). In the underdetermined case (i.e., $K < L$), the matrix $\mathbf{U}$ (see Section 25.4.1.2) is nonsquare (with more columns than rows), and consequently, $\mathbf{U}^H\mathbf{U} \neq \mathbf{I}$

[14] Subspace analysis of the STFDs vs. the covariance matrix is provided in [82].

represents the projection matrix onto the row space of **U**. Therefore, Equation 25.71 becomes only an approximation. An alternative solution consists of exploiting the existence of only one source at some autoterm points. At such points, each autoterm STFD matrix is of rank one, or at least has one large eigenvalue compared to its other eigenvalues. Therefore, one can use rank selection criteria, such as MDL (minimum description length) or AIC (Akaike information criterion) [91], to select autoterm points as those corresponding to STFD matrices of selected rank equal to one. For simplicity, the following criterion can be used:

$$\text{if } \left| \frac{\lambda \text{max}\{\hat{\mathbf{D}}_{\mathbf{zz}}(n,k)\}}{\text{norm}\{\hat{\mathbf{D}}_{\mathbf{zz}}(n,k)\}} - 1 \right| > \epsilon \rightarrow \text{ decide that } (n,k) \text{ is a cross-term point}$$

where ϵ is a small positive scalar (typically, $\epsilon = 1\text{E-}4$) and $\lambda\text{max}\{\hat{\mathbf{D}}_{\mathbf{zz}}(n,k)\}$ represents the largest eigenvalue of $\hat{\mathbf{D}}_{\mathbf{zz}}(n,k)$.

A statistical test to decide whether a time–frequency point is dominated by auto- or cross-terms is proposed in [92]. The latter consider the following test statistic:

$$\frac{\text{Trace}\{\underline{\hat{\mathbf{D}}}_{\mathbf{zz}}(n,k)\}}{\text{norm}\{\underline{\hat{\mathbf{D}}}_{\mathbf{zz}}(n,k)\}} \tag{25.72}$$

To discriminate between noise and either auto- or cross-term, the variance of the test statistic is used. Because only a single value of the test statistic is known at the time–frequency point under test, the variance is estimated using a bootstrap resampling technique [88, 92]. Once the noise regions in the time–frequency domain are identified, a threshold is set to distinguish the autoterms from the cross-terms. In [93], array averaging of the STFDs is used to reduce the cross-terms without smearing the autoterms, allowing the autoterms to be more pronounced and easier to detect in the time–frequency plane.

In [94], it is shown that for real-valued signals, the imaginary parts of the STFDs, when not equal to zero, only correspond to cross-terms whatever the considered point in the time–frequency plane. This result was exploited to derive a criterion for the auto- and cross-term selection. In the case of noisy signals, reference [85] describes a detection criteria of cross- and auto-terms in the time–frequency plane by introducing two thresholds based on the Bayesian and Neyman–Pearson approaches.

The selection of autoterms in the time–frequency domain is still an open problem. And the success of any STFD-based technique depends highly on the performance of the employed autoterm selection procedure.

25.4.5 Time–Frequency Direction-of-Arrival Estimation

In order to obtain the mobile users' spatial information and achieve the space-division multiple access (SDMA), the DOA estimation of far field sources from the multiantenna outputs is one of the important issues in next-generation wireless communications. Thanks to their super resolution and robustness, the subspace-based methods, such as MUSIC [95] and ESPRIT [96], are considered the most popular techniques in traditional array processing. However, all these subspace methods assume the signals impinged on the antennae stationary, while typical signals in wireless communications, such as frequency-hopping signals or frequency-modulated signals, are nonstationary with some *a priori* known information on their time-varying frequency content. In addition, several nonspatial features such as time and frequency signatures of the signals are ignored in conventional methods. These defects may result in unaffordable estimation error.

In most wireless communications systems, the signals are man-made and hence much information contained in these signals is known or can be obtained *a priori*. One can exploit this information not only in the spatial domain but also in the time–frequency domain in order to improve the performance. One of these techniques is the STFD-based DOA estimation method. Recently, several traditional DOA estimation techniques have been extended to nonstationary signals thanks to the use of STFD instead of the covariance matrix. Hence, time–frequency MUSIC (TF-MUSIC) was first introduced in [79]; then time–frequency maximum likelihood (TF-ML), time–frequency signal subspace fitting (TF-SSF), and time–frequency

ESPRIT (TF-ESPRIT) were introduced in [97], [86], and [89], respectively. Below, we describe only TF-MUSIC as an illustrative example on how the STFDs can be exploited for DOA estimation.

25.4.5.1 Data Model

Consider again Equation 25.61, which is often encountered in wireless communications:

$$\mathbf{z}(n) = \mathbf{A}(\theta)\mathbf{s}(n) + \mathbf{n}(n) \tag{25.73}$$

Herein, the propagation matrix $\mathbf{A}(\theta) = [\mathbf{a}(\theta_1), \ldots, \mathbf{a}(\theta_L)]^T$, also known as steering matrix, is parameterized by the parameter vector $\theta = [\theta_1, \ldots, \theta_L]^T$, where $\mathbf{a}(\theta_k)$ and θ_k define the steering vector and the DOA of the kth user, respectively.

Assuming a noise-free environment, the structure of the STFD associated with the above model is given by

$$\mathbf{D}_{\mathbf{zz}}(n, k) = \mathbf{A}(\theta)\mathbf{D}_{\mathbf{ss}}(n, k)\mathbf{A}(\theta)^H \tag{25.74}$$

25.4.5.2 TF-MUSIC

By performing the singular value decomposition (SVD) of the steering matrix,

$$\mathbf{A}(\theta) = [\mathbf{E}_s \mathbf{E}_n][\mathbf{D} \quad \mathbf{0}]^T \mathbf{V}^H \tag{25.75}$$

and incorporating the results in Equation 25.74, it is easily shown that

$$\mathbf{D}_{\mathbf{zz}}(n, k) = [\mathbf{E}_s \mathbf{E}_n]\mathbf{D}(n, k)[\mathbf{E}_s \mathbf{E}_n]^H \tag{25.76}$$

where $\mathbf{D}(n, k)$ is a block-diagonal matrix given by

$$\mathbf{D}(n, k) = \text{diag}[\mathbf{D}\mathbf{V}^H \mathbf{D}_{\mathbf{ss}}(n, k)\mathbf{V}\mathbf{D} \quad \mathbf{0}] \tag{25.77}$$

Since $\mathbf{E}_s$ and $\mathbf{E}_n$, which span the signal subspace and noise subspace, respectively, are fixed and independent of the time–frequency point (n, k), Equation 25.76 reveals that any matrix $\mathbf{D}_{\mathbf{zz}}(n, k)$ is block-diagonalized by the unitary transform $\mathbf{E} = [\mathbf{E}_s \; \mathbf{E}_n]$.

A simple way to estimate $\mathbf{E}_s$ and $\mathbf{E}_n$ is to perform the SVD on a single matrix $\mathbf{D}_{\mathbf{zz}}(n, k)$ or an averaged version of $\mathbf{D}_{\mathbf{zz}}(n, k)$ over the source signatures of interest. But indeterminacy arises in the case where $\mathbf{D}_{\mathbf{ss}}(n, k)$ is singular. To avoid this problem, a joint block diagonalization (JBD) of the combined set of $\{\mathbf{D}_{\mathbf{zz}}(n_l, k_l) | l = 1, \ldots, P\}$ can be performed by exploiting the joint structure (Equation 25.76) of the STFDs. This JBD is achieved by the maximization under unitary transform of the following criterion:

$$C(\mathbf{U}) \triangleq \sum_{l=1}^{P} \sum_{i,j=1}^{L} \left| \mathbf{u}_i^H \mathbf{D}_{\mathbf{zz}}(n_l, k_l) \mathbf{u}_j \right|^2 \tag{25.78}$$

over the set of unitary matrices $\mathbf{U} = [\mathbf{u}_1, \ldots, \mathbf{u}_K]$. Note that in [98], an efficient algorithm for solving Equation 25.78 exists. Once the signal and noise subspaces are estimated, one can use any subspace-based technique to estimate the DOAs. The MUSIC algorithm [95] is then applied to the noise subspace matrix $\hat{\mathbf{E}}_n$ estimated from Equation 25.78. Hence, the TF-MUSIC algorithm estimates the DOAs by finding the L largest peaks of the localization function

$$f(\theta) = \left| \hat{\mathbf{E}}_n^H \mathbf{a}(\theta) \right|^{-2} \tag{25.79}$$

25.4.6 Time–Frequency Source Separation

Currently, blind source separation is considered one of most promising techniques in wireless communications and more specifically in multiuser detection. The underlying problem consists of recovering the original waveforms of the user-emitted signals without any knowledge on their linear mixture. This mixture can be either instantaneous or convolutive. The problem of blind source separation has two inherent indeterminacies such that source signals can only be identified up to a fixed permutation and some complex factors [99].

So far, the problem of blind source separation has been solved using statistical information available on the source signals. The first solution was based on the cancellation of higher-order moments assuming non-Gaussian and independent and identically distributed (i.i.d.) signals [100]. Other solutions based on minimization of cost functions, such as contrast functions [101] or likelihood function [102], have been proposed. In the case of non-i.i.d. signals and even Gaussian sources, solutions based on second-order statistics were also proposed [99].

When the frequency content of the source signals is time varying, one can exploit the powerful tool of the STFDs to separate and recover the incoming signals. In this context, the underlying problem can be regarded as signal synthesis from the time–frequency plane with the incorporation of the spatial diversity provided by the antennae.

In contrast to conventional blind source separation approaches, the STFD-based signal separation techniques allow separation of Gaussian sources with identical spectral shape provided that the sources have different time–frequency signatures. Below, we describe applications of the STFDs for the separation of both instantaneous and convolutive mixtures.

25.4.6.1 Separation of Instantaneous Mixture

The multiantenna signal $\mathbf{z}(n)$ is assumed to be nonstationary and to obey the linear model in Equation 25.55. The problem under consideration consists of identifying the matrix $\mathbf{A}$ and recovering the source signals $\mathbf{s}(n)$ up to a fixed permutation and some complex factors.

By selecting autoterm points, the whitened auto-STFDs have the structure in Equation 25.59 that we recall herein:

$$\underline{\mathbf{D}}_{\mathbf{zz}}(n,k) = \mathbf{U}\mathbf{D}^{a}_{\mathbf{ss}}(n,k)\mathbf{U}^{H} \tag{25.80}$$

with $\mathbf{D}^{a}_{\mathbf{ss}}(n,k)$ a diagonal matrix. The missing unitary matrix $\mathbf{U}$ is retrieved up to permutation and phase shifts by JD of a combined set $\{\underline{\mathbf{D}}_{\mathbf{zz}}(n_a,k_a)|a=1,\ldots,P\}$ of P auto-STFDs. The incorporation of several autoterm points in the JD reduces the likelihood of having degenerate eigenvalues and increases robustness to a possible additive noise. The above JD is defined as the maximization of the following criterion:

$$C_{JD}(\mathbf{V}) \triangleq \sum_{a=1}^{P}\sum_{i=1}^{L}\left|\mathbf{v}_i^H\underline{\mathbf{D}}_{\mathbf{zz}}(n_a,k_a)\mathbf{v}_i\right|^2 \tag{25.81}$$

over the set of unitary matrices $\mathbf{V}=[\mathbf{v}_1,\ldots,\mathbf{v}_L]$.

The selection of cross-term points leads to the whitened cross STFD (Equation 25.60),

$$\underline{\mathbf{D}}_{\mathbf{zz}}(n,k) = \mathbf{U}\mathbf{D}^{c}_{\mathbf{ss}}(n,k)\mathbf{U}^{H} \tag{25.82}$$

with $\mathbf{D}^{c}_{\mathbf{ss}}(n,k)$ an off-diagonal matrix. The unitary matrix $\mathbf{U}$ is found up to permutation and phase shifts by joint off-diagonalization (JOD) of a combined set $\{\underline{\mathbf{D}}_{\mathbf{zz}}(n_c,k_c)|c=1,\ldots,Q\}$ of Q cross-STFDs. This JOD is defined as the maximization of the following criterion:

$$C_{JOD}(\mathbf{V}) \triangleq -\sum_{c=1}^{Q}\sum_{i=1}^{L}\left|\mathbf{v}_i^H\underline{\mathbf{D}}_{\mathbf{zz}}(n_c,k_c)\mathbf{v}_i\right|^2 \tag{25.83}$$

over the set of unitary matrices $\mathbf{V}=[\mathbf{v}_1,\ldots,\mathbf{v}_L]$.

The unitary matrix $\mathbf{U}$ can also be found up to permutation and phase shifts by a combined JD/JOD of the two sets $\{\underline{\mathbf{D}}_{\mathbf{zz}}(n_a,k_a)|a=1,\ldots,P\}$ and $\{\underline{\mathbf{D}}^{c}_{\mathbf{zz}}(n_c,k_c)|c=1,\ldots,Q\}$.

Once the unitary matrix $\mathbf{U}$ is obtained from either the JD, JOD, or combined JD/JOD, an estimate of the mixing matrix $\mathbf{A}$ can be computed by the product $\mathbf{W}^{\#}\mathbf{U}$, where $\mathbf{W}$ is the whitening matrix (see Section 25.4.1.2). An estimate of the source signals $\mathbf{s}(n)$ can then be obtained by the product $\mathbf{A}^{\#}\mathbf{z}(n)$.

25.4.6.2 Separating More Sources Than Sensors

A challenging problem consists of the blind separation of more sources than sensors (i.e., $L > K$); this problem, also known as the underdetermined blind source separation problem, was pointed out for the first time in [102] while separating discrete sources. Since then, several other works based on *a priori* knowledge of the probability density functions of the sources [103, 104] were conducted. In [105], an approach for the resolution of the aforementioned problem exploits the concept of disjoint orthogonality of short Fourier transforms. Herein, for the resolution of the underdetermined problem, we review a STFD-based blind source separation method [84].

We start by selecting autoterm points where only one source exists, as described in Section 25.4.4. The corresponding STFD then has the following form:

$$\mathbf{D}_{\mathbf{zz}}(n,k) = D_{s_i s_i}(n,k)\mathbf{a}_i\mathbf{a}_i^H, \text{ where } (n,k) \in \Omega_i \tag{25.84}$$

where Ω_i denotes the time–frequency support of the ith source. The idea of the algorithm consists of clustering together the autoterm points associated with the same principal eigenvector of $\mathbf{D}_{\mathbf{zz}}(n,k)$ representing a particular source signal. Once the clustering and classification of the autoterms are done, the estimates of the source signals are obtained from the selected autoterms using a time–frequency synthesis algorithm [20]. Note that the missing autoterms in the classification, often due to intersection points, are automatically interpolated in the synthesis process. An advanced clustering technique of the above autoterms based on gap statistics is proposed in [106].

25.4.6.3 Separation of Convolutive Mixtures

Consider a convolutive multiple-input multiple-output linear time-invariant model given by

$$z_i(n) = \sum_{j=1}^{L}\sum_{c=0}^{C} a_{ij}(c)s_j(n-c) \text{ for } i = 1,\ldots,K \tag{25.85}$$

where $s_j(n)$, $j = 1,\ldots,L$, are the L source signals; $z_i(n)$, $i = 1,\ldots,K$, are the $K > L$ sensor signals; and $a_{ij}(c)$ is the transfer function between the jth source and the ith sensor with an overall extent of $(C+1)$ taps. The sources are assumed to have different time–frequency signatures, and the channel matrix $\mathbf{A}$ defined below in Equation 25.87 is full column rank.

In matrix form, Equation 25.85 becomes

$$\mathbf{z}(n) = \mathbf{A}\mathbf{s}(n) \tag{25.86}$$

where

$$\mathbf{s}(n) = [s_1(n),\ldots,s_1(n-(C+C')+1),\ldots,s_L(n-(C+C')+1)]^T$$
$$\mathbf{z}(n) = [z_1(n),\ldots,z_1(n-C'+1),\ldots,z_K(n-C'+1)]^T$$

$$\mathbf{A} = \begin{bmatrix} \mathbf{A}_{11} & \cdots & \mathbf{A}_{1L} \\ \vdots & \ddots & \vdots \\ \mathbf{A}_{K1} & \cdots & \mathbf{A}_{KL} \end{bmatrix} \tag{25.87}$$

with

$$\mathbf{A}_{ij} = \begin{bmatrix} a_{ij}(0) & \cdots & a_{ij}(C) & \cdots & 0 \\ & \ddots & \ddots & \ddots & \\ 0 & \cdots & a_{ij}(0) & \cdots & a_{ij}(C) \end{bmatrix} \tag{25.88}$$

Note that $\mathbf{A}$ is a $[KC' \times L(C+C')]$ matrix and $\mathbf{A}_{ij}$ are $[C' \times (C+C')]$ matrices. C' is chosen such that $KC' \geq L(C+C')$.

Herein, the same formalism as in the instantaneous mixture case is retrieved and the data STFDs still have the same expression as in Equation 25.56. However, the source auto-STFDs, $\mathbf{D}^a_{\mathbf{ss}}(n,k)$, are not diagonal but block diagonal with diagonal blocks of size $(C+C')\times(C+C')$. Note that the block diagonal structure comes from the fact that the cross-terms between $s_i(n)$ and $s_i(n-d)$, where d is some delay, are not zero and depend on the local correlation structure of the signal. This block diagonal structure is exploited to achieve the separation of the convolutive mixture.

25.4.6.4 STFD-Based Separation

First the data vector $\mathbf{z}(n)$ is whitened. The whitening matrix $\mathbf{W}$ is of size $[L(C'+C)\times KC']$ and verifies

$$\mathbf{W}E\{\mathbf{z}(n)\mathbf{z}(n)^H\}\mathbf{W}^H = \mathbf{W}\mathbf{R}_{\mathbf{zz}}\mathbf{W}^H = (\mathbf{W}\mathbf{A}\mathbf{R}_{\mathbf{ss}}^{\frac{1}{2}})(\mathbf{W}\mathbf{A}\mathbf{R}_{\mathbf{ss}}^{\frac{1}{2}})^H = \mathbf{I} \tag{25.89}$$

where $\mathbf{R}_{\mathbf{zz}}$ and $\mathbf{R}_{\mathbf{ss}}$ denote the covariance matrices of $\mathbf{z}(n)$ and $\mathbf{s}(n)$, respectively. Equation 25.89 shows that if $\mathbf{W}$ is a whitening matrix, then

$$\mathbf{U} = \mathbf{W}\mathbf{A}\mathbf{R}_{\mathbf{ss}}^{\frac{1}{2}} \tag{25.90}$$

is a $L(C'+C)\times L(C'+C)$ unitary matrix where $\mathbf{R}_{\mathbf{ss}}^{\frac{1}{2}}$ (Hermitian square root matrix of $\mathbf{R}_{\mathbf{ss}}$) is block diagonal. The whitening matrix $\mathbf{W}$ can be determined from the eigendecomposition of the data covariance matrix $\mathbf{R}_{\mathbf{zz}}$ as in [78].

Now by considering the whitened STFD matrices $\underline{\mathbf{D}}_{\mathbf{zz}}(n,k)$ and the above relations, we obtain the key relation

$$\underline{\mathbf{D}}_{\mathbf{zz}}(n,k) = \mathbf{U}\mathbf{R}_{\mathbf{ss}}^{-\frac{1}{2}}\mathbf{D}_{\mathbf{ss}}(n,k)\mathbf{R}_{\mathbf{ss}}^{-\frac{1}{2}}\mathbf{U}^H = \mathbf{U}\mathbf{D}(n,k)\mathbf{U}^H \tag{25.91}$$

where $\mathbf{D}(n,k) = \mathbf{R}_{\mathbf{ss}}^{-\frac{1}{2}}\mathbf{D}_{\mathbf{ss}}(n,k)\mathbf{R}_{\mathbf{ss}}^{-\frac{1}{2}}$.

Since the matrix $\mathbf{U}$ is unitary and $\mathbf{D}(n,k)$ is block diagonal, the latter just means that any whitened STFD matrix is block diagonal in the basis of the column vectors of matrix $\mathbf{U}$. The unitary matrix can be retrieved by computing the block diagonalization of some matrix $\underline{\mathbf{D}}_{\mathbf{zz}}(n,k)$. But to reduce the likelihood of indeterminacy and increase the robustness of determining $\mathbf{U}$, we consider the JBD of a set $\{\underline{\mathbf{D}}_{\mathbf{zz}}(n_l,k_l);\ l=1,\ldots,P\}$ of P whitened STFD matrices. This JBD is achieved by the maximization under unitary transform of the following criterion:

$$C(\mathbf{U}) \triangleq \sum_{l=1}^{P}\sum_{m=1}^{L}\sum_{i,j=(C'+C)(m-1)+1}^{(C'+C)m}\left|\mathbf{u}_i^H\underline{\mathbf{D}}_{\mathbf{zz}}(n_l,k_l)\mathbf{u}_j\right|^2 \tag{25.92}$$

over the set of unitary matrices $\mathbf{U} = [\mathbf{u}_1,\ldots,\mathbf{u}_{L(C'+C)}]$. Note that an efficient Jacobi-like algorithm for the minimization of Equation 25.92 exists in [98, 107].

Once the unitary matrix $\mathbf{U}$ is determined up to a block diagonal unitary matrix $\mathbf{D}$ coming from the inherent indeterminacy of the JBD problem [108], the recovered signals are obtained up to a filter by

$$\hat{\mathbf{s}}(n) = \mathbf{U}^H\mathbf{W}\mathbf{z}(n) \tag{25.93}$$

According to Equations 25.86 and 25.90, the recovered signals verify

$$\hat{\mathbf{s}}(n) = \tilde{\mathbf{D}}\mathbf{s}(n) \tag{25.94}$$

with

$$\tilde{\mathbf{D}} = \mathbf{D}\mathbf{R}_{\mathbf{ss}}^{-\frac{1}{2}} \tag{25.95}$$

where we recall that the matrix $\mathbf{R}_{\mathbf{ss}}^{-\frac{1}{2}}$ and $\mathbf{D}$ are the block diagonal matrix and unitary block diagonal matrix, respectively. Consequently, $\tilde{\mathbf{D}}$ is also a block diagonal matrix, and the above STFD-based technique leads to the separation of the convolutive mixture up to a filter instead of a full MIMO deconvolution procedure.

Note that if needed, a SIMO (single-input multi-output) deconvolution/equalization [109] can be applied to the estimated sources of Equation 25.94.

25.5 Other TFSP Applications in Wireless Communications

25.5.1 Precoding for LTV Channels

Linear precoding is a useful signal processing tool for coping with frequency-selective propagation channels encountered with high-rate wireless transmission.

Precoding consists of mapping each incoming block of M symbols onto a P-long vector through a $P \times M$ $(P > M)$ matrix referred to as the precoding matrix. Each received block is then multiplied by a $M \times P$ decoding matrix to retrieve the original symbols under the condition $M > L$ and $P = M + L$, where L is the overall channel length. To avoid interblock interference, guard intervals can be used, as in OFDM, for example. This can be done by forcing either the last L rows of the precoding matrix or the first L columns of the decoding matrix to zero [110].

Precoding of LTV channels can be optimized by *a priori* knowledge of the channel temporal evolution. This knowledge can be provided by a feedback channel such that the receiver estimates periodically the channel parameters, also called channel status information (CSI) [110], and sends them back to the transmitter. The latter uses this CSI to predict the channel evolution within a finite time interval and commutes the optimal precoder. The optimality herein should be understood in the sense of maximizing the information rate over the linear channel affected by additive Gaussian noise. Under the constraint of a fixed average transmit power, the optimal precoder, i.e., the optimal precoding matrix, is obtained from the SVD of the channel matrix[15] [111, 113].

In [110], a physical interpretation of the optimal precoding for time- and frequency-dispersive channels is provided thanks to an approximate analytic model for the eigenfunctions of LTV channels.[16] The approximate model is valid for multipath channels with finite Doppler and delay spread. Under the above model and using a time–frequency representation of the eigenfunctions, the latter are shown to be characterized by an energy distribution along curves, in the time–frequency plane, given by contour lines of the time–frequency representation of the LTV channel. In the same reference, it is also shown under mild conditions often met in practice that the TFDs of the right singular vectors of the LTV channel are mainly concentrated along the curves where the energy in the time–frequency domain of the channel equals the square of the associated singular value. The TFDs of the left singular vectors are simply time- and frequency-shift versions of the TFDs of the right singular vectors. This interpretation clearly establishes the optimal power allocation in the time–frequency domain as a generalization of the well-known water-filling principle [112, 114]. The above interpretation also allows an approximate computation of the channel singular vectors and values directly from the time–frequency representation of the LTV channel, without computing the SVD.

25.5.2 Signaling Using Chirp Modulation

TFSP tools can be used for the receiver design and for optimizing the design parameters of a spreading system using a chirp modulation scheme.

Indeed, chirp signals or, equivalently, linear FM signals have been widely used in sonar applications for range and Doppler estimation, as well as in radar systems for pulse compression. Thanks to their particular time–frequency signatures, these signals provide high interference rejection and inherent immunity against

[15]The channel matrix is the transfer matrix from the transmitted block vector to the received block vector.

[16]$v(t)$ is said to be an eigenfunction of $f(t)$ if and only if $\lambda v(t) = \int_{-\infty}^{+\infty} f(t - t')v(t')dt'$, where λ is known as the associated eigenvalue.

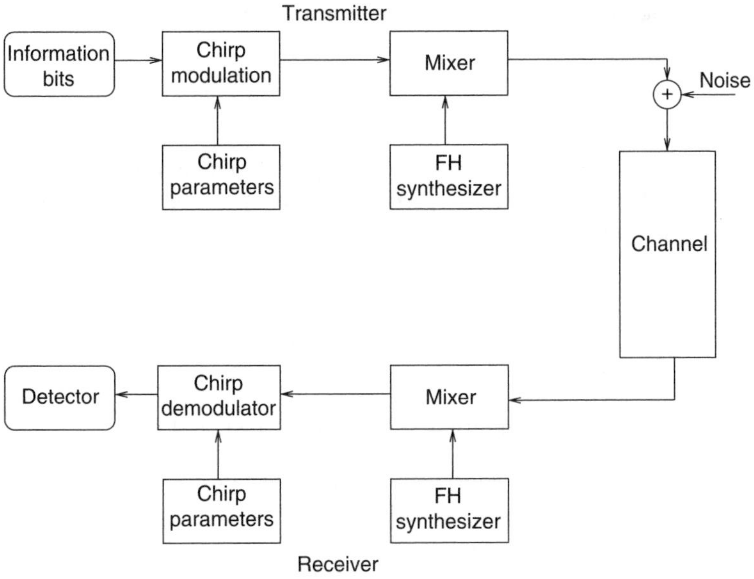

FIGURE 25.11 Chirp modulation signaling and FH communications system.

Doppler shift and multipath fading [115] in wireless communications. In addition, they are bandwidth efficient [117]. It have been shown [118] that for a same SNR and Doppler shift, the chirp signaling outperforms the frequency shift keying (FSK) signaling, thanks to their better cross-coherence properties, compared to FSK.

Signaling using chirp modulation is also seen as a spread-spectrum technique, which is defined as "a mean of transmission in which the signal occupies a bandwidth in excess of the minimum necessary to send the information" [41]. Chirp modulation was first suggested in [117] by using a pair of linear chirps with opposite chirp rates for binary signaling. A system for multiple access within a common frequency band was proposed in [119] by assigning pairs of linear chirps with different chirp rates to several users. In this system, the number of users simultaneously accessing the shared resources is limited by the MAI for a given time–bandwidth product. To reduce this shortcoming, the chirp signals are selected in [116] such that they all have the same power as well as the same bandwidth, offering inherent protection against frequency-selective fading. Further, the combination of chirp modulation signaling with frequency-hopping (FH) multiple access was proposed in [115]. The obtained hybrid system (Figure 25.11) improves communications system performance, especially in multipath fading-dispersive channels. Note that this system was extended to FH-CDMA in [120] and compared to the FSK-FH-CDMA system, leading to the same conclusion as in [115]. In [121], the chirp parameters, to be used in chirp modulation signaling, are selected under the actual time–bandwidth requirements so as to reduce significantly multiple-access interference and bit error rates.

25.5.3 Detection of FM Signals in Rayleigh Fading

Diversity reception is currently one of the most effective techniques for coping with the multipath Rayleigh fading effect in mobile environments [5]. It requires a number of signal transmission paths that carry the same information but have uncorrelated multipath fadings. A circuit to combine the received signals or select one of the paths is necessary. Diversity techniques take advantage of the fact that signals exhibit fades at different places in time, frequency, or space, depending on different situations. However, a diversity scheme normally requires a number of antennae at the transmitter or receiver, resulting in high cost and redundancy of information. Herein, we review a method that can overcome this problem.

Given a transmitted signal $s(t)$ under such an environment, the received signal $\underline{x}(t)$[17] is then considered random and may be modeled as [5]

$$\underline{x}(t) = \sum_{i=1}^{N} \underline{a}_i s(t - \underline{\tau}_i) \exp\{j\underline{\theta}_i\} \tag{25.96}$$

where N is the number of received waves and $\underline{a}_i$, $\underline{\tau}_i$, and $\underline{\theta}_i$ are the random attenuation, multipath delay, and phase shift associated with the ith path, respectively. When considering narrowband communications with frequency or phase modulation, the transmitted signals have the following form:

$$s(t) = \exp\{j[2\pi f_o t + \Psi_s(t)]\} \tag{25.97}$$

where $\Psi_s(t)$ represents the baseband signal and f_o is the carrier frequency. The received signal $\underline{x}(t)$ will then be expressed as

$$\underline{x}(t) = \underline{r}(t) \exp\{j[2\pi f_o t + \Psi_s(t) + \underline{\Psi}_r(t)]\} \tag{25.98}$$

For non-Rician channels (in general, channels with no line-of-sight path), envelope $\underline{r}(t)$ can be assumed to be Rayleigh distributed and hence has an autocorrelation function approximated by [122]

$$R_r(\tau) \approx \alpha(1 + 1/4\eta(\tau)) \tag{25.99}$$

where α is some constant and $\eta(\tau)$ is some particular function. The Fourier transform of the latter, referred to as S_η, exhibits a peak at the zero frequency [5]. The spectrum of the signal envelope is then given by

$$S_R(f) = \alpha(\delta(f) + \frac{1}{4} S_\eta(f)) \tag{25.100}$$

In [124, 125], it is shown that for various types of frequency modulation, the second-order Wigner–Ville spectrum (WVS)[18] (or spectra of other TFDs) of the received FM signal $\underline{x}(t)$ has a delta concentration along the IF of the transmitted signal. The special structure of the envelope spectrum (Equation 25.100) makes the delta concentration possible in the TF plane of the WVS (or spectra of other TFDs). Consequently, the detection of FM signals through the Rayleigh flat fading environment can be achieved in the above time–frequency plane without the use of the conventional diversity techniques or higher-order spectra approach [126].

25.5.4 Mobile Velocity/Doppler Estimation Using Time–Frequency Processing

Many wireless communications systems require prior knowledge of the mobile velocity. This knowledge allows compensation of distortions introduced by the communications channel. In addition, reliable estimates of the mobile velocity are useful for effective dynamic channel assignment and for the optimization of adaptive multiple access. In another chapter of [127], an overview of existing velocity estimators is given, particularly an estimator based on the estimation of the IF of the received signal. In contrast to approaches based on the envelope of the received signal, the IF-based velocity estimators are robust to shadow fading, which is produced by variations of the average of the received signal envelope over few wavelengths. Interested readers can refer to [127] for more details.

---

[17] Random terms are hereafter underlined to distinguish them from deterministic terms.

[18] The second-order time-varying Wigner–Ville spectrum is defined as [123]

$$\mathcal{W}_{\underline{x}}^{(2)}(t, f) \triangleq \int_{-\infty}^{\infty} E\{\underline{x}(t + \tau/2)\underline{x}^*(t - \tau/2)\} e^{-j2\pi f \tau} d\tau$$

25.6 Conclusion

The review of contributions made in applying TFSP to communications in general and wireless communications in particular indicates a wide range of possible uses of TFSP methods and techniques in these areas. Introducing the basics of TFSP and the motivation behind the use of TFSP techniques in wireless communications allowed the obtaining of a deeper understanding of how to match the properties of TFSP to the problems encountered in wireless communications. The application of TFSP in spread-spectrum communications systems includes channel identification, scattering function estimation, and interference (MAI and ISI) mitigation. Improvements of performance can be obtained in all these areas by adapting TFSP methodologies. Similarly, time–frequency array processing techniques are well suited for source localization and blind source separation. Other application issues in wireless communications that we are briefly discussed show that there are great potential benefits in further exploring the use of TFSP techniques for current issues in wireless communications and, more generally, in telecommunications.

In particular, note that the use of TFSP in spread-spectrum and array processing applications attracts more and more attention within the signal processing and communications communities. Indeed, using spatial diversity in conjunction with time and frequency diversity is a very powerful means of exploiting and extracting the received signal information. This is the main motivation behind the increasing number of TFSP-based array processing methods and applications that include source localization and source separation problems.

On the other hand, the spread-spectrum-based applications are driven by the fact that the third and most probably the fourth mobile generation systems will be based on the CDMA/MC-CDMA transmission technique. In such systems, the need to combat interuser interference and interchannel interference leads to the problem of the design of an orthogonal function in the time–frequency domain. Past and present research work has been and is being conducted to optimize the modulation/transmission scheme (e.g., by using chirp modulation) as well as the receiver/detection scheme. However, no final or optimal solution is available yet, and many open problems are still to be solved, for example, by using TFSP theories and methods.

Acknowledgments

The first author acknowledges the funding received from the Australian Research Council. The authors also acknowledge the assistance provided by G. Azemi, A.M. Khalighi, and S.A.C. Alvarez in the review of this chapter.

References

[1] H.V. Poor and G.W. Wornell, Eds., *Wireless Communications: Signal Processing Perspectives*, Englewood Cliffs, NJ: Prentice Hall, 1998.

[2] B. Boashash, Ed., *Time-Frequency Signal Analysis, Methods and Applications*, Melbourne, Australia: Longman Cheshire, 1992.

[3] B. Boashash, *Time-Frequency Signal Analysis and Processing: A Comprehensive Reference*, Oxford: Elsevier, 2003.

[4] L. Cohen, *Time-Frequency Analysis*, Englewood Cliffs, NJ: Prentice Hall, 1995.

[5] W. Jakes, Ed., *Microwave Mobile Communications*, Washington, DC: IEEE Press, 1998 (reprint).

[6] T.S. Rappaport, *Wireless Communications: Principles and Practice*, Englewood Cliffs, NJ: Prentice Hall, 1996.

[7] N. Martins, S. Jesus, C. Gervaise, and A. Quinquis, "A time-frequency approach to blind deconvolution in multipath underwater channels," in *Proceedings of IEEE International Conference on Acoustics, Speech, and Signal Processing (ICASSP)*, Vol. 2, 2002, pp. 1225–1228.

[8] X. Wang and J. Ray Liu, "Channel estimation for multicarrier modulation systems using a time-frequency polynomial model," *IEEE Transactions on Communications*, 50, 1045–1048, 2002.
[9] P. Bello, "Characterization of randomly time-variant linear channels," *IEEE Transactions on Communication Systems*, COM-11, 360–393, 1963.
[10] J.D.D. Parsons and A.S. Bajwa, "Wideband characterisation of fading mobile radio channels," *IEE Proceedings. F. Communications, Radar and Signal Processing*, 129, 95–101, 1982.
[11] K. Pahlavan and A. Levesque, *Wireless Information Networks*, New York: Wiley Interscience, 1995.
[12] R. Steele and L. Hanzo, Eds., *Mobile Radio Communications: Second and Third Generation Cellular and WATM Systems*, 2nd ed., Oxford: John Wiley & Sons, 1999.
[13] J.G. Proakis, *Digital Communications*, 4th ed., New York, McGraw-Hill, 2001.
[14] F. Swarts, P. van Rooyan, I. Oppermann, and M.P. Lotter, Eds., *CDMA Techniques for Third Generation Mobile Systems*, Boston: Kluwer Academic Publishers, 1999.
[15] R. Becher, M. Dillinger, M. Haardt, and W. Mohr, "Broad-band wireless access and future communication networks," *Proceedings of the IEEE*, 89, 58–75, 2001.
[16] L.B. Milstein, "Interference rejection techniques in spread spectrum communications," *Proceedings of the IEEE*, 657–671, 1988.
[17] S. Verdú, *Multiuser Detection*, Cambridge, U.K.: Cambridge University Press, 1998.
[18] V. Sucic, B. Barkat, and B. Boashash, "Performance evaluation of the B distribution," in *Proceedings of the Fifth International Symposium on Signal Processing and Its Applications, ISSPA '99*, Vol. 1, Brisbane, Queensland, Australia, August 1999, pp. 267–270.
[19] B. Boashash and V. Sucic, "Resolution measure criteria for the objective assessment of the performance of quadratic time–frequency distributions," *IEEE Transactions on Signal Processing*, 51, 1253–1263, 2003.
[20] G.F. Boudreaux-Bartels and T.W. Marks, "Time-varying filtering and signal estimation using Wigner distributions," *IEEE Transactions on Acoustics, Speech, and Signal Processing*, 34, 422–430, 1986.
[21] B. Boashash, "Time-frequency signal analysis," in *Advances in Spectrum Analysis and Array Processing*, Vol. I, S. Haykin, Ed., Englewood Cliffs, NJ: Prentice Hall, 1991, chap. 9, pp. 418–517.
[22] T.J. McHale and G.F. Boudreaux-Bartels, "An algorithm for synthesizing signals from partial time-frequency models using the cross Wigner distribution," *IEEE Transactions on Signal Processing*, 41, 1986–1990, 1993.
[23] J.C. Wood and D.T. Barry, "Linear signal synthesis using the Radon-Wigner transform," *IEEE Transactions on Signal Processing*, 42, 2105–2111, 1994.
[24] F. Hlawatsch and W. Krattenthaler, "Signal synthesis algorithms for bilinear time-frequency signal representations," in *The Wigner Distribution: Theory and Applications in Signal Processing*, W. Mecklenbräuker and F. Hlawatsch, Eds., Amsterdam: Elsevier, 1997, pp. 135–209.
[25] A. Francos and M. Porat, "Analysis and synthesis of multicomponent signals using positive time-frequency distributions," *IEEE Transactions on Signal Processing*, 47, 493–504, 1999.
[26] A.M. Sayeed, A. Sendonaris, and B. Aazhang, "Multiuser detection in fast-fading multipath environments," *IEEE Journal on Selected Areas in Communications*, 16, 1691–1701, 1998.
[27] H. Artés and F. Hlawatsch, "Blind multiuser equalization for time-varying channels," in *IEEE Signal Processing Workshop on Signal Proceeding Advances in Wireless Communications, SPAWC '01*, Taoyuan, Taiwan, March 2001, pp. 102–105.
[28] A.M. Sayeed and B. Aazhang, "Exploiting Doppler diversity in mobile wireless communications," in *Proceedings of CISS '97, 1997 Conference on Information Science Systems*, 1997, pp. 287–292.
[29] M.R. Baissas and A.M. Sayeed, "Pilot-based estimation of time-varying multipath channels for CDMA systems," in *International Conference on Acoustics, Speech, and Signal Processing, ICASSP '2000*, Vol. V, Istanbul, Turkey, June 2000, pp. 2657–2660.
[30] M.R. Baissas and A.M. Sayeed, "Channel estimation errors versus Doppler diversity in fast fading channels," in *Proceedings of the Thirty-Fourth Asilomar Conference*, Vol. 2, 2000, pp. 970–974.
[31] N.T. Gaarder, "Scattering function estimation," *IEEE Transactions on Information Theory*, IT-14, 684–693, 1968.

[32] H.L. VanTrees, *Detection, Estimation, and Modulation Theory. Radar-Sonar Signal Processing and Gaussian Signals in Noise*, Malabar, FL, Krieger Publication Co., 1992.

[33] R.A. Altes, "Detection, estimation, and classification with spectrograms," *The Journal of the Acoustical Society of America*, 67, 1232–1246, 1980.

[34] P. Flandrin, "A time-frequency formulation of optimum detection," *IEEE Transactions on Acoustics, Speech, and Signal Processing*, 36, 1377–1384, 1988.

[35] B.L. Johnson, D.W. Ricker, and J.R. Sacha, "The use of iterative deconvolution for scattering function identification," *The Journal of the Acoustical Society of America*, 91, 2790–2798, 1992.

[36] D.W. Ricker and M.J. Gustafson, "A low sidelobe technique for the direct measurement of scattering functions," *IEEE Journal of Oceanic Engineering*, 21, 14–23, 1996.

[37] T. Wang, V.K. Dubey, and J.T. Ong, "Generation of scattering functions for mobile communication channel: a computer simulation approach," *Wireless Information Networks*, 4, 187–204, 1997.

[38] J.S. Sadowsky and V. Kafedziski, "On the correlation and scattering function of the WSSUS channel for mobile communications," *IEEE Transactions on Vehicular Technology*, 47, 270–282, 1998.

[39] L.-T. Nguyen, B. Senadji, and B. Boashash, "Scattering function and time-frequency signal processing," in *International Conference on Acoustics, Speech, and Signal Processing, ICASSP 2001*, Salt Lake City, UT, June 2001.

[40] R. Haas and J.-C. Belfiore, "A time-frequency well-localized pulse for multiple carrier transmission," *Wireless Personal Communications*, 5, 1–18, 1997.

[41] R.L. Pickholtz, L.B. Milstein, and D.L. Schilling, "Spread spectrum for mobile communications," *IEEE Transactions on Vehicular Technology*, 40, 313–322, 1991.

[42] E. Marsy, "Closed-form analytical results for the rejection of narrowband interference in PN spread spectrum systems. Part I. Linear prediction filters," *IEEE Transactions on Communications*, COM–32, 888–896, 1984.

[43] H.V. Poor and L.A. Rusch, "Narrowband interference suppression in spread spectrum CDMA," *IEEE Personal Communications*, Third Quarter, 14–27, 1994.

[44] L.A. Rusch and H.V. Poor, "Multiuser detection techniques for narrowband interference suppression in spread spectrum communications," *IEEE Transactions on Communications*, 43, 1725–1737, 1995.

[45] H.V. Poor and X. Wang, "Blind adaptive suppression of narrowband digital interferers from spread spectrum signals," *Wireless Personal Communications*, 6, 69–96, 1998.

[46] A. Bultan and A.N. Akansu, "A novel time-frequency exciser in spread spectrum communications for chirp-like interference," in *International Conference on Acoustics, Speech, and Signal Processing, ICASSP '98*, Vol. 6, 1998, pp. 3265–3268.

[47] A. Bultan, M.V. Tazebay, and A.N. Akansu, "A comparative performance study of excisers for various interference classes," in *Proceedings of the IEEE-SP International Symposium on Time-Frequency and Time-Scale Analysis*, New York, 1998, pp. 381–384.

[48] S.-H. Hong and B.G. Lee, "Time-frequency localized subspace tracking for narrowband interference suppression of DS/CDMA systems," *Electronics Letters*, 35, 1063–1064, 1999.

[49] J. Horng and R.A. Haddad, "Interference excision in DSSS communication system using time-frequency adaptive block transform," in *Proceedings of the IEEE-SP International Symposium on Time-Frequency and Time-Scale Analysis*, Pittsburgh, PA, October 6–9, 1998, pp. 385–388.

[50] M.V. Tazebay and A.N. Akansu, "Adaptive subband transforms in time-frequency excisers for DSSS communications systems," *IEEE Transactions on Signal Processing*, 43, 2776–2782, 1995.

[51] M.V. Tazebay and A.N. Akansu, "A smart time-frequency exciser for spread spectrum communications," in *International Conference on Acoustics, Speech, and Signal Processing, ICASSP '95*, Vol. 2, New York, 1995, pp. 1209–1212.

[52] M.V. Tazebay and A.N. Akansu, "A performance analysis of interference excision techniques in direct sequence spread spectrum communications," *IEEE Transactions on Signal Processing*, 46, 2530–2535, 1998.

[53] B. Boashash, "Estimating and interpreting the instantaneous frequency of a signal. Part 2. Algorithms and applications," *Proceedings of the IEEE*, 80, 539–569, 1992.

[54] F. Hlawatsch and G.F. Boudreaux-Bartels, "Linear and quadratic time-frequency signal representations," *IEEE Signal Processing Magazine*, 9, 21–67, 1992.

[55] M.G. Amin, "Interference mitigation in spread spectrum communication systems using time-frequency distributions," *IEEE Transactions on Signal Processing*, 45, 90–101, 1997.

[56] M.G. Amin, C. Wang, and A.R. Lindsey, "Optimum interference excision in spread spectrum communications using open-loop adaptive filters," *IEEE Transactions on Signal Processing*, 47, 1966–1976, 1999.

[57] S. Barbarossa and A. Scaglione, "Adaptive time-varying cancellation of wideband interferences in spread-spectrum communications based on time-frequency distributions," *IEEE Transactions on Signal Processing*, 47, 957–965, 1999.

[58] S. Barbarossa, A. Scaglione, S. Spalletta, and S. Votini, "Adaptive suppression of wideband interferences in spread-spectrum communications using the Wigner-Hough transform," in *International Conference on Acoustics, Speech, and Signal Processing, ICASSP '97*, Vol. 5, California, 1997, pp. 3861–3864.

[59] S. Lach, M.G. Amin, and A. Lindsey, "Broadband interference excision for software-radio spread-spectrum communications using time-frequency distribution synthesis," *IEEE Journal on Selected Areas in Communications*, 17, 704–714, 1999.

[60] X. Ouyang and M.G. Amin, "Performance analysis of the DS/SS communications receiver implementing a short time Fourier transform interference excision system," in *Proceedings of the IEEE-SP International Symposium on Time-Frequency and Time-Scale Analysis*, Pittsburgh, PA, October 6–9, 1998, pp. 393–396.

[61] X. Ouyang and M.G. Amin, "Short-time Fourier transform receiver for nonstationary interference excision in direct sequence spread spectrum communications," *IEEE Transactions on Signal Processing*, 49, 851–863, 2001.

[62] S. Barbarossa, "Analysis of multicomponent LFM signals by a combined Wigner-Hough transform," *IEEE Transactions on Signal Processing*, 43, 1511–1515, 1995.

[63] B. Boashash and P. O'Shea, "Use of the cross Wigner-Ville distribution for estimation of instantaneous frequency," *IEEE Transactions on Signal Processing*, 41, 1439–1445, 1993.

[64] R. Prasad, *CDMA for Wireless Personal Communications*, Mobile Communications Series, Artech House, Norwood, MA, 1997.

[65] K. Fazel, "Performance of CDMA/OFDM for mobile communication systems," in *Proceedings of the IEEE 2th ICUPC*, Ottawa, Canada, 1993, pp. 975–979.

[66] S. Hara and R. Prasad, "Overview of multicarrier CDMA," *IEEE Communications Magazine*, 35, 126–133, 1997.

[67] S. Hara and R. Prasad, "Design and performance of multicarrier CDMA system in frequency-selective Rayleigh fading channels," *IEEE Transactions on Vehicular Technology*, 48, 1584–1595, 1999.

[68] N. Yee, J.-P. Linnartz, and G.P. Fettweis, "Multi-carrier CDMA in indoor wireless radio networks," in *Proceedings of the IEEE 4th PIMRC*, Yokohama, Japan, 1993, pp. 109–113.

[69] K.D. Kammeyer, U. Tuisel, H. Schulze, and H. Bochmann, "Digital multicarrier: transmission of audio signals over mobile radio channels," *European Transactions on Telecommunications*, 3, 243–253, 1992.

[70] R.W. Chang, "Synthesis of band-limited orthogonal signals for multichannel data transmission," *Bell System Technical Journal*, 45, 1775–1796, 1966.

[71] J. Kühne, A. Nahler, and G.P. Fettweis, "Multi-user interference evaluation for a one-code multi-carrier spread spectrum CDMA system with imperfect time and frequency synchronization," in *1988 IEEE 5th International Symposium on Spread Spectrum Techniques and Applications — Proceedings. Spread Technology to Africa*, Vol. 2, New York, 1998, pp. 479–483.

[72] V. Aue and G.P. Fettweis, "Multi-carrier spread spectrum modulation with reduced dynamic range," in *Proceedings of the 46th IEEE Vehicular Technology Conference*, April 1996, pp. 914–917.

[73] S. Verdu, *Multiuser Detection*, New York: Cambridge University Press, 1998.

[74] A.H. Sayed and N.R. Yousef, "Wireless location", in *Wiley Encyclopedia of Telecommunications*, J. Proakis, Ed., New York: Wiley & Sons, 2002.

[75] A. Belouchrani and M.G. Amin, "A new approach for blind source separation using time frequency distributions," in *Proceedings SPIE Conference on Advanced Algorithms and Architectures for Signal Processing*, Denver, Colorado, August 1996.

[76] A. Belouchrani and M.G. Amin, "Blind source separation using time-frequency distributions: algorithm and asymptotic performance," in *IEEE Proceeding ICASSP*, Germany, April 1997.

[77] A. Belouchrani and M.G. Amin, "On the Use of Spatial Time Frequency Distributions for Signal Extraction," Special issue of the journal *Multidimensional Systems and Signal Processing*, Boston: Kluwer Academic Publishers, October 1998.

[78] A. Belouchrani and M.G. Amin, "Blind source separation based on time-frequency signal representation," *IEEE Transactions on Signal Proceedings*, 2888–2898, 1998.

[79] A. Belouchrani and M.G. Amin, "Time-frequency MUSIC: a new array signal processing method based on time-frequency signal representation," *IEEE Signal Processing Letters*, 6, 109–110, 1999.

[80] A.S. Kayhan and M.G. Amin, "Spatial evolutionary spectrum for DOA estimation and blind signal separation," *IEEE Transactions on Signal Processing*, 2000.

[81] A.R. Leyman, Z.M. Kamran, and K. Abed-Meraim, "Higher order time frequency based blind source separation technique," *IEEE Signal Processing Letters*, 2000.

[82] Y. Zhang, W. Mu, and M.G. Amin, "Subspace analysis of spatial time frequency distribution matrices," *IEEE Transactions on Signal Processing*, 49, 2001.

[83] A. Belouchrani, K. Abed-Meraim, M.G. Amin, and A. Zoubir, "Joint anti-diagonalization for blind source separation," in *Proceedings ICASSP*, Utah, Vol. 5, May 2001, pp. 2789–2792.

[84] L.-T. Nguyen, A. Belouchrani, K. Abed-Meraim, and B. Boashash, "Separating more sources than sensors using time frequency distributions," in *Proceeding ISSPA*, Kuala-Lampur, Malaysia, Vol. 2, August 2001, pp. 583–586.

[85] L. Giulieri, N. Thirion-Moreau, and P.-Y. Arques, "Blind sources separation based on bilinear time-frequency representations: a performance analysis," in *Proceedings ICASSP*, Vol. 2, 2002, pp. 1649–1652.

[86] L. Jin, Q.-Y. In, and W.-J. Wang, "Time-frequency signal subspace fitting method for direction-of-arrival estimation" in *Proceedings IEEE International Symposium on Circuits and Systems*, ISCAS '2000, Vol. 3, Geneva, May 2000, pp. 375–378.

[87] G. Wang and X.-G. Xia, "Iterative algorithm for direction of arrival estimation with wideband chirp signals," *IEE Proceedings: Radar, Sonar and Navigation*, 147, 233–238, 2000.

[88] L. Cirillo, A. Zoubir, N. Ma, and M.G. Amin, "Automatic classification of auto- and cross-terms of time-frequency distributions in antenna arrays," in *Proceedings of ICASSP*, Orlando, FL, 2002.

[89] A. Hassanien, A.B. Gershman, and M.G. Amin, "Time-frequency ESPRIT for direction-of-arrival estimation of chirp signals," in *Sensor Array and Multichannel Signal Processing Workshop Proceedings*, August 4–6, 2002, pp. 337–341.

[90] Y. Zhang and M.G. Amin, "Blind separation of sources based on their time-frequency signatures," in *Proceedings ICASSP*, Istanbul, Turkey, June 2000.

[91] M. Wax and T. Kailath, "Detection of signals by information theoretic criteria," *IEEE Transactions on Acoustic, Speech and Signal Processing*, ASSP-33, 387–392, 1985.

[92] L. Cirillo, A. Zoubir, and M.G. Amin, "Selection of auto- and cross-terms for blind non-stationary source separation," in *Proceedings of the IEEE International Symposium on Signal Processing and Information Technology (ISSPIT)*, Cairo, Egypt, 2001.

[93] W. Mu, M.G. Amin, and Y. Zhang, "Bilinear signal synthesis in array processing," *IEEE Transactions on Signal Processing*, 51, 90–100, 2003.

[94] L. Giulieri, N. Thirion-Moreau, and P.-Y. Arques, "Blind sources separation using bilinear and quadratic time-frequency representations," in *ICA '2001*, San Diego, December 2001.

[95] R. Schmidt, "Multiple emitter location and signal parameter estimation," *IEEE Transactions on Antennas and Propagation*, 34, 276–280, 1986.

[96] R. Roy and T. Kailath, "ESPRIT: estimation of signal parameters via rotational invariance techniques," *IEEE Transactions Acoustic, Speech, Signal Processing*, 37, 984–995, 1989.

[97] Y. Zhang, W. Mu, and M.G. Amin, "Time-frequency maximum likelihood methods for direction finding," *Journal Franklin Institue*, 337, 483–497, 2000.

[98] A. Belouchrani, M.G. Amin, and K. Abed-Meraim, "Direction finding in correlated noise fields based on joint block-diagonalization of spatio-temporal correlation matrices," *IEEE Signal Processing Letters*, 4, 266–268, 1997.

[99] A. Belouchrani, K. Abed-Meraim, J.F. Cardoso, and E. Moulines, "Blind source separation using second order statistics," *IEEE Transactions on Signal Proceedings*, 434–444, 1997.

[100] C. Jutten and J. Hérault, "Détection de grandeurs primitives dans un message composite par une architecture de calcul neuromimétrique en apprentissage non supervisé," in *Proceedings GRETSI*, Nice, France, 1985.

[101] P. Comon, "Independent component analysis, a new concept?" *Signal Processing*, 36, 287–314, 1994.

[102] A. Belouchrani and J.-F. Cardoso, "Maximum likelihood source separation for discrete sources," in *Proceedings EUSIPCO*, 1994, pp. 768–771.

[103] P. Comon and O. Grellier, "Nonlinear inversion of underdetermined mixtures," in *ICA '99*, Aussois, France, pp. 461–465, January 1999.

[104] K.I. Diamantaras, "Blind separation of multiple binary sources using a single linear mixture," in *International Conference on Acoustics, Speech, and Signal Processing, ICASSP '2000*, Istanbul, Turkey, Vol. V, June 2000, pp. 2657–2660.

[105] A. Jourjine, S. Rickard, and O. Yilmaz, "Blind separation of disjoint orthogonal signals: demixing *n* sources from 2 mixtures," in *International Conference on Acoustics, Speech, and Signal Processing, ICASSP '2000*, Istanbul, Turkey, Vol. 5, June 2000, pp. 2985–2988.

[106] Y. Luo and J. Chambers, "Active source selection using gap statistics for underdetermined blind source separation," in *Proceedings of the 7th International Symposium on Signal Processing and Its Applications (ISSPA)*, Vol. I, Paris, France, July 1–4, 2003, pp. 137–140.

[107] A. Belouchrani, K. Abed-Meraim, and Y. Hua, "Jacobi-like algorithms for joint block diagonalization: application to source localization," *in Proceedings ISPACS*, Melbourne, Australia, November 1998.

[108] H. Bousbia-Saleh, A. Belouchrani, and K. Abed-Meraim, "Jacobi-like algorithm for blind signal separation of convolutive mixtures," *Electronics Letters*, 37, 1049–1050, 2001.

[109] K. Abed-Meraim, W. Qiu, and Y. Hua, "Blind system identification," *Proceedings of the IEEE*, 85, 1310–1322, 1997.

[110] S. Barbarossa and A. Scaglione, "Optimal precoding for transmissions over linear time-varying channels," in *Proceedings of GLOBECOM '99*, Vol. 5, Piscataway, NJ, 1999, pp. 2545–2549.

[111] A. Scaglione, S. Barbarossa, and G.B. Giannakis, "Filterbank transceivers optimizing information rate in block transmissions over dispersive channels," *IEEE Transactions on Information Theory*, 45, 1019–1032, 1999.

[112] R.G. Gallager, *Information Theory and Reliable Communication*, New York: John Wiley & Sons, 1968.

[113] M. Medard, "The Capacity of Time-Varying Multiple User Channels in Wireless Communications," Ph.D. thesis, MIT, Cambridge, MA, September 1995.

[114] A. Goldsmith, "Design and Performance of High-Speed Communication Systems over Time-Varying Radio Channels," Ph.D. thesis, University of California, Berkeley, September 1994.

[115] S.E. El-Khamy, S.E. Shahban, and E.A. Thabet, "Frequency hopped multi-user chirp modulation (FH/MCM) for multipath fading channels," in *IEEE Symposium Antennas and Propagation*, Vol. 1, July 1999, pp. 996–999.

[116] S.E. El-Khamy, S.E. Shahban, and E.A. Thabet, "Efficient multiple access communications using multi-user chirp modulation signals," in *IEEE 4th International Symposium on Spread Spectrum Techniques and Applications*, Vol. 3, September 1996, pp. 1209–1213.

[117] M.R. Winkler, "Chirp Signals for Communications," WESCON convention record, Paper 14.2, 1962.
[118] A.J. Berni and W.D. Gregg, "On the utility of chirp modulation for digital signaling," *IEEE Transactions on Communications*, 1973.
[119] C.E. Cook, "Linear FM signal formats for beacon and communication systems," *IEEE Transactions on Aerospace and Electronic Systems*, 10, 471–478, 1974.
[120] C. Gupta and A. Papandreou-Suppappola, "Wireless CDMA communications using time-varying signals," in *Sixth International Symposium on Signal Processing and Its Applications*, Vol. 1, August 2001, pp. 242–245.
[121] S. Hengstler, D.P. Kasilingam, and A.H. Costa, "A novel chirp modulation spread spectrum technique for multiple access," in *IEEE Seventh International Symposium on Spread Spectrum Techniques and Applications*, Vol. 1, 2002, pp. 73–77.
[122] W.B. Davenport and W.L. Root, *An Introduction to the Theory of Random Signals and Noise*, Washington, DC: IEEE Press, 1987.
[123] B. Boashash and B. Ristic, "Polynomial time-frequency distributions and time-varying higher order spectra: application to the analysis of multicomponent FM signals and to the treatment of multiplicative noise," *Signal Processing*, 67, 1–23, 1998.
[124] L.-T. Nguyen and B. Senadji, "Analysis of nonlinear signals in the presence of Rayleigh fading," in *Proceedings of the Fifth International Symposium on Signal Processing and Its Applications, ISSPA '99*, Brisbane, Queensland, Australia, Vol. 1, August 1999, pp. 411–414.
[125] L.-T. Nguyen and B. Senadji, "Detection of frequency modulated signals in Rayleigh fading channels based on time-frequency distributions," in *International Conference on Acoustics, Speech, and Signal Processing, ICASSP '2000*, Vol. II, Istanbul, Turkey, June 2000, pp. 729–732.
[126] B. Senadji and B. Boashash, "A mobile communications application of time-varying higher order spectra to FM signals affected by multiplicative noise," in *Proceedings of ICICS, 1997 International Conference on Information, Communications, and Signal Processing*, Vol. 3, New York, 1997, pp. 1489–1492.
[127] B. Senadji, G. Azemi, and B. Boashash, "Mobile velocity estimation for wireless communications," chapter in *Signal Processing for Mobile Communications Handbook*, M. Ibnkahla, Ed., Boca Raton, FL: CRC Press, 2003.

26

Monte Carlo Signal Processing for Digital Communications: Principles and Applications

Xiaodong Wang
Columbia University

26.1 Introduction **26**-2
26.2 MCMC Methods **26**-3
General MCMC Algorithms • Applications of MCMC in Digital Communications • Other Applications
26.3 SMC Methods **26**-11
General SMC Algorithms • Resampling Procedures • Applications of SMC in Digital Communications
26.4 Concluding Remarks **26**-23

Abstract

Many statistical signal processing problems found in digital communications involve making inferences about the transmitted information data based on the received signals in the presence of various unknown channel distortions. The *optimal* solutions to these problems are often too computationally complex to implement by conventional signal processing methods. The recently emerged Bayesian Monte Carlo signal processing methods, the relatively simple yet extremely powerful numerical techniques for Bayesian computation, offer a novel paradigm for tackling these problems. These methods fall into two categories: the Markov chain Monte Carlo (MCMC) methods for batch signal processing and the sequential Monte Carlo (SMC) methods for adaptive signal processing. We provide an overview of the theories and applications of both the MCMC and SMC methods. The salient features of these techniques include the following: (1) they are optimal in the sense of achieving minimum symbol error rate; (2) they do not require the knowledge of the channel states, and they do not explicitly estimate the channel by employing training signals or decision feedback; and (3) being soft input soft output in nature, they are well suited for iterative (turbo) processing in coded systems.

0-8493-1657-X/$0.00+$1.50

26.1 Introduction

In digital communications, the transmitter transmits a sequence of symbols $X = (x_1, x_2, \ldots, x_n)$, where $x_i \in \mathcal{A}$ can be either a scaler or a vector and $\mathcal{A}$ is a finite set of all possible transmitted symbols. At the receiver end, the received signal denoted as $Y = r(\tau; X, \Theta)$ is a random continuous function of time τ, which is dependent of the transmitted symbol sequence X and some unknown parameters Θ. Typically Θ is a multidimensional continuous-valued quantity used to describe the channel characteristic. In this context, the following two statistical inference problems based on the maximum likelihood (ML) principle are of interest and have been the subject of intense research in the fields of communications and signal processing:

- *Sequence detection:* Assuming Θ is known, find the ML estimate of X:

$$\hat{X} = \arg\max_{X \in \mathcal{A}^n} p(Y \mid X) \tag{26.1}$$

- *Channel estimation:* Assuming X is known, find the ML estimate of Θ:

$$\hat{\Theta} = \arg\max_{\Theta} p(Y \mid \Theta) \tag{26.2}$$

Typically the transmitter first transmits a sequence of known symbols, based on which channel estimation is performed at the receiver. The information symbol sequence is then transmitted and sequence detection is performed at the receiver using the estimated channel. Such an "estimate-then-plug-in" approach, although popular in engineering practice, is *ad hoc* and bears no theoretical optimality. In fact, the objective of digital communications is to reliably transmit information between the transmitter and receiver, and channel estimation is not necessarily essential for achieving this. Indeed, the channel parameters Θ can be viewed as "hidden" random data and the ultimate symbol sequence detection problem becomes finding

$$\hat{X} = \arg\max_{X \in \mathcal{A}^n} E\{p(Y \mid \Theta, X) \mid X\} \tag{26.3}$$

where $E\{\cdot\}$ denotes the expectation taken with respect to the distribution of Θ. The inference problem in Equation 26.3 can be solved using the expectation–maximization (EM) algorithm [9]. This approach to receiver design in digital communications, although theoretically appealing, suffers from a spectral efficiency problem. This is because in order for the EM algorithm to converge to the ML solution, the initial estimate of the symbol sequence must be close to the ML estimate. Hence, a training symbol sequence has to be employed for this purpose, which results in a loss of spectral efficiency.

An entirely different approach to the design of the optimal digital communications receiver is based on the Bayesian formulation of the problem. In this approach, all unknowns X and Θ are treated as random quantities with some independent prior distributions $p(X)$ and $p(\Theta)$, respectively. We are then interested in computing the *a posteriori* probability distribution of each transmitted symbol:

$$P(x_t = a_\ell \mid Y) = \sum_{X_{[-t]} \in \mathcal{A}^{n-1}} p(X \mid Y) \tag{26.4}$$

$$= \sum_{X_{[-t]} \in \mathcal{A}^{n-1}} \int p(X, \Theta \mid Y) d\Theta, \quad \forall a_\ell \in \mathcal{A} \tag{26.5}$$

where $X_{[-t]} \stackrel{\Delta}{=} X \backslash x_t$. Note that the joint posterior density in Equation 26.5 can be written as

$$p(X, \Theta \mid Y) \propto p(Y \mid X, \Theta)\, p(X)\, p(\Theta) \tag{26.6}$$

In many Bayesian analyses, the computation involved in eliminating the nuisance parameters and missing data is so difficult that one has to resort to some analytical or numerical approximations. These approximations were often case specific and were the bottleneck that prevented the Bayesian method from being widely used. In late 1980s and early 1990s, statisticians discovered that a wide variety of Monte Carlo strategies can be applied to overcome the computational difficulties encountered in almost all likelihood-based inference procedures. Soon afterwards, this rediscovery of the Monte Carlo method as one of the most versatile and powerful computational tools began to invade other quantitative fields such as artificial intelligence, computational biology, engineering, financial modeling, etc. [20].

Suppose we can generate random samples (either independent or dependent)

$$(\boldsymbol{X}^{(1)}, \boldsymbol{\Theta}^{(1)}), (\boldsymbol{X}^{(2)}, \boldsymbol{\Theta}^{(2)}), \ldots, (\boldsymbol{X}^{(\nu)}, \boldsymbol{\Theta}^{(\nu)})$$

from the joint distribution in Equation 26.6. Then we can approximate the marginal distribution in Equation 26.5 by the empirical distribution (i.e., the histogram) based on $\boldsymbol{x}_t^{(1)}, \boldsymbol{x}_t^{(2)}, \ldots, \boldsymbol{x}_t^{(\nu)}$, the component of $\boldsymbol{x}_t$ in $\boldsymbol{X}^{(j)}$,

$$P(\boldsymbol{x}_t = \boldsymbol{a}_\ell \mid \boldsymbol{Y}) \cong \frac{1}{\nu} \sum_{j=1}^{\nu} I\left(\boldsymbol{x}_t^{(j)} = \boldsymbol{a}_\ell\right) \tag{26.7}$$

where $I(\cdot)$ is an indicator function. Depending on how the data are processed and the inference is made, signal processing methods fall into one of two categories: batch processing and adaptive (i.e., sequential) processing. In batch signal processing, the entire data block $\boldsymbol{Y}$ is received and stored before it is processed, and the inference about $\boldsymbol{X}$ is made based on the entire data block $\boldsymbol{Y}$. In adaptive processing, however, inference is made sequentially (i.e., on-line) as the data being received. For example, at time t, after a new sample $\boldsymbol{y}_t$ is received, an update on the inference about some or all elements of $\boldsymbol{X}$ is made. On the other hand, most Monte Carlo techniques fall into one of the following two categories: Markov chain Monte Carlo (MCMC) methods, corresponding to batch processing, and sequential Monte Carlo (SMC) methods, corresponding to adaptive processing.

26.2 MCMC Methods

26.2.1 General MCMC Algorithms

Markov chain Monte Carlo is a class of algorithms that allow one to draw (pseudo-) random samples from an arbitrary target probability distribution, $p(\boldsymbol{x})$, known up to a normalizing constant. The basic idea behind these algorithms is that one can achieve the sampling from p by running a Markov chain whose equilibrium distribution is exactly p. Two basic types of MCMC algorithms, the Metropolis algorithm and the Gibbs sampler, have been widely used in diverse fields. The validity of both algorithms can be proved by the basic Markov chain theory.

26.2.1.1 Metropolis–Hastings Algorithm

Let $p(\boldsymbol{x}) = c \exp\{-f(\boldsymbol{x})\}$ be the target probability distribution from which we want to simulate random draws. The normalizing constant c may be unknown to us. Metropolis et al. [25] first introduced the fundamental idea of evolving a Markov process in Monte Carlo sampling, which was later generalized by Hastings [15]. Starting with any configuration $\boldsymbol{x}^{(0)}$, the algorithm evolves from the current state $\boldsymbol{x}^{(t)} = \boldsymbol{x}$ to the next state $\boldsymbol{x}^{(t+1)}$ as follows.

Algorithm 26.1 (Metropolis–Hastings algorithm)

- Propose a random perturbation of the current state, i.e., $\boldsymbol{x} \to \boldsymbol{x}'$, where $\boldsymbol{x}'$ is generated from a transition function $T(\boldsymbol{x}^{(t)} \to \boldsymbol{x}')$, which is nearly arbitrary (of course, some are better than others in terms of efficiency) and is completely specified by the user.
- Compute the Metropolis ratio

$$r(\boldsymbol{x}, \boldsymbol{x}') = \frac{p(\boldsymbol{x}')T(\boldsymbol{x}' \to \boldsymbol{x})}{p(\boldsymbol{x})T(\boldsymbol{x} \to \boldsymbol{x}')} \tag{26.8}$$

- Generate a random number $u \sim \text{uniform}(0,1)$. Let $\boldsymbol{x}^{(t+1)} = \boldsymbol{x}'$ if $u \leq r(\boldsymbol{x}, \boldsymbol{x}')$, and let $\boldsymbol{x}^{(t+1)} = \boldsymbol{x}^{(t)}$ otherwise.

It is easy to prove that the Metropolis–Hastings (M-H) transition rule results in an actual transition function $A(\boldsymbol{x}, \boldsymbol{y})$ (it is different from T because an acceptance/rejection step is involved) that satisfies the detailed balance condition

$$p(\boldsymbol{x})A(\boldsymbol{x}, \boldsymbol{y}) = p(\boldsymbol{y})A(\boldsymbol{y}, \boldsymbol{x}) \tag{26.9}$$

which necessarily leads to a *reversible* Markov chain with $p(\boldsymbol{x})$ as its invariant distribution.

The Metropolis algorithm has been extensively used in statistical physics over the past 40 years and is the cornerstone of all MCMC techniques recently adopted and generalized in the statistics community. Another class of MCMC algorithms, the Gibbs sampler [8], differs from the Metropolis algorithm in that it uses conditional distributions based on $p(\boldsymbol{x})$ to construct Markov chain moves.

26.2.1.2 Gibbs Sampler

Suppose $\boldsymbol{x} = (x_1, \ldots, x_d)$, where x_i is either a scalar or a vector. In the Gibbs sampler, one systematically or randomly chooses a coordinate, say x_i, and then updates its value with a new sample x_i' drawn from the conditional distribution $p(\cdot \mid \boldsymbol{x}_{[-i]})$. Algorithmically, the Gibbs sampler can be implemented as follows.

Algorithm 26.2 (Gibbs sampler)

Let the current state be $\boldsymbol{x}^{(t)} = (x_1^{(t)}, \ldots, x_d^{(t)})$.

For $i = 1, \ldots, d$, we draw $x_i^{(t+1)}$ from the conditional distribution

$$p\left(x_i \mid x_1^{(t+1)}, \ldots, x_{i-1}^{(t+1)}, x_{i+1}^{(t)}, \ldots, x_d^{(t)}\right)$$

Alternatively, one can randomly scan the coordinate to be updated. Suppose currently $\boldsymbol{x}^{(t)} = (x_1^{(t)}, \ldots, x_d^{(t)})$. Then one can randomly select i from the index set $\{1, \ldots, d\}$ according to a given probability vector $(\pi_1, \ldots, \pi_d)$ and then draw $x_i^{(t+1)}$ from the conditional distribution $p(\cdot \mid \boldsymbol{x}_{[-i]}^{(t)})$, and let $\boldsymbol{x}_{[-i]}^{(t+1)} = \boldsymbol{x}_{[-i]}^{(t)}$.

It is easy to check that *every* individual conditional update leaves p invariant. Suppose currently $\boldsymbol{x}^{(t)} \sim p$. Then $\boldsymbol{x}_{[-i]}^{(t)}$ follows its marginal distribution under p. Thus,

$$p\left(x_i^{(t+1)} \mid \boldsymbol{x}_{[-i]}^{(t)}\right) \cdot p\left(\boldsymbol{x}_{[-i]}^{(t)}\right) = p\left(x_i^{(t+1)}, \boldsymbol{x}_{[-i]}^{(t)}\right) \tag{26.10}$$

which implies that the joint distribution of $(\boldsymbol{x}_{[-i]}^{(t)}, x_i^{(t+1)})$ is unchanged at p after one update.

The Gibbs sampler's popularity in statistics communities stems from its extensive use of *conditional distributions* in each iteration. The data augmentation method [27] first linked the Gibbs sampling structure with missing data problems and EM-type algorithms. The Gibbs sampler was further popularized by [7], where it was pointed out that the conditionals needed in Gibbs iterations are commonly available in many Bayesian and likelihood computations. Under regularity conditions, one can show that the Gibbs sampler chain converges geometrically and its convergence rate is related to how the variables correlate with each other [19]. Therefore, grouping highly correlated variables together in the Gibbs update can greatly speed up the sampler.

26.2.1.3 Other Techniques

A main problem with all the MCMC algorithms is that they may, for some problems, move very slowly in the configuration space or may be trapped in a local mode. This phenomenon is generally called *slow mixing* of the chain. When the chain is slow mixing, estimation based on the resulting Monte Carlo samples becomes very inaccurate. Some recent techniques suitable for designing more efficient MCMC samplers include parallel tempering [10], the multiple-try method [23], and evolutionary Monte Carlo [17].

26.2.2 Applications of MCMC in Digital Communications

In this section, we discuss MCMC-based receiver signal processing algorithms for several typical communication channels, when the channel conditions are unknown *a priori*.

26.2.2.1 MCMC Detectors in AWGN Channels

We start with the simplest channel model in digital communications — the additive white Gaussian noise (AWGN) channel. After filtering and sampling of the continuous-time received waveform, the discrete-time received signal in such a channel is given by

$$y_t = \phi\, x_t + v_t, \quad t = 1, 2, \ldots, n \tag{26.11}$$

where y_t is the received signal at time t, $x_t \in \{+1, -1\}$ is the transmitted binary symbol at time t, $\phi \in \mathbb{R}$ is the received signal amplitude, and v_t is an independent Gaussian noise sample with zero mean and variance σ^2, i.e., $v_t \sim \mathcal{N}(0, \sigma^2)$. Denote $\boldsymbol{X} \triangleq [x_1, \ldots, x_n]$ and $\boldsymbol{Y} \triangleq [y_1, \ldots, y_n]$. Our problem is to estimate the *a posteriori* probability distribution of each symbol based on the received signal $\boldsymbol{Y}$, without knowing the channel parameters (ϕ, σ^2). The solution to this problem based on the Gibbs sampler is as follows. Assuming a uniform prior for ϕ, a uniform prior for $\boldsymbol{X}$ (on $\{-1, +1\}^n$), and an inverse χ^2 prior for σ^2, $\sigma^2 \sim \chi^{-2}(\nu, \lambda)$, the complete posterior distribution is given by

$$p(\boldsymbol{X}, \phi, \sigma^2 \mid \boldsymbol{Y}) \propto p(\boldsymbol{Y} \mid \boldsymbol{X}, \phi, \sigma^2)\, p(\phi)\, p(\sigma^2) p(\boldsymbol{X}) \tag{26.12}$$

The Gibbs sampler starts with arbitrary initial values of $\boldsymbol{X}^{(0)}$ and for $k = 0, 1, \ldots$ iterates between the following two steps.

Algorithm 26.3 (two-component Gibbs detector in AWGN channel)

- Draw a sample $(\phi^{(k+1)}, \sigma^{2(k+1)})$ from the conditional distribution (given $\boldsymbol{X}^{(k)}$)

$$\begin{aligned} p\left(\phi, \sigma^2 \mid \boldsymbol{X}^{(k)}, \boldsymbol{Y}\right) &\propto (\sigma^2)^{-\frac{n}{2}} \exp\left[-\frac{1}{2\sigma^2} \sum_{t=1}^{n} \left(y_t - \phi\, x_t^{(k)}\right)^2\right] \cdot \left(\sigma^2\right)^{-\frac{\nu+2}{2}} \exp\left(-\frac{\nu\lambda}{2\sigma^2}\right) \\ &\propto \pi^{(k+1)}(\phi \mid \sigma^2)\, \pi^{(k+1)}(\sigma^2) \end{aligned} \tag{26.13}$$

where

$$\pi^{(k+1)}(\sigma^2) \sim \chi^{-2}\left(\nu + n - 1,\ \frac{1}{\nu + n - 1}\left[\nu\lambda + \sum_{t=1}^{n} y_t^2 - \frac{1}{n}\left(\sum_{t=1}^{n} y_t\, x_t^{(k)}\right)^2\right]\right) \tag{26.14}$$

and

$$\pi^{(k+1)}(\phi \mid \sigma^2) \sim \mathcal{N}\left(\frac{1}{n}\sum_{t=1}^{n} y_t\, x_t^{(k)},\ \frac{\sigma^2}{n}\right) \tag{26.15}$$

- Draw a sample $\boldsymbol{X}^{(k+1)}$ from the following conditional distribution, given $(\phi^{(k+1)}, \sigma^{2(k+1)})$:

$$\begin{aligned} p(\boldsymbol{X} \mid \phi^{(k+1)}, \sigma^{2(k+1)}, \boldsymbol{Y}) &= \prod_{t=1}^{n} p(x_t \mid y_t, \phi^{(k+1)}, \sigma^{2(k+1)}) \\ &\propto \prod_{t=1}^{n} \exp\left[-\frac{1}{2\sigma^{2(k+1)}}(y_t - \phi^{(k+1)}\, x_t)^2\right] \end{aligned} \tag{26.16}$$

That is, for $t = 1, \ldots, n$ and $b \in \{+1, -1\}$, draw $x_t^{(k+1)}$ from

$$P\left(x_t^{(k+1)} = b\right) = \left[1 + \exp\left(-\frac{2b\,\phi^{(k+1)}\,y_t}{\sigma^{2(k+1)}}\right)\right]^{-1} \tag{26.17}$$

It is worthwhile to note that one can integrate out ϕ and σ^2 in Equation 26.12 analytically to get the marginal target distribution of $\boldsymbol{X}$, which can provides some further insight. More precisely, we have

$$\pi(\boldsymbol{X}) \propto \left[\nu\lambda + \sum_{t=1}^{n} y_t^2 - \frac{1}{n}\left(\sum_{t=1}^{n} x_t y_t\right)^2\right]^{-(n+\nu)/2} \tag{26.18}$$

This defines a distribution on the space of an n-dimensional cube. The mode of this distribution is clearly at $\tilde{\boldsymbol{X}}$, and $-\tilde{\boldsymbol{X}}$, where $\tilde{\boldsymbol{X}} = \text{sign}(\boldsymbol{Y})$. Intuitively, this is the 'obvious solution' in this simple setting but is not easy to generalize. Based on Equation 26.18, we can derive another Gibbs sampling algorithm as follows.

Algorithm 26.4 (one-component Gibbs detector in AWGN channel)

- Choose t from $1, \ldots, n$ by either the random scan (i.e., the t is chosen at random) or the deterministic scan (i.e., one cycle t from 1 to n systematically). Update $\boldsymbol{X}^{(k)}$ to $\boldsymbol{X}^{(k+1)}$, where $x_s^{(k)} = x_s^{(k+1)}$ for $s \neq t$ and $x_t^{(k+1)}$ is drawn from the conditional distribution

$$\pi\left(x_t = b \mid \boldsymbol{X}_{[-t]}^{(k)}\right) = \frac{\pi\left(x_t = b, \boldsymbol{X}_{[-t]}^{(k)}\right)}{\pi\left(x_t = b, \boldsymbol{X}_{[-t]}^{(k)}\right) + \pi\left(x_t = -b, \boldsymbol{X}_{[-t]}^{(k)}\right)} \tag{26.19}$$

where $\pi(\boldsymbol{X})$ is as in Equation 26.18. When the variance σ^2 is known,

$$\pi(\boldsymbol{X}) \propto \exp\left\{\frac{1}{2n\sigma^2}\left(\sum_{t=1}^{n} x_t y_t\right)^2\right\} \tag{26.20}$$

Besides the two Gibbs samplers just described, an attractive alternative is the Metropolis algorithm applied directly to Equation 26.18. Suppose $\boldsymbol{X}^{(k)} = (x_1^{(k)}, \ldots, x_n^{(k)})$. At step $k+1$, the Metropolis algorithm proceeds as follows.

Algorithm 26.5 (Metropolis detector in AWGN channel)

- Choose $t \in \{1, \ldots, n\}$ either by the random scan or by the deterministic scan. Define $\boldsymbol{Z} = (z_1, \ldots, z_n)$ where $z_t = -x_t^{(k)}$ and $z_s = x_s^{(k)}$ for $s \neq t$. Generate independently $U \sim \text{uniform}(0, 1)$. Let $\boldsymbol{X}^{(k+1)} = \boldsymbol{Z}$ if

$$U \leq \min\left\{1, \frac{\pi(\boldsymbol{Z})}{\pi(\boldsymbol{X}^{(k)})}\right\} \tag{26.21}$$

and let $\boldsymbol{X}^{(k+1)} = \boldsymbol{X}^{(k)}$ otherwise.

This Metropolis algorithm differs from the one-component Gibbs detector only slightly in the way of updating $x_t^{(k)}$ to $x_t^{(k+1)}$. That is, the Metropolis algorithm always forces the change (to $-x_t^{(k)}$) unless it is rejected, whereas the Gibbs sampler voluntarily selects whether to make the change so that no rejection is incurred. It is shown in [18] that when the random scan is used, the Metropolis rule always results in a smaller second-largest eigenvalue (not in absolute value) than the corresponding Gibbs sampler. Thus, when the target distribution is relatively peaked (high signal-to-noise ratio (SNR)), the Metropolis algorithm is slightly preferable. However, the Metropolis algorithm may have a large (in absolute value) negative eigenvalue when the target distribution is flatter (low SNR). In practice, however, the large negative eigenvalue is not a serious concern. No clear theory is available when a deterministic scan is used for updating. Simulations suggest that a similar result to that of the random scan samplers seems to hold well.

To overcome the phase ambiguity, one can either restrict ϕ to be positive or, alternatively, use differential encoding. Let the information sequence be $s_t \in \{+1, -1\}, t = 2, \ldots, n$. In differential coding, we construct the transmitted sequence $x_t \in \{+1, -1\}, t = 1, \ldots, n$, such that $x_t = x_{t-1}s_t$. To obtain Monte Carlo draws from the posterior distribution of $p(\boldsymbol{S}, \phi, \sigma^2 \mid \boldsymbol{Y})$, we use one of the MCMC algorithms to generate a Markov chain on $(\boldsymbol{X}, \phi, \sigma^2)$ and then convert the samples of $\boldsymbol{X}$ to $\boldsymbol{S}$ using $s_t^{(k)} = x_t^{(k)} x_{t-1}^{(k)}, t = 2, \ldots, n$. Note that in this way $\boldsymbol{X}$ and $-\boldsymbol{X}$ result in the same $\boldsymbol{S}$. Since $\{\boldsymbol{X}^{(k)}\}$ is a Markov chain, so is $\{\boldsymbol{S}^{(k)}\}$. The transition probability from $\boldsymbol{S}^{(k)}$ to $\boldsymbol{S}^{(k+1)}$ is given by

$$P(\boldsymbol{S}^{(k+1)} \mid \boldsymbol{S}^{(k)}) = P(\boldsymbol{X}^{(k+1)} \mid \boldsymbol{X}^{(k)}) + P(-\boldsymbol{X}^{(k+1)} \mid \boldsymbol{X}^{(k)}) \tag{26.22}$$

where both $\boldsymbol{X}^{(k+1)}$ and $-\boldsymbol{X}^{(k+1)}$ result in $\boldsymbol{S}^{(k+1)}$, and $\boldsymbol{X}^{(k)}$ results in $\boldsymbol{S}^{(k)}$. Note that both $\boldsymbol{X}^{(k)}$ and $-\boldsymbol{X}^{(k)}$ result in $\boldsymbol{S}^{(k)}$, but since $P(-\boldsymbol{X}^{(k+1)} \mid -\boldsymbol{X}^{(k)}) = P(\boldsymbol{X}^{(k+1)} \mid \boldsymbol{X}^{(k)})$, either one can be used.

By denoting $s_1 = x_1$ and $\boldsymbol{S} \triangleq [s_1, s_2, \ldots, s_n]$, we can modify Equation 26.18 to give rise to the marginal target distribution for the s_t:

$$\pi(s_1, \ldots, s_n) \propto \left\{ \nu\lambda + \sum_{t=1}^{n} y_t^2 - \frac{1}{n} \left(\sum_{t=1}^{n} y_t \prod_{i=1}^{t} s_i \right)^2 \right\}^{-(n+\nu)/2} \tag{26.23}$$

Clearly, s_1 is independent of all the other s values and has a uniform marginal distribution.

It is trickier to implement an efficient Gibbs sampler or Metropolis algorithm based on Equation 26.23. For example, the single-site update method (i.e., changing one s_t at a time) may be inefficient because when we propose to change s_t to $-s_t$, all the signs on $y_t, y_{t+1}, \ldots$ have to be changed. This may result in a very small acceptance rate. Since a single update from x_t to $-x_t$ corresponds to changing (s_t, s_{t+1}) $(-s_t, -s_{t+1})$, we can employ proposals

$$(s_t, s_{t+1}) \propto (-s_t, -s_{t+1}), \quad t < n$$

and $s_n \to -s_n$ for the distribution in Equation 26.23.

26.2.2.2 MCMC Equalizers in ISI Channels

Next we consider the Gibbs sampler for blind equalization in an intersymbol interference (ISI) channel [2, 29]. After filtering and sampling the continuous-time received waveform, the discrete-time received signal in such a channel is given by

$$y_t = \sum_{s=0}^{q} \phi_s x_{t-s} + v_t, \quad t = 1, 2, \ldots, n \tag{26.24}$$

where $(q+1)$ is the channel order, $\phi_i \in \mathbb{R}$ is the value of the i-th channel tap, $i = 0, \ldots, q, x_t \in \{+1, -1\}$ is the transmitted binary symbol at time t, and $v_t \sim \mathcal{N}(0, \sigma^2)$ is an independent Gaussian noise sample at time t.

Let $\boldsymbol{X} \triangleq [x_{1-q}, \ldots, x_n]$, $\boldsymbol{Y} \triangleq [y_1, \ldots, y_n]$, and $\boldsymbol{\phi} \triangleq [\phi_0, \ldots, \phi_q]^T$. With a uniform prior for $\boldsymbol{\phi}$, a uniform prior for $\boldsymbol{X}$, and an inverse χ^2 prior for σ^2 (e.g., $\sigma^2 \sim \chi_{\nu,\lambda}^{-2}$), the complete posterior distribution is

$$p(\boldsymbol{X}, \boldsymbol{\phi}, \sigma^2 \mid \boldsymbol{Y}) \sim p(\boldsymbol{Y} \mid \boldsymbol{X}, \boldsymbol{\phi}, \sigma^2)\, p(\boldsymbol{\phi})\, p(\sigma^2) p(\boldsymbol{X}) \tag{26.25}$$

The Gibbs sampler approach to this problem starts with an arbitrary initial value of $\boldsymbol{X}^{(0)}$ and iterates between the following two steps.

Algorithm 26.6 (two-component Gibbs equalizer in ISI channel)

- Draw a sample $(\phi^{(k+1)}, \sigma^{2(k+1)})$ from the conditional distribution (given $X^{(k)}$)

$$p(\phi, \sigma^2 \mid X^{(k)}, Y) \propto (\sigma^2)^{-\frac{n}{2}} \exp\left[-\frac{1}{2\sigma^2}\sum_{t=1}^{n}\left(y_t - \phi^T x_t^{(k)}\right)^2\right](\sigma^2)^{-\frac{\nu+2}{2}} \exp\left(-\frac{\nu\lambda}{2\sigma^2}\right)$$
$$\propto \pi^{(k+1)}(\phi \mid \sigma^2)\,\pi^{(k+1)}(\sigma^2) \tag{26.26}$$

where $x_t^{(k)} \triangleq [x_t^{(k)}, \ldots, x_{t-q}^{(k)}]^T$ for $k = 0, 1, \ldots$ and

$$\pi^{(k+1)}(\sigma^2) \sim \chi^{-2}\left(\nu + n - 1, \frac{\nu\lambda + W^{(k+1)}}{\nu + n - 1}\right) \tag{26.27}$$

$$\pi^{(k+1)}(\phi \mid \sigma^2) \sim \mathcal{N}(\mu^{(k+1)}, \Sigma^{(k+1)}) \tag{26.28}$$

$$W^{(k+1)} = \sum_{t=1}^{n} y_t^2 - \left[\sum_{t=1}^{n} x_t^{(k)} y_t\right]^T \left[\sum_{t=1}^{n} x_t^{(k)} x_t^{(k)T}\right]^{-1} \left[\sum_{t=1}^{n} x_t^{(k)} y_t\right] \tag{26.29}$$

$$\Sigma^{(k+1)} = \left[\frac{1}{\sigma^2}\sum_{t=1}^{n} x_t x_t^T\right]^{-1} \tag{26.30}$$

$$\mu^{(k+1)} = \Sigma^{(k+1)}\left(\frac{1}{\sigma^2}\sum_{t=1}^{n} x_t^{(k)} y_t\right) \tag{26.31}$$

- Draw a sample $X^{(k+1)}$ from the conditional distribution, given $(\phi^{(k+1)}, \sigma^{2(k+1)})$, through the following iterations. For $t = 1 - q, \ldots, n$, generate $x_t^{(k+1)}$ from

$$p\left(x_t \mid \phi^{(k+1)}, \sigma^{2(k+1)}, Y, X_{[-t]}^{(k)}\right) \propto \exp\left[-\frac{1}{2\sigma^{2(k+1)}}\sum_{j=1}^{n}\left(y_j - \phi^{(k+1)T} x_j^{(k)}\right)^2\right] \tag{26.32}$$

where $X_{[-t]}^{(k)} \triangleq [x_{1-q}^{(k+1)}, \ldots, x_{t-1}^{(k+1)}, x_{t+1}^{(k)}, \ldots, x_M^{(k)}]$ and $x_j^{(k)} \triangleq [X_{[-t]}^{(k)}]_{j-q:j}$

Another interesting Gibbs sampling scheme is based on the grouping idea [1]. In particular, a forward–backward algorithm can be employed to sample X jointly, conditional on Y and the parameters. This scheme is shown effective when the X forms a Gaussian Markov model or a Markov chain whose state variable takes on only a few values. In the ISI channel equalization problem, the x_t are independent and identically distributed (i.i.d.) symbols *a priori*, but they are correlated *a posteriori* because of the observed signal Y and the Equation 26.24. The induced correlation among the x_t vanishes after lag q. More precisely, instead of using Equation 26.32 to sample X iteratively, one can draw X altogether.

Algorithm 26.7 (grouping Gibbs equalizer in ISI channel)

- The first few steps are identical to the previous Gibbs equalizer.
- The last step is replaced by the forward–backward scheme. Conditional on ϕ and σ (we suppress the superscript for iteration numbers), we have the joint distribution of X:

$$p(X \mid \phi, \sigma, Y) \propto \exp\left[-\frac{1}{2\sigma^2}\sum_{j=1}^{n}(y_j - \phi^T x_j)^2\right] \equiv \exp\{g_1(x_1) + \cdots + g_n(x_n)\} \tag{26.33}$$

where $x_j = (x_{j-q}, \ldots, x_j)$. Thus, each x_j can take 2^{q+1} possible values. The following two steps produce a sample X from $p\,(X \mid \phi, \sigma, Y)$.

— Forward summation. Define $f_1(\boldsymbol{x}_1) = \exp\{g_1(\boldsymbol{x}_1)\}$ and compute recursively

$$f_{j+1}(\boldsymbol{x}_{j+1}) = \sum_{x_{j-q}=-1}^{1} [f_j(\boldsymbol{x}_j) \exp\{g_{j+1}(\boldsymbol{x}_{j+1})\}] \tag{26.34}$$

— Backward sampling. First draw $\boldsymbol{x}_n = (x_{n-q}, \ldots, x_n)$ from distribution $P(\boldsymbol{x}_n) \propto f_n(\boldsymbol{x}_n)$. Then, for $j = n - q - 1, \ldots, 1$, draw $P(x_j \mid x_{j+1}, \ldots, x_n) \propto f_{j+q}(x_j, \ldots, x_{j+q})$.

Although the grouping idea is attractive for overcoming the channel memory problem, the additional computation cost may offset its advantages. More precisely, the forward–backward procedure needs about 2^q times more memory and about 2^q times more basic operations.

Similar to the previous section, we can integrate out the continuous parameters and write down the marginal target distribution of $\boldsymbol{X}$:

$$\pi(\boldsymbol{X}) \propto [\nu\lambda + W]^{-(n+\nu)/2} \tag{26.35}$$

where

$$W = \sum y_t^2 - \left[\sum \boldsymbol{x}_t y_t\right]^T \left[\sum \boldsymbol{x}_t \boldsymbol{x}_t^T\right]^{-1} \left[\sum \boldsymbol{x}_t y_t\right] \tag{26.36}$$

We can then derive the one-component Gibbs and Metropolis algorithms accordingly. The phase ambiguity (i.e., likelihood unchanged when $\boldsymbol{X}$ is changed to $-\boldsymbol{X}$) can be clearly seen from this joint distribution.

Algorithm 26.8 (one-component Gibb/Metropolis equalizer in ISI channel)

- Choose t from $1, \ldots, n$ by either the random scan or the systematic scan. Let $\boldsymbol{X}^{(k+1)} = \boldsymbol{Z}$, where $z_s = x_s^{(k)}$ for $s \neq t$ and $z_t = -x_t^{(k)}$, with probability

$$\frac{\pi(\boldsymbol{Z})}{\pi(\boldsymbol{X}^{(k)}) + \pi(\boldsymbol{Z})} \tag{26.37}$$

for the Gibbs equalizer, or with probability

$$\min\left\{1, \frac{\pi(\boldsymbol{Z})}{\pi(\boldsymbol{X}^{(k)})}\right\} \tag{26.38}$$

for the Metropolis equalizer, where $\pi(\boldsymbol{X})$ is as in Equation 26.35. Otherwise, let $\boldsymbol{X}^{(k+1)} = \boldsymbol{X}^{(k)}$. When the variance σ^2 is known,

$$\pi(\boldsymbol{X}) \propto \left\{\frac{1}{|\sum \boldsymbol{x}_t \boldsymbol{x}_t^T|^{q/2}}\right\} \exp\left(\frac{1}{2\sigma^2}\left[\sum \boldsymbol{x}_t y_t\right]^T \left[\sum \boldsymbol{x}_t \boldsymbol{x}_t^T\right]^{-1} \left[\sum \boldsymbol{x}_t y_t\right]\right)$$

To overcome the phase ambiguity, we use differential coding in all of our algorithms. Denote $\boldsymbol{S} \triangleq [s_2, \ldots, s_n]$ as the information bits. Let $s_t^{(k)} = x_t^{(k)} x_{t-1}^{(k)}$, $t = 2, \ldots, n$. Since $\boldsymbol{X}^{(k)}$ forms a Markov chain, $\boldsymbol{S}^{(k)}$ is a Markov chain too. The transition probability from $\boldsymbol{S}^{(k)}$ to $\boldsymbol{S}^{(k+1)}$ is

$$P(\boldsymbol{S}^{(k+1)} \mid \boldsymbol{S}^{(k)}) = P(\boldsymbol{X}^{(k+1)} \mid \boldsymbol{X}^{(k)}) + P(-\boldsymbol{X}^{(k+1)} \mid \boldsymbol{X}^{(k)}) \tag{26.39}$$

where both $\boldsymbol{X}^{(k+1)}$ and $-\boldsymbol{X}^{(k+1)}$ result in $\boldsymbol{S}^{(k+1)}$ and $\boldsymbol{X}^{(k)}$ results in $\boldsymbol{S}^{(k)}$.

26.2.2.3 MCMC Multiuser Detector in CDMA Channels

Consider now a CDMA system with K users, employing normalized modulation waveforms $\boldsymbol{h}_1, \ldots, \boldsymbol{h}_K$ and signaling in the presence of additive white Gaussian noise. After filtering and sampling the continuous-time received waveform, the received signal at time t is given by

$$\boldsymbol{r}_t = \boldsymbol{H}\boldsymbol{\Phi}\boldsymbol{x}_t + \boldsymbol{v}_t \tag{26.40}$$

where $\boldsymbol{H} \triangleq [\boldsymbol{h}_1, \boldsymbol{h}_2, \ldots, \boldsymbol{h}_K]$ contains the spreading waveforms of all users, $\boldsymbol{\Phi} \triangleq \text{diag}\{\phi_1, \phi_2, \ldots, \phi_K\}$ contains the real amplitude of each user, $\boldsymbol{x}_t \triangleq [x_{1,t}, x_{2,t}, \ldots, x_{K,t}]^T$ contains the binary symbols sent by all users at time t, and $\boldsymbol{v}_t \sim \mathcal{N}(\boldsymbol{0}, \sigma^2 \boldsymbol{I}_N)$ is the ambient Gaussian noise vector at time t. Clearly, the CDMA system is a generalization of the AWGN in that the multiuser symbol can be viewed as a supersymbol.

At the receiver, the received signal $\boldsymbol{r}_t$ is correlated with each of the spreading waveforms to obtain the following sufficient statistic:

$$\boldsymbol{y}_t \triangleq \boldsymbol{H}^T \boldsymbol{r}_t = \boldsymbol{R}\boldsymbol{\Phi}\boldsymbol{x}_t + \boldsymbol{u}_t \tag{26.41}$$

where

$$\boldsymbol{R} \triangleq \boldsymbol{H}^T \boldsymbol{H} \quad \text{and} \quad \boldsymbol{u}_t \sim \mathcal{N}(\boldsymbol{0}, \sigma^2 \boldsymbol{R})$$

Denote $\boldsymbol{\phi} \triangleq [\phi_1 \cdots \phi_K]^T$ and $\boldsymbol{X}_t = \text{diag}\{x_{1,t}, \ldots, x_{K,t}\}$. Denote also $\mathcal{X} \triangleq [\boldsymbol{x}_1, \ldots, \boldsymbol{x}_n]$ and $\boldsymbol{Y} \triangleq [\boldsymbol{y}_1, \ldots, \boldsymbol{y}_n]$. With a flat prior for $\boldsymbol{\phi}$ and known σ^2, a Gibbs sampler starts with an initial value $\mathcal{X}^{(0)}$ and iterates between the following steps for $j = 0, 1, \ldots$.

Algorithm 26.9 (Gibbs multiuser detector in CDMA channel)

- Draw a sample $\boldsymbol{\phi}^{(j+1)}$ from the following conditional distribution (given $\mathcal{X}^{(j)}$):

$$p(\boldsymbol{\phi} \mid \boldsymbol{X}^{(j)}, \boldsymbol{Y}) \sim \mathcal{N}(\boldsymbol{\mu}^{(j+1)}, \boldsymbol{\Sigma}^{(j+1)}) \tag{26.42}$$

  where

$$\boldsymbol{\Sigma}^{(j+1)} \triangleq \left[\frac{1}{\sigma^2} \sum_{t=1}^{n} \boldsymbol{X}_t^{(j)} \boldsymbol{R} \boldsymbol{X}_t^{(j)} \right]^{-1} \tag{26.43}$$

$$\boldsymbol{\mu}^{(j+1)} \triangleq \boldsymbol{\Sigma}^{(j+1)} \left(\frac{1}{\sigma^2} \sum_{t=1}^{n} \boldsymbol{X}_t^{(j)} \boldsymbol{y}_t \right) \tag{26.44}$$

- For $t = 1, \ldots, n$ and $k = 1, \ldots, K$, draw a sample $x_{k,t}^{(j+1)}$ from the following:

$$\frac{p\left[x_{k,t} = +1 \,\middle|\, \boldsymbol{Y}, \boldsymbol{\phi}^{(j+1)}, \mathcal{X}_{[-k,-t]}^{(j)}\right]}{p\left[x_{k,t} = -1 \,\middle|\, \boldsymbol{Y}, \boldsymbol{\phi}^{(j+1)}, \mathcal{X}_{[-k,-t]}^{(j)}\right]} = \exp\left\{ \frac{2\phi_k^{(j+1)}}{\sigma^2} \left(y_{k,t} - \boldsymbol{e}_k^T \boldsymbol{R} \boldsymbol{\Phi}^{(j+1)} \boldsymbol{x}_{k,t}^{0(j)} \right) \right\} \tag{26.45}$$

  where

$$\boldsymbol{x}_{k,t}^{0(j)} \triangleq \left[x_{1,t}^{(j+1)} \cdots x_{k-1,t}^{(j+1)},\ 0,\ x_{k+1,t}^{(j)} \cdots x_{K,t}^{(j)} \right]^T$$

  and $\boldsymbol{e}_k$ denotes the k-th unit basis vector of $\mathbb{R}^K$.

The one-component Gibbs detector can be derived easily, as in the AWGN cases, and is omitted here. Again, to cope with the phase ambiguity, we use differential encoding for each user. Let the information sequence of the k-th user be $s_{k,t} \in \{+1, -1\}, t = 2, \ldots, n$; then the transmitted sequence for this user is given by $x_{k,t} = x_{k,t-1} s_{k,t}$, $t = 1, \ldots, n$. Denote $\boldsymbol{s}_t \triangleq [s_{1,t} \ \ldots \ s_{K,t}]^T$ and $\mathcal{S} \triangleq [\boldsymbol{s}_2, \ldots, \boldsymbol{s}_n]$, and let $\text{diff}(\mathcal{X}) = \mathcal{S}$.

Another interesting situation is the detection of the multiuser symbols when the channel conditions (i.e., $\boldsymbol{\phi}$ and σ^2) are known to the receiver. In this case, the Equation 26.41 becomes independent for different t values. That is,

$$p(\boldsymbol{X} \mid \boldsymbol{Y}) = \prod_{t=1}^{n} p(\boldsymbol{x}_t \mid \boldsymbol{y}_t)$$

Hence, we can simply consider the case that $n = 1$. The Gibbs algorithms for interference cancellation becomes the following. Given initial values $\boldsymbol{x}^{(0)}$, a Gibbs sampler iterates between the following steps for $j = 1, 2, \ldots$.

Algorithm 26.10 (Gibbs interference canceller in CDMA channel)

- For $k = 1, \ldots, K$, draw a sample $x_k^{(j+1)}$ from the following:

$$\frac{p\left[x_k = +1 \mid \boldsymbol{Y}, \boldsymbol{x}_{[-k]}^{(j)}\right]}{p\left[x_k = -1 \mid \boldsymbol{y}, \boldsymbol{x}_{[-k]}^{(j)}\right]} = \exp\left\{\frac{2\phi_k^{(j+1)}}{\sigma^2}\left(y_k - \boldsymbol{e}_k^T \boldsymbol{R}\boldsymbol{\Phi}\boldsymbol{x}_k^{0(j)}\right)\right\} \tag{26.46}$$

where

$$\boldsymbol{x}_k^{0(j)} \triangleq \left[x_1^{(j+1)} \cdots x_{k-1}^{(j+1)},\ 0,\ x_{k+1}^{(j)} \cdots x_K^{(j)}\right]^T$$

and $\boldsymbol{e}_k$ denotes the k-th unit basis vector of $\mathbb{R}^K$.

26.2.3 Other Applications

In the preceding sections we have illustrated the applications of the Metropolis–Hasting algorithm and the Gibbs sampler to some basic receiver design problems found in digital communications. The convergence behaviors of the above MCMC samplers in these applications are investigated in [4]. Another salient feature of the MCMC-based detector is that since it can incorporate as input the *a priori* symbol probabilities and it produces as output the *a posteriori* probabilities, it can be employed as a soft-input soft-output (SISO) demodulator in a turbo receiver for coded systems. Such a turbo receiver iterates between the demodulation process and channel decoding process by exchanging the so-called extrinsic information between these stages to successively improve the receiver performance. The design and implementation of such a turbo receiver in the presence of an unknown channel remained a challenging open problem until in [28] a blind turbo multiuser receiver was proposed that employs the Gibbs sampler as the SISO demodulator. Following that work, MCMC techniques have been developed to treat more complicated systems, such as coded ISI channels [29], asynchronous code division multiple access (CDMA) with multipath fading [34], nonlinearly modulated CDMA systems [26], multicarrier CDMA systems with space–time coding [31], orthogonal frequency division multiplexing (OFDM) systems with frequency offset and frequency-selective fading [24], and systems with Gaussian minimum-shift keying (GMSK) modulation over multipath fading channels [32]. Finally, we note that for most of these applications, only 100 to 200 MCMC iterations are needed to obtain good estimates of the unknown quantities.

26.3 SMC Methods

26.3.1 General SMC Algorithms

26.3.1.1 Sequential Importance Sampling

Importance sampling is perhaps one of the most elementary, well-known, and versatile Monte Carlo techniques. Suppose we want to estimate $E\{h(\boldsymbol{x})\}$ (with respect to p) using the Monte Carlo method. Since directly sampling from $p(\boldsymbol{x})$ is difficult, we want to find a *trial* distribution, $q(\boldsymbol{x})$, that is reasonably close to p but is easy to draw samples from. Because of the simple identity

$$\begin{aligned} E\{h(\boldsymbol{x})\} &= \int h(\boldsymbol{x})\, p(\boldsymbol{x})\, \mathrm{d}\boldsymbol{x} \\ &= \int h(\boldsymbol{x})\, w(\boldsymbol{x})\, q(\boldsymbol{x})\, \mathrm{d}\boldsymbol{x} \end{aligned} \tag{26.47}$$

where

$$w(\boldsymbol{x}) \triangleq \frac{p(\boldsymbol{x})}{q(\boldsymbol{x})} \tag{26.48}$$

is the importance weight, we can approximate Equation 26.47 by

$$E\{h(\boldsymbol{x})\} \cong \frac{1}{W} \sum_{j=1}^{\nu} h(\boldsymbol{x}^{(j)})\, w(\boldsymbol{x}^{(j)}) \tag{26.49}$$

where $\boldsymbol{x}^{(1)}, \boldsymbol{x}^{(2)}, \ldots, \boldsymbol{x}^{(\nu)}$ are random samples from q and $W = \sum_{j=1}^{n} w(\boldsymbol{x}^{(j)})$. In using this method, we only need to know the expression of $p(\boldsymbol{x})$ up to a normalizing constant, which is the case for many processing problems found in digital communications. Each $\boldsymbol{x}^{(j)}$ is said to be properly weighted by $w(\boldsymbol{x}^{(j)})$ with respect to p.

However, it is usually difficult to design a good trial density function in high-dimensional problems. One of the most useful strategies in these problems is to build up the trial density sequentially. Suppose we can decompose $\boldsymbol{x}$ as $(x_1, \ldots, x_d)$ where each of the x_j may be multidimensional. Then our trial density can be constructed as

$$q(\boldsymbol{x}) = q_1(x_1) q_2(x_2 \mid x_1) \cdots q_d(x_d \mid x_1, \ldots, x_{d-1}) \tag{26.50}$$

by which we hope to obtain some guidance from the target density while building up the trial density. Corresponding to the decomposition of $\boldsymbol{x}$, we can rewrite the target density as

$$p(\boldsymbol{x}) = p(x_1) p(x_2 \mid x_1) \cdots p(x_d \mid x_1, \ldots, x_{d-1}) \tag{26.51}$$

and the importance weight as

$$w(\boldsymbol{x}) = \frac{p(x_1) p(x_2 \mid x_1) \cdots p(x_d \mid x_1, \ldots, x_{d-1})}{q_1(x_1) q_2(x_2 \mid x_1) \cdots q_d(x_d \mid x_1, \ldots, x_{d-1})} \tag{26.52}$$

Equation 26.52 suggests a recursive way of computing and monitoring the importance weight. That is, by denoting $\boldsymbol{x}_t = (x_1, \ldots, x_t)$ (thus, $\boldsymbol{x}_d \equiv \boldsymbol{x}$), we have

$$w_t(\boldsymbol{x}_t) = w_{t-1}(\boldsymbol{x}_{t-1}) \frac{p(x_t \mid \boldsymbol{x}_{t-1})}{q_t(x_t \mid \boldsymbol{x}_{t-1})} \tag{26.53}$$

Then w_d is equal to $w(\boldsymbol{x})$ in Equation 26.52. Potential advantages of this recursion and Equation 26.51 are (1) we can stop generating further components of $\boldsymbol{x}$ if the *partial weight* derived from the sequentially generated *partial sample* is too small; and (2) we can take advantage of $p(x_t|\boldsymbol{x}_{t-1})$ in designing $q_t(x_t|\boldsymbol{x}_{t-1})$. In other words, the marginal distribution $p(\boldsymbol{x}_t)$ can be used to guide the generation of $\boldsymbol{x}$.

Although the idea sounds interesting, the trouble is that Equations 26.51 and 26.52 are not useful at all. The reason is that in order to get Equation 26.51, one needs to have the marginal distribution

$$p(\boldsymbol{x}_t) = \int p(x_1, \ldots, x_d) \mathrm{d}x_{t+1} \cdots \mathrm{d}x_d \tag{26.54}$$

which is perhaps more difficult than the original problem.

In order to carry out the sequential sampling idea, we need to find a sequence of auxiliary distributions, $\pi_1(x_1), \pi_2(\boldsymbol{x}_2), \ldots, \pi_d(\boldsymbol{x})$, so that $\pi_t(\boldsymbol{x}_t)$ is a reasonable approximation to the marginal distribution $p(\boldsymbol{x}_t)$, for $t = 1, \ldots, d-1$, and $\pi_d = p$. We want to emphasize that the π_t are only required to be known up to a normalizing constant and they *only* serve as guides to our construction of the whole sample $\boldsymbol{x} = (x_1, \ldots, x_d)$. The *sequential importance sampling* (SIS) method can then be defined as the following recursive procedure.

Algorithm 26.11 (sequential importance sampling)

For $t = 2, \ldots, d$:

- Draw x_t from $q_t(x_t|\boldsymbol{x}_{t-1})$ and let $\boldsymbol{x}_t = (\boldsymbol{x}_{t-1}, x_t)$.
- Compute

$$u_t = \frac{\pi_t(\boldsymbol{x}_t)}{\pi_{t-1}(\boldsymbol{x}_{t-1})q_t(x_t \mid \boldsymbol{x}_{t-1})} \tag{26.55}$$

  and let $w_t = w_{t-1}u_t$. Here u_t is called an *incremental weight.*

It is easy to show that $\boldsymbol{x}_t$ is properly weighted by w_t with respect to π_t provided that $\boldsymbol{x}_{t-1}$ is properly weighted by w_{t-1} with respect to π_{t-1}. Thus, the whole sample $\boldsymbol{x}$ obtained by SIS is properly weighted by w_d with respect to the target density $p(\boldsymbol{x})$. The auxiliary distributions can also be used to help construct a more efficient trial distribution:

- We can build q_t in light of π_t. For example, one can choose (if possible)

$$q_t(x_t \mid \boldsymbol{x}_{t-1}) = \pi_t(x_t \mid \boldsymbol{x}_{t-1}) \tag{26.56}$$

  Then the incremental weight becomes

$$u_t = \frac{\pi_t(\boldsymbol{x}_t)}{\pi_{t-1}(\boldsymbol{x}_{t-1})} \tag{26.57}$$

  In the same token, we may also want q_t to be $\pi_{t+1}(x_t \mid \boldsymbol{x}_{t-1})$, where the latter involves integrating out x_{t+1}.
- When we observe that w_t is getting too small, we may want to reject the sample halfway and restart. In this way we avoid wasting time on generating samples that are doomed to have little effect in the final estimation. However, as an outright rejection incurs bias, techniques such as the rejection control are needed [22].
- Another problem with the SIS is that the resulting importance weights are often very skewed, especially when d is large. An important recent advance in sequential Monte Carlo to address this problem is the resampling technique [11, 21, 22].

26.3.1.2 SMC for Dynamic Systems

Consider the following dynamic system modeled in a state-space form as

$$\begin{aligned} &\text{state equation} && z_t = f_t(z_{t-1}, \boldsymbol{u}_t) \\ &\text{observation equation} && \boldsymbol{y}_t = g_t(z_t, \boldsymbol{v}_t) \end{aligned} \tag{26.58}$$

where z_t, $\boldsymbol{y}_t$, $\boldsymbol{u}_t$, and $\boldsymbol{v}_t$ are, respectively, the state variable, the observation, the state noise, and the observation noise at time t. They can be either scalars or vectors.

Let $\boldsymbol{Z}_t = (z_0, z_1, \ldots, z_t)$ and let $\boldsymbol{Y}_t = (\boldsymbol{y}_0, \boldsymbol{y}_1, \ldots, \boldsymbol{y}_t)$. Suppose an on-line inference of $\boldsymbol{Z}_t$ is of interest; that is, at current time t we wish to make a timely estimate of a function of the state variable $\boldsymbol{Z}_t$, say $h(\boldsymbol{Z}_t)$, based on the currently available observation $\boldsymbol{Y}_t$. With the Bayes theorem, we realize that the optimal solution to this problem is $E\{h(\boldsymbol{Z}_t)|\boldsymbol{Y}_t\} = \int h(\boldsymbol{Z}_t)p(\boldsymbol{Z}_t|\boldsymbol{Y}_t)\mathrm{d}\boldsymbol{Z}_t$. In most cases an exact evaluation of this expectation is analytically intractable because of the complexity of such a dynamic system. Monte Carlo methods provide us with a viable alternative to the required computation. Specifically, suppose a set of random samples $\{Z_t^{(j)}\}_{j=1}^{\nu}$ is generated from the trial distribution $q(Z_t|Y_t)$. By associating the weight

$$w_t^{(j)} = \frac{p\left(Z_t^{(j)} \mid Y_t\right)}{q\left(Z_t^{(j)} \mid Y_t\right)} \tag{26.59}$$

to the sample $Z_t^{(j)}$, we can approximate the quantity of interest, $E\{h(Z_t)|Y_t\}$, as

$$E\{h(Z_t) \mid Y_t\} \cong \frac{1}{W_t} \sum_{j=1}^{\nu} h\left(Z_t^{(j)}\right) w_t^{(j)} \tag{26.60}$$

where $W_t = \sum_{j=1}^{\nu} w_t^{(j)}$. The pair $(Z_t^{(j)}, w_t^{(j)})$ is a *properly weighted sample* with respect to distribution $p(Z_t|Y_t)$. A trivial but important observation is that the $z_t^{(j)}$ (one of the components of $Z_t^{(j)}$) is also properly weighted by the $w_t^{(j)}$ with respect to the marginal distribution $p(z_t|Y_t)$.

To implement Monte Carlo techniques for a dynamic system, a set of random samples properly weighted with respect to $p(Z_t|Y_t)$ is needed for any time t. Because the state equation in Equation 26.58 possesses a Markovian structure, we can implement a SMC strategy [22]. Suppose a set of properly weighted samples $\{(Z_{t-1}^{(j)}, w_{t-1}^{(j)})\}_{j=1}^{\nu}$ (with respect to $p(Z_{t-1}|Y_{t-1})$) is given at time $(t-1)$. A sequential Monte Carlo filter generates from the set a new one, $\{Z_t^{(j)}, w_t^{(j)}\}_{j=1}^{\nu}$, which is properly weighted at time t with respect to $p(Z_t|Y_t)$, according to the following algorithm.

Algorithm 26.12 (sequential Monte Carlo filter for dynamic systems)

For $j = 1, \ldots, \nu$:

- Draw a sample $z_t^{(j)}$ from a trial distribution $q(z_t|Z_{t-1}^{(j)}, Y_t)$ and let $Z_t^{(j)} = (Z_{t-1}^{(j)}, z_t^{(j)})$.
- Compute the importance weight

$$w_t^{(j)} = w_{t-1}^{(j)} \cdot \frac{p\left(Z_t^{(j)} \mid Y_t\right)}{p\left(Z_{t-1}^{(j)} \mid Y_{t-1}\right) q\left(z_t^{(j)} \mid Z_{t-1}^{(j)}, Y_t\right)} \tag{26.61}$$

The algorithm is initialized by drawing a set of i.i.d. samples $z_0^{(1)}, \ldots, z_0^{(m)}$ from $p(z_0|y_0)$. When y_0 represents the null information, $p(z_0|y_0)$ corresponds to the prior of z_0.

A useful choice of the trial distribution $q(z_t \mid Z_{t-1}^{(j)}, Y_t)$ for the state space model Equation 26.58 is of the form

$$\begin{aligned} q\left(z_t \mid Z_{t-1}^{(j)}, Y_t\right) &= p\left(z_t \mid Z_{t-1}^{(j)}, Y_t\right) \\ &= \frac{p(y_t \mid z_t)\, p\left(z_t \mid z_{t-1}^{(j)}\right)}{p\left(y_t \mid z_{t-1}^{(j)}\right)} \end{aligned} \tag{26.62}$$

For this trial distribution, the importance weight is updated according to

$$w_t^{(j)} \propto w_{t-1}^{(j)} \cdot p\left(y_t \mid z_{t-1}^{(j)}\right) \tag{26.63}$$

26.3.1.3 Mixture Kalman Filter

Many dynamic system models belong to the class of conditional dynamic linear models (CDLMs) of the form

$$\begin{aligned} x_t &= F_{\lambda_t} x_{t-1} + G_{\lambda_t} u_t \\ y_t &= H_{\lambda_t} x_t + K_{\lambda_t} v_t \end{aligned} \tag{26.64}$$

where $\boldsymbol{u}_t \sim \mathcal{N}_c(0, I)$, $\boldsymbol{v}_t \sim \mathcal{N}_c(0, I)$ (here I denotes an identity matrix), and λ_t is a random indicator variable. The matrices F_{λ_t}, G_{λ_t}, H_{λ_t}, and K_{λ_t} are known given λ_t. In this model, the state variable $\boldsymbol{z}_t$ corresponds to $(\boldsymbol{x}_t, \lambda_t)$.

We observe that for a given trajectory of the indicator λ_t in a CDLM, the system is both linear and Gaussian, for which the Kalman filter provides the complete statistical characterization of the system dynamics. The mixture Kalman filter (MKF) [3] can be employed for on-line filtering and prediction of CDLMs. It exploits the conditional Gaussian property and utilizes a marginalization operation to improve the algorithmic efficiency. Instead of dealing with both $\boldsymbol{x}_t$ and λ_t, the MKF draws Monte Carlo samples only in the indicator space and uses a mixture of Gaussian distributions to approximate the target distribution. Compared with the generic SMC method, the MKF is substantially more efficient (e.g., giving more accurate results with the same computing resources).

Let $\boldsymbol{Y}_t = (\boldsymbol{y}_0, \boldsymbol{y}_1, \ldots, \boldsymbol{y}_t)$ and let $\boldsymbol{\Lambda}_t = (\lambda_0, \lambda_1, \ldots, \lambda_t)$. By recursively generating a set of properly weighted random samples $\{(\boldsymbol{\Lambda}_t^{(j)}, w_t^{(j)})\}_{j=1}^{\nu}$ to represent $p(\boldsymbol{\Lambda}_t|\boldsymbol{Y}_t)$, the MKF approximates the target distribution $p(\boldsymbol{x}_t|\boldsymbol{Y}_t)$ by a random mixture of Gaussian distributions

$$\frac{1}{W_t} \sum_{j=1}^{\nu} w_t^{(j)} \mathcal{N}_c\left(\boldsymbol{\mu}_t^{(j)}, \boldsymbol{\Sigma}_t^{(j)}\right) \tag{26.65}$$

where $\boldsymbol{\kappa}_t^{(j)} \triangleq [\boldsymbol{\mu}_t^{(j)}, \boldsymbol{\Sigma}_t^{(j)}]$ is obtained by implementing a Kalman filter for the given indicator trajectory $\boldsymbol{\Lambda}_t^{(j)}$ and $W_t = \sum_{j=1}^{\nu} w_t^{(j)}$. A key step in the MKF is the production at time t of a weighted sample of indicators, $\{(\boldsymbol{\Lambda}_t^{(j)}, \boldsymbol{\kappa}_t^{(j)}, w_t^{(j)})\}_{j=1}^{\nu}$, based on the set of samples, $\{(\boldsymbol{\Lambda}_{t-1}^{(j)}, \boldsymbol{\kappa}_{t-1}^{(j)}, w_{t-1}^{(j)})\}_{j=1}^{\nu}$, at the previous time $(t-1)$ according to the following algorithm.

Algorithm 26.13 (mixture Kalman filter)

For $j = 1, \ldots, \nu$:

- Draw a sample $\lambda_t^{(j)}$ from a trial distribution $q(\lambda_t \mid \boldsymbol{\Lambda}_{t-1}^{(j)}, \boldsymbol{\kappa}_{t-1}^{(j)}, \boldsymbol{Y}_t)$.
- Run a one-step Kalman filter based on $\lambda_t^{(j)}$, $\boldsymbol{\kappa}_{t-1}^{(j)}$, and $\boldsymbol{y}_t$ to obtain $\kappa_t^{(j)}$.
- Compute the weight

$$w_t^{(j)} \propto w_{t-1}^{(j)} \cdot \frac{p\left(\boldsymbol{\Lambda}_{t-1}^{(j)}, \lambda_t^{(j)} \mid \boldsymbol{Y}_t\right)}{p\left(\boldsymbol{\Lambda}_{t-1}^{(j)} \mid \boldsymbol{Y}_{t-1}\right) q\left(\lambda_t^{(j)} \mid \boldsymbol{\Lambda}_{t-1}^{(j)}, \boldsymbol{\kappa}_{t-1}^{(j)}, \boldsymbol{Y}_t\right)} \tag{26.66}$$

26.3.2 Resampling Procedures

The importance sampling weight $w_t^{(j)}$ measures the quality of the corresponding imputed signal sequence $\boldsymbol{Z}_t^{(j)}$. A relatively small weight implies that the sample is drawn far from the main body of the posterior distribution and has a small contribution in the final estimation. Such a sample is said to be ineffective. If there are too many ineffective samples, the Monte Carlo procedure becomes inefficient. This can be detected by observing a large *coefficient of variation* in the importance weight. Suppose $\{w_t^{(j)}\}_{j=1}^{m}$ is a sequence of importance weights. Then the coefficient of variation, υ_t, is defined as

$$\upsilon_t^2 = \frac{\sum_{j=1}^{m} \left(w_t^{(j)} - \bar{w}_t\right)^2 / m}{\bar{w}_t^2} = \frac{1}{m} \sum_{j=1}^{m} \left(\frac{w_t^{(j)}}{\bar{w}_t} - 1\right)^2 \tag{26.67}$$

where $\bar{w}_t = \sum_{j=1}^{m} w_t^{(j)} / m$. Note that if the samples are drawn exactly from the target distribution, then all the weights are equal, implying that $\upsilon_t = 0$. It is shown in [16] that the importance weights resulting from a sequential Monte Carlo filter form a martingale sequence. As more and more data are processed, the coefficient of variation of the weights increases — that is, the number of ineffective samples increases — rapidly.

A useful method for reducing ineffective samples and enhancing effective ones is *resampling* [21]. Roughly speaking, resampling allows those bad samples (with small importance weights) to be discarded and those good ones (with large importance weights) to replicate so as to accommodate the dynamic change of the system. Specifically, let $\{(\boldsymbol{Z}_t^{(j)}, w_t^{(j)})\}_{j=1}^m$ be the original properly weighted samples at time t. A *residual resampling* strategy forms a new set of weighted samples $\{(\tilde{\boldsymbol{Z}}_t^{(j)}, \tilde{w}_t^{(j)})\}_{j=1}^m$ according to the following algorithm (assume that $\sum_{j=1}^m w_t^{(j)} = m$).

Algorithm 26.14 (Resampling Algorithm)

- For $j = 1, \ldots, m$, retain $k_j = \lfloor w_t^{(j)} \rfloor$ copies of the sample $\boldsymbol{Z}_t^{(j)}$. Denote $K_r = m - \sum_{j=1}^m k_j$.
- Obtain K_r i.i.d. draws from the original sample set $\{\boldsymbol{Z}_t^{(j)}\}_{j=1}^m$, with probabilities proportional to $(w_t^{(j)} - k_j), j = 1, \ldots, m$.
- Assign equal weight, i.e., set $\tilde{w}_t^{(j)} = 1$, for each new sample.

The samples drawn by the above residual resampling procedure are properly weighted with respect to $p(\boldsymbol{Z}_t | \boldsymbol{Y}_t)$, provided that m is sufficiently large. In practice, when small to modest m values are used, the resampling procedure can be seen as trading off between bias and variance. That is, the new samples with their weights resulting from the resampling procedure are only approximately proper, which introduces small bias in Monte Carlo estimation. On the other hand, resampling greatly reduces Monte Carlo variance for future samples.

Resampling can be done at any time. However, resampling too often adds computational burden and decreases diversities of the Monte Carlo filter (i.e., it decreases the number of distinctive filters and loses information). On the other hand, resampling too rarely may result in a loss of efficiency. It is thus desirable to give guidance on when to do resampling. A measure of the efficiency of an importance sampling scheme is the *effective sample size* $\bar{m}_t$, defined as

$$\bar{m}_t \triangleq \frac{m}{1 + \upsilon_t^2} \tag{26.68}$$

Heuristically, $\bar{m}_t$ reflects the equivalent size of a set of i.i.d. samples for the set of m weighted ones. It is suggested in [22] that resampling should be performed when the effective sample size becomes small, e.g., $\bar{m}_t \leq \frac{m}{10}$. Alternatively, one can conduct resampling at every fixed-length time interval (say, every five steps).

Instead of the previous resampling scheme suggested in the literature, we may implement a more flexible resampling scheme as follows (assume that $\sum_{j=1}^m w_t^{(j)} = m$):

For $j = 1, \ldots, m$:

1. For $w_t^{(j)} \geq 1$:
   - Retain k_j copies of the sample $\boldsymbol{Z}_t^{(j)}$, where k_j is given in advance (see below).
   - Assign weight $\tilde{w}_t^{(j)} = w_t^{(j)}/k_j$ for each copy.
2. For $w_t^{(j)} < 1$:
   - Kill the sample with probability $1 - f_j$.
   - Assign weight $w_t^{(j)}/f_j$ to the surviving sample.

The advantage of this new resampling method is that we have the flexibility of choosing a proper resampling size k_j as we wish. On one hand, we want to eliminate those hopeless samples and emphasize those promising ones. On the other hand, we do not want to throw away those mediocre ones that may prove important later on (as the dynamical system moves toward them). An empirical choice of the resample size formula is $k_j = \lfloor \sqrt{w_t^{(j)}} \rfloor$ and $f_j = \sqrt{w_t^{(j)}}$. The intuition behind this choice is that it effectively removes those hopeless samples with small weights but still maintains the diversity of the Monte Carlo sample.

26.3.3 Applications of SMC in Digital Communications

In this section we illustrate the application of SMC in the design of blind adaptive receivers for both fading channels and multiple-input multiple-output (MIMO) ISI channels.

26.3.3.1 SMC Receiver in Flat-Fading Channels

Suppose we want to transmit binary symbols $x_1, \ldots, x_n, x_t \in \{+1, -1\}$, through a fading channel whose input–output relationship is given by

$$y_t = \alpha_t x_t + v_t \tag{26.69}$$

where $\{\alpha_t\}_{t=1}^n$ represents the unknown Rayleigh fading process, which can be modeled as the output of a low-pass filter of order r driven by white Gaussian noise,

$$\{\alpha_t\} = \frac{\Psi(D)}{\Phi(D)}\{u_t\} \tag{26.70}$$

where D is the back-shift operator $D^k\, u_t \triangleq u_{t-k}$; $\Phi(z) \triangleq \phi_r z^r + \cdots + \phi_1 z + 1$; $\Psi(z) \triangleq \psi_r z^r + \cdots + \psi_1 z + \psi_0$; and $\{u_t\}_{t=1}^n$ is a white complex Gaussian noise sequence with independent real and imaginary components, $u_t \sim \mathcal{N}_c(0, \sigma^2)$. The inference problem is to estimate the transmitted symbols $\boldsymbol{X} = \{x_t\}_{t=1}^n$, based on the received signals $\boldsymbol{Y} = \{y_t\}_{t=1}^n$. The nuisance parameters are $\boldsymbol{\Theta} = \{\alpha_1, \ldots, \alpha_n, \sigma^2\}$.

Equations 26.69 and 26.70 can be rewritten in the state-space model form, which is instrumental in developing the sequential signal processing algorithm. Define

$$\{z_t\} \triangleq \Psi^{-1}(D)\{\alpha_t\} \Longrightarrow \Phi(D)\{z_t\} = \{u_t\} \tag{26.71}$$

Denote $\boldsymbol{z}_t \triangleq [z_t, \ldots, z_{t-r+1}]^T$. By Equation 26.70 we then have

$$\boldsymbol{z}_t = \boldsymbol{F}\boldsymbol{z}_{t-1} + \boldsymbol{g}u_t, \qquad u_t \overset{\text{i.i.d.}}{\sim} \mathcal{N}_c(0, 1) \tag{26.72}$$

where

$$\boldsymbol{F} \triangleq \begin{pmatrix} -\phi_1 & -\phi_2 & \cdots & -\phi_r & 0 \\ 1 & 0 & \cdots & 0 & 0 \\ 0 & 1 & \cdots & 0 & 0 \\ \vdots & \vdots & \ddots & \vdots & \vdots \\ 0 & 0 & \cdots & 1 & 0 \end{pmatrix} \quad \text{and} \quad \boldsymbol{g} \triangleq \begin{pmatrix} 1 \\ 0 \\ \vdots \\ 0 \end{pmatrix}$$

Because of Equation 26.71, the fading coefficient sequence $\{\alpha_t\}$ can be written as

$$\alpha_t = \boldsymbol{h}^H \boldsymbol{z}_t, \quad \text{where} \quad \boldsymbol{h} \triangleq [\psi_0\ \psi_1\ \cdots\ \psi_r]^H \tag{26.73}$$

Then we have the following state–space model for the system defined by Equations 26.69 and 26.70:

$$\boldsymbol{z}_t = \boldsymbol{F}\boldsymbol{z}_{t-1} + \boldsymbol{g}u_t \tag{26.74}$$

$$y_t = x_t \boldsymbol{h}^H \boldsymbol{z}_t + v_t \tag{26.75}$$

Denote $\boldsymbol{Y}_t \triangleq (y_1, \ldots, y_t)$ and $\boldsymbol{X}_t \triangleq (x_1, \ldots, x_t)$. We are interested in estimating the symbol x_t at time t based on the observation $\boldsymbol{Y}_t$. The Bayes solution to this problem requires the posterior distribution

$$p(\boldsymbol{z}_t, x_t \mid \boldsymbol{Y}_t) = \int p(\boldsymbol{z}_t \mid \boldsymbol{X}_t, \boldsymbol{Y}_t)\, p(\boldsymbol{X}_t \mid \boldsymbol{Y}_t)\, \mathrm{d}\boldsymbol{X}_{t-1} \tag{26.76}$$

Note that with a given $\boldsymbol{X}_t$, the state-space model Equations 26.74 and 26.75 becomes a linear Gaussian system. Hence,

$$p(\boldsymbol{z}_t \mid \boldsymbol{X}_t, \boldsymbol{Y}_t) \sim \mathcal{N}_c(\boldsymbol{\mu}_t(\boldsymbol{X}_t), \boldsymbol{\Sigma}_t(\boldsymbol{X}_t)) \tag{26.77}$$

where the mean $\boldsymbol{\mu}_t(\boldsymbol{X}_t)$ and covariance matrix $\boldsymbol{\Sigma}_t(\boldsymbol{X}_t)$ can be obtained by a Kalman filter with the given $\boldsymbol{X}_t$.

In order to implement the MKF, we need to obtain a set of Monte Carlo samples of the transmitted symbols, $\{(\boldsymbol{X}_t^{(j)}, w_t^{(j)})\}_{j=1}^{\nu}$, properly weighted with respect to the distribution $p(\boldsymbol{X}_t|\boldsymbol{Y}_t)$. Then the *a posteriori* symbol probability can be estimated as

$$P[x_t = +1 \mid \boldsymbol{Y}_t] \cong \frac{1}{W_t} \sum_{j=1}^{\nu} 1\left(x_t^{(j)} = +1\right) w_t^{(j)} \tag{26.78}$$

where $1(\cdot)$ is an indicator function such that $1(x_t = +1) = 1$ if $x_t = +1$ and 0 otherwise.

Hereafter, we let $\boldsymbol{\mu}_t^{(j)} \triangleq \boldsymbol{\mu}_t(\boldsymbol{S}_t^{(j)})$, $\boldsymbol{\Sigma}_t^{(j)} \triangleq \boldsymbol{\Sigma}_t(\boldsymbol{S}_t^{(j)})$, and $\kappa_t^{(j)} \triangleq [\boldsymbol{\mu}_t^{(j)}, \boldsymbol{\Sigma}_t^{(j)}]$. The following algorithm, which is based on the mixture Kalman filter and first appeared in [5], generates properly weighted Monte Carlo samples $\{(\boldsymbol{S}_t^{(j)}, \kappa_t^{(j)}, w_t^{(j)})\}_{j=1}^{\nu}$.

Algorithm 26.15 (SMC receiver in flat-fading channel)

- *Initialization:* Each Kalman filter is initialized as $\kappa_0^{(j)} = [\boldsymbol{\mu}_0^{(j)}, \boldsymbol{\Sigma}_0^{(j)}]$, with $\boldsymbol{\mu}_0^{(j)} = \boldsymbol{0}$, $\boldsymbol{\Sigma}_0^{(j)} = 2\boldsymbol{\Sigma}$, and $j = 1, \ldots, m$, where $\boldsymbol{\Sigma}$ is the stationary covariance of $\boldsymbol{x}_t$ and is computed analytically from Equation 26.6. (The factor 2 is to accommodate the initial uncertainty.) All importance weights are initialized as $w_0^{(j)} = 1$, $j = 1, \ldots, \nu$. Since the data symbols are assumed to be independent, initial symbols are not needed.

  Based on the state-space model Equations 26.74 to 26.75, the following steps are implemented at time t to update each weighted sample.
- For $j = 1, \ldots, \nu$ compute the one-step predictive update of each Kalman filter $\kappa_{t-1}^{(j)}$:

$$\boldsymbol{K}_t^{(j)} = \boldsymbol{F}\boldsymbol{\Sigma}_{t-1}^{(j)}\boldsymbol{F}^H + \boldsymbol{g}\boldsymbol{g}^H \tag{26.79}$$

$$\gamma_t^{(j)} = \boldsymbol{h}^H\boldsymbol{K}_t^{(j)}\boldsymbol{h} + \sigma^2 \tag{26.80}$$

$$\eta_t^{(j)} = \boldsymbol{h}^H\boldsymbol{F}\boldsymbol{\mu}_{t-1}^{(j)} \tag{26.81}$$

- Compute the trial sampling density: For $b \in \{+1, -1\}$, compute:

$$\begin{aligned}\rho_{t,b}^{(j)} &\triangleq P\left[x_t = b \,\middle|\, \boldsymbol{X}_{t-1}^{(j)}, \boldsymbol{Y}_t\right] \\ &\propto p\left(y_t \,\middle|\, x_t = b, \boldsymbol{X}_{t-1}^{(j)}, \boldsymbol{Y}_{t-1}\right) P[x_t = b]\end{aligned} \tag{26.82}$$

  with

$$p\left(y_t \,\middle|\, x_t = b, \boldsymbol{X}_{t-1}^{(j)}, \boldsymbol{Y}_{t-1}\right) \sim \mathcal{N}_c\left(b\eta_t^{(j)}, \gamma_t^{(j)}\right) \tag{26.83}$$

- Impute the symbol x_t: Draw $x_t^{(j)}$ from the set $\{+1, -1\}$ with probability

$$P\left[x_t^{(j)} = b\right] \propto \rho_{t,b}^{(j)}, \quad b \in \{+1, -1\} \tag{26.84}$$

  Append $x_t^{(j)}$ to $\boldsymbol{X}_{t-1}^{(j)}$ and obtain $\boldsymbol{X}_t^{(j)}$.
- Compute the importance weight:

$$\begin{aligned}w_t^{(j)} &= w_{t-1}^{(j)} \cdot p\left(y_t \,\middle|\, \boldsymbol{X}_{t-1}^{(j)}, \boldsymbol{Y}_{t-1}\right) \\ &\propto w_{t-1}^{(j)} \cdot \left[\rho_{t,+1}^{(j)} + \rho_{t,-1}^{(j)}\right]\end{aligned} \tag{26.85}$$

- Compute the one-step filtering update of the Kalman filter $\kappa_{t-1}^{(j)}$: Based on the imputed symbol $x_t^{(j)}$ and the observation y_t, complete the Kalman filter update to obtain $\kappa_t^{(j)} = [\boldsymbol{\mu}_t^{(j)}, \boldsymbol{\Sigma}_t^{(j)}]$, as follows:

$$\boldsymbol{\mu}_t^{(j)} = \boldsymbol{F}\boldsymbol{\mu}_{t-1}^{(j)} + \frac{1}{\gamma_t^{(j)}}\left(y_t - x_t^{(j)}\eta_t^{(j)}\right)\boldsymbol{K}_t^{(j)}\boldsymbol{h}x_t^{(j)} \tag{26.86}$$

$$\boldsymbol{\Sigma}_t^{(j)} = \boldsymbol{K}_t^{(j)} - \frac{1}{\gamma_t^{(j)}}\boldsymbol{K}_t^{(j)}\boldsymbol{h}\boldsymbol{h}^H\boldsymbol{K}_t^{(j)} \tag{26.87}$$

- Perform resampling according to Algorithm 14 in Section 26.3.2, when the effective sample size $\bar{m}_t$ in Equation 26.68 is below a threshold.

In the above algorithm at any time t, the only quantities that need to be stored are $\{\kappa_t^{(j)}, w_t^{(j)}\}_{j=1}^{\nu}$. At each time t, the dominant computation in this receiver involves the ν one-step Kalman filter updates. Since the ν samplers operate independently and in parallel, such a sequential Monte Carlo receiver is well suited for massively parallel implementation.

Since the fading process is highly correlated, the future received signals contain information about current data and channel state. Hence, a delayed estimate is usually more accurate than the concurrent estimate. From the recursive procedure described above, we note by induction that if the set $\{(\boldsymbol{X}_t^{(j)}, w_t^{(j)})\}_{j=1}^{\nu}$ is properly weighted with respect to $p(\boldsymbol{X}_t|\boldsymbol{Y}_t)$, then the set $\{(\boldsymbol{X}_{t+\delta}^{(j)}, w_{t+\delta}^{(j)})\}_{j=1}^{\nu}$ is properly weighted with respect to $p(\boldsymbol{X}_{t+\delta}|\boldsymbol{Y}_{t+\delta})$, for any $\delta > 0$. Hence, if we focus our attention on $\boldsymbol{X}_t$ at time $(t+\delta)$, we obtain the following delayed estimate of the symbol:

$$P[x_t = +1 \mid \boldsymbol{Y}_{t+\delta}] \cong \frac{1}{W_{t+\delta}}\sum_{j=1}^{\nu} 1\left(x_t^{(j)} = +1\right) w_{t+\delta}^{(j)} \tag{26.88}$$

Since the weights $\{w_{t+\delta}^{(j)}\}_{j=1}^{\nu}$ contain information about the signals $(y_{t+1}, \ldots, y_{t+\delta})$, the estimate in Equation 26.88 is usually more accurate. Note that such a delayed estimation method incurs no additional computational cost (i.e., cpu time), but it requires some extra memory for storing $\{(x_{t+1}^{(j)}, \ldots, x_{t+\delta}^{(j)})\}_{j=1}^{\nu}$.

The application of SMC to receiver design in fading channels was first proposed in [5], where it is also shown that the SMC can exploit the code constraint structures when the transmitted symbols are protected by channel codes, with increased computational complexity. In [30], several low-complexity sampling schemes are proposed to reduce the computational complexity when dealing with systems with strong memory. Moreover, in [13, 14], nonparametric SMC receivers based on wavelet transform of the fading process are developed to address the scenario when the fading statistics are unknown *a priori*.

26.3.3.2 SMC Receiver in MIMO ISI Channels

Consider a space-division multiple-access (SDMA) communications system with K users. The k-th user transmits data symbols $\{b_k[i]\}_i$ in the same frequency band at the same time, where $b_k[i] \in \mathcal{A}$ and $\mathcal{A}$ is a signal constellation set. The receiver employs an antenna array consisting of P antenna elements. The received signal at the p-th antenna element is the superposition of the convolutively distorted signals from all users plus the ambient noise, given by

$$\begin{aligned} y_p[i] &= \sum_{k=1}^{K}\sum_{\ell=0}^{L-1} g_{p,\ell,k} b_k[i-\ell] + n_p[i] \\ &= \boldsymbol{g}_p^H \boldsymbol{b}[i] + n_p[i], \quad p = 1, \ldots, P \end{aligned} \tag{26.89}$$

where $n_p[i] \overset{\text{i.i.d.}}{\sim} \mathcal{N}_c(0, \sigma^2)$, L is the length of the channel dispersion in terms of number of symbols, and

$$\boldsymbol{g}_p \triangleq [g_{p,0,1} \cdots g_{p,L-1,1} \cdots g_{p,0,K} \cdots g_{p,L-1,K}]^H$$

$$\boldsymbol{b}[i] \triangleq [b_1[i] \ldots b_1[i-L+1] \ldots b_K[i] \ldots b_K[i-L+1]]^T$$

Denote

$$\boldsymbol{y}[i] \triangleq [y_1[i] \ldots y_P[i]]^T$$
$$\boldsymbol{G} \triangleq \left[\boldsymbol{g}_1^H \ldots \boldsymbol{g}_P^H\right],$$
$$\boldsymbol{n}[i] \triangleq [n_1[i] \ldots n_P[i]]^T$$

Then Equation 26.89 can be written as the following MIMO signal model:

$$\boldsymbol{y}[i] = \boldsymbol{G}\boldsymbol{b}[i] + \boldsymbol{n}[i] \tag{26.90}$$

We now look at the problem of on-line estimation of the multiuser symbols

$$\underline{b}[i] \triangleq [b_1[i] \ldots b_K[i]]^T$$

and the channels $\boldsymbol{H}$ based on the received signals up to time i, $\{\boldsymbol{y}[j]\}_{j=1}^{i}$. Assume that the multiuser symbol streams are independent and identically distributed uniformly *a priori*, i.e., $p(b_k[i] = a_l \in \mathcal{A}) = 1/|\mathcal{A}|$. Denote

$$\boldsymbol{X}[i] \triangleq [\underline{b}[0] \ldots \underline{b}[i]]$$
$$\boldsymbol{Y}[i] \triangleq [\boldsymbol{y}[0] \ldots \boldsymbol{y}[i]]$$

Then the problem becomes one of making Bayesian inference with respect to the posterior density

$$p(\boldsymbol{X}[i], \boldsymbol{H}, \sigma^2 \mid \boldsymbol{Y}[i]) \propto (\pi\sigma^2)^{-(i+1)} \exp\left\{-\frac{1}{\sigma^2}\sum_{j=0}^{i} \|\boldsymbol{y}[j] - \boldsymbol{H}\boldsymbol{b}[j]\|^2\right\} \tag{26.91}$$

For example, an on-line multiuser symbol estimation can be obtained from the marginal posterior distribution $p(\underline{b}[i]|\boldsymbol{Y}[i])$, and an on-line channel state estimation can be obtained from the marginal posterior distribution $p(\boldsymbol{H}|\boldsymbol{Y}[i])$. Although the joint distribution in Equation 26.91 can be written out explicitly up to a normalizing constant, the computation of the corresponding marginal distributions involves very high dimensional integration and is infeasible in practice. Our approach to this problem is the sequential Monte Carlo technique.

For simplicity, assume that the noise variance σ^2 is known. The SMC principle suggests the following basic approach to the blind MIMO signal separation problem discussed above. At time i, draw m random samples

$$\{\underline{b}^{(j)}[i]\}_{j=1}^{m} \sim q(\underline{b}[i] \mid \boldsymbol{X}^{(j)}[i-1], \boldsymbol{Y}[i])$$

from some trial distribution $q(\cdot)$. Then update the important weights $\{w^{(j)}[i]\}_{j=1}^{m}$. The *a posteriori* symbol probability of each user can then be estimated as

$$\begin{aligned} P(b[i] = a_l \mid \boldsymbol{Y}[i]) &= E\{I(b[i] = a_l) \mid \boldsymbol{Y}[i]\} \\ &= \frac{1}{W[i]}\sum_{j=1}^{m} I(b^{(j)}[i] = a_l)\, w^{(j)}[i], \end{aligned} \tag{26.92}$$

$$\text{with} \quad W[i] = \sum_{j=1}^{m} w^{(j)}[i]$$

for $a_l \in \mathcal{A}$, where $I(\cdot)$ is an indicator function such that $I(b[i] = a_l) = 1$ if $b[i] = a_l$ and $I(b[i] = a_l) = 0$ otherwise.

Following the above discussions, the trial distribution is chosen to be

$$q(\underline{b}[i] \mid X^{(j)}[i-1], Y[i]] = p(\underline{b}[i] \mid X^{(j)}[i-1], Y[i]) \tag{26.93}$$

and the importance weight is updated according to

$$w^{(j)}[i] \propto w^{(j)}[i-1] \cdot p(y[i] \mid X^{(j)}[i-1], Y[i-1]) \tag{26.94}$$

We next specify the computation of the two predictive densities in Equations 26.93 and 26.94.

Assume the channel g_p has an *a priori* Gaussian distribution, i.e.,

$$g_p \sim \mathcal{N}_c(\bar{g}_p, \bar{\Sigma}_p) \tag{26.95}$$

Then the conditional distribution of g_p, conditioned on $X[i]$ and $Y[i]$, can be computed as

$$\begin{aligned} p(g_p \mid X[i], Y[i]) &\propto p(X[i], Y[i] \mid g_p)\, p(g_p) \\ &\sim \mathcal{N}_c(g_p[i],\ \Sigma_p[i]) \end{aligned} \tag{26.96}$$

where

$$g_p[i] \triangleq \Sigma_p[i] \left(\bar{\Sigma}_p^{-1} \bar{g}_p + \frac{1}{\sigma^2} \sum_{j=0}^{i} b[j] y_p[j]^* \right) \tag{26.97}$$

$$\Sigma_p[i] \triangleq \left(\bar{\Sigma}_p^{-1} + \frac{1}{\sigma^2} \sum_{j=0}^{n} b[j] b[j]^H \right)^{-1} \tag{26.98}$$

Hence, the predictive density in Equation 26.94 is given by

$$\begin{aligned} p(y[i] \mid X[i-1], Y[i-1]) &\propto \sum_{\underline{a}_l \in \mathcal{A}^K} p(y[i] \mid X[i-1], Y[i-1], \underline{b}[i] = \underline{a}_l) \\ &= \sum_{l} \prod_{p=1}^{P} p(y_p[i] \mid X[i-1], Y[i-1], \underline{b}[i] = \underline{a}_l) \end{aligned} \tag{26.99}$$

where

$$\begin{aligned} &p(y_p[i] \mid X[i-1], Y[i-1], \underline{b}[i] = \underline{a}_l) \\ &= \int p(y_p[i] \mid X[i-1], Y[i-1], \underline{b}[i] = \underline{a}_l, g_p)\, p(g_p \mid X[i-1], Y[i-1]) \mathrm{d}g_p \end{aligned} \tag{26.100}$$

Note that the above is an integral of a Gaussian probability density function (pdf) with respect to another Gaussian pdf. The resulting distribution is still Gaussian, i.e.,

$$p(y_p[i] \mid X[i-1], Y[i-1], \underline{b}[i] = \underline{a}_l) \sim \mathcal{N}_c\left(\mu_{p,l}[i], \sigma_{p,l}^2[i]\right) \tag{26.101}$$

with mean and variance given respectively by

$$\begin{aligned} \mu_{p,l}[i] &\triangleq E\{y_p[i] \mid X[i-1], Y[i-1], \underline{b}[i] = \underline{a}_l\} \\ &= g_p[i-1]^H b[i] \,|_{\underline{b}[i] = \underline{a}_l} \end{aligned} \tag{26.102}$$

$$\begin{aligned} \text{and} \quad \sigma_{p,l}^2[i] &\triangleq \mathrm{Var}\{y_p[i] \mid X[i-1], Y[i-1], \underline{b}[i] = \underline{a}_l\} \\ &= \sigma^2 + b[i]^H \Sigma_p[i-1] b[i] \,|_{\underline{b}[i] = \underline{a}_l} \end{aligned} \tag{26.103}$$

Therefore, Equation 26.99 becomes

$$p(\boldsymbol{y}[i] \mid \boldsymbol{X}[i-1], \boldsymbol{Y}[i-1]) \propto \sum_{l} \prod_{p=1}^{P} \rho_{p,l}[i], \tag{26.104}$$

$$\text{with} \quad \rho_{p,l}[i] \triangleq \frac{1}{\sigma_{p,l}^2[i]} \exp\left\{ -\frac{|y_p[i] - \mu_{p,l}[i]|^2}{\sigma_{p,l}^2[i]} \right\} \tag{26.105}$$

The filtering density in Equation 26.93 can be computed as follows:

$$\begin{aligned} p(\underline{b}[i] = \underline{a}_l \mid \boldsymbol{X}[i-1], \boldsymbol{Y}[i]) &\propto p(\boldsymbol{X}[i-1], \boldsymbol{Y}[i], \underline{b}[i] = \underline{a}_l) \\ &\propto p(\boldsymbol{y}[i] \mid \boldsymbol{X}[i], \boldsymbol{Y}[i-1], \underline{b}[i] = \underline{a}_l) \\ &\propto \prod_{p=1}^{P} \rho_{p,l}[i] \end{aligned} \tag{26.106}$$

Note that the *a posteriori* mean and covariance of the channel in Equations 26.97 and 26.98 can be updated recursively as follows. At time i, after a new sample of $\underline{b}[i]$ is drawn, we combine it with the past samples $\boldsymbol{b}[i-1]$ to form $\boldsymbol{b}[i]$. Let $\mu_p[i]$ and $\sigma_p^2[i]$ be the quantities computed by Equations 26.102 and 26.103 for the imputed $\underline{b}[i]$. It then follows from the matrix inversion lemma that Equations 26.97 and 26.98 become

$$\boldsymbol{g}_p[i] = \boldsymbol{g}_p[i-1] + \left(\frac{y_p[i] - \mu_p[i]}{\sigma_p^2[i]} \right)^* \boldsymbol{\xi}[i] \tag{26.107}$$

$$\text{and} \quad \boldsymbol{\Sigma}_p[i] = \boldsymbol{\Sigma}_p[i-1] - \frac{1}{\sigma_p^2[i]} \boldsymbol{\xi}_p[i] \boldsymbol{\xi}_p[i]^H, \tag{26.108}$$

$$\text{with} \quad \boldsymbol{\xi}_p[i] \triangleq \boldsymbol{\Sigma}_p[i-1] \boldsymbol{b}[i] \tag{26.109}$$

Finally, we summarize the SMC-based blind adaptive equalizer in MIMO channels as follows.

Algorithm 26.16 (SMC receiver in MIMO ISI channel)

- *Initialization:* The initial samples of the channel vectors are drawn from the following *a priori* distribution:

$$\boldsymbol{g}_p^{(j)}[0] \sim \mathcal{N}_c(\boldsymbol{0}, 1000 \boldsymbol{I}_{KL}), \quad j = 1, \ldots, m, \quad p = 1, \ldots, P$$

  All importance weights are initialized as $w_0^{(j)}[0] = 1$, $j = 1, \ldots, m$. Since the data symbols are assumed to be independent, initial symbols are not needed.
  The following steps are implemented at time i to update each weighted sample.
- For $j = 1, \ldots, m$, for each $\underline{a}_l \in \mathcal{A}^K$ and $p = 1, \ldots, P$, compute the following quantities:

$$\mu_{p,l}^{(j)}[i] = \boldsymbol{g}_p^{(j)}[i-1]^H \boldsymbol{b}_l^{(j)}[i] \tag{26.110}$$

$$\sigma_{p,l}^2[i]^{(j)} = \sigma^2 + \boldsymbol{b}_l^{(j)}[i]^H \boldsymbol{\Sigma}_p^{(j)}[i-1] \boldsymbol{b}_l^{(j)}[i] \tag{26.111}$$

$$\rho_{p,l}^{(j)}[i] = \left[\sigma_{p,l}^2[i]^{(j)}\right]^{-1} \exp\left\{ -\frac{\left| y_p[i] - \mu_{p,l}^{(j)}[i] \right|^2}{\sigma_{p,l}^2[i]^{(j)}} \right\} \tag{26.112}$$

  with $\boldsymbol{b}_l^{(j)}[i] \triangleq \boldsymbol{b}^{(j)}[i] \mid_{\underline{b}^{(j)}[i] = \underline{a}_l}$.

- Impute the multiuser symbol $\underline{b}^{(j)}[i]$: Draw $\underline{b}^{(j)}[i]$ from the set $\mathcal{A}^K$ with probability

$$p\left(\underline{b}^{(j)}[i] = \underline{a}_l\right) \propto \prod_{p=1}^{P} \rho_{p,l}^{(j)}[i], \quad a_l \in \mathcal{A}^K \tag{26.113}$$

- Compute the importance weight:

$$w^{(j)}[i] \propto w^{(j)}[i-1] \cdot \sum_{a_l \in \mathcal{A}^K} \prod_{p=1}^{P} \rho_{p,l}^{(j)}[i]$$

  Let $\mu_p^{(j)}[i]$ and $\sigma_p^2[i]^{(j)}$ be the quantities computed in Step 2 with $\underline{a}_l$ corresponding to the imputed symbol $\underline{b}^{(j)}[i]$.
- Update the *a posteriori* mean and covariance of the channels:

$$\boldsymbol{g}_p^{(j)}[i] = \boldsymbol{g}_p^{(j)}[i-1] + \left(\frac{y_p[i] - \mu_p^{(j)}[i]}{\sigma_p^2[i]^{(j)}}\right)^* \boldsymbol{\xi}_p^{(j)}[i]$$

$$\text{and} \quad \boldsymbol{\Sigma}_p^{(j)}[i] = \boldsymbol{\Sigma}_p^{(j)}[i-1] - \frac{1}{\sigma_p^2[i]^{(j)}} \boldsymbol{\xi}_p^{(j)}[i] \boldsymbol{\xi}_p^{(j)}[i]^H,$$

$$\text{with} \quad \boldsymbol{\xi}_p^{(j)}[i] \triangleq \boldsymbol{\Sigma}_p^{(j)}[i-1] \boldsymbol{b}^{(j)}[i]$$

- Perform resampling according to Algorithm 14 in Section 26.3.2, when the effective sample size $\bar{m}_t$ in Equation 26.68 is below a threshold.

The above SMC-based MIMO receiver was first developed in [12]. Note that the complexity of the sampling step in this approach is high since it samples directly from the set $\mathcal{A}^K$. More recently, a new method has been proposed in [6] that has a linear complexity by employing a novel trial sampling density. An SMC-based blind adaptive OFDM receiver that bears a similar structure as above was developed in [33]. Note that in most of these applications, the number of sample streams m is between 50 and 100.

26.4 Concluding Remarks

We have presented an overview on the theories and applications of the emerging field of Monte Carlo Bayesian signal processing. The optimal solutions to many statistical signal processing problems, especially those found in digital communications, are computationally prohibitive to implement by conventional signal processing methods. The Monte Carlo paradigm offers a novel and powerful approach to tackling these problems at a reasonable computational cost. We have outlined two families of Monte Carlo signal processing methodologies—Markov chain Monte Carlo for batch signal processing and sequential Monte Carlo for adaptive signal processing—as well as their applications in several signal processing problems found in digital communications. It is anticipated that the Bayesian Monte Carlo techniques will find wide applications in tackling many challenging signal processing problems in digital communications.

References

[1] C.K. Carter and R. Kohn. On Gibbs sampling for state space models. *Biometrika*, 81:541–553, 1994.

[2] R. Chen and T.-H. Li. Blind restoration of linearly degraded discrete signals by Gibbs sampler. *IEEE Trans. Sig. Proc.*, 43:2410–2413, 1995.

[3] R. Chen and J.S. Liu. Mixture Kalman filters. *J. Roy. Stat. Soc. B*, 62(3):493–509, 2000.

[4] R. Chen, J.S. Liu, and X. Wang. Convergence analyses and comparisons of Markov chain Monte Carlo algorithms in digital communications. *IEEE Trans. Sig. Proc.*, 50(2):255–270, 2002.

[5] R. Chen, X.Wang, and J.S. Liu. Adaptive joint detection and decoding in flat-fading channels via mixture Kalman filtering. *IEEE Trans. Inf. Theory*, 46(6):2079–2094, 2000.
[6] B. Dong, X. Wang, and A. Doucet. A new class of soft MIMO demodulation algorithms. *IEEE Trans. Sig. Proc.*, 51(11): 2003.
[7] A.E. Gelfand and A.F.W. Smith. Sampling-based approaches to calculating marginal densities. *J. Am. Stat. Assoc.*, 85:398–409, 1990.
[8] S. Geman and D. Geman. Stochastic relaxation, Gibbs distribution, and the Bayesian restoration of images. *IEEE Trans. Pattern Anal. Machine Intell.*, PAMI-6(11):721–741, 1984.
[9] C.N. Georghiades and J.C. Han. Sequence estimation in the presence of random parameters via the EM algorithm. *IEEE Trans. Commun.*, COM-45(3):300–308, 1997.
[10] C.J. Geyer. Markov chain Monte Carlo maximum likelihood. In *Computing Science and Statistics: Proceedings of the 23rd Symposium on the Interface*, E.M. Keramigas, Ed. Fairfax: Interface Foundation, 1991, pp. 156–163.
[11] N.J. Gordon, D.J. Salmon, and A.F.M. Smith. A novel approach to nonlinear/non-Gaussian Bayesian state estimation. *IEE Proc. Radar Sig. Proc.*, 140:107–113, 1993.
[12] D. Guo and X. Wang. Blind detection in MIMO systems via sequential Monte Carlo. *IEEE J. Select. Areas Commun.*, 21(3):453–464, 2003.
[13] D. Guo, X. Wang, and R. Chen. Nonparametric adaptive detection based on sequential Monte Carlo and Bayesian model averaging. *Ann. Inst. Stat. Math.*
[14] D. Guo, X. Wang, and R. Chen. Wavelet-based sequential Monte Carlo blind receivers in fading channels with unknown channel statistics. *IEEE Trans. Sig. Proc.*
[15] W.K. Hastings. Monte Carlo sampling methods using Markov chains and their applications. *Biometrika*, 57:97–109, 1970.
[16] A. Kong, J.S. Liu, and W.H. Wong. Sequential imputations and Bayesian missing data problems. *J. Am. Stat. Assoc.*, 89:278–288, 1994.
[17] F. Liang and W.H. Wong. Evolutionary Monte Carlo: applications to c_p model sampling and change point problem. *Statistica Sinica*, 10:317–342, 2000.
[18] H. Liu and G. Xu. A subspace method for signal waveform estimation in synchronous CDMA systems. *IEEE Trans. Commun.*, COM-44(10):1346–1354, 1996.
[19] J.S. Liu. The collapsed Gibbs sampler with applications to a gene regulation problem. *J. Am. Stat. Assoc.*, 89:958–966, 1994.
[20] J.S. Liu. *Monte Carlo Methods for Scientific Computing*. New York: Springer-Verlag, 2001.
[21] J.S. Liu and R. Chen. Blind deconvolution via sequential imputations. *J. Am. Stat. Assoc.*, 90:567–576, 1995.
[22] J.S. Liu and R. Chen. Sequential Monte Carlo methods for dynamic systems. *J. Am. Stat. Assoc.*, 93:1032–1044, 1998.
[23] J.S. Liu, F. Ling, and W.H. Wong. The use of multiple-try method and local optimization in Metropolis sampling. *J. Am. Stat. Assoc.*, 95:121–134, 2000.
[24] B. Lu and X. Wang. Bayesian blind turbo receiver for coded OFDM systems with frequency offset and frequency-selective fading. *IEEE J. Select. Areas Commun.*, 19(12):2516–2527, 2001.
[25] N. Metropolis, A.W. Rosenbluth, A.H. Teller, and E. Teller. Equations of state calculations by fast computing machines. *J. Chem. Phys.*, 21:1087–1091, 1953.
[26] V.D. Phan and X. Wang. Bayesian turbo multiuser detection for nonlinearly modulated CDMA. *Sig. Proc.*, 82(1):42–68, 2002.
[27] M.A. Tanner and W.H. Wong. The calculation of posterior distribution by data augmentation [with discussion]. *J. Am. Stat. Assoc.*, 82:528–550, 1987.
[28] X. Wang and R. Chen. Adaptive Bayesian multiuser detection for synchronous CDMA in Gaussian and impulsive noise. *IEEE Trans. Sig. Proc.*, 48(7):2013–2028, 2000.
[29] X. Wang and R. Chen. Blind turbo equalization in Gaussian and impulsive noise. *IEEE Trans. Veh. Tech.*, 50:1092–1105, 2001.

[30] X. Wang, R. Chen, and D. Guo. Delayed-pilot sampling for mixture Kalman filter with application in fading channels. *IEEE Trans. Sig. Proc.*, 50(2):241–254, 2002.

[31] Z. Yang, B. Lu, and X. Wang. Bayesian Monte Carlo multiuser receiver for space-time coded multi-carrier CDMA systems. *IEEE J. Select. Areas Commun.*, 19(8):1625–1637, 2001.

[32] Z. Yang and X. Wang. Turbo equalization for GMSK signaling over multipath channels based on the Gibbs sampler. *IEEE J. Select. Areas Commun.*, 19(9):1753–1763, 2001.

[33] Z. Yang and X. Wang. Blind detection of OFDM signals in multipath fading channels via sequential Monte Carlo. *IEEE Trans. Sig. Proc.*, 50(2):271–280, 2002.

[34] Z. Yang and X. Wang. Blind turbo multiuser detection for long-code multipath CDMA. *IEEE Trans. Commun.*, 50(1):112–125, 2002.

27

Principles of Chaos Communications

27.1 What Is Chaos? 27-1
Nonlinear Dynamic Systems • Statistical Analysis of Chaotic Signals • Realization of Chaos Generators • Properties of Chaotic Signals
27.2 Communication: Requirements and Resources 27-9
27.3 Chaos in Communications 27-10
The Broadband Aspect • The Complexity Aspect • The Orthogonality Aspect
27.4 Communication Using Broadband Chaotic Carriers ...27-10
Chaos-Based Transmitters • Receiver Design in Chaos Communications
27.5 Chaos for Spreading Code Generation 27-19
27.6 Chaotic vs. Classical Communications 27-20
The Analysis of Chaos Communication Schemes in AWGN • Chaos Communication Methods in Comparison to Classical Solutions

Andreas Abel
ITI GmbH Dresden

Wolfgang Schwarz
Dresden University of Technology

Abstract

In recent decades a newly discovered phenomenon — chaos — attracted much attention among scientists from nonlinear dynamics. In the search for practical applications, communications systems were identified as one of the fields where chaos could turn out to be beneficial. This chapter discusses the application of chaotic signals and systems to communication problems. First, we shortly review the phenomenon of chaos, the properties of chaotic signals, and the means of their generation. Starting from the general communications system structure, the potential application fields of chaos in communications are identified. Then we introduce proposed methods for using chaos in modulation and coding and classify them based on the established point of views. Finally, the performance and usefulness of chaos are analyzed, discussed, and compared to those of the conventional approaches.

27.1 What Is Chaos?

27.1.1 Nonlinear Dynamic Systems

The behavior of a dynamic system is described by state equations. These are difference equations in the discrete-time case:

$$x_{k+1} = g(x_k); x_0 \tag{27.1}$$

0-8493-1657-X/$0.00+$1.50

TABLE 27.1 Asymptotic Behavior of Dynamic Systems

Static	Periodic	Quasi periodic	Chaotic
The state x tends to a constant vector value	The state x tends to a periodic time function	The state–time function is a superposition non harmonic sinusoidal signal	The state evolves in a steady nonperiodic movement
Power Density Spectra of the State Coordinates			
Discrete	Discrete	Discrete	Continuous
One single spectral line at $\omega = 0$	Equidistant spectral lines	Nonequidistant spectral lines	Spectrum spread over a certain frequency range

and ordinary differential equations in the case of continuous time:

$$\frac{dx}{dt} = g(x); x_0 \tag{27.2}$$

The solutions of these equations describe the state evolution of the corresponding system. The sequence $\{x_k, k = 0, 1, 2, \ldots\}$ and respectively the time function $\{x(t), \mathrm{k} \geq 0\}$ are called trajectories of the system. The systems can be represented by block schemes as shown in Figure 27.1. For a system classification the asymptotic behavior, i.e., the evolution of the trajectory for $k \to \infty$ or for $t \to \infty$, is of interest. Here four cases can be distinguished (Table 27.1).

Whereas linear and nonlinear systems can exhibit static, periodic and quasi-periodic behavior, chaotic behavior is only possible if the system, i.e., the function g in Equations 27.1 and 27.2, is nonlinear. Moreover, for chaotic behavior of a continuous-time system the order, i.e., the number of state coordinates, has to be at least three, whereas discrete-time systems of order one can behave chaotically. In this chapter we mainly restrict to this simple case.

In order to generate a chaotic sequence, the map g has to have some special properties:

1. Mapped intervals: The map g has to map a certain state interval to itself. Without loss of generality it can be assumed for g that $[0, 1] \to [0, 1]$. This property guarantees the stability of the state development in the sense that the state values x_k for all k remain in the interval (0,1).
2. Fixed points: A fixed point is mapped by g to itself, i.e., $g(x_f) = x_f$. A fixed point x_f is stable if any trajectory starting in the vicinity of x_f converges to x_f. The condition for this is $|\frac{dg(x_f)}{dx_f}| < 1$. In order to guarantee a steady movement of the state, all fixed points of g have to be unstable. This also refers to any iterate of the map; hence, if one fixed point of the n-th iterate $g^{(n)}$ is stable, the signal x_k for $\mathrm{k} \to \infty$ will be periodic with period n. Thus a necessary condition for the system to exhibit a steady nonperiodic movement is

$$\forall n \forall x_f : g^{(n)}(x_f) = x_f \to \left| \frac{dg^{(n)}(x_f)}{dx_f} \right| > 1$$

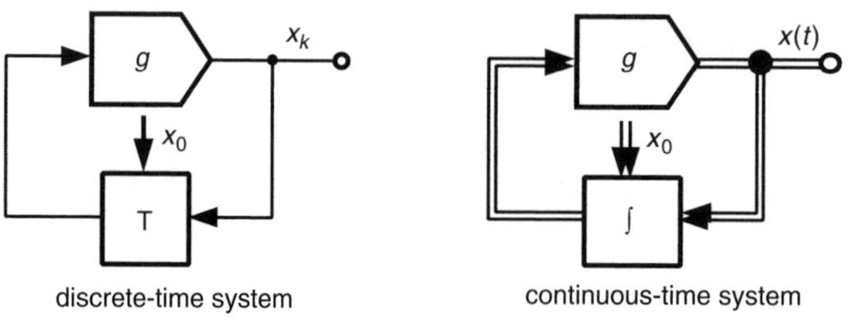

FIGURE 27.1 Block schemes of nonlinear dynamic systems.

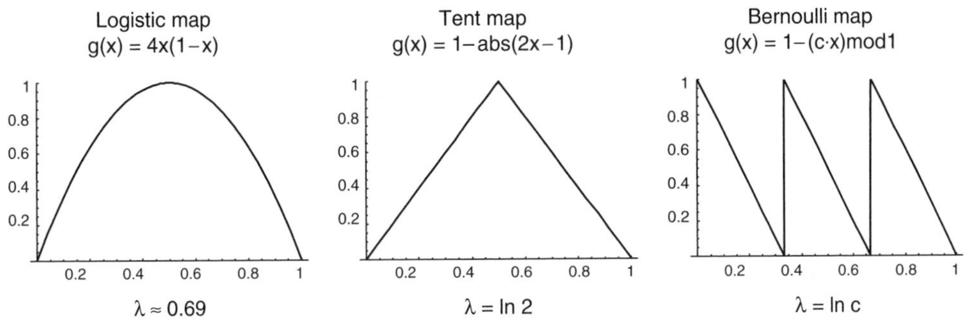

FIGURE 27.2 Examples for chaotic maps.

3. The map g is not invertible.
4. Lyapunov exponent: If at a given state x_0 the derivative g' of g has the property $|g'(x_0)| > 1$ or $\ln|g'(x)| > 0$, two trajectories starting close to x_0 will diverge with increasing k; the map is expanding at x_0. Otherwise, if $\ln|g'(x_0)| < 0$, two neighboring trajectories starting close to x_0 will be coming closer with increasing k and the map is compressing in x_0. A sufficient condition for chaotic behavior is that the map g is expanding in the mean, i.e., that the so-called Lyapunov exponent

$$\lambda = E(|\ln g'(x_k)|) = \int_0^1 \ln|g'(x)| f(x)dx \quad (27.3)$$

where $f(x)$ denotes that the probability density function of the x_k is positive. If the state sequence $\{x_k, k = 0, 1, \ldots\}$ is ergodic, the Lyapunov exponent can also be calculated as time average:

$$\hat{\lambda} = \lim_{K\to\infty} \frac{1}{K} \sum_{k=0}^{K} \ln|g'(x_k)| \quad (27.4)$$

Some examples of chaotic maps are depicted in Figure 27.2.

An important class is the piecewise linear fully stretching maps defined by

$$g(x) = \{g_i(x) = a_i x + b_i,\ x \in J_i, i = 1, 2, \ldots, m\} \quad (27.5)$$

such that $\forall i : g_i : J_i \to X = (0, 1)$. Figure 27.3 shows an example of a map of this kind.

27.1.2 Statistical Analysis of Chaotic Signals

For the analysis and design of communications systems statistical signal properties such as autocorrelation function, power density spectrum or probability distribution functions or densities are used. Chaotic signals are deterministic in principle, but, they also behave like realizations of random signals. In order to obtain proper models for the statistical analysis, chaotic signals have to be randomized. This is done by assuming random initial conditions x_0 in Equation 27.1. Thus x_0 is a random variable and $\{x_k, k = 0, 1, \ldots\}$ is a discrete-time random process.

27.1.2.1 Probability Density Functions

If $f(x;\mathrm{k})$ is the probability density function (pdf) of x_k, then the pdf of the next iterate x_{k+1} is

$$f(x; k+1) = \mathbf{P}(f(x,k)) = \int_0^1 \delta(x - g(u))\, f(u;k)\, du \quad (27.6)$$

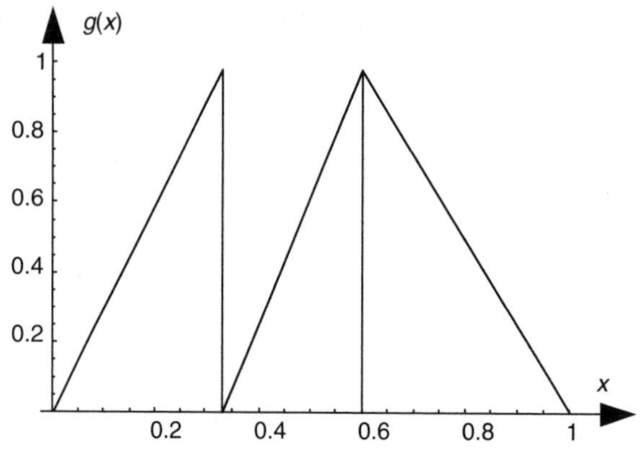

FIGURE 27.3 A piecewise linear fully stretching map.

This relation is the integral representation of the so-called Frobenius–Perron operator. It maps subsequent probability density functions of the process x_k. Thus from known initial distribution $f(x, 0)$, the next distributions $f(x, 1)$, $f(x, 2)$, ... can be calculated and the evolution of the densities can be observed. This process converges if the operator in Equation 27.6 has a stable fixed point, which is then defined by

$$f(x) = \int_0^1 \delta(x - g(u))\, f(u)\, du \tag{27.7}$$

$f(x)$ is called the invariant density of the map g. Of special interest are maps that generate a uniform distribution. The condition for this follows from Equation 27.7:

$$\forall x \in [0, 1] \rightarrow 1 = \int_0^1 \delta(x - g(u))\, du = \sum_{g(u_i)=x} \frac{1}{|g'(u_i)|} \tag{27.8}$$

As an example, we regard the c-adic Renyi map,

$$g(x) = (c \cdot x) \bmod 1, \tag{27.9}$$

the graph of which is depicted in Figure 27.4.

Inserting Equation 27.9 into Equation 27.6 we get the Frobenius–Perron operator for positive values of c:

$$\mathbf{P}(f(x)) = \frac{1}{c} \sum_{n=0}^{c-1} f\left(\frac{x+n}{c}\right) \tag{27.10}$$

It can easily be verified that the uniform distribution is a solution of the equation $\mathbf{P}(f(x)) = f(x)$, and hence an invariant density.

Figure 27.5 shows the action of this operator for the first four time steps. It can be seen how the pdf converges to a uniform distribution. It can be shown that all piecewise linear fully stretching maps fulfill Equation 27.8 and hence generate uniformly distributed signals.

27.1.2.2 Autocovariance and Autocorrelation Function

The autocovariance function of a random sequence $\{x_k, k = 0, 1, \ldots\}$ is defined as

$$c(n) = E(x_k x_{k+n}) \tag{27.11}$$

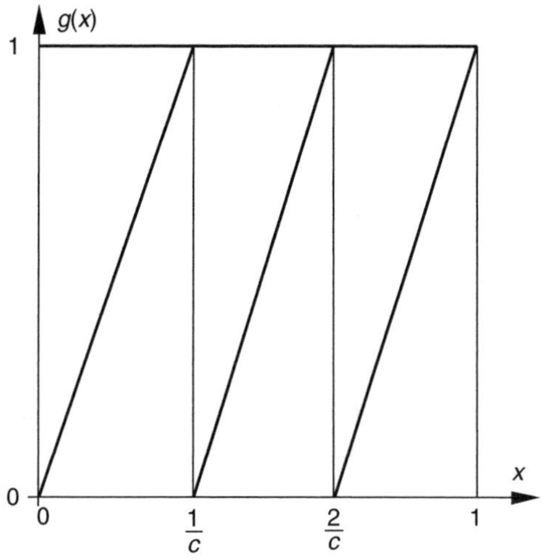

FIGURE 27.4 The 3-adic Renyi map.

For chaotic sequences it can be calculated as ensemble average

$$c(n) = \int_0^1 x g^{(n)}(x) f(x) dx \tag{27.12}$$

where $g^{(n)}$ is the n-th iterate and $f(x)$ denotes the invariant density of the map g. If the map is ergodic, this value coincides with the time average

$$\hat{c}(n) = \lim_{K\to\infty} \frac{1}{K} \sum_{k=0}^{K} x_k x_{k+n} = \lim_{K\to\infty} \frac{1}{K} \sum_{k=0}^{K} x_k g^{(n)}(x) \tag{27.13}$$

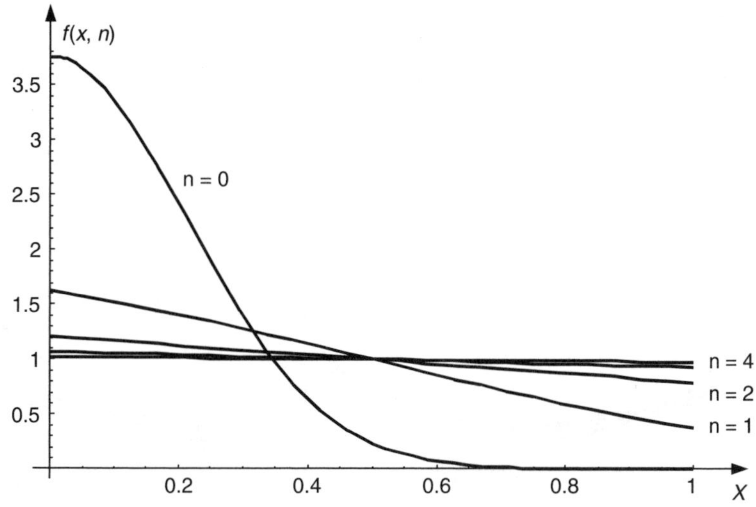

FIGURE 27.5 Subsequent probability functions of the 3-adic Renyi map.

The autocorrelation function (acf) is then

$$r(n) = c(n) - m^2 \tag{27.14}$$

where $m = E(x_k)$ is the average value of the sequence.

The discrete Fourier transform of the autocorrelation function is the power density spectrum (pds)

$$S(\Omega) = \sum_{n=-\infty}^{\infty} r(n)e^{-jn\Omega} \tag{27.15}$$

There are only a few maps for which the acf and hence the pdf can be calculated analytically. The most important class of them is the piecewise linear fully stretching maps (Equation 27.5), which generate sequences with an acf

$$r(n) = r(0)\beta^{-|n|} \tag{27.16}$$

where $r(0) = 1/12$ and β is defined by [2]

$$\beta = \sum_{i=1}^{m} \frac{1}{a_i\,|a_i|} \tag{27.17}$$

So in order to generate an uncorrelated or white noise sequence, the sum in Equation 27.17 has to be zero. This is the case, e.g., for the symmetric tent map (Figure 27.2). Furthermore, if we have a map generating an uncorrelated sequence, any two sequences starting with different initial conditions will be mutually uncorrelated; i.e., their cross-correlation function will be identical to zero.

For our c-adic Renyi map (Equation 27.9) we get the autocorrelation function (Equation 27.16) with $\beta = 1/c$ from Equation 27.17 and the power density spectrum

$$S(\Omega) = \frac{1}{12(1+\beta^2 - 2\beta\cos\Omega)} \tag{27.18}$$

which is depicted in Figure 27.6 for different values of c. For positive values of c and hence for β, we have a monotonously decreasing acf and a low-pass-type spectrum, whereas for negative values of c (this corresponds to the Bernoulli map in Figure 27.2) the acf alternates and we have a high-pass-type spectrum. In order to obtain a flat spectrum we have to choose a high value of c, which makes the realization complicated. For the generation of white noise, the tent map is more suitable than the Renyi map.

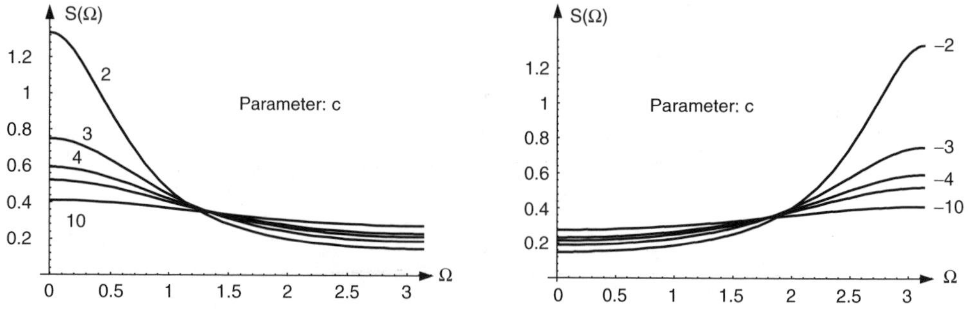

FIGURE 27.6 Power density spectra for the c-adic Renyi map.

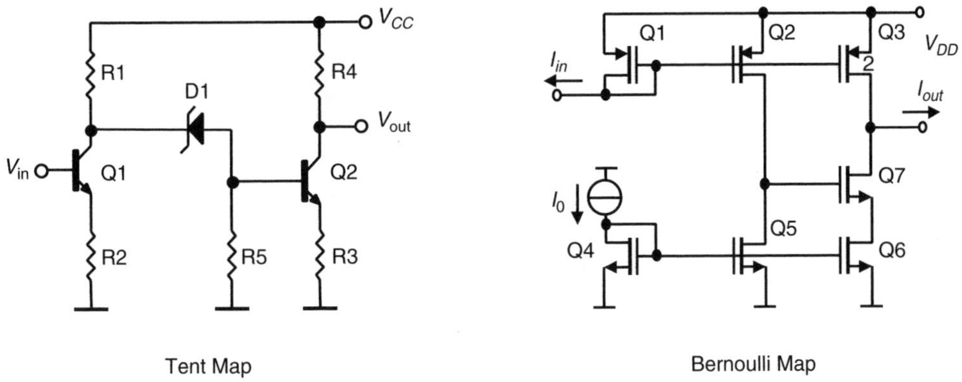

FIGURE 27.7 Circuits realizing chaotic maps.

27.1.3 Realization of Chaos Generators

There are several possibilities to construct electronic circuits that realize chaotic systems and thus can serve as generators for chaotic signals.

27.1.3.1 Discrete-Time Systems

Discrete-time systems are realized on the base of the structure in Figure 27.1. As time-delay block, a sample-and-hold device is used, and for the nonlinear block, a suitable circuit has to be designed. Figure 27.7 shows two example circuits. In the left-hand circuit in the bipolar voltage technique the voltage transfer function is that of the tent map, whereas the right-hand circuit realizes the Bernoulli map in the MOS current technique.

For discrete-time systems methods have been developed for analyzing the statistical properties of the generated signals as well as approaches for a systematic design [7].

27.1.3.2 Bit Stream Generators

Bit stream generators are continuous–discrete systems generating both continuous and discrete value continuous-time signals. In the example circuit in Figure 27.8, which is driven by a periodic clock signal,

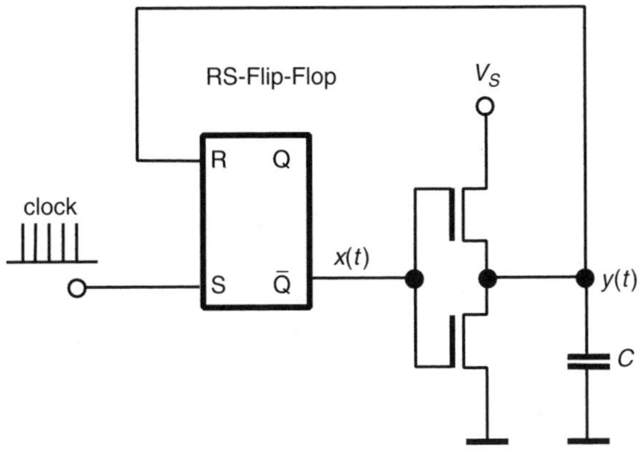

FIGURE 27.8 A bit stream generator.

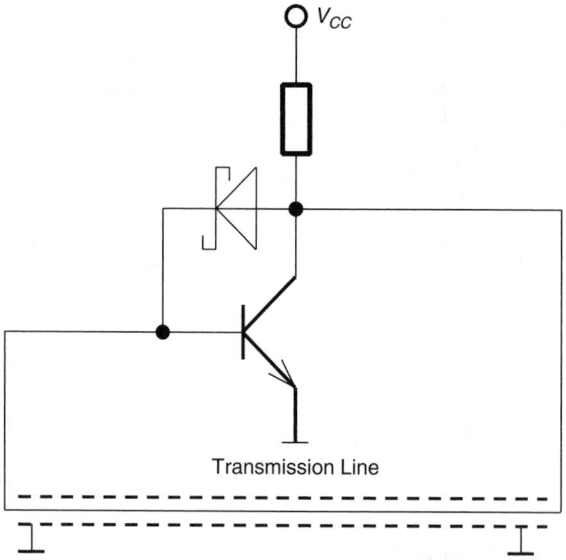

FIGURE 27.9 Chaos generator with transmission line.

$x(t)$ is a chaotic telegraph wave and $y(t)$ is a triangular signal. The time intervals between subsequent switching moments of the RS flip-flop form a chaotic sequence.

27.1.3.3 Continuous-Time Systems

There are many analog systems designed or known to exhibit chaotic behavior. Examples are Chua's circuit, the Colpitts, and other classical oscillator structures. Circuits realizing the well-known Lorenz and Roessler equations have also been designed. The problem is that there exist no analytical methods for a statistical analysis of the signals generated by analog chaotic systems, and so a systematic design is rather complicated. A comprehensive treatment of analog chaotic circuits is given in [3].

27.1.3.4 Distributed Systems

The idea to realize the delay element in Figure 27.1 by means of a transmission line leads to a system with distributed parameters, and thus to a system with infinite dimensional state space. Although very simple in structure, these systems show a wide variety of chaotic behavior and are suitable to generate chaotic signals up to very high frequencies. As an example, Figure 27.9 shows the simplest chaotic circuit known so far [5].

27.1.3.5 Digital Realizations

Digital systems cannot exhibit chaotic behavior in principle. Their state space is always finite, and so an ever continuing state evolution eventually reaches its starting point, and thus the process is periodic. However, it is possible to construct digital systems with very large periods so that the behavior of the system comes close to a chaotic one. An important special case is shift register structures, which can be interpreted as digital realizations of the discrete-time structure in Figure 27.1.

27.1.4 Properties of Chaotic Signals

In concluding this section we summarize some communication-relevant properties of chaotic signals in Table 27.2.

TABLE 27.2 Properties of Chaotic Signals

Property	Description	Application
Deterministic	Generated by deterministic systems (algorithms); can be reproduced in principle	Information coding, measurements
Noise-like	Seemingly random; irregular to an observer not knowing the generation algorithm	Pseudorandom signal generators, information encryption
Broadband	Continuous power density spectrum over an extended frequency range	Broadband stimulation, measurement, and communication
Decorrelated	Signals from: a) the same generator with different initial conditions b) structurally identical systems with slight parameter differences that are not correlated	CDMA, communication and broadband measurements

27.2 Communication: Requirements and Resources

Talking about communication we have the target of this process in mind — the transmission of a message (i.e., information) from one place to another, or from a sender to a recipient. This transmission has to be achieved via some physical medium (or channel) that is a limited resource and imposes restrictions on frequency ranges, bandwidth, and signal power (physical, technical, or administrative limitations) and disturbs the transmission by noise or interfering signals from other sources. Also, other requirements such as security may impose further constraints on the way a message is transmitted.

So the link between the sender and recipient is not simply a one-to-one map from the transmitted to the received message. Since the available resources are always limited, a communications system has to use them as effectively as possible. So the aim of a communications system is not solely the transmission of a message from one point to another, but the achievement of this transmission in an *efficient, robust,* and, if required, *secure* manner using the available channel resources. These three aspects are reflected in the main building blocks of a communication scheme (Figure 27.10).

The *source encoding/decoding* (usually found in digital communications) takes care of removing redundancy from the transmitted message and thus reduces the amount of information to be transmitted.

The *encryption/decryption* prevents unwanted listeners from deciphering the message. Since added redundancy helps deciphering, these blocks do not change the amount of information.

The *channel encoding and modulation/channel decoding and demodulation* serve the adaptation of the message to the specific channel conditions. The message is mapped to a signal in the desired frequency

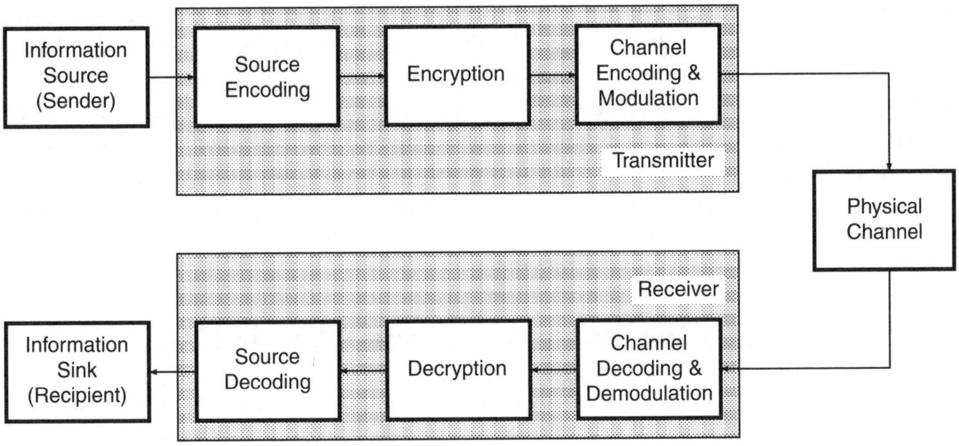

FIGURE 27.10 Principal structure of a communication scheme.

band (modulation) and redundancy is added in a controlled manner (channel encoding), which allows the receiver to detect and eliminate transmission errors.

27.3 Chaos in Communications

The idea to use chaos in communications systems sparked in the early 1990s, when engineers studying the phenomenon of chaos were looking for potential application fields of chaotic signals and systems. From the properties of chaotic signals, which were explained in Section 27.1.4, emerge potential application fields of chaos in communications systems. These applications result from three core aspects of chaos, broadband, complexity, and orthogonality.

27.3.1 The Broadband Aspect

Chaotic signals are the result of complex nonperiodic behavior. Consequently, chaotic signals do not have spectra composed of discrete lines at particular frequencies, but they generate continuous spectra covering whole frequency ranges. This property makes chaotic signals appear similar to noise signals. Often these signals are broadband too. In the progress of research, methods were found that allow the control of the spectral properties of signals from certain classes of chaos generators (see Section 27.1 and [1]), and so there exist means to design noise-like signals with prescribed spectra.

In communications systems broadband signals are used to fight narrowband channel imperfections such as frequency-selective fading or narrowband interferences.

Consequently, chaotic signals became candidates for spread-spectrum communications and are intensively studied in this context (see [6]). Section 27.4 is dedicated to this aspect.

27.3.2 The Complexity Aspect

Chaotic signals, albeit stemming from fairly simple generators, are complexly structured. Irregular behavior is combined with the impossibility to predict a chaotic trajectory over longer periods of time, even if the generator itself is exactly known. The exponential growth of tiny errors (such as from measurements) eventually leads to total divergence and decorrelation between the true trajectory and a prediction.

Complexly structured and hard-to-predict signals are classically used in cryptography. So this field was identified as a further potential application area for chaotic signals and systems (see [7], [8]).

27.3.3 The Orthogonality Aspect

Irregular behavior and sensitivity to initial conditions in chaotic systems imply the exponential growth of microscopic errors to a macroscopic scale. So even two virtually identical chaos generators with virtually the same initial conditions after a while produce totally different signals, which are uncorrelated. This generic ability to produce uncorrelated signals can be used in applications where orthogonal signal sets are of importance. This includes the generation of sequence families for multiuser spread-spectrum applications. Chaos can be used for the assignment of time slots (time hopping, pulse-position modulation), frequency slots (frequency hopping), or codes in code division multiple access (CDMA). Since the generated signals retain much of their uncorrelatedness when truncated to finite length, chaos also provides a means to generate sequences for conventional communication schemes (see [9]). Section 27.5 provides insight into the chaos-based generation of CDMA spreading codes.

27.4 Communication Using Broadband Chaotic Carriers

This section classifies the most common chaos communication methods. Since many approaches share the methods for modulation and reception, various combinations of transmitter and receiver solutions are possible. So we will classify transmitters and receivers separately, which is a commonly used approach in classical communications too [10].

27.4.1 Chaos-Based Transmitters

Chaos is applied in communications solutions especially in the context of channel encoding and modulation. Observing the ongoing research, one observes that the vast majority of the schemes proposed so far concerns modulation in the first place, whereas only a few consider the channel encoding. This is due to the still very early stage in research and exploitation of these methods. Nevertheless, the application of chaos in modulation usually implies the addition of redundancy, as it is found in channel encoding. So from the classification point of view, a strict separation between channel encoding and modulation methods is not useful. Here we will consider the combination of the two as *one* operation, mapping a message signal to a transmission signal, which suits the given physical channel.

For classification, we can follow the traditional approach. In channel encoding we find two main classes of methods [11]:

- Block encoders (static)
- Convolution encoders (dynamic)

The first type is a memoryless device with respect to subsequent message blocks; i.e., the result of the currently performed encoding is independent of the previous ones — it is static with respect to the message. The convolution encoders memorize the previously performed encoding of message symbols and thus encode the message dynamically.

Modulation methods can be classified in essentially the same way [11]:

- Memoryless modulation methods
- Modulation methods with memory

Again, the memory is understood with respect to the message. We can call these methods static or dynamic also. So for a joint consideration of the two methods we find:

- Static encoding/modulation schemes
- Dynamic encoding/modulation schemes

This classification is especially suited for the performance analysis of chaos communication schemes, whose results are discussed in Section 27.6.

27.4.1.1 Static Encoding/Modulation Methods

Static encoding represents a memoryless mapping of the message to be transmitted to a carrier suitable for the given channel. The respective carrier signal is generated independently of the actual message to be transmitted. One can imagine three types of signals to be used:

1. Signals from deterministic signal generators with simple dynamics (such as periodic signals). This is the common approach in the classical communication methods, such as AM, FM, or PM.
2. Signals from deterministic generators with complex dynamics. Such generators are pseudonoise (PN) generators or chaos generators. This type is the focus in this paper.
3. Signals from random processes. These are very similar to the chaos generators with respect to the statistical properties of the signals. Contrary to chaos, there are no means of signal synchronization and coherent reception, but some of the chaos-based schemes will work equivalently when using random noise signals.

27.4.1.1.1 Chaotic Masking

One of the earliest proposals to apply chaos in the transmission of a message is the so-called *chaotic masking* [12]. The main intention here was the achievement of privacy in the message transmission by hiding the message signal in a larger-amplitude chaotic signal:

$$y(t) = x(t) + m(t) \tag{27.19}$$

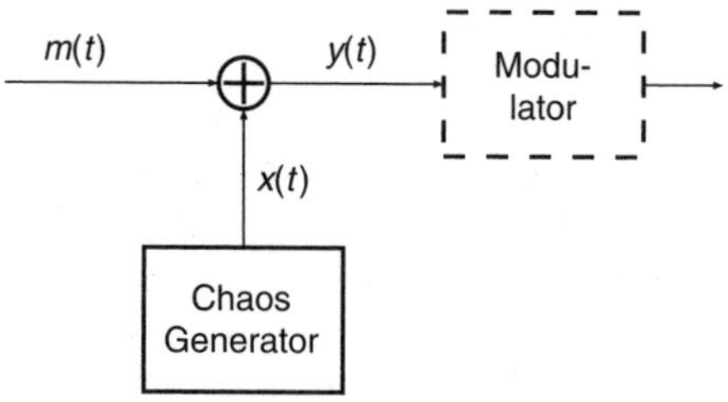

FIGURE 27.11 Chaotic masking: signal block structure.

where y is the transmitted signal, x the chaotic signal, and m the message signal. This method is applicable to any type of message, digital or analog. But it does not adapt the message to a given channel (m and y will be located in the same frequency band — the channel would have to transmit the message frequencies) and thus cannot be considered a modulation method. For this, a further operation such as a classical modulation (e.g., AM, FM) will be required. Figure 27.11 shows the corresponding transmitter structure.

27.4.1.1.2 Chaos Shift Keying

Chaos shift keying (CSK) [13] is an encoding/modulation scheme for digital messages. Depending on the current message symbol (N-ary message) that shall be sent, the transmitter selects one of N chaos generators and transmits the corresponding signal. If the generators produce chaotic signals, which are sufficiently disjoint in a particular property (usually some signal statistics), this modulation operation can be reverted in the receiver. This method is comparable to shift keying methods found in classical communications, such as phase or frequency shift keying. Formally, CSK can be described by

$$y(t) = \begin{cases} x_1(t) & \text{if} \quad m(t) = m_1 \\ x_2(t) & \text{if} \quad m(t) = m_2 \\ \cdots & \\ x_N(t) & \text{if} \quad m(t) = m_N \end{cases} \tag{27.20}$$

where $x(t)$, $y(t)$, and $m(t)$ are the carrier, transmitted signal, and message respectively. The set $\{m_1, m_2, \ldots, m_N\}$ is the domain of the discrete-valued digital message. The corresponding transmitter structure is shown in Figure 27.12.

27.4.1.1.3 Chaotic On-Off Keying

Chaotic on–off keying (cook) is a special case of chaos shift keying for binary message signals. Instead of two chaotic signals transmitted intermittently, in chaotic on–off keying only one of the two carrier signals

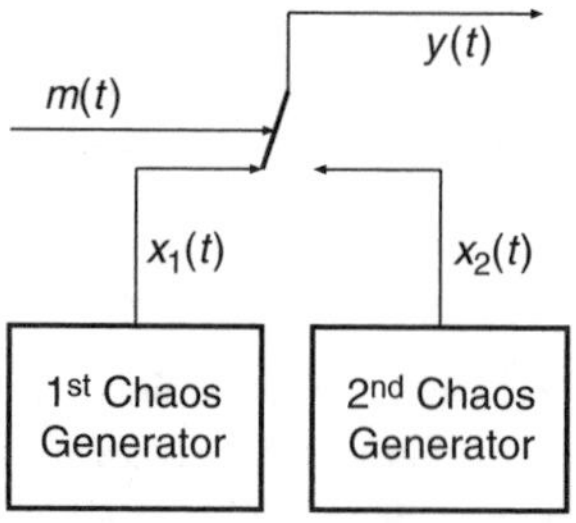

FIGURE 27.12 Chaos shift keying: signal block structure for binary message signals.

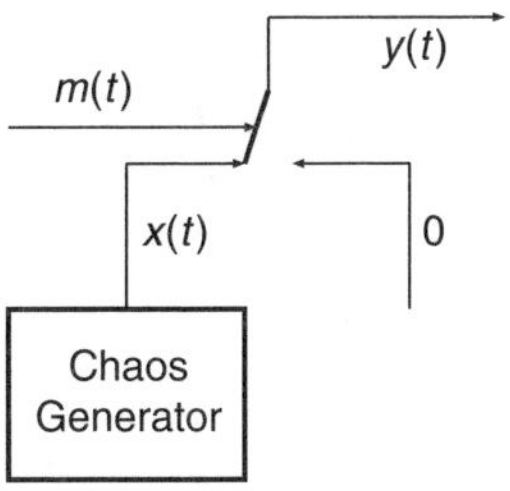

FIGURE 27.13 Chaotic on–off keying: signal block structure.

is chaotic, whereas the other is identical to zero:

$$y(t) = \begin{cases} x(t) & \text{if} \quad m(t) = m_1 \\ 0 & \text{if} \quad m(t) = m_2 \end{cases} \tag{27.21}$$

Figure 27.13 shows the corresponding transmitter structure.

27.4.1.1.4 Transmitted Reference Methods and Differential Chaos Shift Keying

Transmitted reference (TR) methods were originally developed for correlation reception, but they apply to any receiver structure that uses a reference of the transmitted carrier for demodulation. In TR the nonmodulated carrier is transmitted to the receiver via a separate channel realized, e.g., via time or frequency multiplexing. TR methods were developed already during the first attempts to use irregular signals such as natural noise as a message carrier [15]. They allow the application of carrier signals, for which no means for carrier synchronization at the receiver side exist. Their counterpart is the stored-reference (SR) approach, where a receiver is able to generate a replica of the carrier for demodulation purposes. This synchronization problem observed with natural noise eminently exists for chaos too, so the TR approaches reappeared in chaotic communication schemes. The most popular and also most well-studied incarnation of TR in chaos communications is differential chaos shift keying (DCSK) [16].

Differential chaos shift keying uses two channels multiplexed in time. For the transmission of one bit, first the unaltered reference is transmitted in a first time slot, followed by the modulated reference carrying the message symbol in the second time slot. Usually DCSK schemes are binary, but multilevel approaches exist too. Formally, the modulation can be expressed by

$$y(t) = \begin{cases} x(t) & \text{if} \quad n \cdot T_b \le t < \dfrac{2n+1}{2} \cdot T_b \\ x\left(t - \dfrac{T_b}{2}\right) & \text{if} \quad \dfrac{2n+1}{2} \cdot T_b \le t < (n+1) \cdot T_b \text{ and } m(t) = m_1 \\ -x\left(t - \dfrac{T_b}{2}\right) & \text{if} \quad \dfrac{2n+1}{2} \cdot T_b \le t < (n+1) \cdot T_b \text{ and } m(t) = m_2 \end{cases} \tag{27.22}$$

where T_b is the bit duration and n is an integer numbering the currently transmitted bit. A typical DCSK transmitter structure is shown in Figure 27.14.

27.4.1.1.5 Analysis of Schemes with Static Encoding/Modulation

In the analysis of chaotic as well as conventional communication schemes signal space concepts play an important role [11][17]. Since the chaos-based static encoding/modulation schemes map message symbols to waveforms, the well-known principles of orthogonal and antipodal modulation are found among the chaotic schemes too. The classical knowledge about these methods can be used in the assessment of chaos-based schemes in order to establish performance bounds of such schemes [18][19]. However, the exploitation of chaos introduces further peculiarities, which have to be taken into account when studying

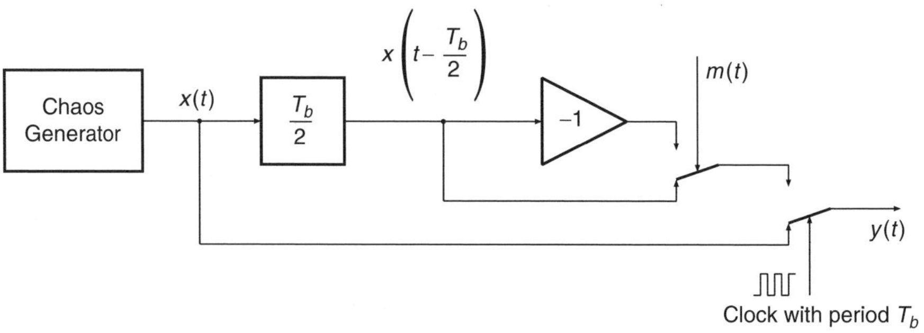

FIGURE 27.14 Differential chaos shift keying transmitter: signal block structure.

the behavior of the schemes and which make the analysis more complex [20]. The underlying approaches of this analysis are sketched and the results are presented in Section 27.6.

27.4.1.2 Dynamic Encoding/Modulation Methods

In the dynamic encoding/modulation method the message is modulated into the message carrier using a discrete-time or a continuous-time dynamical system. Since the state evolution of dynamical systems is influenced by the input signals of the system, the encoder/modulator "memorizes" the prehistory of the message. The currently transmitted signal will depend not only on the current message symbol, but also on a (possibly infinite) number of preceding symbols. The dynamical system, whose behavior is influenced by the message stream, may be an encoder or the generator of the carrier itself.

Such approaches are found in classical communications, e.g., in convolution encoders as well as in differentially encoding modulation schemes such as DPSK.

27.4.1.2.1 Chaotic Modulation/Chaotic Switching

Chaotic modulation (CM) [21] is the general term for dynamic encoding/modulation using chaos. The encoder/modulator is a dynamical system that can be described by a set of state equations (describing the evolution of states x depending on current states and the input signal(s) m) and output equations (mapping current states and input(s) to an output signal y):

$$\begin{aligned} \dot{x}(t) &= g(x(t), m(t)) \\ y(t) &= h(x(t), m(t)) \end{aligned} \tag{27.23}$$

Figure 27.15 shows an example structure for a chaotic modulation transmitter. Here the modulation signal is used to modify the states of the chaos generator. If m is a discrete-valued (e.g., binary) signal, the equations change their structure at the times when the message symbol changes its value. This is equivalent to a switching between different sets of equations with different sets of parameters. Consequently, this particular implementation of CM is termed chaotic switching (CS) [22].

27.4.1.2.2 Encoding Messages into the Symbolic Dynamics of Chaos Generators

This variant of dynamic encoding/modulation forces a chaotic carrier generator to behave in a prescribed way depending on the message to be transmitted. In order to observe the symbolic dynamics of a chaos generator, the state space of the generator (the space of all state vectors x) is partitioned into disjoint subsets, each labeled by a unique symbol. The sequence of labels marking the subset where the state of the system currently resides is termed the symbolic dynamics of the generator. Methods of chaos control allow the influencing of a chaos generator such that its symbolic dynamics follow a prescribed sequence, such as a sequence of message symbols [23]. The knowledge of the partition provides a means for easy message symbol retrieval at the receiver.

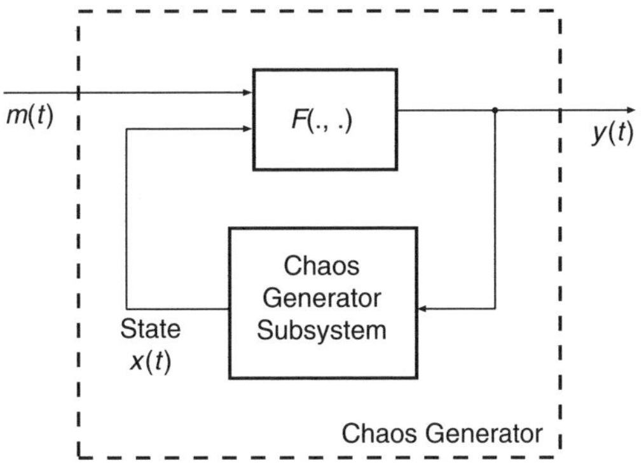

FIGURE 27.15 Chaotic modulation transmitter (example structure).

27.4.2 Receiver Design in Chaos Communications

The reception of a chaotically encoded/modulated message is performed using similar principles as in classical communication schemes. Essentially there are three classes of methods proposed:

- Message reception using a reference signal
- Message reception based on the evaluation of statistical characteristics of the received signal
- Message retrieval using inverse system techniques

27.4.2.1 Message Detection Using a Reference Signal

Reference signals are commonly used in correlation receivers. The reference is generated either by a local, synchronized generator (stored reference) or by a separately transmitted reference signal (transmitted reference).

For generating chaotic reference signals in the receiver from transmitted signals, two solutions have been proposed:

- Drive response synchronization of chaotic systems
- Chaos control methods

27.4.2.1.1 Reference Generation by Drive Response Synchronization

Drive response synchronization was one of the first successful attempts to synchronize chaos generators [24]. Drive response synchronization uses dynamical subsystems of chaos generators for regenerating a chaotic signal from a (maybe distorted) received signal.

The prerequisite for drive response synchronization is the decomposability of the chaotic generator into two subsystems, which are feedback coupled to each other:

$$\begin{aligned}\dot{x}_1(t) &= g_1(x_1(t), x_2(t))\\ \dot{x}_2(t) &= g_2(x_1(t), x_2(t))\end{aligned} \tag{27.24}$$

where x_1 and x_2 represent two vector signals in the general case. For applications, subsystems coupled by one-dimensional signals are favored due to the easier transmission of such signals.

If the two subsystems g_1 and g_2 are stable, they can be expected to reproduce their corresponding state signals x_1 and x_2 when fed with the correct input signal. This is exploited in drive response synchronization. The chaotic drive signal is generated at the transmitter and arrives distorted at the receiver. The stable subsystems are used to restore a copy of the originally transmitted signal. A block diagram illustrating the synchronization principle is shown in Figure 27.16. It has to be noted that this type of synchronization

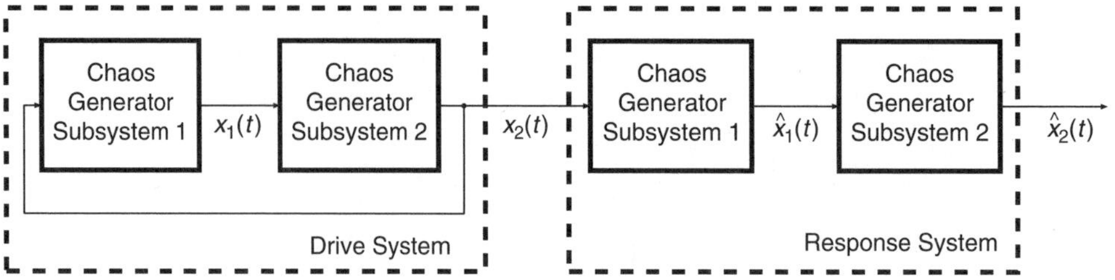

FIGURE 27.16 Drive response synchronization principle.

consists of driven systems; i.e., the behavior strongly depends on the provided input signal. The receiver is not able to generate the desired signal on its own without the correct input. If the input is corrupted, the regenerated signal gets corrupted too; often disturbances like noise in the input produce an even stronger noise at the output.

So far, there does not exist a possibility to generate a fairly clean reference from noisy received signals. This is a significant difference from the synchronizable periodic generators found in classical communications. So as of now there is no possibility to implement a coherent receiver concept based on chaos.

27.4.2.1.2 Reference Generation by Controlling Chaotic Systems

Another method to synchronize chaotic systems is derived from control engineering principles. As for classical systems, for chaotic generators there exist means to force the behavior of the generator to a desired trajectory by applying control input signals derived from the error signal between the generator's current behavior and the desired behavior [25], [26]. This method bears a close resemblance to the classical methods of carrier synchronization. The control may be achieved via some easily accessible quantity, such as a system parameter. Figure 27.17 illustrates the principle.

One remark about the applicability of this method has to be made: chaos generators are very sensitive to parameter changes. So the control of a chaotic generator using such a method will usually fail if the received chaotic signal is significantly distorted. As for the drive response synchronization, this prevents the implementation of a coherent receiver.

27.4.2.1.3 Demodulation Using Reference Signals: Application Examples

The demodulation of a chaotically modulated signal using a reference of the chaotic carrier applies to many of the chaotic encoding/modulation schemes. Typical examples are chaotic masking, chaos shift keying or chaotic switching, and differential chaos shift keying.

In *chaotic masking* the reference can be used for subtraction of the locally generated copy (with the superposed message suppressed) from the message-carrying received signal. Obviously, any additive noise component propagates to the retrieved message signal and any larger disturbance to the original chaotic signal (the message has to be considered a disturbance here too) can spoil the quality of synchronization and thus the quality of the retrieved message. This is another argument against chaotic masking as a communication solution; nevertheless, the approach is included here to give a representative overview

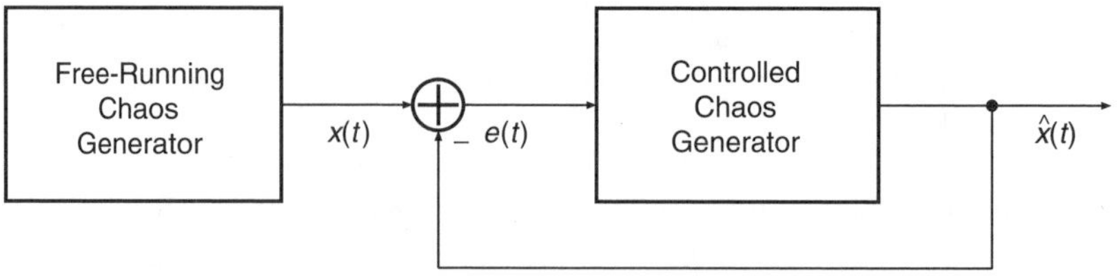

FIGURE 27.17 Illustration of the chaos control approach for reference generation.

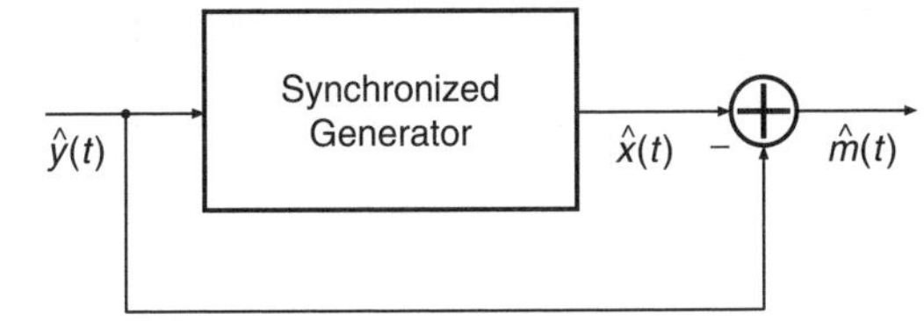

FIGURE 27.18 Chaotic masking receiver using chaos synchronization.

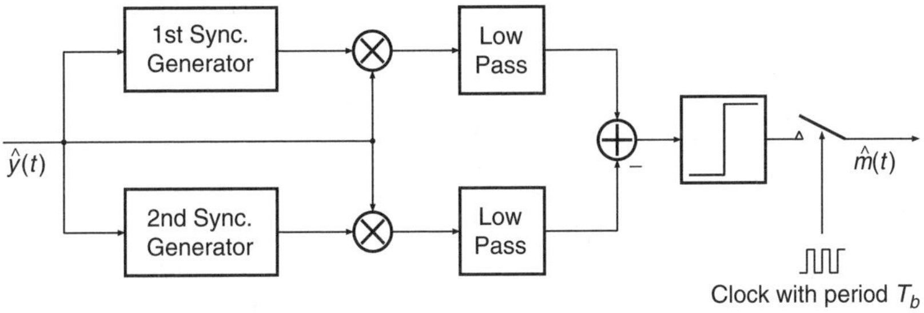

FIGURE 27.19 Chaos shift keying receiver.

of the chaos communications solutions found in the literature. The receiver shown in Figure 27.18 is the one originally proposed with the scheme [12]. The classical principle of correlation reception is a standard solution for demodulation of any modulated signal where references of the carrier are available. Consequently, many solutions proposed for chaos communication receivers are correlator based.

A chaotic transmission using *chaos shift keying* can be received easily by correlation if the receiver is capable of synchronizing to all of the possible chaotic signals corresponding to the transmitted message symbols (see Section 27.4.1.1). A typical receiver setup is shown in Figure 27.19. The better-matching correlation result determines the detected symbol. The shown structure also applies to *chaotic switching*. Since the change in the message symbol always implies a change in the transmitted chaotic signal, synchronization and desynchronization effects take place in such a receiver. Consequently, there will be a transient behavior when switching symbols, which has to be observed in the design of the communications system. A synchronization-free demodulation using a reference signal is found in *differential chaos shift keying*. Since this scheme is transmitted reference, the reference can be taken from the received signal. Reference and message-carrying signals are transmitted in subsequent time slots, so the receiver has to delay the reference part and correlate it with the message-carrying part. The resulting receiver structure is shown in Figure 27.20. This solution has the advantage that the reference and message-carrying signals are equivalently distorted by the transmission channel. The DCSK receiver has essentially the same structure as a suboptimum DPSK receiver (or autocorrelation DPSK receiver) [27]. The DCSK receiver shares the noise

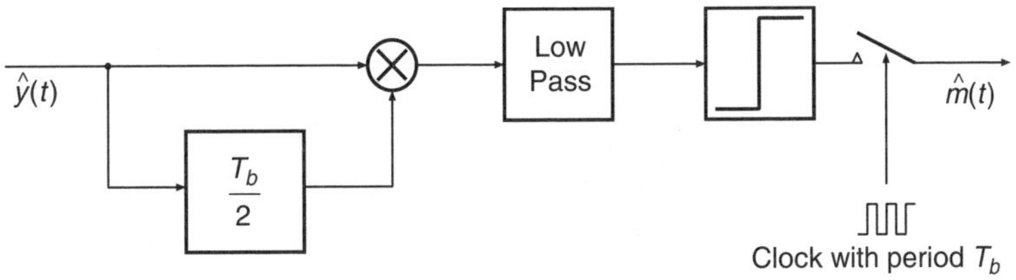

FIGURE 27.20 Differential chaos shift keying receiver.

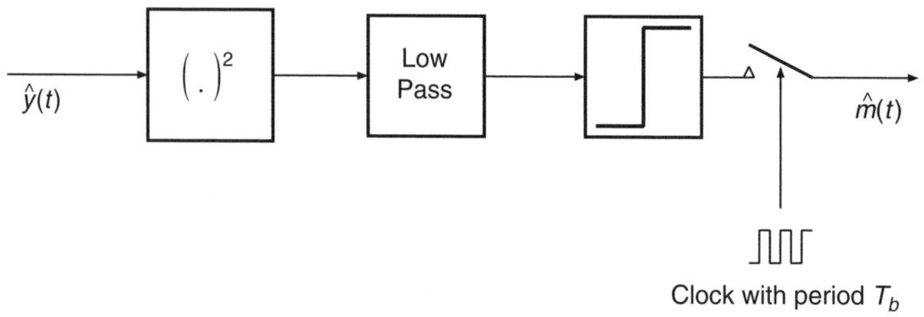

FIGURE 27.21 COOK receiver estimating the transmitted signal power.

robustness problems known for suboptimum reception, which is a common property of all transmitted reference schemes (see Section 27.6).

27.4.2.2 Message Detection Based on Signal Statistics

The observation of particular statistics of the received signal in order to retrieve the message modulated/encoded into it is possible whenever there is a particular statistical property that changes clearly and significantly with the transmitted message symbols. A classical example for this receiver solution is the envelope detectors for AM. However, the estimation of signal statistics is always subject to estimation variance and fluctuations in the estimated values, in particular if the received signals are corrupted by noise or if they are noise-like themselves (as chaos is). So these methods are better suited for digital communications where the outcome of a statistics estimation is expected to represent a finite set of discrete symbols.

A well-suited case for using statistical estimations in the detection of a chaotically encoded/modulated message is *chaotic on–off keying*. The obvious statistical signal property to be observed at the receiver side is the signal power. The corresponding receiver is shown in Figure 27.21.

27.4.2.3 Inverse System Principles

A method particularly applicable to dynamically encoding/modulating schemes using chaos is the *inverse system principle* [28]. It originates from the control theory too. The essential idea is to invert the information (message) processing in the chaos generator during the encoding/modulation operation. This inversion is possible under certain stability constraints [28]. In this case the input and output of the transmitter structure are swapped in order to form the receiver. Figure 27.22 shows this principle applied to the chaotic modulation transmitter displayed in Figure 27.15. The synchronization of the generator subsystem in Figure 27.22 can be understood as a special case of drive response synchronization (see Section 27.4.2.1).

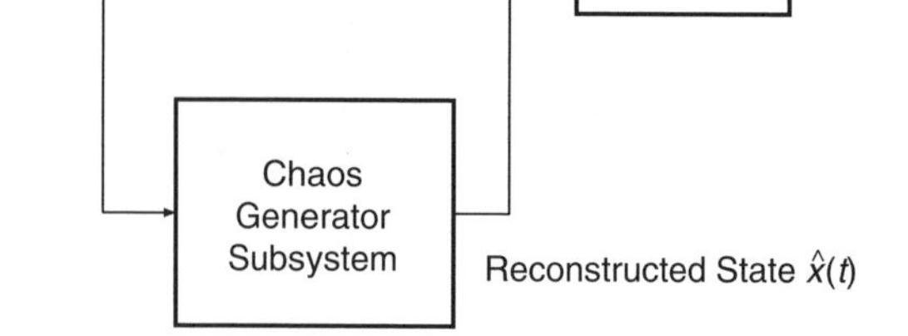

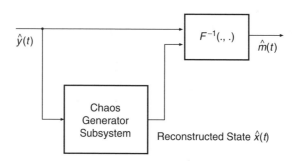

FIGURE 27.22 Chaos communication receiver based on the inverse system principle (transmitter shown in Figure 27.15).

27.4.2.4 Other Detection Principles

There are receiver design solutions for chaos communications that, contrary to the three methods, do not extend classical receiver structures to chaotic communication schemes. These particular receiver designs are of algorithmic nature. Starting from the aim to create *optimum receivers* for chaos, communication solutions are obtained that usually perform better than the methods discussed above at the expense of more extensive signal processing [29].

27.5 Chaos for Spreading Code Generation

CDMA communications systems use code sequences for the transmission of binary messages. In the transmitter the binary message signal is multiplied with the code sequence $x(k)$. The receiver correlates the received signal with a replica of the code sequence. The threshold element estimates the transmitted signal from the result of the correlation. Code sequences are required to have a broadband spectrum; i.e., they should be uncorrelated. In order to avoid crosstalk, code sequences from different transmitter–receiver pairs must not correlate. Since chaotic signals have these properties (see Table 27.2) they were proposed for use in CDMA systems as an alternative to PN sequences [9][32].

Figure 27.23 shows the basic processing scheme of a CDMA system using chaotic code sequences. The two chaos generators in the transmitter and receiver have to generate the same sequence $x(k)$, so they have to be synchronized. If a reliable synchronization channel is not available, this can be done by generating the sequences in advance and storing them in the receiver. This automatically means truncation and periodic repetition, which is an important condition for synchronization. Synchronization is then achieved in the usual way, as it is done with PN sequences, or alternatively, a corresponding matched filter with a chaotically generated sequence of coefficients is used.

The performance of coherent chaos-based CDMA systems can be characterized as follows: as long as a correlation receiver is used for the calculation of the bit error rate, only the second-order characteristics (acf, pds) are taken into account. So the receiver cannot distinguish chaotic and random signals, and the performance of chaos-based systems should compare with that of conventional systems. The benefit of using chaotic sequences consists of two points:

1. With given code length, a larger number of sequences is available compared to the number of PN sequences.
2. Chaotic sequences can be easily generated and their statistical characteristics can be adjusted to their users' needs [31].

The latter point is especially interesting; hence, it was found recently that slightly correlated sequences show a better performance than purely uncorrelated ones [31]. From the desired correlation or spectral properties of the sequence a return map g can be constructed that, from different initial values, generates a variety of code sequences. Another aspect to take into account is the structure of chaotic sequences for

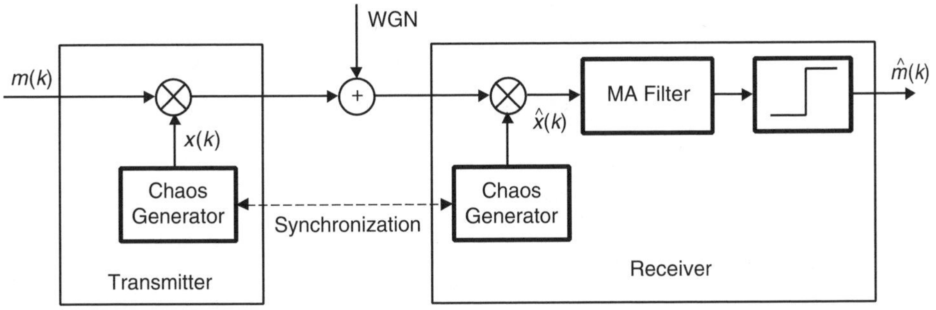

FIGURE 27.23 Baseband model of a CDMA system using chaotic sequences.

the receiver design. This leads to nonlinear receivers, which might perform better than the classical ones. Although first attempts have been made [33], this field of research is still open.

27.6 Chaotic vs. Classical Communications

A very important question to be answered when discussing chaos communications is that of the performance of the proposed schemes. It is now generally accepted that performance studies have to be performed using the commonly accepted measures (such as the bit error rate over the bit energy-to-noise ratio) from classical communications theory. In turn, the results will provide means for an in-depth comparison of chaotic and classical communications solutions, highlighting both the advantages and drawbacks of the methods in this new field.

Since chaos is generated by nonlinear systems, many proposed receiver structures, in particular receivers using references generated by chaos synchronization and inverse system approaches, are nonlinear dynamical systems. The performance analysis of such a communication scheme, even for the comparatively simple situation of additive white gaussian noise (AWGN), quite often turns out to be tedious if not impossible analytically. As a consequence, many of the results known about the performance of chaos communication schemes were obtained by computer simulations. However, there exist analytical solutions (both approximate and exact); see, e.g., [18][20][35]. Some of the results in [20] and [35] concern coherent implementations of chaos communication receivers, where the receiver is assumed to possess a clean reference of the transmitted chaotic signal for correlation. As stated in Section 27.4.2.1, the existing synchronization methods for chaos do not allow the derivation of clean synchronized references of chaotic signals. Thus, so far there exist no means to practically implement a coherent correlation receiver for chaos communications. Some performance studies for chaos communication schemes also present lower performance bounds, which are derived by comparisons with existing classical communication methods (see [18]).

27.6.1 The Analysis of Chaos Communication Schemes in AWGN

27.6.1.1 The Analysis Problem

The analysis of chaos communication schemes is normally more complicated than it is for conventional communications. However, the general structure of the communication setup remains the same, as it is shown in Figure 27.24 for the AWGN situation and binary digital communications. In classical

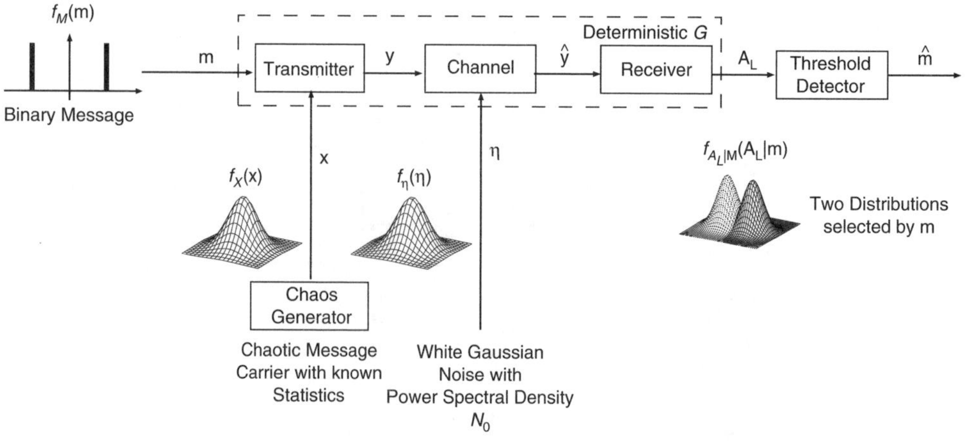

FIGURE 27.24 The analysis problem in a chaos communication scheme.

communications we find two sources of randomness — the message stream and the channel noise. The message carrier itself is fully determined and not random. The analysis of such schemes is performed as a statistical analysis of the deterministic system labeled G in the figure.

In chaos communications the message carrier is still a deterministic signal, but as shown in Sections 27.1.1 and 27.1.2, chaotic signals closely resemble random signals and can be well described and analyzed by statistical means. So from the analysis point of view we arrive at a situation similar to that in the classical communications — we have a deterministic signal processing scheme (G in Figure 27.24), but instead of two, we now find three random input signals: the binary message stream $\mathbf{m}$, the chaotic message carrier $\mathbf{x}$, and the channel noise η:

$$A_L = G\left(\mathbf{m}, \mathbf{x}, \eta\right) \tag{27.25}$$

In order to assess the statistics of the processing result A_L, the statistics of the random input signals have to be known. The statistics can be given in terms of probability density functions or distribution functions, but also, for instance, in terms of moments or cumulants. n-th-order *moments* are the expectations of the nth power of a random variable (e.g., the mean is the first-order moment and the signal power is the second). The *moments* are also the coefficients of a Taylor series expansion of the Fourier transform of a probability density function (the characteristic function) [40]. The complete set of moments gives a full description of the respective random variable. In a similar fashion, the series expansion of the natural logarithm of the characteristic function defines the *cumulants* of a random variable [40]. Well-known and often used cumulants are those of 1st order (mean) and 2nd order (variance or covariance). As for the moments, the complete set of cumulants gives the full statistical description of a random variable or a set of random variables. Contrary to moments, the cumulants show some peculiarities that make them useful quantities for the statistical analysis:

- The cumulants of a sum of independent random variables are calculated as the sum of the respective cumulants of the particular variables.
- Variables with a Gaussian distribution only have cumulants up to 2nd order. So, e.g., the white Gaussian channel noise is very easily described by just two cumulants.
- Joint cumulants of independent random variables are zero. This is a typical situation in chaos communication schemes. This knowledge helps to simplify the statistical analysis of such schemes, since the independence of chaos, message, and noise can usually be assumed.

So the statistical description of the channel noise, as well as of the binary message, is straightforward and appears in the same way as for the classical communication schemes. For the chaotic message carrier we have two options to derive the statistics:

- A statistical analysis based on the Frobenius–Perron operator theory. This requires the theory to be applicable to the used chaos generator, as it is for piecewise linear maps.
- A numerical analysis of the generated chaotic signals in order to estimate the statistics. This is the way to go for the more complex chaos generators, in particular for generators operating in continuous time.

27.6.1.2 Analysis Methods

27.6.1.2.1 Discrete-Time Baseband Modeling

As in classical communications the analysis of a chaos communication scheme with a radio frequency (RF) channel and in continuous time is a tedious and often unsolvable task. In classical communications the solution is the *derivation of equivalent baseband models* (see [10]), where the RF signals of the original scheme are represented by their IQ-demodulated counterparts (spectral shift of the RF band to zero frequency) and the signal processing structures by equivalent models acting on the complex-valued baseband signals. In order to simplify the analysis task further, the models are sampled, resulting in a *discrete-time baseband model* of the communication scheme.

For chaos communications systems this approach is also applicable; however, sometimes one might encounter certain difficulties due to the nonlinearity in the processing schemes. For the general procedure as well as example solutions refer to [36], [37], [38].

27.6.1.2.2 Statistical Analysis of Nonlinear Discrete-Time Baseband Models

An exact statistical analysis of a chaos communication scheme in order to derive exact performance measures (e.g., the bit error rate (BER) as a function of the ratio of the bit energy to the noise power spectral density (E_b/N_0)) is normally a computationally expensive task.

For the analysis two methods are established:

1. The evaluation of particular statistical characteristics of the estimation A_L, such as certain moment or cumulants, which in turn can be used to calculate the BER of the scheme exactly or approximately (as in [20] and [34])
2. The derivation of closed-form expressions for the BER involving the complete statistical description of all input signals (as in [35])

The first approach requires extensive formula manipulations and calculations in order to arrive at the result, which then can be numerically evaluated quite easily.

The second approach is applicable to systems with lower structural complexity and gives a quicker result. However, the derived expressions still require a high computational effort for their evaluation in order to arrive at the performance figures themselves.

In all cases the analytical results give deep insight into the behavior of the analyzed scheme, which cannot be gained from numerical simulations.

The evaluation of particular statistics is made easier if G in Equation 27.25 has a polynomial structure with respect to the input signals, especially chaos and noise, and if A_L can be expressed explicitly in terms of input signal values at various time instants (i.e., the signal processing structure is free of dynamical feedback). Such a structure might be developed using approximation methods such as polynomial Volterra series expansions.

Based on polynomial system structures, two methods become well applicable:

- *Moment calculations using the operator rules for the expectation value operator.* Expectations of sums of random quantities become sums of the expectations of the addends, and the expectation of a product of independent random expressions is the product of the expectations of the respective expressions. In a polynomial dependence this maps any moment (a moment of order n of a random variable z is the expectation of z^n) of the output quantity to functions of moments of the input quantity. The analysis and the results, however, quickly increase in complexity with increasing moment order.
- *Cumulant equations.* Assume a situation where we have a random variable (such as A_L) that depends on other random quantities (such as $\mathbf{x}$, $\mathbf{m}$, and η) by means of a nonlinear function f, as in Equation 27.25. Cumulant equations [39] define the coefficients of a Taylor series expansion of the moments of A_L in terms of the cumulants of $\mathbf{x}$, $\mathbf{m}$, and η. Similar to a Taylor series expansion of the function f in terms of $\mathbf{x}$, $\mathbf{m}$, and η, the cumulant equations contain partial derivatives of the function f by its arguments. If f is a polynomial, the derivatives vanish from certain derivation orders onward, so the expansion becomes finite. This simplifies the analysis and makes the cumulant analysis best applicable to polynomial structures.

The analysis procedures are complicated, so they are not presented here. Refer to other publications covering the topic, in particular [7], but also [20], [34], [36].

As a result of the analysis described here, we obtain the moments or cumulants of the threshold detector input A_L in Figure 27.24. If we know the first and second order moments or cumulants, this will be sufficient to calculate the BER of the scheme if we assume that the probability density function of A_L is close to a Gaussian distribution. If we find the complete set of statistics, we can possibly derive an exact expression for the BER.

27.6.1.3 Analysis Example

In this section we will sketch the analysis procedure for one example scheme — DCSK, as described in Section 27.4.1.1 and shown in Figure 27.14 (transmitter side), with the correlation receiver shown in Figure 27.20.

27.6.1.3.1 *Baseband Model*

The baseband model structure of the transmitter remains as shown in Figure 27.14; we just replace the continuous-time t by the discrete-time k and all signals by their complex-valued baseband counterparts. Now the reference is sent for L samples and is followed by another L samples carrying the message. A bit will take $2L$ samples to be transmitted; hence, we substitute $T_b \rightarrow 2L$. Equation 27.22 becomes

$$y(k) = \begin{cases} x(k) & \text{if} \quad 2n \cdot L \leq k < 2n \cdot L + 1 \\ x(t-L) & \text{if} \quad 2n \cdot L + 1 \leq k < (2n+1) \cdot L \text{ and } m(k) = m_1 \\ -x(t-L) & \text{if} \quad 2n \cdot L + 1 \leq k < (2n+1) \cdot L \text{ and } m(k) = m_2 \end{cases} \tag{27.26}$$

Since the transformation to the baseband preserves linearity, the channel remains the same structurally and the received signal becomes

$$\hat{y}(k) = y(k) + \eta(k) \tag{27.27}$$

The receiver still performs a correlation, but now in the complex domain (Figure 27.25). Formally it is expressed as

$$A_L(k) = \frac{1}{L} \sum_{i=0}^{L-1} \hat{y}(k-i) \cdot \hat{y}^*(k-i-L) \tag{27.28}$$

where the asterisk denotes the complex conjugate.

27.6.1.3.2 *Baseband Signal Models*

The following assumptions will be made about the involved random signals, which will result in particular baseband model properties:

- *The message* is a sequence of balanced and independent binary symbols (each symbol appears with a probability 0.5 and there is no memory in the message). The signal remains unchanged in the baseband.
- *The chaotic message carrier* shall be a chaotically frequency-modulated signal (the scheme is then called FM-DCSK [41]). In the baseband this signal is represented by a chaotic sequence, mapped to the phase angle of points on the complex unit circle (as we find it for the conventional PM/FM modulation).

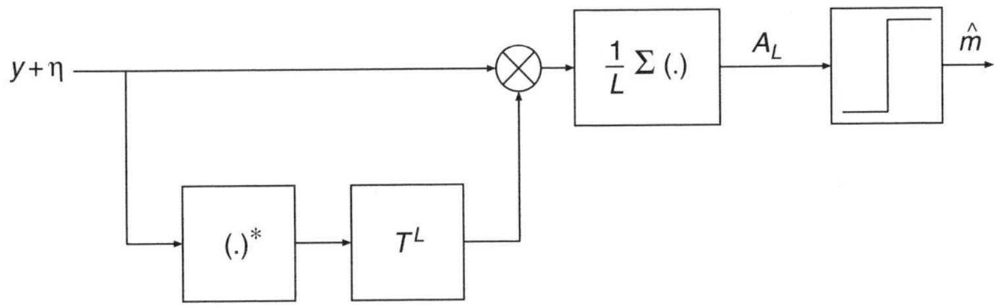

FIGURE 27.25 Baseband model of the DCSK receiver in Figure 27.20.

- *The white Gaussian channel noise* is mapped to a complex white Gaussian noise in discrete time with independent real and imaginary parts (as in the analysis of the classical communications systems).

27.6.1.3.3 *Results of the Statistical Analysis: Gaussian Approximation*

The authors performed an analysis based on cumulant equations in order to assess the performance of the FM-DCSK scheme. In the result of the analysis the first- and second-order cumulants (mean and variance) of the threshold detector input A_L at the time instant of information detection are obtained:

$$\begin{aligned}\chi^{1}_{A_L} &= \frac{m}{2L}\cdot E_b\\ \chi^{1,1}_{A_L,A_L^*} &= \frac{E_b\cdot N_0}{L^2}+\frac{N_0^2}{L}\end{aligned} \tag{27.29}$$

χ denotes the cumulants, the subscript the corresponding random variable(s) (note that the variance of a complex quantity is the joint cumulant of the quantity and its complex conjugate), and the superscript the order with respect to the particular random variables (the cumulant order is the sum of all those). The statistics are represented already in the relevant terms for describing a communications system performance measure: E_b, the energy per transmitted bit (this is preserved in the discrete-time baseband model; i.e., the bit energy can be calculated by summing up the squares of the $2L$ samples carrying one bit); N_0, the power spectral density of the channel noise (which becomes the variance of the complex baseband noise); and m, the value of the transmitted bit (plus or minus 1).

Knowing mean and variance, one can calculate the error rate of the threshold detector in the well-known Bayesian approach, assuming that A_L is Gaussian (which is valid only for large L as well as for small E_b/N_0 [36]):

$$BER=\frac{1}{2}\cdot\operatorname{erfc}\left(\frac{\chi^{1}_{A_L}}{\sqrt{\chi^{1,1}_{A_L,A_L^*}}}\right)=\frac{1}{2}\cdot\operatorname{erfc}\left(\frac{\chi^{1}_{A_L}}{\sqrt{\chi^{1,1}_{A_L,A_L^*}}}\right)=\frac{1}{2}\cdot\operatorname{erfc}\left(\frac{1}{\sqrt{4\frac{N_0}{E_b}+4L\cdot\left(\frac{N_0}{E_b}\right)^2}}\right) \tag{27.30}$$

27.6.1.3.4 *Results of the Statistical Analysis: Exact Solutions*

Continuing the analysis further, one can arrive at a general expression for the cumulants of A_L. Since the threshold detector only evaluates the real part of A_L, we can restrict the cumulants of this quantity. The result of this analysis is

$$\chi^{q}_{\mathrm{Re}(A_L)}=m^q\cdot\left(q!\cdot\left(\frac{N_0}{E_b}\right)^{q-1}+(1+(-1)^q)\cdot L\cdot(q-1)!\cdot\left(\frac{N_0}{E_b}\right)^{q}\right) \tag{27.31}$$

Via the reconstruction of the cumulant-generating function, the derivation of the probability density function, and the calculation of the threshold detector error, one obtains a closed-form expression for the BER [36]:

$$BER=\frac{1}{2^L}\cdot\exp\left(\frac{E_b}{2N_0}\right)\sum_{i=0}^{L-1}\frac{\left(\frac{E_b}{2N_0}\right)^i}{i!}\sum_{j=i}^{L-1}\frac{1}{2^j}\binom{j+L-1}{j-i} \tag{27.32}$$

Figure 27.26 shows the exact and approximate solutions. The performance of the standard classical modulation schemes is given for comparison.

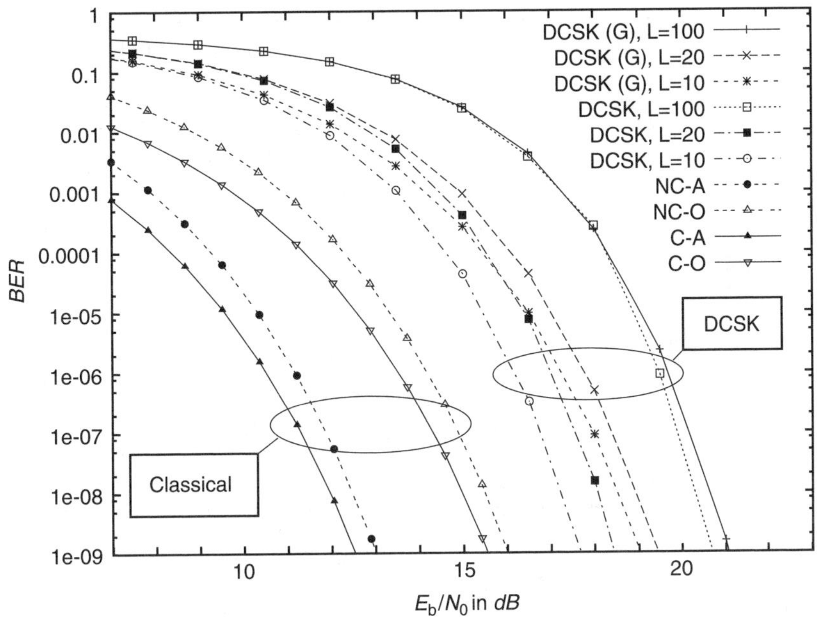

FIGURE 27.26 FM-DCSK performance in AWGN: approximate and exact solutions. G, Gaussian approximations; NC, classical noncoherent receiver for orthogonal (NC-O) and antipodal (NC-A) modulation; C, classical coherent receiver for orthogonal (C-O) and antipodal (C-A) modulation.

27.6.2 Chaos Communication Methods in Comparison to Classical Solutions

The interesting question now is, How do the chaos communication schemes perform compared with the classical approaches?

We shall start with the application of chaos generators to the creation of CDMA sequence families, as it was described in Section 27.5. Here the application of chaos provides a true benefit in terms of performance. The sequence families created by chaos generators are designed to have advantageous correlation properties on average, and thus chaos-based sequence families perform better than classical sequence families [32]. The advantage is usually only a few percent, but this can mean squeezing a few more users into a channel that would otherwise be at its limit. The chaos communication schemes in the current stage of development cannot outperform the classical approaches, but they can in some cases reach a comparable performance. Typical problems arise from the nonlinearity of the generators and their sensitivity to disturbances, as well as from the nonperiodicity of the generated chaotic signals:

1. Proposals based on chaos synchronization or inverse systems in the receiver reproduce sensitive dynamical (sub)systems at the receiver side, which do not exactly match in parameters and already react sensibly on small disturbances. Often they do not withstand a channel noise of typical scale. Consequently, these implementations have a significantly worse performance than the classical solutions (the difference might well reach 15 to 20 dB in terms of the E_b/N_0 ratio in order to achieve the same BER) and are not yet feasible for practical implementations.
2. The nonperiodicity of the signals and the sensitivity of the generators do not allow implemention of a robust synchronized generator, which is tuned to the incoming signal (this is no problem for a periodic generator). So the problem of coherent reception is not yet solved, which implies an inevitable performance lack of chaotic solutions that are effectively noncoherent so far.
3. Many chaos communication proposals show a lack of noise robustness. In classical communications the BER is usually solely described as a function of the ratio E_b/N_0. The bit energy is the product of the average signal power and the bit duration. It does not matter how the bit energy is

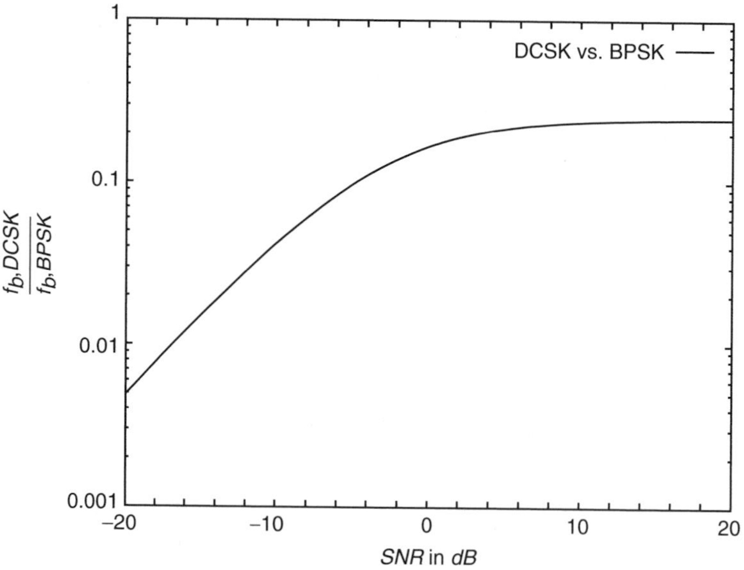

FIGURE 27.27 Bit Rate of FM-DCSK compared to BPSK in the same SNR on the channel and achieving the same BER.

transmitted — either with high power and short duration (high bit rate) or with low power and long duration — the achieved BER in an AWGN situation will remain the same. This is an important prerequisite for CDMA, since it allows the scheme to operate significantly below the noise floor. In many chaos communications configurations the BER depends on further quantities apart from E_b/N_0, such as bit duration or signal power. In the example from Section 27.6.1.3 this additional parameter is L, which is proportional to the bit duration. With increasing L the BER becomes worse for the same E_b/N_0. In other words, a low signal-to-noise ratio (SNR) on the channel spoils the performance. The scheme operates best in a high-transmission-power low-noise environment. This lack of noise robustness is manifested by a shift of the curves to the right in the performance plots (BER as function of E_b/N_0), as it is obvious for the FM-DCSK example in Figure 27.26. In [20] the performance drop in low SNR conditions was calculated using the first- and second- order cumulant analysis result (Equation 27.23). The diagram shows how much slower an FM-DCSK scheme with the discussed receiver has to be operated in order to achieve the same BER under the same SNR on the transmission channel compared to binary phase shift keying (BPSK) (antipodal modulation, coherent receiver).

Looking at the above comparisons it becomes obvious that chaos communications still have to advance in order to grow into real competitors of classical communications. This, however, is an obvious situation, since the classical solutions are about 50 years ahead in research and applications. The upspring of ideas observed in this field should encourage researchers to have a closer look at these topics.

References

[1] A.L. Baranovski, W. Schwarz, and A. Mögel. Statistical Analysis and Design of Chaotic Switched Dynamical Systems. In *Proceedings of ISCAS*, Vol. V, Orlando, FL, May/June 1999, pp. 467–470.

[2] A.L. Baranovski and D. Daems. Design of 1-D Chaotic Maps with Prescribed Statistical Properties. *Int. J. Bifurcation Chaos*, 5, 1585–1598, 1995.

[3] G. Chen and T. Ueta, Ed. *Chaos in Circuits and Systems*. World Scientific Publishing Co., 2002.

[4] M. Götz and W. Schwarz. Bit Energy Efficient Non-Coherent Chaotic Spreading Technique. NDES, 2000.

[5] A. Mögel. Design of Chaotic Broadband Generators, Technical Report. INSPECT Meeting, TU Dresden, October 2, 1997.

[6] M.P. Kennedy, G. Kolumbán, and Z. Jákó. Chaotic Modulation Schemes. In *Application of Chaotic Electronics to Telecommunications*, M.P. Kennedy, R. Rovatti, and G. Setti, Eds. CRC Press, Boca Raton, FL, 2000, chapter 6, pp. 151–183.

[7] W. Schwarz, M. Götz, K. Kelber, A. Abel, T. Falk, and F. Dachselt. Statistical Analysis and Design of Chaotic Systems. In *Application of Chaotic Electronics to Telecommunications*, M.P. Kennedy, R. Rovatti, and G. Setti, Eds. CRC Press, Boca Raton, FL, 2000, chapter 9, pp. 253–305.

[8] F. Dachselt, K. Kelber, and W. Schwarz. Discrete-Time Chaotic Encryption Systems. Part III. Cryptographical Analysis. *IEEE Trans. CAS-I*, 45, 883–888, 1998.

[9] G. Mazzini, R. Rovatti, and G. Setti. Chaos-Based Asynchronous DS-CDMA Systems. In *Application of Chaotic Electronics to Telecommunications*, M.P. Kennedy, R. Rovatti, and G. Setti, Eds. CRC Press, Boca Raton, FL, 2000, chapter 4, pp. 33–79.

[10] S.S. Haykin. *Communication Systems*, 3rd ed. John Wiley & Sons, New York, 1994.

[11] J.G. Proakis. *Digital Communications*, 3rd ed. McGraw-Hill, Singapore, 1995.

[12] L.J. Kocarev, K.S. Halle, K. Eckert, L.O. Chua, and U. Parlitz. Experimental Demonstration of Secure Communications via Chaotic Synchronization. *Int. J. Bifurcation Chaos*, 2(3), 709–713, 1992.

[13] M.P. Kennedy and H. Dedieu. Experimental Demonstration of Binary Chaos-Shift-Keying Using Self-Synchronising Chua's Circuits. In *Proceedings of NDES'94*, Dresden, Germany, World Scientific, Singapore, 1994, pp. 262–275.

[14] G. Kolumbán, M.P. Kennedy, and G. Kis. Performance Improvement of Chaotic Communications Systems. In *Proceedings of ECCTD*, Budapest, Hungary, 1997, pp. 284–289.

[15] B.L. Basore. Noise-Like Signals and Their Detection by Correlation. Ph.D. thesis, MIT, Cambridge, MA, 1952.

[16] G. Kolumbán, B. Vizvari, W. Schwarz, and A. Abel. Differential Chaos Shift Keying: A Robust Coding for Chaos Communication. In *Proceedings of NDES*, Seville, Spain, 1996, pp. 87–92.

[17] J.M. Wozencraft and I.M. Jacobs. *Principles of Communication Engineering*. John Wiley & Sons, New York, 1965.

[18] G. Kolumbán. Theoretical Noise Performance of Correlator-Based Chaotic Communication Schemes. *IEEE Trans. Circuits Syst. 1*, 47(12), 1692–1701, 2000.

[19] R. Rovatti, G. Setti, G. Mazzini, and S. Callegari. Performance Upper Bound for Full-E_b Bilinear FM-DCSK Receivers. In *Proceedings of NDES*, Catania, Italy, 2000, pp. 170–173.

[20] A. Abel, W. Schwarz, and M. Götz. Noise Performance of Chaotic Communication Systems. *IEEE Trans. Circuits Syst. I*, 47(12), 1726–1732, 2000.

[21] K.S. Halle, C.-W. Wu, M. Itoh, and L.O. Chua. Spread Spectrum Communication through Modulation of Chaos. *Int. J. Bifurcation Chaos*, 3(2), 469–477, 1993.

[22] U. Parlitz, L.O. Chua, L.J. Kocarev, K.S. Halle, and A. Chang. Transmission of Digital Signals by Chaotic Synchronization. *Int. J. Bifurcation Chaos*, 2(4), 973–977, 1992.

[23] T. Schimming and J. Schweizer. Chaos Communication from a Maximum Likelihood Perspective. In *Proceedings of NOLTA*, Vol. 1, Crans Montana, Switzerland, 1998, pp. 77–80.

[24] L.M. Pecora and T.L. Carroll. Synchronization in Chaotic Systems. *Phys. Rev. Lett.*, 64(8), 821–824, 1990.

[25] E. Ott, C. Grebogi, and J.A. Yorke. Controlling Chaos. *Phys. Rev. Lett.*, 64(8), 1196–1199, 1990.

[26] H.D.I. Abarbanel, L. Korzinov, A.I. Mees, and I.M. Starobinets. Optimal Control of Nonlinear Systems to Given Orbits. *Syst. Control Lett.*, 31, 263–276, 1997.

[27] Y.U. Okunev. *Phase and Phase-Difference Modulation in Digital Communications*. Artech House, Norwood, MA, 1997.

[28] U. Feldmann, M. Hasler, and W. Schwarz. Communication by Chaotic Signals: The Inverse System Approach. *Int. J. Circuit Theory Appl.*, 24(5), 551–579, 1996.

[29] M. Hasler and T. Schimming. Optimal and Suboptimal Chaos-Receivers. *Proc. IEEE*, 90(5), 733–746, 2002.

[30] G. Mazzini, R. Rovatti, and G. Setti. Capacity of Chaos-Based Asynchronous DS-CDMA Systems with Exponentially Vanishing Autocorrelations. *IEE Electron. Lett.*, 38(25), 1717–1718, 2002.

[31] G. Setti, G. Mazzini, and R. Rovatti. Chaos-Based Spreading Sequence Optimization for DS-CDMA Synchronization. In *Proceedings of NOLTA '98*, Crans Montana, Switzerland, 1998.

[32] R. Rovatti, G. Setti, and G. Mazzini. Chaos-Based Spreading Compared to m-Sequences and Gold Spreading in Asynchronous CDMA Communication Systems. In *Proceedings of ECCTD '97*, Budapest, August 1997.

[33] M. Götz and W. Schwarz. Bit Energy Efficient Non-Coherent Chaotic Spreading Technique. In *Proceedings of NDES*, Catania, Italy, May 2000, pp. 47–51.

[34] A. Abel and W. Schwarz. Chaos Comunications: Principles, Schemes, and System Analysis. *Proc. IEEE*, 90(5), 691–710, 2002.

[35] A.J. Lawrance and G. Ohama. Exact and Analytically Approximate Bit Error Rates in Chaos Communication. In *Proceedings of NOLTA*, Xi'an, People's Republic of China, 2002.

[36] A. Abel, M. Götz, and W. Schwarz. Statistical Analysis of Chaotic Communication Schemes. In *Proceedings of ISCAS*, Vol. 4, Monterey, 1998, pp. IV-465–IV-468.

[37] G. Kolumbán. Performance Evaluation of Chaotic Communications Systems: Determination of Low-Pass Equivalent Model. In *Proceedings of NDES*, Budapest, 1998, pp. 41–51.

[38] M. Hasler. Chaotc Communications over a Noisy Channel [tutorial talk]. In *Proceedings of ECCTD*, Streza, Italy, 1999, p. 256.

[39] A.N. Malakhov. *Cumulant Analysis of Random Non-Gaussian Processes and Their Transformations.* Sovetskoe Radio, Moscow, 1978 (in Russian).

[40] C.L. Nikias and A.P. Petropulu. *Higher-Order Spectra Analysis.* Prentice Hall, Englewood Cliffs, NJ, 1993.

[41] G. Kolumbán, G. Kis, M.P. Kennedy, and Z. Jákó. FM-DCSK: A New and Robust Solution for Chaotic Communications. In *Proceedings of NOLTA*, Vol. 1, Hawaii, 1997, pp. 117–120.

28

Adaptation Techniques and Enabling Parameter Estimation Algorithms for Wireless Communications Systems

28.1 Introduction **28**-2

28.2 Overview of Adaptation Schemes **28**-3
Link and Transmitter Adaptation • Adaptive System Resource Allocation • Receiver Adaptation

28.3 Parameter Measurements **28**-6
Channel Selectivity Estimation • Channel Quality Measurements

28.4 Applications of Adaptive Algorithms: Case Studies **28**-14
Examples for Adaptive Receiver Algorithms • Examples for Link Adaptation and Adaptive Resource Allocation

28.5 Future Research for Adaptation **28**-20

28.6 Conclusion **28**-22

Hüseyin Arslan
University of South Florida

Abstract

Today's wireless services have come a long way since the roll-out of the conventional voice-centric cellular systems. The demand for wireless access in voice and high-rate-data multimedia applications has been increasing. New-generation wireless communications systems are aimed at accommodating this demand through better resource management and improved transmission technologies. The widespread use of the adaptation techniques for the evolution of wireless mobile radio systems is one way to achieve this goal.

0-8493-1657-X/$0.00+$1.50

In this chapter, adaptation algorithms are discussed for improving wireless mobile radio system performance and capacity. First, an overview of adaptive resource management and adaptive transmission technologies is given. Then, parameter measurements that make many of these adaptation techniques possible are discussed in detail. Finally, several examples that demonstrate the application of adaptation techniques, along with the required parameters, are presented.

28.1 Introduction

Wireless communications systems have evolved substantially over the last two decades. The explosive growth of the wireless communication market is expected to continue in the future, as the demand for all types of wireless services is increasing. There is no doubt that the second generation of cellular wireless communications systems was a success. However, these systems were designed to provide good coverage for voice services so that a minimum required signal quality can be ensured over the coverage area. If the received signal quality is well above the minimum required level, the receivers do not exploit this. The speech quality does not improve much, as the quality is mostly dominated by the speech coder. On the other hand, if the signal quality is below the minimum required level, a call drop will be observed. Therefore, such a design requires the use of strong forward error correction (FEC) schemes, low-order modulations, and many other redundancies at the transmission and reception. In essence, the mobile receivers and transmitters are designed for the worst-case channel and received signal conditions. As a result, many users experience unnecessarily high signal quality from which they cannot benefit. While reliable communication is achieved, the system resources are not used efficiently.

New generations of wireless mobile radio systems aim to provide higher data rates and a wide variety of applications (like video, data, etc.) to mobile users while serving as many users as possible. However, this goal must be achieved under spectrum and power constraints. Given the high price of spectrum and its scarcity, the systems must provide higher system capacity and performance through better use of the available resources. Therefore, *adaptation techniques* have been becoming popular for optimizing mobile radio system transmission and reception at the physical layer as well as at the higher layers of the protocol stack.

Traditional system designs focus on allocating fixed resources to the user. Adaptive design methodologies typically identify the user's requirements and then allocate just enough resources, thus enabling more efficient utilization of system resources and consequently increasing capacity. Adaptive channel allocation and adaptive cell assignment algorithms have been studied since the early days of cellular systems. As the demand in wireless access for speech and data has increased, link and system adaptation algorithms have become more important.

For a given average transmit power, adaptation allows the users to experience better signal qualities. Adaptation reduces the average interference observed from other users, as they do not transmit extra power unnecessarily. As a result, the received signal quality will be improved over a large portion of the coverage area. These higher-quality signal levels can be exploited to provide increased data rates through rate adaptation. For a desired received signal quality, this might also translate into less transmit power, leading to improved power efficiency for longer battery life. On the other hand, for a desired minimum signal quality, this might lead to an increased coverage area or better frequency reuse. In addition, adaptive receiver designs allow the receiver to work with reduced signal quality values; i.e., a desired bit-error-rate (BER) or frame-error-rate (FER) performance can be achieved with a lower signal quality. Adaptive receivers can also enable reduced average computational complexities for the same quality of service, which again implies less power consumption. As can be seen, adaptation algorithms lead to improved performance, increased capacity, lower power consumption, increased radio coverage area, and eventually better overall wireless communications system design.

Many adaptation schemes require a form of measurement (or estimation) of various quantities (parameters) that might change over time. These estimates are then used to trigger or perform a multitude of functions, like the adaptation of the transmission and reception. For example, Doppler spread and delay spread estimations, signal-to-noise ratio (SNR) estimation, channel estimation, BER estimation, cyclic redundancy check (CRC) information, and received signal strength measurement are some of the commonly used measurements for adaptive algorithms. As the interest in the adaptation schemes increases, so does the research on improved (fast and accurate) parameter estimation techniques.

In this chapter, an overview of commonly used adaptation techniques and their applications for wireless mobile radio systems is given. Some of the commonly used parameters and their estimation using baseband signal processing techniques are explained in detail. Also, the current and future research issues regarding the improved parameter estimation and extensive use of adaptation techniques are discussed throughout the chapter. Note that there has been a significant amount of research on adaptation of wireless communications systems. This chapter is not intended to cover all these developments, but rather, it is intended to provide the readers an overview and conceptual understanding of adaptation techniques and related parameter estimation algorithms. More emphasis is given on signal processing perspectives of the adaptation of wireless communications systems.

28.2 Overview of Adaptation Schemes

In wireless mobile communications systems, information is transmitted through a radio channel. Unlike other guided media, the radio channel is highly dynamic. The transmitted signal reaches the receiver by undergoing many effects, corrupting the signal, and often placing limitations on the performance of the system.

Figure 28.1 illustrates a wireless communications system that includes some of the effects of the radio channel. The received signal strength varies depending on the distance relative to the transmitter, shadowing

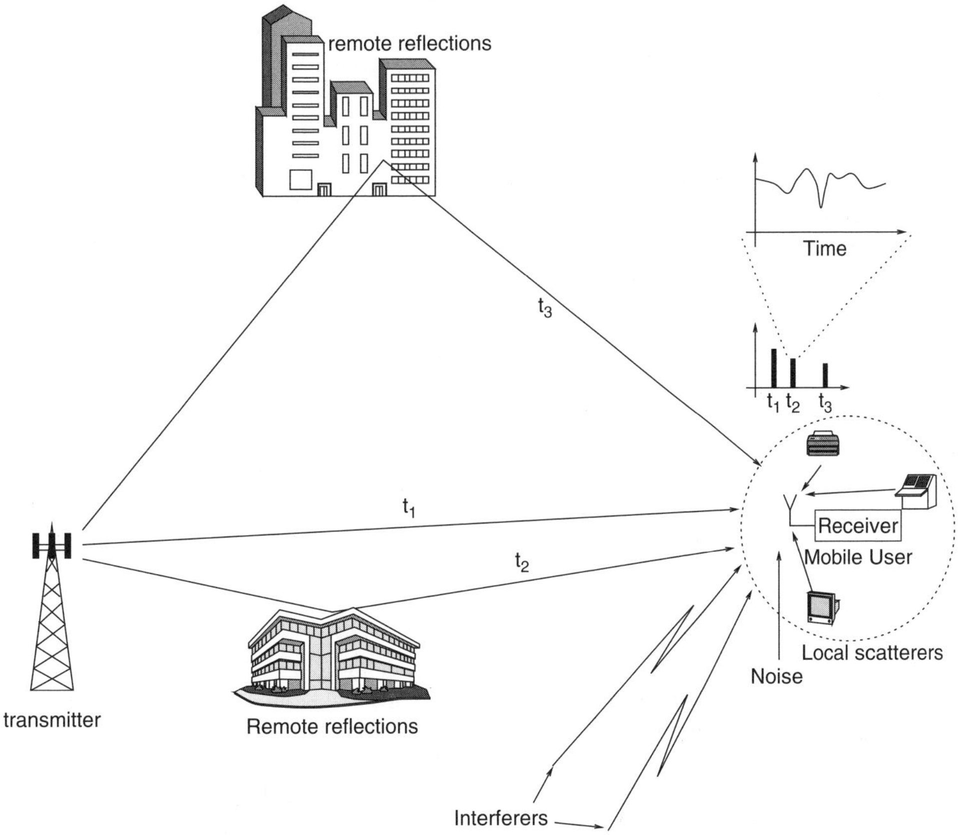

FIGURE 28.1 Illustration of some of the effects of radio channel. Local scatterers cause fading; remote reflectors cause multipath and time dispersion, leading to ISI; mobility of user or scatterers cause time varying channel; reuse of frequencies, adjacent carriers cause interference.

caused by large obstructions, and fading due to reflection, diffraction, and scattering. Mobility of the transmitter, receiver, or scattering objects causes the channel to change over time. Moreover, the interference conditions in the system change rapidly. Most important of all, the radio channel is highly random and the statistical characteristics of the channel are environment dependent. In addition to these changes, the traffic load, type of services, and mobile user characteristics and requirements might also vary in time. Adaptive techniques can be used to address all these changing conditions.

The adaptation strategy can be different depending on the application and services. Constant BER constraint for a given fixed transmission bandwidth and constant throughput constraint are two of the most popular criteria for adaptation. In constant BER, a desired average or instantaneous BER is defined to satisfy the acceptable quality of service. Then the system is adapted to the varying channel and interference conditions so that the BER is maintained below the target value. In order to ensure this for all types of channel and interference conditions, the system changes power, modulation order, coding rate, spreading factor, etc. Note that this changes the throughput as the channel quality changes. On the other hand, for constant throughput case, the adaptations are done to make sure that the effective throughput is constant, where the BER might change.

In general, it is possible to classify the adaptation algorithms as link and transmitter adaptation, adaptation of system resource allocation, and receiver adaptation. In the following sections, brief discussions of these adaptation techniques will be given.

28.2.1 Link and Transmitter Adaptation

A reliable link must ensure that the receiver is able to capture and reproduce the transmitted information bits. Therefore, the target link quality must be maintained all the time in spite of the changes in the channel and interference conditions. As mentioned earlier, one way to achieve this is to design the system for the worst-case scenario so that the target link quality can always be achieved.

If the transmitter sends more power for a specific user, the user benefits from it by having a better link quality, but the level of interference for the other users increases accordingly. On the other hand, if the user does not receive enough power, a reliable link cannot be established. In order to establish a reliable link while minimizing interference to other users, the transmitter should continuously control the transmitted power level. Power control is a simple form of adaptation that compensates for the variation of the received signal level due to path loss, shadowing, and sometimes fading. Numerous studies on power control schemes have been performed for various radio communications systems (see [1] and the references listed therein). In code division multiple-access (CDMA) systems, signals having widely different power levels at the receiver cause strong signals to swamp out weaker ones in a phenomenon known as the near–far effect. Power control mitigates the near–far problem by controlling the transmitted power.

It is possible to trade off power for bandwidth efficiency; i.e., a desired BER (or FER) can be achieved by increasing the power level or by reducing the bandwidth efficiency. One way of establishing a reliable link is to add redundancy to the information bits through FEC techniques. With no other changes, this would normally reduce the information rate (or bandwidth efficiency) of the communication. In the same way, high-quality links can be obtained by transmitting the signals with spectrally less efficient modulation schemes, like binary phase shift keying (BPSK) and quaternary PSK (QPSK). On the other hand, new-generation wireless systems aim for higher data rates made possible through spectrally efficient higher-order modulations. Therefore, a reliable link with higher information rates can be accomplished by continuously controlling the coding and modulation levels. Higher modulation orders with less powerful coding rates are assigned to users that experience good link qualities, so that the excess signal quality can be used to obtain higher data rates. Recent designs have exploited this with adaptive modulation techniques that change the order of the modulation [1], [2], as well as with adaptive coding schemes that change the coding rate [3], [4]. For example, the Enhanced General Packet Radio Service (EGPRS) standard introduces both Gaussian minimum shift keying (GMSK) and 8-PSK modulations with different coding rates through link adaptation and hybrid automatic repeat request (ARQ) [5]. The channel quality is estimated at the receiver, and the information is passed to the transmitter through appropriately defined messages. The transmitter adapts the coding and modulation based on this channel quality feedback. Similarly, variable spreading and coding techniques are present in third-generation CDMA-based systems [3], cdma2000 and wideband CDMA (WCDMA, or UMTS — Universal Mobile Telecommunications System). Higher data rates can be achieved by changing the spreading factor and coding rate, depending on the perceived communication link qualities.

Adaptive antennas and adaptive beam-forming techniques have also been studied extensively to increase the capacity and to improve the performance of wireless communications systems [6]. The adaptive antenna systems shape the radiation pattern in such a way that the information is transmitted (for example, from a base station) directly to the mobile user in narrow beams. This reduces the probability of another user experiencing interference in the network, resulting in improved link quality, which can also be translated into increased network capacity. Although adaptive beam forming is an excellent way to utilize multiple-antenna systems to enhance the link quality, recently different flavors of the usage of multiantenna systems have gained significant interest. Space–time processing and multiple-input multiple-output (MIMO) antenna systems are some new developments that will allow further usage of multiple-antenna systems in wireless communications. Adaptive implementation of these technologies is important for successful and efficient integration of them into wireless communications systems.

28.2.2 Adaptive System Resource Allocation

In addition to physical link adaptation, system resources can also be allocated adaptively to reduce the interference and to improve the overall system quality. This includes adaptive power control, adaptive channel allocation, adaptive cell assignment, adaptive resource scheduling, adaptive spectrum management, congestion, handoff (mobility), admission, and load control strategies. Adaptive system resource allocation considers the current traffic load, as well as the channel and interference conditions. For example, the system could assign more resources to the mobiles that have better link quality to increase the throughput. Alternatively, the system could assign the resources to the user in such a way that the user experiences better quality for the current traffic condition.

Adaptive channel allocation and adaptive cell assignment in hierarchical cellular systems have been studied since the early days of cellular systems. Adaptive channel allocation increases the system capacity through efficient channel utilization and decreased probability of blocked calls [7]. Unlike fixed channel allocation, where the channels are assigned to the cells permanently and the assignment is done based on the worst-case scenario, in adaptive channel assignment, a common pool of channels is shared by many cells, and the channels are assigned with regard to the interference and traffic conditions.

Adaptive cell assignment can increase capacity without increasing the handoff rate. The cells can be assigned to the users depending on their mobility level. Fast-moving mobiles can be assigned to larger umbrella cells (to reduce the number of handoffs), while slow-moving mobiles are assigned to microcells (to increase capacity) [8].

Recently, research on increasing the average throughput of the system through water-filling-based resource allocation has gained significant interest [9], [10], [11]. The main idea is to allocate more resources to the users that experience better link quality, resulting in very efficient use of the available resources. The high-data-rate (HDR) system, which is based on a best-effort radio packet protocol, uses a water-filling-based approach in allocating system resources. Algorithms that deal with compromising the throughput to achieve fairness have also been studied [10], [11].

28.2.3 Receiver Adaptation

Digital wireless communication receiver performance is related to the required value of the signal-to-interference-plus-noise ratio (SINR) so that the BER (or FER) performance can be kept below a certain threshold for reliable communication. For a given complexity, if receiver A requires lower SINR than receiver B to satisfy the same error rate, receiver A is considered to perform better than receiver B.

Receiver adaptation techniques can increase the performance of the receiver, hence reducing the minimum required SINR. As mentioned before, this can be used to increase the coverage area for a fixed transmitted power, or it can be used to reduce the transmitted power requirement for a given coverage area. Moreover, receiver adaptation can reduce the average receiver complexity and the power drain from the battery for the same quality of service. In order to satisfy the desired BER performance, instead of running a computationally complex algorithm for all channel conditions, the receiver can choose the most appropriate algorithm given the system and channel conditions.

Advanced baseband signal processing techniques play a significant role in receiver adaptation. Baseband algorithms used for time and frequency synchronization, baseband filtering, channel estimation and tracking, demodulation and equalization, interference cancellation, soft information calculation, antenna selection and combining, decoding, etc., can be made adaptive depending on the channel and interference conditions.

Conventional receiver algorithms are designed for the worst-case channel and interferer conditions. For example, the channel estimation and tracking algorithms assume the worst-case mobile speed; the channel equalizers assume the worst-case channel dispersion; the interference cancellation algorithms assume that the interferer is always active and constant; and so on. Adaptive receiver design measures the current channel and interferer conditions and tunes the specific receiver function that is most appropriate for the current conditions. For example, a specific demodulation technique may work well in some channel conditions, but might not provide good performance in others. Hence, a receiver might include a variety of demodulators that are individually tuned to a set of channel classes. If the receiver could demodulate the data reliably with a simpler and less complex receiver algorithm under the given conditions, then it is desired to use that algorithm for demodulation.

28.3 Parameter Measurements

Many adaptation techniques require estimation of various quantities like channel selectivity, link quality, network load and congestion, etc. Here, we focus more on physical-layer measurements from a digital signal processing perspective. As discussed earlier, link quality measures have many applications for various adaptation strategies. In addition, information on channel selectivity in time, frequency, and space is very useful for adaptation of wireless communications systems. In this section, these important parameters and their estimation techniques will be discussed.

28.3.1 Channel Selectivity Estimation

In wireless communications, the transmitted signal reaches the receiver through a number of different paths. Multipath propagation causes the signal to be spread in time, frequency, and angle. These spreads, which are related to the selectivity of the channel, have significant implications on the received signal. A channel is considered to be selective if it varies as a function of time, frequency, or space. The information on the variation of the channel in time, frequency, and space is very crucial in adaptation of wireless communications systems.

28.3.1.1 Time Selectivity Measure: Doppler Spread

Doppler shift is the frequency shift experienced by the radio signal when either the transmitter or receiver is in motion, and Doppler spread is a measure of the spectral broadening caused by the temporal rate of change of the mobile radio channel. Therefore, time-selective fading and Doppler spread are directly related. The coherence time of the channel can be used to characterize the time variation of the time-selective channel. It represents the statistical measure of the time window over which the two signal components have strong correlation, and it is inversely proportional to the Doppler spread. Figure 28.2 shows the effect of mobile speed on channel variation and channel correlation in time, as well as the corresponding Doppler spread values in frequency domain.

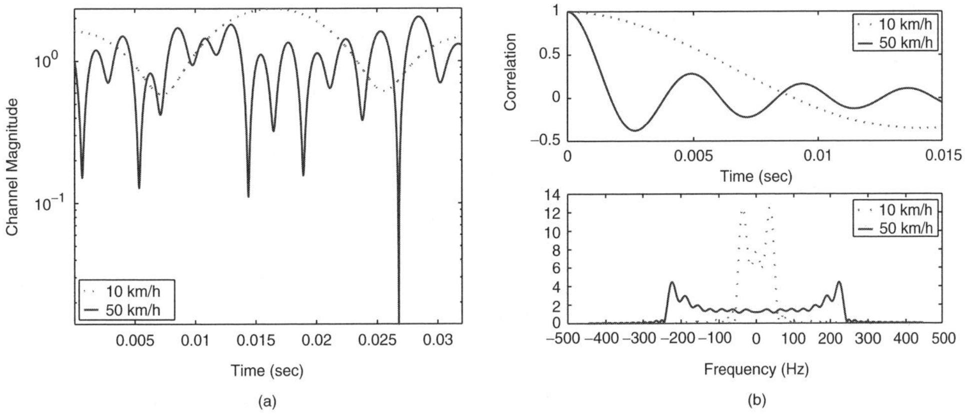

FIGURE 28.2 Illustration of the effect of mobile speed on time variation, time correlation, and Doppler spread of radio channel. (a) Channel time variation for different mobile speeds. (b) Time correlation of channel as a function of the time difference (separation in time) between the samples, and the corresponding Doppler spectrum in frequency.

In an adaptive receiver, Doppler information can be used to improve performance or reduce complexity. For example, in channel estimation algorithms, whether using channel trackers or channel interpolators, instead of fixing the tracker or interpolation parameters for the worst-case Doppler spread value (as commonly done in practice), the parameters can be optimized adaptively based on Doppler spread information [12], [13]. Similarly, Doppler information could be used to control the receiver or transmitter adaptively for different mobile speeds, like variable coding and interleaving schemes [14]. Also, radio network control algorithms, such as handoff, cell assignment, and channel allocation in cellular systems, can utilize the Doppler information [8]. For example, as will be described later, in a hierarchical cell structure, the users are assigned to cells based on their speeds (mobility).

Doppler spread estimation has been studied for several applications in wireless mobile radio systems. Correlation and variation of channel estimates as well as correlation and variation of the signal envelope have been used for Doppler spread estimation [12]. One simple method for Doppler spread estimation is to use *differentials* of the complex channel estimates [15]. The differentials of the channel estimates are very noisy, which require low-pass filtering. The bandwidth of the low-pass filter is also a function of the Doppler estimate. Therefore, such approaches require adaptive receivers that continuously change the filter bandwidth depending on the previously obtained Doppler value. A Doppler estimation scheme based on the autocorrelation of complex channel estimates is described in [16]. Also, a maximum likelihood estimation-based approach, given the channel autocorrelation estimate, is utilized for Doppler spread estimation in [17]. Channel autocorrelation is calculated using the channel estimates over the known field of the transmitted data.

Instead of using channel estimates, the received signal can also be used directly in estimating Doppler spread information. In [18], the Doppler frequency is extracted from the samples of the received signal envelope. Doppler information is calculated as a function of the squared deviation of the signal envelope. Similarly, in [19] the mobile speed is estimated as a function of the deviation of the averaged signal envelope in flat fading channels. For dispersive channels, pattern recognition, using the variation of pattern mean, can be used to quantify the deviation of signal envelope. In [20], the filtered received signal is used to calculate the channel autocorrelation values over each slot. Then, the autocorrelation estimate is used for identification of high- and low-speed mobiles. In [21], multiple antennas are exploited, where a linear relation between the switching rate of the antenna branches and Doppler frequency is given. Also, the level crossing rate of the average signal level has been used in estimating velocity[22], [23].

28.3.1.2 Frequency Selectivity Measure: Delay Spread

The multipath signals that reach the receiver have different delays as the paths that the signals travel through have different lengths. When the relative path delays are on the order of a symbol period or more, images of different transmitted symbols arrive at the same time, causing intersymbol interference (ISI). Delay spread is one of the most commonly used parameters that describes the time dispersiveness of the channel, and it is related to frequency selectivity of the channel. The frequency selectivity can be described in terms of coherence bandwidth, which is a measure of range of frequencies over which the two frequency components have a strong correlation. The coherence bandwidth is inversely proportional to the delay spread [24]. Figure 28.3 shows the effect of time dispersion on channel frequency variation and channel frequency correlation, as well as the corresponding power delay profiles.

Like time selectivity, the information about the frequency selectivity of the channel can be very useful for improving the performance of the adaptive wireless radio systems. For example, in a time division multiple-access (TDMA)-based Global System for Mobile Communications (GSM), the number of channel taps needed for equalization might vary depending on channel dispersion. Instead of fixing the number of channel taps for the worst-case channel condition, we can change them adaptively [25], allowing simpler receivers with reduced battery consumption and improved performance. Similarly, in [26], a TDMA receiver with adaptive demodulator is proposed, using the measurement about the dispersiveness of the channel. Dispersion estimation can also be used for other parts of transmitters and receivers. For example, in frequency domain channel estimation using channel interpolators, instead of fixing the interpolation parameters for the worst expected channel dispersion, we can change the parameters adaptively depending on the dispersion information [27].

Although dispersion estimation can be very useful for many wireless communications systems, it is particularly crucial for orthogonal frequency division multiplexing (OFDM)-based wireless communications systems. OFDM, which is a multicarrier modulation technique, handles the ISI problem due to high-bit-rate communication by splitting the high-rate symbol stream into several lower-rate streams and transmitting them on different orthogonal carriers. The OFDM symbols with increased duration might still be affected by the previous OFDM symbols due to multipath dispersion. Cyclic prefix extension of the OFDM symbol avoids ISI from the previous OFDM symbols if the cyclic prefix length is greater than the maximum excess delay of the channel. Since the maximum excess delay depends on the radio environment, the cyclic prefix length needs to be designed for the worst-case channel condition. This makes the cyclic prefix a significant portion of the transmitted data, thereby reducing spectral efficiency. One way to increase spectral efficiency is to adapt the length of the cyclic prefix depending on the radio environment [28].

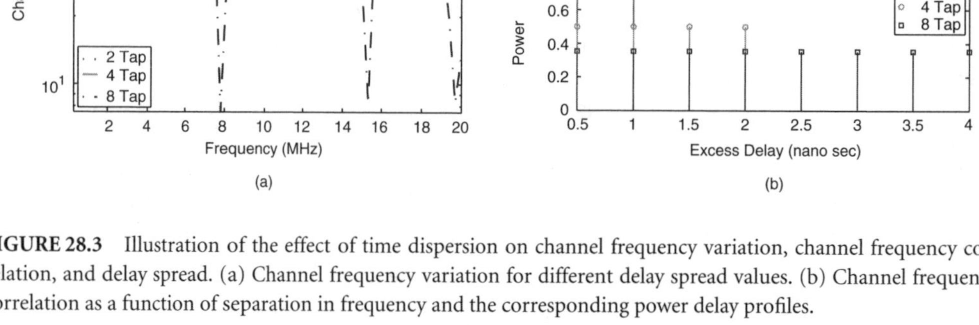

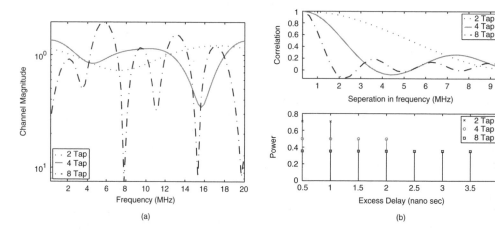

FIGURE 28.3 Illustration of the effect of time dispersion on channel frequency variation, channel frequency correlation, and delay spread. (a) Channel frequency variation for different delay spread values. (b) Channel frequency correlation as a function of separation in frequency and the corresponding power delay profiles.

The adaptation requires estimation of maximum excess delay of the radio channel, which is also related to the frequency selectivity of the channel. In HiperLAN2, which is a wireless local area network (WLAN) standard, a cyclic prefix duration of 800 ns, which is sufficient to allow good performance for channels with delay spread up to 250 ns, is used. Optionally, a short cyclic prefix with 400-ns duration may be used for short-range indoor applications. Delay spread estimation allows adaptation of these various options to optimize the spectral efficiency. Other OFDM parameters that could be changed adaptively using the knowledge of the dispersion include OFDM symbol duration and OFDM subcarrier bandwidth.

Characterization of the frequency selectivity of the radio channel is studied in [29], [30], [31] using the level crossing rate (LCR) of the channel in frequency domain. Frequency domain LCR gives the average number of crossings per Hertz at which the measured amplitude crosses a threshold level. An analytical expression between LCR and the time domain parameters corresponding to a specific multipath power delay profile (PDP) is given. LCR is very sensitive to noise, which increases the number of level crossing and severely deteriorates the performance of the LCR measurement [31]. Filtering the channel frequency response reduces the noise effect, but finding the appropriate filter parameters is an issue. If the filter is not designed properly, one might end up smoothing the actual variation of frequency domain channel response. In [27], instantaneous root mean square (rms) delay spread, which provides information about local (small-scale) channel dispersion, is obtained by estimating the channel impulse response (CIR) in the time domain. The detected symbols in the frequency domain are used to regenerate the time domain signal through inverse fast Fourier transform (IFFT). This signal is then used to correlate the actual received signal to obtain CIR, which is then used for delay spread estimation. Since the detected symbols are random, they might not have good autocorrelation properties, which can be a problem, especially when the number of carriers is low. In addition, the use of detected symbols for correlating the received samples to obtain CIR provides poor results for low SNR values. In [28], the delay spread is also calculated from the instantaneous time domain CIR, wherein the CIR is obtained by taking IFFT of the frequency domain channel estimate. Channel frequency selectivity and delay spread information are calculated using the channel frequency correlation estimates in [24], [32]. An analytical expression between delay spread and coherence bandwidth is also given.

The level of time dispersion can be obtained by using known training sequences and a maximum likelihood-based algorithm. The channel can be modeled with different levels of dispersion. Using these various channel models, the corresponding channel estimates and the residual error can be calculated. From these residual error terms, a decision can be made about the level of dispersion. Note that when the channel is overmodeled, the residual error also becomes smaller. Hence, it is not necessarily true that the model that provides the smaller residual error is the most suitable one. The most appropriate model can be found by several information criteria algorithms, like Bayesian information criteria (BIC) or Akaike information criteria (AIC) [33].

28.3.1.3 Spatial Selectivity Measure: Angle Spread

Angle spread is a measure of how multipath signals are arriving (or departing) with respect to the mean arrival (departure) angle. Therefore, angle spread refers to the spread of angles of arrival (or departure) of the multipaths at the receiving (transmitting) antenna array [34]. Angle spread is related to the spatial selectivity of the channel, which is measured by coherence distance. Like coherence time and frequency, coherence distance provides the measure of the maximum spatial separation over which the signal amplitudes have strong correlation, and it is inversely proportional to angular spread, i.e., the larger the angle spread, the shorter the coherence distance. Figure 28.4 shows the effect of local scattering on angle of arrival. The local scattering in the vicinity of Receiver-2 results in larger angular spreads, as the received signals come from many different directions due to a richer local scattering environment. For a given receiver antenna spacing, this leads to less antenna correlations between the received antenna elements than the correlation of antennas in Receiver-1. Note that although the angular spread is described independent of the other channel selectivity values for the sake of simplicity, in reality the angle of arrival can be related to the path delay. The multipath components that arrive at the receiver earlier (with shorter delays) are expected to have similar angles of arrival (lower angle spread values).

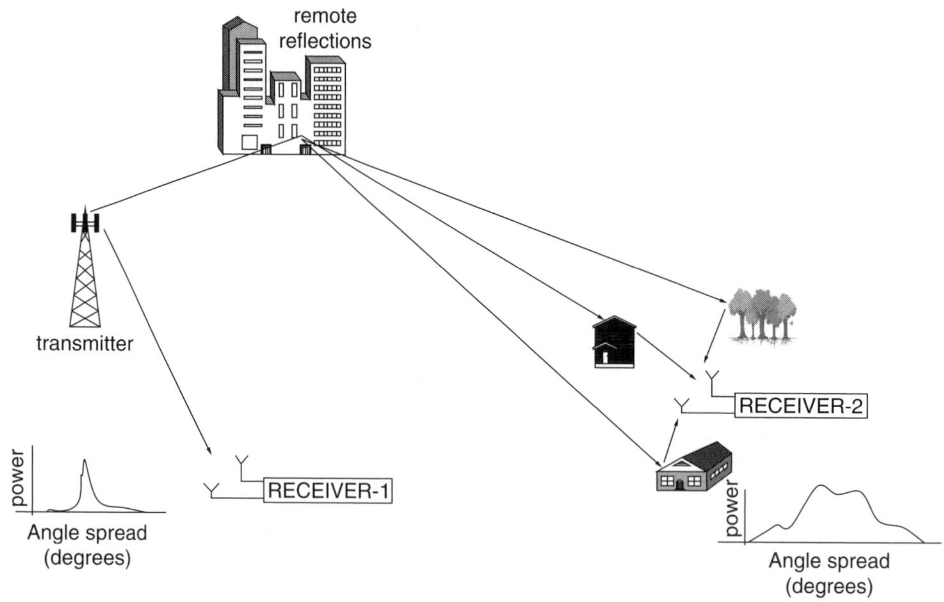

FIGURE 28.4 Illustration of the effect of different local scattering in angle of arrivals. Receiver-1 observes less angle spread compared to Receiver-2. Therefore, receiver antennas in Receiver-1 will have more correlations.

Compared to time and frequency selectivity, spatial selectivity has not been studied widely in the past. However, recently there has been a significant amount of work in multiantenna systems. With the widespread application of multiantenna systems, it is expected that the need for understanding spatial selectivity and related parameter estimation techniques will gain momentum. Spatial selectivity will especially be useful when the requirement for placing antennas close to each other increases, as in the case of multiple antennas at the mobile units.

Spatial correlation between multiple-antenna elements is related to the spatial selectivity, antenna distance, mutual coupling between antenna elements, antenna patterns, etc. [35], [36]. Spatial correlation has significant effects on multiantenna systems. Full capacity and performance gains of multiantenna systems can only be achieved with low antenna correlation values. However, when this is not possible, maximum capacity can be achieved by employing efficient adaptation techniques. Adaptive power allocation is one way to exploit the knowledge of the spatial correlation to improve the performance of multiantenna systems [37]. Similarly, adaptive modulation and coding, which employs different modulation and coding schemes across multiantenna elements depending on the channel correlation, is possible [38], [39]. In MIMO systems, adaptive power allocation has been studied by using the knowledge of channel matrix estimate and eigenvalue analysis [40] [41].

28.3.2 Channel Quality Measurements

Channel quality estimation is by far the most important measurement that can be used in adaptive receivers and transmitters [3]. Different ways of measuring the quality of radio channel are possible, and many of these measurements are done in the physical layer using baseband signal processing techniques. In most of the adaptation algorithms, the target quality measure is the FER or BER, as these are closely related to higher-level quality-of-service parameters like speech and video quality. However, reliable measurement of these qualities requires many measurements, and this causes delays in the adaptation as the process could be very long. Therefore, other types of channel quality measurements that are related to these might be preferred. When the received signal is impaired only by white Gaussian noise, analytical expressions can be found relating the BER to other measurements. For other impairment cases, like colored interferers,

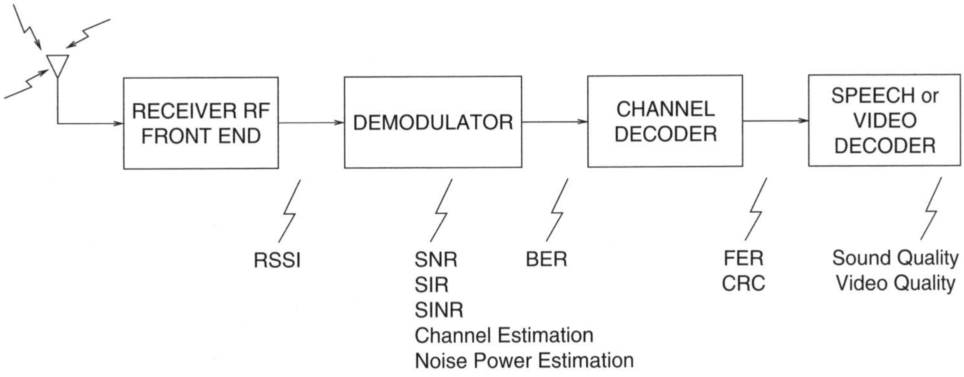

FIGURE 28.5 A simple wireless receiver that shows the estimation points of commonly used parameters.

numerical calculations and computer simulations that relate these measurements to BER can be performed. Therefore, depending on the system, a channel quality is related to the BER. Then, for a target BER (or FER), a required signal quality threshold can be calculated to be used with the adaptation algorithm.

The measurements can be performed at various points of a receiver, depending on the complexity, reliability, and delay requirements. There are trade-offs in achieving these requirements at the same time. Figure 28.5 shows a simple example where some of these measurements can take place. In the following sections, these measurements will be discussed briefly.

28.3.2.1 Measures before Demodulation

Received signal strength (RSS) estimation provides a simple indication of the fading and path loss, and provides the information about how strong the signal is at the receiver front end. If the received signal strength is stronger than the threshold value, then the link is considered to be good. Measuring the signal strength of the available radio channels can be used as part of the scanning and intelligent roaming process in cellular systems. Also, other adaptation algorithms, like power control and handoff, can use this information. The RSS measurement is simply done by reading samples from a channel and averaging them [42]. Compared to other measurements, RSS estimation is simple and computationally less complex, as it does not require the processing and demodulation of the received samples. However, the received signal includes noise, interference, and other channel impairments. Therefore, receiving a good signal strength does not tell much about the channel and signal quality. Instead, it gives an indication of whether a strong signal is present in the channel of interest. For the measurement of RSS, the transmitter might send a pilot signal continuously, as in the WCDMA cellular system, or a link layer beacon can be transmitted at discrete time intervals, as in IEEE 802.11 WLANs.

Since the received signal power fluctuates rapidly due to fading, in order to obtain reliable estimates, the signal needs to be averaged over a time window to compensate for short-term fluctuations. The averaging window size depends on the system, application, variation of the channel, etc. For example, if multiple receiver antennas are involved at the receiver, the window can be shorter than that for a single-antenna receiver.

28.3.2.2 Measures during and after Demodulation

The signal-to-interference ratio (SIR), SNR, and SINR are the most common ways of measuring the channel quality during (or just after) the demodulation of the received signal. SIR (or SNR or SINR) provides information on how strong the desired signal is compared to the interferer (or noise or interference plus noise). Most wireless communications systems are interference limited; therefore, SIR and SINR are more commonly used. Compared to RSS, these measurements provide more accurate and reliable estimates at the expense of computational complexity and with additional delay.

There are many adaptation schemes where these measurements can be exploited. Link adaptation (adaptive modulation and coding, rate adaptation, etc.), adaptive channel assignment, power control, adaptive channel estimation, and adaptive demodulation are only a few of many applications.

SIR estimation can be employed by estimating signal power and interference power separately and then taking the ratio of these two. In many new-generation wireless communications systems, coherent detection, which requires estimation of channel parameters, is employed. These channel parameter estimates can also be used to calculate the signal power. The training (or pilot) sequences can be used to obtain the estimate of SIR. Instead of the training sequences, the data symbols can also be used for this purpose. For example, in [43], where SNR information is used as a channel quality indicator for rate adaptation, the cumulative Euclidean metric corresponding to the decoded trellis path is exploited for channel quality information. Another method for channel quality measurement is the use of the difference between the maximum likelihood decoder metrics for the best path and second-best path, as described in [44]. In a sense, in this technique, some sort of soft information is used for the channel quality indicator. However, this approach does not tell much about the strength of the interferer or the desired signal. There are several other ways of SNR measurement that are based on subspace projection techniques. These approaches can be found in [45] and in the references cited therein.

Often, in obtaining the estimates, the impairment (noise or interference) is assumed to be white and Gaussian distributed to simplify the estimation process. However, in wireless communications systems, the impairment might be caused by a strong interferer, which is colored. For example, in OFDM systems, where the channel bandwidth is wide and the interference is not constant over the whole band, it is very likely that some part of the spectrum is affected more by the interferer than the other parts. Figure 28.6 shows the OFDM frequency spectrum and two types of noise over this spectrum: colored and white. Hence, when the impairment is colored, estimates that take the color of the impairment into account might be needed [46].

Note that since both the desired signal's channel and interferer conditions change rapidly, depending on the application, both short-term and long-term estimates are desirable. Long-term estimates provide information on long-term fading statistics due to shadowing and lognormal fading as well as average interference conditions. On the other hand, short-term estimates provide measurements of instantaneous channel and interference conditions. Applications like adaptive channel assignment and handoff prefer long-term statistics, whereas applications like adaptive demodulation, adaptive interference cancellation, etc., prefer short-term statistics.

For some applications, a direct measure of channel quality from channel estimates would be sufficient for adaptation. As mentioned above, channel estimates only provide information about the desired

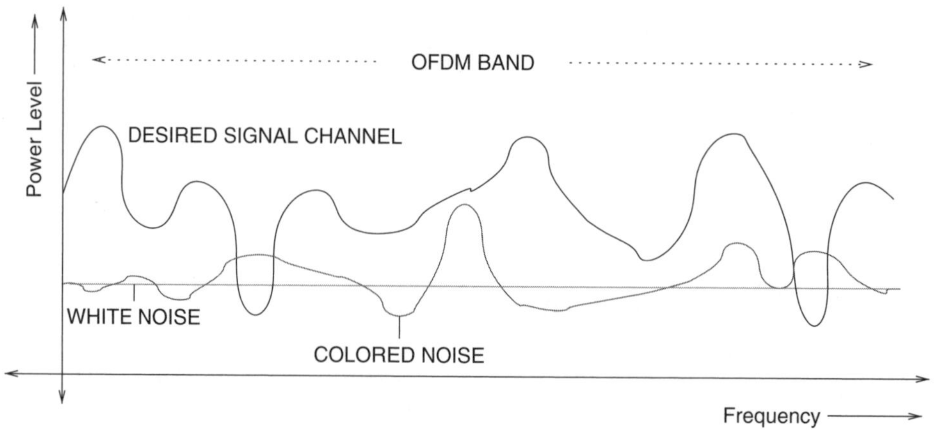

FIGURE 28.6 Representation of OFDM frequency channel response and noise spectrum. Spectrum for both white and colored noise is shown.

signal's power. It is a much more reliable estimate than RSS information, as it does not include the other impairments as part of the desired signal power. However, it is less reliable than SNR (or SINR) estimates, since it does not provide information about the noise or interference powers with respect to the desired signal's power.

Channel estimation for wireless communications systems has a very rich history. A significant amount of work has been done for various systems. In many systems, known information (like pilot symbols, pilot channels, pilot tones, training sequences, etc.) is transmitted along with the unknown data to help the channel estimation process. Blind channel estimation techniques that do not require known information transmission have also been studied extensively. For details on channel estimation for wireless communications systems, refer to [47], [48] and the references listed therein.

28.3.2.3 Measures after Channel Decoding

Channel quality measurements can also be based on postprocessing of the data (after demodulation and decoding). BER, symbol-error-rate (SER), FER, and CRC information are some of the examples of the measurements in this category. BER (or FER) is the ratio of the bits (or frames) that have errors relative to the total number of bits (or frames) received during the transmission. The CRC indicates the quality of a frame, which can be calculated using parity check bits through a known cyclic generator polynomial. FER can be obtained by averaging the CRC information over a number of frames. In order to calculate the BER, the receiver needs to know the actual transmitted bits, which is not possible in practice. Instead, BER can be calculated by comparing the bits before and after the decoder. Assuming that the decoder corrects the bit errors that appear before decoding, this difference can be related to BER. Note that the comparison makes sense only if the frame is error-free (good frame), which is obtained from the CRC information.

As mentioned earlier, although these estimates provide excellent link quality measures, reliable estimates of these parameters require observations over a large number of frames. Especially for low BER and FER measurements, extremely long transmission intervals will be needed. Therefore, for some applications these measures might not be appropriate. Note also that these measurements provide information about the actual operating condition of the receiver. For example, for a given RSS or SINR measure, two different receivers that have different performances will have different BER or FER measurements. Therefore, BER and FER measurements also provide information on the receiver capability as well as the link quality.

28.3.2.4 Measures after Speech or Video Decoding

The speech and video quality, the delays on data reception, and network congestion are some of the parameters that are related to user's perception. Essentially, these are the ultimate quality measures that need to be used for adaptive algorithms. However, these parameters are not easy to measure, and in many cases, real-time measurement might not be possible. On the other hand, these measures are often related to the other measures mentioned above. For example, speech quality for a given speech coder can be related to FER of a specific system under certain assumptions [49]. However, as discussed in [49], some frame errors cause more audible damage than others. Therefore, it is still desired to find ways to measure the speech quality more reliably (and timely) and adapt the system parameters accordingly. Speech (or video) quality measures that take the human perception of the speech (or video) into account are highly desirable.

Perceptual speech quality measurements have been studied in the past. Both subjective and objective measurements are available [50]. Subjective measurements are obtained from a group of people who rate the quality of the speech after listening to the original and received speech. Then a mean opinion score (MOS) is obtained from their feedback. Although these measurements reflect the exact human perception that is desired for adaptation, they are not suitable for adaptation purposes because the measurements are not obtained in real time. On the other hand, the objective measurements can be implemented at the receiver in real time [51]. However, these measurements require a sample of the original speech at the receiver to compare the received voice with the original undistorted voice. Therefore, they are also not applicable for many scenarios.

28.4 Applications of Adaptive Algorithms: Case Studies

28.4.1 Examples for Adaptive Receiver Algorithms

In this section, some representative examples for adaptive receiver algorithms will be discussed briefly. These algorithms can be employed in both base stations and mobile terminals, as well as in many other wireless receivers.

28.4.1.1 Channel Estimation with *A Priori* Information

Channel estimation is an integral part of standard adaptive receiver designs used in digital wireless communications systems. For conventional, coherent receivers, the effect of the channel on the transmitted signal must be estimated to recover the transmitted information. As long as the receiver estimates what the channel did to the transmitted signal, it can accurately recover the information sent.

The estimation of time-varying channel parameters is often based on an approximate underlying model of the radio channel. In fading environments, the coefficients of a channel model exhibit typical trends or quasi-periodic behavior in time, frequency, and space. The ability to track channel variation depends on how fast the channel changes in time, frequency, and space. As mentioned before, this is related to Doppler spread (time variation), delay spread (frequency variation), and angle spread (space variation). By utilizing *a priori* information about the channel variation, adaptive algorithms with larger memories can be designed without sacrificing tracking capability [15]. In contrast to the algorithms that do not exploit this information, adaptive algorithms provide a means of extrapolation of the channel coefficients in time, frequency, and space [13], [52]. For example, in [53], the step size of a simple least mean square (LMS) channel tracker is changed using the Doppler spread information. Similarly, the window size of a sliding window (moving average filtering)-based channel tracking algorithm can be adapted depending on Doppler spread and SNR information [54]. Wiener filtering, which is one of the most popular techniques for channel estimation using interpolation, is an excellent example in exploiting *a priori* information, as the optimal Wiener filter design requires knowledge of Doppler spread and noise power. In most conventional Wiener filtering designs, the worst-case expected Doppler spread values are used, degrading the performance of the algorithm for other Doppler spread values [55]. Recently, two-dimensional interpolation using Wiener filtering for OFDM-based wireless communications systems gained significant interest [28]. In this case, both Doppler spread and delay spread information, as well as noise variance estimates, can be used to optimize the channel tracker performance. Although we have mentioned a few examples, the usage of *a priori* information in channel estimation has been considered by many other authors. Further information can be found in [47], [48].

Figure 28.7 shows a simple coherent receiver structure with an adaptive channel tracker. The receiver includes a parameter measurement block that estimates the necessary parameters for the adaptation of

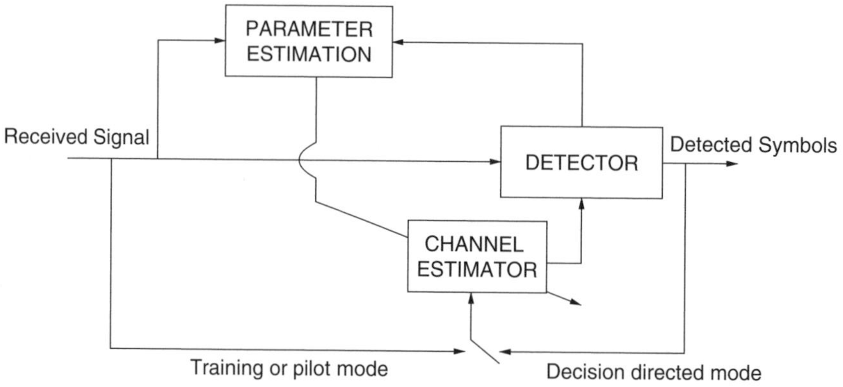

FIGURE 28.7 A simple adaptive channel estimation receiver.

the channel tracker. The necessary parameters can be estimated using the received signal and the output of the detector as described before. The detector requires the channel estimates that can be obtained from the channel tracker.

28.4.1.2 Adaptive Channel Length Truncation for Equalization

Time dispersion in wireless systems can cause ISI, which degrades the performance, often severely. Equalization is a technique used to counter the effects of ISI. In the Telecommunications Industry Association/ Electronics Industry Association/Interim Standard 136 (TIA/EIA/IS-136 or simply IS-136) system, the channel can be assumed to be flat (nondispersive) with respect to the symbol duration most of the time. Equalization does not help much in nondispersive environments and in fact hurts performance by trying to model dispersion that does not exist. However, in hilly terrain channel conditions, the channel is dispersive and requires equalization. Therefore, to design the receiver for the worst-case condition, equalization needs to be used for all the geographical conditions unnecessarily, resulting in a loss due to the mismatch of the implemented receiver to the fading scenario. An adaptive receiver can, on the other hand, have an algorithm that measures the dispersiveness of the channel and uses the appropriate demodulator based on the measurement [26]. This also results in conserving battery power.

In another cellular communications system, GSM, the symbol duration is relatively short compared to that in IS-136. Also, the pulse shaping itself introduces intentional ISI, so that equalization is required even in nondispersive channels. However, the number of channel taps needed for equalization might vary depending on the dispersion (the geographical area). Instead of fixing the number of channel taps for the worst-case condition, the number can be made adaptive [25], allowing simpler receivers with reduced battery consumption and improved performance. Again, the point emphasized here is to avoid overmodeling the signal. Figure 28.8 shows a simple example of an adaptive receiver that measures the level of dispersion and adapts the equalizer number of taps accordingly.

28.4.1.3 Adaptive Interference Cancellation Receivers

The impairment sources in wireless mobile radio systems are numerous. Co-channel interference, which is caused by the reuse of carrier frequencies in nearby cells, is one of the major contributors. Another major interference source is adjacent channel interference, which is caused by the spectral overlap between adjacent channel users. Also, thermal noise and other impairment sources that are commonly modeled as additive white Gaussian noise (AWGN) degrade the performance of a receiver. The statistics of these disturbance sources are different. Conventional receivers commonly assume that the impairment at the receiver is white, which causes performance loss if the actual impairment is colored. By exploiting the statistics of the impairments, better receivers can be designed. For example, *interference whitening* is one such technique that partially suppresses the interference and optimizes the demodulator performance. However, at any given time, the kind of disturbance that is dominant at the receiver is not known before. In order to achieve the best possible performance in all situations, the receiver should estimate the possible disturbance source and adapt the receiver to the second-order statistics of the impairment.

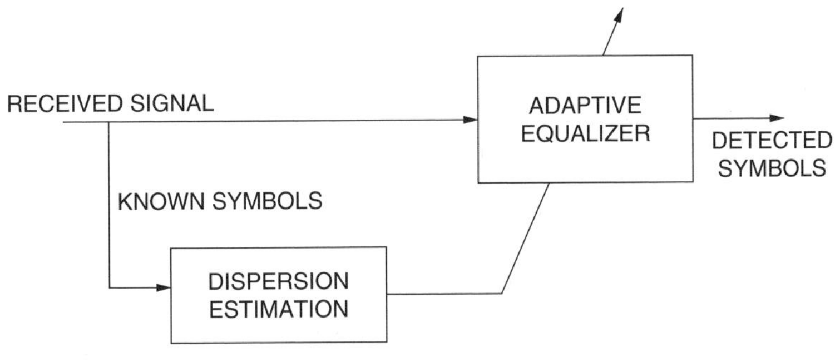

FIGURE 28.8 An adaptive receiver that uses the delay spread (time dispersion) estimate to adjust the equalizer.

Such an adaptive receiver described in [56] improves the performance of the maximum likelihood based receiver.

The interference can also be suppressed by employing interference cancellation techniques in the receivers. For example, joint demodulation (JD) of co-channel signals is a powerful technique for cancelling co-channel interference. In [57], it was shown that the capacity of the IS-136 system can be increased significantly by using a JD receiver. However, the JD receiver given in [57] works well only when there is a single dominant interferer, the mobile speed is low, and the channel is nondispersive. Otherwise, the conventional single-user demodulator (CD) works better than joint demodulation at the targeted operating SINR level. A simple and efficient solution to the above problem is an adaptive receiver that adapts the detector to the system conditions. Figure 28.9 illustrates the schematic of such an adaptive receiver. It contains the conventional detector, the joint detector, and a control unit to control the two detectors. For each slot, the control unit determines which of the two demodulators to be used to recover the data symbols of the desired user. The control unit makes this decision on the basis of certain information obtained from conventional and joint acquisitions. The demodulator selected by the control unit outputs an estimate for the data symbols of the desired user. The details regarding the two demodulators can be found in [58].

The choice for the demodulator can be based on several criteria. Ideally, one would like to know the SNR, SIR, dominant interferer ratio ($\frac{I_1}{I-I_1}$, where I_1 is the dominant interferer and I is the total impairment, including the dominant interferer), and extent of ISI present in the system, among other parameters. Although these quantities are not generally available at the receiver, they can be estimated. For example, carrier and dominant interferer powers are estimated by averaging the corresponding channel tap strengths over multiple slots. The unmodeled impairment power is estimated from the accumulated Euclidean distance metric during the acquisition process (joint or conventional) over the training sequence of the desired signal.

28.4.1.4 Adaptive Soft Information Generation and Decoding

In digital wireless communications systems, forward error correction encoding is commonly used to provide a robust communication link. At the receiver, the decoder performance is optimized when the demodulator provides soft information for the encoded bits. The better soft information generation schemes require the knowledge of the noise covariance, and often the noise covariance changes across the interleaving length. Therefore, a receiver should continuously measure the noise covariance and use these estimates for the improvement of soft bit values.

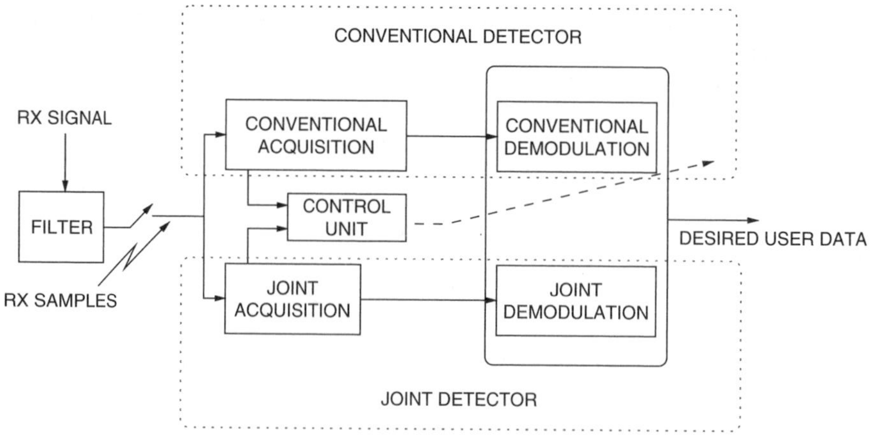

FIGURE 28.9 Example for adaptive interference cancellation receiver. A complex joint demodulation and a less complex single user demodulation used adaptively based on the measured parameters.

28.4.2 Examples for Link Adaptation and Adaptive Resource Allocation

In this section, some examples for adaptation of radio link and adaptive resource allocation will be discussed briefly. Examples in this area are numerous, and there has been a significant amount of research on this area.

28.4.2.1 Adaptive Power Control

Power control has a long and rich history in wireless communications systems [59], [60], [61]. Specifically, for CDMA-based cellular systems, adaptive power control has a significant role, as the performance and capacity of the CDMA systems are normally interference limited. Without power control, an interfering transmitter that is closer to the receiver than the desired signal's transmitter will cause a significant degradation, and this phenomenon is commonly referred to as the *near–far problem*. Power control handles this problem by adaptively controlling the user's power depending on the link quality and desired quality of service (QoS). As a result, the interference observed by other users due to this user will be less, which in turn reduces the average interference observed at the receivers. This results in a high-capacity system with improved battery life for the mobile terminals.

In voice-dominated cellular systems, the objective of the power control was mainly to maintain the minimal (target) link quality at a constant level for individual users. The data rates for all users are constant in this case, and each user experiences roughly the same quality of service. While this was appropriate for voice, recently, with the increased demand for multimedia services and high-speed data access, different objectives and cost functions to optimize the use of power resources have been developed. In mixed-traffic environments, the cost function for each service will be different, leading to different power allocation strategies [62]. Use of constant power along with variable coding, modulation, and spreading that adapts the data rate to the channel variations is one objective that some new-generation wireless systems have been adopting (rate control or rate adaptation) [43]. Also, water-filling types of power assignments, which assign more power to the users that have favorable channels, are being studied extensively [63].

In adaptive power control mechanisms, estimation of the link quality parameters is the key factor. Typical parameters used for adaptation include SIR, FER, and RSS. Doppler spread estimate can also be used to adjust the adaptation rate. Depending on the adaptation rate, power control can be classified as *fast power control* and *slow power control*. Fast power control compensates the changes in power level due to Rayleigh fading (small-scale fading), while slow power control is used for lognormal fading (shadowing) and path loss. The parameters that are used for them can also be different. For example, for fast power control, instantaneous SIR, SNR, SINR, and RSS can be more suitable than FER and BER, which might suit better slow power control. As mentioned in the parameter estimation section, parameter selection depends on the delay, complexity, and accuracy requirements. The estimation errors and delays, between measurements and adaptation of power, limit the efficient application of power control schemes. Therefore, more accurate and practical algorithms that estimate and predict the parameters to be used in adaptation are needed.

28.4.2.2 Adaptive Modulation and Channel Coding

Given the high price of spectrum and its scarcity, it is in the interest of operators to continue evolving their networks toward higher capacity and quality. Adaptive modulation and coding provide a framework to adjust modulation level and FEC coding rate depending on the link quality. Higher-order modulations (HOMs) allow more bits to be transmitted for a given symbol rate. On the other hand, HOM is less power efficient, requiring higher energy per bit for a given BER. Therefore, HOMs should be used only when the link quality is high, as they are less robust to channel impairments. Similarly, strong FEC and interleaving provide robustness against channel impairments at the expense of lower data rate and spectral efficiency, suggesting adaptation of coding rate based on the link quality. Figure 28.10 illustrates the capacity gain that can be achieved by employing adaptive modulation only. First, the BER performances of different modulations as a function of SNR are given in Figure 28.10(a). As can be seen, a desired BER can be achieved with low-order modulations for lower SNRs. Higher-order modulations need better link quality (higher

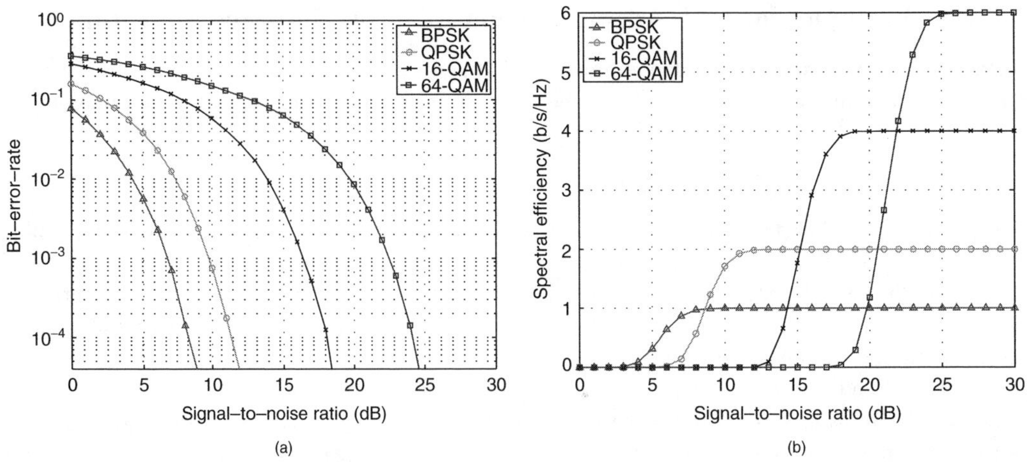

FIGURE 28.10 Illustration of the BER and spectral efficiency of several modulation options. (a) BER plots of different modulations as a function of SNR. (b) Spectral efficiency of different modulation as a function of SNR.

SNRs) in order to obtain the same BER performance. Figure 28.10(b) shows the spectral efficiencies of different uncoded modulations, where an arbitrary packet size of 200 bits is used. Notice that the optimal spectral efficiency for different SNR regions can be obtained through the use of different modulations depending on the SNR.

Link adaptation using adaptive coding and modulation is deployed in some of the new-generation wireless communications systems. For example, EGPRS, which is the evolution of the second-generation GSM, employs two different modulation options (GMSK and 8-PSK) along with different coding rates, resulting in nine different modulation/coding options, as shown in Table 28.1 [5], [43]. In addition, EGPRS introduces the use of a Type II Hybrid ARQ system, commonly known within the specification as incremental redundancy. In link adaptation, the link quality is measured regularly and the most appropriate modulation and coding scheme is assigned for the next transmission interval. On the other hand, in the incremental redundancy scheme, information is first sent with low coding power (high coding rate). This results in a high bit rate if decoding is successful with this rate. However, if decoding fails with such a high rate, additional coded bits (redundancy) should be send so that the transmitted bits can be decoded successfully. However, sending extra coded bits incrementally reduces the resulting bit rate and introduces undesired extra delay. Therefore, the initial code rate and modulation for the incremental redundancy scheme should be based on measurements of the link quality, instead of starting with any arbitrary rate [5]. As a result, by combining incremental redundancy with adaptive initial code rate, lower delays with lower memory requirements, and high data rates can be achieved. The different initial code rates are obtained by puncturing a different number of bits from a common convolution code (rate 1/3). Incremental

TABLE 28.1 EGPRS Modulation and Coding Schemes and Peak Data Rates

Scheme	Modulation	Maximum Rate per Slot (kb/s)	Code Rate
MCS-1	GMSK	8.8	0.53
MCS-2	GMSK	11.2	0.66
MCS-3	GMSK	14.8	0.80
MCS-4	GMSK	17.6	1.00
MCS-5	8-PSK	22.4	0.37
MCS-7	8-PSK	44.8	0.76
MCS-6	8-PSK	29.6	0.49
MCS-8	8-PSK	54.5	0.92
MCS-9	8-PSK	59.2	1.00

redundancy operation is enabled by puncturing a different set of bits each time a block is retransmitted, whereby the code rate is gradually decreased toward 1/3 for every new transmission of the block.

Recent studies introduce new modulation and coding options together with other capacity enhancement techniques to further increase the data rate and throughput of EGPRS [64], [65]. Higher-order modulations like 16-QAM and 64-QAM are being proposed along with some more coding options to optimize the performance.[1]

Adaptive modulation and coding are also successfully employed for new-generation WLAN systems. HiperLAN2 and IEEE 802.11a, both of which use OFDM technology at the physical layer, allow four different modulation options (BPSK, QPSK, 16-QAM, and 64-QAM) with different coding rates. The coding rates are obtained with different puncturing patterns to a mother convolutional code, resulting in eight different modulation and coding options [66]. Similar to link adaptation in EGPRS, an appropriate modulation and coding scheme is used depending on the link quality. Therefore, a data rate ranging from 6 to 54 Mbit/s can be obtained by using various modes. BPSK, QPSK, and 16-QAM are used as mandatory modulation formats, whereas 64-QAM is applied as an optional mode.

Although only a couple of cases are given above, adaptive modulation and coding have attracted many new-generation wireless standards to consider them as options to increase the data rates, and there has been a significant amount of research in this area. Especially in conjunction with the advanced receiver algorithms that reduce the required SINR to lower values, the better link quality values can be exploited to increase the data rates further. Combining adaptive modulation with multiantenna transmitter and MIMO schemes based on the feedback-related channel estimates, channel quality, channel correlation, etc., is one of these interesting research areas. Based on the channel feedback information, the modulation type on multiantenna transmitters can be varied. Similarly, adapting the source coding with the channel coding or modulation is another interesting area of focus for link adaptation. For example, adaptive multirate (AMR) codec allows changing of the compression rate of speech depending on the link quality, as in GSM AMR. For weak link conditions, where heavy FEC is required, AMR has the ability to decrease the codec rate (more speech compression) to allocate more bits for FEC [49].

28.4.2.3 Adaptive Cell and Frequency Assignment

As mentioned before, radio spectrum is very expensive and limited. Efficient use of radio spectrum is very important to maximize the system capacity. The introduction of cellular technology was a major step toward efficient usage of finite spectrum through a concept called *frequency reuse*. The capacity of cellular systems is interference limited, dominated by co-channel interference (CCI) and adjacent channel interference (ACI). Early cellular systems aimed to avoid these major interference sources by designing systems for the worst-case interference conditions along with fixed channel allocation. This is often achieved by employing higher-frequency reuse and by allowing enough carrier spacing between adjacent channels. Both of these reduce the spectral efficiency. Later, more efficient spectrum usage strategies were developed that dynamically assign frequencies relative to current interference, propagation, and traffic conditions. In traditional cellular system designs, the allocation of frequency channels to cells is fixed, which means that each cell can use only a set of frequencies. Even if the other cells are not fully loaded, the cell that does not have any available frequency (fully loaded cell) can not take advantage of it. In dynamic channel allocation, all the channels belong to a global pool and the channels are assigned according to a cost function that considers the CCI and ACI [67]. As a result, for nonuniform traffic conditions, the available channels can be used more efficiently.

Resource utilization is also evolved by employing a concept called hierarchical cellular structures (HCS) [68]. The use of HCS has become a major component in third-generation mobile systems such as UMTS and IMT-2000. In an HCS, various cell sizes are deployed and small cell clusters are overlaid by larger cells. For example, Figure 28.11 shows a two-layer (e.g., microcell and macrocell) hierarchical system. Microcells

[1] 16-QAM and 64-QAM stand for 16-level and 64-level quadrature amplitude modulation, respectively.

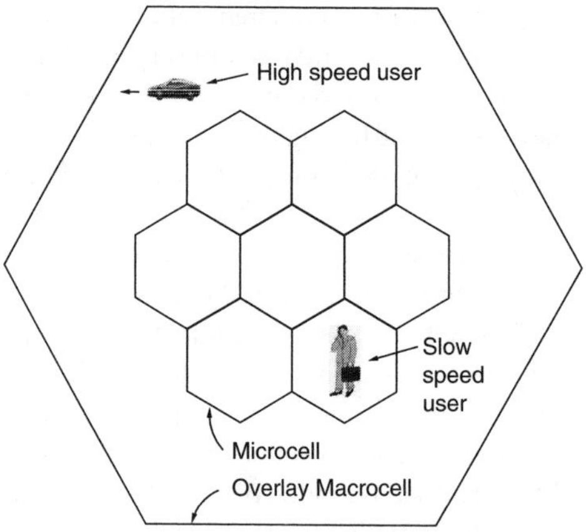

FIGURE 28.11 Illustration of two-layer hierarchical cell structure. High speed mobiles assigned to large cells, low speed mobiles are assigned to smaller cells.

increase capacity within a coverage area, but radio resource management becomes more difficult. The number of handoffs per cell are increased by an order of magnitude, and the time available to make a handoff is decreased. HCS handle this by assigning cells to the mobiles depending on their speeds (Doppler spread estimate). For example, in the two-layer structure given in Figure 28.11, low-speed mobiles are assigned to microcells, whereas high-speed mobiles are assigned to macrocells. Hence, the macrocell/microcell overlay architecture provides a balance between maximizing the capacity per unit area and minimizing the number of handoffs [69]. As a result, the risk of call dropping is reduced, and there are other benefits, like lower handover delays, reduced switching load, and increased QoS. The HCS can be more than two layers (multilayer HCS). For example, picocellular layers can also be included in multilayer HCS. Similarly, communication satellite beams can overlay all the terrestrial layers at the highest hierarchical level.

Recently, dynamic allocation and multitiered design strategies are further generalized to take power control, cell handoff, traffic classes (like multimedia), and user priorities into account. Also, there are several studies toward combining link adaptation schemes with adaptive resource allocation. For example, adaptive modulation (and coding) can be combined with dynamic channel allocation. Similarly, adaptive modulation (and coding) can be combined with handover algorithms to introduce more intelligent handover strategies. All these developments require more sophisticated adaptation of the network, and they are based on many parameter measurements.

28.5 Future Research for Adaptation

Most of the techniques and issues that were described in the previous sections still need further research for the development of more efficient adaptation and parameter estimation algorithms. At the same time, there is a significant amount of effort in evolving current wireless communications systems to provide higher data rates, higher capacity, and better performance. New technologies are being introduced to accommodate these goals, like multicarrier wireless communications, MIMO, and ultrawideband (UWB). Adaptation techniques will be a significant factor in efficient and successful deployment of these technologies.

UWB is a promising technology for future data communications systems, high accuracy (indoor) geolocation devices, sensor applications, etc. Any signal that occupies more than 500 MHz of bandwidth and meets the spectrum mask requirements enforced by spectrum regulation agencies is considered a UWB signal [70]. For example, in the U.S., the Federal Communications Commission (FCC) has allocated 7.5 GHz of spectrum (between 3.1 and 10.6 GHz) for unlicensed use of UWB devices. One of the most popular UWB systems, which is based on impulse radio (IR), utilizes carrierless transmission with very low power spectral density. IR-based UWB techniques are based on the transmission of nanosecond-level short pulses that generate extremely wide spectrums. This results in a covert noise-like signal in a radio channel. Note that within the transmission band of UWB, other technologies also coexist. For example, the OFDM-based WLAN technology at 5-GHz U-NII band is a big concern for UWB signals, as it might create significant interference for UWB signals. In order to provide robustness against narrowband interference, adaptive implementation of UWB systems is very important. For this purpose, several strategies have been developed recently. Multiband UWB is one of the techniques proposed to reduce the effect of narrowband interference. In such techniques, the whole 7.5-GHz bandwidth is divided into several narrower bands that are still wider than 500 MHz. The information is transmitted in these bands depending on the narrowband interference situation. Several versions of multiband schemes are available, some of which can be found in [70]. Estimations of the existence and level of narrowband interference are interesting research topics that fall into the parameter estimation algorithms described before.

The wide bandwidth of UWB offers a capacity much higher than the current narrowband systems. Short-range data transmission rates of over 500 Mbps have been theoretically shown [71], [72]. However, these high data rates are only possible with excellent signal quality values and for short-range communications. High date rates can be traded off with longer ranges and for lower link quality values. Depending on the link quality and distance between transmitter and receiver, the rate can be changed through adaptation of the processing gain. UWB achieves processing gain due to pulse repetition, i.e., transmitting more than one pulse within a bit. For example, by transmitting 100 pulses per bit, a processing gain of 20 dB is obtained. Additional processing gain is obtained due to the low duty cycle, which is the ratio of the pulse repetition interval and the pulse width. Adaptation of processing gain is a research topic that needs to be explored for UWB systems. Similarly, multiple-access capability of UWB systems, which is primarily determined by the processing gain, needs to be explored. Adaptive multiaccess code design depending on delay spread of the channel is one of the interesting research areas. Also, adaptive multiuser detection techniques need to be studied for cancelling multiaccess interference. In summary, adaptation algorithms and related parameter estimation techniques for jointly optimizing the multiaccess capability, power consumption, data rate, and range of UWB systems are needed.

Adaptation of multicarrier systems has already gained some momentum. In multicarrier systems, the transmission bandwidth is much wider than the coherence bandwidth of the channel, resulting in frequency-selective fading channels. Therefore, different carriers experience different channel qualities. This leads to adaptation of each subcarrier individually. Adaptive bit/power loading can be used as an effective tool to get the highest capacity from a multicarrier system provided that the transmitter has the link quality information for each carrier. For example, adaptation of the modulation level for OFDM-based multicarrier systems has been studied recently [73]. The modulation level on different carriers can be changed depending on the link quality observed at the carriers. Transmitting different modulations on each carrier requires a large overhead for signaling. Therefore, approaches that group the neighboring carriers into subsets and use the same modulation in each subset are preferred. The signaling can also be avoided in time division duplexing (TDD) systems under certain assumptions. Unlike frequency division duplexing (FDD) systems, where the channels on the downlink and uplink are different, in TDD systems, using the assumption of the reciprocal and slowly varying channel, the transmitter and receiver can be assumed to experience the same channel response. Therefore, this might eliminate the need for signaling of the channel state information to the transmitter, if the channel estimates are used as the link quality measures. However, in this case, the receiver needs to know which modulation is used at the transmitter for each group. Blind modulation detection techniques can be used for this purpose [74]. Note that although the channels could be the same, the interferences observed in the transmitter and receiver are not necessarily the same. Therefore, the observed link qualities would be different at each end. Issues like these need to be studied further for successful implementation of adaptation techniques in multicarrier systems. Multicarrier CDMA (MC-CDMA), which combines OFDM modulation with CDMA-type multiple accessing, is a technology that is being pushed for fourth-generation cellular networks. The previous adaptation algorithms proposed for CDMA and OFDM technologies need to be revisited and optimized, and new adaptation algorithms need to be developed for MC-CDMA.

MIMO and multiantenna systems bring about a new dimension to wireless channels. The spatial dimension will be used in future communications systems for further improvement of the bandwidth and power efficiency. However, this dimension and the related parameter estimates need to be understood better. Research on parameter estimation for fast and accurate calculation of spatial selectivity, angular spread, antenna correlation, etc., is needed. Also, further research is required on the effect of mutual coupling between antenna elements, the effect of near-field scatterers on antenna patterns and antenna correlations, exploitation of pattern selectivity when the spatial selectivity is not enough, and generation of a desired pattern selectivity between antenna elements adaptively.

As described earlier, MIMO systems, which employ multiple-transmit and multiple-receive antennas, can provide huge capacity and improved performance gains by exploiting spatial selectivity of the channel. However, these gains, in reality, depend heavily on the statistical properties of the channel and the correlations between antenna elements. Among the factors that affect the antenna correlation are the characteristics of the scattering environment. Therefore, an optimal way of using multiple-antenna systems depends on the situation awareness. If the transmitter knows the instantaneous channel gains (the MIMO channel matrix), it can adapt the transmission to maximize the capacity of the MIMO system [40] [41]. Similarly, the instantaneous antenna correlation values can be exploited to adapt the transmission. In many cases, estimation of the perfect instantaneous channel state and antenna correlation information, and feeding this information back to the transmitter might not be possible. This is the case especially when the mobility is high. Instead, other parameter measures like partial (statistical) channel information, average channel selectivity, or angular spread would be useful for adapting the transmitter and receiver. Advanced signal processing techniques to calculate this partial channel and correlation information are needed.

Most of the previous adaptation techniques take place in physical and medium access control (MAC) layers. Future-generation wireless systems will also allow adaptive strategies at the higher networking layers. The higher layers will be more aware of the situation in lower layers. Cross-layer optimization algorithms and cost functions that involve many layers will be developed. This will also create the need for the development of new adaptation parameters and algorithms for the estimation of these parameters.

Current wireless communications systems are based on layered protocol design, and each layer is often designed and operated independently. For example, the channel variation is addressed by adapting the link in the physical or MAC layer using signal processing techniques as described before. The traffic load and delays are adapted by changing the routing tables, by adaptive channel and cell assignment techniques. The layered structure and adaptation of layers locally (and independently) simplify the network design. But the performance and capacity of the network is suboptimal, especially for addressing the requirements of future multimedia wireless services. The future applications will have different data rate, delay, power, and QoS requirements. Cross-layer adaptation could address these requirements by jointly optimizing multidimensional cost functions that involve all protocol layers [75], [76], [77]. As a result, networks with improved end-to-end performance subject to constraints in link quality and available network resources can be obtained, while being aware of application trade-offs.

28.6 Conclusion

Recently, the use of adaptation algorithms for better utilization of the available resources, like power and spectrum, has grown significantly. Several adaptation strategies to increase performance, data rate, capacity, and QoS of wireless communications systems have been introduced. Many of these adaptation techniques depend on accurate estimation of the various parameters. Therefore, further research on efficient parameter estimation techniques is still needed.

There is a significant amount of work needed for the evolution of the current wireless communications systems to accommodate the future demands. Ultrawideband, MIMO, and multicarrier wireless communications are some of the technologies that are being studied extensively. All of them have a common point: their capability to adapt the changing radio channel conditions. Adaptation of multicarrier communications and MIMO schemes has already gained some momentum. Adaptation algorithms for UWB still need to be explored. The flexibility of UWB makes it very attractive for employing successful adaptation schemes.

Acknowledgment

The author thanks T. Yücek for his help and the anonymous reviewers for their comments.

References

[1] S. Sampei, *Applications of Digital Wireless Technologies to Global Wireless Communications*, Englewood Cliffs, NJ: Prentice Hall, 1997.

[2] T. Ikeda, S. Sampei, and N. Morinaga, "TDMA-based adaptive modulation with dynamic channel assignment for high-capacity communication systems," *IEEE Transactions on Communications*, 49(2), 404–412, 2000.

[3] S. Nanda, K. Balachandran, and S. Kumar, "Adaptation techniques in wireless packet data services," *IEEE Transactions on Communications*, 38(1), 54–64, 2000.

[4] R.V. Nobelen, N. Seshadri, J. Whitehead, and S. Timiri, "An adaptive radio link protocol with enhanced data rates for GSM evolution," *IEEE Communications Magazine*, 54–63, 1999.

[5] A. Furuskar, S. Mazur, F. Muller, and H. Olofsson, "EDGE: enhanced data rates for GSM and TDMA/136 evolution," *IEEE Personal Communications Magazine*, 6(3), 56–66, 1999.

[6] S. Anderson, H. Dam, U. Forssen, J. Karlsson, F. Kronestedt, S. Mazur, and K.J. Molnar, "Adaptive antennas for GSM and TDMA systems," *IEEE Personal Communications Magazine*, 6, 74–86, 1999.

[7] W.C. Lee, *Mobile Cellular Telecommunications*, New York: McGraw-Hill, 1995.

[8] G. Pollini, "Trends in handover design," *IEEE Communications Magazine*, 34(3), 82–90, 1996.

[9] P. Bender, P. Black, M. Grob, R. Padavoni, N. Sindhushyana, and S. Viterbi, "CDMA/HDR: a bandwidth efficient high speed wireless data servive for nomadic users," *IEEE Communications Magazine*, 38, 70–77, 2000.

[10] A. Jalali, R. Padovani, and R. Pankaj, "Data throughput of CDMA-HDR: a high efficiency-high data rate personal communication wireless system," in *Proceedings of IEEE Vehicular Technology Conference*, Vol. 3, Tokyo, Japan, May 2000, pp. 1854–1858.

[11] J. Holtzman, "CDMA forward link waterfilling power control," in *Proceedings of IEEE Vehicular Technology Conference*, Vol. 3, Tokyo, Japan, May 2000, pp. 1663–1667.

[12] H. Arslan, L. Krasny, D. Koilpillai, and S. Channakeshu, "Doppler spread estimation for wireless mobile radio systems," in *Proceedings of IEEE WCNC Conference*, Vol. 3, Chicago, IL, September 2000, pp. 1075–1079.

[13] M. Sakamoto, J. Huoponen, and I. Niva, "Adaptive channel estimation with velocity estimator for W-CDMA receiver," in *Proceedings of IEEE Vehicular Technology Conference*, Vol. 3, Tokyo, Japan, May 2000, pp. 2024–2028.

[14] D. Mottier and D. Castelain, "A doppler estimation for UMTS-FDD based on channel power statistics," in *Proceedings of IEEE Vehicular Technology Conference*, Vol. 5, Amsterdam, Netherlands, September 1999, pp. 3052–3056.

[15] L. Lindbom, "Adaptive equalization for fading mobile radio channels," Licentiate thesis, Technology Department, Uppasala University, Uppasala, Sweden, 1992.

[16] M. Morelli, U. Mengali, and G. Vitetta, "Further results in carrier frequency estimation for transmissions over flat fading channels," *IEEE Communications Letters*, 2, 327–330, 1998.

[17] L. Krasny, H. Arslan, D. Koilpillai, and S. Channakeshu, "Doppler spread estimation in mobile radio systems," in *Proceedings of IEEE WCNC Conference*, Vol. 5, no. 5, May 2001, pp. 197–199.

[18] J.H.A. Sampath, "Estimation of maximum Doppler frequency for handoff decisions," in *Proceedings of IEEE Vehicular Technology Conference*, Secaucus, NJ, May 1993, pp. 859–862.

[19] L. Wang, M. Silventoinen, and Z. Honkasalo, "A new algorithm for estimating mobile speed at the TDMA-based cellular system," in *Proceedings of IEEE Vehicular Technology Conference*, Vol. 2, Atlanta, GA, May 1996, pp. 1145–1149.

[20] C. Xiao, K. Mann, and J. Olivier, "Mobile speed estimation for TDMA-based hierarchical cellular systems," in *Proceedings of IEEE Vehicular Technology Conference*, Vol. 2, Amsterdam, Netherlands, September 1999, pp. 2456–2460.

[21] K. Kawabata, T. Nakamura, and E. Fukuda, "Estimating velocity using diversity reception," in *Proceedings of IEEE Vehicular Technology Conference*, Vol. 1, Stockholm, Sweden, June 1994, pp. 371–374.

[22] W. Lee, *Mobile Communications Engineering*, New York: McGraw-Hill, 1998.

[23] M. Austin and G. Stüber, "Eigen-based Doppler estimation for differentially coherent CPM," *IEEE Transactions on Vehicular Technology*, 43, 781–785, 1994.

[24] H. Arslan and T. Yücek, "Delay spread estimation for wireless communication systems," in *Proceedings of the eighth IEEE Symposium on Computers and Communications (ISCC 2003)*, Antalya, Turkey, July 2003, pp. 282–287.

[25] J.-T. Chen, J. Liang, H.-S. Tsai, and Y.-K. Chen, "Joint MLSE receiver with dynamic channel description," *IEEE Journal on Selected Areas in Communications*, 16, 1604–1615, 1998.

[26] L. Husson and J.-C. Dany, "A new method for reducing the power consumption of portable handsets in TDMA mobile systems: conditional equalization," *IEEE Transactions on Vehicular Technology*, 48(6), 1936–1945, 1999.

[27] H. Schober and F. Jondral, "Delay spread estimation for OFDM based mobile communication systems," in *Proceedings of European Wireless Conference*, Florence, Italy, February 2002, pp. 625–628.

[28] F. Sanzi and J. Speidel, "An adaptive two-dimensional channel estimator for wireless OFDM with application to mobile DVB-T," *IEEE Transactions on Broadcasting*, 46(2), 128–133, 2000.

[29] K. Witrisal, Y.-H. Kim, and R. Prasad, "RMS delay spread estimation technique using non-coherent channel measurements," *IEE Electronics Letters*, 34(20), 1918–1919, 1998.

[30] K. Witrisal, Y.-H. Kim, and R. Prasad, "A new method to measure parameters of frequency selective radio channel using power measurements," *IEEE Transactions on Communications*, 49, 1788–1800, 2001.

[31] K. Witrisal and A. Bohdanowicz, "Influence of noise on a novel RMS delay spread estimation method," in *Proceedings of IEEE PIMRC Conference*, Vol. 1, London, U.K., September 2000, pp. 560–566.

[32] H. Arslan and T. Yücek, "Estimation of frequency selectivity for OFDM based new generation wireless communication systems," in *Proceedings of 2003 World Wireless Congress*, San Francisco, CA, May 2003.

[33] T. Söderström and P. Stoica, *Applications of Digital Wireless Technologies to Global Wireless Communications*, Englewood Cliffs, NJ: Prentice Hall, 1997.

[34] A. Paulraj and B. Ng, "Space-time modems for wireless personal communications," *IEEE Personal Communications Magazine*, 5(1), 36–48, 1998.

[35] M.K. Özdemir, H. Arslan, and E. Arvas, "Mutual coupling effect in multi-antenna wireless communication systems," in *IEEE GlobeCom Conference*, San Francisco, CA, December 2003.

[36] M.K. Özdemir, H. Arslan, and E. Arvas, "Dynamics of spatial correlation and implications on MIMO systems," to appear in *IEEE Commun. Mag.*, June 2004.

[37] D.S. Shiu, G.J. Foschini, M.J. Gans, and J.M. Kahn, "Fading correlation and its effects on the capacity of multielement antenna systems," *IEEE Transactions on Communications*, 48(3), 502–513, 2000.

[38] M. Ivrlac, T. Kurpjuhn, C. Brunner, and W. Utschick, "Efficient use of fading correlations in MIMO systems," in *Proceedings of IEEE Vehicular Technology Conference*, Vol. 4, Atlantic City, NJ, October 2001, pp. 2763–2767.

[39] S. Catreux, V. Erceg, D. Gesbert, J. Heath, and "Adaptive modulation and MIMO coding for broadband wireless data networks," *IEEE Communications Magazine*, 40(6), 108–115, 2002.

[40] E. Telatar, "Capacity of multiantenna Gaussian channels," Technical Report, AT&T Bell Laboratories, June 1995.

[41] G.J. Foschini and M.J. Gans, "On limits of wireless communications in a fading environment when using multiple antennas," *Wireless Personal Communications*, 6(3), 311–335, 1998.

[42] TIA/EIA 136-131-B, *TDMA Third Generation Wireless: Digital Traffic Channel Layer 1*, March 2000.
[43] K. Balachandran, S. Kabada, and S. Nanda, "Rate adaptation over mobile radio channels using channel quality information," in *Proceedings of IEEE Globecom'98 Communication Theory Mini Conference Record*, 1998, pp. 46–52.
[44] J. Jacobsmeyer, "Adaptive data rate modem," U.S. Patent 5541955, July 1996.
[45] M. Türkboylari and G.L. Stüber, "An efficient algorithm for estimating the signal-to-interference ratio in TDMA cellular systems," *IEEE Transactions on Communications*, 46(6), 728–731, 1998.
[46] H. Arslan and S. Reddy, "Noise variance and SNR estimation for OFDM based wireless communication systems," in *Proceedings of the 3rd International Conference on Wireless and Optical Communications*, Banff, Alberta, Canada, July 2003.
[47] G.B.H. Arslan, "Channel estimation in narrowband wireless communication systems," *Wireless Communications and Mobile Computing (WCMC) Journal*, 1(2), 201–219, 2001.
[48] G. Bottomley and H. Arslan, "Channel estimation for time-varying channels in wireless communication systems," in *The Wiley Encyclopedia of Telecommunications*, New York: Wiley, 2003.
[49] K. Homayounfar, "Rate adaptive speech coding for universal multimedia access," *IEEE Signal Processing Magazine*, 20(2), 30–39, 2003.
[50] S. Wolf, C. Dvorak, R. Kubichek, and C. South, "How will we rate telecommunications system performance?" *IEEE Communications Magazine*, 29, 23–29, 1991.
[51] S. Voran, "Objective estimation of perceived speech quality: part II," *IEEE Transactions Speech and Audio Processing*, 7, 383–390, 1999.
[52] H. Arslan, R. Ramesh, and A. Mostafa, "Interpolation and channel tracking based receivers for coherent Mary-PSK modulations," in *Proceedings of IEEE Vehicular Technology Conference*, Vol. 3, Houston, TX, May 1999, pp. 2194–2199.
[53] W. Liu, "Performance of joint data and channel estimation using tap variable step size LMS for multipath fast fading channel," in *Proceedings of IEEE Globecom Conference*, Vol. 2, San Francisco, CA, December 1994, pp. 973–978.
[54] M. Benthin and K.-D. Kammeyer, "Influence of channel estimation on the performance of a coherent DS-CDMA system," *IEEE Transactions on Vehicular Technology*, 46, 262–268, 1997.
[55] P. Schramm, "Differentially coherent demodulation for differential BPSK in spread spectrum systems," *IEEE Transactions on Vehicular Technology*, 48, 1650–1656, 1999.
[56] D. Hui and K. Zangi, "An adaptive maximum-likelihood receiver for colored noise and interference," in *Proceedings of IEEE Vehicular Technology Conference*, Vol. 4, Atlantic City, NJ, October 2001, pp. 2257–2261.
[57] A. Hafeez, K. Molnar, and G. Bottomley, "Co-channel interference cancellation for D-AMPS handsets," in *Proceedings of IEEE Vehicular Technology Conference*, Vol. 2, Houston, TX, May 1999, pp. 1026–1030.
[58] A. Hafeez, H. Arslan, and K. Molnar, "Adaptive joint detection of co-channel signals for ANSI-136 handsets," in *Proceedings of IEEE PIMRC Conference*, Vol. 2, San Diego, CA, October 2001, pp. E-105–E-110.
[59] J. Zander, "Performance of optimum transmitter power control in cellular radio systems," *IEEE Transactions on Vehicular Technology*, 41(1), 57–62, 1992.
[60] Z. Rosberg and J. Zander, "Toward a framework for power control in cellular systems," *Wireless Networks*, 4, 215–222, 1998.
[61] S. Ulukus and R.D. Yates, "Stochastic power control for cellular radio systems," *IEEE Transactions on Communications*, 46(6), 784–798, 1998.
[62] S.V. Hanly and D.N. Tse, "Power control and capacity of spread-spectrum wireless networks," *Automatica*, 35(12), 1987–2012, 1999.
[63] A. Goldsmith, "The capacity of downlink fading channels with variable rate and power," *IEEE Transactions on Vehicular Technology*, 46(3), 569–580, 1997.
[64] H. Arslan, T.J.-F. Cheng, and K. Balachandran, "Physical layer evolution for GSM/EDGE," in *Proceedings of IEEE Globecom Conference*, Vol. 5, San Antonio, TX, November 2001.

[65] H. Arslan, T.J.-F. Cheng, and K. Balachandran, "Evolution of EDGE to higher data rates using QAM," in *Proceedings of IEEE Vehicular Technology Conference*, Vol. 4, Atlantic City, NJ, October 2001, pp. 2267–2271.

[66] A. Doufexi, S. Armour, M. Butler, A. Nix, D. Bull, J. McGeehan, and P. Karlsson, "A comparison of the HIPERLAN/2 and IEEE 802.11a wireless LAN standards," *IEEE Communications Magazine*, 40(5), 172–180, 2002.

[67] I. Katzela and M. Naghshineh, "Channel assignment schemes for cellular mobile telecommunication systems: a comprehensive survey," *IEEE Personal Communications Magazine*, 3(3), 10–31, 1996.

[68] N.D. Tripathi, N.J.H. Reed, and H.F.V. Landingham, "Handoff in cellular systems," *IEEE Personal Communications Magazine*, 5(6), 26–37, 1998.

[69] G. Pollini, "Trends in handover design," *IEEE Communications Magazine*, 3, 82–90, 1996.

[70] G. Aiello and G. Rogerson, "Ultrawideband wireless systems," *IEEE Microwave Magazine*, 36–47, 2003.

[71] J. Forrester, G. Evan, D. Leeper, and S. Srinivasa, "Ultra-wideband technology for short or medium-range wireless communications," *Intel Technology Journal*, 1–7, 2001.

[72] P. Mannion, "Ultrawideband radio set to redefine wireless signaling," *EE Times*, September 2002, pp. 71–84.

[73] L. Hanzo, C.H. Wong, and M.S. Yee, *Adaptive Wireless Transceivers: Turbo-Coded, Turbo-Equalized and Space-Time Coded TDMA, CDMA, and OFDM Systems*, New York: John Wiley & Sons, 2002.

[74] S. Reddy, T. Yucek, and H. Arslan, "An efficient blind modulation detection algorithm for adaptive OFDM systems," in *Proceedings of IEEE Vehicular Technology Conference*, Orlando, FL, October 2003.

[75] A. Goldsmith and S. Wicker, "Design challenges for energy-constrained ad hoc wireless networks," *IEEE Wireless Communications* 8–27, 2002.

[76] T. Rappaport, A. Annamalai, R. Buehrer, and W. Tranter, "Wireless communications: past events and a future perspective," *IEEE Communications Magazine*, 40(5), 148–161, 2002.

[77] Z. Haas, "Design methodologies for adaptive and multimedia networks," *IEEE Communications Magazine*, 39(11), 106–107, 2001.

Index

A

Active link protection (ALP), **23**-8, **23**-9, **23**-10
Adaptation techniques and enabling parameter estimation, **28**-1 to **28**-26
adaptation schemes, **28**-3 to **28**-6
applications of adaptive algorithms, **28**-14 to **28**-20
future research for adaptation, **28**-20 to **28**-22
parameter measurements, **28**-6 to **28**-13
Adaptive differential pulse code modulation (ADPCM), **20**-10
Adaptive equalizers **19**-2
Adaptive multirate (AMR), **20**-7, **20**-21, **28**-19
Adaptive resource allocation, examples, **28**-17
Adaptive signaling methodology, **6**-1
Additive Gaussian noise (AGN) covariance, **14**-7
Additive white Gaussian noise (AWGN), **1**-14, **6**-2 to **6**-3, **7**-1
analysis of chaos communication schemes in, **27**-20
channel(s), **4**-26
despreading and, **10**-9
environment, detectors optimum in, **10**-25
error probability, **4**-20
GPS signal model, **22**-5
IF-based estimator and, **5**-13, **5**-14
power spectral density of, **4**-5
receiver performance and, **28**-15
vector, **15**-3
Advanced forward link trilateration (AFLT), **21**-18
Advanced Mobile Phone System (AMPS), **1**-29, **2**-2, **10**-2, **21**-17
Akaike information criterion (AIC), **25**-27
Alamouti scheme, **12**-18, **13**-21
Algebraic code excited linear prediction (ACELP), **20**-10, **20**-14
ALOHA, **24**-11, **24**-12, **24**-16
Alphabet independent optimization, **14**-6
Alternating projection, **21**-12
Ambiguity function, **25**-9
Amplifier(s)
bias in, **3**-20
input back-off, **19**-14
Amplitude
-to-amplitude (AM/AM) conversion, **19**-4, **19**-8, **19**-11
-to-phase (AM/PM) conversion, **19**-4, **19**-8, **19**-11
statistics, **2**-18
Analog-to-digital conversion (ADC), **10**-9
Analysis-by-synthesis (AbS) coders, **20**-5
Angles of arrival (AOAs), **21**-3, **21**-12
Angular delay power spectrum (ADPS), **2**-17, **2**-22
Angular power spectrum (APS), **2**-9
Angular spread, **13**-4
Antenna
array, carrier-phase estimation using, **22**-22
correlation, **12**-16, **28**-22
diversity, **1**-35
selection, **6**-15
spacing, **12**-7
systems, multiple, *see* Multiple antenna systems, performance analysis of
A posteriori probabilities (APPs), **18**-9
Array
geometry, **12**-3
processing, **25**-2
Assisted GPS (AGPS), **21**-2, **21**-18
Asymmetric Digital Subscriber Line (ADSL), **10**-28
Asymptotic multiuser efficiency (AME), **11**-19
Asymptotic smoothing error, **24**-6
Asynchronous CDMA (A-CDMA), **10**-11, **26**-11
Asynchronous transfer mode (ATM), **1**-28, **20**-11
Autocorrelation function (ACF), **2**-6
Automatic repeat request (ARQ), **28**-4
Automatic speech recognition (ASR), **20**-1, **20**-16
Average fade duration (AFD), **4**-13
Average transmit power, adaptation and, **28**-2
Azimuthal power spectrum, **2**-21

B

Background noise, **20**-2, **20**-12
Backlog pressure, **23**-18
Backpropagation (BP) algorithm, **19**-1
Band-limited spread-spectrum (BL-SS) signal, **10**-6
Bandpass noise, **5**-18

Bandwidth, coherence, **2**-7
Barts and Stutzman model, **4**-16
Baseband
  filtering, **28**-6
  model, **27**-23
Base station (BS), **5**-2, **5**-6, **10**-4
Basis expansion model (BEM), **16**-2, **16**-5
  FIR channel parameters, **16**-12
  FIR serial equalizer, **16**-27
Baud rate sampling, **3**-2
Bayesian inference, **26**-20
Beam forming (BF), **6**-15, **15**-19, **15**-20
  scheme, modeling of, **15**-17
  techniques, adaptive, **28**-5
  transmitting strategy, **15**-16
Bell Labs Layered Space—Time (BLAST), **13**-7
Bernoulli map, **27**-3, **27**-6
Bessel function, zero-order modified, **1**-11, **2**-19, **4**-5
Bhattacharyya bound, **4**-27
Bias
  amplifier, **3**-20
  variance and, **26**-16
Binary differential phase shift keying (BDPSK), **17**-10
Binary message stream, **27**-21
Binary phase shift keying (BPSK), **1**-13, **4**-23, **6**-3, **11**-33, **18**-4
  average BER of, **17**-12, **17**-19
  code, **12**-18
  link adaptation and, **28**-4
  signaling scheme, **25**-17
  transmission, isolated, **11**-19
Bipartition algorithm, **2**-13
Birth—death Markov process, **4**-19
Bit error probability, **11**-22
Bit error rate (BER), **3**-19, **6**-10, **8**-6, **10**-10, **11**-17
  BPSK, **17**-12, **17**-19
  chaos communications, **27**-26
  combiner output and, **17**-10
  maintenance of, **28**-4
  performance, SSC scheme, **17**-11
  plotting of, **18**-4
  signal quality and, **28**-2
  target, **28**-11
Bit-interleaved coded modulation (BICM), **6**-12
Bit stream generators, **27**-7
Black box, **2**-4
Blind channel estimation, **3**-2, **3**-10, **28**-13
Blind detection
  code-aided, **11**-33
  two-stage, **11**-35
Blind source separation (BSS), **25**-2
Block codes, **1**-22 to **1**-24
  BCH codes, **1**-23
  coding gain, **1**-22
  Hamming distance, **1**-22
  implementation complexity, **1**-22 to **1**-23
  interleaving, **1**-23 to **1**-24
  linear block code, **1**-22
  Reed—Solomon codes, **1**-23
  vector space and subspace, **1**-22
Block decision feedback equalizers (BDFEs), **16**-2, **18**-5
Block equalization, **16**-7, **18**-17
Block linear equalizers (BLEs), **16**-8, **18**-2, **18**-4
BLUE estimator, **8**-10
Bluetooth, **10**-28, **10**-29
Boosted SIR target, **23**-10
Branch switching, **17**-13, **17**-14
Broadband, *see also* Space—time coding and signal processing for broadband wireless communications
  channel, **13**-4
  chaotic carriers, **27**-10
  MIMO-OFDM systems, capacity of, **12**-4

C

Call dropping, risk of, **28**-20
Canonical channel models, **25**-15
Capacitor coupling, **3**-20
Capacity
  achieved, **12**-6
  CCDF, **15**-2, **15**-10, **15**-27
  distribution, **14**-16
  expressions of mean and variance of, **15**-27
  impact of propagation parameters on, **12**-6
  mean, **15**-18, **15**-30
  MGF, **15**-25
  statistics, **14**-14
  threshold, **15**-11
  variance formulas, **15**-31
Carrier frequency offset (CFO), **8**-1
  acquisition range for, **8**-14
  effect of on system performance, **8**-6
  estimation, **8**-2, **8**-7, **8**-8, **8**-10
Carrier phase
  estimation, **22**-24
  recovery, **10**-20
Cayley differential unitary space—time codes, **14**-2
cdmaOne channels, **10**-17
CDMA 2000 system, implementation of, **5**-28
Cellular network with frequency reuse, **10**-14
Cellular systems, North American, **20**-14
Channel(s), *see also* Mobile channels, modeling and estimation of access, **23**-3
  AGN, **14**-8
  broadband, **13**-4
  capacity, optimum combining, **15**-24
  cdmaOne, **10**-17
  characterizations, **1**-2 to **1**-13, **2**-6
  coding, **1**-21, **20**-11, **28**-17
  coherence bandwidth, **4**-2
  conditional, **6**-6
  covariance matrix, **11**-22, **14**-18
  decorrelated, **10**-27
  delay spread, **13**-4
  demodulation, **27**-9
  discrete-time, **3**-22
  encoding, **27**-9
  energy profiles, **18**-4
  equalization, **16**-1, **19**-21
  frequency-selective, **10**-23, **13**-4

impulse response (CIR), **8**-4, **11**-3, **13**-4, **13**-11, **28**-9
interpolators, **28**-8
inversion, **6**-6
LMS propagation, **4**-2
memoryless, **4**-27
modeling, **3**-1, **25**-13
MSE, normalized, **3**-24
Nakagami fading, **1**-7, **1**-37
overspread, **2**-6
pilot, **10**-15
power, covariance function of, **5**-4
prediction error, **6**-12
probing, noninvasive, **23**-14
quality, **20**-12
radio, **28**-4
Rayleigh fading, **1**-7, **1**-37
reception matrix, **24**-8
Ricean fading, **1**-7, **1**-37
selectivity estimation, **28**-5
sounders, **2**-10
statistical properties of, **2**-2
status information, **14**-2
time-invariant, **3**-6
time-selective, **13**-4
time-variant, **3**-3, **3**-9, **3**-21
tracker, least mean square, **28**-14
transition probability, **18**-10
wideband, **1**-6
wireless propagation, **2**-2
Channel estimation, **3**-8, **16**-23, **26**-2
algorithms, **14**-20, **28**-5
based on least squares criterion, **11**-28
blind, **3**-2
errors in, **6**-10
procedure, centralized, **11**-29
quasi-static channels, **13**-14
SIMO, **3**-13
simulation examples, **3**-22
SISO, **3**-12
training-based, **3**-9
Channel state information (CSI), **8**-6, **11**-2
acquisition of without ambiguity, **14**-19
availability of at receiver, **6**-3
available strategies to cope with missing, **11**-28
feedback, **15**-25
multiuser detection with known, **11**-6
multiuser detection with unknown, **11**-25
perfect, **6**-11, **15**-1, **15**-17, **15**-25
Chaos communications, **27**-1 to **27**-28
chaos in communications, **27**-10
chaotic vs. classical communications, **27**-20 to **27**-26
communication using broadband chaotic carriers, **27**-10 to **27**-19
definition of chaos, **27**-1 to **27**-9
spreading code generation, **27**-19 to **27**-20
Chaos shift keying (CSK), **27**-12, **27**-13, **27**-17
Chaotic masking, **27**-11, **27**-16
Chaotic modulation (CM), **27**-14
Chaotic on—off keying, **27**-12
Chernoff bound, **4**-27
Chip
-matched filter (CMF), **10**-9
rate, **10**-8, **10**-12
-timing error detectors (CEDs), **10**-20
Chirp
modulation, **25**-32
signal, **25**-18, **25**-33
Choi—Williams distribution (CWD), **25**-12
Cholesky decompositions, **14**-12
Cholesky factorization, **11**-36
Chua's circuit, **27**-8
Circulant matrix, **8**-5
Cluster angle spread, **12**-7, **12**-14
Co-channel interference (CCI), **15**-1, **15**-2, **15**-24, **28**-19
Code(s)
-aided techniques, **11**-28
block, **1**-22 to **1**-24
BPSK, **12**-18
convolutional, **12**-20
division multiplexed pilot (CDMP), **4**-27
domain, **10**-10
Hamming, **1**-22
performance parameter, **4**-27
PRN, **22**-2, **22**-20
Reed—Solomon, **1**-22
scrambling, **10**-14
space—frequency, **12**-12
space—time, **1**-22, **13**-11
synchronization, **10**-18
tracking, **10**-19
Code division multiple access (CDMA), **1**-30
asynchronous, **10**-11
capacity, **1**-34
direct-sequence, **25**-18
interference mitigation in, **25**-2
multicarrier, **10**-28, **11**-2, **25**-2, **25**-21
multitone, **1**-31, **1**-32
multiuser, equalizers proposed for, **18**-2
networks, transmitted packets in, **24**-2
radio networks, **10**-2
receiver performance, **10**-25
satellite network, **4**-27
scheme, **1**-31
sequence families, chaos generators and, **27**-25
signal amplitudes, unbalanced, **11**-7
spreading sequence, **10**-10
time division, **1**-31
transmissions, frequency recovery for, **10**-21
wideband, **13**-2
Code division multiple access system(s)
asymptotic studies, **15**-2
asynchronous, **26**-11
capacity of, **10**-13
chaos-based, **27**-19
design of reception structures for, **11**-2
estimation of LTV channels for, **25**-16
FSK-FH, **25**-33
handover processes specific to, **5**-25
multicarrier, **26**-11
power control in, **23**-2

pseudonoise tracking in, **5**-4
wideband, **17**-3
Code excited linear prediction (CELP), **20**-6
algebraic, **20**-10, **20**-14
algorithm, **20**-6
relaxed, **20**-14
speech coder, **20**-14
Coding
distortion, ASR over IP, **20**-19
gain, definition of, **1**-22
linear predictive, **20**-4
source-controlled channel, **20**-15
space—frequency, **12**-9
space—time, **12**-1, **13**-9
tandem source–channel, **20**-14
techniques, **1**-20 to **1**-28
Coherence
bandwidth, **1**-5, **2**-7, **4**-2, **13**-4, **13**-13
distance, **13**-4
time, **1**-5, **4**-2, **13**-4
Coin flips, **23**-13
Collision channel model, **24**-1
Combination logic unit (CLU), **10**-24
Combination unit (CU), **10**-24
Combiner output
MGF of, **17**-21
statistics of, **17**-8, **17**-17
Complementary cumulative distribution function (CCDF), **15**-2
capacity, **15**-2, **15**-10, **15**-13
Gaussian-approximated, **15**-16
Complementary error function, **15**-10
Complementary metal-oxide semiconductor (CMOS), **10**-3
Complex spreading (CS), **10**-6
Composite spreading signature code, **10**-14
Conditional CRB (CCRB), **8**-15, **8**-19
Conditional dynamic linear models (CDLMs), **26**-14
Conditional Fisher information matrix (CFIM), **8**-15
Constant modulus algorithm (CMA), **3**-13, **24**-8
Constellation
diagram, geometry of, **7**-26
IBO used for, **7**-28
Continuous-phase FSK (CPFSK), **1**-17
Continuous-phase modulation (CPM), **1**-13, **1**-18, **10**-4
Continuous-time received signal, **13**-3
Convolutional codes, **1**-24 to **1**-26, **4**-26, **28**-18
Convolutive mixtures, separation of, **25**-30
Corner effect, **5**-3
Corollaries, **14**-4
Correlation
matrix, **12**-12
receiver (CR), **10**-10, **10**-22
Coupling terms, **19**-23
Covariance (COV)
-based estimator, **5**-12
-based method, **5**-10, **5**-16
feedback, **15**-16
matrix, advantages of STFDs over, **25**-25
Craig's approximation, **19**-10
Craig's POE calculation method, **7**-4
Cramer—Rao bound (CRB), **3**-9, **8**-1, **14**-20, **14**-22
conditional, **8**-15, **8**-19
unconditional, **8**-16
Cross-layer optimization algorithms, **28**-22
Cross-polarization discrimination (XPD), **2**-23
Cross-SCORE algorithm, **22**-14
Cross-terms reduction, **25**-11
Cumulant(s)
equations, **27**-22, **27**-24
random variable, **27**-21
Cumulative distribution function (CDF), **17**-3
combiner output, **17**-9
conditional, **17**-8
state-dependent, **17**-8
Cyclic prefix (CP), **8**-3, **12**-5, **13**-4, **13**-13, **28**-8
Cyclic redundancy check (CRC), **28**-2, **28**-13
Cyclostationarity, **13**-21

D

Data
block, **22**-14
predistortion, **7**-16, **7**-28, **7**-29
dc offset-to-signal ac component (DCAC) power ratio, **3**-22, **3**-23
Decision feedback equalizer (DFE), **3**-14, **13**-7, **16**-21, **19**-3
Decoding, soft bit source, **20**-15
Deconvolution, **25**-18
Decorrelating detector, **10**-26, **11**-10
Delay
cross-power spectrum, **2**-7
diversity, **12**-20, **13**-9
-Doppler coefficients, **25**-15
lock loop (DLL), **21**-13, **22**-3
power density spectrum, **2**-7
spread, **1**-5, **2**-3, **2**-4, **4**-2, **28**-8
vs. power trade-off, **23**-3
Demodulation, **25**-20, **27**-9, **27**-16
Despreading operation, **10**-9
Differential chaos shift keying, **27**-17
Differential phase shift keying (DPSK), **1**-13
Differential QPSK (DQPSK), **1**-15
Diffuse scattering, **2**-12
Digital-to-analog conversion (DAC), **10**-9
Digital audio broadcasting (DAB), **1**-19, **8**-2, **10**-28
Digital communications, *see* Monte Carlo signal processing for digital communications
Digital delay-lock loop (DDLL), **10**-19
Digital signal processing (DSP), **10**-2, **13**-23
Digital subscriber line (DSL), **1**-19
Digital Video Broadcasting (DVB), **8**-2, **10**-28
Dirac deltas, **14**-17
Direct equalizer design, **16**-23
Direction-of-arrival (DOA) estimation, **25**-22, **25**-24, **25**-27
Direction of departure (DOD), **2**-8
Direct-sequence spread spectrum (DS/SS), **10**-5

Direct-sequence (DS) technique, **25**-13
Discontinuous transmission (DTX) mechanism, **20**-11
Discrete Fourier transform (DFT), **8**-3, **9**-3
  autocorrelation function, **27**-6
  operation of, **14**-18
Discrete multitone (DMT), **8**-6, **9**-9
Discrete time
  baseband modeling, **27**-21
  systems, **27**-7
Discrete Wavelet Multitone (DWMT) modulation, **9**-16
Distress signal, **23**-13
Distributed power control (DPC) algorithm, **23**-7, **23**-9
Distributed speech recognition (DSR), **20**-17, **20**-18
Diversity, **13**-1
  delay, **13**-9
  gains, **12**-9, **13**-23
  multibranch switched, **17**-14
  order, **13**-2, **13**-5
  paths, GSEC and, **17**-28
  receive antenna, **16**-3
  techniques, classifications of, **1**-35
  transmit, **18**-3
  vs. throughput trade-off, **13**-12
DLR wideband model, **4**-18
Doppler estimation, **25**-34
Doppler frequency, **5**-8, **6**-5, **13**-15
Doppler-lag representations, **25**-9
Doppler shift, **1**-4
Doppler spectrum, **4**-9, **4**-11, **4**-13
Doppler spread, **1**-5, **2**-6, **3**-5, **4**-2, **11**-3, **13**-5, **16**-4, **28**-2, **28**-14
Dot-product mode, **22**-20
Down-link (DL), **10**-4, **15**-7
DPC/ALP, admission control under, **23**-10
Dual-branch switch and stay combining, **17**-4
Durkin's model, **1**-4
Dynamic random access memories (DRAMs), **10**-3
Dynamic systems, asymptotic behavior of, **27**-2

E

Early-minus-late mode, **22**-20
EDGE cellular technology, *see* Enhanced data rate for GSM evolution cellular technology
Eigenvalue(s)
  constraint, **14**-7
  decomposition (EVD), **14**-5
  distribution, **12**-8
  empirical distribution of, **14**-14
  maximum modulus, **23**-6
Elastic traffic, voice over IP, **20**-8
Electromagnetic propagation characteristics, modeling of, **4**-2
Encoding
  dynamic, **27**-14
  rule, SC FDE-STBC, **13**-17
  scheme, Alamouti, **13**-21
  static, **27**-11, **27**-13
End-to-end delay, **20**-8
Energy concentration
  MIMO channels, **18**-14
  SISO channels, **18**-15
Enhanced data rate for GSM evolution (EDGE) cellular technology, **1**-15
Enhanced General Packet Radio Service (EGPRS), **28**-4
  ink adaptation in, **28**-19
  modulation options, **28**-18
Enhanced observed time difference (E-OTD), **21**-18
Enhanced variable-rate coder (EVRC), **20**-14
Envelope
  -based estimators, **5**-22
  first moment of, **5**-29
  function, **25**-20
Equal gain combiner (EGC), **1**-36, **10**-24
Equalization, **3**-21
  adaptive, **19**-1
  block, **16**-7
  frequency-domain, **16**-17
  performance, **3**-23, **3**-24
  serial linear, **16**-10
  trellis, **18**-1
Equalizer(s)
  adaptive, **19**-2
  BEM FIR serial, **16**-27
  block linear, **18**-2, **18**-4
  decision feedback, **3**-14, **16**-21, **19**-3
  design, **16**-13, **16**-16
  Gibbs, **26**-8
  linear, **16**-21
  list-type, **18**-2, **18**-8
  maximum *a posteriori*, **16**-2, **18**-8
  MCMC, **26**-7
  Metropolis, **26**-9
  MMSE, **3**-22, **3**-23, **18**-7
  neural network, **19**-3
  output, expression of, **19**-22
  serial linear, **16**-2
Equation error, **3**-12
Error
  asymptotic smoothing, **24**-6
  average probability of, **4**-22
  covariance, CRB matrix and, **14**-24
  probability, **11**-17, **11**-37, **12**-15
  propagation, **13**-7
  rate performance, **12**-11, **12**-18
  vector, **11**-32
ESPRIT, **2**-11, **21**-12, **25**-27
European Telecommunications Standards Institute (ETSI), **20**-13
Expansion models, **3**-5
Expectation
  —maximization (EM) algorithm, **3**-10, **5**-21, **26**-2
  value operator, **27**-22
Exterior differential calculus, **14**-10
Exterior product, concept of, **14**-10

F

Fading, **1**-4, **21**-14
  coefficient sequence, **26**-17
  fast, **1**-6, **4**-27, **11**-4

flat, **1**-6
frequency-selective, **1**-6, **3**-2
large-scale, **4**-2
multipath, **5**-7, **6**-2, **6**-4, **25**-2
phenomenon, MIMO systems and, **15**-2
Rayleigh, **6**-5, **25**-33
Rice, **4**-24
shadow, **5**-5
slow, **1**-6, **4**-27
small-scale, **4**-2
statistics, **12**-3
time nonselective, **4**-4
time selective, **4**-4
types of, **1**-5
wide-sense stationary, **11**-4
Fading channels, adaptive coded modulation for transmission over, **6**-1 to **6**-19
adaptive system model, **6**-4 to **6**-8
adaptivity in multiantenna systems, **6**-14 to **6**-16
adaptivity in single-input single-output systems, **6**-8 to **6**-14
Fading channels, equalization techniques for, **16**-1 to **16**-29
block equalization, **16**-7 to **16**-10
channel estimation and direct equalizer design, **16**-23
complexity, **16**-22 to **16**-23
existence of zero-forcing solution, **16**-21 to **16**-22
frequency-domain equalization for TIV channels, **16**-17 to **16**-21
performance results, **16**-24 to **16**-27
serial decision feedback equalization, **16**-14 to **16**-17
serial linear equalization, **16**-10 to **16**-14
system model, **16**-6 to **16**-7
wireless channel model, **16**-2 to **16**-6
Fading channels, multiuser detection for, **11**-1 to **11**-39
multiuser detection with known CSI, **11**-6 to **11**-25
multiuser detection with unknown CSI, **11**-25 to **11**-38
signal and channel model, **11**-2 to **11**-6
False alarm probability, **10**-19
Far echoes, **4**-19
Fast fading, **1**-6, **4**-27, **11**-4
Fast Fourier transform (FFT), **1**-18, **9**-3, **16**-17
Fast power control, **28**-17
FD decision feedback equalizers (FDDFEs), **16**-2
FD linear equalizers (FDLEs), **16**-2
Federal Communications Commission (FCC), **21**-2
Field of view (FOV), **22**-4
Filter
FIR prototype, **9**-23
impulse response, prototype, **9**-7
Kalman, **19**-6
Nyquist, **9**-13, **9**-21
prototype, **9**-13
whitened matched, **18**-2, **18**-12, **18**-17
Filter bank(s), **9**-6, **9**-7
communications system, **9**-22
cosine-modulated, **9**-16
critically-sampled, **9**-8
design, **9**-21
O-QAM OFDM, **9**-16
transceivers, **9**-3
transmit, **9**-15
uniform, **9**-7
Filter-bank modulation techniques for transmission over frequency-selective channels, **9**-1 to **9**-29
critically sampled filter banks, **9**-4 to **9**-9
discrete multitone modulation, **9**-9 to **9**-11
discrete wavelet multitone modulation, **9**-16 to **9**-17
filtered multitone modulation, **9**-17 to **9**-27
O-QAM OFDM modulation, **9**-11 to **9**-16
Filtered multitone (FMT)
demodulator, block diagram of, **9**-20
modulation, **9**-1, **9**-4, **9**-17
prototype filter design algorithm for, **9**-22
system receiver, **9**-26
Finite difference time domain (FDTD) method, **2**-12
Finite element method (FEM), **2**-12
Finite impulse response (FIR)
channel, **16**-2
differentiator, **5**-18
filter, **8**-4, **9**-5, **10**-8, **19**-15, **22**-6
Finite-state machines, **13**-16
Finite-state Markov chain (FSMC) model, **1**-9, **1**-10
First-come first-serve (FCFS) algorithm, **24**-2
Fisher information matrix (FIM), **14**-22
calculation of inverse of, **19**-6
inverse of, **19**-16
Fixed bit rate (FBR), **20**-6
Flat-fading channels
SMC receiver in, **26**-17, **26**-18
space—time codes over, **13**-6
Flexible precoding, **9**-25
FMDA scheme, **1**-29
Fontan et al. model, **4**-1
Forced dropout (FDO), **23**-13
Forgetting factor, **11**-31
Forward error correction (FEC), **12**-16, **20**-10
encoding, **28**-16
schemes, **28**-2
Forward-link, **10**-4
Fourier transform, **25**-14
Fractional sampling, **16**-3
Frame-error-rate (FER), **28**-2
Fraud protection, **25**-2
Fraunhofer distance, **1**-3
Free-space propagation mode, **1**-2
Frequency
diversity, **1**-35
division duplexing (FDD), **21**-17, **28**-21
division multiple access (FDMA), **1**-29, **11**-2
division multiplexing (FDM), **9**-3
domain duplexing, **2**-15
-domain (FD) equalization, **16**-2, **16**-17
-hopping (FH) multiple access, **25**-33
-hopping spread spectrum (FH/SS), **10**-5
instantaneous, **25**-4, **25**-6
-modulated (FM) signals, **25**-3
reuse, **28**-19

-selective channels, *see* Filter-bank modulation techniques for transmission over frequency-selective channels
shift keying (FSK), **1**-13, **1**-16
synchronization, **8**-11, **28**-6
time division duplexing (F/TDD), **10**-16
Frobenius—Perron operator, **27**-4
Future mobile communications systems, signal processing for, **1**-1 to **1**-42
channel characterizations, **1**-2 to **1**-13
coding techniques, **1**-20 to **1**-28
diversity technique, **1**-35 to **1**-38
modulation techniques, **1**-13 to **1**-20
multiple access techniques, **1**-29 to **1**-34

G

Gallager bound, **4**-27, **4**-31
Gaussian Markov model, **26**-8
Gaussian minimum shift keying (GMSK), **1**-15, **10**-4, **26**-11, **28**-4
Gaussian noise source, **11**-2
Gaussian random process, **6**-7
Gauss–Seidel iterations, **11**-15
Generalized likelihood ratio test (GLRT), **21**-13
Generalized selection combining (GSC), **17**-2, **17**-3, **17**-26
Generalized switch and examine combining (GSEC), **17**-2
average number of channel estimates needed by, **17**-27
diversity paths and, **17**-28
error performance of, **17**-25
mode of operation, **17**-20
outage probability of, **17**-23, **17**-24
Generalized Viterbi algorithm (GVA), **18**-9
Generalized Wishart density, **14**-13
Generic channel modeling approaches, **2**-3
Geolocation techniques for mobile radio systems, **21**-1 to **21**-23
FCC regulations, **21**-2
geolocation algorithms, **21**-5 to **21**-11
geolocation methods, **21**-3 to **21**-4
impairments to accuracy, **21**-14 to **21**-17
location-based services, **21**-2 to **21**-3
location parameter estimation, **21**-12 to **21**-14
provisions in standards, **21**-17 to **21**-18
Geometrical theory of diffraction (GTD), **2**-12
Geometric dilution of precision (GDOP), **21**-11
Geometry-based stochastic channel model (GSCM), **2**-16
Gibbs interference canceller, **26**-11
Gibbs sampler, **26**-4, **26**-6
Global Positioning System (GPS), **10**-6, **21**-2, **22**-1, *see also* GPS receivers, adaptive arrays for
assisted, **21**-2, **21**-18
signal model, **22**-5
user segments, **22**-2
Global System for Mobile Communications (GSM), **2**-2, **18**-3, **28**-8
Communications Evolution, **13**-2
NLOS propagation and, **21**-15
radios, **5**-4
symbol duration in, **28**-15
voice over wireless and, **20**-11
Goddard algorithms, **3**-13
GPS receivers, adaptive arrays for, **22**-1 to **22**-26
GPS signal model, **22**-5 to **22**-6
interference suppression techniques in GPS, **22**-6 to **22**-19
multipath mitigation in GPS, **22**-19 to **22**-24
Grassmann algebra, **14**-10
GRLN distribution, **4**-9

H

Haar distribution, **14**-12, **14**-13
Hamming codes, **1**-22
Hamming distance, **4**-26
Handoff algorithm design, **5**-2
Handover
decision algorithms, **5**-25, **5**-27
delay, **5**-28
Harmonic retrieval (HR) problem, **8**-7
Harmonic and stochastic excitation (HSX), **20**-15
Harmonic vector excitation (HVXC) coders, **20**-6
Hashemi—Suzuki—Turin model, **2**-14
Hearability, **21**-2, **21**-15
Heisenberg's uncertainty principle, **25**-18
Hermitian nonnegative-definite matrix, **15**-28
Hermitian positive-definite matrix, **15**-7, **15**-8
Hermitian square root matrix, **25**-31
Hidden Markov model (HMM), **3**-10, **3**-19, **20**-17
Hidden random data, **26**-2
Hierarchical cellular structures (HCS), **28**-19
High-data-rate (HDR) system, **28**-5
High-definition TV (HDTV), **1**-19
Higher-order modulations (HOMs), **28**-17
Higher-order statistics (HOS), **3**-2
High-power amplifier (HPA), **7**-2, **7**-12, **19**-1, **19**-4
AM/AM conversion, **19**-20
measurements, **19**-5
modeling errors, **19**-4, **19**-7
response, approximated, **19**-10
saturation point, **19**-12
High-resolution correlators (HRCs), **22**-4
Hostile jamming, **10**-5
Hough transform, **25**-19
Human-to-machine scenarios, **20**-1
Hybrid coders, **20**-5
Hybrid location methods, **21**-9

I

IEEE 802.11i standard, **20**-16
Importance
sampling, **26**-11, **26**-13
weight, **26**-15, **26**-23
Impulse response(s), **2**-2
channel, **28**-9
finite, **5**-18
infinite, **3**-18
point-to-point, **2**-13

prototype filter, **9**-7, **9**-15
time-invariant, **2**-5
Incremental weight, **26**-13
Inelastic traffic, voice over IP, **20**-8
Infinite impulse response (IIR), **3**-18
Information-bearing signal, **3**-20
Input back-off (IBO), **7**-1, **7**-2
definition of, **7**-12
minimal, **7**-14, **7**-28
Instantaneous frequency (IF), **5**-4, **25**-4, **25**-6
-based estimator, **5**-12, **5**-19
-based method, **5**-11
-based velocity estimator, **5**-15
estimator, SNR improvement in, **5**-30
noise, PSD of, **5**-30
vs. baseband sampling, **10**-21
Instantaneous rms delay spread, **2**-7
Interantenna interference, **13**-17
Interblock interference (IBI), **8**-3
Interchip interference (ICI), **10**-24
Interference
background, **11**-32
cancellation, **13**-7, **13**-18, **26**-11
distributions, reduced, **25**-12
interantenna, **13**-17
intersymbol, **1**-6, **8**-2, **13**-17
mitigations, **25**-18
nonresponsive, **23**-16
optimal power response to, **23**-18
per-slot-independent, **23**-17
responsive, **23**-19
-to-signal ratio (ISR), **11**-22
whitening, **28**-15
Interleaving, **12**-6
Intermediate frequency (IF), **10**-9
Intermediate Module Repeaters (IMRs), **4**-19
International standards organizations, channel models adopted by, **2**-2
International Telecommunications Union (ITU), **2**-14, **10**-3
Internet
availability, **20**-9
multimedia subsystem (IMS), **20**-9, **20**-16
protocol (IP), **20**-1, **20**-2
Intersymbol interference (ISI), **1**-6, **1**-29, **8**-2, **13**-15, **25**-35
adaptive equalizers and, **19**-2
channel(s), **13**-4
mitigation, signal modulation design for, **25**-19
multipath propagation and, **3**-2
OFDM and, **28**-8
time dispersion and, **28**-15
Inverse discrete Fourier transform (IDFT), **8**-3, **9**-3, **9**-8
output, **9**-9, **9**-19
processor, **9**-14
Inverse fast Fourier transform (IFFT), **12**-5, **14**-15
Inverse system principles, **27**-18
IP, *see* Internet protocol
Isotropic Doppler spectrum, **4**-11, **4**-13
Isotropic scattering, **5**-9, **5**-14, **5**-21, **5**-29

J

Jammer-to-signal ratio, **22**-11
Jammer subspace, **22**-9
Jamming, hostile, **10**-5
Joint block diagonalization (JBD), **25**-28
Joint cumulants, **27**-21
Joint demodulation (JD), **28**-16
Joint off-diagonalization (JOD), **25**-29
Joint transmit—receive correlation, **12**-13

K

Kalman filter, **19**-6, **19**-23
Kalman processing, **19**-17
Kernel filter, **25**-10
Keyhole effect, **2**-16
Kronecker delta, **3**-9, **9**-5
Kronecker product, **14**-2, **15**-5, **16**-2

L

Lagrange polynomials, **14**-17
Laguerre polynomials, **14**-14
Land mobile satellite, **4**-1
LANs, *see* Local area networks
Laptops, **10**-27
Large-scale fading, **4**-2
Large-scale propagation models, **1**-2 to 4
deterministic approach, **1**-2 to **1**-2
stochastic approach, **1**-3 to **1**-4
Least mean squares (LMS), **13**-20, **28**-14
Least squares
criterion, channel estimation based on, **11**-28
estimation, **21**-7
Lemmas, **14**-4
Level crossing rate (LCR), **4**-12, **28**-9
definition of, **5**-9
estimator, **5**-11
expression of, **4**-13
Likelihood function, **3**-8
Linear equalizers, **16**-21
Linear feedback shift register (LFSR), **10**-8
Linear MMSE (LMMSE), **14**-10
Linear precoding methods, **14**-1
Linear prediction, WMF implementation using, **18**-13
Linear predictive coding (LPC), **20**-4
Linear serial interference cancellation, **11**-15
Linear time-variant channel(s)
delay-Doppler spread function of, **25**-14
estimation of, **25**-16
model, **25**-14
precoding for, **25**-32
Linear time-variant (LTV) system, **11**-2, **25**-2
Line of position (LOP), **21**-3
generation of, **21**-6
intersection of, **21**-5
Line of sight (LOS), **1**-11, **2**-8, **21**-2
component, **5**-6, **5**-7
propagation, **21**-10

Link(s)
initial power level of new, **23**-14
operational, **23**-9
protection, **23**-8, **23**-9, **23**-10
reliability, improved, **12**-1
state, **6**-2
List-type equalizers, **18**-2, **18**-8
List-type MAP algorithm, **18**-9, **18**-10
Local area networks (LANs), **2**-2, **8**-1, **10**-2, **23**-2
Location
algorithms, **21**-9
-based services (LBS), **21**-2
of channel zeros (LOCZ), **8**-13
parameter estimation, **21**-12
Log-distance path loss model, **1**-3
Lognormal shadowing model, **1**-3
Long code, **10**-7
Longley—Rice model, **1**-4
Loo model, **4**-7
Low-complexity diversity combining schemes, **17**-1 to **17**-29
dual-branch switch and stay combining, **17**-4 to **17**-14
generalized switch and examine combining, **17**-20 to **17**-28
multibranch switched diversity, **17**-14 to **17**-20
system and channel models, **17**-3 to **17**-4
Low-density parity check (LDPC), **4**-26, **6**-6
Low-power regime, **14**-15
Lutz model, **4**-14, **4**-15, **4**-16

M

Markov chain
-based analysis, **17**-6
model, **4**-14
Monte Carlo (MCMC) methods, **26**-1, **26**-3
reversible, **26**-4
theory, **17**-8
Matrix inversion lemma, **11**-31
Maximum *a posteriori* (MAP) equalizers, **16**-2, **18**-8
Maximum likelihood
decision rule, **11**-8
detection, constrained, **11**-16
sequence detector, **3**-14
sequence estimation (MLSE), **7**-3, **18**-1, **19**-1
Maximum modulus eigenvalue, **23**-6
Maximum ratio combining (MRC), **1**-36, **4**-24, **10**-24, **17**-2
MBI suppression, **25**-18
McLaurin expansion, **11**-18
Mean excess delay, **1**-4
Mean opinion score (MOS), **20**-5, **28**-13
Mean output energy (MOE), **11**-35
Mean square error (MSE), **5**-25
estimator, **5**-25
matrix, minimized, **14**-4
Media access control (MAC)
layer, **23**-2
methods, capacity of, **1**-33
protocol, **24**-1, **24**-11
schemes, challenges in, **1**-34
Medium-altitude Earth orbits (MEOs), **4**-2
Memory effects, **4**-10
Memoryless nonlinear amplifiers, **19**-4
Methods of moments, **3**-12
Metropolis equalizer, **26**-9
Metropolis—Hastings algorithm, **26**-3
Metropolis—Hastings (M-H) transition rule, **26**-4
Metropolitan area networks (MANs), **8**-2
Millimeter-wave systems, **17**-3
MIMO, *see* Multiple-input multiple-output
MIMO fading channels, equalization techniques for, **18**-1 to **18**-21
block equalizers vs. prefiltered list-type MAP, **18**-17 to **18**-19
block linear and decision-feedback equalizers, **18**-4 to **18**-8
frequency-selective MIMO channel model, **18**-2 to **18**-4
list-type equalizers, **18**-8 to **18**-12
multidimensional whitened matched filter, **18**-12 to **18**-17
MIMO-OFDM wireless systems, **12**-1 to **12**-22
broadband MIMO fading channel, **12**-2 to **12**-4
capacity of broadband MIMO-OFDM systems, **12**-4 to **12**-9
glossary, **12**-21
impact of propagation parameters on space—frequency codes, **12**-14 to **12**-21
space—frequency coded MIMO-OFDM, **12**-9 to **12**-14
MIMO systems, linear precoding for, **14**-1 to **14**-26
channel estimation using precoding techniques, **14**-19 to **14**-24
optimum precoding, **14**-3 to **14**-9
performance analysis and random matrices, **14**-10 to **14**-19
system model, **14**-2 to **14**-3
Minimum description length (MDL), **25**-27
Minimum mean squared error (MMSE), **1**-36
-BDFE, number of multiplications required for, **18**-18
block decision-feedback equalizer, **18**-4
channel estimate, **25**-17
detection, subspace-based blind, **11**-34
detector, **10**-27, **11**-10, **11**-15
equalizer, **3**-22, **3**-23, **3**-25, **8**-6, **18**-7
estimation problem, linear, **25**-17
front end, spatial signals and, **12**-18
linear, **14**-10
predictor, prediction error of, **6**-12
receivers, **11**-21, **11**-32
Minimum output power (MOP), **22**-4, **22**-8
Minimum phase factorization, **18**-13
Minimum shift keying (MSK), **1**-18
Minimum variance
blind detection, **11**-35
unbiased estimator (MVUE), **11**-29
Missed detection probability, **10**-19
Mixed excitation linear prediction (MELP), **20**-5
Mixture Kalman filter (MKF), **26**-14, **26**-15, **26**-18

M-level quadrature amplitude modulation (M-QAM), **7**-1, **7**-2
Mobile channels, modeling and estimation of, **3**-1 to **3**-28
  channel estimation, **3**-8 to **3**-22
  channels models, **3**-3 to **3**-7
  simulation examples, **3**-22 to **3**-35
Mobile satellite channels, **4**-1 to **4**-35
  detection performance analysis, **4**-20 to **4**-31
  statistical propagation models, **4**-2 to **4**-20
Mobile station (MS), **5**-2, **21**-1
  real signal received by, **5**-6
  velocity, **5**-18, **5**-20, **5**-21
Mobile terminal (MT), **10**-4
Mobile velocity, **5**-2
Model(s)
  Barts and Stutzman, **4**-16
  baseband, **27**-23
  basis expansion, **16**-5
  broadband wireless channel, **13**-3
  Canonical channel, **25**-15
  collision channel, **24**-1
  conditional dynamic linear, **26**-14
  continuous-time received signal, **11**-5
  COST 207, **2**-24
  COST 259, **2**-30
  derivation of equivalent baseband, **27**-21
  DLR wideband, **4**-18
  downlink, **11**-5
  DS/SS signal, **10**-6
  Durkin's, **1**-4
  expansion, **3**-5
  fading channel, **11**-2
  finite-state Markov chain, **1**-9, **1**-10
  Fontan et al., **4**-17
  free-space propagation, **1**-2
  Gaussian Markov, **26**-8
  GPS signal, **22**-5
  Hashemi—Suzuki—Turin, **2**-14
  hidden Markov, **3**-10, **3**-19, **20**-17
  IEEE 802.11/HIPERLAN, **2**-27
  ITU, **2**-25, **4**-4
  Karasawa et al., **4**-17
  large-scale propagation, **1**-2
  log-distance path loss, **1**-3
  lognormal shadowing, **1**-3
  Longley—Rice, **1**-4
  Loo, **4**-7
  LTV channel, **25**-14
  Lutz, **4**-14, **4**-15, **4**-16
  Markov chain, **4**-14
  MIMO, **1**-11, **1**-12
  motif, **1**-9
  multipacket reception, **24**-10
  multipath propagation, *see* Multipath propagation models, broadband wireless system
  multistate statistical, **4**-14
  narrowband statistical, **4**-4
  NN phase, **19**-11
  path loss, **1**-3
  physical scattering, **1**-12
  Rayleigh fading channel, **11**-3
  received signal, **5**-5
  Rice and Humphreys, **4**-17
  Rice—lognormal, **4**-7
  Saleh, **7**-30
  Saleh—Valenzuela, **2**-19, **2**-27
  satellite channel, **1**-10
  SIMO, **3**-3
  single-input—single-output, **2**-29
  SISO, **3**-7
  small-scale propagation, **1**-4
  source—filter, **20**-5
  tapped delay line, **3**-3
  3GPP–3GPP2, **2**-29
  transmitted signal, **11**-4
  two-ray fading channel, **1**-8
  Vucetic and Du, **4**-16
  Wakana, **4**-17
  wideband statistical, **4**-18
  Wiener—Hammerstein, **19**-2
Modeling error (ME), **19**-7
Modem, DS/SS, **10**-8
Modulation
  block coded, **1**-27
  chaotic, **27**-14
  chirp, **25**-32
  continuous-phase, **1**-13, **1**-18
  discrete multitone, **9**-9
  Discrete Wavelet Multitone, **9**-16
  filtered multitone, **9**-1, **9**-4, **9**-17
  higher-order, **28**-17
  interval, **9**-15
  multicarrier, **10**-28
  multilevel coded, **1**-27
  O-QAM OFDM, **9**-11, **9**-12
  orthogonal functions, **10**-15
  pulse code, **20**-10
  quadrature amplitude, **1**-15
  single sideband, **9**-17
  structures, coded, **6**-12
  trellis coded, **1**-22, **1**-27, **9**-25
  turbo coded, **1**-27
  vestigial sideband, **9**-17
Moment-generating function (MGF), **15**-3
Monte Carlo estimations, **19**-10
Monte Carlo runs, results of averaging, **3**-22
Monte Carlo signal processing for digital communications, **26**-1 to **26**-25
  MCMC methods, **26**-3 to **26**-11
  SMC methods, **26**-11 to **26**-23
Monte Carlo simulation, **5**-17, **7**-16
Moore—Penrose generalized inverse, **11**-21
Moore—Penrose pseudoinverse, **3**-9
Moore's law, **10**-3
Motif model, **1**-9
Multiaccess communication networks, signal processing for, **24**-1 to **24**-18
  high performance from physical and MAC layers, **24**-16 to 17
  impact of MPR on performance of existing MAC protocols, **24**-11 to **24**-12

interface between physical layer and MAC layer, **24**-8 to **24**-10
MAC layer design for networks with MPR, **24**-13 to **24**-16
MPR at physical layer, **24**-2 to **24**-8
Multiantenna systems, adaptivity in, **6**-14
Multibranch switched diversity, **17**-14
Multicarrier
CDMA, **10**-28, **25**-2, **26**-11, **28**-21
modulation, **10**-28, **28**-8
transmission, **5**-4
Multilayer perceptron (MLP) structures, **19**-2
Multilevel coded modulation (MCM), **1**-27
Multimedia Mobile Access Communications (MMAC), **8**-2
Multipacket reception (MPR), **24**-2
impact of on MAC protocol performance, **24**-11
MAC layer design for networks with, **24**-13
model, **24**-10
Multipath components (MPCs), **2**-3, **5**-6
Multipath elimination
delay lock loop (MEDLL), **22**-4, **2**-22
technology (MET), **22**-4
Multipath fading, **5**-7, **6**-2, **6**-4, **25**-2
Multipath mitigation, GPS, **22**-19
Multipath propagation models, broadband wireless system, **2**-1 to **2**-44
abbreviations, **2**-43 to **2**-44
definition and symbols, **2**-42 to **2**-43
modeling methods for multipath channels, **2**-10 to **2**-18
narrowband, wideband, and directional channel modeling, **2**-3 to **2**-10
propagation aspects and parameterization, **2**-17 to **2**-24
standard models, **2**-24 to **2**-31
Multiple-access interference (MAI), **1**-37, **10**-13, **11**-2, **24**-5, **25**-35
error floor due to, **11**-22
filtered, **11**-13
problem, use of BSS techniques to solve, **25**-2
strength, **11**-21
Multiple antenna systems, performance analysis of, **15**-1 to **15**-32
integral identities, **15**-30 to **15**-31
MIMO systems in presence of co-channel interference, **15**-19 to **15**-31
MIMO systems without co-channel interference, **15**-3 to **15**-19
Multiple-input multiple-output (MIMO), **10**-28
antenna techniques, **13**-23
broadband fading channels, model for, **12**-2
delay spread channels, **12**-9
detectors, decision-feedback, **9**-11
double-directional channels used for, **2**-11
flat-fading channels, **12**-9
ISI channels, SMC receiver in, **26**-19, **26**-22
linear AGN channel, **14**-8
model(s), **1**-11, **1**-12
MRC systems, capacity/outage probability of, **15**-12
prefilter, **13**-18
random matrices, **14**-10
Rayleigh channels, correlated, **15**-14
signal model, **26**-20
system(s), **1**-26
Multiprotocol label switching (MPLS), **20**-11
Multistep linear prediction, **3**-17, **3**-18
Multitone CDMA (MT-CDMA), **1**-31, **1**-32
Multiuser data detection (MUD), **10**-25
MUSIC, **2**-11, **21**-12, **25**-1, **25**-27, **25**-28

N

Nakagami distribution, **4**-6
Nakagami fading
channel, **1**-7, **1**-37
environment, **17**-7
Nakagami—lognormal (NLN) model, **4**-10
NARMA representations, *see* Nonlinear autoregressive moving average representations
Narrowband
channels, modeling of, **2**-3
flat-fading channels, **13**-2
FM interference suppression, **22**-9
interference (NBI), **25**-2
MIMO system, **6**-14
modulated signals, popular, **10**-4
statistical models, **4**-4
transmissions, early—late timing detector for, **10**-20
NATO Narrow Band Voice Coder, **20**-15
Natural gradient (NG) descent learning, **19**-1, **19**-5
Navigation symbols, **22**-2
Near echoes, **4**-19
Near—far problem, **11**-2, **28**-17
Near—far resistance expression, **11**-20
Near—far situation, **10**-19
Network
admission, **23**-3
analyzer, **2**-11
-based speech recognition (NSR), **20**-17, **20**-18
performance comparison, **24**-11
Neural network (NN), **19**-2, *see also* Nonlinear channels, neural networks for transmission over
equalizers, **19**-3
output, **19**-16
phase model, **19**-11
structure, **19**-21
weight, adjusted, **19**-6
Noise
background, **20**-2, **20**-12
bandpass, **5**-18
Gaussian, **11**-2, **11**-7
man-made, **10**-5
masking, **20**-4
power spectral density, **1**-21, **4**-21
thermal, **28**-15
vector, **22**-10
whitening, **11**-10, **11**-11
Noncoherent transmission schemes, **13**-21
Nonisotropic scattering, **5**-15
Nonlinear amplifier (NA), **7**-2

Nonlinear autoregressive moving average (NARMA) representations, **19**-2
Nonlinear channels, neural networks for transmission over, **19**-1 to **19**-28
  channel equalization, **19**-21 to **19**-24
  identification of memoryless nonlinear amplifiers, **19**-4 to **19**-15
  modeling and identification of nonlinear channels with memory, **19**-15 to **19**-21
Nonlinear channels, signaling constellations for transmission over, **7**-1 to **7**-31
  Craig's method, **7**-4 to **7**-5
  probability of symbol error for circular **32**-ary QAM format, **7**-20 to **7**-29
  probability of symbol error for **16**-ary QAM format, **7**-5 to **7**-19
  system model, **7**-3
  TWT based on Saleh model, **7**-30
Nonlinear communications system, **7**-2
Nonlinear dynamic systems, **27**-1
Nonlinear least squares (NLS), **8**-8
Non-line-of-sight (NLOS) propagation, **21**-14, **21**-15
Nonorthogonal inner code, **12**-18
Nonresponsive interference, **23**-16
No-return-to-zero (NRZ), **10**-5
North American cellular systems, **20**-14
Norton distribution, **4**-6
Nuisance parameters, **8**-16
Null subcarriers (NSCs), **8**-4
Nyquist bandwidth, **10**-29
Nyquist filter, **9**-13, **9**-21
Nyquist rate, **13**-3

O

OFDM, *see* Orthogonal frequency division multiplexing
OFDM systems, carrier frequency synchronization for, **8**-1 to **8**-22
  basics of OFDM, **8**-2 to **8**-6
  carrier frequency offset estimation, **8**-7 to **8**-8
  conditional CRB, **8**-19 to **8**-20
  effect of CFO on system performance, **8**-6
  identifiability, **8**-13 to **8**-15
  null-subcarrier-based CFO estimation, **8**-10 to **8**-13
  performance analysis, **8**-15 to **8**-17
  repetitive slots-based CFO estimation, **8**-8 to **8**-10
  simulation results, **8**-17 to **8**-19
OFDM/time division multiple access (OFDM/TDMA), **1**-29, **1**-33
Offset QPSK (OQPSK), **1**-15
Offset quadrature amplitude modulation (O-QAM), **9**-1, **9**-3, **9**-11, **9**-12
  scheme, group band data modem based on, **9**-16
  system, digital implementation of, **9**-13
Optimal power
  allocation, **23**-4
  vector, **23**-6
Orthogonal Alamouti-like matrices, **13**-19
Orthogonal frequency division multiplexing (OFDM), **1**-18, **8**-1, **9**-1, **9**-3, **11**-2, **13**-5, **25**-19
  attractiveness of, **13**-13
  Doufexi space—time coded, **1**-28
  frequency spectrum, **28**-12
  ISI and, **28**-8
  modulation, O-QAM, **9**-11, **9**-12
  multicarrier transmission scheme, **13**-13
  receiver
  symbol
  system(s)
  vector, **1**-19
  versions, **1**-19
  wideband, **1**-19
Orthogonal functions modulation, **10**-15
Orthogonal variable spreading factor (OVSF), **10**-12
  codes, **10**-16, **10**-17
  synchronous, **10**-16
Oscillator frequency stability, **10**-5
Ostrowski—Reich theorem, **11**-15
Outage probability, **15**-12, **15**-21, **17**-15, **17**-23
Overlap factor, **9**-21
Oversampling factor, **11**-24, **11**-38
Overspread channel, **2**-6

P

Pacific digital cellular (PDC) services, **1**-15
Packet loss, **20**-8
  ASR over IP, **20**-19
  recovery techniques, **20**-10
Pairwise error probability (PEP), **12**-11, **12**-13, **13**-5
Palmtops, **10**-27
Parametric coders, **20**-5
Pareto-optimal power vector, **23**-5
Path loss, **5**-6, **5**-26, **6**-2, **6**-4
  adaptation in response to, **6**-5
  feedback, **6**-6
  model, **1**-3
Path synchronization, ideal, **25**-21
PCMA algorithms
  design of, **23**-19
  performance of, **23**-20
Peak-to-average power ratio (PAPR), **1**-18, **3**-20
Perfect root of unity sequences (PRUSs), **13**-14
Per-slot-independent interference, **23**-17
Per survivor processing (PSP), **18**-2, **18**-8
Phase
  ambiguity, **26**-7
  difference (PD), **7**-2
  lock loop (PLL), **22**-3
  shift keying (PSK), **1**-13, **10**-4
  sweep transmit diversity (PSTD), **1**-36
  synchronization, **10**-20
Physical dedicated control channel (PCCH), **10**-16
Physical dedicated data channel (PDCH), **10**-16
Physical scattering model, **1**-12
Pilot
  channel, **10**-15
  filtering, **4**-29

symbol assisted modulation (PSAM), **6**-7, **14**-19
tones, **13**-15
Ping-pong effect, **5**-26
Poisson-distributed random variable, **4**-19
Polarization, broadband wireless networks and, **2**-23
Polyphase
decomposition, **9**-14
filter-bank structures, **9**-6
Power
constraint (PC), **14**-5, **14**-7
control, **23**-2, **28**-17
-controlled multiple access (PCMA), **23**-15
delay profile (PDP), **1**-4, **2**-7, **28**-9
space, **23**-4
spectral density (PSD), **4**-21, **7**-7, **10**-5, **11**-5
vs. delay trade-off, **23**-15
Precoding
flexible, **9**-25
optimum, **14**-2
techniques, MIMO channel estimation using, **14**-19
trellis-augmented, **9**-25, **9**-26
trellis-enhanced, **9**-25
Prediction error statistics, **6**-12, **6**-13
Predistortion, **7**-12, **19**-10
Probability density function (PDF), **7**-4, **15**-7
prescribed shape of, **2**-8
Rayleigh, **6**-8
TOA, **21**-17
Probability distribution functions, chaotic signal, **27**-3
Probability of error (POE), **7**-1, **7**-2
analysis, **7**-6, **7**-7, **7**-11
best-approximated, **7**-7, **7**-9
calculation method, **7**-4
curves, **7**-28
performance, effect of ring ratio on, **7**-15
predefined, **7**-19
QAM circular constellations, **7**-5
Processing gain, **10**-7
Propagation
channels, formal description methods for, **2**-3
delays, unknown, **11**-27
distance, **6**-5
multipath, **2**-4, **3**-1, **10**-22
parameters, impact of on space—frequency codes, **12**-14
paths, **4**-19, **10**-23
three-dimensional, **4**-11
Properly weighted sample, **26**-14
Prototype filter, **9**-13
FIR, **9**-23
impulse response, **9**-15
polyphase component of, **9**-19
stopband mean energy of, **9**-24
Pseudocoherent BPSK detection, **4**-23
Pseudonoise (PN), **5**-4, **10**-8, **27**-11
Pseudorandom binary sequence (PRBS), **10**-8
Pseudorandom noise (PRN), **22**-2, **22**-20
Pseudorange, **22**-2
Public safety answering points (PSAPs), **21**-2
Public Switched Telephone Network (PSTN), **20**-7
Pulse amplitude modulation (PAM), **1**-13, **1**-15
Pulse code modulation (PCM), **20**-10
Pulse roll-off factor, **10**-5

Q

Quadrature amplitude modulation (QAM), **1**-15, **6**-10, **10**-4
constellation(s), **7**-16, **19**-18
M-level, **7**-1, **7**-2
NN equalizers and, **19**-3
offset, **9**-1
Quadrature mirror filter (QMF), **9**-2
Quadrature phase shift keying (QPSK), **1**-13, **16**-24
-based spatial multiplexing, **12**-16
coherently decoded, **6**-11
link adaptation and, **28**-4
modulation, **8**-17
trellis code, **12**-16
Quadrature real spreading (Q-RS), **10**-6
Quality of service (QoS), **1**-2, **5**-26, **13**-13
adaptive power control, **28**-17
communication link and, **23**-2
degradation, **10**-13
requirements, **8**-2, **19**-2
voice over IP, **20**-9
Quasi-static channels, channel estimation, **13**-14

R

Radio channel
random, **28**-4
standard models for mobile, **2**-24
Radio frequency (RF), **10**-5, **17**-2
Radiolocation, **21**-1
Radios, software, **10**-21
Radio systems, *see* Geolocation techniques for mobile radio systems
RAKE receiver, **1**-31, **1**-37, **10**-22, **11**-18, **25**-17, **25**-21
Range-differencing methods, **21**-4
Rank-deficient matrix, **14**-2
Rank selection criteria, **25**-27
Rate of maxima (ROM), **5**-9
Ray
absorption probability, **1**-9
launching, **2**-13
tracing, **2**-13, **2**-16
Rayleigh channel(s), **1**-7, **1**-37, **4**-26, **15**-4
correlated, **15**-5
CRB and, **8**-16
model, **11**-3
two satellite diversity, **4**-30
uncorrelated, **8**-17
Rayleigh distribution, **4**-5
Rayleigh fading, **6**-5, **15**-21
detection of FM signals in, **25**-33
process, unknown, **26**-17
pure, **12**-19
scenarios, **15**-13
Rayleigh—lognormal channel, probability of error, **4**-22

Rayleigh probability density function, **6**-8
Real spreading (RS), **10**-6
Real-time transport protocol/real-time transport control protocol (RTP/RTCP), **20**-9
Receive antenna diversity, **16**-3
Receive correlation matrix, **12**-4
Receive diversity, **12**-1, **13**-13
Received signal
  model, **5**-5
  simulations of, **5**-17
  strength (RSS) estimation, **28**-11
Receiver
  adaptation, **28**-5
  combiner output, **15**-17
  complexity, **12**-2
  correlation, **10**-10, **10**-22
  creation of optimum, **27**-19
  CSI available at, **13**-8
  decorrelating, **11**-12
  errors in channel estimation at, **6**-10
  FMT system, **9**-26
  GPS, **22**-4, **22**-13
  MEDLL, **22**-22
  MLSE, **19**-3
  MMSE, **11**-21, **11**-32
  OFDM, **26**-23
  parameter, Doppler information and, **5**-3
  RAKE, **10**-22, **11**-18, **25**-17, **25**-21
  semiblind least squares smoothing, **24**-5
  signal processing, **13**-23
  sliding-window one-shot multiuser, **11**-24
  SMC, **26**-17, **26**-18
  synchronization between satellite and, **22**-18
  training-based ZF, **24**-4
Reception matrix, **24**-10
Recursive least squares (RLS), **11**-1, **13**-20
Reed—Solomon codes, **1**-22
Reference
  block, **22**-14
  generation, **27**-16
  signal, message detection using, **27**-15
Relaxation code excited linear predictive (RCELP), **20**-14
Renyi map, **27**-6
Resampling
  algorithm, **26**-16
  procedures, **26**-15
  residual, **26**-16
Residual resampling strategy, **26**-16
Responsive interference, **23**-19
Return-link, **10**-4
Rice distribution
  expression of, **4**-6
  modeling of fading according to, **4**-5
Rice factor, **2**-19, **4**-7, **4**-15, **5**-1
Rice fading, **4**-24, **12**-13, **15**-1
  channel, **1**-7, **1**-37, **4**-26
  distribution, **4**-25
Rice and Humphreys model, **4**-17
Rice—lognormal model, **4**-7
Rician/Rayleigh fading scenarios, **15**-28
Ring ratio (RR), **7**-1
  effect of on POE performance, **7**-15
  optimal, **7**-15, **7**-16

S

SAGE, **2**-11
Saleh model, **7**-30
Saleh—Valenzuela model, **2**-19, **2**-27
Sampling, IF vs. baseband, **10**-21
Satellite(s)
  channels, *see* Mobile satellite channels
  communications, low SNRs for, **1**-20
  diversity combining technique, **1**-37
  role of in 4G mobile communications systems, **19**-1
  synchronization between receiver and, **22**-18
Scatterer(s)
  cluster, **12**-2
  location of, **2**-3
Scattering
  distribution, **5**-7, **5**-29
  uncorrelated, **11**-4
Scrambling code, **10**-14
Selection combining (SC), **1**-36, **10**-24, **17**-2
Self-organizing maps (SOMs), **19**-3
Semiblind algorithms, **8**-7
Sensors, blind separation of, **25**-30
Sequential importance sampling, **26**-11, **26**-13
Sequential Monte Carlo (SMC) methods, **26**-1, **26**-3, **26**-11
  applications of in digital communications, **26**-17
  dynamic systems, **26**-13
  filter, **26**-14
  receiver, **26**-17, **26**-18
Serial decision feedback equalizers (SDFEs), **16**-2
Serial linear equalizers (SLEs), **16**-2
Session initiation protocol (SIP), **20**-9
Shadowing, **4**-2, **5**-5, **6**-2
  adaptation in response to, **6**-5
  coherence distance, **4**-4
  correlation function, **4**-14
  effect of, **5**-13
  time-share of, **4**-15
Shannon capacity, **1**-20, **13**-2, **13**-5
Short code, **10**-7, **10**-19, **10**-26
Signal(s)
  additive noise, **16**-3
  band-limited spread-spectrum, **10**-6
  chaotic, **27**-3, **27**-8
  chirp, **25**-18
  continuous-time received, **13**-3
  distortion, **22**-8
  distress, **23**-13
  energy, distribution of, **25**-4
  envelope, modeled, **1**-11
  FM, **25**-3
  GPS, **22**-3, **22**-5
  interpolation, general expression for, **9**-18
  model, DS/SS, **10**-6
  monocomponent analytical, **25**-7
  multicomponent, **25**-6

multipath, bias due to, **22**-19
random, **27**-23
reference, message detection using, **27**-15
separation, STFD-based, **25**-29
simulations of received, **5**-17
statistics, message detection based on, **27**-18
strengths, measurement of, **5**-26
synchronization, **10**-18
synthesis, time—frequency, **25**-13
transmission, **13**-7
Signal-to-interference-plus-noise ratio (SINR), **15**-20
gains, **13**-15
outage probability of, **15**-21
Signal-to-noise ratio (SNR), **1**-14, **1**-31, **5**-13
adaptation and, **6**-2
conditions, *M*-ary transmission under high, **4**-21
definition of, **16**-24
estimation, **28**-2
GPS signal, **22**-3
high, **13**-5
improvement, **5**-30
increased, **8**-3, **25**-22
loss, **3**-13
PDF of received, **17**-22
Rake output, **10**-24
satellite communication, **1**-20
scaled transmitting, **15**-5
signal-dependent, **20**-4
Signature waveform, **11**-4, **25**-21
SIMO, *see* Single-input multiple output
Single-carrier frequency-domain equalizer (SC FDE), **13**-16
Single-input multiple output (SIMO), **3**-2
channel estimation, **3**-13
model, **3**-3
system, **16**-1
Single-input—single-output (SISO)
ergodic capacity in, **12**-7
models, **2**-29, **3**-7
Single sideband (SSB) modulation, **9**-17
Sliding correlator, **10**-19
Slow fading, **1**-6, **4**-27
Slow power control, **28**-17
Small-scale fading, **1**-5, **4**-2
Small-scale propagation model, **1**-4 to **1**-13
parameters of mobile multipath channel, **1**-4 to **1**-5
statistical models for multipath fading channels, **1**-8 to **1**-13
statistical representation of small-scale propagation channel, **1**-7 to **1**-8
types of small-scale fading, **1**-5 to **1**-6
Soft bit source decoding (SBSD), **20**-15
Soft information generation schemes, **28**-16
Software radios, **10**-21
Solid-state power amplifiers (SSPAs), **19**-2
Source
—channel coding, **20**-14
—controlled channel coding (SCCD), **20**-15
encoding, **27**-9
—filter model, **20**-5
Space
division multiple access (SDMA), **1**-31
—frequency code, **12**-12, **12**-14
segment, **22**-2
Space-time coding and signal processing for broadband wireless communications, **13**-1 to **13**-27
broadband wireless channel model, **13**-3 to **13**-5
information-theoretic considerations, **13**-5 to **13**-7
signal transmission issues, **13**-7 to **13**-22
Spatial fading correlation, **12**-3
Spatial multiplexing, **12**-2, **12**-20, **18**-2
gain, **12**-4
QPSK-based, **12**-16
Spatial selectivity, **28**-10
Spatial time—frequency distributions (STFDs), **25**-22, **25**-23
advantages of, **25**-25
-based signal separation, **25**-29
cross-source, **25**-26
Speaker identification, **20**-17
Spectral efficiency, **12**-1
Spectral shaping, **13**-4
Specular reflection, **2**-12
Speech
coding algorithms, **20**-3
decoding, **28**-13
recognition, **20**-17, **20**-20
synthesis, **20**-16
Spreading
code, **10**-6
factor, definition of, **10**-7
function, **2**-5
sequence, **10**-6
Spread-spectrum (SS) techniques, **10**-1 to **10**-31
architecture of DSP-based DS/SS and CDMA receivers, **10**-21 to **10**-24
brief history of wireless communications, **10**-2 to **10**-4
code-division multiple access, **10**-10 to **10**-14
fundamentals of digital spread-spectrum signaling, **10**-4 to **10**-10
multiuser detection, **10**-24 to **10**-27
perspectives, **10**-27 to **10**-29
review of 2G and 3G standards for CDMA mobile communications, **10**-14 to **10**-18
synchronization for spread-spectrum and CDMA signals, **10**-18 to **10**-21
Square envelope detectors (SEDs), **4**-25
Square root raised cosine (SRRC), **10**-5
Stationary probabilities, **17**-10
Steering vectors, **2**-9
Stiefel manifold, **14**-12
Suzuki distribution, **4**-6
Switch and examine combining (SEC), **17**-1
diversity reception, **17**-17
error performance of, **17**-18
L-branch, **17**-15
Switch and stay combining (SSC), **17**-1, **17**-2
alternative discrete-time realization of, **17**-5
dual-branch, **17**-4
L-branch, **17**-14

schemes, dual-branch, **17**-5
strategies, Markov chain built for, **17**-13
Symbol(s)
decision error, **7**-4
detection, **11**-7
duration, **25**-20, **28**-15
energies, **7**-22
energy-to-noise PSD ratio, **4**-25
repetition, **10**-15
time, **10**-19
—vector sequences, **9**-1
Synchronous code-division multiplexing (S-CDM), **10**-10
System
infeasible, **23**-5
near—far resistance, **11**-20
overlap factor, **9**-21
quality of service, **5**-26
resource allocation, adaptive, **28**-5
Szëgo theorem, **14**-17

T

Tall matrix, definition of, **14**-2
Tandeming, **20**-12
Tapped delay lines, **2**-14, **3**-3
Taylor series expansion, **27**-21, **27**-22
Telecommunications Industry Association (TIA), **20**-15, **21**-17
Terminal-based speech recognition (TSR), **20**-17
Text-to-speech conversion (TTS) systems, **20**-16
Thermal noise, **28**-15
Third-generation (3G) systems, **10**-2, **21**-1
Time
delay, **1**-36, **22**-21, **22**-24
dispersion, ISI and, **28**-15
diversity, **1**-35, **12**-1
Time of arrival (TOA), **21**-3
estimation, **22**-4
one-way, **22**-2
probability density function, **21**-17
Time differences of arrival (TDOAs), **21**-3
algorithms, **21**-8, **21**-11
error, NLOS, **21**-16
measurements, **21**-9, **21**-15
Time division
CDMA (TCDMA), **1**-31
duplexing (TDD) systems, **28**-21
frequency hopping (TDFH), **1**-31
multiple access (TDMA), **1**-30, **11**-2, **28**-8
Time—frequency
correlation function, **2**-7
distribution (TFD), **25**-2
source separation, **25**-1
synthesis algorithm, **25**-30
Time—frequency signal processing (TFSP), **25**-1 to **25**-41
other TFSP applications, **25**-32 to **25**-34
spread-spectrum communications systems, **25**-13 to **25**-22
time—frequency array signal processing, **25**-22 to **25**-32
time—frequency signal processing tools, **25**-3 to **25**-13
Time-invariant (TIV) channels, **3**-6, **16**-2, **16**-4
frequency-domain equalization for, **16**-17
matrix, **16**-7
performance results, **16**-24
serial decision feedback equalization, **16**-15
serial linear equalization, **16**-10
Time-variant (TV) channels, **3**-9, **3**-21, **16**-2, **16**-4
matrix, **16**-7
performance results, **16**-26
serial decision feedback equalization, **16**-15
serial linear equalization, **16**-12
Time-variant impulse response, **2**-5
Total angle spread, **12**-7, **12**-15
Total degradation (TD), **7**-1, **7**-14, **19**-13
Total spread, **2**-6
Training-to-information sequence power ratio (TIR), **3**-22
Training sequence (TS), **19**-16
Transfer function, time-variant, **2**-5
Transition probability matrix, **17**-7
Transmission
errors, **20**-2
periods (TPs), **24**-14
Transmit
correlation, **12**-12
diversity, **1**-35, **13**-8, **18**-2
power, MMSE criterion under, **14**-5
—receive correlation, **12**-13
Transmitted reference (TR) methods, **27**-13
Transmitter(s)
buffer, optimally emptying of, **23**-16
chaos-based, **27**-11
Transmitter power control, **23**-1 to **23**-21
packetized data traffic, **23**-15 to **23**-20
streamed continuous traffic, **23**-3 to **23**-15
Transmitting power, normalized, **15**-30
Transport control protocol (TCP), **20**-8
Traveling wave tube (TWT), **7**-2
AM/AM characteristic, **19**-10
amplifiers, **7**-30, **19**-2
Trellis-augmented precoding, **9**-25, **9**-26
Trellis-coded modulation (TCM), **1**-22, **1**-27, **9**-25
Trellis-enhanced precoding, **9**-25
Truncated channel inversion, definition of, **6**-6
Turbo coding (TC), **1**-26, **4**-30
Two-ray fading channel model, **1**-8

U

Ultrawideband (UWB)
geolocation devices and, **28**-20
signaling, **10**-29, **28**-21
systems, **2**-24
Uncertainty principle, Heisenberg's, **25**-18
Uncoded systems, design for, **6**-9
Unconditional CRB (UCRB), **8**-16
Uncorrelated scattering (US), **11**-4
Unequal error protection (UEP), **20**-12
Uniform interleaver, **4**-30
Uniform theory of diffraction (UTD), **2**-12
Unitary matrix, **25**-31
Unitary transformation, **15**-16
Universal frequency reuse, **10**-14

Universal Mobile Telecommunications System (UMTS), **1**-28, **4**-27, **10**-3, **10**-27, **20**-9
networks, first commercial deployment of, **10**-29
terrestrial radio access (UTRA), **10**-15 to **10**-16
Up-link (UL), **10**-4, **10**-24, **15**-7
U.S. Department of Defense (DoD), **22**-1
User datagram protocol (UDP), **20**-8

V

Value-added services, **21**-2
Vandermonde determinant, **15**-29
Vandermonde matrix, **14**-16
Variable bit rate (VBR), **20**-7
Variable spreading factor, **10**-12
V-BLAST, **13**-7, **15**-2
VDSL, **10**-28
Vector
error, **11**-32
minimum variance unbiased estimator of, **11**-29
optimal power, **23**-6
signal, linear model of, **25**-24
sum excited linear prediction (VSELP) paradigm, **20**-14
Velocity estimation for wireless communications, mobile, **5**-1 to **5**-34
application on handover performance, **5**-25 to **5**-28
chapter structure, **5**-5
derivation of equation,
effect of scattering distribution on ZCR method, **5**-29 to **5**-30
existing velocity estimators, **5**-4
importance of velocity estimation, **5**-2 to **5**-4
performance analysis using simulations, **5**-17 to **5**-21
performance analysis of velocity estimators, **5**-12 to **5**-16
principles of mobile velocity estimation, **5**-9 to **5**-12
received signal model and statistics, **5**-5 to **5**-8
Rice factor estimation, **5**-21 to **5**-25
SNR improvement in IF estimator, **5**-30
Very large scale integrated (VLSI) chips, **10**-2, **10**-3
Vestigial sideband (VSB) modulation, **9**-17
Video
decoding, **28**-13
terminals, **10**-27
Virtual cell deployment areas (VCDAs), **2**-30
Virtual subcarriers, **8**-12, **8**-17
Viterbi algorithm (VA), **3**-11, **4**-27, **10**-25, **11**-9, **11**-10, **18**-8, **18**-19
Viterbi convolutional decoder, **1**-24
Viterbi decoder, **4**-26, **19**-17
VLSI chips, *see* Very large scale integrated chips
Voice activity detection (VAD), **20**-11, **20**-12, **20**-21
Voice over IP and wireless, **20**-1 to **20**-25
acronyms and abbreviations, **20**-21 to **20**-23
speech coding for IP and wireless, **20**-2 to **20**-7
voice-enabled services over IP and wireless, **20**-16 to **20**-21
voice over IP, **20**-7 to **20**-11
voice over IP over wireless, **20**-15 to **20**-20-16
voice over wireless, **20**-11 to **20**-15
Voice transport, packet-based alternatives for, **20**-11
Volterra series, **19**-10, **19**-14, **27**-22
Voluntary dropout (VDO), **23**-13
Vucetic and Du model, **4**-16

W

Wakana model, **4**-17
Water filling, **14**-6
capacity, **15**-14, **15**-15
principle, **15**-25
Waveform
coders, **20**-5
interpolation (WI) vocoders, **20**-5
Weight-enumerating function (WEF), **4**-26
Whitened matched filter (WMF), **18**-2, **18**-12, **18**-17
Whitening matrix, **25**-29, **25**-31
Wideband
channels, **1**-6
code division multiple access (WCDMA), **1**-29, **13**-2, **20**-11, **28**-11
speech coding, **20**-2
statistical models, **4**-18
Wide-sense stationary (WSS)
fading, **11**-4
processes, **18**-13
uncorrelated scattering (WSSUS), **2**-1, **25**-2
Wiener filtering, **28**-14
Wiener–Hammerstein models, **19**-2
Wiener–Khintchine theorem, **25**-7
Wigner–Ville distribution (WVD), **25**-3, **25**-7, **25**-12
Wigner–Ville spectrum (WVS), **25**-34
Wireless fidelity (WiFi), **10**-2
Wireless link adaptation, definition of, **6**-2
Wireless local area network (WLAN), **10**-27, **20**-16, **28**-9
Wireless networks, **10**-27
Wireless propagation channel, **2**-2
Wireless revolution, **10**-2
Wold decomposition, **18**-13

X

Xie and Fang distribution, **4**-9
XPD, *see* Cross-polarization discrimination

Z

ZAM distribution, **25**-19
Zero crossing rate (ZCR), **5**-10
-based velocity estimator, **5**-15
method, effect of scattering distribution on, **5**-29
Zero-forcing (ZF), **16**-2
constraint, **14**-7
criterion, **9**-10
MMSE optimization criterion, **18**-2
receiver, training-based, **24**-4
SLE, **16**-14
solution, **16**-21
Zero-order modified Bessel function, **1**-11, **2**-19, **4**-5
Zero padding, **16**-7, **16**-17
Zero stuffing, **13**-4